INTERNATIONAL
CODE COUNCIL®

INTERNATIONAL
BUILDING
CODE®

COMMENTARY - Vol. I

2003

First Printing: January 2004
Second Printing: September 2004
Third Printing: February 2005

ISBN # 1-58001-127-6

COPYRIGHT © 2004
by
INTERNATIONAL CODE COUNCIL, INC.

PRINTED IN THE U.S.A.

PREFACE

The principal purpose of the Commentary is to provide a basic volume of knowledge and facts relating to building construction as it pertains to the regulations set forth in the 2003 *International Building Code*. The person who is serious about effectively designing, constructing and regulating buildings and structures will find the Commentary to be a reliable data source and reference to almost all components of the built environment

As a follow-up to the *International Building Code*, we offer a companion document, the *International Building Code Commentary—Volume I*. Volume I covers Chapters 1 through 15 of the 2003 *International Building Code*. The basic appeal of the Commentary is thus: it provides in a small package and at reasonable cost thorough coverage of many issues likely to be dealt with when using the *International Building Code* — and then supplements that coverage with historical and technical background. Reference lists, information sources and bibliographies are also included.

Throughout all of this, strenuous effort has been made to keep the vast quantity of material accessible and its method of presentation useful. With a comprehensive yet concise summary of each section, the Commentary provides a convenient reference for regulations applicable to the construction of buildings and structures. In the chapters that follow, discussions focus on the full meaning and implications of the code text. Guidelines suggest the most effective method of application, and the consequences of not adhering to the code text. Illustrations are provided to aid understanding; they do not necessarily illustrate the only methods of achieving code compliance.

The format of the Commentary includes the full text of each section, table and figure in the code, followed immediately by the commentary applicable to that text. At the time of printing, the Commentary reflects the most up-to-date text of the 2003 *International Building Code*. As stated in the preface to the *International Building Code,* the content of sections in the code which begin with a letter designation (i.e., Section 307) are maintained by another code development committee. Each section's narrative includes a statement of its objective and intent, and usually includes a discussion about why the requirement commands the conditions set forth. Code text and commentary text are easily distinguished from each other. All code text is shown as it appears in the *International Building Code*, and all commentary is indented below the code text and begins with the symbol ❖.

Readers should note that the Commentary is to be used in conjunction with the *International Building Code* and not as a substitute for the code. **The Commentary is advisory only;** the code official alone possesses the authority and responsibility for interpreting the code.

Comments and recommendations are encouraged, for through your input, we can improve future editions. Please direct your comments to the Codes and Standards Development Department at the Chicago District Office.

TABLE OF CONTENTS

Chapter 1
Administration

General Comments

This chapter contains provisions for the application, enforcement and administration of subsequent requirements of the code. In addition to establishing the scope of the code, Chapter 1 identifies which buildings and structures come under its purview. Section 101 addresses the scope of the code and references the other *International Codes®* that are mentioned elsewhere in the code. Section 102 establishes the applicability of the code and addresses existing structures.

Section 103 establishes the department of building safety and the appointment of department personnel. Section 104 outlines the duties and authority of the building official with regard to permits, inspections and right of entry. It also establishes the authority of the building official to approve alternative materials, used materials and modifications. Section 105 states when permits are required and establishes the procedures for the review of applications and the issuance of permits. Section 106 describes the information that must be included on the construction documents submitted with the application. Section 107 authorizes the building official to issue permits for temporary structures and uses. Section 108 establishes requirements for a fee schedule. Section 109 includes inspection duties of the building official or an inspection agency that has been approved by the building official. Provisions for the issuance of certificates of occupancy are detailed in Section 110. Section 111 gives the building official the authority to approve utility connections. Section 112 establishes the board of appeals and the criteria for making applications for appeal. Administrative provisions for violations are addressed in Section 113, including provisions for unlawful acts, violation notices, prosecution and penalties. Section 114 describes procedures for stop work orders. Section 115 establishes the criteria for unsafe structures and equipment and the procedures to be followed by the building official for abatement and for notification to the responsible party.

Each state's building code enabling legislation, which is grounded within the police power of the state, is the source of all authority to enact building codes. In terms of how it is used, police power is the power of the state to legislate for the general welfare of its citizens. This power enables passage of such laws as building codes. If the state legislature has limited this power in any way, the municipality may not exceed these limitations. While the municipality may not further delegate its police power (e.g., by delegating the burden of determining code compliance to the building owner, contractor or architect), it may turn over the administration of the building code to a municipal official, such as a building official, provided that sufficient criteria are given to establish clearly the basis for decisions as to whether or not a proposed building conforms to the code.

Chapter 1 is largely concerned with maintaining "due process of law" in enforcing the building performance criteria contained in the body of the code. Only through careful observation of the administrative provisions can the building official reasonably hope to demonstrate that "equal protection under the law" has been provided. While it is generally assumed that the administration and enforcement section of a code is geared toward a building official, this is not entirely true. The provisions also establish the rights and privileges of the design professional, contractor and building owner. The position of the building official is merely to review the proposed and completed work and to determine if the construction conforms to the code requirements. The design professional is responsible for the design of a safe structure. The contractor is responsible for constructing the structure in conformance with the plans.

During the course of construction, the building official reviews the activity to ascertain that the spirit and intent of the law are being met and that the safety, health and welfare of the public will be protected. As a public servant, the building official enforces the code in an unbiased, proper manner. Every individual is guaranteed equal enforcement of the provisions of the code. Furthermore, design professionals, contractors and building owners have the right of due process for any requirement in the code.

Purpose

A building code, as with any other code, is intended to be adopted as a legally enforceable document to safeguard health, safety, property and public welfare. A building code cannot be effective without adequate provisions for its administration and enforcement. The official charged with the administration and enforcement of building regulations has a great responsibility, and with this responsibility goes authority. No matter how detailed the building code may be, the building official must, to some extent, exercise his or her own judgement in determining code compliance. The building official has the responsibility to establish that the homes in which the citizens of the community reside and the buildings in which they work are designed and constructed to be structurally stable, with adequate means of egress, light and ventilation and to provide a minimum acceptable level of protection to life and property from fire.

SECTION 101
GENERAL

101.1 Title. These regulations shall be known as the *Building Code* of [NAME OF JURISDICTION], hereinafter referred to as "this code."

❖ The purpose of this section is to identify the adopted regulations by inserting the name of the adopting jurisdiction into the code.

101.2 Scope. The provisions of this code shall apply to the construction, alteration, movement, enlargement, replacement, repair, equipment, use and occupancy, location, maintenance, removal and demolition of every building or structure or any appurtenances connected or attached to such buildings or structures.

Exceptions:

1. Detached one- and two-family dwellings and multiple single-family dwellings (town houses) not more than three stories above grade plane in height with a separate means of egress and their accessory structures shall comply with the *International Residential Code*.

2. Existing buildings undergoing repair, alterations or additions and change of occupancy shall be permitted to comply with the *International Existing Building Code*.

❖ This section establishes when the regulations contained in the code must be followed, whether all or in part. Something must happen (construction of a new building, modification to an existing one or allowing an existing building or structure to become unsafe) for the code to be applicable. While such activity may not be as significant as a new building, a fence is considered a structure and, therefore, its erection is within the scope of the code. The building code is not a maintenance document requiring periodic inspections that will, in turn, result in an enforcement action, although periodic inspections are addressed by the *International Fire Code*® (IFC®).

The first exception mandates that detached one- and two-family dwellings and townhouses that are not more than three stories above grade and have separate means of egress are to comply with the *International Residential Code*® (IRC®) and are not required to comply with the code. This applies to all such structures, whether or not there are lot lines separating them and also to their accessory structures, such as garages and pools. Such structures four stories or more in height are beyond the scope of the IRC and must comply with the provisions of the IBC and its referenced codes.

The second exception provides an option to use the *International Existing Building Code*® (IEBC™) for alterations, repairs, additions or change of occupancy, rather than the provisions of the code, which are contained mainly in Chapter 34.

101.2.1 Appendices. Provisions in the appendices shall not apply unless specifically adopted.

❖ The provisions contained in Appendices A through J are not considered part of the code and are, therefore, not enforceable unless they are specifically included in the ordinance or other adopting law or regulation of the jurisdiction. See Section 1 of the sample ordinance on page v of the code for where the appendices to be adopted are to be specified in the adoption ordinance.

101.3 Intent. The purpose of this code is to establish the minimum requirements to safeguard the public health, safety and general welfare through structural strength, means of egress facilities, stability, sanitation, adequate light and ventilation, energy conservation, and safety to life and property from fire and other hazards attributed to the built environment and to provide safety to fire fighters and emergency responders during emergency operations.

❖ The intent of the code is to set forth regulations that establish the minimum acceptable level to safeguard public health, safety and welfare and to provide protection for fire fighters and emergency responders in building emergencies. The intent becomes important in the application of such sections as Sections 102, 104.11 and 113, as well as any enforcement-oriented interpretive action or judgement. Like any code, the written text is subject to interpretation. Interpretations should not be affected by economics or the potential impact on any party. The only considerations should be protection of public health, safety and welfare and emergency responder safety.

101.4 Referenced codes. The other codes listed in Sections 101.4.1 through 101.4.7 and referenced elsewhere in this code shall be considered part of the requirements of this code to the prescribed extent of each such reference.

❖ The International Code Council® (ICC®) promulgates a complete set of codes to regulate the built environment. These codes are coordinated with each other so as not to contain conflicting provisions. When the code is adopted by a jurisdiction, the codes that regulate a building's electrical, fuel gas, mechanical and plumbing systems are also included in the adoption and are considered a part of the code. The *International Property Maintenance Code*® (IPMC®) and the IFC are also referenced and enable the building official to address unsafe conditions in existing structures. Various other sections of the code also specifically refer to these codes. Note that these codes are listed in Chapter 35 and further identified by the specific year of issue. Only that edition of the code is legally adopted and any future editions are not enforceable. The issuance of new editions of all the *International Codes* occurs concurrently and new editions of the referenced codes are adopted with each new edition of the building code. Adoption is done in this manner so that there are not conflicting provisions in these codes.

101.4.1 Electrical. The provisions of the ICC *Electrical Code* shall apply to the installation of electrical systems, including alterations, repairs, replacement, equipment, appliances, fixtures, fittings and appurtenances thereto.

❖ The ICC *Electrical Code*® (ICC EC™) regulates all aspects of electrical systems and is adopted by reference in this section, as well as Section 2701.1, as the enforceable document for regulating electrical systems. Note that the ICC EC contains the necessary administrative provisions for enforcing the requirements of NFPA 70, the *National Electrical Code*®.

101.4.2 Gas. The provisions of the *International Fuel Gas Code* shall apply to the installation of gas piping from the point of delivery, gas appliances and related accessories as covered in this code. These requirements apply to gas piping systems extending from the point of delivery to the inlet connections of appliances and the installation and operation of residential and commercial gas appliances and related accessories.

❖ The *International Fuel Gas Code*® (IFGC®) regulates gas piping and appliances and is adopted by reference from this section, as well as Section 2801.1, as the enforceable document for regulating gas systems. This section also establishes the scope of the IFGC as extending from the point of delivery to the inlet connections of each gas appliance. The "Point of delivery" is defined in the IFGC as the outlet of the service meter, regulator or shutoff valve.

101.4.3 Mechanical. The provisions of the *International Mechanical Code* shall apply to the installation, alterations, repairs and replacement of mechanical systems, including equipment, appliances, fixtures, fittings and/or appurtenances, including ventilating, heating, cooling, air-conditioning and refrigeration systems, incinerators and other energy-related systems.

❖ The *International Mechanical Code*® (IMC®) regulates all aspects of a building's mechanical systems, including ventilating, heating, cooling, air-conditioning and refrigeration systems, incinerators and other energy-related systems and is adopted by reference from this section, as well as Section 2801.1, as the enforceable document for regulating these systems.

101.4.4 Plumbing. The provisions of the *International Plumbing Code* shall apply to the installation, alteration, repair and replacement of plumbing systems, including equipment, appliances, fixtures, fittings and appurtenances, and where connected to a water or sewage system and all aspects of a medical gas system. The provisions of the *International Private Sewage Disposal Code* shall apply to private sewage disposal systems.

❖ The *International Plumbing Code*® (IPC®) regulates the components of a building's plumbing system, including water supply and distribution piping; sanitary and storm drainage systems; the fixtures and appliances connected thereto and medical gas and oxygen systems and is adopted by reference from this section, as well as Section 2901.1, as the enforceable document for regulating these systems. The *International Private Sewage*

Disposal Code® (IPSDC®) is also adopted as the enforceable document for regulating on-site sewage disposal systems.

101.4.5 Property maintenance. The provisions of the *International Property Maintenance Code* shall apply to existing structures and premises; equipment and facilities; light, ventilation, space heating, sanitation, life and fire safety hazards; responsibilities of owners, operators and occupants; and occupancy of existing premises and structures.

❖ The applicability of the code to existing structures is set forth in Section 101.2 and Chapter 34 and is generally limited to new work or changes in use that occur in these buildings. The IPMC, however, is specifically intended to apply to existing structures and their premises and provides a jurisdiction with an enforceable document for public health, safety and welfare when occupying all buildings, including those that were constructed prior to the adoption of the current building code.

101.4.6 Fire prevention. The provisions of the *International Fire Code* shall apply to matters affecting or relating to structures, processes and premises from the hazard of fire and explosion arising from the storage, handling or use of structures, materials or devices; from conditions hazardous to life, property or public welfare in the occupancy of structures or premises; and from the construction, extension, repair, alteration or removal of fire suppression and alarm systems or fire hazards in the structure or on the premises from occupancy or operation.

❖ The IFC contains provisions for safeguarding structures and premises from the hazards of fire and explosion that result from the materials, substances and operations that may be present in a structure; from circumstances that endanger life, property or public welfare and from the modification or removal of fire suppression and alarm systems. Many of the provisions contained in the code, especially in Chapters 9 and 10, also appear in the IFC. So that all the *International Codes* contain consistent provisions, only one development committee is responsible for considering proposed changes to such provisions and that committee is identified by a letter designation in brackets that appears at the beginning of affected sections. This is described more fully in the preface to the codes. The IFC also contains provisions that are specifically applicable to existing structures and uses and, like the IPMC, provides a jurisdiction with an enforceable document for public health, safety and welfare in all buildings.

101.4.7 Energy. The provisions of the *International Energy Conservation Code* shall apply to all matters governing the design and construction of buildings for energy efficiency.

❖ The *International Energy Conservation Code*® (IECC®) contains provisions for the efficient use of energy in building construction by regulating the design of building envelopes for thermal resistance and low air leakage and the design and selection of mechanical systems for effective use of energy and is adopted by reference in this section, as well as Section 1301.1.1, as the enforceable document for regulating these systems.

SECTION 102
APPLICABILITY

102.1 General. Where, in any specific case, different sections of this code specify different materials, methods of construction or other requirements, the most restrictive shall govern. Where there is a conflict between a general requirement and a specific requirement, the specific requirement shall be applicable.

❖ The most restrictive requirement is to apply where there may be different requirements in the code for a specific installation. In cases where the code establishes a specific requirement for a certain condition, that requirement is applicable even if it is less restrictive than a general requirement elsewhere in the code. For instance, specific requirements for certain uses and occupancies are located in Chapter 4 and take precedence over general requirements found in other chapters of the code. As an example, the requirements contained in Section 402.4 for means of egress in a covered mall building would govern over any differing requirements located in Chapter 10, whether the requirements in Section 402.4 are more or less restrictive.

102.2 Other laws. The provisions of this code shall not be deemed to nullify any provisions of local, state or federal law.

❖ In some cases, other laws enacted by the jurisdiction or the state or federal government may be applicable to a condition that is also governed by a requirement in the code. In such circumstances, the requirements of the code are in addition to that other law that is still in effect, although the building official may not be responsible for its enforcement.

102.3 Application of references. References to chapter or section numbers, or to provisions not specifically identified by number, shall be construed to refer to such chapter, section or provision of this code.

❖ In a situation where the code may make reference to a chapter or section number or to another code provision without specifically identifying its location in the code, assume that the referenced section, chapter or provision is in the code and not in a referenced code or standard.

102.4 Referenced codes and standards. The codes and standards referenced in this code shall be considered part of the requirements of this code to the prescribed extent of each such reference. Where differences occur between provisions of this code and referenced codes and standards, the provisions of this code shall apply.

❖ A referenced code, standard or portion thereof is an enforceable extension of the code as if the content of the standard were included in the body of the code. For example, Section 905.2 references NFPA 14 in its entirety

for the installation of standpipe systems. In those cases where the code references only portions of a standard, the use and application of the referenced standard is limited to those portions that are specifically identified. For example, Section 412.2.6 requires that aircraft hangars must be provided with fire suppression systems as required in NFPA 409. Section 412.2.6 cannot be construed to require compliance with NFPA 409 in its entirety. It is the intent of the code to be in harmony with the referenced standards. If conflicts occur because of scope or purpose, the code text governs.

102.5 Partial invalidity. In the event that any part or provision of this code is held to be illegal or void, this shall not have the effect of making void or illegal any of the other parts or provisions.

❖ Only invalid sections of the code (as established by the court of jurisdiction) can be set aside. This is essential to safeguard the application of the code text to situations whereby a provision is declared illegal or unconstitutional. This section preserves the legislative action that put the legal provisions in place.

102.6 Existing structures. The legal occupancy of any structure existing on the date of adoption of this code shall be permitted to continue without change, except as is specifically covered in this code, the *International Property Maintenance Code* or the *International Fire Code*, or as is deemed necessary by the building official for the general safety and welfare of the occupants and the public.

❖ An existing structure is generally "grandfathered" to be considered approved with code adoption, provided that the building meets a minimum level of safety. Frequently, the criteria for this level are the regulations (or code) under which the existing building was originally constructed. If there are no previous code criteria to apply, the building official must apply those provisions that are reasonably applicable to existing buildings. A specific level of safety is dictated by provisions dealing with hazard abatement in existing buildings and maintenance provisions, as contained in the code, the IPMC and the IFC. These codes are referenced (see Sections 101.4.5 and 101.4.6) and are applicable to existing buildings. Additionally, Chapter 34 comprehensively identifies the pertinent requirements for existing buildings on which a construction operation is intended or that undergoes a change of occupancy.

SECTION 103
DEPARTMENT OF BUILDING SAFETY

103.1 Creation of enforcement agency. The Department of Building Safety is hereby created and the official in charge thereof shall be known as the building official.

❖ This section creates the building department and describes its composition (see Section 109 for a discussion of the inspection duties of the department). Appendix A contains qualifications for the employees of the building department involved in the enforcement of the code. If a jurisdiction desires to establish these qualifications for its employees, Appendix A must be specifically referenced in the adopting ordinance.

The executive official in charge of the building department is named the "building official" by this section. In actuality, the person who is in charge of the department may hold a different title, such as building commissioner, building inspector or construction official. For the purpose of the code, that person is referred to as the "building official."

103.2 Appointment. The building official shall be appointed by the chief appointing authority of the jurisdiction.

❖ This section establishes the building official as an appointed position of the jurisdiction.

103.3 Deputies. In accordance with the prescribed procedures of this jurisdiction and with the concurrence of the appointing authority, the building official shall have the authority to appoint a deputy building official, the related technical officers, inspectors, plan examiners and other employees. Such employees shall have powers as delegated by the building official. For the maintenance of existing properties, see the *International Property Maintenance Code.*

❖ This section provides the building official with the authority to appoint other individuals to assist with the administration and enforcement of the code. These individuals have the authority and responsibility as designated by the building official. Such appointments, however, may be exercised only with the authorization of the chief appointing authority.

SECTION 104
DUTIES AND POWERS OF BUILDING OFFICIAL

104.1 General. The building official is hereby authorized and directed to enforce the provisions of this code. The building official shall have the authority to render interpretations of this code and to adopt policies and procedures in order to clarify the application of its provisions. Such interpretations, policies and procedures shall be in compliance with the intent and purpose of this code. Such policies and procedures shall not have the effect of waiving requirements specifically provided for in this code.

❖ The duty of the building official is to enforce the code and he or she is the "authority having jurisdiction" for all matters relating to the code and its enforcement. It is the duty of the building official to interpret the code and to determine compliance. Code compliance will not always be easy to determine and will require judgement and expertise, particularly when enforcing the provisions of Sections 104.10 and 104.11. In exercising this

authority, however, the building official cannot set aside or ignore any provision of the code.

104.2 Applications and permits. The building official shall receive applications, review construction documents and issue permits for the erection, and alteration, demolition and moving of buildings and structures, inspect the premises for which such permits have been issued and enforce compliance with the provisions of this code.

❖ The code enforcement process is normally initiated with an application for a permit. The building official is responsible for processing applications and issuing permits for the construction or modification of buildings in accordance with the code.

104.3 Notices and orders. The building official shall issue all necessary notices or orders to ensure compliance with this code.

❖ An important element of code enforcement is the necessary advisement of deficiencies and corrections, which is accomplished through written notices and orders. The building official is required to issue orders to abate illegal or unsafe conditions. Section 115.3 contains additional information for these notices.

104.4 Inspections. The building official shall make all of the required inspections, or the building official shall have the authority to accept reports of inspection by approved agencies or individuals. Reports of such inspections shall be in writing and be certified by a responsible officer of such approved agency or by the responsible individual. The building official is authorized to engage such expert opinion as deemed necessary to report upon unusual technical issues that arise, subject to the approval of the appointing authority.

❖ The building official is required to make inspections as necessary to determine compliance with the code or to accept written reports of inspections by an approved agency. The inspection of the work in progress or accomplished is another significant element in determining code compliance. While a department does not have the resources to inspect every aspect of all work, the required inspections are those that are dictated by administrative rules and procedures based on many parameters, including available inspection resources. In order to expand the available resources for inspection purposes, the building official may approve an agency that, in his or her opinion, complies with the criteria set forth in Section 1703. When unusual, extraordinary or complex technical issues arise relative to building safety, the building official has the authority to seek the opinion and advice of experts. Since this usually involves the expenditure of funds, the approval of the jurisdiction's chief executive (or similar position) is required. A technical report from an expert requested by the building official can be used to assist in the approval process (also see Section 1704 for special inspection requirements).

104.5 Identification. The building official shall carry proper identification when inspecting structures or premises in the performance of duties under this code.

❖ This section requires the building official (including by definition all authorized designees) to carry identification in the course of conducting the duties of the position. This removes any question as to the purpose and authority of the inspector.

104.6 Right of entry. Where it is necessary to make an inspection to enforce the provisions of this code, or where the building official has reasonable cause to believe that there exists in a structure or upon a premises a condition which is contrary to or in violation of this code which makes the structure or premises unsafe, dangerous or hazardous, the building official is authorized to enter the structure or premises at reasonable times to inspect or to perform the duties imposed by this code, provided that if such structure or premises be occupied that credentials be presented to the occupant and entry requested. If such structure or premises is unoccupied, the building official shall first make a reasonable effort to locate the owner or other person having charge or control of the structure or premises and request entry. If entry is refused, the building official shall have recourse to the remedies provided by law to secure entry.

❖ The first part of this section establishes the right of the building official to enter the premises in order to make the permit inspections required by Section 109.3. Permit application forms typically include a statement in the certification signed by the applicant (who is the owner or owner's agent) granting the building official the authority to enter areas covered by the permit in order to enforce code provisions related to the permit. The right to enter other structures or premises is more limited. First, to protect the right of privacy, the owner or occupant must grant the building official permission before an interior inspection of the property can be conducted. Permission is not required for inspections that can be accomplished from within the public right-of-way. Second, such access may be denied by the owner or occupant. Unless the inspector has reasonable cause to believe that a violation of the code exists, access may be unattainable. Third, building officials must present proper identification (see Section 104.5) and request admittance during reasonable hours—usually the normal business hours of the establishment—to be admitted. Fourth, inspections must be aimed at securing or determining compliance with the provisions and intent of the regulations that are specifically within the established scope of the building official's authority.

Searches to gather information for the purpose of enforcing the other codes, ordinances or regulations are considered unreasonable and are prohibited by the Fourth Amendment to the U.S. Constitution. "Reasonable cause" in the context of this section must be distinguished from "probable cause," which is required to gain access to property in criminal cases. The burden of proof establishing reasonable cause may vary among jurisdictions. Usually, an inspector must show that the

property is subject to inspection under the provisions of the code; that the interests of the public health, safety and welfare outweigh the individual's right to maintain privacy and that such an inspection is required solely to determine compliance with the provisions of the code.

Many jurisdictions do not recognize the concept of an administrative warrant and may require the building official to prove probable cause in order to gain access upon refusal. This burden of proof is usually more substantial, often requiring the building official to stipulate in advance why access is needed (usually access is restricted to gathering evidence for seeking an indictment or making an arrest); what specific items or information is sought; its relevance to the case against the individual subject; how knowledge of the relevance of the information or items sought was obtained and how the evidence sought will be used. In all such cases, the right to privacy must always be weighed against the right of the building official to conduct an inspection to verify that public health, safety and welfare are not in jeopardy. Such important and complex constitutional issues should be discussed with the jurisdiction's legal counsel. Jurisdictions should establish procedures for securing the necessary court orders when an inspection is deemed necessary following a refusal.

104.7 Department records. The building official shall keep official records of applications received, permits and certificates issued, fees collected, reports of inspections, and notices and orders issued. Such records shall be retained in the official records for the period required for retention of public records.

❖ In keeping with the need for an efficiently conducted business practice, the building official must keep official records pertaining to permit applications, permits, fees collected, inspections, notices and orders issued. Such documentation provides a valuable resource of information if questions arise regarding the department's actions with respect to a building. The code does not require that construction documents be kept after the project is complete. It requires that other documents be kept for the length of time mandated by a jurisdiction's, or its state's, laws or administrative rules for retaining public records.

104.8 Liability. The building official, member of the board of appeals or employee charged with the enforcement of this code, while acting for the jurisdiction in good faith and without malice in the discharge of the duties required by this code or other pertinent law or ordinance, shall not thereby be rendered liable personally and is hereby relieved from personal liability for any damage accruing to persons or property as a result of any act or by reason of an act or omission in the discharge of official duties. Any suit instituted against an officer or employee because of an act performed by that officer or employee in the lawful discharge of duties and under the provisions of this code shall be defended by legal representative of the jurisdiction until the final termination of the proceedings. The building official or any subordinate shall not be liable for cost in any action, suit or proceeding that is instituted in pursuance of the provisions of this code.

❖ The building official, other department employees and members of the appeals board are not intended to be held liable for those actions performed in accordance with the code in a reasonable and lawful manner. The responsibility of the building official in this regard is subject to local, state and federal laws that may supersede this provision. This section further establishes that building officials (or subordinates) must not be liable for costs in any legal action instituted in response to the performance of lawful duties. These costs are to be borne by the state, county or municipality. The best way to be certain that the building official's action is a "lawful duty" is always to cite the applicable code section on which the enforcement action is based.

104.9 Approved materials and equipment. Materials, equipment and devices approved by the building official shall be constructed and installed in accordance with such approval.

❖ The code is a compilation of criteria with which materials, equipment, devices and systems must comply to be suitable for a particular application. The building official has a duty to evaluate such materials, equipment, devices and systems for code compliance and, when compliance is determined, approve the same for use. The materials, equipment, devices and systems must be constructed and installed in compliance with, and all conditions and limitations considered as a basis for, that approval. For example, the manufacturer's instructions and recommendations are to be followed if the approval of the material was based even in part on those instructions and recommendations. The approval authority given to the building official is a significant responsibility and is a key to code compliance. The approval process is first technical and then administrative and must be approached as such. For example, if data to determine code compliance are required, such data should be in the form of test reports or engineering analysis and not simply taken from a sales brochure.

104.9.1 Used materials and equipment. The use of used materials which meet the requirements of this code for new materials is permitted. Used equipment and devices shall not be reused unless approved by the building official.

❖ The code criteria for materials and equipment have changed over the years. Evaluation of testing and materials technology has permitted the development of new criteria that the old materials may not satisfy. As a result, used materials are required to be evaluated in the same manner as new materials. Used materials, equipment and devices must be equivalent to that required by the code if they are to be used again in a new installation.

104.10 Modifications. Wherever there are practical difficulties involved in carrying out the provisions of this code, the building official shall have the authority to grant modifications for individual cases, upon application of the owner or owner's representative, provided the building official shall first find that special individual reason makes the strict letter of this code impractical and the modification is in compliance with the intent and purpose of this code and that such modification does not lessen health, accessibility, life and fire safety, or structural requirements. The details of action granting modifications shall be recorded and entered in the files of the department of building safety.

❖ The building official may amend or make exceptions to the code as needed where strict compliance is impractical. Only the building official has authority to grant modifications. Consideration of a particular difficulty is to be based on the application of the owner and a demonstration that the intent of the code is accomplished. This section is not intended to permit setting aside or ignoring a code provision; rather, it is intended to provide for the acceptance of equivalent protection. Such modifications do not, however, extend to actions that are necessary to correct violations of the code. In other words, a code violation or the expense of correcting one cannot constitute a practical difficulty.

104.11 Alternative materials, design and methods of construction and equipment. The provisions of this code are not intended to prevent the installation of any material or to prohibit any design or method of construction not specifically prescribed by this code, provided that any such alternative has been approved. An alternative material, design or method of construction shall be approved where the building official finds that the proposed design is satisfactory and complies with the intent of the provisions of this code, and that the material, method or work offered is, for the purpose intended, at least the equivalent of that prescribed in this code in quality, strength, effectiveness, fire resistance, durability and safety.

❖ The code is not intended to inhibit innovative ideas or technological advances. A comprehensive regulatory document such as a building code cannot envision and then address all future innovations in the industry. As a result, a performance code must be applicable to and provide a basis for the approval of an increasing number of newly developed, innovative materials, systems and methods for which no code text or referenced standards yet exist. The fact that a material, product or method of construction is not addressed in the code is not an indication that such material, product or method is intended to be prohibited. The building official is expected to apply sound technical judgement in accepting materials, systems or methods that, while not anticipated by the drafters of the current code text, can be demonstrated to offer equivalent performance. By virtue of its text, the code regulates new and innovative construction practices while addressing the relative safety of building occupants. The building official is responsible for determining if a requested alternative provides the equivalent level of protection of public health, safety and welfare as required by the code.

104.11.1 Research reports. Supporting data, where necessary to assist in the approval of materials or assemblies not specifically provided for in this code, shall consist of valid research reports from approved sources.

❖ When an alternative material or method is proposed for construction, it is incumbent upon the building official to determine whether this alternative is, in fact, an equivalent to the methods prescribed by the code. Reports providing evidence of this equivalency are required to be supplied by an approved source, meaning a source that the building official finds to be reliable and accurate. The ICC Evaluation Service is an example of an agency that provides research reports for alternative materials and methods.

104.11.2 Tests. Whenever there is insufficient evidence of compliance with the provisions of this code, or evidence that a material or method does not conform to the requirements of this code, or in order to substantiate claims for alternative materials or methods, the building official shall have the authority to require tests as evidence of compliance to be made at no expense to the jurisdiction. Test methods shall be as specified in this code or by other recognized test standards. In the absence of recognized and accepted test methods, the building official shall approve the testing procedures. Tests shall be performed by an approved agency. Reports of such tests shall be retained by the building official for the period required for retention of public records.

❖ To provide the basis on which the building official can make a decision regarding an alternative material or method, sufficient technical data, test reports and documentation must be provided for evaluation. If evidence satisfactory to the building official indicates that the alternative material or construction method is equivalent to that required by the code, he or she may approve it. Any such approval cannot have the effect of waiving any requirements of the code. The burden of proof of equivalence lies with the applicant who proposes the use of alternative materials or methods.

The building official must require the submission of any appropriate information and data to assist in the determination of equivalency. This information must be submitted before a permit can be issued. The type of information required includes test data in accordance with referenced standards, evidence of compliance with the referenced standard specifications and design calculations. A research report issued by an authoritative agency is particularly useful in providing the building official with the technical basis for evaluation and approval of new and innovative materials and methods of construction. The use of authoritative research reports can greatly assist the building official by reducing the time-consuming engineering analysis necessary to review these materials and methods. Failure to substantiate adequately a request for the use of an alternative is a valid reason for the building official to deny a request. Any tests submitted in support of an application must have been performed by an agency approved by the building official based on evidence that the agency has

the technical expertise, test equipment and quality assurance to properly conduct and report the necessary testing. The test reports submitted to the building official must be retained in accordance with the requirements of Section 104.7.

SECTION 105
PERMITS

105.1 Required. Any owner or authorized agent who intends to construct, enlarge, alter, repair, move, demolish, or change the occupancy of a building or structure, or to erect, install, enlarge, alter, repair, remove, convert or replace any electrical, gas, mechanical or plumbing system, the installation of which is regulated by this code, or to cause any such work to be done, shall first make application to the building official and obtain the required permit.

❖ This section contains the administrative rules governing the issuance, suspension, revocation or modification of building permits. It also establishes how and by whom the application for a building permit is to be made, how it is to be processed, fees and what information it must contain or have attached to it.

In general, a permit is required for all activities that are regulated by the code or its referenced codes (see Section 101.4), and these activities cannot begin until the permit is issued, unless the activity is specifically exempted by Section 105.2. Only the owner or a person authorized by the owner can apply for the permit. Note that this section indicates a need for a permit for a change in occupancy, even if no work is contemplated. Although the occupancy of a building or portion thereof may change and the new activity is still classified in the same group, different code provisions may be applicable. The means of egress, structural loads and light and ventilation provisions are examples of requirements that are occupancy sensitive. The purpose of the permit is to cause the work to be reviewed, approved and inspected to determine compliance with the code.

105.1.1 Annual permit. In lieu of an individual permit for each alteration to an already approved electrical, gas, mechanical or plumbing installation, the building official is authorized to issue an annual permit upon application therefor to any person, firm or corporation regularly employing one or more qualified tradepersons in the building, structure or on the premises owned or operated by the applicant for the permit.

❖ In some instances, such as large buildings or industrial facilities, the repair, replacement or alteration of electrical, gas, mechanical or plumbing systems occurs on a frequent basis, and this section allows the building official to issue an annual permit for this work. This relieves both the building department and the owners of such facilities from the burden of filing and processing individual applications for this activity; however, there are restrictions on who is entitled to these permits. They can be issued only for work on a previously approved installation and only to an individual or corporation that employs persons specifically qualified in the trade for which the permit is issued. If tradespeople who perform the work involved are required to be licensed in the jurisdiction, then only those persons would be permitted to perform the work. If trade licensing is not required, then the building official needs to review and approve the qualifications of the persons who will be performing the work. The annual permit can apply only to the individual property that is owned or operated by the applicant.

105.1.2 Annual permit records. The person to whom an annual permit is issued shall keep a detailed record of alterations made under such annual permit. The building official shall have access to such records at all times or such records shall be filed with the building official as designated.

❖ The work performed in accordance with an annual permit must be inspected by the building official, so it is necessary to know the location of such work and when it was performed. This can be accomplished by having records of the work available to the building official either at the premises or in the official's office, as determined by the official.

105.2 Work exempt from permit. Exemptions from permit requirements of this code shall not be deemed to grant authorization for any work to be done in any manner in violation of the provisions of this code or any other laws or ordinances of this jurisdiction. Permits shall not be required for the following:

Building:

1. One-story detached accessory structures used as tool and storage sheds, playhouses and similar uses, provided the floor area does not exceed 120 square feet (11.15 m²).

2. Fences not over 6 feet (1829 mm) high.

3. Oil derricks.

4. Retaining walls which are not over 4 feet (1219 mm) in height measured from the bottom of the footing to the top of the wall, unless supporting a surcharge or impounding Class I, II or III-A liquids.

5. Water tanks supported directly on grade if the capacity does not exceed 5,000 gallons (18 925 L) and the ratio of height to diameter or width does not exceed 2 to 1.

6. Sidewalks and driveways not more than 30 inches (762 mm) above grade and not over any basement or story below and which are not part of an accessible route.

7. Painting, papering, tiling, carpeting, cabinets, counter tops and similar finish work.

8. Temporary motion picture, television and theater stage sets and scenery.

9. Prefabricated swimming pools accessory to a Group R-3 occupancy, as applicable in Section 101.2, which are less than 24 inches (610 mm) deep, do not exceed 5,000 gallons (18 925 L) and are installed entirely above ground.

10. Shade cloth structures constructed for nursery or agricultural purposes and not including service systems.

11. Swings and other playground equipment accessory to detached one- and two-family dwellings.

12. Window awnings supported by an exterior wall which do not project more than 54 inches (1372 mm) from the exterior wall and do not require additional support of Group R-3, as applicable in Section 101.2, and Group U occupancies.

13. Movable cases, counters and partitions not over 5 feet 9 inches (1753 mm) in height.

Electrical:

Repairs and maintenance: Minor repair work, including the replacement of lamps or the connection of approved portable electrical equipment to approved permanently installed receptacles.

Radio and television transmitting stations: The provisions of this code shall not apply to electrical equipment used for radio and television transmissions, but do apply to equipment and wiring for power supply, the installations of towers and antennas.

Temporary testing systems: A permit shall not be required for the installation of any temporary system required for the testing or servicing of electrical equipment or apparatus.

Gas:

1. Portable heating appliance.

2. Replacement of any minor part that does not alter approval of equipment or make such equipment unsafe.

Mechanical:

1. Portable heating appliance.

2. Portable ventilation equipment.

3. Portable cooling unit.

4. Steam, hot or chilled water piping within any heating or cooling equipment regulated by this code.

5. Replacement of any part which does not alter its approval or make it unsafe.

6. Portable evaporative cooler.

7. Self-contained refrigeration system containing 10 pounds (4.54 kg) or less of refrigerant and actuated by motors of 1 horsepower (746 W) or less.

Plumbing:

1. The stopping of leaks in drains, water, soil, waste or vent pipe provided, however, that if any concealed trap, drain pipe, water, soil, waste or vent pipe becomes defective and it becomes necessary to remove and replace the same with new material, such work shall be considered as new work and a permit shall be obtained and inspection made as provided in this code.

2. The clearing of stoppages or the repairing of leaks in pipes, valves or fixtures, and the removal and reinstallation of water closets, provided such repairs do not involve or require the replacement or rearrangement of valves, pipes or fixtures.

❖ Section 105.1 essentially requires a permit for any activity involving work on a building and its systems and other structures. This section lists those activities that are permitted to take place without first obtaining a permit from the building department. Note that in some cases, such as Items 9, 10, 11 and 12, the work is exempt only for certain occupancies. It is further the intent of the code that even though work may be exempted for permit purposes, it must still comply with the code and the owner is responsible for proper and safe construction for all work being done. Work exempted by the codes adopted by reference in Section 101.4 is also included here.

105.2.1 Emergency repairs. Where equipment replacements and repairs must be performed in an emergency situation, the permit application shall be submitted within the next working business day to the building official.

❖ This section recognizes that in some cases, emergency replacement and repair work must be done as quickly as possible, so it is not practical to take the necessary time to apply for and obtain approval. A permit for the work must be obtained the next day that the building department is open for business. Any work performed before the permit is issued must be done in accordance with the code and corrected if not approved by the building official.

105.2.2 Repairs. Application or notice to the building official is not required for ordinary repairs to structures, replacement of lamps or the connection of approved portable electrical equipment to approved permanently installed receptacles. Such repairs shall not include the cutting away of any wall, partition or portion thereof, the removal or cutting of any structural beam or load-bearing support, or the removal or change of any required means of egress, or rearrangement of parts of a structure affecting the egress requirements; nor shall ordinary repairs include addition to, alteration of, replacement or relocation of any standpipe, water supply, sewer, drainage, drain leader, gas, soil, waste, vent or similar piping, electric wiring or mechanical or other work affecting public health or general safety.

❖ This section distinguishes between what might be termed by some as repairs but are in fact alterations, wherein the code is to be applicable, and ordinary repairs, which are maintenance activities that do not require a permit.

105.2.3 Public service agencies. A permit shall not be required for the installation, alteration or repair of generation, transmission, distribution or metering or other related equipment that is under the ownership and control of public service agencies by established right.

❖ Utilities that supply electricity, gas, water, telephone, television cable, etc., do not require permits for work involving the transmission lines and metering equipment that they own and control; that is, to their point of delivery. Utilities are typically regulated by other laws that give them specific rights and authority in this area. Any equipment or appliances installed or serviced by such

agencies that are not owned by them and under their full control are not exempt from a permit.

105.3 Application for permit. To obtain a permit, the applicant shall first file an application therefor in writing on a form furnished by the department of building safety for that purpose. Such application shall:

1. Identify and describe the work to be covered by the permit for which application is made.
2. Describe the land on which the proposed work is to be done by legal description, street address or similar description that will readily identify and definitely locate the proposed building or work.
3. Indicate the use and occupancy for which the proposed work is intended.
4. Be accompanied by construction documents and other information as required in Section 106.3.
5. State the valuation of the proposed work.
6. Be signed by the applicant, or the applicant's authorized agent.
7. Give such other data and information as required by the building official.

❖ This section requires that a written application for a permit be filed on forms provided by the building department and details the information required on the application. Permit forms will typically have sufficient space to write a very brief description of the work to be accomplished, which is sufficient for only small jobs. For larger projects, the description will be augmented by construction documents as indicated in Item 4. As required by Section 105.1, the applicant must be the owner of the property or an authorized agent of the owner, such as an engineer, architect, contractor, tenant or other. The applicant must sign the application, and permit forms typically include a statement that if the applicant is not the owner, he or she has permission from the owner to make the application.

105.3.1 Action on application. The building official shall examine or cause to be examined applications for permits and amendments thereto within a reasonable time after filing. If the application or the construction documents do not conform to the requirements of pertinent laws, the building official shall reject such application in writing, stating the reasons therefor. If the building official is satisfied that the proposed work conforms to the requirements of this code and laws and ordinances applicable thereto, the building official shall issue a permit therefor as soon as practicable.

❖ This section requires the building official to act with reasonable speed on a permit application. In some instances, this time period is set by state or local law. The building official must refuse to issue a permit when the application and accompanying documents do not conform to the code. In order to ensure effective communication and due process of law, the reasons for denial of an application for a permit are required to be in writing. Once the building official determines that the work described conforms with the code and other applicable laws, the permit must be issued upon payment of the fees required by Section 108.

105.3.2 Time limitation of application. An application for a permit for any proposed work shall be deemed to have been abandoned 180 days after the date of filing, unless such application has been pursued in good faith or a permit has been issued; except that the building official is authorized to grant one or more extensions of time for additional periods not exceeding 90 days each. The extension shall be requested in writing and justifiable cause demonstrated.

❖ Typically, an application for a permit is submitted and goes through a review process that ends with the issuance of a permit. If a permit has not been issued 180 days after the date of filing, however, the application is considered abandoned, unless the applicant was diligent in efforts to obtain the permit. The building official has the authority to extend this time limitation (in increments of 90 days), provided there is reasonable cause. This would cover delays beyond the applicant's control, such as prerequisite permits or approvals from other authorities within the jurisdiction or state. The intent of this section is to limit the time between the review process and the issuance of a permit.

105.4 Validity of permit. The issuance or granting of a permit shall not be construed to be a permit for, or an approval of, any violation of any of the provisions of this code or of any other ordinance of the jurisdiction. Permits presuming to give authority to violate or cancel the provisions of this code or other ordinances of the jurisdiction shall not be valid. The issuance of a permit based on construction documents and other data shall not prevent the building official from requiring the correction of errors in the construction documents and other data. The building official is also authorized to prevent occupancy or use of a structure where in violation of this code or of any other ordinances of this jurisdiction.

❖ This section states the fundamental premise that the permit is only a license to proceed with the work. It is not a license to violate, cancel or set aside any provisions of this code. This is significant because it means that despite any errors or oversights in the approval process, the permit applicant, not the building official, is responsible for code compliance. Also, the permit can be suspended or revoked in accordance with Section 105.6.

105.5 Expiration. Every permit issued shall become invalid unless the work on the site authorized by such permit is commenced within 180 days after its issuance, or if the work authorized on the site by such permit is suspended or abandoned for a period of 180 days after the time the work is commenced. The building official is authorized to grant, in writing, one or more extensions of time, for periods not more than 180 days each. The extension shall be requested in writing and justifiable cause demonstrated.

❖ The permit becomes invalid under two distinct situations—both based on a 180-day period. The first situa-

tion is when no work was ever started 180 days from issuance of a permit. The second situation is when the authorized work has stopped for 180 days. The person who was issued the permit should be notified, in writing, that it is invalid and what steps must be taken to reinstate it and restart the work. The building official has the authority to extend this time limitation (in increments of 180 days), provided the extension is requested in writing and there is reasonable cause, which typically includes events beyond the permit holder's control.

105.6 Suspension or revocation. The building official is authorized to suspend or revoke a permit issued under the provisions of this code wherever the permit is issued in error or on the basis of incorrect, inaccurate or incomplete information, or in violation of any ordinance or regulation or any of the provisions of this code.

❖ A permit is in reality a license to proceed with the work. The building official, however, can suspend or revoke permits shown to be based, all or in part, on any false statement or misrepresentation of fact. A permit can also be suspended or revoked if it was issued in error, such as an omitted prerequisite approval or code violation indicated on the construction documents. An applicant may subsequently apply for a reinstatement of the permit with the appropriate corrections or modifications made to the application and construction documents.

105.7 Placement of permit. The building permit or copy shall be kept on the site of the work until the completion of the project.

❖ The permit, or copy thereof, is to be kept on the job site until the work is complete and made available to the building official or representative to conveniently make required entries thereon.

SECTION 106
CONSTRUCTION DOCUMENTS

106.1 Submittal documents. Construction documents, special inspection and structural observation programs, and other data shall be submitted in one or more sets with each application for a permit. The construction documents shall be prepared by a registered design professional where required by the statutes of the jurisdiction in which the project is to be constructed. Where special conditions exist, the building official is authorized to require additional construction documents to be prepared by a registered design professional.

Exception: The building official is authorized to waive the submission of construction documents and other data not required to be prepared by a registered design professional if it is found that the nature of the work applied for is such that review of construction documents is not necessary to obtain compliance with this code.

❖ This section establishes the requirement to provide the building official with construction drawings, specifications and other documents that describe the structure or system for which a permit is sought (see Section 202 for a complete definition). It describes the information that must be included in the documents, who must prepare them and procedures for approving them.

A detailed description of the work for which an application is made must be submitted. When the work can be briefly described on the application form and the services of a registered design professional are not required, the building official may utilize judgement in determining the need for detailed documents. An example of work that may not involve the submission of detailed construction documents is the replacement of an existing 60-amp electrical service with a 200-amp service. Other sections of the code also contain specific requirements for construction documents, such as Sections 1603, 1901.4, 2101.3 and 3103.2. These provisions are intended to reflect the minimum scope of information needed to determine code compliance. Although this section specifies that "one or more" sets of construction documents be submitted, note that Section 106.3.1 requires one set of approved documents to be retained by the building official and one set to be returned to the applicant, essentially requiring at least two sets of construction documents. The building official should establish a consistent policy of the number of sets required by the jurisdiction and make this information readily available to applicants.

This section also requires the building official to determine that any state professional registration laws be complied with as they apply to the preparation of construction documents.

106.1.1 Information on construction documents. Construction documents shall be dimensioned and drawn upon suitable material. Electronic media documents are permitted to be submitted when approved by the building official. Construction documents shall be of sufficient clarity to indicate the location, nature and extent of the work proposed and show in detail that it will conform to the provisions of this code and relevant laws, ordinances, rules and regulations, as determined by the building official.

❖ The construction documents are required to be of a quality and detail such that the building official can determine that the work conforms to the code and other applicable laws and regulations. General statements on the documents, such as "all work must comply with the *International Building Code,*" are not an acceptable substitute for showing the required information. The following subsections and sections in other chapters indicated in the commentary to Section 106.1 specify the detailed information that must be shown on the submitted documents. When specifically allowed by the building official, documents can be submitted in electronic form.

106.1.1.1 Fire protection system shop drawings. Shop drawings for the fire protection system(s) shall be submitted to indicate conformance with this code and the construction documents and shall be approved prior to the start of system in-

stallation. Shop drawings shall contain all information as required by the referenced installation standards in Chapter 9.

❖ Since the fire protection contractor(s) may not have been selected at the time a permit is issued for construction of a building, detailed shop drawings for fire protection systems are not available. Because they provide the information necessary to determine code compliance, as specified in the appropriate referenced standard in Chapter 9, they must be submitted and approved by the building official before the contractor can begin installing the system. For example, the professional responsible for the design of an automatic sprinkler system should determine that the water supply is adequate, but will not be able to prepare a final set of hydraulic calculations if the specific materials and pipe sizes, lengths and arrangements have not been identified. Once the installing contractor is selected, specific hydraulic calculations can be prepared. Factors such as classification of the hazard, amount of water supply available and the density or concentration to be achieved by the system are to be included with the submission of the shop drawings. Specific data sheets identifying sprinklers, pipe dimensions, power requirements for smoke detectors, etc., should also be included with the submission.

106.1.2 Means of egress. The construction documents shall show in sufficient detail the location, construction, size and character of all portions of the means of egress in compliance with the provisions of this code. In other than occupancies in Groups R-2, R-3, as applicable in Section 101.2, and I-1, the construction documents shall designate the number of occupants to be accommodated on every floor, and in all rooms and spaces.

❖ The complete means of egress system is required to be indicated on the plans to permit the building official to initiate a review and identify pertinent code requirements for each component. Additionally, requiring such information to be reflected in the construction documents requires the designer not only to become familiar with the code, but also to be aware of egress principles, concepts and purposes. The need to ensure that the means of egress leads to a public way is also a consideration during the plan review. Such an evaluation cannot be made without the inclusion of a site plan, as required by Section 106.2.

Information essential for determining the required capacity of the egress components (see Section 1005) and the number of egress components required from a space (see Sections 1014.1 and 1018.1) must be provided. The designer must be aware of the occupancy of a space and properly identify that, along with its resultant occupant load, on the construction documents. In occupancies in Groups I-1, R-2 and R-3, the occupant load can be readily determined with little difference in the number so that the designation of the occupant load on the construction documents is not required.

106.1.3 Exterior wall envelope. Construction documents for all buildings shall describe the exterior wall envelope in sufficient detail to determine compliance with this code. The construction documents shall provide details of the exterior wall envelope as required, including flashing, intersections with dissimilar materials, corners, end details, control joints, intersections at roof, eaves or parapets, means of drainage, water-resistive membrane and details around openings.

The construction documents shall include manufacturer's installation instructions that provide supporting documentation that the proposed penetration and opening details described in the construction documents maintain the weather resistance of the exterior wall envelope. The supporting documentation shall fully describe the exterior wall system which was tested, where applicable, as well as the test procedure used.

❖ This section specifically identifies details of exterior wall construction that are critical to the weather resistance of the wall and requires those details to be provided on the construction documents. Where the weather resistance of the exterior wall assembly is based on tests, the submitted documentation is to describe the details of the wall envelope and the test procedure that was used. This provides the building official with the information necessary to determine code compliance.

106.2 Site plan. The construction documents submitted with the application for permit shall be accompanied by a site plan showing to scale the size and location of new construction and existing structures on the site, distances from lot lines, the established street grades and the proposed finished grades and, as applicable, flood hazard areas, floodways, and design flood elevations; and it shall be drawn in accordance with an accurate boundary line survey. In the case of demolition, the site plan shall show construction to be demolished and the location and size of existing structures and construction that are to remain on the site or plot. The building official is authorized to waive or modify the requirement for a site plan when the application for permit is for alteration or repair or when otherwise warranted.

❖ Certain code requirements are dependent on the structure's location on the lot (see Sections 506.2, 507, 704, 1023 and 1205) and the topography of the site (see Sections 1104, 1107.7.4 and 1803.3). As a result, a scaled site plan containing the data listed in this section is required to permit review for compliance. The building official can waive the requirement for a site plan when it is not required to determine code compliance, such as work involving only interior alterations or repairs.

106.3 Examination of documents. The building official shall examine or cause to be examined the accompanying construction documents and shall ascertain by such examinations whether the construction indicated and described is in accordance with the requirements of this code and other pertinent laws or ordinances.

❖ The requirements of this section are related to those found in Section 105.3.1 regarding the action of the building official in response to a permit application. The building official can delegate review of the construction

documents to subordinates as provided for in Section 103.3.

106.3.1 Approval of construction documents. When the building official issues a permit, the construction documents shall be approved, in writing or by stamp, as "Reviewed for Code Compliance." One set of construction documents so reviewed shall be retained by the building official. The other set shall be returned to the applicant, shall be kept at the site of work and shall be open to inspection by the building official or a duly authorized representative.

❖ The building official must stamp or otherwise endorse as "Reviewed for Code Compliance" the construction documents on which the permit is based. One set of approved construction documents must be kept on the construction site to serve as the basis for all subsequent inspections. To avoid confusion, the construction documents on the site must be the documents that were approved and stamped. This is because inspections are to be performed with regard to the approved documents, not the code itself. Additionally, the contractor cannot determine compliance with the approved construction documents unless they are readily available. Unless the approved construction documents are available, the inspection should be postponed and work on the project halted.

106.3.2 Previous approvals. This code shall not require changes in the construction documents, construction or designated occupancy of a structure for which a lawful permit has been heretofore issued or otherwise lawfully authorized, and the construction of which has been pursued in good faith within 180 days after the effective date of this code and has not been abandoned.

❖ If a permit is issued and construction proceeds at a normal pace and a new edition of the code is adopted by the legislative body, requiring that the building be constructed to conform to the new code is unreasonable. This section provides for the continuity of permits issued under previous codes, as long as such permits are being "actively prosecuted" subsequent to the effective date of the ordinance adopting this edition of the code.

106.3.3 Phased approval. The building official is authorized to issue a permit for the construction of foundations or any other part of a building or structure before the construction documents for the whole building or structure have been submitted, provided that adequate information and detailed statements have been filed complying with pertinent requirements of this code. The holder of such permit for the foundation or other parts of a building or structure shall proceed at the holder's own risk with the building operation and without assurance that a permit for the entire structure will be granted.

❖ The building official has the authority to issue a partial permit to allow for the practice of "fast tracking" a job. Any construction under a partial permit is "at the holder's own risk" and "without assurance that a permit for the entire structure will be granted." The building official is under no obligation to accept work or issue a complete permit in violation of the code, ordinances or statutes simply because a partial permit had been issued. Fast tracking places an unusual administrative and technical burden on the building official. The purpose is to proceed with construction while the design continues for other aspects of the work. Coordinating and correlating the code aspects into the project in phases requires attention to detail and project tracking so that all code issues are addressed. The coordination of these submittals is the responsibility of the registered design professional in responsible charge described in Section 106.3.4.

106.3.4 Design professional in responsible charge.

106.3.4.1 General. When it is required that documents be prepared by a registered design professional, the building official shall be authorized to require the owner to engage and designate on the building permit application a registered design professional who shall act as the registered design professional in responsible charge. If the circumstances require, the owner shall designate a substitute registered design professional in responsible charge who shall perform the duties required of the original registered design professional in responsible charge. The building official shall be notified in writing by the owner if the registered design professional in responsible charge is changed or is unable to continue to perform the duties.

The registered design professional in responsible charge shall be responsible for reviewing and coordinating submittal documents prepared by others, including phased and deferred submittal items, for compatibility with the design of the building.

Where structural observation is required by Section 1709, the inspection program shall name the individual or firms who are to perform structural observation and describe the stages of construction at which structural observation is to occur (see also duties specified in Section 1704).

❖ At the time of permit application and at various intervals during a project, the code requires detailed technical information to be submitted to the building official. This will vary depending on the complexity of the project, but typically includes the construction documents with supporting information, applications utilizing the phased approval procedure in Section 106.3.3 and reports from engineers, inspectors and testing agencies required in Chapter 17. This section specifically references the special inspection requirements in Section 1704 and the structural observation requirements in Section 1709. Since these documents and reports are prepared by numerous individuals, firms and agencies, it is necessary to have a single person charged with responsibility for coordinating their submittal to the building official. This person is the point of contact for the building official for all information relating to the project. Otherwise, the building official could waste time and effort attempting to locate the source of accurate information

when trying to resolve an issue such as a discrepancy in plans submitted by different designers. The requirement that the owner engage a person to act as the design professional in responsible charge is applicable to projects where the construction documents are required by law to be prepared by a registered design professional (see Section 106.1) and when required by the building official. Small projects that do not involve phased approvals, deferred submittals or special inspection reports (see Section 1704) that cannot be readily administered by the building official and staff does not require the designation of a design professional in responsible charge. The person employed by the owner to act as the design professional in responsible charge must be identified on the permit application, but the owner can change the designated person at any time during the course of the review process or work, provided the building official is so notified in writing.

106.3.4.2 Deferred submittals. For the purposes of this section, deferred submittals are defined as those portions of the design that are not submitted at the time of the application and that are to be submitted to the building official within a specified period.

Deferral of any submittal items shall have the prior approval of the building official. The registered design professional in responsible charge shall list the deferred submittals on the construction documents for review by the building official.

Documents for deferred submittal items shall be submitted to the registered design professional in responsible charge who shall review them and forward them to the building official with a notation indicating that the deferred submittal documents have been reviewed and been found to be in general conformance to the design of the building. The deferred submittal items shall not be installed until the design and submittal documents have been approved by the building official.

❖ Often, especially on larger projects, details of certain building parts are not available at the time of permit issuance because they have not yet been designed; for example, exterior cladding, prefabricated items such as trusses and stairs and the components of fire protection systems (see Section 106.1.1.1). The design professional in responsible charge must identify on the construction documents the items to be included in any deferred submittals. Documents required for the approval of deferred items must be reviewed by the design professional in responsible charge for compatibility with the design of the building, forwarded to the building official with a notation that this is the case and approved by the building official before installation of the items. Sufficient time must be allowed for the approval process. Note that deferred submittals differ from the phased permits described in Section 106.3.3 in that they occur after the permit for the building is issued and are not for work covered by separate permits.

106.4 Amended construction documents. Work shall be installed in accordance with the approved construction documents, and any changes made during construction that are not in compliance with the approved construction documents shall be resubmitted for approval as an amended set of construction documents.

❖ Any amendments to the approved construction documents must be filed before constructing the amended item. In the broadest sense, amendments include all addenda, change orders, revised drawings and marked-up shop drawings. Building officials should maintain a policy that all amendments be submitted for review. Otherwise, a significant amendment may not be submitted because of misinterpretation, resulting in an activity that is not approved and that causes a needless delay in obtaining approval of the finished work.

106.5 Retention of construction documents. One set of approved construction documents shall be retained by the building official for a period of not less than 180 days from date of completion of the permitted work, or as required by state or local laws.

❖ A set of the approved construction documents must be kept by the building official as may be required by state or local laws, but for a period of no less than 180 days after the work is complete. Questions regarding an item shown on the approved documents may arise in the period immediately following completion of the work and the documents should be available for review. See Section 104.7 for requirements to retain other records that are generated as a result of the work.

SECTION 107
TEMPORARY STRUCTURES AND USES

107.1 General. The building official is authorized to issue a permit for temporary structures and temporary uses. Such permits shall be limited as to time of service, but shall not be permitted for more than 180 days. The building official is authorized to grant extensions for demonstrated cause.

❖ In the course of construction or other activities, structures that have a limited service life are often necessary. This section contains the administrative provisions that permit such temporary structures without full compliance with the code requirements for permanently occupied structures. This section should not be confused with the scope of Section 3103, which regulates temporary structures larger than 120 square feet (11 m²) in area.
  This section allows the building official to issue permits for temporary structures or uses. The applicant must specify the time period desired for the temporary structure or use, but the approval period cannot exceed 180 days. Structures or uses that are temporary but are anticipated to be in existence for more than 180 days are required to conform to code requirements for permanent structures and uses. The section also autho-

rizes the building official to grant extensions to this time period if the applicant can provide a valid reason for the extension, which typically includes circumstances beyond the applicant's control. This provision is not intended to be used to circumvent the 180-day limitation.

107.2 Conformance. Temporary structures and uses shall conform to the structural strength, fire safety, means of egress, accessibility, light, ventilation and sanitary requirements of this code as necessary to ensure the public health, safety and general welfare.

❖ This section prescribes those categories of the code that must be complied with, despite the fact that the structure will be removed or the use discontinued at some time in the future. These criteria are essential for measuring the safety of any structure or use, temporary or permanent; therefore, the application of these criteria to a temporary structure cannot be waived.
  "Structural strength" refers to the ability of the temporary structure to resist anticipated live, environmental and dead loads (see Chapter 16). It also applies to anticipated live and dead loads imposed by a temporary use in an existing structure.
  "Fire safety" provisions are those required by Chapters 7, 8 and 9 invoked by virtue of the structure's size, use or location on the property.
  "Means of egress" refers to full compliance with Chapter 10.
  "Accessibility" refers to full compliance with Chapter 11 for making buildings accessible to physically disabled persons, a requirement that is repeated in Section 1103.1.
  "Light, ventilation and sanitary" requirements are those imposed by Chapter 12 of the code or applicable sections of the IPC or IMC.

107.3 Temporary power. The building official is authorized to give permission to temporarily supply and use power in part of an electric installation before such installation has been fully completed and the final certificate of completion has been issued. The part covered by the temporary certificate shall comply with the requirements specified for temporary lighting, heat or power in the ICC *Electrical Code*.

❖ Commonly, the electrical service on most construction sites is installed and energized long before all of the wiring is completed. This procedure allows the power supply to be increased as construction demands; however, temporary permission is not intended to waive the requirements set forth in the ICC EC or NFPA 70. Construction power from the permanent wiring of the building does not require the installation of temporary ground-fault circuit-interrupter (GFCI) protection or the assured equipment grounding program, because the building wiring installed as required by the code should be as safe for use during construction as it would be for use after completion of the building.

107.4 Termination of approval. The building official is authorized to terminate such permit for a temporary structure or use and to order the temporary structure or use to be discontinued.

❖ This section provides the building official with the necessary authority to terminate the permit for a temporary structure or use. The building official can order that a temporary structure be removed or a temporary use be discontinued if conditions of the permit have been violated or the structure or use poses an imminent hazard to the public, in which case the provisions of Section 115 become applicable. This text is important because it allows the building official to act quickly when time is of the essence in order to protect public health, safety and welfare.

SECTION 108
FEES

108.1 Payment of fees. A permit shall not be valid until the fees prescribed by law have been paid, nor shall an amendment to a permit be released until the additional fee, if any, has been paid.

❖ The code anticipates that jurisdictions will establish their own fee schedules. It is the intent that the fees collected by the department for building permit issuance, plan review and inspection be adequate to cover the costs to the department in these areas. If the department has additional duties, then its budget will need to be supplemented from the general fund. This section requires that all fees be paid prior to permit issuance or release of an amendment to a permit. Since department operations are intended to be supported by fees paid by the user of department activities, it is important that these fees are received before incurring any expense. This philosophy has resulted in some departments having fees paid prior to the performance of two areas of work: plan review and inspection.

108.2 Schedule of permit fees. On buildings, structures, electrical, gas, mechanical, and plumbing systems or alterations requiring a permit, a fee for each permit shall be paid as required, in accordance with the schedule as established by the applicable governing authority.

❖ The jurisdiction inserts its desired fee schedule at this location. The fees are established by law, such as in an ordinance adopting the code (see page v of the code for a sample), a separate ordinance or legally promulgated regulation, as required by state or local law. Fee schedules are often based on a valuation of the work to be performed. This concept is based on the proposition that the valuation of a project is related to the amount of work to be expended in plan review, inspections and administering the permit, plus an excess to cover the department overhead.

To assist jurisdictions in establishing some uniformity in fees, building evaluation data are published periodically in ICC's *Building Safety Journal*.

108.3 Building permit valuations. The applicant for a permit shall provide an estimated permit value at time of application. Permit valuations shall include total value of work, including materials and labor, for which the permit is being issued, such as electrical, gas, mechanical, plumbing equipment and permanent systems. If, in the opinion of the building official, the valuation is underestimated on the application, the permit shall be denied, unless the applicant can show detailed estimates to meet the approval of the building official. Final building permit valuation shall be set by the building official.

❖ As indicated in Section 108.2, jurisdictions usually base their fees on the value of the work being performed. This section, therefore, requires the applicant to provide this figure, which is to include the total value of the work, including materials and labor, for which the permit is sought. If the building official believes that the value provided by the applicant is underestimated, the permit is to be denied unless the applicant can substantiate the value by providing detailed estimates of the work to the satisfaction of the building official. For the construction of new buildings, the building valuation data referred to in Section 108.2 can be used by the building official as a yardstick against which to compare the applicant's estimate.

108.4 Work commencing before permit issuance. Any person who commences any work on a building, structure, electrical, gas, mechanical or plumbing system before obtaining the necessary permits shall be subject to a fee established by the building official that shall be in addition to the required permit fees.

❖ The building official will incur certain costs (i.e., inspection time and administrative) when investigating and citing a person who has commenced work without having obtained a permit. The building official is, therefore, entitled to recover these costs by establishing a fee, in addition to that collected when the required permit is issued, to be imposed on the responsible party. Note that this is not a penalty, as described in Section 113.4, for which the person can also be liable.

108.5 Related fees. The payment of the fee for the construction, alteration, removal or demolition for work done in connection to or concurrently with the work authorized by a building permit shall not relieve the applicant or holder of the permit from the payment of other fees that are prescribed by law.

❖ The fees for a building permit may be in addition to other fees required by the jurisdiction or others for related items, such as sewer connections, water service taps, driveways and signs. It cannot be construed that the building permit fee includes these other items.

108.6 Refunds. The building official is authorized to establish a refund policy.

❖ This section allows for a refund of fees, which may be full or partial, typically resulting from the revocation, abandonment or discontinuance of a building project for which a permit has been issued and fees have been collected. The refund of fees should be related to the cost of enforcement services not provided because of the termination of the project. The building official, when authorizing a fee refund, is authorizing the disbursement of public funds; therefore, the request for a refund must be in writing and for good cause.

SECTION 109
INSPECTIONS

109.1 General. Construction or work for which a permit is required shall be subject to inspection by the building official and such construction or work shall remain accessible and exposed for inspection purposes until approved. Approval as a result of an inspection shall not be construed to be an approval of a violation of the provisions of this code or of other ordinances of the jurisdiction. Inspections presuming to give authority to violate or cancel the provisions of this code or of other ordinances of the jurisdiction shall not be valid. It shall be the duty of the permit applicant to cause the work to remain accessible and exposed for inspection purposes. Neither the building official nor the jurisdiction shall be liable for expense entailed in the removal or replacement of any material required to allow inspection.

❖ The inspection function is one of the more important aspects of building department operations. This section authorizes the building official to inspect the work for which a permit has been issued and requires that the work to be inspected remain accessible to the building official until inspected and approved. Any expense incurred in removing or replacing material that conceals an item to be inspected is not the responsibility of the building official or the jurisdiction. As with the issuance of permits (see Section 105.4), approval as a result of an inspection is not a license to violate the code and an approval in violation of the code does not relieve the applicant from complying with the code and is not valid.

109.2 Preliminary inspection. Before issuing a permit, the building official is authorized to examine or cause to be examined buildings, structures and sites for which an application has been filed.

❖ The building official is granted authority to inspect the site before permit issuance. This may be necessary to verify existing conditions that impact the plan review and permit approval. This section provides the building official with the right-of-entry authority that otherwise does not occur until after the permit is issued (see Section 104.6).

109.3 Required inspections. The building official, upon notification, shall make the inspections set forth in Sections 109.3.1 through 109.3.10.

❖ The building official is required to verify that the building is constructed in accordance with the approved construction documents. It is the responsibility of the permit holder to notify the building official when the item is ready for inspection. The inspections that are necessary to provide such verification are listed in the following sections, with the caveat in Section 109.3.8 that inspections in addition to those listed here may be required depending on the work involved.

109.3.1 Footing and foundation inspection. Footing and foundation inspections shall be made after excavations for footings are complete and any required reinforcing steel is in place. For concrete foundations, any required forms shall be in place prior to inspection. Materials for the foundation shall be on the job, except where concrete is ready mixed in accordance with ASTM C 94, the concrete need not be on the job.

❖ It is necessary for the building official to inspect the soil upon which the footing or foundation is to be placed. This inspection also includes any reinforcing steel, concrete forms and materials to be used in the foundation, except for ready-mixed concrete that is prepared off site.

109.3.2 Concrete slab and under-floor inspection. Concrete slab and under-floor inspections shall be made after in-slab or under-floor reinforcing steel and building service equipment, conduit, piping accessories and other ancillary equipment items are in place, but before any concrete is placed or floor sheathing installed, including the subfloor.

❖ The building official must be able to inspect the soil and any required under-slab drainage, waterproofing or dampproofing material, as well as reinforcing steel, conduit, piping and other service equipment embedded in or installed below a slab prior to placing the concrete. Similarly, items installed below a floor system other than concrete must be inspected before they are concealed by the floor sheathing or subfloor.

109.3.3 Lowest floor elevation. In flood hazard areas, upon placement of the lowest floor, including the basement, and prior to further vertical construction, the elevation certification required in Section 1612.5 shall be submitted to the building official.

❖ Where a structure is located in a flood hazard area, as established in Section 1612.3, the building official must be provided with certification that either the lowest floor elevation (for structures located in flood hazard areas not subject to high-velocity wave action) or the elevation of the lowest horizontal structural member (for structures located in flood hazard areas subject to high-velocity wave action) is in compliance with Section 1612. This certification must be submitted prior to any construction proceeding above this level.

109.3.4 Frame inspection. Framing inspections shall be made after the roof deck or sheathing, all framing, fireblocking and bracing are in place and pipes, chimneys and vents to be concealed are complete and the rough electrical, plumbing, heating wires, pipes and ducts are approved.

❖ This section requires that the building official be able to inspect the framing members, such as studs, joists, rafters and girders and other items, such as vents and chimneys, that will be concealed by wall construction.

Rough electrical work, plumbing, heating wires, pipes and ducts must have already been approved in accordance with the applicable codes prior to this inspection.

109.3.5 Lath and gypsum board inspection. Lath and gypsum board inspections shall be made after lathing and gypsum board, interior and exterior, is in place, but before any plastering is applied or gypsum board joints and fasteners are taped and finished.

> **Exception:** Gypsum board that is not part of a fire-resistance-rated assembly or a shear assembly.

❖ In order to verify that lath and gypsum board is properly attached to framing members, it is necessary for the building official to be able to conduct an inspection before the plaster or joint finish material is applied. This is required only for gypsum board that is part of either a fire-resistant assembly or a shear wall.

109.3.6 Fire-resistant penetrations. Protection of joints and penetrations in fire-resistance-rated assemblies shall not be concealed from view until inspected and approved.

❖ The building official must have an opportunity to inspect joint protection required by Section 713 and penetration protection required by Section 712 for fire-resistance-rated assemblies before concealed from view.

109.3.7 Energy efficiency inspections. Inspections shall be made to determine compliance with Chapter 13 and shall include, but not be limited to, inspections for: envelope insulation R and U values, fenestration U value, duct system R value, and HVAC and water-heating equipment efficiency.

❖ Items installed in a building that are required by the IECC to comply with certain criteria, such as insulation material, windows, HVAC and water-heating equipment, must be inspected and approved.

109.3.8 Other inspections. In addition to the inspections specified above, the building official is authorized to make or require other inspections of any construction work to ascertain compliance with the provisions of this code and other laws that are enforced by the department of building safety.

❖ Any item regulated by the code is subject to inspection by the building official to determine compliance with the applicable code provision, and no list can include all items in a given building. This section, therefore, gives the building official the authority to inspect any regulated items.

109.3.9 Special inspections. For special inspections, see Section 1704.

❖ Special inspections are to be provided by the owner for the types of work required in Section 1704. The building official is to approve special inspectors and verify that the required special inspections have been conducted. See the commentary to Section 1704 for a complete discussion of this topic.

109.3.10 Final inspection. The final inspection shall be made after all work required by the building permit is completed.

❖ Upon completion of the work for which the permit has been issued and before issuance of the certificate of occupancy required by Section 110.3, a final inspection is to be made. All violations of the approved construction documents and permit are to be noted and the holder of the permit is to be notified of the discrepancies.

109.4 Inspection agencies. The building official is authorized to accept reports of approved inspection agencies, provided such agencies satisfy the requirements as to qualifications and reliability.

❖ As an alternative to the building official conducting the inspection, he or she is permitted to accept inspections of and reports by approved inspection agencies. Appropriate criteria on which to base approval of inspection agencies can be found in Section 1703.

109.5 Inspection requests. It shall be the duty of the holder of the building permit or their duly authorized agent to notify the building official when work is ready for inspection. It shall be the duty of the permit holder to provide access to and means for inspections of such work that are required by this code.

❖ It is the responsibility of the permit holder or other authorized person, such as the contractor performing the work, to arrange for the required inspections when completed work is ready and to allow for sufficient time for the building official to schedule a visit to the site to prevent work from being concealed prior to being inspected. Access to the work to be inspected must be provided, including any special means such as a ladder.

109.6 Approval required. Work shall not be done beyond the point indicated in each successive inspection without first obtaining the approval of the building official. The building official, upon notification, shall make the requested inspections and shall either indicate the portion of the construction that is satisfactory as completed, or notify the permit holder or his or her agent wherein the same fails to comply with this code. Any portions that do not comply shall be corrected and such portion shall not be covered or concealed until authorized by the building official.

❖ This section establishes that work cannot progress beyond the point of a required inspection without the building official's approval. Upon making the inspection, the building official must either approve the completed work or notify the permit holder or other responsible party of that which does not comply with the code. Approvals and notices of noncompliance must be in writing, as required by Section 104.4, to avoid any misunderstanding as to what is required. Any item not approved cannot be concealed until it has been corrected and approved by the building official.

SECTION 110
CERTIFICATE OF OCCUPANCY

110.1 Use and occupancy. No building or structure shall be used or occupied, and no change in the existing occupancy classification of a building or structure or portion thereof shall be made until the building official has issued a certificate of occupancy therefor as provided herein. Issuance of a certificate of occupancy shall not be construed as an approval of a violation of the provisions of this code or of other ordinances of the jurisdiction.

❖ This section establishes that a new building or structure cannot be occupied until a certificate of occupancy is issued by the building official, which reflects the conclusion of the work allowed by the building permit. Also, no change in occupancy of an existing building is permitted without first obtaining a certificate of occupancy for the new use.

The tool that the building official uses to control the uses and occupancies of various buildings and structures within the jurisdiction is the certificate of occupancy. It is unlawful to use or occupy a building or structure unless a certificate of occupancy has been issued. Its issuance does not relieve the building owner from the responsibility for correcting any code violation that may exist.

110.2 Certificate issued. After the building official inspects the building or structure and finds no violations of the provisions of this code or other laws that are enforced by the department of building safety, the building official shall issue a certificate of occupancy that contains the following:

1. The building permit number.
2. The address of the structure.
3. The name and address of the owner.
4. A description of that portion of the structure for which the certificate is issued.
5. A statement that the described portion of the structure has been inspected for compliance with the requirements of this code for the occupancy and division of occupancy and the use for which the proposed occupancy is classified.
6. The name of the building official.
7. The edition of the code under which the permit was issued.
8. The use and occupancy, in accordance with the provisions of Chapter 3.
9. The type of construction as defined in Chapter 6.
10. The design occupant load.
11. If an automatic sprinkler system is provided, whether the sprinkler system is required.
12. Any special stipulations and conditions of the building permit.

❖ The building official is required to issue a certificate of occupancy after a successful final inspection has been completed and all deficiencies and violations have been resolved. This section lists the information that must be included on the certificate. This information is useful to both the building official and the owner because it indicates the criteria under which the structure was evaluated and approved at the time the certificate was issued. This is important when applying Chapter 34 to existing buildings.

110.3 Temporary occupancy. The building official is authorized to issue a temporary certificate of occupancy before the completion of the entire work covered by the permit, provided that such portion or portions shall be occupied safely. The building official shall set a time period during which the temporary certificate of occupancy is valid.

❖ The building official is permitted to issue a temporary certificate of occupancy for all or a portion of a building prior to the completion of all work. Such certification is to be issued only when the building or portion in question can be safely occupied prior to full completion. The certification is intended to acknowledge that some building features may not be completed even though the building is safe for occupancy, or that a portion of the building can be safely occupied while work continues in another area. This provision precludes the occupancy of a building or structure that does not contain all of the required fire protection systems and means of egress. Temporary certificates should be issued only when incidental construction remains, such as site work and interior work that is not regulated by the code and exterior decoration not necessary to the integrity of the building envelope. The building official should view the issuance of a temporary certificate of occupancy as substantial an act as the issuance of the final certificate. Indeed, the issuance of a temporary certificate of occupancy offers a greater potential for conflict because once the building or structure is occupied, it is very difficult to remove the occupants through legal means. The certificate must specify the time period for which it is valid.

110.4 Revocation. The building official is authorized to, in writing, suspend or revoke a certificate of occupancy or completion issued under the provisions of this code wherever the certificate is issued in error, or on the basis of incorrect information supplied, or where it is determined that the building or structure or portion thereof is in violation of any ordinance or regulation or any of the provisions of this code.

❖ The building official is authorized to, in writing, suspend or revoke a certificate of occupancy or completion issued under the provisions of this code wherever the certificate is issued in error, on the basis of incorrect information supplied or where it is determined that the building or structure or portion thereof is in violation of any ordinance, regulation or any of the provisions of this code.

This section is needed to give the building official the authority to revoke a certificate of occupancy for the reasons indicated in the code text. The building official may also suspend the certificate of occupancy until all of the code violations are corrected.

SECTION 111
SERVICE UTILITIES

111.1 Connection of service utilities. No person shall make connections from a utility, source of energy, fuel or power to any building or system that is regulated by this code for which a permit is required, until released by the building official.

❖ This section establishes the authority of the building official to approve utility connections to a building for items such as water, sewer, electricity, gas and steam, and to require their disconnection when hazardous conditions or emergencies exist.

The approval of the building official is required before a connection can be made from a utility to a building system that is regulated by the code, including those referenced in Section 101.4. This includes utilities supplying water, sewer, electricity, gas and steam services. For the protection of building occupants, including workers, such systems must have had final inspection approvals, except as allowed by Section 111.2 for temporary connections.

111.2 Temporary connection. The building official shall have the authority to authorize the temporary connection of the building or system to the utility source of energy, fuel or power.

❖ The building official is permitted to issue temporary authorization to make connections to the public utility system prior to the completion of all work. This acknowledges that, because of seasonal limitations, time constraints or the need for testing or partial operation of equipment, some building systems may be safely connected even though the building is not suitable for final occupancy. The temporary connection and utilization of connected equipment should be approved when the requesting permit holder has demonstrated to the building official's satisfaction that public health, safety and welfare will not be endangered.

111.3 Authority to disconnect service utilities. The building official shall have the authority to authorize disconnection of utility service to the building, structure or system regulated by this code and the codes referenced in case of emergency where necessary to eliminate an immediate hazard to life or property. The building official shall notify the serving utility, and wherever possible the owner and occupant of the building, structure or service system of the decision to disconnect prior to taking such action. If not notified prior to disconnecting, the owner or occupant of the building, structure or service system shall be notified in writing, as soon as practical thereafter.

❖ Disconnection of one or more of a building's utility services is the most radical method of hazard abatement available to the building official and should be reserved for cases in which all other lesser remedies have proven ineffective. Such an action must be preceded by written notice to the utility and the owner and occupants of the building. Disconnection must be accomplished within the time frame established by the building official in the notice. When the hazard to the public health, safety or welfare is so imminent as to mandate immediate disconnection, the building official has the authority and even the obligation to cause disconnection without notice. In such cases, the owner or occupants must be given written notice as soon as possible.

SECTION 112
BOARD OF APPEALS

112.1 General. In order to hear and decide appeals of orders, decisions or determinations made by the building official relative to the application and interpretation of this code, there shall be and is hereby created a board of appeals. The board of appeals shall be appointed by the governing body and shall hold office at its pleasure. The board shall adopt rules of procedure for conducting its business.

❖ This section provides an aggrieved party with a material interest in the decision of the building official a process to appeal such a decision before a board of appeals. This provides a forum, other than the court of jurisdiction, in which to review the building official's actions.

This section literally allows any person to appeal a decision of the building official. In practice, this section has been interpreted to permit appeals only by those aggrieved parties with a material or definitive interest in the decision of the building official. An aggrieved party may not appeal a code requirement per se. The intent of the appeal process is not to waive or set aside a code requirement; rather it is intended to provide a means of reviewing a building official's decision on an interpretation or application of the code or to review the equivalency of protection to the code requirements. The members of the appeals board are appointed by the "governing body" of the jurisdiction, typically a council or administrator, such as a mayor or city manager, and remain members until removed from office. The board must establish procedures for electing a chairperson, scheduling and conducting meetings and administration. Note that Appendix B contains complete, detailed requirements for creating an appeals board, including number of members, qualifications and administrative procedures. Jurisdictions desiring to utilize these requirements must include Appendix B in their adopting ordinance.

112.2 Limitations on authority. An application for appeal shall be based on a claim that the true intent of this code or the rules legally adopted thereunder have been incorrectly interpreted, the provisions of this code do not fully apply or an equally good or better form of construction is proposed. The board shall have no authority to waive requirements of this code.

❖ This section establishes the grounds for an appeal, which claims that the building official has misinterpreted or misapplied a code provision. The board is not allowed to set aside any of the technical requirements of the code; however, it is allowed to consider alternative methods of compliance with the technical requirements (see Section 104.11).

112.3 Qualifications. The board of appeals shall consist of members who are qualified by experience and training to pass on matters pertaining to building construction and are not employees of the jurisdiction.

❖ It is important that the decisions of the appeals board are based purely on the technical merits involved in an appeal. It is not the place for policy or political deliberations. The members of the appeals board are, therefore, expected to have experience in building construction matters. Appendix B provides more detailed qualifications for appeals board members and can be adopted by jurisdictions desiring that level of expertise.

SECTION 113
VIOLATIONS

113.1 Unlawful acts. It shall be unlawful for any person, firm or corporation to erect, construct, alter, extend, repair, move, remove, demolish or occupy any building, structure or equipment regulated by this code, or cause same to be done, in conflict with or in violation of any of the provisions of this code.

❖ Violations of the code are prohibited, and form the basis for all citations and correction notices.

113.2 Notice of violation. The building official is authorized to serve a notice of violation or order on the person responsible for the erection, construction, alteration, extension, repair, moving, removal, demolition or occupancy of a building or structure in violation of the provisions of this code, or in violation of a permit or certificate issued under the provisions of this code. Such order shall direct the discontinuance of the illegal action or condition and the abatement of the violation.

❖ The building official is required to notify the person responsible for the erection or use of a building found to be in violation of the code. The section that is allegedly being violated must be cited so that the responsible party can respond to the notice.

113.3 Prosecution of violation. If the notice of violation is not complied with promptly, the building official is authorized to request the legal counsel of the jurisdiction to institute the appropriate proceeding at law or in equity to restrain, correct or abate such violation, or to require the removal or termination of the unlawful occupancy of the building or structure in violation of the provisions of this code or of the order or direction made pursuant thereto.

❖ The building official must pursue, through the use of legal counsel of the jurisdiction, legal means to correct the violation. This is not optional.

Any extensions of time so that the violations may be corrected voluntarily must be for a reasonable, bona fide cause or the building official may be subject to criticism for "arbitrary and capricious" actions. In general, it is better to have a standard time limitation for correction of violations. Departures from this standard must be for a clear and reasonable purpose, usually stated in writing by the violator.

113.4 Violation penalties. Any person who violates a provision of this code or fails to comply with any of the requirements thereof or who erects, constructs, alters or repairs a building or structure in violation of the approved construction documents or directive of the building official, or of a permit or certificate issued under the provisions of this code, shall be subject to penalties as prescribed by law.

❖ Penalties for violating provisions of the code are typically contained in state law, particularly if the code is adopted at that level, and the building department must follow those procedures. If there is no such procedure already in effect, one must be established with the aid of legal counsel.

SECTION 114
STOP WORK ORDER

114.1 Authority. Whenever the building official finds any work regulated by this code being performed in a manner either contrary to the provisions of this code or dangerous or unsafe, the building official is authorized to issue a stop work order.

❖ This section provides for the suspension of work for which a permit was issued, pending the removal or correction of a severe violation or unsafe condition identified by the building official.

Normally, correction notices, issued in accordance with Section 109.6, are used to inform the permit holder of code violations. Stop work orders are issued when enforcement can be accomplished no other way or when a dangerous condition exists.

114.2 Issuance. The stop work order shall be in writing and shall be given to the owner of the property involved, or to the owner's agent, or to the person doing the work. Upon issuance of a stop work order, the cited work shall immediately cease. The stop work order shall state the reason for the order, and the conditions under which the cited work will be permitted to resume.

❖ Upon receipt of a violation notice from the building official, all construction activities identified in the notice must immediately cease, except as expressly permitted to correct the violation.

114.3 Unlawful continuance. Any person who shall continue any work after having been served with a stop work order, except such work as that person is directed to perform to remove a violation or unsafe condition, shall be subject to penalties as prescribed by law.

❖ This section states that the work in violation must terminate and that all other work, except that which is necessary to correct the violation or unsafe condition, must cease as well. As determined by the municipality or state, a penalty may be assessed for failure to comply with this section.

SECTION 115
UNSAFE STRUCTURES AND EQUIPMENT

115.1 Conditions. Structures or existing equipment that are or hereafter become unsafe, insanitary or deficient because of inadequate means of egress facilities, inadequate light and ventilation, or which constitute a fire hazard, or are otherwise dangerous to human life or the public welfare, or that involve illegal or improper occupancy or inadequate maintenance, shall be deemed an unsafe condition. Unsafe structures shall be taken down and removed or made safe, as the building official deems necessary and as provided for in this section. A vacant structure that is not secured against entry shall be deemed unsafe.

❖ This section describes the responsibility of the building official to investigate reports of unsafe structures and equipment and provides criteria for such determination.

   Unsafe structures are defined as buildings or structures that are insanitary, deficient in light and ventilation or adequate exit facilities, constitute a fire hazard or are otherwise dangerous to human life.

   This section establishes that unsafe buildings can result from illegal or improper occupancies. For example, prima facie evidence of an unsafe structure is an unsecured (open at door or window) vacant building. All unsafe buildings must either be demolished or made safe and secure as deemed appropriate by the building official.

115.2 Record. The building official shall cause a report to be filed on an unsafe condition. The report shall state the occupancy of the structure and the nature of the unsafe condition.

❖ The building official must file a report on each investigation of unsafe conditions, stating the occupancy of the structure and the nature of the unsafe condition. This report provides the basis for the notice described in Section 115.3.

115.3 Notice. If an unsafe condition is found, the building official shall serve on the owner, agent or person in control of the structure, a written notice that describes the condition deemed unsafe and specifies the required repairs or improvements to be made to abate the unsafe condition, or that requires the unsafe structure to be demolished within a stipulated time. Such notice shall require the person thus notified to declare immediately to the building official acceptance or rejection of the terms of the order.

❖ When a building or structure is deemed unsafe, the building official is required to notify the owner or agent of the building as the first step in correcting the problem. Such notice must describe the necessary repairs and improvements to correct the deficiency or must require the unsafe building or structure to be demolished in a specified time in order to provide for public health, safety and welfare. Additionally, such notice requires the immediate response of the owner or agent. If the owner or agent is not available, public notice of such declaration should suffice for the purposes of complying with this section (see Section 115.4). The building offi-

cial may also determine that immediate work is necessary to correct an unsafe condition and seek a lien against the building or structure to compensate the municipality for the cost of remedial action.

115.4 Method of service. Such notice shall be deemed properly served if a copy thereof is (a) delivered to the owner personally; (b) sent by certified or registered mail addressed to the owner at the last known address with the return receipt requested; or (c) delivered in any other manner as prescribed by local law. If the certified or registered letter is returned showing that the letter was not delivered, a copy thereof shall be posted in a conspicuous place in or about the structure affected by such notice. Service of such notice in the foregoing manner upon the owner's agent or upon the person responsible for the structure shall constitute service of notice upon the owner.

❖ The notice must be delivered personally to the owner. If the owner or agent cannot be located, additional procedures are established, including posting the unsafe notice on the premises in question. Such action may be considered the equivalent of personal notice; however, it may or may not be deemed by the courts as representing a "good faith" effort to notify. In addition to complying with this section, therefore, public notice through the use of newspapers and other postings in a prominent location at the government center should be used.

115.5 Restoration. The structure or equipment determined to be unsafe by the building official is permitted to be restored to a safe condition. To the extent that repairs, alterations or additions are made or a change of occupancy occurs during the restoration of the structure, such repairs, alterations, additions or change of occupancy shall comply with the requirements of Section 105.2.2 and Chapter 34.

❖ This section provides that unsafe structures may be restored to a safe condition. This means that the cause of the unsafe structure notice can be abated without the structure being required to comply fully with the provisions for new construction. Any work done to eliminate the unsafe condition, as well as any change in occupancy that may occur, must comply with the code.

Bibliography

The following resource materials are referenced in this chapter or are relevant to the subject matter addressed in this chapter.

Legal Aspects of Code Administration. Country Club Hills, IL: International Code Council, 2002.

NFPA 14-00, *Standpipe and Hose Systems.* Quincy, MA: National Fire Protection Association, 2000.

NFPA 70-02, *National Electrical Code.* Quincy, MA: National Fire Protection Association, 2002.

NFPA 409-95, *Standard on Aircraft.* Quincy, MA: National Fire Protection Association, 1995.

Readings in Code Administration, Volume 1: History/Philosophy/Law. Country Club Hills, IL: Building Officials and Code Administrators International, Inc., 1974.

Rhyne, Charles S. *Survey of the Law and Building Codes*. The American Institute of Architects and the National Association of Home Builders.

Chapter 2:
Definitions

General Comments

All terms defined in the code are listed alphabetically in Chapter 2. The actual definitions of the terms are located as follows:

Where a term is used in more than one chapter, its definition appears in Chapter 2. Of the more than 500 words, terms and phrases defined in the code, over 40 are defined in Chapter 2.

Where a term is unique or primarily pertains to a single chapter, its definition appears within that chapter. In many chapters, the second section is devoted to definitions. For example, definitions applicable to means of egress are found in Section 1002.

Where a term is unique to a single section or subsection of a chapter, its definition appears within that section or subsection. For example, definitions applicable to stages and platforms are found in Section 410.2 (see Section 410).

Purpose

Codes, by their very nature, are technical documents. As such, literally every word, term and punctuation mark can add to or change the meaning of the intended result. This is even more so with a performance-based code where the desired result often takes on more importance than the specific words. Furthermore, the code, with its broad scope of applicability, includes terms inherent in a variety of construction disciplines. These terms often have multiple meanings depending on the context or discipline being used at the time. For these reasons, it is necessary to maintain a consensus on the specific meaning of terms contained in the code. Chapter 2 performs this function by stating clearly what specific terms mean for the purpose of the code.

SECTION 201
GENERAL

❖ This section contains language and provisions that are supplemental to the use of Chapter 2. It gives guidance to the use of the defined words relevant to tense, gender and plurality. Finally, this section provides direction on how to apply terms that are not defined in the code.

201.1 Scope. Unless otherwise expressly stated, the following words and terms shall, for the purposes of this code, have the meanings shown in this chapter.

❖ The use of words and terms in the code is governed by the provisions of this section. This includes code-defined terms as well as those terms that are not.

201.2 Interchangeability. Words used in the present tense include the future; words stated in the masculine gender include the feminine and neuter; the singular number includes the plural and the plural, the singular.

❖ While the definitions contained or referenced in Chapter 2 are to be taken literally, gender and tense are interchangeable.

201.3 Terms defined in other codes. Where terms are not defined in this code and are defined in the *International Fuel Gas Code, International Fire Code, International Mechanical Code*

or *International Plumbing Code,* such terms shall have the meanings ascribed to them as in those codes.

❖ Definitions that are applicable in other *International Codes®* are applicable everywhere the term is used in the code. Definitions of terms can help in the understanding and application of code requirements.

201.4 Terms not defined. Where terms are not defined through the methods authorized by this section, such terms shall have ordinarily accepted meanings such as the context implies.

❖ Words or terms not defined within the *International Code* series are intended to be applied based on their "ordinarily accepted meanings." The intent of this statement is that a dictionary definition may suffice, provided it is in context. Oftentimes, construction terms used throughout the code are not specifically defined in the code or even in a dictionary. In such a case, the definitions contained in the referenced standards (see Chapter 35) and published textbooks on the subject in question are good resources.

SECTION 202
DEFINITIONS

❖ This portion of the commentary addresses only those terms whose definitions appear in Chapter 2. The commentary for definitions that are located elsewhere in the code can be found in the indicated sections that contain those definitions.

ACCESSIBLE. See Section 1102.1.

ACCESSIBLE MEANS OF EGRESS. See Section 1002.1.

ACCESSIBLE ROUTE. See Section 1102.1.

ACCESSIBLE UNIT. See Section 1102.

ACCREDITATION BODY. See Section 2302.1.

ACTIVE FAULT/ACTIVE FAULT TRACE. See Section 1613.1.

ADDITION. An extension or increase in floor area or height of a building or structure.

❖ An extension or increase in floor area or height of a building or structure.
   This term is used to describe the condition when the floor area or height of an existing building or structure is increased (see Chapter 34). This term is only applicable to existing buildings, never new ones.

ADHERED MASONRY VENEER. See Section 1402.1.

ADJUSTED SHEAR RESISTANCE. (Steel Construction). See Section 2202.1.

ADJUSTED SHEAR RESISTANCE. (Wood Construction). See Section 2302.1.

ADMIXTURE. See Section 1902.1.

ADOBE CONSTRUCTION. See Section 2102.1.

   Stabilized adobe. See Section 2102.1.

   Unstabilized adobe. See Section 2102.1.

[F] AEROSOL. See Section 307.2.

   Level 1 aerosol products. See Section 307.2.

   Level 2 aerosol products. See Section 307.2.

   Level 3 aerosol products. See Section 307.2.

[F] AEROSOL CONTAINER. See Section 307.2.

AGGREGATE. See Section 1902.1.

AGGREGATE, LIGHTWEIGHT. See Section 1902.1.

AGRICULTURAL, BUILDING. A structure designed and constructed to house farm implements, hay, grain, poultry, livestock or other horticultural products. This structure shall not be a place of human habitation or a place of employment where ag-

ricultural products are processed, treated or packaged, nor shall it be a place used by the public.

❖ This definition is needed for the proper application of the utility and miscellaneous occupancy group provisions. The use of the building is quite restricted such that buildings that include habitable or public spaces are not agricultural buildings by definition.

AIR-INFLATED STRUCTURE. See Section 3102.2.

AIR-SUPPORTED STRUCTURE. See Section 3102.2.

   Double skin. See Section 3102.2.

   Single skin. See Section 3102.2.

AISLE ACCESSWAY. See Section 1002.1.

[F] ALARM NOTIFICATION APPLIANCE. See Section 902.1.

[F] ALARM SIGNAL. See Section 902.1.

[F] ALARM VERIFICATION FEATURE. See Section 902.1.

ALLEY. See "Public way.

ALLOWABLE STRESS DESIGN. See Section 1602.1.

ALTERATION. Any construction or renovation to an existing structure other than repair or addition.

❖ The code utilizes this term to reflect construction operations intended for an existing building (see Chapter 34), but not within the scope of an addition or repair (see the definitions of "Addition" and "Repair").

ALTERNATING TREAD DEVICE. See Section 1002.1.

ANCHOR. See Section 2102.1.

ANCHOR BUILDING. See Section 402.2.

ANCHORED MASONRY VENEER. See Section 1402.1.

ANNULAR SPACE. See Section 702.1.

[F] ANNUNCIATOR. See Section 902.1.

APPROVED. Acceptable to the building official.

❖ As related to the process of acceptance of building installations, including materials, equipment and construction systems, this definition identifies where ultimate authority rests. Whenever this term is used, it intends that only the enforcing authority can accept a specific installation or component as complying with the code.

APPROVED AGENCY. See Section 1702.1.

APPROVED FABRICATOR. See Section 1702.1.

APPROVED SOURCE. An independent person, firm or corporation, approved by the building official, who is competent and experienced in the application of engineering principles to materials, methods or systems analyses.

❖ The building official sometimes needs to rely on evaluation reports, analyses or other types of reports that purport to validate the use of a material, system or method as complying with the code. This definition establishes that the building official needs to rely on independent, competent individuals or agencies as the source of these reports.

ARCHITECTURAL TERRA COTTA. See Section 2102.1.

AREA. See Section 2102.1.

Bedded. See Section 2102.1.

Gross cross-sectional. See Section 2102.1.

Net cross-sectional. See Section 2102.1.

AREA, BUILDING. See Section 502.1.

AREA OF REFUGE. See Section 1002.1.

AREAWAY. A subsurface space adjacent to a building open at the top or protected at the top by a grating or guard.

❖ Areaways are often constructed to provide access to below-grade building services, including transformers, ventilation shafts and pipe tunnels.

ATRIUM. See Section 404.1.1.

ATTACHMENTS, SEISMIC. See Section 1613.1.

ATTIC. The space between the ceiling beams of the top story and the roof rafters.

❖ The definition of "Attic" identifies the specific portion of a building or structure for the purposes of determining the applicability of requirements that are specific to attics, such as ventilation (see Section 1202) and draftstopping (see Section 716). Additionally, the code has access requirements (see Section 1208) and uniformly distributed live load requirements (see Table 1607.1) for attics. An attic is considered the space or area located immediately below the roof sheathing within the roof framing system of a building. Pitched roof systems, such as gabled, hip, sawtoothed or curved roofs, all create spaces between the roof sheathing and ceiling membrane, which are considered attics.

[F] AUDIBLE ALARM NOTIFICATION APPLIANCE. See Section 902.1.

[F] AUTOMATIC. See Section 902.1.

[F] AUTOMATIC FIRE-EXTINGUISHING SYSTEM. See Section 902.1.

[F] AUTOMATIC SPRINKLER SYSTEM. See Section 902.1.

[F] AVERAGE AMBIENT SOUND LEVEL. See Section 902.1

AWNING. An architectural projection that provides weather protection, identity or decoration and is wholly supported by the building to which it is attached. An awning is comprised of a lightweight, rigid skeleton structure over which a covering is attached.

❖ Similar to a canopy, an awning typically provides weather protection, signage or decoration. But unlike a canopy, an awning relies solely on the building to which it is attached for its means of support. See Section 3105 for general requirements, Section 1607.11 for awning design loads and Section 3202 for encroachment requirements.

BACKING. See Section 1402.1.

BALCONY, EXTERIOR. See Section 1602.1.

[F] BARRICADE. See Section 307.2.

Artificial barricade. See Section 307.2.

Natural barricade. See Section 307.2.

BASE. See Section 1613.1.

BASE FLOOD. See Section 1612.2.

BASE FLOOD ELEVATION. See Section 1612.2.

BASEMENT. That portion of a building that is partly or completely below grade (see "Story above grade plane" and Sections 502.1 and 1612.2).

❖ A basement is a level within a building that has its floor surface below the adjoining ground level. Often due to grading conditions, a basement will also be considered as a story above grade, thereby contributing to the building height (see the commentary to the definition of "Story above grade plane").

BASE SHEAR. See Section 1602.1.

BASIC SEISMIC-FORCE-RESISTING SYSTEMS. See Section 1602.1.

Bearing wall system. See Section 1602.1.

Building frame system. See Section 1602.1.

Dual system. See Section 1602.1.

Inverted pendulum system. See Section 1602.1.

Moment-resisting frame system. See Section 1602.1.

Shear wall-frame interactive system. See Section 1602.1.

BED JOINT. See Section 2102.1.

BLEACHERS. See Section 1002.1.

BOARDING HOUSE. See Section 310.2.

[F] BOILING POINT. See Section 307.2.

BOND BEAM. See Section 2102.1.

BOND REINFORCING. See Section 2102.1.

BOUNDARY ELEMENT. See Sections 1602.1 and 1613.1.

BOUNDARY MEMBERS. See Section 1602.1.

BRACED WALL LINE. See Section 2302.1.

BRACED WALL PANEL. See Section 2302.1.

BRICK. See Section 2102.1.

Calcium silicate (sand lime brick). See Section 2102.1.

Clay or shale. See Section 2102.1.

Concrete. See Section 2102.1.

BRITTLE. See Section 1613.1.

BUILDING. Any structure used or intended for supporting or sheltering any use or occupancy.

❖ The code uses this term to identify those structures that provide shelter for a function or activity.

BUILDING, ENCLOSED. See Section 1609.2.

BUILDING LINE. The line established by law, beyond which a building shall not extend, except as specifically provided by law.

❖ This term defines the limitations or boundaries for construction of a building This line is typically established by a zoning statute or rights-of-way dedication and is not specified in the code.

BUILDING, LOW-RISE. See Section 1609.2.

BUILDING OFFICIAL. The officer or other designated authority charged with the administration and enforcement of this code, or a duly authorized representative.

❖ The statutory power to enforce the code is normally vested in a building department (or the like) of a state, county or municipality whose designated enforcement officer is termed the "building official" (see Section 103.1).

BUILDING, OPEN. See Section 1609.2.

BUILDING, PARTIALLY ENCLOSED. See Section 1609.2.

BUILDING, SIMPLE DIAPHRAGM. See Section 1609.2.

BUILT-UP ROOF COVERING. See Section 1502.1.

BUTTRESS. See Section 2102.1.

CABLE-RESTRAINED, AIR-SUPPORTED STRUCTURE. See Section 3102.2.

CANOPY. An architectural projection that provides weather protection, identity or decoration and is supported by the building to which it is attached and at the outer end by not less than one stanchion. A canopy is comprised of a rigid structure over which a covering is attached.

❖ A canopy is an architectural projection comprised of a rigid structure over which a membrane covering is typically attached providing weather protection, a means of identity or decoration. It is supported by both the building to which it is attached and no less than one stanchion. See Section 3105 for general requirements, Section 1607.11 for design loads and Section 3202 for encroachment requirements.

CANTILEVERED COLUMN SYSTEM. See Section 1602.1.

[F] CARBON DIOXIDE EXTINGUISHING SYSTEMS. See Section 902.1.

CAST STONE. See Section 2102.1.

[F] CEILING LIMIT. See Section 902.1.

CEILING RADIATION DAMPER. See Section 702.1.

CELL. See Section 2102.1.

CEMENT PLASTER. See Section 2502.1.

CEMENTITIOUS MATERIALS. See Section 1902.1.

CERAMIC FIBER BLANKET. See Section 721.1.1.

CERTIFICATE OF COMPLIANCE. See Section 1702.1.

CHIMNEY. See Section 2102.1.

CHIMNEY TYPES. See Section 2102.1.

High-heat appliance type. See Section 2102.1.

Low-heat appliance type. See Section 2102.1.

Masonry type. See Section 2102.1.

Medium-heat appliance type. See Section 2102.1.

CIRCULATION PATH. See Section 1102.1.

CLADDING. See "Components and cladding."

[F] CLEAN AGENT. See Section 902.1.

CLEANOUT. See Section 2102.1.

[F] CLOSED SYSTEM. See Section 307.2.

COLLAR JOINT. See Section 2102.1.

COLLECTOR. See Sections 1613.1 and 2302.1.

COLLECTOR ELEMENTS. See Section 1602.1.

COLUMN. See Section 1902.1.

COLUMN, MASONRY. See Section 2102.1.

COMBINATION FIRE/SMOKE DAMPER. See Section 702.1.

[F] COMBUSTIBLE DUST. See Section 307.2.

[F] COMBUSTIBLE FIBERS. See Section 307.2.

[F] COMBUSTIBLE LIQUID. See Section 307.2.

   Class II. See Section 307.2.

   Class IIIA. See Section 307.2.

   Class IIIB. See Section 307.2.

COMMON PATH OF EGRESS TRAVEL. See Section 1002.1.

COMPONENT. See Section 1613.1.

   Component equipment. See Section 1613.1.

   Component, flexible. See Section 1613.1.

   Component, rigid. See Section 1613.1.

COMPONENTS AND CLADDING. See Section 1609.2.

COMPOSITE MASONRY. See Section 2102.1.

[F] COMPRESSED GAS. See Section 307.2.

COMPRESSIVE STRENGTH OF MASONRY. See Section 2102.1.

CONCRETE. See Section 1902.1.

CONCRETE CARBONATE AGGREGATE. See Section 721.1.1.

CONCRETE, CELLULAR. See Section 721.1.1.

CONCRETE, LIGHTWEIGHT AGGREGATE. See Section 721.1.1.

CONCRETE, PERLITE. See Section 721.1.1.

CONCRETE, SAND-LIGHTWEIGHT. See Section 721.1.1.

CONCRETE, SILICEOUS AGGREGATE. See Section 721.1.1.

CONCRETE (F'_c), SPECIFIED COMPRESSIVE STRENGTH OF. See Section 1902.1.

CONCRETE, VERMICULITE. See Section 721.1.1.

CONFINED REGION. See Section 1602.1.

CONNECTOR. See Section 2102.1.

[F] CONSTANTLY ATTENDED LOCATION. See Section 902.1.

CONSTRUCTION DOCUMENTS. Written, graphic and pictorial documents prepared or assembled for describing the design, location and physical characteristics of the elements of a project necessary for obtaining a building permit.

❖ To determine whether or not proposed construction is in compliance with code requirements, it is necessary that sufficient information be submitted to the building official for review. This typically consists of the drawings (floor plans, elevations, sections, details, etc.), specifications and product information describing the proposed work.

CONSTRUCTION TYPES. See Section 602.

   Type I. See Section 602.2.

   Type II. See Section 602.2.

   Type III. See Section 602.3.

   Type IV. See Section 602.4.

   Type V. See Section 602.5.

[F] CONTINUOUS GAS-DETECTION SYSTEM. See Section 415.2.

CONTRACTION JOINT. See Section 1902.1.

[F] CONTROL AREA. See Section 307.2.

CONTROLLED LOW-STRENGTH MATERIAL. A self-compacted, cementitious material used primarily as a backfill in place of compacted fill.

❖ The definition provided is from ACI 229R-99, *Controlled Low-Strength Materials*. This type of material is known by many "local" names (e.g., flowable fill) and is commonly used in lieu of a compacted backfill. Requirements for its use under the code are outlined in Section 1803.6.

CONVENTIONAL LIGHT-FRAME WOOD CONSTRUCTION. See Section 2302.1.

CORRIDOR. See Section 1002.1.

CORROSION RESISTANCE. The ability of a material to withstand deterioration of its surface or its properties when exposed to its environment.

❖ There are different environments that contain different types of materials to which building construction materials are exposed. "Corrosion resistance" is not an absolute term; it is relative to the building material, and where it is being used. For instance, certain types of plastic polymers might resist corrosion when used in an exterior environment, but might not be resistant to corrosion from certain chemical gases that could be present in a laboratory.

CORROSION RESISTANT. See Section 1502.1.

[F] CORROSIVE. See Section 307.2.

COURT. An open, uncovered space, unobstructed to the sky, bounded on three or more sides by exterior building walls or other enclosing devices.

❖ Though not specifically identified in the definition, the provisions in the code for courts are only applicable to those areas created by the arrangement of exterior walls and used to provide natural light or ventilation (see Section 1205.3 and the definition of "Yard" at the end of this section).

COVER. See Section 2102.1.

COVERED MALL BUILDING. See Section 402.2.

CRIPPLE WALL. See Section 2302.1.

CRYOGENIC FLUID. See Section 307.2.

DALLE GLASS. See Section 2402.1.

DAMPER. See Section 702.1.

DEAD LOADS. See Section 1602.1.

DECK. See Section 1602.1.

DECORATIVE GLASS. See Section 2402.1.

[F] DEFLAGRATION. See Section 307.2.

DEFORMABILITY. See Section 1602.1.
    High deformability element. See Section 1602.1.
    Limited deformability element. See Section 1602.1.
    Low deformability element. See Section 1602.1.

DEFORMATION. See Section 1602.1.
    Limited deformation. See Section 1602.1.
    Ultimate deformation. See Section 1602.1.

DEFORMED REINFORCEMENT. See Section 1902.1.

[F] DELUGE SYSTEM. See Section 902.1.

DESIGN EARTHQUAKE. See Section 1613.1.

DESIGN FLOOD. See Section 1612.2.

DESIGN FLOOD ELEVATION. See Section 1612.2.

DESIGN STRENGTH. See Section 1602.1.

DESIGNATED SEISMIC SYSTEM. See Section 1613.1.

[F] DETACHED STORAGE BUILDING. See Section 307.2.

DETECTABLE WARNING. See Section 1102.1.

[F] DETECTOR, HEAT. See Section 902.1.

[F] DETONATION. See Section 307.2.

DIAPHRAGM. See Sections 1602.1 and 2102.1.
    Diaphragm, blocked. See Sections 1602.1 and 2102.1.
    Diaphragm, boundary. See Section 1602.1.
    Diaphragm, chord. See Section 1602.1.
    Diaphragm, flexible. See Section 1602.1.
    Diaphragm, rigid. See Section 1602.1.
    Diaphragm, unblocked. See Section 2302.1.

DIMENSIONS. See Section 2102.1.
    Actual. See Section 2102.1.
    Nominal. See Section 2102.1.
    Specified. See Section 2102.1.

DISPENSING. See Section 307.2.

DISPLACEMENT. See Section 1613.1.
    Design displacement. See Section 1613.1.

Total design displacement. See Section 1613.1.

Total maximum displacement. See Section 1613.1.

DISPLACEMENT RESTRAINT SYSTEM. See Section 1613.1

DOOR, BALANCED. See Section 1002.1.

DORMITORY. See Section 310.2.

DRAFTSTOP. See Section 702.1.

DRAG STRUT. See Section 2302.1.

[F] DRY-CHEMICAL EXTINGUISHING AGENT. See Section 902.1.

DRY FLOODPROOFING. See Section 1612.2.

DURATION OF LOAD. See Section 1602.1.

DWELLING. A building that contains one or two dwelling units used, intended or designed to be used, rented, leased, let or hired out to be occupied for living purposes.

❖ Dwellings are buildings intended to serve as residences for one or two families. Dwellings can be owner occupied or rented. The term "dwelling," which refers to the building itself, is defined to distinguish it from the term "dwelling unit," which is a single living unit within a building. It is important to recognize that the code is not intended to regulate detached one- and two-family dwellings and townhouses that are no more than three stories in height. These dwellings are regulated by the *International Residential Code®* (IRC®) (see Section 101.2).

DWELLING UNIT. See Section 310.2.

DWELLING UNIT OR SLEEPING UNIT, MULTISTORY. See Section 1102.

DWELLING UNIT OR SLEEPING UNIT, TYPE A. See Section 1102.

DWELLING UNIT OR SLEEPING UNIT, TYPE B. See Section 1102.

EFFECTIVE DAMPING. See Section 1613.1.

EFFECTIVE DEPTH OF SECTION (*d*). See Section 1902.1.

EFFECTIVE HEIGHT. See Section 2102.1.

EFFECTIVE STIFFNESS. See Section 1613.1.

EFFECTIVE WIND AREA. See Section 1609.2.

EGRESS COURT. See Section 1002.1.

ELEMENT. See Section 1602.1.

Ductile element. See Section 1602.1.

Limited ductile element. See Section 1602.1.

Nonductile element. See Section 1602.1.

[F] EMERGENCY ALARM SYSTEM. See Section 902.1.

[F] EMERGENCY CONTROL STATION. See Section 415.2.

EMERGENCY ESCAPE AND RESCUE OPENING. See Section 1002.1.

[F] EMERGENCY VOICE/ALARM COMMUNICATIONS. See Section 902.1.

EQUIPMENT SUPPORT. See Section 1602.1.

ESSENTIAL FACILITIES. See Section 1602.1.

[F] EXHAUSTED ENCLOSURE. See Section 415.2.

EXISTING CONSTRUCTION. See Section 1612.2.

EXISTING STRUCTURE. A structure erected prior to the date of adoption of the appropriate code, or one for which a legal building permit has been issued.

❖ This term is used to identify those structures or buildings that were constructed before the current edition of the code was adopted by the jurisdiction. Often erected under the provisions of an earlier edition of the code, the buildings are exempt from compliance with current code provisions, unless otherwise stated, a hazardous condition is present or when significant alterations or changes in building heights and areas are being made. Applicability of the code to existing buildings is clearly identified in Chapter 34 (see commentary, Chapter 34).

EXIT. See Section 1002.1.

EXIT ACCESS. See Section 1002.1.

EXIT DISCHARGE. See Section 1002.1.

EXIT DISCHARGE, LEVEL OF. See Section 1002.1.

EXIT ENCLOSURE. See Section 1002.1.

EXIT PASSAGEWAY. See Section 1002.1.

EXPANDED VINYL WALL COVERING. See Section 802.1.

[F] EXPLOSION. See Section 902.1.

[F] EXPLOSIVE. See Section 307.2.

    High explosive. See Section 307.2.

    Low explosive. See Section 307.2.

    Mass detonating explosives. See Section 307.2.

    UN/DOTn Class 1 Explosives. See Section 307.2.

        Division 1.1. See Section 307.2.

        Division 1.2. See Section 307.2.

        Division 1.3. See Section 307.2.

        Division 1.4. See Section 307.2.

        Division 1.5. See Section 307.2.

        Division 1.6. See Section 307.2.

EXTERIOR SURFACES. See Section 2502.1.

EXTERIOR WALL. See Section 1402.1.

EXTERIOR WALL COVERING. See Section 1402.1.

EXTERIOR WALL ENVELOPE. See Section 1402.1.

F RATING. See Section 702.1.

FABRICATED ITEM. See Section 1702.1.

[F] FABRICATION AREA. See Section 415.2.

FACILITY. See Section 1102.1.

FACTORED LOAD. See Section 1602.1.

FIBERBOARD. See Section 2302.1.

[F] FIRE ALARM CONTROL UNIT. See Section 902.1.

[F] FIRE ALARM SIGNAL. See Section 902.1.

[F] FIRE ALARM SYSTEM. See Section 902.1.

FIRE AREA. See Section 702.1.

FIRE BARRIER. See Section 702.1.

[F] FIRE COMMAND CENTER. See Section 902.1.

FIRE DAMPER. See Section 702.1.

[F] FIRE DETECTOR, AUTOMATIC. See Section 902.1.

FIRE DOOR. See Section 702.1.

FIRE DOOR ASSEMBLY. See Section 702.1.

FIRE EXIT HARDWARE. See Section 1002.1.

FIRE PARTITION. See Section 702.1.

FIRE PROTECTION RATING. See Section 702.1.

[F] FIRE PROTECTION SYSTEM. See Section 902.1.

FIRE RESISTANCE. See Section 702.1.

FIRE-RESISTANCE RATING. See Section 702.1.

FIRE-RESISTANT JOINT SYSTEM. See Section 702.1.

[F] FIRE SAFETY FUNCTIONS. See Section 902.1.

FIRE SEPARATION DISTANCE. See Section 702.1.

FIRE WALL. See Section 702.1.

FIRE WINDOW ASSEMBLY. See Section 702.1.

FIREBLOCKING. See Section 702.1.

FIREPLACE. See Section 2102.1.

FIREPLACE THROAT. See Section 2102.1.

FIREWORKS. See Section 307.2.

FIREWORKS, 1.3G. See Section 307.2.

FIREWORKS, 1.4G. See Section 307.2.

FLAME RESISTANCE. See Section 802.1.

FLAME SPREAD. See Section 802.1.

FLAME SPREAD INDEX. See Section 802.1.

[F] FLAMMABLE GAS. See Section 307.2.

[F] FLAMMABLE LIQUEFIED GAS. See Section 307.2.

[F] FLAMMABLE LIQUID. See Section 307.2.

    Class IA. See Section 307.2.

    Class IB. See Section 307.2.

    Class IC. See Section 307.2.

[F] FLAMMABLE MATERIAL. See Section 307.2.

[F] FLAMMABLE SOLID. See Section 307.2.

[F] FLAMMABLE VAPORS OR FUMES. See Section 415.2.

[F] FLASH POINT. See Section 307.2.

FLEXIBLE BUILDINGS AND OTHER STRUCTURES. See Section 1609.2.

FLEXIBLE EQUIPMENT CONNECTIONS. See Section 1602.1.

FLEXURAL LENGTH. See Section 1808.1.

FLOOD OR FLOODING. See Section 1612.2.

FLOOD DAMAGE-RESISTANT MATERIALS. See Section 1612.2.

FLOOD HAZARD AREA. See Section 1612.2.

FLOOD HAZARD AREA SUBJECT TO HIGH VELOCITY WAVE ACTION. See Section 1612.2.

FLOOD INSURANCE RATE MAP (FIRM). See Section 1612.2.

FLOOD INSURANCE STUDY. See Section 1612.2.

FLOODWAY. See Section 1612.2.

FLOOR AREA, GROSS. See Section 1002.1.

FLOOR AREA, NET. See Section 1002.1.

FLOOR FIRE DOOR ASSEMBLY. See Section 702.1.

FLY GALLERY. See Section 410.2.

[F] FOAM-EXTINGUISHING SYSTEMS. See Section 902.1.

FOAM PLASTIC INSULATION. See Section 2602.1.

FOLDING AND TELESCOPIC SEATING. See Section 1002.1.

FOOD COURT. See Section 402.2.

FRAME. See Section 1602.1.

    Braced frame. See Section 1602.1.

    Concentrically braced frame (CBF). See Section 1602.1.

    Eccentrically braced frame (EBF). See Section 1602.1.

    Moment frame. See Section 1602.1.

    Ordinary concentrically braced frame (OCBF). See Section 1602.1.

    Special concentrically braced frame (SCBF). See Section 1602.1.

[F] GAS CABINET. See Section 415.2.

[F] GAS ROOM. See Section 415.2.

GLASS FIBERBOARD. See Section 721.1.1.

GLUED BUILT-UP MEMBER. See Section 2302.1.

GRADE FLOOR OPENING. A window or other opening located such that the sill height of the opening is not more than 44 inches (1118 mm) above or below the finished ground level adjacent to the opening.

❖ Openings used for emergency escape or emergency rescue are clearly easier to use the closer they are to grade. This definition specifies that the maximum sill height above the exterior adjacent grade must be no more than 44 inches (1118 mm) for an opening to qualify as a grade floor opening.

GRADE (LUMBER). See Section 2302.1.

GRADE PLANE. See Section 502.1.

GRANDSTAND. See Section 1002.1.

GRAVITY LOAD. See Section 1613.1.

GRIDIRON. See Section 410.2.

GROSS LEASABLE AREA. See Section 402.2.

GROUTED MASONRY. See Section 2102.1.

    Grouted hollow-unit masonry. See Section 2102.1.

    Grouted multiwythe masonry. See Section 2102.1.

GUARD. See Section 1002.1.

GYPSUM BOARD. See Section 2502.1.

GYPSUM PLASTER. See Section 2502.1.

GYPSUM VENEER PLASTER. See Section 2502.1.

HABITABLE SPACE. A space in a building for living, sleeping, eating or cooking. Bathrooms, toilet rooms, closets, halls, storage or utility spaces and similar areas are not considered habitable spaces.

❖ These spaces are normally considered inhabited in the course of residential living and provide the four basic characteristics associated with it: living, sleeping, eating and cooking. All habitable spaces are considered occupiable spaces, though other occupiable spaces, such as halls or utility rooms, are not considered habitable (see the definition of "Occupiable space" in this chapter).

[F] HALOGENATED EXTINGUISHING SYSTEMS. See Section 902.1.

[F] HANDLING. See Section 307.2.

HANDRAIL. See Section 1002.1.

HARDBOARD. See Section 2302.1.

HAZARDOUS CONTENTS. See Section 1613.1.

[F] HAZARDOUS MATERIALS. See Section 307.2.

[F] HAZARDOUS PRODUCTION MATERIAL (HPM). See Section 415.2.

HEAD JOINT. See Section 2102.1.

HEADER (Bonder). See Section 2102.1.

[F] HEALTH HAZARD. See Section 307.2.

HEIGHT, BUILDING. See Section 502.1.

HEIGHT, STORY. See Section 502.1.

HEIGHT, WALLS. See Section 2102.1.

HELIPORT. See Section 412.5.2.

HELISTOP. See Section 412.5.2.

[F] HIGHLY TOXIC. See Section 307.2.

HISTORIC BUILDINGS. Buildings that are listed in or eligible for listing in the National Register of Historic Places, or designated as historic under an appropriate state or local law (see Section 3406).

❖ The code provides a subjective blanket exception for compliance in Section 3406 to those "true" historic buildings that are receiving improvements or undergoing changes of any kind. This definition specifies the criteria for consideration as a historic building.

HORIZONTAL EXIT. See Section 1002.1.

[F] HPM FLAMMABLE LIQUID. See Section 415.2.

[F] HPM ROOM. See Section 415.2.

HURRICANE-PRONE REGIONS. See Section 1609.2.

IMMEDIATELY DANGEROUS TO LIFE AND HEALTH (IDLH). See Section 415.2.

IMPACT LOAD. See Section 1602.1.

IMPORTANCE FACTOR, *I*. See Section 1609.2.

INCOMPATIBLE MATERIALS. See Section 307.2.

INDUSTRIAL EQUIPMENT PLATFORM. See Section 502.1.

[F] INITIATING DEVICE. See Section 902.1.

INSPECTION CERTIFICATE. See Section 1702.1.

INTENDED TO BE OCCUPIED AS A RESIDENCE. See Section 1102.

INTERIOR FINISH. See Section 802.1.

INTERIOR FLOOR FINISH. See Section 802.1.

INTERIOR SURFACES. See Section 2502.1.

INTERIOR WALL AND CEILING FINISH. See Section 802.1.

INTERLAYMENT. See Section 1502.1.

INVERTED PENDULUM-TYPE STRUCTURES. See Section 1613.1.

ISOLATION INTERFACE. See Section 1613.1.

ISOLATION JOINT. See Section 1902.1.

ISOLATION SYSTEM. See Section 1613.1.

ISOLATOR UNIT. See Section 1613.1.

JOINT. See Sections 702.1 and 1602.1.

JURISDICTION. The governmental unit that has adopted this code under due legislative authority.

❖ The governmental unit adopting the code has the legal authority to do so under state statutes.

LABEL. See Section 1702.1.

LIGHT-DIFFUSING SYSTEM. See Section 2602.1.

LIGHT-FRAME CONSTRUCTION. A type of construction whose vertical and horizontal structural elements are primarily formed by a system of repetitive wood or light gage steel framing members.

❖ The code uses the term "light frame" to distinguish this unique type of framing system from other structural systems. The structural integrity of light-frame construction is dependent upon numerous connections or frequent bracing. Other framing systems or terms commonly used in the building industry that are considered as light-frame construction include: "stick built," "post and beam," "pole barn," "platform frame," "western frame" and "balloon frame." Sections 2205 and 2211 pertain to light-frame cold-formed steel construction. Section 2308 defines a specific subcategory of light-frame construction called "conventional light-frame wood construction," which is limited to wood materials.

LIGHT-TRANSMITTING PLASTIC ROOF PANELS. See Section 2602.1.

LIGHT-TRANSMITTING PLASTIC WALL PANELS. See Section 2602.1.

LIMIT STATE. See Section 1602.1.

[F] LIQUID. See Section 415.2.

[F] LIQUID STORAGE ROOM. See Section 415.2.

[F] LIQUID USE, DISPENSING AND MIXING ROOMS. See Section 415.2.

LISTED. See Section 902.1.

LIVE LOADS. See Section 1602.1.

LIVE LOADS (ROOF). See Section 1602.1.

LOAD. See Section 1613.1.

 Gravity load (W). See Section 1613.1.

LOAD AND RESISTANCE FACTOR DESIGN (LRFD). See Section 1602.1.

LOAD FACTOR. See Section 1602.1.

LOADS. See Section 1602.1.

LOADS EFFECTS. See Section 1602.1.

LOT. A portion or parcel of land considered as a unit.

❖ A lot is a legally recorded parcel of land, the boundaries of which are described on a deed. When code requirements are based on some element of a lot (such as yard area or lot line location), it is the physical attributes of the parcel of land that the code is addressing, not issues of ownership. Adjacent lots owned by the same party are treated as if they were owned by different parties because ownership can change at any time. A condominium form of building ownership does not create separate lots (i.e., parcels of land) and such unit owners are treated as separate tenants, not separate lot owners.

LOT LINE. A line dividing one lot from another, or from a street or any public place.

❖ Lot lines are legally recorded divisions between two adjacent land parcels or lots. They are the reference point for the location of buildings for exterior separation and other code purposes (see the definition of "Lot" above).

[F] LOWER FLAMMABLE LIMIT (LFL). See Section 415.2.

LOWEST FLOOR. See Section 1612.2.

MAIN WINDFORCE-RESISTING SYSTEM. See Section 1609.2.

MALL. See Section 402.2.

[F] MANUAL FIRE ALARM BOX. See Section 902.1.

MANUFACTURER'S DESIGNATION. See Section 1702.1.

MARK. See Section 1702.1.

MARQUEE. A permanent roofed structure attached to and supported by the building and that projects into the public right-of-way.

❖ Marquees, unlike canopies and awnings, are fixed, permanent structures that justify sufficiently different requirements from those for other projections (see Section 3106 for code requirements for marquees).

MASONRY. See Section 2102.1.

 Ashlar masonry. See Section 2102.1.

 Coursed ashlar. See Section 2102.1.

 Glass unit masonry. See Section 2102.1.

 Plain masonry. See Section 2102.1.

 Random ashlar. See Section 2102.1.

 Reinforced masonry. See Section 2102.1.

 Solid masonry. See Section 2102.1.

MASONRY UNIT. See Section 2102.1.

 Clay. See Section 2102.1.

 Concrete. See Section 2102.1.

 Hollow. See Section 2102.1.

 Solid. See Section 2102.1.

MAXIMUM CONSIDERED EARTHQUAKE. See Section 1613.1.

MEAN DAILY TEMPERATURE. See Section 2102.1.

MEAN ROOF HEIGHT. See Section 1609.2.

MEANS OF EGRESS. See Section 1002.1.

MECHANICAL-ACCESS OPEN PARKING GARAGES. See Section 406.3.2.

MECHANICAL EQUIPMENT SCREEN. See Section 1502.1.

MEMBRANE-COVERED CABLE STRUCTURE. See Section 3102.2.

MEMBRANE-COVERED FRAME STRUCTURE. See Section 3102.2.

MEMBRANE PENETRATION. See Section 702.1.

MEMBRANE-PENETRATION FIRESTOP. See Section 702.1.

METAL COMPOSITE MATERIAL (MCM). See Section 1402.

METAL COMPOSITE MATERIAL SYSTEM. See Section 1402.

METAL ROOF PANEL. See Section 1502.1.

METAL ROOF SHINGLE. See Section 1502.1.

MEZZANINE. See Section 502.1.

MINERAL BOARD. See Section 721.1.1.

MODIFIED BITUMEN ROOF COVERING. See Section 1502.1.

MORTAR. See Section 2102.1.

MORTAR, SURFACE-BONDING. See Section 2102.1.

[F] MULTIPLE-STATION ALARM DEVICE. See Section 902.1.

[F] MULTIPLE-STATION SMOKE ALARM. See Section 902.1.

NAILING, BOUNDARY. See Section 2302.1.

NAILING, EDGE. See Section 2302.1.

NAILING, FIELD. See Section 2302.1.

NATURALLY DURABLE WOOD. See Section 2302.1.

   Decay resistant. See Section 2302.1.

   Termite resistant. See Section 2302.1.

NOMINAL LOADS. See Section 1602.1.

NOMINAL SIZE (LUMBER). See Section 2302.1.

NONBUILDING STRUCTURE. See Section 1613.1.

NONCOMBUSTIBLE MEMBRANE STRUCTURE. See Section 3102.2.

[F] NORMAL TEMPERATURE AND PRESSURE (NTP). See Section 415.2.

NOSING. See Section 1002.1.

[F] NUISANCE ALARM. See Section 902.1.

OCCUPANCY IMPORTANCE FACTOR. See Section 1613.1.

OCCUPANT LOAD. See Section 1002.1.

OCCUPIABLE SPACE. A room or enclosed space designed for human occupancy in which individuals congregate for amusement, educational or similar purposes or in which occupants are engaged at labor, and which is equipped with means of egress and light and ventilation facilities meeting the requirements of this code.

❖ Occupiable spaces are those areas designed for human occupancy. Based on the nature of the occupancy, various code sections apply. All habitable spaces are also considered occupiable (see the definition of "Habitable space"); however, all occupiable spaces are not habitable. Additionally, some spaces are neither habitable nor occupiable, such as closets, toilet rooms and mechanical equipment rooms.

OPEN PARKING GARAGE. See Section 406.3.2.

[F] OPEN SYSTEM. See Section 307.2.

OPERATING BUILDING. See Section 307.2.

[F] ORGANIC PEROXIDE. See Section 307.2.

   Class I. See Section 307.2.

   Class II. See Section 307.2.

   Class III. See Section 307.2.

   Class IV. See Section 307.2.

   Class V. See Section 307.2.

   Unclassified detonable. See Section 307.2.

OTHER STRUCTURES. See Section 1602.1.

OWNER. Any person, agent, firm or corporation having a legal or equitable interest in the property.

❖ This term defines the person or other legal entity who is responsible for a building and its compliance with the code requirements.

[F] OXIDIZER. See Section 307.2.

   Class 4. See Section 307.2.

   Class 3. See Section 307.2.

   Class 2. See Section 307.2.

   Class 1. See Section 307.2.

[F] OXIDIZING GAS. See Section 307.2.

P-DELTA EFFECT. See Section 1602.1.

PANEL (PART OF A STRUCTURE). See Section 1602.1.

PANIC HARDWARE. See Section 1002.1.

PARTICLEBOARD. See Section 2302.1.

PEDESTAL. See Section 1902.1.

PENETRATION FIRESTOP. See Section 702.1.

PENTHOUSE. See Section 1502.1.

PERMIT. An official document or certificate issued by the authority having jurisdiction which authorizes performance of a specified activity.

❖ The permit constitutes a license issued by the building official to proceed with a specific activity, such as construction of a building, in accordance with all applicable laws.

PERSON. An individual, heirs, executors, administrators or assigns, and also includes a firm, partnership or corporation, its or their successors or assigns, or the agent of any of the aforesaid.

❖ Corporations and other organizations listed in the definition are treated as persons under the law. Also, when the code provides for a penalty (see Section 113.4), the definition makes it clear that the individuals responsible for administering the activities of these various organizations are subject to these penalties.

PERSONAL CARE SERVICE. See Section 310.2.

[F] PHYSICAL HAZARD. See Section 307.2.

PIER FOUNDATIONS. See Section 1808.1.

    Belled piers. See Section 1808.1.

PILE FOUNDATIONS. See Section 1808.1.

    Auger uncased piles. See Section 1808.1.

    Caisson piles. See Section 1808.1.

    Concrete-filled steel pipe and tube piles. See Section 1808.1.

    Driven uncased piles. See Section 1808.1.

    Enlarged base piles. See Section 1808.1.

    Piles. See Section 1808.1.

    Steel-cased piles. See Section 1808.1.

PINRAIL. See Section 410.2.

PLAIN CONCRETE. See Section 1902.1.

PLAIN REINFORCEMENT. See Section 1902.1.

PLASTIC, APPROVED. See Section 2602.1.

PLASTIC GLAZING. See Section 2602.1.

PLASTIC HINGE. See Section 2102.1.

PLATFORM. See Section 410.2.

POSITIVE ROOF DRAINAGE. See Section 1502.1.

PRECAST CONCRETE. See Section 1902.1.

PRESERVATIVE-TREATED WOOD. See Section 2302.1.

PRESTRESSED CONCRETE. See Section 1902.1.

PRESTRESSED MASONRY. See Section 2102.1.

    Prestressed masonry shear wall. See Section 2102.1.

    Ordinary plain prestressed masonry shear wall. See Section 2102.1.

    Special prestressed masonry shear wall. See Section 2102.1.

    Special reinforced masonry shear wall. See Section 2102.1.

PRISM. See Section 2102.1.

PROSCENIUM WALL. See Section 410.2.

PUBLIC ENTRANCE. See Section 1102.1.

PUBLIC-USE AREAS. See Section 1102.1.

PUBLIC WAY. See Section 1002.1.

[F] PYROPHORIC. See Section 307.2.

[F] PYROTECHNIC COMPOSITION. See Section 307.2.

QUALITY ASSURANCE PLAN. A written procedure complying with the requirements of Section 1705.

❖ A quality assurance plan is required to be submitted for certain buildings required to resist seismic forces. Components of the plan are listed in Section 1705. Quality assurance plans are also required for wind requirements for certain buildings (see Section 1706).

RAMP. See Section 1002.1.

RAMP-ACCESS OPEN PARKING GARAGES. See Section 406.3.2.

[F] RECORD DRAWINGS. See Section 902.1.

REFERENCE RESISTANCE (*D*). See Section 2302.1.

REGISTERED DESIGN PROFESSIONAL. An individual who is registered or licensed to practice their respective design profession as defined by the statutory requirements of the professional registration laws of the state or jurisdiction in which the project is to be constructed.

❖ Legal qualifications for engineers and architects are established by the state having jurisdiction. Licensing and registration of engineers and architects are accomplished by written or oral examinations offered by states or by reciprocity (licensing in other states).

REINFORCED CONCRETE. See Section 1902.1.

REINFORCED PLASTIC, GLASS FIBER. See Section 2602.1.

REINFORCEMENT. See Section 1902.1.

REPAIR. The reconstruction or renewal of any part of an existing building for the purpose of its maintenance.

❖ As indicated in Section 105.2.2, the repair of an item typically does not require a permit. This definition makes it clear that repair is limited to work on the item, and does not include complete or substantial replacement or other new work.

REQUIRED STRENGTH. See Sections 1602.1 and 2102.1.

REROOFING. See Section 1502.1.

RESHORES. See Section 1902.1.

RESIDENTIAL AIRCRAFT HANGAR. See Section 412.3.1.

RESIDENTIAL CARE/ASSISTED LIVING FACILITIES. See Section 310.2.

RESISTANCE FACTOR. See Section 1602.1.

RESTRICTED ENTRANCE. See Section 1102.1.

RETRACTABLE AWNING. See Section 3105.2.

ROOF ASSEMBLY. See Section 1502.1.

ROOF COVERING. See Section 1502.1.

ROOF COVERING SYSTEM. See Section 1502.1.

ROOF DECK. See Section 1502.1.

ROOF RECOVER. See Section 1502.1.

ROOF REPAIR. See Section 1502.1.

ROOF REPLACEMENT. See Section 1502.1.

ROOF VENTILATION. See Section 1502.1.

ROOFTOP STRUCTURE. See Section 1502.1.

RUBBLE MASONRY. See Section 2102.1.

　Coursed rubble. See Section 2102.1.

　Random rubble. See Section 2102.1.

　Rough or ordinary rubble. See Section 2102.1.

RUNNING BOND. See Section 2102.1.

SCISSOR STAIR See Section 1002.1.

SCUPPER. See Section 1502.1.

SEISMIC DESIGN CATEGORY. See Section 1613.1.

SEISMIC-FORCE-RESISTING SYSTEM. See Section 1613.1.

SEISMIC FORCES. See Section 1613.1.

SEISMIC RESPONSE COEFFICIENT. See Section 1613.1.

SEISMIC USE GROUP. See Section 1613.1.

SELF-CLOSING. See Section 702.1.

SELF-SERVICE STORAGE FACILITY. See Section 1102.1.

[F] SERVICE CORRIDOR. See Section 415.2.

SERVICE ENTRANCE. See Section 1102.1.

SHAFT. See Section 702.1.

SHAFT ENCLOSURE. See Section 702.1.

SHALLOW ANCHORS. See Section 1602.1.

SHEAR PANEL. See Section 1602.1.

SHEAR WALL. See Sections 1602.1, 1613.1 and 2102.1.

　Detailed plain masonry shear wall. See Section 2102.1.

　Intermediate reinforced masonry shear wall. See Section 2102.1.

　Ordinary plain masonry shear wall. See Section 2102.1.

　Ordinary reinforced masonry shear wall. See Section 2102.1.

　Perforated shear wall. See Section 2302.1.

　Perforated shear wall segment. See Section 2302.1.

　Special reinforced masonry shear wall. See Section 2102.1.

Type I shear wall. See Section 2202.1.

Type II shear wall. See Section 2202.1.

Type II shear wall segment. See Section 2202.1.

SHEAR WALL-FRAME INTERACTIVE SYSTEM. See Section 1613.1.

SHELL. See Section 2102.1.

SHORES. See Section 1902.1.

SHOTCRETE. See Section 1914.1.

SINGLE-PLY MEMBRANE. See Section 1502.1.

[F] SINGLE-STATION SMOKE ALARM. See Section 902.1.

SITE. See Section 1102.1.

SITE CLASS. See Section 1613.1.

SITE COEFFICIENTS. See Section 1613.1.

SKYLIGHT, UNIT. A factory-assembled, glazed fenestration unit, containing one panel of glazing material that allows for natural lighting through an opening in the roof assembly while preserving the weather-resistant barrier of the roof.

❖ This is a specific type of sloped glazing assembly that is factory assembled. The code and the IRC contain specific provisions that are appropriate for this type of building component. Factory assembled units, as opposed to site-built skylights, can be designed, tested and rated as one component that incorporates both glazing and framing, if applicable. The individual components of site-built glazing must be designed to resist the design loads of the codes individually, and are not usually rated as an assembly.

SKYLIGHTS AND SLOPED GLAZING. Glass or other transparent or translucent glazing material installed at a slope of 15 degrees (0.26 rad) or more from vertical. Glazing material in skylights, including unit skylights, solariums, sunrooms, roofs and sloped walls, are included in this definition.

❖ The code regulates skylights and sloped glazing since their failure could result in injury and building damage (see Section 2405 for the code requirements).

SLEEPING UNIT. A room or space in which people sleep, which can also include permanent provisions for living, eating, and either sanitation or kitchen facilities but not both. Such rooms and spaces that are also part of a dwelling unit are not sleeping units.

❖ This definition is included to coordinate the *Fair Housing Act Guidelines* with the code. The definition for "Sleeping unit" is needed to clarify the differences between sleeping units and dwelling units. Some examples would be a hotel guestroom, a dormitory, a boarding house, etc. Another example would be an addition to a studio apartment with a kitchenette (i.e., microwave, sink, refrigerator). Since the cooking arrangements were not permanent, this configuration would be considered a sleeping unit, not a dwelling unit. As already defined in the code, a dwelling unit must contain permanent facilities for living, sleeping, eating, cooking and sanitation.

[F] SMOKE ALARM. See Section 902.1.

SMOKE BARRIER. See Section 702.1.

SMOKE COMPARTMENT. See Section 702.1.

SMOKE DAMPER. See Section 702.1.

[F] SMOKE DETECTOR. See Section 902.1.

SMOKE-DEVELOPED INDEX. See Section 802.1.

SMOKE-PROTECTED ASSEMBLY SEATING. See Section 1002.1.

SMOKEPROOF ENCLOSURE. See Section 902.1.

[F] SOLID. See Section 415.2.

SPACE FRAME. See Section 1602.1.

SPECIAL AMUSEMENT BUILDING. See Section 411.2.

SPECIAL FLOOD HAZARD AREA. See Section 1612.2.

SPECIAL INSPECTION. See Section 1702.1.

Special continuous inspection. See Section 1702.1.

Special periodic inspection. See Section 1702.1.

SPECIAL TRANSVERSE REINFORCEMENT. See Section 1602.1.

SPECIFIED. See Section 2102.1.

SPECIFIED COMPRESSIVE STRENGTH OF MASONRY (f'_m). See Section 2102.1.

SPIRAL REINFORCEMENT. See Section 1902.1.

SPLICE. See Section 702.1.

SPRAYED FIRE-RESISTANT MATERIALS. See Section 1702.1.

STACK BOND. See Section 2102.1.

STAGE. See Section 410.2.

STAIR. See Section 1002.1.

STAIRWAY. See Section 1002.1.

STAIRWAY, EXTERIOR. See Section 1002.1.

STAIRWAY, INTERIOR. See Section 1002.1.

STAIRWAY, SPIRAL. See Section 1002.1.

[F] STANDPIPE SYSTEM, CLASSES OF. See Section 902.1.

    Class I system. See Section 902.1.

    Class II system. See Section 902.1.

    Class III system. See Section 902.1.

[F] STANDPIPE, TYPES OF. See Section 902.1.

    Automatic dry. See Section 902.1.

    Automatic wet. See Section 902.1.

    Manual dry. See Section 902.1.

    Manual wet. See Section 902.1.

    Semiautomatic dry. See Section 902.1.

START OF CONSTRUCTION. See Section 1612.2.

STEEL CONSTRUCTION, COLD-FORMED. See Section 2202.1.

STEEL JOIST. See Section 2202.1.

STEEL MEMBER, STRUCTURAL. See Section 2202.1.

STEEP SLOPE. A roof slope greater than two units vertical in 12 units horizontal (17-percent slope).

❖ This is the general criterion for roof slope that is used throughout the code. Slope requirements for specific roof covering materials are specified in Chapter 15.

STONE MASONRY. See Section 2102.1.

    Ashlar stone masonry. See Section 2102.1.

    Rubble stone masonry. See Section 2102.1.

[F] STORAGE, HAZARDOUS MATERIALS. See Section 415.2.

STORY. That portion of a building included between the upper surface of a floor and the upper surface of the floor or roof next above (also see "Basement," "Mezzanine" and Section 502.1). It is measured as the vertical distance from top to top of two successive tiers of beams or finished floor surfaces and, for the topmost story, from the top of the floor finish to the top of the ceil-ing joists or, where there is not a ceiling, to the top of the roof rafters.

❖ All levels in a building that conform to this description are stories, including basements. A mezzanine is considered part of the story in which it is located. See Chapter 5 for code requirements regarding limitations on the number of stories in a building as a function of the type of construction. See Section 1617 for limits on story drift from earthquake effects.

STORY ABOVE GRADE PLANE. Any story having its finished floor surface entirely above grade plane, except that a basement shall be considered as a story above grade plane where the finished surface of the floor above the basement is:

    1. More than 6 feet (1829 mm) above grade plane;

    2. More than 6 feet (1829 mm) above the finished ground level for more than 50 percent of the total building perimeter; or

    3. More than 12 feet (3658 mm) above the finished ground level at any point.

❖ The determination of a story above grade is important because it contributes to the height of a building for the purpose of applying the allowable building height in stories from Tables 503 and 1018.2. Every story with the finished floor entirely above grade (finished ground level) is a story above grade; however, a story with any portion of the finished floor level below grade is by definition a basement, and must be evaluated in conformance to the three criteria for story above grade. These three criteria are intended to deal with unusual grading of ground adjacent to exterior walls. Without such a consideration, the resulting building height can be reduced because of a berm or other landscaping technique that may be artificially created to reduce the apparent building height. The specific criteria establish the point at which a basement extends far enough above ground that it contributes to the regulated height of the building in number of stories.

STORY DRIFT RATIO. See Section 1613.1.

STRENGTH. See Section 2102.1.

    Design strength. See Section 2102.1.

    Nominal strength. See Sections 1602.1 and 2102.1.

STRENGTH DESIGN. See Section 1602.1.

STRUCTURAL CONCRETE. See Section 1902.1.

STRUCTURAL GLUED-LAMINATED TIMBER. See Section 2302.1.

STRUCTURAL OBSERVATION. See Section 1702.1.

STRUCTURE. That which is built or constructed.

❖ This definition is intentionally broad so as to include within its scope, and therefore the scope of the code (see Section 101.2), everything that is built as an improvement to real property.

SUBDIAPHRAGM. See Section 2302.1.

SUBSTANTIAL DAMAGE. See Section 1612.2.

SUBSTANTIAL IMPROVEMENT. See Section 1612.2.

[F] SUPERVISING STATION. See Section 902.1.

[F] SUPERVISORY SERVICE. See Section 902.1.

[F] SUPERVISORY SIGNAL. See Section 902.1.

[F] SUPERVISORY SIGNAL-INITIATING DEVICE. See Section 902.1.

SWIMMING POOLS. See Section 3109.2.

T RATING. See Section 702.1.

TECHNICALLY INFEASIBLE. See Section 3402.

TENDON. See Section 1902.1.

TENT. Any structure, enclosure or shelter which is constructed of canvas or pliable material supported in any manner except by air or the contents it protects.

❖ Tents can be temporary or permanent structures. When permanent, they are considered membrane-covered structures and are regulated by Section 3102. When erected as temporary enclosures, they are regulated by Section 3103.

THERMOPLASTIC MATERIAL. See Section 2602.1.

THERMOSETTING MATERIAL. See Section 2602.1.

THROUGH PENETRATION. See Section 702.1.

THROUGH-PENETRATION FIRESTOP SYSTEM. See Section 702.1.

TIE-DOWN (HOLD-DOWN). See Section 2302.1.

TIE, LATERAL. See Section 2102.1.

TIE, WALL. See Section 2102.1.

TILE. See Section 2102.1.

TILE, STRUCTURAL CLAY. See Section 2102.1.

[F] TIRES, BULK STORAGE OF. See Section 902.1.

TORSIONAL FORCE DISTRIBUTION. See Section 1613.1.

TOUGHNESS. See Section 1613.1.

[F] TOXIC. See Section 307.2.

TREATED WOOD. See Section 2302.1.

TRIM. See Section 802.1.

[F] TROUBLE SIGNAL. See Section 902.1.

UNADJUSTED SHEAR RESISTANCE. See Section 2202.1.

UNDERLAYMENT. See Section 1502.1.

[F] UNSTABLE (REACTIVE) MATERIAL. See Section 307.2.

Class 4. See Section 307.2.

Class 3. See Section 307.2.

Class 2. See Section 307.2.

Class 1. See Section 307.2.

[F] USE (MATERIAL). See Section 415.2.

VAPOR-PERMEABLE MEMBRANE. A material or covering having a permeance rating of 5 perms (52.9×10^{-10} kg/Pa · s · m²) or greater, when tested in accordance with the dessicant method using Procedure A of ASTM E 96. A vapor-permeable material permits the passage of moisture vapor.

❖ Greater demands on the building envelope due to energy considerations now dictate the need for an outer membrane that reduces wind infiltration. The membranes used in this application may need to allow vapor through, given that a vapor barrier would be needed on the inside of the wall and would be undesirable on the outside of the wall. In such cases, a vapor-permeable membrane would be used.

VAPOR RETARDER. A vapor-resistant material, membrane or covering such as foil, plastic sheeting or insulation facing having a permeance rating of 1 perm (5.7×10^{-11} kg/Pa · s · m²) or less, when tested in accordance with the dessicant method using Procedure A of ASTM E 96. Vapor retarders limit the amount of moisture vapor that passes through a material or wall assembly.

❖ This definition establishes the acceptance criteria and purpose of membranes used in building construction. This definition is important since the use of vapor retarders allows the required ventilation areas for attics and crawl spaces to be reduced. The transfer of less moisture from the interior building spaces into attics and crawl spaces decreases the need to ventilate the moisture in the attics or crawl spaces directly to the outside.

VENEER. See Section 1402.1.

VENTILATION. The natural or mechanical process of supplying conditioned or unconditioned air to, or removing such air from, any space.

❖ Ventilation is the process of moving air to or from building spaces. This definition is used in this chapter to establish minimum levels of air movement within a building for the purposes of providing a healthful interior environment. Ventilation would include both natural (openable exterior windows and doors for wind movement) and mechanical (forced air with mechanical equipment) methods.

[F] VISIBLE ALARM NOTIFICATION APPLIANCE. See Section 902.1.

WALKWAY, PEDESTRIAN. A walkway used exclusively as a pedestrian trafficway.

❖ A pedestrian walkway is an enclosed passageway external to, and not considered part of, the buildings it connects. Intended only for pedestrian use, it can be at grade, below grade or elevated above grade.

WALL. See Section 2102.1.

Cavity wall. See Section 2102.1.

Composite wall. See Section 2102.1.

Dry-stacked, surface-bonded wall. See Section 2102.1.

Masonry-bonded hollow wall. See Section 2102.1.

Parapet wall. See Section 2102.1.

WALL, LOAD-BEARING. See Section 1602.1.

WALL, NONLOAD-BEARING. See Section 1602.1.

[F] WATER-REACTIVE MATERIAL. See Section 307.2.

Class 3. See Section 307.2.

Class 2. See Section 307.2.

Class 1. See Section 307.2.

WEATHER-EXPOSED SURFACES. See Section 2502.1.

WEB. See Section 2102.1.

[F] WET-CHEMICAL EXTINGUISHING SYSTEM. See Section 902.1.

WHEELCHAIR SPACE. See Section 1102.1.

WHEELCHAIR SPACE CLUSTER. See Section 1102.1.

WIND-BORNE DEBRIS REGION. See Section 1609.2.

WIND-RESTRAINT SEISMIC SYSTEM. See Section 1613.

WIRE BACKING. See Section 2502.1.

[F] WIRELESS PROTECTION SYSTEM. See Section 902.1.

WOOD SHEAR PANEL. See Section 2302.1.

WOOD STRUCTURAL PANEL. See Section 2302.1.

Composite panels. See Section 2302.1.

Oriented strand board (OSB). See Section 2302.1.

Plywood. See Section 2302.1.

[F] WORKSTATION. See Section 415.2.

WYTHE. See Section 2102.1.

YARD. An open space, other than a court, unobstructed from the ground to the sky, except where specifically provided by this code, on the lot on which a building is situated.

❖ This definition is used, similar to the definition of "Court," to establish the applicability of code requirements when yards are utilized for natural light or natural ventilation purposes (see Section 1205.2). Whereas a court is bounded on three or more sides with the building or structure, a yard is bounded on two or less sides by the building or structure.

[F] ZONE. See Section 902.1.

Chapter 3:
Use and Occupancy Classification

General Comments

Chapter 3 provides for the classification of buildings, structures and parts thereof based on the purpose or purposes for which they are used.

Section 302 identifies the groups into which all buildings, structures and parts thereof must be classified. It also identifies special performance requirements for incidental use areas located in particular group classifications and the criteria that qualify portions of a structure as an accessory use area. Finally, the section sets forth the design alternatives for a structure that has multiple occupancies not of the same group classification (i.e., mixed occupancies).

Sections 303 through 312 identify the occupancy characteristics of each group classification. In some sections, specific group classifications having requirements in common are collectively organized such that one term applies to all. For example, Groups A-1, A-2, A-3, A-4 and A-5 are individual groups. The general term "Group A," however, includes each of these individual groups. For this reason, each specific assembly group classification is included in Section 303.

In the early years of building code development, the essence of regulatory safeguards from fire was to provide a reasonable level of protection to property. The idea was that if property was adequately protected from fire, then the building occupants would also be protected.

From this outlook on fire safety, the concept of equivalent risk has evolved in the code. This concept maintains that, in part, an acceptable level of risk against the damages of fire respective to a particular occupancy type (group) can be achieved by limiting the height and area of buildings containing such occupancies according to the building's construction type (i.e., its relative fire endurance).

The concept of equivalent risk involves three interdependent considerations: (1) the level of fire hazard associated with the specific occupancy of the facility; (2) the reduction of fire hazard by limiting the floor area(s) and the height of the building based on the fuel load (combustible contents and burnable building components) and (3) the level of overall fire resistance provided by the type of construction used for the building.

The interdependence of these fire safety considerations can be seen by first looking at Tables 601 and 602, which show the fire-resistance ratings of the principal structural elements comprising a building in relation to the five classifications for types of construction. Type I construction is the classification that generally requires the highest fire-resistance ratings for structural elements, whereas Type V construction, which is designated as a combustible type of construction, generally requires the least amount of fire-resistance-rated structural elements. If one then looks at Table 503, the relationship among group classification, allowable heights and areas and types of construction becomes apparent. Respective to each group classification, the greater the fire-resistance rating of structural elements, as represented by the type of construction, the greater the floor area and height allowances. The greater the potential fire hazards indicated as a function of the group, the lesser the height and area allowances for a particular construction type.

As a result of extensive research and advancements in fire technology, today's building codes are more comprehensive and complex regulatory instruments than they were in the earlier years of code development. While the principle of equivalent risk remains an important component in building codes, perspectives have changed and life safety is now the paramount fire issue. Even so, occupancy classification still plays a key part in organizing and prescribing the appropriate protection measures. As such, threshold requirements for fire protection and means of egress systems are based on occupancy classification (see Chapters 9 and 10).

Other sections of the code also contain requirements respective to the classification of building groups. For example, Section 102.6 deals with applicability of the code to existing structures when there is a change of occupancy; Section 704 deals with requirements for exterior wall fire-resistance ratings that are tied to the occupancy classification of a building and Section 803.4 contains interior finish requirements that are dependent upon the occupancy classification.

Purpose

The purpose of this chapter is to classify a building, structure or part thereof into a group based on the specific purpose for which it is designed or occupied. Throughout the code, group classifications are considered a fundamental principle in organizing and prescribing the appropriate features of construction and occupant safety requirements for buildings, especially general building limitations, means of egress, fire protection systems and interior finishes.

SECTION 301
GENERAL

301.1 Scope. The provisions of this chapter shall control the classification of all buildings and structures as to use and occupancy.

❖ As used throughout the code, the classification of an occupancy into a group is established by the requirements of this chapter. The purpose of these provisions is to provide rational criteria for the classification of various occupancies into groups based on their relative fire hazard and life safety properties. This is necessary because the code utilizes group classification as a fundamental principle for differentiating requirements in other parts of the code related to fire and life safety protection.

SECTION 302
CLASSIFICATION

302.1 General. Structures or portions of structures shall be classified with respect to occupancy in one or more of the groups listed below. Structures with multiple uses shall be classified according to Section 302.3. Where a structure is proposed for a purpose which is not specifically provided for in this code, such structure shall be classified in the group which the occupancy most nearly resembles, according to the fire safety and relative hazard involved.

1. Assembly (see Section 303): Groups A-1, A-2, A-3, A-4 and A-5

2. Business (see Section 304): Group B

3. Educational (see Section 305): Group E

4. Factory and Industrial (see Section 306): Groups F-1 and F-2

5. High Hazard (see Section 307): Groups H-1, H-2, H-3, H-4 and H-5

6. Institutional (see Section 308): Groups I-1, I-2, I-3 and I-4

7. Mercantile (see Section 309): Group M

8. Residential (see Section 310): Groups R-1, R-2, R-3 as applicable in Section 101.2, and R-4

9. Storage (see Section 311): Groups S-1 and S-2

10. Utility and Miscellaneous (see Section 312): Group U

❖ This section requires all structures to be classified in one or more of the groups listed according to the structure's purpose and function (i.e., its occupancy). By organizing occupancies with similar fire hazard and life safety properties into groups, the code has adopted the means to differentiate occupancies such that various fire protection and life safety requirements can be rationally organized and applied. Each specific group has an individual classification. Each represents a different characteristic and level of fire hazard that requires special code provisions to lessen the associated risks.

There are some group classifications that are very closely related to other specific groups and, therefore, are collectively referred to as a single group (e.g., Group F applies to Groups F-1 and F-2). In these cases, there are requirements within the code that are common to each specific group classification. These common requirements are applicable based on the reference to the collective classification. For example, the requirements of Section 1024 apply to each specific group classification listed under the term "Group A."

Example 1: Both a restaurant (Group A-2) and a church (Group A-3) are included in Group A, but they have different specific group classifications. Both Groups A-3 and A-4 are subject to the same travel distance limitations (see Table 1015.1) and corridor fire-resistance ratings (see Table 1016.1), although automatic sprinkler systems (see Section 903) are different.

Buildings that contain occupancies classified for more than one use are mixed-use buildings. Buildings with mixed occupancies must comply with one of the design options contained in Section 302.3. Accessory areas, incidental use areas or any combination of these areas are not considered as mixed use. This condition is permitted because such areas either do not represent a significant change in the performance characteristics of the structure or are otherwise appropriately protected by the provisions contained in this section. Where the provisions of this section are exceeded or otherwise not complied with, the occupancy is designated a mixed-use building and must comply with Section 302.3.

In cases where a structure has a purpose that is not specifically identified within any particular group classification, that structure is to be classified in the group that it most closely resembles. Before an accurate classification can be made, however, a detailed description of the activities or processes taking place inside the building, the occupant load and the materials and equipment used and stored therein must be submitted to the building official. The building official must then compare this information to the various group classifications, determine which one the building most closely resembles and classify the building as such.

Example 2: A designer presents the building official with a building occupancy needing a group classification. The building official is informed that the building is to be used as an indoor shooting gallery, open to the public but used mostly by police officers. After reviewing the code, he or she cannot find a specific reference to a shooting gallery in Sections 303 through 312 or in the associated tables. The building official asks the designer for additional information about the activities to be conducted in the building and is told that there will be a small sign-in booth, patron waiting/viewing area and the actual shooting area. Based on this information, the building official can determine that the most logical classification of the building is Group A-3, assembly. This classification is based on the fact that the building

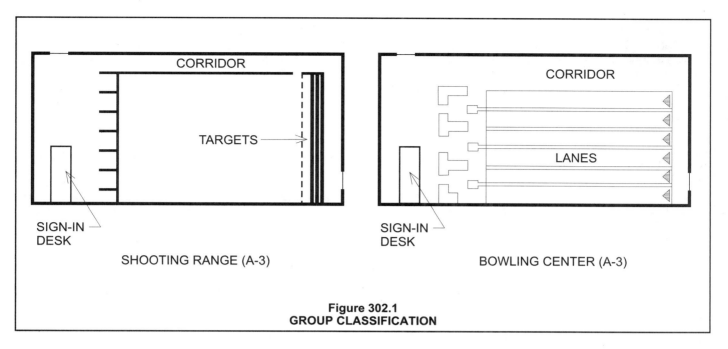

**Figure 302.1
GROUP CLASSIFICATION**

is used for the congregation of people for recreation. A shooting gallery is similar in many respects to a bowling center which is classified as Group A-3 (see Figure 302.1).

302.1.1 Incidental use areas. Spaces which are incidental to the main occupancy shall be separated or protected, or both, in accordance with Table 302.1.1 or the building shall be classified as a mixed occupancy and comply with Section 302.3. Areas that are incidental to the main occupancy shall be classified in accordance with the main occupancy of the portion of the building in which the incidental use area is located.

Exception: Incidental use areas within and serving a dwelling unit are not required to comply with this section.

❖ Incidental use areas are rooms or areas that constitute special hazards or risks to life safety beyond the protection provided by the general code requirements for the occupancy in which they are located. The requirements applicable to the main group classification are not adequate enough in order to protect against the hazards that are unique to these areas. Table 302.1.1 is applicable where an incidental use is not classified separately as to its occupancy (i.e., a mixed occupancy condition). The code intends the protection for the rooms or areas identified in Table 302.1.1 to be mandatory. These protection requirements, however, are not applicable to incidental use areas that are located within and serve a dwelling unit.

Even though incidental use areas are not considered to be mixed-use conditions, such locations may be classified separately from their occupancy and the building designed in accordance with Section 302.3 (i.e., mixed occupancies). In such cases, the presence of the higher relative fire hazard is accounted for by protection features that result from considering such areas as one of the building's group classifications. The application of

Section 302.3 will result in certain building features being governed by code requirements that are commensurate with the higher relative fire hazard of these areas.

For example, an office building (Group B) contains storage areas greater than 100 square feet (9.3 m^2) in area. If the building is classified as a single occupancy of Group B, the storage areas would be regulated as incidental use areas and would be required to be separated from the remainder of the building by 1-hour fire-resistance-rated fire barriers. Otherwise, the building is classified as a mixed occupancy building of Groups B and S. The application of the requirements of Section 302.1.1 for an incidental use area assigned to a Group S occupancy is not required and a 1-hour fire-resistance rating (in accordance with Table 302.1.1) would not be necessary.

The purpose of Table 302.1.1 is to identify the incidental use areas that require special protection and to indicate the required protection. The requirements of this table do not apply to incidental use areas that are located within or that serve a dwelling unit, nor do they apply to occupancies within the building that are classified separately from their occupancy (i.e., in accordance with Section 302.3). In the latter case, the presence of the higher relative fire hazard is accounted for by protection features that result from considering such areas as one of the building's group classifications.

For example, the intent of the code for storage rooms is to mitigate the risk of rooms with a high fuel load (boxes of paper, plastics, etc.). No supervision and the potential for unsupervised storage placed too close to furnaces or boilers characterize the fire hazard associated with these spaces.

The protection requirements identified in Table 302.1.1 vary depending on the incidental use area. In some cases, a specific type of protection is required, while in others there is an option. As indicated by Note a to Table 302.1.1, the requirement for an automatic fire-extinguishing system applies only to the incidental use room or area, not the entire building.

**TABLE 302.1.1
INCIDENTAL USE AREAS**

ROOM OR AREA	SEPARATION[a]
Furnace room where any piece of equipment is over 400,000 Btu per hour input	1 hour or provide automatic fire-extinguishing system
Rooms with any boiler over 15 psi and 10 horsepower	1 hour or provide automatic fire-extinguishing system
Refrigerant machinery rooms	1 hour or provide automatic sprinkler system
Parking garage (Section 406.2)	2 hours; or 1 hour and provide automatic fire-extinguishing system
Hydrogen cut-off rooms	1-hour fire barriers and floor/ceiling assemblies in Group B, F, H, M, S and U occupancies. 2-hour fire barriers and floor/ceiling assemblies in Group A, E, I and R occupancies.
Incinerator rooms	2 hours and automatic sprinkler system
Paint shops, not classified as Group H, located in occupancies other than Group F	2 hours; or 1 hour and provide automatic fire-extinguishing system
Laboratories and vocational shops, not classified as Group H, located in Group E or I-2 occupancies	1 hour or provide automatic fire-extinguishing system
Laundry rooms over 100 square feet	1 hour or provide automatic fire-extinguishing system
Storage rooms over 100 square feet	1 hour or provide automatic fire-extinguishing system
Group I-3 cells equipped with padded surfaces	1 hour
Group I-2 waste and linen collection rooms	1 hour
Waste and linen collection rooms over 100 square feet	1 hour or provide automatic fire-extinguishing system
Stationary lead-acid battery systems having a liquid capacity of more than 100 gallons used for facility standby power, emergency power or uninterrupted power supplies	1-hour fire barriers and floor/ceiling assemblies in Group B, F, H, M, S and U occupancies. 2-hour fire barriers and floor/ceiling assemblies in Group A, E, I and R occupancies

For SI: 1 square foot = 0.0929 m², 1 pound per square inch = 6.9 kPa,
              1 British thermal unit = 0.293 watts, 1 horsepower = 746 watts,
              1 gallon = 3.785 L.

a. Where an automatic fire-extinguishing system is provided, it need only be
   provided in the incidental use room or area.

302.1.1.1 Separation. Where Table 302.1.1 requires a fire-resistance-rated separation, the incidental use area shall be separated

rated from the remainder of the building with a fire barrier. Where Table 302.1.1 permits an automatic fire-extinguishing system without a fire barrier, the incidental use area shall be separated by construction capable of resisting the passage of smoke. The partitions shall extend from the floor to the underside of the fire-resistance-rated floor/ceiling assembly or fire-resistance-rated roof/ceiling assembly or to the underside of the floor or roof deck above. Doors shall be self-closing or automatic-closing upon detection of smoke. Doors shall not have air transfer openings and shall not be undercut in excess of the clearance permitted in accordance with NFPA 80.

❖ Table 302.1.1 identifies incidental use areas and the required separation or other protection to be provided. Where a fire-resistance rating is required, the incidental use area must be separated from other occupancies with fire barriers that comply with Section 706. Where Table 302.1.1 permits protection by an automatic fire-extinguishing system without fire barriers, the walls enclosing the incidental use area must simply resist the passage of smoke.

302.2 Accessory use areas. A fire barrier shall be required to separate accessory use areas classified as Group H in accordance with Section 302.3.2, and incidental use areas in accordance with Section 302.1.1. Any other accessory use area shall not be required to be separated by a fire barrier provided the accessory use area occupies an area not more than 10 percent of the area of the story in which it is located and does not exceed the tabular values in Table 503 for the allowable height or area for such use.

❖ Buildings often have rooms or spaces with an occupancy that is different from, but accessory to, the principal occupancy of the building. When such accessory areas are limited in size, they will not ordinarily represent a significantly different life safety hazard. This principle does not apply, however, to the incidental use areas indicated in Section 302.1.1 or where otherwise indicated in Section 302.3.1 for areas classified as Group H.

For the purposes of determining the required construction and the automatic fire-extinguishing requirements applicable to the occupancy of an accessory area, Section 302.2 permits the occupancy of an accessory area to be treated as if it were the same group classification as the principal occupancy of the fire area in which it is located. Other code requirements applicable to the occupancy of an accessory area, however, must be based on the actual occupancy of that area (e.g., means of egress, design occupant load, etc.).

Accessory use areas in conformance with Section 302.2 are not subject to the provisions for mixed occupancies (see Section 302.3). A designer, however, is not precluded from applying the provisions for mixed occupancies, provided that the requirements set forth in Section 302.3 are met. Occasionally, this approach is useful where there are multiple accessory uses within a single fire area, and some can be treated as nonseparated principal uses (i.e. they are a lesser hazard than the principal occupancy). In order for an occupancy to qualify for evaluation as an accessory use area, all of the following requirements must be satisfied.:

1. The occupancy must be ancillary to the principal purpose for which the structure is occupied. This means that the purpose and function of the area is subordinate and secondary to the structure's primary function. As such, the activities that occur in accessory use areas are necessary for the principal occupancy to properly function and would not otherwise reasonably exist apart from the principal occupancy.

2. The aggregate area within a floor devoted to the occupancies that are designated as accessory use areas must not be greater than 10 percent of the area of that floor [see Figure 302.2(1)].

3. The area devoted to an accessory occupancy is less than the tabular building height and area permitted by Table 503, based on the group classification that most nearly resembles the accessory occupancy under consideration. Height increases for automatic sprinkler protection and area increases for excess street frontage and automatic sprinkler protection based on the provisions of Sections 504 and 506 are not allowed [see Figure 302.2(2)].

Variations (exceptions) to the requirements for an area to be evaluated as an accessory use area are contained in Sections 302.2.1, 303.1 and 305.1. Section 303.1 provides that assembly areas having an occupant load of less than 50 that are accessory to another occupancy are classified as part of the main occupancy. Sections 301.1 and 302.2.1 provide that accessory assembly areas with a floor area of no greater than 750 square feet (70 m²) (regardless of the percentage of the floor area) are not required to be considered sep-

arate occupancies. Additionally, Sections 303.1 and 302.2.1 provide that accessory assembly spaces (i.e., cafeteria, gymnasium, library, etc.) in a Group E building are not to be regulated as separate occupancies. Finally, Sections 303.1 and 302.2.1 provide that accessory religious educational rooms and religious auditoriums need not be considered separate occupancies as long as the occupant load in these spaces is not greater than 100 people, regardless of the percentage of the floor area.

302.2.1 Assembly areas. Accessory assembly areas are not considered separate occupancies if the floor area is equal to or less than 750 square feet (69.7 m²). Assembly areas that are accessory to Group E are not considered separate occupancies. Accessory religious educational rooms and religious auditoriums with occupant loads of less than 100 are not considered separate occupancies.

❖ Recognizing the need for flexibility in the use of buildings with small spaces that are technically defined as assembly, the code stipulates that assembly areas that are accessory to any occupancy and have a floor area of no greater than 750 square feet (70 m²) (regardless of the percentage of the floor area) are not required to be considered separate occupancies. Some examples include lunch rooms or conference rooms in a business office, small meeting rooms, etc.

A Group E occupancy invariably contains many types of assembly spaces other than classrooms, such as auditoriums, cafeterias, gymnasiums, libraries, etc. These spaces that are accessory to a Group E building are not required to be regulated as separate occupancies.

In churches, religious educational rooms are often provided in the same building complex. These religious educational rooms are not to be considered separate occupancies from the church (i.e., the religious auditorium) as long as the occupant load in these spaces is no

FACTORY (F-1)
9,600 SQ.FT.

OFFICE (B)
400 SQ.FT.

OFFICES (B)
800 SQ.FT.

• TOTAL AREA OF FLOOR = 10,800 SQ.FT.

• ACCESSORY USE AREAS (AGGREGATE) = 1,200 SQ.FT.

• MAXIMUM ALLOWABLE AREA OF ACCESSORY USES
  (AGGREGATE) = (.10) (10,800) = 1,080 SQ.FT.

• ACCESSORY USE AREAS EXCEED THE AREA PERMITTED
  (I.E., 1,200 > 1,080)

• BUILDING MUST BE REEVALUATED AS A MIXED OCCUPANCY
  OF OFFICES (B) AND FACTORY (F-1).

For SI: 1 square foot = 0.0929 m².

Figure 302.2(1)
ACCESSORY USES LIMITED BY FLOOR AREA

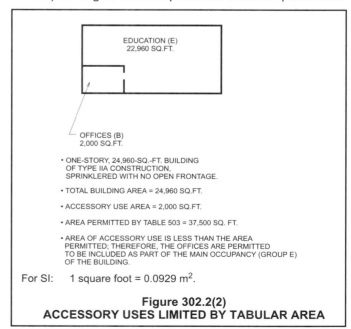

EDUCATION (E)
22,960 SQ.FT.

OFFICES (B)
2,000 SQ.FT.

• ONE-STORY, 24,960-SQ.-FT. BUILDING
  OF TYPE IIA CONSTRUCTION,
  SPRINKLERED WITH NO OPEN FRONTAGE.

• TOTAL BUILDING AREA = 24,960 SQ.FT.

• ACCESSORY USE AREA = 2,000 SQ.FT.

• AREA PERMITTED BY TABLE 503 = 37,500 SQ. FT.

• AREA OF ACCESSORY USE IS LESS THAN THE AREA
  PERMITTED; THEREFORE, THE OFFICES ARE PERMITTED
  TO BE INCLUDED AS PART OF THE MAIN OCCUPANCY (GROUP E)
  OF THE BUILDING.

For SI: 1 square foot = 0.0929 m².

Figure 302.2(2)
ACCESSORY USES LIMITED BY TABULAR AREA

greater than 100 people, regardless of the percentage of the floor area.

302.3 Mixed occupancies. Where a building is occupied by two or more uses not included in the same occupancy classification, the building or portion thereof shall comply with Section 302.3.1 or 302.3.2 or a combination of these sections.

Exceptions:

1. Occupancies separated in accordance with Section 508.

2. Areas of Group H-2, H-3, H-4 or H-5 occupancies shall be separated from any other occupancy in accordance with Section 302.3.2.

3. Where required by Table 415.3.2, areas of Group H-1, H-2 or H-3 occupancy shall be located in a separate and detached building or structure.

4. Accessory use areas in accordance with Section 302.2.

5. Incidental use areas in accordance with Section 302.1.1.

❖ This section describes the provisions governing the condition when a building contains more than one of the groups identified in Sections 303 through 312.

Buildings very often contain more than one group classification. For example, a facility that appears to be used solely for storage often has a small office area for bookkeeping and other purposes. Because the office would be classified as a business occupancy, the building would be identified as a mixed occupancy, storage (Group S-1) and business (Group B), unless the accessory use area provisions of Section 302.2 apply.

Most buildings containing mixed occupancies are clearly identifiable, such as a car dealership that consists of an office/salesroom area and a service garage. There is little question that this building would be considered a mixed occupancy (Groups B and S-1), subject to the provisions of Section 302.3.

Once a building is determined to contain more than one group, the code requires that the combination of potential fire and life safety hazards be addressed. For example, if an occupancy such as a theater is located in the same building where large quantities of paper products are stored, the large occupant load in the theater could be endangered by a fire emergency in the storage area. This is what Section 302.3 is about: recognizing that each group carries distinctive risks and hazards; therefore, in the presence of more than one group, more than one hazard will exist. Accordingly, this section sets forth the requirements for buildings containing more than one group classification so that the appropriate relative code provisions can be applied.

The term "mixed occupancy" is not, in itself, a group classification. It is stating a fact that there exists a combination of different group classifications within one building.

For a building to be subject to the mixed occupancy provisions, it must contain two or more group classifications that are considered to be principal occupancies. It is possible (but not probable) for a building to contain

some spaces dedicated to 26 different occupancies.

Mixed occupancy conditions do not always occur adjacent to each other on the same floor level; they commonly occur in a vertical or stacked manner. For example, consider a two-story building that has the first story devoted to a department store (mercantile occupancy) and the second story devoted to office space (business occupancy). This building is a mixed occupancy condition subject to the same provisions as if the occupancies were located side by side. When applying the mixed occupancy provisions, it does not matter if the occupancies are located horizontally or vertically [see Figures 302.3(1) and 302.3(2)].

This section contains two options that can be used to address mixed occupancy situations: nonseparated uses (see Section 302.3.1) and separated uses (see Section 302.3.2).

When more than two occupancies are present in a building and except for occupancies classified in Group H, the designer may use either or both of the two options available to mixed occupancies. This decision depends on how he or she wishes to provide for the different group classifications present [see Figure 302.3(3)] and the building configuration.

Each design option, when fully complied with, provides a building with an acceptable level of safety. An owner/designer may find one of the options more attractive than another based on the building's purpose, function or construction cost. The selection of which option to use remains with the permit applicant and not the building official [see Figure 302.3(4)].

Exception 1 references Section 508, which contains special provisions for certain buildings with mixed occupancies, including enclosed parking garages below Group A, B, M or R; enclosed parking garage below an open parking garage and open parking structures beneath Groups A, B, I, M and R.

Exception 2 states that areas classified in Group H-2, H-3, H-4 or H-5 are not permitted to be designed as nonseparated mixed occupancies (see Section 301.3.1). As such, Groups H-2, H-3, H-4 and H-5 must be separated from all other occupancies in accordance with Section 302.3.2 or 705. While not a specific requirement, it should be noted that fire barriers between H-2, H-3, H-4 or H-5 occupancies may also be appropriate where incompatible materials are stored or used. The need to separate Groups H-2, H-3, H-4 and H-5 from other occupancies takes into consideration the control area concept for hazardous materials and other exceptions contained in Section 307.9, which allows increased quantities of hazardous materials in a building in specific circumstances without the building or occupancy being classified in Group H (see commentary, Sections 307.9 and 414.2).

Exception 3 stipulates that the design options contained in Sections 302.3.1 and 302.3.2 are not applicable to Group H-1, H-2 or H-3 where Table 415.3.2 requires detached structures for certain materials, such as explosives. These occupancies with such materials must always be located in a building or structure that is detached and separate from any other occupancy, including Groups H-4 and H-5 and other portions of Group H-1, H-2 or H-3 where materials other than those specified in Table 415.3.2 are contained.

Section 415.4 contains special provisions for H-1 occupancies, including the requirement that H-1 occupancies must be in buildings that are occupied for no other use (see commentary, Section 415.4).

Exceptions 4 and 5 relate to Sections 302.1.1 and 302.2, which provide for two conditions under which certain dissimilar occupancies are not treated as mixed occupancies. In consideration of the special protection afforded incidental use areas described in Section 302.1.1, such areas are intended to be allowed without affecting the group classification. Section 302.2 provides for occupancies that are permitted to be considered as accessory use areas when such occupancies meet the specified relative area limitations. For both incidental use areas and accessory use areas, the building is intended to be classified in accordance with the main occupancy. Hence, the provisions of this section do not apply to incidental use areas or accessory use areas.

302.3.1 Nonseparated uses. Each portion of the building shall be individually classified as to use. The required type of construction for the building shall be determined by applying the height and area limitations for each of the applicable occupancies to the entire building. The most restrictive type of construction, so determined, shall apply to the entire building. All other code requirements shall apply to each portion of the building based on the use of that space except that the most restrictive applicable provisions of Section 403 and Chapter 9 shall apply to these nonseparated uses. Fire separations are not required between uses, except as required by other provisions.

❖ This section describes the first option a designer may choose to apply to a building that contains more than one group classification. Except with respect to occupancies classified in Group H, this option differs from the other one in that there is no requirement for various occupancies to be physically separated by any type of fire-resistance-rated assembly. Note, however, in some cases it could happen that there is a wall between two different occupancies that is required by another section of the code to be fire-resistance rated (e.g., a separation wall is provided that also serves as a corridor wall that is required to be fire-resistance rated).

The principle behind nonseparated uses is that the different occupancies within the same building do not have to be separated by fire-resistance-rated assemblies if the building complies throughout with the more restrictive code requirements for minimum construction type high-rise buildings (Section 403) and fire protection systems (Chapter 9). Although each occupancy is separately classified as to its group, a fire-resistance-rated assembly is not required by the nonseparated uses option. A designer may choose to physically separate the occupancies; however, a fire-resistance rating of these separations would not be required by this section.

There are four basic steps to follow when applying the nonseparated uses concept:

Step 1: Determine which group classifications are present in the building.

Step 2: Determine the minimum type of construction based on the height and area of the building for each group in accordance with Chapter 5 and Table 503. Apply the requirement for the highest type to the entire building.

For example, the tabular area for Group S-1 is less than the corresponding tabular area for Group B; therefore, Group S-1 results in a requirement for a higher type of construction and will, therefore, determine the minimum construction type of the building.

Step 3: Apply the most restrictive provisions contained in Section 403 and Chapter 9 throughout all nonseparated uses. It is important to note the threshold requirements for fire protection systems contained in Chapter 9. In some cases, they are simply based on the occupancy and in other cases they are based on height or area criteria. Also, the requirements to be met in some cases only apply to the fire area in which a given occupancy is contained, while in others they apply to the entire building.

For example, if the business occupancy as described in Figure 302.3.1 required a fire alarm system with manual fire alarm boxes, then one would be required for the entire building—even though Chapter 9 does not require a fire alarm system for a storage occupancy.

Step 4: Apply all other requirements of the code, except for Section 403 and Chapter 9, to each occupancy individually based on the specific occupancy of each space (e.g., means of egress see commentary, Chapter 10).

For example, if a building contains both Groups B and M, each portion of the building must comply with the code requirements for its respective group classification. The occupant load for the area classified

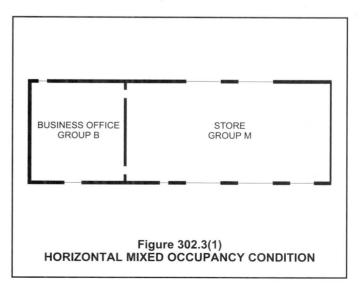

Figure 302.3(1)
HORIZONTAL MIXED OCCUPANCY CONDITION

FIGURE 302.3(2) – FIGURE 302.3(4)

USE AND OCCUPANCY CLASSIFICATION

as Group B is based on the code requirements applicable to the business occupancy. Similarly, the occupant load for the Group M area is based on requirements applicable to the mercantile occupancy. Live loads for the Group B area are applied to the area of business occupancy. Group M live loads are applied to the area of mercantile occupancy. Exterior wall ratings based on fire separation distance (see Table 602) apply to exterior walls of the business occupancy for Group B buildings; the same is true for walls in Group M occupancies.

Example: A building (see Figure 302.3.1) contains a 46,200-square-foot (4292 m²) general warehouse and an accompanying 30,000-square-foot (2787 m²) regional dispatch office for a total area of 76,200 square feet (7079 m²). The building is a one-story, pre-engineered metal building. Assume that there is no allowable area increase for frontage. In applying the four-step procedure for nonseparated uses, the following holds true:

Step 1: The warehouse is classified in Group S-1 and the offices are classified in Group B.

Step 2: The minimum type of construction is based on the lesser allowable area of the building for each occupancy classification. First, determine if any increases can be obtained (no frontage increase—check for installation of a sprinkler system). In Step 3, the building is shown to be sprinklered throughout; therefore, a 300-percent increase can be applied to the tabular areas (see Section 506.3). For the purpose of analysis, assume the building is of Type IIB construction. The Group B requirements would allow the building to be 92,000 square feet (8547 m²) (23,000 square feet + 300 percent of 23,000 square feet). The Group S-1 requirements, however, only allow a building of Type IIB construction to be 70,000 square feet (6503 m²) (17,500 square feet + 300 percent of 17,500 square feet). The most restrictive is the Group S-1 requirement;

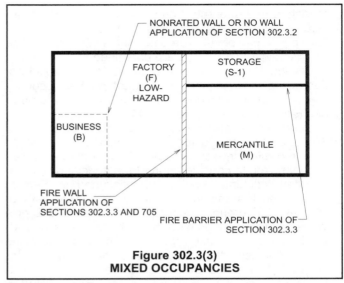

Figure 302.3(3)
MIXED OCCUPANCIES

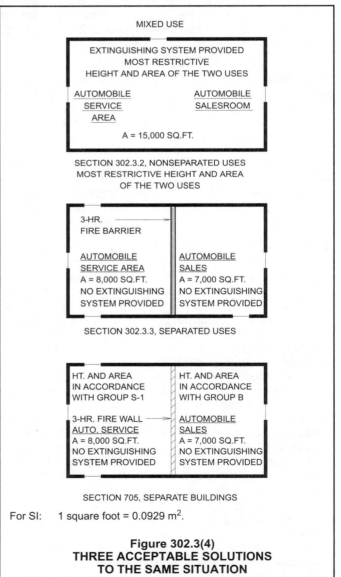

For SI: 1 square foot = 0.0929 m².

Figure 302.3(4)
THREE ACCEPTABLE SOLUTIONS
TO THE SAME SITUATION

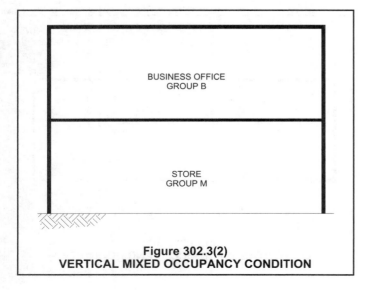

Figure 302.3(2)
VERTICAL MIXED OCCUPANCY CONDITION

therefore, it governs. Hence, the minimum type of construction is required to be Type IIA.

Step 3: The fire area containing the occupancy classified in Group S-1 is more than 12,000 square feet (1115 m²); thus, in accordance with Section 903.2.10, this area must be sprinklered. The area classified as Group B ordinarily is not required to be equipped with sprinklers by Section 903; however, since the Group S-1 provision is more restrictive than the requirement for Group B under Chapter 9, the entire building must be sprinklered.

Step 4: The occupant load of each area should be calculated separately (based on the occupancy of that area) using Section 1004.1.2 and Table 1004.1.2.

Business: 30,000 square feet (2787 m²) divided by 100 square feet (9 m²) per person = 300 people.

Storage: 46,200 square feet (4292 m²) divided by

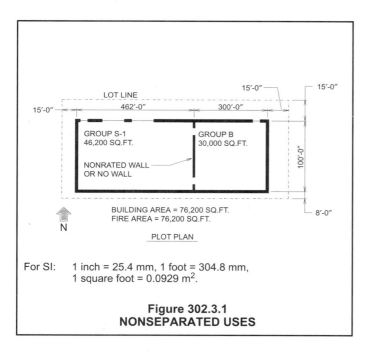

For SI: 1 inch = 25.4 mm, 1 foot = 304.8 mm,
1 square foot = 0.0929 m².

**Figure 302.3.1
NONSEPARATED USES**

500 square feet (46 m²) per person = 93 people.

Total Occupant Load: 393 people.

The evaluation of means of egress in accordance with Chapter 10 is then based on these occupant loads.

The load-bearing members in the north, east and west exterior walls of the building must provide a minimum 1-hour fire-resistance rating as required for Type IIA construction in Table 601. The nonload-bearing walls are also required to have a 1-hour fire-resistance rating because of the 15-foot (4572 m²) fire separation distance (see Table 602).

The fire separation distance on the south side of the mixed occupancy building is 8 feet (2438 mm). Table 602 requires the Group S-1 and B portions of this wall to

have a 1-hour fire-resistance rating. The ratings required by this table apply to both load-bearing and nonload-bearing exterior walls. The substitution of an automatic sprinkler system in lieu of 1-hour fire-resistance-rated construction (as indicated in Note d of Table 601) does not apply to exterior walls; therefore, all exterior walls (whether load bearing or nonload bearing) in this building are required to have a 1-hour fire-resistance rating.

This analysis is continued for other aspects of code compliance for each respective occupancy. Once all four steps have been satisfied, the building is deemed to have an equivalent level of safety as a building with only one occupancy. Any future changes in the occupancy would require review and approval by the local jurisdiction in accordance with the provisions contained in Sections 105.1 and 3405.

302.3.2 Separated uses. Each portion of the building shall be individually classified as to use and shall be completely separated from adjacent areas by fire barrier walls or horizontal assemblies or both having a fire-resistance rating determined in accordance with Table 302.3.2 for uses being separated. Each fire area shall comply with this code based on the use of that space. Each fire area shall comply with the height limitations based on the use of that space and the type of construction classification. In each story, the building area shall be such that the sum of the ratios of the floor area of each use divided by the allowable area for each use shall not exceed one.

Exception: Except for Group H and I-2 areas, where the building is equipped throughout with an automatic sprinkler system, installed in accordance with Section 903.3.1.1, the fire-resistance ratings in Table 302.3.2 shall be reduced by 1 hour but to not less than 1 hour and to not less than that required for floor construction according to the type of construction.

❖ This section describes the second (separate uses) option a designer may choose to apply when constructing a building that contains more than one occupancy classification. The second option differs from the first alternative (i.e., nonseparated uses) (see Section 302.3.1) in three ways:

- Occupancies in different uses that are to be evaluated as separated uses must be located in separate fire areas. As such, they must be separated completely, both horizontally and vertically, with each area having fire barrier walls and horizontal assemblies (see Sections 706 and 710) with a fire-resistance rating as prescribed in Table 302.3.2. It is the designer's option to use a combination of nonseparated and separated uses. In such cases, fire areas with nonseparated uses are evaluated based on the most restrictive code provisions applicable to the occupancies within that fire area.

- The determination of sprinkler requirements in Section 903 is based on the occupancies present, respective to the square footage of the fire area in which they are located.

TABLE 302.3.2
REQUIRED SEPARATION OF OCCUPANCIES (HOURS)[a]

USE	A-1	A-2[c]	A-3	A-4	A-5	B[b]	E	F-1	F-2	H-1	H-2	H-3	H-4	H-5	I-1	I-2	I-3	I-4	M[b]	R-1	R-2	R-3, R-4	S-1	S-2[c]	U
A-1	—	2	2	2	2	2	2	3	2	NP	4	3	2	4	2	2	2	2	2	2	2	2	3	2	1
A-2[c]		—	2	2	2	2	2	3	2	NP	4	3	2	4	2	2	2	2	2	2	2	2	3	2	1
A-3			—	2	2	2	2	3	2	NP	4	3	2	4	2	2	2	2	2	2	2	2	3	2	1
A-4				—	2	2	2	3	2	NP	4	3	2	4	2	2	2	2	2	2	2	2	3	2	1
A-5					—	2	2	3	2	NP	2	1	1	1	2	2	2	2	2	2	2	2	3	2	1
B[b]						—	2	3	2	NP	2	1	1	1	2	2	2	2	2	2	2	2	3	2	1
E							—	3	3	NP	4	3	1	1	3	3	2	3	3	3	3	3	3	3	3
F-1								—	3	NP	2	1	1	1	2	2	2	2	1	2	2	2	2	2	1
F-2									—	NP	2	1	1	1	2	2	2	2	1	2	2	2	2	1	3
H-1										—	NP	NP	NP	NP	NP	NP	NP	NP	NP	NP	NP	NP	NP	NP	NP
H-2											—	1	2	2	4	4	4	4	2	4	4	4	2	2	2
H-3												—	1	1	4	3	3	3	1	3	3	3	1	1	1
H-4													—	1	4	4	4	4	1	4	4	4	1	1	1
H-5														—	4	4	4	3	1	4	4	4	1	1	3
I-1															—	2	2	2	2	2	2	2	4	3	2
I-2																—	2	2	2	2	2	2	3	2	1
I-3																	—	2	2	2	2	2	3	2	1
I-4																		—	2	2	2	2	3	2	1
M[b]																			—	2	2	2	3	2	1
R-1																				—	2	2	3	2	1
R-2																					—	2	3	2	1
R-3, R-4																						—	3	2[d]	1[d]
S-1																							—	1	1
S-2[c]																								—	1
U																									—

For SI: 1 square foot = 0.0929 m².

NP = Not permitted.

a. See exception to Section 302.3.2 for reductions permitted.

b. Occupancy separation need not be provided for storage areas within Groups B and M if the:
   1. Area is less than 10 percent of the floor area;
   2. Area is provided with an automatic fire-extinguishing system and is less than 3,000 square feet; or
   3. Area is less than 1,000 square feet.

c. Areas used only for private or pleasure vehicles shall be allowed to reduce separation by 1 hour.

d. See Section 406.1.4.

e. Commercial kitchens need not be separated from the restaurant seating areas that they serve.

• The determination of the minimum type of construction is based on both the height of each fire area relative to the grade plane and the areas of each occupancy relative to the total floor area per story.

There are six basic steps to follow when using the separated uses option:

Step 1: Determine which occupancies are present in the building.

Step 2: Separate the building into fire areas with fire barrier walls and horizontal assemblies in accordance with Sections 706 and 710.

Step 3: Determine the fire-resistance rating of the fire barrier walls and horizontal assemblies from Table 302.3.2.

Pursuant to the Exception 4, the rating determined from the table may be reduced by 1 hour but not to less than 1 hour or the fire-resistance rating required by Table 601 for floor construction, if the entire building (not just a fire area) is equipped throughout with an automatic sprinkler system in accordance with NFPA 13. Group H and I-2 areas are not allowed this additional reduction even if the entire building is sprinklered.

For example, a completely sprinklered building of Type VB construction contains areas devoted to business and mercantile occupancies. The designer has chosen the separated uses option and has completely separated the areas containing the two different occupancies by fire barrier walls and horizontal assemblies having a minimum 2-hour fire-resistance rating in accordance with Table 302.3.2. Because the entire building is sprinklered, the fire-resistance rating of the fire barrier walls and horizontal assemblies is permitted to be reduced to 1 hour.

It is important to realize that the 1-hour reduction based on the presence of sprinklers can only be utilized if the entire building is sprinklered, not just a fire area.

Step 4: Determine which fire protection code provisions contained in Section 403 and Chapter 9 apply to each fire area. For the application of code provisions, each fire area is taken into consideration separately. If a code section identifies requirements as applicable to a mercantile occupancy, it is only applied to the fire area of the building classified in Group M. The same holds true for the business occupancy fire area. Unlike the first option (nonseparated uses), the provisions of Section 403 and Chapter 9 are applied only to the appropriate fire areas unless otherwise indicated in Chapter 9. For example, Section 903.2.5 requires an automatic sprinkler system to be provided throughout all buildings that contain a Group I fire area.

Step 5: Determine the minimum type of construction of a building based on the height limitations of Sections 503 and 504 and the area limitations of Sections 503 and 506.

Part A: Determine the minimum type of construction required based on the height of each fire area relative to the grade plane. This is accomplished by comparing the height of each fire area respective to the most restrictive use contained within that fire area to the height allowed by Table 503 as modified by Section 504 for buildings in that group.

Part B: Determine the minimum type of construction based on a weighted average of areas occupied by the various occupancies. For each story, determine that the sum of the ratios of the actual floor area of each fire area, respective to the most restrictive occupancy contained therein as compared to the floor areas allowed by Table 503 and as modified by Section 506 for each respective occupancy, does not exceed one. In the evaluation of allowable area, intervening fire barrier walls between different fire areas containing the same occupancy are not a consideration. In determining the floor area per occupancy, all fire areas of the same occupancy are added together.

In determining the allowable areas for each occupancy, the tabular areas from Table 503 are permitted to be modified in accordance with the provisions of Section 506; thus, the allowable areas are intended to include the increases permitted for sprinklers and open perimeter. For determination of the allowable perimeter increase, use the entire building perimeter—not the fire area perimeter. If considering the sprinkler increase, the entire building must be sprinklered, not just particular fire areas.

Step 6: Similar to Step 4 for nonseparated uses described in Section 302.3.1, apply all other code requirements for each fire area individually based on the occupancy or occupancies present (i.e., design occupant load, means of egress elements, exterior wall requirements, etc.) (see commentary, Section 302.3.1).

Example 1: A four-story building contains a retail store on the first floor, a lecture hall on the second floor and offices on the top two floors. Each story is 10,500 square feet (975 m^2). A sprinkler system is not provided. Assume that there is no open perimeter increase [see Figure 302.3.2(1a)]. In applying the six-step procedure for separated uses, the following holds true:

Step 1: Fire area 1 is Group M throughout.
Fire area 2 is Group A-3 throughout.
Fire area 3 is Group B throughout. Note fire area 3 includes both the third and fourth floors, because there is not a rated horizontal assembly located between them.

FIGURE 302.3.2(1a) – FIGURE 302.3.2(1b)   USE AND OCCUPANCY CLASSIFICATION

Step 2: Each occupancy is separated horizontally with an approved fire-resistance-rated assembly. Fire area 1 is Group M, 2 is Group A-3 and 3 is Group B. Section 710 for fire-resistance-rated floor/ceiling assemblies is applicable.

Group M fire area = 10,500 square feet (975 m²).
Group A-3 fire area = 10,500 square feet (975 m²).
Group B fire area = 21,000 square feet (1951 m²) [combined area for two stories; each 10,500 square feet (975 m²)].

Step 3: In accordance with Table 302.3.2, a 2-hour fire-resistance rating is required for the horizontal assembly between fire area 1 (Group M) and fire area 2 (Group A-3). Likewise, a 2-hour fire-resistance-rated horizontal assembly is required between fire area #2 (Group A-3) and fire area 3 (Group B). Since the building is not equipped throughout with an automatic sprinkler system, no reduction in the required rating is allowed.

Step 4: Since the fire areas containing Groups A-3 and M are less than 12,000 square feet (1115 m²) and Group M is no more than three stories in height, neither fire area 1 (Group M) nor fire area 2 (Group A-3) requires sprinklers (see Section 903). Fire area 3 (Group B) does not require sprinklers because neither the provisions in Section 903 nor the high-rise building requirements (Section 403) apply; therefore, no sprinklers are required in any of the fire areas based on occupancy classification. Nevertheless, the requirements for incidental use areas would still be applicable within each fire area (see commentary, Section 302.1.1).

Step 5:
Part A: Review for the minimum types of construction based on the height limitations for each of the three fire areas as follows [see Figure 302.3.2(1b)]:

1. Fire area 1: (Group M), one story in height; the minimum type of construction is Type VB [one story and 40 feet (12 192 mm) in accordance with Table 503].
2. Fire area 2: (Group A-3), two stories in height; the minimum type of construction is Type VA [two stories and 50 feet (15 240 mm) in accordance with Table 503].
3. Fire area 3: (Group B), four stories in height; the minimum type of construction is Type IIIB [four stories and 55 feet (16 764 mm) in accordance with Table 503].

The highest type of construction for any one of the fire areas governs for the entire building; thus, the minimum allowable type of construction permitted for the building based on height is Type IIIB. The designer is permitted to choose any one of the construction types other than Types VA and VB for the design of the building.

If the building were completely sprinklered, Section 504.2 would permit a height increase. Accordingly, under Table 503, the minimum type of construction allowed for the building would be Type VA. Exception 1 to Section 302.3.2 would also allow the 2-hour fire-resistance-rated horizontal assembly (identified in Step 3) to be reduced to 1 hour.

NO FIRE-RESISTANCE RATING

2-HR. FIRE-RESISTANCE RATED IN ACCORDANCE WITH TABLE 302.3.2

GROUP B	FIRE AREA 3
GROUP B	
GROUP A-3	FIRE AREA 2
GROUP M	FIRE AREA 1

GRADE

10,500 SQ.FT. PER FLOOR
NO OPEN PERIMETER INCREASE
NO SPRINKLER SYSTEM

For SI: 1 square foot = 0.0929 m².

Figure 302.3.2(1a)
SEPARATED USES: FIRE AREAS

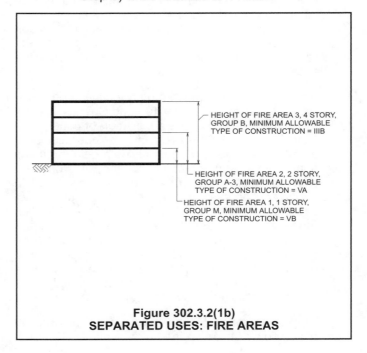

HEIGHT OF FIRE AREA 3, 4 STORY, GROUP B, MINIMUM ALLOWABLE TYPE OF CONSTRUCTION = IIIB

HEIGHT OF FIRE AREA 2, 2 STORY, GROUP A-3, MINIMUM ALLOWABLE TYPE OF CONSTRUCTION = VA

HEIGHT OF FIRE AREA 1, 1 STORY, GROUP M, MINIMUM ALLOWABLE TYPE OF CONSTRUCTION = VB

Figure 302.3.2(1b)
SEPARATED USES: FIRE AREAS

Part B: Since each floor has only one occupancy, the maximum allowable area for a story of the building is derived directly from Table 503. Based on the allowable construction type determined, Type IIIB construction is acceptable for Groups B and M; however, it is not acceptable for Group A-3. Based on the given area of 10,500 square feet (975 m²) and without any area increases, the allowable types of construction permitted for Group A-3 based on area are Types IA, IB, IIA, IIIA, IV and VA.

From Part A it was determined that Type VA construction is not permitted for the height of the building due to Group B; therefore, the allowable types of construction for the building based on height and area are Types IA, IB, IIA, IIIA and IV.

Step 6: Utilize the provisions in other parts of the code for each fire area respective to means of egress, exterior wall requirements, etc., based on the occupancy contained therein. For example, the design occupant load of fire area 1 is based on Group M, while fire area 2 is based on Group A-3.

Example 2: Consider a 76,200-square-foot (7079 m²) building identical to that used in the example for Section 302.3.1, except that the wall separating the 46,200-square-foot (4292 m²) warehouse from the 30,000-square-foot (2787 m²) office space is a fire barrier wall [see Figure 302.3.2(2)].

Step 1: The warehouse fire area is classified in Group S-1. The office fire area is classified in Group B.

Step 2: A fire barrier wall is located between the two occupancies and extends continuously through all concealed spaces to the underside of the roof deck in accordance with Section 706.

Step 3: The fire-resistance rating of the wall is required to be 3 hours by Table 302.3.2. The exception to Section 302.3.2, however, allows this to be reduced by 1 hour when the building is sprinklered throughout in accordance with NFPA 13. In Step 4, we will find that sprinklers are required throughout; hence, the required fire-resistance rating is only 2 hours.

Step 4: The fire area for the occupancy of Group B is one story and 30,000 square feet (2787 m²). A sprinkler system is not required by the code for a one-story Group B fire area. The Group S-1 fire area is 46,200 square feet (4292 m²), which in accordance with Section 903.2.8, requires the entire building to be sprinklered. Note that even though the uses are separated, Section 903.2.8 requires sprinklers throughout all buildings containing a Group S-1 fire area greater than 12,000 square feet (1115 m²).

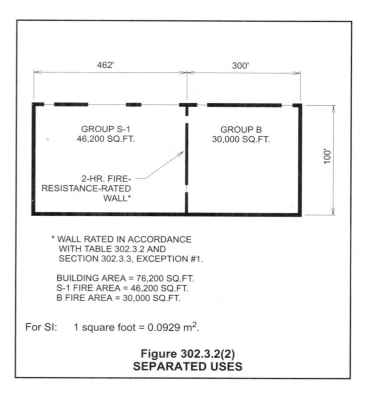

* WALL RATED IN ACCORDANCE WITH TABLE 302.3.2 AND SECTION 302.3.3, EXCEPTION #1.

BUILDING AREA = 76,200 SQ.FT.
S-1 FIRE AREA = 46,200 SQ.FT.
B FIRE AREA = 30,000 SQ.FT.

For SI: 1 square foot = 0.0929 m².

**Figure 302.3.2(2)
SEPARATED USES**

Step 5: Part A: Each fire area is only one story in height. Based on height, any type of construction is acceptable for the occupancies under consideration.

Part B: Make a trial evaluation of the building assuming it to be Type IIB construction.

Actual area of Group S-1 = 46,200 square feet (4292 m²) and actual area of Group B = 30,000 square feet (2787 m²).

Group S-1 maximum allowable area, including allowable increases (allowable tabular area plus sprinkler increase) = 17,500 + (17,500)(3.0) = 70,000 square feet (6503 m²).

The maximum allowable area for Group B including allowable increases (allowable tabular area plus sprinkler increase) = 23,000 + (23,000)(3.0) = 92,000 square feet (8547 m²).

$$\frac{\text{Actual Area (S - 1)}}{\text{Allowable Area (S - 1)}} + \frac{\text{Actual Area(B)}}{\text{Allowable Area(B)}}$$

$$\frac{46,200}{70,000} + \frac{30,000}{92,000} = 0.66 + .326 = .986$$

(which is ≤ 1.0)

Therefore, Type IIB construction is an acceptable minimum type of construction. In the case of multistory buildings, an evaluation of the allowable area must be performed for every story of the building that contains separated

FIGURE 302.3.2(3) – FIGURE 302.3.2(4)   USE AND OCCUPANCY CLASSIFICATION

uses.

When the factor arrived at is substantially below 1.0, it indicates that the building might qualify as a lower type of construction or as nonseparated uses.

Step 6: Evaluate each fire area in the same manner as in Step 4 of the example described above for Section 302.3.1.

The building is deemed to comply with the code after each of the code requirements, as determined by Steps 1 through 6, is satisfied.

Other examples of horizontal separations are detailed in Figures 302.3.2(3) through 302.3.2(5). Also, an example of vertical separation is detailed in Figure 302.3.2(6).

Since Section 302.3.1 (nonseparated uses) makes a reference to applying the most restrictive requirements of Section 403 and Chapter 9, there may be the assumption that with separated uses, requirements can be applied separately to each occupancy in all cases. This is not true. Section 403.3.1 permits a modification to the minimum type of construction for all but certain

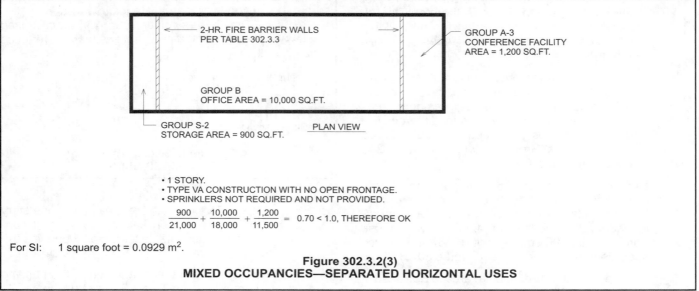

2-HR. FIRE BARRIER WALLS PER TABLE 302.3.3

GROUP A-3 CONFERENCE FACILITY AREA = 1,200 SQ.FT.

GROUP B OFFICE AREA = 10,000 SQ.FT.

GROUP S-2 STORAGE AREA = 900 SQ.FT.

PLAN VIEW

- 1 STORY.
- TYPE VA CONSTRUCTION WITH NO OPEN FRONTAGE.
- SPRINKLERS NOT REQUIRED AND NOT PROVIDED.

$$\frac{900}{21,000} + \frac{10,000}{18,000} + \frac{1,200}{11,500} = 0.70 < 1.0, \text{ THEREFORE OK}$$

For SI: 1 square foot = 0.0929 m^2.

Figure 302.3.2(3)
MIXED OCCUPANCIES—SEPARATED HORIZONTAL USES

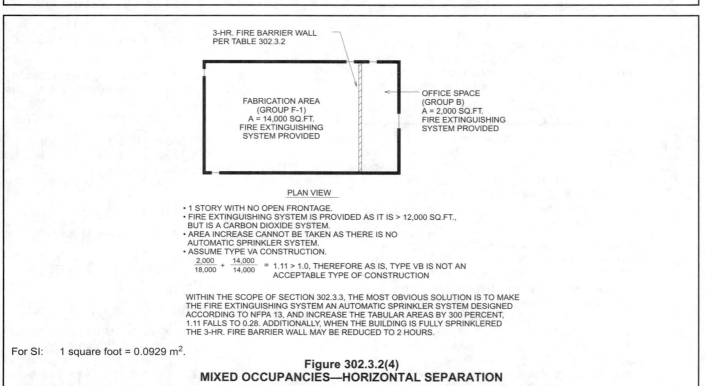

3-HR. FIRE BARRIER WALL PER TABLE 302.3.2

FABRICATION AREA (GROUP F-1) A = 14,000 SQ.FT. FIRE EXTINGUISHING SYSTEM PROVIDED

OFFICE SPACE (GROUP B) A = 2,000 SQ.FT. FIRE EXTINGUISHING SYSTEM PROVIDED

PLAN VIEW

- 1 STORY WITH NO OPEN FRONTAGE.
- FIRE EXTINGUISHING SYSTEM IS PROVIDED AS IT IS > 12,000 SQ.FT., BUT IS A CARBON DIOXIDE SYSTEM.
- AREA INCREASE CANNOT BE TAKEN AS THERE IS NO AUTOMATIC SPRINKLER SYSTEM.
- ASSUME TYPE VA CONSTRUCTION.

$$\frac{2,000}{18,000} + \frac{14,000}{14,000} = 1.11 > 1.0, \text{ THEREFORE AS IS, TYPE VB IS NOT AN}$$
ACCEPTABLE TYPE OF CONSTRUCTION

WITHIN THE SCOPE OF SECTION 302.3.3, THE MOST OBVIOUS SOLUTION IS TO MAKE THE FIRE EXTINGUISHING SYSTEM AN AUTOMATIC SPRINKLER SYSTEM DESIGNED ACCORDING TO NFPA 13, AND INCREASE THE TABULAR AREAS BY 300 PERCENT, 1.11 FALLS TO 0.28. ADDITIONALLY, WHEN THE BUILDING IS FULLY SPRINKLERED THE 3-HR. FIRE BARRIER WALL MAY BE REDUCED TO 2 HOURS.

For SI: 1 square foot = 0.0929 m^2.

Figure 302.3.2(4)
MIXED OCCUPANCIES—HORIZONTAL SEPARATION

occupancies of Type I construction. If one of these occupancies is located within a high-rise building and the separated uses option is selected, modification of the type of construction for that occupancy would not be permitted, since the building can only be classified as one type of construction (see Section 602.1). Additionally, there are sections of Chapter 9 such as Sections 903.25, 903.2.7 or 903.2.8, that require sprinklers throughout the entire building when a particular occupancy is present.

Some sections in Chapter 9, however, do permit separate application of requirements, even though the term "fire area" is not used. For example, a warehouse building with an office would be categorized as Groups B (office) and S-1 (warehouse). If the separated uses option is selected and the office has an occupant load of at least 500 people, then a fire alarm system is only required in the Group B area and not the Group S-1 area (see Section 907.2).

A third option available to the designer of structures that contain more than one occupancy requires that the occupancies are completely separated by fire walls (see Section 705) and is the simplest option to apply and analyze. Then, for code application purposes, two or more separate and independent buildings (see the definition of "Building" in Section 202.2) are created, and each building is reviewed individually.

The purpose of Table 302.3.2 is to set forth the fire-resistance rating required for fire barrier walls and horizontal assemblies used to create separate fire areas in different classified occupancies. The fire-resistance rating of the separation between different occupancies is based on the relative anticipated fire severity of the occupancies. For application of Table 302.3.2, refer to the commentary to Sections 302.3.2, 706.3.6 and 711.3.

Note a: See the commentary to the exception for Section 302.3.2.

Note b: Many business and mercantile occupancies are very similar in operation and to the types of storage that may occur. Storage rooms that are greater than 100 square feet (9 m²) in area and satisfy any of the three options listed are not required to be separated from the business or mercantile occupancy.

Note c: The fire-resistance rating of spaces used solely for private or pleasure vehicles may be reduced by 1 hour.

Note d: See the commentary to the exceptions for Section 302.3.2.

Note e: Kitchens and the restaurants they serve need not be separated from each other by fire barrier walls.

302.4 Spaces used for different purposes. A room or space that is intended to be occupied at different times for different purposes shall comply with all the requirements that are applicable to each of the purposes for which the room or space will be occupied.

❖ Occasionally, a building or space is intended to be occupied for completely different purposes at different times.

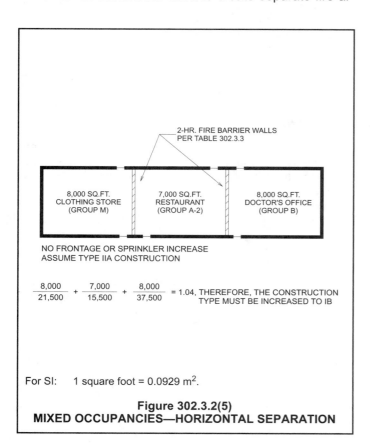

For SI: 1 square foot = 0.0929 m².

Figure 302.3.2(5)
MIXED OCCUPANCIES—HORIZONTAL SEPARATION

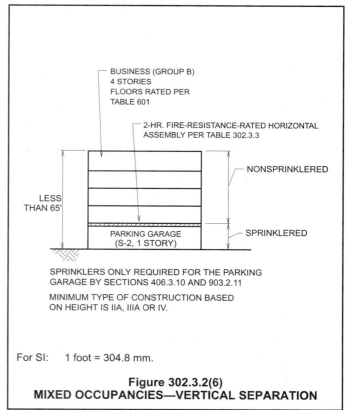

For SI: 1 foot = 304.8 mm.

Figure 302.3.2(6)
MIXED OCCUPANCIES—VERTICAL SEPARATION

For instance, a church hall might be used as a day care center during weekdays and as a reception hall for weddings and other similar events at other times. In these cases, the code provisions for each occupancy must be complied with.

SECTION 303
ASSEMBLY GROUP A

303.1 Assembly Group A. Assembly Group A occupancy includes, among others, the use of a building or structure, or a portion thereof, for the gathering together of persons for purposes such as civic, social or religious functions, recreation, food or drink consumption or awaiting transportation. A room or space used for assembly purposes by less than 50 persons and accessory to another occupancy shall be included as a part of that occupancy. Assembly areas with less than 750 square feet (69.7 m²) and which are accessory to another occupancy according to Section 302.2.1 are not assembly occupancies. Assembly occupancies which are accessory to Group E in accordance with Section 302.2 are not considered assembly occupancies. Religious educational rooms and religious auditoriums which are accessory to churches in accordance with Section 302.2 and which have occupant loads of less than 100 shall be classified as A-3.

Assembly occupancies shall include the following:

A-1 Assembly uses, usually with fixed seating, intended for the production and viewing of the performing arts or motion pictures including, but not limited to:

> Motion picture theaters
> Symphony and concert halls
> Television and radio studios admitting an audience
> Theaters

A-2 Assembly uses intended for food and/or drink consumption including, but not limited to:

> Banquet halls
> Night clubs
> Restaurants
> Taverns and bars

A-3 Assembly uses intended for worship, recreation or amusement and other assembly uses not classified elsewhere in Group A including, but not limited to:

> Amusement arcades
> Art galleries
> Bowling alleys
> Churches
> Community halls
> Courtrooms
> Dance halls (not including food or drink consumption)
> Exhibition halls
> Funeral parlors
> Gymnasiums (without spectator seating)
> Indoor swimming pools (without spectator seating)
> Indoor tennis courts (without spectator seating)
> Lecture halls
> Libraries
> Museums

> Waiting areas in transportation terminals
> Pool and billiard parlors

A-4 Assembly uses intended for viewing of indoor sporting events and activities with spectator seating including, but not limited to:

> Arenas
> Skating rinks
> Swimming pools
> Tennis courts

A-5 Assembly uses intended for participation in or viewing outdoor activities including, but not limited to:

> Amusement park structures
> Bleachers
> Grandstands
> Stadiums

❖ Because of the arrangement and density of the occupant load associated with occupancies classified in the assembly group category, the potential for multiple fatalities and injuries from fire is comparatively high. For example, no other use listed in Section 302.1 contemplates occupant loads as dense as 5 square feet (0.46 m²) per person (see Table 1004.1.2). Darkened spaces in theaters, nightclubs and the like serve to increase hazards. In sudden emergencies, the congestion caused by large numbers of people rushing to exits can cause panic conditions. For these and many other reasons, there is a relatively high degree of hazard to life safety in assembly facilities. The relative hazards of assembly occupancies are reflected in the height and area limitations of Table 503 imposed on buildings in the assembly group that are, in comparison, generally more restrictive than for buildings in other group classifications.

There are five specific assembly group classifications, Groups A-1 through A-5, described in this section. Where used in the code, the general term "Group A" is intended to include all five classifications.

The fundamental characteristics of all assembly occupancies are identified in this section. Structures that are designed or occupied for assembly purposes must be classified in one of the assembly group classifications. Exceptions to this rule are accessory use areas (see Section 302.2) and assembly rooms or spaces in nonassembly buildings meeting the following conditions:

- The occupant load of the assembly space is less than 50; and
- The purpose of the assembly space is accessory to the principal occupancy of the structure (i.e., the activities in the assembly space are subordinate and secondary to the primary occupancy).

The special exception given to assembly spaces in nonassembly buildings is a practical code consideration that permits a mixed occupancy condition to exist without requiring compliance with the provisions for mixed occupancies (see Section 302.3) or accessory use areas (see Section 302.2). All other code requirements applicable to assembly occupancies, however, still apply to the assembly area (e.g., the design occupant load, means of egress design, height, ventilation, floor live loads, etc.). This is because these requirements are most appropriately based on the occupancy of the space and not on the occupancy classification of the building.

Section 302.2.1 contains additional considerations regarding exceptions to the Occupancy Group A classification, including assembly areas less than 750 square feet (69.7 m²) that are accessory to another use (regardless of the size of the main use), assembly areas accessory to a Group E occupancy and religious educational rooms and auditoriums that are accessory to a church. In each of these cases, these rooms are not considered separate occupancies.

Example: An office building, Group B, has a conference room used for staff meetings [see Figure 303.1(1)]. The occupancy of a conference room is classified as Group A-3. Because the floor area of the conference room exceeds 10 percent of the floor area in which it is located, the room does not qualify as an accessory use area; however, because the occupant load of the conference room is less than 50 and its function is clearly accessory to the business area, the provisions of Section 303.1 permit the room to be included in the main occupancy, Group B.

If the accessory use area described in the example above was a public coffee shop, it would not qualify for the exception to the requirements for accessory use areas because the purpose of a coffee shop is not functionally related to the business occupancy [see Figure 303.1(2)]. In such a case, the coffee shop must be classified as an A-3 occupancy and the building evaluated as a mixed occupancy.

A-1: Some of the characteristics of Group A-1 occupancies are large, concentrated occupant loads, low lighting levels, above-normal sound levels and a moderate fuel load.

Group A-1 is characterized by two basic types of activities. The first type is one in which the facility is occupied for the production and viewing of theatrical or operatic performances. Facilities of this type ordinarily have fixed seating; a permanent raised stage; a proscenium wall and curtain; fixed or portable scenery drops; lighting devices; dressing rooms; mechanical appliances or other theatrical accessories and equipment [see Figure 303.1(3)].

The second type is one in which the structure is primarily occupied for the viewing of motion pictures. Facilities of this type ordinarily have fixed seating, no stage, a viewing screen, motion picture projection booth(s) and equipment [see Figure 303.1(4)].

Group A-1 presents a significant potential life safety hazard because of the large occupant loads and the concentration of people within confined spaces. The means of egress is an important factor in the design of such facilities. Theaters for the performing arts that require stages are considered particularly hazardous because of the amount of combustibles such as curtains, drops, scenery, construction materials and other accessories normally associated with stage operation. As such, special protection requirements applicable to stages and platforms are provided in Section 410 (see commentary, Section 410).

A-2: Group A-2 includes occupancies in which people congregate in high densities for social entertainment, such as drinking and dancing (e.g., nightclubs, dance halls, banquet halls, cabarets, etc.) and food and drink consumption (e.g., restaurants). The uniqueness of these occupancies is characterized by some or all of the following:

- Low lighting levels;
- Entertainment by a live band or recorded music generating above-normal sound levels;
- No theatrical stage accessories;
- Later-than-average operating hours;

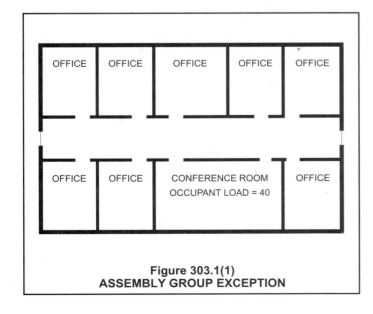

Figure 303.1(1)
ASSEMBLY GROUP EXCEPTION

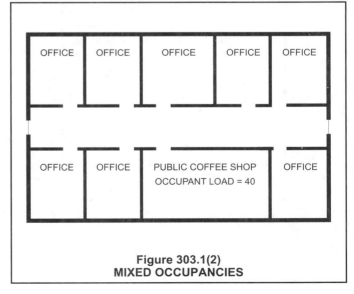

Figure 303.1(2)
MIXED OCCUPANCIES

FIGURE 303.1(3) – FIGURE 303.1(4)

USE AND OCCUPANCY CLASSIFICATION

- Tables and seating arranged or positioned so as to create ill-defined aisles;
- A specific area designated for dancing;
- Service facilities for alcoholic beverages and food; and
- High occupant load density.

The fire records are very clear in identifying that the characteristics listed above often cause a delayed awareness of a fire situation and confuse the appropriate response, resulting in an increased egress time and sometimes panic. Together, these factors may result in extensive life and property losses. These characteristics are only advisory in determining whether Group A-2 is the appropriate classification. Often there are additional characteristics that are unique to a project, which also must be taken into consideration when a classification is made.

Example: The Downtown Club, a popular local night-club/dance hall, features a different band every weekend [see Figure 303.1(5)]. It is equipped with a bar and basic kitchen facilities so that beverages and appetizers can be served. There is a platform for a band to perform on, a dance floor in front of the platform and numerous cocktail tables and chairs. The tables and chairs are not fixed, resulting in a haphazard arrangement such that there are no distinct aisles. When the band performs, the house lights are dimmed and spot-

lights are keyed in on the performers. The club is equipped with a sound system that is used at loud levels. The club is open until 3:00 a.m.—the latest time the local jurisdiction will allow.

From this description of the Downtown Club, one can readily see that the appropriate classification is Group A-2. Sometimes, however, it is not this easy to determine the appropriate classification. In such cases, the building official must seek additional information regarding the function(s) of the building and each area within the building.

A-3: Structures in which people assemble for the purpose of social activities (such as entertainment, recreation and amusement) that are neither classified in Group A-1 or A-2 nor appropriately classified in Group A-4 or A-5 are to be classified in Group A-3. Exhibition halls, libraries, museums, gymnasiums, recreation centers, health clubs, fellowship halls, indoor shooting galleries, bowling centers, billiard halls and the like are among the facilities often classified in Group A-3. Also, since they most nearly resemble this occupancy classification, public and private spaces used for assembly are often classified in Group A-3. These include large courtrooms, meeting rooms and conference centers. Similarly, lecture halls, which are large lecture rooms located in colleges, universities or in schools for students up to 12th grade and that have an occupant load of 50 or more, are also classified in Group A-3. [Note

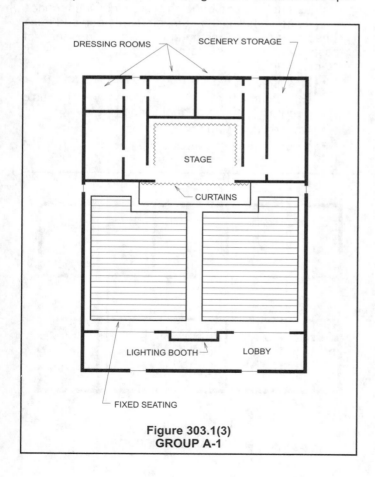

Figure 303.1(3)
GROUP A-1

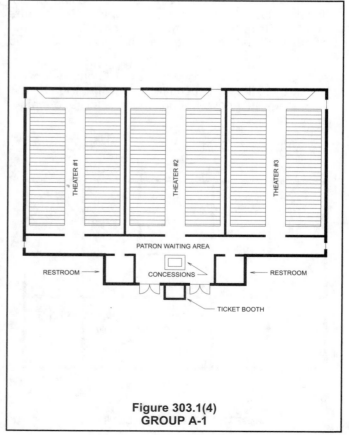

Figure 303.1(4)
GROUP A-1

that assembly uses that are accessory to Group E occupancies are not required to be regulated as separate occupancies (see Section 302.2.1)]. Structures in which people gather for worship and other religious purposes are also classified as Group A-3.

Structures in which people gather exclusively for worship and other religious purposes are also classified as Group A-3. Although such worship and religious purposes are without restriction to any particular sect or creed, the intent of the code is to limit Group A-3 classification to occupancies that are specifically related to worship services, devotions and religious rituals.

Religious facilities differ from other assembly occupancies in that the activity is, by nature, more orderly and their use tends to be most often limited in duration and frequency. Futhermore, the occupants of such facilities usually are very familiar with the facility and are well oriented to its egress pattern. Religious buildings typically contain a vestibule (narthex), a seating area (nave), an alter area (sanctuary) and a chancel (the area around and including the alter, including areas for clergy and choir). This is illustrated in Figure 303.1(6).

Frequently, other occupancies are located within the same structure where religious services (Group A-3) are performed [e.g., classrooms (Group E), care for infants (Group I-2) and staff offices (Group B)]. When this occurs, and depending on their size, these occupancies must be considered as either accessory use areas or other principal occupancies. As indicated in Sections 302.2.1 and 303.1, Group E classrooms with occupant loads less than 100 would not be required to be sepa-

rated from the A-3 portion of the building. Any area that does not qualify as an accessory use area must be classified in another occupancy. Accordingly, the structure then contains multiple occupancies and is subject to the provisions of Section 302.3.

The fire hazard in terms of combustible contents (fuel load) in structures classified in Group A-3 is most often expected to be moderate to low. Because structures classified in Group A-3 vary widely as to the purpose for which they are used, the range of fuel load varies widely. For example, the fuel load in a library or an exhibition hall usually is considerably greater than that normally found in a gymnasium.

A-4: Structures provided with spectator seating in which people assemble to watch an indoor sporting event are to be classified as Group A-4. Arenas, skating rinks, swimming pools and tennis courts are among the facilities often classified as Group A-4. The distinguishing factor between Group A-4 and A-5 structures is whether the event is indoors or outdoors. Group A-4 facilities are limited to indoor structures only. The distinguishing factor between Group A-4 and A-3 facilities is the presence of a defined seating area. While A-3 facilities are indoors (i.e., tennis courts, swimming pools, etc.), they typically do not have a defined seating area in which to view the event. Only facilities that are both indoors and have a defined seating area are to be classified as Group A-4.

A-5: Structures classified in Group A-5 are outdoor facilities where people assemble to review or participate

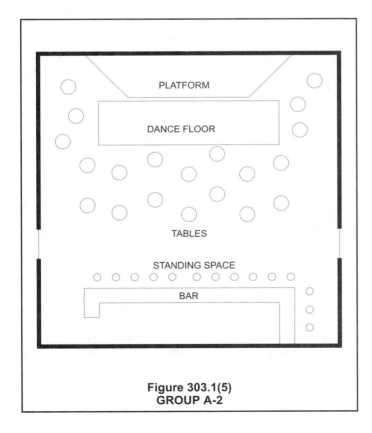

Figure 303.1(5)
GROUP A-2

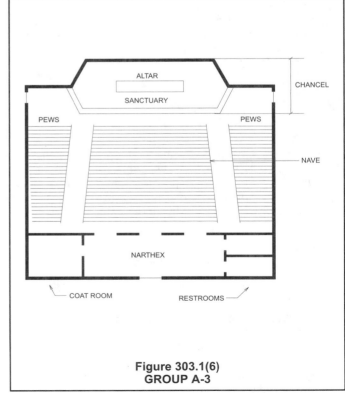

Figure 303.1(6)
GROUP A-3

in social and recreational activities (e.g., stadiums, grandstands, bleachers, coliseums, etc.). In order to qualify as an outdoor facility, the structure must be one where the products of combustion are freely and rapidly vented to the atmosphere (i.e., a structure without enclosures that would prevent the free movement of smoke from the occupied area to the outside). Any recreation facility that has exterior walls that enclose the facility and a roof that fully covers the area would not be classified in Group A-5, but rather in Group A-3 or A-4 depending on whether a seating area has been provided.

Since occupancies classified in Group A-5 are primarily viewing and sports participation areas, the fuel load associated with them is very low (i.e., the structure itself and seats). Because the fuel load present is relatively low and the expectation is that smoke will be quickly evacuated from the structure, the relative fire hazard of occupancies classified in Group A-5 is expected to be low. The life safety hazard from panic that might occur in an emergency, however, is a serious concern; hence, the capability of large crowds to exit the structure quickly and orderly during emergencies is an important design consideration (see Section 1024).

303.1.1 Nonaccessory assembly use. A building or tenant space used for assembly purposes by less than 50 persons shall be considered a Group B occupancy.

❖ The code recognizes that there are often small establishments that typically serve food and have a few seats in them that technically meet the definition of an assembly but pose a lower risk than a typical assembly. These types of spaces are to be considered to be an Occupancy Group B when they are not accessory to another occupancy. Examples of this include small "fast food" establishments, and small "mom-and-pop" restaurants or coffee shops.

SECTION 304
BUSINESS GROUP B

304.1 Business Group B. Business Group B occupancy includes, among others, the use of a building or structure, or a portion thereof, for office, professional or service-type transactions, including storage of records and accounts. Business occupancies shall include, but not be limited to, the following:

Airport traffic control towers
Animal hospitals, kennels and pounds
Banks
Barber and beauty shops
Car wash
Civic administration
Clinic—outpatient
Dry cleaning and laundries; pick-up and delivery stations and self-service
Educational occupancies above the 12th grade
Electronic data processing

Laboratories; testing and research
Motor vehicle showrooms
Post offices
Print shops
Professional services (architects, attorneys, dentists, physicians, engineers, etc.)
Radio and television stations
Telephone exchanges

❖ The risks to life safety in the business occupancy classification are relatively low. Exposure to the potential effects of fire is limited because business-type facilities most often have low fuel loads, are normally occupied only during the daytime and, with some exceptions, are usually occupied for a set number of hours. The occupants, because of the nature of the use, are alert, ambulatory, conscious, aware of their surroundings and generally familiar with the building's features, particularly the means of egress. Historically, this occupancy has one of the better fire safety records for the protection of life and property.

This section identifies the general characteristics and lists examples of occupancies that are classified in Group B. Note that the description recognizes the need for limited storage spaces that are incidental to office occupancies. Classrooms and laboratories that are located in colleges, universities and academies for educating students above the 12th grade and that have an occupant load of less than 50 are classified in Group B. Classrooms with an occupant load greater than 50 are classified in Group A-3 (see Section 303.1). Classrooms for children through the 12th grade with an occupant load of less than 50 are classified in Group E (see Section 305.1). When lecture facilities for large groups (i.e., occupant load greater than 50) are located within the same building where classrooms (i.e., occupant load less than 50) are found, the building is a mixed occupancy (Groups A and B) and is subject to the provisions of Section 302.3.

SECTION 305
EDUCATIONAL GROUP E

305.1 Educational Group E. Educational Group E occupancy includes, among others, the use of a building or structure, or a portion thereof, by six or more persons at any one time for educational purposes through the 12th grade. Religious educational rooms and religious auditoriums, which are accessory to churches in accordance with Section 302.2 and have occupant loads of less than 100, shall be classified as A-3 occupancies.

❖ The risks to life safety in this occupancy vary with the composition of the facilities and also with the ages of the occupants. In general, children require more safeguards than do older, more mature persons. Day care centers are a special problem since they are generally occupied by preschool children who are less capable of responding to an emergency. The hazards found in a day care center are far greater than in normal educational facilities, not so much because of the occupant or fuel load, but because of the inability of the occupants to respond.

This section identifies the criteria for classification of a building in Group E. The two fundamental characteristics of a Group E facility are as follows:

- The facility is occupied by more than five persons (including the instructor); and
- The purpose of the facility is for educating persons at the 12th-grade level and below, but not including more than five occupants less than $2^1/_2$ years of age.

Occupants who are older than $2^1/_2$ years of age are typically capable of walking and talking and have an instinct for self-preservation. As such, it is expected that these occupants are able to recognize an emergency situation and respond to instructions in an appropriate manner. If more than five occupants under the age of $2^1/_2$ years are being cared for, then the facility is considered a child care facility (Group I-4, see Section 308.5.2).

Occupancies used for the education of persons above the 12th-grade level are not included in Group E. These facilities are occupied by adults who are not expected to require special supervision, direction or instruction in a fire or other emergency. By the same measure, however, they also are not closely supervised; therefore, classrooms and laboratories located in colleges, universities and academies for students above the 12th grade are classified in Group B, because the occupancy characteristics and potential hazards to life safety present in these facilities more nearly resemble those of a business occupancy than educational occupancy. Similarly, lecture halls for students above the 12th grade with an occupant load of greater than 50 are classified in Group A-3 (see Section 303.1).

It is common for a school to also have gymnasiums (Group A-3), auditoriums (Group A-1), libraries (Group A-3), offices (Group B) and storage rooms (Group S-1). When this occurs, the building is a mixed occupancy and is subject to the provisions of Section 302.3. In accordance with Section 302.2.1, assembly spaces, such as the gymnasium, auditorium, library and cafeteria, do not have to be considered separate occupancies.

305.2 Day care. The use of a building or structure, or portion thereof, for educational, supervision or personal care services for more than five children older than $2^1/_2$ years of age, shall be classified as a Group E occupancy.

❖ Day care occupancies include facilities intended to be used for the care and supervision of more than five children for a period of less than 24 hours per day and that do not contain at any time more than five children who are $2^1/_2$ years of age or less. Facilities that provide care for more than five occupants greater than the 12th grade are to be classified as adult care facilities (Group I-4, Section 308.5.1).

Children under the age of $2^1/_2$ years usually are not

able to recognize an emergency situation, may not respond appropriately or simply may not be able to egress without assistance; thus, facilities that have more than five children $2^1/_2$ years of age or less are classified as child care facilities and considered to be Group I-4 (see Section 308.5.2). Figure 308.3.1 summarizes the appropriate occupancy classifications for day care, adult care (more than five adults, less than 24 hours) and child care facilities.

SECTION 306
FACTORY GROUP F

306.1 Factory Industrial Group F. Factory Industrial Group F occupancy includes, among others, the use of a building or structure, or a portion thereof, for assembling, disassembling, fabricating, finishing, manufacturing, packaging, repair or processing operations that are not classified as a Group H hazardous or Group S storage occupancy.

❖ The purpose of this section is to identify the characteristics of occupancies that are classified in factory and industrial occupancies and to differentiate Groups F-1 and F-2.

Because of the vast number of diverse manufacturing and processing operations in the industrial community, it is more practical to classify such facilities by their level of hazard rather than their function. In industrial facilities, experience has shown that the loss of life or property is most directly related to fire hazards, particularly the fuel load contributed by the materials being fabricated, assembled or processed.

Statistics show that property losses are comparatively high in factory and industrial occupancies, but the record of fatalities and injuries from fire has been remarkably low. This excellent life safety record can, in part, be attributed to fire protection requirements of the code.

This section requires that all structures that are used for fabricating, finishing, manufacturing, packaging, assembling or processing products or materials are to be classified in either Group F-1 (moderate hazard) or F-2 (low hazard). These classifications are based on the relative level of hazard for the types of materials that are fabricated, assembled or processed. Where the products and materials in a factory present an extreme fire, explosion or health hazard, such facilities are classified in Group H (see Section 307). It should be noted that the term "Group F" is not a specific occupancy, but is a term that collectively applies to Groups F-1 and F-2.

306.2 Factory Industrial F-1 Moderate-Hazard Occupancy. Factory industrial uses which are not classified as Factory Industrial F-2 Low Hazard shall be classified as F-1 Moderate Hazard and shall include, but not be limited to, the following:

Aircraft
Appliances

Athletic equipment
Automobiles and other motor vehicles
Bakeries
Beverages; over 12-percent alcohol content
Bicycles
Boats
Brooms or brushes
Business machines
Cameras and photo equipment
Canvas or similar fabric
Carpets and rugs (includes cleaning)
Clothing
Construction and agricultural machinery
Disinfectants
Dry cleaning and dyeing
Electric generation plants
Electronics
Engines (including rebuilding)
Food processing
Furniture
Hemp products
Jute products
Laundries
Leather products
Machinery
Metals
Millwork (sash & door)
Motion pictures and television filming (without spectators)
Musical instruments
Optical goods
Paper mills or products
Photographic film
Plastic products
Printing or publishing
Recreational vehicles
Refuse incineration
Shoes
Soaps and detergents
Textiles
Tobacco
Trailers
Upholstering
Wood; distillation
Woodworking (cabinet)

❖ Structures classified in Group F-1 (moderate hazard) are occupied for the purpose of fabrication, finishing, manufacturing, packaging, assembly or processing of materials that are combustible or that use combustible products in the production process.

306.3 Factory Industrial F-2 Low-Hazard Occupancy. Factory industrial uses that involve the fabrication or manufacturing of noncombustible materials which during finishing, packing or processing do not involve a significant fire hazard shall be classified as F-2 occupancies and shall include, but not be limited to, the following:

Beverages; up to and including 12-percent alcohol content
Brick and masonry
Ceramic products
Foundries
Glass products
Gypsum
Ice
Metal products (fabrication and assembly)

❖ Structures classified in Group F-2 (low hazard) are occupied for the purpose of fabrication, manufacturing or processing of noncombustible materials. It is acceptable for noncombustible products to be packaged in a combustible material, provided that the fuel load contributed by the packaging is negligible when compared to the amount of noncombustible product. The use of a significant amount of combustible material to package or finish a noncombustible product, however, will result in a Group F-1 classification.

To distinguish when the presence of combustible packaging constitutes a significant fuel load, possibly requiring the reclassification of the building or structure as Group F-1 (moderate-hazard factory and industrial), a reasonable guideline to follow is the "single thickness" rule, which is when a noncombustible product is put in one layer of packaging material.

Examples of acceptable conditions in Group F-2 are:

• Vehicle engines placed on wood pallets for transportation after assembly;

• Washing machines in corrugated cardboard boxes; and

• Soft-drink glass bottles packaged in pressed paper boxes.

Occupancies involving noncombustible items packaged in more than one layer of combustible packaging material are most appropriately classified in Group F-1. Typical examples of packaging that would result in a Group F-1 classification are:

• Chinaware wrapped in corrugated paper and placed in cardboard boxes;

• Glassware set in expanded foam forms and placed in cardboard boxes; and

• Fuel filters individually packed in pressed paper boxes, placed by the gross in a cardboard box and stacked on a pallet for transportation.

Factories and industrial facilities often have offices and areas where large quantities of materials are kept in the same building as manufacturing operations, fabrication processes and assembly processes. The stock areas that do not qualify as accessory use areas (see Section 302.2) are classified as either Group S-1 or S-2, depending on the combustibility of the materials stored. Areas used for offices that do not qualify as accessory use areas are classified in Group B. When these combinations of occupancies occur, as well as other combinations of occupancies, the building is subject to the mixed occupancy provisions in Section 302.3.

SECTION 307
HIGH-HAZARD GROUP H

[F] 307.1 High-Hazard Group H. High-Hazard Group H occupancy includes, among others, the use of a building or structure, or a portion thereof, that involves the manufacturing, processing, generation or storage of materials that constitute a physical or health hazard in quantities in excess of those found in Tables 307.7(1) and 307.7(2) (see also definition of "Control area").

❖ This section identifies the various types of facilities contained in the high-hazard occupancy. This occupancy classification relates to those facilities where the storage of materials or the operations are deemed to be extremely hazardous to life and property, especially when they involve the use of significant amounts of highly combustible, flammable or explosive materials, regardless of their composition (i.e., solids, liquids, gases or dust). Although they are not explosive or highly flammable, other hazardous materials, such as corrosive liquids, highly toxic materials and poisonous gases, still present an extreme hazard to life. Many materials possess multiple hazards, whether physical or health related.

There is a wide range of high-hazard operations in the industrial community; therefore, it is more practical to categorize such facilities in terms of the degree of hazard they present, rather than attempt to define a facility in terms of its function. This method is similar to that used to categorize factory (Section 306) and storage (Section 311) occupancies.

Group H is handled as a separate classification because it represents an unusually high degree of hazard that is not found in the other occupancies. It is important to isolate those industrial or storage operations that pose the greatest dangers to life and property and to reduce such hazards by providing systems or elements of protection through the regulatory provisions of building codes.

Operations that, because of the materials utilized or stored, cause a building or portion of a building to be classified as a high-hazard occupancy are identified in this section. While buildings classified as Group H may not have a large occupant load, the unstable chemical properties of the materials contained on the premises constitute an above-average fuel load and serve as a potential danger to the surrounding area.

The dangers created by the high-hazard materials require special consideration for the abatement of the danger. The classification of a material as high hazard is based on information derived from National Fire Protection Association (NFPA) standards and the Code of Federal Regulations (DOL 29 CFR).

The wide range of materials utilized or stored in buildings creates an equally wide range of hazards to the occupants of the building, the building proper and the surrounding area. Since these hazards range from explosive to corrosive conditions, the high-hazard occupancy has been broken into four subclassifications: Groups H-1 through H-4. A fifth category, Group H-5, is used to represent structures that contain hazardous production material (HPM) facilities. Each of these subclassifications addresses materials that have similar characteristics and the protection requirements attempt to address the hazard involved. These subclassifications are defined by the properties of the materials involved with only occasional reference to specific materials. This performance-based criterion may involve additional research to identify a hazard, but it is the only way to remain current in a rapidly changing field. Material Safety Data Sheets (MSDS) will be a major source for information.

Additional information on hazardous materials can be found in the commentary to the *International Fire Code®* (IFC®).

[F] 307.2 Definitions. The following words and terms shall, for the purposes of this section and as used elsewhere in this code, have the meanings shown herein.

❖ Definitions of terms that are associated with the content of this section are contained herein. These definitions can help in the understanding and application of the code requirements. It is important to emphasize that these terms are not exclusively related to this section but are applicable everywhere the term is used in the code. The purpose for including these definitions within this section is to provide more convenient access to them without having to refer back to Chapter 2.

For convenience, these terms are also listed in Chapter 2 with a cross reference to this section. The use and application of all defined terms, including those defined herein, are set forth in Section 201.

AEROSOL. A product that is dispensed from an aerosol container by a propellant.

Aerosol products shall be classified by means of the calculation of their chemical heats of combustion and shall be designated Level 1, 2 or 3.

Level 1 aerosol products. Those with a total chemical heat of combustion that is less than or equal to 8,600 British thermal units per pound (Btu/lb) (20 kJ/g).

Level 2 aerosol products. Those with a total chemical heat of combustion that is greater than 8,600 Btu/lb (20 kJ/g), but less than or equal to 13,000 Btu/lb (30 kJ/g).

Level 3 aerosol products. Those with a total chemical heat combustion that is greater than 13,000 Btu/lb (30 kJ/g).

❖ The intent of the code is to regulate those aerosols that contain a flammable propellant, such as butane, isobutane or propane. An aerosol product such as whipped cream is a water-based material with a nonflammable propellant (nitrous oxide) and would, therefore, not be regulated as a hazardous material. The contents of the aerosol container may be dispensed in the form of a mist spray, foam, gel or aerated powder.

Because of the wide range of flammability of aerosol products, a classification system was established to determine the required level of fire protection. Categories are defined according to the aerosol's chemical heat of combustion expressed in Btus per pound (Btu/lb). Aerosol category classifications of Levels 1, 2 and 3 are used to avoid confusion with flammable liquid classifications.

Examples of Level 1 aerosol products are shaving gel, whipped cream and air fresheners. Level 1 aerosols are not regulated as a hazardous material and are essentially exempt from the requirements of this chapter. Examples of Level 2 aerosols include some hair sprays and insect repellents. Level 3 aerosols include carburetor cleaner and other petroleum-based aerosols.

AEROSOL CONTAINER. A metal can or a glass or plastic bottle designed to dispense an aerosol. Metal cans shall be limited to a maximum size of 33.8 fluid ounces (1,000 ml). Glass or plastic bottles shall be limited to a maximum size of 4 fluid ounces (118 ml).

❖ All design criteria for the aerosol container, including the maximum size and minimum strength, are set by the U.S. Department of Transportation (DOTn 49 CFR). These container regulations are essential for safe transportation of aerosol products.

BARRICADE. A structure that consists of a combination of walls, floor and roof, which is designed to withstand the rapid release of energy in an explosion and which is fully confined, partially vented or fully vented; or other effective method of shielding from explosive materials by a natural or artificial barrier.

Artificial barricade. An artificial mound or revetment a minimum thickness of 3 feet (914 mm).

Natural barricade. Natural features of the ground, such as hills, or timber of sufficient density that the surrounding exposures that require protection cannot be seen from the magazine or building containing explosives when the trees are bare of leaves.

❖ The use of barricades provides an alternative method of explosion control by minimizing the potential damage due to blast effects and flying debris in the event of an explosion (see Section 414.5.1). Depending on the detonable hazard involved, an effective barricade may be a blast-resistant structure or natural or artificial barrier as provided for in Chapter 34 of the IFC for the storage of explosives.

BOILING POINT. The temperature at which the vapor pressure of a liquid equals the atmospheric pressure of 14.7 pounds per square inch (psi) (101 kPa) gage or 760 mm of mercury. Where an accurate boiling point is unavailable for the material in question, or for mixtures which do not have a constant boiling point, for the purposes of this classification, the 20-percent evaporated point of a distillation performed in accordance with ASTM D 86 shall be used as the boiling point of the liquid.

❖ The boiling point of a liquid is significant in determining the appropriate division for Class I flammable liquids.

Temperatures above the established boiling point for a given liquid would result in the atmospheric pressure no longer being able to keep the liquid in a liquid state. Liquids with low boiling points present a greater fire hazard because of the increased vapor pressure at normal ambient temperatures.

CLOSED SYSTEM. The use of a solid or liquid hazardous material involving a closed vessel or system that remains closed during normal operations where vapors emitted by the product are not liberated outside of the vessel or system and the product is not exposed to the atmosphere during normal operations; and all uses of compressed gases. Examples of closed systems for solids and liquids include product conveyed through a piping system into a closed vessel, system or piece of equipment.

❖ The difference between a closed system and an open system is whether the hazardous material involved in the process is exposed to the atmosphere. While not specific in the definition, certain gases are also allowed in closed systems, as indicated in Tables 307.7(1) and 307.7(2). Materials in closed or open systems are assumed to be "in use" as opposed to "in storage." Gases are always assumed to be in closed systems, since they would be immediately dispersed in an open system if exposed to the atmosphere without some means of containment (see the definition of "Open system").

COMBUSTIBLE DUST. Finely divided solid material that is 420 microns or less in diameter and which, when dispersed in air in the proper proportions, could be ignited by a flame, spark or other source of ignition. Combustible dust will pass through a U.S. No. 40 standard sieve.

❖ These types of materials are combustible solids in a finely divided state. A hazard exists when the concentration of the combustible dust dispersed into the air is within the explosive limitations and is exposed to an ignition source of sufficient energy and duration to initiate combustion. Sawdust from sawing operations is not normally considered to be combustible dust; however, sawdust produced from sanding operations can contain fines that are indeed combustible. Combustible dusts are classified as Group H-2 materials because of their severe deflagration hazard. The means of mitigation for the explosion potential associated with these materials must also be addressed.

COMBUSTIBLE FIBERS. Readily ignitable and free-burning fibers, such as cocoa fiber, cloth, cotton, excelsior, hay, hemp, henequen, istle, jute, kapok, oakum, rags, sisal, Spanish moss, straw, tow, wastepaper, certain synthetic fibers or other like materials.

❖ Operations involving combustible fibers are typically associated with paper milling, recycling, cloth manufacturing, carpet and textile mills and agricultural operations, among others. The primary hazards associated with such operations involve the abundance of materials and their ready ignitability. Many organic fibers are prone to spontaneous ignition if improperly dried and kept in areas without sufficient ventilation.

COMBUSTIBLE LIQUID. A liquid having a closed cup flash point at or above 100°F (38°C). Combustible liquids shall be subdivided as follows:

Class II. Liquids having a closed cup flash point at or above 100°F (38°C) and below 140°F (60°C).

Class IIIA. Liquids having a closed cup flash point at or above 140°F (60°C) and below 200°F (93°C).

Class IIIB. Liquids having a closed cup flash point at or above 200°F (93°C).

The category of combustible liquids does not include compressed gases or cryogenic fluids.

❖ Combustible liquids differ from flammable liquids in that the closed cup flash point of all combustible liquids is at or above 100°F (38°C) (see the definition of "Flash point"). The range of flash point dictates the class of combustible liquid. The flash point range of 100° F (38°C) to 140°F (60°C) for Class II liquids is based on a possible indoor ambient temperature exceeding 100° F (38°C). Only a moderate degree of heating would be required to bring the liquid to its flash point in this type of condition. Class III liquids, which have flash points higher than 140°F (38°C), would require a significant heat source besides ambient temperature conditions to reach their flash point (see the definition of "Flammable liquid"). Combustible liquids are primarily considered Group H-2 materials except for Class IIIB liquids, which are considered Group H-3 because of their relatively high flash point, and Class II and IIIA liquids which are considered Group H-3 when used or stored in normally closed containers or systems pressurized at less than 15 psi (103.4 kPa) gage. Motor oil is a typical example of a Class IIIB combustible liquid.

While cryogenic fluids and compressed gases may be combustible, they are to be regulated separately from combustible liquids.

COMPRESSED GAS. A material, or mixture of materials which:

1. Is a gas at 68°F (20°C) or less at 14.7 pounds per square inch atmosphere (psia) (101 kPa) of pressure; and

2. Has a boiling point of 68°F (20°C) or less at 14.7 psia (101 kPa) which is either liquefied, nonliquefied or in solution, except those gases which have no other health- or physical-hazard properties are not considered to be compressed until the pressure in the packaging exceeds 41 psia (282 kPa) at 68°F (20°C).

The states of a compressed gas are categorized as follows:

1. Nonliquefied compressed gases are gases, other than those in solution, which are in a packaging under the charged pressure and are entirely gaseous at a temperature of 68°F (20°C).

2. Liquefied compressed gases are gases that, in a packaging under the charged pressure, are partially liquid at a temperature of 68°F (20°C).

3. Compressed gases in solution are nonliquefied gases that are dissolved in a solvent.

4. Compressed gas mixtures consist of a mixture of two or more compressed gases contained in a packaging, the hazard properties of which are represented by the properties of the mixture as a whole.

❖ Gases are materials that boil at a temperature of 68°F (20°C) or less at a pressure of 14.7 psia (101.3 kPa). Liquefied and nonliquefied compressed gases are determined by the state of the gas at a temperature of 68°F (20°C). Nonliquefied gases are entirely gaseous, while liquefied gases are partially liquid.

CONTROL AREA. Spaces within a building that are enclosed and bounded by exterior walls, fire walls, fire barriers and roofs, or a combination thereof, where quantities of hazardous materials not exceeding the maximum allowable quantities per control area are stored, dispensed, used or handled.

❖ The use of control areas provides an alternative method for the use and storage of hazardous materials without classifying the building or structure as a high-hazard occupancy (Group H). This concept is based on regulating the allowable quantities of hazardous materials per control area, rather than per building area, by giving credit for further compartmentation through the use of fire barrier walls and horizontal assemblies having a minimum fire-resistance rating of 1 hour. The maximum quantities of hazardous materials within a given control area cannot exceed the amounts for a given material listed in either Table 307.7(1) or 307.7(2) (see commentary, Section 414.2).

CORROSIVE. A chemical that causes visible destruction of, or irreversible alterations in, living tissue by chemical action at the point of contact. A chemical shall be considered corrosive if, when tested on the intact skin of albino rabbits by the method described in DOTn 49 CFR, Part 173.137, such a chemical destroys or changes irreversibly the structure of the tissue at the point of contact following an exposure period of 4 hours. This term does not refer to action on inanimate surfaces.

❖ This definition is derived from DOL 29 CFR; Part 1910.1200. While corrosive materials do not present a fire, explosion or reactivity hazard, they do pose a handling and storage problem. Corrosive materials, therefore, are primarily considered a health hazard and are classified as a Group H-4 material. Many corrosive chemicals are also strong oxidizing agents that require classification as a multiple hazard in accordance with Section 307.8.

CRYOGENIC FLUID. A liquid having a boiling point lower than -150°F (-101°C) at 14.7 pounds per square inch atmosphere (psia) (an absolute pressure of 101 kPa).

❖ Cryogenic fluids present a hazard because they are extremely cold. Should a spill occur, their extremely cold temperature affects other compounds exposed to the spilled cryogenic fluid. Cryogenic fluids may be flammable or nonflammable; however, nonflammable cryogenics may possess properties that cause them to support combustion or react severely with other materials. The code is only intended to classify flammable or oxidizing cryogenic fluids as a hazardous material.

DEFLAGRATION. An exothermic reaction, such as the extremely rapid oxidation of a flammable dust or vapor in air, in which the reaction progresses through the unburned material at a rate less than the velocity of sound. A deflagration can have an explosive effect.

❖ Materials that present a deflagration hazard usually burn very rapidly with the release of energy from a chemical reaction in the form of intense heat. Confined deflagration hazards under pressure can result in an explosion. Most hazardous materials that pose a severe deflagration hazard are classified as Group H-2 in accordance with Section 307.4 (see the definition of "Detonation").

DETACHED BUILDING. A separate single-story building, without a basement or crawl space, used for the storage or use of hazardous materials and located an approved distance from all structures.

❖ This term is used to define the type of structure the code recognizes for the use and storage of hazardous materials in excess of the maximum allowable quantity per control area. While the definition addresses all hazardous materials, a detached storage building is only required for Group H-1, H-2 and H-3 structures as indicated in Sections 302.3.1, 415.3.2 and 415.4 and Table 415.3.2. The location of the structure is regulated by Section 415.3.1 and Table 415.3.1.

DETONATION. An exothermic reaction characterized by the presence of a shock wave in the material which establishes and maintains the reaction. The reaction zone progresses through the material at a rate greater than the velocity of sound. The principal heating mechanism is one of shock compression. Detonations have an explosive effect.

❖ Detonations are distinguished from deflagrations (which are produced by explosive gases, dusts, vapors and mists) by the speed with which they propagate a blast effect. Detonations occur much faster than deflagrations, since they propagate a combustion zone at a velocity greater than the speed of sound. Deflagrations propagate a combustion zone at a velocity less than the speed of sound. The speed of sound is approximately 1,100 feet per second (336 m/s) at sea level. Both detonations and deflagrations may produce explosive results when they occur in a confined space. Materials that are considered a detonation hazard are classified

as Group H-1 materials in accordance with Section 307.3.

DISPENSING. The pouring or transferring of any material from a container, tank or similar vessel, whereby vapors, dusts, fumes, mists or gases are liberated to the atmosphere.

❖ This term refers to a specific operation whereby the act of transferring a material occurs and that has a hazard associated with the liberation of the material in the forms listed in the definition. It is not "handling" and should not be confused with that term (see the definitions of "Closed system" and "Handling").

EXPLOSIVE. Any chemical compound, mixture or device, the primary or common purpose of which is to function by explosion. The term includes, but is not limited to, dynamite, black powder, pellet powder, initiating explosives, detonators, safety fuses, squibs, detonating cord, igniter cord, igniters and display fireworks, 1.3G (Class B, Special).

The term "explosive" includes any material determined to be within the scope of USC Title 18: Chapter 40 and also includes any material classified as an explosive other than consumer fireworks, 1.4G (Class C, Common) by the hazardous materials regulations of DOTn 49 CFR.

> **High explosive.** Explosive material, such as dynamite, which can be caused to detonate by means of a No. 8 test blasting cap when unconfined.

> **Low explosive.** Explosive material that will burn or deflagrate when ignited. It is characterized by a rate of reaction that is less than the speed of sound. Examples of low explosives include, but are not limited to, black powder; safety fuse; igniters; igniter cord; fuse lighters; fireworks, 1.3G (Class B, Special) and propellants, 1.3C.

> **Mass-detonating explosives.** Division 1.1, 1.2 and 1.5 explosives alone or in combination, or loaded into various types of ammunition or containers, most of which can be expected to explode virtually instantaneously when a small portion is subjected to fire, severe concussion, impact, the impulse of an initiating agent or the effect of a considerable discharge of energy from without. Materials that react in this manner represent a mass explosion hazard. Such an explosive will normally cause severe structural damage to adjacent objects. Explosive propagation could occur immediately to other items of ammunition and explosives stored sufficiently close to and not adequately protected from the initially exploding pile with a time interval short enough so that two or more quantities must be considered as one for quantity-distance purposes.

> **UN/DOTn Class 1 explosives.** The former classification system used by DOTn included the terms "high" and "low" explosives as defined herein. The following terms further define explosives under the current system applied by DOTn for all explosive materials defined as hazard Class 1 materials. Compatibility group letters are used in concert with the

division to specify further limitations on each division noted (i.e., the letter G identifies the material as a pyrotechnic substance or article containing a pyrotechnic substance and similar materials).

Division 1.1. Explosives that have a mass explosion hazard. A mass explosion is one which affects almost the entire load instantaneously.

Division 1.2. Explosives that have a projection hazard but not a mass explosion hazard.

Division 1.3. Explosives that have a fire hazard and either a minor blast hazard or a minor projection hazard or both, but not a mass explosion hazard.

Division 1.4. Explosives that pose a minor explosion hazard. The explosive effects are largely confined to the package and no projection of fragments of appreciable size or range is to be expected. An external fire must not cause virtually instantaneous explosion of almost the entire contents of the package.

Division 1.5. Very insensitive explosives. This division is comprised of substances that have a mass explosion hazard, but that are so insensitive there is very little probability of initiation or of transition from burning to detonation under normal conditions of transport.

Division 1.6. Extremely insensitive articles which do not have a mass explosion hazard. This division is comprised of articles that contain only extremely insensitive detonating substances and which demonstrate a negligible probability of accidental initiation or propagation.

❖ Explosives either detonate or deflagrate when initiated by either heat, shock or electric current. While these materials are normally designed and intended to be initiated by detonators under controlled conditions, heat, shock and electric current from uncontrolled sources may initiate these materials to produce an explosion. DOTn classifies explosives in six classes according to the degree of hazard posed by the material. The most dangerous of these materials is capable of almost simultaneous detonation of all of the material in a single load or store. The least-sensitive explosives produce blasts limited to the packages in which they are transported. This definition of explosives includes materials such as detonators, blasting agents and water gels; examples of these materials are listed in DOTy 27 CFR 55.23.

Explosive materials are subdivided into high, low, mass detonating and UN/DOTn Class 1 explosives. High explosives and mass detonating explosives are typically classified as Group H-1 and present a detonation hazard. Low explosives more commonly are classified as Group H-2, as they tend to deflagrate or burn upon ignition. Mass detonating devices present a greater threat to adjacent objects and structures. The IFC, therefore, contains provisions in the form of Table 3305.3 to deal with the separation distances for mass explosion hazards.

The definitions cited in this section are consistent with DOTn 49 CFR; Section 173.50. The hazards of this group of materials vary with the nature of the material,

with some explosives being very sensitive and others less sensitive. Some explosives detonate, others deflagrate and the hazards of others are limited to intense burning. The classification system was designed to correlate with the system of classification developed under recommendations of the United Nations (UN), wherein all explosive materials are placed into a hazard class of Class 1. This class is further divided into six divisions: Divisions 1.1 through 1.6.

FIREWORKS. Any composition or device for the purpose of producing a visible or audible effect for entertainment purposes by combustion, deflagration or detonation that meets the definition of 1.4G fireworks or 1.3G fireworks as set forth herein.

❖ Any device containing an explosive material that produces an audible or visible effect through combustion, deflagration, detonation or explosion is considered a firework. Fireworks are divided into two categories, 1.4G and 1.3G, based on the amount of pyrotechnic composition present.

FIREWORKS, 1.3G. (Formerly Class B, Special Fireworks.) Large fireworks devices, which are explosive materials, intended for use in fireworks displays and designed to produce audible or visible effects by combustion, deflagration or detonation. Such 1.3G fireworks include, but are not limited to, firecrackers containing more than 130 milligrams (2 grains) of explosive composition, aerial shells containing more than 40 grams of pyrotechnic composition, and other display pieces which exceed the limits for classification as 1.4G fireworks. Such 1.3G fireworks are also described as fireworks, 49 CFR (172) by the DOTn.

❖ The definitions of "1.4G fireworks" and "1.3G fireworks" are derived from the U.S. Department of Transportation (DOTn 49 CFR) clarification system for transporting explosives and from NFPA 1124. The amount of pyrotechnic composition is the distinguishing factor between the two types of fireworks (see commentary to the definition of "Pyrotechnic composition"). Fireworks that contain a limited amount of pyrotechnic composition are classified as 1.4G fireworks. 1.4G fireworks represent a physical hazard (Group H-3), while display fireworks represent a detonation hazard (Group H-1).

FIREWORKS, 1.4G. (Formerly Class C, Common Fireworks.) Small fireworks devices containing restricted amounts of pyrotechnic composition designed primarily to produce visible or audible effects by combustion. Such 1.4G fireworks which comply with the construction, chemical composition and labeling regulations of the DOTn for fireworks, 49 CFR (172), and the U.S. Consumer Product Safety Commission (CPSC) as set forth in CPSC 16 CFR: Parts 1500 and 1507, are not explosive materials for the purpose of this code.

❖ The requirements for storage, display and labeling depend on the correct application of this definition. This definition reflects the construction, chemical composition and labeling requirements of the U.S. Consumer Product Safety Commission found in Title 16, Code of

Federal Regulations, Parts 1500 and 1507. Also, 1.4G fireworks are not considered to be explosives in accordance with the provisions of Chapter 33 of the IFC.

FLAMMABLE GAS. A material that is a gas at 68°F (20°C) or less at 14.7 pounds per square inch atmosphere (psia) (101 kPa) of pressure [a material that has a boiling point of 68°F (20°C) or less at 14.7 psia (101 kPa)] which:

1. Is ignitable at 14.7 psia (101 kPa) when in a mixture of 13 percent or less by volume with air; or

2. Has a flammable range at 14.7 psia (101 kPa) with air of at least 12 percent, regardless of the lower limit.

The limits specified shall be determined at 14.7 psi (101 kPa) of pressure and a temperature of 68°F (20°C) in accordance with ASTM E 681.

❖ This term essentially refers to any type of compressed gas that burns in normal concentrations of oxygen in the air (see the definition of "Compressed gas"). The definition is consistent with the provisions of ASTM E 681.

FLAMMABLE LIQUEFIED GAS. A liquefied compressed gas which, under a charged pressure, is partially liquid at a temperature of 68°F (20°C) and which is flammable.

❖ This term essentially refers to any type of liquefied compressed gas that burns in normal concentrations of oxygen in the air (see the definition of "Compressed gas").

FLAMMABLE LIQUID. A liquid having a closed cup flash point below 100°F (38°C). Flammable liquids are further categorized into a group known as Class I liquids. The Class I category is subdivided as follows:

Class IA. Liquids having a flash point below 73°F (23°C) and a boiling point below 100°F (38°C).

Class IB. Liquids having a flash point below 73°F (23°C) and a boiling point at or above 100°F (38°C).

Class IC. Liquids having a flash point at or above 73°F (23°C) and below 100°F (38°C).

The category of flammable liquids does not include compressed gases or cryogenic fluids.

❖ While all flammable liquids have a closed cup flash point less than 100°F (38°C), the further classification of the Class I liquid is dependent on the boiling point (see the definition of "Boiling point"). The 100°F (38°C) flash point limitation for flammable liquids assumes possible indoor ambient temperature conditions of 100°F (38°C).

FLAMMABLE MATERIAL. A material capable of being readily ignited from common sources of heat or at a temperature of 600°F (316°C) or less.

❖ Many standardized tests, such as ASTM E 136 and NFPA 701a, have been developed to assess the flammability and fire hazards of materials. Both of these tests include objective criteria for evaluating the combustibility of different materials; however, great care

must be taken in conducting and evaluating the results of such tests.

FLAMMABLE SOLID. A solid, other than a blasting agent or explosive, that is capable of causing fire through friction, absorption or moisture, spontaneous chemical change, or retained heat from manufacturing or processing, or which has an ignition temperature below 212°F (100°C) or which burns so vigorously and persistently when ignited as to create a serious hazard. A chemical shall be considered a flammable solid as determined in accordance with the test method of CPSC 16 CFR; Part 1500.44, if it ignites and burns with a self-sustained flame at a rate greater than 0.1 inch (2.5 mm) per second along its major axis.

❖ Flammable solids are combustible materials that ignite easily and burn rapidly. Solids that may cause a fire due to friction are considered flammable solids as well as metal powders that can be readily ignited. Examples of flammable solids include nitrocellulose and combustible metals, such as magnesium and titanium.

FLASH POINT. The minimum temperature in degrees Fahrenheit at which a liquid will give off sufficient vapors to form an ignitable mixture with air near the surface or in the container, but will not sustain combustion. The flash point of a liquid shall be determined by appropriate test procedure and apparatus as specified in ASTM D 56, ASTM D 93 or ASTM D 3278.

❖ The flash point is the characteristic used in the classification of flammable and combustible liquids. The Tag Closed Tester (ASTM D 56), the Pensky-Martens Closed Cup Tester (ASTM D 93) and the Small Scale Closed-Cup Apparatus (ASTM D 3278) are the referenced test procedures for determining the flash points of liquids. The applicability of the respective test method is dependent on the viscosity of the test liquid and the expected flash point.

HANDLING. The deliberate transport by any means to a point of storage or use.

❖ The term "handling" is concerned with the transporting or movement of hazardous materials within a building. Handling presents a level of hazard that is of a lesser degree than that of use or dispensing operations but greater than storage. Material is handled only when it is transported from one point to another; it is the act of conveyance. The definition provides the means to determine proper controls necessary to provide safety in the transport mode. Specific handling requirements for various hazardous materials are contained in the IFC.

HAZARDOUS MATERIALS. Those chemicals or substances that are physical hazards or health hazards as defined and classified in this section and the *International Fire Code*, whether the materials are in usable or waste condition.

❖ The term "hazardous materials" refers to those materials that present either a physical or health hazard. A specific listing of hazardous materials is indicated in Sections 307.3, 307.4, 307.5 and 307.6. An occupancy containing greater than the maximum allowable quantity per control area of these materials as indicated in Table 307.7(1) or 307.7(2) is classified in one of the four high-hazard occupancy classifications.

HEALTH HAZARD. A classification of a chemical for which there is statistically significant evidence that acute or chronic health effects are capable of occurring in exposed persons. The term "health hazard" includes chemicals that are toxic or highly toxic, and corrosive.

❖ Materials that present risks to people from handling or exposure are considered health hazards. Examples of these types of materials are indicated in Section 307.6. Buildings and structures containing materials that present a health hazard in excess of the maximum allowable quantity per control areas would be classified as Group H-4. Materials that present a health hazard may also present a physical hazard (see the definition of "Physical hazard") and must comply with the requirements of the code applicable to both hazards.

HIGHLY TOXIC. A material which produces a lethal dose or lethal concentration that falls within any of the following categories:

1. A chemical that has a median lethal dose (LD50) of 50 milligrams or less per kilogram of body weight when administered orally to albino rats weighing between 200 and 300 grams each.

2. A chemical that has a median lethal dose (LD50) of 200 milligrams or less per kilogram of body weight when administered by continuous contact for 24 hours (or less if death occurs within 24 hours) with the bare skin of albino rabbits weighing between 2 and 3 kilograms each.

3. A chemical that has a median lethal concentration (LC50) in air of 200 parts per million by volume or less of gas or vapor, or 2 milligrams per liter or less of mist, fume or dust, when administered by continuous inhalation for 1 hour (or less if death occurs within 1 hour) to albino rats weighing between 200 and 300 grams each.

Mixtures of these materials with ordinary materials, such as water, might not warrant classification as highly toxic. While this system is basically simple in application, any hazard evaluation that is required for the precise categorization of this type of material shall be performed by experienced, technically competent persons.

❖ The definition is derived from DOL 29 CFR; Part 1910.1200. These materials are considered dangerous to humans when either inhaled, absorbed or injected through the skin or ingested orally. Highly toxic materials present a health hazard and are subsequently listed as Group H-4 in Section 307.6. Examples of highly toxic materials include gases such as arsine, fluorine and hydrogen cyanide, liquid acrylic acid and calcium cyanide in solid form.

Mixtures of these materials with ordinary materials, such as water, might not warrant a highly toxic classification. While this system is basically simple in application, any hazard evaluation that is required for the precise categorization of this type of material is to be performed by experienced, technically competent persons.

INCOMPATIBLE MATERIALS. Materials that, when mixed, have the potential to react in a manner that generates heat, fumes, gases or byproducts which are hazardous to life or property.

❖ These materials, whether in storage or in use, constitute a dangerous chemical combination. Determination of which chemicals in combination present a hazard is a difficult situation for the building official. MSDS alone may not provide all of the necessary information. When in doubt, the building official should seek additional information from the manufacturer of the chemicals involved, the building owner or from experts who are knowledgeable in industrial hygiene or chemistry.

OPEN SYSTEM. The use of a solid or liquid hazardous material involving a vessel or system that is continuously open to the atmosphere during normal operations and where vapors are liberated, or the product is exposed to the atmosphere during normal operations. Examples of open systems for solids and liquids include dispensing from or into open beakers or containers, dip tank and plating tank operations.

❖ See the commentary to the definition of "Closed system."

OPERATING BUILDING. A building occupied in conjunction with the manufacture, transportation or use of explosive materials. Operating buildings are separated from one another with the use of intraplant or intraline distances.

❖ Buildings used for storage of explosives are not magazines. This definition is included in this section to address that fact.

ORGANIC PEROXIDE. An organic compound that contains the bivalent -O-O- structure and which may be considered to be a structural derivative of hydrogen peroxide where one or both of the hydrogen atoms have been replaced by an organic radical. Organic peroxides can pose an explosion hazard (detonation or deflagration) or they can be shock sensitive. They can also decompose into various unstable compounds over an extended period of time.

Class I. Those formulations that are capable of deflagration but not detonation.

Class II. Those formulations that burn very rapidly and that pose a moderate reactivity hazard.

Class III. Those formulations that burn rapidly and that pose a moderate reactivity hazard.

Class IV. Those formulations that burn in the same manner as ordinary combustibles and that pose a minimal reactivity hazard.

Class V. Those formulations that burn with less intensity than ordinary combustibles or do not sustain combustion and that pose no reactivity hazard.

Unclassified detonable. Organic peroxides that are capable of detonation. These peroxides pose an extremely high explosion hazard through rapid explosive decomposition.

❖ The chemical structure of organic peroxides differs from that of hydrogen peroxide (an oxidizer) in that an organic radical replaces the hydrogen atoms. Organic chemicals are all carbon based. As a result, organic peroxides pose varying degrees of fire or explosion hazard in addition to their oxidizing properties. The classification of organic peroxides is based on the provisions of NFPA 43B. Proper material classification of organic peroxides is essential to determining the appropriate occupancy classification of the structure. Examples of organic peroxides include acetyl cyclohexane, sulfonyl peroxide and benzoyl peroxide. The actual class of these materials is dependent on the percentage of concentration by weight. Most organic peroxides are available as liquids, pastes or solids in a powder form.

OXIDIZER. A material that readily yields oxygen or other oxidizing gas, or that readily reacts to promote or initiate combustion of combustible materials. Examples of other oxidizing gases include bromine, chlorine and fluorine.

Class 4. An oxidizer that can undergo an explosive reaction due to contamination or exposure to thermal or physical shock. Additionally, the oxidizer will enhance the burning rate and can cause spontaneous ignition of combustibles.

Class 3. An oxidizer that will cause a severe increase in the burning rate of combustible materials with which it comes in contact or that will undergo vigorous self-sustained decomposition due to contamination or exposure to heat.

Class 2. An oxidizer that will cause a moderate increase in the burning rate or that causes spontaneous ignition of combustible materials with which it comes in contact.

Class 1. An oxidizer whose primary hazard is that it slightly increases the burning rate but which does not cause spontaneous ignition when it comes in contact with combustible materials.

❖ The classification of oxidizers is based on the provisions of NFPA 430. Oxidizers, whether a solid, liquid or gas, yield oxygen or another oxidizing gas during a chemical reaction or readily react to oxidize combustibles. The rate of reaction varies with the class of oxidizer. Specific classification of oxidizers is important because of the varying degree of hazard. Example of oxidizers include liquid hydrogen peroxide, nitric acid, sulfuric acid and solids such as sodium chlorite, chromic acid and calcium hypochlorite. Many commercially available swimming pool chemicals are indicative of Class 2 or 3 oxidizers.

OXIDIZING GAS. A gas that can support and accelerate combustion of other materials.

❖ Oxidizers sometimes yield oxidizing gases during a chemical reaction. These gases are capable of supporting and accelerating the combustion of other materials.

Examples of oxidizing gases include bromine, chlorine and fluorine.

PHYSICAL HAZARD. A chemical for which there is evidence that it is a combustible liquid, compressed gas, cryogenic, explosive, flammable gas, flammable liquid, flammable solid, organic peroxide, oxidizer, pyrophoric or unstable (reactive) or water-reactive material.

❖ Those materials that present a detonation hazard, deflagration hazard or readily support combustion are considered physical hazards. Examples of the types of materials that present a physical hazard are included in the definition. Buildings and structures containing materials that present a physical hazard in excess of the maximum allowable quantity per control areas would be classified in Group H-1, H-2 or H-3. Materials that present a physical hazard may also present a health hazard (see the definition of "Health hazard").

PYROPHORIC. A chemical with an autoignition temperature in air, at or below a temperature of 130°F (54.4°C).

❖ The definition is derived from DOL 29 CFR; Part 1910.1200. Pyrophoric materials, whether in a gas, liquid or solid form, are capable of spontaneous ignition at low temperatures. Examples of pyrophoric materials include silane and phosphine gas; liquid diethylaluminum chloride and inert solids such as cesium, plutonium, potassium and robidium.

PYROTECHNIC COMPOSITION. A chemical mixture that produces visible light displays or sounds through a self-propagating, heat-releasing chemical reaction which is initiated by ignition.

❖ Pyrotechnic composition consists of those chemical components, including oxidizers, that cause fireworks to make noise or display light when ignited. The definition is derived from NFPA 1124. The amount of pyrotechnic composition is the determining factor in whether the storage area for consumer fireworks is classified as Group H-3. The pyrotechnic content of consumer fireworks is contained within a significant amount of packaging and nonexplosive materials used in their manufacture, which constitute the bulk of the weight of the fireworks devices.

TOXIC. A chemical falling within any of the following categories:

1. A chemical that has a median lethal dose (LD50) of more than 50 milligrams per kilogram, but not more than 500 milligrams per kilogram of body weight when administered orally to albino rats weighing between 200 and 300 grams each.

2. A chemical that has a median lethal dose (LD50) of more than 200 milligrams per kilogram but not more than 1,000 milligrams per kilogram of body weight when administered by continuous contact for 24 hours (or less if death occurs within 24 hours) with the bare skin of albino rabbits weighing between 2 and 3 kilograms each.

3. A chemical that has a median lethal concentration (LC50) in air of more than 200 parts per million but not more than 2,000 parts per million by volume of gas or vapor, or more than 2 milligrams per liter but not more than 20 milligrams per liter of mist, fume or dust, when administered by continuous inhalation for 1 hour (or less if death occurs within 1 hour) to albino rats weighing between 200 and 300 grams each.

❖ The definition is derived from DOL 29 CFR; Part 1910.1200. These materials are considered dangerous to humans when either inhaled, absorbed or injected through the skin or when orally ingested. Toxic materials differ from highly toxic materials with regard to the specified median lethal dose or concentration of a given chemical. Toxic materials present a health hazard and are subsequently listed as a Group H-4 material in Section 307.6. Examples of toxic materials include gases such as chlorine; phosgene and hydrogen fluoride; liquid alkyl alcohol; methyl isocyanide and phosphorous chloride and barium chloride, benzidine and sodium fluoride in solid form.

UNSTABLE (REACTIVE) MATERIAL. A material, other than an explosive, which in the pure state or as commercially produced, will vigorously polymerize, decompose, condense or become self-reactive and undergo other violent chemical changes, including explosion, when exposed to heat, friction or shock, or in the absence of an inhibitor, or in the presence of contaminants, or in contact with incompatible materials. Unstable (reactive) materials are subdivided as follows:

Class 4. Materials that in themselves are readily capable of detonation or explosive decomposition or explosive reaction at normal temperatures and pressures. This class includes materials that are sensitive to mechanical or localized thermal shock at normal temperatures and pressures.

Class 3. Materials that in themselves are capable of detonation or of explosive decomposition or explosive reaction but which require a strong initiating source or which must be heated under confinement before initiation. This class includes materials that are sensitive to thermal or mechanical shock at elevated temperatures and pressures.

Class 2. Materials that in themselves are normally unstable and readily undergo violent chemical change but do not detonate. This class includes materials that can undergo chemical change with rapid release of energy at normal temperatures and pressures, and that can undergo violent chemical change at elevated temperatures and pressures.

Class 1. Materials that in themselves are normally stable but which can become unstable at elevated temperatures and pressure.

❖ The classification of unstable (reactive) materials is based on provisions in NFPA 704. The different classes of unstable (reactive) materials reflect the degree of susceptibility of the materials to release energy. Unstable (reactive) materials polymerize, decompose or become self-reactive when exposed to heat, air, moisture, pressure or shock. Separation from incompatible materials is essential to minimizing the hazards. Examples of

unstable (reactive) materials include nitromethane, perchloric acid, sodium perchlorate, vinyl acetate and acetic acid.

WATER-REACTIVE MATERIAL. A material that explodes; violently reacts; produces flammable, toxic or other hazardous gases; or evolves enough heat to cause self-ignition or ignition of nearby combustibles upon exposure to water or moisture. Water-reactive materials are subdivided as follows:

Class 3. Materials that react explosively with water without requiring heat or confinement.

Class 2. Materials that may form potentially explosive mixtures with water.

Class 1. Materials that may react with water with some release of energy, but not violently.

❖ These materials liberate significant quantities of heat when reacting with water. Combustible water-reactive materials are capable of self-ignition. Even noncombustible water-reactive materials present a hazard because of the heat liberated during their reaction with water, which may be sufficient to ignite surrounding combustible materials. While a definition for Class 1 water-reactive materials is provided for informational purposes, the maximum allowable quantity per control area of these materials in accordance with Table 307.7(1) is not limited.

[F] 307.3 High-Hazard Group H-1. Buildings and structures containing materials that pose a detonation hazard shall be classified as Group H-1. Such materials shall include, but not be limited to, the following:

Explosives:

Division 1.1
Division 1.2
Division 1.3

Exception: Materials that are used and maintained in a form where either confinement or configuration will not elevate the hazard from a mass fire to mass explosion hazard shall be allowed in H-2 occupancies.

Division 1.4

Exception: Articles, including articles packaged for shipment, that are not regulated as an explosive under Bureau of Alcohol, Tobacco and Firearms regulations, or unpackaged articles used in process operations that do not propagate a detonation or deflagration between articles shall be allowed in H-3 occupancies.

Division 1.5
Division 1.6

Organic peroxides, unclassified detonable
Oxidizers, Class 4
Unstable (reactive) materials, Class 3 detonable and Class 4
Detonable pyrophoric materials

❖ The contents of occupancies in Group H-1 present a detonation hazard. Examples of materials that create this hazard are listed in the section. The definitions for

Group H-1 materials are contained in Section 307.2. Because of the explosion hazard potential associated with Group H-1 materials, occupancies in Group H-1, which contain an excess of the allowable amounts indicated in Table 307.7(1) per control area, are required to be located in detached one-story buildings without basements (see commentary, Sections 302.3.1 and 415.4).

[F] 307.4 High-Hazard Group H-2. Buildings and structures containing materials that pose a deflagration hazard or a hazard from accelerated burning shall be classified as Group H-2. Such materials shall include, but not be limited to, the following:

Class I, II or IIIA flammable or combustible liquids which are used or stored in normally open containers or systems, or in closed containers or systems pressurized at more than 15 psi (103.4 kPa) gage.

  Combustible dusts
  Cryogenic fluids, flammable
  Flammable gases
  Organic peroxides, Class I
  Oxidizers, Class 3, that are used or stored in normally open
    containers or systems, or in closed containers or systems
    pressurized at more than 15 psi (103 kPa) gage
  Pyrophoric liquids, solids and gases, nondetonable
  Unstable (reactive) materials, Class 3, nondetonable
  Water-reactive materials, Class 3

❖ The contents of occupancies in Group H-2 present a deflagration or accelerated burning hazard. Examples of materials that create this hazard are listed. The definitions for Group H-2 materials are contained in Section 307.2. Because of the severe fire or reactivity hazard associated with these types of materials, proper classification is essential in determining the applicable requirements with regard to the mitigation of these hazards.

[F] 307.5 High-Hazard Group H-3. Buildings and structures containing materials that readily support combustion or that pose a physical hazard shall be classified as Group H-3. Such materials shall include, but not be limited to, the following:

Class I, II or IIIA flammable or combustible liquids which
  are used or stored in normally closed containers or systems
  pressurized at less than 15 psi (103.4 kPa) gage.
Combustible fibers
Consumer fireworks, 1.4G (Class C Common)
Cryogenic fluids, oxidizing
Flammable solids
Organic peroxides, Classes II and III
Oxidizers, Class 2
Oxidizers, Class 3, that are used or stored in normally closed
  containers or systems pressurized at less than 15 pounds
  per square inch (103 kPa) gauge
Oxidizing gases
Unstable (reactive) materials, Class 2
Water-reactive materials, Class 2

❖ The contents of occupancies in Group H-3 present a hazard inasmuch as they contain materials that readily support combustion or that present a physical hazard.

Examples of materials that create this hazard are listed. The definitions for Group H-3 materials are contained in Section 307.2. While Group H-3 materials are generally less of a fire or reactivity hazard than Group H-2 materials, they still present a greater physical hazard than materials not currently regulated as high hazard.

[F] 307.6 High-Hazard Group H-4. Buildings and structures which contain materials that are health hazards shall be classified as Group H-4. Such materials shall include, but not be limited to, the following:

  Corrosives
  Highly toxic materials
  Toxic materials

❖ The contents of occupancies in Group H-4 present a hazard inasmuch as they contain materials that are health hazards. Examples of these hazards are listed in this section. The definitions for Group H-4 materials are contained in Section 307.2. While reference is made to chemicals that cause these hazards, the data sheets for these chemicals, which are furnished by the applicant, will need considerable subjective evaluation. Some materials falling into the category of health hazard may also present a physical hazard and would, therefore, require the structure to be designed for multiple hazards in accordance with Section 307.8.

[F] 307.7 Group H-5 structures. Semiconductor fabrication facilities and comparable research and development areas in which hazardous production materials (HPM) are used and the aggregate quantity of materials is in excess of those listed in Tables 307.7(1) and 307.7(2). Such facilities and areas shall be designed and constructed in accordance with Section 415.9.

❖ Hazardous production materials (HPM) include flammable liquids and gases, corrosives, oxidizers and, in many instances, highly toxic materials (see definition for "Hazardous production material" in Section 415.2). In determining the applicable requirements of other sections of the code, HPM facilities are considered to be Group H-5 occupancies. The provisions of Sections 307 and 506.3 as well as Table 1003.2.3.1, however, are not intended to be applicable based on a high-hazard occupancy classification. It is intended that the quantities of materials permitted in Table 415.9.2.1.1 will take precedence over Tables 307.7(1) and 307.7(2).

Table 307.7(1). See page 3-34.

❖ The maximum allowable quantities of high-hazard materials allowed in each control area before having to classify a part of the (or the entire) building as a high-hazard occupancy are given in the table. This table is referenced in Exception 1 of Section 307.9. The materials listed in this table are classified according to their specific occupancy in Sections 307.3 through 307.5 and defined in Section 307.2. This table only contains materials applicable to Groups H-1, H-2 and H-3. The maximum allowable quantities per control area for Group H-4 materials are listed in Table 307.7(2).

The presence of any one or more of the materials listed in Table 307.7(1) in an amount greater than allowed requires that the building or area in which the material is contained be classified as a high-hazard occupancy.

If a building or area contains only the materials listed in either Table 307.7(1) or 307.7(2) in the maximum allowable quantity per control areas or less, then that building or area would not be classified as a high-hazard occupancy. The possible increase in overall danger that might exist should this occur because of the storage and use of incompatible materials is an issue that the code does not specifically address. In such situations, the building official can seek the advice of chemical engineers, fire protection engineers, fire service personnel or other experts in the use of hazardous materials. Based on their advice, the building official can deem the building a high-hazard occupancy.

The maximum allowable quantity per control area listed in Table 307.7(1) based on the concept of control areas as further regulated in Section 414. The quantities listed apply per control area. While every building area also represents a single control area, a given building may have multiple control areas, provided that the allowable amount within each control area is not exceeded and adequate fire-resistance-rated separation is provided between control areas. As indicated in Exception 2 of Section 307.9, a building that utilizes multiple control areas and complies with the applicable provisions of Section 414 is not classified as Group H. The number, degree of separation and location of control areas are indicated in Section 414 (see commentary, Section 414).

Table 307.7(1) is subdivided based on whether the material is in storage or in use in a closed or open system. Definitions of both closed and open systems are found in Section 307.2. Within these subdivisions, the appropriate maximum allowable quantity per control area is listed in accordance with the physical state (solid, liquid or gas) of the material. A column for gas in open systems is not indicated because hazardous gaseous materials should not be allowed in a system that is continuously open to the atmosphere. While hazardous materials within a closed or open system are considered to be in use, Note b clearly indicates that the aggregate quantity of hazardous materials in use and storage within a given control area should not exceed the quantity listed in Table 307.7(1) for storage. Notes d and e of Table 307.7(1) are significant in that, for certain materials, the maximum allowable amount may be increased due to the use of approved hazardous material storage cabinets, sprinklers or both. The notes are intended to be cumulative in that up to four times the base maximum quantity may be allowed per control area, if both sprinklered and in cabinets, without classifying the building as Group H. While the use of cabinets is not always a feasible or practical method of storage, they do provide additional protection to warrant an increase if provided. Construction requirements for hazardous material storage cabinets are contained in the IFC.

While classified as a hazardous material, the code recognizes the relative hazard of Class IIIB liquids as compared to that of other flammable and combustible liquids by classifying them as Group H-3, instead of Group H-2, and by establishing a base maximum allowable quantity per control area of 13,200 gallons (49 962 L). As indicated in Note f, the quantity of Class I oxidizers and Class IIIB liquids, without classifying the occupancy as Group H-3, would not be limited, provided the building is fully sprinklered in accordance with NFPA 13. The hazard presented by Class I oxidizers is that they slightly increase the burning rate of combustible materials that they may come into contact with during a fire. Class IIIB combustible liquids have flash points at or above 200°F (93°C). Motor oil is a typical example of a Class IIIB combustible liquid.

Note i is a specific exception for maximum 660-gallon (2498 L) inside storage tanks of combustible liquids that are connected to a fuel-oil piping system. This exception applies to most oil-fired stationary equipment, whether in industrial, commercial or residential occupancies. NFPA 31 provides further guidance on the type of installations this exception is intending to permit. Note i provides consistency with Sections 603.3.2 and 3401.2 of the IFC for inside storage.

Note k permits a larger amount of Class 3 oxidizers in a building when used for maintenance and health purposes. The quantities proposed are reasonable for occupancies such as the health care industry where Class 3 oxidizers are used for maintenance purposes, sterilization and sanitation of equipment and operation sanitation. The method used to store the oxidizers is subject to the evaluation and approval of the building official. Note k also provides consistency with Note k of Table 2703.1.1(1) of the IFC.

Note l clarifies that the 125 pounds (57 kg) of storage permitted for consumer fireworks represents the net weight of the pyrotechnic composition of the fireworks in a nonsprinklered building. This amount represents approximately $12^1/_2$ shipping cases (less than one and a half pallet loads) of fireworks in a nonsprinklered storage condition. In cases where the net weight of the pyrotechnic composition of the fireworks is unknown, 25 percent of the gross weight of the fireworks is to be used. The gross weight is to include the weight of the packaging.

Table 307.7(2). See page 3-36.

❖ Table 307.7(2), similar to Table 307.7(1), specifies the maximum quantities of hazardous materials, liquids or chemicals allowed per control area before having to classify a part of the (or the entire) building as a high-hazard occupancy. Table 307.7(2), as referenced in Exception 1 of Section 307.9, contains materials classified as Group H-4 in accordance with Section 307.6. While the materials listed in this table are considered health hazards, some materials may also possess physical hazard characteristics more indicative of materials classified as Group H-1, H-2 or H-3.

The maximum quantities per control area listed in Ta-

TABLE 307.7(1) USE AND OCCUPANCY CLASSIFICATION

[F] TABLE 307.7(1)
MAXIMUM ALLOWABLE QUANTITY PER CONTROL AREA OF HAZARDOUS MATERIALS POSING A PHYSICAL HAZARD[a, j, m, n]

MATERIAL	CLASS	GROUP WHEN THE MAXIMUM ALLOWABLE QUANTITY IS EXCEEDED	STORAGE[b] Solid pounds (cubic feet)	STORAGE[b] Liquid gallons (pounds)	STORAGE[b] Gas (cubic feet at NTP)	USE-CLOSED SYSTEMS[b] Solid pounds (cubic feet)	USE-CLOSED SYSTEMS[b] Liquid gallons (pounds)	USE-CLOSED SYSTEMS[b] Gas (cubic feet at NTP)	USE-OPEN SYSTEMS[b] Solid pounds (cubic feet)	USE-OPEN SYSTEMS[b] Liquid gallons (pounds)
Combustible liquid[c, i]	II	H-2 or H-3	N/A	120[d,e]	N/A	N/A	120[d]	N/A	N/A	30[d]
	IIIA	H-2 or H-3	N/A	330[d,e]	N/A	N/A	330[e,f]	N/A	N/A	80[d]
	IIIB	N/A	N/A	13,200[e,f]	N/A	N/A	13,200[e,f]	N/A	N/A	3,300[f]
Combustible fiber	Loose	H-3	(100)	N/A	N/A	(100)	N/A	N/A	(20)	N/A
	Baled	H-3	(1,000)	N/A	N/A	(1,000)	N/A	N/A	(200)	N/A
Consumer fireworks (Class C, Common)	1.4G	H-3	125[d,e,l]	N/A	N/A	N/A	N/A	N/A	N/A	N/A
Cryogenics flammable	N/A	H-2	N/A	45[d]	N/A	N/A	45[d]	N/A	N/A	10[d]
Cryogenics, oxidizing	N/A	H-3	N/A	45[d]	N/A	N/A	45[d]	N/A	N/A	10[d]
Explosives	Division 1.1	H-1	1[c,g]	(1)[c,g]	N/A	0.25[g]	(0.25)[g]	N/A	0.25[g]	(0.25)[g]
	Division 1.2	H-1	1[c,g]	(1)[c,g]	N/A	0.25[g]	(0.25)[g]	N/A	0.25[g]	(0.25)[g]
	Division 1.3	H-1 or 2	5[c,g]	(5)[c,g]	N/A	1[g]	(1)[g]	N/A	1[g]	(1)[g]
	Division 1.4	H-3	50[c,g]	(50)[c,g]	N/A	50[g]	(50)[g]	N/A	N/A	N/A
	Division 1.4G	H-3	125[d,e,l]	N/A	N/A	N/A	N/A	N/A	N/A	N/A
	Division 1.5	H-1	1[c,g]	(1)[c,g]	N/A	0.25[g]	(0.25)[g]	N/A	0.25[g]	(0.25)[g]
	Division 1.6	H-1	1[d,e,g]	N/A	N/A	N/A	N/A	N/A	N/A	N/A
Flammable gas	Gaseous	H-2	N/A	N/A	1,000[d,e]	N/A	N/A	1,000[d,e]	N/A	N/A
	liquefied	H-2	N/A	30[d,e]	N/A	N/A	30[d,e]	N/A	N/A	N/A
Flammable liquid[c]	1A	H-2 or H-3	N/A	30[d,e]	N/A	N/A	30[d]	N/A	N/A	10[d]
	1B and 1C	H-2 or H-3	N/A	120[d,e]	N/A	N/A	120[d]	N/A	N/A	30[d]
Combination flammable liquid (1A, 1B, 1C)	N/A	H-2 or H-3	N/A	120[d,e,h]	N/A	N/A	120[d,h]	N/A	N/A	30[d,h]
Flammable solid	N/A	H-3	125[d,e]	N/A	N/A	125[d,e]	N/A	N/A	25[d]	N/A
Organic peroxide	UD	H-1	1[c,g]	(1)[c,g]	N/A	0.25[g]	(0.25)[g]	N/A	0.25[g]	(0.25)[g]
	I	H-2	5[d,e]	(5)[d,e]	N/A	1[d]	(1)	N/A	1[d]	(1)[d]
	II	H-3	50[d,e]	(50)[d,e]	N/A	50[d]	(50)[d]	N/A	10[d]	(10)[d]
	III	H-3	125[d,e]	(125)[d,e]	N/A	125[d]	(125)[d]	N/A	25[d]	(25)[d]
	IV	N/A	NL	NL	N/A	NL	NL	N/A	NL	NL
	V	N/A	NL	NL	N/A	NL	NL	N/A	NL	NL
Oxidizer	4	H-1	1[c,g]	(1)[c,g]	N/A	0.25[g]	(0.25)[g]	N/A	0.25[g]	(0.25)[g]
	3[k]	H-2	10[d,e]	(10)[d,e]	N/A	2[d]	(2)	N/A	2[d]	(2)[d]
	2	H-3	250[d,e]	(250)[d,e]	N/A	250[d]	(250)[d]	N/A	50[d]	(50)[d]
	1	H-3	4,000[e,f]	(4,000)[e,f]	N/A	4,000[f]	(4,000)[f]	N/A	1,000[f]	(1,000)[f]
Oxidizing gas	Gaseous	H-3	N/A	N/A	1,500[d,e]	N/A	N/A	1,500[d,e]	N/A	N/A
	liquefied	H-3	N/A	15[d,e]	N/A	N/A	15[d,e]	N/A	N/A	N/A

(continued)

[F] TABLE 307.7(1)—continued

MAXIMUM ALLOWABLE QUANTITY PER CONTROL AREA OF HAZARDOUS MATERIALS POSING A PHYSICAL HAZARD[a, j, m, n]

MATERIAL	CLASS	GROUP WHEN THE MAXIMUM ALLOWABLE QUANTITY IS EXCEEDED	STORAGE[b]			USE-CLOSED SYSTEMS[b]			USE-OPEN SYSTEMS[b]	
			Solid pounds (cubic feet)	Liquid gallons (pounds)	Gas (cubic feet at NTP)	Solid pounds (cubic feet)	Liquid gallons (pounds)	Gas (cubic feet at NTP)	Solid pounds (cubic feet)	Liquid gallons (pounds)
Pyrophoric material	N/A	H-2	4[e,g]	(4)[c,g]	50[c,g]	1[g]	(1)[g]	10[c,g]	0	0
Unstable (reactive)	4	H-1	1[e,g]	(1)[c,g]	10[d,g]	0.25[g]	(0.25)[g]	2[c,g]	0.25[g]	(0.25)[g]
	3	H-1 or H-2	5[d,e]	(5)[d,e]	50[d,e]	1[d]	(1)	10[d,e]	1[d]	(1)[d]
	2	H-3	50[d,e]	(50)[d,e]	250[d,e]	50[d]	(50)[d]	250[d,e]	10[d]	(10)[d]
	1	N/A	NL	NL	N/A	NL	N/A	NL	NL	NL
Water reactive	3	H-2	5[d,e]	(5)[d,e]	N/A	5[d]	(5)[d]	N/A	1[d]	(1)[d]
	2	H-3	50[d,e]	(50)[d,e]	N/A	50[d]	(50)[d]	N/A	10[d]	(10)[d]
	1	N/A	NL	NL	N/A	NL	NL	N/A	NL	NL

For SI: 1 cubic foot = 0.023 m^3, 1 pound = 0.454 kg, 1 gallon = 3.785 L.

NL = Not Limited; N/A = Not Applicable; UD = Unclassified Detonable

a. For use of control areas, see Section 414.2.

b. The aggregate quantity in use and storage shall not exceed the quantity listed for storage.

c. The quantities of alcoholic beverages in retail and wholesale sales occupancies shall not be limited providing the liquids are packaged in individual containers not exceeding 1.3 gallons. In retail and wholesale sales occupancies, the quantities of medicines, foodstuffs, consumer or industrial products, and cosmetics containing not more than 50 percent by volume of water-miscible liquids with the remainder of the solutions not being flammable, shall not be limited, provided that such materials are packaged in individual containers not exceeding 1.3 gallons.

d. Maximum allowable quantities shall be increased 100 percent in buildings equipped throughout with an automatic sprinkler system in accordance with Section 903.3.1.1. Where Note e also applies, the increase for both notes shall be applied accumulatively.

e. Maximum allowable quantities shall be increased 100 percent when stored in approved storage cabinets, gas cabinets, exhausted enclosures or safety cans as specified in the *International Fire Code*. Where Note d also applies, the increase for both notes shall be applied accumulatively.

f. The permitted quantities shall not be limited in a building equipped throughout with an automatic sprinkler system in accordance with Section 903.3.1.1.

g. Permitted only in buildings equipped throughout with an automatic sprinkler system in accordance with Section 903.3.1.1.

h. Containing not more than the maximum allowable quantity per control area of Class IA, IB or IC flammable liquids.

i. Inside a building, the maximum capacity of a combustible liquid storage system that is connected to a fuel-oil piping system shall be 660 gallons provided such system conforms to the *International Fire Code*.

j. Quantities in parenthesis indicate quantity units in parenthesis at the head of each column.

k. A maximum quantity of 200 pounds of solid or 20 gallons of liquid Class 3 oxidizers is allowed when such materials are necessary for maintenance purposes, operation or sanitation of equipment. Storage containers and the manner of storage shall be approved.

l. Net weight of the pyrotechnic composition of the fireworks. Where the net weight of the pyrotechnic composition of the fireworks is not known, 25 percent of the gross weight of the fireworks, including packaging, shall be used.

m. For gallons of liquids, divide the amount in pounds by 10 in accordance with Section 2703.1.2 of the *International Fire Code*.

n. For storage and display quantities in Group M and storage quantities in Group S occupancies complying with Section 414.2.4, see Table 414.2.4.

[F] TABLE 307.7(2)
[F] TABLE 307.7(2)
MAXIMUM ALLOWABLE QUANTITY PER CONTROL AREA OF HAZARDOUS MATERIAL POSING A HEALTH HAZARD[a, b, c]

MATERIAL	STORAGE[d]			USE-CLOSED SYSTEMS[d]			USE-OPEN SYSTEMS[d]	
	Solid pounds[e, f]	Liquid gallons (pounds)[e, f]	Gas (cubic feet at NTP)[e]	Solid pounds[e]	Liquid gallons (pounds)[e]	Gas (cubic feet at NTP)[e]	Solid pounds[e]	Liquid gallons (pounds)[e]
Corrosive	5,000	500	810[f, g]	5,000	500	810[f, g]	1,000	100
Highly toxic	10	(10)[i]	20[h]	10	(10)[i]	20[h]	3	(3)[i]
Toxic	500	(500)[i]	810[f]	500	(500)[i]	810[f]	125	(125)[i]

For SI: 1 cubic foot = 0.028 m^3, 1 pound = 0.454 kg, 1 gallon = 3.785 L.

a. For use of control areas, see Section 414.2.

b. In retail and wholesale sales occupancies, the quantities of medicines, foodstuffs, consumer or industrial products, and cosmetics, containing not more than 50 percent by volume of water-miscible liquids and with the remainder of the solutions not being flammable, shall not be limited, provided that such materials are packaged in individual containers not exceeding 1.3 gallons.

c. For storage and display quantities in Group M and storage quantities in Group S occupancies complying with Section 414.2.4, see Table 414.2.4.

d. The aggregate quantity in use and storage shall not exceed the quantity listed for storage.

e. Quantities shall be increased 100 percent in buildings equipped throughout with an approved automatic sprinkler system in accordance with Section 903.3.1.1. Where Note f also applies, the increase for both notes shall be applied accumulatively.

f. Quantities shall be increased 100 percent when stored in approved storage cabinets, gas cabinets or exhausted enclosures as specified in the *International Fire Code*. Where Note e also applies, the increase for both notes shall be applied accumulatively.

g. A single cylinder containing 150 pounds or less of anhydrous ammonia in a single control area in a nonsprinklered building shall be considered a maximum allowable quantity. Two cylinders, each containing 150 pounds or less in a single control area, shall be considered a maximum allowable quantity provided the building is equipped throughout with an automatic sprinkler system in accordance with Section 903.3.1.1.

h. Allowed only when stored in approved exhausted gas cabinets or exhausted enclosures as specified in the *International Fire Code*.

i. Quantities in parenthesis indicate quantity units in parenthesis at the head of each column.

ble 307.7(2) are indicative of industry practice and assume the materials are properly stored and handled in accordance with the IFC. Group H-4 materials, while indeed hazardous, are primarily considered a handling problem and do not possess the same fire, explosion or reactivity hazard associated with other hazardous materials. The base maximum allowable quantity per control area of 810 cubic feet (23 m^3) for gases that are either corrosive or toxic is based on a standard-size chlorine cylinder. Where applicable, Notes e and f provide an increase in the base maximum allowable amount, similar to that in Table 307.7(1) [see commentary, Table 307.7(1)].

The storage or use of irritants and sensitizers is exempt from a Group H-4 occupancy classification. Irritants and sensitizers do not under normal circumstances present significant acute or chronic exposure effects. These materials, however, must still be properly stored and handled in accordance with the IFC.

Note g exempts a building from a Group H-4 classification that contains no more than a single 150-pound (68 kg) cylinder per control area of anhydrous ammonia in an unsprinklered building and no more than two cylinders each containing 150 pounds (68 kg) or less in a single control area of a fully sprinklered building. Anhydrous ammonia, which has a very low density when unconfined, is approximately one-fifth the density of chlorine. Based on their respective *LC*50 values, chlorine is also approximately 80 times more toxic than anhydrous ammonia. The use of cabinets is not recognized as a realistic method of storing cylinders of anhydrous ammonia.

[F] 307.8 Multiple hazards. Buildings and structures containing a material or materials representing hazards that are classified in one

or more of Groups H-1, H-2, H-3 and H-4 shall conform to the code requirements for each of the occupancies so classified.

❖ If materials are present that possess characteristics of more than one high-hazard occupancy, then the structure must be designed to protect against the hazards of each relevant high-hazard occupancy classification. For example, a material could be classified as both a Class 2 oxidizer (Group H-3) and a corrosive (Group H-4). If the given quantity exceeded the maximum allowable quantity per control area individually for both a Class 2 oxidizer and a corrosive, the structure is required to conform to the applicable requirements of both Groups H-3 and H-4.

[F] 307.9 Exceptions: The following shall not be classified in Group H, but shall be classified in the occupancy that they most nearly resemble. Hazardous materials in any quantity shall conform to the requirements of this code, including Section 414, and the *International Fire Code*.

1. Buildings and structures that contain not more than the maximum allowable quantities per control area of hazardous materials as shown in Tables 307.7(1) and 307.7(2) provided that such buildings are maintained in accordance with the *International Fire Code*.

2. Buildings utilizing control areas in accordance with Section 414.2 that contain not more than the maximum allowable quantities per control area of hazardous materials as shown in Tables 307.7(1) and 307.7(2).

3. Buildings and structures occupied for the application of flammable finishes, provided that such buildings or areas conform to the requirements of Section 416 and the *International Fire Code*.

4. Wholesale and retail sales and storage of flammable and combustible liquids in mercantile occupancies conforming to the *International Fire Code*.

5. Closed systems housing flammable or combustible liquids or gases utilized for the operation of machinery or equipment.

6. Cleaning establishments that utilize combustible liquid solvents having a flash point of 140°F (60°C) or higher in closed systems employing equipment listed by an approved testing agency, provided that this occupancy is separated from all other areas of the building by 1-hour fire-resistance-rated fire barrier walls or horizontal assemblies or both.

7. Cleaning establishments which utilize a liquid solvent having a flash point at or above 200°F (93°C).

8. Liquor stores and distributors without bulk storage.

9. Refrigeration systems.

10. The storage or utilization of materials for agricultural purposes on the premises.

11. Stationary batteries utilized for facility emergency power, uninterrupted power supply or telecommunication facilities provided that the batteries are provided with safety venting caps and ventilation is provided in accordance with the *International Mechanical Code*.

12. Corrosives shall not include personal or household products in their original packaging used in retail display or commonly used building materials.

13. Buildings and structures occupied for aerosol storage shall be classified as Group S-1, provided that such buildings conform to the requirements of the *International Fire Code*.

14. Display and storage of nonflammable solid and nonflammable or noncombustible liquid hazardous materials in quantities not exceeding the maximum allowable quantity per control area in Group M or S occupancies complying with Section 414.2.4.

15. The storage of black powder, smokeless propellant and small arms primers in Groups M and R-3 and special industrial explosive devices in Groups B, F, M and S, provided such storage conforms to the quantity limits and requirements prescribed in the *International Fire Code*.

❖ This section lists conditions that are exempt from a high-hazard classification because of the building's construction or use; the packaging of materials, the quantity of materials or the precautions taken to prevent fire. Even if a high-hazard material meets one of the exceptions, its storage and use must comply with the applicable provisions of Section 414.0 and the IFC.

Exception 1 allows specific quantities of hazardous materials to be located in a building without it being classified as a high-hazard occupancy. These quantities have been determined to be relatively safe in amounts no greater than those prescribed in Tables 307.7(1) and 307.7(2) when maintained in accordance with the IFC. While the building may be exempt from a high-hazard occupancy classification, any potential hazard with regard to the use or storage of any hazardous material, regardless of quantity, must be abated.

Exception 2 provides an alternative use of control areas as indicated in Section 414.2 in order to exceed the maximum allowable quantity per control area of Tables 307.7(1) and 307.7(2) within a given building without classifying the building as a high-hazard occupancy. Maximum allowable quantities per control area are not specified with respect to the entire building area since this would, in effect, limit the quantities of hazardous materials within a building before classification as Group H, regardless of building area, degree of fire separation or additional methods of mitigation (see commentary, Section 414.2).

Exception 3 exempts spray painting and similar operations within buildings from being classified as a high-hazard occupancy. This exception requires that all such operations, as well as the handling of flammable finishes, are in accordance with the provisions of Section 416 and the IFC; therefore, an adequately protected typical paint spray booth in a factory (Group F-1) would not result in a high-hazard occupancy classification for either the building or the paint spray area.

Exception 4 relies on the provisions of NFPA 30 and the IFC to regulate the storage of flammable and combustible liquids in wholesale and retail occupancies. The overall permitted amount of flammable and combustible liquids is dependent on the level of sprinkler protection provided in accordance with Section 4.5.6 of NFPA 30. For unsprinklered buildings, the publicly accessible display areas are limited to the aggregate quantity of Class IB, IC, II and IIIA liquids not to exceed 2 gallons per square foot (0.05 L/m²) of gross floor area with a maximum total quantity of 7,500 gallons (28 388 L). Class IA liquids are limited to a total of 120 gallons (454 L). If the display areas were equipped with an NFPA 13 automatic sprinkler system with a minimum design density adequate for an Ordinary Hazard Group 2 occupancy, the density of the stored liquids could be increased to 4 gallons per square foot (0.1 L/m²) of floor area. The aggregate total would still be limited to 7,500 gallons (28 388 L) of Class IB, IC, II and IIIA liquids with a maximum total of 120 gallons (454 L) of Class IA liquids. NFPA 30 would also permit the maximum total quantities listed above to be doubled when enhanced sprinkler protection based on substantiated fire test data is provided. The maximum quantity of liquid that can be stored is based on the floor area actually being used for display of the liquids plus the aisle spaces immediately adjacent thereto.

Example: A local department store has an open area of 25,000 square feet (2323 m²) per floor (i.e., each floor is one room). The area assigned to flammable paint sales is 2,000 square feet (186 m²). The building, including the paint sales area, is protected by an Ordinary Hazard Group 2 sprinkler system. Based on this exception, the facility is permitted to store 7,500 gallons (28 388 L) of oil-based or other types of paint, classified as a high-hazard material, in the 2,000-square-foot (186 m²) area devoted to retail paint sales. While a density of 4 gallons per square foot (0.1 L/m²) of retail paint sales is permitted in

FIGURE 307.9   **USE AND OCCUPANCY CLASSIFICATION**

sprinklered occupancies, the total permitted quantity is limited to 7,500 gallons (28 388 L), as illustrated in Figure 307.9. For calculation purposes, the floor area used to determine the maximum allowable quantity is the entire floor area [2,000 square feet (186 m²)] used for that purpose.

Exception 5 exempts closed systems that are used exclusively for the operation of machinery or equipment. The closed systems, which are essentially not open to the atmosphere, keep flammable or combustible liquids from direct exposure to external sources of ignition as well as prevent the users from coming in direct contact with liquids or harmful vapors. This exception would include systems such as oil-burning equipment, piping for diesel fuel generators and LP-gas cylinders for use in forklift trucks.

Exception 6 exempts cleaning establishments that utilize a closed system for all combustible liquid solvents with a flash point at or above 140°F (60°C). The reference to using equipment listed by an approved testing laboratory does not mean that the entire system needs to be approved but rather the individual pieces of equipment. As with any mechanical equipment or appliance, it should bear the label of an approved agency and be installed in accordance with the manufacturer's installation instructions [see the *International Mechanical Code®* (IMC®)].

Exception 7 covers cleaning establishments that use solvents that have very high flash points [greater than 200°F (93°C)] and that are exceedingly difficult to ignite. Such liquids can be used openly, but with due care.

Exception 8 exempts all retail liquor stores and liquor distribution facilities from the high-hazard occupancy classification, even though most of the contents are considered combustible liquids. The exception takes into account that alcoholic beverages are packaged in individual containers of limited size.

Exception 9 refers to refrigeration systems that utilize refrigerants that may be flammable or toxic. Refrigeration systems do not alter the occupancy classification of the building, provided they are installed in accordance with the IMC. The IMC has specific limitations on the quantity and type of refrigerants that can be used, depending on the occupancy classification of the building.

Exception 10 exempts materials that are used for agricultural purposes, such as fertilizers, pesticides, fungicides, etc., when used on the premises. Agricultural materials stored for direct or immediate use are not usually of such quantities that would constitute a large fuel load or an exceptionally hazardous condition.

Exception 11 addresses battery storage rooms when used as part of an operating system, such as for providing standby power. The batteries used in installations of this type do not represent a significant health, safety or fire hazard. The electrolyte and battery casing contribute little fuel load to an existing fire. The release of hydrogen gas during the operation of battery systems is minimal. Ventilation in accordance with the IMC will disperse the small amounts of liberated hydrogen.

Without Exception 12 certain products that technically are corrosive could cause grocery stores and other mercantile occupancies to be inappropriately classified as Group H-4. This exception allows the maximum allowable quantity per control area in Table 307.7(2) for corrosives to be exceeded in the retail display area. This would include such things as bleaches, detergents and other household cleaning supplies in normal-size containers. The exception also exempts the storage or manufacture of commonly used building materials, such as portland cement, from being inappropriately classified as Group H.

Exception 13 exempts buildings and structures used

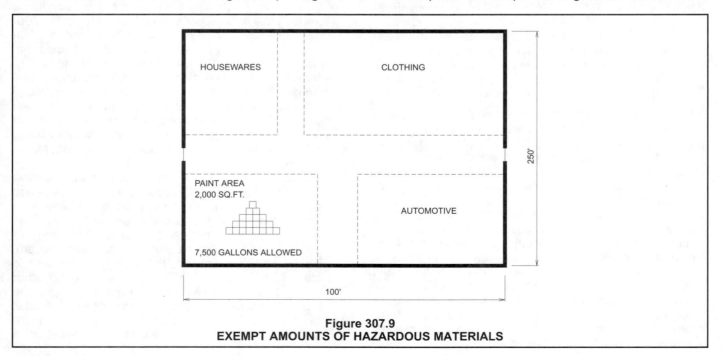

Figure 307.9
EXEMPT AMOUNTS OF HAZARDOUS MATERIALS

for the storage and manufacture of aerosol products, provided they are protected in accordance with the provisions of NFPA 30B and the IFC. The aerosol storage requirements in the IFC, referred to in this exception, are based on the provisions of NFPA 30B. Compliance with the exception exempts buildings from complying with the code provisions for Group H-3 (see Section 307.5), provided the storage and manufacturing of aerosol products comply with the applicable separation, storage limitations and sprinkler design requirements specified in the IFC and NFPA 30B.

Exception 14 permits certain products found in mercantile and storage occupancies, which may be comprised of hazardous materials, to exceed the maximum allowable quantity per control area of Tables 307.7(1) and 307.7(2). The products, however, must be comprised of nonflammable solids, liquids that are nonflammable or noncombustible or health hazard gases. Materials could include swimming pool chemicals, which are typically Class 2 or 3 oxidizers or industrial corrosive cleaning agents (see commentary, Section 414.2.4).

Exception 15 permits the base maximum allowable quantity per control area of black powder, smokeless propellant and small arms primers in Group M and R-3 occupancies to be exceeded, provided the material is stored in accordance with Chapter 33 of the IFC. The requirements are based on the provisions in NFPA 495.

Similarly, special industrial explosive devices are found in a number of occupancies other than Group H (Groups B, F, M and S). Storage of these devices in accordance with the IFC is not required to have a high-hazard occupancy classification. Power drivers are commonly used in the construction industry, and there are stocks of these materials maintained for sale and use by the trade. The automotive airbag industry has evolved with the use of these devices, and they are located in automotive dealerships and personal use vehicles throughout society. The IFC currently exempts up to 50 pounds (23 kg) of these materials from regulation under Chapter 33 (explosives).

SECTION 308
INSTITUTIONAL GROUP I

308.1 Institutional Group I. Institutional Group I occupancy includes, among others, the use of a building or structure, or a portion thereof, in which people are cared for or live in a supervised environment, having physical limitations because of health or age are harbored for medical treatment or other care or treatment, or in which people are detained for penal or correctional purposes or in which the liberty of the occupants is restricted. Institutional occupancies shall be classified as Group I-1, I-2, I-3 or I-4.

❖ Institutional occupancies are comprised of two basic types. The first relates to health care facilities that are intended to provide medical care or treatment for people who have physical or mental disabilities or diseases and other infirmities. This includes persons who are ambula-

tory and capable of self-preservation as well as those who are restricted in their mobility or totally immobile and need assistance to escape a life-threatening situation, such as a fire (i.e., children $2^{1}/_{2}$ years or less, the infirm and the aged). The second type of occupancy relates primarily to detention and correctional facilities. Since security is the major operational consideration in these kinds of facilities, the occupants (inmates) are under some form of restraint and may be rendered incapable of self-preservation without assistance in emergency situations.

The degree of hazards in each type of institutional facility identified in this section varies respective to each kind of occupancy. The code addresses each occupancy separately and the regulatory provisions throughout the code provide the proper means of protection so as to produce an acceptable level of safety to life and property.

This section identifies the occupancies that are included in the general term "Group I." Institutional occupancies are divided into four individual occupancy classifications: Groups I-1, I-2, I-3 and I-4. These classifications are based on the degree of detention and physical mobility of the occupants. The term "Group I" includes each of the individual institutional occupancy classifications.

308.2 Group I-1. This occupancy shall include buildings, structures or parts thereof housing more than 16 persons, on a 24-hour basis, who because of age, mental disability or other reasons, live in a supervised residential environment that provides personal care services. The occupants are capable of responding to an emergency situation without physical assistance from staff. This group shall include, but not be limited to, the following:

Residential board and care facilities
Assisted living facilities
Halfway houses
Group homes
Congregate care facilities
Social rehabilitation facilities
Alcohol and drug centers
Convalescent facilities

A facility such as the above with five or fewer persons shall be classified as a Group R-3 or shall comply with the *International Residential Code* in accordance with Section 101.2. A facility such as above, housing at least six and not more than 16 persons, shall be classified as Group R-4.

❖ An occupancy classified in Group I-1 is characterized by four conditions: it is a health care facility, the occupant load is greater than 16 (in order to be consistent with the definition of "Residential care/assisted living facility"), there is 24-hour-a-day supervision and the occupants are capable of reaching safety in an emergency situation without the need of physical assistance by staff or others.

Any building that has these characteristics but that contains an occupant load of more than five and not more than 16 is classified as Group R-4 (see Section 310.1). Any building that has these characteristics but contains an occupant load of five or less is classified as Group R-3 (see Section 310.1), or shall be constructed in accordance with the *International Residential Code*® (IRC®).

For clarification purposes, a dormitory or apartment complex that houses only elderly people and has a nonmedically trained live-in manager is not classified as an institutional occupancy but rather as a residential occupancy (see Section 310.1). A critical phrase in the code to consider when evaluating this type of facility is "live in a supervised residential environment."

308.3 Group I-2. This occupancy shall include buildings and structures used for medical, surgical, psychiatric, nursing or custodial care on a 24-hour basis of more than five persons who are not capable of self-preservation. This group shall include, but not be limited to, the following:

Hospitals
Nursing homes (both intermediate-care facilities and skilled nursing facilities)
Mental hospitals
Detoxification facilities

A facility such as the above with five or fewer persons shall be classified as Group R-3 or shall comply with the IRC in accordance with Section 101.2.

❖ An occupancy classified in Group I-2 is characterized by four conditions: it is a health care facility, the occupant load is greater than five, there is 24-hour-a-day supervision and there are occupants who are incapable of self-preservation (i.e., some or all of the occupants require physical assistance by staff or others to reach safety in an emergency situation).

As with Group I-1, any building that has these characteristics but that contains an occupant load of five or less is classified as a residential occupancy, Group R-3 (see Section 310.1), or shall be constructed in accordance with the IRC.

The most common examples of facilities classified in Group I-2 are hospitals and nursing homes [see Figures 308.3(1) and 308.3(2)].

It is not uncommon to find dining rooms (Group A-2), staff offices (Group B), gift shops (Group M) and other nonmedically related areas in buildings classified as Group I-2. Unless the area of the nonmedical occupancy qualifies as an accessory use area (see commentary, Section 302.2), the buildings are considered as a mixed occupancy and subject to the provisions of Sections 302.3.

In addition to the general requirements contained in this section, Section 407 contains specific requirements for Group I-2.

308.3.1 Child care facility. A child care facility that provides care on a 24-hour basis to more than five children $2^{1}/_{2}$ years of age or less shall be classified as Group I-2.

❖ Child care facilities housing more than five children $2^{1}/_{2}$ years old and younger on a 24-hour basis are classified as Group I-2 because children that age are not generally capable of responding to an emergency but must be led or carried to safety. Under such circumstances, the occupants are considered nonambulatory. The distinguishing factor between Groups I-2 and I-4 is the amount of time the facility provides care; Group I-2 facilities provide care on a 24-hour basis, while Group I-4 facilities must be less than 24 hours. It is also assumed that medical supervision is present in Group I-2 facilities. Figure 308.3.1 summarizes the different occupancy classifications for care facilities.

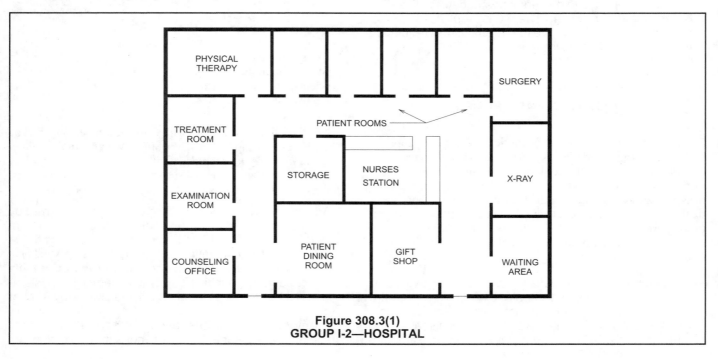

**Figure 308.3(1)
GROUP I-2—HOSPITAL**

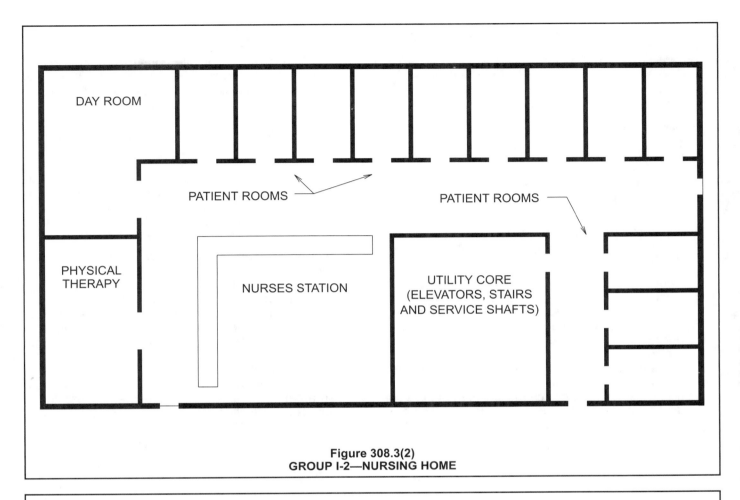

Figure 308.3(2)
GROUP I-2—NURSING HOME

24-HOUR CARE			
Age and capability of residents	0-5 occupants	6-16 occupants	Over 16 occupants
2^1/$_2$ years of age or less	R-3	I-2	I-2
Over 2^1/$_2$ years of age and capable of self-preservation	R-3	R-4	I-1
Over 2^1/$_2$ years of age and not capable of self-preservation	R-3	I-2	I-2

LESS THAN 24-HOUR CARE—DAY CARE		
Age and capability of residents	0-5 occupants	Over 5 occupants
2^1/$_2$ years of age or less	R-3	I-4 (Exception permits E)
Over 2^1/$_2$ years of age thru 12th grade and capable of self-preservation	R-3	E
Over 12th grade and capable of self-preservation	R-3	A-3
Over 2^1/$_2$ years of age and not capable of self-preservation	R-3	I-4

Figure 308.3.1
OCCUPANCY CLASSIFICATION OF CARE FACILITIES

308.4 Group I-3. This occupancy shall include buildings and structures that are inhabited by more than five persons who are under restraint or security. An I-3 facility is occupied by persons who are generally incapable of self-preservation due to security measures not under the occupants' control. This group shall include, but not be limited to, the following:

> Prisons
> Jails
> Reformatories
> Detention centers
> Correctional centers
> Prerelease centers

Buildings of Group I-3 shall be classified as one of the occupancy conditions indicated in Sections 308.4.1 through 308.4.5 (see Section 408.1).

❖ An occupancy classified in Group I-3 is characterized by three conditions: it is a location where the occupants are under restraint or where security is closely supervised; there are more than five occupants and the occupants are not capable of self-preservation because the conditions of confinement are not under their control (i.e., they require assistance by the facilities' staff to reach safety in an emergency situation).

Buildings that have these characteristics but that contain an occupant load of five or less are classified as a residential occupancy, but are still subject to special hardware requirements (see Chapter 10).

It is recognized that not all Group I-3 occupancies exercise the same level of restraint; thus, to distinguish these different levels, the code defines five different conditions of occupancy based on the degree of access to the exit discharge.

The five occupancy conditions are summarized in Figure 308.4.

308.4.1 Condition 1. This occupancy condition shall include buildings in which free movement is allowed from sleeping areas, and other spaces where access or occupancy is permitted, to the exterior via means of egress without restraint. A Condition 1 facility is permitted to be constructed as Group R.

❖ Condition 1 areas are those where the occupants have unrestrained access to the exterior of the building. As such, a key or remote control release device is not needed for any occupant to reach the exterior of the building (exit discharge) at any time. These types of buildings are referred to as low-security facilities. A work-release center is a typical Condition 1 facility (see Figure 308.4.1).

Because of the lack of restraint associated with a Condition 1 building, it resembles a residential use more than a detention facility and, therefore, is classified in Group R (see Section 310).

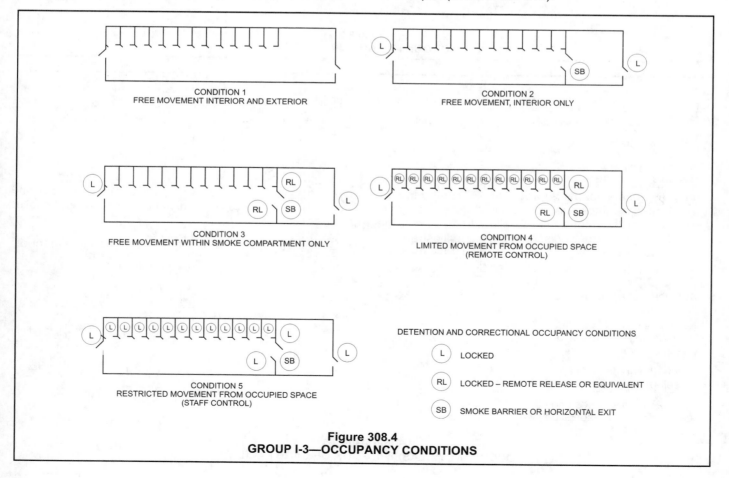

**Figure 308.4
GROUP I-3—OCCUPANCY CONDITIONS**

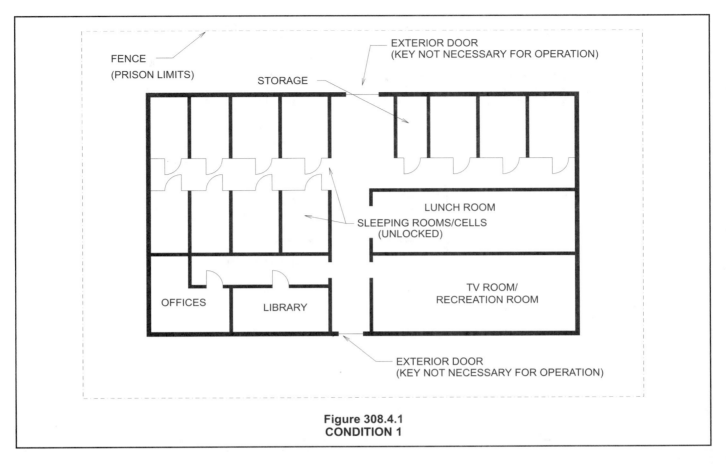

Figure 308.4.1
CONDITION 1

308.4.2 Condition 2. This occupancy condition shall include buildings in which free movement is allowed from sleeping areas and any other occupied smoke compartment to one or more other smoke compartments. Egress to the exterior is impeded by locked exits.

❖ Condition 2 areas are those in which the movement of occupants is not controlled within the exterior walls of the building (i.e., the occupants have unrestrained access within the building). As such, there is free movement by the occupants between smoke compartments (as created by smoke barriers) however, the occupants must rely on someone else to allow them to exit the building to the area of discharge. This is illustrated in Figure 308.4.2.

308.4.3 Condition 3. This occupancy condition shall include buildings in which free movement is allowed within individual smoke compartments, such as within a residential unit comprised of individual sleeping units and group activity spaces, where egress is impeded by remote-controlled release of means of egress from such a smoke compartment to another smoke compartment.

❖ Condition 3 areas are those in which free movement by the occupants is permitted within an individual smoke compartment; however, movement of occupants from one smoke compartment (as created by smoke barriers) to another smoke compartment and from within the building to the exterior (exit discharge) is controlled by

remote release locking devices. As such, the occupants in the facility are dependent on the staff for their release from each smoke compartment or to the exterior (exit discharge). This condition is illustrated in Figure 308.4.3.

308.4.4 Condition 4. This occupancy condition shall include buildings in which free movement is restricted from an occupied space. Remote-controlled release is provided to permit movement from sleeping units, activity spaces and other occupied areas within the smoke compartment to other smoke compartments.

❖ Condition 4 areas are those in which the movement of occupants from any room or space within a smoke compartment (as created by smoke barriers) to another smoke compartment or to the exterior (exit discharge) is controlled by remote release locking devices. Any movement within the facility requires activation of a remote control lock system to release the designated area (see Figure 308.4.4). The occupants within a Condition 4 area must rely on an activation system in the event of an emergency in order to evacuate the area.

Condition 4 facilities most often are penal facilities where the occupants are considered relatively safe to handle in large groups. As such, many occupants can be released simultaneously from their individual sleeping areas when they need to travel to dining areas or move to another area.

FIGURE 308.4.2 – FIGURE 308.4.3

USE AND OCCUPANCY CLASSIFICATION

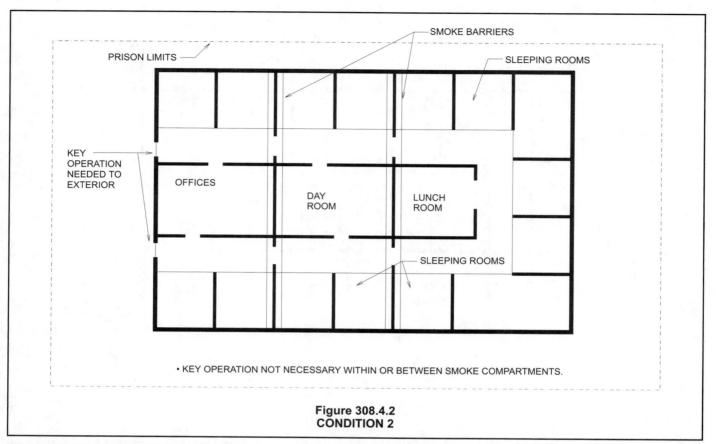

• KEY OPERATION NOT NECESSARY WITHIN OR BETWEEN SMOKE COMPARTMENTS.

Figure 308.4.2
CONDITION 2

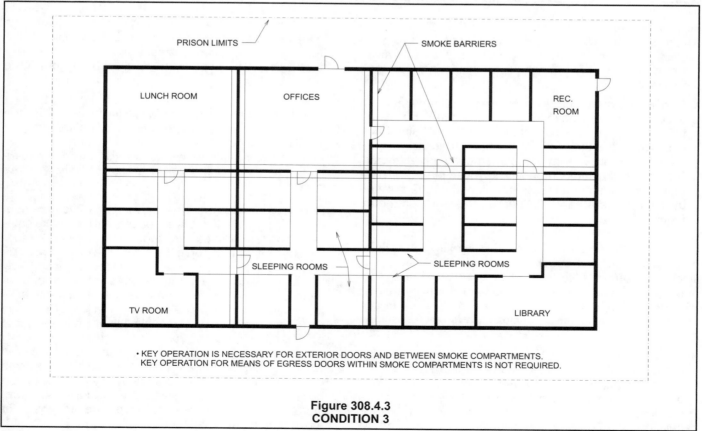

• KEY OPERATION IS NECESSARY FOR EXTERIOR DOORS AND BETWEEN SMOKE COMPARTMENTS.
KEY OPERATION FOR MEANS OF EGRESS DOORS WITHIN SMOKE COMPARTMENTS IS NOT REQUIRED.

Figure 308.4.3
CONDITION 3

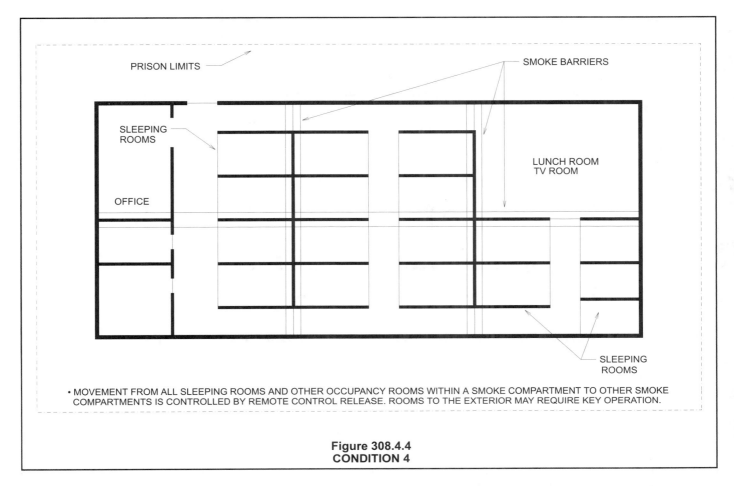

PRISON LIMITS

SMOKE BARRIERS

SLEEPING ROOMS

LUNCH ROOM TV ROOM

OFFICE

SLEEPING ROOMS

• MOVEMENT FROM ALL SLEEPING ROOMS AND OTHER OCCUPANCY ROOMS WITHIN A SMOKE COMPARTMENT TO OTHER SMOKE COMPARTMENTS IS CONTROLLED BY REMOTE CONTROL RELEASE. ROOMS TO THE EXTERIOR MAY REQUIRE KEY OPERATION.

Figure 308.4.4
CONDITION 4

308.4.5 Condition 5. This occupancy condition shall include buildings in which free movement is restricted from an occupied space. Staff-controlled manual release is provided to permit movement from sleeping units, activity spaces and other occupied areas within the smoke compartment to other smoke compartments.

❖ Condition 5 are those areas in which the occupants are not allowed free movement to any other room or space within a smoke compartment (as created by smoke barriers), to another smoke compartment or to the exterior (exit discharge) unless the locking device controlling their area of confinement is manually released by a staff member. Once released from an individual space, a staff member is responsible for unlocking all doors from that location to the next smoke compartment. This is the most restrictive occupancy condition, as each occupant must be released on an individual basis and escorted to other areas.

Condition 5 facilities are most often used for maximum security or solitary confinement areas where the occupants are considered to be dangerous to others, including staff members, and cannot safely be handled in large groups (see Figure 308.4.5).

308.5 Group I-4, day care facilities. This group shall include buildings and structures occupied by persons of any age who re-

ceive custodial care for less than 24 hours by individuals other than parents or guardians, relatives by blood, marriage or adoption, and in a place other than the home of the person cared for. A facility such as the above with five or fewer persons shall be classified as a Group R-3 or shall comply with the *International Residential Code* in accordance with Section 101.2. Places of worship during religious functions are not included.

❖ Facilities that contain provisions for the care of more than five adults (greater than the 12th grade) or more than five children ($2^1/_2$ years of age or less) are classified as Group I-4. Group I-4 facilities are less restrictive in some of the requirements (e.g., height and area) than the other Group I occupancies. Group I-4 facilities are intended to be used for less than 24 hours and are not intended to provide medical supervision. Day care facilities are not intended to be a residence for the people cared for. The staff members are assumed to not be related to the individuals in the day care facilities. In order to be consistent with the other Group I facilities, Group I-4 buildings that have five or less people are also to be classified as Group R-3, or shall be constructed in accordance with the IRC. The definition for Group I-4 facilities could be construed to include places of worship; therefore, the last sentence of Section 308.5 is added to clarify that I-4 provisions would not supply to places of worship unless the space is also used for day care facilities.

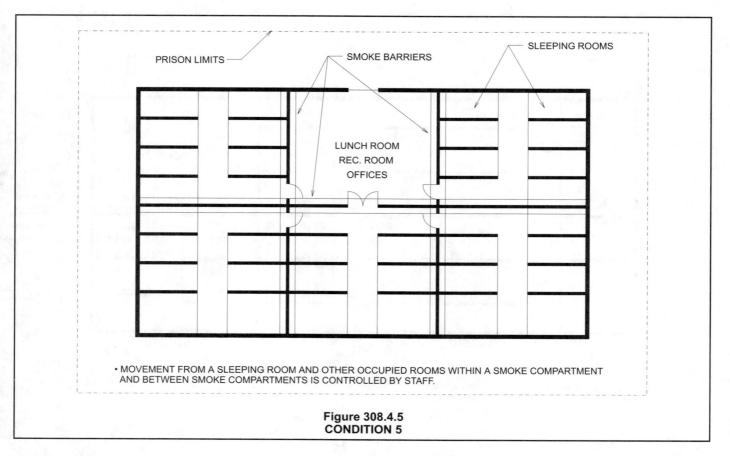

Figure 308.4.5
CONDITION 5

308.5.1 Adult care facility. A facility that provides accommodations for less than 24 hours for more than five unrelated adults and provides supervision and personal care services shall be classified as Group I-4.

> **Exception:** A facility where occupants are capable of responding to an emergency situation without physical assistance from the staff shall be classified as Group A-3.

❖ Adult care facilities are assumed to be for people passed the 12th grade. For people below 12th grade, the building must be classified as Group I-2 (occupants are 2$\frac{1}{2}$ years of age or less) or as Group E (occupants greater than 2$\frac{1}{2}$ years of age but below 12th grade). In addition, there must be more than five adults in the facility and they must not be related in any manner. Facilities in which the adults are related would be more appropriately classified in Group R. The exception clarifies that the classification of Group I-4 for an adult day care facility does not apply to facilities that provide services for adults who are capable of responding to an emergency unassisted. In that case, the facility is simply a place of assembly, an A-3 occupancy.

308.5.2 Child care facility. A facility that provides supervision and personal care on less than a 24-hour basis for more than five children 2$\frac{1}{2}$ years of age or less shall be classified as Group I-4.

> **Exception:** A child day care facility that provides care for more than five but no more than 100 children 2$\frac{1}{2}$ years or less of age, when the rooms where such children are cared for are

located on the level of exit discharge and each of these child care rooms has an exit door directly to the exterior, shall be classified as Group E.

❖ As with Group I-2 child care facilities, the occupants of Group I-4 child care facilities are limited to 2$\frac{1}{2}$ years of age or less. The distinguishing factor between the two occupancies is the amount of time the facility provides care; Group I-2 facilities provide care on a 24-hour basis while Group I-4 facilities must be less than 24 hours. It is also assumed that medical supervision is not present in Group I-4 facilities. Occupants younger than 2$\frac{1}{2}$ years of age are not typically capable of independently responding to an emergency, but must be led or carried to safety. Under such circumstances, the occupants are considered nonambulatory.

A child care facility in which the number of occupants is greater than five but not more than 100 is permitted to be classified as Group E, provided the children are all located in rooms on the level of exit discharge and all of the rooms have exit doors directly to the exterior. This exception is only applicable to rooms and spaces used for child care and is not intended to apply to accessory spaces such as restrooms, offices and kitchens.

By permitting the facility to be classified as Group E, the building would not be required to be sprinklered unless the fire area was greater than 20,000 square feet (1858 m²). A Group I-4 would be required to be sprinklered regardless of the area.

SECTION 309
MERCANTILE GROUP M

309.1 Mercantile Group M. Mercantile Group M occupancy includes, among others, buildings and structures or a portion thereof, for the display and sale of merchandise, and involves stocks of goods, wares or merchandise incidental to such purposes and accessible to the public. Mercantile occupancies shall include, but not be limited to, the following:

Department stores
Drug stores
Markets
Motor fuel-dispensing facilities
Retail or wholesale stores
Sales rooms

❖ The characteristics of occupancies classified in Group M are contained in this section. Because mercantile occupancies normally involve the display and sale of large quantities of combustible merchandise, the fuel load in such facilities can be relatively high, potentially exposing the occupants (customers and sales personnel) to a high degree of fire hazard. Mercantile operations often attract large crowds (particularly in large department stores and covered malls and especially during weekends and holidays). There are two factors that alleviate the risks to life safety: the occupant load normally has a low-to-moderate density and the occupants are alert, mobile and able to respond in an emergency situation. The degree of openness and the organization of the retail display found in most mercantile occupancies is generally orderly and does not present an unusual difficulty for occupant evacuation.

Contained herein are general descriptions of the kinds of occupancies that are classified in Group M. Mercantile buildings most often have both a moderate occupant load and a high fuel load, which is in the form of furnishings and the goods being displayed, stored and sold [see Figure 309.1(1)].

The key characteristics that differentiate occupancies classified in Group M from those classified in Group B (see Section 304) are the larger quantity of goods or merchandise available for sale and the lack of familiarity of the occupants with the building, particularly its means of egress. To be classified in Group M, the goods that are on display must be accessible to the public. If a patron sees an item for sale, then that item is generally available for purchase at that time (i.e., there is a large stock of goods). If a store allows people to see the merchandise but it is not available on the premises, such as an automobile showroom, then the occupancy classification of business (Group B) should be considered. A mercantile building is accessible to the public, many of whom may not be regular visitors. A business building, however, is primarily occupied by regular employees who are familiar with the building arrangement and, most importantly, the exits. This awareness of the building and the exits can be an important factor in a fire emergency.

Automotive, fleet-vehicle, marine and self-service fuel-dispensing facilities, as defined in the IFC, are classified in the mercantile occupancy, as are the conve-

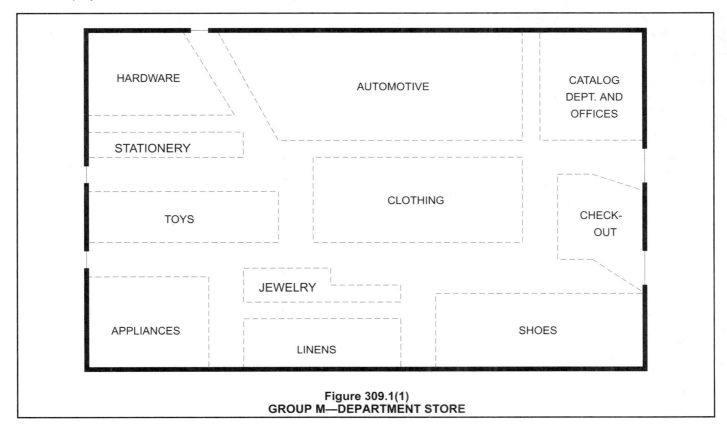

Figure 309.1(1)
GROUP M—DEPARTMENT STORE

nience stores often associated with such occupancies [(see Figure 309.1(2)]. Quick-lube, tune-up, muffler and tire shops are not included in this classification. Those facilities that typically conduct automotive service and minor repair work are treated as a repair garage (Group S-1, also defined in the IFC) [see Figure 309.1(2)].

309.2 Quantity of hazardous materials. The aggregate quantity of nonflammable solid and nonflammable or noncombustible liquid hazardous materials stored or displayed in a single control area of a Group M occupancy shall not exceed the quantities in Table 414.2.4.

❖ This section addresses an exception for control areas of mercantile occupancies containing certain nonflammable or noncombustible materials and health hazard gases that are stored in accordance with Table 414.2.4. This exception allows Group H-4 materials, which present a health hazard rather than a physical hazard, as well as limited Group H-2 and H-3 materials, to be stored in both the retail display and stock areas of regulated mercantile occupancies in excess of the maximum allowable quantity per control area of Tables 307.7(1) and 307.7(2) without classifying the building as Group H. To correctly classify a building where products that have hazardous properties are stored and sold, the code user must also be aware of the provisions contained in Section 307.9, Notes b and c of Table 307.7(1) and Note b of Table 307.7(2). These provisions give the quantity limitations for specific high-hazard products in mercantile display areas, including medicines, foodstuffs, cosmetics and alcoholic beverages.

Without this option, many mercantile occupancies could technically be classified as Group H. The in-creased quantities of certain hazardous materials are based on the recognition that, while there is limited risk in mercantile occupancies, the packaging and storage arrangements can be controlled. For further information on the storage limitations required for these types of materials in mercantile occupancies, see Section 2703.11 of the IFC.

SECTION 310
RESIDENTIAL GROUP R

310.1 Residential Group R. Residential Group R includes, among others, the use of a building or structure, or a portion thereof, for sleeping purposes when not classified as an Institutional Group I. Residential occupancies shall include the following:

R-1 Residential occupancies where the occupants are primarily transient in nature, including:

    Boarding houses (transient)
    Hotels (transient)
    Motels (transient)

R-2 Residential occupancies containing sleeping units or more than two dwelling units where the occupants are primarily permanent in nature, including:

    Apartment houses
    Boarding houses (not transient)
    Convents
    Dormitories
    Fraternities and sororities
    Monasteries
    Vacation timeshare properties

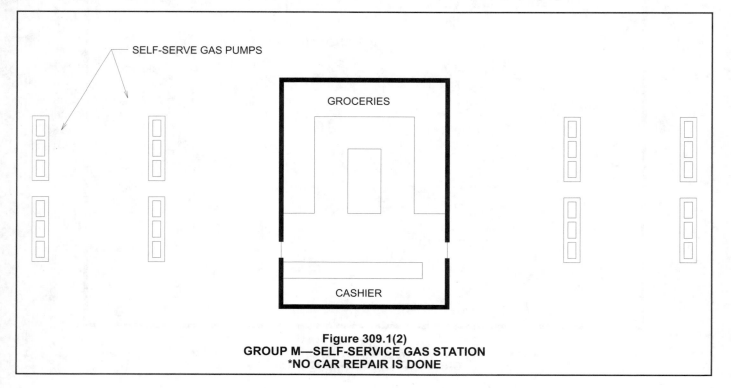

Figure 309.1(2)
GROUP M—SELF-SERVICE GAS STATION
*NO CAR REPAIR IS DONE

Hotels (nontransient)
Motels (nontransient)

R-3 Residential occupancies where the occupants are primarily permanent in nature and not classified as R-1, R-2, R-4 or I and where buildings do not contain more than two dwelling units as applicable in Section 101.2, or adult and child care facilities that provide accommodations for five or fewer persons of any age for less than 24 hours. Adult and child care facilities that are within a single-family home are permitted to comply with the *International Residential Code* in accordance with Section 101.2.

R-4 Residential occupancies shall include buildings arranged for occupancy as residential care/assisted living facilities including more than five but not more than 16 occupants, excluding staff.

Group R-4 occupancies shall meet the requirements for construction as defined for Group R-3 except as otherwise provided for in this code or shall comply with the *International Residential Code* in accordance with Section 101.2.

❖ Residential occupancies represent some of the highest fire safety risks of any of the occupancies listed in Chapter 3. There are several reasons for this condition:

- Structures in the residential occupancy house the widest range of occupant types, i.e., infants to the aged, for the longest periods of time. As such, residential occupancies are more susceptible to the frequency of careless acts of the occupants; therefore, the consequences of exposure to the effects of fire are the most serious.

- Most residential occupants are asleep approximately one-third of every 24-hour period. When sleeping, they are not likely to become immediately aware of a developing fire. Also, if awakened from sleep by the presence of fire, the residents often may not immediately react in a rational manner and delay their evacuation.

- The fuel load in residential occupancies is often quite high, both in quantity and variety. Also, in the construction of residential buildings, it is common to use extensive amounts of combustible materials.

- Another portion of the fire problem in residential occupancies relates to the occupants' lack of vigilance in the prevention of fire hazards. In their own domicile or residence, people tend to relax and are often prone to allow fire hazards to go unabated, thus, in residential occupancies, fire hazards tend to accrue over an extended period of time and go unnoticed or are ignored.

Most of the nation's fire problems occur in Group R buildings and, in particular, one- and two-family dwellings, which account for more than 80 percent of all deaths from fire in residential occupancies and about two-thirds of all fire fatalities in all occupancies. One- and two-family dwellings also account for more than 80 percent of residential property losses from fire and more than one-half of all property losses from fire.

Because of the relatively high fire risk and potential for loss of life in buildings classified in Groups R-1 (hotels and motels) and R-2 (apartments and dormitories), the code has stringent provisions for the protection of life in these occupancies. Group R-3 occupancies, however, are not generally considered to be in the same domain and, thus, are not subject to the same level of regulatory control as is provided in other occupancies. Group R-3 facilities are one- or two-family dwellings where the occupants are generally more familiar with their surroundings, and, because they are single units or duplexes, tend to pose a lower risk of injury or death.

Due to the growing trend to care for people in a residential environment, residential care/assisted living facilities are also classified as Group R. Specifically, these facilities are classified as Group R-4. "Mainstreaming" people who are recovering from alcohol or drug addiction and people who are developmentally disabled is reported to have therapeutic and social benefits. A residential environment often fosters this mainstreaming.

Buildings in Group R are described herein. A building or part of a building is considered to be a residential occupancy if it is intended to be used for sleeping accommodations (including residential care/assisted living facilities) and is not an institutional occupancy. Institutional occupancies are similar to residential occupancies in many ways; however, they differentiated from each other in that institutional occupants are in a supervised environment, and, in the case of Groups I-2 and I-3 occupancies, are under some form of restraint or physical limitation that makes them incapable of complete self-preservation. The number of these occupants who are under supervision or are incapable of self-preservation is the distinguishing factor for being classified as an institutional or residential occupancy.

The term "Group R" refers collectively to the four individual residential occupancy classifications: Groups R-1, R-2, R-3 and R-4. These classifications are differentiated in the code based on the following criteria: (1) whether the occupants are transient or nontransient in nature; (2) the type and number of dwellings contained in a single building and (3) the number of occupants in the facility.

R-1: The key characteristic of Group R-1 that differentiates it from other Group R occupancies is the number of transient occupants (i.e., those whose length of stay is less than 30 days).

The most common building types classified in Group R-1 are hotels, motels and boarding houses. Group R-1 occupancies do not typically have cooking facilities in the unit. When a unit is not equipped with cooking facilities, it does not meet the definition of a "Dwelling unit" in Section 310.2. When this occurs, such units are treated as guestrooms for the application of code provisions [see Figure 310.1(1)]. Guestrooms are required to be separated from each other by fire partitions and horizontal assemblies (see Section 310.3).

FIGURE 310.1(1)   USE AND OCCUPANCY CLASSIFICATION

Other occupancies are often found in buildings classified in Group R-1. These occupancies include nightclubs (Group A-2), restaurants (Group A-2), gift shops (Group M), health clubs (Group A-3) and storage facilities (Group S-1). When this occurs, the building is a mixed occupancy and is subject to the provisions of Section 302.3.

R-2: The length of the occupants' stay plus the arrangement of the facilities provided are the basic factors that differentiate occupancies classified in Group R-2 from other occupancies in Group R. The occupants of facilities or areas classified in Group R-2 are primarily not transient, capable of self-preservation and share their means of egress in whole or in part with other occupants outside of their sleeping area or dwelling unit. The separation between dwelling units must, as a minimum, meet the requirements contained in Sections 310.3, 708.1 and 711.3. Building types ordinarily classified in Group R-2 include apartments, boarding houses (when the occupants are not transient) and dormitories [see Figures 310.1(2) and 310.1(3)].

Individual dwelling units in Group R-2 are either rented by tenants or owned by the occupants. The code does not make a distinction between either type of tenancy unless the dwelling unit is located on a separate parcel of land. When this occurs, lot lines defining the land parcel exist and the requirements for fire separation must be met.

Dormitories are generally associated with university or college campuses for use as student housing, but this is changing rapidly. Many dormitories are now being built as housing for elderly people who wish to live with other people their own age and who do not need 24-hour-a-day medical supervision. The only difference between the dormitory that has just been described and the dormitory found on a college campus is the age of its occupants. If the elderly people must have 24-hour-a-day supervision (i.e., a nurse or doctor on the premises), the building is no longer considered a residential occupancy but an institutional occupancy (Group I-2 assuming greater than five occupants) and would have to comply with the applicable provisions of the code.

Similar to Group R-1, individual rooms in dormitories are also required to be separated from each other by fire partitions and horizontal assemblies in accordance with Sections 310.3, 708.1 and 711.3. When college classes are not in session, the rooms in dormitories are sometimes rented out for periods less than 30 days to convention attendees and other visitors. When dormitories undergo this type of transient use, they more closely resemble Group R-1.

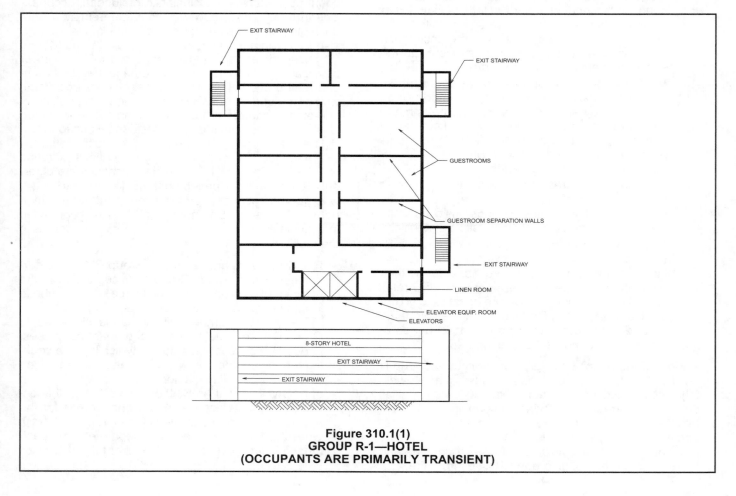

Figure 310.1(1)
GROUP R-1—HOTEL
(OCCUPANTS ARE PRIMARILY TRANSIENT)

Buildings containing dormitories often contain other occupancies, such as cafeterias or dining rooms (Group A-2), recreation rooms (Group A-3) and office (Group B) or meeting rooms (Group A-3). When this occurs, the building is considered a mixed occupancy and is subject to the provisions of Section 302.3 [see Figure 310.1(4)].

R-3: Group R-3 facilities include all detached one- and two-family dwellings and multiple (three or more) single-family dwellings (townhouses) more than three stories in height. Those buildings three or less stories in height are not classified as Group R-3 and must be regulated by the IRC (see Section 101.2). Each pair of dwelling units in multiple single-family dwellings greater than three stories in height must be separated with fire walls (see Section 705) or by two exterior walls (see Table 602) in order to be classified as Group R-3.

Buildings that are one- and two-family dwellings and multiple single-family dwellings less than three stories in height and which contain another occupancy (e.g., Groups R-1, R-2, I-4, etc.) must be regulated as a mixed occupancy in accordance with the code and are not required to comply with the provisions of the IRC [see Figures 310.1(5) and 310.1(6)]. In addition, all institutional facilities that accommodate five or less people are to be classified as Group R-3.

Buildings that are classified as Group R-3, while limited in height, are not limited in the allowable area per floor as indicated in Table 503.

All dwelling units must be separated from each other by fire partitions and horizontal assemblies (in accordance with Sections 310.3, 708.1 and 711.3) unless required to be separated by fire walls.

R-4: When a limited number of people who require personal care live in a residential environment, a facility is no longer classified as Group I-1 but as a residential care/assisted living facility, Group R-4. Ninety-eight percent of households in the U.S. have less than 16 occupants, thus the limit of 16 would allow equal access for the disabled. The number of occupants is those that receive care and is not intended to include staff. With the exception of height and area limitations, Group R-4 facilities must satisfy the construction requirements of Group R-3.

310.2 Definitions. The following words and terms shall, for the purposes of this section and as used elsewhere in this code, have the meanings shown herein.

❖ Definitions of terms that are associated with the content of this section are contained herein. These definitions can help in the understanding and application of the code requirements. One should keep in mind, however, that in many cases, terms defined in the code may also be defined by ordinances and statutes of local and state governments. In such cases, code users must focus on the specific features that define the term relative to the code and not its generally held meaning.

It is important to emphasize that these terms are not exclusively related to this section, but are applicable everywhere the term is used in the code. The purpose for

Figure 310.1(2)
GROUP R-2—BOARDING HOUSE
(OCCUPANTS PRIMARILY NOT TRANSIENT)

FIGURE 310.1(3) – FIGURE 310.1(4)

USE AND OCCUPANCY CLASSIFICATION

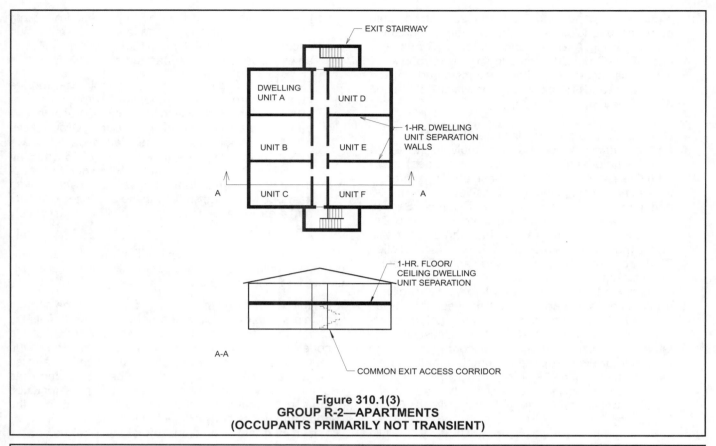

Figure 310.1(3)
GROUP R-2—APARTMENTS
(OCCUPANTS PRIMARILY NOT TRANSIENT)

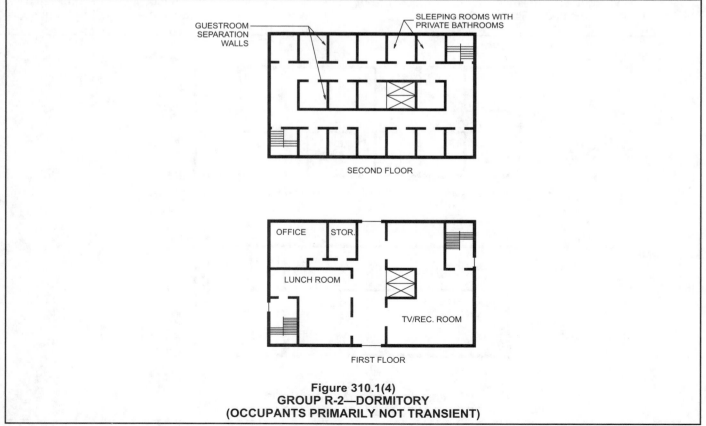

Figure 310.1(4)
GROUP R-2—DORMITORY
(OCCUPANTS PRIMARILY NOT TRANSIENT)

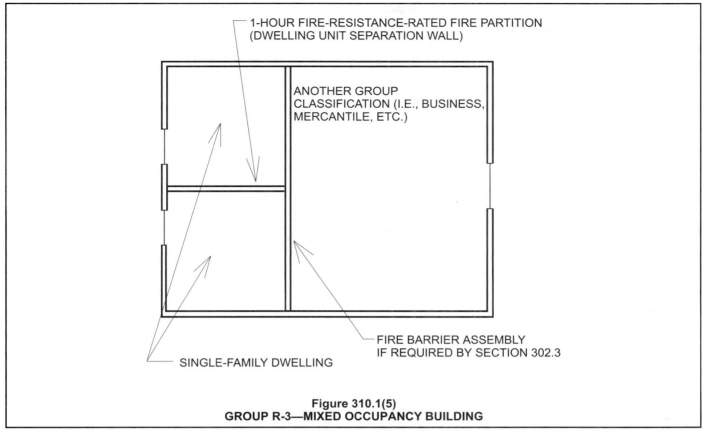

Figure 310.1(5)
GROUP R-3—MIXED OCCUPANCY BUILDING

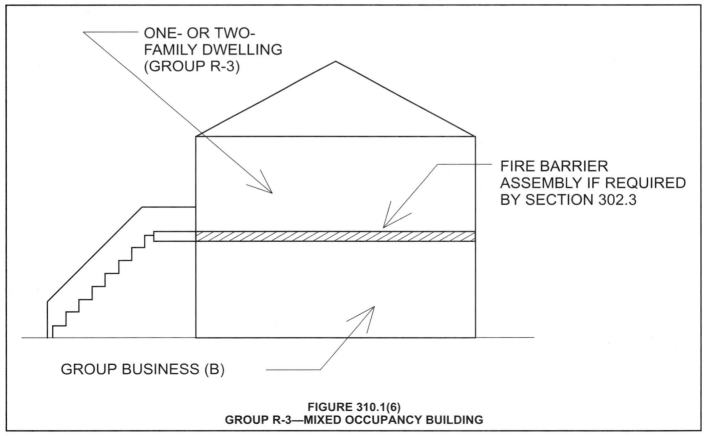

FIGURE 310.1(6)
GROUP R-3—MIXED OCCUPANCY BUILDING

including these definitions within this section is to provide more convenient access to them without having to refer back to Chapter 2. For convenience, these terms are also listed in Chapter 2 with a cross reference to this section.

The use and application of all defined terms, including those defined herein, are set forth in Section 201.

BOARDING HOUSE. A building arranged or used for lodging for compensation, with or without meals, and not occupied as a single-family unit.

❖ A boarding house is a structure housing lodgers or boarders in which the occupants are provided lodging or meals and lodging for a fee. The individual rooms used for lodging usually do not contain all of the permanent living provisions of a dwelling unit (e.g., permanent cooking facilities). Most often, the term "boarding house" describes a facility that is primarily for transient occupants; however, these facilities might also be used for nontransient purposes. Depending on the extent of transiency, a boarding house could be classified as Group R-1 when an occupant typically stays for less than 30 days or Group R-2 when the length of stay is 30 days or more [see commentary, Section 310.1 and Figure 310.1(2)].

DORMITORY. A space in a building where group sleeping accommodations are provided in one room, or in a series of closely associated rooms, for persons not members of the same family group, under joint occupancy and single management, as in college dormitories or fraternity houses.

❖ Dormitories typically consist of a large room serving as a community sleeping room or many smaller rooms grouped together and serving as private or semiprivate sleeping rooms. A typical setting for dormitories is on college campuses; however, sleeping areas of a fire station and similar lodging facilities for occupants not of the same family group are also considered dormitories. Dormitories most often are not the permanent residence of the occupants. They are typically occupied only for a designated period of time, such as a school year. Though limited, the period of occupancy is usually more than 30 days, which provides the occupant with a familiarity of the structure such that the occupancy is not considered transient. A dormitory is classified as Group R-2 (see commentary, Section 310.1).

Structures containing a dormitory often have a cafeteria or central eating area and common recreational areas. When such conditions exist, the structure must comply with the mixed occupancy provisions of the code [see Section 302.3 and Figure 310.1(4)].

DWELLING UNIT. A single unit providing complete, independent living facilities for one or more persons, including permanent provisions for living, sleeping, eating, cooking and sanitation.

❖ A dwelling unit contains elements necessary for independent living, including provisions for living spaces (family rooms, living rooms, dens, etc.); sleeping quarters; food preparation and eating spaces and personal hygiene, cleanliness and sanitation facilities. The allowable minimum room sizes for habitable rooms in a dwelling unit, as well as the minimum required levels of light and ventilation for such spaces, are provided in Chapter 12.

A dwelling unit is occupied in one of two ways, either through renting or by ownership. The code requirements are applied consistently to all dwellings, regardless of the type of ownership. Both owner-occupied and rented or leased dwellings must comply with the requirements of the code.

A dwelling unit can exist singularly as a one-family dwelling or in combination with other dwelling units. When two dwelling units are grouped together in the same structure, the structure is considered a two-family dwelling. Three or more dwelling units in the same structure are considered as a multiple-family dwelling or multiple single-family dwellings.

PERSONAL CARE SERVICE. The care of residents who do not require chronic or convalescent medical or nursing care. Personal care involves responsibility for the safety of the resident while inside the building.

❖ Used in conjunction with Group I-1 and R-4 facilities, this item refers to personal care for residents in a supervised environment who do not require medical supervision. The purpose of personal care service is to distinguish between residents who require medical supervision and those who require solely physical supervision (i.e., to provide safety for the residents while inside the building).

RESIDENTIAL CARE/ASSISTED LIVING FACILITIES. A building or part thereof housing persons, on a 24-hour basis, who because of age, mental disability or other reasons, live in a supervised residential environment which provides personal care services. The occupants are capable of responding to an emergency situation without physical assistance from staff. This classification shall include, but not be limited to, the following: residential board and care facilities, assisted living facilities, halfway houses, group homes, congregate care facilities, social rehabilitation facilities, alcohol and drug abuse centers and convalescent facilities.

❖ Residential care/assisted living facilities are essentially Group I-1 facilities with a smaller number of occupants. The same provisions that are applicable to Group I-1 facilities are applicable to residential care/assisted living facilities (e.g., occupants are there on a 24-hour basis, are capable of responding to an emergency situation without physical assistance, etc.).

SECTION 311
STORAGE GROUP S

311.1 Storage Group S. Storage Group S occupancy includes, among others, the use of a building or structure, or a portion thereof, for storage that is not classified as a hazardous occupancy.

❖ This section requires that all structures (or parts thereof) designed or occupied for the storage of moderate- and low-hazard materials are to be classified in either Group S-1 (moderate hazard) or S-2 (low hazard).

Even though a storage facility may be part of another occupancy or stand alone as a separate building or operation, the characteristics of the occupancy and the level of hazards present in such facilities require that storage occupancies be treated as separate and distinct considerations in the code enforcement process.

The life safety problems in structures used for storage of moderate- and low-hazard materials are minimal because the number of people involved in a storage operation is usually small and normal work patterns require the occupants to be dispersed throughout the facility. The problems of fire safety, particularly as they relate to the protection of stored contents, are directly associated with the amount and combustibility of the materials (including packaging) that are housed on the premises.

Storage facilities typically contain significant amounts of combustible or noncombustible materials that are kept in a common area. Because of the combustion or explosive characteristics of certain materials (see Section 307), a structure (or portion thereof) that is used to store high-hazard materials, which does not meet one of the exceptions identified in Section 307.9, may not be classified as Group S. A building that has hazardous materials in storage in quantities that exceed the limitations of Section 307.9 is to be classified as Group H, high-hazard uses, and is to comply with Section 307.

Storage occupancies consist of two basic types: Groups S-1 and S-2, which are based on the properties of the materials being stored. The distinction between Group S-1 and S-2 is similar to that between Group F-1 and F-2, as outlined in Section 306.

311.2 Moderate-hazard storage, Group S-1. Buildings occupied for storage uses which are not classified as Group S-2 including, but not limited to, storage of the following:

Aerosols, Levels 2 and 3
Aircraft repair hangar
Bags; cloth, burlap and paper
Bamboos and rattan
Baskets
Belting; canvas and leather
Books and paper in rolls or packs
Boots and shoes
Buttons, including cloth covered, pearl or bone
Cardboard and cardboard boxes
Clothing, woolen wearing apparel
Cordage
Furniture
Furs
Glues, mucilage, pastes and size
Grains
Horns and combs, other than celluloid
Leather

Linoleum
Lumber
Motor vehicle repair garages complying with the maximum allowable quantities of hazardous materials listed in Table 307.7(1) (see Section 406.6)
Photo engravings
Resilient flooring
Silks
Soaps
Sugar
Tires, bulk storage of
Tobacco, cigars, cigarettes and snuff
Upholstery and mattresses
Wax candles

❖ Buildings in which combustible materials are stored and that burn with ease are classified in Group S-1, moderate-hazard storage occupancies. Examples of the kinds of materials that, when stored, are representative of occupancies classified in Group S-1 are also listed in this section.

Repair garages (e.g., painting, body work, engine overhaul, etc.) are also classified as Group S-1 (see Figure 311.2) and must be in compliance with Section 406.6. In addition, to avoid a Group H classification, the amounts of hazardous materials in the garage must be less than the maximum allowable quantity per control area permitted in Table 307.7(1).

311.3 Low-hazard storage, Group S-2. Includes, among others, buildings used for the storage of noncombustible materials such as products on wood pallets or in paper cartons with or without single thickness divisions; or in paper wrappings. Such products are permitted to have a negligible amount of plastic trim, such as knobs, handles or film wrapping. Storage uses shall include, but not be limited to, storage of the following:

Aircraft hangar
Asbestos
Beverages up to and including 12-percent alcohol in metal, glass or ceramic containers
Cement in bags
Chalk and crayons
Dairy products in nonwaxed coated paper containers
Dry cell batteries
Electrical coils
Electrical motors
Empty cans
Food products
Foods in noncombustible containers
Fresh fruits and vegetables in nonplastic trays or containers
Frozen foods
Glass
Glass bottles, empty or filled with noncombustible liquids
Gypsum board
Inert pigments
Ivory
Meats
Metal cabinets
Metal desks with plastic tops and trim
Metal parts

FIGURE 311.2 – FIGURE 311.3

USE AND OCCUPANCY CLASSIFICATION

Metals
Mirrors
Oil-filled and other types of distribution transformers
Parking garages, open or enclosed
Porcelain and pottery
Stoves
Talc and soapstones
Washers and dryers

❖ Buildings in which noncombustible materials are stored are classified as Group S-2, low-hazard storage occupancies (see Figure 311.3). It is acceptable for stored noncombustible products to be packaged in combustible materials as long as the quantity of packaging is kept to an insignificant level.

As seen in Group F-1 and F-2 classifications, it is important to be able to distinguish when the presence of combustible packaging constitutes a significant fuel load; as such, a fuel load might require the building to be classified in Group S-1, moderate-hazard storage. A simple guideline to follow is the "single thickness" rule, which is when a noncombustible product is put in one layer of packaging material.

Examples of materials qualified for storage in Group S-2 storage facilities are as follows:

- Vehicle engines placed on wood pallets for transportation after assembly;
- Washing machines in corrugated cardboard boxes; and
- Soft-drink glass bottles packaged in pressed paper boxes.

Structures used to store noncombustible materials packaged in more than one layer of combustible packaging material are to be classified in Group S-1. Examples of materials that, because of packaging, do not qualify for classification in Group S-2 are:

- Chinaware wrapped in corrugated paper and placed in cardboard boxes;

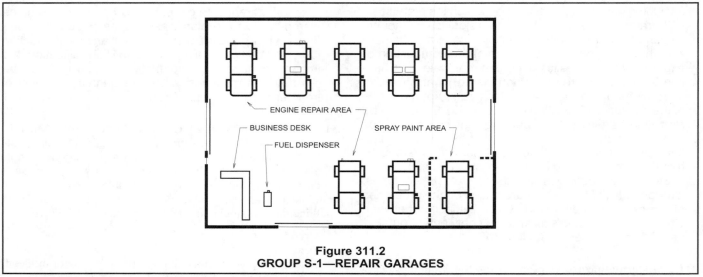

Figure 311.2
GROUP S-1—REPAIR GARAGES

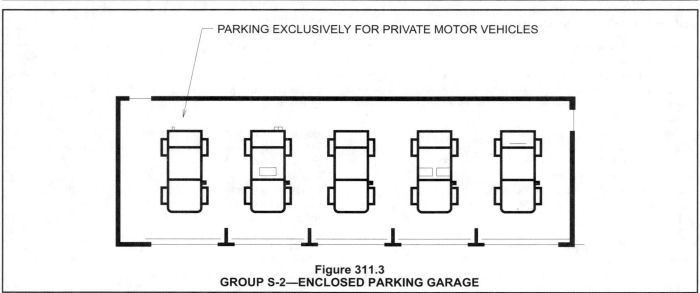

Figure 311.3
GROUP S-2—ENCLOSED PARKING GARAGE

- Glassware set in expanded foam forms and placed in a cardboard box; and
- Fuel filters individually packed in pressed paper boxes, placed by the gross in a cardboard box and then stacked on a wood pallet for transportation.

SECTION 312
UTILITY AND MISCELLANEOUS GROUP U

312.1 General. Buildings and structures of an accessory character and miscellaneous structures not classified in any specific occupancy shall be constructed, equipped and maintained to conform to the requirements of this code commensurate with the fire and life hazard incidental to their occupancy. Group U shall include, but not be limited to, the following:

Agricultural buildings
Aircraft hangars, accessory to a one- or two-family residence (see Section 412.3)
Barns
Carports
Fences more than 6 feet (1829 mm) high
Grain silos, accessory to a residential occupancy
Greenhouses
Livestock shelters
Private garages
Retaining walls
Sheds
Stables
Tanks
Towers

❖ This section identifies the characteristics of occupancies classified in Group U. Structures that are classified in Group U are typically accessory to another building or structure and are not more appropriately classified in another occupancy. Miscellaneous storage buildings accessory to detached one- and two-family dwellings and multiple single-family dwellings (townhouses) no more than three stories in height, however, are intended to be designed and built in accordance with the IRC (see Section 101.2).

Structures classified as Group U, such as fences, equipment, foundations, retaining walls, etc., are somewhat outside the primary scope of the code (i.e., means of egress, fire resistance, etc.). They are not usually considered to be habitable or occupiable. Nevertheless, many code provisions do apply and need to be enforced (e.g., structural design and material performance).

Bibliography

The following resource materials are referenced in this chapter or are relevant to the subject matter addressed in this chapter.

ASTM D 56-01, *Test Method for Flash Point by Tag Closed Tester.* West Conshohocken, PA: ASTM International, 2001.

ASTM D 86-01, *Test Method for Distillation of Petroleum Products.* West Conshohocken, PA: ASTM International, 2001.

ASTM D 93-00, *Test Methods for Flash Point by Pensky-Martens Closed Cup Tester.* West Conshohocken, PA: ASTM International, 2000.

ASTM D 3278-96, *Test Methods for Flash Point of Liquids by Small Scale Closed Cup Apparatus.* West Conshohocken, PA: ASTM International, 1996.

ASTM E 136-99, *Test Method for Behavior of Materials in a Vertical Tube Furnace at 750° C.* West Conshohocken, PA: ASTM International, 1999.

ASTM E 681-01, *Test Method for Concentration Limits of Flammability of Chemicals (Vapors and Gases).* West Conshohocken, PA: ASTM International, 2001.

CPSC 16 CFR; Part 1500-98, *Hazardous Substances and Articles; Administration and Enforcement Regulations.* Bethesda, MD: Consumer Product Safety Commission, 1998.

DOL 29 CFR; Part 1910-92, *Occupational Safety and Health Standards.* Washington, DC: United States Department of Labor, 1992.

DOTn 49 CFR; Part 100-199-94, *Specification for Transportation of Explosive and Other Dangerous Articles, Shipping Containers.* Washington, DC: United States Department of Transportation, 1994.

DOTy 27 CFR; Part 1-92-92, *Alcohol, Tobacco and Firearms.* Washington, DC: United States Department of Treasury, 1992.

IFC-2003; *International Fire Code.* Falls Church, VA: International Code Council, 2003.

IMC-2003; *International Mechanical Code.* Falls Church, VA: International Code Council, 2000.

IRC-2003; *International Residential Code.* Falls Church, VA: International Code Council, 2003.

NFPA 13-99, *Installation of Sprinkler Systems.* Quincy, MA: National Fire Protection Association, 1999.

NFPA 30-00, *Flammable and Combustible Liquids Code.* Quincy, MA: National Fire Protection Association, 1996.

NFPA 30B-98, *Manufacture and Storage of Aerosol Products.* Quincy, MA: National Fire Protection Association, 1998.

NFPA 31-01, *Installation of Oil-Burning Equipment.* Quincy, MA: National Fire Protection Association, 2001.

NFPA 33-00, *Spray Application Using Flammable and Combustible Materials.* Quincy, MA: National Fire Protection Association, 2000.

NFPA 34-00, *Dipping and Coating Processes Using Flammable or Combustible Liquids.* Quincy, MA: National Fire Protection Association, 2000.

NFPA 43B-93, *Storage of Organic Peroxide Formulations*. Quincy, MA: National Fire Protection Association, 1993.

NFPA 231D-98, *Storage of Rubber Tires.* Quincy, MA: National Fire Protection Association, 1998.

NFPA 430-95, *Storage of Liquid and Solid Oxidizers*. Quincy, MA: National Fire Protection Association, 1995.

NFPA 495-96, *Explosive Materials Code*. Quincy, MA: National Fire Protection Association, 1996.

NFPA 701-99, *Methods of Fire Tests for Flame-Resistant Textiles and Films*. Quincy, MA: National Fire Protection Association, 1999.

NFPA 704-96, *Identification of the Fire Hazards of Materials*. Quincy, MA: National Fire Protection Association, 1996.

NFPA 1124-98, *Code for the Manufacture, Transportation and Storage of Fireworks*. Quincy, MA: National Fire Protection Association, 1998.

USC Title 18; Chapter 40-70, *Importation, Manufacture, Distribution and Storage of Explosive Materials.* Washington, DC: United States Code, 1970.

Chapter 4:
Special Detailed Requirements
Based On Use And Occupancy

General Comments

The provisions of Chapter 4 are supplemental to the remainder of the code. Chapter 4 contains provisions that may alter requirements found elsewhere in the code; however, the general requirements of the code still apply unless modified within the chapter. For example, the height and area limitations established in Chapter 5 apply to all special occupancies unless Chapter 4 contains height and area limitations. In this case, the limitations in Chapter 4 supersede those in other sections. An example of this is the height and area limitations given in Section 406.3.5, which supersede the limitations given in Table 503 and Section 503 for open parking garages.

The *International Fire Code®* (IFC®) contains provisions applicable to the storage, handling and use of hazardous substances, materials or devices and, therefore, must also be complied with when dealing with such occupancies as those involving flammable and combustible liquids. Similarly, the *International Mechanical Code®* (IMC®) and the *International Plumbing Code®* (IPC®) include provisions for specific applications, such as hazardous exhaust systems and hazardous material piping.

In some instances, it may not be necessary to apply the provisions of Chapter 4. For example, if a covered mall building complies with the provisions of the code for Group M, Section 402 does not apply; however, other sections that deal with a use, process or operation must be applied to that specific occupancy, such as Sections 410, 411 and 414.

Purpose

The purpose of Chapter 4 is to combine in one chapter the provisions of the code applicable to special uses and occupancies. Hazardous occupancies and operations may occur in more than one group; therefore, the applicable provisions for the specific hazardous occupancy or operation apply to multiple groups. Also, while the provisions for all structures are interrelated to form an overall protection system by providing requirements for specific occupancies in Chapter 4, the package of protection features is more easily identified.

Chapter 4 contains the requirements for protecting special uses and occupancies. The provisions in this chapter reflect those occupancies and groups that require special consideration and are not addressed elsewhere in the code. The chapter includes requirements for buildings and conditions that apply to one or more groups, such as high-rise buildings or atriums. Special uses may also imply specific occupancies and operations, such as for Groups H-1, H-2, H-3, H-4 and H-5, application of flammable finishes and combustible storage or for a specific occupancy within a much larger occupancy, such as covered mall buildings, motor-vehicle-related occupancies, special amusement buildings and aircraft-related occupancies. Finally, in order that the overall package of protection features can be easily understood, occupancies such as Groups I-2 and I-3 and underground buildings are addressed.

SECTION 401
SCOPE

401.1 Detailed use and occupancy requirements. In addition to the occupancy and construction requirements in this code, the provisions of this chapter apply to the special uses and occupancies described herein.

❖ This section provides guidance on how Chapter 4 is to be applied with respect to other sections of the code. Section 401.1 indicates that all other provisions of the code apply except as modified by Chapter 4.

The requirements contained in Chapter 4 are intended to apply to special uses and occupancies as defined by the various sections in this chapter. These requirements are applicable in addition to other chapters of the code.

SECTION 402
COVERED MALL BUILDINGS

402.1 Scope. The provisions of this section shall apply to buildings or structures defined herein as covered mall buildings not exceeding three floor levels at any point nor more than three stories above grade. Except as specifically required by this section, covered mall buildings shall meet applicable provisions of this code.

Exceptions:

1. Foyers and lobbies of Groups B, R-1 and R-2 are not required to comply with this section.

2. Buildings need not comply with the provisions of this section where they totally comply with other applicable provisions of this code.

❖ This section primarily addresses shopping centers with a maximum of three levels that consist of one or more sizeable department stores, known as anchor buildings, and numerous smaller retail stores, all of which are interconnected by means of a covered, climate-controlled, public pedestrian way. The complex may include movie theaters, bowling lanes, ice arenas, offices and dining and drinking establishments. The complex may also include single- or multiple-level buildings, with a vast majority of shopping centers being one or two levels and with an occasional mezzanine.

This section addresses the special considerations associated with covered mall buildings, including construction, egress and fire protection systems. It does not, in general, apply to the anchor buildings, which are usually large department stores along the perimeter of the covered mall building. To be considered an anchor building for purposes of applying the code, the building must have complete egress facilities, including the required number and capacity of exits, independent of the covered mall building.

Originally, the provisions of this section applied to the typical covered mall building: a one- to three-level structure consisting primarily of retail space and a covered pedestrian way. More recently, other types of covered mall buildings, such as airport passenger terminals and office centers, have also been constructed in accordance with this section. The use of Section 402 for other than shopping malls may require some variation in the egress provisions. The design occupant load in Section 402.4.1.1 is intended to apply to shopping malls. When the section is used for other covered mall buildings, such as an office center, the means of egress provisions of Chapter 10 will apply.

Because of the differences in occupant load, this section is not intended to apply to the foyers and lobbies of business and residential occupancies that may or may not contain retail space. Such areas are usually constructed in accordance with Section 404.

Covered mall buildings that comply with all other applicable provisions of the code need not comply with the provisions of this section; however, covered mall buildings that are designed and constructed to comply with the provisions of this section must also comply with the provisions of the code that are not otherwise modified in this section.

For example, the egress provisions in Section 402.4 are applicable to covered mall buildings designed in accordance with this section. The provisions in Chapter 10 related to door swing and stairways are also applicable, however, since Section 402 contains no similar provisions.

402.2 Definitions. The following words and terms shall, for the purposes of this chapter and as used elsewhere in this code, have the meanings shown herein.

❖ Definitions of terms that are associated with the content of this section are contained herein. These definitions can help in the understanding and application of the code requirements. It is important to emphasize that these terms are not exclusively related to this section but are applicable everywhere the term is used in the code. The purpose for including these definitions within this section is to provide more convenient access to them without having to refer back to Chapter 2.

For convenience, these terms are also listed in Chapter 2 with a cross reference to this section. The use and application of all defined terms, including those defined herein, are set forth in Section 201.

ANCHOR BUILDING. An exterior perimeter building of a group other than H having direct access to a covered mall building but having required means of egress independent of the mall.

❖ A key to understanding what distinguishes an anchor building from a tenant space is that anchor buildings are typically retail establishments (Group M), although this may not always be the case. The anchor building is typically some facility that, by its nature, draws a considerable number of people. The tenants in the adjoining covered mall building then seek to capitalize on this traffic generated by the anchor building. The scale or size of the building is not a primary factor in determining whether it is an anchor building; rather, its function is such that it draws people to the site in sufficient numbers so that other facilities can benefit from being located in the same facility (see Figure 402.2).

Generally, the anchor building will have its own identity and there is a high probability that it will have separate management and its own hours of operation. This will necessitate a means of egress that does not rely on the mall being open for its patrons to enter or leave. Similarly, the means of egress from the covered mall building cannot rely on the anchor building being open for patrons to exit through; hence, egress facilities for the anchor building must be independent of those for the covered mall building. An anchor building is a separate building from the covered mall building and must comply with the provisions of the code for its own identity, except as modified by this section (see Sections 402.7.3.1 and 402.6).

COVERED MALL BUILDING. A single building enclosing a number of tenants and occupants such as retail stores, drinking and dining establishments, entertainment and amusement facilities, passenger transportation terminals, offices, and other similar uses wherein two or more tenants have a main entrance into one or more malls. For the purpose of this chapter, anchor build-

ings shall not be considered as a part of the covered mall building.

❖ The covered mall building is the entire area of the building (area of mall plus gross leasable area), excluding the anchor buildings. Passenger transportation terminals frequently are developed as wide concourses with small shops along the sides. For this reason, passenger transportation facilities are included. Transportation facilities used for freight or other purposes are not to be considered a covered mall building (see Figure 402.2).

FOOD COURT. A public seating area located in the mall that serves adjacent food preparation tenant spaces.

❖ Typical covered mall building layouts include a central gathering area for food and drink consumption. These areas are usually located in the mall itself. Tables and chairs are provided for the public's use to consume the food and drink. This public area is usually surrounded by numerous tenant spaces where food is prepared and sold over the counter. A separate design occupant load is required to be calculated in accordance with Section 402.4.1.4.

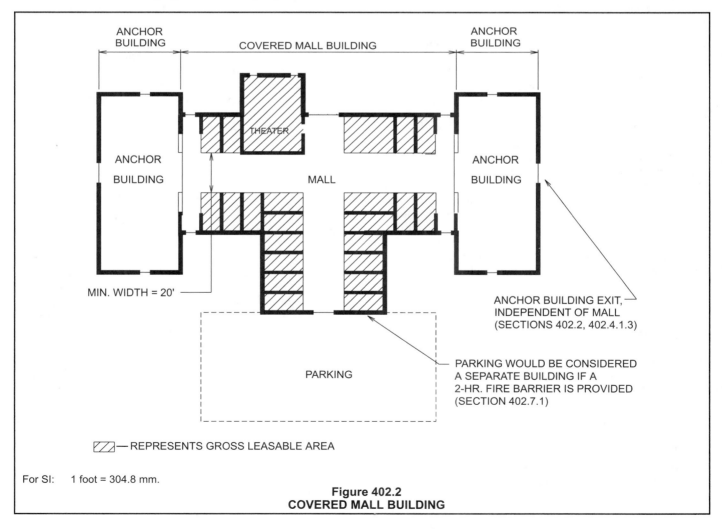

For SI: 1 foot = 304.8 mm.

Figure 402.2
COVERED MALL BUILDING

GROSS LEASABLE AREA. The total floor area designed for tenant occupancy and exclusive use. The area of tenant occupancy is measured from the centerlines of joint partitions to the outside of the tenant walls. All tenant areas, including areas used for storage, shall be included in calculating gross leasable area.

❖ The gross leasable area represents the aggregate area available in the building for tenant occupancy. It does not include the area of the mall, unless portions of it are leased for the purposes of setting up separate tenant spaces. The area is used to determine the design occupant load in accordance with Section 402.4.1.1.

MALL. A roofed or covered common pedestrian area within a covered mall building that serves as access for two or more tenants and not to exceed three levels that are open to each other.

❖ The mall is an interior, climate-controlled pedestrian way that is open to the tenant spaces within the building and typically connects to the anchor buildings.

402.3 Lease plan. Each covered mall building owner shall provide both the building and fire departments with a lease plan showing the location of each occupancy and its exits after the certificate of occupancy has been issued. No modifications or changes in occupancy or use shall be made from that shown on the lease plan without prior approval of the building official.

❖ The required lease plan is permitted to be submitted after the certificate of occupancy has been issued because many times the developer does not have the information at the time of construction. The location of tenant separations may not be known until leases are negotiated with prospective tenants. This may occur after the mall has opened.

During initial construction, it is anticipated that the building department will require tenant improvements to be submitted through the building permit process. After tenant spaces are prepared for occupancy and the lease plan is developed, subsequent modifications and changes must be submitted for approval prior to commencing construction or changing the use. It is important that the fire department receives copies of current lease plans, since not only does the fire department perform fire prevention inspections, but also the lease plans assist in fire department response to an emergency.

402.4 Means of egress. Each tenant space and the covered mall building shall be provided with means of egress as required by this section and this code. Where there is a conflict between the requirements of this code and the requirements of this section, the requirements of this section shall apply.

❖ Each individual occupancy or tenant space within the covered mall building is required to have a means of egress that complies with other provisions of the code—especially Chapter 10. Keep in mind that the requirements of Section 402.4 will supersede some of the provisions of Chapter 10. In order to comply, travel through the mall area is considered as exit access and must comply with Section 402.4.4. Travel distance

within a tenant space need only be measured to the entrance of the mall from the space (see Figure 402.4). As such, two distinct criteria must be met with regard to exit access travel distance: travel within the mall and travel within the tenant space.

402.4.1 Determination of occupant load. The occupant load permitted in any individual tenant space in a covered mall building shall be determined as required by this code. Means of egress requirements for individual tenant spaces shall be based on the occupant load thus determined.

❖ Since the tenant spaces of covered mall buildings can be used for varied occupancies, the design occupant loads will also vary. Although each tenant space contributes to the gross leasable area of Section 402.4.1.1, each must have its own occupant load calculated along with the applicable means of egress requirements. For example, a restaurant with its own dining area may occupy a tenant space. The design occupant load for this space is calculated in accordance with Table 1004.1.2 based on the assembly occupancy of the dining areas and on the commercial kitchen occupancy of the food preparation areas. Door swing direction and panic and fire exit hardware requirements would be based on this occupant load.

402.4.1.1 Occupant formula. In determining required means of egress of the mall, the number of occupants for whom means of egress are to be provided shall be based on gross leasable area of the covered mall building (excluding anchor buildings) and the occupant load factor as determined by the following equation.

$$OLF = (0.00007)(GLA) + 25 \qquad \textbf{(Equation 4-1)}$$

where:

OLF = The occupant load factor (square feet per person).

GLA = The gross leasable area (square feet).

❖ The capacity of the exits serving the covered mall buildings must satisfy the calculated occupant load based on the gross leasable area. The occupant load factors (OLF) were determined empirically by surveying more than 270 covered mall shopping centers, studying mercantile occupancy parking requirements [5.0 cars per 1,000 square feet (93 m²) of gross leasable area] and observing the number of occupants per vehicle during peak seasons (4.0 per car).

The formula used for determining occupant load results in a smooth transition between the occupant load and the gross leasable area with a range of 30 to 50.

For example, if the gross leaseable area of the covered mall building is 400,000 square feet (37 160 m²), the calculated OLF is determined as follows:

OLF = (0.00007)(400,000) + 25

OLF = 53 square feet (4.9 m²) per person

Since the design occupant load factor cannot exceed 50 in accordance with Section 402.4.1.2, the required OLF would be 50 square feet (4.6 m²) per person. If the gross leasable area of the covered mall building is 55,000 square feet (5109 m²), the calculated OLF is determined as follows:

$$OLF = (0.00007)(55,000) + 25$$

$$OLF = 29 \text{ square feet } (2.7 \text{ m}^2) \text{ per person}$$

Since the OLF need not be less than 30 in accordance with Section 402.4.1.2, the required occupant load factor would be 30 square feet (2.8 m²) per person. The code would permit the use of an occupant load factor of 29, since it would result in a higher design occupant load and consequently increased width of exits.

402.4.1.2 OLF range. The occupant load factor (*OLF*) is not required to be less than 30 and shall not exceed 50.

❖ Although not mandatory, an occupant load factor of less than 30 may be used at the designer's discretion (see Section 1004.2). In no case should an OLF of greater than 50 be used, since it would result in too small a design occupant load.

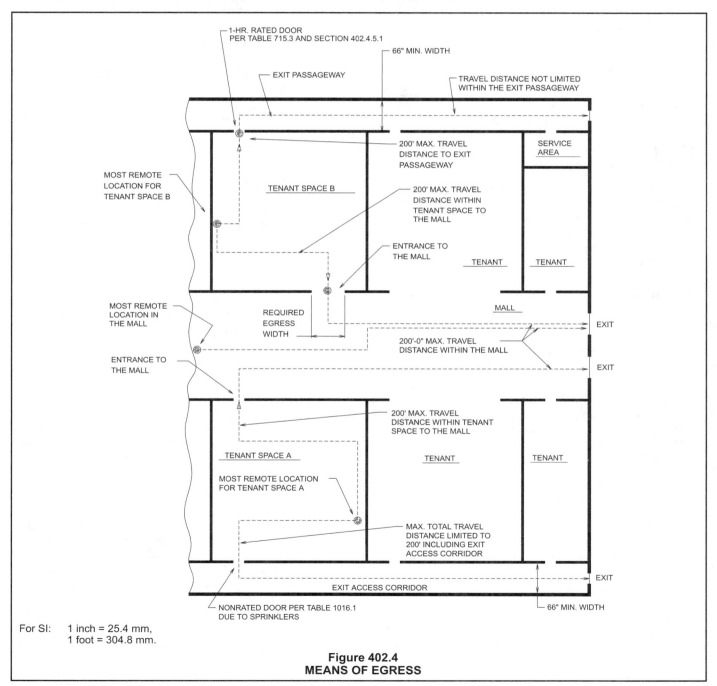

For SI: 1 inch = 25.4 mm,
           1 foot = 304.8 mm.

Figure 402.4
MEANS OF EGRESS

402.4.1.3 Anchor buildings. The occupant load of anchor buildings opening into the mall shall not be included in computing the total number of occupants for the mall.

❖ Although anchor buildings are allowed access to the mall portions of the covered mall building, the occupant load of the mall does not have to be increased for this access. People may certainly discharge from the anchor building into the mall on their own accord; however, the anchor building is required to provide sufficient exits for its occupant load.

402.4.1.4 Food courts. The occupant load of a food court shall be determined in accordance with Section 1004. For the purposes of determining the means of egress requirements for the mall, the food court occupant load shall be added to the occupant load of the covered mall building as calculated above.

❖ The occupant load for food courts must be determined in accordance with Section 1004. Typically, tables and chairs are provided for the public's use in a food court. As such, the design occupant load for a food court is usually calculated at a rate of one person for every 15 square feet (1.4 m²) of net floor space for an unconcentrated assembly occupancy. The occupant loads for all food courts are added to the mall occupant load calculated in accordance with Section 402.4.1.1 and Equation 4-1. Food courts are not considered as part of the gross leasable area of the mall; therefore, the occupant loads of the food courts must be calculated separately and then added to the overall mall building occupant load.

402.4.2 Number of means of egress. Wherever the distance of travel to the mall from any location within a tenant space used by persons other than employees exceeds 75 feet (22 860 mm) or the tenant space exceeds an occupant load of 50, not less than two means of egress shall be provided.

❖ This section dictates the minimum number of paths of travel an occupant is to have available to avoid a fire incident in the occupied tenant space. While providing multiple egress doorways from every tenant space is unrealistic, a point does exist where alternative egress paths must be provided, based on the number of occupants at risk, the distance any one occupant must travel to reach the mall and the relative hazards associated with the occupancy of the space.

The 75-foot (22 860) travel distance and the 50-person occupant load limitations represent an empirical judgment of the risks associated with a single means of egress from a tenant space based on the inherent risks associated with the occupancy (such as occupant mobility, occupant familiarity with the mall, occupant response and the fire growth rate).

402.4.3 Arrangements of means of egress. Assembly occupancies with an occupant load of 500 or more shall be so located in the covered mall building that their entrance will be immedi-

ately adjacent to a principal entrance to the mall and shall have not less than one-half of their required means of egress opening directly to the exterior of the covered mall building.

❖ Large tenant spaces with assembly occupant loads of 500 or more must be carefully located to minimize the hazards of egress. Movie theaters, nightclubs, etc., must be located on the perimeter of the covered mall building and adjacent to the mall's exits if their occupant loads exceed 499. A maximum of 50 percent of the exits from these assembly occupancies are permitted to discharge into the mall. Other exits from these tenant spaces must discharge directly to the exterior with a capacity of at least one-half the occupant load.

402.4.3.1 Anchor building means of egress. Required means of egress for anchor buildings shall be provided independently from the mall means of egress system. The occupant load of anchor buildings opening into the mall shall not be included in determining means of egress requirements for the mall. The path of egress travel of malls shall not exit through anchor buildings. Malls terminating at an anchor building where no other means of egress has been provided shall be considered as a dead-end mall.

❖ As included in the definition, the required exits and exit capacities for an anchor building must be provided independently of the mall or mall exits. Since independent exits are provided, the occupant load of anchor buildings is not included in determining the exit requirements for the mall. If independent exits are not provided, the space cannot be considered an anchor building and is to be treated as any other tenant space.

402.4.4 Distance to exits. Within each individual tenant space in a covered mall building, the maximum distance of travel from any point to an exit or entrance to the mall shall not exceed 200 feet (60 960 mm).

The maximum distance of travel from any point within a mall to an exit shall not exceed 200 feet (60 960 mm).

❖ The maximum permissible travel distance from any point in the tenant space to the mall or from anywhere in the mall to an exit is 200 feet (60 960 mm). As such, the mall is treated like an aisle in a large store, the large store being the covered mall building. Although the distance is less than that permitted in Group M with an automatic sprinkler system, it must be recognized that this travel distance is measured from a point within the mall and not from the most remote point within the covered mall building (i.e., within a tenant space). Because of the uncertainty with respect to tenant improvements, it is more reliable to regulate travel distance within the mall instead of throughout the covered mall building.

402.4.5 Access to exits. Where more than one exit is required, they shall be so arranged that it is possible to travel in either direction from any point in a mall to separate exits. The minimum width of an exit passageway or corridor from a mall shall be 66 inches (1676 mm).

> **Exception:** Dead ends not exceeding a length equal to twice the width of the mall measured at the narrowest location within the dead-end portion of the mall.

❖ In order to accommodate the many occupants anticipated in a covered mall building, exits are to be distributed equally throughout the mall. If exits were congregated in certain areas, egress would be compounded by the convergence of a large number of people to one or more areas of the building. The corridors and exit access passageways from a mall must be at least 66 inches (1676 mm) wide.

The maximum dead end permitted in a mall is twice the width of the mall. For a mall with a minimum width of 20 feet (6096 mm), dead ends are permitted to be 40 feet (12 192 mm) or less in length. The allowance for dead ends is in recognition of the reduced hazard represented by a dead end in a mall and because the relatively large minimum width provides alternative paths of travel. Additionally, dead ends are not as critical in a mall, since the volume of the space and the automatic sprinkler system minimize the potential for the space to become untenable in a fire condition.

402.4.5.1 Exit passageway enclosures. Where exit passageway enclosures provide a secondary means of egress from a tenant space, doors to the exit passageway enclosures shall be 1-hour fire doors. Such doors shall be self-closing and be so maintained or shall be automatic-closing by smoke detection.

❖ If exit passageways must be provided because of either travel distance limitations or the number of means of egress required from a tenant space, the passageways must be enclosed as required by the code. Additionally, the doors opening from the tenant spaces into the exit enclosures must be 1-hour rated. These doors must be self-closing or automatic-closing by smoke detection to provide the necessary protection as the occupants of a tenant space flee a fire situation.

402.4.6 Service areas fronting on exit passageways. Mechanical rooms, electrical rooms, building service areas and service elevators are permitted to open directly into exit passageways provided that the exit passageway is separated from such rooms with 1-hour fire-resistance-rated walls and 1-hour opening protectives.

❖ In a covered mall building, it is necessary to provide for services to the tenant spaces that are maintained by the mall management (e.g., water, electricity, telephone and fire protection). These services must be located in a common space controlled by the mall management and, therefore, cannot be located within the tenant spaces. Frequently, these services are logically located with direct access to exit passageways at the rear of the tenant spaces.

Exit passageways are generally treated similar to exit stairs in that only openings from normally occupied spaces are permitted. This would prohibit doors or utility penetrations to such mechanical/electrical rooms. In the case of covered malls the code allows an exception to that general rule, provided that the fire-resistance rating of the exit enclosure is maintained by appropriate opening protection, such as fire doors, fire dampers and through-penetration firestopping.

402.5 Mall width. For the purpose of providing required egress, malls are permitted to be considered as corridors but need not comply with the requirements of Section 1005.1 of this code where the width of the mall is as specified in this section.

❖ The pedestrian malls that serve as the exit access routes for the tenant space occupants are not required to comply with Section 1003.2.3. The design capacity of the mall width is already established by Section 402.5.1. No further egress calculations are required by the code for mall corridors.

402.5.1 Minimum width. The minimum width of the mall shall be 20 feet (6096 mm). The mall width shall be sufficient to accommodate the occupant load served. There shall be a minimum of 10 feet (3048 mm) clear exit width to a height of 8 feet (2438 mm) between any projection of a tenant space bordering the mall and the nearest kiosk, vending machine, bench, display opening, food court or other obstruction to means of egress travel.

❖ The minimum width of a mall [20 feet (6096 mm)] is based on the need to provide adequate exit access. Together with the automatic sprinkler system, the physical separation further reduces the need for a separation between tenant spaces and the mall.

So that an aggregate clear width of 20 feet (6096 mm) is always provided in the mall, a minimum 10-foot (3048 mm) clear and unobstructed space is to be maintained to a height of 8 feet (2438 mm) in front of, adjacent to and parallel to the store fronts. The requirement applies to kiosks, vending machines, benches, small stands, merchandise displays and any other potential obstruction to egress (see Figure 402.5.1).

402.6 Types of construction. The area of any covered mall building, including anchor buildings, of Type I, II, III and IV construction, shall not be limited provided the covered mall building and attached anchor buildings and parking garages are surrounded on all sides by a permanent open space of not less than 60 feet (18 288 mm) and the anchor buildings do not exceed three stories in height. The allowable height and area of anchor buildings greater than three stores in height shall comply with Section 503, as modified by Sections 504 and 506. The construction type of open parking garages and enclosed parking garages shall comply with Sections 406.3 and 406.4, respectively.

❖ Covered mall buildings are considered to be special types of unlimited area buildings and are exempt from the area limitations of Table 503 when they are limited in height to three floor levels at any point and not more than three stories above grade. It should be noted that the height limitations in feet, specified in Table 503 based on the type of construction classification, are applicable to covered mall buildings. The allowance of an unlimited area anchor building is based on the restriction of construction types to noncombustible (Types I and II), noncombustible/combustible (Type III) and heavy timber (Type IV) and the effectiveness of the automatic sprinkler system. When the anchor building is over three stories high, then it must comply with the general provisions of Chapter 5, such as Sections 503, 504 and 506. The last sentence of the code text serves as a reminder that the construction of parking garages is regulated by the general provisions found in Section 406 and that they are not regulated by the covered mall building's type of construction. It also clarifies that the garage is not included in the unlimited area as stated in the first sentence of the code text.

402.7 Fire-resistance-rated separation. Fire-resistance-rated separation is not required between tenant spaces and the mall. Fire-resistance-rated separation is not required between a food court and adjacent tenant spaces or the mall.

❖ From an operational point of view, separating the tenant space from the mall is not practical in that customer flow

would be restricted and, therefore, detract from the merchandising purpose of covered malls. Historical experience and reliability data on automatic sprinkler performance indicate that a separation between the tenant space and the mall is not warranted.

402.7.1 Attached garage. An attached garage for the storage of passenger vehicles having a capacity of not more than nine persons and open parking garages shall be considered as a separate building where it is separated from the covered mall building by a fire barrier having a fire-resistance rating of at least 2 hours.

Exception: Where an open parking garage or enclosed parking garage is separated from the covered mall building or anchor building a distance greater than 10 feet (3048 mm), the provisions of Table 602 shall apply. Pedestrian walkways and tunnels which attach the open parking garage or enclosed parking garage to the covered mall building or anchor building shall be constructed in accordance with Section 3104.

❖ In accordance with Section 402.1, adjacent buildings may be considered separate buildings only if they are separated by exterior walls or fire walls. In recognition of the limited hazard presented by attached garages and

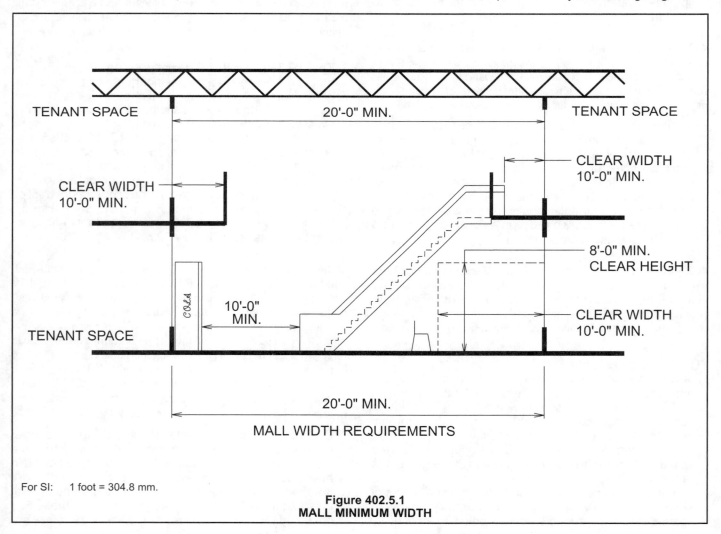

For SI:　1 foot = 304.8 mm.

Figure 402.5.1
MALL MINIMUM WIDTH

open parking garages, this section permits such structures to be considered separate buildings, provided that a fire barrier having a fire-resistance rating of at least 2 hours is provided. See the commentary to Section 706 relative to the construction of fire barriers.

The 2-hour fire-resistance rating is consistent with the requirements of Table 302.3.2 for Groups M and S-2. If the 2-hour fire barrier is not provided, the parking structure can be considered as part of the covered mall building and be limited. See the commentary to Section 3104 for the pedestrian walkways exception.

402.7.2 Tenant separations. Each tenant space shall be separated from other tenant spaces by a fire partition complying with

Section 708. A tenant separation wall is not required between any tenant space and the mall.

❖ In order to limit the spread of smoke, tenant separation walls are required to be fire partitions (see Section 708) having a fire-resistance rating of at least 1 hour and extending from the floor to the underside of the ceiling (see Figure 402.7.2). Extending tenant separations to the floor slab or roof deck above is not always practical or possible because of operation of the heating, ventilating and air-conditioning (HVAC) system. The effectiveness of the automatic sprinkler system is also a reason for not requiring tenant separations to extend above the ceiling, including attic spaces (see Exception 4 to Section 708.4).

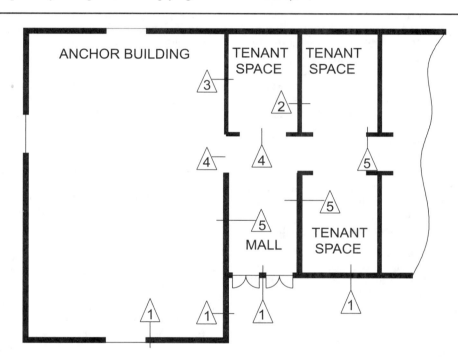

△1 WALL CONSTRUCTED AS REQUIRED FOR
AN EXTERIOR WALL

△2 WALL CONSTRUCTED AS REQUIRED FOR A
TENANT SEPARATION BY SECTION 402.7.2

△3 WALL CONSTRUCTED AS REQUIRED BY
SECTION 402.7.3 (FIRE WALL OR FIRE BARRIER)

△4 UNPROTECTED OPENING

△5 WALL CONSTRUCTED AS REQUIRED BY
SECTION 402.7

Figure 402.7.2
TENANT SPACE AND ANCHOR BUILDING SEPARATIONS

402.7.3 Anchor building separation. An anchor building shall be separated from the covered mall building by fire walls complying with Section 705.

> **Exception:** Anchor buildings of not more than three stories above grade which have an occupancy classification of the same uses permitted as tenants of the covered mall building shall be separated by 2-hour fire-barriers complying with Section 705.

❖ In general, anchor buildings are typically viewed as being separate from the covered mall building (see commentary, Sections 402.1 and 402.6). As separate buildings, the code requires that a fire wall complying with Section 705 be constructed. The exception to this section may be used in situations where the occupancy and size limitations of the anchor building are the same as what would be permitted for any other tenant within the covered mall building. In those situations, the code permits the separation to be constructed as a fire barrier instead of as a fire wall. The exception would not be applicable if the anchor building was over three stories in height or if it contained an occupancy that was not normally permitted within the mall. An example would be a high-rise office building or a hotel that was connected with the covered mall building (see Figure 402.7.2 and commentary, Section 402.7.3.1).

402.7.3.1 Openings between anchor building and mall. Except for the separation between Group R-1 sleeping units and the mall, openings between anchor buildings of Type IA, IB, IIA and IIB construction and the mall need not be protected.

❖ As with tenant separations, openings between the anchor building and the pedestrian area of the mall need not be protected. Such separations would defeat the merchandising purpose of covered malls by restricting customer flow and visual access. While the openings may be unprotected, the separations between the anchor building and all tenant spaces in the covered mall building are to be constructed as required in Section 402.7.3 for a fire wall (see commentary, Section 705) or fire barrier (see commentary, Section 706) since the anchor building is considered a separate building (see Figure 402.7.2).

[F] 402.8 Automatic sprinkler system. The covered mall building and buildings connected shall be equipped throughout with an automatic sprinkler system in accordance with Section 903.3.1.1, which shall comply with the following:

1. The automatic sprinkler system shall be complete and operative throughout occupied space in the covered mall building prior to occupancy of any of the tenant spaces. Unoccupied tenant spaces shall be similarly protected unless provided with approved alternate protection.

2. Sprinkler protection for the mall shall be independent from that provided for tenant spaces or anchors. Where tenant spaces are supplied by the same system, they shall be independently controlled.

> **Exception:** An automatic sprinkler system shall not be required in spaces or areas of open parking garages constructed in accordance with Section 406.2.

❖ The covered mall building and connected buildings, such as anchor buildings, must be protected with an automatic sprinkler system to protect life and property effectively. As has been discussed throughout the section, numerous allowances (such as reduced tenant separations and elimination of area limitations) are based on the effectiveness of the automatic sprinkler system.

The sprinkler system is to be designed, installed, tested and maintained in accordance with Chapter 9, the IFC and NFPA 13. Additionally, the system must be installed such that any portion serving tenant spaces may be shut down independently without affecting the operation of the systems protecting the mall area. This special feature is in recognition of the frequent need to shut down the system so that changes can be made to it as a result of tenant improvements and modifications.

Section 909.12.3 requires operation of the sprinkler system to activate automatically the mechanical smoke control system (where an automatic control system is utilized). It is imperative that the zoning of the sprinkler system match the zoning of the smoke control system. This is necessary so that the area where water flow has occurred will also be the area from which smoke is removed.

402.8.1 Standpipe system. The covered mall building shall be equipped throughout with a standpipe system as required by Section 905.3.3.

❖ This section provides a reminder to the code user and a direct reference to the section that regulates the need for standpipes or hose connections for covered mall buildings (see commentary, Section 905.3.3).

402.9 Smoke control. A smoke control system shall be provided where required for atriums in Section 404.

❖ All covered mall buildings that are two or three stories in height must be provided with a smoke control system. Malls inherently contain unenclosed floor openings that may allow for the migration of smoke and hot gases from a fire condition. See Sections 404.4 and 909 for a discussion of the requirements of the smoke control system, noting the exceptions listed for floor openings.

402.10 Kiosks. Kiosks and similar structures (temporary or permanent) shall meet the following requirements:

1. Combustible kiosks or other structures shall not be located within the mall unless constructed of any of the following materials:

1.1. Fire-retardant-treated wood complying with Section 2303.2.

1.2. Foam plastics having a maximum heat release rate not greater than 100kW (105 Btu/h) when tested in accordance with the exhibit booth protocol in UL 1975.

1.3. Aluminum composite material (ACM) having a flame spread index of not more than 25 and a smoke-developed index of not more than 450 when tested as an assembly in the maximum thickness intended for use in accordance with ASTM E 84.

2. Kiosks or similar structures located within the mall shall be provided with approved fire suppression and detection devices.

3. The minimum horizontal separation between kiosks or groupings thereof and other structures within the mall shall be 20 feet (6096 mm).

4. Each kiosk or similar structure or groupings thereof shall have a maximum area of 300 square feet (28 m²).

❖ Other potential sources of combustibles within the mall area are kiosks and similar structures. As with the restrictions on plastic signs, in order to maintain the mall as a viable means of egress, the amount of combustibles within the mall must be minimized. The restriction on kiosk construction and location is also intended to minimize the potential for fire spread through the mall area. These restrictions apply to both permanent and temporary structures.

Kiosks and similar structures are permitted to be of noncombustible construction or may be of combustible construction where the provisions of Item 1 are followed. The code allows combustible kiosks to be made of fire-retardant-treated wood or certain foam plastics and aluminum composites that provide performance comparable to that of the fire-retardant-treated wood. As such, the amount of combustibles within the mall and the potential for fire spread through the mall are minimized.

If the kiosk or similar structure has a cover, the automatic sprinkler system within the covered mall building will not be able to control effectively a fire within the kiosk (or similar structure) in the early stages of development. Suppression and detection devices must, therefore, be installed within such structures when there is anything that shields the mall sprinkler system from areas within the kiosk.

In order to minimize a fire exposure hazard, kiosks and similar structures must be located at least 20 feet (6096 mm) from each other or be situated in groups that are appropriately separated from other kiosks. Although the structure itself is often noncombustible or is of the limited types of acceptable combustible materials, it is recognized that combustibles may be displayed within the structure and, therefore, the separation minimizes the potential for fire spread from kiosk to kiosk. As required by Section 402.5.1, the kiosk must also be located at least 10 feet (3048) from any projection of a

tenant space.

To further control the use of kiosks and similar structures within the mall, their size is restricted to 300 square feet (28 m²) or less. This area restriction attempts to minimize the potential for a significant fire within the mall area. A kiosk larger than 300 square feet (28 m²) must be considered as a tenant space.

402.11 Security grilles and doors. Horizontal sliding or vertical security grilles or doors that are a part of a required means of egress shall conform to the following:

1. They shall remain in the full open position during the period of occupancy by the general public.

2. Doors or grilles shall not be brought to the closed position when there are more than 10 persons occupying spaces served by a single exit or 50 persons occupying spaces served by more than one exit.

3. The doors or grilles shall be openable from within without the use of any special knowledge or effort where the space is occupied.

4. Where two or more exits are required, not more than one-half of the exits shall be permitted to include either a horizontal sliding or vertical rolling grille or doors.

❖ When improperly installed or when their use is not properly supervised, security grilles and doors can prohibit or delay egress beyond an acceptable time period. Such security devices are common in covered mall buildings. In every case, the grille must be openable to permit egress from the inside without the use of tools, keys or special knowledge or effort when the grille spans across a means of egress. For example, during business hours in a mercantile use, the grille must remain in its full, open position. The grille may be taken to a partially closed position when it provides the sole means of egress and no more than 10 persons occupy the space. If a second means of egress is required, the grille may not be closed, as long as more than 50 persons occupy the space.

Because security grilles represent such a threat to prompt egress, the number of occupants exposed to the risk of a lack of supervision of the devices is limited. Except where a single means of egress is permitted, alternative means of egress must be available and not equipped with such devices. Security grilles must not be used on more than 50 percent of the exits.

402.12 Standby power. Covered mall buildings exceeding 50,000 square feet (4645 m²) shall be provided with standby power systems that are capable of operating the emergency voice/alarm communication system.

❖ Covered mall buildings of a substantial size are required to have a standby power system for the emergency voice/alarm communication system. See Section 2702.2.13 of the commentary for further discussion of when a covered mall building exceeds 50,000 square feet (4645 m²).

[F] 402.13 Emergency voice/alarm communication system. Covered mall buildings exceeding 50,000 square feet (4645 m²) in total floor area shall be provided with an emergency voice/alarm communication system. Emergency voice/alarm communication systems serving a mall, required or otherwise, shall be accessible to the fire department. The system shall be provided in accordance with Section 907.2.12.2.

❖ An emergency voice/alarm communication system is required to allow the fire department full control during an emergency situation. Such a system is required when the covered mall building exceeds an aggregate area of 50,000 square feet (4645 m²) (see commentary, Sections 907.2.12.2 and 907.2.20).

402.14 Plastic signs. Within every store or level and from sidewall to sidewall of each tenant space facing the mall, plastic signs shall be limited as specified in Sections 402.14.1 through 402.14.5.

❖ In order that the mall can be used as an exit access, combustible materials in the mall must be restricted; therefore, the size, area, location and amount of exposed plastic signs are restricted. This section restricts only plastic signs that are along the wall separating the tenant space from the mall (see Figure 402.14). If the sign housing and face panels are noncombustible or

glass, there are no limitations. Light-transmitting plastic interior signs not located in the mall, such as those located within an individual tenant space, must comply with the requirements of Chapter 26.

402.14.1 Area. Plastic signs shall not exceed 20 percent of the wall area facing the mall.

❖ The signs located on the wall facing the mall must not exceed 20 percent of the wall area. The restrictions on size, coupled with the restrictions on location and height, are intended to prevent fire spread via the plastic signs.

402.14.2 Height and width. Plastic signs shall not exceed a height of 36 inches (914 mm), except if the sign is vertical, the height shall not exceed 96 inches (2438 mm) and the width shall not exceed 36 inches (914 mm).

❖ Signs that are wider than they are high (horizontal panels) must not be more than 36 inches (914 mm) in height. The width of signs is controlled by the maximum area (see Section 402.14.1) and the location restrictions with respect to adjacent tenant spaces (see Section 402.14.3). Vertical signs must not be more than 36 inches (914 mm) in width and 96 inches (2438 mm) in height and are also subject to the area and location re-

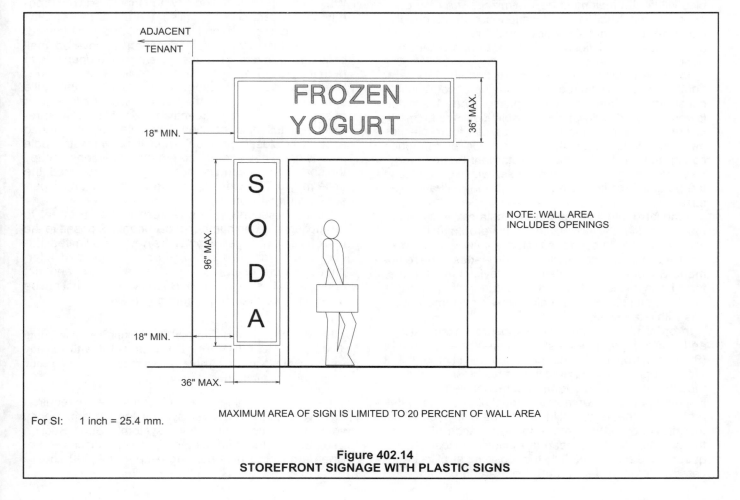

For SI: 1 inch = 25.4 mm.

MAXIMUM AREA OF SIGN IS LIMITED TO 20 PERCENT OF WALL AREA

Figure 402.14
STOREFRONT SIGNAGE WITH PLASTIC SIGNS

strictions of other sections. The restrictions are intended to minimize horizontal and vertical fire spread through the mall area via the signs.

402.14.3 Location. Plastic signs shall be located a minimum distance of 18 inches (457 mm) from adjacent tenants.

❖ To minimize the potential for fire spread, plastic signs must be located at least 18 inches (457 mm) from adjacent tenant spaces.

402.14.4 Plastics other than foam plastics. Plastics other than foam plastics used in signs shall be light-transmitting plastics complying with Section 2606.4 or shall have a self-ignition temperature of 650°F (343°C) or greater when tested in accordance with ASTM D 1929, and a flame spread index not greater than 75 and smoke-developed index not greater than 450 when tested in the manner intended for use in accordance with ASTM E 84 or meet the acceptance criteria of Section 803.2.1 when tested in accordance with NFPA 286.

❖ To minimize the potential fire hazard, light-transmitting plastic signs located in a mall must comply with the specific material requirements of Section 2606.4 (see commentary, Section 2606.4) or must meet one of the alternate test methods that are permitted. Additionally, the backs and edges of the sign must be encased to reduce the likelihood of ignition and fire spread.

402.14.4.1 Encasement. Edges and backs of plastic signs in the mall shall be fully encased in metal.

❖ In order that the amount of exposed plastic materials in the mall is minimized, as well as to reduce the potential for ignition of a plastic panel, the edges and backs of plastic signs must be encased in metal. To conceptually understand this requirement, consider how easily a piece of paper ignites when a flame is held to its edge versus being held in the middle of the paper.

402.14.5 Foam plastics. Foam plastics used in signs shall have flame-retardant characteristics such that the sign has a maximum heat-release rate of 150 kilowatts when tested in accordance with UL 1975 and the foam plastics shall have the physical characteristics specified in this section. Foam plastics used in signs installed in accordance with Section 402.14 shall not be required to comply with the flame spread and smoke-developed indexes specified in Section 2603.3.

❖ Foam plastics that are used in signs located in a mall must comply with the provisions of Sections 402.14.5.1 and 402.14.5.2. The provisions of the code include specific material requirements for the foam plastics, such as the allowable heat-release rate, density and thickness permitted.

402.14.5.1 Density. The minimum density of foam plastics used in signs shall not be less than 20 pounds per cubic foot (pcf) (320 kg/M³).

❖ Foam plastics used in signs are limited to a minimum density of no less than 20 pounds per cubic foot

(pcf)(320 kg/m³). This limitation lessens the likelihood that the sign will add a significant fuel load to the tenant space entrance.

402.14.5.2 Thickness. The thickness of foam plastic signs shall not be greater than $^1/_2$-inch (12.7 mm).

❖ The thickness of the foam plastic sign is also limited to lessen the likelihood that the sign will add a significant fuel load to the tenant space entrance.

402.15 Fire department access to equipment. Rooms or areas containing controls for air-conditioning systems, automatic fire-extinguishing systems or other detection, suppression or control elements shall be identified for use by the fire department.

❖ Consideration should be given to fire department access during an emergency. Fire department personnel may need to either determine that the fire protection systems are functioning properly or override the automatic features to manually activate or shut down a particular system. For this reason, the fire department must have access to controls for the air-conditioning and fire protection systems. The controls are to be clearly identified so that fire department personnel can properly operate them.

SECTION 403
HIGH-RISE BUILDINGS

403.1 Applicability. The provisions of this section shall apply to buildings having occupied floors located more than 75 feet (22 860 mm) above the lowest level of fire department vehicle access.

Exception: The provisions of this section shall not apply to the following buildings and structures:

1. Airport traffic control towers in accordance with Section 412.

2. Open parking garages in accordance with Section 406.3.

3. Buildings with an occupancy in Group A-5 in accordance with Section 303.1.

4. Low-hazard special industrial occupancies in accordance with Section 503.1.2.

5. Buildings with an occupancy in Group H-1, H-2 or H-3 in accordance with Section 415.

❖ High-rise buildings are defined as buildings with occupied floors located more than 75 feet (22 860 mm) above the lowest level of fire department vehicle access. Such buildings require special consideration relative to fire protection because of the difficulties associated with smoke movement (stack effect), egress time and access by fire department personnel. This section contains provisions for high-rise buildings to address these special considerations.

In accordance with Section 403.2, a basic requirement for all high-rise buildings regulated by this section

FIGURE 403.1(1)   SPECIAL DETAILED REQUIREMENTS BASED ON USE AND OCCUPANCY

(see Sections 403.1 and 403.2 for exceptions) is that an automatic sprinkler system is provided throughout the building. Additionally, the sprinkler system must comply with the provisions of Section 403.3 if the modifications identified in Sections 403.3.1 and 403.3.2 are to be utilized. Earlier editions of the codes permitted a compartmentation option for high-rise buildings with occupancies in Groups B, R-1 and R-2, the only groups previously governed by the high-rise provisions. Since the compartmentation option did not appear to be economically viable for the owner or for the governmental entity expected to provide resources to suppress a fire in a compartmented building, the option was eliminated. This is not intended to be a reflection on the value of properly designed and constructed passive fire protection.

Sections 403.5, 403.6 and 403.7 address fire alarm and communication system requirements to aid in fire detection, occupant notification and fire department notification and communication. The remaining provisions of Section 403 address required building systems and equipment features related to the building's fire protection systems.

The provisions of Section 403 apply to all high-rise buildings except those identified in the five exceptions. Exception 1 addresses airport traffic control towers and is based on the limited fuel load and the limited number of persons occupying the tower (see Section 412). Places of outdoor assembly (Group A-5) and open parking garages are exempted because of the free ventilation to the outside that exists in such structures. Low-hazard special industrial occupancies as defined in Section 503.1.2 may be exempted when approved by the building official. Such buildings should be evaluated based on the occupant load and the hazards of the occupancy and its contents to determine if the protection features required by Section 403 are necessary. Build-

ings with occupancies in Groups H-1, H-2 and H-3 are excluded from the provisions of this section because the fire hazard characteristics of such occupancies have not yet been considered.

The provisions are applicable to all buildings when the highest occupied floor is more than 75 feet (22 860 mm) above the lowest level of fire department vehicle access [see Figure 403.1(1)]. The lowest level of fire department vehicle access refers to the lowest ground level at which the fire department vehicle could be staged at the exterior of the building for carrying out fire-fighting operations. The definition of a "High-rise building" comes from the International Symposium on Fire Safety in High-Rise Buildings sponsored by the General Services Administration. The definition developed at the symposium, which was attended by representatives from England, France, Canada, Sweden and the United States, is as follows:

"A high-rise building is one in which emergency evacuation is not practical and in which fire must be fought internally because of height. The usual characteristics of such a building are:

A-1: Beyond the reach of fire department equipment;
A-2: Poses a potential for significant stack effect; and
A-3: Requires unreasonable evacuation time."

The 75-foot (22 860 mm) height limitation was determined from the effective reach of a 100-foot (30 480 mm) aerial apparatus based on: buildings set back from the curb and access restrictions such as parked vehicles and obstructions, such as utility lines. The applicability of this section, however, is not based on the availability of such apparatus within the community.

Stack effect is illustrated in Figure 403.1(2), while typical evacuation times for buildings are shown in Figure 403.1(3).

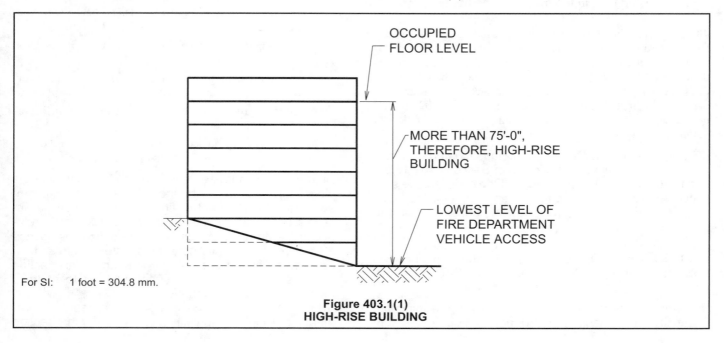

For SI: 1 foot = 304.8 mm.

Figure 403.1(1)
HIGH-RISE BUILDING

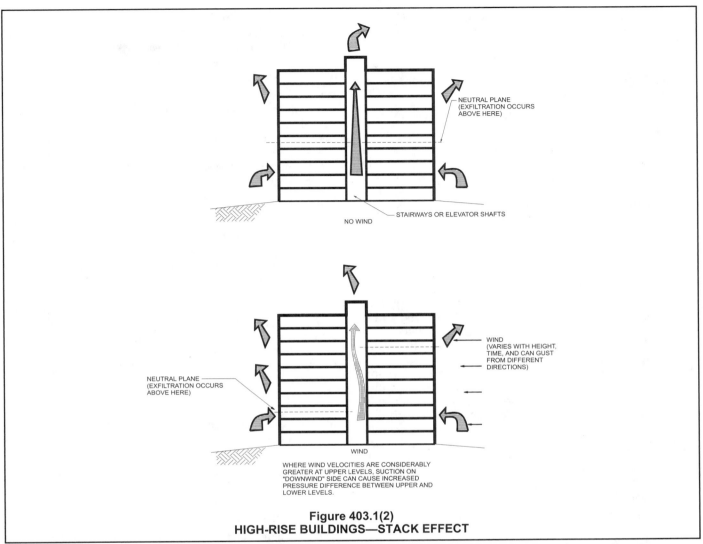

Figure 403.1(2)
HIGH-RISE BUILDINGS—STACK EFFECT

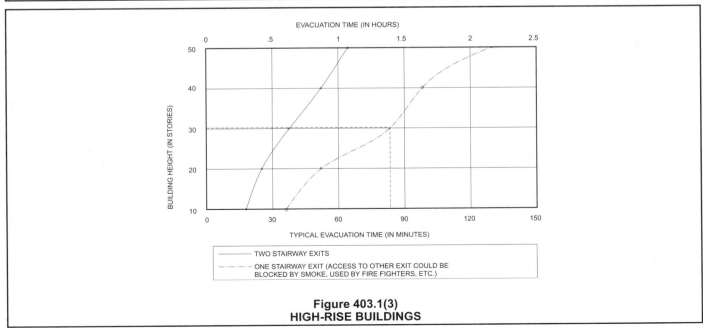

Figure 403.1(3)
HIGH-RISE BUILDINGS

[F] 403.2 Automatic sprinkler system. Buildings and structures shall be equipped throughout with an automatic sprinkler system in accordance with Section 903.3.1.1 and a secondary water supply where required by Section 903.3.5.2.

> **Exception:** An automatic sprinkler system shall not be required in spaces or areas of:
>
> 1. Open parking garages in accordance with Section 406.3.
>
> 2. Telecommunications equipment buildings used exclusively for telecommunications equipment, associated electrical power distribution equipment, batteries and standby engines, provided that those spaces or areas are equipped throughout with an automatic fire detection system in accordance with Section 907.2 and are separated from the remainder of the building with fire barriers consisting of 1-hour fire-resistance-rated walls and 2-hour fire-resistance-rated floor/ceiling assemblies.

❖ Because of the difficulties associated with manual suppression of a fire in a high-rise building, all high-rise buildings regulated by this section are required to be protected throughout with an automatic sprinkler system. The sprinkler system is to be in accordance with Section 903.3.1.1, which requires compliance with NFPA 13. The section does not reference Section 903.3.1.2, since NFPA 13R is limited to buildings that are four stories or less in height and is not a suitable sprinkler system for a high-rise building. If a high-rise building contains an open parking garage along with Group B occupancies above, then the open parking garage portion of the building need not be protected with an automatic sprinkler system. See the commentary to Section 406.3 for a discussion of the fire hazards associated with open parking garages.

If a telecommunications equipment facility is part of a high-rise building, automatic sprinkler protection can be eliminated from certain areas if an automatic sprinkler fire detection system is provided and the telecommunication equipment areas are separated from other building areas with fire-resistance-rated construction of the specified ratings. This same exception is provided in Section 903.3.1.1.1, but is restated here to emphasize its application to all buildings, including high-rise buildings. This exception is based on the need to maintain uninterrupted operation of telecommunications equipment, which frequently includes emergency telephone and similar communication services. The presence of an automatic sprinkler system or other automatic fire suppression system may be detrimental to public safety should disruption of emergency services occur.

In order to increase the reliability of the sprinkler system should an earthquake disable the primary water supply, a secondary water supply is required for buildings located at sites with the specified seismic design category listed in Section 903.3.5.2. The secondary water supply must be equal to the hydraulically calculated sprinkler demand and must have a duration of no less than 30 minutes.

403.3 Reduction in fire-resistance rating. The fire-resistance-rating reductions listed in Sections 403.3.1 and 403.3.2 shall be allowed in buildings that have sprinkler control valves equipped with supervisory initiating devices and water-flow initiating devices for each floor.

❖ Since the overall reliability of the sprinkler system is greatly improved by the required control valves and water-flow devices, certain code modifications are allowed. These reductions may be utilized, provided that the building is equipped with an automatic sprinkler system in accordance with Section 403.2 and NFPA 13.

Sprinkler control valves with supervisory initiating devices and water-flow initiating devices must be provided for each floor in order to realize the reduction in rating. Annunciation by floor is intended to allow rapid identification of the fire location by the fire department. Sprinkler control valves on each floor are intended to permit servicing activated systems without impairing the water supply to large portions of the building. It is, therefore, necessary to equip the valves with supervisory initiating devices. NFPA 13 does not require annunciation per floor, since some systems may take on configurations that would not lend themselves to this type of zoning. In accordance with Section 102.4, the criteria in this section take precedence over the permissive language of NFPA 13 for the purposes of utilizing the modification permitted by Section 403.3.

403.3.1 Type of construction. The following reductions in the minimum construction type allowed in Table 601 shall be allowed as provided in Section 403.3:

1. Type IA construction shall be allowed to be reduced to Type IB.

2. In other than Groups F-1, M and S-1, Type IB construction shall be allowed to be reduced to Type IIA.

3. The height and area limitations of the reduced construction type shall be allowed to be the same as for the original construction type.

❖ One of the reductions in fire-resistance ratings allowed by Section 403.3 is an across-the-board decrease in the building's type of construction. All high-rise buildings that initially must be of Type IA construction for height and area purposes are allowed to be reclassified as Type IB. Additionally, high-rise buildings that initially must be of Type IB can be reclassified to Type IIA construction for occupancies other than Groups F-1, M and S-1. This reduction is not applicable to moderate-hazard buildings because of their customarily higher fuel loads. In determining the minimum allowable type of construction for a high-rise building, Table 503 is applied in the usual fashion (see Section 503 and Table 503). Item 3 in this section reminds the code user that even though the reductions are taken, the allowable area for the building may still use the original areas specified for the building from Table 503 before the type of construction is reduced. Section 403.3.1 is applied once the minimum type of construction has been determined. This results in the reduction of construction by one type. Generally, when Tables 601 and 602 are then applied to the new type of construction, a reduction in fire-resistance ratings for most structural elements is achieved.

For example, an eight-story, high-rise office building is to be constructed and equipped with an automatic sprinkler system in accordance with Sections 403.2 and 403.3. The minimum allowable construction type is Type IB. In accordance with Section 403.3.1, however, the required fire-resistance ratings for the structural elements would be determined using the column for Type IIA construction in Tables 601 and 602. It should be noted that Table 503 does not permit the use of unprotected noncombustible construction in a high-rise building.

403.3.2 Shaft enclosures. The required fire-resistance rating of the fire barrier walls enclosing vertical shafts, other than exit enclosures and elevator hoistway enclosures, shall be reduced to 1 hour where automatic sprinklers are installed within the shafts at the top and at alternate floor levels.

❖ As with previous sections, based on the effectiveness and reliability of a complete automatic sprinkler system in accordance with Sections 403.2 and 403.3, the fire-resistance rating of shaft enclosures other than exit enclosures and elevator hoistway enclosures may be reduced to 1 hour, provided that sprinklers are also installed within the shafts at the top and at alternate floor levels.

The reduced fire-resistance rating does not apply to exit and elevator shafts since they may continue to be used for egress or fire-fighter access during a fire and, therefore, the integrity of the shaft must be maintained. If, for some reason, the fire penetrates the 1-hour shaft, it will affect the conditions on other floors; hence, the need to maintain the protection for the exit or elevator shaft. By the time the fire penetrates the 1-hour fire-resistance-rated shaft, the affected floors should be evacuated.

403.4 Emergency escape and rescue. Emergency escape and rescue openings required by Section 1025 are not required.

❖ This section of the code reinforces other tradeoffs or allowances permitted by the code for buildings equipped throughout with automatic sprinkler systems. Exception 1 to Section 1025.1 does not require emergency escape and rescue openings for a building protected with an automatic sprinkler system. The high-rise building provisions are consistent with this allowance.

[F] 403.5 Automatic fire detection. Smoke detection shall be provided in accordance with Section 907.2.12.1.

❖ Automatic smoke detectors are required in all high-rise buildings in certain locations so that a fire will be detected in its early stages of development. Smoke detectors must be installed in rooms that are not typically occupied. Spaces specifically identified as requiring

automatic fire detection are: mechanical equipment; electrical and transformer and telephone equipment rooms as required by Section 907.2.12.1. See Section 907.2 of the commentary for further requirements of the automatic fire detection system based on the occupancy.

[F] 403.6 Emergency voice/alarm communication systems. An emergency voice/alarm communication system shall be provided in accordance with Section 907.2.12.2.

❖ By definition, one characteristic of a high-rise building is high evacuation times. As such, the traditional fire alarm system, which usually results in simultaneous total building evacuation, is not practical; therefore, the alarm system should be able to:

- Direct the occupants of the fire zone to an area of refuge or exit;
- Notify the fire department of the existence of the fire and
- Sound no alarms outside the fire zone until deemed desirable.

The alarm signal to the fire zone may include continuous sounding devices (bells, horns, chimes, etc.) as well as voice direction, which may momentarily silence the continuous sounding devices so as to be heard clearly. In accordance with Section 907.9.1, visible alarm notification appliances are required in the common and public areas of the high-rise building and certain guestrooms and dwelling units. The message of the voice/alarm is to be predetermined but need not be a recorded message. It would typically indicate that a fire has been reported at a specific location and that occupants should await further instructions or evacuate in accordance with the building's fire emergency plan. The voice/alarm signal will usually commence with a 3- to 10-second alert signal followed by the message. See Section 907.2.12.1 of the commentary for further requirements of the emergency voice/alarm communication system.

[F] 403.7 Fire department communications system. A two-way fire department communications system shall be provided for fire department use in accordance with Section 907.2.12.3.

❖ High-rise buildings have also posed a challenge to the traditional communication systems used by the fire service for fire-to-ground communications; therefore, a fire department communications system must be provided to assist fireground officers in communicating with the fire fighters working in various areas of the building. The system must be capable of operating between the fire command center and every elevator; elevator lobby; emergency and standby power rooms; fire pump rooms; areas of refuge and exit stairway. Additional guidance on a two-way fire department communication system can be found in Section 907.2.12.3 and in NFPA 72.

[F] 403.8 Fire command. A fire command center complying with Section 911 shall be provided in a location approved by the fire department.

❖ Fireground operations usually involve establishing an incident command post where the fire officer can observe what is happening, control arriving personnel and equipment and direct resources and fire-fighting operations effectively. Because of the difficulties in controlling a fire in a high-rise building, a separate room (enclosed in 1-hour-rated construction) within the building must be established to assist the fire officer. The room must be provided at a location that is acceptable to the fire department, usually along the front of the building or near the main entrance. The room must contain equipment necessary to monitor or control fire protection and other building service systems (see Section 911 for further information).

403.9 Elevators. Elevator operation and installation shall be in accordance with Chapter 30.

❖ The primary purpose of elevator requirements for high-rise buildings is to make elevators available for use by the fire department. Section 3002.4 provides minimum size requirements for elevators that allows persons requiring emergency medical services to be moved on an ambulance stretcher. The minimum elevator size and the provisions relating to elevator lobbies (see Sections 707.14.1 and 1007.6) also permit relocation of physically disabled individuals who may not be able to use the stairs.

This section also explicitly states that elevator operation and installation are to be in accordance with Chapter 30, which contains provisions for the design, construction, installation, operation and maintenance of elevators.

403.10 Standby power. A standby power system complying with Section 2702 shall be provided for standby power loads specified in Section 403.10.2.

❖ Standby power is required to increase the probability that the fire protection systems, exit illumination and elevators will continue to function in the event of failure of normal building service. By referencing Section 2702, these provisions pick up the requirements of not only the electrical code, but also NFPA 110 and 111.

403.10.1 Special requirements for standby power systems. If the standby system is a generator set inside a building, the system shall be located in a separate room enclosed with 2-hour fire-resistance-rated fire barrier assemblies. System supervision with manual start and transfer features shall be provided at the fire command center.

❖ Standby power systems are intended to supply power automatically to selected loads in the event of failure of the normal power source. The system is to be capable of supplying the connected loads within 60 seconds of normal power failure. If the standby power system is a generator set in the building, the generator must be located in a room that is separated from the remaining parts of the building by fire barrier assemblies having a fire-resistance rating of at least 2 hours. The purpose of the enclosure is to decrease the probability that a fire can affect both the normal and standby power systems. It is important to note that this requirement is more restrictive than the provisions found in the ICC *Electrical Code*® (ICC EC™). The ICC EC would only require either the 2-hour fire protection or a sprinkler system, but not both. Because high-rise buildings are required to be sprinklered, these standby systems end up with both the sprinkler protection and the separation. Although not specifically stated, consideration should also be given to the routing of the conduit to minimize the impact of a single fire on both primary and standby electrical systems. An acceptable alternative, therefore, would be to locate the generator external to the building or embed the conduits in concrete columns, slabs or other fireproofing materials subject to the limitations of Section 714.3. The provisions of Section 2702 are applicable to standby power systems.

403.10.2 Standby power loads. The following are classified as standby power loads:

1. Power and lighting for the fire command center required by Section 403.8;

2. Electrically powered fire pumps;

3. Ventilation and automatic fire detection equipment for smokeproof enclosures.

Standby power shall be provided for elevators in accordance with Section 3003.

❖ All fire protection and emergency systems required in Sections 403.8 and 403.9 must be connected to the standby power system. Additionally, equipment that is necessary for the proper operation of a fire protection system, such as an electrically powered fire pump, must also be transferable to the standby source. An arrangement must be made such that any elevator may be connected to the standby power with one elevator designated as the primary recipient of power once the standby system is activated. This primary elevator must be capable of serving all floors of a building. In extremely tall buildings, it may therefore, be necessary to designate multiple elevators as being primary recipients in order to permit usage of both local and express elevators. Based on language in the ICC EC, the capacity of the standby system must be such that all equipment that has to be operational at the same time will be able to function. The system need not, however, be capable of supplying the load for all connected equipment simultaneously if automatic load shedding is provided. For example, if multiple elevators are connected to the system, the standby power system need only be capable of supplying one elevator such that the elevator provides access to all floors and that central controls restrict operation of other elevators at the same time.

403.11 Emergency power systems. An emergency power system complying with Section 2702 shall be provided for emergency power loads specified in Section 403.11.1.

❖ Emergency power is required to ensure that fire protection systems and exit illumination systems will continue to function with only a very minor interruption in the event of a failure of normal building service. By referencing Section 2702, these provisions pick up the requirements of not only the ICC EC, but also NFPA 110 and 111.

403.11.1 Emergency power loads. The following are classified as emergency power loads:

1. Exit signs and means of egress illumination required by Chapter 10;

2. Elevator car lighting;

3. Emergency voice/alarm communications systems;

4. Automatic fire detection systems; and

5. Fire alarm systems.

❖ The illumination of exit signs and the means of egress as required by Sections 1006 and 1011, alarm and detection systems and elevator car lighting must be considered emergency electrical systems. As such, the systems are to comply with the provisions of Section 2702, the ICC EC and the referenced standards. Because of the critical nature of the systems and the potential for panic in the areas covered, the load must be picked up within 10 seconds after failure of the normal power supply.

403.12 Stairway door operation. Stairway doors other than the exit discharge doors shall be permitted to be locked from stairway side. Stairway doors that are locked from the stairway side shall be capable of being unlocked simultaneously without unlatching upon a signal from the fire command center.

❖ Section 1008.1.8.7 requires that all egress doors for interior stairways be readily openable from both sides. It is often desirable to control movement of persons within a building and to provide additional security from external threats; therefore, this section permits locking of stairway doors from the stair side when all doors are capable of being simultaneously unlocked. Since high-rise buildings are difficult to evacuate and people are often relocated to another floor level, access from the stairway to a floor could be essential in a fire emergency; therefore, all stairway doors that are to be locked from the stairway side must have the capability of being unlocked simultaneously on a signal from the fire command center. The unlocking of the door must not negate the latching feature, which is essential to the operation of the door as a fire door. Section 403.10.2, Item 1, by its reference to Section 403.8, requires the locking feature to be connected to the standby power system. When the door is unlocked during an emergency, it should not automatically relock on closure. Electrically powered locks should be designed such that when power to the locking device is interrupted, the lock is released. This is intended to enable door locks to be operable from inside the stairway if power to the lock is interrupted. The building official should review the emergency release operation of stairway doors to determine that they remain unlocked.

403.12.1 Stairway communications system. A telephone or other two-way communications system connected to an approved constantly attended station shall be provided at not less than every fifth floor in each required stairway where the doors to the stairway are locked.

❖ If the stairway doors are locked to restrict reentry as permitted in Section 403.12, a two-way communications system must be provided at no less than every fifth floor and must be connected to the standby power system. This system is required to be connected to a constantly attended location, which could be within the building, or to a central station that monitors fire alarms and is manned 24 hours a day, 7 days a week. Use of the fire command center is not recommended, since it may not be constantly attended. The system will permit occupants in the stairway to notify the attended location that the stairway doors need to be unlocked to access another floor or because conditions in the stairway prevent its continued use.

403.13 Smokeproof exit enclosures. Every required stairway serving floors more than 75 feet (22 860 mm) above the lowest level of fire department vehicle access shall comply with Sections 909.20 and 1019.1.8.

❖ This section serves as a reminder that the smokeproof enclosure provisions of Section 1019.1.8 are applicable to high-rise buildings. It is important to understand that this requirement is only applicable in the portions of the building where the stack effect and the spread of smoke will be greatest. For example, if the high-rise building consists of a tower portion with a larger area on the bottom that is only a couple of stories in height, these requirements would only be applicable to stairways within the tower portion and not for stairs that only serve the low-rise portion of the building.

403.14 Seismic considerations. For seismic considerations, see Chapter 16.

❖ A reminder is provided in this section of the code that high-rise buildings are required to be designed for the effects of seismic activity. The intent of this section is to note that mechanical and electrical components are required to be designed for forces determined by Section 1621. This would also require that elevators be designed to resist the effects of the seismic forces. These elements are critical for the safety of high-rise buildings so that all of the required fire protection systems and the electrical components that secure them will be designed for the applicable seismic loads.

SECTION 404
ATRIUMS

404.1 General. Vertical openings meeting the requirements of this section are not required to be enclosed in other than Group H occupancies.

❖ Unprotected vertical openings are often identified as the factor responsible for fire spread in incidents involving fire fatalities or extensive property damage. Section 404 addresses the need for protection of these specific building features in lieu of providing a complete floor separation. This section addresses one particular type of vertical opening that is permissible in all buildings other than Group H. Atriums are regulated by this section and are considered acceptable when in compliance with the criteria of this section. Additionally, the code permits other types of vertical openings that are not addressed by this section. For example, Section 707.2 permits openings within residential dwelling units to be unprotected. Other unprotected vertical openings that are permitted by the code include covered malls (Section 402); communicating spaces in buildings of Group I-3 (Section 408.5); mezzanines (Section 505); escalator or supplement stair openings (Section 707.2); open parking garages (Section 406.3); enclosed parking garages (Section 406.4) or as otherwise noted in Section 707.2.

An atrium is a space within a building that extends vertically and connects two or more stories. Atriums are not to be considered unprotected vertical openings; rather, the vertical openings are protected by means other than enclosure of the shaft or a complete floor assembly. This section does not apply to floor openings permitted to be unenclosed by Section 707.2.

Section 404 is applicable when a shaft enclosure would normally be required by Section 707.2, but because of the nature, use or design of the space, the shaft enclosure is not provided and, therefore, does not comply with the provisions of Section 707. Spaces that are separated from the atrium by shaft enclosure assemblies complying with the provisions of Section 707 are not considered as part of the atrium. Section 404 applies only to the spaces that are contained within the atrium. Atriums are permitted in all buildings except Group H.

404.1.1 Definition. The following word and term shall, for the purposes of this chapter and as used elsewhere in this code, have the meaning shown herein.

❖ Definitions of terms that are associated with the content of this section are contained herein. These definitions can help in the understanding and application of the code requirements. It is important to emphasize that these terms are not exclusively related to this section but are applicable everywhere the term is used in the code. The purpose for including this definition within this section is to provide more convenient access to it without having to refer back to Chapter 2.

For convenience, this term is also listed in Chapter 2 with a cross reference to this section.

The use and application of all defined terms, including that defined herein, are set forth in Section 201.

ATRIUM. An opening connecting two or more stories other than enclosed stairways, elevators, hoistways, escalators, plumbing, electrical, air-conditioning or other equipment, which is closed at the top and not defined as a mall. Stories, as used in this definition, do not include balconies within assembly groups or mezzanines that comply with Section 505.

❖ The definition identifies that an atrium is a floor opening or a series of floor openings that connect the environments of adjacent stories. Building features such as stairways, elevators, hoistways, escalators, plumbing, electrical, air-conditioning or other equipment openings are required to be enclosed in fire-resistance-rated shafts in accordance with Section 707.2 or must be protected with one of the methods specified in the exceptions to Section 707.2, 712.4 or 716.6. These items are exempted from the atrium provisions. An atrium is not defined by size or use. A series of floor openings that are enclosed with exterior walls, yet open at the roof, would be considered a court and would be exempt from the requirements of Section 404. Balconies and mezzanines are not individual stories and are not subject to being called an atrium.

404.2 Use. The floor of the atrium shall not be used for other than low fire hazard uses and only approved materials and decorations in accordance with the *International Fire Code* shall be used in the atrium space.

> **Exception:** The atrium floor area is permitted to be used for any approved use where the individual space is provided with an automatic sprinkler system in accordance with Section 903.3.1.1.

❖ Because an automatic sprinkler system at the ceiling of an atrium may not be effective for a fire on the floor of the atrium due to the atrium height or obstructions to the sprinkler discharge, the use and activities of the floor level and the types of materials in the atrium space must be controlled. This section applies to all atriums regardless of their height or area. Low fire-hazard uses would limit the atrium floor to such functions as pedestrian walk-through areas, security desks, reception areas, etc. Storage areas, fabrication areas, office areas, etc., would not be low fire-hazard uses. Chapter 8 of the IFC regulates the use of decorative materials and furnishings.

If the floor area is equipped with an automatic sprinkler system that can provide the required protection, then its use is not restricted. The exception stipulates that such areas must be equipped with an automatic sprinkler system as is required throughout the remainder of the atrium.

[F] 404.3 Automatic sprinkler protection. An approved automatic sprinkler system shall be installed throughout the entire building.

Exceptions:

1. That area of a building adjacent to or above the atrium need not be sprinklered provided that portion of the building is separated from the atrium portion by a 2-hour fire-resistance-rated fire barrier wall or horizontal assembly or both.

2. Where the ceiling of the atrium is more than 55 feet (16 764 mm) above the floor, sprinkler protection at the ceiling of the atrium is not required.

❖ One means of controlling the spread of fire and smoke through vertical openings is to control and extinguish the fire as early as possible; therefore, all floor areas that are connected by the atrium are to be protected with an electrically supervised automatic sprinkler system, including the atrium space itself. The system is to be designed, installed, tested and maintained in accordance with the provisions of Section 903.3.1.1 and the IFC. The sprinkler system must comply with Section 903.3.1.1 and NFPA 13. Although this sprinkler system is not necessarily required throughout the entire building, in accordance with Exception 1, Section 903.3.1.1 indicates that it is not the intent that this be considered a limited area system, but that it otherwise conform to the requirements of NFPA 13 (e.g., fire department connection and alarms are required). It should be noted that even in residential occupancies in which the code would permit an NFPA 13R system, an NFPA 13 system must be used. Since Chapter 9 requires that the sprinkler system be supervised, the reliability of the sprinkler system is improved. Exception 1 clarifies that since the atrium protection is an alternative to the shaft enclosure requirements, the sprinkler system is not required in areas that are separated from the atrium by 2-hour fire barrier walls or horizontal assemblies that would otherwise be required for a shaft enclosure. Such fire barrier walls must conform to Section 706, while the horizontal assemblies must comply with Section 711.

Exception 2 permits the required sprinkler system to be deleted from the ceiling areas of atriums where the vertical distance between the atrium floor and atrium ceiling is greater than 55 feet (16 764 mm). A ceiling height of more than 55 feet (16 764 mm) may not only delay the response of the sprinkler system but also magnify the design problems associated with obtaining the correct density of the sprinkler heads. This exception does not alter the use limitations of the atrium as stated in Section 404.2, nor does it exempt any adjacent floor areas with smaller ceiling heights that are included in the atrium boundary in accordance with Exception 1.

404.4 Smoke control. A smoke control system shall be installed in accordance with Section 909.

Exceptions:

1. Smoke control is not required for floor openings meeting the requirements of Section 707.2, Exception 2, 7, 8 or 9.

2. Smoke control is not required for floor openings meeting the requirements of Section 1019.1, Exception 8 or 9.

❖ In order to prevent the migration of smoke throughout interconnected levels of a building via the atrium, a mechanical smoke control system is to be installed in atriums connecting two or more stories in accordance with the provisions of Section 909. Other methods of smoke control may be used if they provide the same or a higher level of protection, as indicated in Section 909.6 or 909.7. The smoke control system for the atrium is required to be connected to a standby source of power in accordance with Sections 404.6 and 909.11.

This section also requires two-story covered mall buildings to have a smoke control system, since Section 402.9 refers to the atrium provisions for smoke control purposes. The only exceptions to the requirements for smoke control are those specified in Section 707.2 and in the open stairways serving only one adjacent floor. Section 404 is only one method of addressing unenclosed floor openings. Other exceptions in Section 707.2 direct the code user to other possible design methods. The building designer may choose Exception 2, 7, 8 or 9 as an alternative to the atrium requirements. (see commentary, Section 707.2).

404.5 Enclosure of atriums. Atrium spaces shall be separated from adjacent spaces by a 1-hour fire barrier wall.

Exceptions:

1. A glass wall forming a smoke partition where automatic sprinklers are spaced 6 feet (1829 mm) or less along both sides of the separation wall, or on the room side only if there is not a walkway on the atrium side, and between 4 inches and 12 inches (102 mm and 305 mm) away from the glass and so designed that the entire surface of the glass is wet upon activation of the sprinkler system. The glass shall be installed in a gasketed frame so that the framing system deflects without breaking (loading) the glass before the sprinkler system operates.

2. A glass-block wall assembly in accordance with Section 2110 and having a $^3/_4$-hour fire protection rating.

3. The adjacent spaces of any three floors of the atrium shall not be required to be separated from the atrium where such spaces are included in computing the atrium volume for the design of the smoke control system.

❖ One of the basic premises of atrium requirements is that an engineered smoke control system combined with an

FIGURE 404.5   SPECIAL DETAILED REQUIREMENTS BASED ON USE AND OCCUPANCY

automatic fire sprinkler system that is properly supervised provide an adequate alternative to the fire-resistance rating of a shaft enclosure. It is also recognized that some form of boundary is required to assist the smoke control system in containing smoke to just the atrium area. The basic requirement, therefore, is that the atrium space be separated from adjacent areas by fire barriers having a fire-resistance rating of at least 1 hour.

The reference to fire barriers is intended to utilize the provisions of Section 706 that relate to fire-resistance-rated construction. Also, openings in the wall are required to be protected, in accordance with Section 706.7.

In accordance with Section 707.4, shafts are required to have a fire-resistance rating of at least 2 hours if connecting more than three stories, and 1 hour when connecting two or three stories. The basis for the 1-hour requirement in Section 404.5 is that an automatic sprinkler system can be substituted for 1 hour of fire resistance of a shaft enclosure. The allowance is consistent with the 1-hour fire-resistance rating reduction permitted in high-rise buildings (see Section 403.3.2).

In lieu of a 1-hour fire-resistance-rated separation, Exception 1 allows adjacent spaces to be separated by glass walls where automatic sprinklers have been installed to protect the glass. The sprinklers are to be prescriptively located so as to wet the entire surface of the glass wall. If there is a floor surface on each side of the wall, both sides of the glass must be protected. The glass must be in a gasketed frame such that the framing system can deflect without breaking the glass.

Although this exception does not address obstructions or other window treatments, consideration must be given to locating such items to avoid interference with the required sprinkler heads. Without specific test evidence, curtain rods, traverse rods, curtains, draperies, etc., must be located at least 12 inches (305 mm) from the window surface (see Figure 404.5). Any doors through the required 1-hour fire barrier wall must be $^3/_4$-hour rated in accordance with Table 715.3.

Exception 2 allows a glass block wall assembly conforming to Section 2110. It is important to note that these glass block assemblies do not require the sprinkler protection that is required by Exception 1.

Exception 3 recognizes the desire to have at least some floors open to the atrium, permitting a maximum of three. The three-floor restriction is consistent with the basic premise that the life safety hazard becomes significant when more than three floors are open. It should be noted that the three floor levels may be at any height and need not be consecutive floor levels. To compensate for the increased hazard to occupants, the smoke control system must be sized for the volume of the atrium and all floors open to the atrium.

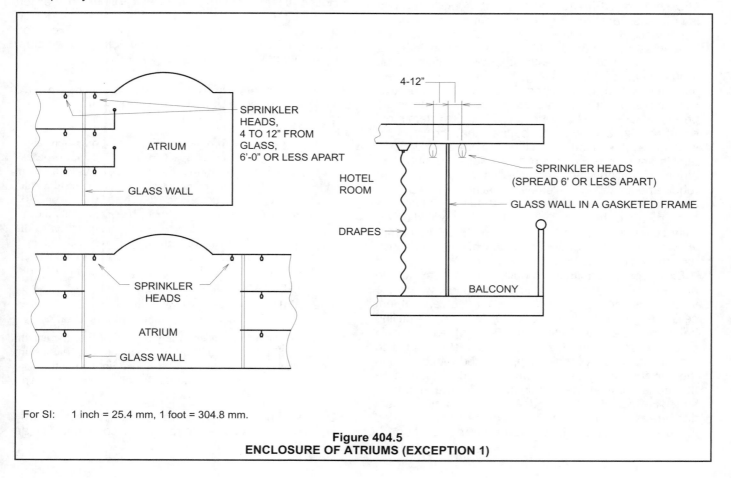

For SI: 1 inch = 25.4 mm, 1 foot = 304.8 mm.

Figure 404.5
ENCLOSURE OF ATRIUMS (EXCEPTION 1)

404.6 Standby power. Equipment required to provide smoke control shall be connected to a standby power system in accordance with Section 909.11.

❖ To enhance the reliability of the required smoke control system, this section of the code mandates that the mechanical exhaust must be connected to a standby power system. The reference to Section 909.11 allows the atrium's equipment to receive its primary power from the building's power system. A secondary standby power source is required should the building's primary system fail.

404.7 Interior finish. The interior finish of walls and ceilings of the atrium shall not be less than Class B with no reduction in class for sprinkler protection.

❖ Although Chapter 8 contains provisions governing the use of interior finishes, trim and decorative materials, specific limitations are provided for an atrium. Similar to the use limitations contained in Section 404.2, a minimum interior finish classification is specified to limit the fuel load within the atrium spaces. At least a Class B finish is mandated for the walls and ceilings of the atrium. Such a finish classification limits the material's flame spread index to a range of 26 through 75 and a smoke-developed index rating of 450 or less. The sprinkler reductions shown in Table 803.5 are not applicable to the interior finishes of an atrium space.

404.8 Travel distance. In other than the lowest level of the atrium, where the required means of egress is through the atrium space, the portion of exit access travel distance within the atrium space shall not exceed 200 feet (60 960 mm).

❖ On all floors with direct communication to the atrium for exit access (other than the floor level of the atrium), the total exit access travel distance through the atrium is limited to 200 feet (60 960 mm). Since smoke is being drawn to the atrium, the time allowed to reach an exit through the atrium is limited. Travel distance at the atrium floor level and the levels that do not communicate with the atrium for exit access in the building is regulated by Table 1015.1.

SECTION 405
UNDERGROUND BUILDINGS

405.1 General. The provisions of this section apply to building spaces having a floor level used for human occupancy more than 30 feet (9144 mm) below the lowest level of exit discharge.

Exceptions:

1. One- and two-family dwellings, sprinklered in accordance with Section 903.3.1.3.

2. Parking garages with automatic fire suppression systems in compliance with Section 405.3.

3. Fixed guideway transit systems.

4. Grandstands, bleachers, stadiums, arenas and similar facilities.

5. Where the lowest story is the only story that would qualify the building as an underground building and has an area not exceeding 1,500 square feet (139 m²) and has an occupant load less than 10.

❖ An underground building presents a unique hazard to life safety. Due to its isolation and inaccessibility, occupants within the structure and fire fighters attempting to locate and suppress a fire are presented with a unique fire protection challenge.

To egress the structure, occupants must travel in an upward direction. The direction of occupant travel is the same as the direction that the products of combustion travel. As such, the occupants are potentially exposed to the products of combustion along the entire means of egress.

Fire fighters are also confronted by constant exposure to the products of combustion. Beginning their descent above the actual location of the fire source, fire fighters encounter an increasing amount of smoke, heat and flame as they attempt to locate and extinguish the fire source. These extreme conditions could significantly hinder the effectiveness of the fire department if not offset by appropriate fire protection requirements. The requirements for underground buildings are, in some ways, similar to those for high-rise structures. Both types of structures present an unusual hazard since they are virtually inaccessible to exterior fire department suppression and rescue operations with the increased potential to trap occupants inside. To counteract these hazards, such structures are required by Section 405.2 to be built of noncombustible, fire-resistance-rated construction. Additionally, they are required by Section 405.3 to be equipped with an automatic sprinkler system and a smoke control system in accordance with Section 405.5. Standby and emergency power systems are also required in these structures by Sections 405.9 and 405.10.

Underground buildings that require the occupants of the lowest floor level to travel upwards for more than 30 feet (9144 mm) to reach the level of exit discharge present a significant hazard to the occupants. As such, Section 405 is applicable to buildings with a floor level more than 30 feet (9144 mm) below the lowest level of exit discharge (see Figure 405.1).

Structures regulated by Section 405 are also subject to all other applicable code provisions. Additionally, underground buildings to which this section does not apply are still subject to all other code provisions, including fire suppression (Section 903); standpipe systems (Section 905); fire alarm and detection (Section 907) and emergency escape (Section 1025).

There are five exceptions to the applicability of Section 405. These exceptions are in consideration of specific types of structures to which the requirements of this section are impractical, unnecessary or to which alternative provisions apply.

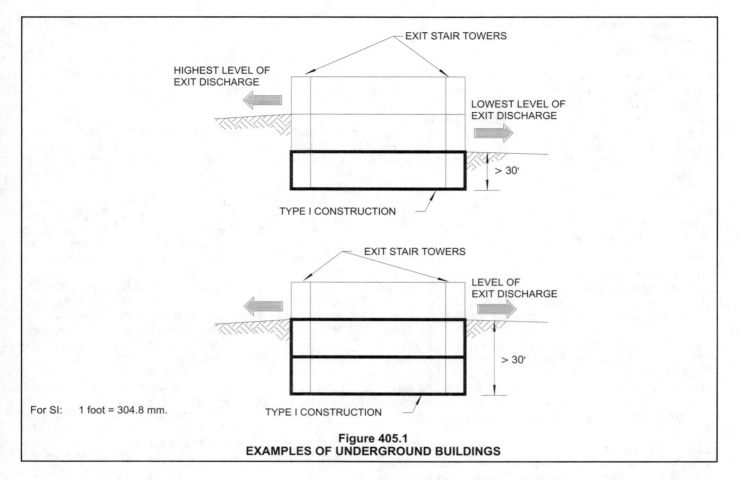

Figure 405.1
EXAMPLES OF UNDERGROUND BUILDINGS

For SI: 1 foot = 304.8 mm.

405.2 Construction requirements. The underground portion of the building shall be of Type I construction.

❖ Two of the key concerns with underground construction are to limit the fuel load and to increase the fire resistance of underground buildings. To reasonably ensure that the building elements will remain structurally sound during exposure to fire, all structural elements of the underground portions are required to be of noncombustible, protected construction; therefore, any portion of the structure located below grade level is required to be of Type I construction (see Figure 405.1).

Portions of the structure above grade, however, are permitted to be of any type of construction.

Any portion of the structure above ground level must conform to the height and area limitations of Table 503. This table does not regulate the maximum depth of an underground structure; therefore, the height, area and depth of most Type I structures are not limited.

[F] 405.3 Automatic sprinkler system. The highest level of exit discharge serving the underground portions of the building and all levels below shall be equipped with an automatic sprinkler system installed in accordance with Section 903.3.1.1. Water-flow switches and control valves shall be supervised in accordance with Section 903.4.

❖ One of the most effective preventive measures to fire growth is the installation of an automatic sprinkler sys-

tem. Because of the unique conditions for occupant egress and fire department access in the underground portion of a building, automatic sprinkler protection is required. The level of exit discharge and all floor levels below are required to be sprinklered throughout in accordance with Section 903.3.1.1. This section does permit a portion of a building to extend above the level of exit discharge and not be equipped with an automatic sprinkler system. If, however, another code section (Section 403.2 or 903) requires an automatic sprinkler system in the above-ground portion, such a requirement would still be applicable. Note that a smoke control system is required. Automatic sprinkler systems are essential elements of any smoke control system. Without suppression, the size of the fire or the resulting products of combustion will rapidly overwhelm most mechanical smoke control systems.

405.4 Compartmentation. Compartmentation shall be in accordance with Sections 405.4.1 through 405.4.3.

❖ Compartmentation is a key element in the egress and fire access plan for floor areas in an underground building. Subdivision into separate compartments through the use of smoke barriers (see Section 706) permits occupants to travel horizontally to escape the fire condition and provides a staging area for the fire service (see Figure 405.4).

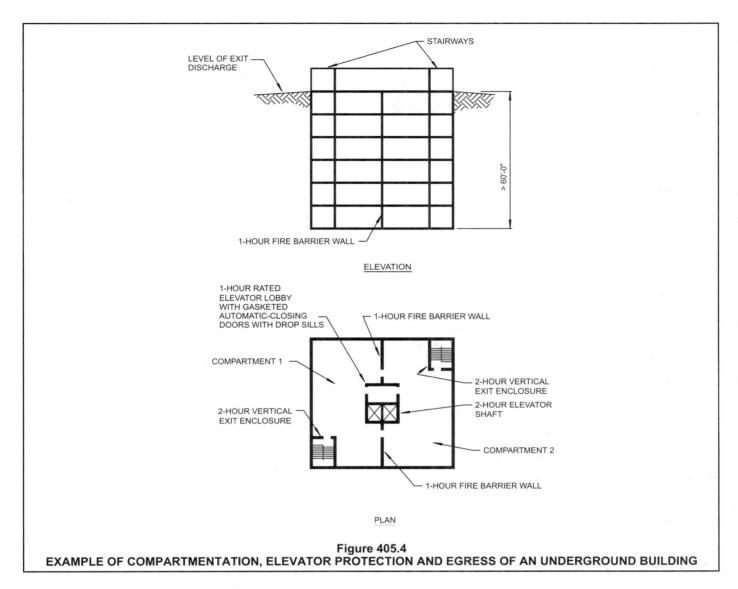

Figure 405.4
EXAMPLE OF COMPARTMENTATION, ELEVATOR PROTECTION AND EGRESS OF AN UNDERGROUND BUILDING

405.4.1 Number of compartments. A building having a floor level more than 60 feet (18 288 mm) below the lowest level of exit discharge shall be divided into a minimum of two compartments of approximately equal size. Such compartmentation shall extend through the highest level of exit discharge serving the underground portions of the building and all levels below.

> **Exception:** The lowest story need not be compartmented where the area does not exceed 1,500 square feet (139 m²) and has an occupant load of less than 10.

❖ The 60-foot (18 288 mm) threshold is based on establishing a reasonable limitation on the required vertical travel distance for the occupants and fire service before the added protection of compartmentation is beneficial.

It is important to realize that the code requires, at a minimum, two compartments of approximately equal size. The maximum size of each compartment is not limited and is, therefore, a design decision subject to the approval of the building official.

An exception is permitted at the lowest level of the un-derground building as long as that level is relatively small in area [less than or equal to 1,500 square feet (139 m²)] and serves a low occupant load (less than 10), since an evacuation to the next higher level would not be expected to adversely affect that level.

405.4.2 Smoke barrier penetration. The separation between the two compartments shall be of minimum 1-hour fire barrier wall construction that shall extend from floor slab to floor deck above. Openings between the two compartments shall be limited to plumbing and electrical piping and conduit penetrations firestopped in accordance with Section 712. Doorways shall be protected by fire door assemblies that are automatic-closing by smoke detection in accordance with Section 715.3 and shall be provided with gasketing and a drop sill to minimize smoke leakage. Where provided, each compartment shall have an air supply and an exhaust system independent of the other compartments.

❖ The fire barrier walls that create the compartments must be fire-resistance rated for 1 hour and must meet the continuity provisions specified in Section 706.4. Penetrations through the fire barrier wall are limited to plumbing, piping and penetrations that are vital to the fire protection systems (i.e., sprinkler piping and electrical raceways). Special fire-resistance-rated doors are required to maintain the intended load of compartmentation, including gasketing and drop sills. Separate air distribution systems are required for each compartment. Components of the air distribution system are not permitted to penetrate the fire barrier walls. These requirements are intended to help provide for the independence of the compartments separated by the fire barrier walls by minimizing the penetrations therein.

405.4.3 Elevators. Where elevators are provided, each compartment shall have direct access to an elevator. Where an elevator serves more than one compartment, an elevator lobby shall be provided and shall be separated from each compartment by a 1-hour fire barrier wall. Doors shall be gasketed, have a drop sill, and be automatic-closing by smoke detection installed in accordance with Section 907.10.

❖ Elevators are permitted to serve more than one compartment. When different compartments utilize a common elevator, a 1-hour fire-resistance-rated elevator lobby is required. This lobby provides additional separation from floor slab to floor slab between adjacent compartments, thus helping to create their independence from the immediate effects of fire (see Figure 405.4).

405.5 Smoke control system. A smoke control system shall be provided in accordance with Sections 405.5.1 and 405.5.2.

❖ A smoke control system is required for all underground buildings. The smoke control system is an integral part of the required fire protection systems for underground buildings. Specific details and requirements are listed in Sections 405.5.1 and 405.5.2.

405.5.1 Control system. A smoke control system is required to control the migration of products of combustion in accordance with Section 909 and the provisions of this section. Smoke control shall restrict movement of smoke to the general area of fire origin and maintain means of egress in a usable condition.

❖ The general design requirements, special inspection and test requirements along with the applicable exhaust methods are contained in Section 909. The performance goal of the smoke control system is to contain the smoke and hot gases generated by a fire condition to that immediate area. This allows the building occupants to access their required exits and evacuate the building before smoke movement traps them below grade.

405.5.2 Smoke exhaust system. Where compartmentation is required, each compartment shall have an independent smoke control system. The system shall be automatically activated and capable of manual operation in accordance with Section 907.2.18.

❖ Each compartment of underground buildings with a floor level more than 60 feet (18 288 mm) below the

level of exit discharge is required to be equipped with an independent smoke exhaust system. This system is required not only to facilitate egress during a fire, but also to improve fire department access to the fire source by maintaining visibility that is otherwise impossible given the inability of the fire service to ventilate manually the underground portion of the building.

[F] 405.6 Fire alarm systems. A fire alarm system shall be provided where required by Section 907.2.19.

❖ The ability to communicate and offer warning of a fire scenario can increase the time available for egress from the building. Underground buildings with a floor level greater than 60 feet (18 288 mm) below the level of exit discharge are, therefore, required to be provided with a manual fire alarm system. An emergency voice/alarm communication system is required as part of this system.

[F] 405.7 Public address. A public address system shall be provided where required by Section 907.2.19.1.

❖ In underground buildings where a fire alarm system is not required [lowest floor level is 60 feet (18 288 mm) or less below exit discharge], a public address system must be provided. This communication between the highest level of discharge and underground levels can be a vital link between emergency personnel and building occupants. This is especially important in an environment in which visibility is likely to be impaired in order to determine the status of emergency evacuation from a fire scene below grade.

405.8 Means of egress. Means of egress shall be in accordance with Sections 405.8.1 and 405.8.2.

❖ This section is simply the the introduction for means of egress in underground buildings.

405.8.1 Number of exits. Each floor level shall be provided with a minimum of two exits. Where compartmentation is required by Section 405.4, each compartment shall have a minimum of one exit and shall also have an exit access doorway into the adjoining compartment.

❖ The means of egress from underground buildings is an integral part of the safety precautions necessary to abate the hazards of people being located more than 30 feet (9144 mm) below the level of exit discharge. Sections 405.8.1 and 405.8.2 provide the minimum number of exits and requirements for smokeproof enclosures. These provisions create safe and usable elements that the building occupants can enter to flee the immediate effects of a fire located well below grade.

The arrangement of exits and exit access doors for an underground building is regulated by this section (see Figure 405.4).

Based on the requirements of this section and Section 405.4, an underground building having a floor level more than 30 feet (9144 mm) below the level of exit discharge requires a minimum of two exits, and cannot qualify as a single-exit building. Additional exits may be required in accordance with Section 1018.1. Underground buildings requiring compartmentation must have an exit in each and every compartment plus a doorway into the adjoining compartment.

405.8.2 Smokeproof enclosure. Every required stairway serving floor levels more than 30 feet (9144 mm) below its level of exit discharge shall comply with the requirements for a smokeproof enclosure as provided in Section 1019.1.8.

❖ In order for exit stairways to provide the necessary means of egress, all stairways of an underground building must be constructed as smokeproof enclosures or pressurized stairways. The requirements for smokeproof enclosures are contained in Section 1019.1.8 and include a 2-hour enclosure along with limited access via a vestibule. The requirements for pressurized stairways are listed in Section 909.20. These provisions not only will maintain a protected path of travel for building occupants to egress the underground building but also will allow the fire department to enter for rescue and fire-fighting operations.

[F] 405.9 Standby power. A standby power system complying with Section 2702 shall be provided standby power loads specified in Section 405.9.1.

❖ All underground buildings regulated by this section are required to be provided with a standby power system to increase the probability that critical systems will be operational in the event of a loss of normal power supply. The purpose of the standby power system is to provide an alternative means of supplying power to selected building systems should the normal electrical source fail.

405.9.1 Standby power loads. The following loads are classified as standby power loads.

1. Smoke control system.
2. Ventilation and automatic fire detection equipment for smokeproof enclosures.
3. Fire pumps.

Standby power shall be provided for elevators in accordance with Section 3003.

❖ The standby power system is required to supply electrical power to equipment that is essential to emergency egress, fire-fighting and rescue operations.

405.9.2 Pick-up time. The standby power system shall pick up its connected loads within 60 seconds of failure of the normal power supply.

❖ This section of the code specifies how quickly the standby power system must be activated to provide the necessary power for the loads specified in Section 405.9.1. The standby power systems must be capable of providing full power to those items within 60 seconds of when the main electrical power source goes down.

This maintains the level of safety necessary to evacuate the building and begin fire-fighting operations.

[F] 405.10 Emergency power. An emergency power system complying with Section 2702 shall be provided for emergency power loads specified in Section 405.10.1.

❖ In addition to the standby power system, all underground buildings regulated by this section are also required to be provided with an emergency power system. This system is required to provide electrical power to fire warning devices, lighting and other equipment required in the underground building.

405.10.1 Emergency power loads. The following loads are classified as emergency power loads:

1. Emergency voice/alarm communications systems.
2. Fire alarm systems.
3. Automatic fire detection systems.
4. Elevator car lighting.
5. Means of egress and exit sign illumination as required by Chapter 10.

❖ The emergency power system is required to supply electrical power to equipment that is essential to detecting and warning others of a fire condition and to provide for illuminated evacuation of the underground building. Because of the critical nature of the systems and the potential for panic in the areas covered, the loads must be picked up within 10 seconds after failure of the normal power supply.

405.11 Standpipe system. The underground building shall be equipped throughout with a standpipe system in accordance with Section 905.

❖ Just as high-rise buildings or other large buildings with special features are required to be provided with a standpipe system, so are underground buildings. A Class I automatic wet or manual wet standpipe system must be provided. Standpipe systems allow the fire department or other trained personnel to fight the fires at basement levels where it is impractical to run supply lines from the fire department trucks or outside hydrants. A quick and convenient water source for fire department use is essential to containing an underground fire.

SECTION 406
MOTOR-VEHICLE-RELATED OCCUPANCIES

❖ Included in this section of the code are all the special use and occupancy requirements for those buildings and structures that house motor vehicles. Corresponding requirements for aircraft-related occupancies can be found in Section 412. By definition, all structures that provide services or are used for the storage or parking of motor vehicles are regulated by these provisions. These requirements are applicable regardless of the types of fuel source used to power the vehicles' motors, the number of vehicles present, whether the vehicles are privately or commercially owned and whether they are used strictly for passengers or for freight. Typically, fire hazards for motor-vehicle occupancies are low because of the limited amounts of combustibles present along with the steel frame and metal clad bodies. Motor vehicle structures normally have sufficient space around each vehicle for parking purposes, which serves to further limit the fuel load. The design occupant loads are usually low and pose little if any life safety risks.

This section contains requirements for five different motor-vehicle occupancies along with general requirements for parking garages. Section 406.1 addresses private garages and carports that are accessory to residential occupancies. All other parking garages are divided up into either open parking garages or enclosed parking garages. These Group S-2 structures represent a slightly higher hazard than private garages or carports. As such, the heights and areas of these structures along with their type of construction classification are rigorously controlled. Requirements are also provided for those buildings and structures that provide service, care and repair of the vehicles. Stations where motor fuels are sold and dispensed are classified as Group M occupancies. The requirements for such facilities are contained in Section 406.5. Garages that provide repair services for vehicles also represent a slightly higher hazard; therefore, Section 406.6 identifies those code requirements necessary to abate the hazards of repair garages and their Group S-1 occupancy classification.

406.1 Private garages and carports.

406.1.1 Classification. Buildings or parts of buildings classified as Group U occupancies because of the use or character of the occupancy shall not exceed 1,000 square feet (93 m²) in area or one story in height except as provided in Section 406.1.2. Any building or portion thereof that exceeds the limitations specified in this section shall be classified in the occupancy group other than Group U that it most nearly resembles.

❖ Whereas private garages are smaller than parking garages by definition, the impact of the fuel load represented by the vehicle is somewhat higher. Such garages are frequently attached to residential-type occupancies, with the potential for fire exposure to the residential area and the occupants therein. The code does not specifically prohibit the servicing of vehicles in private garages, since it would be impractical to enforce. For example, the occupant of a residential dwelling unit will occasionally work on a vehicle in an attached private garage. For this reason, there are special provisions for private garages.

Private garages and carports are usually considered accessory to the dwelling units they serve. Section 312 classifies these utility and miscellaneous uses as Group U. The code limits these structures to 1,000 square feet (93 m²) in area and one story in height. Those limitations

correspond to a fuel load associated with the passenger motor vehicles that is matched by the specified separation requirements. Separation and construction details for private garages are provided in Section 406.1.4. This separation includes the $^1/_2$-inch (12.7 mm) regular or $^5/_8$-inch (15.9 mm) Type X gypsum board applied to the garage side (see Section 406.1.4 for further details).

Any garage that exceeds the height and area limits of Section 406.1.1 or the area increase of Section 406.1.2 must be reclassified. Once the specified height and area limits have been exceeded, the garage must be classified for the increased hazards. This results in the garage being classified as Group S-2, and the provisions of Sections 406.2 and 406.4 must now be met.

406.1.2 Area increase. Group U occupancies used for the storage of private or pleasure-type motor vehicles where no repair work is done or fuel dispensed are permitted to be 3,000 square feet (279 m²), when the following provisions are met:

1. For a mixed occupancy building, the exterior wall and opening protection for the Group U portion of the building shall be as required for the major occupancy of the building. For such mixed occupancy building, the allowable floor area of the building shall be as permitted for the major occupancy contained therein.

2. For a building containing only a Group U occupancy, the exterior wall and opening protection shall be as required for a Group R-1 or R-2 occupancy.

More than one 3,000-square-foot (279 m²) Group U occupancy shall be permitted to be in the same building, provided each 3,000-square-foot (279 m²) area is separated by fire walls complying with Section 705.

❖ The area of a private garage can be increased to 3,000 square feet (279 m²) when all of the conditions of this section are met. If the garage will only be used for the parking and storage of private or pleasure-type motor vehicles, then the area increase may be used. Care must be taken such that no repair work or fuel dispensing is done within these areas. The limitations to private- or pleasure-type motor vehicles prevent commercial motor vehicles and their increased hazards from being located in this oversized private garage.

As part of the increased area provisions for private garages, the exterior walls and any openings therein must be properly protected. If the garage is part of another occupancy, then the exterior wall ratings for that occupancy apply to the garage. For example, an automobile insurance office may have several parking bays attached to it. These bays will be used to bring damaged cars in for the purposes of evaluating the damage and estimating the cost of repairs for the vehicle owner. The insurance office spaces are usually classified as Group B. In this case, the allowable building area is based on the Group B occupancy, the garage area is limited to 3,000 square feet (279 m²) and the exterior walls and openings in the garages are rated as part of the Group B occupancy.

For other applications, where the buildings only will

be used to park vehicles, the building is classified as Group U. The exterior walls and openings therein are then rated as required for a Group R-1 and R-2 occupancy.

Additional private garages can be added to a Group U building, but they must be separated with fire walls. The 3,000-square-foot (279 m²) limit is a maximum. To reduce the risk of a fire from involving other adjacent garage spaces, fire walls are required to compartmentalize and further isolate the hazards.

406.1.3 Garages and carports. Carports shall be open on at least two sides. Carport floor surfaces shall be of approved noncombustible material. Carports not open on at least two sides shall be considered a garage and shall comply with the provisions of this section for garages.

Exception: Asphalt surfaces shall be permitted at ground level in carports.

The area of floor used for parking of automobiles or other vehicles shall be sloped to facilitate the movement of liquids to a drain or toward the main vehicle entry doorway.

❖ Carports are covered structures, either free-standing or attached to the side of a building, for the purposes of providing shelter for motor vehicles. Carports must be open on at least two sides. This allows a fire condition to vent quickly to the outside. This further prevents a fire from going undetected and usually limits the amount of other incidental storage that may be included in a fully enclosed garage. Attached carports that are not open on at least two sides are considered a garage and must then be separated from the residence with the required distance (see Section 406.1.4 for further details).

The use of a combustible floor finish that could absorb flammable and combustible liquids commonly used and stored within garages and carports presents a potential safety hazard to the occupants of an attached dwelling. The floors of private garages must be sloped towards the main vehicle doors. Although not specified, the floor must be positively sloped [$^1/_4$:12 (2-percent slope)] to prevent the accumulation of any spilled flammable and combustible liquids and their vapors. This minimizes the risk of the vapors building up to a point where a fire condition could result.

406.1.4 Separation. Separations shall comply with the following:

1. The private garage shall be separated from the dwelling unit and its attic area by means of a minimum $^1/_2$-inch (12.7 mm) gypsum board applied to the garage side. Garages beneath habitable rooms shall be separated from all habitable rooms above by not less than $^5/_8$-inch Type X gypsum board or equivalent. Door openings between a private garage and the dwelling unit shall be equipped with either solid wood doors, or solid or honeycomb core steel doors not less than $1^3/_8$ inches (34.9 mm) thick, or doors in compliance with Section 715.3.3. Openings from a private garage directly into a room used for sleeping purposes shall not be permitted.

2. Ducts in a private garage and ducts penetrating the walls or ceilings separating the dwelling unit from the garage shall be constructed of a minimum 0.019-inch (0.48 mm) sheet steel and shall have no openings into the garage.

3. A separation is not required between a Group R-3 and U carport provided the carport is entirely open on two or more sides and there are not enclosed areas above.

❖ As indicated in Item 1, when a private garage is attached to a residence, the adjacent areas (including attics) are to be separated to provide a minimum level of protection. This separation is to be constructed of at least 2-inch (12.3 mm) gypsum wallboard applied to the garage side. In a location where the garage is beneath any type of habitable space, the separation requirement is increased by requiring that $^5/_8$-inch (15.9 mm) Type X gypsum board or some other material that would provide an equivalent level of protection be used. Doors opening within the adjacent wall are required to be protected with a minimum 1 $^3/_8$-inch-thick (34 mm) solid core wood or 1 $^3/_8$-inch-thick (34 mm) solid or honeycomb steel door. Although the 1 $^3/_8$-inch-thick (34 mm) solid core wood and honeycomb steel doors are not listed as fire doors, they still provide adequate protection. Alternatively, doors with a 20-minute fire protection rating in accordance with Section 715.3.3 may be used. All door openings from the garage, whether solid core wood, honeycomb steel or 20-minute rated, are prohibited from opening directly into a bedroom or any other room used for sleeping purposes. It is important to note that this separation requirement, including the protection and the limitation on the openings, is only applicable to garages and not to carports.

With regards to Item 2, ducts in the walls and ceilings that do not penetrate into the garage are already protected by the gypsum wallboard; therefore, additional separation is not necessary. If the ducts do penetrate through the separation, then the specified protection must be provided.

In accordance with Item 3, a roofed structure open on two or more sides without any enclosed uses above poses no special hazard to the occupants of the dwelling. If a fire were to start under the carport roof, the smoke, hot gases and flames would be able to escape out the open sides, providing the structure with an adequate amount of protection. See the commentary to Section 406.1.3 for additional related discussion.

406.2 Parking garages.

406.2.1 Classification. Parking garages shall be classified as either open, as defined in Section 406.3, or enclosed and shall meet the appropriate criteria in Section 406.4. Also see Section 508 for special provisions for parking garages.

❖ Parking garages are considered to be storage occupancies (Group S-2). In addition to the provisions of Sections 406.2, 406.3 and 406.4, parking garages must also comply with the code provisions for the storage group.

The overall building fire loading in parking garages is

typically low because of the considerable amount of metal in vehicles, which absorbs heat and the average weight of combustibles per square foot being relatively low [approximately 2 pounds per square foot (9.8 kg/m² psf)]. Still, a vehicle fire may be quite extensive.

Parking garages are divided into one of two categories. Those parking garages that contain sufficient clear openings in their exterior walls that meet the requirements of Section 406.3.3.1 can be classified as open parking garages. All other parking garages must be considered as enclosed parking garages and must comply with Section 406.4. Special height and area provisions for parking garages located below other uses and occupancies are contained in Sections 508.2, 508.3, 508.4 and 508.7.

406.2.2 Clear height. The clear height of each floor level in vehicle and pedestrian traffic areas shall not be less than 7 feet (2134 mm). Vehicle and pedestrian areas accommodating van-accessible parking required by Section 1106.5 shall conform to ICC A117.1.

❖ Previously, some model building codes did not specify ceiling heights for nonoccupiable spaces such as parking garages. This section requires a clear height of 7 feet (2134 mm). This minimum height permits free and unobstructed egress around the vehicles and exit access areas and allows for smoke and hot gases from a fire to accumulate above the building occupants. Van-accessible parking areas must comply with ICC A117.1.

406.2.3 Guards. Guards shall be provided in accordance with Section 1012 at exterior and interior vertical openings on floor and roof areas where vehicles are parked or moved and where the vertical distance to the ground or surface directly below exceeds 30 inches (762 mm).

❖ For the same reasons as identified in the commentary to Section 1012, guards are required around all vertical openings in the floors and roofs of parking garages where the vertical distance between adjacent levels exceeds 30 inches (762 mm). Where required, the guards are to be installed in accordance with the provisions of Section 1012.

406.2.4 Vehicle barriers. Parking areas shall be provided with exterior or interior walls or vehicle barriers, except at pedestrian or vehicular accesses, designed in accordance with Section 1607.7. Vehicle barriers not less than 2 feet (607 mm) high shall be placed at the ends of drive lanes, and at the end of parking spaces where the difference in adjacent floor elevation is greater than 1 foot (305 mm).

❖ This section requires wheel guards, vehicle barriers, structural walls, etc., at specific locations where a 1-foot (305 mm) dropoff exists. Vehicle barriers are to be designed in accordance with Section 1607.7.3 and should be of noncombustible material in order to further control the combustible fuel load within parking garages.

406.2.5 Ramps. Vehicle ramps shall not serve as an exit element.

❖ Since the vehicular ramps of parking garages are open to all levels, they are directly exposed to the effects of smoke and hot gases. In addition, the vehicle ramps are often sloped at a rate greater than 1:12 (8-percent slope). These ramps, therefore, cannot be counted as part of the required exits from each level or tier. Certainly occupants can travel from their vehicles on sloped parking levels to access the required exit stairways; however, occupants cannot continuously travel down the vehicle ramps to reach the exit discharge.

406.2.6 Floor surface. Parking surfaces shall be of concrete or similar noncombustible and nonabsorbent materials.

Exception: Asphalt parking surfaces are permitted at ground level.

The area of floor used for parking of automobiles or other vehicles shall be sloped to facilitate the movement of liquids to a drain or toward the main vehicle entry doorway.

❖ To avoid acquiring a buildup of flammable liquids on the floor of parking garages, the floor finish is required to be noncombustible and nonabsorbent. Because of pollution concerns, however, garage floors are not automatically required to be drained into the building drainage system. This is consistent with requirements of the Environmental Protection Agency (EPA) for storm water drainage, as well as the philosophy of hazardous materials (i.e., localize spills and treat them). If floor drains are provided, however, they must be installed in accordance with the *International Plumbing Code®* (IPC®).

The exception recognizes that many states have allowed the use of asphalt paving surfaces at grade levels of parking garages with no record of fire hazards. This exception does not allow asphalt paving to be used on stories above or below grade. Although not specified, the floor must be positively sloped at some rate such as $1/_4$:12 (2-percent slope) to prevent the accumulation of any spilled flammable and combustible liquids and their vapors. This minimizes the risk of the vapors building up to a point where a fire condition could result.

406.2.7 Mixed separation. Parking garages shall be separated from other occupancies in accordance with Section 302.3.2.

❖ If a building or structure consists not only of parking garages but also other uses and occupancies, then the building is treated as any other mixed occupancy. The building designer must choose one of the options available in Section 302.3 to address the mixed use and occupancy issues. Keep in mind that Sections 508.2, 508.3, 508.4 and 508.7 contain special provisions for mixed separations when the parking garages are located beneath other groups.

406.2.8 Special hazards. Connection of a parking garage with any room in which there is a fuel-fired appliance shall be by means of a vestibule providing a two-doorway separation.

> **Exception:** A single door shall be allowed provided the sources of ignition in the appliance are at least 18 inches (457 mm) above the floor.

❖ As part of the special use and occupancy requirements for parking garages, all possible ignition sources must be controlled and isolated. Specifically, all heating equipment must be located in rooms that are separated from the main areas where the vehicles are parked. Doors connecting the heating equipment rooms and the main parking area must be done with a vestibule or air-lock arrangement such that one must pass through two doors prior to entering the other room. Again, this is done to minimize the possibility of any spilled flammable liquids and the resulting vapors from coming in contact with the ignition sources of the heating equipment.

An allowance is made if the heating equipment is located at least 18 inches (457 mm) above the floor of the separated room. In such a case, the vestibule/airlock arrangement with a double-door system is not required and can be done with just a single door. If this exception is used, care must be taken by the building and fire officials that these special stipulations and conditions are part of the certificate of occupancy.

406.2.9 Attached to rooms. Openings from a parking garage directly into a room used for sleeping purposes shall not be permitted.

❖ Just as a private garage cannot have openings into bedrooms (see Item 1 in Section 406.1.4), other parking garages are likewise limited. The risks of a fire quickly spreading into adjacent areas where occupants may be sleeping are too great.

406.3 Open parking garages.

406.3.1 Scope. Except where specific provisions are made in the following subsections, other requirements of this code shall apply.

❖ The code defines four types of garages: private, enclosed parking, repair garages and open parking. With the exception of private garages, all such structures are classified as Group S. Unlike other storage occupancies in which the fuel load is evenly distributed, garages represent a different fire hazard.

Because of the generally fast-burning upholstery and the gasoline or diesel fuel content, fires in individual vehicles can be quite extensive. Because of both the overall combustible loading within a parking garage, which yields a low combustible fuel load [approximately 2 pound per square foot (psf) (9.8 kg/m^2)], and the considerable amount of metal in vehicles, which absorbs heat, the overall fire hazards are low. Section 406.3 provides requirements that are unique to open parking garages, while Section 406.1 addresses private garages and Section 406.4 addresses enclosed parking garages.

Because of the permanently open exterior walls of open parking garages, which permit the dissipation of heated gases, special provisions are made for heights and areas of such structures.

Open parking garages are classified as Group S-2 and all of the provisions for this group are applicable except as modified herein.

406.3.2 Definitions. The following words and terms shall, for the purposes of this chapter and as used elsewhere in this code, have the meanings shown herein.

❖ Definitions of terms that are associated with the content of this section are contained herein. These definitions can help in the understanding and application of the code requirements. It is important to emphasize that these terms are not exclusively related to this section but are applicable everywhere the term is used in the code. The purpose for including these definitions within this section is to provide more convenient access to them without having to refer back to Chapter 2.

For convenience, these terms are also listed in Chapter 2 with a cross reference to this section.

The use and application of all defined terms, including those defined herein, are set forth in Section 201.

MECHANICAL-ACCESS OPEN PARKING GARAGES. Open parking garages employing parking machines, lifts, elevators or other mechanical devices for vehicles moving from and to street level and in which public occupancy is prohibited above the street level.

❖ These types of garages are constructed with most of the same attributes as other garages. They have numerous parking levels that house the motor vehicles; however, public access to these upper levels is not permitted. Employees take control of the vehicles at the street level and drive into some sort of a vertical mechanical device, which conveys them to the upper parking levels. These structures are still provided with the required openings of Section 406.3.3.1.

OPEN PARKING GARAGE. A structure or portion of a structure with the openings as described in Section 406.3.3.1 on two or more sides that is used for the parking or storage of private motor vehicles as described in Section 406.3.4.

❖ Open parking garages are defined as having uniformly distributed openings on no less than two sides totaling no less than 40 percent of the building perimeter. The aggregate area of the openings is to be a minimum of 20 percent of the total wall area of all perimeter walls (see Figure 406.3.3.1).

RAMP-ACCESS OPEN PARKING GARAGES. Open parking garages employing a series of continuously rising floors or a series of interconnecting ramps between floors permitting the movement of vehicles under their own power from and to the street level.

❖ These types of garages employ vehicular ramps or sloped tiers that connect all of the levels of the structure.

Whether the vehicles are self-parked or employee parked is immaterial.

406.3.3 Construction. Open parking garages shall be of Type I, II or IV construction. Open parking garages shall meet the design requirements of Chapter 16. For vehicle barriers, see Section 406.2.4.

❖ One of the basic premises of the provisions for open parking garages is that the overall fuel load is low and that the average fuel load per square foot is low. The construction of open parking garages, therefore, is restricted to Type II or IV, so that the combustible loading of cars is not exceeded by that of the structure.

406.3.3.1 Openings. For natural ventilation purposes, the exterior side of the structure shall have uniformly distributed openings on two or more sides. The area of such openings in exterior walls on a tier must be at least 20 percent of the total perimeter wall area of each tier. The aggregate length of the openings considered to be providing natural ventilation shall constitute a minimum of 40 percent of the perimeter of the tier. Interior walls shall be at least 20 percent open with uniformly distributed openings.

   Exception: Openings are not required to be distributed over 40 percent of the building perimeter where the required openings are uniformly distributed over two opposing sides of the building.

❖ Key attributes of an open parking garage, based on the fire records of such facilities, include the exterior wall openings and ventilation of the structure. This section requires that 40 percent of the building perimeter have openings that are uniformly distributed on no less than two sides of the structure. In addition to providing for adequate distribution, the section also requires that a minimum of 20 percent of the total perimeter wall area at each level must be open. The openings in the exterior wall must be free so that natural ventilation will occur without interior wall obstructions.

In instances where the open parking garage is provided with openings on two opposite sides of the structure, thereby providing cross ventilation, the 40-percent criterion does not apply; however, the openings must still meet the 20-percent criterion of the total perimeter wall area.

In every case, interior walls are to have distributed openings totaling 20 percent of the wall area to allow ventilation of all spaces (see Figure 406.3.3.1).

406.3.4 Uses. Mixed uses shall be allowed in the same building as an open parking garage subject to the provisions of Sections 302.3, 402.7.1, 406.3.13, 508.3, 508.4 and 508.7.

❖ Because of the increased height and area limitations provided for open parking garages, these structures are limited to just private motor vehicles. Facilities used for the parking of trucks or buses must be classified as an enclosed parking garage even if the opening requirements of Section 406.3.3.1 are met. Open parking garages or buildings that contain open parking garages

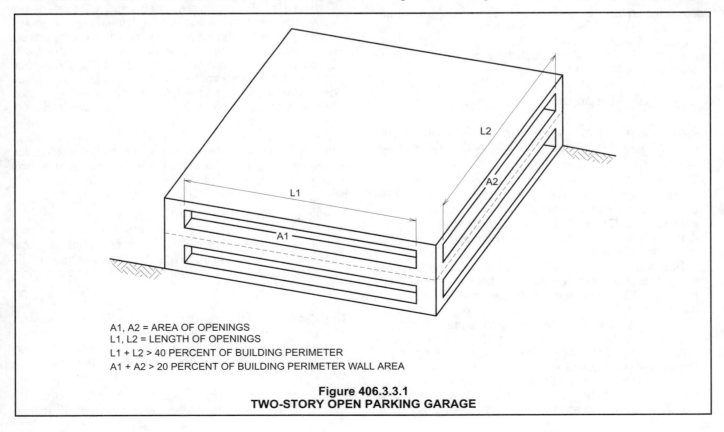

A1, A2 = AREA OF OPENINGS
L1, L2 = LENGTH OF OPENINGS
L1 + L2 > 40 PERCENT OF BUILDING PERIMETER
A1 + A2 > 20 PERCENT OF BUILDING PERIMETER WALL AREA

Figure 406.3.3.1
TWO-STORY OPEN PARKING GARAGE

may contain other occupancies as permitted by the other general provisions of the code.

406.3.5 Area and height. Area and height of open parking garages shall be limited as set forth in Chapter 5 for Group S-2 occupancies and as further provided for in Section 302.3.

❖ When an open parking garage is located within a building containing other occupancies, the code requires that the height and area limitations for the open parking garage be determined by using the normal provisions found in Chapter 5 for a Group S-2 occupancy. This essentially means that the garage would receive the same allowable areas of an enclosed parking garage or a storage building that was classified in the same occupancy.

406.3.5.1 Single use. When the open parking garage is used exclusively for the parking or storage of private motor vehicles, with no other uses in the building, the area and height shall be permitted to comply with Table 406.3.5, along with increases allowed by Section 406.3.6.

> **Exception:** The grade-level tier is permitted to contain an office, waiting and toilet rooms having a total combined area of not more than 1,000 square feet (93 m²). Such area need not be separated from the open parking garage.

In open parking garages having a spiral or sloping floor, the horizontal projection of the structure at any cross section shall not exceed the allowable area per parking tier. In the case of an open parking garage having a continuous spiral floor, each 9 feet 6 inches (2896 mm) of height, or portion thereof, shall be considered a tier.

The clear height of a parking tier shall not be less than 7 feet (2134 mm), except that a lower clear height is permitted in mechanical-access open parking garages where approved by the building official.

❖ When an open parking garage is located in a building that is only used as a parking garage, the type of hazard that is created by the occupancy is greatly reduced. Because of the openness requirements in Section 406.3.3.1 and the dissipation of heated gases as well as

the overall low fuel load, the height and area limitations for single-use open parking garages are not the same as for other buildings of Group S-2. The height requirements are based on actual fire experience and full-scale fire tests conducted in the United States, Great Britain, Japan and Switzerland. For these single-use buildings that comply with the other provisions of this section, Table 406.3.5 replaces Table 503 for the determination of heights and areas. Although these open parking garages are considered as being a single occupancy, the exception permits a limited amount of support-type spaces to be included in the structure.

TABLE 406.3.5. See below.

❖ Table 406.3.5 is used in the same manner as Table 503. For a given construction type, the height and area limitations can be determined from the table. For example, an open parking garage of Type IIB construction may be a maximum of eight tiers in height for a ramp access garage and the area would then be limited to 50,000 square feet (4545 m²).

The height restrictions in the table are measured as defined in Section 202. In this instance, parking may be permitted on the highest level of the structure, which also constitutes the roof of the structure. The areas provided are per tier.

406.3.6 Area and height increases. The allowable area and height of open parking garages shall be increased in accordance with the provisions of this section. Garages with sides open on three-fourths of the building perimeter are permitted to be increased by 25 percent in area and one tier in height. Garages with sides open around the entire building perimeter are permitted to be increased 50 percent in area and one tier in height. For a side to be considered open under the above provisions, the total area of openings along the side shall not be less than 50 percent of the interior area of the side at each tier, and such openings shall be equally distributed along the length of the tier.

Allowable tier areas in Table 406.3.5 shall be increased for open parking garages constructed to heights less than the table maximum. The gross tier area of the garage shall not exceed that permitted for the higher structure. At least three sides of each

TABLE 406.3.5
OPEN PARKING GARAGES AREA AND HEIGHT

TYPE OF CONSTRUCTION	AREA PER TIER (square feet)	HEIGHT (in tiers)		
		Ramp access	Mechanical access	
			Automatic sprinkler system	
			No	Yes
IA	Unlimited	Unlimited	Unlimited	Unlimited
IB	Unlimited	12 tiers	12 tiers	18 tiers
IIA	50,000	10 tiers	10 tiers	15 tiers
IIB	50,000	8 tiers	8 tiers	12 tiers
IV	50,000	4 tiers	4 tiers	4 tiers

For SI: 1 square foot = 0.0929 m².

such larger tier shall have continuous horizontal openings not less than 30 inches (762 mm) in clear height extending for at least 80 percent of the length of the sides, and no part of such larger tier shall be more than 200 feet (60 960 mm) horizontally from such an opening. In addition, each such opening shall face a street or yard accessible to a street with a width of at least 30 feet (9144 mm) for the full length of the opening, and standpipes shall be provided in each such tier.

Open parking garages of Type IB and II construction, with all sides open, shall be unlimited in allowable area where the height does not exceed 75 feet (22 860 mm). For a side to be considered open, the total area of openings along the side shall not be less than 50 percent of the interior area of the side at each tier, and such openings shall be equally distributed along the length of the tier. All portions of tiers shall be within 200 feet (60 960 mm) horizontally from such openings.

❖ The area modifications permitted by Section 406.3.6 for open building perimeter similar to Section 506.2 are also applicable to the area, therefore, restrictions in Table 406.3.5. The allowable area for a four-story open parking garage of Type IIB construction, which has open sides on 75 percent of the building's perimeter, may be increased by 25 percent of 50,000 square feet (4645 m²) [or 12,500 square feet (1161 m²)] yielding a maximum permitted area per tier of 62,500 square feet (5806 m²).

406.3.7 Location on property. Exterior walls and openings in exterior walls shall comply with Tables 601 and 602. The distance from an adjacent lot line shall be determined in accordance with Table 602 and Section 704.

❖ The required fire separation is intended to provide sufficient open space adjacent to the exterior wall openings to allow for free ventilation and to reduce the probability for fire spread to adjacent structures. The exterior wall openings in an open parking garage are not subject to the opening limitations with a fire separation distance greater than 10 feet (3048 mm) (see Table 704.8, Note c).

406.3.8 Means of egress. Where persons other than parking attendants are permitted, open parking garages shall meet the means of egress requirements of Chapter 10. Where no persons other than parking attendants are permitted, there shall not be less than two 36-inch-wide (914 mm) exit stairways. Lifts shall be permitted to be installed for use of employees only, provided they are completely enclosed by noncombustible materials.

❖ Typical means of egress elements are required for open parking garages based on a design occupant load of one person per 200 square feet (19 m²) of gross floor area (see Table 1004.1.2). For mechanical-access open parking garages or for attendant-only ramp-access parking garages, the code would permit two exit stairs, each 36 inches (914 mm) wide.

406.3.9 Standpipes. Standpipes shall be installed where required by the provisions of Chapter 9.

❖ Depending on the height of an open parking garage, Section 905.3.1 may require a standpipe system. The standpipe system would aid the fire department in on-site fire-fighting operations, since Table 406.3.5 permits buildings that may not allow for conventional hoses and apparatus to be carried from ground operations to the fire floor.

406.3.10 Sprinkler systems. Where required by other provisions or this code, automatic sprinkler systems and standpipes shall be installed in accordance with the provisions of Chapter 9.

❖ Although sprinkler systems are not mandated by Section 903.2 for open parking garages, sprinkler systems may be required for other occupancies based on the special provisions of Sections 508.3 and 508.7. As such, any sprinkler systems required for those other occupancies may have to be extended into the open parking garage levels to achieve the required coverage.

406.3.11 Enclosure of vertical openings. Enclosure shall not be required for vertical openings except as specified in Section 406.3.8.

❖ Based on the use of open parking garages requiring convenient and frequent access between floor levels, enclosure of most vertical openings is not required or wanted due to the opening requirements of Section 406.3.3.1. This would include the interior exit stairways as permitted by Exception 5 to Section 1019.1.

406.3.12 Ventilation. Ventilation, other than the percentage of openings specified in Section 406.3.3.1, shall not be required.

❖ Mechanical ventilation systems are not required in a structure that is inherently open to the exterior atmosphere.

406.3.13 Prohibitions. The following uses and alterations are not permitted:

1. Vehicle repair work.

2. Parking of buses, trucks and similar vehicles.

3. Partial or complete closing of required openings in exterior walls by tarpaulins or any other means.

4. Dispensing of fuel.

❖ This section reinforces the concepts and requirements of Section 406.3.3.1 and the definition of "Open parking garage." Any other use or alteration of an open parking garage is specifically prohibited.

406.4 Enclosed parking garages.

406.4.1 Heights and areas. Enclosed vehicle parking garages and portions thereof that do not meet the definition of open parking garages shall be limited to the allowable heights and areas specified in Table 503. Roof parking is permitted.

❖ Enclosed parking garages are considered to be storage occupancies. In addition to the provisions of Sections 406.2 and 406.4, enclosed parking garages must also comply with the code provisions for Group S-2 low-hazard storage occupancies.

   The overall building fire load in enclosed parking garages is typically low because of the considerable amount of metal in vehicles that absorbs heat and the average weight of combustibles per square foot being relatively low [approximately 2 psf (9.8 kg/m²)]. Still, a vehicle fire may be quite extensive.

   Special height and area limitations are not provided for enclosed parking garages, which is similar to what is done in Table 406.3.5 for open parking garages. The typical allowable building heights and areas of Table 503 are to be used for these Group S-2 structures. It should be noted that Sections 508.2, 508.3 and 508.4 contain special provisions for enclosed parking garages located beneath other groups and occupancies. See those sections for further discussion of these special building arrangements.

406.4.2 Ventilation. A mechanical ventilation system shall be provided in accordance with the *International Mechanical Code*.

❖ Enclosed parking garages are required to be ventilated in accordance with the provisions of the IMC. If motor vehicles will be operated for a period of time exceeding 10 seconds, the ventilation return air for the enclosed parking garage ventilation system must be exhausted (see Section 404 of the IMC).

406.5 Motor fuel-dispensing facilities.

406.5.1 Construction. Motor fuel-dispensing facilities shall be constructed in accordance with the *International Fire Code* and this section.

❖ Motor fuel-dispensing facilities are those areas where motor fuels are stored, sold or otherwise dispensed to vehicles. These buildings and structures are classified as Group M. Motor fuel-dispensing facilities must be constructed in accordance with the applicable requirements of Chapter 22 of the IFC. Specific details are provided for the special hazards relative to fuel-dispensing systems and areas.

406.5.2 Canopies. Canopies under which fuels are dispensed shall have a clear, unobstructed height of not less than 13 feet 6 inches (4115 mm) to the lowest projecting element in the vehicle drive-through area. Canopies and their supports over pumps shall be of noncombustible materials, fire-retardant-treated wood complying with Chapter 23, wood of Type IV sizes or of construction providing 1-hour fire resistance. Combustible materials used in or on a canopy shall comply with one of the following:

1. Shielded from the pumps by a noncombustible element of the canopy, or wood of Type IV sizes;

2. Plastics covered by aluminum facing having a minimum thickness of 0.010 inch (0.30 mm) or corrosion-resistant steel having a minimum base metal thickness of 0.016 inch (0.41 mm). The plastic shall have a flame spread index of 25 or less and a smoke-developed index of 450 or less when tested in the form intended for use in accordance with ASTM E 84 and a self-ignition temperature of 650°F (343°C) or greater when tested in accordance with ASTM D 1929; or

3. Panels constructed of light-transmitting plastic materials shall be permitted to be installed in canopies erected over motor vehicle fuel-dispensing station fuel dispensers, provided the panels are located at least 10 feet (3048 mm) from any building on the same property and face yards or streets not less than 40 feet (12 192 mm) in width on the other sides. The aggregate areas of plastics shall not exceed 1,000 square feet (93 m²). The maximum area of any individual panel shall not exceed 100 square feet (9.3 m²).

❖ Almost all motor fuel-dispensing facilities are provided with a flat canopy-style roof structure that is column supported. Obviously these structures must be designed to meet the applicable wind and snow loads of Chapter 16 for life safety purposes. For fire safety purposes, the canopies must pose little fire risk and be located above any immediate fire hazards from the fuel-dispensing areas. Limitations are provided for all combustible materials used in the canopy construction.

406.6 Repair garages.

406.6.1 General. Repair garages shall be constructed in accordance with the *International Fire Code* and this section. This occupancy shall not include motor fuel-dispensing facilities, as regulated in Section 406.5.

❖ Repair garages are those in which provisions are made for the care, repair and painting of vehicles. Repair garages have an inherently higher fire hazard represented by numerous vehicles in some state of repair, with possible spray painting. As such, these buildings or structures are classified as Group S-1. Repair garages must also meet the applicable requirements of Chapter 22 of the IFC and Section 416 if there are paint-spraying operations.

   All motor vehicle work that involves repairs or reconstruction of damaged or nonfunctioning vehicles is considered a higher hazard. These types of facilities provide a variety of services, including tune-ups, oil changes, engine or transmission overhauls and body work, and are all considered as repair garages. Additional protection and detection systems are required to match the increased hazards.

406.6.2 Mixed uses. Mixed uses shall be allowed in the same building as a repair garage subject to the provisions of Section 302.3.

❖ Repair garages usually have integral office and administration areas or are part of an automobile dealership. To minimize the hazards and functions housed in the repair areas, the code reminds the user that these spaces are considered as different occupancies and must therefore comply with the general code requirements for mixed uses. The actual requirements will depend on the size of the spaces and how the designer chooses to deal with the different uses.

406.6.3 Ventilation. Repair garages shall be mechanically ventilated in accordance with the *International Mechanical Code*. The ventilation system shall be controlled at the entrance to the garage.

❖ Repair garages are required to be ventilated in accordance with the provisions of the IMC. If motor vehicles will be operated for a period of time exceeding 10 seconds, the ventilation return air for the repair garage ventilation system must be exhausted (see Sections 403 and 404 of the IMC).

406.6.4 Floor surface. Repair garage floors shall be of concrete or similar noncombustible and nonabsorbent materials.

Exception: Slip-resistant, nonabsorbent, interior floor finishes having a critical radiant flux not more than 0.45 W/cm2, as determined by NFPA 253, shall be permitted.

❖ To avoid a buildup of flammable liquids on the floor of repair garages, the floor finish is in general required to be noncombustible and nonabsorbent. Because of pollution concerns, however, garage floors are not automatically required to be drained into the building drainage system. This is consistent with requirements of the EPA for storm water drainage, as well as the philosophy of hazardous materials (i.e., localize spills and treat them). If floor drains are provided, however, they must be installed in accordance with the IPC.

406.6.5 Heating equipment. Heating equipment shall be installed in accordance with the *International Mechanical Code*.

❖ As part of the special use and occupancy requirements for repair garages, all possible ignition sources must be controlled and isolated. Specifically, all heating equipment must be installed in accordance with the IMC, which contains provisions that address items such as the elevation of the appliance above the floor and the use of overhead unit heaters.

[F] 406.6.6 Gas detection system. Repair garages used for repair of vehicles fueled by nonodorized gases, such as hydrogen and nonodorized LNG, shall be provided with an approved flammable gas-detection system.

❖ Specially fueled motor vehicles pose a higher level of risk if they are powered by hydrogen, nonodorized liquified natural gas (LNG) or similar types of fuels. Any spills of these fuel cells or tanks may not be recognized by repairmen. Repair garages servicing these vehicles must be provided with a detection system that will identify the presence of these flammable gases.

[F] 406.6.6.1 System design. The flammable gas-detection system shall be calibrated to the types of fuels or gases used by vehicles to be repaired. The gas detection system shall be designed to activate when the level of flammable gas exceeds 25 percent of the lower explosive limit. Gas detection shall also be provided in lubrication or chassis repair pits of garages used for repairing nonodorized LNG-fueled vehicles.

❖ The required flammable gas-detection system must be able to alert the occupants to the presence of the nonodorized fuels prior to reaching concentration levels that pose an explosive fire risk. Complete details and installation information for these systems must be submitted as part of the permit process to verify compliance with the specified calibration requirements and detection levels.

[F] 406.6.6.2 Operation. Activation of the gas detection system shall result in all of the following:

1. Initiation of distinct audible and visual alarm signals in the repair garage.

2. Deactivation of all heating systems located in the repair garage.

3. Activation of the mechanical ventilation system, where the system is interlocked with gas detection.

❖ Once the required flammable gas detection system identifies a gas level exceeding 25 percent of the lower explosive limit (LEL), it must do three things: alert the occupants of the emergency, cut power to all possible ignition sources and dilute the concentration level with additional ventilation air.

[F] 406.6.6.3 Failure of the gas detection system. Failure of the gas detection system shall result in the deactivation of the heating system, activation of the mechanical ventilation system when the system is interlocked with the gas detection system and cause a trouble signal to sound in an approved location.

❖ The required flammable gas-detection system must include a default mode of operation. If the system malfunctions or loses power, it must in effect do the same operational requirements referenced Section 406.6.6.2. This redundancy will lessen the likelihood of an explosive fire risk for the specialized motor fuels.

SECTION 407
GROUP I-2

407.1 General. Occupancies in Group I-2 shall comply with the provisions of this section and other applicable provisions of this code.

❖ The provisions of Section 407 are largely based on a report by what was the Council of American Building Official's (CABO) Board for the Coordination of Model

Codes (BCMC), entitled "Health Care Facilities." One of the purposes of that report was to obtain national uniformity; as such, the provisions contained in Section 407 reflect previous requirements of the other former model codes, full-scale fire tests and historical fire experience. Features unique to health care facilities, such as a low fuel load, the presence of staff personnel, operating practices and procedures and the regulatory process, were also included in developing the provisions. One of the concerns addressed by this section is the need to provide an acceptable level of protection without interfering with the operation of the facility.

Underlying the protection requirements for Group I-2 is a "defend-in-place" philosophy based on the difficulties associated with evacuating such facilities.

The first level of protection established is the individual room (typically a patient sleeping room). Horizontal evacuation to an adjacent smoke compartment provides the second level of protection. If necessary, the third level of protection involves evacuation of the floor or entire building.

This section has been created not only because of considerations unique to health care facilities, but also to facilitate the consideration of the entire package of protection features. The basic protection features include early detection, fire containment, horizontal evacuation and fire extinguishment. Additionally, Chapter 4 of the IFC contains criteria for emergency planning and preparedness.

As previously noted, the provisions of this section are based in part on a relatively low fuel load. Certain areas within Group I-2, however, will have a higher concentration of combustibles. These areas, as well as other areas that represent a specific hazard within Group I-2, are addressed in Section 302.1.1 as incidental use areas. Because of the nature of the activity, the incidental use area is required to be protected with an automatic fire-extinguishing system, separated with a fire barrier assembly or both (see Table 302.1.1).

These provisions apply to all occupancies classified as Group I-2. The requirements of Section 407 are intended to be additional special use and occupancy provisions for facilities that provide care for more than five persons who are not capable of self-preservation. Other requirements for Group I-2 occupancies are provided throughout the code.

407.2 Corridors. Corridors in occupancies in Group I-2 shall be continuous to the exits and separated from other areas in accordance with Section 407.3 except spaces conforming to Sections 407.2.1 through 407.2.4.

❖ Since the occupants of areas classified as Group I-2 are incapable of self-preservation, the means of egress serving such areas must facilitate the evacuation of disabled persons. Accordingly, exit access corridors serving Group I-2 occupancies are required to be continuous to the exits and, except as provided for in Sections 407.2.1 through 407.2.4, separated from other use ar-

eas. As such, corridors in Group I-2 occupancies are intended to provide a protected path of travel that once entered leads directly and continuously to the exits (see also commentary, Sections 1013.2.2 and 1016.5). Because of various operational requirements of occupancies classified as Group I-2, it is desirable to have certain areas open to exit access corridors. Recognizing these needs and the fact that such areas are equipped throughout with an automatic sprinkler system, the code permits certain areas to be open to corridors provided they are properly arranged and equipped with compensating protection features (e.g., automatic fire detection). Sections 407.2.1 through 407.2.4 identify those areas and the protection features required to maintain the safety of the corridor.

In addition to practical reasons, such as in the case of nurses' stations, it is also desirable to designate certain areas being open to the corridor for treatment purposes. In many instances, the patient is encouraged to leave the patient room for both physical as well as social benefit. So that patients can easily be encouraged to socialize, the areas need to be open to the corridor. If the area were separated with only a door opening, many patients would not be willing to enter the room and benefit from group activities and social interaction.

Additionally, such waiting areas serve a useful purpose during a fire emergency. If evacuation becomes necessary, patients usually will be relocated to an adjacent smoke compartment in which the areas open to the corridor will serve as an area of refuge. Section 407.4.1 contains provisions for the net area that must be available for the relocation of patients. Waiting areas provide a space in which patients can be relocated such that treatment can continue even during a fire emergency.

407.2.1 Spaces of unlimited area. Waiting areas and similar spaces constructed as required for corridors shall be permitted to be open to a corridor, only where all of the following criteria are met:

1. The spaces are not occupied for patient sleeping units, treatment rooms, hazardous or incidental use areas as defined in Section 302.1.1.

2. The open space is protected by an automatic fire detection system installed in accordance with Section 907.

3. The corridors onto which the spaces open, in the same smoke compartment, are protected by an automatic fire detection system installed in accordance with Section 907, or the smoke compartment in which the spaces are located is equipped throughout with quick-response sprinklers in accordance with Section 903.3.2.

4. The space is arranged so as not to obstruct access to the required exits.

❖ In the design of health care facilities, it is desirable to have small waiting areas and similar spaces open to the corridor. Such seating areas or gathering spaces provide residents and others a place to visit outside of the patient rooms. This provides physical and social benefits to the residents by encouraging them to leave their room and socialize with others. When approved by the building official, these areas also provide a designated area for smoking, which otherwise is prohibited by Section 310.2 of the IFC. Designated smoking areas provide better supervision of smoking activities and a "safer" alternative to occupants who otherwise might attempt to smoke in patient rooms. For these reasons and in consideration of the additional protection features specified, small waiting areas and similar spaces are permitted to be open to corridors in Group I-2 occupancies within each smoke compartment.

Other than patient sleeping rooms, treatment rooms, hazardous or incidental use areas and spaces of unlimited area, such as offices or dining rooms, are permitted to be open to exit access corridors (see Figure 407.2.1).

To reduce the likelihood that a fire within such an open space could develop beyond the incipient stage, thereby jeopardizing the integrity of the corridor, the area is to be equipped with an automatic fire detection system (see Section 907). Whereas the areas also often serve as designated smoking areas, arrangements should be made to minimize the potential for unwanted alarms caused by activating the fire detection system. The detectors are to be located so as to provide the appropriate coverage to the space.

Due to the unlimited size of the space and consequently the potential for a significant number of people in the area, an automatic fire detection system is also required in all corridors open to the spaces within the

smoke compartment or the smoke compartment in which such spaces are located must be protected throughout with quick-response sprinklers.

Waiting areas and similar spaces open to corridors must be arranged so as to not obstruct access to exits. There must be a clear path of travel, of the minimum required width for the corridor, which should be maintained at all times. As such, furniture layouts must be arranged in a manner that obstruction of the exits does not occur under normal use.

407.2.2 Nurses' stations. Spaces for doctors' and nurses' charting, communications and related clerical areas shall be permitted to be open to the corridor, when such spaces are constructed as required for corridors.

❖ Nurses' stations need to be located to provide quick access to patients and to permit visual or audible monitoring of the patient rooms. For these and other practical reasons, nurses' stations are permitted to be open to the corridor provided the walls surrounding the area are constructed as required for corridor walls (see Figure 407.2.2). Except as required in Section 407.6, the additional protection of automatic fire detection is not required because of the reduced risk to life safety and attendance by the staff (see Section 407.6). This section is not intended to apply to nurses' offices, administrative supply storage areas, drug distribution stations or similar areas with a higher fuel load and from which continual observation of the nursing unit is not essential.

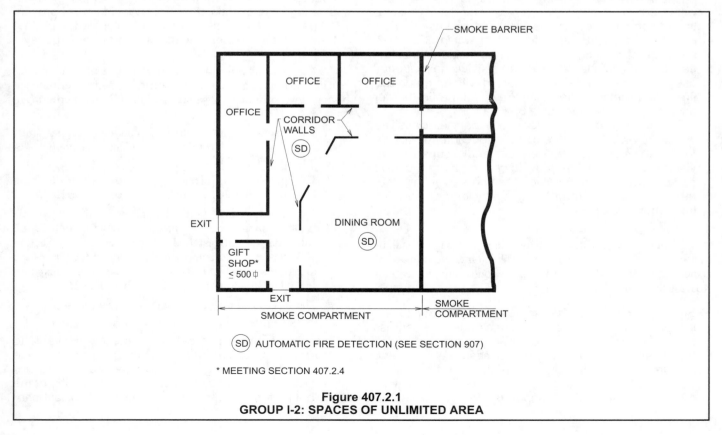

SD AUTOMATIC FIRE DETECTION (SEE SECTION 907)

* MEETING SECTION 407.2.4

Figure 407.2.1
GROUP I-2: SPACES OF UNLIMITED AREA

407.2.3 Mental health treatment areas. Areas wherein mental health patients who are not capable of self-preservation are housed, or group meeting or multipurpose therapeutic spaces other than incidental use areas as defined in Section 302.1.1, under continuous supervision by facility staff, shall be permitted to be open to the corridor, where the following criteria are met:

1. Each area does not exceed 1,500 square feet (140 m²).

2. The area is located to permit supervision by the facility staff.

3. The area is arranged so as not to obstruct any access to the required exits.

4. The area is equipped with an automatic fire detection system installed in accordance with Section 907.2.

5. Not more than one such space is permitted in any one smoke compartment.

6. The walls and ceilings of the space are constructed as required for corridors.

❖ To encourage residents of mental health facilities to participate in group meetings and therapeutic activities, treatment areas may be open to the corridor, provided that the area does not exceed 1,500 square feet (139 m²) and only one such area is provided within a smoke compartment. Other protection features that must be provided to permit the area to be open include: staff supervision; unobstructed access to exits; automatic fire detection in accordance with Section 907.2 within the area and the surrounding walls and ceilings constructed

as required for corridors. These provisions are based on the increased mobility of the occupants and direct supervision by staff. This section applies to all areas of a building classified as Group I-2, including smoke compartments that contain patient sleeping rooms that house mental health patients who are not physically capable of self-preservation.

As mentioned in the commentary to Section 407.2, one of the primary reasons for permitting areas to be open to the corridor is to encourage and facilitate participation by the residents in group activities. If the space is not used for such treatment or therapeutic purposes, therefore, the need to have the area open to the corridor is reduced and the section is not applicable.

407.2.4 Gift shops. Gift shops less than 500 square feet (46.5 m²) in area shall be permitted to be open to the corridor provided the gift shop and storage areas are fully sprinklered and storage areas are protected in accordance with Section 302.1.1.

❖ Gift shops are not permitted to be open to a corridor in occupancies classified in Group I-2 unless they are less than 500 square feet (46 m²) in area and protected in accordance with Section 302.1.1. Although Table 302.1.1 does not contain a specific entry for gift shops in Group I-2 facilities, the intent of this section is to provide some level of separation for these incidental use areas. A comparable level of protection would be the 1-hour separation with a fire barrier specified for storage rooms.

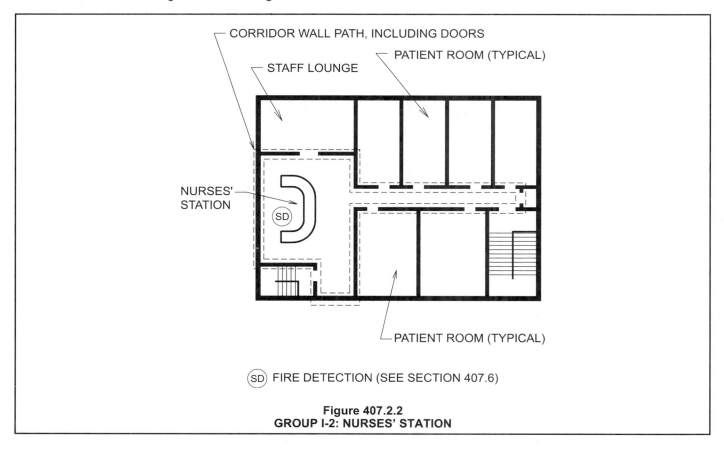

CORRIDOR WALL PATH, INCLUDING DOORS

PATIENT ROOM (TYPICAL)

STAFF LOUNGE

NURSES' STATION

SD

PATIENT ROOM (TYPICAL)

SD FIRE DETECTION (SEE SECTION 407.6)

**Figure 407.2.2
GROUP I-2: NURSES' STATION**

407.3 Corridor walls. Corridor walls shall be constructed as smoke partitions.

❖ Corridor walls are required to form a barrier to limit smoke transfer, thus providing protection for exit access routes, to separate areas that contain combustible materials that could produce smoke and to provide adequate separation of the patient sleeping rooms. This section essentially provides a reference to the requirements of Section 710 for smoke partitions. In a building protected throughout with an automatic sprinkler system, the probability that a fire will develop that is life threatening to persons outside the room of origin is reduced; therefore, corridor walls need only be able to resist the passage of smoke.

The corridor wall is not required to have a fire-resistance rating, but the wall is to terminate at the underside of the floor or roof deck above or to the underside of the ceiling membrane if it is capable of resisting the passage of smoke. The walls must be of materials consistent with the building's type of construction classification.

Note that the provisions of Section 302.1.1 must be considered in determining the fire-resistance rating of any corridor walls that also enclose incidental use areas (see Figure 407.3).

407.3.1 Corridor doors. Corridor doors, other than those in a wall required to be rated by Section 302.1.1 or for the enclosure of a vertical opening or an exit, shall not have a required fire pro-

tection rating and shall not be required to be equipped with self-closing or automatic-closing devices, but shall provide an effective barrier to limit the transfer of smoke and shall be equipped with positive latching. Roller latches are not permitted. Other doors shall conform to Section 715.3.

❖ As with corridor wall construction (see commentary, Section 407.3), the door assemblies need only be capable of resisting the passage of smoke. Corridor doors in occupancies classified as Group I-2 are not required to be self-closing or automatic-closing except for doors serving incidental use areas as required in Section 302.1.1, doors in walls that enclose a vertical opening and exit doors. This provision is primarily intended to apply to patient room corridor doors. Doors in walls that separate incidental use areas in accordance with Section 302.1.1.1 are required to be self-closing to minimize the possibility that corridors will be affected by a fire originating in those adjacent areas. Self-closing or automatic-closing doors in enclosures of vertical openings are still required to avoid a breech in vertical opening protection that would enable rapid fire spread from story to story and for adequate protection of the exits. It is important to recognize that this section contains a specific set of requirements for the corridor doors in this occupancy and that the provisions of this section must be followed versus the provisions of Section 710.5 (see code and commentary, Section 102.1). Full-scale fire tests and historical experience indicate that the automatic

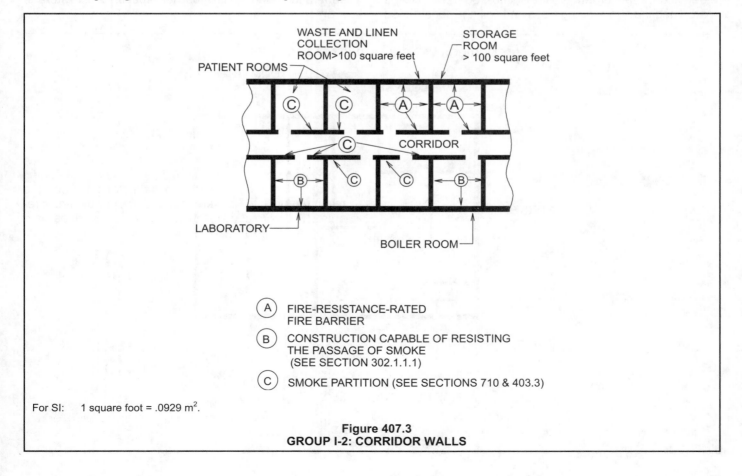

For SI: 1 square foot = .0929 m².

Figure 407.3
GROUP I-2: CORRIDOR WALLS

sprinkler system will provide adequate protection to persons outside of the room of origin. It is also expected that staff will be adequately trained and will close the doors to patient sleeping rooms. The presence of the automatic sprinkler system, particularly one equipped with quick-response heads, increases the probability that staff will be able to close the door to the room of origin and thereby limit the size of the fire. Roller latches are not permitted on these doors, as they are not regarded as providing positive latching that will allow the door to act as a reliable smoke barrier.

407.3.2 Locking devices. Locking devices that restrict access to the patient room from the corridor, and that are operable only by staff from the corridor side, shall not restrict the means of egress from the patient room except for patient rooms in mental health facilities.

❖ Locking devices are permitted on patient room doors, provided egress from the patient room is not restricted. Such locking devices enable residents in long-term care facilities to secure their rooms and staff to secure unoccupied rooms. In such cases, it is expected that the staff still has access, via key or other opening device.

The need to lock corridor doors to restrain patients in mental health facilities, however, is recognized. When such locking is necessary, provisions must be made for continuous supervision and prompt release. Additional guidance on locking arrangements can be found in Section 408.4, which applies to occupancies in Group I-3. A decision needs to be made to determine if the building more closely represents an occupancy in Group I-2 or I-3. In occupancies in Group I-3, it is generally assumed that the occupants are capable of self-preservation once the doors are unlocked. This may not be the case in some mental health facilities and, therefore, the provision applicable to occupancies in Group I-2 may be more appropriate.

Similar protection may also be necessary in certain portions of other occupancies in Group I-2 where the movement of the residents must be controlled (i.e., patients with Alzheimer's disease).

407.4 Smoke barriers. Smoke barriers shall be provided to subdivide every story used by patients for sleeping or treatment and to divide other stories with an occupant load of 50 or more persons, into at least two smoke compartments. Such stories shall be divided into smoke compartments with an area of not more than 22,500 square feet (2092 m²) and the travel distance from any point in a smoke compartment to a smoke barrier door shall not exceed 200 feet (60 960 mm). The smoke barrier shall be in accordance with Section 709.

❖ One of the basic premises of this section is that vertical evacuation of occupancies in Group I-2 is extremely difficult. It is essential, therefore, to provide an area of refuge on every story used by patients for sleeping or treatment. The area of refuge is intended to provide a protected area to which patients may be relocated if it becomes necessary to evacuate them. For this reason, each story used by patients for sleeping or treatment

and all other stories with an occupant load of 50 or more must be divided into at least two smoke compartments with an area of no more than 22,500 square feet (2092 m²) (see the definition of "Smoke compartment" in Section 702). The travel distance from any point in a smoke compartment to a smoke barrier door is limited to 200 feet (60 960 mm) (see Figure 407.4).

Smoke barriers used to divide one smoke compartment from the next are to be constructed in accordance with Section 709. The smoke barrier is different from that required for corridor walls in Section 407.3. Among other things, a smoke barrier is required to have a fire-resistance rating of 1 hour and, except as otherwise provided for in the exception to Section 709.4, provide a continuous separation through all concealed spaces to resist the passage of fire and smoke. The maximum area of the smoke compartment provides two protection features. First, the number of persons that are exposed to a fire will be restricted. Typical designs indicate that the number of persons exposed will usually be less than 100 and, more likely, 50 or less. Second, the overall dimensions of the smoke compartment must result in a maximum travel distance to the smoke barrier door of 200 feet (60 960 mm) for all design configurations.

407.4.1 Refuge area. At least 30 net square feet (2.8 m²) per patient shall be provided within the aggregate area of corridors, patient rooms, treatment rooms, lounge or dining areas and other low-hazard areas on each side of each smoke barrier. On floors not housing patients confined to a bed or litter, at least 6 net square feet (0.56 m²) per occupant shall be provided on each side of each smoke barrier for the total number of occupants in adjoining smoke compartments.

❖ To provide adequate space for occupants who might be relocated to an adjacent smoke compartment, minimum net areas per occupant are required. In determining the area available, it should be noted that the facility may still need to function as a medical care facility even while the patients are relocated. On floors housing nonambulatory patients (i.e., patient sleeping floors), a minimum of 30 net square feet (2.8 m²) per patient is required to be provided in low-hazard areas, such as waiting areas, corridors or patient rooms. This is to permit the relocation of the patient on a bed or litter to an area of safety on the same floor. The 30 net square feet (2.8 m²) is intended only to apply to the number of patients and not the calculated occupant load as determined using Table 1004.1.2.

For floors classified as Group I-2, which are not used for patient sleeping rooms (i.e., treatment floors), the net area required is 6 square feet (0.56 m²) per occupant. In this instance, the area required is based on the calculated occupant load as determined by Table 1004.1.2. The 6-square-foot (0.56 m²) minimum is based on an assumption that some occupants will be using walkers, wheelchairs and possibly a few litters for patients who are being transported through the area at the time of the fire.

FIGURE 407.4   SPECIAL DETAILED REQUIREMENTS BASED ON USE AND OCCUPANCY

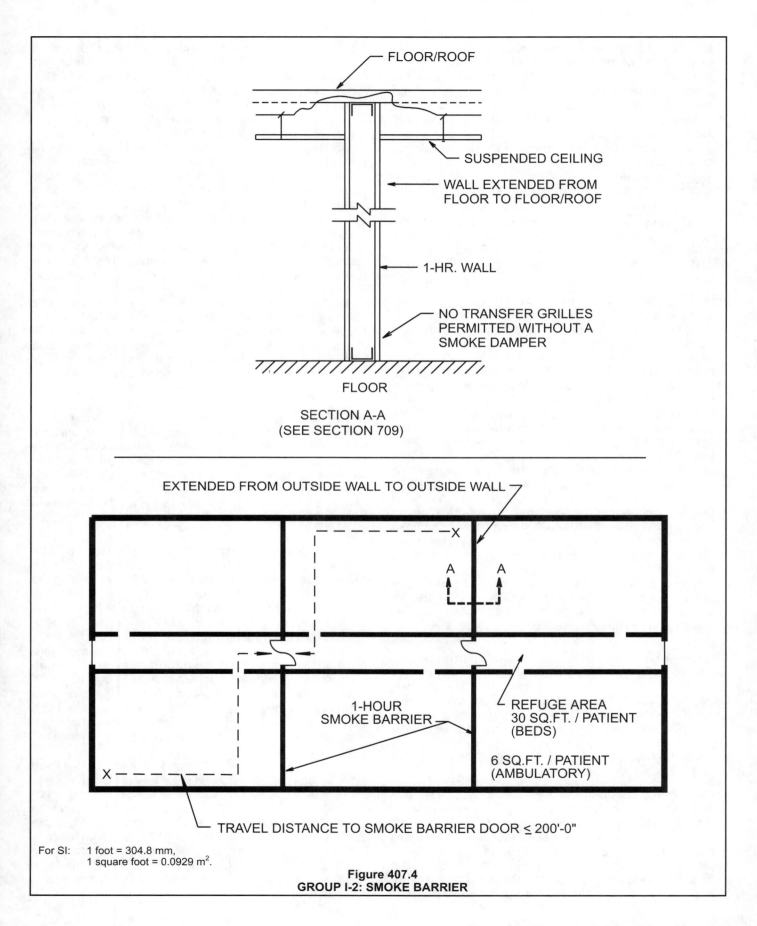

SECTION A-A
(SEE SECTION 709)

For SI: 1 foot = 304.8 mm,
1 square foot = 0.0929 m².

Figure 407.4
GROUP I-2: SMOKE BARRIER

407.4.2 Independent egress. A means of egress shall be provided from each smoke compartment created by smoke barriers without having to return through the smoke compartment from which means of egress originated.

❖ To prevent creation of a dead-end smoke compartment, exits are to be arranged so as to permit access without returning through a smoke compartment from which egress originated. This section does not require an exit from within each smoke compartment. See Figures 407.4.2(1) and 407.4.2(2) for acceptable and unacceptable egress arrangements.

[F] 407.5 Automatic sprinkler system. Smoke compartments containing patient sleeping units shall be equipped throughout with an automatic fire sprinkler system in accordance with Section 903.3.1.1. The smoke compartments shall be equipped with approved quick-response or residential sprinklers in accordance with Section 903.3.2.

❖ Recognizing the effectiveness of an automatic sprinkler system, the reference to Section 903.3.1.1 clarifies that the suppression system required by Section 903.2.5 must be a sprinkler system in accordance with the code and NFPA 13. Furthermore, in response to recent proven technology, smoke compartments containing patient rooms are required to be equipped throughout with quick-response or residential sprinklers.

The majority of fire injuries in hospitals begin with the ignition of clothing on a person, a mattress, a pillow, bedding or linen. The Building and Fire Research Laboratory of the National Institute of Standards and Technology (NIST) conducted fire tests in patient rooms using quick-response sprinklers and smoke detectors. The test results were published in a July 1993 report entitled "Measurement of Room Conditions and Response of Sprinklers and Smoke Detectors During a Simulated Two-Bed Hospital Patient Room Fire." The report concluded that quick-response sprinklers actuated before the patient's life would be threatened by a fire in the his or her room.

Prior to the development of quick-response sprinkler technology, smoke detection in patient rooms classified as Group I-2 occupancies was typically required by codes, due to the slower response time of standard sprinklers. While properly operating standard sprinklers are effective, the extent of fire growth and smoke production that can occur before sprinkler activation creates the need for early warning to enable faster response by staff and initiation of egress that is critical in occupancies containing persons incapable of self-preservation. With quick-response or residential sprinklers installed throughout the smoke compartment and additional protection provided as indicated for spaces open to corridors; incidental use areas and certain specific health care facilities, smoke detectors in patient sleeping units are no longer required in Group I-2 occupancies.

[F] 407.6 Automatic fire detection. Corridors in nursing homes (both intermediate-care and skilled nursing facilities), detoxification facilities and spaces permitted to be open to corridors by

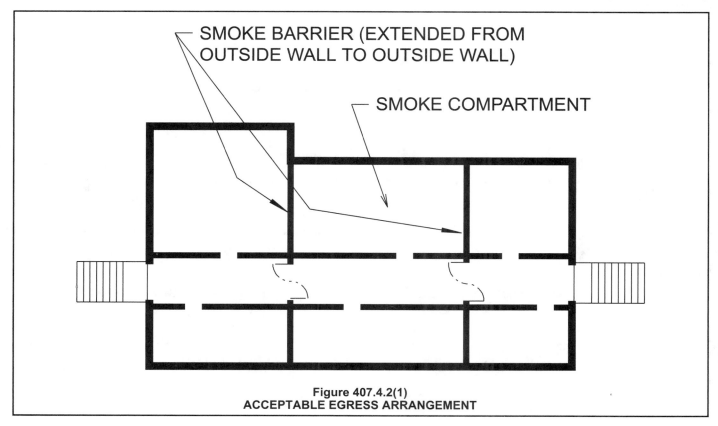

Figure 407.4.2(1)
ACCEPTABLE EGRESS ARRANGEMENT

FIGURE 407.4.2(2)   SPECIAL DETAILED REQUIREMENTS BASED ON USE AND OCCUPANCY

Section 407.2 shall be protected by an automatic fire detection system installed in accordance with Section 907.

Exceptions:

1. Corridor smoke detection is not required where patient sleeping units are provided with smoke detectors that comply with UL 268. Such detectors shall provide a visual display on the corridor side of each patient sleeping unit and an audible and visual alarm at the nursing station attending each unit.

2. Corridor smoke detection is not required where patient sleeping unit doors are equipped with automatic door-closing devices with integral smoke detectors on the unit sides installed in accordance with their listing, provided that the integral detectors perform the required alerting function.

❖ Automatic fire detection is required in areas open to corridors in occupancies classified as Group I-2 and corridors in nursing homes and detoxification facilities. In recognition of quick-response sprinkler technology and the fact that the sprinkler system is electronically supervised and doors to patient rooms are supervised by staff on a continual basis when in the open position, it is now believed that patient room smoke detectors are not required for adequate fire safety in patient sleeping units. In nursing homes and detoxification facilities, however, some redundancy is appropriate because such facilities typically have less control over furnishings and personal items and, thereby, result in a less predictable and usu-

ally higher fire hazard load than other Group I-2 occupancies. Also, there is generally less staff supervision in these facilities than in other health care facilities and, thus, less, control over patient smoking and other fire causes. To provide additional protection against fires spreading from the room of origin, therefore, automatic fire detection is required in corridors of nursing homes and detoxification facilities. It should be noted that fire detection is not required in corridors of other Group I-2 occupancies except where otherwise specifically required in the code (see Section 907.2.6). Similarly, since areas open to the corridor very often are the room of fire origin and because such areas are no longer required by the code to be under visual supervision by staff, some redundancy to protection by the sprinkler system is requested. Accordingly, all areas open to corridors must be protected by an automatic fire detection system. This requirement provides an additional level of protection against sprinkler system failures or lapses in staff supervision. There are two exceptions to the requirement for an automatic fire detection system in corridors of nursing homes and detoxification facilities. In both exceptions, the alternative methods of protection specifically provide an equivalent level of safety to what is required in patient sleeping rooms.

Exception 1 requires smoke detectors to be located in the patient's room, which activate both a visual display on the corridor side of the patient room and a visual and audible alarm at the nurses' station serving or attending the room. Detectors complying with UL 268 are in-

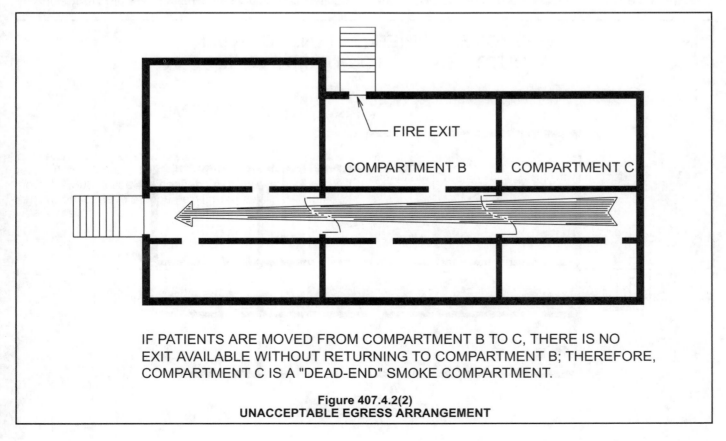

FIRE EXIT

COMPARTMENT B   COMPARTMENT C

IF PATIENTS ARE MOVED FROM COMPARTMENT B TO C, THERE IS NO EXIT AVAILABLE WITHOUT RETURNING TO COMPARTMENT B; THEREFORE, COMPARTMENT C IS A "DEAD-END" SMOKE COMPARTMENT.

Figure 407.4.2(2)
UNACCEPTABLE EGRESS ARRANGEMENT

tended for open area protection and for connection to a normal power supply or as part of a fire alarm system.

This exception, however, is specifically designed so as not to require the detectors to activate the building's fire alarm system where approved patient room smoke detectors are installed and where visual and audible alarms are provided. This is in response to the concern over unwanted alarms. It should be noted that the required alarm signals will not necessarily indicate to staff that a fire emergency exists. The nursing call system may typically be used to identify numerous conditions within the room.

Exception 2 addresses the situation where smoke detectors are incorporated within automatic door-closing devices. The units are acceptable as long as the required alarm functions are still provided. Such units are usually listed as combination door closer and hold-open devices.

407.7 Secured yards. Grounds are permitted to be fenced and gates therein are permitted to be equipped with locks, provided that safe dispersal areas having 30 net square feet (2.8 m²) for bed and litter patients and 6 net square feet (0.56 m²) for ambulatory patients and other occupants are located between the building and the fence. Such provided safe dispersal areas shall not be located less than 50 feet (15 240 mm) from the building they serve.

❖ Group I-2 occupancies specializing in the treatment of mental disabilities, such as Alzheimer's, often provide a secured yard for patient use. During an emergency, it may be difficult to control residents if they are allowed to have direct access to a public way. This section essentially creates a safe area where people may move from the building without having unrestricted access to the public way.

SECTION 408
GROUP I-3

408.1 General. Occupancies in Group I-3 shall comply with the provisions of this section and other applicable provisions of this code (see Section 308.4).

❖ The provisions of Section 408 address the unique features of detention and correctional occupancies in Group I-3. The provisions are based on full-scale fire tests, fire experience (in particular several multiple-death fires that occurred between 1974 and 1979) and the provisions for occupancies in Group I-2. With respect to evacuation, occupancies in Groups I-2 and I-3 are similar in that the occupants are typically not capable of self-preservation and, therefore, staff must assist in evacuation.

The general approach is a defend-in-place philosophy based on the difficulties associated with evacuation, especially in medium- and high-security facilities. The first level of protection is established as the room that is typically a sleeping room or cell. Horizontal evacuation to an adjacent smoke compartment provides the second level of protection, and vertical evacuation is the third level of protection. The evacuation difficulty associated with occupancies in Group I-3 is not occupant mobility; rather, it is the need to maintain security and, in some instances, to keep the residents segregated. There are also instances such as protected witness facilities in which the identity of the occupant must not be revealed to other occupants or the public.

It is recognized that the broad classification of Group I-3 includes a variety of locking arrangements based on the various levels of security required. As such, Section 308.4 contains definitions for five different occupancy conditions that are intended to represent the various locking arrangements and different levels of security. Regardless of the security level, fire protection and life safety features must be provided in order to achieve an acceptable level of protection without interfering with the operation of the facility and the need to maintain security.

This section has been developed out of considerations unique to detention and correctional facilities and to facilitate the consideration of the entire package of protection features. The basic protection features provided include early detection, fire containment, evacuation and fire extinguishment. Section 408.7 of the IFC contains provisions for emergency preparedness, including staff training, staffing requirements and the need to be able to obtain and identify emergency keys. Additional consideration should be given to restricting the fuel load within resident housing areas. It should be also noted that Section 302.1.1 contains provisions for incidental use areas within Group I-3. The incidental use areas represent different degrees of hazard than are associated with the hazards of the primary building area—in this case, the housing of occupants under physical restraint. The different hazard is typically related to the basic nature of the contents or the quantity of combustibles.

The provisions of Section 408 apply to all occupancies in Group I-3 and those portions of occupancies that are considered Group I-3, except Condition 1 facilities (see Figure 408.1). As defined in Section 308.4.1, Condition 1 facilities permit free egress even to the exterior and, therefore, such occupancies are subject to the same provisions applicable to Group R.

The elimination of Condition 1 from the provisions of Section 408 may result in some interesting applications of the code. For example, the corridor separation requirements in Section 1016.1 for occupancies in Group R are more restrictive than the corridor requirements for occupancies in Group I-3, Condition 2. This may result in a request to consider a facility as Condition 2 even though the building is really Condition 1. Such a request could be considered acceptable only if the additional requirements in the code and in the IFC for Condition 2 are met, including automatic fire suppression and staffing (see Section 408.7.2 of the IFC).

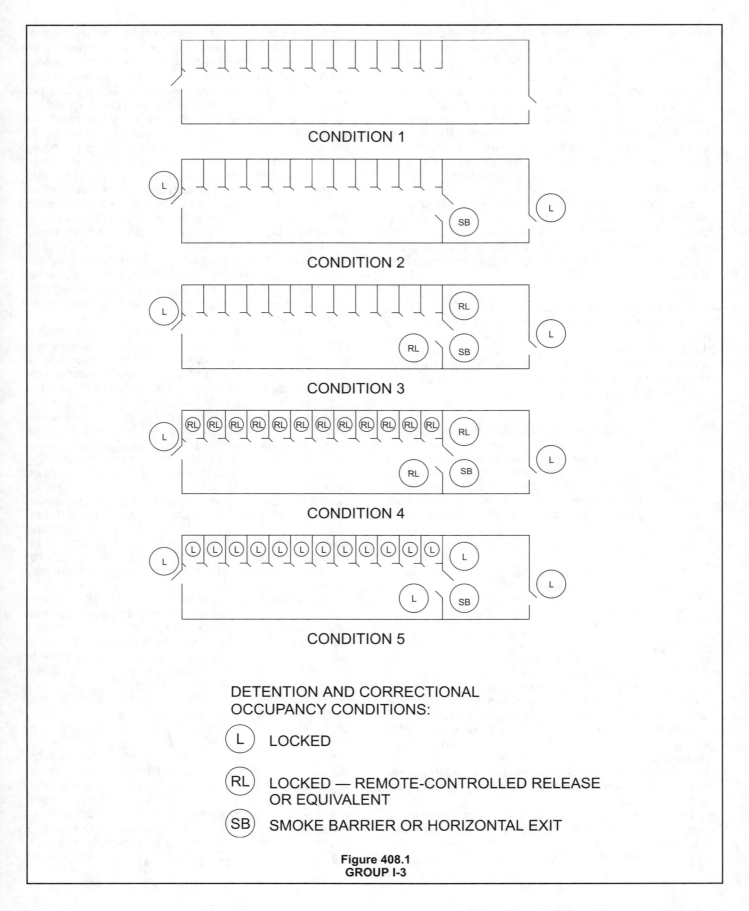

Figure 408.1
GROUP I-3

408.2 Mixed occupancies. Portions of buildings with an occupancy in Group I-3 that are classified as a different occupancy shall meet the applicable requirements of this code for such occupancies. Where security operations necessitate the locking of required means of egress, provisions shall be made for the release of occupants at all times.

Means of egress from detention and correctional occupancies that traverse other use areas shall, as a minimum, conform to requirements for detention and correctional occupancies.

> **Exception:** It is permissible to exit through a horizontal exit into other contiguous occupancies that do not conform to detention and correctional occupancy egress provisions but that do comply with requirements set forth in the appropriate occupancy, as long as the occupancy is not a high-hazard use.

❖ In accordance with the provisions of Section 302.3, portions of an occupancy in Group I-3 may be classified as a separate occupancy and meet the provisions of the code for that occupancy. Since the area may be used by the residents, however, a need may exist for the other occupancy to also contain security provisions, such as the locking of egress doors. This section specifically permits such a condition as long as arrangements have been made for release of the occupants within these areas at any time they are occupied. Acceptable methods include having either staff operate the locks or remote release of the locks, similar to that which is provided in the housing areas.

Although it is indicated that the section applies to portions of occupancies in Group I-3, consideration should be given to allowing the necessary security to be provided in separate buildings that are part of a detention or correctional facility. Applications of this section should be restricted to the buildings that must be secure because they are occupied by residents, and arrangements for quick release of the locks must be provided at all times the building is occupied. If residents are to be permitted into an area or building without staff supervision, the residents should be able to initiate their own evacuation.

The means of egress provisions from this section and Chapter 10 that are applicable to Group I-3 are also applicable to the path through different occupancies. For example, if the means of egress from a cell block wing traverses through a Group A area, then the requirements of the code for the Group I-3 egress are also applicable to the path through the assembly occupancy.

The exception in the code permits egress through a different occupancy (nonhigh-hazard use) that is separated from the Group I-3 areas with a horizontal exit. This separation voids the requirements that Group I-3 egress provisions are also applicable to the different occupancies. The reason for this, of course, is that Group I-3 occupants have fled the immediate effects of the fire condition in the original detention or correctional areas and are now harbored in an area of refuge.

408.3 Means of egress. Except as modified or as provided for in this section, the provisions of Chapter 10 shall apply.

❖ Because of the need to provide security in occupancies in Group I-3, many of the means of egress requirements of Chapter 10 have been modified for Group I-3. Except as modified in Section 408, however, all other provisions of Chapter 10 are applicable.

408.3.1 Door width. Doors to resident sleeping units shall have a clear width of not less than 28 inches (711 mm).

❖ Except for sleeping units that are provided for physically disabled individuals, sleeping unit doors may be a minimum of 28 inches (711 mm) in width. This section is not intended to negate the need to provide a 32-inch (813 mm) door if the room is intended for use by physically disabled persons. In an occupancy in Group I-3, most sleeping units need not be accessible.

408.3.2 Sliding doors. Where doors in a means of egress are of the horizontal-sliding type, the force to slide the door to its fully open position shall not exceed 50 pounds (220 N) with a perpendicular force against the door of 50 pounds (220 N).

❖ Swinging doors in occupancies in Group I-3 may not be acceptable for security reasons. If the door swings into a room, the door can easily be blocked and staff access can be prevented. If the door swings out, staff does not have control of the door and the residents can use the door as a means to overcome a staff member; therefore, horizontal sliding doors are used quite extensively and are permitted in the means of egress. The doors must be capable of being opened with a force of 50 pounds (222 N) even if a 50-pound (222 N) force is being applied perpendicular to the door. The maximum force permitted to open the door exceeds the restrictions in Section 1008.1.3.3 in recognition of the physical capabilities of staff members.

408.3.3 Spiral stairs. Spiral stairs that conform to the requirements of Section 1009.9 are permitted for access to and between staff locations.

❖ Recognizing the physical capabilities of the staff, spiral stairs are permitted for access to and between staff locations. In multistory facilities, spiral stairs are often used between staff control areas on different levels. As such, staff can move between the areas without entering adjacent housing areas and without losing the space that would be required for a traditional stair. Spiral stairs provide an option for vertical circulation in guard towers and similar observation areas where the floor area is limited and the desire is to have the least amount of obstruction of floor area possible.

408.3.4 Exit discharge. Exits are permitted to discharge into a fenced or walled courtyard. Enclosed yards or courts shall be of a size to accommodate all occupants, a minimum of 50 feet (15 240 mm) from the building with a net area of 15 square feet (1.4 m²) per person.

❖ For security purposes, exits from occupancies in Group I-3 do not normally provide access to a public way. In some instances, the building is located in a complex that has perimeter walls preventing access to a public way but that permit occupants to move away from the building. In other instances or in complexes where the mixing of residents must be controlled, a yard or court is provided for exit discharge. Such arrangements are acceptable provided that the occupants have an area that is at least 15 square feet (1.4 m²) per person and is located at least 50 feet (15 240 mm) from the building.

This 50-foot (15 240 mm) distance enables residents to move away from the building and should be adequate spatial separation—especially since the building is required to be protected with an automatic fire suppression system in accordance with Section 903.2.5. The 15-square-foot (1.4 m²) measurement is provided so that the residents have adequate space in which to stand and to move around since they may need to remain in the yard or court for some time.

408.3.5 Sallyports. A sallyport shall be permitted in a means of egress where there are provisions for continuous and unobstructed passage through the sallyport during an emergency egress condition.

❖ A sallyport is a compartment established for security purposes that restricts the movement of individuals. A sallyport consists of two or more security doors that do not normally operate simultaneously. One door will not normally open until the other doors are closed and locked. Sallyports are provided for security purposes to control movement and passage of residents from one area to another. The sallyport is to be arranged such that during an emergency, the doors may be opened simultaneously so as to permit free and unobstructed movement from the area. In many instances, the sallyport doors contain a manual release that can be operated by staff during an emergency to override the normal electric operation.

Although not addressed by this section, if the sallyport is in the path that the fire department or fire brigade will use to bring fire hoses, additional consideration may need to be given to permit the hose to go through the sallyport. An alternative would be to provide a standpipe connection on the housing-area side of the sallyport.

408.3.6 Vertical exit enclosures. One of the required vertical exit enclosures in each building shall be permitted to have glazing installed in doors and interior walls at each landing level providing access to the enclosure, provided that the following conditions are met:

1. The vertical exit enclosure shall not serve more than four floor levels.

2. Vertical exit enclosure doors shall not be less than ³/₄-hour fire doors complying with Section 715.3.

3. The total area of glazing at each floor level shall not exceed 5,000 square inches (3.23 m²) and individual panels of glazing shall not exceed 1,296 square inches (0.84 m²).

4. The glazing shall be protected on both sides by an automatic fire sprinkler system. The sprinkler system shall be designed to wet completely the entire surface of any glazing affected by fire when actuated.

5. The glazing shall be in a gasketed frame and installed in such a manner that the framing system will deflect without breaking (loading) the glass before the sprinkler system operates.

6. Obstructions, such as curtain rods, drapery traverse rods, curtains, drapes or similar materials shall not be installed between the automatic sprinklers and the glazing.

❖ In addressing the security needs in Group I-3, the limitation on openings in the enclosures for vertical exits given in Sections 1019.1 and 1019.1.1 is modified to facilitate visibility into the exit.

A vertical exit that is open to view in a correctional facility is useful and necessary for two reasons. First, the movement of the inmates is observable, thus reducing the potential for concealment, physical attacks on other inmates and other undesirable activities that could otherwise take place in enclosed spaces that are not observable.

Second, the movement of inmates who are under physical restraint is observable from the exterior, thus increasing the level of safety for correctional officers. The alternative of closed-circuit television within the exits is not as desirable because the potential for malfunction of or intentional damage to the equipment makes this method a less reliable form of observation.

The conditions specified that permit glazing require full enclosure of the floor opening so that there is no direct communication between stories. These criteria provide a measure of fire integrity, if somewhat less than that otherwise required of a vertical exit enclosure.

408.4 Locks. Egress doors are permitted to be locked in accordance with the applicable use condition. Doors from an area of refuge to the exterior are permitted to be locked with a key in lieu of locking methods described in Section 408.4.1. The keys to unlock the exterior doors shall be available at all times and the locks shall be operable from both sides of the door.

❖ The locking arrangements for egress doors depend on the occupancy condition (1, 2, 3, 4 or 5) given in Section 308.4. Except for Condition 1 buildings, the doors to the exterior from an occupancy in Group I-3 may be locked and need not be remotely controlled. The door locks are to be arranged so that they can be released from the exterior of the building, as well as the interior. This feature accomplishes two purposes. First, since the door can be unlocked from the outside, staff need not enter the housing area in which the fire may be located to release the locks. Second, many facilities do not permit staff members in the housing area to carry the keys for the exterior doors; therefore, there may be no benefit in requiring the door lock to be arranged to permit operation from the interior, although such an arrangement is not prohibited.

The provisions of Section 408.7.4 of the IFC concerning the marking of keys applies to both exterior and interior door locks. For this reason, it is usually desirable to minimize the number of keys required for an emergency; therefore, the exterior doors will most likely be keyed alike. Unfortunately, many facilities do not take this into consideration and the number of emergency keys becomes excessive to the extent that it is virtually impossible to identify each key by sight and touch.

408.4.1 Remote release. Remote release of locks on doors in a means of egress shall be provided with reliable means of operation, remote from the resident living areas, to release locks on all required doors. In Occupancy Conditions 3 or 4, the arrangement, accessibility and security of the release mechanism(s) required for egress shall be such that with the minimum available staff at any time, the lock mechanisms are capable of being released within 2 minutes.

> **Exception:** Provisions for remote locking and unlocking of occupied rooms in Occupancy Condition 4 are not required provided that not more than 10 locks are necessary to be unlocked in order to move occupants from one smoke compartment to a refuge area within 3 minutes. The opening of necessary locks shall be accomplished with not more than two separate keys.

❖ The provisions of this section do not mandate remote release locks; rather, the facility determines the level

and manner in which security is to be provided. Door-locking arrangements fall into one of the conditions defined in Section 308.4. If remote release locks are provided, the means of operating them must be external to the resident housing area so the staff is not required to enter the housing area to release the locks. Clearly, in Conditions 3 and 4, excessive delay in releasing locks puts the occupants at additional risk. Locks in Conditions 3 and 4 must be able to be released promptly (within 2 minutes) with the minimum staff that will be available. A control center is typically provided for two or more housing areas from which the locks are controlled. In evaluating the use and arrangement of the remote release, consideration must be given to the intended staffing levels at different times of the day, as well as the location(s) of the remote release.

The condition definitions of Section 308.4 are intended to relate to the time necessary to evacuate the residents to an area of refuge, such as a smoke compartment; therefore, in Condition 4, manual release locks may be used only when no more than 10 locks have to be unlocked in order to move all occupants from one smoke compartment to another and provided that this can be accomplished within 3 minutes. The time period of 3 minutes is presumed to be a reasonable staff response time to a fire emergency. So that the time to release the locks is kept to a minimum, it cannot take more than two separate keys to release the 10 locks. Again, the intended staff levels of the facility need to be considered (see Figure 408.4.1).

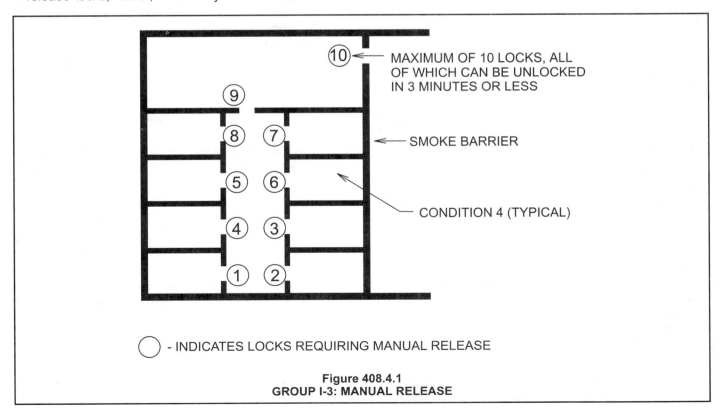

Figure 408.4.1
GROUP I-3: MANUAL RELEASE

[F] 408.4.2 Power-operated doors and locks. Power-operated sliding doors or power-operated locks for swinging doors shall be operable by a manual release mechanism at the door, and either emergency power or a remote mechanical operating release shall be provided.

> **Exception:** Emergency power is not required in facilities with 10 locks or less complying with the exception to Section 408.4.1.

❖ To increase the likelihood that the doors in the means of egress can be operated even during a power failure, two alternative locking arrangements are required whenever the doors or locks are electrically operated. First, a manual release mechanism is to be provided at the door. Also, a remote backup (second) mechanical operating door release system must be provided or the electrical door control system must be provided with an emergency power source. If provided, the mechanical operating door release system must be capable of being operated from outside the residential housing area.

This section does not apply to manually operated doors and locks that meet the requirements of the exception to Section 408.4.1.

408.4.3 Redundant operation. Remote release, mechanically operated sliding doors or remote release, mechanically operated locks shall be provided with a mechanically operated release mechanism at each door, or shall be provided with a redundant remote release control.

❖ If a mechanically operated remote release mechanism is provided as the remote release system, either a second, redundant system is required or each remote release door or lock is to be provided with a mechanically operated device at the door. The redundant operation is essential to increase the likelihood that the doors will continue to operate in the event of failure of the primary locking system. If provided, the redundant mechanical operating door release system must also be operable from outside the residential housing area.

This provision does not apply to power-operated doors and locks that are required to have alternative systems in accordance with Section 408.4.2.

408.4.4 Relock capability. Doors remotely unlocked under emergency conditions shall not automatically relock when closed unless specific action is taken at the remote location to enable doors to relock.

❖ So as to maintain means of egress doors in an open position during an emergency condition, once a door in a means of egress is unlocked via an emergency unlocking system, the door must not automatically relock when closed unless a specific action is taken at the remote location. The normal operation of such doors and locks is typically arranged so that the door will automatically relock upon closure.

Many electronic-locking systems operate such that the door automatically relocks upon closure for security reasons. To comply with the provisions of this section, however, a specific switch is usually provided that

serves as an emergency control and does not permit the door to relock without operating it. This section does not require that doors automatically release upon activation of the fire alarm system.

408.5 Vertical openings. Vertical openings shall be enclosed in accordance with Section 707.

> **Exception:** A floor opening between floor levels of residential housing areas is permitted without enclosure protection between the levels, provided that both of the following conditions are met:
>
> 1. The entire normally occupied areas so interconnected are open and unobstructed so as to enable observation of the areas by supervisory personnel.
>
> 2. Means of egress capacity is sufficient to provide simultaneous egress for all occupants from all interconnected levels and areas.

The height difference between the highest and lowest finished floor levels shall not exceed 23 feet (7010 mm). Each story, considered separately, has at least one-half of its individual required means of egress capacity provided by exits leading directly out of that story without traversing another story within the interconnected area.

❖ Ordinarily, vertical openings are required to be enclosed by fire-resistance-rated construction in accordance with Section 707. This section contains provisions that allow communicating floor levels of occupancies in Group I-3 to be open without enclosure protection between levels when these special protection features are provided. This provision applies regardless of whether or not the communicating floor levels are a mezzanine.

The following is a discussion of the conditions that must be met:

1. The areas must be sufficiently open and unobstructed so that a fire in one part will be immediately obvious to the occupants and supervisory personnel in the area. The intent of the provision is not to require open cell fronts; rather, it is to enable the staff to observe events occurring on both levels from one location. This arrangement is also usually desirable from a security standpoint [see Figure 408.5(1)].

2. The exit capacity required is determined based on the occupant load of all floor levels exiting simultaneously. It is not unusual for most of the residents to be on the main level in the common space during periods when they are not required to be in the sleeping rooms; therefore, this level must be able to handle the potential occupant load of all levels. To minimize the potential that a fire in the day room would block access to all exits, each level must have at least one-half of its individual exit capacity accessible directly from that level [see Figures 408.5.(1) and 408.5(2)].

The height differential between the top and bottom floor levels must not exceed 23 feet (7010 mm) and a minimum of one-half of the required exit capacity of each level must be provided at the level without entering another story within the interconnected area.

The limitation of 23 feet (7010 mm) expands the general code limitations on unprotected floor openings beyond two stories, because in Group I-3 it is not uncommon to have more than two interconnected levels in this open arrangement; however, the effective limitation in number of stories is three. As with atriums (see Section 404), an automatic sprinkler system is required to control the spread of smoke and fire through vertical openings. Beyond the specified dimension of 23 feet (7010 mm), effective unobstructed sight becomes less attainable. Additionally, the limitation is in lieu of a smoke control system requirement for this open area that would otherwise be required for atriums complying with Section 404. Note that floor openings meeting the requirements of an atrium complying with Section 404 are allowed with no height limitations. Such openings would require smoke control and special smoke alarms. Vertical exits would not be permitted in the atrium.

408.6 Smoke barrier. Occupancies in Group I-3 shall have smoke barriers complying with Section 709 to divide every

story occupied by residents for sleeping, or any other story having an occupant load of 50 or more persons, into at least two smoke compartments.

Exception: Spaces having direct exit to one of the following, provided that the locking arrangement of the doors involved complies with the requirements for doors at the compartment barrier for the use condition involved:

1. A public way.
2. A building separated from the resident housing area by a 2-hour fire-resistance-rated assembly or 50 feet (15 240 mm) of open space.
3. A secured yard or court having a holding space 50 feet (15 240 mm) from the housing area that provides 6 square feet (0.56 m^2) or more of refuge area per occupant, including residents, staff and visitors.

❖ One of the basic premises of this section is that evacuation of a Group I-3 facility is impractical because of the need to maintain security. The security measures may also result in a delayed evacuation as compared to residential occupancies in which free egress is provided; therefore, it is essential to provide an area of refuge to which residents can be relocated when it becomes necessary to evacuate them from the sleeping areas. For this reason, each sleeping floor and each floor with an occupant load of 50 or more are required to be divided

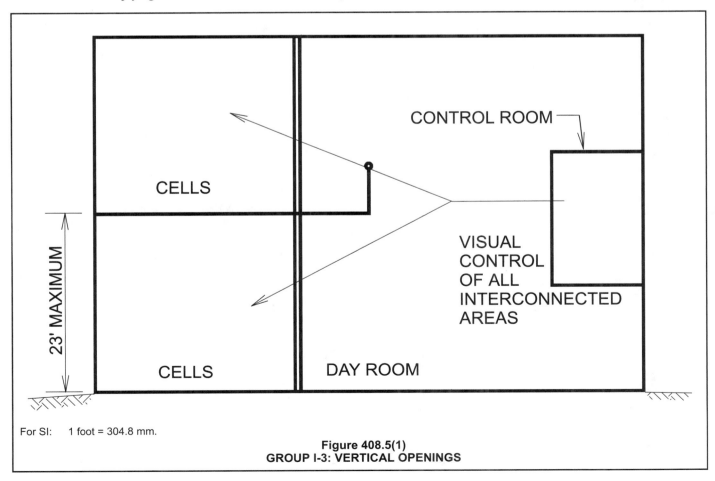

For SI: 1 foot = 304.8 mm.

Figure 408.5(1)
GROUP I-3: VERTICAL OPENINGS

into at least two smoke compartments using smoke barriers that comply with Section 709 (see commentary, Section 709). It should be noted that Section 709 provides several exceptions to smoke barriers used in Group I-3 occupancies (e.g., construction, doors and opening protectives).

The three exceptions are common-sense alternatives to movement into a smoke compartment. If the particular occupancy condition affords free, unobstructed access with no delay to the exterior, to a separate building or through a horizontal exit, then defend-in-place provisions are not needed.

408.6.1 Smoke compartments. The maximum number of residents in any smoke compartment shall be 200. The travel distance to a door in a smoke barrier from any room door required as exit access shall not exceed 150 feet (45 720 mm). The travel distance to a door in a smoke barrier from any point in a room shall not exceed 200 feet (60 960 mm).

❖ This section defines the maximum allowable size of smoke compartments by limiting the number of residents in any single smoke compartment to 200, the maximum door-to-door distance within the compartment to 150 feet (45 720 mm) and the maximum total travel distance to 200 feet (60 960 mm). These limita-

tions are intended to enable evacuation of a smoke compartment for all of the occupants (see Figure 408.6.1).

408.6.2 Refuge area. At least 6 net square feet (0.56 m^2) per occupant shall be provided on each side of each smoke barrier for the total number of occupants in adjoining smoke compartments. This space shall be readily available wherever the occupants are moved across the smoke barrier in a fire emergency.

❖ To provide adequate space for the residents evacuated to an adjacent smoke compartment, an additional 6 square feet (0.56 m^2) per person must be readily available for the total number of occupants in the adjoining smoke compartment (see Figure 408.6.2). The 6-square-foot (0.56 m^2) criterion enables the occupants to be evacuated to the adjacent smoke compartment without delay from crowding.

In a fire emergency, any action that must be taken to unlock the space will result in a delay in moving occupants into the area of refuge. As such, the code requires the refuge area to be readily available in order to preclude consideration (as an area of refuge) of a space that may be normally locked and, therefore, not immediately usable.

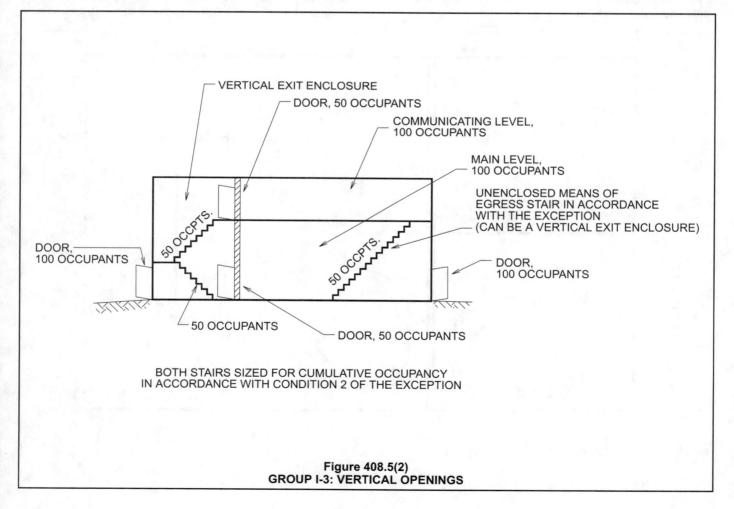

Figure 408.5(2)
GROUP I-3: VERTICAL OPENINGS

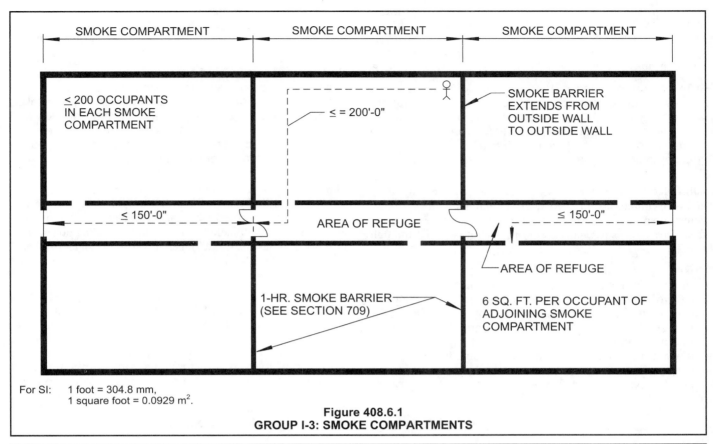

For SI: 1 foot = 304.8 mm,
            1 square foot = 0.0929 m^2.

Figure 408.6.1
GROUP I-3: SMOKE COMPARTMENTS

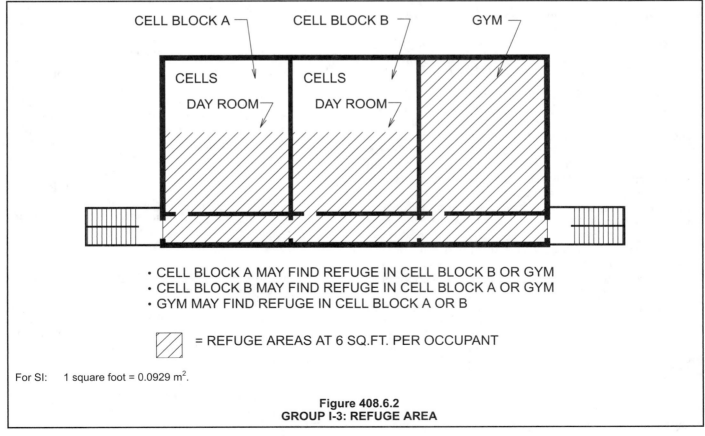

• CELL BLOCK A MAY FIND REFUGE IN CELL BLOCK B OR GYM
• CELL BLOCK B MAY FIND REFUGE IN CELL BLOCK A OR GYM
• GYM MAY FIND REFUGE IN CELL BLOCK A OR B

= REFUGE AREAS AT 6 SQ.FT. PER OCCUPANT

For SI: 1 square foot = 0.0929 m^2.

Figure 408.6.2
GROUP I-3: REFUGE AREA

408.6.3 Independent egress. A means of egress shall be provided from each smoke compartment created by smoke barriers without having to return through the smoke compartment from which means of egress originates.

❖ To prevent creating a dead-end smoke compartment, exits are to be arranged so as to permit access without returning through a compartment from which egress originated. An exit is not required from within each smoke compartment [see Figures 407.4.2(1) and (2) for acceptable and unacceptable exit arrangements].

408.7 Subdivision of resident housing areas. Sleeping areas and any contiguous day room, group activity space or other common spaces where residents are housed shall be separated from other spaces in accordance with Sections 408.7.1 through 408.7.4.

❖ Resident housing areas usually consist of sleeping areas and other contiguous common area spaces, such as day rooms or activity areas onto which sleeping areas open. The separation required by this section is intended to separate housing areas, where the occupants in those areas may be asleep, from other activity areas since people asleep are more vulnerable to a fire emergency. The separation requirements are based on the relative evacuation difficulty as determined by the occupancy condition. Sections 408.7.1 through 408.7.4 give the separation requirements for the various occupancy conditions for which separation is necessary.

408.7.1 Occupancy Conditions 3 and 4. Each sleeping area in Occupancy Conditions 3 and 4 shall be separated from the adjacent common spaces by a smoke-tight partition where the travel distance from the sleeping area through the common space to the exit access corridor exceeds 50 feet (15 240 mm).

❖ A smoke-tight partition is required between the sleeping areas and common areas in Conditions 3 and 4 when travel distance from the sleeping area through the common area to an exit access corridor exceeds 50 feet (15 240 mm). This requirement recognizes the need to protect sleeping areas from nonsleeping areas due to the increased hazards attendant with common areas, delayed recognition of fire hazards by sleeping persons and the restrictions on movement for these two occupancy conditions. Moreover, release from the sleeping area may be delayed and as such, evacuation with extended travel through a common space may not be safe. In such cases, the sleeping area will be an area of refuge until the fire is controlled and the occupants in the sleeping area can be safely evacuated.

It should be noted that the provisions contained in Section 1013.2 are applicable to this circumstance, and as such, sleeping areas, including those in Group I-3 occupancies, are precluded from egressing through another sleeping area. Additionally, egress is not allowed from sleeping areas through toilet rooms.

408.7.2 Occupancy Condition 5. Each sleeping area in Occupancy Condition 5 shall be separated from adjacent sleeping areas, corridors and common spaces by a smoke-tight partition. Additionally, common spaces shall be separated from the exit access corridor by a smoke-tight partition.

❖ Given that movement from an occupied space in Condition 5 is even more restricted than for the other conditions, smoke-tight partitions are required to be installed between areas, sleeping areas, common spaces and the exit access corridor. Protection is provided because the staff-controlled release from the sleeping area is not as immediate as in other conditions.

408.7.3 Openings in room face. The aggregate area of openings in a solid sleeping room face in Occupancy Conditions 2, 3, 4 and 5 shall not exceed 120 square inches (77 419 mm²). The aggregate area shall include all openings including door undercuts, food passes and grilles. Openings shall be not more than 36 inches (914 mm) above the floor. In Occupancy Condition 5, the openings shall be closeable from the room side.

❖ Food pass-throughs, door undercuts and grilles are a functional necessity; therefore, openings are limited in size and close to the floor. The openings are required to be close to the floor to minimize the potential for the smoke to pass through the opening. By placing the opening close to the floor, the smoke will not pass through a smoke-tight barrier initially; rather, there will be some time delay as the smoke layer descends to 36 inches (914 mm) above the floor before smoke is transmitted through such openings.

408.7.4 Smoke-tight doors. Doors in openings in partitions required to be smoke tight by Section 408.7 shall be substantial doors, of construction that will resist the passage of smoke. Latches and door closures are not required on cell doors.

❖ If the partition is required to be capable of resisting the passage of smoke, door openings within the enclosure are also to be protected with an assembly that is capable of resisting the passage of smoke. As with other sections of the code that deal with doors and walls capable of resisting the passage of smoke, application of the section requires judgment by the building official. The intent is a solid door surface without louvered openings. Door closers and latches are not required. It is anticipated that the security lock will secure the door in the closed position. A door closer can interfere with the removal of the occupants not only in an emergency but also during normal operations.

408.8 Windowless buildings. For the purposes of this section, a windowless building or portion of a building is one with nonopenable windows, windows not readily breakable or without windows. Windowless buildings shall be provided with an engineered smoke control system to provide ventilation (mechanical or natural) in accordance with Section 909 for each windowless smoke compartment.

❖ An engineered smoke control system is required for smoke compartments in which there are no openings through which the products of combustion can be vented. As defined for this section only, a windowless building is a building or portion thereof with only nonoperable windows, windows that are not readily breakable or without windows. The intent of this section is that staff must have some means to ventilate the products of combustion; therefore, if the window cannot be broken by items readily available to the staff, the area is considered windowless.

SECTION 409
MOTION PICTURE PROJECTION ROOMS

409.1 General. The provisions of this section shall apply to rooms in which ribbon-type cellulose acetate or other safety film is utilized in conjunction with electric arc, xenon or other light-source projection equipment that develops hazardous gases, dust or radiation. Where cellulose nitrate film is utilized or stored, such rooms shall comply with NFPA 40.

❖ Cellulose acetate or other safety film used in the presence of potential ignition sources, including projectors or related equipment, creates a potential hazard. The provisions of this section are intended to minimize the potential for exposure of the audience and other occupants to the hazards associated with the presence of ignition sources in close proximity to the fuel load.

Provisions are established for projection rooms in recognition of the hazards attendant to the projection of ribbon-type cellulose acetate or other safety film in such spaces.

Safety film is typically made of cellulose acetate, polyester or triacetate film. In the early 1950s, safety film began to replace cellulose nitrate film. Although the use and manufacture of cellulose nitrate film has virtually ceased in the United States, large quantities of such film still exist, primarily for archival purposes. Cellulose nitrate film presents a serious hazard because degradation under external heat below ignition temperatures causes a chemical reaction. This chemical reaction releases heat of sufficient quantity to raise the substance to ignition temperatures, resulting in spontaneous combustion. After ignition, rapid burning and the production of toxic and flammable gases takes place. If such film is used, the provisions of NPFA 40 are applicable as are the provisions of Section 306 of the IFC.

The provisions of this section apply only where the previously mentioned hazards exist. The use of ribbon-type cellulose acetate or other safety film in conjunction with electric arc, xenon or other light-source projection equipment that develops hazardous gases, dust or radiation invokes these provisions. The provisions of this section do not apply where nonprofessional projection equipment is used, except that statements specific to cellulose nitrate film apply if such film is present regardless of the nature of the equipment. The safety film used by the motion picture industry since 1951 has fire hazard characteristics similar to ordinary paper of the same thickness and form.

If cellulose nitrate film is used, the provisions of NFPA 40 apply. NFPA 40 contains minimum requirements to provide a reasonable level of protection for the storage and handling of cellulose nitrate film. The standard does not contain provisions for the manufacture of cellulose nitrate film, since it has not been manufactured in the United States since 1951.

409.1.1 Projection room required. Every motion picture machine projecting film as mentioned within the scope of this section shall be enclosed in a projection room. Appurtenant electrical equipment, such as rheostats, transformers and generators, shall be within the projection room or in an adjacent room of equivalent construction.

❖ The projection room is to be enclosed in accordance with Section 409.2 to isolate the spaces housing the potential hazard. To minimize the potential for arcs or sparks to come in contact with the film, electrical equipment such as rheostats, transformers and generators should be located in a separate room. This section permits such equipment to be in the same room, but it should be arranged to minimize the potential for arcs or sparks to come in contact with the film, reducing the potential for ignition. Additional requirements for electrical equipment permitted in the projection room and requirements for nonprofessional equipment are given in NFPA 70.

In order to remind the operator that the room is not protected as required for cellulose nitrate film use, signs must be posted indicating that only safety film is to be used in the projection room. The signs are to be located on the outside of each projection room door and in the room itself. The minimum letter size ensures that the sign is legible.

409.2 Construction of projection rooms. Every projection room shall be of permanent construction consistent with the construction requirements for the type of building in which the projection room is located. Openings are not required to be protected.

The room shall have a floor area of not less than 80 square feet (7.44 m²) for a single machine and at least 40 square feet (3.7 m²) for each additional machine. Each motion picture projector, floodlight, spotlight or similar piece of equipment shall have a clear working space of not less than 30 inches by 30 inches (762 mm by 762 mm) on each side and at the rear thereof, but only one such space shall be required between two adjacent projectors. The projection room and the rooms appurtenant thereto shall have a ceiling height of not less than 7 feet 6 inches (2286 mm). The aggregate of openings for projection equipment shall not exceed 25 percent of the area of the wall between the projection room and the auditorium. Openings shall be provided with glass or other approved material, so as to close completely the opening.

❖ The projection room is to be constructed as required by the construction type of the building. Since the room dimensions are small, ventilation is provided to reduce heat buildup and accumulation of dust particles. A minimum fire-resistance rating is not required unless by other provisions of the code. Limited openings are permitted and are not required to be protected so as to facilitate the intended function of projecting the image onto the auditorium screen. Openings to the auditorium must not exceed 25 percent of the area of the wall between the projection room and the auditorium. Glass or other approved material must be provided that will completely close the opening and prevent a sharing of atmospheres with the audience or seating area.

In order to minimize the buildup of heat in any area, adequate floor space must be available. The space is required to be at least 80 square feet (7.44 m²) in size for a single projector and an additional 40 square feet (3.7 m²) for each additional projector. The room must be of adequate size so that no less than 30 inches (762 mm) of clear working space is available on each side and behind the projectors, floodlights, spotlights and similar equipment. For adjacent equipment, a clear space of 30 inches (762 mm) is adequate.

Whereas the projection room is considered an occupiable room, the ceiling height must not be less than 7 feet 6 inches (2286 mm) (see commentary, Section 120 .2).

409.3 Projection room and equipment ventilation. Ventilation shall be provided in accordance with the *International Mechanical Code*.

❖ All projection rooms and the equipment located within the rooms must be properly ventilated to abate the hazards of any hot gases and smoke generated by the equipment. Further, any intense heat given off by the equipment must be directly exhausted to the exterior rather than be absorbed by the building's ventilation system. Sections 502.11.1 and 502.11.2 of the IMC contain requirements for projectors provided with an exhaust discharge and those without an exhaust connection.

409.3.1 Projection room.

409.3.1.1 Supply air. Each projection room shall be provided with adequate air supply inlets so arranged as to provide well-distributed air throughout the room. Air inlet ducts shall provide an amount of air equivalent to the amount of air being exhausted by projection equipment. Air is permitted to be taken from the outside; from adjacent spaces within the building, provided the volume and infiltration rate is sufficient; or from the building air-conditioning system, provided it is so arranged as to provide sufficient air when other systems are not in operation.

❖ Separate supply (Section 409.3.1.1) and exhaust (Section 409.3.1.2) requirements are provided for projection rooms. Exhaust rates are specified in Section 502.11.2 of the IMC for projectors without exhaust connection.

Sufficient air must be introduced into each projection room to allow the effects of hot gases, smoke and heat to be diluted and to provide a make-up air source for the exhausted air. Although the code does not prescriptively state where the air-supply inlets are to be located, the intent is to create a well-ventilated room. Several air-supply inlets should be provided so that the entire room has adequate air changes. The amount of air supplied to each projection room must be approximately equal to the volume of air that is exhausted according to the IMC. Make-up supply air can be obtained from any source as long as sufficient quantities are provided at all times that the exhaust system is operating.

409.3.1.2 Exhaust air. Projection rooms are permitted to be exhausted through the lamp exhaust system. The lamp exhaust system shall be positively interconnected with the lamp so that the lamp will not operate unless there is the required airflow. Exhaust air ducts shall terminate at the exterior of the building in such a location that the exhaust air cannot be readily recirculated into any air supply system. The projection room ventilation system is permitted to also serve appurtenant rooms, such as the generator and rewind rooms.

Each projection machine shall be provided with an exhaust duct that will draw air from each lamp and exhaust it directly to the outside of the building. The lamp exhaust is permitted to serve to exhaust air from the projection room to provide room air circulation. Such ducts shall be of rigid materials, except for a flexible connector approved for the purpose. The projection lamp or projection room exhaust system, or both, is permitted to be combined but shall not be interconnected with any other exhaust or return system, or both, within the building.

❖ Exhaust air can be achieved by either of two methods. Projectors with an integral lamp exhaust system use air to cool the equipment during operation. The amount of air exhausted with these systems is part of the manufacturers' installation instructions. These integral exhaust systems must be directly connected to the mechanical exhaust systems. Further, the lamp exhaust system must be interconnected to the projector's power supply such that when the projector is operating, its exhaust system will be automatically activated.

The other method of providing exhaust is to exhaust the room itself. This method can only be used when the projectors do not have an integral lamp exhaust system. Exhaust rates for this type of system are specified in the IMC based on the type of projector. The exhaust ducts for the projection room must be terminated at the exterior of the building and cannot be combined with any other exhaust or return system present in the building. The intent of this requirement is so that the hot gases, smoke and heat will not be allowed to be disbursed elsewhere in the building.

409.4 Lighting control. Provisions shall be made for control of the auditorium lighting and the means of egress lighting systems of theaters from inside the projection room and from at least one other convenient point in the building.

❖ The projection room is usually an attended location, but it is also a likely area for a fire to originate. Lighting controls that provide adequate normal and emergency auditorium lighting must be provided in the projection room. Upon detection of a fire anywhere in the building, the operator can immediately provide adequate illumination for the occupants egressing the auditorium. A second location for the lighting controls is also required somewhere else in the building, so that entry into the projection room is not required to provide the necessary illumination, as it could be the location of fire origin. This second control is required to be conveniently located for the staff yet inaccessible to unauthorized persons.

409.5 Miscellaneous equipment. Each projection room shall be provided with rewind and film storage facilities.

❖ In order to keep all operations related to projection activities in the protected enclosure, rewind and film storage facilities are to be located in the projection room or in an adjacent room in accordance with Section 409.3.1.2.

SECTION 410
STAGES AND PLATFORMS

410.1 Applicability. The provisions of this section shall apply to all parts of buildings and structures that contain stages or platforms and similar appurtenances as herein defined.

❖ Historically, most significant theater fires originated on the stage. The 1903 Iroquois Theater fire in Chicago serves as a vivid example of a stage fire and its potentially tragic effects— 602 people lost their lives. Hazards associated with stages include: combustible scenery and lighting suspended overhead; scenic elements, contents and acoustical treatment on the back and sides of the stage; workshops, scene docks and dressing rooms located around the stage perimeter and storage areas and property rooms located underneath the stage.

The protection requirements set forth in this section are intended to limit the threat from stage fires to an audience and reduce the likelihood of a large fire in the stage area. These provisions include construction restrictions, automatic sprinkler systems, ventilation, separation of the stage from the audience and compartmentalization of appurtenant rooms to the stage.

The CABO Board for the Coordination of Model Codes (BCMC), May 1992 report, entitled "Stages, Platforms and Sound Stages," provided recommendations for national uniformity relative to stage and platform construction. Subsequent to this report, a comprehensive review of the requirements in the model codes pertaining to stages, platforms and sound stages was undertaken by the board. Representatives from a broad range of experts, including fire protection consultants, art design consultants and the American Society of Theater Consultants (ASTC), were heard in numerous public hearings. Based on this report, various principles were developed that subsequently were used in the requirements contained in this section. These principles are summarized as follows:

1. The fire hazards of stages are not necessarily a function of type (legitimate, regular or thrust) but rather area and height. Accordingly, a stage area in excess of 1,000 square feet (93 m²) or a height

in excess of 50 feet (15 240 mm) is the likely threshold that represents a significant potential for fuel load due to scenery, drops, etc.

2. Stages that exceed 1,000 square feet (93 m²) or the 50-foot (15 240 mm) height threshold require a means of emergency ventilation.

3. Stages that exceed 1,000 square feet (93 m²) and the 50-foot (15 240 mm) height threshold require adequate sprinkler protection.

4. Stages and platforms are similar to floor construction; therefore, the type of construction should be consistent with that of the floor construction in the building.

5. Separation of the stage from the seating area and dressing rooms, property rooms, etc., is critical in order to provide a degree of fire containment.

Section 410.3 contains provisions for the construction of the stage and gallery. Stage openings, decorations, equipment and scenery are addressed in Sections 410.3.3 through 410.3.6. Stage ventilation is required in accordance with Section 410.3.7. Platforms must be constructed in accordance with Section 410.4. The construction of auxiliary stage spaces is addressed in Section 410.5. Sections 410.6 and 410.7 contain the requirements for automatic sprinkler systems and standpipes, respectively.

All parts of a building or structure that contain a stage, platform or similar appurtenance are required to comply with the provisions of this section.

410.2 Definitions. The following words and terms shall, for the purposes of this section and as used elsewhere in this code, have the meanings shown herein.

❖ Definitions of terms that are associated with the content of this section are contained herein. These definitions can help in the understanding and application of the code requirements. It is important to emphasize that these terms are not exclusively related to this section but are applicable everywhere the term is used in the code. The purpose for including these definitions within this section is to provide more convenient access to them without having to refer back to Chapter 2.

For convenience, these terms are also listed in Chapter 2 with a cross reference to this section.

The use and application of all defined terms, including those defined herein, are set forth in Section 201.

FLY GALLERY. A raised floor area above a stage from which the movement of scenery and operation of other stage effects are controlled.

❖ The fly gallery is an integral part of the stage construction. This building feature provides an elevated, limited-size working area where the stagehands can operate and control the movement of suspended scenery or other visual effects.

GRIDIRON. The structural framing over a stage supporting equipment for hanging or flying scenery and other stage effects.

❖ The gridiron is directly tied to the functions of the fly gallery. This structural element usually consists of lightweight trusses that are used to suspend not only scenery but also behind-the-scenes equipment.

PINRAIL. A rail on or above a stage through which belaying pins are inserted and to which lines are fastened.

❖ This structural beam provides support for the tying off of various ropes and cables that are used to raise and lower suspended scenery.

PLATFORM. A raised area within a building used for worship, the presentation of music, plays or other entertainment; the head table for special guests; the raised area for lecturers and speakers; boxing and wrestling rings; theater-in-the-round stages; and similar purposes wherein there are no overhead hanging curtains, drops, scenery or stage effects other than lighting and sound. A temporary platform is one installed for not more than 30 days.

❖ Platforms are raised areas that are used for public performances and presentations that, except for lighting, do not incorporate any overhead hanging curtains, drops, scenery or stage effects. Additionally, while it is not specified, the fuel load on platforms (e.g., a podium for a lecturer or a boxing ring) is anticipated to be low.

Thus, since the fuel load on platforms is ordinarily low and there is no fuel load overhead in areas that would be difficult to access, the code requirements for platforms are less stringent than for stages.

Since many platforms are installed and used on a temporary basis, this definition specifically defines temporary use as 30 days or less. Temporary use is less likely to accumulate additional fuel loads, storage or waste materials on, beneath or above the platform.

PROSCENIUM WALL. The wall that separates the stage from the auditorium or assembly seating area.

❖ A proscenium wall separates the stage area from the audience. The performance on the stage is viewed through a large opening known as the main proscenium opening. Where the stage height is greater than 50 feet (15 240 mm), a proscenium wall of approved construction is required by the code (see Section 410.3.4).

STAGE. A space within a building utilized for entertainment or presentations, which includes overhead hanging curtains, drops, scenery or stage effects other than lighting and sound. Stage area shall be measured to include the entire performance area and adjacent backstage and support areas not separated from the performance area by fire-resistance-rated construction. Stage height shall be measured from the lowest point on the stage floor to the highest point of the roof or floor deck above the stage.

❖ For the purpose of applying the code, areas used for entertainment and presentations are defined to be on stage. The fuel load differences between various stage

types are typically due to size or height; hence, a single definition is all that is needed in conjunction with the code that provides the thresholds at which protection requirements are mandated.

410.3 Stages. Stage construction shall comply with Sections 410.3.1 through 410.3.7.

❖ This section indicates the scope of the code provisions that apply to stage construction.

410.3.1 Stage construction. Stages shall be constructed of materials as required for floors for the type of construction of the building in which such stages are located.

Exceptions:

1. Stages of Type IIB or IV construction with a nominal 2-inch (51 mm) wood deck, provided that the stage is separated from other areas in accordance with Section 410.3.4.

2. In buildings of Type IIA, IIIA and VA construction, a fire-resistance-rated floor is not required, provided the space below the stage is equipped with an automatic fire-extinguishing system in accordance with Section 903 or 904.

3. In all types of construction, the finished floor shall be constructed of wood or approved noncombustible materials. Openings through stage floors shall be equipped with tight-fitting, solid wood trap doors with approved safety locks.

❖ Stages, most simply, are floor construction, and as such, must conform to the requirements for all other floors in the building so that the building's designated construction classification will be consistent throughout.

Since wood is the material of choice for stage floors because it allows for nailing of sets and props and provides the best acoustics for dance and other performances, the code allows three exceptions to the requirement for stage construction.

Exception 1 permits stages in buildings of any construction classification to be constructed with a nominal 2-inch (51 mm) wood deck supported by either unprotected noncombustible construction (Type IIB) or heavy timber construction (Type IV) where the stage is separated from the audience by a proscenium wall built in accordance with Section 410.3.4.

Under Exception 2, stage floors are not required to be fire-resistance rated in buildings of Types IIA, IIIA and VA where the space below the stage is equipped with an automatic fire-extinguishing system installed in accordance with Section 903 or 904. In this case, the requirement for fire-resistance-rated floors is offset by the protection supported by the automatic fire-extinguishing system.

Exception 3 recognizes that stages are predominantly constructed with a finished floor of wood and identifies wood as an acceptable finish material for all construction classifications. It also permits openings in stage floors to be protected with tight-fitting solid wood doors with safety locks to secure them. The normal operation of stage trap doors does not typically allow them to have a fire-resistance rating.

410.3.1.1 Stage height and area. Stage areas shall be measured to include the entire performance area and adjacent backstage and support areas not separated from the performance area by fire-resistance-rated construction. Stage height shall be measured from the lowest point on the stage floor to the highest point of the roof or floor deck above the stage.

❖ This section simply repeats the information that is found in the definition of "Stage" in Section 410.2. It is appropriate that these technical requirements be located in the code, rather than only in the definition.

410.3.2 Galleries, gridirons, catwalks and pinrails. Beams designed only for the attachment of portable or fixed theater equipment, gridirons, galleries and catwalks shall be constructed of approved materials consistent with the requirements for the type of construction of the building; and a fire-resistance rating shall not be required. These areas shall not be considered to be floors, stories, mezzanines or levels in applying this code.

Exception: Floors of fly galleries and catwalks shall be constructed of any approved material.

❖ This section identifies the components of the areas located above the stage that are used to conceal moveable scenery from the audience and where the operation of such scenery and other stage effects is controlled.

The normal use and operations of a stage requires that equipment, normally rigging, be installed and rearranged as production requirements vary. This equipment is normally installed by clamping or welding to the structural framing over and around the stage, making protecting the framing by encasement or membranes infeasible; hence, a fire-resistance rating is not required for these elements. In order to control the fuel load in this area, however, the code requires that all of the elements must be of approved materials consistent with the requirements of the building's construction type. The exception to this requirement is that in all cases the floors of fly galleries and catwalks may be constructed of any approved material regardless of the building's construction type. Such spaces are generally limited in area, access is restricted to authorized personnel and they require floor materials that will deaden sound so as not to disrupt the performance below. These incidental areas do not have to meet the code's requirements as another story, mezzanine level or equipment platforms.

410.3.3 Exterior stage doors. Where protection of openings is required, exterior exit doors shall be protected with fire doors that comply with Section 715. Exterior openings that are located on the stage for means of egress or loading and unloading purposes, and that are likely to be open during occupancy of the theater, shall be constructed with vestibules to prevent air drafts into the auditorium.

❖ If exterior opening protectives are required, exit discharge doors directly from the stage to the outside must

be fire door assemblies in accordance with Section 715.

Since one of the major concerns is containing a stage fire to the stage area, any exterior opening from the stage that is likely to be opened during a performance is to be constructed with a vestibule to prevent air drafts into the auditorium. This vestibule requirement applies to all exterior openings that are likely to be open regardless of their intended use.

410.3.4 Proscenium wall. Where the stage height is greater than 50 feet (15 240 mm), all portions of the stage shall be completely separated from the seating area by a proscenium wall with not less than a 2-hour fire-resistance rating extending continuously from the foundation to the roof.

❖ The protection afforded by a 2-hour fire separation and opening protection (see Section 410.3.5) is required where the stage height is greater than 50 feet (15 240 mm). Stages with a height greater than 50 feet (15 240 mm) permit multiple settings and large amounts of scenery and scenic elements in dense configurations. The height may reduce the effectiveness of the suppression system and the multiple settings hung over the stage may further obstruct the suppression system and impede access to a fire originating high above the stage. Stages with a height less than 50 feet (15 240 mm) do not require separation from the audience since they represent a limited fuel load potential caused by scenery, drops, etc.

410.3.5 Proscenium curtain. The proscenium opening of every stage with a height greater than 50 feet (15 240 mm) shall be provided with a curtain of approved material or an approved water curtain complying with Section 903.3.1.1. The curtain shall be designed and installed to intercept hot gases, flames and smoke, and to prevent a glow from a severe fire on the stage from showing on the auditorium side for a period of 20 minutes. The closing of the curtain from the full open position shall be effected in less than 30 seconds, but the last 8 feet (2438 mm) of travel shall require not less than 5 seconds.

❖ Fuel loads in a stage with a height greater than 50 feet (15 240 mm) most often are high and the resulting fire is severe. In such cases, to permit the audience to evacuate the seating area without being threatened directly or indirectly by a stage fire, the proscenium opening is required to be protected with an approved fire curtain or water curtain.

A fire curtain must be capable of preventing the passage of flame or smoke for a period of 20 minutes. Also, since one of the goals is to protect against panic, the curtain should be able to prevent the glow produced by a stage fire from showing on the audience side for 20 minutes. Protection for a period of 20 minutes is based upon evacuation time. Typically, if all exits serving the audience are used, the space should be evacuated in 3 to 4 minutes. If half of the exits were blocked, the evacuation time for the space is still estimated to be only 6 to 7 minutes.

The curtain must close quickly upon detection of a fire so that containment offered by the curtain will be achieved as early in the fire as possible. The entire opening should, therefore, be closed in 30 seconds. The speed of descent must be slowed for the last 8 feet (2438 mm) of travel in order to allow individuals on the stage to move away from the curtain and to avoid injury from being hit by the batten in the bottom pocket of the curtain. It should be noted that only stages with a height greater than 50 feet (15 240 mm) are required to have a proscenium wall and curtain.

Water curtains conforming with Section 903.3.1.1 are viewed as providing an equivalent degree of protection to that of a fire curtain. This, in part, recognizes that emergency ventilation is also required in accordance with Section 410.3.7, which will assist in controlling smoke movement. The duration of protection is, of course, continuous as long as the sprinkler system is in operation.

410.3.5.1 Activation. The curtain shall be activated by rate-of-rise heat detection installed in accordance with Section 907.10 operating at a rate of temperature rise of 15 to 20°F per minute (8 to 11°C per minute), and by an auxiliary manual control.

❖ The proscenium curtain is required to close automatically upon detection of a temperature rate increase of 15 to 20°F (8 to 11°C) per minute. Typical rate-of-rise heat detectors meet this performance criteria. The detection system must be installed in accordance with Section 907.10 to perform the intended fire safety functions. In addition to the automatic-closing requirement, a means of manually closing the curtain (e.g., rolling ball, knife, lever release) is required both as an alternative closing mechanism and as a means of closing the curtain if a fire is detected prior to activation of the detectors.

410.3.5.2 Fire test. A sample curtain with a minimum of two vertical seams shall be subjected to the standard fire test specified in ASTM E 119 for a period of 30 minutes. The curtain shall overlap the furnace edges by an amount that is appropriate to seal the top and sides. The curtain shall have a bottom pocket containing a minimum of 4 pounds per linear foot (58 N/m) of batten. The exposed surface of the curtain shall not glow, and flame or smoke shall not penetrate the curtain during the test period. Unexposed surface temperature and hose stream test requirements are not applicable to the proscenium fire safety curtain test.

❖ Fabric materials capable of passing the ASTM E 119 test for a 30-minute period are often reinforced with wire to strengthen the fabric. This reinforcement becomes discontinuous at the seams. Also, the threads used to stitch the seams and the manner in which the seams are stitched are normally the weaker part of the curtain. If there is a failure, the seams will ordinarily fail first. For these reasons, at least two vertical seams must be included in the test specimen.

The overlay and batten in the bottom pocket is intended to simulate field conditions. If a fire were to start on the stage, the initial pressure of the heated air would cause the curtain to move outward. When the roof vents open (similar to the condition in the test chamber), the relative pressure on the curtain reverses, swinging the curtain in toward the stage and away from the audience. Under these conditions, gaps can develop with a flexible material. In order to arrest this condition, proscenium curtains are weighted down at the bottom with a batten. During the test there must be no gaps, either around the opening to the chamber or in the seams, which would allow smoke or flames to penetrate. Similarly, if a gap were to open, the glow of the fire would be visible.

Since the intent of the code is only to form a temporary barrier in order to increase the time available for the audience to exit, the curtain is not intended to form a barrier such as would normally be required in a 2-hour fire-resistance-rated wall; hence, the unexposed surface temperature and hose stream criteria of the ASTM E 119 test are not applicable.

410.3.5.3 Smoke test. Curtain fabrics shall have a smoke-developed rating of 25 or less when tested in accordance with ASTM E 84.

❖ It is important that proscenium curtain fabric not produce smoke under fire conditions so that smoke is not introduced into the area where the audience is located. For this reason, the treated fabric must be tested in accordance with referenced standard ASTM E 84 and must demonstrate a maximum smoke-developed rating of 25. In many cases, the fire-resistance rating of the proscenium curtain is obtained by treating the fabric with a chemical solution. During a fire, this solution serves to protect the fabric and, through chemical reaction, provides a barrier against the passage of smoke and flames.

410.3.5.4 Tests. The completed proscenium curtain shall be subjected to operating tests prior to the issuance of a certificate of occupancy.

❖ Since the effectiveness of the proscenium curtain is dependent on it functioning properly, it must be operationally tested before it is approved. The activation by automatic and manual means and the rate of descent for closing the curtain must be checked for proper operation.

410.3.6 Scenery. Combustible materials used in sets and scenery shall be rendered flame resistant in accordance with Section 805 and the *International Fire Code*. Foam plastics and materials containing foam plastics shall comply with Section 2603 and the *International Fire Code*.

❖ This section recognizes that scenery and sets are decorative materials. As such, combustible materials used in sets and scenery must be rendered flame resistant in accordance with the provisions of Section 805 and the IFC. It should be noted that Section 2603 and the IFC also provide the appropriate rules for regulating decorative materials made of or containing foamed plastics.

410.3.7 Stage ventilation. Emergency ventilation shall be provided for stages larger than 1,000 square feet (93 m²) in floor area, or with a stage height greater than 50 feet (15 240 mm). Such ventilation shall comply with Section 410.3.7.1 or 410.3.7.2.

❖ In addition to an automatic sprinkler system, emergency ventilation is one of the key fire protection components for stages that contain large fuel loads. Stages with large areas or heights permit multiple settings and large amounts of scenery and scenic elements in dense configurations. Increased height also permits multiple settings to be hung over the stage that may reduce the effectiveness of the suppression system or impede access to a fire originating high above the stage. Based on these factors and the potential hazard presented by them, stages larger than 1,000 square feet (93 m²) in area or with a height greater than 50 feet (15 240 mm) are required to be equipped with emergency ventilation to control smoke movement and minimize the potential for fire and smoke to spread to the seating area.

410.3.7.1 Roof vents. Two or more vents constructed to open automatically by approved heat-activated devices and with an aggregate clear opening area of not less than 5 percent of the area of the stage shall be located near the center and above the highest part of the stage area. Supplemental means shall be provided for manual operation of the ventilator. Curbs shall be provided as required for skylights in Section 2610.2. Vents shall be labeled.

❖ Where vents are used to ventilate a stage, the code requires that there be a minimum of two vents provided for redundancy. Those vents are to be located near the center and above the highest part of the stage, since this is the point at which smoke and hot gases are likely to accumulate. The minimum required aggregate clear opening area of the vents is required to be 5 percent of the floor area of the stage. The vents must open automatically by approved heat-activated devices, often fusible links, and must also be operable by a supplemental manual means. The manual means serves as a backup to the automatic detectors and permits the vent to be opened as a precaution prior to achieving the heat required to activate the heat device. The vents are to be provided with curbs as required for skylights in Section 2610.2. In addition, the vents must be labeled by an approved agency (see Section 1703.5 for further labeling requirements).

410.3.7.2 Smoke control. Smoke control in accordance with Section 909 shall be provided to maintain the smoke layer interface not less than 6 feet (1829 mm) above the highest level of the assembly seating or above the top of the proscenium opening where a proscenium wall is provided in compliance with Section 410.3.4.

❖ In addition to roof vents, another option for emergency ventilation of stages is provided in this section (i.e., smoke control). This section, in concert with the provisions contained in Section 909, provides the criteria for smoke control for stages larger than 1,000 square feet (93 m²) in floor area or with a stage height greater than 50 feet (15 240 mm). In such cases, the smoke layer interface must be maintained 6 feet (1829 mm) above the highest level of the assembly seating or above the top of the proscenium opening when one is provided. The smoke layer is maintained 6 feet (1829 mm) above the proscenium opening in order to prevent smoke from entering into the audience area.

410.4 Platform construction. Permanent platforms shall be constructed of materials as required for the type of construction of the building in which the permanent platform is located. Permanent platforms are permitted to be constructed of fire-retardant-treated wood for Type I, II, and IV construction where the platforms are not more than 30 inches (762 mm) above the main floor, and not more than one-third of the room floor area and not more than 3,000 square feet (279 m²) in area. Where the space beneath the permanent platform is used for storage or any other purpose other than equipment, wiring or plumbing, the floor construction shall not be less than 1-hour fire-resistance-rated construction. Where the space beneath the permanent platform is used only for equipment, wiring or plumbing, the underside of the permanent platform need not be protected.

❖ This section establishes the scope of the code requirements for platform construction. A permanent platform is basically raised floor construction. As such, the construction must be consistent with that of all the floors in the building.

Permanent platforms can be of fire-retardant-treated wood in limited applications. In buildings of Type I, II or IV construction, platforms can be of fire-retardant-treated wood if they are no more than 30 inches (762 mm) in height, are no more than one-third of the room floor area and are 3,000 square feet (279 m²) or less in area. A platform that meets these three size limits poses a small fire risk relative to the normal fuel load in the room. As such, the platform deck and its supporting construction can be of fire-retardant-treated wood materials. If the platform exceeds any one of the three size limits, then it must be constructed of materials consistent with the building's type of construction classification. A basic premise of Section 410.4 is that the platform must not significantly increase the fire hazard of the space or building. As such, the space beneath platforms is regulated to appropriately abate the risk. Where the space beneath a permanent platform is utilized for any purpose other than electrical wiring or plumbing, the platform is to provide a 1-hour fire-resistance rating. The purpose of the fire-resistance-rating requirement is to provide some structural integrity to the platform should a fire occur in the concealed space. In the case of permanent platforms where the space beneath the platform is only used for plumbing or electrical wiring, no further protection is required since the fire risk will be minimal.

410.4.1 Temporary platforms. Platforms installed for a period of not more than 30 days are permitted to be constructed of any materials permitted by the code. The space between the floor and the platform above shall only be used for plumbing and electrical wiring to platform equipment.

❖ Temporary platforms may be constructed of any material regardless of the building's construction classification due to their limited time of use and their normally limited size and fuel load. Because temporary platforms are permitted to be constructed of any approved material, the space beneath a temporary platform may never be used for any purpose other than electrical wiring or plumbing to the platform equipment. Such lines serving other areas may not be located beneath a temporary platform.

410.5 Dressing and appurtenant rooms. Dressing and appurtenant rooms shall comply with Sections 410.5.1 through 410.5.4.

❖ Because stages are open to the viewing audience and typically contain a substantial fuel load, it is essential to contain fires in accessory rooms around the stage to the room of origin; therefore, this section contains provisions for the separation of such areas from the stage and from one another. Section 410.5.4 also contains provisions for adequate egress so that an occupant on the stage and in areas above and below it may be able to egress the area safely should a fire occur in the stage area.

410.5.1 Separation from stage. Where the stage height is greater than 50 feet (15 240 mm), the stage shall be separated from dressing rooms, scene docks, property rooms, workshops, storerooms and compartments appurtenant to the stage and other parts of the building by a fire barrier wall and horizontal assemblies, or both, with not less than a 2-hour fire-resistance rating with approved opening protectives. For stage heights of 50 feet (15 240 mm) or less, the required stage separation shall be a fire barrier wall and horizontal assemblies, or both, with not less a 1-hour fire-resistance rating with approved opening protectives.

❖ Separation of the stage from appurtenant rooms provides a significant level of fire containment. Such containment is fundamental to limiting the spread of fire from adjacent spaces to the stage area as well as the growth of fires in the stage area itself. The 2-hour fire-resistance rating for stages with a height greater than 50 feet (15 240 mm), such as large theatrical stages, acknowledges the significant potential for a large fuel load and is consistent with the 2-hour rating of Section 410.3.4. Although stages with a height of 50 feet (15 240 mm) or less represent a limited potential fuel load, a 1-hour fire-resistance rating will provide an additional level of compartmentation in the event a fire originates at the stage (which may not be sprinklered if the stage complies with Exception 2 of Section 410.6). Such stages are likely to be found in educational occupancies or meeting halls.

410.5.2 Separation from each other. Dressing rooms, scene docks, property rooms, workshops, storerooms and compart-

ments appurtenant to the stage shall be separated from each other by fire barrier wall and horizontal assemblies, or both, with not less than a 1-hour fire-resistance rating with approved opening protectives.

❖ As an additional level of protection, rooms and compartments appurtenant to the stage must be separated from one another by approved 1-hour fire-resistance-rated fire barrier walls and horizontal assemblies. Rooms appurtenant to the stage often have very high fuel loads (i.e., property rooms and activities with a history of fire incidents, such as scenery workshops). This additional level of fire containment is intended to minimize fire growth and limit fires in these areas to the room of origin.

410.5.3 Opening protectives. Openings other than to trunk rooms and the necessary doorways at stage level shall not connect such rooms with the stage, and such openings shall be protected with fire door assemblies that comply with Section 715.

❖ Other than the doorways required at stage level, openings directly into the stage from dressing rooms and other appurtenant rooms are not permitted pursuant to Section 410.5.1 and this section. Since the separation from the stage is required to have a fire-resistance rating, all door openings therein must be protected in accordance with Section 715. As with the requirements for the separation of such spaces from the stage, the restriction on openings is intended to minimize the exposure of fire risk.

410.5.4 Stage exits. At least one approved means of egress shall be provided from each side of the stage; and from each side of the space under the stage. At least one means of escape shall be provided from each fly gallery and from the gridiron. A steel ladder, alternating tread stairway or spiral stairway is permitted to be provided from the gridiron to a scuttle in the stage roof.

❖ Because of the relative fire hazards associated with stages, this section requires that a minimum of one approved means of egress be provided from each side of the stage and from each side of the occupied space under the stage. This is intended to provide the occupants of a stage ready access to evacuate the stage area in the event of a fire. It should be noted that in accordance with Exception 7 to Section 1019.1, means of egress are not required to be enclosed if they are within the stage area. Fly galleries and gridirons are ordinarily occupied by a few persons at any one time who are likely to be very familiar with the stage operations. As such, these areas need only be provided with one approved means of egress. In these areas, ladders, alternating tread devices or spiral stairs are permitted (see Exception 5, Section 1014.6.1). In all cases, interior means of egress stairs serving these areas are not required to be enclosed (see Exception 7, Section 1019.1). A steel ladder, alternating tread stairways or spiral stairways from the gridiron to the roof is permitted.

[F] 410.6 Automatic sprinkler system. Stages shall be equipped with an automatic fire-extinguishing system in accordance with Chapter 9. The system shall be installed under the roof and gridiron, in the tie and fly galleries, in places behind the proscenium wall of the stage, and in dressing rooms, lounges, workshops and storerooms accessory to such stages.

Exceptions:

1. Sprinklers are not required under stage areas less than 4 feet (1219 mm) in clear height utilized exclusively for storage of tables and chairs, provided the concealed space is separated from the adjacent spaces by not less than $^5/_8$-inch (15.9 mm) Type X gypsum board.

2. Sprinklers are not required for stages 1,000 square feet (93 m^2) or less in area and 50 feet (15 240 mm) or less in height where curtains, scenery or other combustible hangings are not retractable vertically. Combustible hangings shall be limited to a single main curtain, borders, legs and a single backdrop.

❖ Stages contain significant quantities of combustible materials stored in, around and above the stage that are located in close proximity to large quantities of lighting equipment (i.e., scenery and lighting above the stage). There also is scenery on the sides and rear of the stage; shops located along the back and sides of the stage and storage, props, trap doors and lifts under the stage floor. This combination of fuel load and ignition sources increases the potential for a fire. As such, stages and accessory areas, such as dressing rooms, workshops and storerooms, are required to be protected with an automatic fire-extinguishing system.

The references to Chapter 9 indicate that the fire-extinguishing system may be designed in accordance with NFPA 13 and Section 903.3.1.1 or, if not more than 20 sprinklers are required on any single connection, a limited area sprinkler system may be used (see Section 903.3.5.1.1).

Exception 1 to the requirements set forth in Section 410.6 applies to areas less than 4 feet (1219 mm) in clear height under stages that are used only for the storage of tables and chairs. Because of the limited fuel load present, such areas are not required to be protected with sprinklers, provided that the concealed space is separated from all adjacent spaces by no less than $^5/_8$-inch (15.9 mm) Type X gypsum board. This level of separation is intended to provide fire containment in the absence of sprinkler protection. This arrangement is often found in educational buildings where the room is used as a multipurpose room.

Exception 2 to Section 410.6 recognizes that stages 1,000 square feet (93 m^2) or less in area and 50 feet (15 240 mm) or less in height represent a limited potential for combustibles that does not warrant the requirements for an automatic sprinkler system. The code further limits the amounts of combustible materials by not allowing any vertical retractable curtains, hangings, etc. It should be noted, however, that although sprinkler protection is not required, the requirements of Section 410.5 still apply.

[F] 410.7 Standpipes. Standpipe systems shall be provided in accordance with Section 905.

❖ Because of the potentially large fuel load and three-dimensional aspect of the fire hazard associated with stages greater than 1,000 square feet (93 m^2) in area, a Class III wet standpipe system is required on each side of such stages. The standpipes are required to be equipped with both a 1$^1/_2$-inch (38 mm) and 2$^1/_2$-inch (64 mm) hose connection. The required design criteria for the standpipe system, including hoses and cabinets, is specified in Section 905.3.4.

SECTION 411
SPECIAL AMUSEMENT BUILDINGS

411.1 General. Special amusement buildings having an occupant load of 50 or more shall comply with the requirements for the appropriate Group A occupancy and this section. Amusement buildings having an occupant load of less than 50 shall comply with the requirements for a Group B occupancy and this section.

Exception: Amusement buildings or portions thereof that are without walls or a roof and constructed to prevent the accumulation of smoke.

For flammable decorative materials, see the *International Fire Code*.

❖ The provisions for special amusement buildings are based on a CABO Board for the Coordination of the Model Codes (BCMC) report dated May 1988. The subject of special amusement buildings was placed on the agenda of the BCMC as a result of the May 11, 1984 fire in the "Haunted Castle" at the Six Flags Great Adventure Park in Jackson Township, New Jersey. At the time of this fire, there were 28 to 34 visitors and three employees in the Haunted Castle. Eight of the visitors, unable to exit the structure immediately, died in the fire. The major factors contributing to the loss of life in this fire were:

1. The failure to detect and extinguish the fire in its incipient stage by means of fixed fire detection and suppression systems.

2. The ignition of synthetic foam material and subsequent fire and smoke spread involving combustible interior finishes and contents.

3. The difficulty of escape by occupants based on fire conditions in the haunted-house type of environment.

A special amusement building is one in which the egress is not readily apparent, is intentionally confounded or is not readily available (see Section 411.2). This section addresses the hazards associated with such use by requiring automatic fire detection (see Section 411.3), automatic sprinkler protection (see Section 411.4), an emergency voice/alarm communication system (see Section 411.6) and specific means of egress lighting and marking (see Section 411.7). Additionally, only interior finish materials that meet the most stringent flame spread classification, Class A, are permitted in special amusement buildings.

In addition to the provisions of this section, other applicable requirements such as means of egress (i.e., occupant load, travel distance, etc.) and building construction (i.e., type of construction, fire-resistance ratings, etc.) apply in accordance with the appropriate assembly group classification (see Section 411.1).

Special amusement buildings are considered as assembly uses based on the definition in Section 303.1. This section further specifies that a Group A classification is only applicable when the building's design occupant load is 50 or more. The provisions of this section apply in addition to the other requirements for the appropriate assembly use, usually Group A-1 or A-3. Very small special amusement buildings are not required to be classified as Group A. If the design occupant load is less than 50, then the building is classified as Group B. This acknowledges the lesser hazards associated with smaller groups of people. Although smaller buildings are classified as Group B, they must still meet the requirements of this section as special amusement buildings.

Section 411 does not apply to a facility that is constructed to permit the free and immediate ventilation of the products of combustion to the outside. Such free and immediate ventilation addresses the hazard associated with many special amusement buildings relative to the rapid accumulation of smoke in a building in which the egress is not readily apparent, confounded or not readily available.

All flammable decorative materials used in special amusement buildings are required to follow the provisions of the IFC. The use of flammable and combustible materials in these types of buildings must be strictly regulated to limit the fire hazards to the building occupants (refer to Sections 805 and 806 of the IFC for those restrictions).

411.2 Special amusement building. A special amusement building is any temporary or permanent building or portion thereof that is occupied for amusement, entertainment or educational purposes and that contains a device or system that conveys passengers or provides a walkway along, around or over a course in any direction so arranged that the means of egress path is not readily apparent due to visual or audio distractions or is intentionally confounded or is not readily available because of the nature of the attraction or mode of conveyance through the building or structure.

❖ This section provides a definition for a special amusement building. In general, a special amusement building is a building or portion thereof in which people gather (thus, an assembly group) and in which egress is either not readily apparent due to distractions, is intentionally confounded (i.e., maze) or is not readily available. The definition includes all such facilities, including portable and temporary structures. The hazard associated with such buildings is not related to the permanence or

length of use; therefore, seasonal uses (such as haunted houses at Halloween) and portable uses (carnival rides) are included if they meet the criteria in the definition.

[F] 411.3 Automatic fire detection. Special amusement buildings shall be equipped with an automatic fire detection system in accordance with Section 907.

❖ The automatic fire detection system is required to provide early warning of fire and must comply with Section 907. The detection system is required regardless of the presence of staff in the building.

[F] 411.4 Automatic sprinkler system. Special amusement buildings shall be equipped throughout with an automatic sprinkler system in accordance with Section 903.3.1.1. Where the special amusement building is temporary, the sprinkler water supply shall be of an approved temporary means.

Exception: Automatic fire sprinklers are not required where the total floor area of a temporary special amusement building is less than 1,000 square feet (93 m^2) and the travel distance from any point to an exit is less than 50 feet (15 240 mm).

❖ One protection strategy to minimize the potential hazard to occupants is to control fire development. As such, special amusement buildings are required to be protected with an automatic sprinkler system. If the building is small [less than 1,000 square feet (93 m²)] and the travel distance to exits is short [less than 50 feet (15 240 mm)] and only used on a temporary basis (such as at Halloween), automatic sprinklers are not required. In such buildings, it is anticipated that automatic fire detection and the resulting alarm (see Section 411.5) will provide additional egress time for the limited number of occupants.

Since many special amusement buildings are portable or temporary, it is not practical to require a permanent, automatic sprinkler water supply as required in Chapter 9. Instead, the building official may allow a reliable, temporary water supply. There are other unique design considerations for the sprinkler system, such as drainage and pipe and sprinkler support, which may be necessary because of the movement of the structures from one location to another.

[F] 411.5 Alarm. Actuation of a single smoke detector, the automatic sprinkler system or other automatic fire detection device shall immediately sound an alarm at the building at a constantly attended location from which emergency action can be initiated including the capability of manual initiation of requirements in Section 907.2.11.2.

❖ Upon activation of either the automatic fire detection or the automatic sprinkler systems, an alarm is required to be sounded at a constantly attended location. The staff at the location is expected to be capable of then providing the required egress illumination, stopping the conflicting or confusing sounds and distractions and activating the exit marking required by Section 411.7. It is also anticipated that the staff would be capable of preventing additional people from entering the building.

[F] 411.6 Emergency voice/alarm communications system. An emergency voice/alarm communications system shall be provided in accordance with Sections 907.2.11 and 907.2.12.2, which is also permitted to serve as a public address system and shall be audible throughout the entire special amusement building.

❖ An integral part of the fire protection systems required for special amusement buildings is an emergency voice/alarm communications system. This system can serve as a public address system to alert the building occupants of the fire emergency and provide them with the proper emergency instructions. The system must be installed in accordance with NFPA 72 and must be heard throughout the entire special amusement building. Upon activation, the system must sound an alert tone followed by the necessary voice instructions. These instructions can save valuable time in directing the building occupants to quickly and safely egress.

411.7 Exit marking. Exit signs shall be installed at the required exit or exit access doorways of amusement buildings. Approved directional exit markings shall also be provided. Where mirrors, mazes or other designs are utilized that disguise the path of egress travel such that they are not apparent, approved low-level exit signs and directional path markings shall be provided and located not more than 8 inches (203 mm) above the walking surface and on or near the path of egress travel. Such markings shall become visible in an emergency. The directional exit marking shall be activated by the automatic fire detection system and the automatic sprinkler system in accordance with Section 907.2.11.2.

❖ During normal operation, the illumination and marking of the egress path may not be adequate to allow for prompt egress from the building. Activation of the automatic fire detection system or automatic sprinkler system must activate the directional exit marking. Where obstructions and confusion may still exist because of the nature of the facility, low-level exit signs and directional path markings are required in order to direct the occupants toward the exits.

411.8 Interior finish. The interior finish shall be Class A in accordance with Section 803.1.

❖ All interior finish materials must be tested for surface-burning performance in accordance with ASTM E 84. Due to the potential for fire to spread quickly in the relatively confined spaces in these structures, only Class A materials are permitted to be used as interior finish in a special amusement building. These special amusement buildings are not permitted the one classification reduction that would normally be allowed by Table 803.5 in a sprinklered building.

SECTION 412
AIRCRAFT-RELATED OCCUPANCIES

❖ This section contains code requirements for some very specialized occupancies dealing with aircraft. Aircraft pose some of the same hazards associated with motor vehicles; therefore, some of the same requirements are applicable for life and fire safety issues. Section 412.1 addresses airport traffic control towers. Although these structures do not house aircraft, they are included in this section for consistency. These structures pose unique hazards to occupants because of their extreme height and limited routes of escape. This section contains requirements governing the permitted types of construction and necessary egress along with the needed fire protection systems.

Sections 412.2 and 412.3 are composed of code provisions that regulate both commercial and residential aircraft hangars. The overall building fire load for hangars is typically low because of the considerable amounts of metal in the aircraft that can absorb heat and provide limited combustibility to sustain a fire. An aircraft fire, however, can be quite severe because of the flammability of the fuel contained in its tank(s). Provisions for regulating the exterior walls, basements and floor surfaces of commercial aircraft hangars are specified along with separation and fire suppression requirements.

Section 412.4 addresses specialized hangars that are used for the painting of aircraft with flammable materials. Just as Section 416 contains limitations and requirements for the application of flammable finishes, so does Section 412.4 for aircraft paint hangars. Use and occupancy requirements commensurate with the hazards are provided, including construction, fire protection systems and necessary ventilation.

Lastly, Section 412.5 identifies those code requirements applicable to helicopter landing and service ports. Although the code relies heavily on the requirements of NFPA 418 for heliports and helistops located on the roofs of buildings and structures, there are provisions applicable for all locations. This section contains requirements for size limitations, design and means of egress.

412.1 Airport traffic control towers.

412.1.1 General. The provisions of this section shall apply to airport traffic control towers not exceeding 1,500 square feet (140 m²) per floor occupied only for the following uses:

1. Airport traffic control cab.
2. Electrical and mechanical equipment rooms.
3. Airport terminal radar and electronics rooms.
4. Office spaces incidental to the tower operation.
5. Lounges for employees, including sanitary facilities.

❖ Airport traffic control towers are structures designed for highly specific functions. These functions include the housing of vital electronic equipment, providing an elevated structure for electronic communication, such as radar, and providing an observation area that allows an unobstructed view of the ground and airspace for air traffic controllers. This functional configuration creates special hazards, including limited means of egress, limited fire-fighting accessibility to upper floors and vulnerability to exposure fires. The code provisions are based on providing adequate fire protection and life safety to a small number of occupants within a limited area [1,500 square feet (139 m²) per floor maximum], while allowing a structure configuration that accommodates the intended function. The intent of the code is to restrict the fuel load by limiting the structure construction to mostly noncombustible materials; minimizing combustible contents and potential ignition sources by limiting the use of the structure; providing automatic fire detection systems, adequate and reliable egress and fire department access and providing standby power in towers over 65 feet (19 812 mm) in height. These measures result in prompt evacuation and fire department accessibility early in a fire event.

This section applies only to airport traffic control towers that do not exceed 1,500 square feet (139 m²) per floor and are limited to uses exclusively related to air traffic control purposes. Rooms and spaces incidental to the air traffic control function, such as equipment rooms, offices, lounges and restrooms, are permitted in the tower. The section does not apply to air traffic control towers containing other uses, including assembly areas, observation decks, restaurants and terminal operations. The 1,500-square-foot (139 m²) restriction is based, in part, on a survey of existing towers that would fall within the above restrictions.

412.1.2 Type of construction. Airport traffic control towers shall be constructed to conform to the height and area limitations of Table 412.1.2.

❖ The limited area per floor, low fuel load, adequate egress provisions and low occupant load present a relatively low potential fire hazard. The provisions of Table 503 are, therefore, modified by Table 412.1.2. The allowable height and area modifications permitted by Sections 504 and 506 do not apply to Table 412.1.2.

TABLE 412.1.2
HEIGHT AND AREA LIMITATIONS FOR AIRPORT TRAFFIC CONTROL TOWERS

TYPE OF CONSTRUCTION	HEIGHT[a] (feet)	MAXIMUM AREA (square feet)
IA	Unlimited	1,500
IB	240	1,500
IIA	100	1,500
IIB	85	1,500
IIIA	65	1,500

For SI: 1 foot = 304.8 mm, 1 square foot = 0.093 m².
a. Height to be measured from grade to cab floor.

❖ Table 412.1.2 functions the same as Table 503, except that the height restriction is based on feet only and not on stories, and the actual height is measured from

grade plane to the finished floor of the cab or the highest occupied level. The number of stories is not a criterion, since many of the towers do not have occupied stories for the entire height between the ground and the occupiable level at the top. If the type of construction is known, the allowable height is given by the corresponding figure in the second column. If the height is known, the permitted construction types are determined from the first column. The third column restates the area-per-floor restriction given in Section 412.1.1.

412.1.3 Egress. A minimum of one exit stairway shall be permitted for airport traffic control towers of any height provided that the occupant load per floor does not exceed 15. The stairway shall conform to the requirements of Section 1009. The stairway shall be separated from elevators by a minimum distance of one-half of the diagonal of the area served measured in a straight line. The exit stairway and elevator hoistway are permitted to be located in the same shaft enclosure, provided they are separated from each other by a 4-hour separation having no openings. Such stairway shall be pressurized to a minimum of 0.15 inch of water column (43 Pa) and a maximum of 0.35 inch of water column (101 Pa) in the shaft relative to the building with stairway doors closed. Stairways need not extend to the roof as specified in Section 1009.12. The provisions of Section 403 do not apply.

Exception: Smokeproof enclosures as set forth in Section 1019.1.8 are not required where required stairways are pressurized.

❖ The benefit of a second exit is greatly reduced when the two exits would be in such close proximity. With the low fuel load, limited area per floor and limited number of occupants, only one exit is required per floor provided that the occupant load of each floor is 15 or less. The occupant load restriction of 15 persons is based on the calculated occupant load for a business use of 1,500 square feet (139 m²) (see Table 1004.1.2).

Since only one exit stairway is provided, elevators must be remotely located from the stairway a distance of one-half of the diagonal of the area served, as if they were the second required exit. If conditions are such that the stairway is not usable but the elevators are, the elevators provide an alternative escape route from the tower and fire department access to each floor level. The elevators are to be considered an egress route only if the stairway is not usable.

The code does permit the required exit stairway and the normally provided elevator to be in the same shaft enclosure; however, to maintain their independence and usability, a 4-hour separation is required between the two building features. No intercommunicating openings are permitted between these two elements to increase the likelihood that one will remain operational during a fire situation. Additionally, an exit stairway located in the same shaft enclosure as the elevator hoistway must be pressurized in accordance with Section 909.20.5.

The exit stairway for an airport traffic control tower does not need to provide roof access. Because of the limited area of these buildings, the fire department most likely will not need access to the roof for fire-fighting operations. Roof access for repair and mechanical equipment maintenance is not mandated by the code. This section of the code also notes that the requirements for high-rise buildings in regards to stairway door operations and communication are also not applicable because of the limited occupant load.

Finally, an exception is provided for smokeproof enclosures. Assuming that the exit stairway is not in the same shaft enclosure as the elevator hoistway, the stairway must still be a smokeproof enclosure in accordance with Section 1019.1.8. Alternatively, the required exit stairway may be pressurized in accordance with Section 909.20.5. These provisions serve to maintain the required means of egress as the building occupants travel large vertical distances to the discharge doors and the public way.

[F] 412.1.4 Automatic fire detection systems. Airport traffic control towers shall be provided with an automatic fire detection system installed in accordance with Section 907.2.

❖ To ensure early fire detection and to alert the occupants to egress the building during the incipient stage of a fire event, an automatic fire detection system is required in accordance with Section 907.2.22. The code does not require an automatic sprinkler system because of the limited fuel load, limited occupant load and protection afforded by early detection.

[F] 412.1.5 Standby power. A standby power system that conforms to Section 2702 shall be provided in airport traffic control towers more than 65 feet (19 812 mm) in height. Power shall be provided to the following equipment:

1. Pressurization equipment, mechanical equipment and lighting.

2. Elevator operating equipment.

3. Fire alarm and smoke detection systems.

❖ To increase the reliability of the protection afforded by equipment provided for egress and fire department accessibility, a standby power system is required in airport traffic control towers that exceed 65 feet (19 812 mm) in height. The system must support the pressurization and mechanical equipment, egress illumination, elevator operation, fire alarm system and the automatic fire detection system. Height is measured in accordance with the definition found in Section 202. If the pressurized stairway alternatives in Section 412.1.3 are utilized, the mechanical equipment required to pressurize the stairway must be connected to the standby power system to provide the same degree of reliability required for the smokeproof enclosure. The reference to Section 2702 incorporates some of the same standby power requirements contained under the high-rise provisions based on the premise that towers over 65 feet (19 812 mm) in height represent similar hazards to those of high-rise buildings.

412.1.6 Accessibility. Airport traffic control towers need not be accessible as specified in the provisions of Chapter 11.

❖ Airport traffic control towers are not required to meet the accessibility requirements of Chapter 11 or ICC A117.1. Most likely, the employees of an airport traffic control tower will not be persons with physical disabilities based on the necessary work. Further, meeting the accessibility requirements for all services and facilities would most likely result in the building exceeding the 1,500-square-foot (139 m^2) limitation.

412.2 Aircraft hangar.

❖ The requirements of Section 412.2 address commercial (or nonresidential) aircraft hangars. It should be noted that most commercial aircraft hangars will not be limited in height in terms of feet (see Section 504.1) or in area (see Section 507) based on the presence of an automatic sprinkler system. All commercial aircraft hangars, however, must be regulated in regards to exterior wall fire-resistance ratings, basement limitations, floor surface construction requirements, heating equipment separation and finishing restrictions. All of these provisions serve to abate the hazards associated with large aircraft and their integral fuel tanks to acceptable fire safety levels.

412.2.1 Exterior walls. Exterior walls located less than 30 feet (9 144 mm) from property lines, lot lines or a public way shall have a fire-resistance rating not less than 2 hours.

❖ To abate the hazards of a fire condition from spreading from a commercial aircraft hangar to adjacent buildings and structures, the code requires 2-hour fire-resistance-rated exterior walls that are located less than 30 feet (9144 mm) from lot lines, property lines or public ways. Since aircraft hangars are classified as moderate-hazard storage facilities, Group S-1, the provisions of Table 602 are further modified regardless of the hangar's type of construction classification. Fire-resistance-rated exterior walls permit the fire department additional time and protection as it attempts to take control of a fire situation in a hangar located less than 30 feet (9144 mm) from the lot lines, property lines or public ways.

412.2.2 Basements. Where hangars have basements, the floor over the basement shall be of Type IA construction and shall be made tight against seepage of water, oil or vapors. There shall be no opening or communication between the basement and the hangar. Access to the basement shall be from outside only.

❖ As part of the hangar requirements, the use and separation of any basement levels are rigidly controlled. A fire in the main level of a hangar could pose a severe hazard not only to the occupants of the basement but also to the fire department itself. The floor/ceiling assembly located between the hangar and the basement must be of Type IA construction. With a floor of 2-hour fire-resistance-rated construction along with the supporting construction, a fire on the main level should be contained. Not only is the floor construction required to be rated, but it must be properly sealed or otherwise made waterproof. This requirement prevents the chances of a fire or other liquids and vapors from seeping into the floor construction and leaking into the basement level. Further, the code prohibits any openings or access between the main level of the hangar and the basement level. This would include fire-resistance-rated shafts. Any opening into the basement has to be made from the exterior of the structure via areaway construction.

412.2.3 Floor surface. Floors shall be graded and drained to prevent water or fuel from remaining on the floor. Floor drains shall discharge through an oil separator to the sewer or to an outside vented sump.

❖ These provisions go hand in hand with the waterproof requirements of Section 412.2.2. The floors of all hangars must be positively sloped to prevent standing liquids. This requirement is not only for personnel safety but also to minimize the effects of any spilled flammable liquids. Floor drains, if provided, must discharge their contents into an oil separator or an outside sump. This prevents the flammable liquids from entering the jurisdiction's sewer system and causing additional hazards.

412.2.4 Heating equipment. Heating equipment shall be placed in another room separated by 2-hour fire-resistance-rated construction. Entrance shall be from the outside or by means of a vestibule providing a two-doorway separation.

Exceptions:

1. Unit heaters suspended at least 10 feet (3048 mm) above the upper surface of wings or engine enclosures of the highest aircraft that are permitted to be housed in the hangar and at least 8 feet (2438 mm) above the floor in shops, offices and other sections of the hangar communicating with storage or service areas.

2. A single interior door shall be allowed, provided the sources of ignition in the appliances are at least 18 inches (457 mm) above the floor.

❖ As part of the special use and occupancy requirements for commercial aircraft hangars, all possible ignition sources must be controlled and isolated. Specifically, all heating equipment must be located in rooms that are separated from the main areas where the aircraft are parked. This separation must be 2-hour fire-resistance-rated construction. Although not explicitly stated, all openings through the rated walls must be protected. Doors connecting the heating equipment rooms and the main hangar area must be done with a vestibule or airlock arrangement such that one must pass through two doors prior to entering the other room. Again, this is done to minimize the possibility of any spilled flammable liquids and the resulting vapors from coming in contact with the ignition sources of the heating equipment.

Two exceptions to the separation requirements are provided. Unit heaters that are carefully located high above not only the floor surfaces but also the fuel tanks and engine compartments of the aircraft pose little risk. An allowance is also made if the heating equipment is located at least 18 inches (457 mm) above the floor of the separated room. In such a case, the vestibule/airlock arrangement with double door system is not required and can be done with just a single door. If either exception is used, care must be taken by the building and fire officials that these special stipulations and conditions are part of the certificate of occupancy.

412.2.5 Finishing. The process of "doping," involving use of a volatile flammable solvent, or of painting, shall be carried on in a separate detached building equipped with automatic fire-extinguishing equipment in accordance with Section 903.

❖ Any application of spraying or "doping" of flammable finishes or solvent treatments to aircraft is prohibited within the hangar. These types of operations must be done in a separate detached building that is provided with an automatic fire-extinguishing system. Although not specifically stated here, the intent of the code is to treat those types of buildings as an aircraft paint hangar in accordance with Section 412.4.

[F] 412.2.6 Fire suppression. Aircraft hangars shall be provided with fire suppression as required in NFPA 409.

Exception: Group II hangars as defined in NFPA 409 storing private aircraft without major maintenance or overhaul are exempt from foam suppression requirements.

❖ To minimize the fire hazards associated with commercial aircraft hangars, all such buildings are required to be protected with a fire suppression system. This requirement is applicable regardless of the size of the hangar in terms of height or area or the types and quantities of aircraft that are stored. The fire suppression system must be designed and installed in accordance with NFPA 409. This standard contains specific requirements for the suppression systems needed to properly protect hangars. Also contained within the standard are specific definitions. A Group II hangar intended only to store private aircraft is exempt from the standard's requirements for a foam suppression system. Such a hangar cannot be used for major maintenance or aircraft overhaul.

412.3 Residential aircraft hangars. Residential aircraft hangars as defined in Section 412.3.1 shall comply with Sections 412.3.2 through 412.3.6.

❖ This section of the code contains provisions that account for small, limited-size aircraft hangars that are truly accessory and incidental to a dwelling unit. Housing developments located along or adjacent to small-scale airports are the most obvious application of these code requirements. As part of the scoping requirement of this section, the hangar must meet all of the criteria listed in its definition in Section 412.3.1. A hangar that exceeds the limitations must then meet the provisions of Section 412.2. Included in the code requirements are fire separation from the adjacent residence; adequate means of egress; smoke detection and alarms; independent mechanical and plumbing systems and height and area limitations.

412.3.1 Definition. The following word and term shall, for the purposes of this chapter and as used elsewhere in this code, have the meaning shown herein.

❖ The definition of a term that is associated with the content of this section is contained herein. This definition can help in the understanding and application of the code requirements. It is important to emphasize that this term is not exclusively related to this section but is applicable everywhere the term is used in the code. The purpose for including this definition within this section is to provide more convenient access to it without having to refer back to Chapter 2.

For convenience, this term is also listed in Chapter 2 with a cross reference to this section.

The use and application of all defined terms, including that defined herein, are set forth in Section 201.

RESIDENTIAL AIRCRAFT HANGAR. An accessory building less than 2,000 square feet (186 m²) and 20 feet (6096 mm) in height, constructed on a one- or two-family residential property where aircraft are stored. Such use will be considered as a residential accessory use incidental to the dwelling.

❖ A residential aircraft hangar is considered an accessory or incidental structure to a residential house similar to any other shed or detached garage. One- or two-family dwellings are required to be constructed in accordance with the *International Residential Code®* (IRC®) (see Section 101.2). Although that code is also applicable to accessory structures associated with one- or two-family dwellings, it is clear that this section of the code was intended to control aircraft hangars since the IRC is silent on the subject.

A residential aircraft hangar is limited to 2,000 square feet (186 m²) in area and 20 feet (6096 mm) in height. The hangar must be integral to a residential home on the same property. As such, the hangar would be considered the same as the private garage used to store the home's motor vehicles.

412.3.2 Fire separation. A hangar shall not be attached to a dwelling unless separated by walls having a fire-resistance rating of not less than 1 hour. Such separation shall be continuous from the foundation to the underside of the roof and unpierced except for doors leading to the dwelling unit. Doors into the dwelling unit must be equipped with self-closing devices and conform to the requirements of Section 715 with at least a 4-inch (102 mm) noncombustible raised sill. Openings from a hanger directly into a room used for sleeping purposes shall not be permitted.

❖ A residential aircraft hangar can either be a stand-alone detached structure or attached to the residential dwelling. If the hangar is in close proximity or attached to the dwelling, it must then be separated with 1-hour fire-resistance-rated walls. This fire separation must be continuous from the floor slab up to the roof sheathing. The

only openings permitted in the fire separation are normal access doors. Windows or other vent openings are prohibited as are any openings between the hangar and the bedrooms of the dwelling. The doors must have a ³/₄-hour opening protective fire protection rating in accordance with Table 715.3. A 4-inch-high (102 mm) noncombustible step is required at the door along with a self-closing device. All of these fire separation requirements serve to isolate the fire hazards associated with the hangar from the occupants of the dwelling, similar to a private garage.

412.3.3 Egress. A hangar shall provide two means of egress. One of the doors into the dwelling shall be considered as meeting only one of the two means of egress.

❖ The hangar must be provided with two separate and remotely located means of egress. This redundancy provides persons with an alternative means of escaping a fire in the hangar. One of the means of egress can be the access door into the adjacent dwelling unit. The other means of egress could be the aircraft entrance door subject to the requirements of Sections 1008.1.1 and 1008.1.2.

[F] 412.3.4 Smoke detection. Smoke alarms shall be provided within the hangar in accordance with Section 907.2.21.

❖ A smoke alarm is required in the hangar space in accordance with Section 907.2.21. Similar to the smoke alarms required in each and every bedroom, in the immediate vicinity of the bedrooms and in each and every story of the dwelling, the hangar smoke alarm must be interconnected such that one alarm will activate all other alarms. An early detection and warning alert is essential to abate the hazards of an aircraft hangar located adjacent to a residence.

412.3.5 Independent systems. Mechanical and plumbing drain, waste and vent (DWV) systems installed within the hangar shall be independent of the systems installed within the dwelling. Building sewer lines may connect outside the structures.

Exception: Smoke detector wiring and feed for electrical subpanels in the hangar.

❖ To maintain the fire separation requirements of Section 412.3.2, the only openings permitted are normal access doors. Likewise, the mechanical and plumbing systems for the hangar must be independent of the systems within the residential house. The plumbing drain or waste line could discharge into its own building sewer line that then connects to the house's building sewer line. This connection must be done outside of both the house and the hangar. Electrical wiring serving as the feed for the subpanels in the hangar is permitted to penetrate the fire separation along with the wiring for the smoke alarm required by Section 412.3.4 and the necessary interconnection with the other interior alarms.

412.3.6 Height and area limits. Residential aircraft hangars shall not exceed 2,000 square feet (186 m²) in area and 20 feet (6096 mm) in height.

❖ As previously stated in the definition in Section 412.3.1, the residential aircraft hangar is limited to 2,000 square feet (186 m²) in area and 20 feet (6096 mm) in height. These limits control the fire hazard associated with such a use. The type and number of aircraft stored in the hangar are not limited. Since the hangar is considered as part of the residential use, it can be constructed of any materials that are permitted for the house, including wood-frame construction.

412.4 Aircraft paint hangars. Aircraft painting operations where flammable liquids are used in excess of the maximum allowable quantities per control area listed in Table 307.7(1) shall be conducted in an aircraft paint hangar that complies with the provisions of Section 412.4.

❖ This section provides requirements for aircraft-related structures that exceed normal hangar storage purposes. The painting or cleaning of all aircraft with flammable liquids must be carefully controlled in an aircraft paint hangar. To determine the applicability of these requirements, the building owner must provide a complete list of all flammable liquids intended to be used in the building along with their anticipated quantities. If the amounts exceed the maximum allowable quantities per control area, then the building must be classified as an aircraft paint hangar. See Table 307.7(1) and Section 414.2 for further discussion of the maximum allowable quantities per control area of hazardous materials.

412.4.1 Occupancy group. Aircraft paint hangars shall be classified as Group H-2. Aircraft paint hangars shall comply with the applicable requirements of this code and the *International Fire Code* for such occupancy.

❖ Similar to any other building containing hazardous materials in excess of the maximum allowable quantities per control area, the building or structure must be classified as Group H. Because of the flammable liquids present, the aircraft paint hangar is classified as Group H-2. Based on the equivalent risk theory, the requirements of Section 412.4, and other applicable portions of the *International Building Code*® (IBC®) and the IFC must be followed for this Group H-2 occupancy.

412.4.2 Construction. The aircraft paint hangar shall be of Type I or II construction.

❖ All aircraft paint hangars must be constructed as a Type I or II building. Special height modifications are provided in Section 504.1 for aircraft paint hangars along with special area limitations in Section 507.7. If these modifications are not applicable, then the height and area limitations of Table 503 for a Group H-2 structure must be used. The requirement for noncombustible construction limits the fuel load that is added to the occupancy contents by the structure.

412.4.3 Operations. Only those flammable liquids necessary for painting operations shall be permitted in quantities less than the maximum allowable quantities per control area in Table 307.7(1). Spray equipment cleaning operations shall be conducted in a liquid use, dispensing and mixing room.

❖ To lessen the likelihood of a fire in the actual painting or aircraft cleaning areas, only flammable liquids necessary for those operations are permitted. The cleaning and maintenance of spray equipment is further limited to a liquid use, dispensing and mixing room. This requirement for separation and segregation of the different operations reduces the fire risks associated with the painting and cleaning services.

412.4.4 Storage. Storage of flammable liquids shall be in a liquid storage room.

❖ The storage of all flammable liquids must be limited to a liquid storage room, which is further defined in Section 415.2. The applicable requirements of Section 415 and the IFC must be followed for liquid storage rooms to reduce the fire risks to the rest of the hangar operation.

412.4.5 Fire suppression. Aircraft paint hangars shall be provided with fire suppression as required in NFPA 409.

❖ To minimize the fire hazards associated with aircraft paint hangars, all such buildings are required to be protected with a fire suppression system. This requirement is applicable regardless of the size of the hangar in terms of height or area or the types and quantities of aircraft that are being cleaned or painted. The fire suppression system must be designed and installed in accordance with NFPA 409. This standard contains specific requirements for the suppression systems needed to properly protect paint hangars.

412.4.6 Ventilation. Aircraft paint hangars shall be provided with ventilation as required in the *International Mechanical Code*.

❖ Integral to the requirements for aircraft paint hangars are the ventilation provisions. Hazardous exhaust systems for controlling the overspray of painting and aircraft cleaning operations are required. These systems along with the necessary ventilation of the occupiable spaces must be in accordance with the IMC.

412.5 Heliports and helistops.

❖ This section contains special use and occupancy requirements for a very unique and specialized aircraft occupancy—those related to helicopters. Because of the limited space requirements necessary for the landing and taking off of helicopters, these areas are more likely to be incorporated into other buildings and structures, such as hospitals or large office buildings than facilities for fixed-wing aircraft.

Included in these requirements are provisions for minimum clearance sizes, structural use and design, means of egress and a referenced standard for rooftop locations. Heliports and helistops pose less fire hazards than the storage of these aircraft in hangars, but increased life safety risks.

412.5.1 General. Heliports and helistops may be erected on buildings or other locations where they are constructed in accordance with this section.

❖ Heliports and helistops are permitted to be located anywhere as long as they meet the requirements of Section 412.5. Certainly federal, state and local governments may have restrictions on the locations of heliports and helistops for general aviation purposes; however, the code only addresses those fire and life safety hazards associated with their locations to other buildings and structures and the means of egress from the same.

412.5.2 Definitions. The following words and terms shall, for the purposes of this chapter and as used elsewhere in this code, have the meanings shown herein.

❖ These definitions can help in the understanding and application of the code requirements. It is important to emphasize that these terms are not exclusively related to this section but are applicable everywhere the term is used in the code. The purpose for including these definitions within this section is to provide more convenient access to them without having to refer back to Chapter 2.

For convenience, these terms are also listed in Chapter 2 with a cross reference to this section.

The use and application of all defined terms, including those defined herein, are set forth in Section 201.

HELIPORT. An area of land or water or a structural surface that is used, or intended for use, for the landing and taking off of helicopters, and any appurtenant areas that are used, or intended for use, for heliport buildings and other heliport facilities.

❖ A heliport includes not only the immediate landing and take-off pad, but also all other adjacent service areas. The fueling, maintenance, repairs or storage of helicopters may be done within or outside of a building or structure. These outside areas or enclosed spaces are considered as part of the heliport.

HELISTOP. The same as a "Heliport," except that no fueling, defueling, maintenance, repairs or storage of helicopters is permitted.

❖ A helistop, by definition, is limited only to the immediate landing and take-off pad. Examples of helistops would be the pad located on top of a hospital for the unloading of emergency room patients, a pad for discharging commuters outside of an office building or the pad used to load and unload tourists at a sight-seeing attraction.

412.5.3 Size. The touchdown or landing area for helicopters of less than 3,500 pounds (1588 kg) shall be a minimum of 20 feet (6096 mm) in length and width. The touchdown area shall be surrounded on all sides by a clear area having a minimum aver-

age width at roof level of 15 feet (4572 mm) but with no width less than 5 feet (1524 mm).

❖ The landing pad for small helicopters [less than 3,500 pounds (1588 kg) in weight] must be a 20-foot-diameter (6096 mm) circle. In addition, a concentric circle must be provided around the landing pad, which provides a clear area with an average width of 15 feet (4572 mm) but with the least dimension no less than 5 feet (1524 mm). This additional clear space provides an increased landing area during windy conditions when pinpoint landing is not possible. Further, this clear area maintains the necessary separation between the rotating blades and all adjacent construction.

412.5.4 Design. Helicopter landing areas and the supports thereof on the roof of a building shall be noncombustible construction. Landing areas shall be designed to confine any flammable liquid spillage to the landing area itself and provisions shall be made to drain such spillage away from any exit or stairway serving the helicopter landing area or from a structure housing such exit or stairway. For structural design requirements, see Section 1605.5.

❖ Landing areas (helistops) located on the roofs of buildings must be of noncombustible materials, including the supporting construction. This requirement is necessary to provide a structurally sound support for the additional weight of the helicopter and its loads. This section refers to Section 1605.5 for further structural design requirements. Section 1605.5 requires that the building designer must account for the increased roof loads, including impact loads in the structural design.

Rooftop landing areas must be sloped or diked to prevent any spillage of flammable fuel from the helicopter from entering the building. This is most important since penthouse doors or exit stairs could allow spilled hazardous materials and vapors to enter the building, which would pose an unacceptable fire hazard.

412.5.5 Means of egress. The means of egress from heliports and helistops shall comply with the provisions of Chapter 10. Landing areas located on buildings or structures shall have two or more means of egress. For landing platforms or roof areas less than 60 feet (18 288 mm) in length, or less than 2,000 square feet (187 m²) in area, the second means of egress may be a fire escape or ladder leading to the floor below.

❖ As with all means of egress, the required egress paths from heliports and helistops must be in accordance with Chapter 10. Rooftop landing areas must be provided with at least two remotely located means of egress so that the helicopter occupants have redundant means to leave the landing area and enter the building. For very small landing areas, the code would permit one exit back into the building while the other could take the form of a fire escape or ladder to the next lower floor level. These requirements are reprinted in Section 1018.1.2 for consistency purposes (see that section for further means of egress discussion).

412.5.6 Rooftop heliports and helistops. Rooftop heliports and helistops shall comply with NFPA 418.

❖ In addition to the specific requirements of Section 412.5, rooftop heliports and helistops must comply with NFPA 418. That standard provides further life safety and fire safety requirements associated with rooftop landing areas.

SECTION 413
COMBUSTIBLE STORAGE

413.1 General. High-piled stock or rack storage in any occupancy group shall comply with the *International Fire Code*.

❖ This section alerts the code user to the specific high-piled combustible storage requirements contained in Chapter 23 of the IFC. High-piled storage, whether solid piled or in racks, in excess of 12 feet (3658 mm) in height, requires specific storage considerations, including fire protection design features, in order to be adequately protected. This section also prescribes the criteria needed to protect combustible storage in concealed spaces and similar locations.

Chapter 23 of the IFC provides requirements for the high-piled storage of combustible materials regardless of occupancy. High-piled storage of combustible materials could include solid-piled, palletized, shelf or rack storage where the top of storage is in excess of 12 feet (3658 mm) in height. Commodity classifications for all types of products as well as fire protection requirements based on NFPA 231 and 231C are contained in Chapter 23.

413.2 Attic, under-floor and concealed spaces. Attic, under-floor and concealed spaces used for storage of combustible materials shall be protected on the storage side as required for 1-hour fire-resistance-rated construction. Openings shall be protected by assemblies that are self-closing and are of noncombustible construction or solid wood core not less than $1^3/_4$ inch (45 mm) in thickness.

Exceptions:

1. Areas protected by approved automatic sprinkler systems.

2. Group R-3 and U occupancies.

❖ The severity of a potential fire hazard increases when combustibles are located within concealed spaces and similar areas that provide limited access to manual fire fighting. The areas offer low supervision and, therefore, increase the potential for a fire to develop and spread undetected through the building. This section regulates the minimum level of separation required between storage areas and the main occupiable area in nonsprinklered buildings. Since the intent is to protect against a fire in the storage area from endangering the other occupied areas of the building, the required 1-hour fire-resistance rating need only be achieved from the storage side. While any access openings in the 1-hour fire-resistant construction need not be rated, they must be self-closing and of either noncombustible construction or a minimum $1^3/_4$ inch (44 mm) thickness of solid wood core.

Exception 1 exempts the storage area from being separated by 1-hour fire-resistance-rated construction provided the area is protected by an approved automatic sprinkler system. This exception only requires the sprinkler system in the attic, under-floor or concealed space. Complete sprinkler protection throughout the building is not required in order to be in compliance with the exception.

Exception 2 clarifies that storage in residential occupancies consisting of not more than two dwelling units (Group R-3) and utility structures are exempt from the separation requirement.

SECTION 414
HAZARDOUS MATERIALS

[F] 414.1 General. The provisions of this section shall apply to buildings and structures occupied for the manufacturing, processing, dispensing, use or storage of hazardous materials.

❖ This section, along with Sections 307 (High-Hazard Group H) and 415 (Groups H-1, H-2, H-3, H-4 and H-5), and the IFC, are intended to be companion provisions for the treatment of occupancies that contain hazardous materials. Any building or structure utilizing hazardous materials, regardless of quantity, is to comply with all of the applicable provisions of both the code and the IFC. This section also contains design alternatives for the use and storage of hazardous materials without classifying the building as a high hazard Group H occupancy through the use of control areas (Section 414.2) or the mercantile display option (Section 414.2.4). While Section 414 contains general construction-related requirements for high-hazard occupancies, they are not indicative of a specific Group H occupancy classification but are dictated by hazardous material requirements in the IFC. Construction-related provisions for specific Group H occupancies are contained in Section 415.

The provisions of Section 414 apply to the use and storage of hazardous materials whether or not the building is classified as Group H. Requirements for specific materials are contained in the IFC.

[F] 414.1.1 Other provisions. Buildings and structures with an occupancy in Group H shall also comply with the applicable provisions of Section 415 and the *International Fire Code*.

❖ Section 415 is referenced for specific provisions applicable to occupancies classified as Groups H-1, H-2, H-3, H-4 and H-5. Regardless of the actual quantity of hazardous materials present, the use and storage of all such materials are required to comply with the applicable provisions of the IFC.

[F] 414.1.2 Materials. The safe design of hazardous material occupancies is material dependent. Individual material requirements are also found in Sections 307 and 415, and in the *International Mechanical Code* and the *International Fire Code.*

❖ This section emphasizes that high-hazard occupancies are different than other occupancies in that they are material dependent. This section alerts the code user to companion provisions in both the IMC and the IFC. Section 307 contains specific parameters for when a high-hazard occupancy classification is warranted. Section 415 contains specific building requirements dependent on the actual Group H occupancy classification for the building or area.

[F] 414.1.2.1 Aerosols. Level 2 and 3 aerosol products shall be stored and displayed in accordance with the *International Fire Code*. See Section 311.2 and the *International Fire Code* for occupancy group requirements.

❖ When Level 2 and 3 aerosol products are stored or displayed in accordance with Chapter 28 of the IFC, they may be classified as a Group S-1 occupancy as stated in Section 311.2. The protection required by the IFC is important so that the hazards created by these aerosols are addressed. The reference to the IFC will also address those locations where limited quantities of aerosol products are allowed in other occupancies.

[F] 414.1.3 Information required. Separate floor plans shall be submitted for buildings and structures with an occupancy in Group H, identifying the locations of anticipated contents and processes so as to reflect the nature of each occupied portion of every building and structure. A report identifying hazardous materials including, but not limited to, materials representing hazards that are classified in Group H to be stored or used, shall be submitted and the methods of protection from such hazards shall be indicated on the construction documents. The opinion and report shall be prepared by a qualified person, firm or corporation approved by the building official and shall be provided without charge to the enforcing agency.

❖ A detailed storage plan, other than an architectural floor plan, is essential for assisting fire department and other emergency response personnel in hazardous materials situations. A report such as a Hazardous Materials Management Plan (HMMP), as indicated in the IFC, or other approved plan should be submitted to aid fire department personnel in the building design preplanning phase.

[F] 414.2 Control areas. Control areas shall be those spaces within a building where quantities of hazardous materials not exceeding the maximum quantities allowed by this code are stored, dispensed, used or handled.

❖ This section, in conjunction with the maximum allowable quantity tables in Section 307, utilizes a density concept for hazardous materials through the use of control areas. The intent of the control area concept is to provide an alternative method for the handling of hazardous materials without classifying the occupancy as

Group H. In order to not be considered Group H, the amount of hazardous materials within any single control area bounded by fire barriers, fire walls and exterior walls cannot exceed the maximum allowable quantity for a specific material listed in Table 307.7(1) or 307.7(2) (see Figure 414.2). A control area may be an entire building or a portion thereof. Note that when an entire building is the control area, the entire maximum allowable quantity of material from Table 307.7(1) or 307.7(2) may be located anywhere in the building (see commentary, Section 307.9, Exception 1).

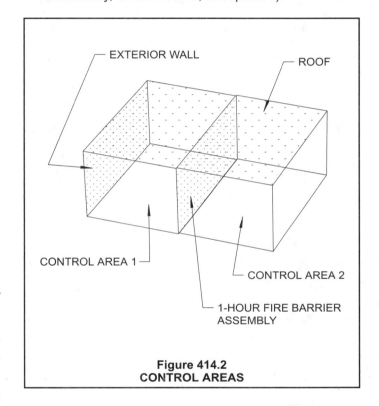

**Figure 414.2
CONTROL AREAS**

[F] 414.2.1 Construction requirements. Control areas shall be separated from each other by not less than a 1-hour fire barrier constructed in accordance with Chapter 7.

❖ Control areas are compartments of a building surrounded by fire barrier walls and fire-resistance-rated fire floor/ceiling assemblies. If there are no fire barriers or fire-resistance-rated floor/ceiling assemblies, the entire building is a single control area, for the purpose of applying these code provisions. Therefore, if more than the permitted maximum allowable quantities of Table 307.7(1) or 307.7(2) are anticipated in the building, additional control areas with minimum 1-hour fire barrier wall construction (2 hours where more than three stories) must be provided in order to not warrant a high-hazard occupancy classification. The provisions for required fire barriers also minimize the possibility of simultaneous involvement of multiple control areas due to a single fire condition. A fire in a single control area would involve only the amount of hazardous materials as limited by the maximum allowable quantities.

[F] 414.2.2 Number. The maximum number of control areas within a building shall be in accordance with Table 414.2.2.

❖ The maximum quantity of hazardous materials, therefore, which are permitted in a building without classifying it as a high-hazard occupancy, is regulated per control area and not per building area. The quantity limitation for the entire building would be established based on the number of permitted control areas on each floor of the building in accordance with Table 414.2.2. Based on the table, the first floor could contain four control areas with up to 100 percent of the maximum allowable quantity of hazardous materials per control area. For example, a single control area in a nonsprinklered building could contain up to 30 gallons (113 L) of Class 1A flammable liquids, 125 pounds (57 kg) of Class III organic peroxides, 250 pounds (113 kg) of Class 2 oxidizers and 500 gallons (1893 L) of corrosive liquids based on the maximum allowable quantities of Tables 307.7(1) and 307.7(2). Those quantities could be contained in each of four different control areas, provided that all control areas are separated from each other with minimum 1-hour fire barriers. Please note that in order to have more control areas per floor than indicated in Table 414.2.2, a fire wall in accordance with Section 705 would be required in order to create separate additional building areas.

TABLE 414.2.2. See below.

❖ The purpose of Table 414.2.2 is to establish the maximum allowable quantity of hazardous materials permitted in a building without classifying the building as a high-hazard occupancy based on the use of control areas. The number of control areas and permitted quantities of hazardous materials per control area are reduced when stored or used above the first floor level. This table also sets forth the minimum vertical fire barrier assemblies between adjacent control areas on the same floor. For floor levels above the third floor, a minimum 2-hour fire barrier is required from adjacent areas to aid

fire department response due to the additional time necessary for them to access the hazardous material storage areas on upper floors. The required fire-resistance rating of the floor/ceiling assembly above the control area is dictated by the required fire-resistance rating of the enclosure walls of the control area in order to maintain continuity and integrity. Special attention needs to be given to the rating of the floor for the control area, especially for levels below the fourth level. Section 414.2.3 requires that the floor of the control area and its supporting construction have a minimum 2-hour fire-resistance rating. In general, this will not affect floors above the third level since the continuity provisions of Section 706.4 when combined with the 2-hour separation requirement from Table 414.2.2 will ensure the floor that supports the fire barrier has an equivalent rating; however, in situations where Table 414.2.2 only requires the walls separating control areas to be of 1-hour fire-resistance-rated construction, Section 414.2.3 still would require a 2-hour fire-resistance-rated floor. This could greatly affect the design of a two-story building that was originally intended to be of Type IIB construction (see commentary, Section 414.2.3).

The percentage of quantities of hazardous materials per control area per floor area is intended to be cumulative. A two-story building, therefore, could contain three control areas with each having 75 percent of the maximum allowable quantity on the second floor in addition to the four control areas containing 100 percent each of the maximum allowable quantity on the first floor. This condition would require that adequate fire-resistance-rated separation be provided.

Note a clarifies that the maximum allowable quantity of hazardous materials per control area is based on Tables 307.7(1) and 307.7(2). The maximum permitted amount includes the increases allowed by either an automatic sprinkler system in accordance with NFPA 13, approved hazardous material storage cabinets or both where applicable.

Note b limits the number of control areas in mercan-

[F] TABLE 414.2.2
DESIGN AND NUMBER OF CONTROL AREAS

FLOOR LEVEL		PERCENTAGE OF THE MAXIMUM ALLOWABLE QUANTITY PER CONTROL AREA[a]	NUMBER OF CONTROL AREAS PER FLOOR[b]	FIRE-RESISTANCE RATING FOR FIRE BARRIERS IN HOURS[c]
Above grade	Higher than 9	5	1	2
	7-9	5	2	2
	6	12.5	2	2
	5	12.5	2	2
	4	12.5	2	2
	3	50	2	1
	2	75	3	1
	1	100	4	1
Below grade	1	75	3	1
	2	50	2	1
	Lower than 2	Not Allowed	Not Allowed	Not Allowed

a. Percentages shall be of the maximum allowable quantity per control area shown in Tables 307.7(1) and 307.7(2), with all increases allowed in the notes to those tables.

b. There shall be a maximum of two control areas per floor in Group M occupancies and in buildings or portions of buildings having Group S occupancies with storage conditions and quantities in accordance with Section 414.2.4.

c. Fire barriers shall include walls and floors as necessary to provide separation from other portions of the building.

tile and storage occupancies to two. The limitation is due to the increased quantities of certain hazardous materials that are allowed above the maximum allowable quantities of Tables 307.7(1) and 307.7(2). The two permitted control areas in the mercantile occupancy are assumed to be the retail sales area and the adjoining stockroom. Note, however, that unless adequate separation (1-hour fire barrier with approved opening protectives) is provided between the retail sales area and the stockroom, the entire mercantile occupancy would be considered a single control area (see commentary, Section 414.2.4).

Note c clarifies the fire barrier separation needed to establish the boundaries of the control area that include not only the vertical wall assemblies but also the floor/ceiling assemblies in order to be adequately separated from all adjacent interior spaces.

Example: Determine the maximum amount of Class IB flammable liquids that can be stored within a single-story, 10,000-square-foot (929 m²) nonsprinklered Group F-1 occupancy (see Figure 414.2.2) of Type IIB construction without classifying the storage area as Group H-2. Based on a maximum allowable quantity of 120 gallons (454 L) for Class IB flammable liquids from Table 307.7(1), a maximum of 120 gallons (454 L) can be stored in each of the four control areas; therefore, while the building may actually contain a total of 480 gallons (1817 L), a maximum of 120 gallons (454 L) is permitted in each control area that is separated from all adjacent control areas by minimum 1-hour fire barriers in accordance with Section 706. The building, in this case, could still be classified as Group F-1. An automatic fire suppression system would not be required,

since the 12,000-square-foot (1115 m²) threshold for suppression of Group F-1 fire areas is not exceeded (Section 903.2.3) and there are no control areas containing hazardous materials that exceed the maximum allowable quantities. Notes d and e of Table 307.7(1) would allow the base quantity of Class IB flammable liquids to be increased 100 percent in buildings protected with an automatic sprinkler system or when the material is stored in approved hazardous material storage cabinets. In this example, this would result in increasing the maximum allowable quantity of Class IB flammable liquids by a factor of two; therefore, the building could now contain a total of 960 gallons (3634 L) with a maximum of 240 gallons (908 L) in each of the four control areas, separated as required by the code, and still maintain a Group F-1 classification. If both an automatic sprinkler system and hazardous material storage cabinets are used to protect Class IB flammable liquids, then the base quantity of Table 307.7(1) could be increased by a factor of four. The building in this example, therefore, with both sprinkler protection and approved cabinets, could contain a total of 1920 gallons (7267 L) with a maximum of 480 gallons (1817 L) in each of the four control areas, separated as required by the code, and still maintain a Group F-1 classification. The allowable increase in the maximum allowable quantities is offset by the additional level or levels of protection. The use of control areas provides a tradeoff based on building compartmentation. Fire protection (automatic sprinkler systems) and controlled storage through the use of approved hazardous material storage cabinets also adds a degree of protection, justifying the increased allowable quantities.

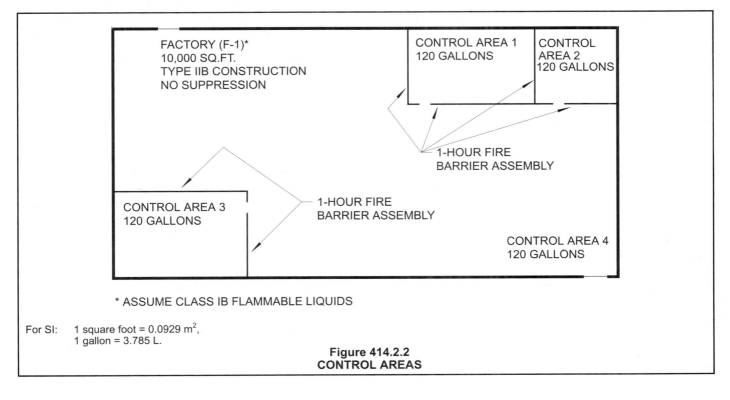

FACTORY (F-1)*
10,000 SQ.FT.
TYPE IIB CONSTRUCTION
NO SUPPRESSION

CONTROL AREA 1
120 GALLONS

CONTROL AREA 2
120 GALLONS

1-HOUR FIRE
BARRIER ASSEMBLY

CONTROL AREA 3
120 GALLONS

1-HOUR FIRE
BARRIER ASSEMBLY

CONTROL AREA 4
120 GALLONS

* ASSUME CLASS IB FLAMMABLE LIQUIDS

For SI: 1 square foot = 0.0929 m²,
          1 gallon = 3.785 L.

Figure 414.2.2
CONTROL AREAS

[F] 414.2.3 Separation. The required fire-resistance rating for fire barrier assemblies shall be in accordance with Table 414.2.2. The floor construction of the control area, and the construction supporting the floor of the control area, shall have a minimum 2-hour fire-resistance rating.

❖ The fire separation requirements for control areas, both horizontal and vertical, is dependent on their location in a building in accordance with Table 414.2.2. The amount of hazardous materials per control area, as well as the number of control areas per floor, are reduced if stored or used above the first floor.

Where the control area is located above the first floor, the floor assembly and all supporting construction for the control area would require a minimum 2-hour fire-resistance rating. The required 2-hour fire-resistance rating of the floor construction only refers to the floor of the control area. The increased fire-resistance rating and reduced quantities are intended to aid fire department personnel. The use of control areas on upper floors provides an alternative method for multistory research and laboratory-type facilities that may need to use a limited amount of hazardous materials throughout various portions of the building. Without control areas, the maximum allowable quantity for a hazardous material would be limited to a single building area regardless of the overall size or height of the building. For example, if control areas are not utilized, a 50,000-square-foot (4645 m²) single-story building would be limited to the same quantity of hazardous materials as a two-story building with 5,000 square feet (464 m²) per floor.

[F] 414.2.4 Hazardous material in Group M display and storage areas and in Group S storage areas. The aggregate quantity of nonflammable solid and nonflammable or noncombustible liquid hazardous materials permitted within a single control area of a Group M or S occupancy or an outdoor control area is permitted to exceed the maximum allowable quantities per control area specified in Tables 307.7(1) and 307.7(2) without classifying the building or use as a Group H occupancy, provided that the materials are displayed and stored in accordance with the *International Fire Code* and quantities do not exceed the maximum allowable specified in Table 414.2.4.

❖ This section addresses an option for control areas containing certain nonflammable or noncombustible hazardous materials that are stored in mercantile and storage occupancies, including outdoor control areas. This option would allow Group H-4 materials, which present a health hazard rather than a physical hazard, as well as limited Group H-2 and H-3 materials, such as oxidizers, to be stored in both retail display and stock areas of regulated mercantile occupancies and in storage-related occupancies in excess of the maximum allowable quantities of Tables 307.7(1) and 307.7(2) without classifying the building as Group H. Without this option, many mercantile and storage occupancies could be classified technically as Group H. The increased quantities of certain hazardous materials are based on the recognition that while there is limited risk in these occupancies, the

packaging and storage arrangements can be controlled. For further information on the storage limitations required for these types of materials, see Section 2703.11.3 of the IFC.

This section, in conjunction with Table 414.2.4, establishes the maximum quantity of the indicated hazardous materials permitted within a single control area of a mercantile occupancy. As indicated in Table 414.2.4, this section only applies to certain nonflammable solids and nonflammable or noncombustible liquids. Please note that this option is not applicable to mercantile and storage occupancies containing hazardous materials other than those indicated in Table 414.2.4. For example, the retail sale of products containing flammable/combustible liquids or products classified as aerosols would require compliance in accordance with Section 307.8, the IFC and the applicable provisions of either NFPA 30 or 30B.

Table 414.2.4. See page 4-77.

❖ Table 414.2.4 lists the hazardous materials eligible for the mercantile and storage occupancy option and the corresponding maximum permitted quantities per control area depending on the extent of protection provided. The permitted quantities of each listed material are independent of each other as well as the various classes or physical state of a specific material. For example, a given control area could contain up to the permitted maximum quantity of Class 2 solid oxidixers, Class 3 solid oxidizers and Class 2 liquid oxidizers, in addition to the permitted quantities of corrosive materials.

Notes b and c would allow the listed maximum quantity in Table 414.2.4 to be increased due to the use of sprinklers, approved hazardous materials storage cabinets or both. The notes are intended to be cumulative in that up to four times the listed amount may be allowed per control area, if the building is fully sprinklered and approved cabinets are utilized, without classifying the building as Group H.

Note d simply refers to Table 414.2.2 for the design and permitted number of control areas. Note b of Table 414.2.2 limits mercantile and storage occupancies utilizing this option to two control areas.

The 100-percent increase in maximum quantities for outdoor control areas permitted by Note f is based on the reduced exposure hazard to the building and its occupants. The increase encourages exterior storage applications without mandating sprinkler protection or approved hazardous material storage cabinets.

Notes g and h recognize that Class 2 and 3 solid oxidizers include several disinfectants that are commonly used in recreational, potable and wastewater treatment. Without these exceptions, the tabular maximum allowable quantities allowed in Group M and S occupancies would not be sufficient to sustain trade demand during times of peak usage. Because small containers of these materials have not been involved in losses, the exceptions permit additional containers of 10 pounds (10 kg) or less. Note that IFC Section 2703.11.3.6 limits the tabular quantities to individual containers of 100 pounds (45 kg) or less, whereas these exceptions give the retailer/wholesaler the option of increasing quantities on the shelves when the packaging sizes are reduced and limited to 10 pounds (5 kg) or less.

Note i recognizes the inherently higher level of protection and safety afforded by a sprinkler system and that, by definition, the only hazard presented by Class 1 oxidizers is that they slightly increase the burning rate of combustible materials with which they may come into contact during a fire. Materials with such properties present nowhere near the level of hazard of many ordinary commodities that might be found in a Group M or S occupancy, such as foam plastics. To put this matter into perspective, Class 1 oxidizers are materials with a de-

gree of hazard similar to that of toilet bowl cleaner crystals. Note i also correlates with Table 307.7(1), Note f.

Note j further recognizes the lesser hazard of Class 1 oxidizers and the inherent safety of storing hazardous materials outdoors by allowing quantities to be unlimited in outdoor control areas. Note j also correlates with IFC Table 2703.1.1(3).

[F] 414.3 Ventilation. Rooms, areas or spaces of Group H in which explosive, corrosive, combustible, flammable or highly toxic dusts, mists, fumes, vapors or gases are or may be emitted due to the processing, use, handling or storage of materials shall be mechanically ventilated as required by the *International Fire Code* and the *International Mechanical Code*.

Ducts conveying explosives or flammable vapors, fumes or

[F] TABLE 414.2.4
MAXIMUM ALLOWABLE QUANTITY PER INDOOR AND OUTDOOR CONTROL AREA IN GROUP M AND S OCCUPANCIES
NONFLAMMABLE SOLIDS AND NONFLAMMABLE AND NONCOMBUSTIBLE LIQUIDS [d,e,f]

CONDITION		MAXIMUM ALLOWABLE QUANTITY PER CONTROL AREA	
Material[a]	Class	Solids pounds	Liquids gallons
A. Health-hazard materials—nonflammable and noncombustible solids and liquids			
1. Corrosives[b, c]	Not Applicable	9,750	975
2. Highly toxics	Not Applicable	20[b, c]	2[b, c]
3. Toxics[b, c]	Not Applicable	1,000	100
B. Physical-hazard materials—nonflammable and noncombustible solids and liquids			
1. Oxidizers[b, c]	4	Not Allowed	Not Allowed
	3	1,150[g]	115
	2	2,250[h]	225
	1	18,000[i, j]	1,800[i, j]
2. Unstable (reactives)[b, c]	4	Not Allowed	Not Allowed
	3	550	55
	2	1,150	115
	1	Not Limited	Not Limited
3. Water (reactives)	3[b, c]	550	55
	2[b, c]	1,150	115
	1	Not Limited	Not Limited

For SI: 1 pound = 0.454 kg, 1 gallon = 3.785 L.

a. Hazard categories are as specified in the *International Fire Code*.

b. Maximum allowable quantities shall be increased 100 percent in buildings that are sprinklered in accordance with Section 903.3.1.1. When Note c also applies, the increase for both notes shall be applied accumulatively.

c. Maximum allowable quantities shall be increased 100 percent when stored in approved storage cabinets, in accordance with the *International Fire Code*. When Note b also applies, the increase for both notes shall be applied accumulatively.

d. See Table 414.2.2 for design and number of control areas.

e. Allowable quantities for other hazardous material categories shall be in accordance with Section 307.

f. Maximum quantities shall be increased 100 percent in outdoor control areas.

g. Maximum amounts are permitted to be increased to 2,250 pounds when individual packages are in the original sealed containers from the manufacturer or packager and do not exceed 10 pounds each.

h. Maximum amounts are permitted to be increased to 4,500 pounds when individual packages are in the original sealed containers from the manufacturer or packager and do not exceed 10 pounds each.

i. The permitted quantities shall not be limited in a building equipped throughout with an automatic sprinkler system in accordance with Section 903.3.1.1.

j. Quantities are unlimited in an outdoor control area.

dusts shall extend directly to the exterior of the building without entering other spaces. Exhaust ducts shall not extend into or through ducts and plenums.

> **Exception:** Ducts conveying vapor or fumes having flammable constituents less than 25 percent of their lower flammable limit (LFL) are permitted to pass through other spaces.

Emissions generated at workstations shall be confined to the area in which they are generated as specified in the *International Fire Code* and the *International Mechanical Code*.

The location of supply and exhaust openings shall be in accordance with the *International Mechanical Code*. Exhaust air contaminated by highly toxic material shall be treated in accordance with the *International Fire Code*.

A manual shutoff control for ventilation equipment required by this section shall be provided outside the room adjacent to the principal access door to the room. The switch shall be of the break-glass type and shall be labeled: VENTILATION SYSTEM EMERGENCY SHUTOFF.

❖ This section requires mechanical ventilation for occupancies utilizing hazardous materials when required by the IFC. The specific ventilation requirements are required to be in accordance with the applicable provisions in the IMC.

With regard to ducts conveying hazardous exhaust, the intent of this section is to minimize the potential for spreading hazardous exhaust to other parts of the building as a result of duct leakage or failure. In the event of a duct fire or explosion, other areas of the building could be jeopardized. Section 414.3 requires the exhaust flow to be maintained at concentrations below the lower flammability limit (LFL) of the contaminant. In all cases, ducts conveying hazardous exhaust must not extend into or through other ducts or plenum spaces, unless the flammable vapor-air mixtures are less than 25 percent of the LFL of the vapor being generated. The reference to work stations addresses spaces within a Group H-5 occupancy that utilize hazardous production materials within a fabrication area. The required control of emissions at work stations is intended to prevent exposure of the work station operator to hazardous fumes or vapors and to prevent hazardous concentrations of the materials used in the manufacturing process.

The exhaust system required by this section must be provided with an emergency manual shutoff control (kill switch) that will permit the exhaust system to be shut down without requiring personnel to enter the storage room. In the event of an emergency, it would be desirable to shut off an exhaust fan that is a source of ignition or is exhausting significant quantities of hazardous substances as a result of leaking storage containers. Under some circumstances, continued operation of an exhaust system could increase the level of hazard or cause the spread of fire.

To prevent tampering and unauthorized use of the shutoff control, such controls must be of the type that requires a seal to be broken before they can be actuated. The control must be clearly identified as to its purpose.

[F] 414.4 Hazardous material systems. Systems involving hazardous materials shall be suitable for the intended application. Controls shall be designed to prevent materials from entering or leaving process or reaction systems at other than the intended time, rate or path. Automatic controls, where provided, shall be designed to be fail safe.

❖ Process-type systems involving the use of hazardous materials generally involve many design variables. Many times, the building official may not be aware of the potentially dangerous chemical combination or use of incompatible materials that might be inherent in a given system. This section simply requires that all systems involving the use of hazardous materials should be designed to prevent an unwanted mixture of hazardous materials from occurring.

[F] 414.5 Inside storage, dispensing and use. The inside storage, dispensing and use of hazardous materials in excess of the maximum allowable quantities per control area of Tables 307.7(1) and 307.7(2) shall be in accordance with Sections 414.5.1 through 414.5.5 of this code and the *International Fire Code*.

❖ Sections 414.5.1 through 414.5.5 contain construction-related items for the inside storage and use of hazardous materials in excess of the maximum allowable quantities of Tables 307.7(1) and 307.7(2). The applicability of the provisions of Section 414.5 is dependent on the specific hazardous material requirements in the IFC.

[F] 414.5.1 Explosion control. Explosion control shall be provided in accordance with the *International Fire Code* as required by Table 414.5.1 where quantities of hazardous materials specified in that table exceed the maximum allowable quantities in Table 307.7(1) or where a structure, room or space is occupied for purposes involving explosion hazards as required by Section 415 or the *International Fire Code*.

❖ It is usually impractical to design a building to withstand the pressure created by an explosion; therefore, this section mandates an explosion relief system for all structures, rooms or spaces with occupancies involving explosion hazards. Explosions may result from the overpressurization of a containing structure, by physical/chemical means or by a chemical reaction. During an explosion, a sudden release of a high-pressure gas occurs and the energy is dissipated in the form of a shock wave.

All structures, rooms or spaces with occupancies involving explosion hazards must be equipped with some method of explosion control as required by Section 415 or the material-specific requirements in the IFC. Table 414.5.1 also specifies when explosion control is required based on certain materials or occupancies where the quantities of hazardous materials involved exceed the maximum allowable quantities in Table 307.7(1). Section 911 of the IFC recognizes explosion (deflagration) venting and explosion (deflagration) prevention systems as acceptable methods of explosion control where appropriate. The use of barricades or other explosion protective devices, such as magazines, may be permitted as the means of explosion control where indicated in the IFC as an acceptable alternative and approved by the building official.

TABLE 414.5.1. See below.

❖ This table designates when some methods of explosion control are required for specific material or special use conditions. The applicability of this table assumes the

quantities of hazardous materials involved exceed the maximum allowable quantities in Table 307.7(1). Section 911 of the IFC provides design criteria for explosion (deflagration) venting. Explosion prevention (suppression) systems, where utilized, must comply with NFPA 69. Barricade construction must be designed and installed in accordance with NFPA 495. As indicated in Note b to Table 414.5.1, the IFC provides additional guidance as to the applicability and design criteria for explosion control methods.

[F] TABLE 414.5.1
EXPLOSION CONTROL REQUIREMENTS[a]

MATERIAL	CLASS	EXPLOSION CONTROL METHODS	
		Barricade construction	Explosion (deflagration) venting or explosion (deflagration) prevention systems[b]
HAZARD CATEGORY			
Combustible dusts[c]	—	Not Required	Required
Cryogenic flammables	—	Not Required	Required
Explosives	Division 1.1	Required	Not Required
	Division 1.2	Required	Not Required
	Division 1.3	Not Required	Required
	Division 1.4	Not Required	Required
	Division 1.5	Required	Not Required
	Division 1.6	Required	Not Required
Flammable gas	Gaseous	Not Required	Required
	Liquefied	Not Required	Required
Flammable liquid	IA[d]	Not Required	Required
	IB[e]	Not Required	Required
Organic peroxides	U	Required	Not Permitted
	I	Required	Not Permitted
Oxidizer liquids and solids	4	Required	Not Permitted
Pyrophoric gas	—	Not Required	Required
Unstable (reactive)	4	Required	Not Permitted
	3 Detonable	Required	Not Permitted
	3 Nondetonable	Not Required	Required
Water-reactive liquids and solids	3	Not Required	Required
	2[g]	Not Required	Required
SPECIAL USES			
Acetylene generator rooms	—	Not Required	Required
Grain processing	—	Not Required	Required
Liquefied petroleum gas-distribution facilities	—	Not Required	Required
Where explosion hazards exist[f]	Detonation	Required	Not Permitted
	Deflagration	Not Required	Required

a. See Section 414.1.3.
b. See the *International Fire Code*.
c. As generated during manufacturing or processing. See definition of "Combustible dust" in Chapter 3.
d. Storage or use.
e. In open use or dispensing.
f. Rooms containing dispensing and use of hazardous materials when an explosive environment can occur because of the characteristics or nature of the hazardous materials or as a result of the dispensing or use process.
g. A method of explosion control shall be provided when Class 2 water-reactive materials can form potentially explosive mixtures.

[F] 414.5.2 Monitor control equipment. Monitor control equipment shall be provided where required by the *International Fire Code.*

❖ Monitor control equipment consists of limit controls such as liquid level controls for atmospheric tanks, temperature level controls, controls for hazardous materials required to be stored at other than ambient temperature and pressure relief devices for stationary tanks. The provisions for limit controls are located in Sections 2704.8 and 2705.1.4 of the IFC. This section provides a correlative cross reference to limit control requirements that may be dictated by a specific hazardous material in the IFC.

[F] 414.5.3 Automatic fire detection systems. Group H occupancies shall be provided with an automatic fire detection system in accordance with Section 907.2.

❖ This section requires an automatic fire (smoke) detection system for certain high-hazard occupancies. In accordance with Section 907.2.5, occupancies utilizing highly toxic gases, organic peroxides or oxidizers in excess of the maximum allowable quantities from Tables 307.7(1) and 307.7(2) would require an automatic smoke detection system installed in accordance with NFPA 72. Chapters 37, 38 and 39 of the IFC contain the specific design parameters as to when the automatic smoke detection system is required.

[F] 414.5.4 Standby or emergency power. Where mechanical ventilation, treatment systems, temperature control, alarm, detection or other electrically operated systems are required, such systems shall be provided with an emergency or standby power system in accordance with the ICC *Electrical Code.*

Exceptions:

1. Storage areas for Class I and II oxidizers.
2. Storage areas for Class III, IV and V organic peroxides.
3. Storage, use and handling areas for highly toxic or toxic materials as provided for in the *International Fire Code.*
4. Standby power for mechanical ventilation, treatment systems and temperature control systems shall not be required where an approved fail-safe engineered system is installed.

❖ A backup emergency power source is considered essential for required systems monitoring hazardous materials; therefore, when limit controls, detection systems or mechanical ventilation is required for a specific hazardous material, an emergency electrical system or standby power system is required.

Exceptions 1 and 2 address low-hazard oxidizers and organic peroxides that do not present a severe fire or reactivity hazard. Highly toxic and toxic materials (see Exception 3) must conform to applicable requirements of Chapter 37 of the IFC. For example, emergency power may be required for treatment systems utilized to process the accidental release of highly toxic or toxic com-

pressed gases caused by a leak or rupture in storage cylinders or tanks. Without emergency power, all required monitoring systems, including the treatment system for neutralizing potential leaking gas, would be rendered inoperative if a power failure or other electrical system failure occurred.

Exception 4 recognizes the use of an engineered system that is designed to always fail in the appropriate design mode without human intervention in lieu of the emergency power system. The intent of the exception is to permit alternative systems that are not subject to power interruptions. The exception, as noted, does not apply to detection and alarm systems, but addresses those systems essential to the removal of hazardous fumes and vapors from potentially occupied areas.

[F] 414.5.5 Spill control, drainage and containment. Rooms, buildings or areas occupied for the storage of solid and liquid hazardous materials shall be provided with a means to control spillage and to contain or drain off spillage and fire protection water discharged in the storage area where required in the *International Fire Code.* The methods of spill control shall be in accordance with the *International Fire Code.*

❖ This section references the IFC for material-specific occupancies containing materials in excess of the maximum allowable quantities, which would require some method of spill control, drainage and containment. The specific provisions for providing adequate spill control, drainage and containment, when required, are located in Section 2704.2 of the IFC.

[F] 414.6 Outdoor storage, dispensing and use. The outdoor storage, dispensing and use of hazardous materials shall be in accordance with the *International Fire Code.*

❖ This section requires the outdoor storage, dispensing and use of hazardous materials to be in accordance with the provisions of the IFC, regardless of the quantity. Certain provisions in the IFC, however, are only applicable when the maximum allowable quantities are exceeded. In general, the permitted quantity per outdoor control area exceeds that permitted for the inside storage or use of the same material. The outdoor storage or use of hazardous materials in excess of the maximum allowable quantities does not result in a high-hazard occupancy classification but simply dictates the need for additional requirements.

[F] 414.6.1 Weather protection. Where weather protection is provided for sheltering outdoor hazardous material storage or use areas, such storage or use shall be considered outdoor storage or use, provided that all of the following conditions are met:

1. Structure supports and walls shall not obstruct more than one side nor more than 25 percent of the perimeter of the storage or use area.
2. The distance from the structure and the structure supports to buildings, lot lines, public ways or means of egress to a public way shall not be less than the distance required for

an outside hazardous material storage or use area without weather protection.

3. The overhead structure shall be of approved noncombustible construction with a maximum area of 1,500 square feet (140 m²).

> **Exception:** The increases permitted by Section 506 apply.

❖ This section provides the minimum construction requirements for outdoor storage areas of hazardous materials that require protection from the elements. The need for weather protection is dependent on the specific material requirements in the IFC. The structure should be sufficiently open to allow for adequate cross ventilation and should be located with respect to lot lines, public ways or the means of egress to the public way as required for the specific hazardous material in the IFC. The overhead structure is required to be of noncombustible construction to eliminate the possibility of adding to the fuel load in a fire condition. For ready access to materials, many such structures are commonly constructed as an attached canopy. Depending on the hazardous material involved, additional protection, such as fire suppression of the outside storage area, fire-resistance-rated exterior wall construction or a limitation on exterior wall openings, may be necessary. Structures not in compliance with the provisions of this section would be regulated as inside storage, and as such, would be considered a high-hazard occupancy if the maximum allowable quantities of a specific hazardous material were exceeded.

The exception permits the area of the overhead structure to exceed the 1,500-square-foot (139 m²) area limitation if either excess frontage (see Section 506.2) or an automatic sprinkler system (see Section 506.3) is provided.

[F] 414.7 Emergency alarms. Emergency alarms for the detection and notification of an emergency condition in Group H occupancies shall be provided as set forth herein.

❖ An emergency alarm is required in all areas utilized for the storage, dispensing, use and handling of hazardous materials in accordance with Sections 414.7.1 through 414.7.3. This section assumes the area in question utilizes hazardous materials in excess of the maximum allowable quantities indicated in Tables 307.7(1) and 307.7(2).

[F] 414.7.1 Storage. An approved manual emergency alarm system shall be provided in buildings, rooms or areas used for storage of hazardous materials. Emergency alarm-initiating devices shall be installed outside of each interior exit or exit access door of storage buildings, rooms or areas. Activation of an emergency alarm-initiating device shall sound a local alarm to alert occupants of an emergency situation involving hazardous materials.

❖ A manual pull station or other emergency signal device approved by the building official must be provided outside of each egress door to hazardous material storage

areas. Activation of the device is intended to warn the building occupants of a potential dangerous condition within the hazardous material storage area. The alarm signal required by this section is intended only to be a local alarm. A complete evacuation alarm system, excluding a local trouble alarm located within the immediate high-hazard area, is not required.

[F] 414.7.2 Dispensing, use and handling. Where hazardous materials having a hazard ranking of 3 or 4 in accordance with NFPA 704 are transported through corridors or exit enclosures, there shall be an emergency telephone system, a local manual alarm station or an approved alarm-initiating device at not more than 150-foot (45 720 mm) intervals and at each exit and exit access doorway throughout the transport route. The signal shall be relayed to an approved central, proprietary or remote station service or constantly attended on-site location and shall also initiate a local audible alarm.

❖ This section requires access to an approved supervised signaling device along the transport route when hazardous materials must be transported through corridors or exit enclosures. A spill or other incident involving hazardous materials within a corridor or exit may render it unusable for egress. A loud audible alarm, which relays a signal to a remote station, allows a quick response by emergency responders.

[F] 414.7.3 Supervision. Emergency alarm systems shall be supervised by an approved central, proprietary or remote station service or shall initiate an audible and visual signal at a constantly attended on-site location.

❖ This section requires an approved method of electrical supervision for the emergency alarm systems for the hazardous material storage and use areas required by Sections 414.7.1 and 414.7.2. The method of supervision should also be in compliance with the applicable provisions of NFPA 72.

[F] SECTION 415
GROUPS H-1, H-2, H-3, H-4 AND H-5

415.1 Scope. The provisions of this section shall apply to the storage and use of hazardous materials in excess of the maximum allowable quantities per control area listed in Section 307.9. Buildings and structures with an occupancy in Group H shall also comply with the applicable provisions of Section 414 and the *International Fire Code*.

❖ This section establishes the application of Section 415 and references the IFC for additional specific hazardous material requirements. Section 415 is only applicable when the maximum allowable quantity of a hazardous material listed in either Table 307.7(1) or 307.7(2) is exceeded. The provisions of Section 414, however, are applicable wherever hazardous materials are stored or used, regardless of quantity.

415.2 Definitions. The following words and terms shall, for the purposes of this chapter and as used elsewhere in the code, have the meanings shown herein.

CONTINUOUS GAS-DETECTION SYSTEM. A gas detection system where the analytical instrument is maintained in continuous operation and sampling is performed without interruption. Analysis is allowed to be performed on a cyclical basis at intervals not to exceed 30 minutes.

❖ This term refers to a system that is capable of constantly monitoring the presence of highly toxic or toxic compressed gases at or below the permissible exposure limit (PEL) for the gas. A continuous gas detection system will provide notification of a leak or rupture in a compressed gas cylinder or tank in a storage or use condition.

EMERGENCY CONTROL STATION. An approved location on the premises where signals from emergency equipment are received and which is staffed by trained personnel.

❖ This definition identifies the room or area located in the hazardous production materials (HPM) facility that is utilized for the purpose of receiving various alarms and signals. The smoke detectors located in the building's recirculation ventilation ducts, the gas-monitoring/detection system and the telephone/fire protective signaling systems located outside of HPM storage rooms are all required to be connected to the emergency control station. The location of the emergency control station must be approved by the building official. An approved location should be based on personnel being able to adequately monitor the necessary alarms and signals and on the fire department being able to gain access quickly when responding to emergency situations. Additionally, the room must be occupied by persons who are trained to respond to the various alarms and signals in the appropriate fashion.

EXHAUSTED ENCLOSURE. An appliance or piece of equipment that consists of a top, a back and two sides providing a means of local exhaust for capturing gases, fumes, vapors and mists. Such enclosures include laboratory hoods, exhaust fume hoods and similar appliances and equipment used to locally retain and exhaust the gases, fumes, vapors and mists that could be released. Rooms or areas provided with general ventilation, in themselves, are not exhausted enclosures.

❖ Exhausted enclosures, such as laboratory hoods or exhaust fume hoods, are utilized to contain hazardous fumes and vapors. The use of an approved exhausted enclosure may allow an increase in the maximum allowable quantity per control area of a given hazardous material. Exhausted enclosures are typically utilized when highly toxic or toxic compressed gases are involved. The exhausted enclosures are required to be of noncombustible construction, have specific ventilation criteria and be protected by an approved fire-extinguishing system.

FABRICATION AREA. An area within a semiconductor fabrication facility and related research and development areas in which there are processes using hazardous production materials. Such areas are allowed to include ancillary rooms or areas such as dressing rooms and offices that are directly related to the fabrication area processes.

❖ This definition describes the basic component of an HPM facility. The code uses this definition to provide certain material limitations on both a quantity and density basis, and to require enclosure of the fabrication areas with fire barrier assemblies. The fabrication area of an HPM facility is the area where the hazardous materials are actively handled and processed. The fabrication area includes accessory rooms and spaces, such as work stations and employee dressing rooms.

FLAMMABLE VAPORS OR FUMES. The concentration of flammable constituents in air that exceed 10 percent of their lower flammable limit (LFL).

❖ Vapors or fumes are considered to be flammable when they exceed 10 percent of their lower flammable limit (LFL). The LFL of a given vapor in air is the concentration at which flame propagation could occur in the presence of an ignition source (see the definition of "Lower flammable limit").

GAS CABINET. A fully enclosed, noncombustible enclosure used to provide an isolated environment for compressed gas cylinders in storage or use. Doors and access ports for exchanging cylinders and accessing pressure-regulating controls are allowed to be included.

❖ Gas cabinets are used to provide adequate control for escaping gas in the event of a leaking cylinder of compressed gases. Gas cabinets are commonly used when dealing with highly toxic and toxic compressed gases. Sections 2703.8.6 and 3704.1.2 of the IFC provide additional construction and ventilation requirements for gas cabinets.

GAS ROOM. A separately ventilated, fully enclosed room in which only compressed gases and associated equipment and supplies are stored or used.

❖ Gas rooms are used exclusively for the storage or use of hazardous gases in excess of the maximum allowable quantities per control area permitted by Tables 307.7(1) and 307.7(2). Gas rooms are commonly used as an alternative storage area for HPM gases in a Group H-5 facility.

HAZARDOUS PRODUCTION MATERIAL (HPM). A solid, liquid or gas associated with semiconductor manufacturing that has a degree-of-hazard rating in health, flammability or reactivity of Class 3 or 4 as ranked by NFPA 704 and which is used directly in research, laboratory or production processes that have as their end product materials that are not hazardous.

❖ This definition identifies those specific materials that can be contained within an hazardous production materials (HPM) facility. The restriction in the definition for only hazardous materials with a Class 3 or 4 rating is not intended to exclude materials that are less hazardous, but to clarify that materials of the indicated higher ranking are still permitted in an HPM facility without classifying the building as Group H. NFPA 704 is referenced in order to establish the degree of hazard ratings for all materials as related to health, flammabiity and reactivity risks.

HPM FLAMMABLE LIQUID. An HPM liquid that is defined as either a Class I flammable liquid or a Class II or Class IIIA combustible liquid.

❖ This definition clarifies that an HPM liquid is essentially all classes of flammable or combustible liquids except Class IIIB combustible liquids, which have a flash point at or above 200°F (93°C). Class IIIB liquids, therefore, are not considered a hazardous production material.

HPM ROOM. A room used in conjunction with or serving a Group H-5 occupancy, where HPM is stored or used and which is classified as a Group H-2, H-3 or H-4 occupancy.

❖ An HPM room in a Group H-5 facility is utilized for the storage and use of hazardous production materials in excess of the maximum allowable quantities permitted in Table 307.7(1) or 307.7(2). The rooms are, therefore, considered a Group H-2, H-3 or H-4 occupancy depending on the type of hazardous material.

IMMEDIATELY DANGEROUS TO LIFE AND HEALTH (IDLH). The concentration of air-borne contaminants which poses a threat of death, immediate or delayed permanent adverse health effects, or effects that could prevent escape from such an environment. This contaminant concentration level is established by the National Institute of Occupational Safety and Health (NIOSH) based on both toxicity and flammability. It generally is expressed in parts per million by volume (ppm v/v) or milligrams per cubic meter (mg/m³). If adequate data do not exist for precise establishment of IDLH concentrations, an independent certified industrial hygienist, industrial toxicologist, appropriate regulatory agency or other source approved by the code official shall make such determination.

❖ The definition of "Immediately dangerous to life and health (IDLH)" is the minimum concentration of an air-borne contaminant, such as a highly toxic compressed gas, that a person could be exposed to before the risk of permanent adverse side effects. The IDLH is important in determining the design of treatment systems for highly toxic or toxic compressed gases.

LIQUID. A material that has a melting point that is equal to or less than 68°F (20°C) and a boiling point that is greater than 68°F (20°C) at 14.7 pounds per square inch absolute (psia) (101 kPa). When not otherwise identified, the term "liquid" includes both flammable and combustible liquids.

❖ This definition specifies the criteria to establish when material is considered a liquid based on its melting and boiling points. When the term "liquid" is referred to, it is

intended to include both flammable and combustible liquids.

LIQUID STORAGE ROOM. A room classified as a Group H-3 occupancy used for the storage of flammable or combustible liquids in a closed condition.

❖ Liquid storage rooms are utilized in Group H-5 facilities. These rooms are utilized exclusively for the storage of flammable and combustible liquids in closed containers in excess of the maximum allowable quantities per control area permitted by Tables 307.7(1) and 307.7(2). The storage room itself is considered a Group H-3 occupancy in accordance with Section 307.5.

LIQUID USE, DISPENSING AND MIXING ROOMS. Rooms in which Class I, II and IIIA flammable or combustible liquids are used, dispensed or mixed in open containers.

❖ This term refers to all nonstorage rooms utilized exclusively for flammable and combustible liquids other than Class IIIB liquids. Class IIIB liquids have a flash point in excess of 200°F (93°C) and are not considered hazardous.

LOWER FLAMMABLE LIMIT (LFL). The minimum concentration of vapor in air at which propagation of flame will occur in the presence of an ignition source. The LFL is sometimes referred to as "LEL" or "lower explosive limit."

❖ When the vapor-to-air ratio is somewhere between the LFL and the upper flammable limit (UFL), fires and explosions can occur upon introduction of an ignition source. The UFL is the maximum vapor-to-air concentration above which propagation of flame will not occur. If a vapor-to-air mixture is below the LFL, it is described as being "too lean" to burn, and if it is above the UFL, it is "too rich" to burn.

NORMAL TEMPERATURE AND PRESSURE (NTP). A temperature of 70°F (21°C) and a pressure of 1 atmosphere [14.7 psia (101 kPa)].

❖ This term refers to the standard room temperature at atmospheric pressure. Atmospheric pressure results from the weight of air elevated above the earth's surface. At sea level, the atmosphere exerts a pressure of 14.7 pounds per square inch (psi) (101kPa). The properties of commercially available compressed gases are indicated at their normal temperature and pressure (NTP).

SERVICE CORRIDOR. A fully enclosed passage used for transporting HPM and purposes other than required means of egress.

❖ Though HPM facility occupants may be exposed to limited HPM quantities during the course of their employment, their means of egress are protected from the HPM hazards by confining the HPM being transferred to its own passageway. A service corridor is only required when the HPM must be carried from a storage room or external area to a fabrication area through a passageway.

SOLID. A material that has a melting point, decomposes or sublimes at a temperature greater than 68°F (20°C).

❖ The temperature at which a solid melts is the melting point. A material will begin to melt as heat is added to it and, thus, eventually change to a liquid state.

STORAGE, HAZARDOUS MATERIALS.

1. The keeping, retention or leaving of hazardous materials in closed containers, tanks, cylinders or similar vessels, or

2. Vessels supplying operations through closed connections to the vessel.

❖ This term refers to all hazardous materials that are essentially being stored in a static condition. The material is considered in use once it is placed into action by either handling or transport or in a closed or open system.

USE (MATERIAL). Placing a material into action, including solids, liquids and gases.

❖ This definition describes the active utilization mode of a material as opposed to the inactive nature of storage or the limited movement or transport involved in handling. This mode tends to be more hazardous in that the material is exposed to human or mechanical contact rather than being confined to a closed container, thus increasing the potential for spills or other releases of liquids, vapors, gases or solids.

WORKSTATION. A defined space or an independent principal piece of equipment using HPM within a fabrication area where a specific function, laboratory procedure or research activity occurs. Approved or listed hazardous materials storage cabinets, flammable liquid storage cabinets or gas cabinets serving a workstation are included as part of the workstation. A workstation is allowed to contain ventilation equipment, fire protection devices, detection devices, electrical devices and other processing and scientific equipment.

❖ Workstations further subdivide a fabrication area and provide relatively self-contained, specialized areas where HPM processes are conducted. Workstation controls limit the quantity of materials and impose limitations on the design of these processes to include, but not be limited to, protection by local exhaust; sprinklers; automatic and emergency shutoffs; construction materials and HPM compatibility. Excess materials are prohibited and must be contained in storage rooms designed to accommodate such hazards.

415.3 Location on property. Group H shall be located on property in accordance with the other provisions of this chapter. In Group H-2 or H-3, not less than 25 percent of the perimeter wall of the occupancy shall be an exterior wall.

Exceptions:

1. Liquid use, dispensing and mixing rooms having a floor area of not more than 500 square feet (46.5 m2)

need not be located on the outer perimeter of the building where they are in accordance with the *International Fire Code* and NFPA 30.

2. Liquid storage rooms having a floor area of not more than 1,000 square feet (93 m2) need not be located on the outer perimeter where they are in accordance with the *International Fire Code* and NFPA 30.

3. Spray paint booths that comply with the *International Fire Code* need not be located on the outer perimeter.

❖ This section specifies the location of Group H storage areas within a building. In order to provide adequate access for fire-fighting operations and venting of the products of combustion, Group H-2 and H-3 storage areas within a building must be located along an exterior wall.

Exception 1 recognizes the use of inside storage rooms that are utilized for operations involving flammable and combustible liquids. These types of rooms, which have no perimeter access, are permitted by the IFC and NFPA 30. The size of the room as well as the quantity of liquids, however, is restricted due to the lack of perimeter access and ventilation options.

Exception 2 is similar to Exception 1 except that a larger room area is permitted since the flammable and combustible liquids are in a static storage condition.

Spray paint booths are typically power-ventilated structures within a building that have their own enclosure separate from the exterior wall construction. The location and fire separation requirements for the spray paint booth must be in accordance with Section 416 and NFPA 33.

415.3.1 Group H minimum distance to lot lines. Regardless of any other provisions, buildings containing Group H occupancies shall be set back a minimum distance from lot lines as set forth in Items 1 through 4 below. Distances shall be measured from the walls enclosing the occupancy to lot lines, including those on a public way. Distances to assumed property lines drawn for the purposes of determination of exterior wall and opening protection are not to be used to establish the minimum distance for separation of buildings on sites where explosives are manufactured or used when separation is provided in accordance with the quantity distance tables specified for explosive materials in the *International Fire Code.*

1. Group H-1. Not less than 75 feet (22 860 mm) and not less than required by the *International Fire Code.*

   Exceptions:

   1. Fireworks manufacturing buildings separated in accordance with NFPA 1124.

   2. Buildings containing the following materials when separated in accordance with Table 415.3.1:

      2.1. Organic peroxides, unclassified detonable.

      2.2. Unstable reactive materials Class 4.

      2.3. Unstable reactive materials, Class 3 detonable.

      2.4. Detonable pyrophoric materials.

2. Group H-2. Not less than 30 feet (9144 mm) where the area of the occupancy exceeds 1,000 square feet (93 m²) and it is not required to be located in a detached building.

3. Groups H-2 and H-3. Not less than 50 feet (15 240 mm) where a detached building is required (see Table 415.3.2).

4. Groups H-2 and H-3. Occupancies containing materials with explosive characteristics shall be separated as required by the *International Fire Code*. Where separations are not specified, the distances required shall not be less than the distances required by Table 415.3.1.

❖ Due to the potentially volatile nature of hazardous materials, specific setback requirements are necessary for Group H occupancies. These provisions take precedence over provisions in the code that may specify a minimum fire separation distance based on building construction type and exposure (see Table 602). The listed conditions are dependent on the type of materials that are indicative of the specified Group H occupancies, the size of the hazardous material storage area and whether a detached building is required by Table 415.3.2. Buildings with explosive materials must comply with Table 415.3.1.

Exception 1 in Item 1 recognizes that fireworks manufacturing buildings present unique hazards due to the potential volume (net weight) of fireworks in any single building. NFPA 1124 specifies the minimum separation distances between all process buildings, public highways and other inhabited buildings.

Exception 2 addresses the fact that these specific materials have a different explosive hazard and, therefore, permits distances to be established using Table 415.3.1 and the modification that is found in Note a of the table.

When dealing with explosives, it is important to note that the base paragraph makes a distinction between the assumed property lines (see Section 702) that are used for determining exterior wall and opening protection and those that are used for the separation of buildings where explosives are involved. Where explosives are involved, the separation distances are measured between structures and not to some imaginary property line that is assumed to be between them. It is reasonable to make this distinction since the separation distances for explosives far exceed the normal exterior wall and opening separation requirements.

TABLE 415.3.1. See page 4-86.

❖ Table 415.3.1 establishes minimum separation distances for the permanent storage of explosives from selected property classes. The intent of the table is to place explosive storage in a sufficiently remote location from occupied buildings, lot lines and other magazines in a manner to reduce exposure of such properties from damage if a detonation of a magazine occurs. The storage of explosives must also comply with the applicable provisions of Chapter 33 of the IFC and NFPA 495.

TABLE 415.3.2. See page 4-88.

❖ Table 415.3.2 establishes when hazardous materials must be stored in detached structures. The need for detached storage is a function of the type, physical state and quantity of material.

415.3.2 Group H-1 and H-2 or H-3 detached buildings. Where a detached building is required by Table 415.3.2, there are no requirements for wall and opening protection based on location on property.

❖ This section recognizes that detached buildings are required to be adequately separated from lot lines and other important buildings. As such, additional exposure protection for exterior walls is not needed. Table 415.3.2 requires detached buildings for some of the materials found in Groups H-2 and H-3 due to the large quantities involved. Even though these materials are less of a hazard when compared to those found in Group H-1, they do create a significant hazard in sufficiently large quantities.

415.4 Special provisions for Group H-1 occupancies. Group H-1 occupancies shall be in buildings used for no other purpose, shall not exceed one story in height and be without basement, crawl spaces or other under-floor spaces. Roofs shall be of lightweight construction with suitable thermal insulation to prevent sensitive material from reaching its decomposition temperature.

Group H-1 occupancies containing materials which are in themselves both physical and health hazards in quantities exceeding the maximum allowable quantities per control area in Table 307.7.(2) shall comply with requirements for both Group H-1 and H-4 occupancies.

❖ Due to the explosion hazard potential associated with Group H-1 materials, Group H-1 occupancies are required to be in separate detached structures. The limitation of one story is based on the need to exit a building with a detonation hazard as soon as possible. Exiting from a second story or basement, even within a fire-resistance-rated stairway enclosure, may not offer sufficient protection where an explosion hazard exists. The one-story limitation is also intended to limit the volume of detonable materials that could be stored in any one structure.

Group H-1 occupancies that contain materials that present a health as well as a detonation hazard must also comply with the applicable requirements for a Group H-4 occupancy (see commentary, Section 307.8).

415.4.1 Floors in storage rooms. Floors in storage areas for organic peroxides, pyrophoric materials and unstable (reactive) materials shall be of liquid-tight, noncombustible construction.

❖ Noncombustible floors are required to prevent the structure from contributing to a fire scenario. The floors are also required to be liquid tight to prevent the spread of hazardous materials to areas outside the storage room.

TABLE 415.3.1

SPECIAL DETAILED REQUIREMENTS BASED ON USE AND OCCUPANCY

TABLE 415.3.1
MINIMUM SEPARATION DISTANCES FOR BUILDINGS CONTAINING EXPLOSIVE MATERIALS

QUANTITY OF EXPLOSIVE MATERIAL[a]		MINIMUM DISTANCE (feet)		
		Lot lines[b] and inhabited buildings[c]		
Pounds over	Pounds not over	Barricaded[d]	Unbarricaded	Separation of magazines[d, e, f]
2	5	70	140	12
5	10	90	180	16
10	20	110	220	20
20	30	125	250	22
30	40	140	280	24
40	50	150	300	28
50	75	170	340	30
75	100	190	380	32
100	125	200	400	36
125	150	215	430	38
150	200	235	470	42
200	250	255	510	46
250	300	270	540	48
300	400	295	590	54
400	500	320	640	58
500	600	340	680	62
600	700	355	710	64
700	800	375	750	66
800	900	390	780	70
900	1,000	400	800	72
1,000	1,200	425	850	78
1,200	1,400	450	900	82
1,400	1,600	470	940	86
1,600	1,800	490	980	88
1,800	2,000	505	1,010	90
2,000	2,500	545	1,090	98
2,500	3,000	580	1,160	104
3,000	4,000	635	1,270	116
4,000	5,000	685	1,370	122
5,000	6,000	730	1,460	130
6,000	7,000	770	1,540	136
7,000	8,000	800	1,600	144
8,000	9,000	835	1,670	150
9,000	10,000	865	1,730	156
10,000	12,000	875	1,750	164
12,000	14,000	885	1,770	174
14,000	16,000	900	1,800	180
16,000	18,000	940	1,880	188
18,000	20,000	975	1,950	196

(continued)

TABLE 415.3.1—continued
MINIMUM SEPARATION DISTANCES FOR BUILDINGS CONTAINING EXPLOSIVE MATERIALS

QUANTITY OF EXPLOSIVE MATERIAL[a]		MINIMUM DISTANCE (feet)		
		Lot lines[b] and inhabited buildings[c]		
Pounds over	Pounds not over	Barricaded[d]	Unbarricaded	Separation of magazines[d, e, f]
20,000	25,000	1,055	2,000	210
25,000	30,000	1,130	2,000	224
30,000	35,000	1,205	2,000	238
35,000	40,000	1,275	2,000	248
40,000	45,000	1,340	2,000	258
45,000	50,000	1,400	2,000	270
50,000	55,000	1,460	2,000	280
55,000	60,000	1,515	2,000	290
60,000	65,000	1,565	2,000	300
65,000	70,000	1,610	2,000	310
70,000	75,000	1,655	2,000	320
75,000	80,000	1,695	2,000	330
80,000	85,000	1,730	2,000	340
85,000	90,000	1,760	2,000	350
90,000	95,000	1,790	2,000	360
95,000	100,000	1,815	2,000	370
100,000	110,000	1,835	2,000	390
110,000	120,000	1,855	2,000	410
120,000	130,000	1,875	2,000	430
130,000	140,000	1,890	2,000	450
140,000	150,000	1,900	2,000	470
150,000	160,000	1,935	2,000	490
160,000	170,000	1,965	2,000	510
170,000	180,000	1,990	2,000	530
180,000	190,000	2,010	2,010	550
190,000	200,000	2,030	2,030	570
200,000	210,000	2,055	2,055	590
210,000	230,000	2,100	2,100	630
230,000	250,000	2,155	2,155	670
250,000	275,000	2,215	2,215	720
275,000	300,000	2,275	2,275	770

For SI: 1 pound = 0.454 kg, 1 foot = 304.8 mm.

a. The number of pounds of explosives listed is the number of pounds of trinitrotoluene (TNT) or the equivalent pounds of other explosive.

b. The distance listed is the distance to lot line, including lot lines at public ways.

c. For the purpose of this table, an inhabited building is any building on the same property that is regularly occupied by people. Where two or more buildings containing explosives or magazines are located on the same property, each building or magazine shall comply with the minimum distances specified from inhabited buildings and, in addition, they shall be separated from each other by not less than the distance shown for "Separation of magazines," except that the quantity of explosive materials contained in detonator buildings or magazines shall govern in regard to the spacing of said buildings or magazines from buildings or magazines, as a group, shall be considered as one building or magazine, and the total quantity of explosive materials stored in such group shall be treated as if the explosive were in a single building or magazine located on the site of any building or magazine of the group, and shall comply with the minimum distance specified from other magazines or inhabited buildings.

d. Barricades shall effectively screen the building containing explosives from other buildings, public ways or magazines. Where mounds or revetted walls of earth are used for barricades, they shall not be less than 3 feet in thickness. A straight line from the top of any side wall of the building containing explosive materials to the eave line of any other building, magazine or a point 12 feet above the centerline of a public way shall pass through the barricades.

e. Magazine is a building or structure, other than an operating building, approved for storage of explosive materials. Portable or mobile magazines not exceeding 120 square feet (11 m^2) in area need not comply with the requirements of this code, however, all magazines shall comply with the *International Fire Code*.

f. The distance listed is permitted be reduced by 50 percent where approved natural or artificial barriers are provided in accordance with the requirements in Note d.

TABLE 415.3.2
REQUIRED DETACHED STORAGE

DETACHED STORAGE IS REQUIRED WHEN THE QUANTITY OF MATERIAL EXCEEDS THAT LISTED HEREIN			
Material	**Class**	**Solids and Liquids (tons)[a,b]**	**Gases (cubic feet)[a,b]**
Explosives	Division 1.1 Division 1.2 Division 1.3 Division 1.4 Division 1.4[c] Division 1.5 Division 1.6	Maximum Allowable Quantity Maximum Allowable Quantity Maximum Allowable Quantity Maximum Allowable Quantity 1 Maximum Allowable Quantity Maximum Allowable Quantity	Not Applicable
Oxidizers	Class 4	Maximum Allowable Quantity	Maximum Allowable Quantity
Unstable (reactives) detonable	Class 3 or 4	Maximum Allowable Quantity	Maximum Allowable Quantity
Oxidizer, liquids and solids	Class 3 Class 2	1,200 2,000	Not Applicable Not Applicable
Organic peroxides	Detonable Class I Class II Class III	Maximum Allowable Quantity Maximum Allowable Quantity 25 50	Not Applicable Not Applicable Not Applicable Not Applicable
Unstable (reactives) nondetonable	Class 3 Class 2	1 25	2,000 10,000
Water reactives	Class 3 Class 2	1 25	Not Applicable Not Applicable
Pyrphoric gases	Not Applicable	Not Applicable	2,000

For SI: 1 ton = 906 kg, 1 cubic foot = 0.02832 M^3.

a. For materials that are detonable, the distance to other buildings or lot lines shall be as specified in Table 415.3.1 based on trinitrotoluene (TNT) equivalence of the material. For materials classified as explosives, see Chapter 33 the *International Fire Code*. For all other materials, the distance shall be as indicated in Section 415.3.1.

b. "Maximum Allowable Quantity" means the maximum allowable quantity per control area set forth in Table 307.7(1).

c. Limited to Division 1.4 materials and articles, including articles packaged for shipment, that are not regulated as an explosive under Bureau of Alcohol, Tobacco and Firearms (BATF) regulations or unpackaged articles used in process operations that do not propagate a detonation or deflagration between articles, providing the net explosive weight of individual articles does not exceed 1 pound.

415.5 Special provisions for Group H-2 and H-3 occupancies. Group H-2 and H-3 occupancies containing quantities of hazardous materials in excess of those set forth in Table 415.3.2 shall be in buildings used for no other purpose, shall not exceed one story in height and shall be without basements, crawl spaces or other under-floor spaces.

Group H-2 and H-3 occupancies containing water-reactive materials shall be resistant to water penetration. Piping for conveying liquids shall not be over or through areas containing water reactives, unless isolated by approved liquid-tight construction.

Exception: Fire protection piping.

❖ This section in conjunction with Table 415.3.2 specifies when a Group H-2 or H-3 occupancy must be in a detached structure. Detached structures used for the storage of hazardous materials are not intended to be mixed-use occupancies.

The design and construction of buildings used for storing water-reactive materials must be such that water will not be permitted to come in contact with the stored materials. The building materials should be designed to resist the passage of flowing water. Piping conveying liquids, other than sprinkler system piping, is prohibited in any area containing water-reactive materials.

415.5.1 Floors in storage rooms. Floors in storage areas for organic peroxides, oxidizers, pyrophoric materials, unstable (reactive) materials and water-reactive solids and liquids shall be of liquid-tight, noncombustible construction.

❖ Noncombustible floors are required to prevent the structure from contributing to a fire scenario. The floors are also required to be liquid tight to prevent the spread of hazardous materials to areas outside the storage room.

415.5.2 Waterproof room. Rooms or areas used for the storage of water-reactive solids and liquids shall be constructed in a manner that resists the penetration of water through the use of waterproof materials. Piping carrying water for other than approved automatic fire sprinkler systems shall not be within such rooms or areas.

❖ Similar to Section 415.5.1, storage rooms containing water-reactive materials must be waterproofed. Though water piping may not be run into or through such rooms, the code recognizes that automatic sprinkler systems are a more regulated type of water piping system and have a low leakage and failure rate when properly installed and maintained.

415.6 Smoke and heat venting. Smoke and heat vents complying with Section 910 shall be installed in the following locations:

1. In occupancies classified as Group H-2 or H-3, any of which are over 15,000 square feet (1394 m²) in single floor area.

   Exception: Buildings of noncombustible construction containing only noncombustible materials.

2. In areas of buildings in Group H used for storing Class 2, 3 and 4 liquid and solid oxidizers, Class 1 and unclassified detonable organic peroxides, Class 3 and 4 unstable (reactive) materials, or Class 2 or 3 water-reactive materials as required for a Class V hazard classification.

   Exception: Buildings of noncombustible construction containing only noncombustible materials.

❖ This section requires smoke and heat vents in roofs of one-story buildings or portions thereof for the specified high-hazard occupancy classifications and storage conditions (see Section 910.2 for the one-story limitation). The use of an automatic venting system provides a mechanism for relieving smoke from the building to not only aid in fire-fighting operations but help provide better visibility for those occupants still trying to egress the building. Section 910 provides the design requirements for curtain boards and smoke vents. Where approved by the building official, an engineered smoke exhaust system may be used in lieu of smoke and heat vents.

Condition 1 of Section 415.6 requires the venting based on the occupancy being a single large floor area greater than 15,000 square feet (1393 m²) for Group H-2 and H-3 occupancies. Due to the potential types and quantities of hazardous materials that could be present, venting is required even if the materials present are not those indicated in Condition 2. Condition 2 requires venting based only on the presence of certain hazardous materials. This condition, however, assumes the quantity of materials listed is in excess of the maximum allowable quantities per control area permitted in Table 307.7(1); therefore, the occupancy is classified as Group H. The exceptions address a fairly limited situation where the buildings and the contents are noncombustible. An example would be an H-3 storage and packaging facility for oxidizing gases that is located in a Type II building and involves only noncombustible contents. The oxidizing gases do not burn (they support and accelerate the combustion of combustible materials); therefore, if the building and contents are noncombustible, there would be no benefit in providing smoke and heat venting.

415.7 Group H-2. Occupancies in Group H-2 shall be constructed in accordance with Sections 415.7.1 through 415.7.4 and the *International Fire Code*.

❖ Sections 415.7.1 through 415.7.4 contain specific construction requirements for occupancies that contain Group H-2 materials in excess of the maximum allowable quantities. Such occupancies are also required to comply with all material-specific related provisions in the IFC.

415.7.1 Combustible dusts, grain processing and storage. The provisions of Sections 415.7.1.1 through 415.7.1.5 shall apply to buildings in which materials that produce combustible dusts are stored or handled. Buildings that store or handle combustible dusts shall comply with the applicable provisions of NFPA 61, NFPA 120, NFPA 651, NFPA 654, NFPA 655, NFPA 664 and NFPA 85, and the *International Fire Code*.

❖ Combustible dusts can generate high-pressure gas by combustion in air. The pressure created can destroy process equipment and structures. While there is a minimum concentration of dust required, there is no reliable upper concentration limitation beyond which combustion will not occur. The ignition source is often the process equipment itself. The provisions of this section primarily address the construction of a building that contains the hazard. The construction requirements are primarily intended to reduce the exposure hazard should a fire or explosion occur. To minimize the impact of the explosion on the structure, explosion control in accordance with Section 414.5.1 and the IFC is required. The provisions of this section apply to all buildings in which combustible dusts and particles are of sufficient quantity to be readily ignited and subject to explosion. The concentration that would be required to constitute such a hazard is dependent on the particle size. Additional guidance on the relative fire risk associated with various combustible dusts can be found in the referenced standards. In addition to the provisions of this section, compliance with the provisions of the IFC and the referenced standards is essential to minimize the potential ignition and the fire and explosion hazards. Chapter 13 of the IFC contains specific safety precautions for combustible dust-producing operations and references additional standards for explosion protection based on the type of process involved. Essentially, each of the referenced standards prescribes reasonable requirements for safety to life and property from fire and explosion. The standards also minimize the resulting damage should a fire or explosion occur. More specifically, they contain provisions for construction, ventilation, explosion venting, equipment, heating devices, dust control, fire protection and supplemental requirements related to electrical wiring and equipment, provisions concerning protection from sparks, cutting and welding and smoking and signage regulations.

415.7.1.1 Type of construction and height exceptions. Buildings shall be constructed in compliance with the height and area limitations of Table 503 for Group H-2; except that where erected of Type I or II construction, the heights and areas of grain elevators and similar structures shall be unlimited, and where of Type IV construction, the maximum height shall be 65 feet (19 812 mm) and except further that, in isolated areas, the maximum height of Type IV structures shall be increased to 85 feet (25 908 mm).

❖ The construction of buildings in which combustible dusts can be readily ignitable and subject to an explosion hazard is restricted to the height and area limits in Table 503 for Group H-2. The limitation on construction is intended to reduce the available fuel source that could ignite and serve as the source of a dust explosion. Grain elevators and similar structures that require a height in excess of that permitted in Table 503 may be of unlimited height if the structures are of Type I or II construction; or such structures may be 65 feet (19 812 mm) if of Type IV construction. If the structure is of Type IV construction and located so as not to constitute an explosion hazard, the allowable height may be increased to 85 feet (25 908 mm). Although the buildings are required to have an automatic sprinkler system in accordance with Section 903.2.4, the speed at which the combustion process or explosion could occur may reduce the effectiveness of the sprinkler system. Additional guidance on methods to prevent explosions can be found in Section 911 of the IFC and applicable referenced standards, such as NFPA 69, for explosion prevention systems.

415.7.1.2 Grinding rooms. Every room or space occupied for grinding or other operations that produce combustible dusts shall be enclosed with fire barriers and horizontal assemblies or both that have not less than a 2-hour fire-resistance rating where the area is not more than 3,000 square feet (279 m²), and not less than a 4-hour fire-resistance rating where the area is greater than 3,000 square feet (279 m²).

❖ Grinding rooms are to be separated from remaining parts of the building by construction having a fire-resistance rating of at least 2 hours. If the area of the room is greater than 3,000 square feet (279 m²), the fire-resistance rating of the enclosure must be at least 4 hours. Not only is it the intent to protect the room from an explosion fire but also, assuming the explosion venting results in minimal structural damage, for the enclosure to contain the resulting fire, if any, to the room of origin. Without venting, it is doubtful whether the enclosure could withstand the forces exerted on the wall by an explosion.

415.7.1.3 Conveyors. Conveyors, chutes, piping and similar equipment passing through the enclosures of rooms or spaces shall be constructed dirt tight and vapor tight, and be of approved noncombustible materials complying with Chapter 30.

❖ The intent of this section is to restrict conveyors that pass through walls so that they do not serve as an ignition source or avenue of fire spread to adjacent areas.

These conveyors and similar equipment must be constructed of noncombustible materials and be dirt and vapor tight.

415.7.1.4 Explosion control. Explosion control shall be provided as specified in the *International Fire Code*, or spaces shall be equipped with the equivalent mechanical ventilation complying with the *International Mechanical Code*.

❖ The pressure exerted by a combustible dust explosion typically ranges from 13 to 89 psi (89 to 614 kPa). It is impractical to construct a building that will withstand such pressures; therefore, either a means of explosion control must be provided in accordance with Section 911 of the IFC or an equivalent mechanical ventilation system must be provided. The mechanical ventilation system must be in accordance with the IMC and must be designed to control the dust concentration below hazardous levels. The mechanical ventilation is dependent on the concentration at which a dust explosion could occur. Additional guidance on the relative fire risk associated with various combustible dusts can be found in the referenced standards.

415.7.1.5 Grain elevators. Grain elevators, malt houses and buildings for similar occupancies shall not be located within 30 feet (9144 mm) of interior lot lines or structures on the same lot, except where erected along a railroad right-of-way.

❖ The 30-foot (9144 mm) spatial separation between grain elevators (or similar structures with regard to lot lines) and other structures is intended to reduce the exposure hazard. It is intended not only to reduce the damage to the adjacent structure but also to minimize the potential that a fire in the adjacent structure would affect the grain elevator. Grain elevators are singled out because of the additional height permitted in Section 415.7.1.1 and the typical openness of the structure. The 30-foot (9144 mm) spatial separation is not required if the grain elevator is located along a railroad right-of-way. The railroad right-of-way will offer some spatial separation, except when a car is being loaded from the grain elevator, in which case the 30-foot (9144 mm) criterion may not be practical.

415.7.1.6 Coal pockets. Coal pockets located less than 30 feet (9144 mm) from interior lot lines or from structures on the same lot shall be constructed of not less than Type IB construction. Where more than 30 feet (9144 mm) from interior lot lines, or where erected along a railroad right-of-way, the minimum type of construction of such structures not more than 65 feet (19 812 mm) in height shall be Type IV.

❖ Like grain elevators, coal pockets are typically open structures and there is a need to provide spatial separation between the coal pocket and the adjacent structures to reduce the exposure fire risk. Type IB construction is the required minimum if a coal pocket is within 30 feet (9144 mm) of another structure or lot line. Like grain elevators, if the spacial separation is in excess of 30 feet

(9144 mm), Type IV construction is permitted to a maximum height of 65 feet (19 812 mm).

415.7.2 Flammable and combustible liquids. The storage, handling, processing and transporting of flammable and combustible liquids shall be in accordance with the *International Mechanical Code* and the *International Fire Code.*

❖ The storage of flammable and combustible liquids constitutes a special hazard because of the potential fire severity and ease of ignition related to such liquids. The flash point of a liquid is the most important criterion of the hazards associated with flammable and combustible liquids because it represents the lowest temperature at which sufficient vapor will be given off to form an ignitable or flammable mixture with air. Although it is the flammable vapors that burn or explode, flammable liquids (by definition) are normally stored above their flash point. Since flammable and combustible liquids are found in virtually every industrial plant and many other occupancies, minimum requirements for the storage of such liquids are set forth in the provisions of Chapter 34 of the IFC and NFPA 30.

Additionally, the provisions of Sections 415.7.2.1 through 415.7.2.10 permit the inside tank storage of flammable and combustible liquids. It is intended that all of these provisions, where applicable, be complied with in order to allow the installation of large tanks of flammable and combustible liquids within the building. These provisions were developed in part due to the cost of complying with environmental considerations regarding the installation of underground storage tanks. In downtown areas, above-ground inside storage tanks are commonly used in conjunction with emergency power generators. While not specifically defined, these provisions are intended for tanks proposed for fixed installation as opposed to containers and portable tanks, such as 55-gallon (208 L) drums.

415.7.2.1 Mixed occupancies. Where the storage tank area is located in a building of two or more occupancies, and the quantity of liquid exceeds the maximum allowable quantity for one control area, the use shall be completely separated from adjacent fire areas in accordance with the requirements of Section 302.3.2.

❖ This section requires mandatory fire-resistance-rated separation of the Group H storage tank area from adjacent fire areas of other groups. By referencing Section 302.3.2 for separated mixed occupancies, the storage tank area would require a minimum fire barrier construction, both horizontally and vertically, whenever a mixed occupancy condition occurs.

415.7.2.1.1 Height exception. Where storage tanks are located within only a single-story building, the height limitation of Section 503 shall not apply for Group H.

❖ The intent of this exception is to allow Group H storage tank areas to be located on an upper floor in a multistory building. As long as the storage tank area is located

within a single story, the height limitations of Table 503, with respect to both height in feet and number of stories, would not apply to the Group H fire area. The height limitation for the building would be dictated by the other occupancy groups in the building and the actual construction type. The height exception assumes the storage tank area complies with all of the applicable provisions of Section 415.7.2 and Chapter 34 of the IFC.

415.7.2.2 Tank protection. Storage tanks shall be noncombustible and protected from physical damage. A fire barrier wall or horizontal assemblies or both around the storage tank(s) shall be permitted as the method of protection from physical damage.

❖ The intent of this section is to reduce the potential for industrial accidents involving storage tanks by some type of physical barrier, such as a pipe bollard or barricade. This section would also allow the fire barrier wall assembly that is enclosing the storage tank area to qualify as the means of protection from physical damage. Any method of physical protection, other than a fire barrier wall assembly, must be approved by building official.

415.7.2.3 Tanks. Storage tanks shall be approved tanks conforming to the requirements of the *International Fire Code.*

❖ The design, construction and installation of all storage tanks for flammable and combustible liquids must comply with the applicable provisions of Chapter 34 of the IFC.

415.7.2.4 Suppression. Group H shall be equipped throughout with an approved automatic sprinkler system, installed in accordance with Section 903.

❖ As with any high-hazard fire area, the storage tank area is required to be protected with an automatic sprinkler system. The use of an alternative extinguishing agent in accordance with Section 904 would be subject to the approval of the building official. A foam-extinguishing system in lieu of an automatic sprinkler system may be a more viable option when protecting a localized flammable liquid storage tank.

415.7.2.5 Leakage containment. A liquid-tight containment area compatible with the stored liquid shall be provided. The method of spill control, drainage control and secondary containment shall be in accordance with the *International Fire Code.*

Exception: Rooms where only double-wall storage tanks conforming to Section 415.7.2.3 are used to store Class I, II and IIIA flammable and combustible liquids shall not be required to have a leakage containment area.

❖ In order to prevent the spread of flammable and combustible liquids to adjacent nonstorage areas in the event of a leak in a storage tank, an adequately sized containment area is required. The sizing of the containment area should be designed to contain a spill from the largest vessel plus the fire protection water for the required duration. It is also important that the drainage system be sufficient to drain not only the flammable or combustible liquid but also the drainage water from the sprinkler system as well as hose streams. Since the outer wall of a double-walled storage tank acts as the secondary means of containment, the exception would allow the omission of the leakage containment area for all flammable and combustible liquids.

415.7.2.6 Leakage alarm. An approved automatic alarm shall be provided to indicate a leak in a storage tank and room. The alarm shall sound an audible signal, 15 dBa above the ambient sound level, at every point of entry into the room in which the leaking storage tank is located. An approved sign shall be posted on every entry door to the tank storage room indicating the potential hazard of the interior room environment, or the sign shall state: WARNING, WHEN ALARM SOUNDS, THE ENVIRONMENT WITHIN THE ROOM MAY BE HAZARDOUS. The leakage alarm shall also be supervised in accordance with Chapter 9 to transmit a trouble signal.

❖ This section requires an alarm system to indicate a leak in the storage tank. The alarm must be supervised as well as sound a local signal. Automatic electrical supervision of the leakage alarm is required in accordance with Section 901.6.3.

415.7.2.7 Tank vent. Storage tank vents for Class I, II or IIIA liquids shall terminate to the outdoor air in accordance with the *International Fire Code*.

❖ Section 3404.2.7.3.3 of the IFC specifies the proper location of the tank vent to enable flammable vapors to be released to the outside air and dissipate without creating a hazard and requires the tank vent to terminate a minimum of 12 feet (3658 mm) above grade and 5 feet (1524 mm) from building openings and lot lines to reduce the possibility of ignitable concentrations of the vapors collecting near the discharge end of the vent pipe, entering building openings or exposing adjoining property.

415.7.2.8 Room ventilation. Storage tank areas storing Class I, II or IIIA liquids shall be provided with mechanical ventilation. The mechanical ventilation system shall be in accordance with the *International Mechanical Code* and the *International Fire Code*.

❖ This section requires mechanical ventilation for storage tank areas in order to maintain any vapor buildup at safe levels until the hazard of a spill or leak can be abated. The ventilation requirements for flammable or combustible liquid storage tanks must be in accordance with the applicable provisions of the IMC and the IFC.

415.7.2.9 Explosion venting. Where Class I liquids are being stored, explosion venting shall be provided in accordance with the *International Fire Code*.

❖ Class I flammable liquids, when stored in sufficient quantities, can readily produce vapors within the explosive limitations. Explosion venting must be in accordance with the provisions of Sections 414.5.1 and 911 and Chapter 34 of the IFC.

415.7.2.10 Tank openings other than vents. Tank openings other than vents from tanks inside buildings shall be designed to ensure that liquids or vapor concentrations are not released inside the building.

❖ Additional provisions are necessary to minimize the possibility of an accidental release of flammable or combustible liquids or their vapors via any connection to the tank other than a vent. This includes requirements for liquid-tight openings below the liquid level in the tank, overflow protection and vapor recovery connections as indicated in Section 3404.2.7.5 of the IFC.

415.7.3 Liquefied petroleum gas-distribution facilities. The design and construction of propane, butane, propylene, butylene and other liquefied petroleum gas-distribution facilities shall conform to the applicable provisions of Sections 415.7.3.1 through 415.7.3.5.2. The storage and handling of liquefied petroleum gas systems shall conform to the *International Fire Code*. The design and installation of piping, equipment and systems that utilize liquefied petroleum gas shall be in accordance with the *International Fuel Gas Code*. Liquefied petroleum gas-distribution facilities shall be ventilated in accordance with the *International Mechanical Code* and Section 415.7.3.1.

❖ Requirements for the design and construction of liquefied petroleum gas distribution and storage facilities are provided. The requirements are intended to minimize the potential for an uncontrolled discharge of the product. Although not referenced in this section, it should be noted that NFPA 58 is adopted by reference through the IMC and the IFC. This section distinguishes between liquefied petroleum gas facilities in separate buildings (Section 415.7.3.3), attached buildings (Section 415.7.3.4) and rooms within buildings (Section 415.7.3.5). The exposure hazard increases as the distance between the liquefied petroleum gas facility and the proximity decreases. It should be noted that the location, floor construction and venting restrictions are based on the fact that liquefied petroleum gas will spread along the floor since it is heavier than air.

This section applies to the design and construction of liquefied petroleum gas distribution facilities regardless of the composition of the gas. Chapter 38 of the IFC contains provisions for equipment, processes and operations for the storage and handling of liquefied petroleum gases. The IMC contains provisions for piping systems and references NFPA 58 for ventilation provisions of liquefied petroleum gas distribution facilities. The IMC provides performance requirements for the mechanical ventilation system when natural ventilation is not provided. In either case, the openings must not be more than 6 inches (152 mm) above the floor since liquefied petroleum gas will spread across the floor.

415.7.3.1 Air movement. Liquefied petroleum gas- distribution facilities shall be provided with air inlets and outlets arranged so that air movement across the floor of the facility will be uniform. The total area of both inlet and outlet openings shall be at least 1 square inch (645 mm²) for each 1 square foot (0.093 m²) of floor area. The bottom of such openings shall not be more than 6 inches (152 mm) above the floor.

❖ Air inlet and outlet openings are intended to prevent liquefied petroleum gas from accumulating in the facility, thereby minimizing the potential for ignition to occur. The openings must be within 6 inches (152 mm) of the floor since liquefied petroleum gas is heavier than air.

415.7.3.2 Construction. Liquefied petroleum gas-distribution facilities shall be constructed in accordance with Section 415.7.3.3 for separate buildings, Section 415.7.3.4 for attached buildings or Section 415.7.3.5 for rooms within buildings.

❖ Construction requirements vary depending on the location of the liquefied petroleum gas facility with respect to exposures. Separate buildings are required to comply with the provisions of Section 415.7.3.3, attached buildings are required to comply with Section 415.7.3.4 and rooms within buildings are required to comply with Section 415.7.3.5.

415.7.3.3 Separate buildings. Where located in separate buildings, liquefied petroleum gas-distribution facilities shall be occupied exclusively for that purpose or for other purposes having similar hazards. Such buildings shall be limited to one story in height and shall conform to Sections 415.7.3.3.1 through 415.7.3.3.3.

❖ This section requires the liquefied petroleum gas facility to be in a separate single-story building used exclusively for the purpose intended. As such, potential ignition and fuel sources are minimized to only those involved directly with liquefied petroleum gas distribution or building services, such as electricity for lighting circuits. Additionally, the requirements of Chapter 5 establish the height and area limitations that are applicable for this high-hazard occupancy.

415.7.3.3.1 Floors. The floor shall not be located below ground level and any spaces beneath the floor shall be solidly filled or shall be unenclosed.

❖ Since liquefied petroleum gas is heavier than air, the floor level must not be below ground level. Furthermore, there must not be any void spaces (such as crawl spaces beneath the floor) that could serve as cavities for the collection of liquefied petroleum gas vapors.

415.7.3.3.2 Materials. Walls, floors, ceilings, columns and roofs shall be constructed of noncombustible materials.

❖ To minimize additional fuel sources, the building is required to be constructed of noncombustible materials.

415.7.3.3.3 Explosion venting. Explosion venting shall be provided in accordance with the *International Fire Code*.

❖ This section requires explosion venting in accordance with Section 911 of the IFC. NFPA 58, as referenced in Chapter 38 of the IFC, contains specific venting requirements for separate buildings or structures depending on the construction materials involved.

415.7.3.4 Attached buildings. Where liquefied petroleum gas-distribution facilities are located in an attached structure, the attached perimeter shall not exceed 50 percent of the perimeter of the space enclosed and the facility shall comply with Sections 415.7.3.3 and 415.7.3.4.1. Where the attached perimeter exceeds 50 percent, such facilities shall comply with Section 415.7.3.5.

❖ For the purposes of this section, an attached building is defined as a building with at least 50 percent of its perimeter not attached to an adjacent structure. If the attached perimeter exceeds 50 percent, the facility is required to comply with the provisions for rooms within buildings (see Section 415.7.3.5).

415.7.3.4.1 Fire separation assemblies. Separation of the attached structures shall be provided by fire barrier walls and horizontal assemblies, or both, having a fire-resistance rating of not less than 1 hour and shall not have openings. Fire barrier walls and horizontal assemblies, or both, between attached structures occupied only for the storage of LP-gas are permitted to have fire doors that comply with Section 715. Such fire barrier walls and horizontal assemblies, or both, shall be designed to withstand a static pressure of at least 100 pounds per square foot (psf) (4788 Pa), except where the building to which the structure is attached is occupied by operations or processes having a similar hazard.

❖ To prevent the spread of fire and liquefied petroleum gas vapors, common walls are required to be fire barriers with a fire-resistance rating of at least 1 hour and there must not be any openings therein. If the building is used for storage only, thereby reducing the potential for unwanted discharge, the common wall may have openings that are protected in accordance with Section 715. The common wall is required to withstand a static pressure of at least 100 psf (4788 Pa) to minimize the damage should an explosion occur. This requirement does not apply if the common wall separates different liquefied petroleum gas processes within the attached building and does not separate the liquefied petroleum gas facility from another hazard. If the hazard in an exposed structure represents a potential fire severity or duration in excess of that intended for a 1-hour fire separation, consideration should be given to providing a higher fire-resistance rating to protect the liquefied petroleum gas facility from a fire in the attached structure.

415.7.3.5 Rooms within buildings. Where liquefied petroleum gas-distribution facilities are located in rooms within buildings, such rooms shall be located in the first story above grade plane and shall have at least one exterior wall with sufficient exposed area to provide explosion venting as required in the *International Fire Code*. The building in which the room is located shall not have a basement or unventilated crawl space and the room shall comply with Sections 415.7.3.5.1 and 415.7.3.5.2.

❖ To minimize the potential exposure to adjacent areas, rooms containing liquefied petroleum gas facilities must be provided with explosion venting in at least one exterior wall. The explosion venting is to be in accordance with Section 414.5.1 and Section 911 of the IFC. In order to restrict liquefied petroleum gas from spreading to other areas, the liquefied petroleum gas facility must not be in a building with a basement or unventilated crawl space.

415.7.3.5.1 Materials. Walls, floors, ceilings and roofs of such rooms shall be constructed of approved noncombustible materials.

❖ To minimize additional fuel sources, the building is required to be constructed of noncombustible materials. The type of approved noncombustible materials utilized should be based on the explosion venting design criteria.

415.7.3.5.2 Common construction. Walls and floor/ceiling assemblies common to the room and to the building within which the room is located shall have a fire barrier wall and horizontal assembly or both of not less than a 1-hour fire-resistance rating and without openings. Common walls for rooms occupied only for storage of LP-gas are permitted to have opening protectives complying with Section 715. Such walls and ceiling shall be designed to withstand a static pressure of at least 100 psf (4788 Pa).

> **Exception:** Where the building, within which the room is located, is occupied by operations or processes having a similar hazard.

❖ To prevent the spread of fire and liquefied petroleum gas vapors, common walls must have a fire-resistance rating of at least 1 hour and there must not be any openings therein. If the room is used for storage only, thereby reducing the potential for unwanted discharge, the common wall may have openings that are protected in accordance with Section 715. Section 715 would require the opening protectives to have a fire protection rating of at least $^3/_4$ hour. If the hazard in an adjacent area represents a potential fire severity or duration in excess of that intended for a 1-hour fire separation, consideration should be given to providing a higher fire-resistance rating to protect the liquefied petroleum gas facility from fire in the adjacent areas. The common walls are required to be able to withstand a static pressure of at least 100 pounds psf (4788 Pa) to minimize the damage should an explosion occur. This requirement does not apply to common walls that separate different liquefied petroleum gas processes within the same room and that do not separate the liquefied petroleum gas facility from another hazard.

415.7.4 Dry cleaning plants. The construction and installation of dry cleaning plants shall be in accordance with the require-

ments of this code, the *International Mechanical Code*, the *International Plumbing Code* and NFPA 32. Dry cleaning solvents and systems shall be classified in accordance with the *International Fire Code*.

❖ Dry cleaning plants are required to comply with the provisions of the IPC, the IMC and NFPA 32. It should be noted that Chapter 12 of the IFC also contains provisions for dry cleaning plants and references NFPA 32 for their maintenance. NFPA 32 contains provisions for the prevention and control of fire and explosion hazards incidental to dry cleaning operations and for the protection of employees and the public. This standard addresses such issues as location, construction, building services, processes and equipment and fire control.

415.8 Groups H-3 and H-4. Groups H-3 and H-4 shall be constructed in accordance with the applicable provisions of this code and the *International Fire Code*.

❖ This section contains provisions that are applicable for all Group H-3 and H-4 occupancies in addition to material-specific requirements in the IFC. These requirements assume that the Group H-3 and H-4 occupancies contain more than the maximum allowable quantities of hazardous materials per control area permitted by Tables 307.7(1) and 307.7(2). Sections 307.5 and 307.6 contain a list of those materials that could be present in Group H-3 and H-4 occupancies.

415.8.1 Gas rooms. When gas rooms are provided, such rooms shall be separated from other areas by not less than a 1-hour fire barrier.

❖ In order to restrict the potential fire involvement of a storage room of hazardous gases within a Group H-3 or H-4 occupancy, a 1-hour fire barrier separation is required from other adjacent areas. The required 1-hour fire barrier separation is not due to a mixed occupancy condition but rather further compartmentation in a Group H-3 or H-4 occupancy.

415.8.2 Floors in storage rooms. Floors in storage areas for corrosive liquids and highly toxic or toxic materials shall be of liquid-tight, noncombustible construction.

❖ Storage rooms of corrosive liquids and highly toxic and toxic materials would be classified as a Group H-4 occupancy. Noncombustible floors are required to prevent the structure from contributing to a fire scenario. The floors are also required to be liquid tight to prevent the spread of liquids from a spill to areas outside the storage room.

415.8.3 Separation—highly toxic solids and liquids. Highly toxic solids and liquids not stored in approved hazardous materials storage cabinets shall be isolated from other hazardous materials storage by construction having a 1-hour fire-resistance rating.

❖ The potential involvement of materials that may be incompatible in storage situations with highly toxic solids and liquids must be addressed. The required degree of separation, either by approved cabinets or 1-hour fire barrier construction, is also intended to restrict the highly toxic materials from other combustibles.

415.9 Group H-5.

415.9.1 General. In addition to the requirements set forth elsewhere in this code, Group H-5 shall comply with the provisions of Section 415.9 and the *International Fire Code.*

❖ This section contains the requirements for Group H-5 facilities that utilize HPM. Section 415.9 was intended to be a design option for buildings that utilize hazardous materials in excess of the maximum allowable quantities permitted by Tables 307.7(1) and 307.7(2). HPM facilities are considered unique high-hazard occupancies. The HPM occupancy classification assumed that while the manufacturing process involved the use of hazardous materials, the end product by itself was not hazardous.

The provisions of Section 415.9 resulted from nonuniform regulation of semiconductor fabrication facilities and are based on code changes submitted by groups associated with the semiconductor industry. These industries include electronic high-tech industries, such as semiconductor microchip fabrication, as well as the floppy disk and telecommunication industries. Additionally, there are other comparable activities involving similar technology that also use high-hazard materials. HPM include flammable liquids and gases, corrosives, oxidizers and, in many instances, highly toxic materials. The HPM may be a solid, liquid or gas with a degree of hazard rating in health, flammability or reactivity of Class 3 or 4 in accordance with NFPA 704, which is used directly in research, laboratory or production processes.

Structures that contain HPM facilities are required to meet the provisions of Section 415.9 as well as any other code provisions for a Group H-5 occupancy classification unless specifically modified in Section 415.9. Additionally, Chapter 18 of the IFC provides complementary fire safety controls that reduce the risk of hazard to an acceptable level in existing HPM buildings and fabrication areas.

415.9.2 Fabrication areas.

❖ The fabrication area of an HPM facility is where the hazardous materials are actively handled and processed. The fabrication area includes accessory rooms and spaces such as workstations and employee dressing rooms.

415.9.2.1 Hazardous materials in fabrication areas.

❖ Sections 415.9.2.1.1 and 415.9.2.1.2 regulate the quantities and densities of HPM for each fabrication area of a Group H-5 facility.

415.9.2.1.1 Aggregate quantities. The aggregate quantities of hazardous materials stored and used in a single fabrication area shall not exceed the quantities set forth in Table 415.9.2.1.1.

Exception: The quantity limitations for any hazard category in Table 415.9.2.1.1 shall not apply where the fabrication area contains quantities of hazardous materials not exceeding the maximum allowable quantities per control area established by Tables 307.7(1) and 307.7(2).

❖ This section regulates the total amount of hazardous materials, whether in use or storage, within a single fabrication area based on the density/quantity of material specified in Table 415.9.2.1.1. The exception permits a fabrication area to have a total quantity of HPM of either the quantity specified in Table 415.9.2.1.1 or the maximum allowable quantities per control area specified in Table 307.7(1) or 307.7(2), whichever is greater. For example, a small fabrication area may have a quantity in accordance with Tables 307.7(1) and 307.7(2) even though it may exceed the quantities specified in Table 415.9.2.1.1. Conversely, a large fabrication area may have an aggregate quantity that exceeds those specified in Tables 307.7(1) and 307.7(2) if it does not exceed the quantities specified in Table 415.9.2.1.1.

It should be reiterated that this section refers to an aggregate quantity of hazardous materials that are in use and storage within a single fabrication area. Section 415.9.2.1.2 further limits the amount of HPM that is being stored within a fabrication area.

415.9.2.1.2 Hazardous production materials. The maximum quantities of hazardous production materials stored in a single fabrication area shall not exceed the maximum allowable quantities per control area established by Tables 307.7(1) and 307.7(2).

❖ This section provides the overall limit of HPM that can be stored within a single fabrication area based on Tables 307.7(1) and 307.7(2). Any storage in excess of the permitted maximum allowable quantities would be required to be stored in an HPM storage room, liquid storage room or gas room. The aggregate quantities of hazardous materials in both a storage and use condition within a single fabrication area are regulated by Section 415.9.2.1.1.

TABLE 415.9.2.1.1. See page 4-96.

❖ Table 415.9.2.1.1 controls the permitted densities/quantities of hazardous materials in a single fabrication area. In accordance with Section 415.9.2.1.1, the quantities permitted in Tables 307.7(1) and 307.7(2) may be exceeded, provided the quantities in Table 415.9.2.1.1 are maintained subject to the storage limitations of Section 415.9.2.1.2. Conversely, the maximum quantities of Table 415.9.2.1.1 may be exceeded, provided the maximum allowable quantities of Tables 307.7(1) and 307.7(2) are not exceeded.

In accordance with Note a, hazardous materials in piping need not be included in evaluating the permitted quantity. Note b establishes that a specific quantity limitation, rather than a calculated density, is required for certain hazardous materials.

TABLE 415.9.2.1.1
QUANTITY LIMITS FOR HAZARDOUS MATERIALS IN A SINGLE FABRICATION AREA IN GROUP H-5[a]

HAZARD CATEGORY		SOLIDS (pounds per square feet)	LIQUIDS (gallons per square feet)	GAS (feet^3 @ NTP/square feet)
PHYSICAL-HAZARD MATERIALS				
Combustible dust		Note b	Not Applicable	Not Applicable
Combustible fiber	Loose	Note b	Not Applicable	Not Applicable
	Baled	Note b		
Combustible liquid	II		0.01	
	IIIA		0.02	
	IIIB	Not Applicable	Not Limited	Not Applicable
Combination Class I, II and IIIA			0.04	
Cryogenic gas	Flammable	Not Applicable	Not Applicable	Note c
	Oxidizing			1.25
Explosives		Note b	Note b	Note b
Flammable gas	Gaseous	Not Applicable	Not Applicable	Note b
	Liquefied			Note c
Flammable liquid	IA		0.0025	
	IB		0.025	
	IC	Not Applicable	0.025	Not Applicable
Combination Class IA, IB and IC			0.025	
Combination Class I, II and IIIA			0.04	
Flammable solid		0.001	Not Applicable	Not Applicable
Organic peroxide				
Unclassified detonable		Note b		
Class I		Note b		
Class II		0.025	Not Applicable	Not Applicable
Class III		0.1		
Class IV		Not Limited		
Class V		Not limited		
Oxidizing gas	Gaseous			1.25
	Liquefied			1.25
Combination of gaseous and liquefied		Not Applicable	Not Applicable	1.25
Oxidizer	Class 4	Note b	Note b	
	Class 3	0.003	0.003	
	Class 2	0.003	0.003	Not Applicable
	Class 1	0.003	0.003	
Combination	Class 1, 2, 3	0.003	0.003	
Pyrophoric material		Note b	0.00125	Notes c and d
Unstable reactive	Class 4	Note b	Note b	Note b
	Class 3	0.025	0.0025	Note b
	Class 2	0.1	0.01	Note b
	Class 1	Not Limited	Not Limited	Not Limited
Water reactive	Class 3	Note b	0.00125	
	Class 2	0.25	0.025	Not Applicable
	Class 1	Not Limited	Not Limited	
HEALTH-HAZARD MATERIALS				
Corrosives		Not Limited	Not Limited	Not Limited
Highly toxic		Not Limited	Not Limited	Note c
Toxics		Not Limited	Not Limited	Note c

For SI: 1 pound per square foot = 4.882 kg/m^2, 1 gallon per square foot = 0.025 L/m^2, 1 cubic foot @ NTP/square foot = 0.305 M^3 @ NTP/m^2, 1 cubic foot = 0.02832 M^3.

a. Hazardous materials within piping shall not be included in the calculated quantities.

b. Quantity of hazardous materials in a single fabrication shall not exceed the maximum allowable quantities per control area in Tables 307.7(1) and 307.7(2).

c. The aggregate quantity of flammable, pyrophoric, toxic and highly toxic gases shall not exceed 9,000 cubic feet at NTP.

d. The aggregate quantity of pyrophoric gases in the building shall not exceed the amounts set forth in Table 415.3.2.

415.9.2.2 Separation. Fabrication areas, whose sizes are limited by the quantity of hazardous materials allowed by Table 415.9.2.1.1, shall be separated from each other, from exit access corridors, and from other parts of the building by not less than 1-hour fire barriers.

Exceptions:

1. Doors within such fire barrier walls, including doors to corridors, shall be only self-closing fire assemblies having a fire-protection rating of not less than $^3/_4$ hour.

2. Windows between fabrication areas and exit access corridors are permitted to be fixed glazing listed and labeled for a fire protection rating of at least $^3/_4$ hour in accordance with Section 715.

❖ Fabrication areas must be separated from other parts of the building, egress corridors and other fabrication areas by fire barriers having a fire-resistance rating of at least 1 hour. While the potential fire exposure could warrant a higher fire-resistance rating, the additional protection features indicative of a Group H-5 facility result in a 1-hour fire-resistance rating being acceptable. The fire barriers must be constructed in accordance with Section 706. One reason for separating one fabrication area from another is the limitation of HPM quantities in a single fabrication area. If the need exists for quantities in excess of those permitted by Section 415.9.2.1.1, the area can be divided into two fabrication areas with 1-hour fire barriers.

Exception 1 clarifies that all doors within the fire barrier walls that comprise the fabrication area must be self-closing and have a fire protection rating of $^3/_4$ hour; therefore, if a corridor wall was also part of the fabrication area, a $^3/_4$-hour opening protective would be required for the door opening. Openings in rated corridors that are not part of a fabrication area typically are required to only have a 20-minute fire protection rating. The doors are required to be self-closing since fabrication areas are "clean rooms" under positive pressure.

The $^3/_4$-hour requirement of Exception 2 will permit the use of fire protection rated glazing or wired glass windows in accordance with Section 715.4. The use of windows is often desirable to facilitate the security of activities in the fabrication area as well as emergency response in the event of an accident.

415.9.2.3 Location of occupied levels. Occupied levels of fabrication areas shall be located at or above the first story above grade plane.

❖ Due to the potential extensive use of hazardous materials, the occupiable levels of fabrication areas are not permitted in basements or other areas below grade. Maintaining fabrication areas at or above grade aids in the egress of the occupants and access for fire department operations.

415.9.2.4 Floors. Except for surfacing, floors within fabrication areas shall be of noncombustible construction.

Openings through floors of fabrication areas are permitted to be unprotected where the interconnected levels are used solely

for mechanical equipment directly related to such fabrication areas (see also Section 415.9.2.5).

Floors forming a part of an occupancy separation shall be liquid tight.

❖ Fabrication area floors are required to be noncombustible. The intent of the requirement for noncombustible floors is to prevent the structure from being involved in the fire from a liquid that may seep into the floor system. The requirement is not intended to apply to floor-covering material.

This section permits the interconnection of the operation floor and a mechanical floor, provided that the mechanical floor is essentially unoccupied. The intent is to prevent leakage past a required separation and to allow multiple-level fabrication areas where the space above or below the operation level is used for mechanical equipment, ducts and similar service equipment. Where a rated separation is required, the floors must be liquid tight. Whereas the hazards often involve liquids, floors are required to be liquid tight to prevent the spread of liquids from a spill to areas outside the fabrication area. "Liquid tight" requires that floors be sealed where they intersect with walls. This can be accomplished by a cove base of the same material as the floor.

415.9.2.5 Shafts and openings through floors. Elevator shafts, vent shafts and other openings through floors shall be enclosed when required by Section 707. Mechanical, duct and piping penetrations within a fabrication area shall not extend through more than two floors. The annular space around penetrations for cables, cable trays, tubing, piping, conduit or ducts shall be sealed at the floor level to restrict the movement of air. The fabrication area, including the areas through which the ductwork and piping extend, shall be considered a single conditioned environment.

❖ Section 707.2 specifies the conditions when a shaft enclosure is required. This section provides the requirements for specific opening penetrations through a floor within a fabrication area. The maximum number of stories that are allowed to be interconnected is three, since penetrations are limited to two floors. The penetrations for duct and piping are to have the annular space protected. When such penetrations exist, the fire separation required by Section 415.9.2.2 is to separate the entire interconnected volume.

415.9.2.6 Ventilation. Mechanical exhaust ventilation shall be provided throughout the fabrication area at the rate of not less than 1 cubic foot per minute per square foot (0.044 L/S/m^2) of floor area. The exhaust air duct system of one fabrication area shall not connect to another duct system outside that fabrication area within the building.

A ventilation system shall be provided to capture and exhaust fumes and vapors at workstations.

Two or more operations at a workstation shall not be connected to the same exhaust system where either one or the combination of the substances removed could constitute a fire, explosion or hazardous chemical reaction within the exhaust duct system.

Exhaust ducts penetrating occupancy separations shall be

contained in a shaft of equivalent fire-resistance construction. Exhaust ducts shall not penetrate fire walls.

Fire dampers shall not be installed in exhaust ducts.

❖ A major design factor is the required minimum ventilation. The minimum rate is 1 cubic foot per minute (cfm) per square foot (0.044 L/S/m²), or about eight air changes per hour for an 8-foot-high (2438 mm) fabrication area. Typically, at least 24 air changes per hour are provided for operational needs and where unclassified electrical systems are to be used. In order to prevent the spread of gases or fumes, the exhaust air duct system of any fabrication area must not be connected to any other ventilation system.

With regard to workstations, the required ventilation system is intended to prevent exposure of the work station operator to hazardous fumes or vapors and prevent hazardous concentrations of materials used in the manufacturing process. Additionally, ventilation systems must be designed to reduce the possibility of a chemical reaction occurring between two or more HPM or a material's and exhaust system's components. Where two or more chemicals may react dangerously if exhausted through a common duct, they must be removed separately. NFPA 491, which is included in the *NFPA Fire Protection Guide on Hazardous Materials*, is a good first reference when trying to determine if any two agents are reactive.

Ventilation should not be interrupted by a fire damper when a fire or other emergency occurs involving a workstation. This helps reduce the likelihood that hazardous combustion byproducts or hazardous concentrations of HPM are not forced back into the workstation or clean room. Continuous ventilation through a duct enclosed in a fire-resistance-rated shaft is required. Fire walls define separations between buildings. Ducts must never penetrate a barrier common to another building or occupancy. This reduces the likelihood of tampering with or interrupting the duct integrity.

415.9.2.7 Transporting hazardous production materials to fabrication areas. Hazardous production materials shall be transported to fabrication areas through enclosed piping or tubing systems that comply with Section 415.9.6.1, through service corridors complying with Section 415.9.4, or in exit access corridors as permitted in the exception to Section 415.9.3. The handling or transporting of hazardous production materials within service corridors shall comply with the *International Fire Code*.

❖ Except as permitted in the exception to Section 415.9.3 for existing facilities, all HPM must be transported to the fabrication area by service corridors or piping (see Sections 415.9.4 and 415.9.6). This reduces the chance that an egress route will be affected by an accidental leak or spill. Chapter 18 of the IFC contains additional requirements for the transport of HPM through all egress components as well as service corridors.

415.9.2.8 Electrical.

❖ Sections 415.9.2.8.1 and 415.9.2.8.2 provide specific electrical requirements with respect to ventilation systems in fabrication areas and at workstations.

415.9.2.8.1 General. Electrical equipment and devices within the fabrication area shall comply with the ICC *Electrical Code*. The requirements for hazardous locations need not be applied where the average air change is at least four times that set forth in Section 415.9.2.6 and where the number of air changes at any location is not less than three times that required by Section 415.9.2.6. The use of recirculated air shall be permitted.

❖ Consistent with Chapter 27, electrical installations are required to comply with the ICC EC and subsequently NFPA 70. This section specifically permits the dilution of concentrations of hazardous areas such that the hazardous location provisions of NFPA 70 need not apply. This is because some of the equipment used in HPM facilities cannot comply with the hazardous location provisions of NFPA 70. The nature of the fabrication process also dictates the use of the dilution concept. Typically, such facilities have a ventilation system designed to provide at least 24 air changes per hour.

At least 3 cfm per square foot (0.132 L/S/m²) or 24 air changes in an 8-foot-high (2438 mm) fabrication area is required at all locations in order not to apply the hazardous location provisions of NFPA 70. Also, an average rate of 4 cfm per square foot (0.176 L/S/m²) or 32 air changes in an 8-foot-high (2438 mm) fabrication area are required if the hazardous location provisions of NFPA 70 are not applied.

415.9.2.8.2 Workstations. Workstations shall not be energized without adequate exhaust ventilation. See Section 415.9.2.6 for workstation exhaust ventilation requirements.

❖ Incidental exposure to flammable fumes or vapors can occur at workstations where hazardous materials are used. Either a mechanical or electrical interlock must be provided to engage the required exhaust ventilation system before HPM enters the workstation. This reduces the likelihood of gas or vapor exposure.

415.9.3 Exit access corridors. Exit access corridors shall comply with Chapter 10 and shall be separated from fabrication areas as specified in Section 415.9.2.2. Exit access corridors shall not contain HPM and shall not be used for transporting such materials, except through closed piping systems as provided in Section 415.9.6.3.

Exception: Where existing fabrication areas are altered or modified, HPM is allowed to be transported in existing exit access corridors, subject to the following conditions:

1. Corridors. Exit access corridors adjacent to the fabrication area where the alteration work is to be done shall comply with Section 1016 for a length determined as follows:

    1.1. The length of the common wall of the corridor and the fabrication area; and

    1.2. For the distance along the exit access corridor to the point of entry of HPM into the exit access corridor serving that fabrication area.

2. Emergency alarm system. There shall be an emergency telephone system, a local manual alarm station or other

approved alarm-initiating device within exit access corridors at not more than 150-foot (45 720 mm) intervals and at each exit and exit access doorway. The signal shall be relayed to an approved central, proprietary or remote station service or the emergency control station and shall also initiate a local audible alarm.

3. Pass-throughs. Self-closing doors having a fire-protection rating of not less than 1 hour shall separate pass-throughs from existing exit access corridors. Pass-throughs shall be constructed as required for the exit access corridors, and protected by an approved automatic fire-extinguishing system.

❖ Exit access corridors are required to comply with the requirements of Section 1016 and must be separated from fabrication areas in accordance with Section 415.9.2.2. Additionally, exit access corridors in new buildings are not to be used for transporting HPM to the fabrication area. In accordance with Section 415.9.2.7, HPM must be transported by service corridors or piping (see Sections 415.9.4 and 415.9.6 and Figure 415.9.3).

The exception addresses HPM facilities that existed before the adoption and enforcement of this section. It permits the transport of HPM in exit access corridors in existing buildings under the conditions specified in Items 1, 2 and 3. When alterations are made to a fabrication area, such corridors must be upgraded. Chapter 18 of the IFC provides additional requirements for the storage, handling and transporting of HPM materials in existing HPM facilities. Item 1 of the exception requires that corridors be upgraded to a 1-hour fire-resistance rating for the entire length of the wall common to the fabrication area. Additionally, the entire length of the corridor that is used for transporting HPM to the fabrication area being renovated must also be upgraded to have a 1-hour fire-resistance rating. Item 2 requires a means to notify trained personnel of an emergency condition in an exit access corridor. A local alarm is required to alert the occupants of a potential hazardous condition. The intent of Item 3 is to permit a "pass-through" for providing HPM to a fabrication area. A pass-through, such as a storage cabinet, is used to store and receive HPM for the fabrication area. The pass-through must be separated from the exit access corridor by 1-hour fire-resistance-rated construction, including a 1-hour rated self-closing fire door, and be suppressed.

415.9.4 Service corridors.

❖ This section regulates the corridor that may be used for the transport of HPM in a Group H-5 facility of new construction. For the transport of HPM via piping systems, see Section 415.9.6.

415.9.4.1 Occupancy. Service corridors shall be classified as Group H-5.

❖ The intent of this section is to clarify that a potential separation due to the mixed occupancy provisions of Section 302.3 does not apply, since service corridors are considered part of the Group H-5 occupancy.

415.9.4.2 Use conditions. Service corridors shall be separated from exit access corridors as required by Section 415.9.2.2. Service corridors shall not be used as a required exit access corridor.

❖ Service corridors are used for transporting HPM in lieu of piping. The separation between a service corridor and exit access corridor must be in accordance with Section 415.9.2.2. As such, a minimum 1-hour fire separation is required. Service corridors are also not intended to be used as a portion of the required means of egress.

415.9.4.3 Mechanical ventilation. Service corridors shall be mechanically ventilated as required by Section 415.9.2.6 or at not less than six air changes per hour, whichever is greater.

❖ Since the same HPM that may be present in the fabrication area may be transported in the service corridors, an adequate means of ventilation is required. This section requires a ventilation rate of either 1cfm per square foot (0.044 L/S/m²) of floor area in accordance with Section 415.9.2.6 or six air changes per hour, whichever is greater depending on the size and volume of the service corridor.

415.9.4.4 Means of egress. The maximum distance of travel from any point in a service corridor to an exit, exit access corridor or door into a fabrication area shall not exceed 75 feet (22 860 mm). Dead ends shall not exceed 4 feet (1219 mm) in length. There shall be not less than two exits, and not more than one-half of the required means of egress shall require travel into a fabrication area. Doors from service corridors shall swing in the direction of egress travel and shall be self-closing.

❖ The exit access travel distance is limited to 75 feet (22 860 mm), except that the travel distance may be measured from a door to a fabrication area. In order to minimize the potential for an occupant to be trapped in a service corridor, dead ends must not exceed 4 feet (1219 mm). As such, the occupant would most likely be intimate with the source of the problem or located in an adjacent fabrication area from which egress is provided via exit access, not service, corridors.

An occupant at any location in a service corridor is required to have access to at least two means of egress. No more than 50 percent of the means of egress may be through a fabrication area that, in most cases, will be one of the means of egress. This assumes that, should an incident or problem occur, it will be in either the service corridor or the fabrication area—not both. Because of the potential for a hazardous condition, all doors to service corridors must be self-closing and swing in the direction of egress travel. Smoke-detector-actuated hold-open devices are not acceptable because smoke detectors may not respond to hazardous conditions that do not involving products of combustion.

FIGURE 415.9.3   SPECIAL DETAILED REQUIREMENTS BASED ON USE AND OCCUPANCY

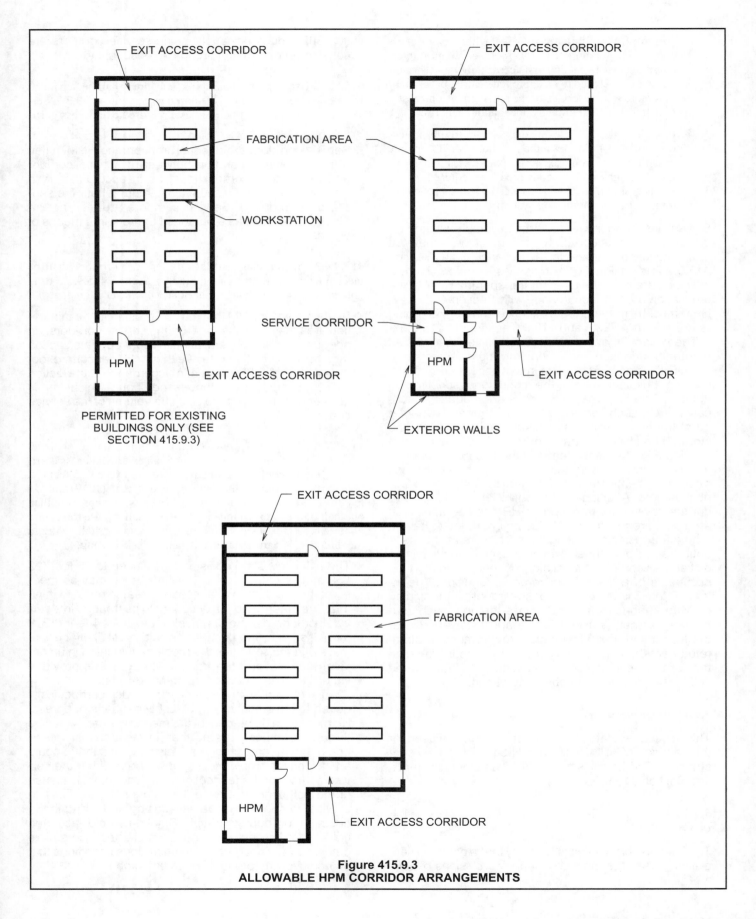

Figure 415.9.3
ALLOWABLE HPM CORRIDOR ARRANGEMENTS

415.9.4.5 Minimum width. The minimum clear width of a service corridor shall be 5 feet (1524 mm), or 33 inches (838 mm) wider than the widest cart or truck used in the corridor, whichever is greater.

❖ The minimum width requirement for the service corridor is to provide adequate clearance to get around the device used to transport the HPM in the event it was involved in an emergency incident in the service corridor. The size of the transport carts and trucks depends on the type and quantity of materials utilized in the facility.

415.9.4.6 Emergency alarm system. Emergency alarm systems shall be provided in accordance with this section and Sections 414.7.1 and 414.7.2. The maximum allowable quantity per control area provisions shall not apply to emergency alarm systems required for HPM.

❖ This section requires an emergency alarm system in all areas where HPM is transported or stored. The applicability of either Section 414.7.1 or 414.7.2 depends on whether the HPM material is in a storage or use condition. This section also clarifies that the requirement for an emergency alarm system in a Group H-5 facility in the locations identified in Sections 415.9.4.6.1 through 415.9.6.4.3 is not dependent on whether the maximum allowable quantities per control area of Tables 307.7(1) or 307.7(2) are exceeded.

415.9.4.6.1 Service corridors. An emergency alarm system shall be provided in service corridors, with at least one alarm device in each service corridor.

❖ An emergency telephone system or manual alarm pull station is required in service corridors. Such devices must initiate an alarm at the emergency control station as well as activate a local audible device.

415.9.4.6.2 Exit access corridors and exit enclosures. Emergency alarms for exit access corridors and exit enclosures shall comply with Section 414.7.2.

❖ Since HPM materials would not be in exit access corridors or exit enclosures unless they were being transported to another approved area, the emergency alarm requirements of Section 414.7.2 for dispensing, use and handling must be complied with.

415.9.4.6.3 Liquid storage rooms, HPM rooms and gas rooms. Emergency alarms for liquid storage rooms, HPM rooms and gas rooms shall comply with Section 414.7.1.

❖ This section requires compliance with the emergency alarm requirements of Section 414.7.1 for hazardous materials in a storage condition. This section addresses storage areas, which by their designation contain HPM in quantities greater than those listed in Table 307.7(1) or 307.7(2).

415.9.4.6.4 Alarm-initiating devices. An approved emergency telephone system, local alarm manual pull stations, or other approved alarm-initiating devices are allowed to be used as emergency alarm-initiating devices.

❖ This section classifies what constitutes an approved alarm-initiating device (see commentary, Section 414.7).

415.9.4.6.5 Alarm signals. Activation of the emergency alarm system shall sound a local alarm and transmit a signal to the emergency control station.

❖ The alarm signal is required to be transmitted to the emergency control station in order to notify trained personnel of an emergency condition. A local alarm is required to alert the occupants of a potentially hazardous condition.

415.9.5 Storage of hazardous production materials.

❖ Section 415.9.5 addresses general storage provisions for HPM materials in a Group H-5 facility.

415.9.5.1 General. Storage of HPM in fabrication areas shall be within approved or listed storage cabinets or gas cabinets, or within a workstation. The storage of hazardous production materials in quantities greater than those listed in Tables 307.7(1) or 307.7(2) shall be in liquid storage rooms, HPM rooms or gas rooms as appropriate for the materials stored. The storage of other hazardous materials shall be in accordance with other applicable provisions of this code and the *International Fire Code*.

❖ Even though the amount of HPM in a fabrication area is controlled, it still must be within approved cabinets or a workstation. This requirement is intended to limit the exposure to occupants of the fabrication area to only the material in use within that area.

The larger amounts of HPM typically stored in separate areas present a hazard comparable to other Group H facilities; therefore, such storage rooms need to meet similar code requirements. Storage rooms containing HPM in quantities greater than permitted by Tables 307.7(1) and 307.7(2) are required to comply with the applicable provisions of Section 415.9.5.2, depending on the state of the material.

415.9.5.2 Construction.

❖ This section deals with construction requirements for two types of storage rooms: those classified as HPM rooms or gas rooms and those utilized as liquid storage rooms. The size and separation of such rooms is dependent on the type of materials stored.

415.9.5.2.1 HPM rooms and gas rooms. HPM rooms and gas rooms shall be separated from other areas by not less than a 2-hour fire barrier where the area is 300 square feet (27.9 m²) or more and not less than a 1-hour fire barrier where the area is less than 300 square feet (27.9 m²).

❖ The amount of hazardous material in an HPM room or gas room is only limited by room size; therefore, since the storage rooms in excess of 300 square feet (28 m²) are not limited by size or quantity of hazardous material, a minimum 2-hour fire barrier separation is required. It should be noted that this section assumes the storage areas have at least one exterior wall with a minimum 30-foot (9144 mm) fire separation distance (see Section 415.9.5.3).

415.9.5.2.2 Liquid storage rooms. Liquid storage rooms shall be constructed in accordance with the following requirements:

1. Rooms in excess of 500 square feet (46.5 m²) shall have at least one exterior door approved for fire department access.

2. Rooms shall be separated from other areas by fire barriers having a fire-resistance rating of not less than 1-hour for rooms up to 150 square feet (13.9 m²) in area and not less than 2 hours where the room is more than 150 square feet (13.9 m²) in area.

3. Shelving, racks and wainscoting in such areas shall be of noncombustible construction or wood of not less than 1inch (25 mm) nominal thickness.

4. Rooms used for the storage of Class I flammable liquids shall not be located in a basement.

❖ This section assumes storage rooms are utilized for the storage of flammable and combustible liquids in closed containers. The size of the rooms is severely restricted when only a 1-hour fire barrier is required to limit the size of a potential fire. For rooms greater than 150 square feet (14 m²) in area, a minimum 2-hour fire barrier is required because the size of the room or quantity of hazardous materials is not limited. Due to the need for readily available access in a fire scenario, all liquid storage rooms in excess of 500 square feet (46 m²) must have at least one exterior door that provides access to the room. In accordance with Section 415.9.5.3, all liquid storage rooms, regardless of size, must have at least one exterior wall.

The construction limitations on the shelving racks and wainscotting are intended to limit their potential involvement in a fire scenario. Due to the volatile nature of Class I flammable liquids and the need for quick fire department access, they must be located in liquid storage rooms at or above grade.

415.9.5.2.3 Floors. Except for surfacing, floors of HPM rooms and liquid storage rooms shall be of noncombustible liquid-tight construction. Raised grating over floors shall be of noncombustible materials.

❖ Similar to the floor construction of fabrication areas, the floors of HPM rooms and liquid storage rooms are required to be noncombustible and liquid tight (see commentary, Section 415.9.2.4).

415.9.5.3 Location. Where HPM rooms, liquid storage rooms and gas rooms are provided, they shall have at least one exterior wall and such wall shall be not less than 30 feet (9144 mm) from property lines, including property lines adjacent to public ways.

❖ Access from the exterior for fire-fighting operations is essential for storage rooms of hazardous materials. These rooms by their designation contain more than the maximum allowable quantities per control area listed in Tables 307.7(1) and 307.7(2). The 30-foot (9144 mm) limitation is based in part on the fact that the exterior wall need not have a fire-resistance rating depending on the construction type of the building. Such walls may also be used for explosion venting when such venting is required.

415.9.5.4 Explosion control. Explosion control shall be provided where required by Section 414.5.1.

❖ This section requires a method of explosion control only when an explosion hazard exists. Not all storage rooms are required to have explosion control. Section 414.5.1 requires a method of explosion control, such as explosion venting or barricade construction, depending on the specific hazardous material involved in accordance with the IFC. Table 911.1 of the IFC provides guidance as to when explosion control is required for various hazardous materials.

415.9.5.5 Exits. Where two exits are required from HPM rooms, liquid storage rooms and gas rooms, one shall be directly to the outside of the building.

❖ Depending on the common path of egress travel available in accordance with Section 1013.3, two exits may be required from all HPM rooms, liquid storage rooms and gas rooms. For example, a liquid storage room containing Class IB flammable liquids in normally closed containers would be considered a Group H-3 occupancy. As such, a maximum common path of travel of 25 feet (7620 mm) would be permitted before a second exit, one of which was directly to the exterior, is required. Storage rooms are required to be located along an exterior wall. In addition to the requirements of this section, all liquid storage rooms in excess of 500 square feet (46 m²) are required to have at least one exterior door regardless of travel distance.

415.9.5.6 Doors. Doors in a fire barrier wall, including doors to corridors, shall be self-closing fire assemblies having a fire-protection rating of not less than ³/₄ hour.

❖ Doors that are in an interior wall that comprise part of the storage room enclosure are required to have a minimum ³/₄-hour fire protection rating. The actual rating is dependent on the fire-resistance rating of the fire barrier separation required. While a 1-hour fire barrier would permit a door with a ³/₄-hour fire protection rating, a 2-hour fire barrier would require a door to have a 1 ¹/₂-hour fire protection rating in accordance with Table 715.3.

415.9.5.7 Ventilation. Mechanical exhaust ventilation shall be provided in liquid storage rooms, HPM rooms and gas rooms at the rate of not less than 1 cubic foot per minute per square foot (0.044 L/S/m2) of floor area or six air changes per hour, whichever is greater, for categories of material.

Exhaust ventilation for gas rooms shall be designed to operate at a negative pressure in relation to the surrounding areas and direct the exhaust ventilation to an exhaust system.

❖ Similar to fabrication areas and service corridors, mechanical exhaust ventilation is required (see commentary, Sections 415.9.2.6 and 415.9.4.3).

415.9.5.8 Emergency alarm system. An approved emergency alarm system shall be provided for HPM rooms, liquid storage rooms and gas rooms.

Emergency alarm-initiating devices shall be installed outside of each interior exit door of such rooms.

Activation of an emergency alarm-initiating device shall sound a local alarm and transmit a signal to the emergency control station.

An approved emergency telephone system, local alarm manual pull stations or other approved alarm-initiating devices are allowed to be used as emergency alarm-initiating devices.

❖ An emergency alarm system is required for all storage rooms containing hazardous materials in excess of the quantities permitted in Tables 307.7(1) and 307.7(2). See the commentary to Section 415.9.4.6 for similar emergency alarm requirements.

415.9.6 Piping and tubing.

❖ This section addresses the requirements for all piping and tubing utilized to transport HPM throughout a Group H-5 facility.

415.9.6.1 General. Hazardous production materials piping and tubing shall comply with this section and ANSI B31.3.

❖ The nature of the HPM involved requires strict control of both the materials used in piping and the joint methods used to avoid leakage. ANSI B 31.3 provides criteria for piping installation applicable to HPM facilities.

415.9.6.2 Supply piping and tubing.

❖ This section regulates the design requirements for all piping and tubing, including where it can be located within a Group H-5 facility.

415.9.6.2.1 HPM having a health-hazard ranking of 3 or 4. Systems supplying HPM liquids or gases having a health-hazard ranking of 3 or 4 shall be welded throughout, except for connections, to the systems that are within a ventilated enclosure if the material is a gas, or an approved method of drainage or containment is provided for the connections if the material is a liquid.

❖ The primary purpose of this section is to minimize the potential for a leak of HPM. The use of mechanical com-

pression-type fittings or other nonwelded joints in areas of health hazard 3 or 4 must be limited to areas where any leaks will be vented and the materials exhausted. If the material is a liquid, then a method of drainage or containment is required to minimize the hazard.

415.9.6.2.2 Location in service corridors. Hazardous production materials supply piping or tubing in service corridors shall be exposed to view.

❖ Concealment of piping in service corridors must be avoided to provide a means of monitoring for leaks.

415.9.6.2.3 Excess flow control. Where HPM gases or liquids are carried in pressurized piping above 15 pounds per square inch gauge (psig) (103.4 kPa), excess flow control shall be provided. Where the piping originates from within a liquid storage room, HPM room or gas room, the excess flow control shall be located within the liquid storage room, HPM room or gas room. Where the piping originates from a bulk source, the excess flow control shall be located as close to the bulk source as practical.

❖ Excess flow control valves regulate the flow of hazardous materials within the piping system. The valves are designed to shut off in the event the predetermined flow is exceeded.

415.9.6.3 Installations in exit access corridors and above other occupancies. The installation of hazardous production material piping and tubing within the space defined by the walls of exit access corridors and the floor or roof above or in concealed spaces above other occupancies shall be in accordance with Section 415.9.6.2 and the following conditions:

1. Automatic sprinklers shall be installed within the space unless the space is less than 6 inches (152 mm) in the least dimension.

2. Ventilation not less than six air changes per hour shall be provided. The space shall not be used to convey air from any other area.

3. Where the piping or tubing is used to transport HPM liquids, a receptor shall be installed below such piping or tubing. The receptor shall be designed to collect any discharge or leakage and drain it to an approved location. The 1-hour enclosure shall not be used as part of the receptor.

4. HPM supply piping and tubing and HPM nonmetallic waste lines shall be separated from the exit access corridor and from occupancies other than Group H-5 by construction as required for walls or partitions that have a fire protection rating of not less than 1 hour. Where gypsum wallboard is used, joints on the piping side of the enclosure are not required to be taped, provided the joints occur over framing members. Access openings into the enclosure shall be protected by approved fire-resistance-rated assemblies.

5. Readily accessible manual or automatic remotely activated fail-safe emergency shutoff valves shall be installed

on piping and tubing other than waste lines at the following locations:

5.1. At branch connections into the fabrication area.

5.2. At entries into exit access corridors.

Exception: Transverse crossings of the corridors by supply piping that is enclosed within a ferrous pipe or tube for the width of corridor need not comply with Items 1 through 5.

❖ The installation of HPM piping in the space above an exit access corridor or another occupancy, as well as the cavity of the egress corridor wall, present a potential source of hazard to the building's occupants. The five requirements provided for such installations serve to mitigate the hazard by suppression, ventilation, containment and ignition control. In order to address the potential hazards, the containment must be a fire barrier with at least a 1-hour fire-resistance rating, drainage receptors, excess flow control and shutoff valves. The elimination of the taping of the wallboard joints on the piping side of a rated assembly is in recognition of actual installation difficulties and the reduced likelihood of a fire on the interior of the wall cavity. To eliminate the taping of joints, however, the joints must occur over framing members.

When the piping traverses a corridor, the use of a co-axial enclosed pipe around the HPM piping is considered acceptable for providing the required separation and containment of a potential leak. The assumption is that the open ends of that pipe are in an HPM facility and, therefore, a leak into the outer casing can be monitored. If the adjacent areas that contain the open ends are not in an HPM facility, then the outer-jacket method cannot be used.

415.9.6.4 Identification. Piping, tubing and HPM waste lines shall be identified in accordance with ANSI A13.1 to indicate the material being transported.

❖ To facilitate monitoring, detecting and controlling any possible leaks, all piping, tubing and HPM waste lines must be identified in accordance with ANSI A13.1. This standard contains criteria for properly identifying the piping system, including labeling and frequency or distribution of the signs or labels.

415.9.7 Continuous gas-detection systems. A continuous gas-detection system shall be provided for HPM gases when the physiological warning properties of the gas are at a higher level than the accepted permissible exposure limit (PEL) for the gas and for flammable gases in accordance with this section.

❖ A gas detection system in the room or area utilized for the storage or use of HPM gases provides early notification of a leak that is occurring before the escaping gas reaches hazardous concentration levels.

415.9.7.1 Where required. A continuous gas-detection system shall be provided in the areas identified in Sections 415.9.7.1.1 through 415.9.7.1.4.

❖ Sections 415.9.7.1.1 through 415.9.7.1.4 prescribe the locations in a Group H-5 facility when a gas detection system is required.

415.9.7.1.1 Fabrication areas. A continuous gas-detection system shall be provided in fabrication areas when gas is used in the fabrication area.

❖ All fabrication areas that utilize HPM gases must have a gas detection system. It should be noted that gas detection is often installed within workstations as a means of early detection of leaks. Such detection is generally not acceptable as an alternative to gas detection for the fabrication area, since a leak may occur remote from the workstation.

415.9.7.1.2 HPM rooms. A continuous gas-detection system shall be provided in HPM rooms when gas is used in the room.

❖ HPM rooms, which by definition contain more than the quantities of hazardous materials per control area permitted by Tables 307.7(1) and 307.7(2), are required to have a gas detection system.

415.9.7.1.3 Gas cabinets, exhausted enclosures and gas rooms. A continuous gas-detection system shall be provided in gas cabinets and exhausted enclosures. A continuous gas-detection system shall be provided in gas rooms when gases are not located in gas cabinets or exhausted enclosures.

❖ In the potential event of a leaking cylinder of a hazardous gas, gas cabinets, exhausted enclosures and gas rooms must have a gas detection system.

415.9.7.1.4 Exit access corridors. When gases are transported in piping placed within the space defined by the walls of an exit access corridor, and the floor or roof above the exit access corridor, a continuous gas-detection system shall be provided where piping is located and in the exit access corridor.

Exception: A continuous gas-detection system is not required for occasional transverse crossings of the corridors by supply piping that is enclosed in a ferrous pipe or tube for the width of the corridor.

❖ In addition to the requirements of Section 415.9.6.3 for HPM piping and tubing in an exit access corridor, a gas detection system is required for early notification of a potential leak of an HPM gas.

The exception is similar to the exception in Section 415.9.6.3. In essence, the gas detection system required by this section as well as the additional provisions in Section 415.9.6.3 are not required if the exception is met (see commentary, Section 415.9.6.3).

415.9.7.2 Gas-detection system operation. The continuous gas-detection system shall be capable of monitoring the room, area or equipment in which the gas is located at or below the PEL or ceiling limit of the gas for which detection is provided. For flammable gases, the monitoring detection threshold level shall be vapor concentrations in excess of 20 percent of the lower explosive limit (LEL). Monitoring for highly toxic and toxic gases shall also comply with the requirements for such material in the *International Fire Code*.

❖ Gas detection is required based on the potential for health-threatening levels as established by nationally accepted health standards, such as those used by the Occupational Safety and Health Administration (OSHA). The permissible exposure limitation of a gas is the legal limitation for long-term exposure (8 to 10 hours normally). The American Conference of Government Industrial Hygienists (ACGIH) publishes threshold limitation values based on a time-weighted average. It should be noted that state and local laws may also contain limitations.

Additionally, gas detection is to be provided when dispensing occurs that may result in vapor concentrations in excess of 20 percent of the LEL. Lower explosive limitations can be obtained from suppliers or other printed sources, such as NFPA 325M. Chapter 37 of the IFC contains additional requirements for the monitoring of highly toxic and toxic compressed gases.

415.9.7.2.1 Alarms. The gas detection system shall initiate a local alarm and transmit a signal to the emergency control station when a short-term hazard condition is detected. The alarm shall be both visual and audible and shall provide warning both inside and outside the area where the gas is detected. The audible alarm shall be distinct from all other alarms.

❖ The required local alarm is intended to alert occupants to a hazardous condition in the vicinity of where the HPM gases are being stored or used. The alarm is not intended to be an evacuation alarm; however, it is required to be monitored to hasten emergency personnel response.

415.9.7.2.2 Shutoff of gas supply. The gas detection system shall automatically close the shutoff valve at the source on gas supply piping and tubing related to the system being monitored for which gas is detected when a short-term hazard condition is detected. Automatic closure of shutoff valves shall comply with the following:

1. Where the gas-detection sampling point initiating the gas detection system alarm is within a gas cabinet or exhausted enclosure, the shutoff valve in the gas cabinet or exhausted enclosure for the specific gas detected shall automatically close.

2. Where the gas-detection sampling point initiating the gas detection system alarm is within a room and compressed gas containers are not in gas cabinets or an exhausted enclosure, the shutoff valves on all gas lines for the specific gas detected shall automatically close.

3. Where the gas-detection sampling point initiating the gas detection system alarm is within a piping distribution manifold enclosure, the shutoff valve supplying the mani-

fold for the compressed gas container of the specific gas detected shall automatically close.

Exception: Where the gas-detection sampling point initiating the gas detection system alarm is at the use location or within a gas valve enclosure of a branch line downstream of a piping distribution manifold, the shutoff valve for the branch line located in the piping distribution manifold enclosure shall automatically close.

❖ Where gas detection systems are required, automatic emergency shutoff valves are required to stop the flow of hazardous material from possibly deteriorating further in an emergency.

415.9.8 Manual fire alarm system. An approved manual fire alarm system shall be provided throughout buildings containing Group H-5. Activation of the alarm system shall initiate a local alarm and transmit a signal to the emergency control station. The fire alarm system shall be designed and installed in accordance with Section 907.

❖ Due to the type and potential quantities of hazardous materials that could be present in a Group H-5 facility, a manual means of activating an evacuation alarm is essential for the safety of the occupants in an emergency situation. The alarm signal must be transmitted to the emergency control station. The fire alarm must also be in accordance with the applicable provisions of Section 907 and NFPA 72.

415.9.9 Emergency control station. An emergency control station shall be provided on the premises at an approved location, outside of the fabrication area and shall be continuously staffed by trained personnel. The emergency control station shall receive signals from emergency equipment and alarm and detection systems. Such emergency equipment and alarm and detection systems shall include, but not necessarily be limited to, the following where such equipment or systems are required to be provided either in Section 415.9 or elsewhere in this code:

1. Automatic fire sprinkler system alarm and monitoring systems.

2. Manual fire alarm systems.

3. Emergency alarm systems.

4. Continuous gas-detection systems.

5. Smoke detection systems.

6. Emergency power system.

❖ This section specifies the systems that are to be monitored by an emergency control station. The fire alarm system signals are received at the emergency control station that must be located in an area approved by the building official. The emergency control station should not be located in an area where hazardous materials are stored, used or transported, such as a fabrication area. See the commentary to Section 415.2 for the definition of "Emergency control station."

415.9.10 Emergency power system. An emergency power system shall be provided in Group H-5 occupancies where required in Section 415.9.10.1. The emergency power system shall be designed to supply power automatically to required electrical systems when the normal electrical supply system is interrupted.

❖ A backup emergency power source is considered essential for systems that are monitoring and protecting hazardous materials in a Group H-5 occupancy. Without an emergency power system, all required electrical controls or equipment monitoring hazardous materials would be rendered inoperative if a power failure or other electrical system failure were to occur.

415.9.10.1 Where required. Emergency power shall be provided for electrically operated equipment and connected control circuits for the following systems:

1. HPM exhaust ventilation systems.

2. HPM gas cabinet ventilation systems.

3. HPM exhausted enclosure ventilation systems.

4. HPM gas room ventilation systems.

5. HPM gas detection systems.

6. Emergency alarm systems.

7. Manual fire alarm systems.

8. Automatic sprinkler system monitoring and alarm systems.

9. Electrically operated systems required elsewhere in this code applicable to the use, storage or handling of HPM.

❖ This section specifies the types of systems within a Group H-5 occupancy that are required to be connected to an approved emergency power system. As indicated in Section 2702, all emergency power systems must be installed in accordance with the applicable requirements of the ICC EC and NFPA 110 and 111.

415.9.10.2 Exhaust ventilation systems. Exhaust ventilation systems are allowed to be designed to operate at not less than one-half the normal fan speed on the emergency power system where it is demonstrated that the level of exhaust will maintain a safe atmosphere.

❖ Emergency power for exhaust ventilation is required to prevent hazardous concentrations of HPM fumes or vapors in areas such as workstations or fabrication areas. Fans for exhaust ventilation draw a considerable amount of current when operating. Running exhaust fans at a reduced speed may be desirable when it will not endanger the operator or result in a hazardous condition; however, exhaust fans must not be run at a speed less than 50 percent of its rating, even if a slower speed will not produce a serious hazard.

415.9.11 Fire sprinkler system protection in exhaust ducts for HPM.

❖ This section prescribes the sprinkler system requirements for exhaust ducts for HPM. The requirements depend on the construction materials of the exhaust duct.

415.9.11.1 General. Automatic fire sprinkler system protection shall be provided in exhaust ducts conveying vapors, fumes, mists or dusts generated from HPM in accordance with this section and the *International Mechanical Code.*

❖ An exhaust duct for HPM materials could convey flammable and combustible fumes, vapors or ducts. To provide protection against the spread of fire within the exhaust system and to prevent a duct fire from involving the building, sprinkler protection is required in the exhaust duct. The use of an extinguishing agent other than water based on agent compatibility would be subject to local approval. Section 510 of the IMC contains additional requirements for protecting hazardous exhaust systems.

415.9.11.2 Metallic and noncombustible, nonmetallic exhaust ducts. Automatic fire sprinkler system protection shall be provided in metallic and noncombustible, nonmetallic exhaust ducts where all of the following conditions apply:

1. Where the largest cross-sectional diameter is equal to or greater than 10 inches (254 mm).

2. The ducts are within the building.

3. The ducts are conveying flammable vapors or fumes.

❖ Sprinklers are required within each individual duct when all three of the following conditions exist:

1. Cross-sectional diameter at the widest point is equal to or exceeds 10 inches (254 mm).

2. Ducts are located within the building.

3. Ducts convey gases or vapors within the flammable range.

Figure 415.9.11.2 illustrates how to measure the cross-sectional diameter of various duct shapes. Provisions of this section require the square and rounded ducts to be protected by automatic sprinklers. The round or elliptical ducts depicted on the right side of the diagram are not required to be protected.

415.9.11.3 Combustible nonmetallic exhaust ducts. Automatic fire sprinkler system protection shall be provided in combustible nonmetallic exhaust ducts where the largest cross-sectional diameter of the duct is equal to or greater than 10 inches (254 mm).

Exceptions:

1. Ducts listed or approved for applications without automatic fire sprinkler system protection.

2. Ducts not more than 12 feet (3658 mm) in length installed below ceiling level.

❖ Galvanized steel is not an appropriate duct material for all substances handled in the broad category of hazardous exhaust systems. A duct material compatible with the exhaust must be selected, and factors such as corrosiveness, abrasion resistance, chemical resistance and operating temperatures must be taken into account. Section 510.8 of the IMC contains additional provisions for duct construction that is part of a hazardous exhaust system. As such, automatic sprinkler protection is also required for all combustible nonmetallic ducts with a cross-sectional diameter equal to or greater than 10 inches (254 mm).

Exception 1 provides that automatic sprinklers are not required where the risk to people or property is limited, such as when nonmetallic ducts approved for installation without sprinklers are used. Exception 2 provides that when ducts do not exceed 12 feet (3658 mm) in length and are installed exposed below ceiling level, sprinklers may be omitted. The limited duct length and exposure reduces the potential of a concealed fire hazard.

415.9.11.4 Automatic sprinkler locations. Sprinkler systems shall be installed at 12-foot (3658 mm) intervals in horizontal ducts and at changes in direction. In vertical ducts, sprinklers shall be installed at the top and at alternate floor levels.

❖ Adequate sprinkler coverage needs to be maintained to prevent the spread of fire within the exhaust system. All sprinkler system components must also be in compliance with the applicable provisions of NFPA 13.

[F] SECTION 416
APPLICATION OF FLAMMABLE FINISHES

416.1 General. The provisions of this section shall apply to the construction, installation and use of buildings and structures, or parts thereof, for the spraying of flammable paints, varnishes and lacquers or other flammable materials or mixtures or compounds used for painting, varnishing, staining or similar purposes. Such construction and equipment shall comply with the *International Fire Code.*

❖ The principal hazards associated with paint spraying and spray booths originate from the presence of flammable liquids or powders and their vapors or mists. The purpose of this section is to provide requirements that address the hazards associated with spray applications and dipping or coating applications involving flammable paints, varnishes and lacquers. The requirements listed include such areas as ventilation, automatic sprinklers, control of ignition sources and proper operation of the equipment. In addition to the provisions of this section, the provisions of the IMC and Chapter 15 of the IFC, also apply.

The provisions of this section apply to the indoor use of spray applications and dipping or coating applications involving flammable paints, varnishes and lacquers. Outdoor applications involving flammable paints, varnishes and lacquers are not covered, since overspray deposits are not likely to create hazardous conditions, and flammable vapor-air mixtures are minimized because of atmospheric dilution. Safeguards such as ventilation, ignition control, material storage and waste disposal (see Chapter 15 of the IFC) should still apply. This reference to the IFC and its provisions regarding flammable finishes covers the application of flammable or combustible materials when applied as a spray by compressed air; "airless" or "hydraulic atom-

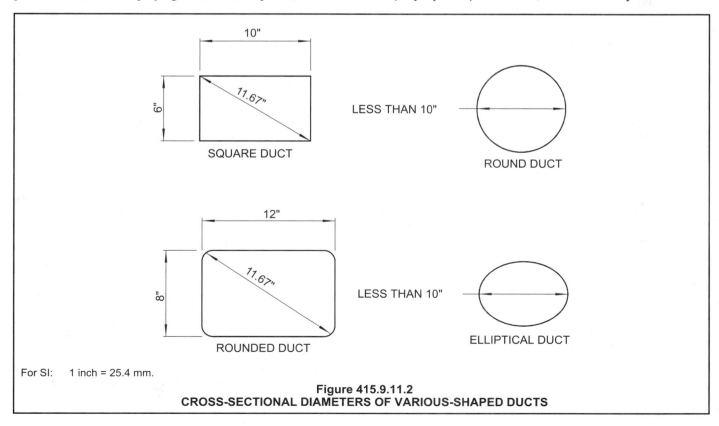

For SI: 1 inch = 25.4 mm.

Figure 415.9.11.2
CROSS-SECTIONAL DIAMETERS OF VARIOUS-SHAPED DUCTS

ization" steam; electrostatically or by any other means in continuous or intermittent processes. They also cover the application of combustible powders when applied by powder spray guns, electrostatic powder spray guns, fluidized beds or electrostatic fluidized beds. The IFC and its referenced standards contain provisions relating to the location of spray areas; ignition sources; ventilation; liquid storage and handling; protection; operation; maintenance and training. The provisions found in Chapter 15 of the IFC also apply to processes in which articles or materials are passed through the contents of tanks, vats or containers of flammable or combustible liquids, including dipping; roll; flow and curtain coating; finishing; treating; cleaning and similar processes. Requirements for the location of dipping and coating processes; ventilation; equipment construction; liquid storage and handling; protection; operation; maintenance and training are also provided.

416.2 Spray rooms. Spray rooms shall be enclosed with fire barrier walls and horizontal assemblies or both with not less than a 1-hour fire-resistance rating. Floors shall be waterproofed and drained in an approved manner.

❖ A spray room is a power-ventilated, fully enclosed room used exclusively for open spraying of flammable and combustible materials. The entire spray room is considered the spray area. The primary difference between a spray room and a spray booth is that spray booths are partially open. A spray room is to be enclosed with fire barrier assemblies having a fire-resistance rating of at least 1 hour. Waterproof floors are to be arranged to drain either to the outside of the building, to internal drains or to other suitable places. Properly designed and guarded drains or scuppers of sufficient number and size to dispose of all surplus water likely to be discharged by automatic sprinklers must be provided.

416.2.1 Surfaces. The interior surfaces of spray rooms shall be smooth and shall be so constructed to permit the free passage of exhaust air from all parts of the interior and to facilitate washing and cleaning, and shall be so designed to confine residues within the room. Aluminum shall not be used.

❖ Rough, corrugated or uneven surfaces are difficult to clean. Periodic cleaning of interior surfaces reduces the fire hazard posed by the accumulation of flammable or combustible coatings. Due to the physical properties of aluminum, it is unsuitable for cleaning and scraping of overspray residue.

416.3 Spraying spaces. Spraying spaces shall be ventilated with an exhaust system to prevent the accumulation of flammable mist or vapors in accordance with the *International Mechanical Code.* Where such spaces are not separately enclosed, noncombustible spray curtains shall be provided to restrict the spread of flammable vapors.

❖ The objective of ventilation is to remove flammable vapors and mists so as to minimize the potential for a flash fire or explosion. The spray area is required to be venti-

lated in accordance with the provisions of the IMC. When the spray process is not separated from the other operations or areas, a noncombustible spray curtain must be provided to restrict the spread of flammable vapors. The IFC requires ventilation systems and enclosures to be interlocked with the spraying equipment to ensure that fans operate and doors remain closed when in use.

416.3.1 Surfaces. The interior surfaces of spraying spaces shall be smooth and continuous without edges, and shall be so constructed to permit the free passage of exhaust air from all parts of the interior and to facilitate washing and cleaning, and shall be so designed to confine residues within the spraying space. Aluminum shall not be used.

❖ See the commentary to Section 416.2.1.

416.4 Fire protection. An automatic fire-extinguishing system shall be provided in all spray, dip and immersing spaces and storage rooms, and shall be installed in accordance with Chapter 9.

❖ Spray application operations are to be protected with an automatic fire suppression system. While an automatic sprinkler system is the most desirable type of suppression system, this section would allow the use of other effective extinguishing agents, such as dry chemical or foam. Because of the ease of ignition of flammable liquids, specifically spray paint applications of a flammable liquid, an automatic fire suppression system is required in rooms in which spray painting is conducted, as well as rooms in which flammable materials are used for painting, brushing, dipping or mixing on a regular basis. Dry sprinklers should not be used for the protection of spray operations, except possibly at the exhaust duct penetrations to the outside. Wet-pipe, preaction or deluge systems should be used so that water is placed on the fire in the shortest possible time. If the entire building is not protected with an automatic sprinkler system, then the system protecting the spray booth or room may be a limited area sprinkler system in accordance with Section 903.3.5.1.1. Locating sprinklers in paint spray booths presents an unusual problem, since the sprinkler may become clogged with paint. The most satisfactory solution is to locate the sprinklers in areas in a way that the paint spray will most likely not reach the sprinkler. Even in the most ideal locations, the sprinkler will need to be cleaned very frequently. A coating of grease with a low melting point, such as petroleum jelly, motor oil or a soft neutral soap, will facilitate cleaning of the sprinkler. Other alternatives that are commonly used include polyethylene or cellophane bags with a thickness of 0.003 inches (0.076 mm) or less or thin paper bags to protect the sprinklers.

Additional guidance on fire protection requirements for the application of flammable finishes can be found in the IFC.

[F] SECTION 417
DRYING ROOMS

417.1 General. A drying room or dry kiln installed within a building shall be constructed entirely of approved noncombustible materials or assemblies of such materials regulated by the approved rules or as required in the general and specific sections of Chapter 4 for special occupancies and where applicable to the general requirements of Chapter 28.

❖ This section establishes specific safety requirements for drying rooms and dry kilns used in conjunction with the drying (accelerated seasoning) of lumber, and dryers or dehydrators used to reduce the moisture content of agricultural products. Included are drying operations for other combustible materials, such as certain building materials, textiles and fabrics. Drying operations associated with the application of flammable finishes are regulated by Section 416. The hazards associated with drying rooms and dry kilns relate to the volume of readily combustible materials being processed at temperatures conducive to their ignition in the event of a system malfunction.

The materials or assemblies used in the construction of drying rooms or dry kilns located within buildings must be noncombustible, regardless of the type of construction of the building itself (see the commentary to Section 703.4 for a discussion of noncombustibility criteria) when required by the approved rules or Chapter 4 or 28. The use of noncombustible materials of construction for drying rooms and dry kilns provides for a measure of confinement of a fire occurring within and prevents the room or kiln from contributing to the fuel load exposed to an unwanted fire.

417.2 Piping clearance. Overhead heating pipes shall have a clearance of not less than 2 inches (51 mm) from combustible contents in the dryer.

❖ Piped steam and hot-water indirect heating systems used in drying rooms or dry kilns can operate at temperatures of up to several hundred degrees Fahrenheit and are considered a safe and appropriate method of providing heat to the drying processes. A minimum clearance of 2 inches (51 mm) between hydronic piping and adjacent combustible contents in the drying room or dry kiln is required to reduce the likelihood of contact ignition of the combustible contents. Though hydronic piping is frequently installed overhead, the clearance is required regardless of the pipes' location within the drying room or kiln.

417.3 Insulation. Where the operating temperature of the dryer is 175°F (79°C) or more, metal enclosures shall be insulated from adjacent combustible materials by not less than 12 inches (305 mm) of airspace, or the metal walls shall be lined with $^1/_4$-inch (6.35 mm) insulating mill board or other approved equivalent insulation.

❖ The insulation of metal drying room and kiln enclosures from adjacent combustible materials is required when the drying apparatus operates at or above 175°F (79°C). Insulation can be accomplished by the use of airspace or by lining the metal dryer walls with insulating material. The insulation will reduce the possibility that combustible materials in proximity to the drying room or dry kiln will be ignited by heat transfer from it. The temperature of 175°F (79°C) has been determined to be the highest allowable temperature without any clearance to combustibles.

417.4 Fire protection. Drying rooms designed for high-hazard materials and processes, including special occupancies as provided for in Chapter 4, shall be protected by an approved automatic fire-extinguishing system conforming to the provisions of Chapter 9.

❖ The possibility of ignition increases in a drying room utilized in conjunction with hazardous materials such as combustible fibers and other finely divided materials. The installation of an approved fire-extinguishing system in accordance with Chapter 9 will contain a fire within the drying room until the fire department (or industrial fire brigade) arrives. In sprinklered buildings, fire suppression can be easily and economically provided by installing sprinklers within the drying room or area. Wet-pipe, preaction or deluge systems should be used so that water is placed on the fire in the shortest possible time. Note that dryers used in conjunction with plywood, veneer and composite board mill operations are required by the IFC to be protected by an approved deluge water-spray system. See the commentary for Chapter 19 of the IFC for further information on lumber and woodworking facilities. In nonsprinklered or partially sprinklered buildings, the system protecting the drying room may be an approved, limited area sprinkler system in accordance with Section 903.3.5.1.1. An alternative fire-extinguishing system consisting of dry chemical, carbon dioxide, clean agent or other approved nonwater-based fire-extinguishing system in accordance with Section 904 can provide an equivalent level of protection, depending on the nature of the drying room contents. See the commentary to Section 904 for a discussion of the various alternative fire-extinguishing systems available.

[F] SECTION 418
ORGANIC COATINGS

❖ The manufacture of organic coatings encompasses operations that produce decorative and protective coatings for architectural uses, industrial products and other specialized purposes. Requirements of this chapter address the hazards associated with the manufacture of solvent-based organic coatings. Water-based materials are exempt from these requirements. These provisions are consistent with Chapter 20 of the IFC, which contains more detailed requirements for the manufacture of organic coatings. NFPA 35 also provides additional guidance on the maintenance of facilities utilized for the manufacture of organic coatings.

418.1 Building features. Manufacturing of organic coatings shall be done only in buildings that do not have pits or basements.

❖ Basements, pits or depressed first-floor construction are prohibited because of the tendency for hazardous material vapors to accumulate in low areas and the difficulty in fighting fires in basements and pits in such occupancies.

418.2 Location. Organic coating manufacturing operations and operations incidental to or connected therewith shall not be located in buildings having other occupancies.

❖ Incidental occupancies involve operations and activities closely related to the primary occupancy and are necessary for the efficient, continuous and safe performance of the manufacture of organic coatings. Administration, storage, shipping and receiving, as well as other related but not indispensable operations, should be located in separate buildings.

418.3 Process mills. Mills operating with close clearances and that process flammable and heat-sensitive materials, such as nitrocellulose, shall be located in a detached building or noncombustible structure.

❖ Milling of heat-sensitive materials, such as nitrocellulose, is an extraordinary hazard. Such operations must be located in separate, single-purpose buildings away from other uses and high-hazard operations. Pebble mills pose a special vapor ignition hazard caused by static electricity. Both the grinding material and inner lining of these mills are made of materials with good insulating characteristics. Because static electricity is produced during milling, it has nowhere to go. Generally, the inside of this type of mill is either partially or totally inerted using nitrogen or carbon dioxide gas to prevent ignition.

418.4 Tank storage. Storage areas for flammable and combustible liquid tanks inside of structures shall be located at or above grade and shall be separated from the processing area by not less than 2-hour fire-resistance-rated fire barriers.

❖ Tank storage located below grade is prohibited. Basements located under grade-level storage areas should be discouraged. Below-grade flammable liquid fires are extremely difficult to fight. Similarly, above-grade spills will flow to lower floors, possibly resulting in spill fires on more than one building level. Tank storage of raw materials must be confined to locations at or above grade level. Tank storage rooms must be separated from process and other storage areas by fire barriers that are a minimum of 2-hour fire-resistance rated. If possible, these rooms should be accessible on at least one side from the outside. Cutoff rooms with access from two or even three sides are considered ideal.

418.5 Nitrocellulose storage. Nitrocellulose storage shall be located on a detached pad or in a separate structure or a room enclosed with no less than 2-hour fire-resistance-rated fire barriers.

❖ Once ignited, nitrocellulose will continue to burn even in the absence of oxygen. Extra precautions must be used to prevent ignition and fire spread; therefore, nitrocellulose storage should be in either a separate detached structure, on a detached exterior pad or fully enclosed by a minimum 2-hour fire barrier.

418.6 Finished products. Storage rooms for finished products that are flammable or combustible liquids shall be separated from the processing area by fire barriers having a fire-resistance rating of at least 2 hours, and openings in the walls shall be protected with approved opening protectives.

❖ Finished products, which are also classified as flammable or combustible liquids, must also comply with the applicable provisions of Chapter 34 of the IFC and NFPA 30. As a minimum, a 2-hour fire barrier with approved opening protectives is required between the processing area and the storage area to eliminate mutual involvement in a single fire scenario. A higher degree of fire separation may be required depending on the building's occupancy classification.

Bibliography

The following resource materials are referenced in this chapter or are relevant to the subject matter addressed in this chapter.

ANSI A13.1-96, *Scheme for the Identification of Piping Systems.* New York: American National Standards Institute, 1996.

ANSI B31.3-99, *Process Piping — with B31a -96 and B31.3b-97 Addenda.* New York: American National Standards Institute, 1999.

ASTM E 84-01, *Test Methods for Surface Burning Characteristics of Building Materials.* West Conshohocken, PA: ASTM International, 2001.

ASTM E 119-00, *Test Methods for Fire Tests of Building Construction and Materials.* West Conshohocken, PA: ASTM International, 2000.

Boring, Delbert F. and others. *Fire Protection Through Modern Building Codes,* 5th ed. Washington, DC: American Iron and Steel Institute, 1981.

Fire Protection Handbook, 19th ed. Quincy, MA: National Fire Protection Association, 2002.

Fothergill, John W. and John H. Klote. *Design of Smoke Control Systems for Buildings.* Atlanta: American Society for Heating Refrigerating and Air-Conditioning Engineers, Inc., 1983.

"Health Care Facilities." (BCMC) Report. CABO Board for the Coordination of the Model Codes. Falls Church, VA: Council of American Building Officials, 1990.

NFPA 13-99, *Installation of Sprinkler Systems.* Quincy, MA: National Fire Protection Association, 1999.

NFPA 13R-99, *Installation of Sprinkler Systems in Residential Occupancies Up To and Including Four Stories in Height.* Quincy, MA: National Fire Protection Association, 1999.

NFPA 30-96, *Flammable and Combustible Liquids Code.* Quincy, MA: National Fire Protection Association, 1996.

NFPA 30A-96, *Automotive and Marine Service Station Code.* Quincy, MA: National Fire Protection Association, 1996.

NFPA 32-96, *Drycleaning Plants.* Quincy, MA: National Fire Protection Association, 1996.

NFPA 33-95, *Spray Application Using Flammable and Combustible Materials.* Quincy, MA: National Fire Protection Association, 1995.

NFPA 34-95, *Dipping and Coating Processes Using Flammable or Combustible Liquids.* Quincy, MA: National Fire Protection Association, 1995.

NFPA 40-97, *Storage and Handling of Cellulose Nitrate Motion Picture Film.* Quincy, MA: National Fire Protection Association, 1997.

NFPA 61-95, *Prevention of Fires and Dust Explosions in Agricultural Food Products Facilities.* Quincy, MA: National Fire Protection Association, 1995.

NFPA 68-94, *Deflagration Venting.* Quincy, MA: National Fire Protection Association, 1994.

NFPA 69-97, *Explosion Prevention Systems.* Quincy, MA: National Fire Protection Association, 1997.

NFPA 70-99, *National Electrical Code.* Quincy, MA: National Fire Protection Association, 1999.

NFPA 72-02, *National Fire Alarm Code.* Quincy, MA: National Fire Protection Association, 2002.

NFPA 120-99, *Coal Preparation Plants.* Quincy, MA: National Fire Protection Association, 1999.

NFPA 231-98, *General Storage.* Quincy, MA: National Fire Protection Association, 1998.

NFPA 231C-98, *Rack Storage of Materials.* Quincy, MA: National Fire Protection Association, 1998.

NFPA 495-96, *Explosive Materials Code.* Quincy, MA: National Fire Protection Association, 1996.

NFPA 651-98, *Manufacture of Aluminum Powder.* Quincy, MA: National Fire Protection Association, 1998.

NFPA 654-00, *Prevention of Fire and Dust Explosions in the Chemical, Dye, Pharmaceutical and Plastics Industries.* Quincy, MA: National Fire Protection Association, 2000.

NFPA 655-93, *Prevention of Sulfur Fires and Explosions.* Quincy, MA: National Fire Protection Association, 1993.

NFPA 664-98, *Prevention of Fires and Explosions in Wood Processing and Woodworking Facilities.* Quincy, MA: National Fire Protection Association, 1998.

NFPA 704-96, *Flammable and Combustible Liquids Code.* Quincy, MA: National Fire Protection Association, 1996.

NIST IR 5240-93, *Measurement of Room Conditions and Response of Sprinklers and Smoke Detectors During a Simulated Two-Bed Hospital Patient Room Fire.* Gaithersburg, MD: National Institute of Standards and Technology, July 1993.

"Special Amusement Buildings." (BCMC) Report. CABO Board for the Coordination of the Model Codes. Falls Church, VA: Council of American Building Officials, 1988.

"Stages, Platforms and Sound Stages." (BCMC) Report. CABO Board for the Coordination of the Model Codes. Falls Church, VA: Council of American Building Officials, 1992.

UL 217-97, *Single and Multiple Station Smoke Detectors—with Revisions thru August 1998.* Northbrook, IL: Underwriters Laboratories Inc., 1997.

UL 268-96, *Smoke Detectors for Fire Protection Signaling Systems—with Revisions through January, 1999.* Northbrook, IL: Underwriters Laboratories Inc., 1996.

Chapter 5:
General Building Heights And Areas

General Comments

Chapter 5 contains the provisions that regulate the minimum type of construction, area modifications, height modifications, mezzanines, unlimited area structures and special provisions for buildings, including parking garages.

Section 501.1 gives the scope of Chapter 5. It specifies that the provisions of Chapter 5 apply not only to new construction, but also to existing building additions.

Section 502 defines terms that are related to the use and application of Chapter 5 and other provisions of the code related to height and area.

Section 503 addresses the allowable height and area limitations of buildings. Table 503 establishes the limits for the height and area of a building as a function of the type of construction. It is a risk/safety table where the risk is based on the occupancy and the safety is related to the type of construction.

Section 504 describes the modifications to height limits in Table 503 that are allowed for buildings that have a sprinkler system installed in accordance with specified standards.

Section 505 provides the criteria and regulations for mezzanines and industrial equipment platforms.

Section 506 describes the allowable modifications to area limits given in Table 503 for buildings with either sprinkler systems or a larger separation distance from adjacent properties or public ways.

Section 507 includes requirements for buildings with no area limit. These provisions are commonly used for very large industrial warehouses or factories.

Section 508 contains requirements for parking garages and other miscellaneous facilities.

Chapter 5 is one of the most important chapters in the code because many other code requirements depend on the establishment of the minimum required type of construction for a building. A building has one minimum required construction type and must conform to those requirements, with certain exceptions for buildings incorporating parking garages (see Section 508). Without the correct determination of the minimum type of construction, misapplication of the code is probable.

Table 503 is the keystone in setting thresholds for building size based on the building's use and the materials with which it is constructed. It is aimed at providing an acceptable level of hazard for building occupants by limiting fire load and fire hazards relating to height and area. In accordance with Table 503, buildings of higher construction types [noncombustible materials and a higher degree of fire-resistance rating of elements (see Chapter 6)] can be larger and higher than buildings of lower construction types (combustible and unprotected with fire-resistance ratings). The area and height thresholds set by Table 503 incorporate the limits that had been established by the three model codes that were "merged" when producing the code. These thresholds are reduced for buildings over three stories in height in accordance with Section 503.3.

The limits of Table 503 are modified by the text of the chapter, particularly the height and area modifications of Sections 504 and 506. Sections 507 and 508 give conditions under which the height and area limitations of Table 503 may not apply, such as in the case of an unlimited area building.

Purpose

The purpose of Chapter 5 is to regulate the size of structures based on the specific hazards associated with their occupancy and the materials of which they are constructed. Chapter 5 also provides for adjustments to the allowable area and height, based on the presence of fire protection systems for building occupants and fire service personnel in the event of a fire.

The code review for the design of any building begins with Chapter 5, with the determination of minimum required type of construction based on the use and size of the building. Misapplication of Chapter 5 can result in a multitude of related errors in subsequent code application, since many code requirements depend on the type of construction that is required for the building.

SECTION 501
GENERAL

501.1 Scope. The provisions of this chapter control the height and area of structures hereafter erected and additions to existing structures.

❖ Chapter 5 is applicable to all new structures and existing structures that are to be enlarged. Allowable height and area are evaluated on the basis of use classification, type of construction, location on the property relative to lot lines and other structures and the presence of an automatic sprinkler system.

In the case of additions to existing buildings, the designer or plan reviewer must evaluate the entire building, with the addition, as if it were a new structure, for purposes of determining allowable area and height. For instance, if an existing Type IIB building is to have an addition, the aggregate area of the new building (existing plus the addition) must be within the limits established by Table 503 for the allowable area and height of a Type IIB building, taking into account area and height modification for open frontage and sprinklers.

If the aggregate area of the existing building and the addition exceeds the allowable area for Type IIB construction in Table 503, the addition is not permitted unless something is done to solve the allowable area problem. The solution could either be adding sprinklers to the building to get an allowable area increase in accordance with Section 506.3, providing a fire wall between the addition and the existing building or creating two buildings in accordance with Section 503.1. If a fire wall is used, the designer must check the allowable area of the existing building again because the perimeter conditions have changed. Specifically, if the allowable area of the existing building results from an increase due to open frontage (see Section 506.2), then the following question must be answered: With the reduced open perimeter (there is no longer an open perimeter where the fire wall and adjacent building will be located), does the existing building exceed the area limitations of Table 503 based on its use and construction type? If so, another solution must be found.

501.2 Premises identification. Approved numbers or addresses shall be provided for new buildings in such a position as to be clearly visible and legible from the street or roadway fronting the property. Letters or numbers shall be a minimum 3 inches (76 mm) in height and stroke of minimum 0.5 inch (12.7 mm) of a contrasting color to the background itself.

❖ Address numbers are critical for emergency responders such as the fire department and ambulance companies. The size and color criteria are intended to aid visibility from the street.

SECTIONS 502
DEFINITIONS

502.1 Definitions. The following words and terms shall, for the purposes of this chapter and as used elsewhere in this code, have the meanings shown herein.

❖ Definitions of terms can help in the understanding and application of the code requirements. The purpose for including these definitions within this chapter is to provide more convenient access to them without having to refer back to Chapter 2. For convenience, these terms are also listed in Chapter 2 with a cross reference to this section.

AREA, BUILDING. The area included within surrounding exterior walls (or exterior walls and fire walls) exclusive of vent shafts and courts. Areas of the building not provided with surrounding walls shall be included in the building area if such areas are included within the horizontal projection of the roof or floor above.

❖ Allowable building areas (as established by the provisions of Chapter 5 and Table 503) are a function of the potential fire hazard and the level of fire endurance of the building's structural elements, as defined by the types of construction in Chapter 6. A building area is the "footprint" of the building; that is, the area measured within the perimeter formed by the inside surface of the exterior walls, which excludes spaces that are inside this perimeter and open to the outside atmosphere at the top, such as open shafts and courts (see Section 1206). When a portion of the building has no exterior walls, the area regulated by Chapter 5 is defined by the projection of the roof or floor above [see Figure 502.1(1)]. The roof overhang on portions of a building where there are exterior enclosure walls does not add to the building area because the area is defined by exterior walls.

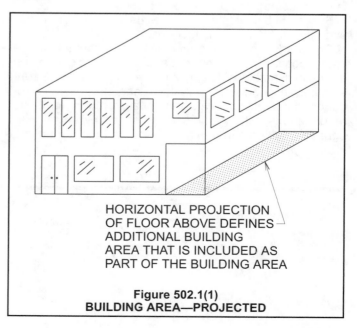

HORIZONTAL PROJECTION
OF FLOOR ABOVE DEFINES
ADDITIONAL BUILDING
AREA THAT IS INCLUDED AS
PART OF THE BUILDING AREA

Figure 502.1(1)
BUILDING AREA—PROJECTED

BASEMENT. That portion of a building that is partly or completely below grade plane (See "Story above grade plane" in Section 202). A basement shall be considered as a story above grade plane where the finished surface of the floor above the basement is:

1. More than 6 feet (1829 mm) above grade plane;

2. More than 6 feet (1829 mm) above the finished ground level for more than 50 percent of the total building perimeter; or

3. More than 12 feet (3658 mm) above the finished ground level at any point.

❖ This definition parallels that of "Story above grade plane" (see Chapter 2). The determination of whether a basement meets the definition of "Story above grade plane" is important because it contributes to the height of a building in regard to Table 503 and the total allowable area of the building in accordance with Sections 503.1.1 and 503.3. Every story with the finished floor entirely above grade (finished ground level) is a story above grade. In addition, three specific criteria in the definition establish the threshold at which a basement extends far enough above ground to contribute to the regulated height of the building in number of stories. Figure 502.1(2) describes the application of these criteria.

GRADE PLANE. A reference plane representing the average of finished ground level adjoining the building at exterior walls. Where the finished ground level slopes away from the exterior walls, the reference plane shall be established by the lowest points within the area between the building and the lot line or, where the lot line is more than 6 feet (1829 mm) from the building, between the building and a point 6 feet (1829 mm) from the building.

❖ This term is used in the definitions of "Basement" and "Story above grade plane." It is critical in determining the height of a building and the number of stories above grade, which are regulated by this chapter. Since the finished ground surface adjacent to the building may vary (depending on site conditions), the mean average taken at various points around the building constitutes the grade plane. One method of determining the grade plane elevation is illustrated in Figure 502.1(3), where the ground slopes uniformly along the length of each exterior wall.

Situations may arise where the ground adjacent to the building slopes away from the building because of site or landscaping considerations. In this case, the lowest finished ground level at any point between the building's exterior wall and a point 6 feet (1829 mm) from the building [or the lot line, if closer than 6 feet (1829 mm)] comes under consideration. These points are used to determine the elevation of the grade plane as illustrated in Figures 502.1(4) and 502.1(5).

In the context of the code, the term "grade" means the finished ground level at the exterior walls. While the grade plane is a hypothetical horizontal plane derived as indicated above, the grade is that which actually ex-

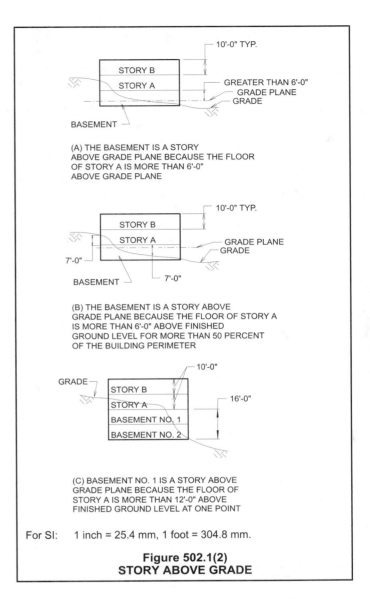

(A) THE BASEMENT IS A STORY ABOVE GRADE PLANE BECAUSE THE FLOOR OF STORY A IS MORE THAN 6'-0" ABOVE GRADE PLANE

(B) THE BASEMENT IS A STORY ABOVE GRADE PLANE BECAUSE THE FLOOR OF STORY A IS MORE THAN 6'-0" ABOVE FINISHED GROUND LEVEL FOR MORE THAN 50 PERCENT OF THE BUILDING PERIMETER

(C) BASEMENT NO. 1 IS A STORY ABOVE GRADE PLANE BECAUSE THE FLOOR OF STORY A IS MORE THAN 12'-0" ABOVE FINISHED GROUND LEVEL AT ONE POINT

For SI: 1 inch = 25.4 mm, 1 foot = 304.8 mm.

Figure 502.1(2)
STORY ABOVE GRADE

ists or is intended to exist at the completion of site work. The only situation where the grade plane and the grade are identical is when the site is perfectly level for a distance of 6 feet (1829 mm) from all exterior walls.

HEIGHT, BUILDING. The vertical distance from grade plane to the average height of the highest roof surface.

❖ This definition establishes the two points of measurement that determine the height of a building in feet. This measurement is used to determine compliance with the building height limitations of Section 503.1 and Table 503, as well as other sections of the code where the height of the building is a factor in the requirements (for example, see Section 1406.2.2).

The lower point of measurement is the grade plane (see the definition of "Grade plane" above). The upper point of measurement is the roof surface of the building, with consideration given to sloped roofs (such as a hip or gable roof). In the case of sloped roofs, the average

FIGURE 502.1(3) – FIGURE 502.1(4)   GENERAL BUILDING HEIGHTS AND AREAS

height would be used as the upper point of measurement, rather than the eave line or the ridge line. The average height of the roof is the mid-height between the roof eave and the roof ridge, regardless of the shape of the roof.

This definition also indicates that building height is measured to the highest roof surface. In the case of a building with multiple roof levels, the highest of the various roof levels must be used to determine the building height. If the highest of the various roof levels is a sloped roof, then the average height of that sloped roof must be used. The average height of multiple roof levels is not to be used to determine the building height.

A penthouse is not intended to affect the measurement of building height. By definition, a "Penthouse" is a structure that is built above the roof of a building (see Section 1502.1).

The distance that a building extends above ground also determines the relative hazards of that building. Simply stated, a taller building presents relatively greater safety hazards than a shorter building for several reasons, including fire service access and time for occupant egress. The code specifically defines how building height is measured to enable various code requirements, such as type of construction and fire suppression, to be consistent with those relative hazards. [see Figure 502.1(6) for the computation of building height in terms of feet and stories].

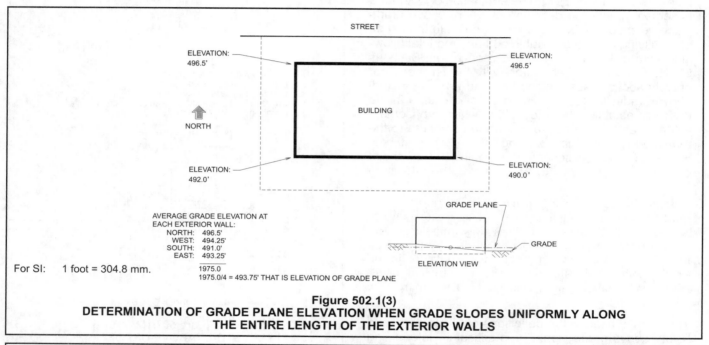

For SI: 1 foot = 304.8 mm.

Figure 502.1(3)
**DETERMINATION OF GRADE PLANE ELEVATION WHEN GRADE SLOPES UNIFORMLY ALONG
THE ENTIRE LENGTH OF THE EXTERIOR WALLS**

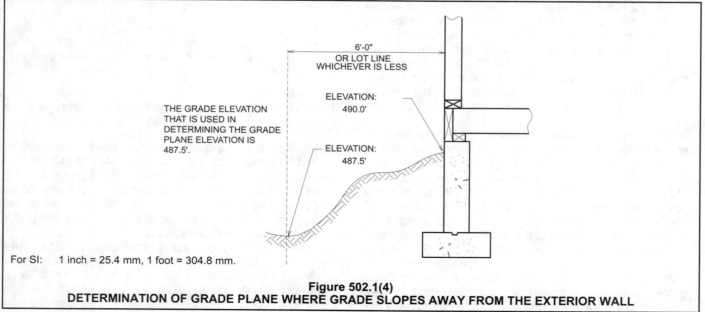

For SI: 1 inch = 25.4 mm, 1 foot = 304.8 mm.

Figure 502.1(4)
DETERMINATION OF GRADE PLANE WHERE GRADE SLOPES AWAY FROM THE EXTERIOR WALL

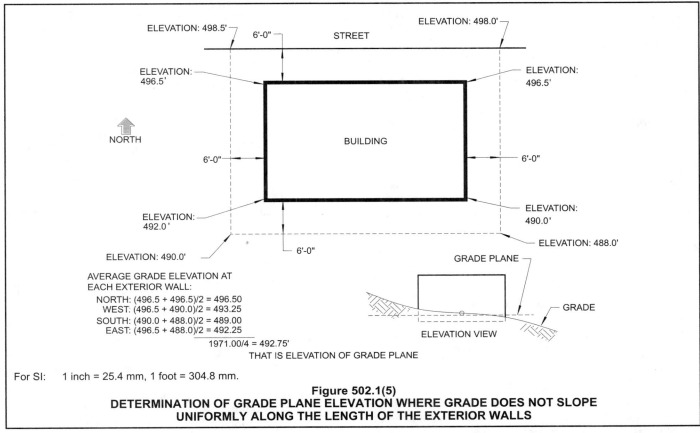

AVERAGE GRADE ELEVATION AT
EACH EXTERIOR WALL:
NORTH: (496.5 + 496.5)/2 = 496.50
WEST. (496.5 + 490.0)/2 = 493.25
SOUTH: (490.0 + 488.0)/2 = 489.00
EAST: (496.5 + 488.0)/2 = 492.25
1971.00/4 = 492.75'
THAT IS ELEVATION OF GRADE PLANE

For SI: 1 inch = 25.4 mm, 1 foot = 304.8 mm.

Figure 502.1(5)
DETERMINATION OF GRADE PLANE ELEVATION WHERE GRADE DOES NOT SLOPE
UNIFORMLY ALONG THE LENGTH OF THE EXTERIOR WALLS

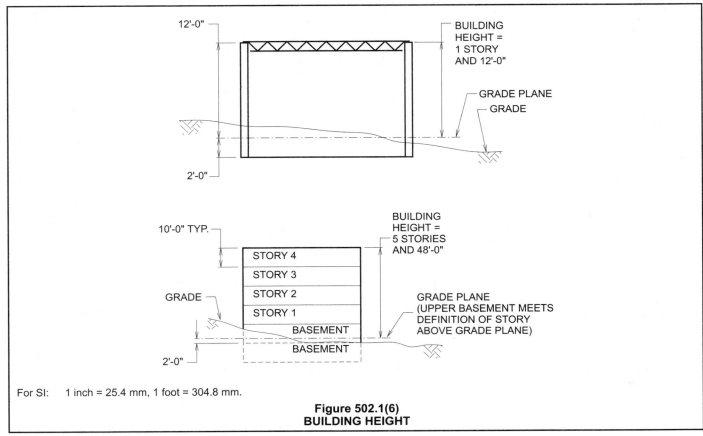

For SI: 1 inch = 25.4 mm, 1 foot = 304.8 mm.

Figure 502.1(6)
BUILDING HEIGHT

HEIGHT, STORY. The vertical distance from top to top of two successive finished floor surfaces; and, for the topmost story, from the top of the floor finish to the top of the ceiling joists or, where there is not a ceiling, to the top of the roof rafters.

❖ This definition is needed for application of seismic provisions in Chapter 16. Specifically, the story height is required in the context of calculating story drift determination and P-delta effects in Section 1617.3, which refers the user to ASCE 7.

INDUSTRIAL EQUIPMENT PLATFORM. An unoccupied, elevated platform in an industrial occupancy used exclusively for mechanical systems or industrial process equipment, including the associated elevated walkways, stairs and ladders necessary to access the platform (see Section 505.5).

❖ A distinction is made between industrial equipment platforms and mezzanines by way of definition. Industrial equipment platforms, covered in Section 505.5, are unoccupied and used exclusively for housing equipment and providing access thereto, and are not subject to the requirements for mezzanines. Their purpose could also be to allow access for maintenance, repair or modification of elevated or very large equipment. Industrial equipment platforms allow efficient use of high bay areas by locating infrequently accessed equipment or processes overhead without the occupant load or increasing the hazard to occupants in the room. Elevated floor areas that do not meet this definition would be subject to the requirements for mezzanines.

MEZZANINE. An intermediate level or levels between the floor and ceiling of any story with an aggregate floor area of not more than one-third of the area of the room or space in which the level or levels are located (see Section 505).

❖ A common design feature in factories, warehouses and mercantile buildings is an intermediate loft, or platform, between the story levels of a building. This type of feature, or mezzanine, can be found in buildings of all occupancies. The code must deal with whether this intermediate level is another story of the building, and whether it can simply be treated as part of the story in which it is contained. The basic rule is that it must be less than one-third of the area of the floor below (of the room in which it is located) in order to be considered a mezzanine. Requirements for mezzanines are found in Section 505.

STORY. That portion of a building included between the upper surface of a floor and the upper surface of the floor or roof next above (also see "Basement" and "Mezzanine").

❖ All levels in a building that conform to this description are stories, including basements. For instance, Section 712.4.3.1 restricts penetrations that connect "more than three stories," which would include basements. Certain code provisions regarding number of stories, however, pertain only to stories above grade (for instance, the

maximum height provisions of Table 503). A mezzanine is considered part of the story in which it is located.

SECTION 503
GENERAL HEIGHT AND AREA LIMITATIONS

503.1 General. The height and area for buildings of different construction types shall be governed by the intended use of the building and shall not exceed the limits in Table 503 except as modified hereafter. Each part of a building included within the exterior walls or the exterior walls and fire walls where provided shall be permitted to be a separate building.

❖ The provisions for governing the height and area of buildings on the basis of occupancy group classification and type of construction are established in this section. It also establishes Table 503 as the primary tool for determining the minimum type of construction. All buildings are subject to these limitations unless more specific code provisions for a building type provide for different height or area limitations. For instance, Section 507 allows certain buildings to be unlimited in area due to lack of exposure, low hazard level, construction type, the presence of fire safety systems or a combination of these characteristics. Given these specific provisions, Table 503 would not apply.

Table 601 is used with Table 503 to determine acceptable risk and fire safety levels for a building. Classification by occupancy, in accordance with the descriptions in Chapter 3, can be considered as establishing the level of "risk" associated with the use of a building. The various construction types, described in Chapter 6 and Table 601, can be thought of as various levels of safety in regard to fire resistance. Table 503 becomes a risk/safety matrix that sets a minimum level of safety (construction type) in accordance with the risk (the occupancy classification). The following is an example for the use of Table 503:

Example: Figure 503.1 shows a two-story factory building of Type IIB construction. Assuming the building is not sprinklered [fire areas < 12,000 square feet (1115 m²)] and does not qualify for any area or height increases, what is the maximum allowable area per story and height of the building? Given the size of the building, is Type IIB construction adequate? What is the minimum required construction type?

Answer: From Table 503, the allowable area per story for an F-1 of Type IIB construction is 15,500 square feet (1440 m²). Since the actual area is 120 × 150 = 18,000 square feet (1672 m²), Type IIB construction is not acceptable. The minimum required type of construction is Type IV, Type IIIA or Type IIA, depending on the desired materials for construction. All of these construction types for an F-1 have allowable per-story areas greater than 18,000 square feet (1672 m²) in accordance with Table 503. The actual height in feet and stories is also within the table limits for these construction types.

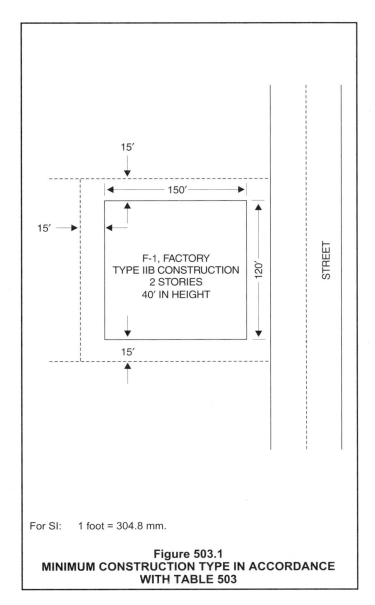

15'

150'

15' →

F-1, FACTORY
TYPE IIB CONSTRUCTION
2 STORIES
40' IN HEIGHT

120'

STREET

15'

For SI: 1 foot = 304.8 mm.

Figure 503.1
MINIMUM CONSTRUCTION TYPE IN ACCORDANCE
WITH TABLE 503

503.1.1 Basements. Basements need not be included in the total allowable area provided they do not exceed the area permitted for a one-story building.

❖ Total allowable area of a building (the aggregate area of all floors) is determined in accordance with Section 503.3. For buildings higher than three stories above grade, the total allowable area must be shared by all floor levels (see commentary, Section 503.3). The floor area of a basement, however, is not required to be counted as part of the total building area when evaluating total allowable area in accordance with Section 503.3; therefore, if the total allowable area of a building is 50,000 square feet (4645 m^2) in accordance with Section 503.3, the total aggregate area of all floors above grade must not exceed 50,000 square feet (4645 m^2). The presence of a basement, however, is permitted to

make the actual total area of the building larger than 50,000 square feet (4645 m^2).

503.1.2 Special industrial occupancies. Buildings and structures designed to house low-hazard industrial processes that require large areas and unusual heights to accommodate craneways or special machinery and equipment including, among others, rolling mills; structural metal fabrication shops and foundries; or the production and distribution of electric, gas or steam power, shall be exempt from the height and area limitations of Table 503.

❖ This section provides an exemption from the limits of Table 503. The occupancies that may use this exemption are quite limited. The term "low-hazard industrial processes" clearly addresses those occupancies in which flammable or combustible materials are not present in the manufacturing, processing or distributing function. Furthermore, the exemption is only applicable when large areas or unusual heights beyond that permitted by Table 503 are necessary to accommodate the specific manufacturing process.

It is the responsibility of the building official to determine when the application of this section is appropriate. The mere cost impact of the application of Table 503 does not dictate an exemption if it is not both a low hazard and in need of an unusual size. The building official, in assessing the proposed construction, may wish to compare what is being proposed to other similar facilities and protection features provided for them.

TABLE 503. See page 5-8.

❖ Table 503 is the foremost code provision table used in establishing equivalent risk (offsetting a building's inherent fire hazard—represented by occupancy—with materials and construction features). Sections 504 and 506 give height and area increases to the limits of Table 503 for buildings with certain features. Section 507 contains provisions by which buildings can in fact be unlimited in area; however, the height limitations of Table 503 would still apply.

Table 503 has three components:

1. The left column represents all of the group classifications defined in Chapter 3.

2. The top row lists all of the types of construction subclassifications explained in Chapter 6. Each cell in this matrix contains the specific height (in stories above grade) and area (per floor) limitations for the group/type of construction combination in question.

3. Maximum building height in feet above grade plane is shown along a top row, just below the construction-type classifications. "Grade plane" is defined in Section 502.1. Buildings must meet both the height in stories (shown in each cell) and the height in feet (shown along the upper row) criteria to be considered in compliance with that type of construction.

TABLE 503
ALLOWABLE HEIGHT AND BUILDING AREAS
Height limitations shown as stories and feet above grade plane.
Area limitations as determined by the definition of "Area, building," per floor.

GROUP	Hgt(feet) Hgt(S)	TYPE I A	TYPE I B	TYPE II A	TYPE II B	TYPE III A	TYPE III B	TYPE IV HT	TYPE V A	TYPE V B
		UL	160	65	55	65	55	65	50	40
A-1	S	UL	5	3	2	3	2	3	2	1
	A	UL	UL	15,500	8,500	14,000	8,500	15,000	11,500	5,500
A-2	S	UL	11	3	2	3	2	3	2	1
	A	UL	UL	15,500	9,500	14,000	9,500	15,000	11,500	6,000
A-3	S	UL	11	3	2	3	2	3	2	1
	A	UL	UL	15,500	9,500	14,000	9,500	15,000	11,500	6,000
A-4	S	UL	11	3	2	3	2	3	2	1
	A	UL	UL	15,500	9,500	14,000	9,500	15,000	11,500	6,000
A-5	S	UL	UL	UL	UL	UL	UL	UL	UL	UL
	A	UL	UL	UL	UL	UL	UL	UL	UL	UL
B	S	UL	11	5	4	5	4	5	3	2
	A	UL	UL	37,500	23,000	28,500	19,000	36,000	18,000	9,000
E	S	UL	5	3	2	3	2	3	1	1
	A	UL	UL	26,500	14,500	23,500	14,500	25,500	18,500	9,500
F-1	S	UL	11	4	2	3	2	4	2	1
	A	UL	UL	25,000	15,500	19,000	12,000	33,500	14,000	8,500
F-2	S	UL	11	5	3	4	3	5	3	2
	A	UL	UL	37,500	23,000	28,500	18,000	50,500	21,000	13,000
H-1	S	1	1	1	1	1	1	1	1	NP
	A	21,000	16,500	11,000	7,000	9,500	7,000	10,500	7,500	NP
H-2	S	UL	3	2	1	2	1	2	1	1
	A	21,000	16,500	11,000	7,000	9,500	7,000	10,500	7,500	3,000
H-3	S	UL	6	4	2	4	2	4	2	1
	A	UL	60,000	26,500	14,000	17,500	13,000	25,500	10,000	5,000
H-4	S	UL	7	5	3	5	3	5	3	2
	A	UL	UL	37,500	17,500	28,500	17,500	36,000	18,000	6,500
H-5	S	3	3	3	3	3	3	3	3	2
	A	UL	UL	37,500	23,000	28,500	19,000	36,000	18,000	9,000
I-1	S	UL	9	4	3	4	3	4	3	2
	A	UL	55,000	19,000	10,000	16,500	10,000	18,000	10,500	4,500
I-2	S	UL	4	2	1	1	NP	1	1	NP
	A	UL	UL	15,000	11,000	12,000	NP	12,000	9,500	NP
I-3	S	UL	4	2	1	2	1	2	2	1
	A	UL	UL	15,000	10,000	10,500	7,500	12,000	7,500	5,000
I-4	S	UL	5	3	2	3	2	3	1	1
	A	UL	60,500	26,500	13,000	23,500	13,000	25,500	18,500	9,000
M	S	UL	11	4	4	4	4	4	3	1
	A	UL	UL	21,500	12,500	18,500	12,500	20,500	14,000	9,000
R-1	S	UL	11	4	4	4	4	4	3	2
	A	UL	UL	24,000	16,000	24,000	16,000	20,500	12,000	7,000
R-2[a]	S	UL	11	4	4	4	4	4	3	2
	A	UL	UL	24,000	16,000	24,000	16,000	20,500	12,000	7,000
R-3[a]	S	UL	11	4	4	4	4	4	3	3
	A	UL	UL	UL	UL	UL	UL	UL	UL	UL
R-4	S	UL	11	4	4	4	4	4	3	2
	A	UL	UL	24,000	16,000	24,000	16,000	20,500	12,000	7,000
S-1	S	UL	11	4	3	4	3	4	3	1
	A	UL	48,000	26,000	17,500	26,000	17,500	25,500	14,000	9,000
S-2[b, c]	S	UL	11	5	4	4	4	5	4	2
	A	UL	79,000	39,000	26,000	39,000	26,000	38,500	21,000	13,500
U[c]	S	UL	5	4	2	3	2	4	2	1
	A	UL	35,500	19,000	8,500	14,000	8,500	18,000	9,000	5,500

For SI: 1 foot = 304.8 mm, 1 square foot = 0.0929 m^2.

UL = Unlimited, NP = Not permitted.

a. As applicable in Section 101.2.

b. For open parking structures, see Section 406.3.

c. For private garages, see Section 406.1.

Example 1: A business building of Type IIIB construction is built to four stories above grade; however, the building height in feet above grade plane is 60 feet (18 288 mm). Type IIIB construction would not be permitted unless the building is fully sprinklered in accordance with Section 504.2, since the height limitation for a Group B, Type IIIB building in Table 503 is 55 feet (16 764 mm).

Example 2: A mercantile building of Type IIB construction is 50 feet (15 240 mm) above grade plane, but five stories above the grade. It would not be permitted to be Type IIB construction unless fully sprinklered in accordance with Section 504.2, since it exceeds the number of stories permitted in Table 503 for a Group M building of Type IIB construction.

The area limitations appearing in each cell represent the absolute maximum square footage per floor for each group/type of construction combination before increases in accordance with Section 506 or adjustments due to total allowable building area in accordance with Section 503.3. The definition of "Area, building" in Section 502.1 indicates the application of the tabular area.

Unprotected combustible construction (Type VB) is not permitted for buildings of Groups H-1 and I-2. This type of construction—in particular, the fire hazard caused by the unprotected wood structure and the concealed spaces—does not provide the minimum protection required by the code to maintain equivalent risk in the presence of these occupancies.

For many groups in Type IA or IIB construction, both the height above grade and the area per floor are not limited by the code because of the required fire-resistance ratings of these construction types. In addition, the area of Group A-5 buildings (outdoor assembly for viewing) is unlimited in any construction type.

Simply put, the application of Table 503 is accomplished by identifying the group classification of the building in question along the left column, as well as identifying the cell in that row that matches the building's height (in stories and feet) and area per floor. The resulting type of construction (at the top of the resulting column) represents the minimum required type of construction for the building before the application of increases in accordance with Sections 504 and 506 and adjustments for total allowable area in accordance with Section 503.3.

Example 3: Assuming no area or height increases for a Group S-1 building [25 feet (7620 mm) above grade plane] of 11,500 square feet (1115 m²) per floor, what is the minimum required construction type of the building in accordance with Table 503?

Answer: In accordance with the Group S-1 row of Table 503, the minimum required construction type is Type VA, which allows 14,000 square feet (1301 m²) per floor and limits the height to three stories and 50 feet (15 240 mm). The building cannot be Type VB, since it exceeds one story and 9,000 square feet (836

m²) per floor. Type IIB (noncombustible, unprotected) is also acceptable.

503.1.3 Buildings on same lot. Two or more buildings on the same lot shall be regulated as separate buildings or shall be considered as portions of one building if the height of each building and the aggregate area of buildings are within the limitations of Table 503 as modified by Sections 504 and 506. The provisions of this code applicable to the aggregate building shall be applicable to each building.

❖ Ordinarily, two buildings on the same lot are subject to code provisions for exterior wall fire-resistance ratings in Table 602 and opening protectives in Tables 704.8, 715.3 and 715.4 based on the "imaginary line" described in the definition of "Fire separation distance" (see Section 702.1). The primary purpose of this section is to eliminate the application of these provisions when two small buildings can be regulated as one larger building on the premise that there is no need to protect a building from itself. In other words, if the two buildings under consideration were actually connected (making one building) and could meet the area and height limitations for one building based on construction type, the connecting interior wall would not be required to have been rated; therefore, it is inconsistent to require the protection of the facing exterior walls simply because there is not a physical connection between two portions of the same building.

For example, in Figure 503.1.3, Buildings A and B are being constructed on the same lot at the same time. If they are considered to be separate buildings, then the code requirements for exterior wall fire-resistance ratings and opening protectives (see Tables 602, 704.8, 715.3 and 715.4) would apply to each on the basis of the placement of an imaginary lot line between the two buildings.

Under Section 503.1.3, however, if they are considered to be one building (for minimum type of construction purposes), then there is no need for protection of the facing exterior walls and Tables 602, 704.8, 715.3 and 715.4 do not apply to the walls between the two buildings. These walls would have to meet exterior wall requirements for the type of construction specified in Table 601 for exterior bearing walls only. If they were nonload-bearing walls, then the required fire-resistance rating would be zero.

Although facing exterior walls between two buildings on the same lot are not subject to fire-resistance rating and opening size limitations by using this provision (except as previously indicated for load-bearing walls), the remainder of the exterior walls in both buildings must comply with Tables 601, 602, 704.8, 715.3 and 715.4. Three of the walls in Building A and three of the walls in Building B, therefore, would still be required to meet the requirements of Tables 602, 704.8, 715.3 and 715.4, and the exemption would only apply to the two walls (one from each building) that face each other.

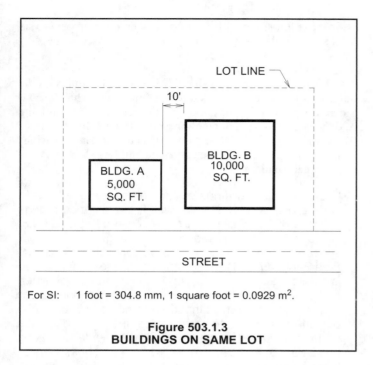

For SI: 1 foot = 304.8 mm, 1 square foot = 0.0929 m².

Figure 503.1.3
BUILDINGS ON SAME LOT

503.1.4 Type I construction. Buildings of Type I construction permitted to be of unlimited tabular heights and areas are not subject to the special requirements that allow unlimited area buildings in Section 507 or unlimited height in Sections 503.1.2 and 504.3 or increased height and areas for other types of construction.

❖ Buildings of Type I construction, because of their superior fire resistance (which are permitted to be of unlimited height and area in accordance with Table 503), do not have to meet the requirements of various code sections that require fire suppression systems or a certain fire separation distance (see Sections 504.2, 506.2 and 507) in order to be unlimited in area and height. These buildings may be of unlimited size based on their type of construction alone. High-rise buildings are required to be sprinklered in accordance with Section 403.2, and this section should not be construed to be a release from that requirement if a building fits the definition of "High rise" in Section 403.1.

503.2 Party walls. Any wall located on a lot line between adjacent buildings, which is used or adapted for joint service between the two buildings, shall be constructed as a fire wall in accordance with Section 705, without openings and shall create separate buildings.

❖ This section defines party walls, requires that they be constructed as fire walls in accordance with Section 705 and specifically prohibits openings. Party walls are walls located on property lines; fire walls may or may not be located on property lines. In other words, all party walls must be fire walls, but not all fire walls are party walls. In accordance with Section 705.8, protected openings of limited size are permitted in fire walls, but

no openings are permitted in party walls.

In accordance with Section 705, fire walls that bisect a building create two separate buildings for code purposes; however, the two buildings share common property. Party walls also create two separate buildings for code purposes, but the two buildings do not share any property except the wall itself; therefore, a higher level of protection is warranted and openings are prohibited.

SECTION 504
HEIGHT MODIFICATIONS

504.1 General. The heights permitted by Table 503 shall only be increased in accordance with this section.

Exception: The height of one-story aircraft hangars, aircraft paint hangars and buildings used for the manufacturing of aircraft shall not be limited if the building is provided with an automatic fire-extinguishing system in accordance with Chapter 9 and is entirely surrounded by public ways or yards not less in width than one and one-half times the height of the building.

❖ Section 504 contains exceptions to the height limitations of Table 503 for certain buildings. These exceptions are made on the basis of the structure's proposed occupancy and of fire safety features included in the design of the structure. The principal exception given is the increase for buildings equipped throughout with an automatic sprinkler system (see Section 504.2). With the exception of Group H (high hazard), no occupancy group and type of construction combination are assumed to have automatic sprinkler protection; therefore, when a sprinkler system is installed either as an option or where required by other sections of the code, an increase in height may be applicable (see the exceptions to Section 504.2) because of the increased protection provided by such a system.

The exception permits fully sprinklered aircraft hangars to exceed the limits of Table 503 as long as they have the specified open area surrounding the building. The exception is necessary to accommodate the size of large aircraft within the building, and the hazard is mitigated by the requirement for sprinklers and a minimum fire separation distance. Aircraft paint hangars may be unlimited in area as well under certain conditions (see commentary, Section 507.7).

504.2 Automatic sprinkler system increase. Where a building is equipped throughout with an approved automatic sprinkler system in accordance with Section 903.3.1.1, the value specified in Table 503 for maximum height is increased by 20 feet (6096 mm) and the maximum number of stories is increased by one story. These increases are permitted in addition to the area increase in accordance with Sections 506.2 and 506.3. For Group R buildings equipped throughout with an approved automatic sprinkler system in accordance with Section 903.3.1.2, the value specified in Table 503 for maximum height is increased by 20 feet (6096 mm) and the maximum number of stories is increased by one story, but shall not exceed four stories or 60 feet (18 288 mm), respectively.

Exceptions:

1. Group I-2 of Type IIB, III, IV or V construction.

2. Group H-1, H-2, H-3 or H-5.

3. Fire-resistance rating substitution in accordance with Table 601, Note d.

❖ This section permits the height limitations of Table 503 to be increased one story and 20 feet (6096 mm) when the building is protected throughout with an approved automatic sprinkler system. When used in this context, the phrase "equipped throughout" (see Section 903.3.1.1) means the entire structure is provided with sprinkler protection, and the only exceptions to this are the specific locations identified in Section 903.3.1.1.1 or in the applicable standard. If a building contains an area that meets the conditions of Section 903.3.1.1.1 and an exception therein, then sprinklers may be omitted in that specific location of the building and the building may be considered protected throughout for purposes of Section 504.2 (see the commentary to Section 903.3.1.1.1 for application of the exceptions).

By referencing Section 903.3.1.1 (NFPA 13 sprinkler system), the code requires a system in accordance with NFPA 13 (not NFPA 13R or NFPA 13D). Only buildings equipped throughout with systems installed in accordance with NFPA 13 qualify for the height increase. The systems and coverage criteria of NFPA 13R and 13D are considered to provide a lower level of protection for certain residential buildings and, therefore, do not qualify for the height increase, except as specifically provided in the sentence regarding Group R occupancies. Buildings of residential occupancies may receive a somewhat limited increase of one story up to a maximum of four stories and 60 feet (18 288 mm) if equipped with a sprinkler system that conforms to Section 903.3.1.2 (an NFPA 13R system). This alternative is intended to address the differences between application and protection provided by an NFPA 13 system versus an NFPA 13R system. If a Group R building is protected throughout with an NFPA 13 system in accordance with Section 903.3.1.1, then the four-story, 60-foot (18 288 mm) limitation would not apply.

It should be noted that the height increase is granted even if the sprinkler system is required by Chapter 9 of the code, based on the use and size of the building. For instance, Section 903.2.8 requires all three-story (and higher) Group R-2 buildings to be sprinklered. Even so, a Type VA, Group R-2 building protected throughout with an NFPA 13R system (in accordance with Section 903.3.1.2) is allowed up to four stories (one story more than the story limitation shown in Table 503 for Group R-2, Type VA construction) and up to 60 feet (18 288 mm) in height.

On the other hand, an NFPA 13R sprinkler system would not yield a height-in-stories increase for a Type IIB building of Group R-2, since Table 503 already per-

mits four stories. The height-in-feet threshold, however, would be 60 feet (18 288 mm), a 5-foot (1524 mm) increase over what is allowed by Table 503 for Type IIB construction. If the R-2 building is protected throughout with a system in accordance with NFPA 13 (see Section 903.3.1.1), then the height thresholds would be five stories and 75 feet (22 860 mm) [the permitted one-story, 20-foot (6096 mm) increase over Table 503 limits in accordance with the general requirements of this section].

Exception 1 indicates that several types of construction are eliminated from the height increase when used for Group I-2. Each of these types of construction either lacks the fire-resistance rating deemed necessary or includes combustible construction materials in a use where the occupants must rely on the fire sprinkler system for defend-in-place protection. Exception 2 indicates that buildings of Groups H-1, H-2, H-3, and H-5 are not eligible for the general height increase because of the higher hazards associated with these occupancies.

Exception 3 states that when an automatic sprinkler system is installed to address the exception allowed for the trade off of sprinklers for 1-hour fire-resistance-rated construction given in Note d of Table 601, the height increase allowed in this section cannot be taken. For example, in a building constructed using Type IIA construction, the required 1-hour fire-resistance rating of all structural members could be reduced to zero if the building were equipped throughout with a sprinkler system in accordance with Section 903.3.1.1 (NFPA 13 system); however, the allowable height for the occupancy classification of, for example, Group B for Type IIA construction could not be increased by 20 feet (6096 mm) to 85 feet (25 908 mm), as would otherwise be allowed in this section.

504.3 Roof structures. Towers, spires, steeples and other roof structures shall be constructed of materials consistent with the required type of construction of the building except where other construction is permitted by Section 1509.2.1. Such structures shall not be used for habitation or storage. The structures shall be unlimited in height if of noncombustible materials and shall not extend more than 20 feet (6096 mm) above the allowable height if of combustible materials (see Chapter 15 for additional requirements).

❖ Certain roof structures may exceed the height limitations of Table 503 in accordance with this section and Chapter 15. These roof structures on a noncombustible building (Types I and II) must be of noncombustible construction, and in some cases must have ratings equivalent to those required for exterior walls (see Section 1509.5.2 for enclosed towers and spires). Section 1509.2.1 contains exceptions for penthouse-type structures, mechanical equipment enclosures and screens when not in proximity to a property line. Structures that are used for habitation or storage must be considered an additional story, unless meeting the requirements for a penthouse in Section 1509.2. The height limitation for noncombustible roof structures in accordance with this section is unlimited; however, Section 1509.5 and its subsections contain additional restrictions for structures of a certain size or use.

SECTION 505
MEZZANINES

505.1 General. A mezzanine or mezzanines in compliance with this section shall be considered a portion of the floor below. Such mezzanines shall not contribute to either the building area or number of stories as regulated by Section 503.1. The area of the mezzanine shall be included in determining the fire area defined in Section 702. The clear height above and below the mezzanine floor construction shall not be less than 7 feet (2134 mm).

❖ Although mezzanines provide an additional or intermediate useable floor level in a building, they are not considered an additional story as long as they comply with the requirements of Section 505. Building height and area limitations are intended to offset the inherent fire hazard associated with specific occupancy groups and with materials and features of a specific construction type. Because of a mezzanine's restricted size and its required openness to the room or space below, a mezzanine does not contribute significantly to a building's inherent fire hazard; therefore, the area of a mezzanine is not considered when applying the provisions of Section 503.1 for building area limitations, and mezzanines are not considered in determining the height in stories of a building as regulated by Table 503. The occupant and fuel load of the mezzanine should be taken into consideration, however, when determining the necessity for fire protection systems. As such, the area of the mezzanine is to be included in the calculation of the size of the fire area for sprinkler thresholds (see the commentary for the definition of "Fire area" in Section 702.1).

This section does not include any requirements for the construction of a mezzanine or for fire-resistance ratings; therefore, the mezzanine is to be constructed of materials consistent with the construction type of the building. Required fire-resistance ratings are determined on the basis of Table 601 for the appropriate construction type.

Mezzanines are required to have a ceiling height of not less than 7 feet (2134 mm), and the ceiling height below the mezzanine must also be 7 feet (2134 mm). Even habitable rooms located in mezzanines may have a ceiling height of 7 feet (2134 mm), in accordance with Exception 4 in Section 1208.2.

505.2 Area limitation. The aggregate area of a mezzanine or mezzanines within a room shall not exceed one-third of the area of that room or space in which they are located. The enclosed portions of rooms shall not be included in a determination of the size of the room in which the mezzanine is located. In determining the allowable mezzanine area, the area of the mezzanine shall not be included in the area of the room.

Exception: The aggregate area of mezzanines in buildings and structures of Type I or II construction for special industrial occupancies in accordance with Section 503.1.2 shall not exceed two-thirds of the area of the room.

❖ So as not to contribute significantly to a building's inherent fire hazard, a mezzanine is restricted to a maximum of one-third of the area of the room with which it shares a common atmosphere. The area may consist of multiple mezzanines at the same or different floor levels, provided that the aggregate area does not exceed the one-third limitation. If the area limitation is exceeded, the provisions of this section do not apply and the level is considered a story.

In determining the allowable area of the mezzanine, the enclosed spaces of the room below are not to be included in calculating the room size. Although the mezzanine area is included in the calculation of fire area size, it is not included in the area of the room when computing the allowable mezzanine area. For example, a room contains 5,000 square feet (465 m²), 500 of which are enclosed and not part of the common atmosphere with the mezzanine. A mezzanine may be provided in the room such that the area of the mezzanine is not more than 1,500 square feet (139 m²). In accordance with Section 505.4, the space at the mezzanine level must be open, except that enclosed spaces are permitted for mezzanines in accordance with the exceptions to Section 505.4.

By definition, in special industrial occupancies (see commentary, Section 503.1.2), the inherent fire hazard is very low; therefore, mezzanines in such occupancies located in buildings of Type I or II construction are permitted to constitute up to two-thirds of the area of the room in which they are located, in accordance with the exception. The limitation on construction types further reduces the fire hazard associated with such occupancies.

Equipment platforms, as defined in Section 502.1, are not considered mezzanines (see Section 505.5 for requirements for industrial equipment platforms).

505.3 Egress. Each occupant of a mezzanine shall have access to at least two independent means of egress where the common path of egress travel exceeds the limitations of Section 1013.3. Where a stairway provides a means of exit access from a mezzanine, the maximum travel distance includes the distance traveled on the stairway measured in the plane of the tread nosing.

Exceptions:

1. A single means of egress shall be permitted in accordance with Section 1014.1.

2. Accessible means of egress shall be provided in accordance with Section 1007.

❖ A mezzanine can be likened to a single room when considering means of egress. As with rooms, if the occupant load of the mezzanine exceeds the limitations of Table 1014.1 for the specific use of the space, at least two independent means of egress must be provided for the mezzanine. For example, if a mezzanine containing office areas (Group B) has an occupant load exceeding

50, a second means of egress from the mezzanine is required. Additionally, if a mezzanine has one means of egress and it is by means of an open stair to the floor below, the travel distance from the most remote point on the mezzanine to the bottom of the stair may not exceed 75 feet (22 860 mm) in accordance with Section 1013.3 for common path of travel [see the exceptions to that section that allow 100 feet (30 480 mm) in some circumstances]. If the travel distance to the bottom of the stair exceeds the limits of Section 1013.3, then a second means of egress must be provided from the mezzanine.

Because mezzanines that comply with Section 505 are considered a portion of the floor below in accordance with Section 505.1, the stair leading from it is not considered an exit stairway and is not required to be a vertical exit enclosure.

This section does not require that an exit be provided directly from the mezzanine level (Exception 2 of Section 505.4, however, requires mezzanines that are not open to the room to have access to an exit directly from the mezzanine level, if two means of egress are provided). When two means of egress are required, however, they are to be located remote from each other in accordance with Section 1014.2, as for any other space, so that if one means of egress is blocked by fire or smoke, then the other will be presumed available.

When exits are not located on the same level as the mezzanine, the occupant load of the mezzanine is added to the room or space below and the required means of egress width for that room is determined accordingly (see Section 1005). For example, if a room (Group B) has an occupant load of 45 and a mezzanine (also Group B) has an occupant load of 15, the total occupant load for the space served is 60; therefore, the room must have two exit access doors in accordance with Table 1014.1. The mezzanine itself, however, needs only one means of egress.

The requirements for accessible means of egress are not in any way intended to be affected by these provisions, as clarified in Exception 2. The requirements for accessibility relate only to the requirements for the occupancy and use of the space as provided in Chapters 10 and 11 of the code.

505.4 Openness. A mezzanine shall be open and unobstructed to the room in which such mezzanine is located except for walls not more than 42 inches (1067 mm) high, columns and posts.

Exceptions:

1. Mezzanines or portions thereof are not required to be open to the room in which the mezzanines are located, provided that the occupant load of the aggregate area of the enclosed space does not exceed 10.

2. A mezzanine having two or more means of egress is not required to be open to the room in which the mezzanine is located, if at least one of the means of egress provides direct access to an exit from the mezzanine level.

3. Mezzanines or portions thereof are not required to be open to the room in which the mezzanines are located,

provided that the aggregate floor area of the enclosed space does not exceed 10 percent of the mezzanine area.

4. In industrial facilities, mezzanines used for control equipment are permitted to be glazed on all sides.

5. In Group F occupancies of unlimited area, meeting the requirements of Section 507.2 or 507.3, mezzanines or portions thereof are not required to be open to the room in which the mezzanines are located, provided that an approved fire alarm system is installed throughout the entire building or structure and notification appliances are installed throughout the mezzanines in accordance with the provisions of NFPA 72. In addition, the fire alarm system shall be initiated by automatic sprinkler water flow.

❖ A mezzanine presents a unique fire threat to the occupant. If a mezzanine is closed off from the larger room, an undetected fire could develop such that it would jeopardize or eliminate the opportunity for occupant escape.

The exceptions address situations where the hazard is reduced. A low occupant load, typical of small mezzanines, would permit a mezzanine to be enclosed in accordance with Exception 1. Occupant load is calculated in accordance with Section 1004.1 for the use of the mezzanine space. Similarly, Exception 3 permits the enclosure of a limited portion of a mezzanine.

Exception 2 permits enclosure of the mezzanine based on the availability of an exit at the mezzanine level, and the mezzanine has at least two means of egress. The definition of "exit" is important for this exception. "Exit" is defined in Section 1002.1 as an element that is separated by fire-resistance-rated construction and opening protectives, and includes exterior doors; therefore, the mezzanine must be served by at least one vertical exit enclosure (Section 1019), a horizontal exit (Section 1021), an exit passageway (Section 1020), an exterior stair or ramp (Section 1022) or an exterior door discharging directly at grade. In addition, another means of egress is required, which may be an open stair to the room below.

Exceptions 4 and 5 address industrial facilities, where enclosure may be necessary for noise reduction or atmospheric control. The term "unlimited area" in Exception 5 refers to buildings that qualify as unlimited area buildings in accordance with Section 507.2 or 507.3, which requires the building to be sprinklered throughout. Although certain Group F occupancies could qualify as unlimited area buildings without sprinklers in accordance with Section 507.1, these buildings are not within the scope of this exception and need to meet the conditions of another exception in order to be enclosed.

505.5 Industrial equipment platforms. Industrial equipment platforms in buildings shall not be considered as a portion of the floor below. Such equipment platforms shall not contribute to either the building area or the number of stories as regulated by Section 503.1. The area of the industrial equipment platform shall not be included in determining the fire area. Industrial equipment platforms shall not be a part of any mezzanine, and such platforms and the walkways, stairs and ladders providing access to an equipment platform shall not serve as a part of the means of egress from the building.

❖ "Industrial equipment platform" is defined in Section 502.1 as an unoccupied, elevated platform in an industrial occupancy used exclusively for supporting mechanical systems or industrial process equipment and providing access to them. If an elevated platform does not meet all the conditions of this definition, then it must be considered either a mezzanine or another story.

Industrial equipment platforms are treated as part of the equipment they support (within the limitations of the subsections to this section), and do not contribute in any way to the area of the building, the number of stories, the area of any mezzanine or any fire area. If industrial equipment platforms are located in the same room as a mezzanine, however, the aggregate area of the platforms and mezzanines is limited by Section 505.5.1.

The definition of "Industrial equipment platform" includes the associated elevated walkways, stairs and ladders necessary to access the platform. Elements that serve an industrial equipment platform are not permitted to serve as a means of egress for occupants of the building unless they meet all the requirements for means of egress in Chapter 10. Because they are not used for means of egress, the elements used to access these platforms could be something other than a stair or ramp. For example, a permanent ladder could be used to access an industrial equipment platform. Since the code does not address ladder construction for this circumstance, other appropriate standards, such as OSHA standards, should be consulted for design.

Guards are required for elevated walking surfaces serving industrial equipment platforms in accordance with Section 1012.1 (see Section 505.5.3). Additionally, if a stair is used to access an equipment platform, it would be subject to all the dimensional requirements for stairs in accordance with Section 1009.1.

505.5.1 Area limitations. The aggregate area of all industrial equipment platforms within a room shall not exceed two-thirds of the area of the room in which they occur. Where an equipment platform is located in the same room as a mezzanine, the area of the mezzanine shall be determined by Section 505.2, and the combined aggregate area of the equipment platforms and mezzanines shall not exceed two-thirds of the room in which they occur.

❖ In determining the allowable area of an equipment platform, the enclosed spaces of the room below are not to be included in calculating the room size, and neither is the area of the equipment platform itself. Whereas the area of mezzanines is included in the calculation of fire area (see Section 505.1), the area of equipment platforms is not in accordance with Section 505.5. The area of mezzanines and equipment platforms, however, is summed when they are in the same room, and the ag-

gregate area is limited to two-thirds of the area of the room. The total area of a mezzanine still cannot exceed one-third of the area of the room.

505.5.2 Fire suppression. Where located in a building that is required to be protected by an automatic sprinkler system, industrial equipment platforms shall be fully protected by sprinklers above and below the platform, where required by the standards referenced in Section 903.3.

❖ In buildings or spaces that are required to be protected with an automatic fire suppression system, fire suppression above and below the equipment platform is needed so the equipment platform will not obstruct sprinkler coverage or delay sprinkler activation if a fire develops below the platform. As stated in Section 903.3.1.1, sprinkler installation is to be in accordance with NFPA 13. This section should not be construed to require sprinkler protection for equipment platforms where such a system is not otherwise required.

505.5.3 Guards. Equipment platforms shall have guards where required by Section 1012.1.

❖ Guards are required for equipment platforms in the same manner that they are required at open walking surfaces in other parts of the building, and are subject to the same requirements for height, design load and configuration. Section 1012.3 contains an exception that allows spacing of guard balusters in Group F occupancies to only be close enough to prevent the passage of a 21-inch (533 mm) sphere.

SECTION 506
AREA MODIFICATIONS

506.1 General. The areas limited by Table 503 shall be permitted to be increased due to frontage (I_f) and automatic sprinkler system protection (I_s) in accordance with the following:

$$A_a = A_t + \left[\frac{A_t I_f}{100}\right] + \left[\frac{A_t I_s}{100}\right] \qquad \textbf{(Equation 5-1)}$$

where:

A_a = Allowable area per floor (square feet).

A_t = Tabular area per floor in accordance with Table 503 (square feet).

I_f = Area increase due to frontage (percent) as calculated in accordance with Section 506.2.

I_s = Area increase due to sprinkler protection (percent) as calculated in accordance with Section 506.3.

❖ Besides increasing the fire resistance (using a higher construction type) of the building structure, in general there are two circumstances that would decrease a building's fire hazard. These are: (1) isolating the building from other structures and (2) equipping it with a fire suppression system. Equation 5-1 takes these circumstances into account. The equation determines the largest single floor area for any building, taking into account the permissible increases due to open frontage and the installation of a fire suppression system. The following examples demonstrate this point:

Example 1: A single-story, Group B building of Type IIB construction does not have a fire suppression system but has a frontage increase of 50 percent in accordance with Section 506.2. In Equation 5-1, the following values would be given to the terms:

A_t = 23,000 square feet (1765 m²) (from Table 503)

I_f = 50 percent

I_s = 0

The equation $A_a = A_t + [(A_t)(I_f)/100] + [(A_t)(I_s)/100]$

becomes A_a = 23,000 + [(23,000)(50)/100] + 0

> = 23,000 + 11,500

> = 34,500 square feet (3205 m²).

The allowable area per floor for the building is 34,500 square feet (3205 m²).

Example 2: A five-story, Group R-2 building of Type IIA construction has a frontage increase of 75 percent in accordance with Section 506.2. The building is fully sprinklered with an NFPA 13 system and the increase yielded by Section 506.3 is 200 percent. In Equation 5-1, the following values would be given to the terms:

A_t = 24,000 square feet (2230 m²) (from Table 503)

I_f = 75 percent

I_s = 200 percent

Equation 5-1 becomes:
A_a = 24,000 + [(24,000)(75)/100] + [(24,000)(200)/100]

> = 24,000 + 18,000 + 48,000.

> = 90,000 maximum square feet (8361 m²) per story [a single largest story may not exceed 90,000 square feet (8361 m²)].

At this point, because the building is more than three stories, an adjustment needs to be made in accordance with Section 503.3, which limits the aggregate area of all floors in the building. In accordance with that section, the total allowable area for the building may be no more than three times the maximum allowable area per floor (A_a). Assuming all stories of the building have equal area, the maximum allowable area of each floor is calculated as follows:

(3)(A_a) = (3)(90,000) = 270,000 square feet (25 083 m²) total allowable area.

270,000/4 stories = 67,500 square feet (6271 m²) per story.

Therefore, if the total allowable area is divided equally among the four stories, the allowable area per story for the building is 67,500 square feet (6271 m²).

506.1.1 Basements. A single basement need not be included in the total allowable area provided such basement does not exceed the area permitted for a one-story building.

❖ The commentary to Section 503.1.1 contains the same requirement. The buildings in the examples of the preceding section would be permitted to have a basement level not exceeding the maximum area per story calculated for the building. Specifically, the building in Example 2 could have a basement of up to 90,000 square feet (8361 m²) in addition to the four stories above grade of 67,500 square feet (6271 m²) each.

In order to not be considered part of the total allowable area for the building, the basement cannot meet the definition of "Story above grade plane" (see the definition of "Basement" in Section 502.1 and its commentary). Additionally, if more than one basement is present, the extra basement levels must be counted as contributing to the total allowable area for the building. In other words, for the purposes of Section 503.3, only one basement level may be considered as not contributing to the total building area.

506.2 Frontage increase. Every building shall adjoin or have access to a public way to receive an area increase for frontage. Where a building has more than 25 percent of its perimeter on a public way or open space having a minimum width of 20 feet (6096 mm), the frontage increase shall be determined in accordance with the following:

$$I_f = 100 \left[\frac{F}{P} - 0.25 \right] \frac{W}{30} \qquad \textbf{(Equation 5-2)}$$

where:

I_f = Area increase due to frontage.

F = Building perimeter which fronts on a public way or open space having 20 feet (6096 mm) open minimum width (feet).

P = Perimeter of entire building (feet).

W = Width of public way or open space (feet) in accordance with Section 506.2.1.

❖ The allowable area of a building is permitted to be increased when it has a certain amount of frontage on streets or open spaces since this provides access to the structure by fire service personnel, a temporary refuge area for occupants as they leave the building in a fire emergency and a reduced exposure to and from adjacent structures.

This section indicates how the frontage increase, or the term "I_f" in Equation 5-1, is calculated. There is no requirement in the code that buildings must have at least 25 percent of their perimeter on a public way or open space; however, in order to qualify for an area increase, a building must have more than 25 percent of its perimeter on a public way or open space having a minimum width of at least 20 feet (6096 mm) in accordance with this section. When the calculations are done, the maximum percent increase for a fully open perimeter (full frontage—the entire perimeter fronts on a public way or open space) is 75 percent.

"Public way" is defined in Section 1002.1, and Section 506.2.2 requires that an open space that is not a public way be on the same lot or dedicated for public use, and it must have access from a street or an approved fire lane in order to contribute to the frontage increase (see commentary, Section 506.2.2).

The term *"W"* in Equation 5-2 represents the minimum width of any open space for the portion of the perimeter that fronts on a public way or open space with a width of at least 20 feet (6096 mm). The following example demonstrates this point:

Example: Refer to Figure 506.2 below. In Equation 5-2, the terms would have the following values:

F = 500 feet (15 240 mm), since all sides of the building front on a public way or open space have 20 feet (6096 mm) minimum open width.

P = 500 feet (15 240 mm), the length of the entire perimeter.

W = 20 feet (6096 mm), the least dimension of the open space around the building. If one side of the building had only 15 feet (4572 mm) of open space, W would still be 20 (although F would be less than 500), since it represents the minimum width of the open space at the portions of the building that front on a public way or open space. Section 506.2.1 limits the value of W to a maximum of 30 feet (9144 mm) for the purposes of this equation.

Equation 5-2, $I_f = 100[(F/P) - 0.25](W/30)$, becomes:

$$I_f = 100[(500/500) - 0.25](20/30)$$
$$= 100(.75)(2/3)$$
$$= 50 \text{ percent}$$

Therefore, the term *"I_f"* equals 50 when plugging into Equation 5-1 for allowable area per floor.

506.2.1 Width limits. W must be at least 20 feet (6096 mm) and the quantity W divided by 30 shall not exceed 1.0. Where the value of W varies along the perimeter of the building, the calculation performed in accordance with Equation 5-2 shall be based on the weighted average of each portion of exterior wall and open space where the value of W is between 20 and 30 feet (6096 and 9144 mm).

Exception: The quantity W divided by 30 shall be permitted to not exceed 2.0 when all of the following conditions exist:

1. The building is permitted to be unlimited in area by Section 507; and

2. The only provision preventing unlimited area is compliant with the 60-foot (18 288 mm) public way or yard requirement, as applicable

❖ The amount of area increase (that is, the value of I_f) will vary depending on the minimum width of the open space used in the frontage increase calculation. The general requirement is that W equal at least 20 feet (6096 mm) and that the term $W/30$ not exceed a value of 1 (in effect, that the value of W not exceed 30). The value of W is the weighted average of the portions of the wall and each open space when the width of the open space is between 20 and 30. See the following two examples which illustrate this requirement:

Example 1: In Figure 506.2.1(1), the value of W would be 26 feet (7925 mm), calculated as follows:
W = [(200 x 25 feet) + (100 x 20 feet) + (200 x 30 feet)] / 500 = 26 feet.

Example 2: In Figure 506.2.1(2), the value of W would be 30, since the minimum width of the open space that qualifies as open frontage is greater than 30. The reason the value of W must be taken at 30 and not 50 is that this section sets an upper limit on the value of the $W/30$ term in Equation 5-2 (that value is limited to 1.0).

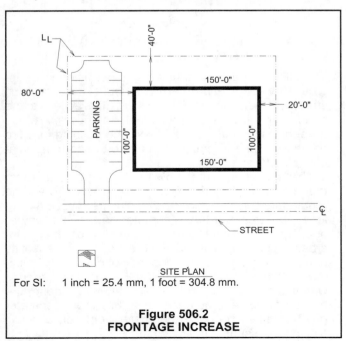

For SI: 1 inch = 25.4 mm, 1 foot = 304.8 mm.

Figure 506.2
FRONTAGE INCREASE

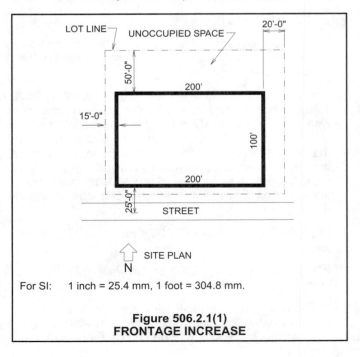

For SI: 1 inch = 25.4 mm, 1 foot = 304.8 mm.

Figure 506.2.1(1)
FRONTAGE INCREASE

The exception to Section 506.2.1 states that for certain buildings, the term "*W*/30" is permitted to have a maximum value of 2.0. This exception applies to buildings that would be allowed to be unlimited in area in accordance with Section 507, save for the fact that the open area of 60 feet (18 288 mm) required by Section 507 is not met. Therefore, the weighted average of *W* would be calculated between 20 feet (6096 mm) and 60 feet (18 288 mm).

506.2.2 Open space limits. Such open space shall be either on the same lot or dedicated for public use and shall be accessed from a street or approved fire lane.

❖ The requirement that the open space be on the same lot is so that the owner or the jurisdiction can control the space that is assumed to be open for purposes of the area increase. It is obvious that one cannot encumber a neighbor's (or a future neighbor's) property with a requirement that the space will always remain unoccupied.

Any part of the perimeter that is not accessible to the fire department by means of a street or fire lane cannot be considered open for the purposes of Section 506.2. For instance, if the back side of a building on a narrow lot cannot be reached by means of a fire lane on one side of the building (and there is no alley or street at the back), that portion of the perimeter is not considered open for purposes of frontage increase, even if there is actual open space exceeding 20 feet (6096 mm) in width.

This section does not require that a fire lane or street extend immediately adjacent to every portion of the perimeter that is considered open for purposes of the increase. Rather, access by a fire lane must be provided up to the open side such that fire department personnel can approach the side and pull hoses across the open area to fight a fire, and no corner of the building will impede the use of hoses and equipment on that side of the building. The following examples demonstrate this point:

Example 1: In Figure 506.2.2(1), the south side of the street can be considered open perimeter (frontage). The north side of the street cannot be considered open perimeter for purposes of the increase, since it is not accessible from the street.

Example 2: In Figure 506.2.2(2), all sides of the building are considered open perimeter (frontage) for purposes of the increase. Access up to each side of the building is provided by means of a fire lane or street.

Section 503 of the *International Fire Code*® (IFC®) specifies that access roads extend to within 150 feet (45 720 mm) of all portions of the exterior walls of a building; however, there are exceptions that would permit the omission of such roads under certain circumstances. One such exception is for buildings equipped throughout with an automatic sprinkler system.

The IFC also stipulates that the access roads must be at least 20 feet (6096 mm) in unobstructed width, although it also gives the building official authority to require greater widths if necessary for effective fire-fighting operations. The type of surface necessary for the approved fire lane is determined by the local building official with input from the fire department, but the road must be capable of supporting the imposed loads of fire apparatus and be surfaced so as to provide all-weather driving capabilities.

For SI: 1 inch = 25.4 mm, 1 foot = 304.8 mm.

Figure 506.2.1(2)
FRONTAGE INCREASE

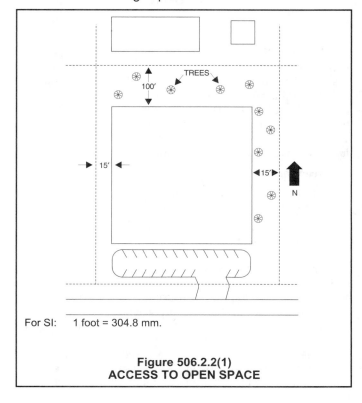

For SI: 1 foot = 304.8 mm.

Figure 506.2.2(1)
ACCESS TO OPEN SPACE

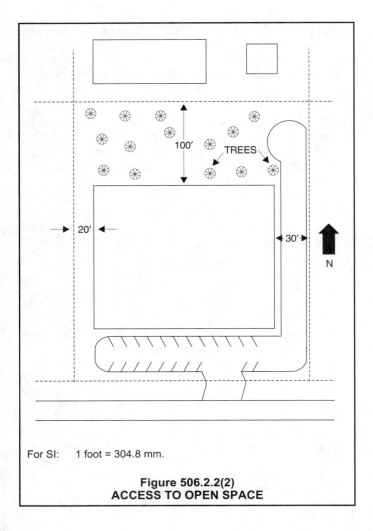

For SI: 1 foot = 304.8 mm.

Figure 506.2.2(2)
ACCESS TO OPEN SPACE

signed and installed in accordance with NFPA 13, as stipulated in Section 903.3.1.1. It is intended that only buildings protected throughout the entire structure with a system designed in accordance with NFPA 13 be eligible for the sprinkler increase permitted by this section, except as specifically modified by the exceptions to Section 903.3.1.1. Those exceptions permit the omission of sprinklers in certain locations within buildings because of conditions that exist in those locations. Even if an exception to Section 903.3.1.1 is utilized in the sprinkler design, the building is still eligible for the Section 506.3 area increase because the exempted locations either have a negligible impact on the fire load of the building or it is likely that other requirements will abate the hazard associated with these rooms. This section also clarifies that these area increases are cumulative with the height/story increases in Section 504.

Exception 1 does not permit the area increase for buildings with Group H-1, H-2 or H-3 occupancy areas, because of the higher level of hazard that these groups represent.

Exception 2 states that when an automatic sprinkler system is installed to address the exception allowed for the tradeoff of sprinklers for 1-hour fire-resistance- rated construction given in Note d of Table 601, the area increases allowed in this section cannot be taken. For example, in a building constructed using Type IIA construction, the required 1-hour fire-resistance rating of all structural members could be reduced to zero if the building were fully equipped with a sprinkler system in accordance with Section 903.3.1.1 (NFPA 13); however, the allowable height for the occupancy classification Group B for Type IIA construction, for example, will not be increased by the percentages that would otherwise be allowed in this section.

506.3 Automatic sprinkler system increase. Where a building is protected throughout with an approved automatic sprinkler system in accordance with Section 903.3.1.1, the area limitation in Table 503 is permitted to be increased by an additional 200 percent (I_s = 200 percent) for multistory buildings and an additional 300 percent (I_s = 300 percent) for single-story buildings. These increases are permitted in addition to the height and story increases in accordance with Section 504.2.

Exceptions:

1. Buildings with an occupancy in Group H-1, H-2 or H-3.

2. Fire-resistance rating substitution in accordance with Table 601, Note d.

❖ This section permits an increase of tabular areas (see Table 503) for each type of construction if the building in question is equipped throughout with an automatic sprinkler system. This section provides for two levels of increase: 200 percent for multistory buildings and 300 percent for single-story buildings.

The scope of the phrase "protected throughout with an automatic sprinkler system" means that the entire structure is to be provided with sprinkler protection de-

506.4 Area determination. The maximum area of a building with more than one story shall be determined by multiplying the allowable area of the first floor (A_a), as determined in Section 506.1, by the number of stories as listed below.

1. For two-story buildings, multiply by 2;

2. For three-story or higher buildings, multiply by 3; and,

3. No story shall exceed the allowable area per floor (A_a), as determined in Section 506.1 for the occupancies on that floor.

Exceptions:

1. Unlimited area buildings in accordance with Section 507.

2. The maximum area of a building equipped throughout with an automatic sprinkler system in accordance with Section 903.3.1.2 shall be determined by multiplying the allowable area per floor (A_a), as determined in Section 506.1 by the number of stories.

❖ This section establishes a total allowable building area (i.e., aggregate area of all floor levels, excluding basements in accordance with Section 503.1.1), in effect reducing the allowable area per floor in Table 503 for buildings exceeding three stories in height. The following examples demonstrate this point:

Two-story building: A building has an allowable area per floor (A_a), determined in accordance with Section 506.1, of 10,000 square feet (929 m^2) per floor. The total allowable area of the building in accordance with Section 506.4 is 10,000 × 2 (stories) = 20,000 square feet (1858 m^2). Each story above grade, therefore, may be 10,000 square feet (929 m^2). In addition, the building may have one or more basements of no more than 10,000 square feet (929 m^2) each.

Four-story building: A building has an allowable area per floor (A_a), determined in accordance with Section 506.1, of 10,000 square feet (929 m^2). The total allowable area of the building in accordance with Section 506.4 is 10,000 × 3 (stories) (the maximum multiplier in accordance with Section 506.4) = 30,000 square feet (2787 m^2). If each story is of uniform size, therefore, each story above grade may be 7,500 square feet (697 m^2) [30,000 square feet (2787 m^2)/4 stories]. In addition, the building may have one or more basements of no more than 10,000 square feet (929 m^2) each.

If under the provisions of this section, the allowable area per floor is not exceeded for any single floor, and the total allowable area in accordance with this section is exceeded, a higher construction type must be used or the conditions of a height or area increase must be met. The following example demonstrates this point:

Building exceeds total allowable area: A four-story Group B building of Type IIB construction is proposed. The actual area per floor is 34,000 square feet (3159 m^2), and the building has a fully open perimeter. The building does not have a fire suppression system.

Section 506.2 yields an increase of 75 percent for full open frontage for this building, resulting in an allowable area per floor of 40,250 square feet (3739 m^2) (23,000 × 1.75; see Section 506.2 for frontage increase calculation). In accordance with Section 506.4, the total allowable area of the building (aggregate of all floors above grade) is 40,250 × 3 stories (the maximum multiplier in accordance with the section) = 120,750 square feet (11 218 m^2). Spread out between four stories having equal areas, each story may have no more than 120,750/4 = 30,187 square feet (2804 m^2); therefore, the building does not comply. Four possible solutions for this problem are:

1. **Change construction type to Type IIA.** Instead of Type IIB, Type IIA could be attempted: The allowable area per floor for Type IIA construction is 37,500 × 1.75 = 65,625 square feet (6097 m^2). The total allowable area of the building (aggregate area of all floors above grade) is 65,625 × 3 (stories) (the maximum multiplier in accordance with Section 503.3) = 196,875 square feet (18 290 m^2). Spread out among four stories, each story may have no more than 196,875/4 = 49,218 square feet (4572 m^2). Since the actual area per

story is 34,000 square feet (3159 m^2), the building complies.

2. **Change construction type to Type IIIA.** Type IIIA construction, which has a greater tabular area than Type IIB construction, would also work: The allowable area per floor for Type IIIA construction is 28,500 × 1.75 = 49,875 square feet (4633 m^2). The total allowable area of the building (aggregate area of all floors above grade) is 49,875 × 3 (stories) (the maximum multiplier in accordance with Section 503.3) = 149,625 square feet (13 900 m^2). Spread out among four stories, each story will have no more than 149,625/4 = 37,406 square feet (3475 m^2). Since the actual area per story is 34,000 square feet (3159 m^2), the building complies.

3. **Provide sprinklers.** Alternatively, if the building is equipped throughout with an automatic fire suppression system, the allowable area per floor for Type IIB construction would be 23,000 × 3.75 = 86,250 square feet (8013 m^2) [represents a 200-percent increase for sprinklers in accordance with Section 506.3, plus a 75-percent increase for frontage in accordance with Section 506.2 (see commentary, Sections 506.2 and 506.3)]. The total allowable area of the building (aggregate area of all floors above grade) is 86,250 × 3 (stories) (the maximum multiplier in accordance with Section 503.3) = 258,750 square feet (24 038 m^2). Spread out among four stories, each story will have no more than 258,750/4 = 64,687 square feet (6009 m^2). Since the actual area per story is 34,000 square feet (3159 m^2), the building complies.

4. **Provide sprinklers and change construction type.** If sprinklers are provided, Type VA construction would also work: The allowable area per floor for Type VA construction is 18,000 × 3.75 = 67,500 square feet (6271 m^2). The total allowable area of the building (aggregate area of all floors above grade) is 67,500 × 3 (stories) (the maximum multiplier in accordance with Section 503.3) = 202,500 square feet (18 812 m^2). Spread out among four stories, each story will have no more than 202,500/4 = 50,625 square feet (4703 m^2). Since the actual area per story is 34,000 square feet (3159 m^2), the building complies.

This section does not require that the total allowable area be equally distributed among the stories of a building, as long as no single story exceeds the allowable area per floor (A_a) determined in accordance with Section 506.1. The following example demonstrates this point:

Building with stories of unequal area: Suppose the maximum allowable area per floor (A_a) for a building is calculated in accordance with Section 506.1 to be 40,250 square feet (3739 m^2) and the building has four stories. In accordance with this section, the total allowable area of the building is 40,250 × 3 (stories) (the maximum multiplier permitted) = 120,750 square feet (11 218 m²). As long as no single story exceeds 40,250 square feet (3739 m²), the maximum allowable area could be divided among the stories unequally. For instance, the first floor could have an area of 40,250 square feet (3739 m²) and the three upper floors could each have an area of (120,750 - 40,250)/3 = 26,833 square feet (2493 m²).

Exception 1 is necessary to establish that the use of the unlimited area provisions in Section 507 would obviate the need for compliance with Section 504. The unlimited area provisions apply to one- or two-story buildings, depending upon the use and sprinkler provisions utilized.

Exception 2 is an increase allowed for residential sprinkler systems installed in accordance with NFPA 13R. The area increases allowed by Section 506.3 are only applicable when the building is to be equipped with a sprinkler system in accordance with NFPA 13— not NFPA 13R. The three-story multiplier limit or "three-story cap" was initially evaluated based upon use of area increases resulting from the installation of an NFPA 13 system. Since buildings equipped with an NFPA 13R system are not allowed an area increase, it was determined that the "three-story cap" had an unintentional detrimental impact on residential buildings. Specifically, for a four-story residential building with an NFPA 13R sprinkler system, the provisions of Section 506.4 would reduce the area per floor and the total building area. Given that the maximum height of buildings with NFPA 13R systems is four stories, this application of an increase in allowable area was determined to be warranted.

SECTION 507
UNLIMITED AREA BUILDINGS

507.1 Nonsprinklered, one story. The area of a one-story, Group F-2 or S-2 building shall not be limited when the building is surrounded and adjoined by public ways or yards not less than 60 feet (18 288 mm) in width.

❖ By definition, single-story occupancies of Groups F-2 and S-2 are not permitted to contain significant combustible materials (see Sections 306.3 and 311.3); therefore, because the fire load of the contents is lower, the hazard is lower. No other structures may be located within the 60-foot (18 288 mm) clear open space required around the building, and the clear open space must be on the same lot dedicated for public use to preclude any reduction of this isolation of the building. The type of construction is not restricted and sprinklers are not required for unlimited area, single-story buildings of Groups F-2 and S-2, in accordance with this section.

507.2 Sprinklered, one story. The area of a one-story, Group B, F, M or S building or a one-story Group A-4 building of other than Type V construction shall not be limited when the building is provided with an automatic sprinkler system throughout in accordance with Section 903.3.1.1, and is surrounded and adjoined by public ways or yards not less than 60 feet (18 288 mm) in width.

Exceptions:

1. Buildings and structures of Type I and II construction for rack storage facilities which do not have access by the public shall not be limited in height provided that such buildings conform to the requirements of Section 507.1 and NFPA 231C.

2. The automatic sprinkler system shall not be required in areas occupied for indoor participant sports, such as tennis, skating, swimming and equestrian activities, in occupancies in Group A-4, provided that:

    2.1. Exit doors directly to the outside are provided for occupants of the participant sports areas, and

    2.2. The building is equipped with a fire alarm system with manual fire alarm boxes installed in accordance with Section 907.

❖ Because of the excellent record in controlling and preventing fires, the installation of a sprinkler system throughout single-story buildings of the listed groups permits them to be unlimited in area, as long as they are surrounded by public ways or yards not less than 60 feet (18 288 mm) in width. This also applies to two-story buildings of these groups, with the exception of Group A-4 (see Section 507.3).

The life safety hazards in buildings of these uses, because of the typical activities of the occupants and their level of awareness, are considered low enough that larger floor areas and increased fire loads can be tolerated. For Groups B, F, M and S, this section may be applied to buildings of all construction types; it is not limited to noncombustible construction only. For A-4 indoor sporting arenas, the unlimited area provision is limited to buildings of Type I, II, III and IV only, given the size of the occupant load in these types of facilities. The sprinkler system is required to be designed and installed in accordance with NFPA 13, in accordance with Section 903.3.1.1. The height of the unlimited area building cannot exceed the limits established in Table 503 and Section 504.2 for the use and type of construction. In addition, the maximum travel distance to exits is limited in accordance with Section 1015.

Exception 1 specifically permits Type I and II buildings with rack storage to be unlimited in height if they are surrounded by 60 feet (18 288 mm) of open space, do not permit access by the public and comply with NFPA 231C. That standard contains provisions for storage configuration, in-rack sprinkler coverage and system requirements that offset the hazard introduced by increased height.

Exception 2 exempts certain areas of unlimited area Group A-4 occupancies where lack of fuel loading and excessive ceiling heights would reduce the need for and effectiveness of the system. Group A-4 occupancies have indoor participant sports areas, such as tennis courts, skating rinks, swimming pools, baseball fields, basketball courts and equestrian activities, with spectator seating usually situated around the perimeter of the sports field or area. These types of indoor recreation areas often require very large, open areas with such high ceilings that the installation of an automatic sprinkler system in the immediate participant sport area would be ineffective. The potential for significant fire involvement in such an area is generally quite low because of the low fuel load; therefore, sprinkler coverage is unnecessary for the playing field in most of these buildings. These areas are, therefore, exempt from the suppression requirement of this section, provided the conditions regarding exiting and the required fire alarm system are met. When an indoor arena or sports facility is built in accordance with Section 507.1 without suppression in the participant sport area, occupants of the playing field or participant sport area must be able to exit the building directly from the playing area, without having to pass through other parts of the building. This eliminates the hazard of having to pass through higher fuel load areas such as locker rooms or concession areas. The building must also be equipped with a manual alarm system that complies with Section 907. This manual alarm system provides an additional and acceptable level of life safety in spite of the omission of sprinklers over the playing field area.

Omission of sprinkler coverage is permitted in the unlimited area building for the participant sport area only. All other areas are required to be equipped with an automatic sprinkler system. This includes all other rooms and spaces in the building, such as the spectator seating areas, locker rooms, restaurants, lounges, shops, arcades, skyboxes and storage areas.

507.3 Two story. The area of a two-story, Group B, F, M or S building shall not be limited when the building is provided with an automatic sprinkler system in accordance with Section 903.3.1.1 throughout, and is surrounded and adjoined by public ways or yards not less than 60 feet (18 288 mm) in width.

❖ The rationale for allowing unlimited area two-story buildings is the same as for one-story structures of these uses. The type of construction is not restricted. Group A is not included in the occupancies that can be two stories and unlimited in area because of the hazards associated with higher occupant loads. The exceptions to Section 507.1 are not applicable to two-story buildings (see commentary, Section 507.2).

507.4 Reduced open space. The permanent open space of 60 feet (18 288 mm) required in Sections 507.1, 507.2 and 507.3 shall be permitted to be reduced to not less than 40 feet (12 192 mm) provided the following requirements are met:

1. The reduced open space shall not be allowed for more than 75 percent of the perimeter of the building.

2. The exterior wall facing the reduced open space shall have a minimum fire-resistance rating of 3 hours.

3. Openings in the exterior wall, facing the reduced open space, shall have opening protectives with a fire-resistance rating of 3 hours.

❖ The minimum clear open space surrounding unlimited area buildings may be reduced if the exterior walls are protected with fire-resistance-rated construction and the openings are protected with rated assemblies. All three conditions of this section must be met in order to reduce the open space to a minimum of 40 feet (12 192 mm). The criteria and standards for wall assemblies and opening protection are contained in Chapter 7. The rated walls and protected openings are required only for those portions of the walls that do not face at least 60 feet (18 288 mm) of unoccupied space.

507.5 Group A-3 buildings. The area of a one-story, Group A-3 building used as a church, community hall, dance hall, exhibition hall, gymnasium, lecture hall, indoor swimming pool or tennis court of Type I or II construction shall not be limited when all of the following criteria are met:

1. The building shall not have a stage other than a platform.

2. The building shall be equipped throughout with an automatic sprinkler system in accordance with Section 903.3.1.1.

3. The assembly floor shall be located at or within 21 inches (533 mm) of street or grade level and all exits are provided with ramps complying with Section 1010.1 to the street or grade level.

4. The building shall be surrounded and adjoined by public ways or yards not less than 60 feet (18 288 mm) in width.

❖ Assembly buildings pose a greater risk in general, primarily because of their relatively large occupant loads and the density of such loads; therefore, the unlimited area provisions are more restrictive for assembly. First, the assembly occupancies included in these provisions include only A-4 (see Section 507.2) and limited types of A-3, as listed in this section. Second, these occupancies must be one story. In the case of the A-3 occupancies of the types listed, the construction classifications allowed to be unlimited area include Types I and II only, which are noncombustible structures. The types of A-3 assemblies listed—churches, dance halls, community halls, exhibition halls, gymnasiums, lecture halls, indoor swimming pools and tennis courts—are typically well-lit, open spaces, containing a low fuel load. Libraries and museums are excluded, as they generally have a higher fuel load, as are bowling alleys and pool halls which can often be poorly lit and have higher levels of activity. Remember that the dance halls included in A-3 are not to be confused with nightclubs, which are an A-2 occupancy.

The unlimited area A-3 occupancy is further restricted, including requirements related to the reduction of fire hazard, and that the building be equipped throughout with an automatic sprinkler system complying with NFPA 13; that large open spaces surround the building, as required for most unlimited area buildings; that issues related to relative ease in exiting the building in an emergency, including no stages, are addressed; and that the assembly floor be located close to grade.

507.6 High-hazard use groups. Group H-2, H-3 and H-4 fire areas shall be permitted in unlimited area buildings having occupancies in Groups F and S, in accordance with the limitations of this section. Fire areas located at the perimeter of the unlimited area building shall not exceed 10 percent of the area of the building nor the area limitations specified in Table 503 as modified by Section 506.2, based upon the percentage of the perimeter of the fire area that fronts on a street or other unoccupied space. Other fire areas shall not exceed 25 percent of the area limitations specified in Table 503. Fire-resistance-rating requirements of fire barrier assemblies shall be in accordance with Table 302.3.2.

❖ This section sets forth provisions for limited areas of Group H-2, H-3 and H-4 occupancies in unlimited area buildings. Group F facilities often inherently contain a certain amount of Group H-2, H-3 and H-4 occupancies when hazardous materials are needed for the industrial processes. Similarly, it is reasonable to provide for limited quantities of hazardous material storage in a storage occupancy.

The allowable area of the Group H-2, H-3 or H-4 occupancy is dependent on the type of construction of the building and the location of the high-hazard occupancy in the building. If the high-hazard occupancy is surrounded on all sides by an unlimited area building, it is difficult for fire department personnel to locate and access that area; therefore, Group H-2, H-3 and H-4 occupancies that are not located at the perimeter of the building are limited to 25 percent of the area limitation specified in Table 503 for the building type of construction, as shown in Figure 507.6(1).

More ready access to the high-hazard occupancy from the exterior of the building provides the fire department with an opportunity to respond more effectively to an incident. As such, Group H-2, H-3 and H-4 occupancies that are located on the perimeter of an unlimited area building are permitted to be a maximum of 10 percent of the total building area, or the maximum area permitted by Table 503 as modified by Section 506.2 for the Group H use, whichever is less.

The increase permitted in Section 506.2 is to be based on the open frontage of Group H-2, H-3 and H-4 fire areas. In Figure 507.6(2), the Group H-2 fire area is located in the corner of the Group F-1 building, so that two walls of the perimeter of the Group H-2 fire area front an unoccupied open space. In accordance with Section 506.2, the area increase due to frontage (I_f) is calculated to be 25 percent, in accordance with the figure. Applying this increase to the area from Table 503, the allowable area for the Group H-2 is $(7,000)(1.25) = 8,750$ square feet (813 m^2). This is less than 10 percent of the total floor area, so 8,250 is the maximum allowable area for the Group H-2 area.

Locating the Group H-2 fire area further away from the perimeter of the Group F-1 building, as occurs in Figure 507.6(3), qualifies the Group H-2 fire area for an even greater allowable area increase [see Figure 507.6(3)].

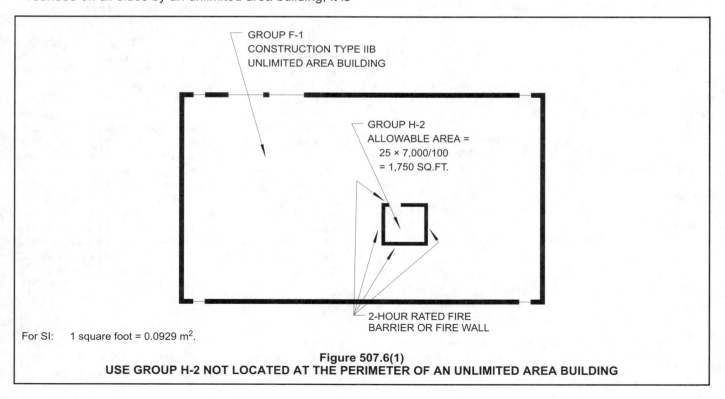

GROUP F-1
CONSTRUCTION TYPE IIB
UNLIMITED AREA BUILDING

GROUP H-2
ALLOWABLE AREA =
25 × 7,000/100
= 1,750 SQ.FT.

2-HOUR RATED FIRE
BARRIER OR FIRE WALL

For SI: 1 square foot = 0.0929 m².

Figure 507.6(1)
USE GROUP H-2 NOT LOCATED AT THE PERIMETER OF AN UNLIMITED AREA BUILDING

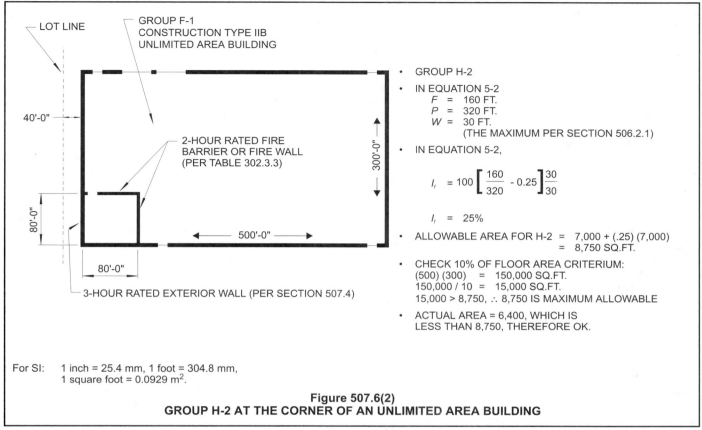

Figure 507.6(2)
GROUP H-2 AT THE CORNER OF AN UNLIMITED AREA BUILDING

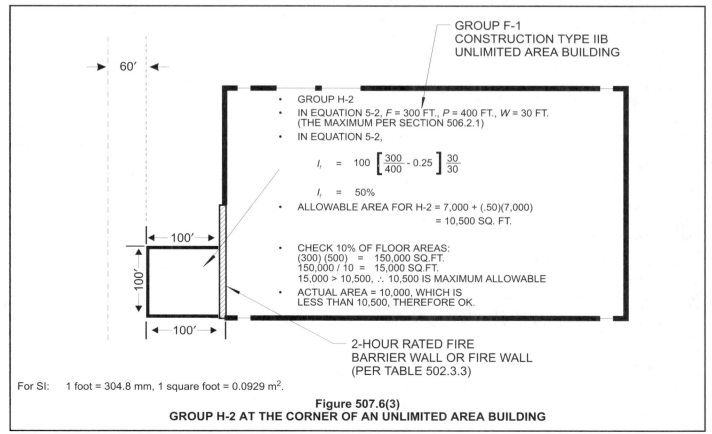

Figure 507.6(3)
GROUP H-2 AT THE CORNER OF AN UNLIMITED AREA BUILDING

507.7 Aircraft paint hangar. The area of a one-story, Group H-2 aircraft paint hangar shall not be limited where such aircraft paint hangar complies with the provisions of Section 412.4 and is entirely surrounded by public ways or yards not less in width than one and one-half times the height of the building.

❖ Because of their specialized nature and the required size of these facilities, Group H-2 aircraft paint hangars are not limited in height (see the exception to Section 504.1) and area in accordance with this section, if they meet all the requirements of Sections 412.4 through 412.4.6. These sections require that the building be of Type I or II construction, and have a fire suppression system in accordance with NFPA 409. This standard requires a foam-water suppression system in the paint areas and water sprinklers in all accessory areas. The criteria for the foam-water system depends on the size (and classification in accordance with NFPA 409) of the hangar. Other requirements of Section 412.4 include compartmentalized storage and use of flammable liquids, compliance with the IFC for spray application of flammable liquids and compliance with the *International Mechanical Code*® (IMC®) for ventilation of flammable-finish application areas (see Section 412.4 and commentary).

507.8 Group E buildings. The area of a one-story Group E building of Type II, IIIA or IV construction shall not be limited when the following criteria are met:

1. Each classroom shall have not less than two means of egress, with one of the means of egress being a direct exit to the outside of the building complying with Section 1017.

2. The building is equipped throughout with an automatic sprinkler system in accordance with Section 903.3.1.1.

3. The building is surrounded and adjoined by public ways or yards not less than 60 feet (18 288 mm) in width.

❖ This section permits one of the previously excluded occupancy groups (educational) to be built as an unlimited area building, if a direct exit to the outside from each classroom is provided in addition to another means of egress. It is clear in the wording that to fulfill this requirement, students must be able to egress the classroom directly to the outside without intervening corridors, exit passageways or exit enclosures. The unlimited area school building must also be fully sprinklered in accordance with NFPA 13, and have open frontage on all sides of at least 60 feet (18 288 mm).

507.9 Motion picture theaters. In buildings of Type I or II construction, the area of one-story motion picture theaters shall not be limited when the building is provided with an automatic sprinkler system throughout in accordance with Section 903.3.1.1 and is surrounded and adjoined by public ways or yards not less than 60 feet (18 288 mm) in width.

❖ Type I or II motion picture theaters may be unlimited in area in accordance with this section by virtue of being of

noncombustible construction, low fire loading, high roof structures and sprinkler protection. Means of egress must be in accordance with the assembly provisions of Chapter 10.

SECTION 508
SPECIAL PROVISIONS

508.1 General. The provisions in this section shall permit the use of special conditions that are exempt from, or modify, the specific requirements of this chapter regarding the allowable heights and areas of buildings based on the occupancy classification and type of construction, provided the special condition complies with the provisions specified in this section for such condition and other applicable requirements of this code.

❖ The subsections of Section 508 are exceptions to the general height and area limitations of Chapter 5. Most of the subsections deal with attached parking structures, and contain conditions by which these can be attached to buildings without creating hardship in regard to allowable area and height for the building. The conditions involve a 3-hour fire-resistance-rated separation between the garage and other parts of the building.

508.2 Group S-2 enclosed parking garage with Group A, B, M or R above. A basement and/or the first story above grade plane of a building shall be considered as a separate and distinct building for the purpose of determining area limitations, continuity of fire walls, limitation of number of stories and type of construction, when all of the following conditions are met:

1. The basement and/or the first story above grade plane is of Type IA construction and is separated from the building above with a horizontal assembly having a minimum 3-hour fire-resistance rating.

2. Shaft, stairway, ramp or escalator enclosures through the horizontal assembly shall have not less than a 2-hour fire-resistance rating with opening protectives in accordance with Table 715.3.

Exception: Where the enclosure walls below the horizontal assembly have not less than a 3-hour fire-resistance rating with opening protectives in accordance with Table 715.3, the enclosure walls extending above the horizontal assembly shall be permitted to have a 1-hour fire-resistance rating provided:

1. The building above the horizontal assembly is not required to be of Type I construction;

2. The enclosure connects less than four stories, and

3. The enclosure opening protectives above the horizontal assembly have a minimum 1-hour fire protection rating.

3. The building above the horizontal assembly contains only Group A having an assembly room with an occupant load of less than 300, or Group B, M or R.

4. The building below the horizontal assembly is a Group S-2 enclosed parking garage, used for the parking and storage of private motor vehicles.

Exceptions:

1. Entry lobbies, mechanical rooms and similar uses incidental to the operation of the building shall be permitted.

2. Group A having an assembly room with an occupant load of less than 300, or Group B or M shall be permitted in addition to those uses incidental to the operation of the building (including storage areas), provided that the entire structure below the horizontal assembly is protected throughout by an approved automatic sprinkler system.

3. The maximum building height in feet shall not exceed the limits set forth in Table 503 for the least restrictive type of construction involved.

❖ Section 508.2 describes the circumstances in which a Type IA enclosed parking structure on the basement level or first story above grade (see the definition of "Basement" in Section 502.1) could be located beneath a building of certain groups without actually being considered a part of the larger building, and therefore is not subject to all the requirements that may be applicable in a mixed-use situation. There are five conditions, which are significantly restrictive: (1) separation by a 3-hour horizontal assembly; (2) with all openings protected by a 2-hour shaft (the shaft openings having $1^1/_2$-hour protectives; see Table 715.3); (3) the use above the parking is restricted to Group A with an occupant load of less than 300, B, M or R; (4) the building below is used exclusively for parking and storage of private motor vehicles and is a Group S-2 enclosed parking garage and (5) the building does not exceed the height limitations of Table 503 for the type of construction of the upper portion of the structure (A, B, M or R).

This is one of the rare circumstances where there could be two different construction types in a single structure without being separated by a fire wall. It is possible that following the conventional provisions for mixed uses in Section 302.3 would be less restrictive than the conditions of this section. Compliance with the general provisions of Section 302.3 is permissible, and this section should be viewed as an alternative means of compliance with Section 302.3, as mentioned in that section. There are two exceptions within the five items: the exception to Item 2 indicates the conditions by which the shaft construction protecting openings in the horizontal separation may be 1-hour rated above the horizontal separation and the exceptions to Item 4 indicate the conditions by which uses other than parking can be located on the same level as the enclosed parking garage.

508.3 Group S-2 enclosed parking garage with Group S-2 open parking garage above. A Group S-2 enclosed parking garage located in the basement or first story below a Group S-2 open parking garage shall be classified as a separate and distinct

building for the purpose of determining the type of construction when the following conditions are met:

1. The allowable area of the structure shall be such that the sum of the ratios of the actual area divided by the allowable area for each separate occupancy shall not exceed 1.0.

2. The Group S-2 enclosed parking garage is of Type I or II construction and is at least equal to the fire-resistance requirements of the Group S-2 open parking garage.

3. The height and the number of the floors above the basement shall be limited as specified in Table 406.3.5.

4. The floor assembly separating the Group S-2 enclosed parking garage and Group S-2 open parking garage shall be protected as required for the floor assembly of the Group S-2 enclosed parking garage. Openings between the Group S-2 enclosed parking garage and Group S-2 open parking garage, except exit openings, shall not be required to be protected.

5. The Group S-2 enclosed parking garage is used exclusively for the parking or storage of private motor vehicles, but shall be permitted to contain an office, waiting room and toilet room having a total area of not more than 1,000 square feet (93 m²), and mechanical equipment rooms incidental to the operation of the building.

❖ Parking garages of both types, enclosed and open, are addressed in Section 406. Special height and area allowances are given in Section 406.3.5 for open parking structures; however, these special height and area provisions are not applicable if any level of the parking garage does not meet the definition of "Open parking garage" by not having the requisite clear open area to the exterior. This would normally preclude having parking levels below grade in an open parking garage.

Section 508.3 contains provisions that would allow an open parking structure to take advantage of the special height and area limits for open parking structures in Section 406.3.5 while incorporating, in the same building, enclosed parking levels below grade. This is an alternative to treating the whole building as an enclosed parking garage.

There are five conditions listed that must be met in order to use this alternative. In considering Item 1, appropriate increases in accordance with Section 506 are to be considered for each portion, and for purposes of frontage increase the same measurement of open perimeter may apply to both the upper and lower garages. Item 2 essentially requires the enclosed parking structure below to meet or exceed the fire resistance of the open parking structure above. At a minimum, the enclosed parking structure must be of Type IIB construction (all noncombustible materials, without fire-resistance-rated protection); however, if the open parking structure above is Type VA, IIIA, IIA or IA, the enclosed parking structure would also have to meet or exceed the required ratings of the building elements from Table 601 for the construction type of the open parking structure. Item 3 limits the entire height (in stories and feet) above grade to the Table 406.3.5 limits; therefore, even if the first story above grade is part of the enclosed parking garage portion of the building, the height of the structure above grade could not exceed what would be permitted by Table 406.3.5 if the entire structure above grade were part of the open parking garage portion. In regard to Item 4, the required protection of the floor assembly of the S-2 would depend on the construction type and Table 601. Item 5 permits certain accessory uses to be present when taking advantage of these alternative provisions.

508.4 Parking beneath Group R. Where a maximum one-story above grade plane Group S-2 parking garage, enclosed or open, or combination thereof, of Type I construction or open of Type IV construction, with grade entrance, is provided under a building of Group R, the number of stories to be used in determining the minimum type of construction shall be measured from the floor above such a parking area. The floor assembly between the parking garage and the Group R above shall comply with the type of construction required for the parking garage and shall also provide a fire-resistance rating not less than the mixed occupancy separation required in Section 302.3.2.

❖ This section permits an extra story (above the limits of Table 503), based on construction type, for Group R buildings with parking on the first level. There are several conditions that must be met: the parking is limited to one story above grade; must be Type IV (if open) or I (open or enclosed) construction; the entrance to the garage must be at grade; and a 2-hour horizontal assembly must be provided between the parking and the Group R occupancy in accordance with Table 302.3.3. The limitation of Table 503 for height above grade plane (in feet) is not changed under this circumstance.

For instance, a fully sprinklered Group R-2 of Type VB construction would normally be permitted to be three stories above grade (Table 503 limit of two stories plus the additional story increase for sprinklers in Section 504.2). If it meets the conditions of this section for parking on the first story above grade, then it could actually be four stories above grade; however, the building height in feet cannot exceed 60 feet (12 192 mm) above grade plane in accordance with Table 503 for Type VB construction and Section 504.2, which allows a height increase for sprinklers.

508.5 Group R-2 buildings of Type IIIA construction. The height limitation for buildings of Type IIIA construction in Group R-2 shall be increased to six stories and 75 feet (22 860 mm) where the first-floor construction above the basement has a fire-resistance rating of not less than 3 hours and the floor area is subdivided by 2-hour fire-resistance-rated fire walls into areas of not more than 3,000 square feet (279 m²).

❖ This section contains special provisions for increasing the height of Type IIIA Group R-2 buildings. The higher rating would apply to the floor structure of the first story above grade, and the fire walls subdividing the building into per-floor areas of not more than 3,000 square feet (279 m²) must extend from foundation to roof in accor-

dance with the definition of "Fire wall" in Section 702.1. In addition, the fire walls must comply with the requirements in Section 705. This section constitutes a tradeoff of building area for extra height, and depends on the materials and rating requirements for Type IIIA construction, as well as the extra fire resistance of the first floor level. Group R-2 buildings of this height are required to be sprinklered in accordance with Section 903.2.8.

508.6 Group R-2 buildings of Type IIA construction. The height limitation for buildings of Type IIA construction in Group R-2 shall be increased to nine stories and 100 feet (30 480 mm) where the building is separated by not less than 50 feet (15 240 mm) from any other building on the lot and from property lines, the exits are segregated in an area enclosed by a 2-hour fire-resistance-rated fire wall and the first-floor construction has a fire-resistance rating of not less than 1¹/₂ hours.

❖ This section contains special provisions for increasing the height of Type IIA Group R-2 buildings. The higher 1¹/₂-hour rating would apply to the floor structure of the first story above grade, and the exits are required to be enclosed with fire walls extending from foundation to roof, otherwise meeting all the requirements of Section 705. A separation distance of at least 50 feet (15 240 mm) from other buildings on the same lot or property lines is also required to use this exception. This section constitutes a tradeoff of extra protection for exits for extra building height, and depends also on the materials and rating requirements for Type IIA construction, as well as the extra fire-resistance rating protecting the first floor from the basement. The building must be sprinklered in accordance with Section 903.2.8.

508.7 Open parking garage beneath Groups A, I, B, M and R. Open parking garages constructed under Groups A, I, B, M and R shall not exceed the height and area limitations permitted under Section 406.3. The height and area of the portion of the building above the open parking garage shall not exceed the limitations in Section 503 for the upper occupancy. The height, in both feet and stories, of the portion of the building above the open parking garage shall be measured from grade plane and shall include both the open parking garage and the portion of the building above the parking garage.

❖ This section addresses a special mixed-use condition, and is another rare circumstance wherein a building can be designated with two different types of construction for determining height and area. This provision is only to be applied when an open parking structure (Group S-2) is to be constructed below a Group A, I, B, M or R occupancy (see Figure 508.7). If an open parking structure is located below any other occupancy group, Section 302.3 must be applied, as for any other mixed use condition. In accordance with Section 508.1, this section is an alternative to the general mixed-use provisions in Section 302.3 that can be applied where advantageous to the design of buildings with open parking on the lower levels.

In the application of Section 508.7, there are two criteria that must be met for code compliance (see Figure 508.7):

1. The height and area of the open parking structure comprising a part of a mixed-use group building must not exceed the limitations for open parking structures permitted in Section 406.3.5 and Table 406.3.5.

2. The allowable height of the occupancy located above the open parking structure is to be determined in accordance with Section 503 and Table 503. The height is the vertical distance (measured in feet and stories) from the grade plane to the top of the average height of the highest roof surface, in accordance with the definition of "Building height" in Section 502.1. Allowable heights and areas may be modified in accordance with Sections 504 and 506.

508.7.1 Fire separation. Fire separation assemblies between the parking occupancy and the upper occupancy shall correspond to the required fire-resistance rating prescribed in Table 302.3.2 for the uses involved. The type of construction shall apply to each occupancy individually, except that structural members, including main bracing within the open parking structure, which is necessary to support the upper occupancy, shall be protected with the more restrictive fire-resistance-rated assemblies of the groups involved as shown in Table 601. Means of egress for the upper occupancy shall conform to Chapter 10 and shall be separated from the parking occupancy by fire barriers having at least a 2-hour fire-resistance rating as required by Section 706, with self-closing doors complying with Section 715.

Means of egress from the open parking garage shall comply with Section 406.3.

❖ This section contains additional conditions for the use of Section 508.7 as an alternative to the general mixed-use provisions in Section 302.3. It contains an additional five criteria:

1. The open parking structure and occupancy located above must be separated, both horizontally and vertically if necessary, by fire separation assemblies having a fire-resistance rating corresponding to that specified in Table 302.3.3

   Example: An open parking structure, Group S-2, is located below an office (Group B); therefore, based on Table 302.3.3, all vertical and horizontal assemblies separating the two groups are required to have a minimum fire-resistance rating of 2 hours.

2. The upper and lower portions of the building may be constructed of a separate type of construction (except as discussed in criterion No. 5 below). The minimum type of construction for an open parking structure is Type IIB or IV, depending on the thresholds established in Table 406.3.5.

   Example: The open parking structure may be of Type IIB construction and the offices located above the open parking structure may be of Type IB construction.

3. Regardless of the construction types being used, all structural members, including main bracing in the

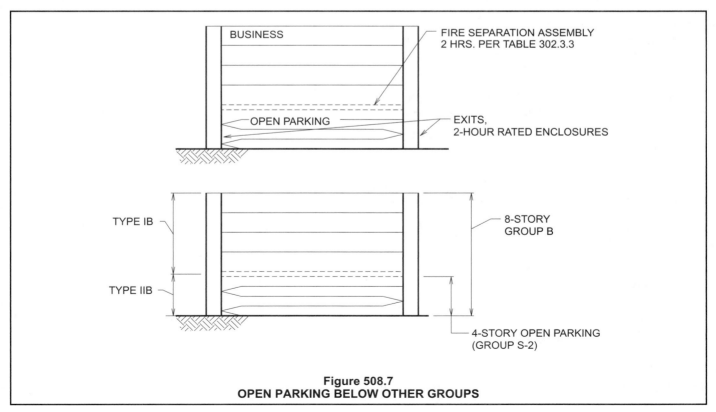

BUSINESS

FIRE SEPARATION ASSEMBLY
2 HRS. PER TABLE 302.3.3

OPEN PARKING

EXITS,
2-HOUR RATED ENCLOSURES

TYPE IB

TYPE IIB

8-STORY
GROUP B

4-STORY OPEN PARKING
(GROUP S-2)

Figure 508.7
OPEN PARKING BELOW OTHER GROUPS

open parking structure for the stability of the upper occupancy, must be rated in accordance with the most restrictive fire-resistance-rating requirement in accordance with Table 601.

Example: Consider a building where the upper occupancy is of Type IB, protected construction, and the open parking structure is of Type IIB, unprotected construction. In accordance with Table 601, all load-bearing walls and structural frames, including columns and girders necessary to support the upper occupancy, must have at least a 2-hour fire-resistance rating, in accordance with the requirements of Type IB construction. This typically applies to the columns and bracing in the entire structure, beams supporting the floor separating the upper occupancy from the open parking structure and any transfer beams located in the open parking structure. The floor of the open parking structure and any members supporting these floors, however, are permitted to have a zero fire-resistance rating in accordance with the requirements for Type IIB construction.

4. Exiting facilities within and from the upper occupancy are to conform to Chapter 10. In addition, the egress facilities from the upper occupancy must be separated from the parking area by fire-resistance-rated wall assemblies of at least 2 hours, which must be fire barriers meeting the requirements of Section 706. These egress facilities are required to maintain the 2-hour protection for their full height and must be continuous to the level of exit discharge. The fire-resistance-rating reduction of vertical exit enclosures for structures less than four stories in height (see Section 1019) is not applicable to an exit passing through the open parking structure in accordance with this section. Door openings to the vertical exit enclosure must have self-closing doors that comply with Section 715, with a fire protection rating of 1 $^1/_2$ hours in accordance with Table 715.2.

5. Exiting facilities within and from the open parking structure are to conform to Section 406.3.8, which also references Chapter 10.

It should be noted that if the building containing the upper occupancy is required to be sprinklered throughout, then the open parking structure below must also be sprinklered in accordance with Section 406.3.10. If the building is required to be sprinklered for a height or area increase in accordance with Section 504.2 or 506.3, then the entire structure, including the parking garage, must be sprinklered.

Bibliography

The following resource materials are referenced in this chapter or are relevant to the subject matter addressed in this chapter.

IFC-2003, *International Fire Code*. Falls Church, VA: International Code Council, 2003.

IMC-2003, *International Mechanical Code*. Falls Church, VA: International Code Council, 2003.

NFPA 13-99, *Installation of Sprinkler Systems*. Quincy, MA: National Fire Protection Association, 1999.

NFPA 13D-99, *Installation of Sprinkler Systems in One- and Two-Family Dwellings and Manufactured Homes*. Quincy, MA: National Fire Protection Association, 1999.

NFPA 13R-99, *Installation of Sprinkler Systems in Residential Occupancies Up to Four Stories in Height*. Quincy, MA: National Fire Protection Association, 1999.

NFPA 231C-98, *Rack Storage of Materials*. Quincy, MA: National Fire Protection Association, 1998.

NFPA 409-95, *Standard on Aircraft Hangars*. Quincy, MA: National Fire Protection Association, 1995.

Chapter 6:
Types Of Construction

General Comments

Chapter 6 contains the requirements to classify buildings into one of five types of construction. Tables 601 and 602 provide the minimum hourly fire-resistance ratings for the structural elements based on the type of construction of the building and fire separation distance. Section 602 describes each construction type in detail. Section 603 describes the use of combustible materials in buildings of noncombustible construction.

Correct classification of a building by its type of construction is essential. Many code requirements applicable to a building, such as allowable height and area (see Chapter 5), are dependent on its type of construction. If a building is placed in an incorrect construction classification (for example, one that is overly restrictive), its owner may be penalized by increased construction costs. On the other hand, when a building is incorrectly classified in an overly lenient type of construction, it will not be constructed in a manner that takes into account the relative risks associated with its size or function. The provisions of this chapter, coupled with Chapter 3 and Tables 601 and 602, establish the basis for the "equivalent risk theory" on which the entire code is based.

Purpose

The purpose of classifying buildings or structures by their type of construction is to account for the response or participation that a building's structure will have in a fire condition originating within the building as a result of its occupancy or fuel load.

The code requires every building to be classified as one of five possible types of construction: Types I, II, III, IV and V. Each type of construction denotes the kinds of materials that are to be used [i.e., noncombustible steel, concrete, masonry, combustible (wood, plastic) or heavy timber (HT)], and the minimum fire-resistance ratings that are associated with the structural elements in a building having that classification, i.e., 0, 1, $1^1/_2$, 2 or 3 hours. Type I and II construction have building elements that are noncombustible. Type III construction has noncombustible exterior walls and combustible or noncombustible interior elements. Type IV construction has noncombustible exterior walls and HT interior elements. Type V construction has building elements that are combustible. Type I, II, III and V construction are further subdivided into two categories (IA and IB, IIA and IIB, IIIA and IIIB, VA and VB).

SECTION 601
GENERAL

601.1 Scope. The provisions of this chapter shall control the classification of buildings as to type of construction.

❖ This section requires that all buildings be assigned a type of construction classification as indicated in the "General Comments" above.

SECTION 602
CONSTRUCTION CLASSIFICATION

602.1 General. Buildings and structures erected or to be erected, altered or extended in height or area shall be classified in one of the five construction types defined in Sections 602.2 through 602.5. The building elements shall have a fire-resistance rating not less than that specified in Table 601 and exterior walls shall have a fire-resistance rating not less than that specified in Table 602.

❖ This section requires that each building or structure be put into one of five possible construction classifications:

Type I, II, III, IV or V. All structural members are required to have a fire-resistance rating in accordance with Table 601. Additionally, the exterior walls of the structure must satisfy the requirements in Table 602, which bases the fire-resistance rating on the fire separation distance.

The use of multiple construction classifications in a single building is very limited and can only be done when specifically called out in the code. An example of combining types of construction is an office building of Type IIA construction located above an open parking structure of Type IIB construction, as described in Sections 508.8 and 508.7.1.

A more common example is where a single structure is divided into two compartments by using a fire wall, resulting in two separate buildings or structures—each of which may be of a different type of construction. While a structure may contain more than one building (for example, separation by a party wall), each building is to be individually assigned a type of construction.

Also, a building may have elements that comply with the requirements of more than one type of construction, in which case the building as a whole must be assigned the less restrictive type of construction. The designer may have intended, however, to comply with a higher type of construction, in which case those elements not in compliance with the intended type of construction are to be brought into compliance. Contact with the designer may be appropriate to eliminate a plan reviewer guessing as to the designer's intention, since the selection of the type of construction remains the prerogative of the permit applicant.

Section 602.1 applies to both new construction and additions. The provisions in Section 3402 on existing structures, Section 503 on general height and area limitations, Chapter 7 on fire-resistant materials and construction, and the applicable portions of the code depend on the requirements of this section.

602.1.1 Minimum requirements. A building or portion thereof shall not be required to conform to the details of a type of construction higher than that type, which meets the minimum requirements based on occupancy even though certain features of such a building actually conform to a higher type of construction.

❖ These requirements permit design flexibility by allowing various building materials and components to be used. A building must, as a minimum, meet all of the requirements of a given type to be classified as such, even though portions of that building meet the criteria of a higher construction type (i.e., greater fire-resistance ratings). This is consistent with the concept that the code is a minimum requirement. For example, a building classified as Type III construction is not prohibited from having construction that is superior, but it could not be reclassified into a higher type of construction unless it met all of the requirements for that construction type. In a normal situation, the design professional has identified the construction classification on the drawings. When this assignment has not been made, the code official is placed in a position of verifying the designer's intent and selecting the least-restrictive type that will meet all of the code requirements.

602.2 Types I and II. Type I and II construction are those types of construction in which the building elements listed in Table 601 are of noncombustible materials.

❖ Buildings of Type I and II construction are required to be constructed of noncombustible materials (see Section 703.4), and, therefore, are frequently referred to as "noncombustible construction." All Type I and II structural members have a fire-resistance rating as required by Tables 601 and 602. A typical example of a building of Type IA, IB or IIA construction would be a high-rise structure or a very large low-rise structure. These buildings are permitted to be relatively large in height and area due to the fire resistance afforded the structure's components. The structural members of a building of Type IIB construction do not have the same fire resistance as structural members in a building of Type IA, IB or IIA construction. As such, the height and area requirements are not as large (see Figure 602.2 for an example of Type I or II construction).

Type I and II construction are divided into four subclassifications: Types IA, IB, IIA and IIB. The difference among the four subclassifications is the degree of fire-resistance rating required for similar elements and

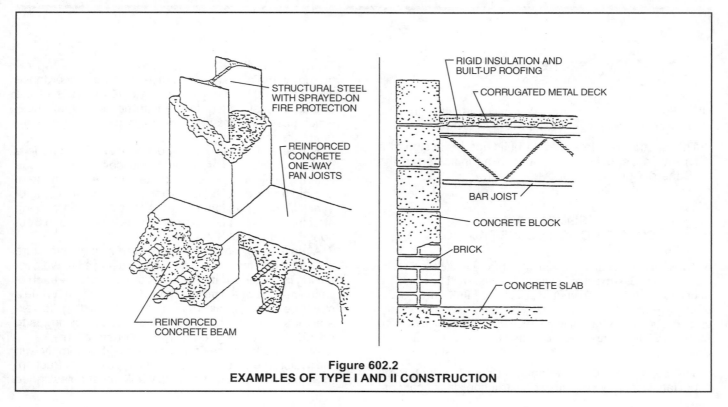

Figure 602.2
EXAMPLES OF TYPE I AND II CONSTRUCTION

assemblies. For example, the required rating for structural frame members in Type IA construction is 3 hours, for Type IB is 2 hours, for Type IIA is 1 hour and for Type IIB is 0 hours. The required fire-resistance ratings of structural elements range from zero for Type IIB construction to 3 hours for most of the elements of Type IA construction. Often, the fire-resistance ratings required by Tables 601 and 602 for structural elements are achieved by "fireproofing" structural members. Fireproofing is typically the process of creating a fire-resistance-rated assembly that incorporates the structural member by encapsulating it, either by boxing it in or by spraying on a coating to achieve the required fire-resistance ratings. It should be noted that when a protective covering is used to provide the fire-resistance rating, it must be a noncombustible material, except as indicated in Section 603.1, Item 18 and in Section 703.2.2. Fire-retardant-treated wood (FRTW), although combustible, is permitted in limited uses in buildings of Type I and II construction (see Section 603 and Table 601, Note C). While FRTW is permitted in certain applications in buildings of Type I and II construction, it is not assumed to be fire-resistance rated, and generally does not afford any higher fire-resistance rating than untreated wood material.

602.3 Type III. Type III construction is that type of construction in which the exterior walls are of noncombustible materials and the interior building elements are of any material permitted by this code. Fire-retardant-treated wood framing complying with Section 2303.2 shall be permitted within exterior wall assemblies of a 2-hour rating or less.

❖ Buildings of Type III construction are made with both combustible and noncombustible materials. The exterior walls are required to be noncombustible with load-bearing exterior walls required to have a minimum 2-hour fire-resistance rating. Exterior nonload-bearing walls are not required by Table 601 to have a fire-resistance rating, but must comply with the provisions of Table 602. The interior elements (i.e., floors, roofs and walls) are permitted to be of combustible materials. An example of a typical building of Type III construction is a structure having its exterior walls constructed of concrete, masonry or other approved noncombustible materials, but with a wood-frame floor and roof construction (see Figure 602.3 for an example of Type III construction). Type III construction is further divided into two subclassifications: Types IIIA and IIIB. An example of a building of Type IIIA construction is one in which the interior load-bearing walls, floors, roofs [those members that are less than 20 feet (6096 mm) to the lowest member] and all structural members are protected to provide a minimum 1-hour fire-resistance rating. The structural members of a building of Type IIIB construction are not required to have a fire-resistance rating with the exception of the exterior load-bearing walls.

Although FRTW does not meet the specifications of the code as a noncombustible material, it is permitted as a substitute for noncombustible materials in exterior wall assemblies of Type III construction. While the exterior walls are permitted to be either nonload bearing or

load bearing, the required fire-resistance rating of the exterior wall must be no greater than 2 hours where FRTW is used. FRTW is required to comply with the provisions in Section 2303.2.

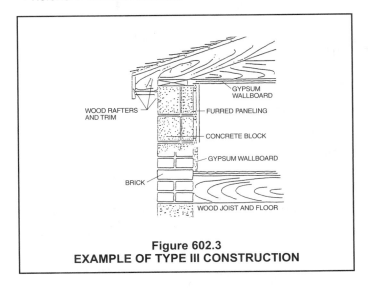

Figure 602.3
EXAMPLE OF TYPE III CONSTRUCTION

602.4 Type IV. Type IV construction (Heavy Timber, HT) is that type of construction in which the exterior walls are of noncombustible materials and the interior building elements are of solid or laminated wood without concealed spaces. The details of Type IV construction shall comply with the provisions of this section. Fire-retardant-treated wood framing complying with Section 2303.2 shall be permitted within exterior wall assemblies with a 2-hour rating or less.

❖ This section provides the general regulations for Type IV HT construction. HT construction requires the exterior walls to be constructed of noncombustible materials. The interior elements are required to be constructed of solid or laminated wood without any concealed spaces. All of the combustible structural elements are permitted to be unprotected because of the massive element sizes and the requirement that there not be any concealed spaces, such as soffits, plenums or suspended ceilings. Sections 602.4.1 through 602.4.7 provide specific requirements for the connection of structural members and minimum dimensions. An examination of Table 503 indicates that the allowable height and area for Type IV construction is greater than that permitted for buildings of Type IIB construction. This distinction is based on testing that demonstrated that HT structural members perform better structurally under fire conditions than comparable unprotected steel structural members because of charring, which insulates the wood mass.

As with Type III construction, FRTW is permitted as a substitute for noncombustible materials in exterior wall assemblies of Type IV construction. While the exterior walls are permitted to be either nonload bearing or load bearing, they must have a fire-resistance rating of no greater than 2 hours where FRTW is used in the exterior wall. FRTW is required to comply with the provisions of

Section 2303.2.

It should be noted that the sizes of specified timber members are nominal. Actual sizes for sawn timbers will generally be $^{1}/_{2}$ inch (12.7 mm) to $^{3}/_{4}$ inch (19.1 mm) less than nominal in both width and thickness as required by the U.S. Department of Commerce Standard PS 20-99. Due to requirements for additional finishing (sanding), glued laminated sizes will be less, as shown in Figure 602.4.

Nominal (inches)	3	4	6	8	10	12	14	16
Actual (inches)	$2^{1}/_{4}$	$3^{1}/_{8}$	$5^{1}/_{8}$	$6^{3}/_{4}$	$8^{3}/_{4}$	$10^{3}/_{4}$	$12^{1}/_{4}$	$14^{1}/_{4}$

For SI: 1 inch = 25.4 mm.

Figure 602.4
NOMINAL VERSUS ACTUAL DIMENSIONS FOR GLUED-LAMINATED MEMBERS

602.4.1 Columns. Wood columns shall be sawn or glued laminated and shall not be less than 8 inches (203 mm), nominal, in any dimension where supporting floor loads and not less than 6 inches (152 mm) nominal in width and not less than 8 inches (203 mm) nominal in depth where supporting roof and ceiling loads only. Columns shall be continuous or superimposed and connected in an approved manner.

❖ Minimum construction requirements and dimensions for timber columns are provided in this section. Columns are required to be a minimum of 8 inches (203 mm) nominal in any dimension if they support floor loads, or a minimum of 6 by 8 inches (152 by 203 mm) nominal if they support a roof and ceiling. Timber columns are required to be continuous or superimposed, positioned on or over each other, through floors for the entire height of the building. The design engineer or architect must provide details of all column connections. As with all structural members, each column must also be adequately fastened to other structural members in order to withstand the loads that will be placed upon the column. Some typical examples include reinforced concrete or metal caps, steel or iron column caps and timber splice plates [see Figures 602.4.1(1) and 602.4.1(2)].

602.4.2 Floor framing. Wood beams and girders shall be of sawn or glued-laminated timber and shall be not less than 6 inches (152 mm) nominal in width and not less than 10 inches (254 mm) nominal in depth. Framed sawn or glued-laminated timber arches, which spring from the floor line and support floor loads, shall be not less than 8 inches (203 mm) nominal in any dimension. Framed timber trusses supporting floor loads shall have members of not less than 8 inches (203 mm) nominal in any dimension.

❖ Minimum construction requirements and dimensions for floor framing are provided in this section. Girders are the principal horizontal structural members that support columns or beams. Beams are the structural members that support a floor or roof. Both girders and beams are required to be a minimum 6 inches (152 mm) wide and 10 inches (254 mm) deep. Both framed timber trusses supporting floor loads and framed sawn or glued laminated timber arches that spring from the floor line and support floor loads are required to be at least 8 inches (203 mm) in any dimension.

602.4.3 Roof framing. Wood-frame or glued-laminated arches for roof construction, which spring from the floor line or from grade and do not support floor loads, shall have members not less than 6 inches (152 mm) nominal in width and have less than 8 inches (203 mm) nominal in depth for the lower half of the height and not less than 6 inches (152 mm) nominal in depth for the upper half. Framed or glued-laminated arches for roof construction that spring from the top of walls or wall abutments, framed timber trusses and other roof framing, which do not support floor loads, shall have members not less than 4 inches (102 mm) nominal in width and not less than 6 inches (152 mm) nominal in depth. Spaced members shall be permitted to be composed of two or more pieces not less than 3 inches (76 mm) nominal in thickness where blocked solidly throughout their intervening spaces or where spaces are tightly closed by a continuous wood cover plate of not less than 2 inches (51 mm) nominal in thickness secured to the underside of the members. Splice plates shall be not less than 3 inches (76 mm) nominal in thickness. Where protected by approved automatic sprinklers under the roof deck, framing members shall be not less than 3 inches (76 mm) nominal in width.

❖ Minimum construction requirements and dimensions for arches and other types of roof framing are provided in this section. Other types of roof framing included in this section are heavy timber trusses with spaced members. When the members of a heavy timber truss are split and placed on either side of a main member, such as a web connecting a chord, each component of the web must be 3 inches (76 mm) or more in nominal thickness. The space between the two web members must be protected with a 2-inch-thick (51 mm) cover plate [see Figure 602.4.3(1)], or solidly filled with blocking [see Figure 602.4.3(2)]. The size of the roof framing members is dependent on the configuration used and is regulated by this section.

If a building of Type IV construction is equipped with approved automatic sprinklers under the roof deck, the minimum size of the roof framing members is reduced to 3 inches (76 mm). Roof framing members of a smaller size will have a lower resistance to fire than the 6-inch by 8-inch (152 mm by 203 mm) or 4-inch by 6-inch (102 mm by 152 mm) members required by this section. The tradeoff allowing smaller roof framing members when the building is equipped with an automatic sprinkler system is consistent with the concept of maintaining "equivalent risk" for the building.

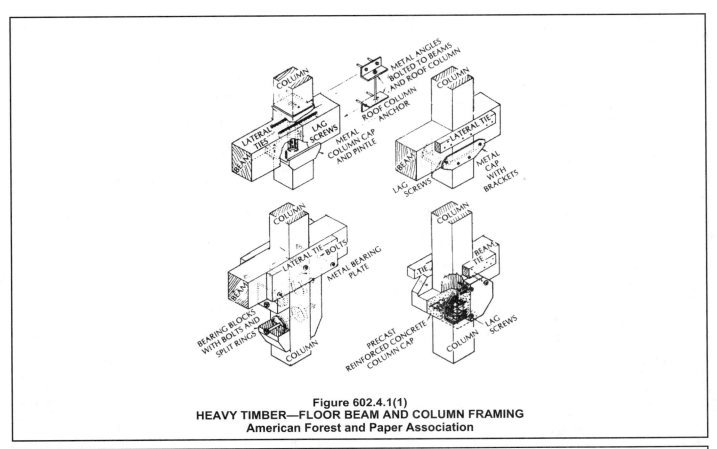

Figure 602.4.1(1)
HEAVY TIMBER—FLOOR BEAM AND COLUMN FRAMING
American Forest and Paper Association

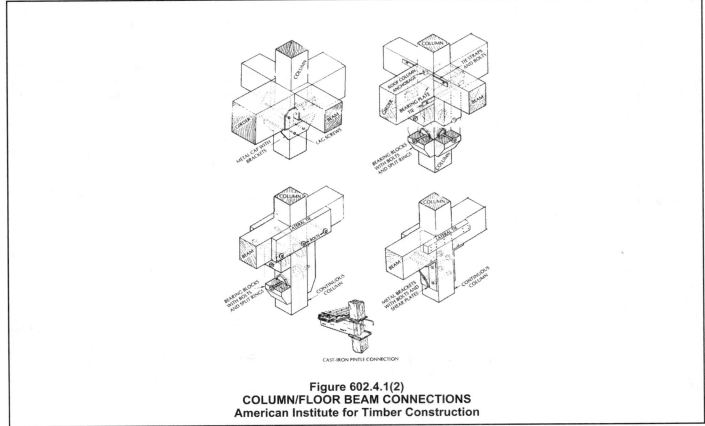

Figure 602.4.1(2)
COLUMN/FLOOR BEAM CONNECTIONS
American Institute for Timber Construction

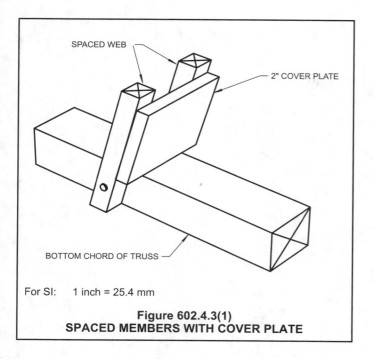

For SI: 1 inch = 25.4 mm

Figure 602.4.3(1)
SPACED MEMBERS WITH COVER PLATE

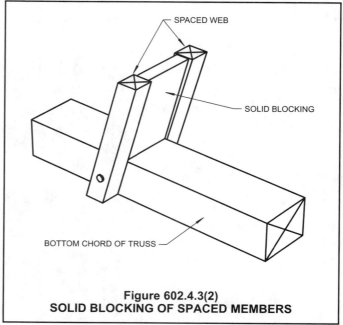

Figure 602.4.3(2)
SOLID BLOCKING OF SPACED MEMBERS

602.4.4 Floors. Floors shall be without concealed spaces. Wood floors shall be of sawn or glued-laminated planks, splined or tongue-and-groove, of not less than 3 inches (76 mm) nominal in thickness covered with 1-inch (25 mm) nominal dimension tongue-and-groove flooring, laid crosswise or diagonally, or 0.5-inch (12.7 mm) particleboard or planks not less than 4 inches (102 mm) nominal in width set on edge close together and well spiked and covered with 1-inch (25 mm) nominal dimension flooring or $^{15}/_{32}$-inch (12 mm) wood structural panel or 0.5-inch (12.7 mm) particleboard. The lumber shall be laid so that no continuous line of joints will occur except at points of support. Floors shall not extend closer than 0.5 inch (12.7 mm) to walls. Such 0.5-inch (12.7 mm) space shall be covered by a molding fastened to the wall and so arranged that it will not obstruct the swelling or shrinkage movements of the floor. Corbeling of masonry walls under the floor shall be permitted to be used in place of molding.

❖ Heavy timber flooring is required to consist of minimum 3-inch-thick (76 mm) sawn or glued-laminated planks, splined floors or tongue-and-groove floors with an overlayment of 1-inch (25 mm) tongue-and-groove flooring, laid crosswise or diagonally. Heavy timber flooring may also consist of $^1/_2$-inch (13 mm) particleboard or planks at least 4 inches (102 mm) in width set on edge and secured together, with an appropriate overlayment such as 1-inch (25 mm) hardwood flooring or a $^{15}/_{32}$-inch (12.7 mm) wood structural panel. Flooring in Type IV construction is not permitted to have concealed spaces because an undetected fire can spread quickly in combustible concealed floor spaces [see Figure 602.4.4(1)]. Because of the support afforded by adjacent members, continuous joints must only occur over supports.
Wood flooring must be fastened to supports that are perpendicular to the planking. Fastening must not be made to beams or girders that are parallel to the planks [see Figure 602.4.4(2)]. This precaution is intended to prevent separation of the planks because of differential movement of the beam relative to the girders and possible expansion/contraction due to differing moisture or humidity levels. This section requires a $^1/_2$-inch (12.7 mm) clearance between the wood flooring and exterior walls. This will prevent damage to the walls if the flooring expands due to rain during construction. This space also creates a potential flue for flames and hot gases. It should be emphasized that the integrity of the floor assembly must be maintained to provide the equivalent of a 1-hour fire-resistance rating. In addition, the ½-inch (12.7 mm) gap must be protected by a molding connected to the wall so that any possible contracting or expanding of the floor is not impeded. If masonry walls are utilized, corbeling of the masonry may be used as an alternate to the molding requirements.

602.4.5 Roofs. Roofs shall be without concealed spaces and wood roof decks shall be sawn or glued laminated, splined or tongue-and-groove plank, not less than 2 inches (51 mm) thick, $1^1/_8$-inch-thick (32 mm) wood structural panel (exterior glue), or of planks not less than 3 inches (76 mm) nominal in width, set on edge close together and laid as required for floors. Other types of decking shall be permitted to be used if providing equivalent fire resistance and structural properties.

❖ Minimum construction requirements and dimensions for roof decks are provided in this section. As required for floors, roofs are not permitted to have concealed spaces [see Figure 602.4.4(1)]. If the materials used in roof construction are different from those described in this section, the roof must have a 1-hour fire-resistance rating and be of the same structural properties.

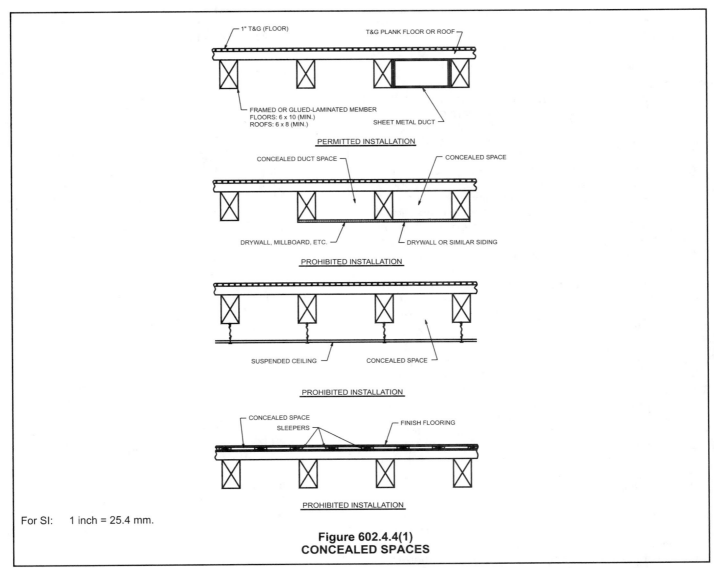

For SI: 1 inch = 25.4 mm.

**Figure 602.4.4(1)
CONCEALED SPACES**

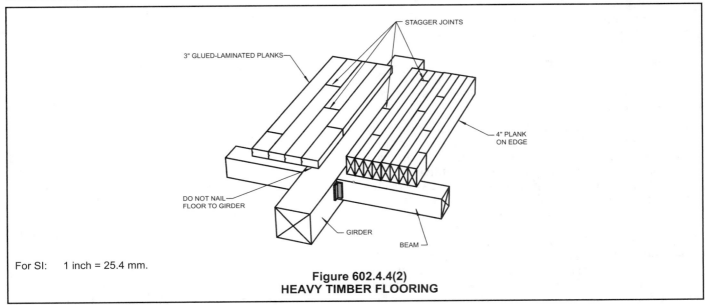

For SI: 1 inch = 25.4 mm.

**Figure 602.4.4(2)
HEAVY TIMBER FLOORING**

602.4.6 Partitions. Partitions shall be of solid wood construction formed by not less than two layers of 1-inch (25 mm) matched boards or laminated construction 4 inches (102 mm) thick, or of 1-hour fire-resistance-rated construction.

❖ Minimum construction requirements and dimensions for partitions are provided in this section. Partitions must either be formed by not less than two layers of 1-inch (25 mm) matched boards or laminated construction 4 inches (102 mm) thick if they are constructed of solid wood. Partitions are permitted to be constructed of materials other than solid wood if they have a 1-hour fire-resistance rating. An example of the use of alternative materials is when a fire-resistance rating for an exit access corridor wall is required. It is common practice to utilize a 1-hour fire-resistance-rated stud and gypsum wallboard assembly between the exposed columns to form the walls of the exit access corridor.

602.4.7 Exterior structural members. Where a horizontal separation of 20 feet (6096 mm) or more is provided, wood columns and arches conforming to heavy timber sizes shall be permitted to be used externally.

❖ Columns and arches that conform to minimum dimensional requirements may be used on the exterior if a fire separation distance of at least 20 feet (6096 mm) is maintained, although the exterior wall itself must be of noncombustible construction. If a fire separation distance of at least 20 feet (6096 mm) is maintained, the risk of exposure of the wood members to fire from an adjacent building is reduced, and the heavy timber columns and arches are permitted to be exposed to the exterior.

If a building of Type IV construction has a fire separation distance of less than 20 feet (6096 mm), the wood columns and arches are to be located on the interior side of the exterior wall. The noncombustible construction of the exterior wall will provide some degree of protection to the interior timber members; therefore, placing the wood structural members inside an exterior wall is preferable to placing them within 20 feet (6096 mm) of a lot line or adjacent building with no exposure protection.

602.5 Type V. Type V construction is that type of construction in which the structural elements, exterior walls and interior walls are of any materials permitted by this code.

❖ Type V construction allows the use of all types of materials, both noncombustible and combustible, but is most commonly constructed of dimensional lumber (see Figure 602.5 for an example of Type V construction). It is divided into two subclassifications: Types VA and VB. An example of a typical building of Type VA construction is a wood frame building in which the interior and exterior load-bearing walls, floors, roofs [those members that are less than 20 feet (6096 mm) to the lowest member] and all structural members are protected to provide a minimum 1-hour fire-resistance rating. An example of a building of Type VB construction is found in the typical single-family home, where a fire-resistance rating is not

required for the structural members. Type V construction is required to comply with Table 601 and Chapter 23.

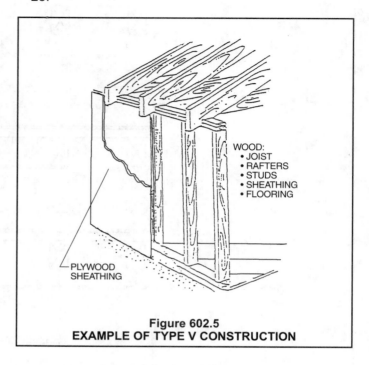

WOOD:
• JOIST
• RAFTERS
• STUDS
• SHEATHING
• FLOORING

PLYWOOD SHEATHING

**Figure 602.5
EXAMPLE OF TYPE V CONSTRUCTION**

SECTION 603
COMBUSTIBLE MATERIAL IN TYPE I AND II CONSTRUCTION

603.1 Allowable materials. Combustible materials shall be permitted in buildings of Type I or II construction in the following applications and in accordance with Sections 603.1.1 through 603.1.3:

1. Fire-retardant-treated wood shall be permitted in:

    1.1. Nonbearing partitions where the required fire-resistance rating is 2 hours or less.

    1.2. Nonbearing exterior walls where no fire rating is required.

    1.3. Roof construction as permitted in Table 601, Note c, Item 3.

2. Thermal and acoustical insulation, other than foam plastics, having a flame spread index of not more than 25.

    Exceptions:

    1. Insulation placed between two layers of noncombustible materials without an intervening airspace shall be allowed to have a flame spread index of not more than 100.

    2. Insulation installed between a finished floor and solid decking without intervening airspace shall be allowed to have a flame spread index of not more than 200.

3. Foam plastics in accordance with Chapter 26.

4. Roof coverings that have an A, B or C classification.

5. Interior floor finish and interior finish, trim and millwork such as doors, door frames, window sashes and frames.

6. Where not installed over 15 feet (4572 mm) above grade, show windows, nailing or furring strips, wooden bulkheads below show windows, their frames, aprons and show cases.

7. Finished flooring applied directly to the floor slab or to wood sleepers that are firestopped in accordance with Section 717.2.7.

8. Partitions dividing portions of stores, offices or similar places occupied by one tenant only and which do not establish a corridor serving an occupant load of 30 or more shall be permitted to be constructed of fire-retardant-treated wood, 1-hour fire-resistance-rated construction or of wood panels or similar light construction up to 6 feet (1829 mm) in height.

9. Platforms as permitted in Section 410.

10. Combustible exterior wall coverings, balconies, bay or oriel windows, or similar appendages in accordance with Chapter 14.

11. Blocking such as for handrails, millwork, cabinets, and window and door frames.

12. Light-transmitting plastics as permitted by Chapter 26.

13. Mastics and caulking materials applied to provide flexible seals between components of exterior wall construction.

14. Exterior plastic veneer installed in accordance with Section 2605.2.

15. Nailing or furring strips as permitted by Section 803.4.

16. Heavy timber as permitted by Note c, Item 2, to Table 601 and Sections 602.4.7 and 1406.3.

17. Aggregates, component materials and admixtures as permitted by Section 703.2.2.

18. Sprayed cementitious and mineral fiber fire-resistance-rated materials installed to comply with Section 1704.11.

19. Materials used to protect penetrations in fire-resistance-rated assemblies in accordance with Section 712.

20. Materials used to protect joints in fire-resistance-rated assemblies in accordance with Section 713.

21. Materials allowed in the concealed spaces of buildings of Type I and II construction in accordance with Section 717.5.

22. Materials exposed within plenums complying with Section 602 of the *International Mechanical Code.*

❖ FRTW does not meet the criteria in the code for a noncombustible material. It is, however, permitted as an alternative to noncombustible materials in specific locations in Type I and II construction. For example, the use of FRTW in walls of Type I and II construction has been limited to nonload-bearing partitions with a fire-resistance rating of no greater than 2 hours and nonload-bearing exterior walls without a fire-resistance rating. Additionally, roofs in buildings of Type I and II are permitted to be constructed of FRTW as long as the

building is no greater than two stories in height.

Other permitted uses of combustible materials in buildings of Type I and II construction include thermal and acoustical insulation having a flame spread rating of no greater than 25 when tested in accordance with ASTM E 84. Foam plastics complying with Chapter 26 are also permitted as elements of noncombustible construction.

In addition to the above, Items 4 through 22 serve as a further listing of the various code sections that regulate the use of combustible materials in noncombustible construction. These items specifically identify where the code addresses combustible materials used in noncombustible construction and are not to be considered blanket exceptions to the requirements of this chapter. All 22 options for the use of combustible materials are controlled by the sections listed and are to be used as described in those specific code sections. These items have been historically permitted and have been determined to not contribute a substantial risk within buildings.

603.1.1 Ducts. The use of nonmetallic ducts shall be permitted when installed in accordance with the limitations of the *International Mechanical Code.*

❖ The *International Mechanical Code®* (IMC®) provides requirements for nonmetallic ducts that deal with the issue of flammability, flame spread, etc. This section clarifies that Chapter 6 is not intended to override those requirements.

603.1.2 Piping. The use of combustible piping materials shall be permitted when installed in accordance with the limitations of the *International Mechanical Code* and the *International Plumbing Code.*

❖ The *International Plumbing Code®* (IPC®) provides requirements for combustible piping materials, such as plastic, that deal with the issue of flammability, flame spread, etc. This section clarifies that Chapter 6 is not intended to override those requirements.

603.1.3 Electrical. The use of electrical wiring methods with combustible insulation, tubing, raceways and related components shall be permitted when installed in accordance with the limitations of the ICC *Electrical Code.*

❖ The *ICC Electrical Code®* (ICC EC™) provides requirements for combustible wiring materials that deal with the issue of flammability, flame spread, etc. This section clarifies that Chapter 6 is not intended to override those requirements.

Table 601. See page 6-10.

❖ Table 601 has three components: the top rows list the types of construction and their subclassifications; the left column lists the various building elements that are regulated by the table and each cell in the table contains the minimum required fire-resistance rating, in hours, for the various elements based on the required type of

TABLE 601   TYPES OF CONSTRUCTION

construction of the building. Notes a through f apply as specifically referenced in the table.

The following describes the items in the left column of the table titled "Building Element":

Row 1. *Structural frame.* This category includes the structural (load-bearing) components of the building frame. To delay vertical load-carrying collapse of a building due to fire exposure for a theoretical amount of time, the structural components are required to maintain a minimum degree of fire resistance. The components listed, with the exception of Type IV construction, must also comply with Section 714, "Fire-Resistance Rating of Structural Members."

Structural members supporting a wall include such components as columns, girders, trusses, lintels and beams. As required for all structural components, the members supporting a wall are required to have a minimum degree of fire resistance to prevent collapse of both the member itself, and the wall it is supporting. Structural members supporting walls, therefore, are required to comply with Table 601 and Section 714. A structural member supporting a wall must have a fire-resistance rating that is equal to or greater than the fire-resistance rating required for the wall it is supporting.

Row 2. *Load-Bearing and Nonload-bearing walls—exterior.* Exterior walls are the outermost walls that enclose the structure. Their required fire-resistance rating is established by the highest of two fire-resistance ratings. To determine the fire-resistance rating, it must first be determined whether the exterior wall supports any structural load other than its own weight (making it load bearing), or if it supports only its own weight (making it nonload bearing). The second fire-resistance rating is based on the exterior wall's fire separation distance from lot lines. Whichever of the two requires the higher fire-resistance rating will dictate the mini-

TABLE 601
FIRE-RESISTANCE RATING REQUIREMENTS FOR BUILDING ELEMENTS (hours)

BUILDING ELEMENT	TYPE I A	TYPE I B	TYPE II A[d]	TYPE II B	TYPE III A[d]	TYPE III B	TYPE IV HT	TYPE V A[d]	TYPE V B
Structural frame[a] Including columns, girders, trusses	3[b]	2[b]	1	0	1	0	HT	1	0
Bearing walls Exterior[f] Interior	3 3[b]	2 2[b]	1 1	0 0	2 1	2 0	2 1/HT	1 1	0 0
Nonbearing walls and partitions Exterior	See Table 602								
Nonbearing walls and partitions Interior[e]	0	0	0	0	0	0	See Section 602.4.6	0	0
Floor construction Including supporting beams and joists	2	2	1	0	1	0	HT	1	0
Roof construction Including supporting beams and joists	$1^{1}/_{2}$[c]	1[c]	1[c]	0	1[c]	0	HT	1[c]	0

For SI: 1 foot = 304.8 mm.

a. The structural frame shall be considered to be the columns and the girders, beams, trusses and spandrels having direct connections to the columns and bracing members designed to carry gravity loads. The members of floor or roof panels which have no connection to the columns shall be considered secondary members and not a part of the structural frame.

b. Roof supports: Fire-resistance ratings of structural frame and bearing walls are permitted to be reduced by 1 hour where supporting a roof only.

c. 1. Except in Factory-Industrial (F-1), Hazardous (H), Mercantile (M) and Moderate-Hazard Storage (S-1) occupancies, fire protection of structural members shall not be required, including protection of roof framing and decking where every part of the roof construction is 20 feet or more above any floor immediately below. Fire-retardant-treated wood members shall be allowed to be used for such unprotected members.

   2. In all occupancies, heavy timber shall be allowed where a 1-hour or less fire-resistance rating is required.

   3. In Type I and II construction, fire-retardant-treated wood shall be allowed in buildings including girders and trusses as part of the roof construction when the building is:
      i. Two stories or less in height;
      ii. Type II construction over two stories; or
      iii. Type I construction over two stories and the vertical distance from the upper floor to the roof is 20 feet or more.

d. An approved automatic sprinkler system in accordance with Section 903.3.1.1 shall be allowed to be substituted for 1-hour fire-resistance-rated construction, provided such system is not otherwise required by other provisions of the code or used for an allowable area increase in accordance with Section 506.3 or an allowable height increase in accordance with Section 504.2. The 1-hour substitution for the fire resistance of exterior walls shall not be permitted.

e. Not less than the fire-resistance rating required by other sections of this code.

f. Not less than the fire-resistance rating based on fire separation distance (see Table 602).

mum required fire-resistance rating of the exterior wall.

In addition to Tables 601 and 602, exterior walls must comply with Section 704 and Chapter 14. Section 1019.4 has specific fire-resistance rating requirements for exterior walls adjacent to an exit stairway.

Row 3. *Load-Bearing and Nonload-bearing walls—interior.* This category includes all nonload-bearing walls and partitions (for example, the common wall separating two offices in the same suite). These walls must comply with all of the material requirements associated with their type of construction classification. Nonload-bearing interior walls and partitions are not required to have a fire-resistance rating. Where they occur in buildings of Type I or II construction and are to be constructed of wood, the wood must be fire-retardant treated as indicated in Section 603.1. Nonload-bearing interior walls may be required to be fire-resistance rated when they separate mixed uses (see Section 302.3.2), dwelling units, tenant spaces, guestrooms and corridor walls (see Sections 706 and 708).

Additionally, this category includes the structural (load bearing) interior walls of a building. To delay vertical load-carrying collapse of a building due to fire exposure for a predetermined amount of time, the structural partitions are required to maintain a minimum degree of fire resistance. Structural frame elements supporting such walls must comply with Table 601 as well as have at least the same degree of fire resistance as the supported wall.

Row 4. *Floor construction.* Floor construction provides a natural fire compartment in a building by means of a horizontal barrier that retards the vertical passage of fire from floor to floor. In order to accomplish this, floor assemblies, including the beams and structural members supporting the floor, must comply with Table 601 and Section 711. Ceilings are included if they are part of the tested assembly.

Row 5. *Roof construction.* Proper roof construction is necessary to prevent collapse from fire as well as potential impingement on adjacent buildings. Roof construction must comply with Table 601 and Section 711. Additionally, when a portion of the roof construction is located below the height limitation, the entire structural member must be protected to meet the minimum fire-resistance rating requirements (see Figure 601). The fire-resistance rating of a column must also be continuous for the full height of the col-

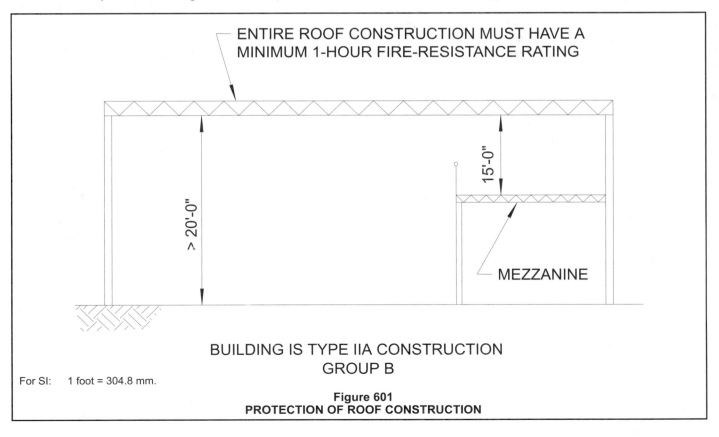

ENTIRE ROOF CONSTRUCTION MUST HAVE A MINIMUM 1-HOUR FIRE-RESISTANCE RATING

15'-0"

> 20'-0"

MEZZANINE

BUILDING IS TYPE IIA CONSTRUCTION
GROUP B

For SI: 1 foot = 304.8 mm.

Figure 601
PROTECTION OF ROOF CONSTRUCTION

TABLE 602

TYPES OF CONSTRUCTION

umn and not reduced or eliminated at a height of 20 feet (6096 mm) and above (see Note c, Item 1).

Note a clarifies the extent of the structural frame of the building or structure. Any structural item that provides direct connections to columns and bracing members that are designed to carry a gravity load is considered part of the structural frame. Secondary members (e.g., floor or roof panels without a connection to the column) are not considered part of the structural frame and as such are not required to be rated in accordance with Table 601.

Note b permits the fire-resistance ratings of structural frame and interior load-bearing walls in buildings of Type I and II construction to be reduced by 1 hour if the members are supporting only the roof.

Note c, Item 1 permits the roof construction to not have a fire-resistance rating when all structural members of the roof are at least 20 feet (6096 mm) above the floor immediately below. This alternative is applicable for all occupancy classifications except Groups F-1, H, M and S-1. In buildings of Type I and II construction, fire-retardant-treated wood may be utilized for unprotected roof members.

Note c, Item 2 permits heavy timber construction to be utilized in the roof construction as an alternative to having a fire-resistance rating of 1 hour or less.

Note c, Item 3 permits roof construction in buildings of Type II construction to include FRTW for any size building, and for Type I construction as long as the building is no greater than two stories in height or for Type I construction of any height when the vertical distance from the upper floor to the roof is 20 feet (6096 mm) or more.

Note d permits buildings of Type IIA, IIIA and VA construction to use an automatic sprinkler system in compliance with NFPA 13 as an alternative to 1-hour fire-resistance-rated construction. In order to utilize the substitution, the sprinkler system must not be required for any other reason, including for increases in allowable height and area. The sprinkler system may not be used as an alternative to the fire-resistance rating for exterior walls. This means that this exception is applicable in very limited circumstances.

Note e references Section 602.4.6, which indicates the minimum construction requirements and dimensions for nonload-bearing interior partitions in heavy timber construction.

Note f is a reminder that exterior load-bearing walls must also comply with the fire-resistance rating listed in Table 602 based on the fire separation distance. Exterior load-bearing walls must satisfy the larger of the two fire-resistance ratings.

TABLE 602. See below.

❖ Table 602 establishes the minimum fire-resistance ratings for all exterior walls. The required ratings are based on the fuel load, probable fire intensity of the various occupancy classifications and the fire separation distance (see Figure 602). In using the table, the occupancy classification of the building and the fire separation distance of the walls must be determined. Once determined, the required fire-resistance rating is obtained by referring to the appropriate row and column corresponding to these parameters. It should be noted that the fire-resistance rating requirements of Table 601, which are based on construction type, also apply. Exterior load-bearing walls must conform to the higher of the fire-resistance ratings specified in Tables 601 and 602. Exterior nonload-bearing walls need only comply with Table 602.

TABLE 602
FIRE-RESISTANCE RATING REQUIREMENTS FOR EXTERIOR WALLS BASED ON FIRE SEPARATION DISTANCE[a]

FIRE SEPARATION DISTANCE (feet)	TYPE OF CONSTRUCTION	GROUP H	GROUP F-1, M, S-1	GROUP A, B, E, F-2, I, R[b], S-2, U
< 5[c]	All	3	2	1
≥ 5	IA	3	2	1
< 10	Others	2	1	1
≥ 10	IA, IB	2	1	1
< 30	IIB, VB	1	0	0
	Others	1	1	1
≥ 30	All	0	0	0

For SI: 1 foot = 304.8 mm.

a. Load-bearing exterior walls shall also comply with the fire-resistance rating requirements of Table 601.

b. Group R-3 and Group U when used as accessory to Group R-3, as applicable in Section 101.2 shall not be required to have a fire-resistance rating where the fire separation distance is 3 feet or more.

c. See Section 503.2 for party walls.

Note b permits Group R-3 and U structures (which are accessory to Group R-3), except for detached one- and two-family dwellings and multiple single-family dwellings (townhouses) no more than three stories high with separate means of egress and their accessory structures (see Exception 1 to Section 101.2), to not have a fire-resistance rating when the fire separation distance is at least 3 feet (914 mm).

Note c requires walls that are located on property lines between adjacent buildings (i.e., zero fire separation distance) to be constructed as fire walls and be rated in accordance with Table 705.4.

Bibliography

The following resource material is referenced in this chapter or is relevant to the subject matter addressed in this chapter.

ASTM E 84-01, *Test Methods for Surface Burning Characteristics of Building Materials.* West Conshohocken, PA: ASTM International, 2001.

Basic Code Enforcement: Building Inspection. Country Club Hills, IL: Building Officials and Code Administrators International, Inc., 1993.

DOC PS 20-99, *American Softwood Lumber Standard*, United States Department of Commerce, 1999.

Fire Protection Through Modern Building Codes. Washington, DC: American Iron and Steel Institute, 1981.

NFPA 13-99, *Standard for the Installation of Sprinkler Systems.* Quincy, MA: National Fire Protection Association, 1999.

Technical Report No. 1: Comparative Fire Test on Wood and Steel Joists. Washington, DC: National Forest Products Association, 1961.

Timber Construction Manual, 2nd edition. American Institute of Timber Construction. New York: John Wiley and Sons Inc., 1974.

Wood Construction Data No. 5: Heavy Timber Construction Details. Washington, DC: National Forest Products Association, 1961.

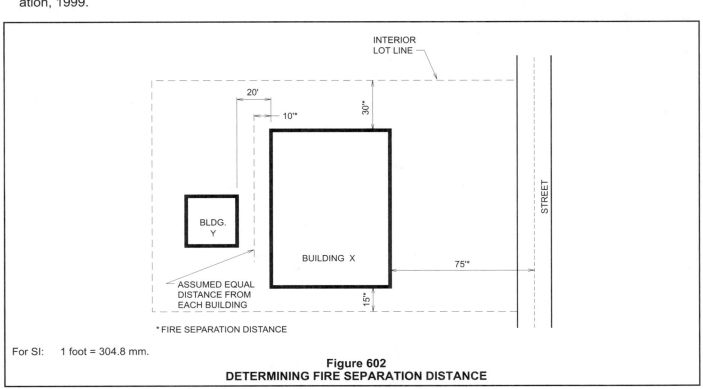

For SI: 1 foot = 304.8 mm.

Figure 602
DETERMINING FIRE SEPARATION DISTANCE

Chapter 7:
Fire-Resistance-Rated Construction

General Comments

The provisions of Chapter 7 represent fundamental concepts of fire performance that all buildings are expected to achieve. These concepts have been addressed in the codes since their inception. Other sections of the code require materials or assemblies to have a fire-resistance rating because of the relative fire hazard associated with the occupancy, type of construction and function associated with the material or assembly. The desired performance and construction of the material or assembly is then evaluated based on test standards and acceptance criteria as set forth in this chapter.

The design professional or permit applicant (see Section 106.1) is responsible for determining and providing documentation on the fire performance characteristics of a material or assembly to support an application for a permit. The construction documents are required to indicate the desired performance characteristics along with substantiating data indicating how this will be achieved. The building official must then evaluate the indicated characteristics to ascertain that compliance with the applicable code provisions has been achieved. He or she is also required to evaluate the details of the material or assembly to determine that the required performance criteria are achieved. Even though the construction documents provide the necessary information, it is critical that the field inspection verify that the assembly or material is installed in accordance with those approved documents.

For example, if a wall assembly is required to have a fire-resistance rating, the construction documents are required to indicate the fire-resistance rating intended to be provided. A detail of the wall assembly could refer to the design number of a specific assembly that has been tested and provides sufficient information to the contractor to verify construction and installation in accordance with the details of the tested assembly. The building official is required to review the design documents to determine that the designated fire-resistance rating is in compliance with the required fire-resistance rating. He or she should also determine that the detail is consistent with those of the tested assembly. Lastly, the design professional, if contracted to perform post-design services, and the building official should determine that the assembly is constructed or installed as designed and tested.

The tests or analytical procedures acceptable in determining the fire-resistance rating of assemblies and combustibility of materials are addressed in Section 703. In addition to the fire-resistance rating requirements of Chapter 7, special fire protection requirements are included in Section 603.1 for certain uses of interior partitions.

This chapter also contains provisions for protection features such as fire walls (Section 705); fire barriers (Section 706); shafts (Section 707); fire partitions (Section 708); smoke barriers (Section 709); smoke partitions (Section 710); floor/ceiling and roof/ceiling assemblies (Section 711); penetration protection (Section 712); fire-resistant joints (Section 713); fire-resistance rating of structural members (Section 714); fire opening protectives (Section 715); duct openings (Section 716) and concealed spaces—fireblocking and draftstopping (Section 717). Provisions for special features such as plaster (Section 718) and insulating materials (Section 719) are also addressed in this chapter. The chapter ends with prescriptive (Section 720) and calculation (Section 721) methods for determining fire-resistance ratings.

Purpose

This chapter identifies the acceptable techniques and methods by which proposed construction can be evaluated to determine its ability to limit the spread and effects of fire. Chapter 7 also contains general provisions, including fire performance requirements for interior and exterior walls, opening protectives, fireblocking and draftstopping. Ideally, fire growth and fire spread will be contained in the building of origin and any adjacent buildings will be protected against fire exposure. Where compartmentation is required by other sections of the code, fire spread will be restricted to the compartment (fire area) of origin.

SECTION 701
GENERAL

701.1 Scope. The provisions of this chapter shall govern the materials and assemblies used for structural fire resistance and fire-resistance-rated construction separation of adjacent spaces to safeguard against the spread of fire and smoke within a building and the spread of fire to or from buildings.

❖ As stated in Section 701.1, the provisions of Chapter 7 apply to materials and assemblies that are required to have a fire-resistance rating. The performance criteria contained in this chapter apply to fire characteristics and do not set aside other provisions in the code, such as structural load-carrying capability, durability and weather resistance.

The use of combustible materials is governed by the type of construction classification in accordance with Chapter 6. The provisions set forth in Chapter 7 may, however, depend on the combustibility of materials or mandate the use of noncombustible materials. For example, the continuity of fire walls (see Section 705.6) depends on the use of noncombustible materials in roof construction. Provisions mandating the use of noncombustible materials in fire walls are contained in Section 705.3.

Section 703.2.2 indicates that combustible aggregates and nailing strips are permitted to be used, provided that the desired fire-resistance rating is not detrimentally affected. Additional information on the use of combustible aggregates can be found in Chapter 19.

Whenever the code mandates the use of materials or assemblies required to be noncombustible or have a fire-resistance rating, the performance of the material or assembly is to be evaluated in accordance with the provisions of this chapter. The required fire-resistance ratings are based on the potential fire hazard of the occupancy in terms of fire severity and fire duration. Two sources of information on the potential fire severity associated with each occupancy classification are Tables 302.3.2 (see Section 302.3.2) and 706.3.7 (see Section 706.3.7).

SECTION 702
DEFINITIONS

702.1 Definitions. The following words and terms shall, for the purposes of this chapter, and as used elsewhere in this code, have the meanings shown herein.

❖ This section contains definitions of terms that are associated with the subject matter of this chapter. It is important to emphasize that these terms are not exclusively related to this chapter but are applicable everywhere the term is used in the code.

Definitions of terms can help in the understanding and application of the code requirements. The purpose for including these definitions within this chapter is to provide more convenient access to them without having to refer back to Chapter 2.

For convenience, these terms are also listed in Chapter 2 with a cross reference to this section.

The use and application of all defined terms, including those defined herein, are set forth in Section 202.

ANNULAR SPACE. The opening around the penetrating item.

❖ This term defines the space created between the outer surface of a penetrating item and the construction penetrated. If left unfilled, this open space provides a means of free passage of smoke, fire and products of combustion [see Figure 702.1(1) for an example of an annular space]. Sections 712.3.1, 712.4.1 and 712.4.2 contain the requirements for protection of annular spaces.

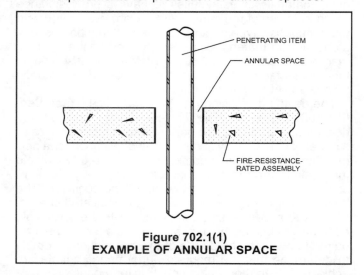

Figure 702.1(1)
EXAMPLE OF ANNULAR SPACE

CEILING RADIATION DAMPER. A listed device installed in a ceiling membrane of a fire-resistance-rated floor/ceiling or roof/ceiling assembly to limit automatically the radiative heat transfer through an air inlet/outlet opening.

❖ See the definition of "Damper."

COMBINATION FIRE/SMOKE DAMPER. A listed device installed in ducts and air transfer openings designed to close automatically upon the detection of heat and to also resist the passage of air and smoke. The device is installed to operate automatically, controlled by a smoke detection system, and where required, is capable of being positioned from a remote command station.

❖ This damper is used when the code requires not only a fire damper but also a damper designed to limit the passage of smoke through the duct penetration. Fire and smoke dampers are required at duct penetrations of shafts in accordance with Section 716.5.3.1 (see the definition of "Damper").

DAMPER. See "Ceiling radiation damper," "Combination fire/smoke damper," "Fire damper" and "Smoke damper."

❖ Fire dampers are used in heating, ventilating and air-conditioning (HVAC) duct systems that pass through fire-resistance-rated walls or floors. Dampers are provided to maintain the fire-resistance rating of the assembly being penetrated. Fire dampers are regulated by UL 555, ceiling dampers by UL 555C and smoke dampers by UL 555S.

DRAFTSTOP. A material, device or construction installed to restrict the movement of air within open spaces of concealed areas of building components such as crawl spaces, floor/ceiling assemblies, roof/ceiling assemblies and attics.

❖ Draftstopping is required in concealed combustible spaces to limit the movement of air, smoke and other products of combustion. Draftstopping materials are permitted to be combustible based on the rationale that a large enough combustible material will act as a deterrent against the movement of air and other products of combustion [see Figures 702.1(2) and 702.1(3) for typical draftstopping applications (also see Section 717)].

F RATING. The time period that the through-penetration firestop system limits the spread of fire through the penetration when tested in accordance with ASTM E 814.

❖ See the definition of "Through-penetration firestop system."

FIRE AREA. The aggregate floor area enclosed and bounded by fire walls, fire barriers, exterior walls or fire-resistance-rated horizontal assemblies of a building.

❖ This term is used to describe a specific and controlled area within a building that may consist of a portion of the floor area within a single story, one entire story or the combined floor area of several stories, depending on how these areas are enclosed and separated from other floor areas. Where a fire barrier wall with a fire-resistance rating in accordance with Section 706.3 and Table 302.3.2 or 706.3.7 divides the floor area of a one-story building, each floor area on each side of the wall constitutes a separate fire area [see Figure 702.1(4)]. If a floor/ceiling assembly separating the two stories in a two-story building is fire-resistance rated in accordance with Section 706.3.6 and Table 302.3.2 or Section 706.3.7 and Table 706.3.7, then each story is a separate fire area [see Figure 702.1(5)]. In cases where mezzanines (see Section 505) are present, the floor area of the mezzanine is included in the fire area calculations, even though the area of the mezzanine does not contribute to the building area calculations (see commentary, Section 505.1).

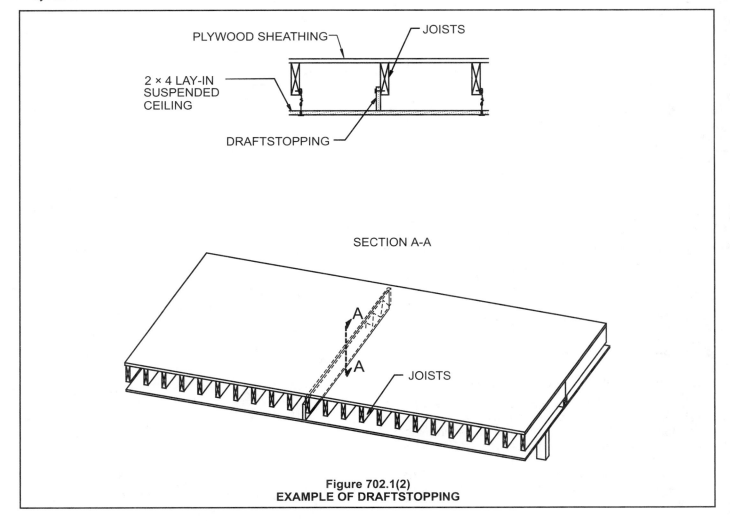

Figure 702.1(2)
EXAMPLE OF DRAFTSTOPPING

FIGURE 702.1(3) – FIGURE 702.1(4)   FIRE-RESISTANCE-RATED CONSTRUCTION

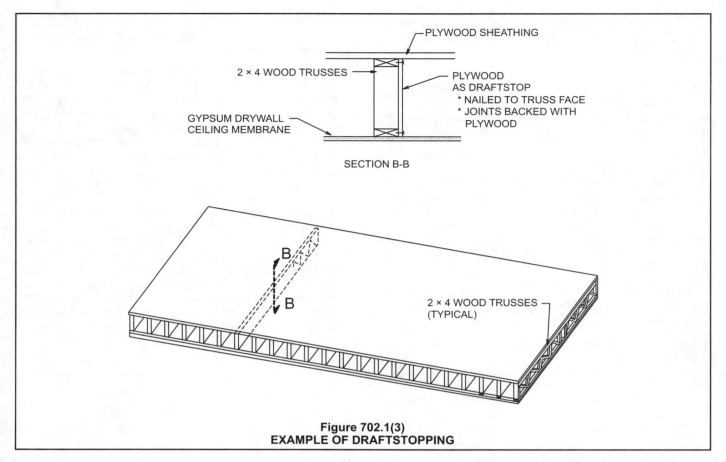

Figure 702.1(3)
EXAMPLE OF DRAFTSTOPPING

FIRE BARRIER. A fire-resistance-rated vertical or horizontal assembly of materials designed to restrict the spread of fire in which openings are protected.

❖ The term represents walls or floors with a fire-resistance rating that are constructed in accordance with Sections 706 and 711.

FIRE DAMPER. A listed device, installed in ducts and air transfer openings of an air distribution system or smoke control system, designed to close automatically upon detection of heat, to interrupt migratory airflow, and to restrict the passage of flame. Fire dampers are classified for use in either static systems that will automatically shut down in the event of a fire, or in dynamic systems that continue to operate during a fire. A dynamic fire damper is tested and rated for closure under airflow.

❖ See the definition of "Damper."

FIRE DOOR. The door component of a fire door assembly.

❖ A fire door is the primary component of a fire door assembly. The fire protection rating assigned to a tested fire door is only valid if the door is installed in a labeled frame with appropriate hardware. Where any of the required components are omitted or otherwise altered, the opening protective is not considered to provide the same fire protection rating as that of the tested assembly.

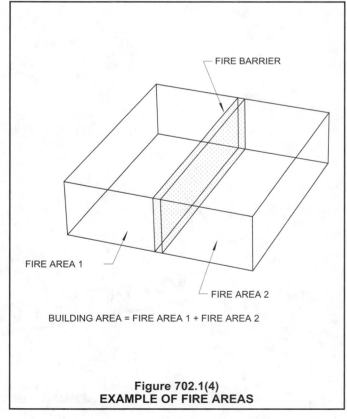

Figure 702.1(4)
EXAMPLE OF FIRE AREAS

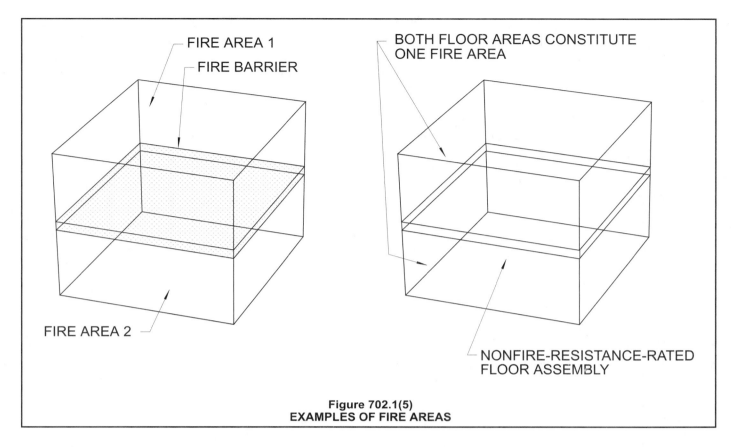

FIRE AREA 1
FIRE BARRIER
FIRE AREA 2

BOTH FLOOR AREAS CONSTITUTE ONE FIRE AREA
NONFIRE-RESISTANCE-RATED FLOOR ASSEMBLY

Figure 702.1(5)
EXAMPLES OF FIRE AREAS

FIRE DOOR ASSEMBLY. Any combination of a fire door, frame, hardware, and other accessories that together provide a specific degree of fire protection to the opening.

❖ Often called "opening protectives," fire door assemblies are required to be fire tested in accordance with NFPA 80. The test must evaluate the door assembly with all appurtenant hardware in place. In order to achieve the tested ratings, the assembly must be installed in the same manner as tested. It is important to note that doors and frames are sometimes manufactured separately; therefore, each must be labeled.

FIRE PARTITION. A vertical assembly of materials designed to restrict the spread of fire in which openings are protected.

❖ Fire partitions are used as wall assemblies to separate adjacent tenant spaces in covered mall buildings, dwelling units, guestrooms and to enclose corridors. Section 708.3 establishes the required fire-resistance ratings of fire partitions.

FIRE PROTECTION RATING. The period of time that an opening protective assembly will maintain the ability to confine a fire as determined by tests prescribed in Section 715. Ratings are stated in hours or minutes.

❖ The term "fire protection rating" applies to the fire performance of an opening protective, such as a fire door, which is determined through tests performed in accordance with NFPA 252 or UL 10C.

FIRE RESISTANCE. That property of materials or their assemblies that prevents or retards the passage of excessive heat, hot gases or flames under conditions of use.

❖ All materials offer some degree of fire resistance. A sheet of plywood has a low level of fire resistance as compared to a concrete block, which has a higher level of fire resistance. The fire resistance of a material or an assembly is evaluated by testing performed in accordance with ASTM E 119. Tested materials will be assigned a fire-resistance rating consistent with the demonstrated performance.

FIRE-RESISTANCE RATING. The period of time a building element, component or assembly maintains the ability to confine a fire, continues to perform a given structural function, or both, as determined by the tests, or the methods based on tests, prescribed in Section 703.

❖ The period of time a building element, component or assembly maintains the ability to confine a fire, continues to perform a given structural function or both as determined by tests or the methods based on tests, prescribed in Section 703.

The fire-resistance rating is developed using standardized test methods (i.e., ASTM E 119, etc). Assemblies rated under these tests are deemed to be able to perform their function for a specified period of time under specific fire conditions (standard time-temperature curve).

FIGURE 702.1(6)   **FIRE-RESISTANCE-RATED CONSTRUCTION**

FIRE-RESISTANT JOINT SYSTEM. An assemblage of specific materials or products that are designed, tested, and fire-resistance rated in accordance with either ASTM E 1966 or UL 2079 to resist for a prescribed period of time the passage of fire through joints made in or between fire-resistance-rated assemblies.

❖ In order to maintain the fire-resistant integrity of fire-resistance-rated assemblies, joints that occur within an assembly or between adjacent assemblies must be protected through an installation that has been tested in accordance with ASTM E 1966 or UL 2079. Some common examples of applications where a fire-resistant joint system would be required are expansion joints in fire-resistance-rated floors or walls and the junction between fire-resistance-rated floors and walls [see Figure 702.1(9) for examples].

FIRE SEPARATION DISTANCE. The distance measured from the building face to the closest interior lot line, to the centerline of a street, alley or public way, or to an imaginary line between two buildings on the lot. The distance shall be measured at right angles from the face of the wall.

❖ Fire separation distance is the distance from the exterior wall of the building to one of the three following property locations, measured perpendicular to the exterior wall face: an interior lot line [see Figure 702.1(6)], the centerline of a street or public way [see Figure 702.1(7)] or an imaginary line between two buildings on the same lot [see Figure 702.1(8)]. The imaginary line can be located anywhere between the two buildings; it is the designer's choice, but

once established, the location of the line applies to both buildings and cannot be revised.

The distance can vary with irregular-shaped lots and buildings, as shown in Figures 702.1(6) and 702.1(7). When applying the exterior wall requirements of Table 602, the required exterior wall fire-resistance rating might vary along a building side; for example, where the lot line is not parallel to the exterior wall.

FIRE WALL. A fire-resistance-rated wall having protected openings, which restricts the spread of fire and extends continuously from the foundation to or through the roof, with sufficient structural stability under fire conditions to allow collapse of construction on either side without collapse of the wall.

❖ In order to be considered a fire wall, a wall must be constructed of noncombustible materials (unless separating buildings of Type V construction) that are fire-resistance rated and constructed so it will remain in place if the construction on either side of it collapses. The wall is not required to remain in place if construction on both sides of it collapses (i.e., the fire wall is not required to be a free-standing cantilever wall). A fire wall is commonly used to divide a structure into separate buildings (see the definition of "Area, building" in Section 502.1) or to separate a new addition from the existing portion of a structure (see Section 3410.2.3). A fire wall also has stringent requirements for continuity and for protection of any openings through it (see Section 705 for fire wall requirements). The code does not recognize a horizontal fire wall, due to the structural stability requirements of Section 705.2.

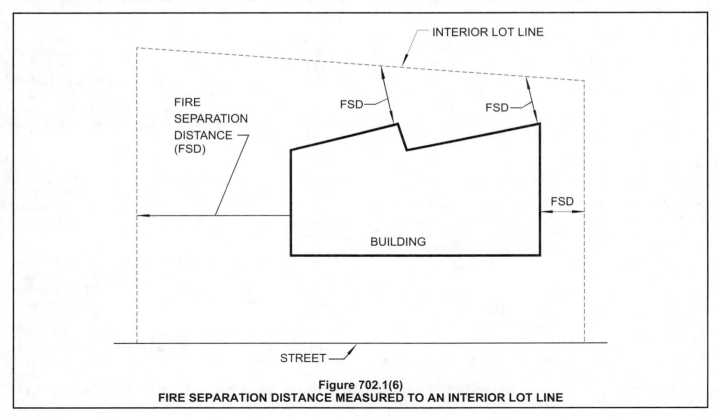

Figure 702.1(6)
FIRE SEPARATION DISTANCE MEASURED TO AN INTERIOR LOT LINE

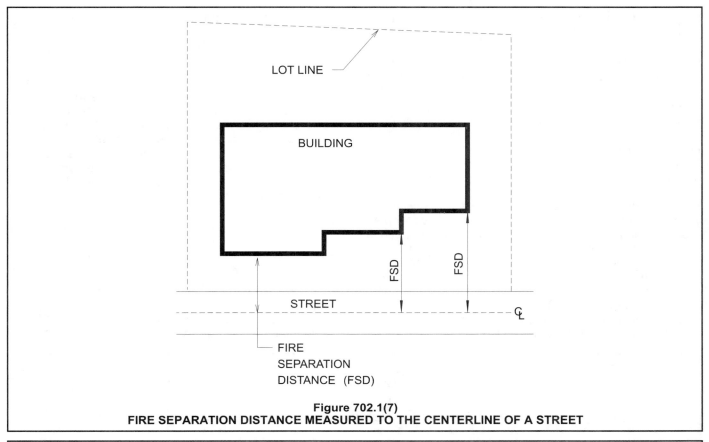

Figure 702.1(7)
FIRE SEPARATION DISTANCE MEASURED TO THE CENTERLINE OF A STREET

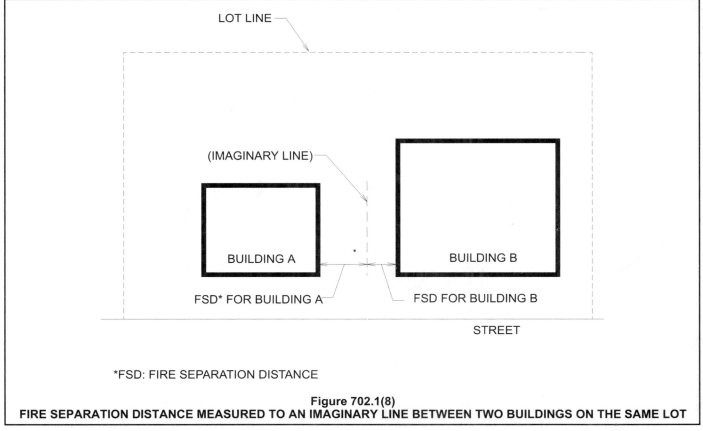

Figure 702.1(8)
FIRE SEPARATION DISTANCE MEASURED TO AN IMAGINARY LINE BETWEEN TWO BUILDINGS ON THE SAME LOT

FIGURE 702.1(9)   FIRE-RESISTANCE-RATED CONSTRUCTION

FIRE WINDOW ASSEMBLY. A window constructed and glazed to give protection against the passage of fire.

❖ Fire windows are more generally called "opening protectives" and are one type of protected opening (see Section 715.4). They are required to be tested in accordance with NFPA 257.

FIREBLOCKING. Building materials installed to resist the free passage of flame to other areas of the building through concealed spaces.

❖ Fireblocking is required to provide a barrier against the passage of flame and hot gases through concealed spaces and to seal the openings around penetrations through walls and floor/ceiling assemblies. This barrier is intended to limit the spread of flame, heat and other products of combustion through the building construction. Some fireblocking materials are permitted to be combustible based on the rationale that a substantial combustible material will provide a barrier adequate to perform the intended function (also see Section 717).

FLOOR FIRE DOOR ASSEMBLY. A combination of a fire door, a frame, hardware and other accessories installed in a horizontal plane, which together provide a specific degree of fire protection to a through opening in a fire-resistance-rated floor (see Section 712.4.6).

❖ The requirements for floor fire doors are found in Section 712.4.6. Floor fire door assemblies are required to be tested in accordance with ASTM E 119 and are used to protect openings in fire-resistance-rated floors. They are one alternative for protecting a floor opening, such as an access opening to mechanical equipment. See the commentary to Section 712.4.6 for additional information on floor fire door assemblies.

JOINT. The linear opening in or between adjacent fire-resistance-rated assemblies that is designed to allow independent movement of the building in any plane caused by thermal, seismic, wind or any other loading.

❖ This term defines the opening created when two fire-resistance-rated assemblies meet and an open space occurs between the assemblies [see Figure 702.1(9) for examples of building joints].

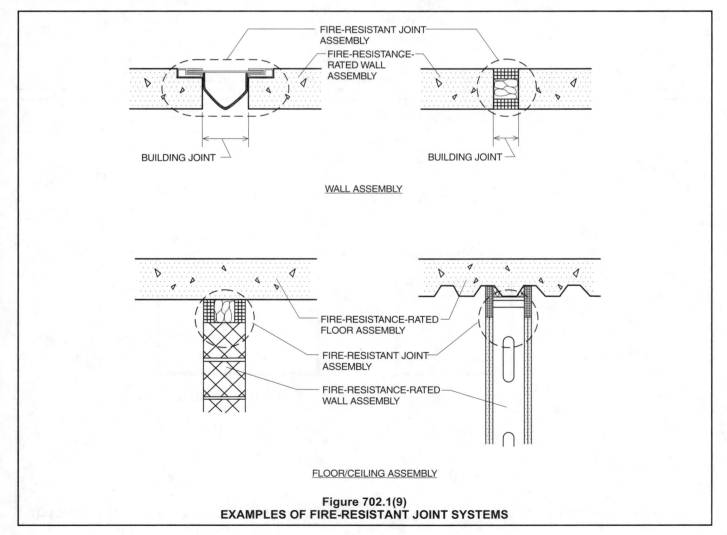

Figure 702.1(9)
EXAMPLES OF FIRE-RESISTANT JOINT SYSTEMS

MEMBRANE PENETRATION. An opening made through one side (wall, floor or ceiling membrane) of an assembly.

❖ This term refers to the situation where a penetration is made of a single layer of a fire-resistance-rated assembly. Sections 712.3.2 and 712.4.2 establish criteria where the penetration of a single membrane of an assembly is permitted while still considering the assembly to have the required fire-resistance rating [see Figure 702.1(10) for an example of a single membrane penetration].

Where a penetration is made completely through a fire-resistance-rated wall or floor/ceiling assembly, it must be protected in accordance with Section 712.3.1, 712.4.1 or 712.4.3 as applicable for the assembly penetrated.

MEMBRANE-PENETRATION FIRESTOP. A material, device or construction installed to resist for a prescribed time period the passage of flame and heat through openings in a protective membrane in order to accommodate cables, cable trays, conduit, tubing, pipes or similar items.

❖ Sections 712.3.2 and 712.4.2 both require such devices to comply as either part of a tested ASTM E 119 assembly or as a through-penetration firestop system.

PENETRATION FIRESTOP. A through-penetration firestop or a membrane-penetration firestop.

❖ See the definitions of "Membrane-penetration firestop" and "Through-penetration firestop system."

SELF-CLOSING. As applied to a fire door or other opening, means equipped with an approved device that will ensure closing after having been opened.

❖ A self-closing opening protective refers to a door that is normally in the closed position. When the door is opened and released, the self-closing feature returns the door to the closed and latched position.

The code also uses the term "automatic-closing" (see Section 715.3.7.2), although the terms "self-closing" and "automatic-closing" apply to different conditions and are not interchangeable. "Automatic-closing" refers to a fire door or fire window that is normally in the open position. These opening protectives are returned to the closed condition upon activation of fire detectors or smoke detectors or loss of power, which automatically releases the hold-open device allowing the door to close.

SHAFT. An enclosed space extending through one or more stories of a building, connecting vertical openings in successive floors, or floors and roof.

❖ Shafts are successive openings in the floors. A shaft is required to be enclosed with fire-resistance-rated assemblies to help prevent the vertical spread of fire from story to story. Stairway and elevator floor openings are examples of shafts. Provisions for vertical shafts are found in Section 707. These provisions are applicable to floor/ceiling openings required to be enclosed as specified in Sections 714.3 and 714.4.

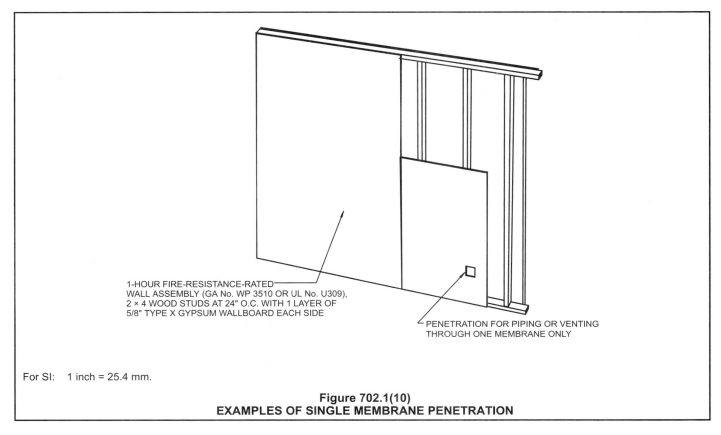

1-HOUR FIRE-RESISTANCE-RATED
WALL ASSEMBLY (GA No. WP 3510 OR UL No. U309),
2 × 4 WOOD STUDS AT 24" O.C. WITH 1 LAYER OF
5/8" TYPE X GYPSUM WALLBOARD EACH SIDE

PENETRATION FOR PIPING OR VENTING
THROUGH ONE MEMBRANE ONLY

For SI: 1 inch = 25.4 mm.

Figure 702.1(10)
EXAMPLES OF SINGLE MEMBRANE PENETRATION

FIGURE 702.1(11)   FIRE-RESISTANCE-RATED CONSTRUCTION

SHAFT ENCLOSURE. The walls or construction forming the boundaries of a shaft.

❖ See the definition of "Shaft."

SMOKE BARRIER. A continuous membrane, either vertical or horizontal, such as a wall, floor, or ceiling assembly, that is designed and constructed to restrict the movement of smoke.

❖ A smoke barrier is a fire-resistance-rated assembly that is different from a fire partition, fire barrier or fire wall. Smoke barriers are required for occupancies in Groups I-2 and I-3 and in underground structures. They include walls and floor/ceiling assemblies that are constructed with a 1-hour fire-resistance rating and are intended to create a smoke compartment. The construction requirements for a smoke barrier provide resistance to the transmission of smoke [see Figure 702.1(11) and Section 709].

SMOKE COMPARTMENT. A space within a building enclosed by smoke barriers on all sides, including the top and bottom.

❖ A smoke compartment is the space bounded on all sides by exterior walls or by 1-hour fire-resistance-rated assemblies constructed to resist the transmission of smoke. Smoke compartments are required by Sections 405.4, 407.4 and 408.6 to create spaces that protect occupants from the products of combustion produced by a fire in an adjacent smoke compartment.

SMOKE DAMPER. A listed device installed in ducts and air transfer openings that is designed to resist the passage of air and smoke. The device is installed to operate automatically, controlled by a smoke detection system, and where required, is capable of being positioned from a remote command station.

❖ See the definition of "Damper."

SPLICE. The result of a factory and/or field method of joining or connecting two or more lengths of a fire-resistant joint system into a continuous entity.

❖ In many instances, the actual length of a joint required to be protected by a fire-resistant joint system exceeds the length of the joint system, especially when a prefabricated system is used. When two or more lengths of a joint system are necessary to protect a joint, the ends of each section must be joined by a splice to maintain the fire-resistive integrity of the assembly. Splices may be either factory or field installed, but in either case, the splice must be tested in accordance with UL 2079.

T RATING. The time period that the penetration firestop system, including the penetrating item, limits the maximum temperature rise to 325°F (163°C) above its initial temperature through the penetration on the nonfire side when tested in accordance with ASTM E 814.

❖ See the definition of "Through-penetration firestop system."

THROUGH PENETRATION. An opening that passes through an entire assembly.

❖ This term is different from single membrane penetration. It defines the situation where an opening is created to accommodate an object that passes completely through an assembly from one side to the other [see Figure 702.1(12) for an example of a through penetration].

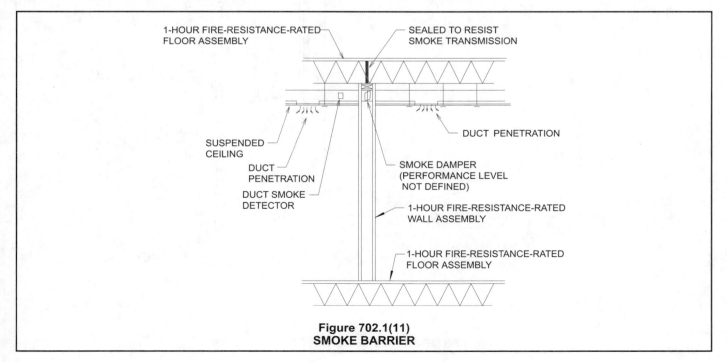

Figure 702.1(11)
SMOKE BARRIER

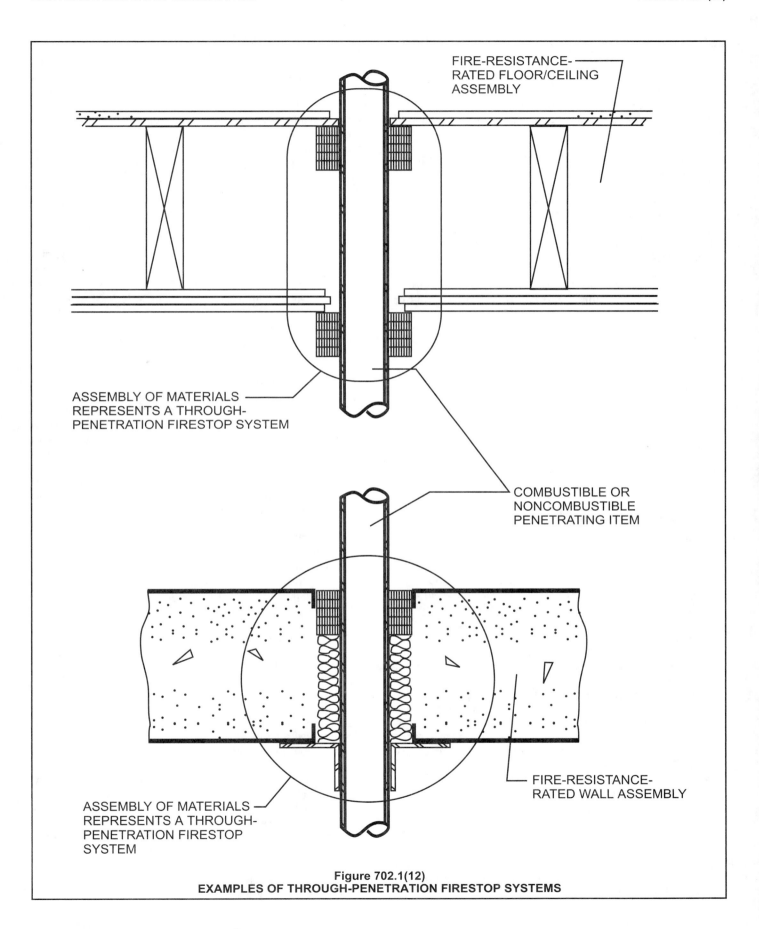

Figure 702.1(12)
EXAMPLES OF THROUGH-PENETRATION FIRESTOP SYSTEMS

THROUGH-PENETRATION FIRESTOP SYSTEM. An assemblage of specific materials or products that are designed, tested and fire-resistance rated to resist for a prescribed period of time the spread of fire through penetrations. The F and T rating criteria for penetration firestop systems shall be in accordance with ASTM E 814. See definitions of "F rating" and "T rating."

❖ One method of protection for penetrations through fire walls, fire barriers, fire partitions and fire-resistance-rated floor/ceiling assemblies is to provide a through-penetration firestop system (TPFS) (see Sections 712.3.1 and 712.4.1). Through-penetration firestop systems maintain the required protection from the spread of fire, the passage of hot gases and the transfer of heat. The protection is often provided by an intumescent material. Upon exposure to high temperatures, this material expands as much as eight to 10 times its original volume, forming a high-strength char. This is one of several types of through-penetration firestop systems available.

This definition is based on information from three sources: Section 3.1.1 of ASTM E 814; a compilation of definitions from ASTM standards and the *Fire Resistance Directory* by Underwriters Laboratories Inc.

SECTION 703
FIRE-RESISTANCE RATINGS AND FIRE TESTS

703.1 Scope. Materials prescribed herein for fire resistance shall conform to the requirements of this chapter.

❖ Standard fire test methods for determining fire-resistance ratings and combustibility of materials are covered in this section. Section 703.2 addresses the acceptable testing method to be used in determining the fire-resistance ratings of building assemblies and structural elements, while Section 703.3 provides alternative methods. Section 703.4 defines methods to determine if a material is combustible or noncombustible. The flame spread rating provisions in Section 703.4.2 are used in evaluating the acceptability of composite materials. The minimum fire-resistance ratings for structural elements are specified in Table 601, based on the type of construction classification. These fire-resistance ratings are one of the elements taken into consideration when determining the allowable height and area of buildings. By using materials and construction methods that result in the required fire-resistance ratings (see commentary, Table 503), the inherent fire hazards associated with different uses are, in part, addressed. Other sections of the code also contain provisions for required fire-resistance ratings in specific applications based on the need to separate uses, processes, fire areas or building areas.

The integrity of structural elements required to have a fire-resistance rating is addressed by other sections of the code and is based, in part, on the intended use and purpose of the assembly.

703.2 Fire-resistance ratings. The fire-resistance rating of building elements shall be determined in accordance with the test procedures set forth in ASTM E 119 or in accordance with Section 703.3. Where materials, systems or devices that have not been tested as part of a fire-resistance-rated assembly are incorporated into the assembly, sufficient data shall be made available to the building official to show that the required fire-resistance rating is not reduced. Materials and methods of construction used to protect joints and penetrations in fire-resistance-rated building elements shall not reduce the required fire-resistance rating.

Exception: In determining the fire-resistance rating of exterior bearing walls, compliance with the ASTM E 119 criteria for unexposed surface temperature rise and ignition of cotton waste due to passage of flame or gases is required only for a period of time corresponding to the required fire-resistance rating of an exterior nonbearing wall with the same fire separation distance, and in a building of the same group. When the fire-resistance rating determined in accordance with this exception exceeds the fire-resistance rating determined in accordance with ASTM E 119, the fire exposure time period, water pressure, and application duration criteria for the hose stream test of ASTM E 119 shall be based upon the fire-resistance rating determined in accordance with this exception.

❖ The fire-resistance properties of materials and assemblies must be measured and specified in accordance with a common test standard. For this reason, the fire-resistance ratings of building assemblies and structural elements are required to be determined in accordance with ASTM E 119. The ASTM E 119 test method evaluates the ability of an assembly to contain a fire, to retain its structural integrity or both in terms of endurance time during the test conditions. The test standard also contains conditions for measuring heat transfer through membrane elements protecting framing or surfaces. The fire exposure is based on the standard time-temperature curve (see Figure 703.2).

The extent to which the test specimen is representative of in-place conditions is determined by the test standard and what is actually tested. For example, the test standard does not provide for the effect on fire endurance of conventional openings in the assembly, such as electrical outlets, plumbing pipes, etc., unless specifically included in the construction assembly tested. The test reports will, however, typically indicate an allowance for openings for steel electrical outlet boxes that do not exceed 16 square inches (0.0103 m^2) in area, provided that the area of such openings does not exceed 100 square inches (0.0645 m^2) for any 100 square feet (9.3 m^2) of wall area. Since, in actuality, assemblies are often penetrated by cables, conduits, ducts, pipes, vents and similar materials, such penetrations that are not part of the tested assembly must be adequately protected and sealed for the assembly to perform as required in resisting the spread of fire. This section, therefore, also mandates that adequate documentation be submitted in order for the building official to verify that the fire-resistance rating of the assembly is not compromised by the penetration. In addition, the materials and methods used to protect joints in or between fire-resistance-rated assemblies must be submitted to determine that the required fire-resistance ratings are maintained. The documentation submitted as evidence that the required fire-resistance ratings are not reduced can be in the form of specific fire tests conducted to evaluate a specific penetration method; however, care must be taken to determine that the materials and methods of the proposed penetration or joint protection correspond to the details of the actual penetration or joint protection assembly that was tested.

ASTM E 119 states that it is not intended to determine the following:

- Contribution of the tested assembly to the fire hazard;
- Measurement of the degree of control or limitation of the passage of smoke or products of combustion;
- Simulation of the fire behavior at joints between building elements, such as floor-to-wall or wall-to-wall connections; or
- Measurement of the flame spread over the surface of the tested element.

Documentation that a component or assembly has been tested to ASTM E 119 is typically in the form of a test report issued by an approved, qualified testing agency. The test report contains certain information, including the time periods of fire resistance (e.g., 1 hour, $1^1/_2$ hours, 2 hours, etc.) and other significant details of the material or assembly. In lieu of an actual copy of a test report, the building official may accept the use of a design number found in a compendium of ASTM E 119 test reports such as, but not limited to, the *Fire Resistance Design Manual* published by the Gypsum Association, the *Fire Resistance Directory* published by Underwriters Laboratories Inc. or *Certification Listings* published by Warnock Hersey. Each publication includes a detailed description of the assembly, its hourly fire-resistance rating and other pertinent details, such as specifications of materials and alternative assembly details.

These publications detail the design of columns, beams, walls, partitions, floor/ceiling and roof/ceiling assemblies that have been tested or analyzed in accordance with ASTM E 119. The designs are classified in accordance with their use and fire-resistance rating. Often, wall, partition, floor/ceiling and roof/ceiling assemblies are further classified with sound transmission class (STC) ratings. Many floor/ceiling assemblies also include impact insulation classification (IIC) ratings. The designs indicate whether an assembly was tested under load-bearing or nonload-bearing conditions, and what structural height limitations there are for nonload-bearing assemblies. Many of the publications have introductory material that provides details on the uses and limitations of the tested assemblies, such as the amount of penetrations permitted in an assembly for electrical outlet boxes.

The exception to the section applies to exterior load-bearing walls that are required to have a fire-resistance rating for structural integrity and not as a fire barrier (e.g., due to fire separation distance in accordance with Section 704). The exception is necessary because ASTM E 119 does not distinguish between structural and fire containment features of a wall. For example, assume that a concrete double-tee wall assembly was tested in accordance with ASTM E 119. The flanges of the double-tees were $1^1/_2$ inches (38 mm) thick and the load applied to the wall was 9,600 pounds per lineal foot (14 150 kg/m). Because the temperature of the unexposed surface rose an average of 250°F (121°C) in 21 minutes, the fire-resistance rating assigned to the assembly was 21 minutes. The fire exposure was continued after the unexposed surface temperature criterion was exceeded and the wall assembly continued to support the applied load for 2 hours, at which time the test was terminated. Thus, the wall should be acceptable where structural performance for 2 hours is required and prevention of fire spread from building to building is not under consideration because of the fire separation distance. The exception, therefore, allows exterior load-bearing walls to be tested after the first failure criterion of ASTM E 119 is reached, if other than failure to sustain the applied load, and allows a structural fire-resistance rating to be determined. It should be noted that the hose stream test procedure is to be based on the fire exposure time period related to the structural fire-resistance rating.

Section 704.7 also addresses the unexposed surface temperatures of exterior walls required to have a fire-resistance rating.

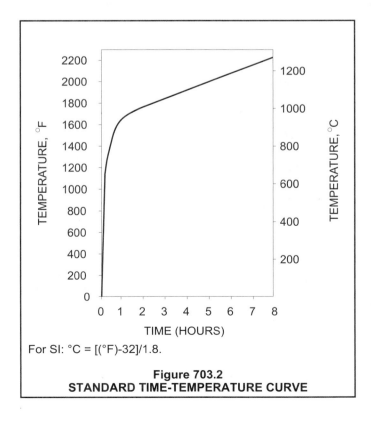

For SI: °C = [(°F)-32]/1.8.

**Figure 703.2
STANDARD TIME-TEMPERATURE CURVE**

703.2.1 Nonsymmetrical wall construction. Interior walls and partitions of nonsymmetrical construction shall be tested with both faces exposed to the furnace, and the assigned fire-resistance rating shall be the shortest duration obtained from the two tests conducted in compliance with ASTM E 119. When evidence is furnished to show that the wall was tested with the least fire-resistant side exposed to the furnace, subject to acceptance of the building official, the wall need not be subjected to tests from the opposite side (see Section 704.5 for exterior walls).

❖ In applying Sections 703.2 and 703.2.1, it is the intent of the code that the required fire-resistance rating of interior walls be based on a fire exposure from either side. For example, it is appropriate to expect the required fire performance of an exit access corridor wall from the corridor side as well as from the room side. If the corridor becomes involved in a fire, the corridor wall must restrict the spread of fire into adjacent spaces; however, Section 704.5 does allow exterior walls with a fire separation distance of more than 5 feet (1524 mm) to be fire-resistance rated from the interior only.

Section 703.2.1 permits testing from only one side of nonsymmetrical assemblies, provided it can be demonstrated that the tested side results in the lower value (i.e., the nontested side, if tested, would result in a higher value).

703.2.2 Combustible components. Combustible aggregates are permitted in gypsum and portland cement concrete mixtures approved for fire-resistance-rated construction. Any approved component material or admixture is permitted in assemblies if the resulting tested assembly meets the fire-resistance test requirements of this code.

❖ The combustibility of materials is regulated by the code when it has been determined that the material or assembly could significantly contribute to a fire hazard; however, the limited use of combustible materials, such as combustible aggregates, is permitted as long as the desired fire-resistance rating can still be obtained. The use of a combustible aggregate in the material or assembly will not significantly contribute to fire growth or severity.

703.2.3 Restrained classification. Fire-resistance-rated assemblies tested under ASTM E 119 shall not be considered to be restrained unless evidence satisfactory to the building official is furnished by the registered design professional showing that the construction qualifies for a restrained classification in accordance with ASTM E 119. Restrained construction shall be identified on the plans.

❖ The effect on the fire-resistance rating of a structural element that is part of an assembly is addressed, to an extent, through the distinction in ASTM E 119 between "restrained" and "unrestrained" conditions. The standard defines this distinction as follows: "A restrained condition in fire tests, as used in this test method, is one in which expansion of the supports of a load-carrying element resulting from the effects of the fire is resisted by forces external to the element. An unrestrained condition is one in which the load-carrying element is free to

expand and rotate at its supports." In evaluating fire-resistance ratings of structural elements, the conditions of support must be consistent with the tested condition as described in the test report.

A restrained classification yields higher fire-resistance ratings than unrestrained; therefore, the code takes the conservative approach, and thus the lesser rating by assuming the in-place conditions to be unrestrained unless structural documentation is provided that supports a restrained condition.

703.3 Alternative methods for determining fire resistance. The application of any of the alternative methods listed in this section shall be based on the fire exposure and acceptance criteria specified in ASTM E 119. The required fire resistance of a building element shall be permitted to be established by any of the following methods or procedures:

1. Fire-resistance designs documented in approved sources.

2. Prescriptive designs of fire-resistance-rated building elements as prescribed in Section 720.

3. Calculations in accordance with Section 721.

4. Engineering analysis based on a comparison of building element designs having fire-resistance ratings as determined by the test procedures set forth in ASTM E 119.

5. Alternative protection methods as allowed by Section 104.11.

❖ When based on the fire exposure and acceptance criteria of ASTM E 119, an analytical method can be used in lieu of an actual ASTM E 119 fire test to establish the fire-resistance rating of a single component or assembly.

1. Methods for determining fire-resistance ratings of concrete or masonry assemblies have been standardized in ACI 216.1/TMS 0216.1 (see Section 720.1 and commentary) as well as in the following:

   • PCI MNL 124, Precast/Prestressed Concrete Institute.

   • *Reinforced Concrete Fire Resistance*, Concrete Reinforcing Steel Institute.

   • CMIFC SR267.01B, Concrete and Masonry Industry Firesafety Committee.

   • *A Compilation of Fire Tests on Concrete Masonry Assemblies*, National Concrete Masonry Association.

   • NCMA TEK 6A, National Concrete Masonry Association.

   • NCMA 35D, National Concrete Masonry Association.

The new standard contains analytical methods for determining fire-resistance ratings of precast concrete, prestressed concrete, reinforced concrete and masonry assemblies. The methods are based on information gathered from actual fire tests of assemblies. This new section requires that assemblies that have not been tested in accordance with ASTM E 119 comply with ACI 216.1/TMS 0216.1. The commentary to Section 703.2 also identifies additional resource materials relative to test reports.

2 & 3. Sections 720 and 721 provide further methods of determining fire-resistance ratings (see the commentary to these sections).

4. An approved engineering analysis allows the fire protection properties of a component or assembly to be determined using information generated from previous fire tests through empirical calculations. There are several analytical methods available to designers for many types of construction materials. An engineering analysis available for calculating fire-resistance ratings of heavy timber construction uses, as a basis of procedure, is an empirical mathematical model that generates a conservative design for the minimum dimensions of a wood beam or column. These dimensions account for the changes that occur in the member due to charring of the wood after any given period of fire exposure. AISI's *Designing Fire Protection for Steel Columns* presents the factors that influence the fire-resistance ratings of steel columns in frequently used sizes and shapes. The publication also contains a discussion of the fire protection materials most often used with steel columns and provides accepted methods for calculating the fire-resistance rating of steel columns predicated on the ASTM E 119 fire exposure standard. Similarly, AISI's *Designing Fire Protection for Steel Beams* presents the factors that influence the fire-resistance ratings of steel beams in frequently used sizes. As typical test reports specify minimum beam sizes only, three different ways of evaluating beam substitutions are provided. Since the ASTM E 119 test furnace is incapable of testing assemblies that incorporate large structural members, AISI's *Designing Fire Protection for Steel Trusses* proposes concepts for determining fire protection designs by applying existing test data. The application of these methods, as well as information on the fire resistance of archaic construction, is summarized in the Building Officials and Code Administrators International® (BOCA®) publication, *Guidelines for Determining Fire Resistance of Building Elements*. It is important to emphasize that all analytical methods of calculating fire-resistance ratings must be substantiated as being based on the fire exposure and acceptance criteria of ASTM E 119. Upon submission of such documentation to the building official, the analytical methods mentioned herein, or any other such analytical methods, can be approved as the basis for showing compliance with the fire-resistance ratings required by the code.

5. As indicated in Section 104.11, alternative protection may be provided if approved by the building official. An example of an alternative protection method is the water-filled steel column system. The system consists of hollow, liquid-filled columns that are interconnected with pipe loops to allow the water to circulate freely. Engineering studies confirmed by tests have shown that critical temperatures are not reached in the steel columns as long as they remain filled with liquid.

703.4 Noncombustibility tests. The tests indicated in Sections 703.4.1 and 703.4.2 shall serve as criteria for acceptance of building materials as set forth in Sections 602.2, 602.3 and 602.4 in Type I, II, III and IV construction. The term "noncombustible" does not apply to the flame spread characteristics of interior finish or trim materials. A material shall not be classified as a noncombustible building construction material if it is subject to an increase in combustibility or flame spread beyond the limitations herein established through the effects of age, moisture or other atmospheric conditions.

❖ The restrictions on the use of combustible materials are primarily found in Chapter 6; however, certain sections of Chapter 7 (such as use of combustibles in Section 703.2.2 and fire walls in Section 705.3) contain restrictions on the use of combustible materials or separate provisions depending on whether the material is combustible or noncombustible. Sections 703.4.1 and 703.4.2 contain the appropriate test criteria by which a material is to be evaluated to ascertain whether it is combustible or noncombustible.

Materials that are considered noncombustible must be capable of maintaining the required performance characteristics (noncombustibility and flame spread ratings) regardless of age, moisture or other atmospheric conditions. If exposure to atmospheric conditions results in an increase in combustibility or flame spread rating beyond the limitations specified, the material is considered combustible.

The criteria established by this section are primarily based on the National Board of Fire Underwriters' (NBFU) *National Building Code*, 1955 edition, which permitted a noncombustible material to contain a limited amount of combustible material, provided that it did not contribute to fire propagation. The requirement for noncombustibility is considered to apply to certain materials used for walls, roofs and other structural elements, but not to surface finish materials. The determination of whether a material is noncombustible from the standpoint of minimum required clearances to heating appliances, flues or other sources of high temperature is based on the ASTM E 136 definition of "Noncombustibility," not the prescriptive definition in Section 703.4.2. Note that the *International Mechanical Code®* (IMC®) contains a definition of "Noncombustible materials," which does not contain provisions for the use of composite materials (defined in Section 703.4.2) as acceptable noncombustible materials.

703.4.1 Elementary materials. Materials required to be noncombustible shall be tested in accordance with ASTM E 136.

❖ Materials intended to be classified as noncombustible are to be tested in accordance with ASTM E 136. In accordance with Section 8 of ASTM E 136, such materials are acceptable as noncombustible materials when at least three of four specimens tested conform to all of the following criteria:

1. The recorded temperature of the surface and interior thermocouples shall not at any time during the test rise more than 54°F (30°C) above the furnace temperature at the beginning of the test.

2. There shall not be flaming from the specimen after the first 30 seconds.

3. If the weight loss of the specimen during testing exceeds 50 percent, the recorded temperature of the surface and interior thermocouples shall not at any time during the test rise above the furnace air temperature at the beginning of the test, and there shall not be flaming of the specimen.

The use of the test standard is limited to elementary materials and excludes laminated and coated materials because of the uncertainties associated with more complex materials and with products that cannot be tested in a realistic configuration. The test standard is also limited to solid materials and does not measure the self-heating tendencies of large masses of materials, such as resin-impregnated mineral fiber insulation.

The defined furnace temperature in the standard, 1,382°F (750°C), is representative of temperatures that are known to exist during building fires, although temperatures between 1,800°F (982°C) and 2,200°F (1204°C) are frequently achieved during intense fires. For most building materials, however, complete burning of the combustible fraction will occur as readily at 1,382°F (750°C) as compared to higher temperatures. The criterion requiring four specimens to be tested recognizes the variable nature of the measurements and the fact that there are difficulties in observing the presence and duration of flaming.

The need to measure and limit the duration of flaming and the rise in temperature recognizes that a brief period of flaming and a small amount of heating are not considered serious limitations on the use of building materials. Test results have shown that such criteria limit the combustible portion of noncombustible materials to a maximum of 3 percent. The 50-percent weight-loss limitation precludes the possibility that combustion of low-density materials will occur so rapidly that the recorded temperature rise and the measured flaming duration will be less than the prescribed limitations. The 50-percent limitation is considered appropriate for materials that contain appreciable quantities of combined water or gaseous components.

703.4.2 Composite materials. Materials having a structural base of noncombustible material as determined in accordance with Section 703.4.1 with a surfacing not more than 0.125 inch (3.18 mm) thick that has a flame spread index not greater than 50 when tested in accordance with ASTM E 84 shall be acceptable as noncombustible materials.

❖ In recognition that an essentially noncombustible material with a thin combustible coating will not contribute appreciably to an ambient fire, this section provides criteria by which a composite material may be determined acceptable as a noncombustible material. The structural base of the composite material must meet the criteria for elementary materials (see Section 703.4.1). The material may have a surface not more than $^1/_8$-inch (3.2 mm) thick applied to the noncombustible base. This surface is required to have a flame spread rating no greater than 50 when tested in accordance with ASTM E 84. For a discussion of the ASTM E 84 test method, refer to the commentary to Section 803.1.

In accordance with this section, material such as gypsum board—which consists of a noncombustible base and a combustible (paper) surface—is considered to be a composite material. As noted in the commentary to Section 703.4, a composite material is considered a noncombustible material in the context of the code. The IMC, however, intentionally includes only limited applications of the criteria of this section in the definition of a noncombustible material; therefore, composite materials are not considered acceptable noncombustible materials from the standpoint of clearances to heating appliances, flues or other sources of high temperature because of their potentially poor response to radiant heat exposure.

SECTION 704
EXTERIOR WALLS

704.1 General. Exterior walls shall be fire-resistance rated and have opening protection as required by this section.

❖ This section addresses the fire protection functions of the exterior enclosures of a building, which are:

- To prevent the spread of fire between buildings; and
- To prevent the spread of fire between stories in a single building.

In order to accomplish these functions, this section regulates the fire protection capability of the exterior wall, including limitations on openings in the walls. The requirements are based on the fire separation distance, the occupancy that represents a relative fuel loading and the percentage of openings in the wall. Section 704.9 also regulates the location of openings in order to prevent ignition of materials in adjacent stories.

All other applicable provisions of the code concerning exterior walls still apply. The provisions of this section refer to the fire-resistance requirements contained in Tables 601 and 602 (see Section 704.5). Table 601 establishes minimum fire-resistance ratings for load-bearing exterior walls based on the type of construction that addresses the structural integrity of the building under fire conditions. Table 602 establishes ratings based on fire separation distance in an effort to address conflagration concerns.

704.2 Projections. Cornices, eave overhangs, exterior balconies and similar architectural appendages extending beyond the floor area shall conform to the requirements of this section and Section 1406. Exterior egress balconies and exterior exit stairways shall also comply with Sections 1013.5 and 1022.1. Projections shall not extend beyond the distance determined by the following two methods, whichever results in the lesser projection:

1. A point one-third the distance to the lot line from an assumed vertical plane located where protected openings are required in accordance with Section 704.8.

2. More than 12 inches (305 mm) into areas where openings are prohibited.

❖ This section is intended to correlate fire protection performance requirements with other required performance characteristics of exterior walls. These attributes include combustibility of exterior finishes (Section 1406) and exterior stairways (Sections 1013.5 and 1022.1).

Section 704.2 further limits the location of combustible projections relative to fire separation distance, since the exterior wall ratings are measured to the wall itself and not the projection.

704.2.1 Type I and II construction. Projections from walls of Type I or II construction shall be of noncombustible materials or combustible materials as allowed by Sections 1406.3 and 1406.4.

❖ The code requires projections from walls of Type I and II materials to be noncombustible. An exception for combustible materials is for balconies (Section 1406.3) and bay windows (Section 1406.4). Buildings of Type I through IV construction require the exterior walls to be noncombustible; however, the code places a higher restriction on projections on the two highest types of construction, while Section 704.2.2 allows combustible projections on buildings of Type III and IV construction. This approach mirrors the general requirements that all elements of Types I and II be noncombustible, while Types III and IV are permitted to be constructed of combustible interior materials.

704.2.2 Type III, IV or V construction. Projections from walls of Type III, IV or V construction shall be of any approved material.

❖ See the commentary to Section 1406.

704.2.3 Combustible projections. Combustible projections located where openings are not permitted or where protection of openings is required shall be of at least 1-hour fire-resistance-rated construction, Type IV construction or as required by Section 1406.3.

Exception: Type V construction shall be allowed for R-3 occupancies, as applicable in Section 101.2.

❖ The provisions of Section 704.2.3 mirror those of Section 704.8, which addresses radiant heat exposure through openings. Where openings are neither permitted nor required to be protected, the code further re-

quires the combustible projections to be either 1-hour rated or of heavy timber. As the fire separation distance decreases, the potential for conflagration between buildings becomes more acute; thus, the need to limit the type of combustibles on the exterior wall.

704.3 Buildings on the same lot. For the purposes of determining the required wall and opening protection and roof-covering requirements, buildings on the same lot shall be assumed to have an imaginary line between them.

Where a new building is to be erected on the same lot as an existing building, the location of the assumed imaginary line with relation to the existing building shall be such that the exterior wall and opening protection of the existing building meet the criteria as set forth in Sections 704.5 and 704.8.

Exception: Two or more buildings on the same lot shall either be regulated as separate buildings or shall be considered as portions of one building if the aggregate area of such buildings is within the limits specified in Chapter 5 for a single building. Where the buildings contain different occupancy groups or are of different types of construction, the area shall be that allowed for the most restrictive occupancy or construction.

❖ This section addresses buildings on the same lot and requires that an imaginary line be established between buildings in order to determine exterior wall fire ratings and opening protectives (see the definition of "Fire separation distance"). This section takes the approach of limiting the conflagration hazard between buildings on the same property.

The exception permits two buildings on the same lot to be exempt from Sections 704.5 and 704.8 when considered as one building in accordance with Section 503.1.3. The provisions of Sections 704.9 and 704.10 would still apply, since a need exists to restrict fire spread between stories within a building. Although not specifically identified in the exception, Section 704.11 would not apply, since it is dependent on the exterior wall being required to have a fire-resistance rating in accordance with Section 704.5, which relates to measurement of fire separation distance.

704.4 Materials. Exterior walls shall be of materials permitted by the building type of construction.

❖ This section merely reinforces the general type of construction requirements of Chapter 6. This provision is found throughout Chapter 7 relative to fire walls (Section 705.3), fire barriers (Section 706.2), fire partitions (Section 708.2), smoke barriers (Section 709.2), smoke partitions (710.2) and horizontal assemblies (Section 711.2).

704.5 Fire-resistance ratings. Exterior walls shall be fire-resistance rated in accordance with Tables 601 and 602. The fire-resistance rating of exterior walls with a fire separation distance of greater than 5 feet (1524 mm) shall be rated for exposure to fire from the inside. The fire-resistance rating of exterior walls with a fire separation distance of 5 feet (1524 mm) or less shall be rated for exposure to fire from both sides.

❖ This section establishes the minimum fire-resistance rating that is required for exterior walls based on occupancy classification and fire separation distance. The required fire-resistance ratings of exterior walls for preventing fire spread between buildings is specified in Table 602 (see Figure 704.5 for determining fire separation distance). The more restrictive provisions of Tables 601 and 602 determine the minimum required fire-resistance rating.

The fire-resistance rating of exterior walls is based on fire exposure from the interior. This prevents radiant heat from a fire in the building from exposing an adjacent building. Additionally, if the fire separation distance is 5 feet (1524 mm) or less, the fire-resistance rating of an exterior wall is required to be based on fire exposure from both sides, due to the possibility of direct flame impingement on the exterior surface from an adjacent building. Beyond 5 feet (1524 mm), the exterior walls are less likely to ignite from exposure from adjacent buildings. The 5-foot (1524 mm) distance is based on a

realistic fire plume from an opening as reported in a Canadian National Research Council paper by J.H. McGuire.

704.6 Structural stability. The wall shall extend to the height required by Section 704.11 and shall have sufficient structural stability such that it will remain in place for the duration of time indicated by the required fire-resistance rating.

❖ This section specifies the performance that is required of an exterior wall. The wall is to extend the full height of the building and be constructed to remain in place for the duration of time indicated by the required fire-resistance rating. As such, any structural element that supports the wall must also be capable of providing the required fire-resistance rating.

704.7 Unexposed surface temperature. Where protected openings are not limited by Section 704.8, the limitation on the rise of temperature on the unexposed surface of exterior walls as required by ASTM E 119 shall not apply. Where protected openings are limited by Section 704.8, the limitation on the rise of temperature on the unexposed surface of exterior walls as required by ASTM E 119 shall not apply provided that a correc-

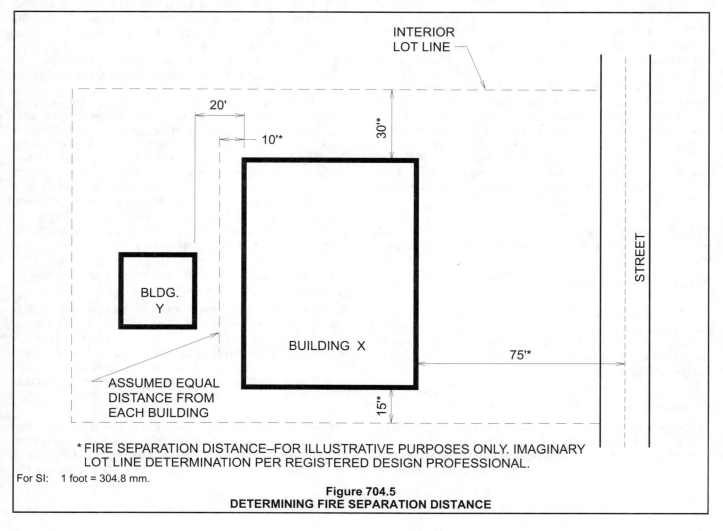

* FIRE SEPARATION DISTANCE–FOR ILLUSTRATIVE PURPOSES ONLY. IMAGINARY LOT LINE DETERMINATION PER REGISTERED DESIGN PROFESSIONAL.

For SI: 1 foot = 304.8 mm.

Figure 704.5
DETERMINING FIRE SEPARATION DISTANCE

tion is made for radiation from the unexposed exterior wall surface in accordance with the following formula:

$$A_e = A + (A_f \times F_{eo}) \qquad \text{(Equation 7-1)}$$

where:

A_e = Equivalent area of protected openings.

A = Actual area of protected openings.

A_f = Area of exterior wall surface in the story under consideration exclusive of openings, on which the temperature limitations of ASTM E 119 for walls are exceeded.

F_{eo} = An "equivalent opening factor" derived from Figure 704.7 based on the average temperature of the unexposed wall surface and the fire-resistance rating of the wall.

❖ The purpose of this provision is to allow the use of an exterior wall assembly that does not meet all of the requirements of ASTM E 119 (see Section 703.2). One of the conditions of acceptance specified in ASTM E 119 for load-bearing and nonload-bearing wall assemblies is that the transmission of heat through the wall during the fire test must not raise the average temperature of the unexposed surface more than 250°F (121°C) above ambient (room) temperature. A wall assembly that passes all aspects of the fire test other than the unexposed surface temperature rise is permitted under the provisions of this section.

The concept for this approach has its basis in the fire test requirements for openings, such as doors and windows. Neither NFPA 252 nor 257 limits the unexposed surface temperature as a condition of acceptance. Any window or door assembly that passes the fire test is acceptable, regardless of the unexposed surface temperature rise. This section of the code acknowledges that the radiant heat exposure to an adjacent building from such an assembly might be significant because of a high unexposed surface temperature. Table 704.8 also illustrates how the code addresses the potential exposure from openings. As the fire separation distance decreases, the permitted amount of unprotected openings also decreases. With a sufficiently large fire separation distance, there is a point at which the amount of protected openings is not limited. Below this point, even protected openings must be limited because of the potential exposure caused by high unexposed surface temperatures.

The first sentence in Section 704.7 indicates that the unexposed surface temperature rise limitation of ASTM E 119 [250°F (121°C) above ambient] does not apply to the wall assembly in circumstances where the code does not limit protected openings [i.e., fire separation distances of greater than 20 feet (6096 mm) in accordance with Table 704.8]. Where protected openings are not limited, virtually 100 percent of the wall is permitted to be comprised of protected openings that do not have limitations on heat transmission characteristics. It is, therefore, consistent to allow the use of a wall assembly without regard to the heat transmission characteristic of

that assembly.

The remainder of Section 704.7 addresses how this approach applies in circumstances where protected openings are limited by Table 704.8. By limiting protected openings, the code is, in effect, limiting the amount of wall area that could have a high temperature transmission and correspondingly high exposure threat. If a given wall is permitted to have only 45-percent protected and 15-percent unprotected openings, 40 percent of the wall would theoretically have to be a rated assembly that complies with all of the parameters of ASTM E 119. Without Section 704.7, a wall assembly that exceeds the temperature transmission endpoint when tested could not be used. Section 704.7 establishes that such an assembly can be used, provided that the excessive heat transmission is accounted and compensated for in accordance with the indicated formula.

This formula effectively results in a reduction of the allowable amount of protected openings that Table 704.8 would otherwise permit. In theory, a wall assembly that exhibits a high temperature transmission characteristic is equivalent to a protected opening assembly that also may exhibit a high temperature transmission. If such a wall assembly is used, the code achieves an equivalent level of protection by a correspondingly lower amount of permitted protected openings. The formula requires that an equivalent area of protected openings (A_e) be determined by adding a certain amount of factored wall area ($A_f \times F_{eo}$) to the actual area of protected openings (A). Thus, the equivalent area (A_e) will be larger than the actual area, but the larger equivalent area must still meet the limitations of Table 704.8. In reality, the actual amount of protected window and door openings must be less than the percentage set forth in Table 704.8 to compensate for that wall area with a higher than normally permitted temperature transmission characteristic.

For example, assume a mercantile building of Type VA construction has an exterior nonload-bearing wall of 1,000 square feet (93 m²) for a given story. This wall has a fire separation distance of greater than 10 feet (3048 mm) but less than 15 feet (4572 mm), and the wall contains 200 square feet (19 m²) of protected openings. The wall is required to be rated for 1 hour, but the fire test for the proposed assembly shows that the average unexposed surface temperature after 1 hour of testing is 1,150°F (621°C). The equivalent opening factor (F_{eo}) is determined from Figure 704.7, and in this case, F_{eo} = 0.3. The equivalent area of protected openings is then calculated as follows:

$$
\begin{aligned}
A_e &= 200 \text{ ft}^2 + (800 \text{ ft}^2 \times 0.3) \\
&= 200 \text{ ft}^2 + 240 \text{ ft}^2 \\
&= 440 \text{ ft}^2
\end{aligned}
$$

Table 704.8 permits 45 percent or 450 square feet (42 m²) of protected openings for this fire separation distance. The wall in this case is acceptable, since it contains the "equivalent" of 440 square feet (41 m²) of protected openings. What can be seen from this is that the excessive heat transmission from this wall assem-

FIGURE 704.7 – 704.8

FIRE-RESISTANCE-RATED CONSTRUCTION

bly has the effect of modifying the actual area of protected openings from 200 square feet (19 m²) to an equivalent area of 440 square feet (41 m²). In this manner, the code accomplishes a comparable level of protection by roughly equating the performance of this wall [with only 200 square feet (19 m²) of protected openings] to a wall that meets the maximum 250°F (121°C) temperature use [which could have up to 450 square feet (42 m²) of protected openings].

Section 704.7 must be used only where a wall assembly that does not meet the temperature endpoint limitations of ASTM E 119 is intended to be used. This is determined from fire test information on the specific assembly under consideration. If such information does not indicate that the temperature endpoint criterion was exceeded, then Section 704.7 is not applicable at the end of the time period equal to the required fire-resistance rating.

In the case of wall assemblies for which fire-resistance ratings are substantiated without specific fire testing, such as cast-in-place, precast or prestressed concrete panels, there are sources of information available for determining the necessary characteristics for use in conjunction with this provision. For example, PCI MNL 124 utilizes previous fire test data to establish this information for a range of concrete formulations.

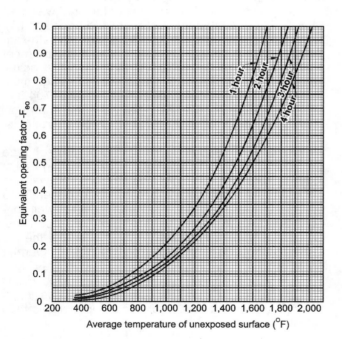

For SI: °C = [(°F) - 32] / 1.8.

FIGURE 704.7
EQUIVALENT OPENING FACTOR

❖ Figure 704.7 illustrates how to determine the equivalent opening factor (F_{eo}) used in Section 704.7. The figure is based on the following formula:

$$F_{eo} = (T_u + 460)^4 / (T_e + 460)^4$$

where:

T_u = Average temperature in degrees Fahrenheit for the unexposed wall surface at the time the required fire-resistance rating is reached during test conditions.

T_e = 1,700°F (927°C) for a 1-hour fire-resistance rating, 1,850°F (1010°C) for a 2-hour fire-resistance rating, 1,925°F (1052°C) for a 3-hour fire-resistance rating and 2,000°F (1093°C) for a 4-hour fire-resistance rating.

The T_e values represent the points on the curve that comprise the standard time-temperature curve used in the ASTM E 119 test procedure.

704.8 Allowable area of openings. The maximum area of unprotected or protected openings permitted in an exterior wall in any story shall not exceed the values set forth in Table 704.8. Where both unprotected and protected openings are located in the exterior wall in any story, the total area of the openings shall comply with the following formula:

$$\frac{A}{a} + \frac{A_u}{a_u} \leq 1.0 \qquad \textbf{(Equation 7-2)}$$

where:

A = Actual area of protected openings, or the equivalent area of protected openings, A_e (see Section 704.7).

a = Allowable area of protected openings.

A_u = Actual area of unprotected openings.

a_u = Allowable area of unprotected openings.

❖ The maximum area of openings, both protected and unprotected, permitted in all exterior walls is established in Table 704.8. To be considered protected, the opening must be provided with opening protection in accordance with Section 715. If the wall area in any story consists of both protected and unprotected openings, the total area must comply with the indicated formula.

The reason for including this parameter is to provide acceptance criteria for a wall with a combination of protected and unprotected openings. The values in Table 704.8 represent two separate levels of performance, protected openings and unprotected openings, that are deemed acceptable from a fire exposure standpoint. For example, in the case of a fire separation distance greater than 10 to 15 feet (3048 to 4572 mm), the criteria used in Table 704.8 results in the position that a fire-resistance-rated wall with 15-percent unprotected openings would be an acceptable exposure risk. It has been separately determined that a fire-resistance-rated wall with 45-percent protected openings would also be an acceptable exposure risk, but there is no determination of any acceptable combinations of protected and unprotected openings. As such, it should not be assumed that Table 704.8 allows both maximum amounts in the same wall.

For example, if an exterior wall consists of 2,000

square feet (189 m²) for a given story and the fire separation distance is 18 feet (5486 mm), a would equal 1,500 square feet (142 m²) and a_u would equal 500 square feet (47 m²). The wall contains 300 square feet (28 m²) of protected openings and 400 square feet (38 m²) of unprotected openings. Using the formula, it can be determined that the proposed arrangement is acceptable. In this instance, the sum is exactly 1 and the maximum total area of opening has been achieved.

As demonstrated in the following chart using the equation, an exterior wall with a fire separation of 10 to 15 feet (3048 to 4572 mm) is not permitted to have openings constituting more than 45 percent of the wall area. To reach that 45 percent, each opening must be protected. Likewise, if a designer desires to have openings constituting 25 percent of the wall area, an alternative is to provide 10 percent of the area in unprotected openings and 15 percent of the area in protected openings.

Unprotected openings (plain glass)	15%	10%	5%	0%
Protected openings (fire windows)	0%	15%	30%	45%
Maximum area of all openings	15%	25%	35%	45%

Note that these provisions apply regardless of whether or not an exterior wall is required to have a fire-resistance rating. This recognizes that even nonfire-resistance-rated exterior walls provide some protection against exposure from adjacent buildings or stories.

Table 704.8. See below.

❖ Table 704.8 contains the limitations for protected and unprotected openings in exterior walls in terms of percentages with respect to fire separation distances. The table is based on a formula that was developed under the following assumptions:

1. With a fire separation distance of greater than 30 feet (9144 mm), an exterior wall may have unlimited openings.

2. The fire plume projects 6 feet (1829 mm) from any opening.

3. The exterior openings are equally spaced and distributed in the wall.

One of the assumptions is uniform distribution of openings. Typically, if the openings are not uniformly distributed, the wall should be divided into areas with a relatively uniform distribution. The table applies to each area individually.

For example, if an exterior wall is 100 feet (30 480 mm) in length, 10 feet (3048 mm) in height and has 300 square feet (28 m²) of protected openings all located within 50 feet (15 240 mm) of the wall length, the wall should be evaluated in two segments, with the 50 feet (15 240 mm) of wall with the openings evaluated as follows:

$$\text{Percent of protected openings} = \frac{300}{(50 \times 10)} = 60 \text{ percent}$$

Therefore, the fire separation distance for that portion of the wall containing the openings would have to be greater than 15 feet (4572 mm) in order to be in compliance with Section 704.8.

Note d prohibits unprotected openings in occupancies of Groups H-2 and H-3 having a fire separation distance of 15 feet (4572 mm) or less. This is because buildings within these occupancies containing hazardous materials require additional exposure protection to be considered an acceptable fire risk.

Note e identifies that for buildings that are separated by a fire wall, where the buildings have differing roof

TABLE 704.8
MAXIMUM AREA OF EXTERIOR WALL OPENINGS[a]

CLASSIFICATION OF OPENING	FIRE SEPARATION DISTANCE (feet)							
	0 to 3[e,h]	Greater than 3 to 5[b]	Greater than 5 to 10[d,f]	Greater than 10 to 15[c,d,f]	Greater than 15 to 20[c,f]	Greater than 20 to 25[c,f]	Greater than 25 to 30[c,f]	Greater than 30
Unprotected	Not Permitted[g]	Not Permitted[b,g]	10%[g]	15%[g]	25%[g]	45%[g]	70%[g]	No Limit
Protected	Not Permitted	15%	25%	45%	75%	No Limit	No Limit	No Limit

For SI: 1 foot = 304.8 mm.

a. Values given are percentage of the area of the exterior wall.

b. For occupancies in Group R-3, as applicable in Section 101.2, the maximum percentage of unprotected and protected exterior wall openings shall be 25 percent.

c. The area of openings in an open parking structure with a fire separation distance of greater than 10 feet shall not be limited.

d. For occupancies in Group H-2 or H-3, unprotected openings shall not be permitted for openings with a fire separation distance of 15 feet or less.

e. For requirements for fire walls for buildings with differing roof heights, see Section 705.6.1.

f. The area of unprotected and protected openings is not limited for occupancies in Group R-3, as applicable in Section 101.2, with a fire separation distance greater than 5 feet.

g. Buildings whose exterior bearing wall, exterior nonbearing wall and exterior structural frame are not required to be fire-resistance rated shall be permitted to have unlimited unprotected openings.

h. Includes accessory buildings to Group R-3 as applicable in Section 101.2.

heights, the provisions for openings in the exterior wall above the fire wall are regulated by Section 705.6.1 (see commentary, Section 705.6.1). Note g is intended to coordinate with the rating requirements of Tables 601 and 602. This assumes that since fire separation distance or structural considerations do require a rating, it is therefore reasonable that openings need not be regulated.

704.8.1 Automatic sprinkler system. In buildings equipped throughout with an automatic sprinkler system in accordance with Section 903.3.1.1, the maximum allowable area of unprotected openings in occupancies other than Groups H-1, H-2 and H-3 shall be the same as the tabulated limitations for protected openings.

❖ One of the assumptions made in the development of Table 704.8 was that unprotected openings in buildings equipped throughout with an automatic sprinkler system are treated the same as protected openings; therefore, in buildings protected with an automatic sprinkler system, the maximum allowable area of unprotected openings is the same as the allowable area of protected openings. The reference to Section 903.3.1.1 clarifies that the system is to be an automatic sprinkler system that is designed and installed in accordance with NFPA 13 as modified by Section 903.3.1.1. This section is not applicable to occupancies in Groups H-1, H-2 and H-3 because of the associated hazards. Note that this section does not exclude sprinklered occupancies in Groups H-4 and H-5 from the allowable increase for unprotected openings, since materials associated with these occupancies do not present the same fire or explosion hazard as materials associated with Group H-1, H-2 or H-3.

704.8.2 First story. In occupancies other than Group H, unlimited unprotected openings are permitted in the first story of exterior walls facing a street that have a fire separation distance of greater than 15 feet (4572 mm), or facing an unoccupied space. The unoccupied space shall be on the same lot or dedicated for public use, shall not be less than 30 feet (9144 mm) in width, and shall have access from a street by a posted fire lane in accordance with the *International Fire Code.*

❖ Since the first story of a building is generally readily available for fire department access and manual suppression efforts, unprotected openings are not restricted in exterior walls facing a street if the fire separation distance is greater than 15 feet (4572 mm) and the building, or portion thereof, is other than Group H. In addition, the amount of unprotected openings in the first story of buildings, other than Group H, is not restricted where the building faces an unoccupied space. In order to allow access by the fire department, the unoccupied space must have a minimum width of 30 feet (9144 mm) and be located on the same lot as the building or be a space dedicated for public use with access to the unoccupied space provided by a posted fire lane having a width of not less than 20 feet (6096 mm) [see Section 503.2.1 of the *International Fire Code*® (IFC®)].

Figure 704.8.2 provides two examples of the' application of the provisions of this section.

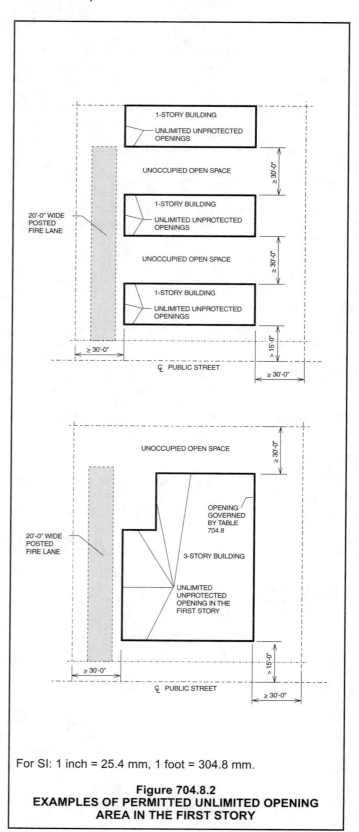

For SI: 1 inch = 25.4 mm, 1 foot = 304.8 mm.

Figure 704.8.2
EXAMPLES OF PERMITTED UNLIMITED OPENING
AREA IN THE FIRST STORY

704.9 Vertical separation of openings. Openings in exterior walls in adjacent stories shall be separated vertically to protect against fire spread on the exterior of the buildings where the openings are within 5 feet (1524 mm) of each other horizontally and the opening in the lower story is not a protected opening in accordance with Section 715.4.8. Such openings shall be separated vertically at least 3 feet (914 mm) by spandrel girders, exterior walls or other similar assemblies that have a fire-resistance rating of at least 1 hour or by flame barriers that extend horizontally at least 30 inches (762 mm) beyond the exterior wall. Flame barriers shall also have a fire-resistance rating of at least 1 hour. The unexposed surface temperature limitations specified in ASTM E 119 shall not apply to the flame barriers or vertical separation unless otherwise required by the provisions of this code.

Exceptions:

1. This section shall not apply to buildings that are three stories or less in height.

2. This section shall not apply to buildings equipped throughout with an automatic sprinkler system in accordance with Section 903.3.1.1 or 903.3.1.2.

3. Open parking garages.

❖ Where unprotected openings occur in adjacent stories, a fire that breaks out of an opening in a lower story can spread vertically to upper stories of the building. A vertical panel or horizontal flame barrier is necessary to minimize the possibility of vertical flame spread. The vertical panel is to be at least 3 feet (914 mm) in height and have a fire-resistance rating of at least 1 hour [see Figure 704.9(1)]. Full-scale fire tests have shown that the vertical panel should extend at least 30 inches (762 mm) above the finished floor level. This is to prevent the transfer of radiant heat to furnishings and other com-

bustibles located in the lower portion of the room in the story above the assumed fire exposure. Horizontal flame barriers are required to extend at least 30 inches (762 mm) beyond the face of the exterior wall and have a fire-resistance rating of at least 1 hour [see Figure 704.9(2)].

This section requires such protection when the opening on the lower level is not protected in accordance with Section 715. The use of glazing is one example of why this section is so specific. Wired glass, for example, has been shown to retain its integrity when exposed to a fire condition for a certain period of time. For this reason, it is a suitable exterior opening protective and, if the opening in the lower story is protected with wired glass, a flame barrier or vertical shield is not required, since the potential for flame propagation through the window prior to fire department arrival is minimized.

Wired glass is not, however, a suitable vertical shield above an unprotected opening. Wired glass permits a considerable amount (approximately 50 percent) of radiant heat energy to pass through. Since the initial concern is heat radiated from the flame plume, wired glass does little to prevent ignition of combustibles in the room on the upper floor.

This section permits the elimination of the unexposed surface temperature limitations prescribed by ASTM E 119 for flame barriers and vertical shields. Fire spread between floors is primarily the result of radiant heat, not conductive heat transfer. Full-scale tests have shown that only a small degree of fire resistance is required for adequate protection, since radiant energy is the primary concern and glazing in the unprotected openings above the story of fire origin may break within 1 minute.

Protection is not required for buildings three stories or less in height. This exception is consistent with Table

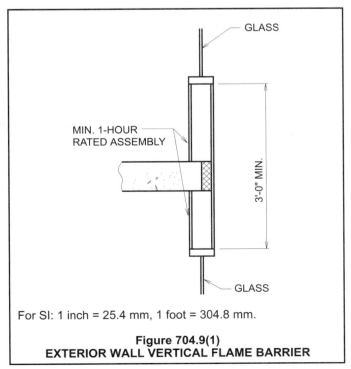

For SI: 1 inch = 25.4 mm, 1 foot = 304.8 mm.

Figure 704.9(1)
EXTERIOR WALL VERTICAL FLAME BARRIER

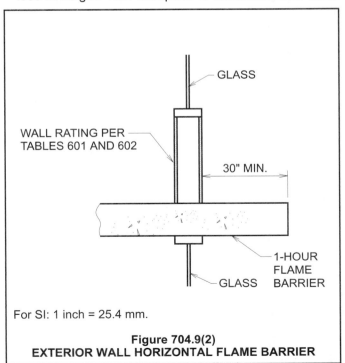

For SI: 1 inch = 25.4 mm.

Figure 704.9(2)
EXTERIOR WALL HORIZONTAL FLAME BARRIER

503, which permits buildings of three stories in height to be of Type IIB, IIIB and VB construction when these buildings do not require a fire-resistance rating for the floor construction.

The exception for buildings protected with an automatic sprinkler system is consistent with the philosophy in the section that the exposure threat of such buildings is considerably reduced. The exception is based on the effectiveness and reliability of automatic sprinkler systems. The reference to Section 903.3.1.1 or 903.3.1.2 clarifies that the system must be an automatic sprinkler system that is designed and installed in accordance with NFPA 13 as modified by Section 903.3.1.1 or in accordance with NFPA 13R as modified by Section 903.3.1.2. Exception 3 acknowledges the practicality of constructing open parking structures and the need to have a significant amount of openings for natural ventilation.

704.10 Vertical exposure. For buildings on the same lot, approved protectives shall be provided in every opening that is less than 15 feet (4572 mm) vertically above the roof of an adjoining building or adjacent structure that is within a horizontal fire separation distance of 15 feet (4572 mm) of the wall in which the opening is located.

Exception: Opening protectives are not required where the roof construction has a fire-resistance rating of not less than 1 hour for a minimum distance of 10 feet (3048 mm) from the adjoining building and the entire length and span of the supporting elements for the fire-resistance-rated roof assembly has a fire-resistance rating of not less than 1 hour.

❖ A fire in a building that is adjacent to a taller building can be the source of fire exposure to openings in the taller building. Although the height of a fire plume is dependent on several factors, this section requires opening protectives in the wall where the openings are within 15 feet vertically (457 mm) above the roof of a building that is within a horizontal fire separation distance of 15 feet (457 mm) (see Figure 704.10). Full-scale tests have indicated that exterior flame plumes may extend higher than 16 feet (4877 mm) above a window using a fuel load of 8 pounds per square foot (psf) (39 kg/m^2). Since this provision is based on a fire exposure from the roof of the lower building, it does not apply where the roof construction is 1-hour fire-resistance rated, which reduces the potential for fire exposure.

704.11 Parapets. Parapets shall be provided on exterior walls of buildings.

Exceptions: A parapet need not be provided on an exterior wall where any of the following conditions exist:

1. The wall is not required to be fire-resistance rated in accordance with Table 602 because of fire separation distance.

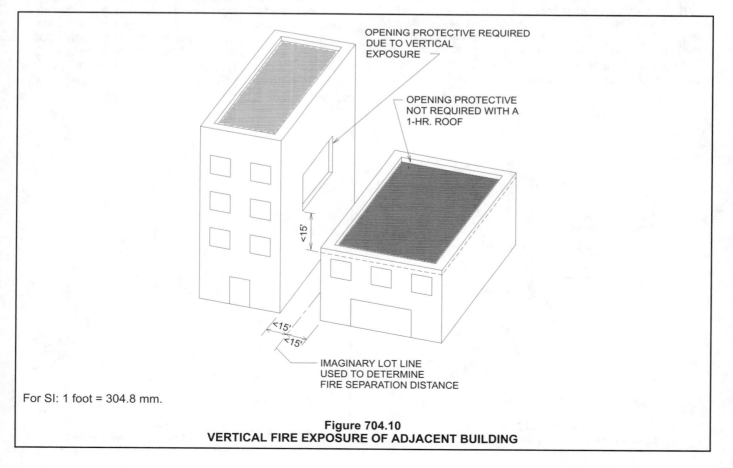

For SI: 1 foot = 304.8 mm.

Figure 704.10
VERTICAL FIRE EXPOSURE OF ADJACENT BUILDING

2. The building has an area of not more than 1,000 square feet (93 m²) on any floor.

3. Walls that terminate at roofs of not less than 2-hour fire-resistance-rated construction or where the roof, including the deck and supporting construction, is constructed entirely of noncombustible materials.

4. One-hour fire-resistance-rated exterior walls that terminate at the underside of the roof sheathing, deck or slab, provided:

    4.1. Where the roof/ceiling framing elements are parallel to the walls, such framing and elements supporting such framing shall not be of less than 1-hour fire-resistance-rated construction for a width of 4 feet (1220 mm) measured from the interior side of the wall for Groups R and U and 10 feet (3048 mm) for other occupancies.

    4.2. Where roof/ceiling framing elements are not parallel to the wall, the entire span of such framing and elements supporting such framing shall not be of less than 1-hour fire-resistance-rated construction.

    4.3. Openings in the roof shall not be located within 5 feet (1524 mm) of the 1-hour fire-resistance-rated exterior wall for Groups R and U and 10 feet (3048 mm) for other occupancies.

    4.4. The entire building shall be provided with not less than a Class B roof covering.

5. In occupancies of Groups R-2 and R-3 as applicable in Section 101.2, both provided with a Class C roof covering, the exterior wall shall be permitted to terminate at the roof sheathing or deck in Type III, IV and V construction provided:

    5.1. The roof sheathing or deck is constructed of approved noncombustible materials or of fire-retardant-treated wood, for a distance of 4 feet (1220 mm); or

    5.2. The roof is protected with 0.625-inch (15.88 mm) Type X gypsum board directly beneath the underside of the roof sheathing or deck, supported by a minimum of nominal 2-inch (51 mm) ledgers attached to the sides of the roof framing members, for a minimum distance of 4 feet (1220 mm).

6. Where the wall is permitted to have at least 25 percent of the exterior wall areas containing unprotected openings based on fire separation distance as determined in accordance with Section 704.8.

❖ Exterior walls that are required to have a fire-resistance rating in accordance with Table 602 (see Exception 1) are to be continuous from the foundation to no less than 30 inches (762 mm) above the roof surface. The minimum height of the parapet [30 inches (762 mm) as referenced in Section 704.11.1] minimizes the risk of direct flame spread and the risk of fire spread by radiant heat from flames above the roof. This provision only applies where a fire-resistance rating is required because of the

fire separation distance; therefore, if a load-bearing wall is required to have a fire-resistance rating by Table 601, but not by Table 602, a parapet would not be required. Both load-bearing and nonload-bearing walls required to have a fire-resistance rating by Table 602 must comply with this section.

Parapets are not required under any circumstances in any building that does not exceed 1,000 square feet (93 m²) in area. Exceptions 3, 4 and 5 address situations where the roof deck or sheathing is constructed of noncombustible materials, fireretardant-treated wood or of fire-resistance-rated construction, in which case the exterior wall may stop at the underside of the roof deck or sheathing (see the commentary to Section 2303.2 for fireretardant-treated wood). Alternatively, the exterior wall may terminate against a single layer of $5/_8$-inch (15.9 mm) Type X gypsum board (see Exception 5) that is attached tight to the underside of the combustible roof deck or sheathing. Not only do the restrictions reduce the likelihood of ignition from an exposing fire but they also reduce the likelihood of flame adjacent to the exterior wall. Exception 6 acknowledges the reduction in hazard associated with increases in fire separation distance. The net effect of this exception is to not reference a parapet where the fire separation distance is greater than 15 feet (4572 mm) (see Table 704.8).

704.11.1 Parapet construction. Parapets shall have the same fire-resistance rating as that required for the supporting wall, and on any side adjacent to a roof surface, shall have noncombustible faces for the uppermost 18 inches (457 mm), including counterflashing and coping materials. The height of the parapet shall not be less than 30 inches (762 mm) above the point where the roof surface and the wall intersect. Where the roof slopes toward a parapet at a slope greater than two units vertical in 12 units horizontal (16.7-percent slope), the parapet shall extend to the same height as any portion of the roof within a fire separation distance where protection of wall openings is required, but in no case shall the height be less than 30 inches (762 mm).

❖ This section requires the inside surface of the top of the parapet to be constructed of noncombustible construction. For sloping roofs, the parapet is required to extend to a height that is at the same elevation of the adjacent roof when the exterior wall is located at a fire separation distance that necessitates protected openings. This provision acknowledges that at higher slopes, the roof configuration is analogous to wall construction.

704.12 Opening protection. Windows required to be protected in accordance with Section 704.8, 704.9, or 704.10 shall comply with Section 715.4.8. Other openings required to be protected with fire doors or shutters in accordance with Sections 704.8, 704.9 and 704.10 shall comply with Section 715.3.

Exception: Fire protective assemblies are not required where the building is protected throughout by an automatic sprinkler system and the exterior openings are protected by an approved water curtain using automatic sprinklers approved for that use. The sprinklers and the water curtain shall be installed in accordance with NFPA 13.

❖ This section describes the manner in which openings in exterior walls must be protected. It also identifies the performance characteristics of the exterior opening protective. The purpose of providing exterior opening protectives is to reduce the fire risk exposure to and from adjacent structures, to reduce the potential for vertical fire spread and to protect exits from other portions of the structure.

The primary need to provide exterior opening protectives is determined by Section 704.8. Exterior opening protectives are required by Section 704.8 based on the fire separation distance and the area of the openings relative to the exterior wall area. Sections 704.9 and 704.10 contain provisions that require exterior opening protectives in certain instances. Exterior walls required by Table 601 to have a fire-resistance rating need not have exterior opening protectives unless also required by Section 704.8, 704.9 or 704.10.

When opening protective assemblies are provided, they are to be self-closing, automatic-closing or be fixed in place (see Sections 715.3 and 715.4.8). The opening protectives may include fire door assemblies and shutters (see Section 715.3); fire windows or wired glass (see Section 715.4).

An alternative method to protect openings in exterior walls is to provide outside automatic sprinklers in accordance with NFPA 13. When outside sprinklers are used, they are to be provided with an automatic water supply and a fire department connection. Additional details on the installation of such systems can be found in Chapter 9 and the referenced standard. In this case, a protected window or door assembly is not required, since the sprinklers will provide the necessary level of protection (see Figure 704.12).

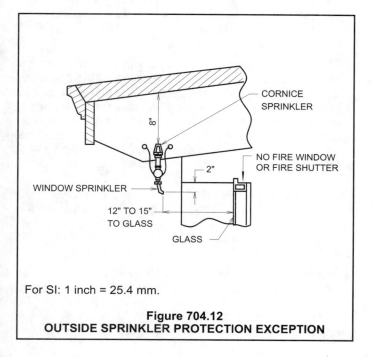

For SI: 1 inch = 25.4 mm.

Figure 704.12
OUTSIDE SPRINKLER PROTECTION EXCEPTION

704.12.1 Unprotected openings. Where protected openings are not required by Section 704, windows and doors shall be constructed of any approved materials. Glazing shall conform to the requirements of Chapters 24 and 26.

❖ This section allows for combustible construction, provided a protected opening is not required.

704.13 Joints. Joints made in or between exterior walls required by this section to have a fire-resistance rating shall comply with Section 713.

Exception: Joints in exterior walls that are permitted to have unprotected openings.

❖ Joints, such as expansion or seismic, are another form of openings in exterior walls, and therefore, must be considered with regard to maintaining the fire-resistance ratings of the exterior walls. This section requires that all joints located in exterior walls required to be fire-resistance rated are to be protected by a joint system that has a fire-resistance rating and complies with the requirements of Section 713 (see commentary, Section 713). The exception to this requirement is for joints located in exterior walls permitted by Table 704.8 to have unprotected openings. In general, this section requires fire-resistant joint systems to protect joints in exterior walls of unsprinklered buildings with a fire separation distance of 5 feet (1524 mm) or less; in sprinklered buildings with a fire separation distance of 3 feet (914 mm) or less (see Section 704.8.1) and in Group H-2 or H-3 buildings with a fire separation distance of 15 feet (4572 mm) or less (see Note d to Table 704.8).

704.13.1 Voids. The void created at the intersection of a floor/ceiling assembly and an exterior curtain wall assembly shall be protected in accordance with Section 713.4.

❖ See Section 713.4.

704.14 Ducts and air transfer openings. Penetrations by air ducts and air transfer openings in fire-resistance-rated exterior walls required to have protected openings shall comply with Section 716.

Exception: Foundation vents installed in accordance with this code are permitted.

❖ The code not only requires protected window and door openings, but also air transfer openings located in exterior walls are required to have protected openings in accordance with Table 704.8. The only exception is for foundation vents installed in accordance with Section 1203.3.

SECTION 705
FIRE WALLS

705.1 General. Each portion of a building separated by one or more fire walls that comply with the provisions of this section shall be considered a separate building. The extent and location of such fire walls shall provide a complete separation. Where a fire wall also separates groups that are required to be separated by a fire barrier wall, the most restrictive requirements of each separation shall apply. Fire walls located on lot lines shall also comply with Section 503.2. Such fire walls (party walls) shall be constructed without openings.

❖ Fire walls serve to create separate buildings (see the definition of "Area, building" in Section 502.1); therefore, all provisions of the code—including height and area limitations, fire protection systems and means of egress—are applied individually to the building on each side of the wall. As such, the fire wall must also provide the same protection afforded by exterior walls, namely: structural integrity, structural independence and adequate fire resistance for exposure protection. The weather resistance provided by exterior walls is not relevant since there is no direct weather exposure of the wall's surfaces, except for the parapet. In accordance with Section 503.2, a party wall is a fire wall that is constructed on a property line.

The basic performance characteristics of fire walls are defined in Section 705.2. Figure 705.1(1) shows an example of noncombustible fire wall construction. Fire walls must provide the same level of structural integrity and independence that is afforded by an exterior wall. Under fire conditions, therefore, the building on either side of the wall is required to be capable of collapse without causing the wall itself to collapse. This would not prohibit a fire wall from also being a structural load-bearing wall as long as this performance characteristic can be achieved. The required fire-resistance rating for a fire wall is a function of the occupancy involved; however, the minimum fire-resistance rating for a fire wall is 2 hours (see Table 705.4).

Continuity of fire walls is essential to their ability to provide protection for the buildings on either side. For buildings where the roof surface is at the same level on each side, the provisions of Section 705.6 require that the fire wall be continuous from the foundation to or through the roof deck or sheathing, depending on the type of roof construction. For buildings where the roof surfaces are at different levels on either side of the fire wall, Section 705.6.1 provides two options for the extension of the fire wall. While Section 705.6 requires that a fire wall be continuous, these requirements do not dictate that it must be constructed in a single vertical plane.

The offsetting of fire walls is not precluded as long as the required fire-resistance rating and structural stability can be provided continuously to or through the roof. It is not intended, however, that fire walls can be provided in the horizontal plane to create separate buildings. There are only a few instances in which two vertical sections of a building can truly act as separate buildings with respect to complying with all of the provisions of the code, including structural independence, means of egress, etc.

Where a fire wall is intended to create separate buildings, it must provide the same level of protection as an exterior wall. An example of a fire wall construction detail is shown in Figures 705.1(1) and 705.1(2).

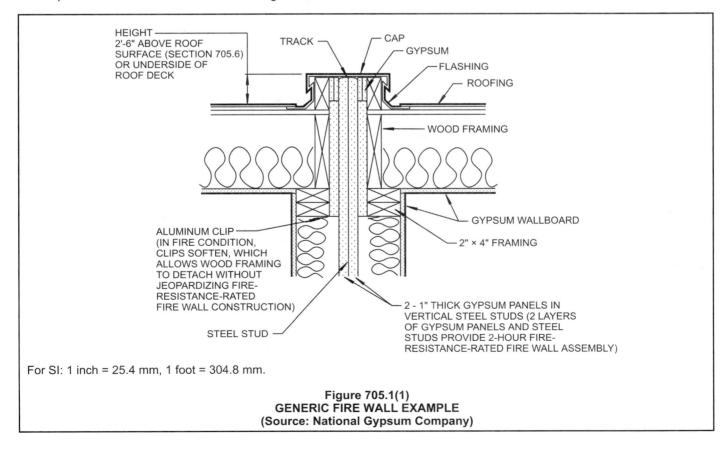

For SI: 1 inch = 25.4 mm, 1 foot = 304.8 mm.

Figure 705.1(1)
GENERIC FIRE WALL EXAMPLE
(Source: National Gypsum Company)

FIGURE 705.1(2) – FIGURE 705.1(4)   **FIRE-RESISTANCE-RATED CONSTRUCTION**

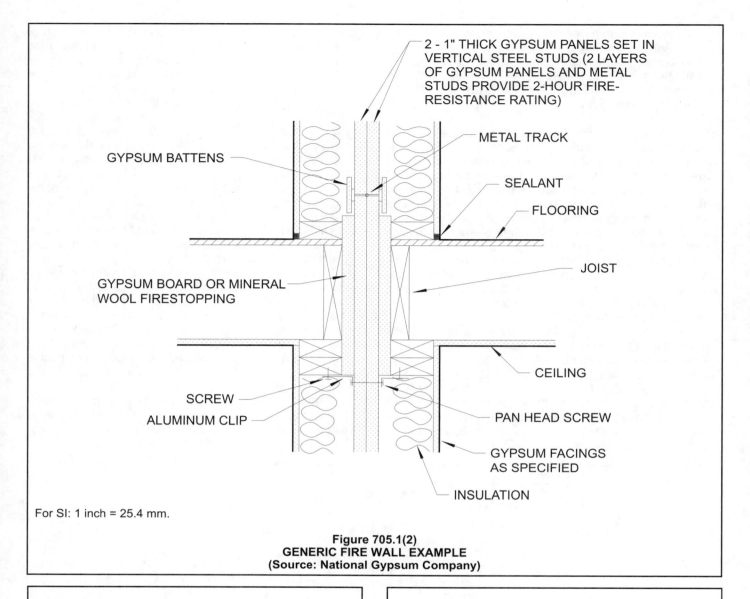

For SI: 1 inch = 25.4 mm.

Figure 705.1(2)
GENERIC FIRE WALL EXAMPLE
(Source: National Gypsum Company)

Average fuel load psf	kg/m²	Equivalent fire endurance (hours)
5	24.4	$^{1}/_{2}$
$7^{1}/_{2}$	36.6	$^{3}/_{4}$
10	48.8	1
15	73.2	$1^{1}/_{2}$
20	97.6	2
30	146.5	3
40	195.3	$4^{1}/_{2}$
50	244.1	6
60	292.9	$7^{1}/_{2}$

Figure 705.1(3)
RELATIONSHIP BETWEEN FUEL LOAD AND
FIRE ENDURANCE

Occupancy	Combustibles in occupancy (psf)	Fire severity (hours)
Assembly	5 to 10	$^{1}/_{2}$ to 1
Business	5 to 10	$^{1}/_{2}$ to 1
Educational	5 to 10	$^{1}/_{2}$ to 1
Factory — Industrial		
Low hazard	0 to 10	0 to 1
Moderate hazard	10 to 25	1 to 3
Hazardous	Variable	Variable
Institutional	5 to 10	$^{1}/_{2}$ to 1
Mercantile	10 to 20	1 to 2
Residential	5 to 10	$^{1}/_{2}$ to 1
Storage		
Low hazard	0 to 10	0 to 1
Moderate hazard	10 to 30	1 to 1

Figure 705.1(4)
OCCUPANCIES—FUEL LOAD—FIRE SEVERITY

705.2 Structural stability. Fire walls shall have sufficient structural stability under fire conditions to allow collapse of construction on either side without collapse of the wall for the duration of time indicated by the required fire-resistance rating.

❖ Since the collapse of one building from fire conditions should not cause the collapse of an adjacent building, a fire wall is required to be capable of withstanding the collapse of the construction on either side of the wall. An exception to the provision is found in Exception 5 to Section 705.6.

705.3 Materials. Fire walls shall be of any approved noncombustible materials.

Exception: Buildings of Type V construction.

❖ This section requires that fire walls be constructed of noncombustible materials unless both buildings are of Type V (combustible) construction. This is consistent with the provisions of Section 602, which require exterior walls of buildings of Type I, II, III and IV to be built of noncombustible construction, while buildings of Type V construction are to be built of combustible exterior walls.

705.4 Fire-resistance rating. Fire walls shall have a fire-resistance rating of not less than that required by Table 705.4.

❖ The required fire-resistance rating must comply with the more restrictive occupancy of Table 705.4. For example, if two buildings of Type II construction with occupancies in Groups S-1 and B are built of any construction type and separated by a fire wall, the fire wall is required to have a fire-resistance rating of 3 hours (in accordance with Table 705.4 based on Group S-1).

TABLE 705.4
FIRE WALL FIRE-RESISTANCE RATINGS

GROUP	FIRE-RESISTANCE RATING (hours)
A, B, E, H-4, I, R-1, R-2, U	3[a]
F-1, H-3[b], H-5, M, S-1	3
H-1, H-2	4[b]
F-2, S-2, R-3, R-4	2

a. Walls shall be not less than 2-hour fire-resistance rated where separating buildings of Type II or V construction.
b. For Group H-1, H-2 or H-3 buildings, also see Sections 415.4 and 415.5.

❖ The fire-resistance ratings represent an approximate relationship between the fuel load and an exposure to fire severity that is equivalent to the standard time-temperature curve. The relationship is based on a number of full-scale fire tests by the U.S. Department of Commerce's National Bureau of Standards (NBS) conducted in the 1920s to determine how actual building fires compared with the temperatures represented by the standard time-temperature curve (see commentary, Section 703.2).

An analysis of tests documented by S.H. Ingberg indicates that the weight per square foot (psf) of ordinary combustibles, such as wood and paper, with a heat of combustion of 7,000 to 8,000 British thermal units (Btu) per pound (16 282 to 18 608 kJ/kg), is related to hourly fire severity as described in Figure 705.1(3). By comparing Table 705.4 to Figure 705.1(4), it shows that the table is not based solely on the average fuel load of the occupancy. Not only are other factors, such as occupant density and evacuation capability, incorporated into the required fire-resistance rating, but it must also be recognized that fuel loads in a building are not evenly distributed. For example, the average fuel load for an occupancy in Group B may be 5 to 10 psf (24 to 49 kg/m²). The fuel load in file and storage rooms, however, may be 10 to 20 psf (49 to 98 Kg/m²). While other sections of the code may address these areas by mandating an automatic fire suppression system with smoke partitions or a fire-resistance-rated enclosure of the incidental use area, the potential for higher fuel loads is also factored into determining the required fire-resistance rating.

The minimum fire-resistance ratings required in Table 705.4 generally correlate with the fire-resistance ratings required by Table 302.3.2; however, Table 705.4 contains provisions that acknowledge the need for more conservative ratings for fire walls due to their unique role in creating separate buildings.

Due to the unique nature of the hazards presented by occupancies of Groups H-1, H-2 and H-3, Table 705.4 includes Note b, which references Sections 415.4 and 415.5 for specific requirements relative to the construction of such buildings (see commentary, Sections 415.4 and 415.5).

705.5 Horizontal continuity. Fire walls shall be continuous from exterior wall to exterior wall and shall extend at least 18 inches (457 mm) beyond the exterior surface of exterior walls.

Exceptions:

1. Fire walls shall be permitted to terminate at the interior surface of combustible exterior sheathing or siding provided the exterior wall has a fire-resistance rating of at least 1 hour for a horizontal distance of at least 4 feet (1220 mm) on both sides of the fire wall. Openings within such exterior walls shall be protected by fire assemblies having a fire protection rating of not less than $^3/_4$ hour.

2. Fire walls shall be permitted to terminate at the interior surface of noncombustible exterior sheathing, exterior siding or other noncombustible exterior finishes provided the sheathing, siding, or other exterior noncombustible finish extends a horizontal distance of at least 4 feet (1220 mm) on both sides of the fire wall.

3. Fire walls shall be permitted to terminate at the interior surface of noncombustible exterior sheathing where the building on each side of the fire wall is protected by an automatic sprinkler system installed in accordance with Section 903.3.1.1 or 903.3.1.2.

❖ Historically, the codes have addressed the hazards of fire exposure at the fire wall from only a vertical perspective; namely, at the roof (see Section 705.6). Section 705.5 addresses a similar hazard from the horizontal perspective; namely, at the intersection of the fire wall and the exterior wall. The horizontal exposure is likewise currently addressed in Section 1019.1.4 relative to stair enclosure exposure.

The 18-inch (457 mm) extension is intended to abate the potential for fire to travel from one building to the other around the fire wall. The 18-inch (457 mm) extension is required to extend the full height of the fire wall. The three exceptions acknowledge the effect certain types of exterior wall construction will have on fire breaching the exterior wall and exposing the adjacent building; namely, fire-resistance-rated construction (see Exception 1), noncombustible finish materials (see Exception 2) and noncombustible sheathing materials coupled with sprinkler protection (see Exception 3) provide the necessary barrier to limit fire spread across the exterior surface. The difference between Exceptions 2 and 3 is that Exception 2 would not permit a combustible exterior finish to be placed over the noncombustible exterior wall construction, while Exception 3 would allow a combustible exterior finish, provided the building is sprinklered with either an NFPA 13 or 13R system.

705.5.1 Exterior walls. Where the fire wall intersects the exterior walls, the fire-resistance rating for the exterior walls on both sides of the fire wall shall have a 1-hour fire-resistance rating with $^3/_4$-hour opening protection where opening protection is required. The fire-resistance rating of the exterior wall shall extend a minimum of 4 feet (1220 mm) on each side of the intersection of the fire wall to exterior wall. Exterior wall intersections at fire walls that form an angle equal to or greater than 180 degrees (3.14 rad) do not need exterior wall protection.

❖ Fire walls create separate buildings with an appropriately rated wall between the buildings; however, as with two separate buildings, opposing exterior walls represent an exposure hazard that is abated by the provisions of Sections 704.5 and 704.8. This same condition occurs with buildings separated by fire walls relative to the exterior walls where they are at an angle of less than 180 degrees (3.14 rad).

Where the walls are at an angle less than 180 degrees (3.14 rad) and the opposing walls are at a fire separation distance that requires protected exterior openings (Section 704.8), the exterior walls are required to be 1-hour fire-resistance rated with $^3/_4$-hour opening protectives. This protective is required to extend for a distance of 4 feet (1219 mm) on both sides of the fire wall.

The 4 feet (1219 mm) of exterior wall, which is 1-hour rated with protected openings, serves two purposes relative to code compliance. First, the walls comply with Exception 1 to Section 705.5, thus eliminating the need for the 18-inch (457 mm) fire wall extension. Second, the wall is required in order to comply with Section 705.5.1. If the exterior wall configuration is such that the fire separation distance is greater than 30 feet (9144 mm); then the exterior wall protection of Section 705.5.1 would not be required since protected openings are not

required in accordance with Table 704.8. The provisions of Section 705.5, however, would still be applicable.

705.5.2 Horizontal projecting elements. Fire walls shall extend to the outer edge of horizontal projecting elements such as balconies, roof overhangs, canopies, marquees and architectural projections that are within 4 feet (1220 mm) of the fire wall.

Exceptions:

1. Horizontal projecting elements without concealed spaces provided the exterior wall behind and below the projecting element has not less than 1-hour fire-resistance-rated construction for a distance not less than the depth of the projecting element on both sides of the fire wall. Openings within such exterior walls shall be protected by fire assemblies having a fire protection rating of not less than $^3/_4$ hour.

2. Noncombustible horizontal projecting elements with concealed spaces, provided a minimum 1-hour fire-resistance-rated wall extends through the concealed space. The projecting element shall be separated from the building by a minimum of 1-hour fire-resistance-rated construction for a distance on each side of the fire wall equal to the depth of the projecting element. The wall is not required to extend under the projecting element where the building exterior wall is not less than 1-hour fire-resistance rated for a distance on each side of the fire wall equal to the depth of the projecting element. Openings within such exterior walls shall be protected by fire assemblies having a fire protection rating of not less than $^3/_4$ hour.

3. For combustible horizontal projecting elements with concealed spaces, the fire wall need only extend through the concealed space to the outer edges of the projecting elements. The exterior wall behind and below the projecting element shall be of not less than 1-hour fire-resistance-rated construction for a distance not less than the depth of the projecting elements on both sides of the fire wall. Openings within such exterior walls shall be protected by fire assemblies having a fire-protection rating of not less than $^3/_4$ hour.

❖ Fire walls are typically used to reduce the building area for purposes of code compliance; however, many structures divided by fire walls are still detailed as a contiguous structure with projecting elements located on the exterior wall that extend across the fire wall. These projecting elements represent a potential conduit for fire to be transferred from one side of the building to the other side of the building. Even if the projecting elements are terminated at a distance in close proximity to the fire wall, the potential for fire spread to the adjacent projecting element still exists. In this section, 4 feet (1219 mm) (measured from the fire wall) is considered the appropriate threshold.

This section requires the fire wall to extend to the outer edge of the projecting element when such element is located within 4 feet (1219 mm) of the fire wall. For example, in a contiguous residential structure with interior lot lines and fire walls between each dwelling unit, and where balconies are provided for each dwelling unit and these balconies are within 4 feet (1219 mm) of the fire wall, then the fire wall must be extended to the outer edge of the balcony.

Exception 1 acknowledges the reduction in hazard due to the lack of a concealed space crossing the fire wall; however, since the projecting element is combustible, there is still a potential for fire to transfer across the wall, thus the need for a rated exterior wall separation between the inside of the building and the combustible projection.

Exceptions 2 and 3 address projecting elements with concealed spaces of noncombustible and combustible construction, respectively.

705.6 Vertical continuity. Fire walls shall extend from the foundation to a termination point at least 30 inches (762 mm) above both adjacent roofs.

Exceptions:

1. Stepped buildings in accordance with Section 705.6.1.

2. Two-hour fire-resistance-rated walls shall be permitted to terminate at the underside of the roof sheathing, deck or slab provided:

    2.1. The lower roof assembly within 4 feet (1220 mm) of the wall has not less than a 1-hour fire-resistance rating and the entire length and span of supporting elements for the rated roof assembly has a fire-resistance rating of not less than 1 hour.

    2.2. Openings in the roof shall not be located within 4 feet (1220 mm) of the fire wall.

    2.3. Each building shall be provided with not less than a Class B roof covering.

3. Walls shall be permitted to terminate at the underside of noncombustible roof sheathing, deck, or slabs where both buildings are provided with not less than a Class B roof covering. Openings in the roof shall not be located within 4 feet (1220 mm) of the fire wall.

4. In buildings of Type III, IV and V construction, walls shall be permitted to terminate at the underside of combustible roof sheathing or decks provided:

    4.1. There are no openings in the roof within 4 feet (1220 mm) of the fire wall,

    4.2. The roof is covered with a minimum Class B roof covering, and

    4.3. The roof sheathing or deck is constructed of fire-retardant-treated wood for a distance of 4 feet (1220 mm) on both sides of the wall or the roof is protected with $^5/_8$ inch (15.9 mm) Type X gypsum board directly beneath the underside of the roof sheathing or deck, supported by a minimum of 2-inch (51 mm) nominal ledgers attached to the sides of the roof framing members for a minimum distance of 4 feet (1220 mm) on both sides of the fire wall.

5. Buildings located above a parking garage designed in accordance with Section 508.2 shall be permitted to

have the fire walls for the buildings located above the parking garage extend from the horizontal separation between the parking garage and the buildings.

❖ Since one of the primary purposes of a fire wall is to provide protection from an exposure fire, the wall must be continuous from the foundation to or through the roof. The fire wall is required to extend to 30 inches (762 mm) above the roof surface [see Figure 705.1(1)], except where other provisions are met that will restrict the spread of fire (see Exceptions 1 through 7). If the two buildings have different roof levels, the fire wall must comply with the provisions of Section 705.6.1.

Exception 2 to this section's requirement for a parapet applies to buildings where: (a) the maximum required fire-resistance rating of the fire wall is 2 hours; (b) the roof assembly located within 4 feet (1219 mm) of the fire wall has a minimum fire-resistance rating of 1 hour (including all supporting elements and structural members); (c) no roof openings are located within 4 feet (1219 mm) of the fire wall and (d) the buildings on either side of the fire wall have a minimum Class B roof covering.

The provisions of Exception 2 are not applicable to buildings with combustible roof construction where a fire wall is required to have a fire-resistance rating of 3 hours or greater. Buildings where the fire wall is required to have a 3-hour or greater fire-resistance rating must comply with the general provisions of Section 705.6, with Exceptions 3 through 5 for either noncombustible or fire-retardant-treated roof construction as applicable or with Exceptions 3 through 5 in order to qualify for the parapet extension exception.

Exception 3 states that, since the intent of the parapet is to prevent fire spread over the roof, it is not required if the entire roof is of noncombustible construction; the fire wall is continuous to the underside of the roof deck, roof slab or roof sheathing; the roof-covering material has a minimum Class B classification in accordance with the provisions of Section 1505.3 (see commentary, Section 1505.3) and there are no roof openings located within 4 feet (1219 mm) on either side of the fire wall.

The intent of the requirement for a Class B roof covering is due to the fact that roof coverings are not considered part of the roof, and therefore, are not subject to the noncombustibility requirements of this section. This section requires that the roof covering be effective against moderate fire test exposure to reduce the potential for fire spread across the roof. The restriction on the location of roof openings is intended to minimize the potential of the spread of fire from one building to an adjacent building via roof openings, such as skylights or roof windows. This fire spread could occur from a fan condition that breaches the roof opening in the involved building and exposes roof openings in the adjacent building to direct fire or to burning brands, which could land on the roof, burn through the roof opening and ignite combustibles in the interior of the adjacent building. The 4-foot (1219 mm) threshold provides a reasonable approximation of a fire plume, beyond which the exposure to fire of the adjacent building is reduced.

Exception 5 to the requirement in Section 705.6 for a parapet is the condition in buildings of Type III, IV and V construction where the sheathing or deck is constructed of noncombustible materials or approved fire retardant-treated wood for a distance of 4 feet (1219 mm) on both sides of the wall (see the commentary to Section 2303.2 for additional information on fire-retardant-treated wood) or where the prescribed application of $^5/_8$-inch (15.9 mm) Type X gypsum board is applied to the underside of the deck for a distance of 4 feet (1219 mm) on both sides of the wall. Combustible materials may not extend through the fire wall; the roof coverings on both buildings must have a minimum Class B rating and no roof openings are to be located within 4 feet (1219 mm) of the fire wall (see commentary, Section 508.2). The intent is the same: to resist the passage of flame beyond the fire wall.

705.6.1 Stepped buildings. Where a fire wall serves as an exterior wall for a building and separates buildings having different roof levels, such wall shall terminate at a point not less than 30 inches (762 mm) above the lower roof level, provided the exterior wall for a height of 15 feet (4572 mm) above the lower roof is not less than 1-hour fire-resistance-rated construction from both sides with openings protected by assemblies having a $^3/_4$-hour fire protection rating.

Exception: Where the fire wall terminates at the underside of the roof sheathing, deck or slab of the lower roof, provided:

1. The lower roof assembly within 10 feet (3048 mm) of the wall has not less than a 1-hour fire-resistance rating and the entire length and span of supporting elements for the rated roof assembly has a fire-resistance rating of not less than 1 hour.

2. Openings in the lower roof shall not be located within 10 feet (3048 mm) of the fire wall.

❖ The basic provisions of Section 705.6 require that a parapet extend a minimum of 30 inches (762 mm) above the roof surfaces on both sides of the fire wall, which would then require fire walls separating buildings with different roof heights to extend 30 inches (762 mm) above the highest roof surface. The provisions of Section 705.6.1 address situations where a fire wall separates adjacent buildings with a difference in roof height of greater than 30 inches (762 mm). These provisions acknowledge that fire exposure to the exterior wall of an adjacent building from the roof of a lower building represents, to a certain extent, a reduced hazard.

This section retains the fire wall extension above the lower roof surface and places a 15-foot (4572 mm) limit on rated opening protectives, while the exception allows the fire wall extension to be eliminated as long as a fire barrier, in the form of 1-hour-rated roof construction, is provided that extends a minimum of 10 feet (3048 mm) from the fire wall.

The option provided by this section permits openings in an exterior wall that extend above the fire wall, where the fire wall extends 30 inches (762 mm) above the lower roof surface and the exterior wall extending above

the fire wall is constructed as a 1-hour fire-resistance-rated wall, rated for exposure from both sides, for a distance of 15 feet (4572 mm) above the lower roof surface. Where the fire wall and exterior wall comply with these requirements, openings are permitted in the exterior wall extension, provided that all openings located within 15 feet (4572 mm) of the lower roof surface are protected with $^3/_4$-hour-rated opening protectives. Openings located more than 15 feet (4572 mm) above the lower roof surface are not required to have opening protectives [see Figure 705.6.1(1)]. The provisions of this section are similar to those contained in Section 704.10 for vertical exposure of exterior openings. They require that when the difference in height between the roof surfaces of the adjacent buildings is less than 15 feet (4572 mm), the exterior wall extension is required to extend to the underside of the upper roof deck [see Figure 705.6.1(2)]. The exception permits openings in the exterior wall that extends above the fire wall, where: (a) the fire wall terminates at the bottom of the roof deck of the lower roof; (b) the lower roof assembly has a minimum 1-hour fire-resistance rating for a minimum distance of 10 feet (3048 mm) from the fire wall and (c) there are no openings located in the lower roof within 10 feet (3048 mm) of the fire wall. Openings located in the exterior wall above the lower roof surface are not required to have opening protectives [see Figure 705.6.1(3)]. It must be noted that all structural elements, including beams, columns and bearing walls that provide support for the fire-resistance-rated roof assembly, must also have a 1-hour fire-resistance rating for their entire length or span in order to maintain the effectiveness of the rated roof assembly. The provisions of this section are similar to those contained in Section 1019.1.4 for the protection of the exterior walls of exit stairways.

705.7 Combustible framing in fire walls. Adjacent combustible members entering into a concrete or masonry fire wall from opposite sides shall not have less than a 4-inch (102 mm) distance between embedded ends. Where combustible members frame into hollow walls or walls of hollow units, hollow spaces shall be solidly filled for the full thickness of the wall and for a distance not less than 4 inches (102 mm) above, below and between the structural members, with noncombustible materials approved for fireblocking.

❖ In order to retain the fire-resistance capability of the wall where combustible members will frame into it, hollow walls or walls of hollow units must be solidly filled for the thickness of the wall and for a distance of not less than 4 inches (102 mm) above, below and between the structural members. Consistent with the construction of the walls, the fireblocking materials are to be noncombustible and approved for fireblocking in accordance with Section 717.2. If combustible members enter both sides of a fire wall, there must be at least 4 inches (102 mm) of masonry between the embedded ends.

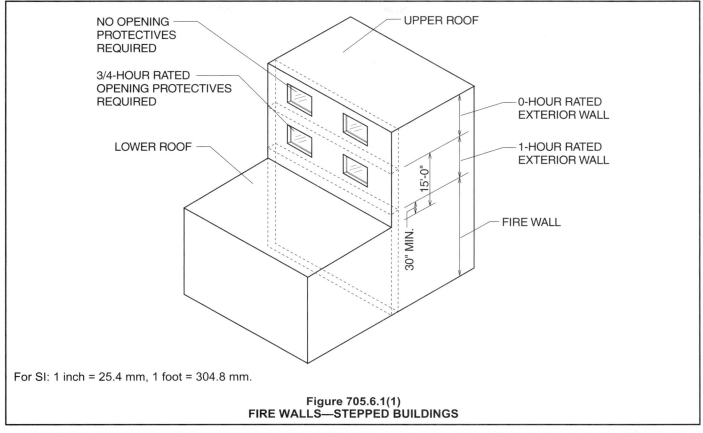

For SI: 1 inch = 25.4 mm, 1 foot = 304.8 mm.

Figure 705.6.1(1)
FIRE WALLS—STEPPED BUILDINGS

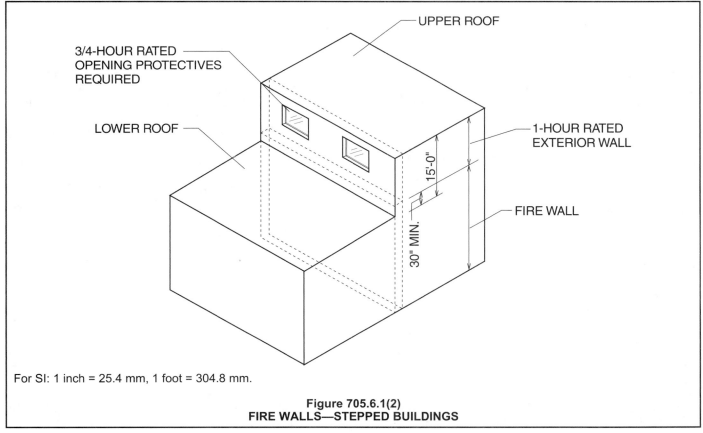

For SI: 1 inch = 25.4 mm, 1 foot = 304.8 mm.

Figure 705.6.1(2)
FIRE WALLS—STEPPED BUILDINGS

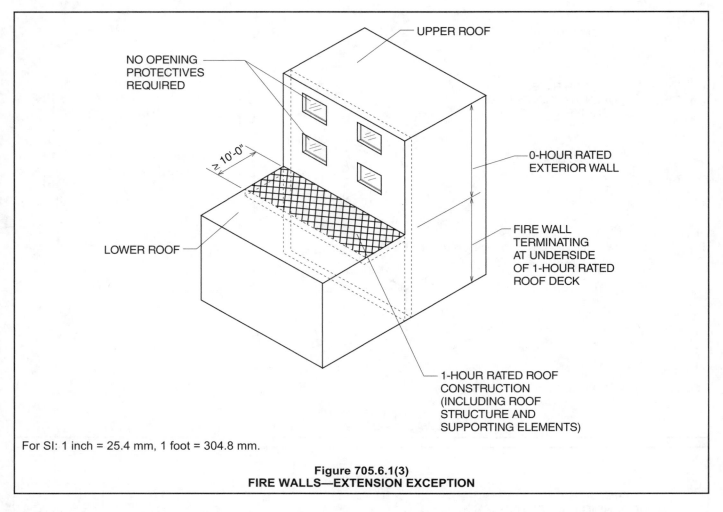

For SI: 1 inch = 25.4 mm, 1 foot = 304.8 mm.

Figure 705.6.1(3)
FIRE WALLS—EXTENSION EXCEPTION

705.8 Openings. Each opening through a fire wall shall be protected in accordance with Section 715.3 and shall not exceed 120 square feet (11 m²). The aggregate width of openings at any floor level shall not exceed 25 percent of the length of the wall.

Exceptions:

1. Openings are not permitted in party walls constructed in accordance with Section 503.2.

2. Openings shall not be limited to 120 square feet (11 m²) where both buildings are equipped throughout with an automatic sprinkler system installed in accordance with Section 903.3.1.1.

❖ In order to maintain the integrity of the fire wall, the maximum area and percent of openings in the wall are restricted. When provided, the openings must be properly protected so that the fire-resistance rating of the wall is maintained. This section prescribes the maximum area and the percent of openings that may be permitted in a fire wall at any one floor level. The provisions must be used in concert with Section 503.2, which limits openings for party walls.

Fire wall openings have restrictive limitations in their size and total area because of the critical function that a fire wall serves. To maintain the required fire perfor-

mance of the fire wall, each opening through a fire wall is restricted in area to 120 square feet (11 m²), and the aggregate width of all openings at any one floor level may not constitute more than 25 percent of the length of the wall. The 120-square-foot (11 m²) limitation provides a reasonable size through which industrial machinery may pass and corresponds with the maximum area limitations of many tested fire doors.

Recognizing the effectiveness of automatic sprinklers, the 120-square-foot (11 m²) opening limitation does not apply where the complete buildings on both sides of the fire wall are sprinklered (see Exception 2); however, the aggregate width of all openings in a fire wall at any one floor level is still limited to 25 percent of the length of the wall.

705.9 Penetrations. Penetrations through fire walls shall comply with Section 712.

❖ In order to maintain the integrity of the required fire-resistance rating, penetrations through the fire wall must be properly protected. Acceptable protection methods for various penetrations of fire walls are identified in Sections 712.2 and 712.3.

705.10 Joints. Joints made in or between fire walls shall comply with Section 713.

❖ Joints, such as expansion or seismic, are another form of openings in fire walls and, therefore, must be considered with regard to maintaining the fire-resistance ratings of fire walls. This section requires all joints that are located in fire walls to be protected by a joint system with a fire-resistance rating and to comply with the requirements of Section 713.

705.11 Ducts and air transfer openings. Ducts and air transfer openings shall not penetrate fire walls.

> **Exception:** Penetrations by ducts and air transfer openings of fire walls that are not on a lot line shall be allowed provided the penetrations comply with Sections 712 and 716. The size and aggregate width of all openings shall not exceed the limitations of Section 705.8.

❖ The general provisions of this section mirror those of Section 503.2 for party walls. The exception further permits duct and transfer openings for fire walls not located on a lot line, provided the maximum aggregate area provisions of Sections 712.3.3 and 716 are met.

SECTION 706
FIRE BARRIERS

706.1 General. Fire barriers used for separation of shafts, exits, exit passageways, horizontal exits or incidental use areas, to separate different occupancies, to separate a single occupancy into different fire areas, or to separate other areas where a fire barrier is required elsewhere in this code or the *International Fire Code*, shall comply with this section.

❖ The provisions of this section apply to assemblies that are required to have a fire-resistance rating and are used for separating exits, incidental use areas, shafts, hazardous materials control areas and fire areas. Fire barriers provide a higher degree of protection than fire partitions (see Section 708). The amount of openings is limited in fire barriers, and fire barrier wall assemblies must be continuous from the top of a fire-resistance-rated floor/ceiling assembly to below to the floor or roof slab/deck above. Unlike fire partitions, there are no circumstances under which a fire barrier wall is permitted to terminate at a ceiling. Fire barriers are used for a variety of purposes, such as mixed occupancies, areas of refuge separations, shafts and exit and floor opening enclosures. Fire barriers also include interior walls that serve to subdivide a space by separating one fire area from an adjacent fire area and for separating incidental use areas (see Section 302.1.1.1). Fire-resistance-rated assemblies used to separate exit access corridors as well as tenant, dwelling unit and guestroom separations are fire partitions (see Section 708). The provisions of this section provide minimum requirements for the fire-resistance rating, continuity, combus-

tibility and protection of openings and penetrations in order to help maintain the reliability of the fire separation assembly. As with any fire-resistance-rated assembly, consideration must be given to the openings and penetrations that are provided within the assembly. The intent, in part, is to maintain the fire-resistance rating of the assembly. These sections recognize that fire spread beyond a fire-resistance-rated compartment is often attributed to the protection given to any opening or penetration in the fire barrier, or the lack thereof.

Since the fire barrier is intended to provide a reliable subdivision of areas, the construction that structurally supports the assembly is required to provide at least the same hourly fire-resistance rating as the fire barrier. This is applicable regardless of the type of construction of the building. Structural stability is regulated by Section 708.4.

706.2 Materials. The walls and floor assemblies shall be of materials permitted by the building type of construction.

❖ The types of materials used in fire barriers are to be consistent with Sections 602 through 602.5 for the type of construction classification of the building. The fire-resistance ratings of fire barriers used to separate mixed occupancies are determined in accordance with Section 302.3 (see commentary, Section 302.3). Fire barriers are permitted to be of combustible materials in Type III, IV and V construction, and are required to be of noncombustible materials in Type I and II construction.

706.3 Fire-resistance rating. The fire-resistance rating of the walls and floor assemblies shall comply with this section.

❖ This section identifies the types of assemblies, both vertical and horizontal (see also Section 711.3), which are required to be fire-resistance rated and constructed as fire barriers.

706.3.1 Shaft enclosures. The fire-resistance rating of the fire barrier separating building areas from a shaft shall comply with Section 707.4.

❖ See Section 707 for information regarding where shaft enclosures are required.

706.3.2 Exit enclosures. The fire-resistance rating of the fire barrier separating building areas from an exit shall comply with Section 1019.1.

❖ See Section 1019.1, which requires enclosure of vertical exit enclosures.

706.3.3 Exit passageway. The fire-resistance rating of the separation between building areas and an exit passageway shall comply with Section 1020.1.

❖ See Section 1020.3, which requires exit passageways (as defined in Section 1002) to be constructed of fire barriers.

706.3.4 Horizontal exit. The fire-resistance rating of the separation between building areas connected by a horizontal exit shall comply with Section 1021.1.

❖ See Section 1021.2, which requires horizontal exits (as defined in Section 1002) to be constructed of fire barriers or fire walls.

706.3.5 Incidental use areas. The fire barrier separating incidental use areas shall have a fire-resistance rating of not less than that indicated in Table 302.1.1.

❖ Table 302.1.1 requires either an automatic fire suppression system, fire barrier or both in order to protect incidental use areas. Section 302.1.1.1 further requires nonfire-resistance-rated construction, which is capable of resisting the passage of smoke where only a fire suppression system, not a rated separation, is required.

706.3.6 Separation of mixed occupancies. Where the provisions of Section 302.3.2 are applicable, the fire barrier separating mixed occupancies shall have a fire-resistance rating of not less than that indicated in Section 302.3.2 based on the occupancies being separated.

❖ When separated occupancies are the chosen option for mixed uses in a building, the required fire separation is either a fire barrier, meeting the requirements of this section, or a horizontal (floor-ceiling) fire separation assembly (see Section 303.3.2). This accomplishes the desired separation into different fire areas.

706.3.7 Single-occupancy fire areas. The fire barrier separating a single occupancy into different fire areas shall have a fire-resistance rating of not less than that indicated in Table 706.3.7.

❖ See the commentary to Table 706.3.7.

TABLE 706.3.7
FIRE-RESISTANCE RATING REQUIREMENTS FOR FIRE BARRIER ASSEMBLIES BETWEEN FIRE AREAS

OCCUPANCY GROUP	FIRE-RESISTANCE RATING (hours)
H-1, H-2	4
F-1, H-3, S-1	3
A, B, E, F-2, H-4, H-5, I, M, R, S-2	2
U	1

❖ One of the alternatives available in addressing fire protection systems in many buildings is to divide the building into separate fire areas. Since many of the fire suppression system thresholds (see Section 903.2) are based upon fire area, separation of a single occupancy into small fire areas would be an acceptable method for avoiding the use of sprinklers. This is a classic type of tradeoff: sprinklers versus compartmentation. If the separation is provided, each fire area may be evaluated separately for purposes of determining the applicable provisions. Table 706.3.7 provides the minimum re-

quired fire-resistance ratings of the fire barrier wall or horizontal assembly separating two fire areas of the same occupancy groups.

Areas separated with fire barriers are not considered separate buildings; they are considered separate fire areas. Two areas must be separated by a fire wall or exterior walls to be considered separate buildings. Two areas separated with fire barriers are still considered as part of a single building. This distinction is critical in determining compliance with allowable height and area and other code provisions.

706.4 Continuity of fire barrier walls. Fire barrier walls shall extend from the top of the floor/ceiling assembly below to the underside of the floor or roof slab or deck above and shall be securely attached thereto. These walls shall be continuous through concealed spaces such as the space above a suspended ceiling. The supporting construction for fire barrier walls shall be protected to afford the required fire-resistance rating of the fire barrier supported except for 1-hour fire-resistance-rated incidental use area separations as required by Table 302.1.1 in buildings of Type IIB, IIIB and VB construction. Hollow vertical spaces within the fire barrier wall shall be firestopped at every floor level.

Exceptions:

1. The maximum required fire-resistance rating for assemblies supporting fire barriers separating tank storage as provided for in Section 415.7.2.1 shall be 2 hours, but not less than required by Table 601 for the building construction type.

2. Shaft enclosure shall be permitted to terminate at a top enclosure complying with Section 707.12.

❖ To minimize the potential for fire spread from the exposed side of a vertical fire barrier (wall) to the unexposed side, such assemblies must be continuous from a fire-resistance-rated floor/ceiling assembly below to the underside of the floor slab or roof deck above (see Figure 706.4). To maintain the efficiency of the fire barrier, it must be continuous through all concealed spaces (such as a space above a suspended ceiling), be constructed tight and securely attached to the underside of the floor slab or roof deck.

Fire barriers must be supported by construction having an equivalent fire-resistance rating. The intent of this requirement is to prevent the effectiveness of the assembly from being circumvented by a fire that threatens the supporting elements. The requirement for structural support applies to all types of construction.

All hollow vertical spaces in the fire separation walls must be fireblocked at the ceiling and floor or roof levels in accordance with Section 717.2.2. This continuity criterion is what distinguishes fire barriers from fire partitions (see commentary, Section 708.4).

Fire barriers are required by Section 302.1.1.1 for incidental use areas. The incidental use area provisions compartmentalize areas of the building with increased risks, thereby providing protection for other areas of the building with comparatively lower risk. By requiring these incidental use fire barriers to be supported by construction with equivalent fire resistance, the code is philosophically protecting the increased risk area from fire in the low risk portion of the building. This would be particularly onerous, for example, in buildings with unprotected types of construction with areas in business buildings of Type IIB, IIIB or VB construction. The location of an incidental use storage area would effectively change the required type of construction in a portion of the building.

This exception maintains the fire-rated separation without triggering the type of construction problems in unrated buildings. The philosophy is similar to that recognized for tenant, guestroom and exit corridor separations in Section 708.4 and smoke barriers in Section 709.4.

706.5 Horizontal fire barriers. Horizontal fire barriers shall be constructed in accordance with Section 711.

❖ The areas being separated by fire barriers are sometimes above and below each other. For instance, when separated occupancies are chosen due to height limitations of the occupancy, it might be necessary for the floor/ceiling assembly between the two occupancies to be a fire-resistance-rated assembly meeting the requirements of Table 302.3.2. Horizontal assemblies are also used to subdivide a fire area into smaller fire areas as discussed in Section 706.3.7. The code references Section 711, which provides detailed requirements for horizontal assemblies.

706.6 Exterior walls. Where exterior walls serve as a part of a required fire-resistance-rated enclosure, such walls shall comply with the requirements of Section 704 for exterior walls and the fire-resistance-rated enclosure requirements shall not apply.

Exception: Exterior walls required to be fire-resistance rated in accordance with Section 1022.6.

❖ If an area is required to be enclosed by fire barriers and an exterior wall constitutes part of the enclosure, the exterior wall is only required to comply with the fire-resistance rating requirements in Section 704, unless the area enclosed is part of an exit stairway (see commentary, Section 1022.6). The intent of the fire barrier requirements is to subdivide or enclose areas to protect them from a fire in the building. The exterior wall need only have a fire-resistance rating if required for structural stability (see Table 601) or because of exterior exposure potential (see Table 602 and Sections 704.5 and 1022.6).

706.7 Openings. Openings in a fire barrier wall shall be protected in accordance with Section 715. Openings shall be limited to a maximum aggregate width of 25 percent of the length of the wall, and the maximum area of any single opening shall not exceed 120 square feet (11 m²). Openings in exit enclosures shall also comply with Section 1019.1.1.

Exceptions:

1. Openings shall not be limited to 120 square feet (11 m²) where adjoining fire areas are equipped throughout

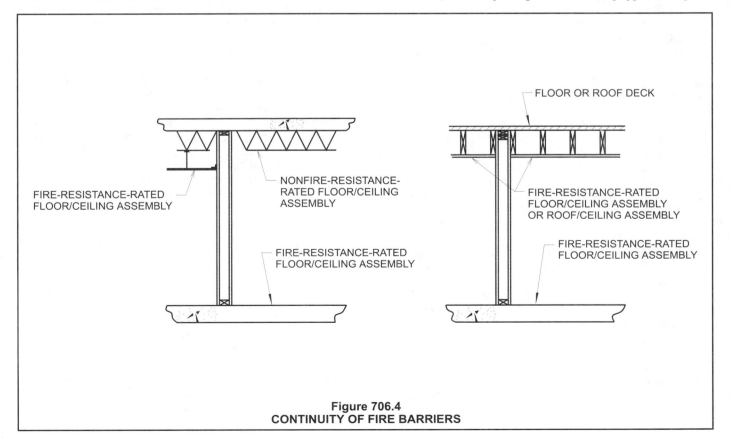

Figure 706.4
CONTINUITY OF FIRE BARRIERS

with an automatic sprinkler system in accordance with Section 903.3.1.1.

2. Fire doors serving an exit enclosure.

3. Openings shall not be limited to 120 square feet (11 m²) or an aggregate width of 25 percent of the length of the wall where the opening protective assembly has been tested in accordance with ASTM E 119 and has a minimum fire-resistance rating not less than the fire-resistance rating of the wall.

❖ To maintain the viability of the fire barrier, the aggregate width of openings is restricted to a maximum of 25 percent of the length of the wall. This limitation is based on the fact that the criteria for opening protectives do not include limitations on unexposed surface temperature or radiant heat transfer. Consistent with typical listing limitations, a single opening protective is limited to a maximum of 120 square feet (11 m²). It should be noted, however, that certain opening protectives, such as fire windows, are often limited to much smaller areas per opening.

The reference to Section 1020.4 specifies that in exit enclosures, only openings for the purpose of egressing from normally occupied spaces are permitted. Spaces that are not normally occupied, such as janitor closets or mechanical and electrical rooms, are not permitted to open directly into the exit enclosure, since fire in those areas may go undetected for a long period of time; therefore, these areas pose a greater fire hazard than occupied spaces.

In order to maintain the required fire-resistance rating of the assembly, opening protectives must have a fire protection rating in accordance with Section 715. A fire door assembly includes all required hardware, anchorage, frames and sills necessary to provide an acceptable opening protective. The reference to Section 715 is intended to identify the required fire protection rating for the opening protective, as indicated in Table 715.3, as well as the applicable test standards.

Openings in fire barriers are not limited in size when all fire areas separated by that assembly are equipped throughout with an automatic sprinkler system (see Exception 1). This exception is similar to one made for fire walls (see Section 705.8), based on the effectiveness of automatic sprinkler systems. Although the openings in the fire barrier are not limited in size under this exception, they are still required to be protected by opening protectives that meet the requirements of Section 715.

Exception 2 acknowledges the practicality of the 25-percent limitation for walls of an exit enclosure. Most exit enclosures are of such limited size that the placement of the fire door in the wall of the enclosure exceeds 25 percent of the wall.

Exception 3 addresses new opening protective products that have been tested to the more rigorous provisions of ASTM E 119 rather than, or in addition to, the opening protective standard NFPA 252. Since the opening protective has been tested to the same standard as

the wall itself, it is then logical to allow such opening protective without restrictions.

706.8 Penetrations. Penetrations through fire barriers shall comply with Section 712.

❖ In order to maintain the integrity of the fire barrier, penetrations into and through it must be properly protected. Acceptable protection methods for various penetrations of fire barriers are identified in Sections 712.3 through 712.3.4, while methods for protection of penetrations of fire-resistance-rated floor/ceiling and roof/ceiling assemblies are identified in Sections 712.4 through 712.4.6.

706.8.1 Prohibited penetrations. Penetrations into an exit enclosure shall only be allowed when permitted by Section 1019.1.2.

❖ See Section 1019.1.2.

706.9 Joints. Joints made in or between fire barriers shall comply with Section 713.

❖ This section regulates joints or linear openings created between building assemblies, which are sometimes referred to as construction, expansion or seismic joints. These joints are most often created where the structural design of a building necessitates a separation between building components in order to accommodate anticipated structural displacements caused by thermal expansion and contraction, seismic activity, wind or other loads. Figure 713.1 illustrates some of the most common locations of these joints.

These linear openings create a "weak link" in fire-resistance-rated assemblies, which can compromise the integrity of the tested assembly by allowing an avenue for the passage of fire and the products of combustion through the assembly. In order to maintain the efficacy of the fire-resistance-rated assembly, these openings must be protected by a joint system with a fire-resistance rating equal to the adjacent assembly. It is not the intent of this section to regulate joints installed in assemblies that are provided to control shrinkage cracking, such as a saw-cut control joint in concrete (see Section 713).

706.10 Ducts and air transfer openings. Penetrations by ducts and air transfer openings shall comply with Sections 712 and 716.

❖ See Sections 712 and 716.

SECTION 707
SHAFT ENCLOSURES

707.1 General. The provisions of this section shall apply to vertical shafts where such shafts are required to protect openings and penetrations through floor/ceiling and roof/ceiling assemblies.

❖ This section applies to all vertical shafts, including those covered by other sections of the code, namely: interior stairways (see Section 1019.1); refuse waste- and linen-handling chutes (see Section 707.13) and elevator and dumbwaiter hoistways (see Section 707.14).

All openings or penetrations in floor/ceiling or roof/ceiling assemblies are required to be protected with a vertical shaft enclosure, unless one of the exceptions provided for in Section 707.2 is applicable.

The key consideration is one of determining whether or not a shaft is required. The requirements for shaft enclosures are found in Section 707.2.

707.2 Shaft enclosure required. Openings through a floor/ceiling assembly shall be protected by a shaft enclosure complying with this section.

Exceptions:

1. A shaft enclosure is not required for openings totally within an individual residential dwelling unit and connecting four stories or less.

2. A shaft enclosure is not required in a building equipped throughout with an automatic sprinkler system in accordance with Section 903.3.1.1 for an escalator opening or stairway which is not a portion of the means of egress protected according to Item 2.1 or 2.2:

    2.1. Where the area of the floor opening between stories does not exceed twice the horizontal projected area of the escalator or stairway and the opening is protected by a draft curtain and closely spaced sprinklers in accordance with NFPA 13. In other than Groups B and M, this application is limited to openings that do not connect more than four stories.

    2.2. Where the opening is protected by approved power-operated automatic shutters at every floor penetrated. The shutters shall be of noncombustible construction and have a fire-resistance rating of not less than 1.5 hours. The shutter shall be so constructed as to close immediately upon the actuation of a smoke detector installed in accordance with Section 907.10 and shall completely shut off the well opening. Escalators shall cease operation when the shutter begins to close. The shutter shall operate at a speed of not more than 30 feet per minute (152.4 mm/s) and shall be equipped with a sensitive leading edge to arrest its progress where in contact with any obstacle, and to continue its progress on release therefrom.

3. A shaft enclosure is not required for penetrations by pipe, tube, conduit, wire, cable, and vents protected in accordance with Section 712.4.

4. A shaft enclosure is not required for penetrations by ducts protected in accordance with Section 712.4.

Grease ducts shall be protected in accordance with the *International Mechanical Code.*

5. A shaft enclosure is not required for floor openings complying with the provisions for covered malls or atriums.

6. A shaft enclosure is not required for approved masonry chimneys, where annular space protection is provided at each floor level in accordance with Section 717.2.5.

7. In other than Groups I-2 and I-3, a shaft enclosure is not required for a floor opening that complies with the following:

    7.1. Does not connect more than two stories.

    7.2. Is not part of the required means of egress system except as permitted in Section 1019.1.

    7.3. Is not concealed within the building construction.

    7.4. Is not open to a corridor in Group I and R occupancies.

    7.5. Is not open to a corridor on nonsprinklered floors in any occupancy.

    7.6. Is separated from floor openings serving other floors by construction conforming to required shaft enclosures.

8. A shaft enclosure is not required for automobile ramps in open parking garages and enclosed parking garages constructed in accordance with Sections 406.3 and 406.4, respectively.

9. A shaft enclosure is not required for floor openings between a mezzanine and the floor below.

10. A shaft enclosure is not required for joints protected by a fire-resistant joint system in accordance with Section 713.

11. Where permitted by other sections of this code.

❖ In general, all floor openings that connect two or more stories are to be enclosed in shafts that are constructed in accordance with this section. The exceptions identify floor openings that are not required to be protected by a shaft enclosure.

Exception 1 identifies that a floor opening connecting no more than four stories within a dwelling unit is permitted without a shaft enclosure. This is consistent with the permitted omission of stairway enclosures in such uses in accordance with Exception 3 to Section 1019.1. Exception 2 addresses stairways and escalators in fully sprinklered buildings that are not part of the required means of egress. Item 2.1 requires a draft curtain to protect the opening, the size of which is limited based on the configuration of the stair or escalator [see Figure 707.2(1)]. This option allows such floor openings to occur for the full height of the building, regardless of the number of stories for Groups B and M. For all other occupancies, there is a four-story limitation (three floors permitted to contain openings). Item 2.2 requires automatic shutters at each floor opening and is not limited as to height or occupancy [see Figure 707.2(2)]. Exceptions 3 through 6 recognize other floor openings that are permitted without a shaft enclosure in accordance with the referenced section.

Exception 7 addresses the issue of floor openings that are not a part of the required means of egress. For example, supplemental stairs are often provided for the convenience of the building's occupants. Such stairways serve as a means of communication between two adjacent floors and are not meant to be used as a required means of egress. To maintain the integrity of the exit access corridor, such stairs may not connect with an exit access corridor in Groups I and R [see Figure 707.2(3)] because occupants can be sleeping and the integrity of the corridor system is especially important under such conditions.

Supplemental stairs are also not permitted to be connected to any other floor opening that connects to an additional floor level. Such stairs must be separated from such floor openings by construction that complies with this section for shaft enclosures. Additionally, such stairs must comply with the other requirements of Chapter 10, such as headroom, handrails, guards and tread and riser relationships.

Exception 8 addresses the practicality of a shaft enclosure for automobile ramps in both open and enclosed parking garages. Since a mezzanine is considered part of the floor below (see Section 505.1), Exception 9 does not require a shaft between levels. A joint protected in accordance with Section 713 is considered a visible method of protection between floor levels and also does not require the additional protective offered by a shaft enclosure (see Exception 10).

707.3 Materials. The shaft enclosure shall be of materials permitted by the building type of construction.

❖ Shaft construction must be of appropriate materials consistent with the type of construction required by Section 602.

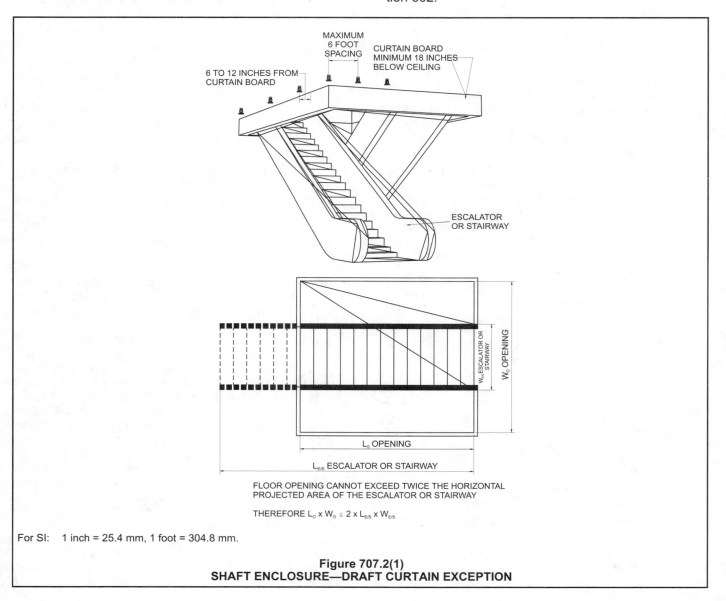

For SI: 1 inch = 25.4 mm, 1 foot = 304.8 mm.

Figure 707.2(1)
SHAFT ENCLOSURE—DRAFT CURTAIN EXCEPTION

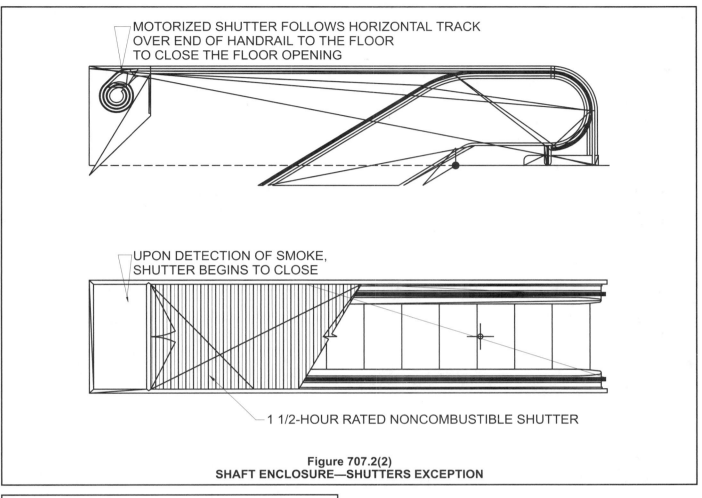

MOTORIZED SHUTTER FOLLOWS HORIZONTAL TRACK
OVER END OF HANDRAIL TO THE FLOOR
TO CLOSE THE FLOOR OPENING

UPON DETECTION OF SMOKE,
SHUTTER BEGINS TO CLOSE

1 1/2-HOUR RATED NONCOMBUSTIBLE SHUTTER

Figure 707.2(2)
SHAFT ENCLOSURE—SHUTTERS EXCEPTION

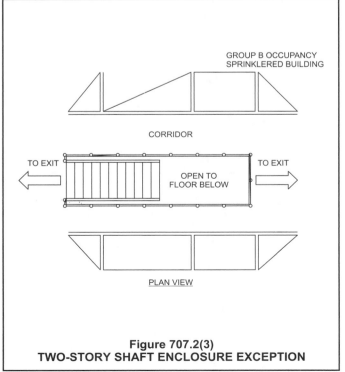

GROUP B OCCUPANCY
SPRINKLERED BUILDING

CORRIDOR

TO EXIT

OPEN TO
FLOOR BELOW

TO EXIT

PLAN VIEW

Figure 707.2(3)
TWO-STORY SHAFT ENCLOSURE EXCEPTION

707.4 Fire-resistance rating. Shaft enclosures shall have a fire-resistance rating of not less than 2 hours where connecting four stories or more and not less than 1 hour where connecting less than four stories. The number of stories connected by the shaft enclosure shall include any basements but not any mezzanines. Shaft enclosures shall be constructed as fire barriers in accordance with Section 706. Shaft enclosures shall have a fire-resistance rating not less than the floor assembly penetrated, but need not exceed 2 hours.

❖ The required fire-resistance rating for a shaft enclosure is related to the fire-resistance rating of the penetrated floor/ceiling or roof/ceiling assembly, since the purpose of the shaft is to maintain the integrity of the floor assembly. For this reason, the shaft must have a fire-resistance rating equivalent to the penetrated floor/ceiling assembly, but not to exceed 2 hours. The maximum rating is based on the historical experience with 2-hour shaft enclosures and a determination that a 2-hour fire-resistance-rated shaft enclosure with properly protected openings will perform as required (see Figure 707.4).

The fire-resistance rating of a shaft enclosure that connects less than four stories may be reduced to 1 hour. With respect to exit enclosures, the reduction recognizes that the limited height of exit travel will result in a shorter period of time required to evacuate the building. The reduction also recognizes that Table 503 mandates a type of construction that requires only a 1-hour fire-resistance-rated floor/ceiling assembly in many three-story buildings. It should be noted, however, that the criteria are based on the number of connected stories and not on the height of the building.

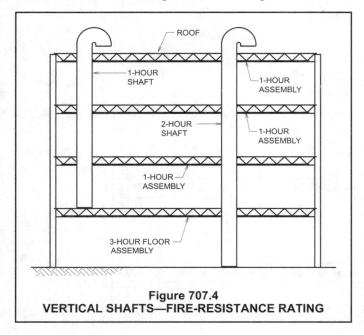

Figure 707.4
VERTICAL SHAFTS—FIRE-RESISTANCE RATING

707.5 Continuity. Shaft enclosure walls shall extend from the top of the floor/ceiling assembly below to the underside of the floor or roof slab or deck above and shall be securely attached thereto. These walls shall be continuous through concealed spaces such as the space above a suspended ceiling. The supporting construction shall be protected to afford the required fire-resistance rating of the element supported. Hollow vertical spaces within the shaft enclosure construction wall shall be firestopped at every floor level.

❖ A shaft wall is a fire barrier (see Section 706.1); therefore, the continuity provisions for shafts mirror those for fire barriers (see Section 706.4).

707.6 Exterior walls. Where exterior walls serve as a part of a required shaft enclosure, such walls shall comply with the requirements of Section 704 for exterior walls and the fire-resistance-rated enclosure requirements shall not apply.

Exception: Exterior walls required to be fire-resistance rated in accordance with Section 1022.6.

❖ If an area is required to be enclosed by fire barriers and an exterior wall constitutes part of the enclosure, the exterior wall is only required to comply with the fire-resistance rating requirements in Section 704, unless the area enclosed is part of an exit stairway (see commentary, Section 1022.6). The intent of the fire barrier requirements is to subdivide or enclose areas to protect them from a fire in the building. The exterior wall need only have a fire-resistance rating if required for struc-

tural stability (see Table 601) or because of an exterior exposure potential (see Table 602 and Sections 704.5 and 1022.6).

707.7 Openings. Openings in a shaft enclosure shall be protected in accordance with Section 715 as required for fire barriers. Such openings shall be self-closing or automatic-closing by smoke detection.

❖ In order to maintain the integrity of the shaft enclosures, all openings must be protected with approved opening protectives (see Section 715). An example of a protected opening is illustrated in Figure 707.7. Pipe, tube and conduit penetrations must be protected in accordance with Section 707.8, and all joints must be protected in accordance with Section 707.9. The provisions in Section 715 are not specific to shaft protection regarding openings. The provisions in Section 715 specific to fire barriers are applicable to shaft enclosures.

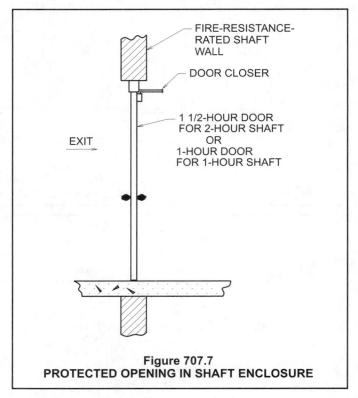

Figure 707.7
PROTECTED OPENING IN SHAFT ENCLOSURE

707.7.1 Prohibited openings. Openings other than those necessary for the purpose of the shaft shall not be permitted in shaft enclosures.

❖ Only openings that are required so that the shaft can serve its intended purpose are permitted in the shaft enclosure.

707.8 Penetrations. Penetrations in a shaft enclosure shall be protected in accordance with Section 712 as required for fire barriers.

❖ See the commentary to Section 712.

707.8.1 Prohibited penetrations. Penetrations other than those necessary for the purpose of the shaft shall not be permitted in shaft enclosures. Ducts shall not penetrate exit shaft enclosures.

Exception: Duct penetrations as permitted in Section 1019.1.2.

❖ The provisions of Section 707.8.1 mirror those of Section 707.7 for openings. The only type of duct penetration that is permitted is one that either ventilates or pressurizes the shaft enclosure. All others are prohibited.

707.9 Joints. Joints in a shaft enclosure shall comply with Section 713.

❖ See Section 713.

707.10 Ducts and air transfer openings. Penetrations of a shaft enclosure by ducts and air transfer openings shall comply with Sections 712 and 716.

❖ These provisions reference the applicable protection features of Sections 712 and 716 for duct penetrations. It is important to note that the type of duct penetration is limited to those in Section 1019.1.2 (see Section 707.8.1).

707.11 Enclosure at the bottom. Shafts that do not extend to the bottom of the building or structure shall:

1. Be enclosed at the lowest level with construction of the same fire-resistance rating as the lowest floor through which the shaft passes, but not less than the rating required for the shaft enclosure;

2. Terminate in a room having a use related to the purpose of the shaft. The room shall be separated from the remainder of the building by construction having a fire-resistance rating and opening protectives at least equal to the protection required for the shaft enclosure; or

3. Be protected by approved fire dampers installed in accordance with their listing at the lowest floor level within the shaft enclosure.

Exceptions:

1. The fire-resistance-rated room separation is not required provided there are no openings in or penetrations of the shaft enclosure to the interior of the building except at the bottom. The bottom of the shaft shall be closed off around the penetrating items with materials permitted by Section 717.3.1 for draftstopping, or the room shall be provided with an approved automatic fire suppression system.

2. A shaft enclosure containing a refuse chute or laundry chute shall not be used for any other purpose and shall terminate in a room protected in accordance with Section 707.13.4.

3. The fire-resistance-rated room separation and the protection at the bottom of the shaft are not required provided there are no combustibles in the shaft and there are no openings or other penetrations through the shaft enclosure to the interior of the building.

❖ Proper shaft enclosures must include all sides as well as the top and bottom, unless the bottom of the shaft is at the bottom of the structure. Where the bottom of the shaft is not at the bottom of the structure, the fire-resistance rating for the bottom of the shaft must be the same as that rating required for the shaft enclosure [see Figure 707.11(1)]. If a duct extends through the bottom of the shaft, a fire damper is required at the penetration (see Item 3). This section also recognizes that the purpose of some shafts cannot be accomplished if the bottom must be enclosed; therefore, a room is permitted at the bottom of the shaft (see Item 2), but it must relate to the shaft's purpose and have the same enclosure as the shaft [see Figure 707.11(2)]. Two of the three exceptions enlarge the "open shaft bottom" concept to include two other traditional shaft systems.

Exception 1 provides for an open shaft bottom where the shaft is to be utilized for a purpose on the bottom floor, such as an exhaust system. The shaft walls are not to be interrupted with protected or unprotected openings or penetrations. Bottom closure by draftstopping is required around items entering the base of the shaft [see Figure 707.11(3)], or the room that contains the shaft bottom is to be equipped throughout with an automatic fire suppression system complying with the requirements of Chapter 9 [see Figure 707.11(4)]. Either protection method will reduce the migration of products of combustion into the shaft. The exception differs from the main text of the section in that the room use is not required to be related to the purpose of the shaft. For example, a toilet room at the bottom of an air shaft would be permitted by this exception, even though the use of the toilet room does not relate to the purpose of the shaft to the same extent as an HVAC room.

Exception 2 is unique to refuse or laundry chute shafts and prohibits the options afforded in this section. This exception mandates compliance with Section 707.13.4.

Exception 3 is intended to deal with a shaft, such as a light well, where the shaft volume can be considered as part of the story in which the bottom of the shaft terminates. The conditions for the exception are based on the fact that the shaft contents will not contribute to the fuel load and that the shaft serves and connects to one story—the story in which it terminates [see Figure 707.11(5)].

FIGURE 707.11(1) – FIGURE 707.11(5)   FIRE-RESISTANCE-RATED CONSTRUCTION

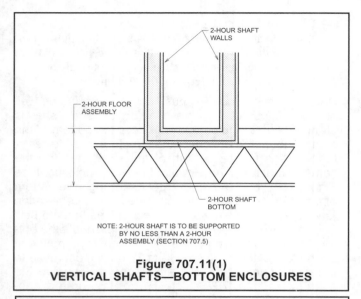

NOTE: 2-HOUR SHAFT IS TO BE SUPPORTED
BY NO LESS THAN A 2-HOUR
ASSEMBLY (SECTION 707.5)

Figure 707.11(1)
VERTICAL SHAFTS—BOTTOM ENCLOSURES

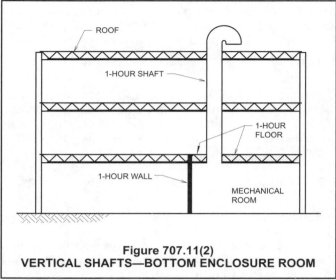

Figure 707.11(2)
VERTICAL SHAFTS—BOTTOM ENCLOSURE ROOM

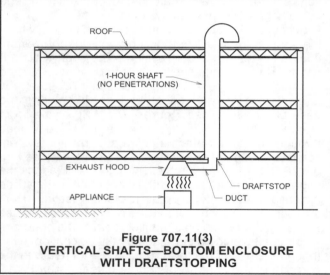

Figure 707.11(3)
VERTICAL SHAFTS—BOTTOM ENCLOSURE
WITH DRAFTSTOPPING

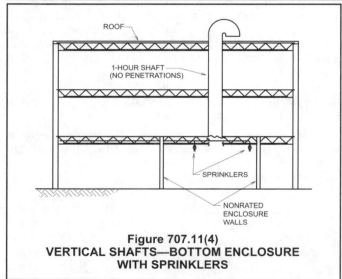

Figure 707.11(4)
VERTICAL SHAFTS—BOTTOM ENCLOSURE
WITH SPRINKLERS

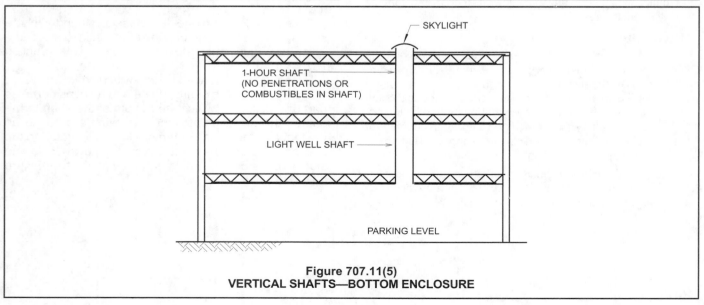

Figure 707.11(5)
VERTICAL SHAFTS—BOTTOM ENCLOSURE

707.12 Enclosure at the top. A shaft enclosure that does not extend to the underside of the roof deck of the building shall be enclosed at the top with construction of the same fire-resistance rating as the topmost floor penetrated by the shaft, but not less than the fire-resistance rating required for the shaft enclosure.

❖ Proper shaft enclosures must include all sides and the top unless the top of the shaft is also the roof of the building. The fire-resistance rating for the top of the shaft must not be less than the required fire-resistance of the shaft enclosure (see Figure 707.12). The required rating for the top of the shaft that extends to the roof deck must be consistent with the requirements of Table 601 and Section 711 for roof construction.

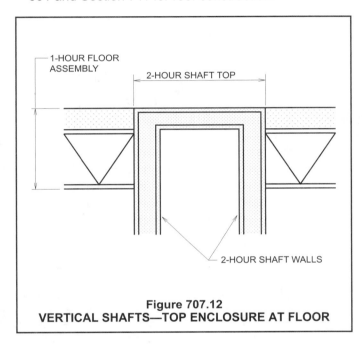

Figure 707.12
VERTICAL SHAFTS—TOP ENCLOSURE AT FLOOR

707.13 Refuse and laundry chutes. Refuse and laundry chutes, access and termination rooms and incinerator rooms shall meet the requirements of Sections 707.13.1 through 707.13.6.

Exception: Chutes serving and contained within a single dwelling unit.

❖ Refuse- and laundry-handling systems represent a rapid fire spread or detonation hazard similar to other mechanical systems, including dust and stock conveying systems and hazardous exhausts. The treatment that such systems receive in this section is similar to that afforded similar mechanical systems.

Most of its emphasis is on the vertical chute systems because such systems interconnect multiple stories and contain combustible material. Additionally, these systems all too often receive an ignition source, such as discarded chemicals and materials capable of producing spontaneous combustion, smoking materials that have not been extinguished and hot ashes, embers and charcoal. Such systems are also subject to acts of vandalism. The construction, fire suppression and termination requirements of refuse- and laundry-handling systems

are set forth herein.

The exception recognizes the small size, limited use and occupant control of refuse and laundry chutes commonly installed within dwelling units.

707.13.1 Refuse and laundry chute enclosures. A shaft enclosure containing a refuse or laundry chute shall not be used for any other purpose and shall be enclosed in accordance with Section 707.4. Openings into the shaft, including those from access rooms and termination rooms, shall be protected in accordance with this section and Section 715. Openings into chutes shall not be located in exit access corridors. Opening protectives shall be self-closing or automatic-closing upon the actuation of a smoke detector installed in accordance with Section 907.10, except that heat-activated closing devices shall be permitted between the shaft and the termination room.

❖ This section requires that the vertical systems or chutes be enclosed in a fire-resistance-rated shaft in accordance with Section 707.4. As with all shafts, the fire-resistance integrity of the shaft must be protected at all penetrations of the shaft enclosure. All access openings and termination points must be treated as shaft enclosure penetrations requiring opening protectives. Refuse and laundry access openings are prohibited from being located in exit access corridors or exits, and fire records support this precaution. While a corridor is a very convenient location for access to these chutes, a relatively minor fire in the chute could render the means of egress elements unusable.

707.13.2 Materials. A shaft enclosure containing a refuse or laundry chute shall be constructed of materials as permitted by the building type of construction.

❖ The types of materials used in fire barriers are to be consistent with Sections 602 through 602.5 for the type of construction classification of the building. The fire-resistance ratings of fire barriers used to separate mixed occupancies are determined in accordance with Section 302.3 (see commentary, Section 302.3). Fire barriers are permitted to be of combustible materials in Type III, IV and V construction and are required to be of noncombustible materials in Type I and II construction.

707.13.3 Refuse and laundry chute access rooms. Access openings for refuse and laundry chutes shall be located in rooms or compartments completely enclosed by construction that has a fire-resistance rating of not less than 1 hour and openings into the access rooms shall be protected by opening protectives having a fire protection rating of not less than $^3/_4$ hour and shall be self-closing or automatic-closing upon the detection of smoke.

❖ Openings into refuse and laundry chutes, in addition to having opening protectives, must be enclosed in a room having a fire-resistance rating. Like all fire-resistance-rated rooms, the egress door into the access room enclosing the shaft access opening must be a fire door (see Figure 707.13.3). This further emphasizes the precautions necessary with systems that connect multiple stories.

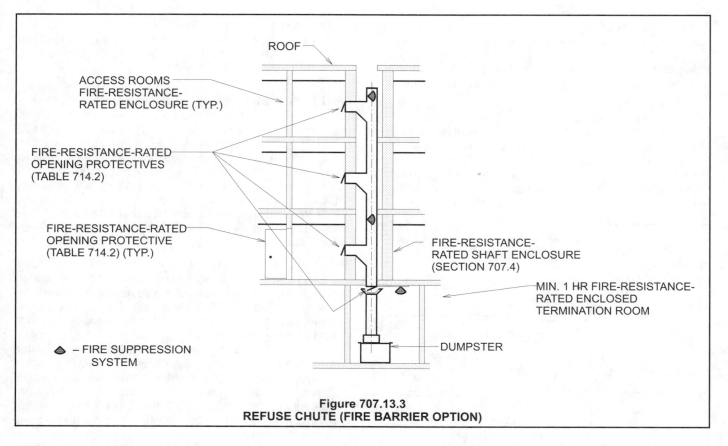

Figure 707.13.3
REFUSE CHUTE (FIRE BARRIER OPTION)

Labels in figure:
- ROOF
- ACCESS ROOMS FIRE-RESISTANCE-RATED ENCLOSURE (TYP.)
- FIRE-RESISTANCE-RATED OPENING PROTECTIVES (TABLE 714.2)
- FIRE-RESISTANCE-RATED OPENING PROTECTIVE (TABLE 714.2) (TYP.)
- FIRE-RESISTANCE-RATED SHAFT ENCLOSURE (SECTION 707.4)
- MIN. 1 HR FIRE-RESISTANCE-RATED ENCLOSED TERMINATION ROOM
- DUMPSTER
- ◆ – FIRE SUPPRESSION SYSTEM

707.13.4 Termination room. Refuse and laundry chutes shall discharge into an enclosed room completely separated from the remainder of the building by construction that has a fire-resistance rating of not less than 1 hour and openings into the termination room shall be protected by opening protectives having a fire protection rating of not less than $^3/_4$ hour and shall be self-closing or automatic-closing upon the detection of smoke. Refuse chutes shall not terminate in an incinerator room. Refuse and laundry rooms that are not provided with chutes need only comply with Table 302.1.1.

❖ Refuse and laundry chutes are required to terminate in rooms enclosed by fire barriers so as to segregate such rooms from all parts of the building. The termination room represents the collection and fuel load concentration point and, as such, must be protected not only from outside ignition sources, but also to retard the spread of fire originating in the termination room. Accordingly, the waste chute must not terminate in a room containing an incinerator. The purpose of waste chutes originally was to collect the refuse for incineration; thus, proximity to the incinerator was thought desirable. The hazards of ignition, however, preclude such convenience.

707.13.5 Incinerator room. Incinerator rooms shall comply with Table 302.1.1.

❖ This section requires that incinerators be enclosed by fire barriers (see Table 302.1.1) for the same reasons that Section 707.13.4 requires termination rooms to be enclosed.

707.13.6 Automatic fire sprinkler system. An approved automatic fire sprinkler system shall be installed in accordance with Section 903.2.10.2.

❖ This section requires that the chute, termination room and incinerator room associated with a waste or linen system be protected with an automatic fire sprinkler. Note that the requirement for suppression is within the chute itself and not within the required shaft that encloses the chute. Section 903.2.10.2 identifies the location of the sprinkler protection.

707.14 Elevator and dumbwaiter shafts. Elevator hoistway and dumbwaiter enclosures shall be constructed in accordance with Section 707.4 and Chapter 30.

❖ The hoistway enclosure is the fixed structure consisting of vertical walls or partitions that isolates the enclosure from all other building areas or from an adjacent enclosure in which the hoistway doors and door assemblies are installed. With the exception of observation elevators, the hoistway is normally enclosed with fire barriers (see Section 707.4). A hoistway enclosure for fire spread purposes may not be required if suitable protection measures are provided. Section 707.2 lists exceptions for shaft enclosures around floor openings and is applicable to all hoistways. In addition, shaft enclosures are not required for elevators located in an atrium, since there is no penetration of floor assemblies. Elevator hoistways are enclosed to ensure that flame, smoke and hot gases from a fire do not have an avenue of travel from one floor to another through a concealed space (see the discussion of stack effect in the commentary to Section 707.14.1). Enclosures are also provided to restrict contact with moving equipment and to protect people from falling.

707.14.1 Elevator lobby. Elevators opening into a fire-resistance-rated corridor as required by Section 1016.1 shall be provided with an elevator lobby at each floor containing such a corridor. The lobby shall separate the elevators from the corridor by fire partitions and the required opening protection. Elevator lobbies shall have at least one means of egress complying with Chapter 10 and other provisions within this code.

Exceptions:

1. In office buildings, separations are not required from a street-floor elevator lobby provided the entire street floor is equipped with an automatic sprinkler system in accordance with Section 903.3.1.1.

2. Elevators not required to be located in a shaft in accordance with Section 707.2.

3. Where additional doors are provided in accordance with Section 3002.6. Such doors shall be tested in accordance with UL 1784 without an artificial bottom seal.

4. In other than Group I-3, and buildings more than four stories above the lowest level of fire department vehicle access, lobby separation is not required where the building, including the lobby and corridors leading to the lobby, is protected by an automatic sprinkler system installed throughout in accordance with Section 903.3.1.1 or 903.3.1.2.

❖ Elevator shafts create a passage for the accumulation and spread of hot smoke and gases from a fire to upper stories of a building. The majority of deaths in fires are a result of smoke. For example, in the fire at the MGM Grand Hotel in Las Vegas, 70 of the 84 deaths occurred on the upper floors where smoke concentration was the greatest, even though the fire was at the first floor. This could be attributed to stack effect, which is a phenomenon that exists in high-rise buildings. Stack effect describes the movement of air inside and outside a building. During a fire, the presence of stack effect generally results in the movement of smoke and combustion products from lower levels to upper levels through shafts in the building. This section requires that an elevator lobby be provided between the elevator and the corridor where the corridor is required to be of fire-resistance-rated construction; therefore, it is important that Table 1016.1 be consulted to determine if a fire-resistance-rated corridor is required. For example, a sprinklered building of Groups A, B, E, F, M, S and U would not require fire-resistance-rated corridors, while an unsprinklered building would require such corridors when the occupant load served by the corridor exceeds 30. Conversely, both sprinklered and unsprinklered buildings of Group R would require fire-resistance-rated corridors when such corridors serve more than 10 occupants. The potential for smoke migration via the stack

effect is reduced by a sprinkler system. A sprinkler system, coupled with a low-rise structure, creates a scenario where smoke movement is not a significant problem that results in an untenable situation. Exception 4 acknowledges this condition and is limited to four-story buildings in all groups other than Group I-3 that are fully sprinklered in accordance with NFPA 13 or 13R (as applicable). These buildings do not require the elevator lobby.

Effectively, the following buildings that contain elevators opening into a corridor would require elevator lobbies:

- All Group I-3 buildings.
- Nonsprinklered buildings of all groups that have required fire-resistance-rated corridors, except office buildings in compliance with Exception 1.
- Five story and higher fully sprinklered buildings of Groups H, I-1 and R.

Exception 3 relates to the special attention required for additional doors in a hoistway. The elevator door is required to meet the provisions for openings in a corridor, which allows that opening to be tested with an artificial seal across the bottom when tested in accordance with UL 1784. Any additional doors need to be tested without the artificial bottom seal. Testing without the artificial bottom seal is a more rigorous test, as it requires the door construction to be tighter to meet the leakage rating requirements of UL 1784.

SECTION 708
FIRE PARTITIONS

708.1 General. The following wall assemblies shall comply with this section.

1. Walls separating dwelling units in the same building.

2. Walls separating sleeping units in occupancies in Group R-1, hotel occupancies, R-2 and I-1.

3. Walls separating tenant spaces in covered mall buildings as required by Section 402.7.2.

4. Corridor walls as required by Section 1016.1.

5. Elevator lobby separation as required by Section 707.14.1.

❖ Fire partitions are wall assemblies that enclose an exit access corridor or that separate tenant spaces in covered malls, dwelling units and guestrooms. Openings in fire partitions must be properly protected, but the amount of openings in a fire partition is not limited. Fire partitions must be continuous from floor slab to the floor slab or roof deck above or to a fire-resistance-rated floor/ceiling or roof/ceiling assembly. This is what distinguishes fire partitions from fire barriers. Although fire partitions must normally be supported by construction having a comparable fire-resistance rating, in buildings of Type IIB, IIIB and VB construction, exit access corridor walls and guestroom separation walls are not required to be supported by fire-resistance-rated construction, according to Section 708.4. Also, tenant separations in covered mall buildings of IIB, IIIB and VB construction are not required to be supported by fire-resistance-rated construction.

This section identifies the types of walls that are considered fire partitions. The term "Fire partition" is defined in Section 702 as a partition designed to restrict the spread of fire with protected openings. As such, a corridor wall that is not required to be fire-resistance rated in accordance with Table 1016.1 would not be considered a fire partition. This is especially important when considering the continuity provisions of Section 708.11, which would not be applicable. This is reinforced in Section 1016.1.

708.2 Materials. The walls shall be of materials permitted by the building type of construction.

❖ The types of materials used in fire partitions are to be consistent with Sections 602.2 through 602.5 for the type of construction classification of the building. Fire partitions are permitted to be of combustible construction in Type III, IV and V construction, and are required to be of noncombustible materials in Type I and II construction, except as permitted in Exception 1 to Section 603.1.

708.3 Fire-resistance rating. The fire-resistance rating of the walls shall be at least 1 hour.

Exceptions:

1. Corridor walls as permitted by Table 1016.1.

2. Dwelling unit and sleeping unit separations in buildings of Type IIB, IIIB and VB construction shall have fire-resistance ratings of not less than $^1/_2$ hour in buildings equipped throughout with an automatic sprinkler system in accordance with Section 903.3.1.1.

❖ The reduction in fire-resistance rating allowed in the exception to Section 708.3 is applicable to all dwelling unit and sleeping unit separation walls. The exception permits dwelling unit and sleeping unit separations in buildings of Type IIB, IIIB and VB construction to be reduced in required fire-resistance rating from 1 hour to $^1/_2$ hour when the building is equipped throughout with an automatic sprinkler system. "Dwelling units" are defined as single units providing complete, independent living facilities, which include permanent provisions for living, sleeping, eating, cooking and sanitation. Sleeping units are intended for use primarily for sleeping, such as those found in Groups R-1, R-2 and I-1, which do not necessarily satisfy the criteria to be considered as a dwelling unit.

708.4 Continuity. Fire partitions shall extend from the top of the floor assembly below to the underside of the floor or roof slab or deck above or to the fire-resistance-rated floor/ceiling or roof/ceiling assembly above, and shall be securely attached thereto. If the partitions are not continuous to the deck, and where constructed of combustible construction, the space between the ceiling and the deck above shall be fireblocked or draftstopped in accordance with Sections 717.2.1 and 717.3.1 at the partition line. The supporting construction shall be protected to afford the required fire-resistance rating of the wall sup-

ported, except for tenant and sleeping unit separation walls and exit access corridor walls in buildings of Type IIB, IIIB and VB construction.

Exceptions:

1. The wall need not be extended into the crawl space below where the floor above the crawl space has a minimum 1-hour fire-resistance rating.

2. Where the room-side fire-resistance-rated membrane of the corridor is carried through to the underside of a fire-resistance-rated floor or roof above, the ceiling of the corridor shall be permitted to be protected by the use of ceiling materials as required for a 1-hour fire-resistance-rated floor or roof system.

3. Where the corridor ceiling is constructed as required for the corridor walls, the walls shall be permitted to terminate at the upper membrane of such ceiling assembly.

4. The fire partition separating tenant spaces in a mall, complying with Section 402.7.2, is not required to extend beyond the underside of a ceiling that is not part of a fire-resistance-rated assembly. A wall is not required in attic or ceiling spaces above tenant separation walls.

5. Fireblocking or draftstopping is not required at the partition line in Group R-2 buildings that do not exceed four stories in height provided the attic space is subdivided by draftstopping into areas not exceeding 3,000 square feet (279 m²) or above every two dwelling units, whichever is smaller.

6. Fireblocking or draftstopping is not required at the partition line in buildings equipped with an automatic sprinkler system installed throughout in accordance with Section 903.3.1.1 or 903.3.1.2 provided that automatic sprinklers are installed in combustible floor/ceiling and roof/ceiling spaces.

❖ To minimize the potential for fire spread from the exposed side of the fire partition to the unexposed side, such partitions must be continuous from the floor assembly to the underside of a fire-resistance-rated floor/ceiling assembly or roof/ceiling assembly. In the absence of a rated floor/ceiling or roof/ceiling assembly, the fire partition is to be continuous to the floor slab or roof deck above (see Figure 708.4). All hollow vertical spaces in the fire partition must be fireblocked at the ceiling and floor or roof levels in accordance with Section 717.2.1, and draftstopped in line with the wall with materials in accordance with Section 717.3.1. Fire partitions are not required to be continuous through concealed spaces whereas fire barriers are to be continuous (see Section 706.4). This is the primary difference between fire barriers and fire partitions.

Fire partitions serving as exit access corridor walls and tenant and guestroom separation walls in buildings of Type IIB, IIIB and VB construction are not required to be supported by structural elements having the same fire-resistance rating. The primary purpose of rated corridor walls is to prevent fire spread from a room to the corridor to maintain the protection of the means of egress. Secondarily, the rated corridor wall also prevents fire spread from a corridor to adjacent rooms should the fire involve the corridor; therefore, the fire-resistance rating of the floor construction supporting the wall is not critical to the performance of the corridor wall. The construction supporting tenants and guestroom separation walls is exempted from the requirements for fire-resistance rating since the compartmentalization provided by these walls is similar to that provided by corridor walls. Additionally, if the supporting construction were required to be fire-resistance rated, this section would effectively result in all buildings containing guestrooms (occupancies in Groups R-1, R-2 and I-1) being built of protected construction only.

The remaining construction types are not included, since the supporting structural elements (floors and columns) are required to have at least a 1-hour rating, which is the maximum fire-resistance rating required for corridor walls. Similarly, tenant separations in covered mall buildings of Type IIB construction are not required to be supported by construction having an equivalent fire-resistance rating.

Exception 2 allows for only one side (the room side) of a fire-resistance-rated wall assembly to be continuous to the rated assembly above. The membrane on the corridor side of the wall is carried up to the underside of the corridor ceilings, provided the corridor ceiling is constructed of materials that have been tested as part of a 1-hour assembly.

Exception 3 allows for a corridor to be constructed in a "tunnel" fashion—with rated walls and a rated top. In these cases, the walls are not required to extend to the underside of a rated assembly.

In order to limit the spread of smoke, tenant separation walls in a mall are required to be fire partitions (see Sections 402.7.2 and 708.1) having a fire-resistance rating of a least 1 hour and extending from the floor to the underside of the ceiling. Exception 4 acknowledges that extending tenant separations to the floor slab or roof deck above is not always practical or possible because of operation of the HVAC system. The effective-

ness of the automatic sprinkler system is also a reason for not requiring tenant separations to extend above the ceiling, including attic spaces.

Exception 5 requires fire walls above every other dwelling unit or at a maximum spacing of 3,000 square feet (279 m²). This design requirement allows for proper ventilation of the attic space above dwelling units. In typical multiple-family construction, with dwelling units along the front and back of the building, trusses run front to rear with soffit vents at each end. If a wall is required at each dwelling unit, the wall will block cross ventilation, eliminating the use of ridge vents. Soffit and ridge venting allow natural air circulation that, in turn, lowers the roof sheathing temperature in the winter, relieving many of the problems associated with ice dams. This exception is the same as permitted for draftstopping in Section 717.4.2, Exception 3.

Exception 6 recognizes the added protection afforded a building that is equipped throughout with an automatic sprinkler system in accordance with Section 903.3.1.1 or 903.3.1.2 and NFPA 13 and 13R, respectively, and permits those fire partitions used for dwelling unit and guestroom separations to terminate at the underside of the ceiling membrane at the top floor of the building. For buildings equipped with a sprinkler system that conforms to the NFPA 13 standard, the attic area is required to be sprinklered, thus providing protection that offsets the fact that the dwelling unit separations do not extend to the roof deck. In buildings equipped throughout with an NFPA 13R system, the attic areas may not be sprinklered. NFPA 13R allows the omission of sprinkler protection in combustible attics, provided the space is not used for living purposes or storage. Sprinkler protection, however, is necessary, since the protection is considered available to control fires in the incipient stage and keep unoccupied attic areas from becoming involved. For these buildings, the exception requires sprinkler protection and the fire partitions do not need to extend to the roof deck.

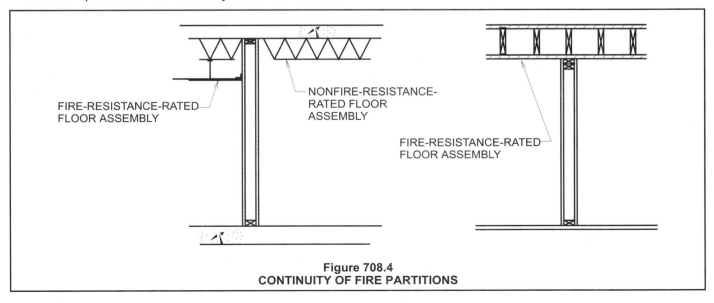

FIRE-RESISTANCE-RATED FLOOR ASSEMBLY

NONFIRE-RESISTANCE-RATED FLOOR ASSEMBLY

FIRE-RESISTANCE-RATED FLOOR ASSEMBLY

Figure 708.4
CONTINUITY OF FIRE PARTITIONS

708.5 Exterior walls. Where exterior walls serve as a part of a required fire-resistance-rated enclosure, such walls shall comply with the requirements of Section 704 for exterior walls and the fire-resistance-rated enclosure requirements shall not apply.

❖ If an area is required to be separated by fire partitions and an exterior wall constitutes part of the separation, the exterior wall is only required to comply with the fire-resistance-rating requirements for exterior walls in Section 704. The intent of the fire partition requirements is to subdivide or separate areas to protect them from a fire in the building.

708.6 Openings. Openings in a fire partition shall be protected in accordance with Section 715.

❖ Section 715.3 includes the requirements for fire doors in fire partitions; namely, $^1/_3$-hour doors are required for corridors and $^3/_4$-hour doors for tenant dwelling unit and guestroom separations in accordance with Table 715.3. It is important to note that Section 715.3.3 requires smoke and draft control doors for corridors. Section 715.4 includes the requirements for windows located in a corridor.

708.7 Penetrations. Penetrations through fire partitions shall comply with Section 712.

❖ This section simply states that penetrations through fire partitions are required to conform to the requirements of Sections 712.2 and 712.3. This is the same as the penetration requirements for fire barriers.

708.8 Joints. Joints made in or between fire partitions shall comply with Section 713.

❖ Joints, such as expansion or seismic, are another form of opening in fire partitions, and, therefore, must be considered with regard to maintaining the fire-resistance ratings of these walls. This section requires all joints that are located in fire partitions to be protected by a joint system that has a fire-resistance rating and complies with the requirements of Section 713 (see commentary, Section 713).

708.9 Ducts and air transfer openings. Penetrations by ducts and air transfer openings shall comply with Sections 712 and 716.

❖ See Sections 712 and 716.

SECTION 709
SMOKE BARRIERS

709.1 General. Smoke barriers shall comply with this section.

❖ Smoke barriers divide areas of a building into separate smoke compartments. A smoke barrier is designed to resist fire and smoke spread so that occupants can be evacuated horizontally to adjacent smoke compartments. This concept has proven effective in Group I-2 and I-3 occupancies. Sections 407.4 and 408.6 identify where smoke barriers are required. Also, while not cross referenced in this section, smoke barriers may be utilized in other applications, such as part of a smoke control system (see Section 909.5) and accessible means of egress (see Section 1007.6.2).

709.2 Materials. Smoke barriers shall be of materials permitted by the building type of construction.

❖ The types of materials used in smoke barriers are to be consistent with Sections 602.2 through 602.5 for the type of construction classification of the building. Smoke barriers are permitted to be of combustible construction in Type III, IV and V construction, and are required to be of noncombustible materials in Type I and II construction, except as permitted in Exception 1 to Section 603.1.

709.3 Fire-resistance rating. A 1-hour fire-resistance rating is required for smoke barriers.

Exception: Smoke barriers constructed of minimum 0.10-inch-thick (2.5 mm) steel in Group I-3 buildings.

❖ Smoke barriers are intended to create an area of refuge; therefore, they are to be capable of resisting the passage of smoke (see Section 709.4). Smoke barriers are also required to have a fire-resistance rating of at least 1 hour. The smoke barrier is not intended or expected to be exposed to fire for extended periods and is, therefore, not required to have a fire-resistance rating exceeding 1 hour. The occupancies in which smoke barriers are required are also generally required to be sprinklered (see Section 903.2.5). The exception of this section allows smoke barriers in occupancies in Group I-3 to be constructed of nominal 0.10-inch (2.5 mm) steel plate. This exception to the 1-hour fire-resistance-rated assembly recognizes the security needs of such facilities and at the same time provides the requisite smoke barrier performance.

709.4 Continuity. Smoke barriers shall form an effective membrane continuous from outside wall to outside wall and from floor slab to floor or roof deck above, including continuity through concealed spaces, such as those found above suspended ceilings, and interstitial structural and mechanical spaces. The supporting construction shall be protected to afford the required fire-resistance rating of the wall or floor supported in buildings of other than Type IIB, IIIB or VB construction.

Exception: Smoke barrier walls are not required in interstitial spaces where such spaces are designed and constructed with ceilings that provide resistance to the passage of fire and smoke equivalent to that provided by the smoke barrier walls.

❖ Smoke barriers are to be continuous from outside wall to outside wall and from floor slab to floor or roof deck above. The provisions require the barrier to be continuous through all concealed and interstitial spaces, such as the space between the ceiling and the floor or roof deck above. Smoke barriers are not required to extend through interstitial spaces if the space is designed and constructed such that fire and smoke will not spread from one smoke compartment to another; therefore, the construction assembly forming the bottom of the interstitial space must provide the required fire-resistance rating and be capable of resisting the passage of smoke from the spaces below.

Since the primary performance of smoke barriers is to achieve protection on the fire floor, the supporting construction is not required to provide the same degree of fire resistance for buildings of Type IIB, IIIB and VB construction. As with fire partitions serving as exit access corridor walls (see Section 708.4), these three construction types are identified because the floor construction is not otherwise required to have a fire-resistance rating, and it is not considered essential to only require fire-resistance-rated floor construction because the floor is supporting a smoke barrier.

When designing for occupancies in Groups I-2 and I-3, it is often desirable to incorporate a horizontal exit and a smoke barrier. The code does not prevent such a combination, but in such cases, the wall construction must meet the provisions of this section as well as Section 706 for fire barrier assemblies and Section 1021 for horizontal exits. The more restrictive provisions of each section would apply. For example, the building would be required to have a fire barrier with a fire-resistance rating of at least 2 hours (see Section 1021.2) and duct penetrations would need to be protected with a combination fire and smoke damper (see Sections 716.5.2 and 716.5.5) or with separate fire and smoke dampers. Furthermore, the supporting construction would then be required to provide at least the same fire-resistance rating as the wall in all types of construction (see Section 706.4).

709.5 Openings. Openings in a smoke barrier shall be protected in accordance with Section 715.

> **Exception:** In Group I-2, where such doors are installed across corridors, a pair of opposite-swinging doors without a center mullion shall be installed having vision panels with approved fire-resistance-rated glazing materials in approved fire-resistance-rated frames, the area of which shall not exceed that tested. The doors shall be close fitting within operational tolerances, and shall not have undercuts, louvers or grilles. The doors shall have head and jamb stops, astragals or rabbets at meeting edges and automatic-closing devices. Positive-latching devices are not required.

❖ In order to maintain the integrity of the smoke barrier, door assemblies and double means of egress corridor doors are required to have a fire protection rating of at least 20 minutes when tested in accordance with Section 716. In occupancies in Group I-2, however, double means of egress cross-corridor doors are not required to have positive latching hardware, since center mullions are prohibited, as they could serve as a possible obstruction in moving beds through the door opening. Without a center mullion, the latching hardware would utilize the floor below and the door frame above for latching purposes. The required stops and door-closing

devices will assist in keeping the doors closed. Doors, other than cross-corridor doors, are required to be provided with positive latches as required for the fire protection rating of the door.

When double egress corridor doors are provided, they are required to have a vision panel consisting of protection-rated glazing, in accordance with Section 715.3.6. Vision panels are required to reduce the likelihood of an injury caused by opening the door into a person standing at or approaching the door from the opposite side. Vision panels also enable an individual to be alerted to a fire or smoke condition on the other side before opening the door. Since the primary purpose of a smoke barrier is to resist smoke spread, the doors are not to have undercuts, louvers or grilles. The only openings permitted in the door assembly are those clearances that are necessary for the proper operation of the door. The opening that occurs at the meeting edges of the door is to be protected with rabbets or astragals (see Figure 709.5). In this manner, the doors will close completely, thus preventing the spread of smoke. Stops are to be provided on the head and jambs of all doors in smoke barriers. Such doors must also comply with UL 1784 in accordance with Section 715.3.3.

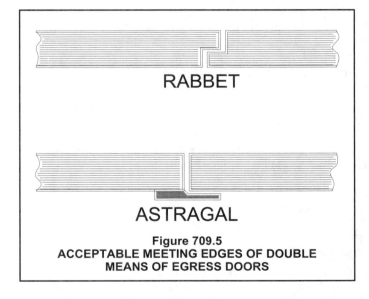

Figure 709.5
ACCEPTABLE MEETING EDGES OF DOUBLE
MEANS OF EGRESS DOORS

709.6 Penetrations. Penetrations through smoke barriers shall comply with Section 712.

❖ The provisions for penetrations of smoke barriers are found in Sections 712.2 and 712.3.

709.7 Joints. Joints made in or between smoke barriers shall comply with Section 713.

❖ Joints, such as expansion or seismic, are another form of openings in fire partitions, and, therefore, must be considered with regard to maintaining the fire-resistance ratings of these walls. This section requires all joints that are located in fire partitions to be protected by a joint system that has a fire-resistance rating and complies with the requirements of Section 713 (see commentary, Section 713).

709.8 Ducts and air transfer openings. Penetrations by ducts and air transfer openings shall comply with Sections 712 and 716.

❖ See Sections 712 and 716.

SECTION 710
SMOKE PARTITIONS

710.1 General. Smoke partitions installed as required elsewhere in the code shall comply with this section.

❖ Section 407.3 requires corridor walls for Group I-2 occupancies to be constructed to resist the passage of smoke. The construction is a smoke partition as described in this section.

710.2 Materials. The walls shall be of materials permitted by the building type of construction.

❖ As with most wall or partition assemblies, except for fire walls, there are no additional requirements or restrictions on the materials used, except that they meet the requirements for the type of construction.

710.3 Fire-resistance rating. Unless required elsewhere in the code, smoke partitions are not required to have a fire-resistance rating.

❖ Smoke partitions are used to prevent the ready and quick passage of smoke into corridors in Group I-2. The automatic sprinkler system obviates the need for a fire-resistance-rated assembly in this use; however, the issue of smoke propagation must be addressed with the use of a smoke partition.

710.4 Continuity. Smoke partitions shall extend from the floor to the underside of the floor or roof deck above or to the underside of the ceiling above where the ceiling membrane is constructed to limit the transfer of smoke.

❖ The continuity provisions for smoke partitions are similar to those for fire partitions (see Section 708.4), except that the issue is smoke spread, not fire; therefore, the allowance for termination at the underside of the ceiling membrane (as opposed to the underside of the floor or roof deck above) relates to the ability of the ceiling membrane to limit the spread of smoke. Typical "lay in" ceiling tiles, for instance, would not serve this function; however, a drywall ceiling that is taped and finished would be an example of construction that could resist the passage of smoke.

710.5 Openings. Windows shall be sealed to resist the free passage of smoke or be automatic-closing upon detection of smoke. Doors in smoke partitions shall comply with this section.

❖ The issue related to openings in smoke partitions is only the passage of smoke.

710.5.1 Louvers. Doors in smoke partitions shall not include louvers.

❖ Louvers in doors are a ready opening to allow smoke to move from an incidental use area to the remainder of the building. This would defeat the sole purpose of a smoke partition.

710.5.2 Smoke and draft-control doors. Where required elsewhere in the code, doors in smoke partitions shall be tested in accordance with UL 1784 with an artificial bottom seal installed across the full width of the bottom of the door assembly. The air leakage rate of the door assembly shall not exceed 3.0 cubic feet per minute per square foot [$ft^3/(min\ ft^2)$] (0.015424 m^3/sm^2) of door opening at 0.10 inch (24.9 Pa) of water for both the ambient temperature test and the elevated temperature exposure test.

❖ The requirements for testing doors for smoke partitions in accordance with UL 1784 are consistent with the requirements for all fire door assemblies (see Section 715.3.3).

710.5.3 Self-closing or automatic-closing doors. Where required elsewhere in the code, doors in smoke partitions shall be self-closing or automatic-closing in accordance with Section 715.3.7.3.

❖ The provisions for a smoke partition are such that doors are required to remain closed when not in use, or otherwise be capable of closing when a fire is detected. This is entirely consistent with the requirements for all doors in fire-resistance-rated assemblies. Without devices to automatically close doors or self-closing doors, the smoke partition as a method of limiting smoke spread is rendered useless.

710.6 Penetrations and joints. The space around penetrating items and in joints shall be filled with an approved material to limit the free passage of smoke.

❖ As with all other elements related to smoke partitions, the sole issue is resisting the passage of smoke. Penetrations need to be made completely tight in the assembly such that smoke will not pass.

710.7 Ducts and air transfer openings. Air transfer openings in smoke partitions shall be provided with a smoke damper complying with Section 716.3.2.

Exception: Where the installation of a smoke damper will interfere with the operation of a required smoke control system in accordance with Section 909, approved alternative protection shall be utilized.

❖ Smoke dampers are required to be provided to maintain the integrity of the smoke partition as a means to prevent the spread of smoke. The exception relating to required operation of smoke control is consistent with the exceptions given elsewhere regarding air transfer openings and ducts with fire dampers or smoke dampers (see Section 716). An effective smoke control system will protect the building against the spread of smoke.

SECTION 711
HORIZONTAL ASSEMBLIES

711.1 General. Floor and roof assemblies required to have a fire-resistance rating shall comply with this section.

❖ The ceiling assembly is often an integral part of a fire-resistance-rated floor/ceiling or roof/ceiling assembly; therefore, the integrity of the ceiling assembly must be maintained in order to reduce the potential for premature failure of the floor or roof of a building.

Much of the information contained in the section is obtained from and based on conditions of testing various assemblies. One focus of the section is on maintaining the integrity of ceiling assemblies, which are an integral component of the fire-resistance rating of floor/ceiling and roof/ceiling assemblies.

711.2 Materials. The floor and roof assemblies shall be of materials permitted by the building type of construction.

❖ The types of materials used in floor and roof assemblies are to be consistent with Sections 602.2 through 602.5 for the type of construction classification of the building. Floors and roofs are permitted to be of combustible construction in Type III, IV and V construction, and are required to be of noncombustible materials in Type I and II construction, except as permitted in Exception 1 to Section 603.1.

711.3 Fire-resistance rating. The fire-resistance rating of floor and roof assemblies shall not be less than that required by the building type of construction. Where the floor assembly separates mixed occupancies, the assembly shall have a fire-resistance rating of not less than that required by Section 302.3.2 based on the occupancies being separated. Where the floor assembly separates a single occupancy into different fire areas, the assembly shall have a fire-resistance rating of not less than that required by Section 706.3.7. Floor assemblies separating dwelling units in the same building or sleeping units in occupancies in Group R-1, hotel occupancies, R-2 and I-1 shall be a minimum of 1-hour fire-resistance-rated construction.

> **Exception:** Dwelling unit and sleeping unit separations in buildings of Type IIB, IIIB, and VB construction shall have fire-resistance ratings of not less than $^1/_2$ hour in buildings equipped throughout with an automatic sprinkler system in accordance with Section 903.3.1.1.

❖ The requirement for the fire-resistance rating of floor and roof assemblies can be found in Table 601 and can be a function of the type of construction of the building or the occupancy of the building. As an example, consider a two-story building of Type IIA construction of Group B, with the basement having a storage occupancy in Group S-1. Assume the designer has chosen Section 302.3.2 for separated occupancies as the option for compliance with the mixed occupancy condition. In this case, even though Table 601 requires 1-hour fire-resistance-rated floor/ceiling assemblies throughout the building, the assembly separating the basement from the first floor of the building is also regulated by

Section 302 for mixed occupancies. Ultimately, Section 302.3.2 references Table 302.3.2, which requires that the floor/ceiling construction separating Group S-1 from Group B will provide a 3-hour fire-resistance-rated separation. Chapter 6 also contains provisions governing materials that may be used in the construction of such assemblies based on the type of construction of the building. The requirement for 1-hour floor assemblies separating dwelling units and sleeping units mirrors that found in Section 708 for fire partitions. There are other sections of the code that require a fire-resistance rating for the floor or roof assembly or where other provisions are dependent on the fire-resistance rating or construction of the assembly (examples of such sections include Sections 704.10, 705.6 and the exceptions to Section 1023).

The exception for dwelling unit and guestroom separations in unprotected types of construction is based on the protection provided by a sprinkler system installed in accordance with Section 903.3.1.1 (NFPA 13). This presumably will reduce the potential fire exposure of the unit separation to that which makes a minimum $^1/_2$-hour fire-resistance rating adequate. The reason that only unprotected types of construction are permitted to take advantage of this tradeoff is because Table 601 would not otherwise require the floor/ceiling assemblies of such structures to be fire-resistance rated, whereas all other types of construction require the floor/ceiling assembly to provide at least a 1-hour fire-resistance rating.

711.3.1 Ceiling panels. Where the weight of lay-in ceiling panels, used as part of fire-resistance-rated floor/ceiling or roof/ceiling assemblies, is not adequate to resist an upward force of 1 lb/ft.^{2} (48 Pa), wire or other approved devices shall be installed above the panels to prevent vertical displacement under such upward force.

❖ When a ceiling membrane constitutes part of a fire-resistance-rated floor/ceiling or roof/ceiling assembly, the ability of the ceiling membrane to remain in place and not be displaced by the upward pressure of a fire condition is necessary to maintain the viability of the fire-resistance rating. Figure 711.3.1 shows a floor/ceiling assembly that uses hold-down clips to keep the ceiling panels in place. When the weight of the ceiling panel is less than 1 pound per square foot (psf) (48 Pa), wire or other approved means must be provided to prevent uplift of the ceiling panels. Manufacturers' literature can be used to determine the weight of the panel.

711.3.2 Access doors. Access doors shall be permitted in ceilings of fire-resistance-rated floor/ceiling and roof/ceiling assemblies provided such doors are tested in accordance with ASTM E 119 as horizontal assemblies and labeled by an approved agency for such purpose.

❖ Access doors are often necessary in order to service mechanical and plumbing systems above the ceiling. This section states that if such doors are used, they must be tested in accordance with ASTM E 119 as a horizontal assembly. This makes it clear that the standard fire test for doors (NFPA 80 or 257) is not acceptable.

711.3.3 Unusable space. In 1-hour fire-resistance-rated floor construction, the ceiling membrane is not required to be installed over unusable crawl spaces. In 1-hour fire-resistance-rated roof construction, the floor membrane is not required to be installed where unusable attic space occurs above.

❖ The section is a special exception for specific assemblies to allow the deletion of floor decking or the ceiling membrane from a required fire-resistance-rated floor/ceiling assembly used in certain applications. In an attic application, the floor sheathing may be deleted from a fire-resistance-rated assembly as long as the joist and ceiling remain identical to the tested assembly and the attic space is not usable space where potential combustible materials or ignition sources may be located [see Figure 711.3.3(1)]. Over a crawl space, the ceiling membrane may be deleted subject to the same conditions [see Figure 711.3.3(2)].

In determining whether a space is "unusable," the building official must verify that combustible materials other than construction elements will not be located therein and ignition sources will be minimal. As such, pipes, conduit and ducts may be permitted in an unusable space.

711.4 Continuity. Assemblies shall be continuous without openings, penetrations or joints except as permitted by this section and Sections 707.2, 712.4 and 713. Skylights and other penetrations through a fire-resistance-rated roof deck are permitted to be unprotected, provided that the structural integrity of the fire-resistance-rated roof construction is maintained. Unprotected skylights shall not be permitted in roof construction required to be fire-resistance rated in accordance with Section

704.10. The supporting construction shall be protected to afford the required fire-resistance rating of the horizontal assembly supported.

❖ All floor/ceiling and roof/ceiling assemblies are to be continuous without openings or penetrations, except as permitted by this section. The continuity of the assembly is critical to its ability to limit fire and smoke spread. The continuity provision applies regardless of whether a fire-resistance rating is required, since floor/ceiling assemblies are also intended to restrict vertical smoke movement [see Figure 711.4(1)]. Penetrations of the roof surface of any roof assembly are permitted, provided that the fire-resistance rating, if required, is maintained as determined by the fire-resistance test or the penetration sealant in accordance with Section 711.5 [see Figure 711.4(2)]. The fire-resistance rating required by Table 601 for roof construction is intended to minimize the threat of premature structural failure of the roof construction under fire conditions. These provisions, with the exception of the requirements of Sections 704.10 and 705.6, are not intended to create a barrier in order to contain the fire within the building. Nonfire-resistance-rated skylight and roof window assemblies are, therefore, permitted to be installed in fire-resistance-rated roof assemblies, provided that the structural integrity of the roof assembly is not reduced and the provisions of Section 704.10 for protection of vertical exposure do not apply (see commentary, Section 704.10). The issue of structural integrity refers to the effect the collapse of a skylight assembly, under fire conditions, would have on the roof structure.

The reference to the fireblocking and draftstopping requirements of Section 717 is intended to alert the user of the code to these requirements, which are in addition to the provisions of this section. As noted in the commentary to Section 717, fireblocking and draftstopping are separate issues from fire-resistance ratings.

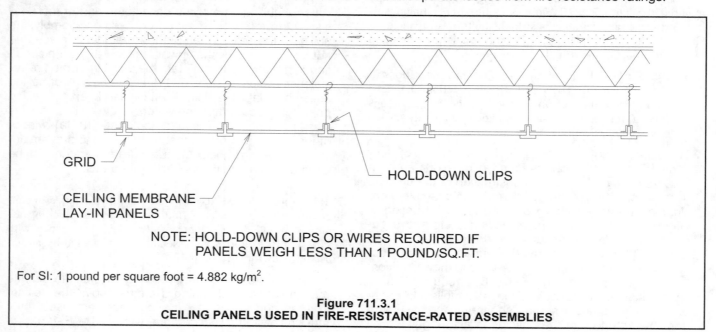

GRID

HOLD-DOWN CLIPS

CEILING MEMBRANE
LAY-IN PANELS

NOTE: HOLD-DOWN CLIPS OR WIRES REQUIRED IF
PANELS WEIGH LESS THAN 1 POUND/SQ.FT.

For SI: 1 pound per square foot = 4.882 kg/m^2.

Figure 711.3.1
CEILING PANELS USED IN FIRE-RESISTANCE-RATED ASSEMBLIES

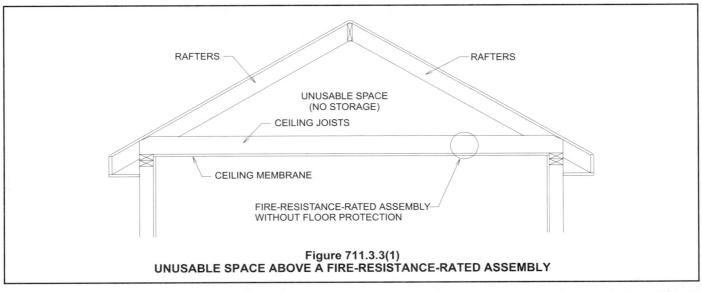

Figure 711.3.3(1)
UNUSABLE SPACE ABOVE A FIRE-RESISTANCE-RATED ASSEMBLY

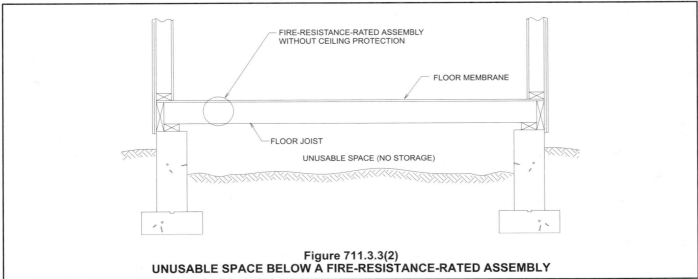

Figure 711.3.3(2)
UNUSABLE SPACE BELOW A FIRE-RESISTANCE-RATED ASSEMBLY

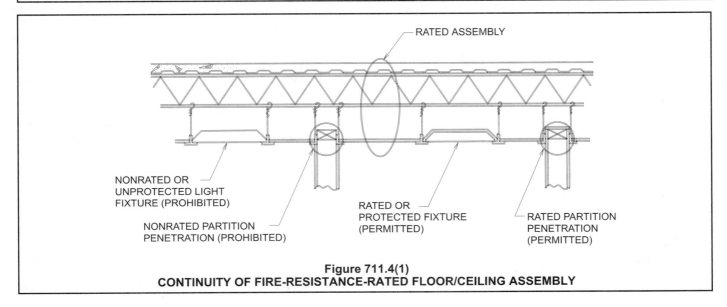

Figure 711.4(1)
CONTINUITY OF FIRE-RESISTANCE-RATED FLOOR/CEILING ASSEMBLY

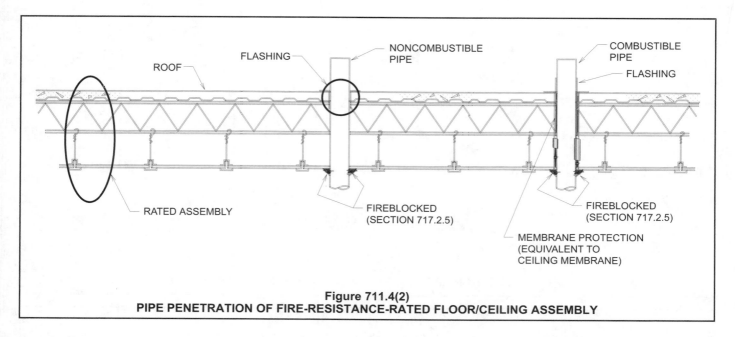

Figure 711.4(2)
PIPE PENETRATION OF FIRE-RESISTANCE-RATED FLOOR/CEILING ASSEMBLY

711.5 Penetrations. Penetrations through fire-resistance-rated horizontal assemblies shall comply with Section 712.

❖ Penetration protection for horizontal assemblies is found in Sections 712.2 and 712.4.

711.6 Joints. Joints made in or between fire-resistance-rated horizontal assemblies shall comply with Section 713. The void created at the intersection of a floor/ceiling assembly and an exterior curtain wall assembly shall be protected in accordance with Section 713.4.

❖ Joints, such as expansion or seismic, are another form of openings in fire partitions and, therefore, must be considered with regard to maintaining the fire-resistance ratings of these walls. This section requires all joints that are located in fire partitions to be protected by a joint system that has a fire-resistance rating and complies with the requirements of Section 713 (see commentary, Section 713).

711.7 Ducts and air transfer openings. Penetrations by ducts and air transfer openings shall comply with Sections 712 and 716.

❖ See Sections 712 and 716.

SECTION 712
PENETRATIONS

712.1 Scope. The provisions of this section shall govern the materials and methods of construction used to protect through penetrations and membrane penetrations.

❖ This section addresses the specific requirements for maintaining the integrity of fire-resistance-rated assemblies at penetrations. The provisions of this section apply to penetrations of fire-resistance-rated walls

(Section 712.3), fire-resistance-rated floor/ceiling and roof/ceiling assemblies (Sections 712.4, 712.4.2 and 712.4.6), nonfire-resistance-rated assemblies (Section 712.4.3) and both rated and nonrated horizontal assemblies (Sections 712.4 and 712.4.5). Penetrations of fire-resistance-rated assemblies range from combustible pipe and tubing to noncombustible wiring with combustible covering to noncombustible items such as pipe, tube, conduit and ductwork.

Each type of penetration requires a specific method of protection, which is based on the type of fire-resistance-rated assembly that is penetrated and the type of penetrating item. To determine the type of penetration protection required, the first step is to identify whether the penetrated assembly is a fire-resistance-rated wall; a fire-resistance-rated floor/ceiling or roof/ceiling assembly or a nonfire-resistance rated floor/ceiling or roof/ceiling assembly. Next, identify the type of penetrating item in the applicable section and determine the applicable method of penetration protection necessary.

712.2 Installation details. Where sleeves are used, they shall be securely fastened to the assembly penetrated. The space between the item contained in the sleeve and the sleeve itself and any space between the sleeve and the assembly penetrated shall be protected in accordance with this section. Insulation and coverings on or in the penetrating item shall not penetrate the assembly unless the specific material used has been tested as part of the assembly in accordance with this section.

❖ Sleeves are typically installed when the opening penetrates an assembly with interior voids, such as a steel-stud-framed wall. The sleeve must be securely fastened to the penetrated assembly to prevent the sleeve from becoming dislodged and adversely affecting the performance of the annular space protection. The space between the sleeve and the penetrating item, as well as between the sleeve and the rated assembly, must be protected in accordance with Section 712.3.1. Without attention to these conditions, a sleeved penetration may not perform as required by the code.

Section 712.3.1 requires that combustible penetrations be protected with assemblies that have been tested in accordance with ASTM E 119 (see Section 712.3.1) or ASTM E 814 (see Section 712.3.1.2). If a sleeve is used in conjunction with the penetrating items, then the annular space within and around the sleeve must also be protected with materials that meet the conditions of the test standards. The purpose of this section is to provide the building official with the necessary text to enforce the conditions of the penetration protection standards.

If combustible insulation material or coverings on the penetrating item could provide a path for fire to travel through the assembly, the method of protection is rendered ineffective. This is in conflict with the conditions of the test standards. This section clarifies that materials that are not part of the tested assembly, whether the assembly was tested under ASTM E 119 or is a through-penetration firestop system tested in accordance with ASTM E 814, are not allowed to pass through the penetration unless they are protected and evaluated in the test standards.

712.3 Fire-resistance-rated walls. Penetrations into or through fire walls, fire barriers, smoke barrier walls, and fire partitions shall comply with this section.

❖ In order to maintain the integrity of the fire-resistance-rated wall assembly (fire walls, fire barriers, smoke barriers and fire partitions), penetrations into and through the rated assembly must be properly protected. Acceptable protection methods for various penetrations of wall assemblies are identified in Sections 712.3.1 through 712.3.2.

712.3.1 Through penetrations. Through penetrations of fire-resistance-rated walls shall comply with Section 712.3.1.1 or 712.3.1.2.

Exception: Where the penetrating items are steel, ferrous or copper pipes or steel conduits, the annular space between the penetrating item and the fire-resistance-rated wall shall be permitted to be protected as follows:

1. In concrete or masonry walls where the penetrating item is a maximum 6-inch (152 mm) nominal diameter and the opening is a maximum 144 square inches (0.0929 m^2), concrete, grout or mortar shall be permitted where installed the full thickness of the wall or the thickness required to maintain the fire-resistance rating; or

2. The material used to fill the annular space shall prevent the passage of flame and hot gases sufficient to ignite cotton waste when subjected to ASTM E 119 time-temperature fire conditions under a minimum positive pressure differential of 0.01 inch (2.49 Pa) of water at the location of the penetration for the time pe-

riod equivalent to the fire-resistance rating of the construction penetrated.

❖ Combustible cables, wires, pipes, tubes and conduits that penetrate fire-resistance-rated walls are required to be properly protected [see Figures 712.3.1(1) and 712.3.1(2)]. These penetrations are required to be tested in accordance with ASTM E 119 (see Section 712.3.1) or ASTM E 814 (see Section 712.3.1.2).

Because combustible materials have a greater propensity to spread fire through a penetration, the requirements for combustible penetrating items are considerably more restrictive than for noncombustible penetrating items. Penetration of fire-resistance-rated walls by noncombustible cables, wires, pipes, tubes, conduits and vents represent a weak link in the continuity of the required fire-resistance rating and are required to be properly protected. These penetrations are to be protected with materials that will maintain the integrity of the wall for the duration of the required fire-resistance rating. In lieu of a system tested in accordance with ASTM E 119 (see Section 712.3.1), ASTM E 814 or UL 1479 (see Section 712.3.1.2), annular space protection in accordance with the exception is acceptable. Annular space protection is permitted for the protection of penetrations of wall assemblies by noncombustible items. The annular space is defined in Section 702.1 as the perimeter space between the penetrating item and the rated assembly. If sleeves are used, the annular space includes the space between the penetrating item and the sleeve, as well as the space between the sleeve and the rated assembly [see Figure 712.3.1(3)]. Item 1 of the exception permits the use of concrete, grout or mortar to provide annular space protection for certain penetrations of concrete and masonry wall assemblies. The concrete, grout or mortar must be provided for the full thickness of the wall, unless evidence can be provided that demonstrates that the required fire-resistance rating can be achieved with a lesser depth.

Concrete, grout and mortar have traditionally been used as protection for the annular space in penetrations of concrete and masonry walls. The presumption has been that experience has shown this form of protection to be viable.

Because annular space protection is not based on specific performance testing of each penetration arrangement and detail, the performance of the protection is dependent on the annular space protection material alone. As such, Item 2 requires that the material be prequalified as to its ability to prevent the passage of flame and hot gases sufficient to ignite cotton waste and when subjected to the time-temperature criteria of ASTM E 119. This is consistent with the criteria required for through-penetration protection systems (ASTM E 814) in which the "T" rating is not limited. Since it is very likely that the penetration in an actual fire will be exposed to a positive pressure, this section specifies that the test fire exposure include a positive pressure of 0.01 inch (0.25 mm) of water column as a further means to verify the performance of this protection method as required by the code.

FIGURE 712.3.1(1) – FIGURE 712.3.1(3)

FIRE-RESISTANCE-RATED CONSTRUCTION

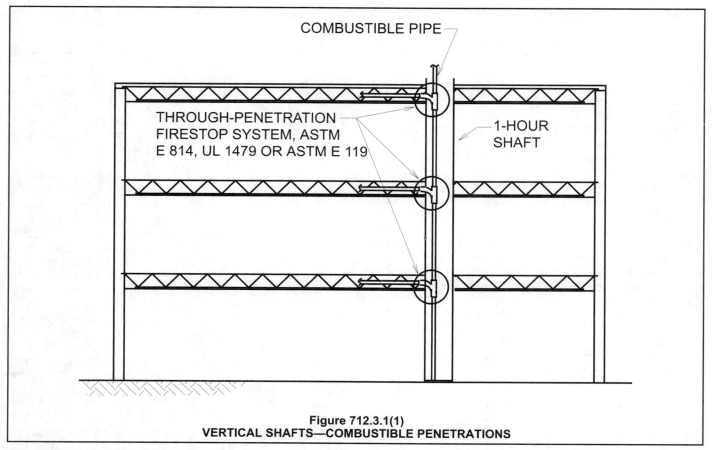

COMBUSTIBLE PIPE

THROUGH-PENETRATION
FIRESTOP SYSTEM, ASTM
E 814, UL 1479 OR ASTM E 119

1-HOUR
SHAFT

Figure 712.3.1(1)
VERTICAL SHAFTS—COMBUSTIBLE PENETRATIONS

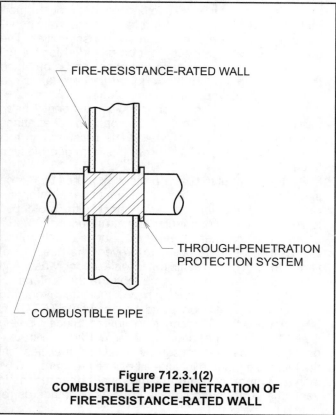

FIRE-RESISTANCE-RATED WALL

THROUGH-PENETRATION
PROTECTION SYSTEM

COMBUSTIBLE PIPE

Figure 712.3.1(2)
COMBUSTIBLE PIPE PENETRATION OF
FIRE-RESISTANCE-RATED WALL

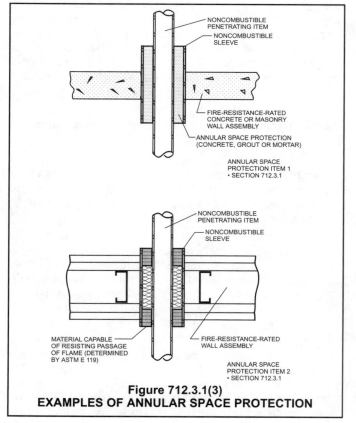

NONCOMBUSTIBLE
PENETRATING ITEM
NONCOMBUSTIBLE
SLEEVE

FIRE-RESISTANCE-RATED
CONCRETE OR MASONRY
WALL ASSEMBLY

ANNULAR SPACE PROTECTION
(CONCRETE, GROUT OR MORTAR)

ANNULAR SPACE
PROTECTION ITEM 1
• SECTION 712.3.1

NONCOMBUSTIBLE
PENETRATING ITEM
NONCOMBUSTIBLE
SLEEVE

MATERIAL CAPABLE
OF RESISTING PASSAGE
OF FLAME (DETERMINED
BY ASTM E 119)

FIRE-RESISTANCE-RATED
WALL ASSEMBLY

ANNULAR SPACE
PROTECTION ITEM 2
• SECTION 712.3.1

Figure 712.3.1(3)
EXAMPLES OF ANNULAR SPACE PROTECTION

712.3.1.1 Fire-resistance-rated assemblies. Penetrations shall be installed as tested in an approved fire-resistance-rated assembly.

❖ This section requires, as an option, that the tested assembly in accordance with ASTM E 119 include the penetrations along with the proposed type of protection—the entire assembly is then tested. As an option, this section is not used frequently due to the limitations placed on the tested assembly relative to its application. Penetration protection is most often provided in accordance with the exception to Section 712.3.1 and the provisions of Section 712.3.1.2.

712.3.1.2 Through-penetration firestop system. Through penetrations shall be protected by an approved penetration firestop system installed as tested in accordance with ASTM E 814 or UL 1479, with a minimum positive pressure differential of 0.01 inch (2.49 Pa) of water and shall have an F rating of not less than the required fire-resistance rating of the wall penetrated.

❖ In order to maintain effective compartmentation of a building to restrict the spread of fire, all penetrations through fire-resistance-rated assemblies must be protected. In recent history, there have been several examples of buildings where extensive fire damage and loss of life was attributed, at least in part, to lack of or improper installation of penetration protection. In the absence of a through-penetration firestop system (TPFS) to protect penetrations of fire-resistance-rated assemblies, the potential exists for fire to spread beyond the initial area of fire origin. One report on the source of origin of fires in buildings indicated that 23 percent of all building fires originate from electrical systems. This fact, coupled with the number of penetrations through walls and floors created by electrical distribution piping, helps to underscore the necessity for the protection of all penetrations through rated assemblies. It must be noted that penetrations are not limited to electrical systems only, but that openings to accommodate plumbing and mechanical systems also contribute to the number of penetrations through any given assembly.

TPFS consist of specific materials or an assembly of materials that are designed to restrict the passage of fire and hot gases for a prescribed period of time through openings made in fire-resistance-rated walls, floor/ceiling or roof/ceiling assemblies. In certain instances, the TPFS is also required to limit the transfer of heat from the fire side to the unexposed side. In order to determine the effectiveness of a TPFS in restricting the passage of fire and the transfer of heat, firestop systems are required to be subjected to fire testing. ASTM E 814 or UL 1479 are the test methods developed specifically for the evaluation of a firestop system's ability to resist the passage of flame and hot gases, withstand thermal stresses and restrict transfer of heat through the penetrated assembly.

The basic provisions of ASTM E 814 and UL 1479 require that a test assembly consisting of a specific wall or floor construction, containing through penetrations of various types and sizes, be constructed. The TPFS to be tested is installed in accordance with the manufacturer's instructions around the penetrations. The test assembly is then exposed to a test fire that corresponds to the time-temperature curve established by ASTM E 119 (see the commentary to Section 703.2 for more discussion on ASTM E 119). After the fire test exposure, the TPFS is subjected to a hose stream, which evaluates the ability of the TPFS to resist the effects of erosion and thermal shock. After completion of the ASTM E 814 or UL 1479 procedure, two ratings for the test subject are established, which indicate how the test specimen withstands exposure to the test fire for a specified period of time.

Ratings for the TPFS are generated based on the results of the testing, and are reported as an "F" (flame) rating and a "T" (temperature) rating. The "F" rating indicates the period of time, in hours, the tested TPFS remained in place without allowing the passage of fire during exposure or water during the hose stream. The "T" rating indicates the time, in hours, it took for the temperature, as recorded by thermocouples placed at specified locations on the unexposed side of the test assembly, to reach 325°F (162°C) above ambient. It must be noted that in order to obtain a "T" rating, the system must first obtain an "F" rating. "F" ratings are required for all TPFS and must be equal to the fire-resistance rating of the assembly penetrated. A "T" rating is not required for wall penetrations, but a minimum of 1 hour is required where the TPFS is installed in a floor assembly where the penetrating item is a pipe, tube or conduit that is in direct contact with a combustible material. This requirement is intended to minimize the potential for ignition of the combustible material on the unexposed side of the assembly due to elevated temperatures transmitted via the pipe, tube or conduit.

Two of the most common types of materials used in TPFS are intumescent and endothermic materials. Intumescent materials expand to approximately eight to 10 times their original volume when exposed to temperatures exceeding 250°F (121°C). The expansion of the material fills the voids or openings within the penetration to resist the passage of flame, while the outer layer of the expanded intumescent material forms an insulating charred layer that assists in limiting the transfer of heat. The expansion properties of intumescent materials allow them to seal openings left by combustible penetrating items that burn away during a fire, but do not retard heat as well as endothermic materials. Intumescent materials are typically used with combustible penetrating items or where a higher "T" rating is not required.

Endothermic materials provide protection through chemically bound water released in the form of steam when exposed to temperatures exceeding 600°F (316°C). This released water provides a cooling of the penetration and retards heat transfer through the penetration. Endothermic materials tend to be superlatively to resistant to heat transfer and have higher "T" ratings, but do not expand to fill voids left by combustible penetrating items that burn away during a fire. Endothermic materials are, therefore, typically used with noncombustible penetrating items, and a higher "T" rating is required.

712.3.2 Membrane penetrations. Membrane penetrations shall comply with Section 712.3.1. Where walls and partitions are required to have a minimum 1-hour fire-resistance rating, recessed fixtures shall be installed such that the required fire resistance will not be reduced.

Exceptions:

1. Steel electrical boxes that do not exceed 16 square inches (0.0103 m²) in area provided the total area of such openings does not exceed 100 square inches (0.0645 m²) for any 100 square feet (9.29 m²) of wall area. Outlet boxes on opposite sides of the wall shall be separated as shown:

    1.1. By a horizontal distance of not less than 24 inches (610 mm);

    1.2. By a horizontal distance of not less than the depth of the wall cavity where the wall cavity is filled with cellulose loose fill, rockwool or slag mineral wool insulation;

    1.3. By solid fireblocking in accordance with Section 717.2.1;

    1.4. By protecting both outlet boxes with listed putty pads; or

    1.5. By other listed materials and methods.

2. Membrane penetrations for listed electrical outlet boxes of any material are permitted provided such boxes have been tested for use in fire-resistance-rated assemblies and are installed in accordance with the instructions included in the listing. Outlet boxes on opposite sides of the wall shall be separated as follows:

    2.1. By a horizontal distance of not less than 24 inches (610 mm);

    2.2. By solid fireblocking in accordance with Section 717.2.1;

    2.3. By protecting both outlet boxes with listed putty pads; or

    2.4. By other listed materials and methods.

3. The annular space created by the penetration of a fire sprinkler provided it is covered by a metal escutcheon plate.

❖ Membrane penetrations are treated similar to through penetrations; namely, noncombustible conduits, pipes and tubes that penetrate only one membrane of a fire-resistance-rated wall assembly are required to have the annular space protected in accordance with Section 712.3.1. Combustible penetrations through only one membrane of a wall assembly would require either a tested assembly (see Section 712.3.1.1) or a through-penetration protection system that complies with Section 712.3.1.2 for the penetrated membrane.

Penetrations, such as electrical outlet boxes, can affect the fire-resistance rating of an assembly. The criteria in Exception 1 [see Figure 712.3.2(1)] limits the size of steel electrical outlet boxes [16 square inches (.0103

m²)] and the amount allowed in a 100-square-foot area (100 square inches). These criteria are consistent with the criteria determined from fire tests that have generally shown that, within these limitations, these penetrations will not adversely affect the fire-resistance rating of the wall. The UL *Fire Resistance Directory* indicates that the opening in a wallboard facing is to be cut such that the clearance between the box and the wallboard does not exceed $1/_8$ inch (3.2 mm).

Electrical outlet boxes of steel or materials other than steel (see Exception 2) can be utilized in the fire-resistance-rated wall if they have been specifically tested as part of a fire-resistance-rated assembly. Steel outlet boxes that exceed the size limitations of this section would also require the same prequalification by testing. Exceptions 1 and 2 also allow electrical boxes on opposite sides of the wall if they meet one of several criteria, such as separation by a horizontal distance of 24 inches (610 mm).

Exception 3 provides an alternative to the annular space protection provisions of Section 712.3.1 for fire sprinkler piping that penetrates a single membrane, provided that the annular space around the pipe is completely covered by an escutcheon plate of noncombustible material [see Figure 712.3.2(2)]. The nature of the hazard posed by single membrane penetrations of sprinkler piping is limited, due to the size of the opening, the potential amount of openings and the presence of a sprinkler system. The installation of a noncombustible escutcheon provides protection against free passage of fire through the annular space, as well as allowing for the movement of sprinkler piping without breaking during a seismic event. These provisions correlate with the requirements of the National

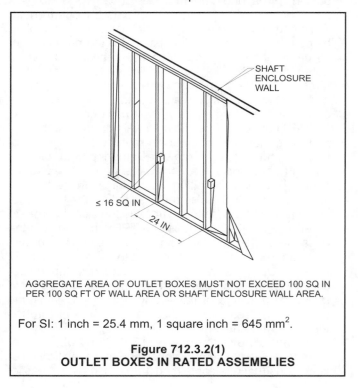

AGGREGATE AREA OF OUTLET BOXES MUST NOT EXCEED 100 SQ IN PER 100 SQ FT OF WALL AREA OR SHAFT ENCLOSURE WALL AREA.

For SI: 1 inch = 25.4 mm, 1 square inch = 645 mm².

Figure 712.3.2(1)
OUTLET BOXES IN RATED ASSEMBLIES

Earthquake Hazard Reduction Program (NEHRP) provisions recommended by the Building Seismic Safety Council (BSSC) for installation of sprinkler systems to resist seismic forces.

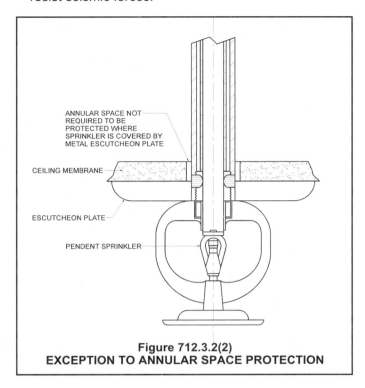

ANNULAR SPACE NOT REQUIRED TO BE PROTECTED WHERE SPRINKLER IS COVERED BY METAL ESCUTCHEON PLATE

CEILING MEMBRANE

ESCUTCHEON PLATE

PENDENT SPRINKLER

Figure 712.3.2(2)
EXCEPTION TO ANNULAR SPACE PROTECTION

712.3.3 Ducts and air transfer openings. Penetrations of fire-resistance-rated walls by ducts and air transfer openings that are not protected with fire dampers shall comply with this section.

❖ Duct penetrations are typically protected with fire dampers in accordance with Section 716.5; however, if the designer doors do not provide such damper, as it may interface with the system design, this section allows for such removal, provided the penetration is protected as part of a tested assembly (see Section 712.3.1.1) or is protected with some type of through-penetration firestop system (see Section 712.3.1.2).

712.3.4 Dissimilar materials. Noncombustible penetrating items shall not connect to combustible items beyond the point of firestopping unless it can be demonstrated that the fire-resistance integrity of the wall is maintained.

❖ This section limits the common practice of using a short metal nipple to penetrate a rated assembly, firestopping for the metal penetration (which is substantially less expensive than firestopping for plastic) and then connecting to plastic pipe or conduit on the room side of the wall. Arguably, there is a distance at which such connection is safe; however, this distance is variable and cannot be specified in the body of the code, hence the requirement for demonstration of fire-resistance integrity. An identical provision is found in Section 712.4.5 regarding penetrations of horizontal assemblies.

712.4 Horizontal assemblies. Penetrations of a floor, floor/ceiling assembly or the ceiling membrane of a roof/ceiling assembly shall be protected in accordance with Section 707. Penetrations permitted by Exceptions 3 and 4 of Section 707.2 shall comply with Sections 712.4.1 through 712.4.4.

Exception: Penetrations located within the same room or undivided area as floor openings are not required to have a shaft enclosure in accordance with Exception 1, 2, 5, 7, 8 or 9 in Section 707.2.

❖ Penetrations of floor/ceiling assemblies and ceiling membranes of roof/ceiling assemblies are required to be protected by a shaft enclosure unless otherwise exempted by Section 707.2 (also see Exceptions 3 and 4 to Section 707.2). The acceptable methods by which a floor or roof assembly can be interrupted by a penetration are identified in Sections 712.4 through 712.4.6. The exception to this section clarifies that penetration protection is not required if shaft protection is not required. This is specifically for cases not already addressed in Exceptions 3 and 4 to Section 707.2.

As noted in the commentary to Section 711, roof construction that is required to have a fire-resistance rating is intended to minimize the threat of premature structural failure of the roof construction under fire conditions and not to create a barrier to contain fire within the building. Section 711.4 permits unprotected penetrations in fire-resistance-rated roof assemblies, provided that the structural integrity of the roof assembly is not reduced.

712.4.1 Through penetrations. Through penetrations of fire-resistance-rated horizontal assemblies shall comply with Section 712.4.1.1 or 712.4.1.2.

Exceptions:

1. Penetrations by steel, ferrous or copper conduits, pipes, tubes, vents, concrete, or masonry through a single fire-resistance-rated floor assembly where the annular space is protected with materials that prevent the passage of flame and hot gases sufficient to ignite cotton waste when subjected to ASTM E 119 time-temperature fire conditions under a minimum positive pressure differential of 0.01 inch (2.49 Pa) of water at the location of the penetration for the time period equivalent to the fire-resistance rating of the construction penetrated. Penetrating items with a maximum 6-inch (152 mm) nominal diameter shall not be limited to the penetration of a single fire-resistance-rated floor assembly provided that the area of the penetration does not exceed 144 square inches (92 900 mm²) in any 100 square feet (9.3 m²) of floor area.

2. Penetrations in a single concrete floor by steel, ferrous or copper conduits, pipes, tubes and vents with a maximum 6-inch (152 mm) nominal diameter provided concrete, grout or mortar is installed the full thickness of the floor or the thickness required to maintain the fire-resistance rating. The penetrating items with a maximum 6-inch (152 mm) nominal diameter shall not be limited to the penetration of a single concrete floor

provided that the area of the penetration does not exceed 144 square inches (0.0929 m²).

3. Electrical outlet boxes of any material are permitted provided that such boxes are tested for use in fire-resistance-rated assemblies and installed in accordance with the tested assembly.

❖ The code addresses the penetration of fire-resistance-rated horizontal separations in much the same way as penetrations through vertical assemblies; namely, by requiring the penetrations to either be part of a total assembly or be protected with a through-penetration firestop system, (see commentary, Section 712.3.1).

Exception 1 is similar to Exception 2 of Section 712.3.1. This exception allows the noncombustible penetrating item to connect two stories (penetrate one floor), regardless of the size of the penetrating item [see Figure 712.4.1(1)]. Where the size of the individual penetration [6-inch diameter (152 mm)] and the aggregate area [144 square inches (0.095 m²)] is limited in any 100 square feet (9.29 m²), the code allows such penetrations of noncombustible construction to be through an unlimited number of floors.

Exception 2 allows a 6-inch-diameter (152 mm) penetration through a concrete floor similar to Exception 1 of Section 712.3.1. Where the area of the penetration is limited to 144 square inches (0.095 mm²), such a penetration is permitted to be through an unlimited number of floors [see Figure 712.4.1(2) for methods of annular space protection for Exceptions 1 and 2].

Exception 3 mirrors Exception 3 to Section 712.3.2 for membrane penetrations, but is applicable to through penetrations of a horizontal assembly, provided the outlet box has been tested.

712.4.1.1 Fire-resistance-rated assemblies. Penetrations shall be installed as tested in an approved fire-resistance-rated assembly.

❖ See the commentary to Section 712.3.1.1.

712.4.1.2 Through-penetration firestop system. Through penetrations shall be protected by an approved through-penetration firestop system installed and tested in accordance with ASTM E 814 or UL 1479, with a minimum positive pressure differential of 0.01 inch (2.49 Pa) of water. The system shall have an F rating and a T rating of not less than 1 hour but not less than the required rating of the floor penetrated.

Exception: Floor penetrations contained and located within the cavity of a wall do not require a T rating.

❖ This section differs slightly from Section 712.3.1.2 for wall penetrations in that a "T" rating is required for penetrations of horizontal assemblies, but is not required for penetrations of wall assemblies (see commentary, Section 712.3.1.2).

When permitted as an alternative to a shaft enclosure, according to Exception 3 of Section 707.2, cables, cable trays, conduits, tubes and pipes may penetrate a floor assembly when an approved through-penetration firestop system is used. An approved through-penetration firestop system is one that has been tested in accordance with ASTM E 814. The test method determines the performance of the protection system with respect to exposure to a standard time-temperature fire test and hose stream test. The performance of the protection system is dependent on the specific assembly of materials tested, including the number, type and size of penetrations and the types of floors or walls in which it is installed. It should also be noted that tests have been conducted at various pressure differentials; however, the current criterion used is 0.01 inch (2.49 Pa) of water gauge, and only tests with such minimum pressure throughout the test period are to be accepted. In evaluating test reports, the building official must determine that the tested assembly is truly representative of the manner in which the system is to be installed. It should be noted that the ASTM E 814 test establishes two ratings: the "F" rating, which identifies the ability of the material to resist the passage of flame, and the "T" rating, which identifies the thermal transmission characteris-

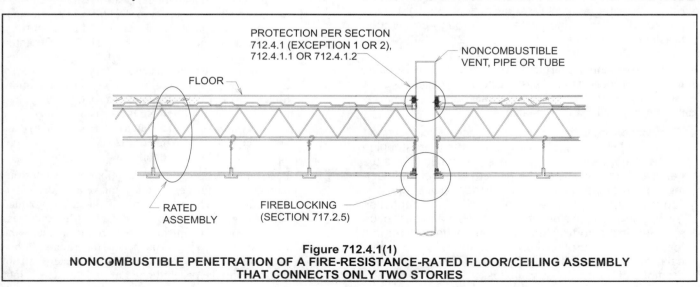

Figure 712.4.1(1)
NONCOMBUSTIBLE PENETRATION OF A FIRE-RESISTANCE-RATED FLOOR/CEILING ASSEMBLY
THAT CONNECTS ONLY TWO STORIES

tics of the material or assembly.

The exception is intended to set aside the "T" rating requirement where the penetrating item is located in a cavity of a wall and is separated from contact with adjacent materials in the occupiable floor area.

712.4.2 Membrane penetrations. Penetrations of membranes that are part of a fire-resistance-rated horizontal assembly shall comply with Section 712.4.1.1 or 712.4.1.2. Where floor/ceiling assemblies are required to have a minimum 1-hour fire-resistance rating, recessed fixtures shall be installed such that the required fire resistance will not be reduced.

Exceptions:

1. Membrane penetrations by steel, ferrous or copper conduits, electrical outlet boxes, pipes, tubes, vents, concrete, or masonry-penetrating items where the annular space is protected either in accordance with Section 712.4.1 or to prevent the free passage of flame and the products of combustion. Such penetrations shall not exceed an aggregate area of 100 square inches (64 500 mm²) in any 100 square feet (9.3 m²) of ceiling area in assemblies tested without penetrations.

2. Membrane penetrations by listed electrical outlet boxes of any material are permitted provided such boxes have been tested for use in fire-resistance-rated assemblies and are installed in accordance with the instructions included in the listing.

3. The annular space created by the penetration of a fire sprinkler provided it is covered by a metal escutcheon plate.

❖ Penetrations of fire-resistance-rated floor/ceiling and roof/ceiling assemblies that have a ceiling membrane as a component of the tested assembly, such as steel or wood joist assemblies using acoustical tiles or gypsum wallboard, are limited to those penetrations that are listed in the tested assembly (see Section 712.4.1.1) or

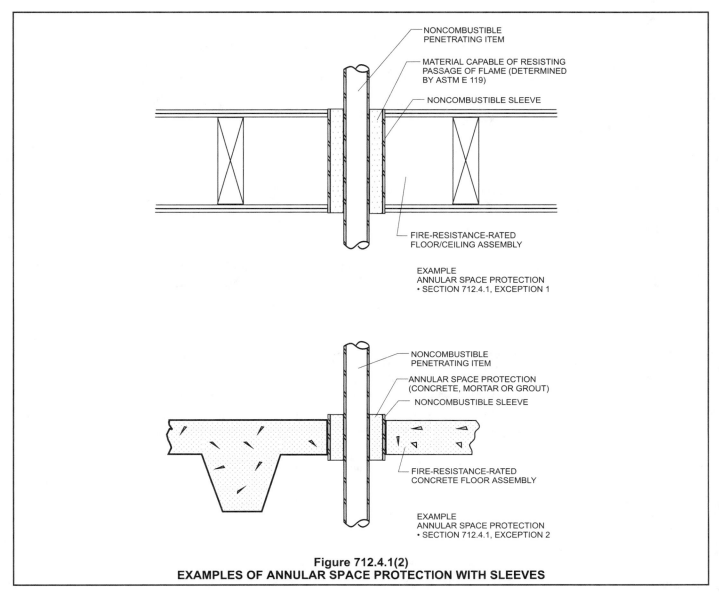

NONCOMBUSTIBLE
PENETRATING ITEM

MATERIAL CAPABLE OF RESISTING
PASSAGE OF FLAME (DETERMINED
BY ASTM E 119)

NONCOMBUSTIBLE SLEEVE

FIRE-RESISTANCE-RATED
FLOOR/CEILING ASSEMBLY

EXAMPLE
ANNULAR SPACE PROTECTION
• SECTION 712.4.1, EXCEPTION 1

NONCOMBUSTIBLE
PENETRATING ITEM

ANNULAR SPACE PROTECTION
(CONCRETE, MORTAR OR GROUT)

NONCOMBUSTIBLE SLEEVE

FIRE-RESISTANCE-RATED
CONCRETE FLOOR ASSEMBLY

EXAMPLE
ANNULAR SPACE PROTECTION
• SECTION 712.4.1, EXCEPTION 2

Figure 712.4.1(2)
EXAMPLES OF ANNULAR SPACE PROTECTION WITH SLEEVES

FIGURE 712.4.2(1) – FIGURE 712.4.2(2)

FIRE-RESISTANCE-RATED CONSTRUCTION

those protected in accordance with ASTM E 814 [see Section 712.4.1.2 and Figure 712.4.2(1)]. Penetrations of a ceiling membrane by items that are not part of a tested assembly create a point for possible fire penetration into the space above the ceiling membrane, which in turn could jeopardize the fire-resistance rating of the assembly. The exceptions contain protection requirements for specific types of penetrations that are not required to be part of the tested assembly.

Exception 1 allows for a maximum aggregate amount of noncombustible pipe, tube, vents and conduit penetrations of a ceiling that is part of a fire-resistance-rated assembly to be 100 square inches (0.0645 m²) per 100 square feet (9.29 m²) of ceiling area. The annular space of the penetration can be protected either by fireblocking (see Section 717.2.5), annular space protection (see Exception 1 of Section 712.4.1) or a through-penetration firestop system tested to ASTM E 814 [see Section 712.4.1.2 and Figure 712.4.2(2)].

Annular space protection is not an option for combustible items because it does not have the ability to protect the resulting void created when the combustible penetrating item burns away.

Exception 1 also allows penetrations of electrical outlet boxes that can affect the fire-resistance rating of an assembly. The criteria of this section limit the size of noncombustible electrical outlet boxes that penetrate a ceiling membrane to an aggregate area that shall not exceed 100 square inches (64 500 mm²) in 100 square feet (9.3 m²) of ceiling area. All of the openings at the penetrations of the outlets are required to be protected as mentioned above. The area limitations are consistent with the criteria determined from fire tests, which have shown that within these limitations, these penetra-

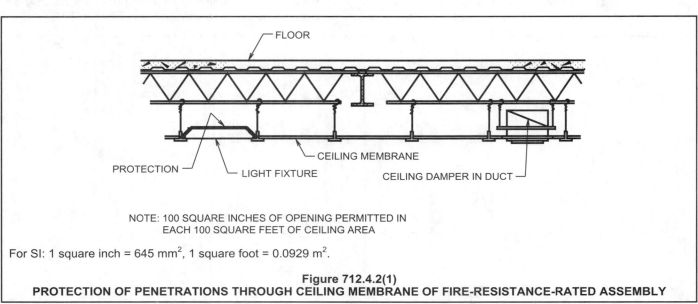

NOTE: 100 SQUARE INCHES OF OPENING PERMITTED IN
EACH 100 SQUARE FEET OF CEILING AREA

For SI: 1 square inch = 645 mm², 1 square foot = 0.0929 m².

Figure 712.4.2(1)
PROTECTION OF PENETRATIONS THROUGH CEILING MEMBRANE OF FIRE-RESISTANCE-RATED ASSEMBLY

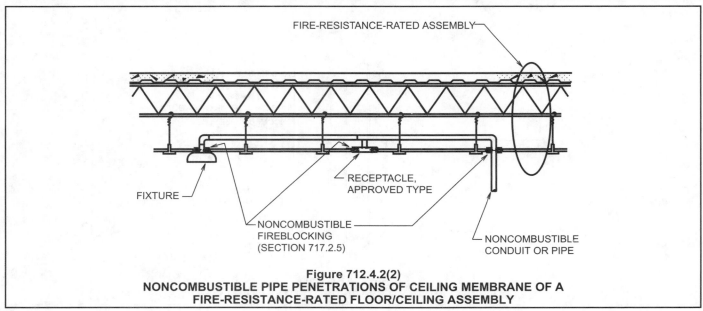

Figure 712.4.2(2)
**NONCOMBUSTIBLE PIPE PENETRATIONS OF CEILING MEMBRANE OF A
FIRE-RESISTANCE-RATED FLOOR/CEILING ASSEMBLY**

tions will not adversely affect the fire-resistance rating of the floor/ceiling assembly.

The exception permits electrical outlet boxes and fittings, which have been evaluated and found suitable for this purpose, to be located in the floor or ceiling of the fire-resistance-rated assembly. Reference sources for fire-resistance-rated assemblies, such as the UL *Fire Resistance Directory*, state that certain tested floor assemblies will permit the installation of electrical and communication connection inserts that appear in the UL listing category "Outlet Boxes and Fittings Classified for Fire Resistance" without reducing the fire-resistance rating of the assembly. However, classified outlet boxes and fittings installed in floor/ceiling assemblies that do not specify their use will jeopardize the rating unless compensating protection is provided.

Exceptions 2 and 3 are identical to those of Section 712.3.2 (see commentary, Section 712.3.2).

712.4.3 Nonfire-resistance-rated assemblies. Penetrations of horizontal assemblies without a required fire-resistance rating shall meet the requirements of Section 707 or shall comply with Sections 712.4.3.1 through 712.4.3.2.

❖ This section limits the penetrations of nonfire-resistance-rated floor assemblies to prevent the migration of smoke through a building, despite being of an unprotected type of construction. Permitted protection methods are outlined in Sections 712.4.3.1 through 712.4.3.2.

Where penetrations of the ceiling membrane of a nonfire-resistance-rated roof/ceiling assembly occur, the penetrations are required to be protected by fireblocking in accordance with the requirements of Section 717.2.5 to restrict the spread of fire through the concealed roof space above the ceiling membrane (see commentary, Section 721.6.4).

712.4.3.1 Noncombustible penetrating items. Noncombustible penetrating items that connect not more than three stories are permitted provided that the annular space is filled with an approved noncombustible material to resist the free passage of flame and the products of combustion.

❖ Noncombustible penetrations connecting three stories or less (penetrating two or less floors) are permitted when the annular space of the penetrating item is fireblocked at each floor line in accordance with Section 717.2.5 (see Figure 712.4.3.1).

712.4.3.2 Penetrating items. Penetrating items that connect not more than two stories are permitted provided that the annular space is filled with an approved material to resist the free passage of flame and the products of combustion.

❖ This section is not limited as to the combustibility of the penetrating item. This section allows both combustible and noncombustible penetrations of a single nonfire-resistance-rated floor (thus connecting two stories), provided the annular space is fireblocked in accordance with Section 717.2.5. This section does not require a

noncombustible firestop to protect the annular space, while Section 712.4.3.1 does require a noncombustible firestop material.

712.4.4 Ducts and air transfer openings. Penetrations of horizontal assemblies by ducts and air transfer openings that are not required to have dampers shall comply with this section. Ducts and air transfer openings that are protected with dampers shall comply with Section 716.

❖ This section mirrors that of Section 712.3.3 for direct penetrations through walls (see commentary, Section 712.3.3).

712.4.5 Dissimilar materials. Noncombustible penetrating items shall not connect to combustible materials beyond the point of firestopping unless it can be demonstrated that the fire-resistance integrity of the horizontal assembly is maintained.

❖ This section mirrors that of Section 712.3.4 for direct penetrations through walls (see commentary, Section 712.3.4).

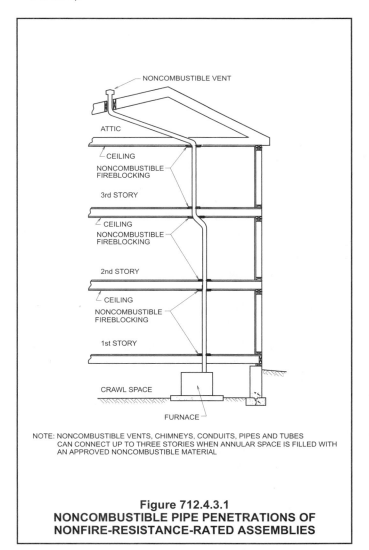

NOTE: NONCOMBUSTIBLE VENTS, CHIMNEYS, CONDUITS, PIPES AND TUBES CAN CONNECT UP TO THREE STORIES WHEN ANNULAR SPACE IS FILLED WITH AN APPROVED NONCOMBUSTIBLE MATERIAL

Figure 712.4.3.1
NONCOMBUSTIBLE PIPE PENETRATIONS OF
NONFIRE-RESISTANCE-RATED ASSEMBLIES

712.4.6 Floor fire doors. Floor fire doors used to protect openings in fire-resistance-rated floors shall be tested in the horizontal position in accordance with ASTM E 119, and shall achieve a fire-resistance rating not less than the assembly being penetrated. Floor fire doors shall be labeled by an approved agency.

❖ Floor fire doors are an alternative means of protecting a floor opening that does not contain penetrating equipment and is not protected by a shaft enclosure in accordance with Section 707. Most horizontal floor openings are the result of common features, such as stairs and atriums, and protection for these floor openings is provided by means of vertical shafts in accordance with Section 707. It was recognized that certain openings, such as small openings between floors for infrequent access to equipment or systems, could be efficiently protected by a fire door assembly in the same way that openings in vertical fire separation assemblies are required to be protected by tested and labeled fire doors in accordance with Section 715. Because a small access opening is similar to openings created for penetrations, the provisions are located as a subsection to Section 712.4, along with the requirements for protection of penetrations. Floor fire doors are treated no differently than the fire-resistance-rated floor in which they reside; namely, they are evaluated against ASTM E 119.

SECTION 713
FIRE-RESISTANT JOINT SYSTEMS

713.1 General. Joints installed in or between fire-resistance-rated walls, floor or floor/ceiling assemblies and roofs or roof/ceiling assemblies shall be protected by an approved fire-resistant joint system designed to resist the passage of fire for a time period not less than the required fire-resistance rating of the wall, floor or roof in or between which it is installed. Fire-resistant joint systems shall be tested in accordance with Section 713.3. The void created at the intersection of a floor/ceiling assembly and an exterior curtain wall assembly shall be protected in accordance with Section 713.4.

Exception: Fire-resistant joint systems shall not be required for joints in all of the following locations:

1. Floors within a single dwelling unit.

2. Floors where the joint is protected by a shaft enclosure in accordance with Section 707.

3. Floors within atriums where the space adjacent to the atrium is included in the volume of the atrium for smoke control purposes.

4. Floors within malls.

5. Floors within open parking structures.

6. Mezzanine floors.

7. Walls that are permitted to have unprotected openings.

8. Roofs where openings are permitted.

9. Control joints not exceeding a maximum width of 0.625 inch (15.9 mm) and tested in accordance with ASTM E 119.

❖ This section regulates joints or linear openings created between building assemblies, which are sometimes referred to as construction, expansion or seismic joints. These joints are most often created where the structural design of a building necessitates a separation between building components in order to accommodate anticipated structural displacements caused by thermal expansion and contraction, seismic activity, wind or other loads. Figure 713.1 illustrates some of the most common locations of these joints.

These linear openings create a "weak link" in fire-resistance-rated assemblies that can compromise the integrity of the tested assembly by allowing an avenue for the passage of fire and the products of combustion through the assembly. In order to maintain the efficacy of the fire-resistance-rated assembly, these openings must be protected by a fire-resistance-rated joint system with a rating equal to the adjacent assembly. It is not the intent of this section to regulate joints installed in assemblies that are provided to control shrinkage cracking, such as a saw-cut control joint in concrete.

This section contains nine locations where a fire-resistant joint system is not required to be installed to protect the joint. Exception 1 states that a fire-resistant joint is not required for joints contained within a single dwelling unit. This exception is similar to Section 707.2, Exception 1. Exception 2 exempts fire-resistant joints completely enclosed in a shaft that complies with Section 707. Exception 3 exempts the requirement for fire-resistant joints that are located in floors in an atrium that complies with Section 404. This exception is similar to Section 707.2, Exception 5. Exception 4 exempts joints located in malls that comply with Section 402 and is similar to Section 707.2, Exception 5. Exception 5 exempts joints located in open parking structures that comply with Section 406.3 and is similar to Section 707.2, Exception 8. Exception 6 exempts joints in mezzanines that comply with Section 505 and is similar to Section 707.2, Exception 9. As a type of opening/penetration, a joint represents a similar hazard to exterior wall fire exposures as other openings. Exception 7 acknowledges that if an exterior wall is permitted to have unprotected openings (see Table 704.8), then the joint does not require protection as well. A joint is a much smaller opening than an unprotected opening. Exception 8 exempts joints located in roof decks that are not required to have protected openings and is similar to that contained in Section 711.4, which allows skylights and other penetrations in rated roof decks, provided the structural integrity is not affected. Exception 9 acknowledges UL 2079, as referenced in Section 713.3, which permits control joints with a maximum joint width of $^5/_8$ inch (15.9 mm) to be tested in accordance with UL 263/ASTM E 119. This allowance is in recognition of fire test data that has existed since the 1960s regarding control joints and perimeter relief joints in fire-rated systems, the performance of these devices and systems in the field and a rational approach regarding minor movement [defined here as a maximum of $^5/_8$ inch (15.9 mm)] where no appreciable damage or fatigue is incurred by the joint system that would have profound impact on the joint's fire performance. Since all buildings move, it is not prudent to have to cycle and test joint systems that only move a few millimeters. Although their movement is limited, control joint systems must still pass the rigors of ASTM E 119 in accordance with UL 2079 and this exception.

713.2 Installation. Fire-resistant joint systems shall be securely installed in or on the joint for its entire length so as not to dislodge, loosen or otherwise impair its ability to accommodate expected building movements and to resist the passage of fire and hot gases.

❖ This section requires that joint systems be installed for its full length or height, due to the fact that the openings to be protected by these types of joints are most often continuous.

713.3 Fire test criteria. Fire-resistant joint systems shall be tested in accordance with the requirements of either ASTM E 1966 or UL 2079. Nonsymmetrical wall joint systems shall be tested with both faces exposed to the furnace, and the assigned fire-resistance rating shall be the shortest duration obtained from the two tests. When evidence is furnished to show that the wall was tested with the least fire-resistant side exposed to the furnace, subject to acceptance of the building official, the wall need not be subjected to tests from the opposite side.

Exception: For exterior walls with a horizontal fire separation distance greater than 5 feet (1524 mm), the joint system shall be required to be tested for interior fire exposure only.

❖ In order to determine the anticipated protection provided by a given assembly, the hourly fire-resistance rating must be determined by testing the joint system in accordance with UL 2079 or ASTM E 1966. This standard includes specific criteria for test specimen preparation, placement, configuration, size and testing conditions of joint systems. Also included in the text protocol is the minimum positive pressure differential to be used for the test, which is intended to assist in evaluating whether or not the joint will remain in place during a fire condition. A joint splice must be tested, since the presence and orientation of a splice in a joint system can affect the fire performance of the joint. These splices may occur where the length of the joint to be protected exceeds the length of a prefabricated joint or where a cold joint occurs in a field-installed system. The maximum joint width must be tested when in a fully expanded or extended condition in order to evaluate the fire-resistance rating. Joints that are designed to transfer structural building loads are required to have a su-

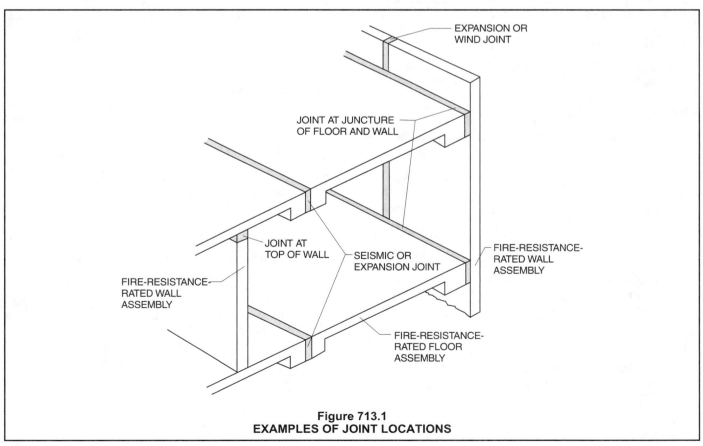

Figure 713.1
EXAMPLES OF JOINT LOCATIONS

perimposed load during the test, which is consistent with the requirements for the testing of load-bearing fire-resistance-rated assemblies. Finally, the test addresses requirements for joints that are intended to accommodate building movement, such as expansion, seismic and wind sway joints, including preconditioning cycling, which is intended to allow for evaluation of the joint's ability to withstand cyclical movement over its anticipated life.

Test data has indicated that the orientation of nonsymmetrical joints can have an effect on the performance of the joint. As a result, in accordance with Item 1, all joint systems must be tested for fire exposure from both sides so that the required protection will be provided regardless of which side of the joint is exposed to fire. The exception for joints in exterior walls correlates with Section 704.5 for fire exposure rating of exterior walls.

713.4 Exterior curtain wall/floor intersection. Where fire-resistance-rated floor or floor/ceiling assemblies are required, voids created at the intersection of the exterior curtain wall assemblies and such floor assemblies shall be sealed with an approved material or system to prevent the interior spread of fire. Such material or systems shall be securely installed and capable of preventing the passage of flame and hot gases sufficient to ignite cotton waste where subjected to ASTM E 119 time-temperature fire conditions under a minimum positive pressure differential of 0.01 inch of water (2.49 Pa) for the time period at least equal to the fire-resistance rating of the floor assembly. Height and fire-resistance requirements for curtain wall spandrels shall comply with Section 704.9.

❖ At this time, there does not exist a standard specifically developed to evaluate the interface between a fire-resistance-rated horizontal assembly and an exterior curtain wall; however, the void created between such assemblies can range anywhere between 1 to 8 inches (25 to 203 mm) or more, which clearly requires sealing to prevent the spread of flames and products of combustion between adjacent stories.

This section merely requires that the interface be sealed with a material or system that will pass the cotton wash ignition parameters of ASTM E 119 under positive pressure. A task group of the ASTM E 05 Subcommittee 011 is developing a standard entitled *ASTM Standard Test Method for Determining the Fire Endurance of Perimeter Fire Barrier Systems Using the Intermediate Scale, Multistory Test Apparatus.*

Compliance with Section 704.9 for curtain wall spandrels is also required (see commentary, Section 704.9).

SECTION 714
FIRE-RESISTANCE RATING OF
STRUCTURAL MEMBERS

714.1 Requirements. The fire-resistance rating of structural members and assemblies shall comply with the requirements for the type of construction and shall not be less than the rating required for the fire-resistance-rated assemblies supported.

Exception: Fire barriers and fire partitions as provided in Sections 706.4 and 708.4, respectively.

❖ This section contains provisions that apply to structural members that are required to have a fire-resistance rating. The required rating is usually based on the type of construction and Table 601. Other sections of the code, such as the continuity provisions of Sections 706.4 and 708.4, may require structural members to have a fire-resistance rating because they support fire-resistance-rated construction. This section addresses issues such as embedment of service facilities (see Section 714.3), impact protection (see Section 714.4) and exterior structural members (see Section 714.5).

The type of construction is one means by which structural members are required to have a fire-resistance rating. The required fire-resistance ratings are identified in Table 601. Additionally, structural members that either support assemblies or are required to have a fire-resistance rating must also have a fire-resistance rating of no less than the rating of the assembly or structural member supported (see Sections 706.4, 708.4 and 711.4). The minimum required fire-resistance rating for the structural member, therefore, is the more restrictive of the above two criteria.

There are instances in which the code does not require the supporting structural members to have a fire-resistance rating. For example, the supporting construction for tenant and guestroom separations, exit access corridors (see Section 708.4) and smoke barriers (see Section 709.4) is not required to have a fire-resistance rating.

714.2 Protection of structural members. Protection of columns, girders, trusses, beams, lintels or other structural members that are required to have a fire-resistance rating shall comply with this section.

❖ This section and its subsections address the type of protection that is required for structural members. The type of protection is typically a function of the type of member (column versus girder) and the number of floors supported.

714.2.1 Individual protection. Columns, girders, trusses, beams, lintels or other structural members that are required to have a fire-resistance rating and that support more than two floors or one floor and roof, or support a load-bearing wall or a nonload-bearing wall more than two stories high, shall be individually protected on all sides for the full length with materials having the required fire-resistance rating. Other structural members required to have a fire-resistance rating shall be protected by individual encasement, by a membrane or ceiling protection as specified in Section 711, or by a combination of both. Columns shall also comply with Section 714.2.2.

❖ This section determines when a required fire-resistance rating can be provided by membrane protection (such as a floor/ceiling assembly) and when individual encasement of the structural member is required. Membrane or ceiling protection (as specified in Section 711) is only permitted for structural members that support two floors or less; a floor and a roof only or load-bearing and nonload-bearing walls two stories or less in height [see Figure 714.2.1(1)]. Also, membrane protection is never permitted for columns (see commentary, Section 714.2.2). In all other instances, structural members must achieve the required fire-resistance rating by individual encasement [see Figure 714.2.1(2)]. Membrane protection methods are limited in this manner to reduce the risk of catastrophic structural failure that might occur because of failure of the membrane or ceiling protection that affects multiple structural elements.

Regardless of the method used–individual encasement or membrane protection–a structural element must be protected for its entire length or height. The risk represented by structural collapse during a fire is significantly increased in multistory buildings, since the occupants require additional time for egress and a single structural element that supports multiple elements requires more conservative protection. There is also a greater probability of the structural element reaching its design load as the tributary area increases.

714.2.2 Column protection above ceilings. Where columns require a fire-resistance rating, the entire column, including its connections to beams or girders, shall be protected. Where the column extends through a ceiling, fire resistance of the column shall be continuous from the top of the floor through the ceiling space to the top of the column.

❖ Membrane or ceiling protection is prohibited for columns required to have a fire-resistance rating; therefore, the materials that "wrap" a required fire-resistance-rated column must continue through the con-

cealed space above a ceiling, even if the ceiling is part of a fire-resistance-rated floor/ceiling or roof/ceiling assembly and regardless of the number of stories supported or the height involved. Current testing equipment cannot accommodate membrane-type protection schemes for columns and, therefore, there have been no tested assemblies employing this type of protection, nor is there any basis for evaluation of membrane-type protection for columns. This section also requires the connection to be adequately protected (see Figure 714.2.2).

714.2.3 Truss protection. The required thickness and construction of fire-resistance-rated assemblies enclosing trusses shall be based on the results of full-scale tests or combinations of tests on truss components or on approved calculations based on such tests that satisfactorily demonstrate that the assembly has the required fire resistance.

❖ Trusses can be of an almost infinite number of configurations and sizes. By nature, they are typically designed and constructed of different structural components. This leads to the impracticality of such trusses being tested. The code acknowledges this by allowing full scale tests, tests on the individual components or some form of calculations that demonstrate that the structural elements that form the truss are provided with the requisite protection to attain the fire-resistance rating.

714.2.4 Attachments to structural members. The edges of lugs, brackets, rivets and bolt heads attached to structural members shall be permitted to extend to within 1 inch (25 mm) of the surface of the fire protection.

❖ ASTM E 119 allows for either the structural element to be tested under maximum load conditions to ascertain

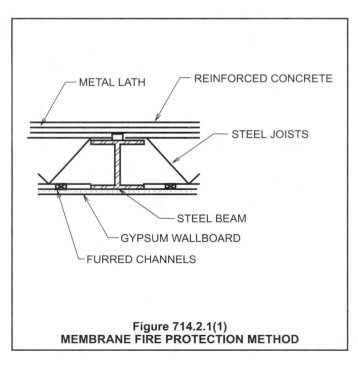

Figure 714.2.1(1)
MEMBRANE FIRE PROTECTION METHOD

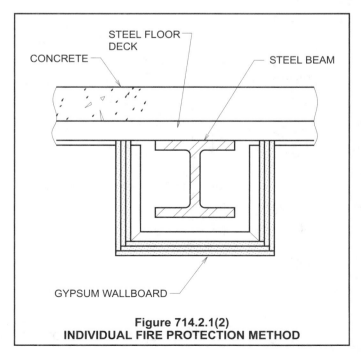

Figure 714.2.1(2)
INDIVIDUAL FIRE PROTECTION METHOD

its load-carrying capability under the prescribed rating time period or an alternative test of the protection. Under the alternative test, one of the key considerations when testing steel structural elements is the heat transfer through the protective covering and the corresponding temperature increase in the range of 1,000 to 1,300°F (538 to 704°C), depending on the type of element. Concrete structural elements have similar restrictions regarding reinforcing steel (see Section 714.2.5). This section is intended to limit the transfer of heat to the structural element from the connection elements by requiring the connection element to be protected with at least 1 inch (25 mm) of some form of protection.

714.2.5 Reinforcing. Thickness of protection for concrete or masonry reinforcement shall be measured to the outside of the reinforcement except that stirrups and spiral reinforcement ties are permitted to project not more than 0.5-inch (12.7 mm) into the protection.

❖ Since load-bearing assemblies must sustain the superimposed load while being subjected to standard time-temperature fire conditions, concrete, masonry and reinforcing steel must be able to provide the required strength at elevated temperatures. This involves providing adequate cover over the steel reinforcement so that the stress induced in the reinforcement is less than its yield stress, which is commonly referred to as "yield strength." Tests of steel reinforcing bars show that at a temperature of approximately 1,100°F (593° C), the

yield strength of steel is reduced to approximately 50 percent of that at ambient temperature conditions. Similar tests of prestressing tendons show that at approximately 800°F (427°C) the tensile strength is approximately 50 percent of that at ambient conditions. Cover requirements, therefore, are typically based on limiting the reinforcing and prestressing steel temperatures to 1,100 and 800°F (593 and 427°C), respectively.

This section allows for stirrups and trusses to project into the required cover by a minimal distance, acknowledging the practicality of construction tolerances while providing adequate cover to the main reinforcing steel, which is located inside of the stirrups and truss. The $1/2$-inch (12.7 mm) dimension reflects a No. 4 reinforcing bar.

714.3 Embedments and enclosures. Pipes, wires, conduits, ducts or other service facilities shall not be embedded in the required fire protective covering of a structural member that is required to be individually encased.

❖ Piping, wires, conduits, ducts and other service facilities are not to be embedded in the fire protective covering of a structural member that is required to be individually encased in accordance with Section 714.2.1. The fire protection performance of encasement materials is critical to achieve the required fire-resistance rating, and would be impaired if the continuity of the encasement is interrupted. This does not, however, prevent the installation if these items are installed within the hollow

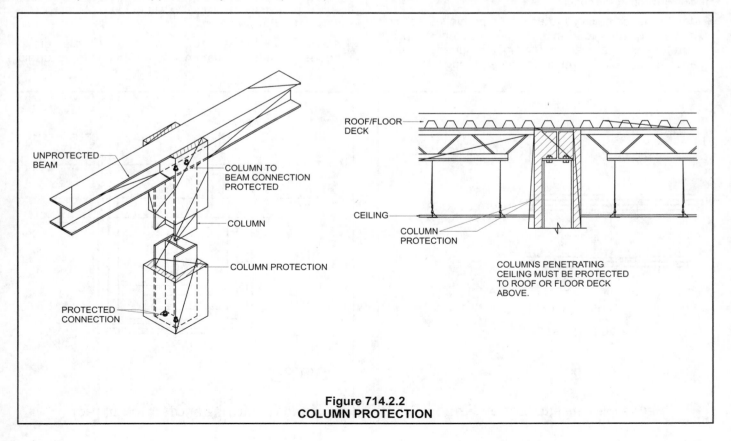

Figure 714.2.2
COLUMN PROTECTION

spaces that may be created by the encasement. When items are installed in such hollow spaces, penetrations of the protective assembly must be properly protected so that the fire-resistance rating is maintained.

714.4 Impact protection. Where the fire protective covering of a structural member is subject to impact damage from moving vehicles, the handling of merchandise or other activity, the fire protective covering shall be protected by corner guards or by a substantial jacket of metal or other noncombustible material to a height adequate to provide full protection, but not less than 5 feet (1524 mm) from the finished floor.

❖ To aid in the reliability of the fire protective covering, impact protection must be provided if the covering is subject to impact from moving vehicles, handling of merchandise or other activities. The impact protection is to be provided to at least 5 feet (1524 mm) above the finished floor or as needed to prevent impact damage.

714.5 Exterior structural members. Load-bearing structural members located within the exterior walls or on the outside of a building or structure shall be provided with the highest fire-resistance rating as determined in accordance with the following:

1. As required by Table 601 for the type of building element based on the type of construction of the building;

2. As required by Table 601 for exterior bearing walls based on the type of construction; and

3. As required by Table 602 for exterior walls based on the fire separation distance.

❖ Exterior load-bearing structural members, such as columns or girders, must have the same fire-resistance rating as is required for exterior load-bearing walls. As such, the required fire-resistance rating is the higher rating of that found in Table 601 for type of construction for structural elements or required in Table 602 based upon separation distance.

714.6 Bottom flange protection. Fire protection is not required at the bottom flange of lintels, shelf angles and plates, spanning not more than 6 feet (1829 mm) whether part of the structural frame or not, and from the bottom flange of lintels, shelf angles and plates not part of the structural frame, regardless of span.

❖ Structural frame elements (see Note a to Table 601) located in an exterior wall are viewed as a critical component to the overall integrity of the exterior wall construction. For this reason, in protected types of construction, such elements are required to be protected from external, as well as internal, fire exposure. This addresses not only exposure from adjacent buildings but fire exposure from the building itself due to a breached opening that exposes the exterior side of the structural element.

The bottom flange of short-span [6 feet (1829 mm) or less] lintels that are designed as part of the building's structural frame system and those lintels that are not designed as part of the structural frame do not require a fire-resistance rating. The risk of a structural failure

caused by leaving the bottom flange unprotected is not great enough to warrant a fire-resistance rating.

714.7 Seismic isolation systems. Fire-resistance ratings for the isolation system shall meet the fire-resistance rating required for the columns, walls, or other structural elements in which the isolation system is installed in accordance with Table 601.

Isolation systems required to have a fire-resistance rating shall be protected with approved materials or construction assemblies designed to provide the same degree of fire resistance as the structural element in which it is installed when tested in accordance with ASTM E 119 (see Section 703.2).

Such isolation system protection applied to isolator units shall be capable of retarding the transfer of heat to the isolator unit in such a manner that the required gravity load-carrying capacity of the isolator unit will not be impaired after exposure to the standard time-temperature curve fire test prescribed in ASTM E 119 for a duration not less than that required for the fire resistance rating of the structure element in which it is installed.

Such isolation system protection applied to isolator units shall be suitably designed and securely installed so as not to dislodge, loosen, sustain damage, or otherwise impair its ability to accommodate the seismic movements for which the isolator unit is designed and to maintain its integrity for the purpose of providing the required fire-resistance protection.

❖ This section states that the fire-resistance requirements for seismic isolation systems are the same as other structural elements as required in Table 601. Additionally the fire-resistance method used must be able to withstand the anticipated movement of the system without the fire-resistance application methods and systems affecting its functionality during a seismic event. Elements of the system can either be individually protected or contained within a tested assembly. Another key element of this section is that, when tested under exposure to the standard time-temperature curve test, the structural elements are still able to carry the gravity load as designed.

SECTION 715
OPENING PROTECTIVES

715.1 General. Opening protectives required by other sections of this code shall comply with the provisions of this section.

❖ This section regulates two types of opening protectives: fire doors (Section 715.2) and fire-protection-rated glazing, including wired glass (Section 715.3).

Fire doors are considered opening protectives and are installed in openings in fire-resistance-rated wall assemblies, including fire walls, fire barriers, fire partitions and exterior walls where the openings are required to be protected (see Section 704.8). Note that where a load-bearing exterior wall requires a fire-resistance rating in accordance with Table 601, it is for structural integrity purposes. The openings are not required to be protected unless an opening protective is required by another section of the code, including Section 704.8.

In addition to fire doors, another form of an opening

protective is fire-protection-rated glazing (commonly referred to as "fire windows"). Fire windows refer to the entire assembly, which may consist of a frame and glazing assembly or a glass-block assembly.

The fire protection ratings in this section are based on the acceptance criteria of NFPA 252, 257 and UL10A, 14B or 14C. The fire protection rating acceptance criteria are, therefore, not the same as that required for a fire-resistance rating for building construction (structural elements). Fire-resistance ratings of building construction are determined by ASTM E 119. Usually, the fire protection rating required for a opening protective is less than the required fire-resistance of the wall (see Table 715.3 and Sections 715.4 and 715.4.8). This is justified by the assumed lack of potential fire exposure in the vicinity of the opening, since clear space on both sides of a door opening is required for operational purposes and for an unobstructed egress path, and the corresponding exposure at a window is also reduced. Sections 715.3 and 715.4 (and subsections) reference the appropriate test standards and require that all opening protectives be labeled (see Sections 715.3.5 and 715.4.9), with the exception being oversized doors that require a certificate of inspection by an approved agency (see Section 715.3.5.2).

715.2 Fire-resistance-rated glazing. Labeled fire-resistance-rated glazing tested as part of a fire-resistance-rated wall assembly in accordance with ASTM E 119 shall not be required to comply with this section.

❖ This section allows any glazing that has passed ASTM E 119 as part of an assembly in lieu of the requirements of Section 715. Section 715 focuses on fire protection-rated glazing, which is concerned only with confining a fire and not providing the thermal protection of a fire-resistance rating (see the definitions of "Fire protection rating" and "Fire- resistance rating" are defined in Section 702.1).

715.3 Fire door and shutter assemblies. Approved fire door and fire shutter assemblies shall be constructed of any material or assembly of component materials that conforms to the test requirements of Section 715.3.1, 715.3.2 or 715.3.3 and the fire protection rating indicated in Table 715.3. Fire door assemblies and shutters shall be installed in accordance with the provisions of this section and NFPA 80.

Exceptions:

1. Labeled protective assemblies that conform to the requirements of this section or UL 10A, UL 14B and UL 14C for tin-clad fire door assemblies.

2. Floor fire doors in accordance with Section 712.4.6.

❖ When the code refers to a fire door, the intent is that the term "fire door" (see the definition in Section 702.31) includes the frame, related hardware and accessories, unless otherwise noted (see the definition of "Fire door assembly"). For example, although there is not a direct mention in the section relative to mandating latching hardware, the referenced test standard, NFPA 80, ad-

dresses the requirements relative to positive latch devices. Labels that indicate compliance with UL 10A, 14B or 14C are acceptable (see Exception 1).

Exception 2 exempts floor fire doors from the requirements of this section, since Section 712.4.6 requires floor fire doors to comply with a different test standard (ASTM E 119) and their rating must not be less than the rating of the horizontal assembly being penetrated (see the commentary to Section 712.4.6 for a discussion of floor fire doors).

TABLE 715.3. See page 7-73.

❖ The table prescribes the minimum fire protection ratings for fire doors relative to the nature and fire-resistance rating of the wall. Once the purpose and fire-resistance rating of the wall are identified, the minimum required fire protection rating of the door is determined using the table. Typically, the minimum permitted fire door fire protection rating is less than the required fire-resistance rating of the wall. This is justified by the fact that under normal conditions of use, the potential fire exposure in the vicinity of the door opening is lessened, since there will be a clear space on both sides of the opening. As indicated in the commentary to Section 715.1, the acceptance criteria for a 1-hour door (fire protection rating) is not the same as the acceptance criteria for a 1-hour wall (fire-resistance rating).

The hourly designation in the table indicates the duration of fire test exposure and is actually called the "fire protection rating." Although not recognized by the code, fire doors are frequently, though inadequately, classified by only an alphabetical letter as follows:

- Class A—Openings in fire walls and in walls that divide a single building into fire areas.
- Class B—Openings in enclosures of vertical communications through buildings and in 2-hour fire-resistance-rated partitions providing horizontal fire separations.
- Class C—Openings in walls or partitions between rooms and corridors having a fire-resistance rating of 1 hour or less.

There are also Class D and E doors for exterior wall applications. These are similar to Class B and C doors, respectively.

Although the code does not refer to the alphabetical classification, many design documents still refer to fire doors by this classification. It should be noted, however, that Class B doors have either a 1-hour or 1¹/₂-hour rating. As noted in Table 715.3, the 1-hour Class B door would be acceptable for a 1-hour shaft enclosure but not in a 2-hour fire barrier or exit enclosure. Likewise, Class C doors have a variety of fire protection ratings; therefore, it is imperative that the hourly fire protection rating be indicated in order to verify compliance with the code, since the letter designation does not specifically relate to the code requirements in all cases.

TABLE 715.3
FIRE DOOR AND FIRE SHUTTER FIRE PROTECTION RATINGS

TYPE OF ASSEMBLY	REQUIRED ASSEMBLY RATING (hours)	MINIMUM FIRE DOOR AND FIRE SHUTTER ASSEMBLY RATING (hours)
Fire walls and fire barriers having a required fire-resistance rating greater than 1 hour	4 3 2 $1^1/_2$	3 3[a] $1^1/_2$ $1^1/_2$

Fire barriers having a required fire-resistance rating of 1 hour:
   Shaft, exit enclosure and exit passageway walls
   Other fire barriers
	
	1
	1
		
		1
		$^3/_4$

Fire partitions:
   Corridor walls
 	
   Other fire partitions
	
	1
	0.5
	1
		
		$^1/_3$[b]
		$^1/_3$[b]
		$^3/_4$

Exterior walls	3 2 1	$1^1/_2$ $1^1/_2$ $^3/_4$

a. Two doors, each with a fire protection rating of $1^1/_2$ hours, installed on opposite sides of the same opening in a fire wall, shall be deemed equivalent in fire protection rating to one 3-hour fire door.
b. For testing requirements, see Section 715.3.3.

715.3.1 Side-hinged or pivoted swinging doors. Side-hinged and pivoted swinging doors shall be tested in accordance with NFPA 252 or UL 10C. After 5 minutes into the NFPA 252 test, the neutral pressure level in the furnace shall be established at 40 inches (1016 mm) or less above the sill.

❖ Fire door assemblies are to be tested in accordance with the referenced test standards. The tests require that an assembly be exposed to a standard fire exposure throughout a specified time period, followed by the application of a specified fire hose stream. As such, the tests provide a relative measure of fire performance of door assemblies under the specified fire exposure conditions. The basic condition of acceptance is that the door remains in the opening such that the movement (deflection) does not exceed prescribed limitations.

In addition to the fire protection ratings in Table 715.3, other sections of the code prescribe additional performance criteria for fire door assemblies. For example, Section 715.3.4 requires doors in exit enclosures to have a maximum temperature increase on the unexposed surface of 450°F (232°C) after 30 minutes. Although temperature rise is not a condition of acceptance, test standards require that unexposed surface temperatures be measured and documented in the test report.

715.3.2 Other types of doors. Other types of doors, including swinging elevator doors, shall be tested in accordance with NFPA 252 or UL 10B. The pressure in the furnace shall be maintained as nearly equal to the atmospheric pressure as possi-

ble. Once established, the pressure shall be maintained during the entire test period.

❖ This section also references NFPA 252 (as does Section 715.3.1 for side-hinged doors) for other types of doors. NFPA 252 is a general test standard that is not specific to any type of door. UL 10B was developed for negative pressure testing of fire door and fire shutter assemblies that is viewed as a viable test method, that provided the pressure is maintained at or nearly at atmospheric pressure.

715.3.3 Door assemblies in corridors and smoke barriers. Fire door assemblies required to have a minimum fire protection rating of 20 minutes where located in corridor walls or smoke barrier walls having a fire-resistance rating in accordance with Table 715.3 shall be tested in accordance with NFPA 252 or UL 10C without the hose stream test. If a 20-minute fire door assembly contains glazing material, the glazing material in the door itself shall have a minimum fire protection rating of 20 minutes and be exempt from the hose stream test. Glazing material in any other part of the door assembly, including transom lites and sidelites, shall be tested in accordance with NFPA 257, including the hose stream test, in accordance with Section 715.4. Fire door assemblies shall also meet the requirements for a smoke- and draft-control door assembly tested in accordance with UL 1784 with an artificial bottom seal installed across the full width of the bottom of the door assembly. The air leakage rate of the door assembly shall not exceed 3.0 cfm per square foot (0.01524 m³/s·m²) of door opening at 0.10 inch (24.9 Pa) of water for both the ambient temperature and elevated temperature tests. Louvers shall be prohibited.

Exceptions:

1. Viewports that require a hole not larger than 1 inch (25 mm) in diameter through the door, have at least an 0.25-inch-thick (6.4 mm) glass disc and the holder is of metal that will not melt out where subject to temperatures of 1,700°F (927°C).

2. Corridor door assemblies in occupancies of Group I-2 shall be in accordance with Section 407.3.1.

3. Unprotected openings shall be permitted for corridors in multitheater complexes where each motion picture auditorium has at least one-half of its required exit or exit access doorways opening directly to the exterior or into an exit passageway.

❖ NFPA 252 and UL 10C standards require a hose stream test on all fire door assemblies. The hose stream test is intended to provide a measurement of structural integrity by evaluating the ability of the assembly to withstand impact. It has been determined, however, that a hose stream is not justified for 20-minute doors, which are only acceptable as corridor or smoke barrier doors. Smoke barriers are intended to simply manage smoke; therefore, impact is not as critical.

The 20-minute fire protection classification was developed in response to a concern about determining the equivalency of other types of doors to a $1^3/_4$-inch (44 mm) solid-bonded wood core door, which provided approximately a 20-minute fire protection rating. Therefore, while a $1^3/_4$-inch (44 mm) solid-bonded wood core door may perform similar to a 20-minute door, one should not consider them as 20-minute doors without a label to substantiate performance (see Section 715.3.5 for labeling requirements).

This section also mandates that glass sidelights and glass transom perform the same as fire windows as regulated by Section 715.4. In other words, the sidelight and transom is not considered part of the door assembly and, therefore, it must be tested with the hose stream test in accordance with NFPA 257.

Fire doors in corridors and smoke barriers are also required to meet the criteria of UL 1784. This standard is intended to address the movement of smoke through the door that is located in a fire-resistance-rated assembly. It is important to note that compliance with this standard as well as NFPA 252 (or UL 10B and 10C), is not required for all doors located in corridors–only corridors required to be fire-resistance rated in accordance with Section 1016.1.

715.3.4 Doors in vertical exit enclosures and exit passageways. Fire door assemblies in vertical exit enclosures and exit passageways shall have a maximum transmitted temperature end point of not more than 450°F (232°C) above ambient at the end of 30 minutes of standard fire test exposure.

Exception: The maximum transmitted temperature end point is not required in buildings equipped throughout with an automatic sprinkler system installed in accordance with Section 903.3.1.1 or 903.3.1.2.

❖ This section requires that door assemblies utilized in vertical exit enclosures and exit passageways as defined in Chapter 10 comply with the requirements of Section 715 and have been tested in accordance with either NFPA 252 or UL 10C. This section adds a requirement that door construction is limited to a temperature rise on the unexposed side of 450°F (232°C) during the first 30 minutes of the standard fire test (NFPA 252). Door labels are required to indicate that the temperature on the unexposed side is 450°F (232°C) or less above ambient temperature; otherwise, it would be possible for a door to pass the test even if the temperature on the unexposed surface exceeds 450°F (232°C), which would comply with this additional requirement. It should be noted that the temperature rise criterion is not otherwise a limitation with respect to a door receiving a fire protection rating. Therefore, simply specifying a $1^1/_2$-hour fire protection rating on the door does not fully reflect the additional requirements of Section 715.3.4.

The reason for limiting the temperature rise of the unexposed surface of exit doors to 450°F (232°C) is that a higher allowable temperature would provide enough radiant heat to discourage or even prevent building occupants from closely approaching or passing by the door assembly during a fire emergency. The intent is to protect the occupants in the vertical exit enclosure and exit passageway from excessive radiant heat. There is an

exception for buildings that are equipped throughout with either an NFPA 13 or 13R automatic sprinkler system. The reason for the exception is that a sprinkler system would likely prevent a door from exceeding approximately 250°F (121°C), thus there is no need for door temperature concern during a fire.

715.3.4.1 Glazing in doors. Fire-protection-rated glazing in excess of 100 square inches (0.065 m²) shall be permitted in fire door assemblies when tested in accordance with NFPA 252 as components of the door assemblies and not as glass lights, and shall have a maximum transmitted temperature end point of 450°F (232°C) in accordance with Section 715.3.4.

Exception: The maximum transmitted temperature end point is not required in buildings equipped throughout with an automatic sprinkler system installed in accordance with Section 903.3.1.1 or 903.3.1.2.

❖ This section requires glazing panels in fire doors, which are to have a maximum temperature rise of 450°F (232°C), to be tested as a component of the door, since NFPA 252 does not require the temperature of the glazing component to be measured. Some glazing materials have the ability to limit temperature rise and have been tested in accordance with ASTM E 119 as wall assemblies. Some of these glazing materials have also been tested in door assemblies.

Glazing in fire doors that are required to limit temperature rise should meet the same temperature rise requirements. This section clarifies that glazing that does not meet temperature rise requirements is not permitted.

The 100-square-inch (0.065 m²) limitation is consistent with the exceptions listed in Section 715.3.6.1.

715.3.5 Labeled protective assemblies. Fire door assemblies shall be labeled by an approved agency. The labels shall comply with NFPA 80, and shall be permanently affixed to the door or frame.

❖ The requirement by Section 1703.5 that fire doors be labeled indicates that the door assembly has not only passed the required fire test but that a follow-up inspection was also performed during production (see Section 1703.5.2). The building official should, therefore, verify that fire door assemblies are properly labeled, and not rely solely on a test report. To provide specific guidance, compliance with NFPA 80 is required. Labels that indicate compliance with NFPA 252, UL 10A, 10B, 10C, 14B or 14C are acceptable (see Sections 715.3, 715.3.1 and 715.3.2)

715.3.5.1 Fire door labeling requirements. Fire doors shall be labeled showing the name of the manufacturer, the name of the third-party inspection agency, the fire protection rating and, where required for fire doors in exit enclosures by Section 715.3.4, the maximum transmitted temperature end point. Smoke and draft control doors complying with UL 1784 shall be labeled as such. Labels shall be approved and permanently affixed. The label shall be applied at the factory or location where fabrication and assembly are performed.

❖ Labels on fire doors apply to the door only. The building official should verify that the door frame, hardware and accessories are also labeled for use with a labeled fire door. In addition to the appropriate fire protection rating, the label for exit doors is required to indicate the temperature rise on the unexposed surface after 30 minutes (see commentary, Section 715.3.4).

A label is also to serve as evidence of both a required fire protection rating and third-party inspection—not solely as a manufacturer's identification (see commentary, Section 1703.5). To ensure appropriate labeling it is required to occur at the actual location where fabrication and assembly occur.

715.3.5.2 Oversized doors. Oversized fire doors shall bear an oversized fire door label by an approved agency or shall be provided with a certificate of inspection furnished by an approved testing agency. When a certificate of inspection is furnished by an approved testing agency, the certificate shall state that the door conforms to the requirements of design, materials and construction, but has not been subjected to the fire test.

❖ Recognizing that doors may exceed the required tested size, the code indicates that oversized doors that have a certificate of inspection from an approved testing agency are acceptable. For example, rolling-steel-type fire doors are normally listed for 120 square feet (11 m²) with no dimension in excess of 12 feet (3658 mm). Rolling-steel-type doors that exceed these dimensions are permitted as long as they have certification indicating that they are constructed of materials of the same grade, thickness, shape, etc., as labeled doors. This is normally indicated with a certificate that reads "Oversized Fire Door Certificate."

715.3.5.3 Smoke and draft control door labeling requirements. Smoke and draft control doors complying with UL 1784 shall be labeled in accordance with Section 715.3.5.1 and shall show the letter "S" on the fire rating label of the door. This marking shall indicate that the door and frame assembly are in compliance when listed or labeled gasketing is also installed.

❖ When doors are required to be approved for smoke and draft control, they require a special label containing the letter "S." Without this label it is very difficult to assess in the field whether a door contains this feature both for construction inspections and future maintenance.

715.3.5.4 Fire door frame labeling requirements. Fire door frames shall be labeled showing the names of the manufacturer and the third-party inspection agency.

❖ It is required that the fire door frame be labeled with the manufacturer's name and third-party testing agency to assist in the assessment of the door in the field. The actual rating is not required to be included. It is likely that a single type of door frame may be used with a variety of

rated doors; however, when a smoke and draft control door is used, it may make a difference which frame is used.

715.3.6 Glazing material. However–Fire-protection-rated glazing conforming to the opening protection requirements in Section 715.3 shall be permitted in fire door assemblies.

❖ This section of the code addresses all types of glass in fire-resistance-rated doors, no matter if tested glass or wired glass is used.

715.3.6.1 Size limitations. Wired glass used in fire doors shall comply with Table 715.4.3. Other fire-protection-rated glazing shall comply with the size limitations of NFPA 80.

Exceptions:

1. Fire-protection-rated glazing in fire doors located in fire walls shall be prohibited except that where serving as a horizontal exit, a self-closing swinging door shall be permitted to have a vision panel of not more than 100 square inches (0.065 m²) without a dimension exceeding 10 inches (254 mm).

2. Fire-protection-rated glazing shall not be installed in fire doors having a $1^1/_2$-hour fire protection rating intended for installation in fire barriers, unless the glazing is not more than 100 square inches (0.065 m²) in area.

❖ Table 715.4.3 limits the quantity of wired glass permitted in fire doors, depending on the fire protection rating of the door. These limitations are based on tests conducted with such panels and doors. In accordance with Sections 715.3.6.3 and 715.4.9, all fire protection-rated glass must be labeled.

NFPA 80 permits fire protection-rated glazing in fire doors with a fire protection rating up to 3 hours. For 1-, $1^1/_2$- and 3-hour doors, the maximum area is 100 square inches (0.065 m²). For $1/_2$-, $1/_3$- and $3/_4$-hour doors, the allowable area is a function of the glazed area tested.

715.3.6.2 Exit and elevator protectives. Approved fire-protection-rated glazing used in fire doors in elevator and stairway shaft enclosures shall be so located as to furnish clear vision of the passageway or approach to the elevator or stairway.

❖ The purpose of vision panels in exit doors and elevator doors is to permit observation of people who may be on the other side of the door before opening it. The limitations on the size of panels in Section 715.3.6.1 still apply.

715.3.6.3 Labeling. Fire-protection-rated glazing shall bear a label or other identification showing the name of the manufacturer, the test standard and the fire protection rating. Such label or other identification shall be issued by an approved agency and shall be permanently affixed.

❖ This provision of the code, which is similar to others, maintains proper labeling for each piece of fire protection-rated glazing. The label must specify the name or fully identifying logo of the manufacturer, the exact test standard that was used to evaluate the glass and the rating established by the test. It is most important to identify to the code user the specific test standard used to properly evaluate the fire performance of the glass. Some doors and glass are tested in accordance with the applicable test standards for doors and windows, while some are tested to the ASTM E 119 standard for walls, columns and floor/ceiling assemblies. A fire protection rating established by NFPA 252 or UL 10C and NFPA 257 for windows is less stringent than a fire-resistance rating as established by the ASTM E 119 test. The essential information contained in the label now provides the necessary data to verify code compliance.

715.3.6.4 Safety glazing. Fire-protection-rated glazing installed in fire doors or fire window assemblies in areas subject to human impact in hazardous locations shall comply with Chapter 24.

❖ This section provides a cross reference to the safety glazing requirements of Chapter 24, specifically Section 2406.2. Not only do glass panels need to be fire-resistance rated for a specified number of hours, but the glass also must pass the test requirements of CPSC 16 CFR, Part 1201. Specific hazardous locations, such as doors and large panels of glass, are identified in Section 2406.2. This requirement reduces the hazards of someone falling into a fire-resistance-rated glass panel.

715.3.7 Door closing. Fire doors shall be self-closing or automatic-closing in accordance with this section.

Exception: Fire doors located in common walls separating sleeping units in Group R-1 shall be permitted without automatic-closing or self-closing devices.

❖ In order for fire doors to be effective, they must be in the closed position; therefore, the preferred arrangement is to install self-closing and self-latching doors that will normally be closed. Recognizing that operational practices often require doors to be open for an extended period of time, automatic-closing doors are permitted as long as this opening will not pose a threat to occupant safety and the doors will self-latch. Automatic-closing devices enable the opening to be protected during a fire condition. The basic requirement for closing devices and specific requirements for automatic-closing and self-closing devices are given in NFPA 80 (see Section 715.3.7.2).

There is an exception for doors in the separation walls of side-by-side sleeping units in Group R-1 occupancies. Because the need for a sleeping unit to have access to the adjacent sleeping unit is minimal, the number of open doors will not pose a threat to the building occupants. Most doors will be closed at night for privacy and will maintain the necessary fire-resistance separation. Since these doors will probably be propped open during the day, the code acknowledges a common-sense solution.

715.3.7.1 Latch required. Unless otherwise specifically permitted, single fire doors and both leaves of pairs of side-hinged swinging fire doors shall be provided with an active latch bolt that will secure the door when it is closed.

❖ This section merely reinforces the acceptance criteria of fire test standards for fire doors. These standards require that the door not move from its original position by more than one and a half times the door thickness in the direction perpendicular to the plane of the door. This criteria basically requires the door to have position latching in order to pass the test.

715.3.7.2 Automatic-closing fire door assemblies. Automatic-closing fire door assemblies shall be self-closing in accordance with NFPA 80.

❖ NFPA 80 requires doors to be automatic-closing with a closing device and a separate hold/release device or hold-open mechanism that closes upon activation of an automatic fire detector that is acceptable to the building official.

715.3.7.3 Smoke-activated doors. Automatic-closing fire doors installed in the following locations shall be automatic-closing by the actuation of smoke detectors installed in accordance with Section 907.10 or by loss of power to the smoke detector or hold-open device. Fire doors that are automatic-closing by smoke detection shall not have more than a 10-second delay before the door starts to close after the smoke detector is actuated.

1. Doors installed across a corridor.

2. Doors that protect openings in horizontal exits, exits or exit access corridors required to be of fire-resistance-rated construction.

3. Doors that protect openings in walls required to be fire-resistance rated by Table 302.1.1.

4. Doors installed in smoke barriers in accordance with Section 709.5.

5. Doors installed in fire partitions in accordance with Section 708.6.

6. Doors installed in a fire wall in accordance with Section 705.8.

❖ Since the integrity of the means of egress can be compromised by smoke, door openings are required to be automatic-closing upon the detection of smoke if the door is not self-closing. The automatic closer is also to activate upon loss of power to the smoke detector and to the hold-open device. This is essential since, if power to the smoke detector is interrupted before the door closes, the fire can still grow and spread through the door opening without the fire door having served its purpose.

NFPA 72, as referenced in Chapter 9, contains criteria relative to the number and location of detectors necessary for door release service. NFPA 80 also provides guidance and criteria for the location and number of closing devices required for a fire door.

715.3.7.4 Doors in pedestrian ways. Vertical sliding or vertical rolling steel fire doors in openings through which pedestrians travel shall be heat activated or activated by smoke detectors with alarm verification.

❖ Where vertical sliding or rolling doors are provided as a required opening protective through which pedestrian travel is intended, automatic-closing actuation is not permitted to be near a smoke detector unless alarm verification is provided. Sudden closing of these doors and the potential of false actuation of smoke detectors create a hazard to occupants egressing through the opening.

Although pedestrian travel is intended in openings regulated by this section, a common oversight is that the vertical rolling or sliding door in the closed position is not permitted to be considered as required egress, as stated in Section 1008.1.2, since the door is not side swinging.

715.3.8 Swinging fire shutters. Where fire shutters of the swinging type are installed in exterior openings, not less than one row in every three vertical rows shall be arranged to be readily opened from the outside, and shall be identified by distinguishing marks or letters not less than 6 inches (152 mm) high.

❖ The fire department needs access to openings for ventilation purposes and entry; therefore, if an extensive amount of swinging shutters is used, at least one in every three vertical rows must be readily openable from the outside. The operable shutter must also be easily recognized for use by the fire department. Since the shutter must be openable from the outside, the identification must also be on the exterior.

715.3.9 Rolling fire shutters. Where fire shutters of the rolling type are installed, such shutters shall include approved automatic-closing devices.

❖ Rolling fire shutters are not self-closing. They must be activated in order to initiate the closing sequence.

715.4 Fire-protection rated glazing. Glazing in fire window assemblies shall be fire protection rated in accordance with this section and Table 715.4. Glazing in fire doors shall comply with Section 715.3.6. Fire-protection-rated glazing installed as an opening protective in fire partitions, smoke barriers and fire barriers shall be tested in accordance with and shall meet the acceptance criteria of NFPA 257 for a fire protection rating of 45 minutes. Fire-protection-rated glazing shall also comply with NFPA 80. Fire-protection-rated glazing required in accordance with Section 704.12 for exterior wall opening protection shall be tested in accordance with and shall meet the acceptance criteria of NFPA 257 for a fire protection rating as required in Section 715.4.8.

Exceptions:

1. Wired glass in accordance with Section 715.4.3.

2. Fire-protection-rated glazing in 0.5-hour fire-resistance-rated partitions is permitted to have an 0.33-hour fire protection rating.

❖ The issue of what constitutes a fire window and the appropriate test standards has been further complicated by recent testing of 1-hour and 1$^1/_2$-hour glazing materials and glass-block assemblies. The test procedure referenced for fire windows (fire protection-rated glazing) is NFPA 257, which is a pass/fail test with a specified time of exposure.

While light-transmitting materials are usually considered fire windows, some materials have been tested in accordance with ASTM E 119 for use as a wall; therefore, it is imperative that the building official carefully review which test procedure is used and how the product is evaluated. A glazing material that is tested in accordance with ASTM E 119 would not require the application of Section 715 (see commentary, Section 715.2).

Modern technology has resulted in developing light-transmitting materials that are capable of obtaining a fire-resistance rating (ASTM E 119) instead of or in addition to a fire protection rating (NFPA 257). The building official needs to review the test reports carefully to determine the intended application and tests conducted. This section only applies to materials that are tested in accordance with NFPA 257 and that are intended as an opening protective in a wall with a fire-resistance rating. This includes fire barriers, fire partitions and smoke barriers.

Other sections of the code restrict the use of fire window assemblies. For example, Section 2110.1.1 restricts glass block to specific applications and Section 1019.1.4 limits the use of fire windows in stair enclosures (see Section 715.4.3 for a discussion on wired glass).

Fire window assemblies are rated in accordance with NFPA 257. The test standard evaluates the effectiveness of windows when used as opening protectives to remain in place during the specified time of exposure. In addition to being subjected to a predetermined fire condition, the assembly is also subjected to a hose stream impact test. The test procedure does not measure or evaluate heat transmission or radiation through the assembly.

The conditions of acceptance require that the assembly remain in place and not be loosened from the frame or develop any openings around the perimeter of the frame or individual glass lights during the fire exposure test. During the hose stream test, the window assembly must remain in place, but is permitted to have glass dislodged from the central portion of each glass light as long as the amount dislodged does not exceed 5 percent of the area. Also, during the hose stream test, openings created by separation of the glass edges from the frame are limited to 30 percent of the perimeter. For glass blocks, at least 70 percent of it must not develop through the openings.

There is an exception that would allow a $^1/_3$-hour rating for openings in $^1/_2$-hour-rated fire partitions. Fire partitions are only allowed to be $^1/_2$-hour rated in accordance with the exceptions to Section 708.3 and Table 1016.1.

TABLE 715.4
FIRE WINDOW ASSEMBLY FIRE PROTECTION RATINGS

TYPE OF ASSEMBLY	REQUIRED ASSEMBLY RATING (hours)	MINIMUM FIRE WINDOW ASSEMBLY RATING (hours)
Interior walls:		
Fire walls	All	NP[a]
Fire barriers and fire partitions	> 1	NP[a]
	1	$^3/_4$
Smoke barriers	1	$^3/_4$
Exterior walls	>1	$1^1/_2$
	1	$^3/_4$
Party walls	All	NP[a]

a. Not permitted except as specified in Section 715.2.

❖ Table 715.4 prescribes the minimum fire protection ratings for fire windows, which are not to be confused with the ratings for fire doors provided in Table 715.3 relative to the nature and fire-resistance rating of the wall. Once the purpose and fire-resistance rating of the wall are identified, the minimum required fire protection rating of windows is determined using the table. Typically, the minimum permitted fire window fire protection rating is less than the required fire-resistance rating of the wall. The hourly designation in the table indicates the duration of fire test exposure and is actually called the "fire protection rating."

715.4.1 Testing under positive pressure. NFPA 257 shall evaluate fire-protection-rated glazing under positive pressure. Within the first 10 minutes of a test, the pressure in the furnace shall be adjusted so at least two-thirds of the test specimen is above the neutral pressure plane, and the neutral pressure plane shall be maintained at that height for the balance of the test.

❖ Under fire conditions there is likely to be positive pressure acting on the window; therefore, this section requires that tested window assemblies be subjected to positive pressure conditions during the test. This requirement is consistent with standards used for testing window assemblies in the U.S. and other countries.

715.4.2 Nonsymmetrical glazing systems. Nonsymmetrical fire-protection-rated glazing systems in fire partitions, fire barriers or in exterior walls with a fire separation of 5 feet (1524 mm) or less pursuant to Section 704 shall be tested with both faces exposed to the furnace, and the assigned fire protection rating shall

be the shortest duration obtained from the two tests conducted in compliance with NFPA 257.

❖ This section requires that glazing systems that by design are not the same on either side be tested (NFPA 257) on both sides. The rating resulting from these tests will be determined from the lower performing side. These requirements are for fire-type barriers (versus smoke) within buildings and when used on an exterior wall with a small separation distance [less than 5 feet (1524 mm)]. The concern here is the same as for nonsymmetrical wall assemblies in Section 703.2.1.

715.4.3 Wired glass. Steel window frame assemblies of 0.125-inch (3.2 mm) minimum solid section or of not less than nominal 0.048-inch-thick (1.2 mm) formed sheet steel members fabricated by pressing, mitering, riveting, interlocking or welding and having provision for glazing with $^1/_4$-inch (6.4 mm) wired glass where securely installed in the building construction and glazed with $^1/_4$-inch (6.4 mm) labeled wired glass shall be deemed to meet the requirements for a $^3/_4$-hour fire window assembly. Wired glass panels shall conform to the size limitations set forth in Table 715.4.3.

❖ An approved fire window assembly normally includes a tested fire window frame and glazing materials. This section permits the use of tested $^1/_4$-inch (6.4 mm) wired glass with a specific steel frame to be considered as a $^3/_4$-hour fire window assembly without the need to pass the NFPA 257 test standard (see the exception to Section 715.4). The impact of this is to exempt the frame from being tested.

This section has a significant impact on wall assemblies that are not of masonry construction. Most fire window frames are tested for use in masonry construction only. The individual evaluation of the frame will indicate whether it may be used in drywall construction. Fire window frames evaluated for drywall construction are further subdivided into those intended to be supported by a noncombustible floor and those intended to be installed above the floor.

Wired glass for fire windows must be installed in either a tested fire window frame (see Section 715.4) or a steel frame that conforms to this section. The maximum size limitations in Table 715.4.3 are based on actual tests of fire windows and fire doors with vision panels of $^1/_4$-inch-thick (6.4 mm) wired glass.

This section requires a fire protection rating of 45 minutes for consistency with criteria previously noted in the model codes (ASTM E 163), even though Table 715.4.3 specifies ratings based on location for other types of assemblies.

While Table 715.4.3 indicates the maximum size of a panel of wired glass, the table does not limit the number of panels. Such limitations are found in Section 715.4.7.2, which limits the total width of all openings in fire barriers and fire partitions to a maximum of 25 percent of the length of the wall. With respect to door assemblies, the fire test results and NFPA 80 will limit the amount of wired glass permitted in certain applications. For example, a 1-hour fire door is permitted to have only one panel of no more than 100 square inches (0.065 m²), whereas a ³/₄-hour fire door is permitted to have more than one panel of no more than 1,296 square inches (0.85 m²) each, provided it has been successfully tested in accordance with NFPA 252. In the case of 20-minute fire doors, both the size of an individual panel and the total amount of wired glass permitted are limited only by the test results.

TABLE 715.4.3
LIMITING SIZES OF WIRED GLASS PANELS

OPENING FIRE PROTECTION RATING	MAXIMUM AREA (square inches)	MAXIMUM HEIGHT (inches)	MAXIMUM WIDTH (inches)
3 hours	0	0	0
1¹/₂-hour doors in exterior walls	0	0	0
1 and 1¹/₂ hours	100	33	10
³/₄ hour	1,296	54	54
20 minutes	Not Limited	Not Limited	Not Limited
Fire window assemblies	1,296	54	54

For SI: 1 inch = 25.4 mm, 1 square inch = 645.2 mm².

❖ Based on the rating of the opening, the maximum area, height and width of wired glass panels are to be in accordance with the table. The limitations are based on maximum sizes of wired glass panels that have been tested in accordance with ASTM E 163, which was referenced in earlier editions of the model codes. The required rating of the opening is determined by Table 715.4.3 for all opening protectives, with limitations on all fire protection-rated glazing, including wired glass, in Sections 715.4.7 and 715.4.8.

715.4.4 Nonwired glass. Glazing other than wired glass in fire window assemblies shall be fire-protection-rated glazing installed in accordance with and complying with the size limitations set forth in NFPA 80.

❖ This section merely reinforces Section 715.4.

715.4.5 Installation. Fire-protection-rated glazing shall be in the fixed position or be automatic-closing and shall be installed in approved frames.

❖ In order to adequately protect the opening, protection must be in place during a fire condition. This is achieved by either having the glazing in a permanently fixed portion or having the protective activated in order to initiate the closing sequence.

715.4.6 Window mullions. Metal mullions that exceed a nominal height of 12 feet (3658 mm) shall be protected with materials

to afford the same fire-resistance rating as required for the wall construction in which the protective is located.

❖ Fire windows are normally tested in panels not exceeding 1,296 square inches (0.85 m²) and 54 inches (1372 mm) in width or length, respectively. The code does, however, permit panels to be installed adjacent to one another. This section requires that mullions higher than 12 feet (3658 mm) meet the criteria for fire resistance (ASTM E 119), not fire protection (NFPA 257).

715.4.7 Interior fire window assemblies. Fire-protection-rated glazing used in fire window assemblies located in fire partitions and fire barriers shall be limited to use in assemblies with a maximum fire-resistance rating of 1 hour in accordance with this section.

❖ Since fire protection-rated glazing, including wired glass, has a ³/₄-hour fire-resistance rating (see Sections 715.4 and 715.4.2, respectively), fire windows may only be used in fire partitions (1-hour rated in accordance with Section 708.3) and fire barriers having a fire-resistance rating of 1 hour.

715.4.7.1 Where permitted. Fire-protection-rated glazing shall be limited to fire partitions designed in accordance with Section 708 and fire barriers utilized in the applications set forth in Sections 706.3.5 and 706.3.6 where the fire-resistance rating does not exceed 1 hour.

❖ This section identifies the specific 1-hour-rated assemblies that are referred to in Section 715.4.7. Fire partitions are 1-hour fire-resistance rated in accordance with Section 708.3. Fire barriers, on the other hand, can have ratings as high as 4 hours (see Table 302.3.3). This section limits the use of fire windows to fire barriers that are 1-hour rated, 1-hour incidental use separations and 1-hour mixed occupancy separations. It must be noted that fire barriers used to separate exit elements (see Sections 706.3.2, 706.3.3 and 706.3.4) are not permitted to have fire windows, even where the separation is 1-hour fire-resistance rated. Since fire walls are intended to separate areas into two or more buildings, the code establishes a higher degree of protection by not allowing fire windows in such walls.

715.4.7.2 Size limitations. The total area of windows shall not exceed 25 percent of the area of a common wall with any room.

❖ This section mirrors that of Section 706.7 for fire barriers, but is applicable to both fire partitions and fire barriers.

715.4.8 Exterior fire window assemblies. Exterior openings, other than doors, required to be protected by Section 704.12, where located in a wall required by Table 602 to have a fire-resistance rating of greater than 1 hour, shall be protected with an assembly having a fire protection rating of not less than 1¹/₂ hours. Exterior openings required to be protected by Section 704.8, where located in a wall required by Table 602 to have a fire-resistance rating of 1 hour, shall be protected with an assembly having a fire protection rating of not less than $^3/_4$ hour. Exterior openings required to be protected by Section 704.9 or 704.10 shall be protected with an assembly having a fire protection rating of not less than $^3/_4$ hour. Openings in nonfire-resistance-rated exterior wall assemblies that require protection in accordance with Section 704.8, 704.9 or 704.10 shall have a fire protection rating of not less than $^3/_4$ hour.

❖ This section prescribes the manner in which openings in exterior walls are to be protected. It identifies the required performance characteristics of the exterior opening protective when the need to provide protected openings exists (see Sections 704.8, 704.9, 704.10 and 704.12). The purpose of providing exterior opening protectives is to reduce the fire risk exposure to and from adjacent structures, to reduce the potential for vertical fire spread and to protect exits adequately when the separation from other portions of the structure involves exterior walls.

The primary need to provide exterior opening protectives is determined by Section 704.8. Exterior opening protectives are required by Section 704.8 based on the fire separation distance and the area of the openings relative to the exterior wall area. Sections 704.9 and 704.10 contain provisions that require exterior opening protectives in certain instances. Section 1005.3.2.1 also requires exterior opening protectives for the protection of exits. Exterior walls required by Table 602 to have a fire-resistance rating need not have exterior opening protectives unless also required by Section 704.8.

Consistent with other opening protective provisions in the code (see Table 715.3), if the opening is required to be protected and is in a wall required to have a fire-resistance rating of more than 1 hour, the opening protective assembly must have a fire protection rating of not less than $1^1/_2$ hours. For walls with a 3- or 4-hour rating, the minimum required rating of the opening protective is still $1^1/_2$ hours. If the opening is required to be protected and is in a wall required to have a fire-resistance rating of 1 hour, the opening protective assembly must have a fire protection rating no less than $^3/_4$ hour. Exterior opening protectives required by Sections 704.9 and 704.10 must have a fire protection rating of no less than $^3/_4$ hour, based on the 1-hour fire-resistance rating afforded by the spandrel (see Section 704.9) and roof (see Section 704.10).

715.4.9 Labeling requirements. Fire-protection-rated glazing shall bear a label or other identification showing the name of the manufacturer, the test standard, and the fire protection rating. Such label or identification shall be issued by an approved agency and shall be permanently affixed.

❖ This section requires fire protection-rated (see Section 715.4) and fire-resistance-rated (see Section 715.2) fire window assemblies to be labeled by an approved test agency. This requirement enables the code user to verify that a fire window has been tested to the correct standard and that the test was properly performed by

experienced people with correct equipment.

This section details the necessary information required on the fire window label and how the label is to be applied. These descriptive requirements are used to verify that fire window assemblies have been correctly tested. The label on each and every fire window assembly must contain the manufacturer's name, the test standard used, the edition of the test standard, the fire protection rating and whether the rating was achieved as an opening or as part of the wall assembly. The label must be applied at the manufacturer's plant by the third-party inspection agency.

As previously discussed, fire protection ratings are specific to windows where fire resistance is specific to building elements, such as walls and structural elements. In some cases, glazing is tested in accordance with ASTM E 119 for fire resistance (see commentary, Section 715.2).

SECTION 716
DUCTS AND AIR TRANSFER OPENINGS

716.1 General. The provisions of this section shall govern the protection of ducts and air transfer openings in fire-resistance-rated assemblies.

❖ Fire dampers, smoke dampers and combination fire/smoke dampers protect openings created by duct penetrations and transfer openings in fire-resistance-rated assemblies. Ceiling radiation dampers protect duct penetrations, which only penetrate the ceiling membrane of a fire-resistance-rated assembly. This section includes installation and testing details in Sections 716.2 and 716.3, respectively. Section 716.3 also indicates that dampers are to bear the label of an approved agency. When dampers are provided, they must be properly maintained and, therefore, must be accessible (see Section 716.4). Section 716.5 indicates the conditions in which dampers are required at penetrations through vertical assemblies. Duct penetrations and transfer openings through horizontal assemblies, including requirements for ceiling dampers, are regulated in Section 716.6. The provisions of Section 716 are duplicated in Section 607 of the IMC.

All the requirements for protection of duct penetrations and transfer openings are found in this section. This section is directly referenced by sections that address vertical and horizontal fire-resistance-rated assemblies, as follows:

Exterior walls:	Section 704.14
Fire walls:	Section 705.11
Fire barriers:	Section 706.10
Shafts:	Section 707.10
Fire partitions:	Section 708.9
Smoke barriers:	Section 709.8
Horizontal assemblies:	Section 711.7

716.1.1 Ducts and air transfer openings without dampers. Ducts and air transfer openings that penetrate fire-resistance-rated assemblies and are not required by this section to have dampers shall comply with the requirements of Section 712.

❖ This section identifies how duct penetrations and air transfer openings are to be properly protected, either in accordance with the requirements of Section 716 using fire or fire/smoke dampers where required or in accordance with Section 712 with materials that have been tested to either ASTM E 814 or ASTM E 119 to maintain the fire-resistance rating of the fire assembly. This section is coordinated with Sections 712.3.3 (fire-resistance-rated walls) and 712.4.4 (horizontal assemblies).

716.2 Installation. Fire dampers, smoke dampers, combination fire/smoke dampers and ceiling dampers located within air distribution and smoke control systems shall be installed in accordance with the requirements of this section, the manufacturer's installation instructions and listing.

❖ This section performs two regulating functions. First, it requires that dampers be installed in accordance with the manufacturer's installation instructions and listing. Such instructions will result in an installation that not only that protects the penetration when the damper is actuated, but that also results in a protected opening should the duct collapse. Second, this section introduces Sections 716.2.1 and 716.2.2, which address the relationship of dampers to smoke control systems and hazardous exhaust ducts.

716.2.1 Smoke control system. Where the installation of a fire damper will interfere with the operation of a required smoke control system in accordance with Section 909, approved alternative protection shall be utilized.

❖ Fire dampers are not permitted where they will interfere with, obstruct or inhibit the proper operation of a required smoke control system. In this situation, fire dampers must be replaced with an alternative means of protection, subject to the approval of the building official. Alternative protection may be in the form of a fire-resistance-rated shaft enclosure, steel subducts extending at least 22 inches (559 mm) vertically in an exhaust shaft having continuous upward airflow or a fire-resistance-rated horizontal duct enclosure. If a smoke control system is installed that is not required by the code, it is still subject to the same fire damper requirements as a required smoke control system; however, such nonrequired systems would be permitted to employ fire dampers instead of the mandatory alternative protection prescribed in this section for required smoke control systems. A required smoke control system is an important life safety system, and the intent is to make it as reliable as possible by eliminating devices or system designs that could render it inoperative.

Of course, it is logical and desirable to design and install all life safety protection systems as if they were required, since their presence instills a sense of security in the building occupants and, regardless of the reasons

for installing the system, it is expected to function as intended.

Note that this section does more than simply allow fire dampers to be omitted in required smoke control systems. This section actually mandates the omission of fire dampers where such system operation could be jeopardized.

716.2.2 Hazardous exhaust ducts. Fire dampers for hazardous exhaust duct systems shall comply with the *International Mechanical Code.*

❖ This section refers the user to the IMC. While not stated in the IMC, duct penetrations are typically not provided with fire dampers, as this may interfere with the exhausting of hazardous materials. Instead of fire damper protection, horizontal ducts are typically placed in a fire-resistance-rated horizontal shaft, which in turn, penetrates the wall.

716.3 Damper testing and ratings. Dampers shall be listed and bear the label of an approved testing agency indicating compliance with the standards in this section. Fire dampers shall comply with the requirements of UL 555. Only fire dampers labeled for use in dynamic systems shall be installed in heating, ventilation and air-conditioning systems designed to operate with fans on during a fire. Smoke dampers shall comply with the requirements of UL 555S. Combination fire/smoke dampers shall comply with the requirements of both UL 555 and UL 555S. Ceiling radiation dampers shall comply with the requirements of UL 555C.

❖ A fire damper is a device designed to close automatically upon detection of heat, to interrupt migratory airflow and to restrict the passage of flame through duct penetrations of rated assemblies. Fire dampers are required to be tested for use in accordance with UL 555. The criteria for acceptance of fire dampers require that they remain in the opening for the fire exposure period for which they are rated; close and latch automatically within 60 seconds and remain in place such that movement (warping) does not exceed prescribed limitations. In addition to being tested, fire dampers must be labeled and, in accordance with Section 1704.3, require a follow-up service.

Fire dampers are classified by UL 555 for use in static and dynamic airflow conditions. Fire dampers installed in air distribution systems that remain in operation after smoke or heat from a fire is detected (a dynamic airflow condition) must be labeled for such use. Static fire dampers may not operate properly under dynamic conditions; therefore, fire dampers used in systems designed with dynamic airflow must be tested and labeled for closure under these conditions.

The manufacturer's installation instructions must be followed. Fire dampers are to be installed in the assembly such that an opening created by a duct collapse is properly protected. Dampers are to be installed in accordance with their tested application, either vertically or horizontally. Although the test criteria of UL 555 are not a function of a horizontal versus vertical application, there are specific tests required for spring-operated dampers that are not required for gravity-operated types; therefore, the type of damper used requires the building official to determine if it is acceptable for the given application.

Fire dampers are required to have the minimum fire protection rating established in Table 716.3.1.

A smoke damper is a device installed in a duct, typically in smoke control systems or to protect smoke barrier penetrations, which operates to seal the duct against smoke leakage. Fire dampers tested to UL 555 are not rated for smoke leakage. Fire dampers are activated using a fusible link or other heat-responsive device that activates the damper at a predetermined temperature. Smoke dampers, on the other hand, are tested for leakage in accordance with the requirements of UL 555S and are activated by a smoke detector in the duct or area to be served.

Ceiling dampers must conform to the requirements of UL 555C, which provides criteria for the construction, performance and testing of ceiling dampers. Ceiling dampers are designed to limit the radiative heat transfer through an air inlet/outlet opening in a fire-resistance-rated floor/ceiling or roof/ceiling assembly. The description of the tested assembly will include a description of the ceiling damper. A fire damper is a device designed to close automatically upon detection of heat, to interrupt migratory airflow and to restrict the passage of flame through duct penetrations of rated assemblies. Since ceiling and fire dampers have different design criteria, a fire damper must not be used in place of a ceiling damper unless a fire damper meets the temperature performance capabilities of a ceiling damper.

716.3.1 Fire protection rating. Fire dampers shall have the minimum fire protection rating specified in Table 716.3.1 for the type of penetration.

❖ See the commentary to Table 716.3.1.

TABLE 716.3.1
FIRE DAMPER RATING

TYPE OF PENETRATION	MINIMUM DAMPER RATING (hours)
Less than 3-hour fire-resistance-rated assemblies	1.5
3-hour or greater fire-resistance-rated assemblies	3

❖ This table summarizes the required hourly ratings for fire dampers based on the fire-resistance-rated assembly that is being penetrated by the air distribution system. The left-hand column lists the associated hourly ratings of either wall or floor/ceiling assemblies. The code user enters the line item in the left-hand column based on the applicable fire-resistance rating of the element penetrated and reads across to the right column for the appropriate fire damper rating. These fire

damper ratings are obtained from the UL 555 test standard and represent the ratings necessary to maintain the integrity of the rated wall or floor/ceiling assembly.

716.3.1.1 Fire damper actuating device. The fire damper actuating device shall meet one of the following requirements:

1. The operating temperature shall be approximately 50°F (10°C) above the normal temperature within the duct system, but not less than 160°F (71°C).

2. The operating temperature shall be not more than 286°F (141°C) where located in a smoke control system complying with Section 909.

3. Where a combination fire/smoke damper is located in a smoke control system complying with Section 909, the operating temperature rating shall be approximately 50°F (10°C) above the maximum smoke control system designed operating temperature, or a maximum temperature of 350°F (177°C). The temperature shall not exceed the UL 555S degradation test temperature rating for a combination fire/smoke damper.

❖ The thresholds for actuation are based on UL 555. For static systems, UL 555 identifies 160°F (71C°) as the minimum and 215°F (102°C) as the maximum temperature (see Item 1). For dynamic systems, the standard includes a minimum value of 160°F (71°C) and a maximum value of 350°F (177°C) (see Items 2 and 3).

716.3.2 Smoke damper ratings. Smoke damper leakage ratings shall not be less than Class II. Elevated temperature ratings shall not be less than 250°F (121°C).

❖ In accordance with UL 555S, smoke dampers are marked with a temperature rating starting at 250°F (121°C) and rising in increments of 100°F (122°C). Leakage ratings range from Classes I to IV in accordance with the standard and are further identified for tested pressure ranging from 1 inch (25 mm) of water to 12 inches (305 mm) of water.

716.3.2.1 Smoke damper actuation methods. The smoke damper shall close upon actuation of a listed smoke detector or detectors installed in accordance with Section 907.10 and one of the following methods, as applicable:

1. Where a damper is installed within a duct, a smoke detector shall be installed in the duct within 5 feet (1524 mm) of the damper with no air outlets or inlets between the detector and the damper. The detector shall be listed for the air velocity, temperature and humidity anticipated at the point where it is installed. Other than in mechanical smoke control systems, dampers shall be closed upon fan shutdown where local smoke detectors require a minimum velocity to operate.

2. Where a damper is installed above smoke barrier doors in a smoke barrier, a spot-type detector listed for releasing service shall be installed on either side of the smoke barrier door opening.

3. Where a damper is installed within an unducted opening in a wall, a spot-type detector listed for releasing service shall be installed within 5 feet (1524 mm) horizontally of the damper.

4. Where a damper is installed in a corridor wall, the damper shall be permitted to be controlled by a smoke detection system installed in the corridor.

5. Where a total-coverage smoke detector system is provided within areas served by a heating, ventilation and air-conditioning (HVAC) system, dampers shall be permitted to be controlled by the smoke detection system.

❖ This section provides specific information in order to properly place a smoke detector to ensure that it will most efficiently detect smoke (see Item 1). Additionally, for passive systems (not part of a smoke control system), the location of detectors is critical to their performance. Smoke dampers are installed in not only duct penetrations, but also transfer openings. A method for properly controlling dampers in such arrangements is provided in Item 3. Where smoke detection is provided throughout a corridor or an entire area, such detection more efficiently detects smoke than does a duct-type detector. Such detection is necessary for controlling dampers (Items 4 and 5). Smoke dampers may also be operated remotely, especially when used as fire dampers and as part of a smoke removal system. Combination fire and smoke dampers can be activated by either the heat or smoke sensor. Once the primary heat-sensing device has been activated, a remote control system can be used to open the damper to permit its use as a smoke damper in the smoke removal system. Then, if the temperature rises to the damper's maximum degradation test temperature, the secondary heat sensor closes the damper again.

716.4 Access and identification. Fire and smoke dampers shall be provided with an approved means of access, large enough to permit inspection and maintenance of the damper and its operating parts. The access shall not affect the integrity of fire-resistance-rated assemblies. The access openings shall not reduce the fire-resistance rating of the assembly. Access points shall be permanently identified on the exterior by a label having letters not less than 0.5 inch (12.7 mm) in height reading: SMOKE DAMPER or FIRE DAMPER. Access doors in ducts shall be tight fitting and suitable for the required duct construction.

❖ Fire and smoke dampers must be properly maintained so that they will operate as intended. The need to maintain dampers, as well as reset them and replace fusible links after operation, requires that they be accessible. The access doors may need to be fire doors in accordance with Section 715, depending on the location of the door, in order to maintain a fire-resistance rating. Access doors in the duct itself need not have a fire protection rating.

716.5 Where required. Fire dampers, smoke dampers, combination fire/smoke dampers and ceiling radiation dampers shall be provided at the locations prescribed in this section. Where an assembly is required to have both fire dampers and smoke dampers, combination fire/smoke dampers or a fire damper and a smoke damper shall be required.

❖ In order to maintain the fire-resistance rating of vertical fire walls, fire barriers and fire partitions, fire dampers must be provided for all duct penetrations. Fire dampers are to be installed in the assembly such that an opening created by a duct collapse is properly protected. Fire dampers are to have a fire-resistance rating as indicated in Table 716.3.1. Section 716.6 addresses the requirements for fire dampers in floor/ceiling assemblies and ceiling membranes of fire-resistance-rated roof/ceiling assemblies. The manufacturer's installation instructions, which should indicate how the damper was installed during the test, must be followed so that the operation and functioning of the installed assembly will be consistent with that required by the code.

Figure 716.5 shows the difference between where a fire damper is required and where one is not. In this example, the designer has chosen to separate mixed occupancies in accordance with the option of Section 302.3.2. The occupancy separation wall (fire barrier) that is required to be fire-resistance rated is also required to have the duct penetration protected by a fire damper, while the nonrated wall requires no such protection.

The code requires that a fire damper not be used as a smoke damper unless the location of the installation is appropriate for the dual purpose. The most likely location for this to occur would be where the air distribution system also serves as a smoke control system and a duct penetrates a fire-resistance-rated assembly. Section 716.2.1 requires alternative protection to be provided where fire dampers will interfere with the operation of a smoke control system. A combination fire and smoke damper, such as that described above, may be approved for use as alternative protection.

716.5.1 Fire walls. Ducts and air transfer openings permitted in fire walls in accordance with Section 705.11 shall be protected with approved fire dampers installed in accordance with their listing.

❖ Fire walls create separate buildings. These multiple buildings may be on a single lot or, in the case of zero lot line construction, the fire wall may be located on the lot line. In the case where the fire wall separates different buildings on different lots, the code does not permit openings (see Sections 503.2 and 705.11). In such instances, a duct penetration is not permitted. Where the fire wall is not located on a lot line, the wall is permitted to be penetrated by a duct or transfer opening provided the opening is protected with a fire damper.

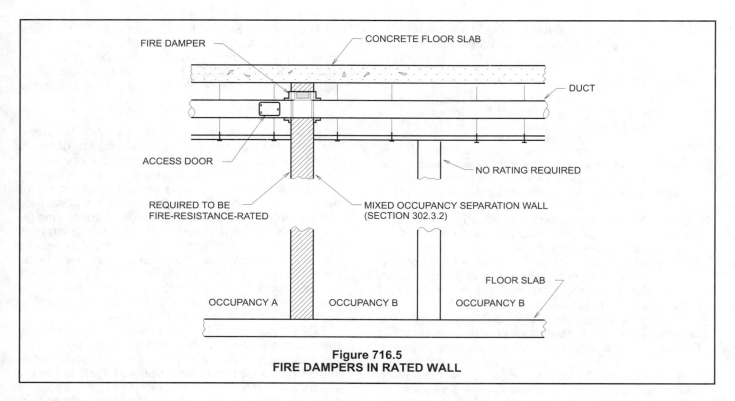

Figure 716.5
FIRE DAMPERS IN RATED WALL

716.5.2 Fire barriers. Ducts and air transfer openings in fire barriers shall be protected with approved fire dampers installed in accordance with their listing.

> **Exception:** Fire dampers are not required at penetrations of fire barriers where any of the following apply:
>
> 1. Penetrations are tested in accordance with ASTM E 119 as part of the fire-resistance-rated assembly.
>
> 2. Ducts are used as part of an approved smoke control system in accordance with Section 909.
>
> 3. Such walls are penetrated by ducted HVAC systems, have a required fire-resistance rating of 1 hour or less, are in areas of other than Group H and are in buildings equipped throughout with an automatic sprinkler system in accordance with Section 903.3.1.1 or 903.3.1.2. For the purposes of this exception, a ducted HVAC system shall be a duct system for conveying supply, return or exhaust air as part of the structure's HVAC system. Such a duct system shall be constructed of sheet steel not less than 26 gage thickness and shall be continuous from the air-handling appliance or equipment to the air outlet and inlet terminals.

❖ This section includes three exceptions to the general requirement on how open duct and air transfer openings in rated construction be protected. Exception 1 states that fire dampers are not required for duct penetrations of fire barriers when the assembly has been tested in accordance with ASTM E 119 without fire dampers and the required fire-resistance rating is obtained. In this case, the penetration assembly has been demonstrated as preserving the fire-resistance rating of the wall assembly.

Exception 2 reinforces the provisions of Section 716.2.1 (see commentary). A fire damper may interfere with the operation of the smoke control system; however, some form of alternative protection must be provided due to the duct penetrating the fire barrier.

Exception 3 states that, in occupancies other than Group H, fire dampers may be omitted in duct penetrations of walls having a fire-resistance rating of 1 hour or less; when part of a ducted HVAC system in a building protected throughout by an automatic sprinkler system. This exception specifies criteria as to what type of HVAC system would qualify for this allowance. Fire dampers would still be required for penetrations of walls requiring a rating greater than 1 hour or for fire-resistance-rated floor/ceiling assemblies. Fire dampers would also still be required for transfer openings (openings without ducts on both sides) through fire barriers. If a plenum system is used, fire dampers must be installed.

716.5.3 Shaft enclosures. Ducts and air transfer openings shall not penetrate a shaft serving as an exit enclosure except as permitted by Section 1019.1.2.

❖ See the commentary to Section 1019.1.2 for additional information.

716.5.3.1 Penetrations of shaft enclosures. Shaft enclosures that are permitted to be penetrated by ducts and air transfer openings shall be protected with approved fire and smoke dampers installed in accordance with their listing.

> **Exceptions:**
>
> 1. Fire dampers are not required at penetrations of shafts where:

1.1. Steel exhaust subducts are extended at least 22 inches (559 mm) vertically in exhaust shafts, provided there is a continuous airflow upward to the outside;

1.2. Penetrations are tested in accordance with ASTM E 119 as part of the rated assembly;

1.3. Ducts are used as part of an approved smoke control system designed and installed in accordance with Section 909, and where the fire damper will interfere with the operation of the smoke control system; or

1.4. The penetrations are in parking garage exhaust or supply shafts that are separated from other building shafts by not less than 2-hour fire-resistance-rated construction.

2. In Group B occupancies, equipped throughout with an automatic sprinkler system in accordance with Section 903.3.1.1, smoke dampers are not required at penetrations of shafts where:

2.1. Bathroom and toilet room exhaust openings with steel exhaust subducts, having a wall thickness of at least 0.019 inches (0.48 mm) that extend at least 22 inches (559 mm) vertically and the exhaust fan at the upper terminus, powered continuously in accordance with the provisions of Section 909.11, maintains airflow upward to the outside; or

2.2. Ducts are used as part of an approved smoke control system, designed and installed in accordance with Section 909, and where the smoke damper will interfere with the operation of the smoke control system.

3. Smoke dampers are not required at penetration of exhaust or supply shafts in parking garages that are separated from other building shafts by not less than 2-hour fire-resistance-rated construction.

❖ This section requires both fire and smoke dampers at duct and air transfer openings in the shaft wall. The fire damper is required due to the penetration of a fire-resistance-rated wall. The smoke damper is required in order to limit the migration of smoke to other parts of the building via the shaft and the chimney effect (see commentary, Section 707.14.1). The exceptions allow the omission of fire dampers and smoke dampers under certain conditions. Exception 1 addresses fire dampers and Exceptions 2 and 3 address smoke dampers.

Exception 1.1 recognizes that the presence of a vertical subduct in a shaft will also offer some degree of fire resistance should the duct outside of the shaft collapse. Steel exhaust ducts that consist of a 22-inch (559 mm) vertical upturn in the shaft need not be protected with a fire damper if there is a continuous upward airflow to the outside. The continuous airflow will create a negative pressure in the shaft as compared to adjacent spaces, thereby minimizing the spread of hot gases from the shaft [see Figure 716.5.3.1(1)].

Exceptions 1.2 and 1.3 are identical to Exceptions 1 and 2 to Section 716.5.2 (see commentary, Section

716.5.2).

Exception 1.4 states that fire dampers may also be omitted in exhaust and supply shafts that serve a garage and are separated from all other shafts in the building by a 2-hour fire separation assembly [see Figure 716.5.3.1(2)]. Requiring fire dampers in garage exhaust and supply shafts would not significantly prevent the spread of smoke and fire within the garage, since the vehicle ramp from floor to floor is a much greater conduit of smoke and fire in a garage.

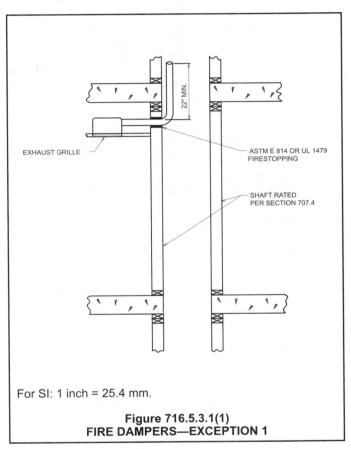

For SI: 1 inch = 25.4 mm.

Figure 716.5.3.1(1)
FIRE DAMPERS—EXCEPTION 1

716.5.4 Fire partitions. Duct penetrations in fire partitions shall be protected with approved fire dampers installed in accordance with their listing.

Exceptions: In occupancies other than Group H, fire dampers are not required where any of the following apply:

1. The partitions are tenant separation and corridor walls in buildings equipped throughout with an automatic sprinkler system in accordance with Section 903.3.1.1 or 903.3.1.2 and the duct is protected as a through penetration in accordance with Section 712.

2. The duct system is constructed of approved materials in accordance with the *International Mechanical Code* and the duct penetrating the wall meets all of the following minimum requirements:

2.1. The duct shall not exceed 100 square inches (0.06 m²).

FIGURE 716.5.3.1(2)   FIRE-RESISTANCE-RATED CONSTRUCTION

2.2. The duct shall be constructed of steel a minimum of 0.0217 inch (0.55 mm) in thickness.

2.3. The duct shall not have openings that communicate the corridor with adjacent spaces or rooms.

2.4. The duct shall be installed above a ceiling.

2.5. The duct shall not terminate at a wall register in the fire-resistance-rated wall.

2.6. A minimum 12-inch-long (0.30 m) by 0.060-inch-thick (1.52 mm) steel sleeve shall be centered in each duct opening. The sleeve shall be secured to both sides of the wall and all four sides of the sleeve with minimum $1^1/_2$-inch by $1^1/_2$-inch by 0.060-inch (0.038 m by 0.038 m by 1.52 mm) steel retaining angles. The retaining angles shall be secured to the sleeve and the wall with No. 10 (M5) screws. The annular space between the steel sleeve and wall opening shall be filled with rock (mineral) wool batting on all sides.

❖ This section includes two exceptions to the general requirement that duct and air transfer openings in rated construction be protected.

Exceptions 1 and 2 specifically address two circumstances wherein fire dampers are not required in ducts penetrating fire partitions, in areas other than Group H. Since an automatic fire sprinkler system reduces the potential for duct collapse by controlling the fire, Exception 1 states that dampers are not required to protect penetrations of tenant separation or corridor walls in buildings protected throughout with an approved automatic sprinkler system and protected as a through penetration [see Figure 716.5.4(1)]. The reference to Sections 903.3.1.1 and 903.3.1.2 establishes that the exception only applies to buildings equipped throughout with an automatic sprinkler system designed and installed in accordance with either NFPA 13 or 13R and does not apply to buildings with an NFPA 13D system. The second exception states that fire dampers are not required where a steel duct penetrates a wall but does not have openings serving adjacent rooms or spaces [see Figure 716.5.4(2)]. This exception has several criteria, one of which is to utilize a sleeve with rock mineral wool batting on all sides. Steel ducts have been shown to have the ability to remain in place during severe fire exposures. With no openings into the corridor to allow for smoke and heat transfer in the protected corridor enclosure, a fire damper is not necessary to attain the required fire performance of the penetration.

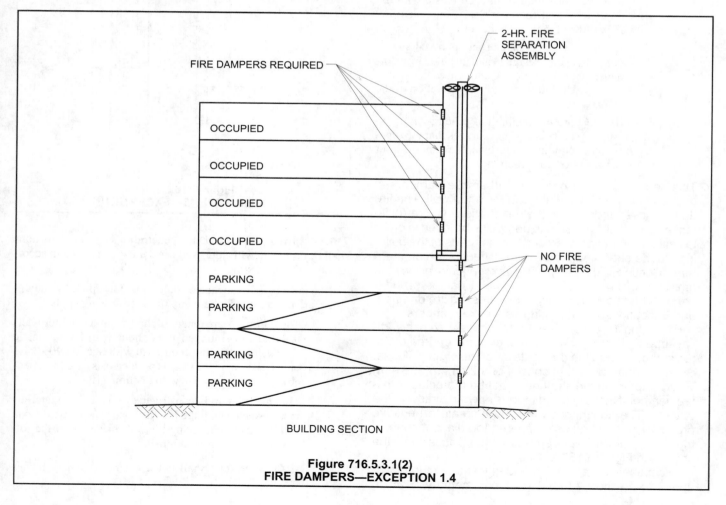

Figure 716.5.3.1(2)
FIRE DAMPERS—EXCEPTION 1.4

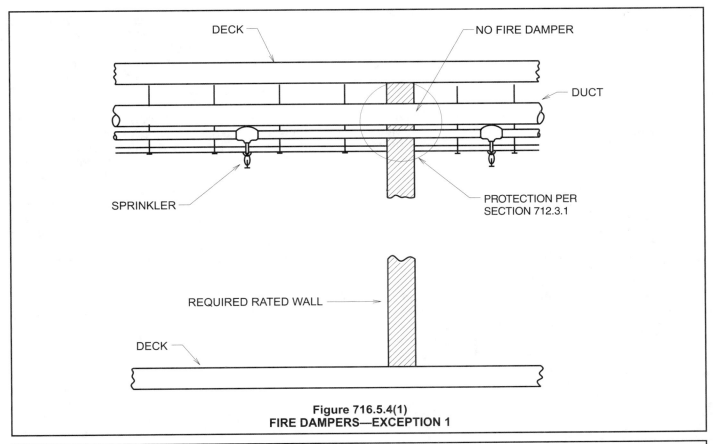

DECK

NO FIRE DAMPER

DUCT

SPRINKLER

PROTECTION PER
SECTION 712.3.1

REQUIRED RATED WALL

DECK

Figure 716.5.4(1)
FIRE DAMPERS—EXCEPTION 1

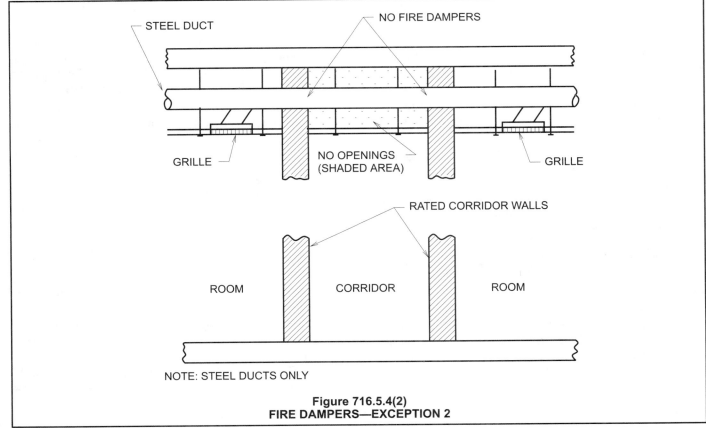

STEEL DUCT

NO FIRE DAMPERS

GRILLE

NO OPENINGS
(SHADED AREA)

GRILLE

RATED CORRIDOR WALLS

ROOM

CORRIDOR

ROOM

NOTE: STEEL DUCTS ONLY

Figure 716.5.4(2)
FIRE DAMPERS—EXCEPTION 2

716.5.4.1 Corridors. A listed smoke damper designed to resist the passage of smoke shall be provided at each point a duct or air transfer opening penetrates a corridor enclosure required to have smoke and draft control doors in accordance with Section 715.3.3.

Exceptions:

1. Smoke dampers are not required where the building is equipped throughout with an approved smoke control system in accordance with Section 909, and smoke dampers are not necessary for the operation and control of the system.

2. Smoke dampers are not required in corridor penetrations where the duct is constructed of steel not less than 0.019-inch (0.48 mm) in thickness and there are no openings serving the corridor.

❖ This section is coordinated with the provisions of Section 715.3.3 regarding opening protection for rated corridor walls. Similar to Section 715.3.3, which requires that the fire doors also be tested for smoke and draft control, this section requires that duct and air transfer openings be protected with smoke dampers. Unless the exceptions of Section 715.5.4 are applicable, this type of opening would also require a fire damper. Exception 1 cites the concern of the damper interfering with the smoke control system (see commentary, Section 716.2.1). Exception 2 is similar to Exceptions 2.2 and 2.3 of Section 716.5.4 for fire dampers. Also see the commentary to Section 716.5.5, which addresses smoke dampers in smoke barrier walls.

716.5.5 Smoke barriers. A listed smoke damper designed to resist the passage of smoke shall be provided at each point a duct or air transfer opening penetrates a smoke barrier. Smoke dampers and smoke damper actuation methods shall comply with Section 716.3.2.1.

Exception: Smoke dampers are not required where the openings in ducts are limited to a single smoke compartment and the ducts are constructed of steel.

❖ Smoke barriers are required for both Group I-2 (Section 407.4) and Group I-3 (Section 408.6) occupancies. The concept of required smoke barriers is to limit the migration of smoke. In fact, Section 709.4 requires that smoke barriers form an effective continuous membrane from outside wall to outside wall and from floor slab to floor or roof deck above. This would include continuity through concealed spaces, such as those found above suspended ceilings and interstitial structural and mechanical spaces.

To prevent smoke migration across the smoke barrier via duct penetrations, an approved damper designed to resist the passage of smoke is to be provided at each point where a duct penetrates a smoke barrier. The damper must close upon detection of smoke by an approved smoke detector within the duct. If the duct penetration is above a smoke-barrier door assembly, smoke detectors

for the doors may also be used to operate the damper and the duct detector may be eliminated. A reference to Section 716.3.2.1 is provided for activation mechanisms.

Additionally, smoke dampers are not required in a fully ducted system that does not contain any vents or openings that would spread smoke across the smoke barrier; in smoke control systems or in duct penetrations of smoke barriers where all supply and return diffusers and registers in the duct are contained in a single smoke compartment. As indicated in the commentary to Section 716.3.2, there are three classes of smoke dampers that are further evaluated based on tested pressure. The designer must evaluate the specific conditions of the installation to determine which class of damper should be installed. If a fire damper is required (i.e., the smoke barrier is also a mixed occupancy separation or corridor wall), a combination fire and smoke damper or separate fire and smoke dampers may be installed.

716.6 Horizontal assemblies. Penetrations by ducts and air transfer openings of a floor, floor/ceiling assembly or the ceiling membrane of a roof/ceiling assembly shall be protected by a shaft enclosure that complies with Section 707 or shall comply with this section.

❖ In general, all floor openings that connect two or more stories are to be enclosed in shafts that are constructed in accordance with Section 707. Exception 4 to Section 707.2 references Section 712.4.4 for duct and air transfer openings (penetration), which further refers to Section 716. Sections 716.6.1 through 716.6.3 identify those conditions where a shaft is not required and either a fire damper (Sections 716.6.1 and 716.6.3) or ceiling radiation damper (Section 716.6.2) is required. There is one exception within Section 716.6.1 that allows the elimination of the fire damper. Through ceiling radiation dampers would still be required in some cases.

716.6.1 Through penetrations. In occupancies other than Groups I-2 and I-3, a duct and air transfer opening system constructed of approved materials in accordance with the *International Mechanical Code* that penetrates a fire-resistance-rated floor/ceiling assembly that connects not more than two stories is permitted without shaft enclosure protection provided a fire damper is installed at the floor line.

Exception: A duct is permitted to penetrate three floors or less without a fire damper at each floor provided it meets all of the following requirements.

1. The duct shall be contained and located within the cavity of a wall and shall be constructed of steel not less than 0.019 inch (0.48 mm) (26 gage) in thickness.

2. The duct shall open into only one dwelling unit or sleeping unit and the duct system shall be continuous from the unit to the exterior of the building.

3. The duct shall not exceed 4-inch (102 mm) nominal diameter and the total area of such ducts shall not exceed

100 square inches (0.065 m²) in any 100 square feet (9.3 m²) of floor area.

4. The annular space around the duct is protected with materials that prevent the passage of flame and hot gases sufficient to ignite cotton waste where subjected to ASTM E 119 time-temperature conditions under a minimum positive pressure differential of 0.01 inch (2.49 Pa) of water at the location of the penetration for the time period equivalent to the fire-resistance rating of the construction penetrated

5. Grille openings located in a ceiling of a fire-resistance-rated floor/ceiling or roof/ceiling assembly shall be protected with a ceiling radiation damper in accordance with Section 716.6.2.

❖ Penetrations of a fire-resistance-rated floor/ceiling assembly by air ducts or plenums are permitted where no more than two stories are connected (i.e., the duct or plenum penetrates a single floor) and a fire damper is provided at the floor line. The fire damper must be tested in accordance with UL 555 [see Figure 716.6.1(1)]. A roof/ceiling assembly penetrated by a duct that is open to the atmosphere is not required to have fire dampers installed at the penetrations because it is considered acceptable for the fire to vent to the outside atmosphere [see Figure 716.6.1(2)]. As noted, there is an exception to this section that allows the elimination of fire dampers at each floor when the ducts extend through a maximum of three floors. There are several criteria that must be followed to allow this, including:

• Minimum duct thickness (increases time for heat to penetrate and provides more durability).

• Limited to a single sleeping unit or dwelling unit (to reduce the likelihood that one dwelling unit or sleeping unit will be able to endanger another).

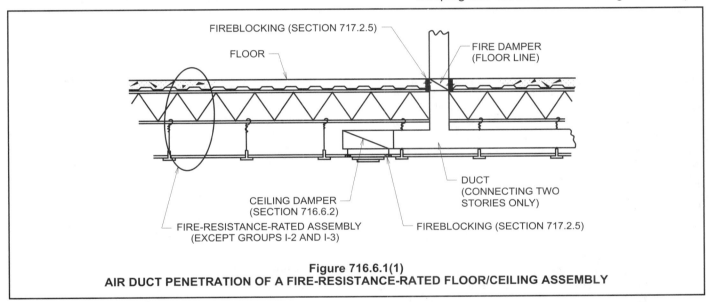

Figure 716.6.1(1)
AIR DUCT PENETRATION OF A FIRE-RESISTANCE-RATED FLOOR/CEILING ASSEMBLY

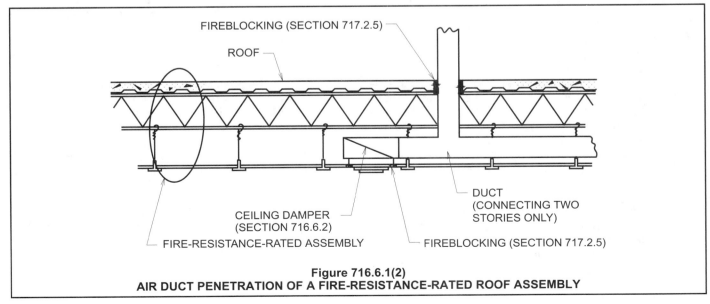

Figure 716.6.1(2)
AIR DUCT PENETRATION OF A FIRE-RESISTANCE-RATED ROOF ASSEMBLY

- Size of the duct opening (smaller opening means smaller amounts of fire effluents).
- Annular protection (reduces the chance of duct leakage from penetrated walls/barriers).
- Ceiling radiation dampers at grills in fire-resistant ceilings (maintains ceiling fire resistance).

All criteria revolve around limiting the impact of hot gases being present and spreading from one area to another.

716.6.2 Membrane penetrations. Where duct systems constructed of approved materials in accordance with the *International Mechanical Code* penetrate a ceiling of a fire-resistance-rated floor/ceiling or roof/ceiling assembly, shaft enclosure protection is not required provided an approved ceiling radiation damper is installed at the ceiling line. Where a duct is not attached to a diffuser that penetrates a ceiling of a fire-resistance-rated floor/ceiling or roof/ceiling assembly, shaft enclosure protection is not required provided an approved ceiling radiation damper is installed at the ceiling line. Ceiling radiation dampers shall be tested in accordance with UL 555C and constructed in accordance with the details listed in a fire-resistance-rated assembly or shall be labeled to function as a heat barrier for air-handling outlet/inlet penetrations in the ceiling of a fire-resistance-rated assembly. Ceiling radiation dampers shall not be required where ASTM E 119 fire tests have shown that ceiling radiation dampers are not necessary in order to maintain the fire-resistance rating of the assembly. Ceiling radiation dampers shall not be required where exhaust duct penetrations are protected in accordance with Section 712.4.2, where exhaust ducts are located within the cavity of a wall, and where exhaust ducts do not pass through another dwelling unit or tenant space.

❖ Unless otherwise tested, a ceiling radiation damper, tested in accordance with UL 555C (see commentary, Section 716.3), is to be installed at the line where the ceiling is both penetrated by a noncombustible air duct and is an integral component of the fire-resistance rating. The requirement for a ceiling radiation damper applies to ductwork that penetrates a ceiling membrane. Section 716.6.1 contains the requirements for fire dampers where the duct penetrates through the entire assembly [see Figure 716.6.1(1) for ceiling radiation damper and fire damper locations].

Ceiling radiation dampers are designed to limit the radiative heat transfer through an air inlet or outlet opening in a fire-resistance-rated floor/ceiling or roof/ceiling assembly. On the other hand, a fire damper is a device designed to close automatically upon detection of heat, to interrupt migratory airflow and to restrict the passage of flame through duct penetrations of rated assemblies.

Ceiling radiation dampers are not required to be installed if they comply with Section 712.4.2, are limited to a single dwelling unit or tenant spaces and are contained within a wall. Section 712.4.2 would allow either assemblies tested to ASTM E 119 or the use of a through-penetration firestop system.

716.6.3 Nonfire-resistance-rated assemblies. Duct systems constructed of approved materials in accordance with the *International Mechanical Code* that penetrate nonfire-resistance-rated floor assemblies and that connect not more than two stories are permitted without shaft enclosure protection provided that the annular space between the assembly and the penetrating duct is filled with an approved noncombustible material to resist the free passage of flame and the products of combustion. Duct systems constructed of approved materials in accordance with the *International Mechanical Code* that penetrate nonfire-resistance-rated floor assemblies and that connect not more than three stories are permitted without shaft enclosure protection provided that the annular space between the assembly and the penetrating duct is filled with an approved noncombustible material to resist the free passage of flame and the products of combustion, and a fire damper is installed at each floor line.

Exception: Fire dampers are not required in ducts within individual residential dwelling units.

❖ Duct systems that penetrate a single nonrated floor/ceiling assembly (thus connect two stories) are permitted, provided the annular space is protected. A fire damper is not required [see Figure 716.6.3(1)]. Air ducts may connect three stories (penetrate two floors) or less where fire dampers are provided at each floor line. A $1^1/_2$-hour-rated fire damper (see Table 716.3.1) is required even though the floor is not required to have a fire-resistance rating because the damper is an alternative to the 1-hour shaft enclosure [see Figure 716.6.3(2)].

716.7 Flexible ducts and air connectors. Flexible ducts and air connectors shall not pass through any fire-resistance-rated assembly. Flexible air connectors shall not pass through any wall, floor or ceiling.

❖ Flexible air ducts are prohibited from penetrating fire-resistance-rated assemblies. Flexible air connectors are not permitted to pass through any wall, floor or ceiling (fire-resistance rated or not). An inadequate seal at the assembly penetration could allow smoke or flame to penetrate the assembly. Flexible air ducts and connectors can be constructed of both combustible and noncombustible components; therefore, the duct's resistance to the passage of fire could be less than the resistance of the penetrated assembly. All construction assemblies, whether fire-resistance rated or not, have some inherent resistance to the spread of fire. For example, duct penetrations can significantly affect fire resistance. The surface contour of the duct or the presence of insulation and vapor barrier materials on the duct exterior makes it difficult to seal effectively (fireblock) the annular spaces around flexible air duct and connector penetrations. For example, flexible ducts cannot be used for risers passing through floor/ceiling assemblies unless enclosed in a fire-resistance-rated shaft.

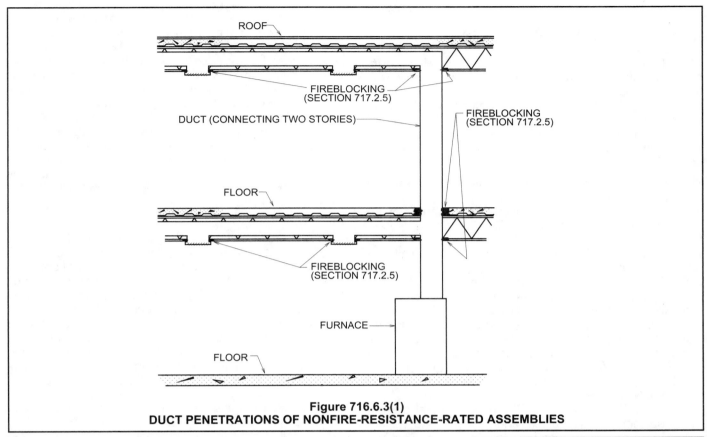

Figure 716.6.3(1)
DUCT PENETRATIONS OF NONFIRE-RESISTANCE-RATED ASSEMBLIES

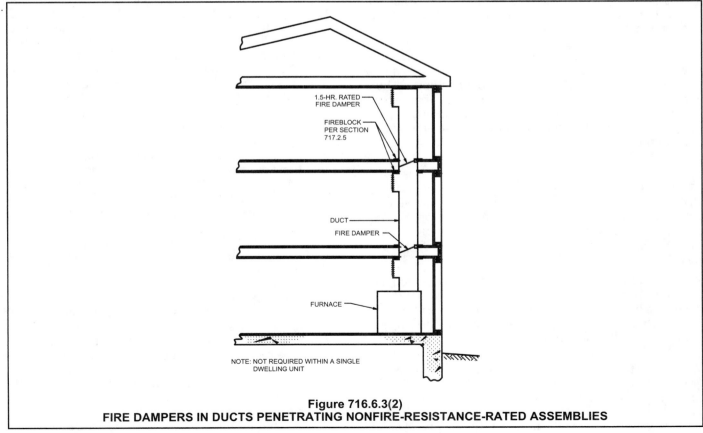

Figure 716.6.3(2)
FIRE DAMPERS IN DUCTS PENETRATING NONFIRE-RESISTANCE-RATED ASSEMBLIES

SECTION 717
CONCEALED SPACES

717.1 General. Fireblocking and draftstopping shall be installed in combustible concealed locations in accordance with this section. Fireblocking shall comply with Section 717.2. Draftstopping in floor/ceiling spaces and attic spaces shall comply with Sections 717.3 and 717.4, respectively. The permitted use of combustible materials in concealed spaces of noncombustible buildings shall be limited to the applications indicated in Section 717.5.

❖ During a fire, flame, smoke and gases will spread via the paths of least resistance. Certain assemblies create void spaces, which will not only affect the spread of fire but, since the voids are concealed, will also make access for suppression more difficult.

The code has established two means by which fire spread within void spaces can be controlled: fireblocking and draftstopping. Fireblocking involves the use of building materials to prevent the movement of flame and gases to other areas through concealed spaces. Draftstopping involves the use of building materials to prevent the movement of air, smoke, gases and flames to other areas through large concealed spaces. For example, the protection in an attic space is draftstopping, while the protection in the cavity of a wall assembly is fireblocking.

Section 717.2.1 identifies the materials that are acceptable for use as fireblocks and Section 717.3.1 specifies the acceptable materials for draftstops. Sections 717.2, 717.3 and 717.4 address where fireblocking and draftstopping are required.

Fireblocks and draftstops are to be installed in combustible concealed spaces to prevent the spread of flame, smoke and gases. If a fire condition exists in a concealed space, the fireblocks and draftstops will help contain the fire until it can be suppressed. This section also includes provisions for permitted combustibles in concealed spaces in buildings of Type I and II construction (see Section 717.5 and Exception 22 of Section 603.1).

717.2 Fireblocking. In combustible construction, fireblocking shall be installed to cut off concealed draft openings (both vertical and horizontal) and shall form an effective barrier between floors, between a top story and a roof or attic space. Fireblocking shall be installed in the locations specified in Sections 717.2.2 through 717.2.7.

❖ This section merely states the performance requirements associated with fireblocking. The intent of fireblocking is to reduce the ability of a fire, smoke and gases from moving to different parts of the building through combustible concealed spaces.

717.2.1 Fireblocking materials. Fireblocking shall consist of 2-inch (51 mm) nominal lumber or two thicknesses of 1-inch (25 mm) nominal lumber with broken lap joints or one thickness of 0.719-inch (18.3 mm) wood structural panel with joints backed by 0.719-inch (18.3 mm) wood structural panel or one

thickness of 0.75-inch (19 mm) particleboard with joints backed by 0.75-inch (19 mm) particleboard. Gypsum board, cement fiber board, batts or blankets of mineral wool or glass fiber or other approved materials installed in such a manner as to be securely retained in place shall be permitted as an acceptable fireblock. Batts or blankets of mineral or glass fiber or other approved nonrigid materials shall be permitted for compliance with the 10-foot (3048 mm) horizontal fireblocking in walls constructed using parallel rows of studs or staggered studs. Loose-fill insulation material shall not be used as a fireblock unless specifically tested in the form and manner intended for use to demonstrate its ability to remain in place and to retard the spread of fire and hot gases. The integrity of fireblocks shall be maintained.

❖ This approved list of fireblocking materials has not been tested for fire resistance, but is simply deemed as acceptable materials for slowing the spread of flame and products of combustion. Their effectiveness depends on their correct installation, something which must be inspected at the time of construction.

Various insulation materials have been evaluated for use as fireblocking and have been shown to perform well if installed in accordance with their specifications. For insulating material to function well as fireblocking, it must completely fill the area where fireblocking is required so that no unblocked passages exist.

Certain insulating materials have been evaluated for use as fireblocks. The evaluation reports that specify the configuration and attachment for the insulating materials in the construction assemblies be fireblocked, and are based on tests performed on the assemblies. Although the code does not specify a test standard for fireblocking, these materials have been tested using the ASTM E 119 standard and their performance during the test is compared to the performance of other conventional fireblocking material. The use of insulation for fireblocking is only for the 10-foot (3048 mm) horizontal fireblocking in walls as required by Item 2 in Section 717.2.2. This is likely related to the fact that Section 717.2.2 only requires fireblocking at floors and ceilings and that fires burn more vigorously vertically than they do horizontally; therefore, more durable materials, such as wood, would be required.

Loose-fill insulation will not remain in place under fire conditions and, therefore, the code requires that it be tested if it is to be used as fireblocking. In the process of performing the test procedure, it is necessary to install the insulation so that it is securely attached and will be retained during the test; thus, specifications for its use as a fireblock are created.

Certain construction designs employing materials for sound transmission control require wider-than-usual concealed wall spaces. Fireblocking of these spaces at locations required by the code can be challenging. Simply because a space is filled with insulation does not guarantee that fireblocking will be accomplished. This section allows the use of batts or blankets of mineral or glass fiber insulation to serve as fireblocking in these spaces, but the installation details must be approved and inspected in the field to show all concealed spaces are effectively fireblocked at the required locations.

717.2.1.1 Double stud walls. Batts or blankets of mineral or glass fiber or other approved nonrigid materials shall be allowed as fireblocking in walls constructed using parallel rows of studs or staggered studs.

❖ See the commentary to Section 717.2.1.

717.2.2 Concealed wall spaces. Fireblocking shall be provided in concealed spaces of stud walls and partitions, including furred spaces, and parallel rows of studs or staggered studs, as follows:

a. Vertically at the ceiling and floor levels.

b. Horizontally at intervals not exceeding 10 feet (3048 mm).

❖ The intent of this section is prevent the spread of fires within walls. Losses in this area have been seen in existing structures of balloon construction. This section requires both vertical and horizontal blocking. Vertical blocking simply requires fireblocking at floors and ceilings; therefore, a tall warehouse-type building may not need any vertical fireblocking. Horizontal blocking, however, would be required every 10 feet (3048 mm). Section 717.2.1 provides the various methods for fireblocking. One method addressed is the use of mineral or glass fiber insulation, which is only allowed for horizontal fireblocking (see commentary, Section 717.2.1).

717.2.3 Connections between horizontal and vertical spaces. Fireblocking shall be provided at interconnections between concealed vertical stud wall or partition spaces and concealed horizontal spaces created by an assembly of floor joists or trusses, and between concealed vertical and horizontal spaces such as occur at soffits, drop ceilings, cove ceilings and similar locations.

❖ To prevent fire and smoke from spreading between vertical and horizontal spaces, fireblocks are required [see Figures 717.2.3(1-3)]. If not provided, a concealed fire could spread throughout the floor level because of the interconnection of the concealed combustible wall and ceiling spaces.

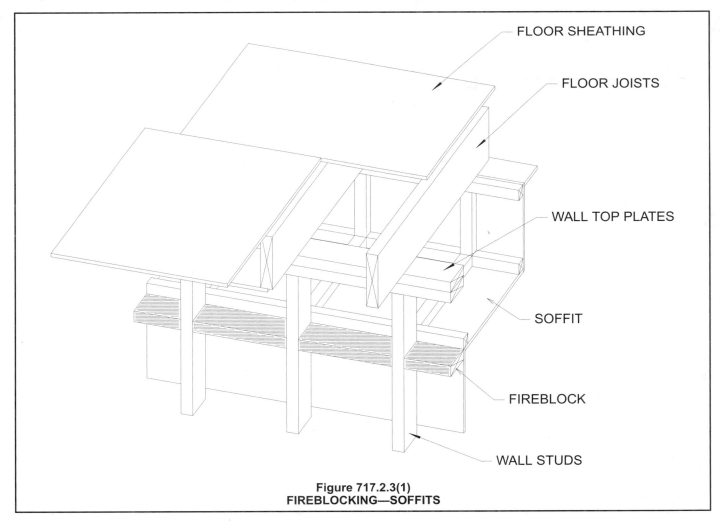

Figure 717.2.3(1)
FIREBLOCKING—SOFFITS

FIGURE 717.2.3(2) – FIGURE 717.2.3(3)

FIRE-RESISTANCE-RATED CONSTRUCTION

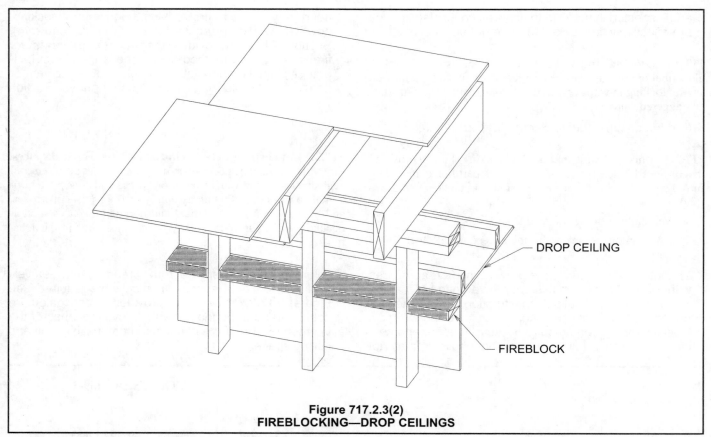

Figure 717.2.3(2)
FIREBLOCKING—DROP CEILINGS

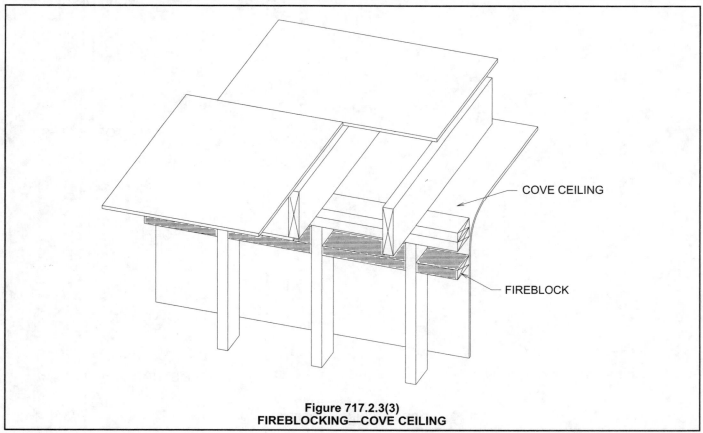

Figure 717.2.3(3)
FIREBLOCKING—COVE CEILING

717.2.4 Stairways. Fireblocking shall be provided in concealed spaces between stair stringers at the top and bottom of the run. Enclosed spaces under stairs shall also comply with Section 1019.1.5.

❖ Fireblocks are required at the top and bottom of concealed spaces between stair stringers (see Figure 717.2.4). Similar to fireblocks between horizontal and vertical spaces, fireblocks at the top and bottom of a run will provide a barrier between the floor below the stair and the floor that the stair serves. Additionally, a reference is made to Section 1019.1.5, which requires the interior of the space under stairs to contain 1-hour fire-resistant-rated construction or the stairway to be fire resistant (whichever is greater).

717.2.5 Ceiling and floor openings. Where annular space protection is provided in accordance with Exception 6 of Section 707.2, Exception 1 of Section 712.4.2, or Section 712.4.3, fireblocking shall be installed at openings around vents, pipes, ducts, chimneys and fireplaces at ceiling and floor levels, with an approved material to resist the free passage of flame and the products of combustion. Factory-built chimneys and fireplaces shall be fireblocked in accordance with UL 103 and UL 127.

❖ Sections 707 and 712 allow penetrations of ceilings and floors to be fireblocked [see Figures 717.2.5(1-4)]. It should be noted that depending on the type of penetration and the number of floors penetrated (or stories connected), a shaft may be required (see Section 707.2). Additional requirements for penetrations through rated floor/ceiling and roof/ceiling assemblies and shaft requirements can be found in Section 712.4.

UL 103 and 127 specify that only metal fireblocks and insulation shields are to be used with factory-built chimneys and fireplaces. The IMC contains a reference similar to that made in the UL standards. When fire-resistance-rated floor/ceiling assemblies are penetrated, the higher degree of protection afforded by tested through-penetration protection systems is required.

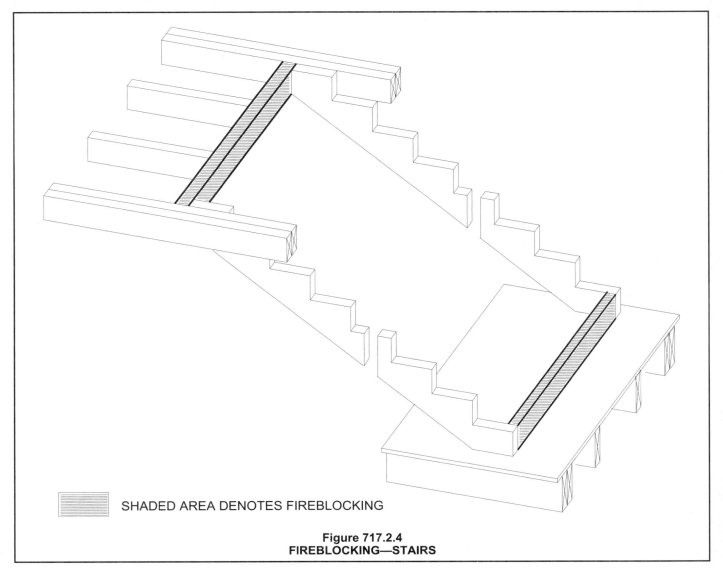

SHADED AREA DENOTES FIREBLOCKING

Figure 717.2.4
FIREBLOCKING—STAIRS

FIGURE 717.2.5(1) – FIGURE 717.2.5(4)

FIRE-RESISTANCE-RATED CONSTRUCTION

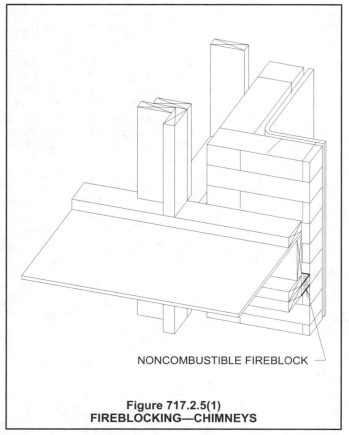

NONCOMBUSTIBLE FIREBLOCK

Figure 717.2.5(1)
FIREBLOCKING—CHIMNEYS

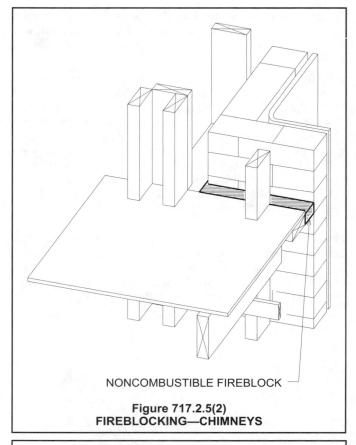

NONCOMBUSTIBLE FIREBLOCK

Figure 717.2.5(2)
FIREBLOCKING—CHIMNEYS

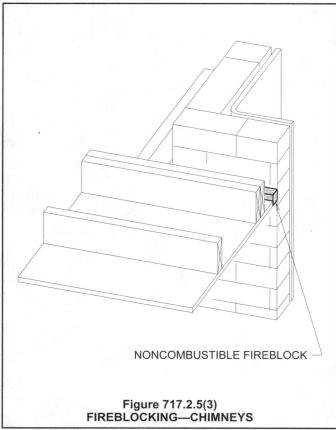

NONCOMBUSTIBLE FIREBLOCK

Figure 717.2.5(3)
FIREBLOCKING—CHIMNEYS

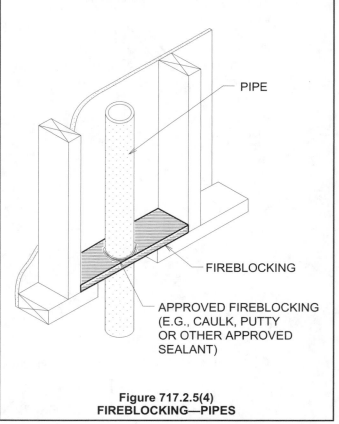

PIPE

FIREBLOCKING

APPROVED FIREBLOCKING
(E.G., CAULK, PUTTY
OR OTHER APPROVED
SEALANT)

Figure 717.2.5(4)
FIREBLOCKING—PIPES

717.2.6 Architectural trim. Fireblocking shall be installed within concealed spaces of exterior wall finish and other exterior architectural elements where permitted to be of combustible construction as specified in Section 1406 or where erected with combustible frames, at maximum intervals of 20 feet (6096 mm). If noncontinuous, such elements shall have closed ends, with at least 4 inches (102 mm) of separation between sections.

Exceptions:

1. Fireblocking of cornices is not required in single-family dwellings, as applicable in Section 101.2. Fireblocking of cornices of a two-family dwelling as applicable in Section 101.2 is required only at the line of dwelling unit separation.

2. Fireblocking shall not be required where installed on noncombustible framing and the face of the exterior wall finish exposed to the concealed space is covered by one of the following materials:

    2.1. Aluminum having a minimum thickness of 0.019 inch (0.5 mm).

    2.2. Corrosion-resistant steel having a base metal thickness not less than 0.016 inch (0.4 mm) at any point.

    2.3. Other approved noncombustible materials.

❖ Combustible exterior wall finish and exterior architectural elements are required to be fireblocked at 20-foot (6096 mm) intervals to prevent the spread of fire or smoke through concealed spaces (see Figure 717.2.6). Adjacent sections of exterior trim need not be considered as contributing to the 20-foot (6096 mm) limitation if there is a minimum 4-inch (102 mm) separation between sections of exterior trim and the ends are closed.

Trim on the building exterior presents the same concern as interior combustible concealed wall spaces and, therefore, requires fireblocking.

717.2.7 Concealed sleeper spaces. Where wood sleepers are used for laying wood flooring on masonry or concrete fire-resistance-rated floors, the space between the floor slab and the underside of the wood flooring shall be filled with an approved material to resist the free passage of flame and products of combustion or fireblocked in such a manner that there will be no open spaces under the flooring that will exceed 100 square feet (9.3 m^2) in area and such space shall be filled solidly under permanent partitions so that there is no communication under the flooring between adjoining rooms.

Exceptions:

1. Fireblocking is not required for slab-on-grade floors in gymnasiums.

2. Fireblocking is required only at the juncture of each alternate lane and at the ends of each lane in a bowling facility.

❖ Concealed spaces created by floor sleepers must be fireblocked into areas no greater than 100 square feet (9.3 m^2) or such spaces must be filled with materials capable of resisting flame, smoke or gases (see Figure 717.2.7).

Since floor sleepers may cover large areas, this fireblock or material will restrict a fire from spreading within the concealed space to other areas of the floor. This section is commonly applied when a raised, wood-finish floor surface is installed in Type I or II construction.

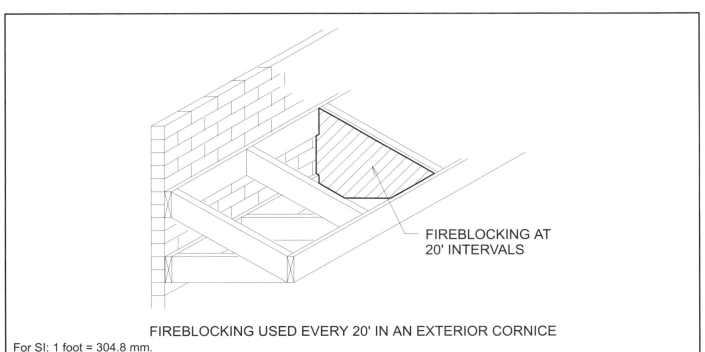

FIREBLOCKING USED EVERY 20' IN AN EXTERIOR CORNICE

For SI: 1 foot = 304.8 mm.

Figure 717.2.6
FIREBLOCKING—ARCHITECTURAL TRIM

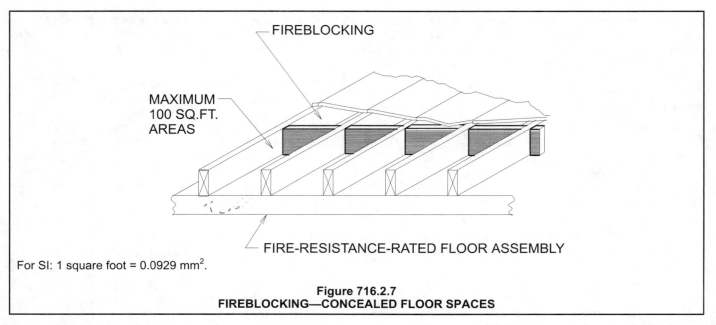

FIREBLOCKING

MAXIMUM 100 SQ.FT. AREAS

FIRE-RESISTANCE-RATED FLOOR ASSEMBLY

For SI: 1 square foot = 0.0929 mm².

Figure 716.2.7
FIREBLOCKING—CONCEALED FLOOR SPACES

717.3 Draftstopping in floors. In combustible construction, draftstopping shall be installed to subdivide floor/ceiling assemblies in the locations prescribed in Sections 717.3.2 through 717.3.3.

❖ Draftstopping is required in combustible concealed floor construction as indicated in Sections 717.3.1 through 717.3.3. Attic draftstopping is regulated in Section 717.4.

717.3.1 Draftstopping materials. Draftstopping materials shall not be less than 0.5-inch (12.7 mm) gypsum board, 0.375-inch (9.5 mm) wood structural panel, 0.375-inch (9.5 mm) particleboard or other approved materials adequately supported. The integrity of draftstops shall be maintained.

❖ Similar to fireblocks in small concealed spaces, draftstops also act as a barrier to smoke and gases. Because of the large areas that require this barrier, however, the code permits certain types of sheathing material such as $1/_2$-inch (12.7 mm) gypsum board, $3/_8$-inch (9.5 mm) plywood and particleboard or other approved materials. The draftstop material must be properly supported and capable of remaining in place when subjected to initial fire exposure.

717.3.2 Groups R-1, R-2, R-3 and R-4. Draftstopping shall be provided in floor/ceiling spaces in Group R-1 buildings, in Group R-2 buildings as applicable in Section 101.2 with three or more dwelling units, in Group R-3 buildings as applicable in Section 101.2 with two dwelling units and in Group R-4 buildings. Draftstopping shall be located above and in line with the dwelling unit and sleeping unit separations.

Exceptions:

1. Draftstopping is not required in buildings equipped throughout with an automatic sprinkler system in accordance with Section 903.3.1.1.

2. Draftstopping is not required in buildings equipped throughout with an automatic sprinkler system in ac-

cordance with Section 903.3.1.2, provided that automatic sprinklers are also installed in the combustible concealed spaces.

❖ To maintain the integrity of dwelling or sleeping unit separation walls in buildings of Groups R-1, R-2, R-3 and R-4, draftstopping is to be provided when the dwelling or sleeping unit separation wall is not continuous to the floor sheathing above. The draftstopping must be installed directly above the dwelling or sleeping unit separation wall (see Figure 717.3.2). Even if the tenant separation walls are not required to be fire-resistance rated, the wall plus the draftstopping above are considered a barrier to the spread of fire and smoke. As such, the draftstop offers some level of protection to the occupants of one dwelling or sleeping unit from a fire occurring in another dwelling or sleeping unit.

The exception indicates that the draftstopping need not be provided if sprinklers are installed above and below the ceiling, since sprinkler activation will control the spread of fire. The sprinkler system must be installed in accordance with Section 903.3.1.1 or 903.3.1.2 and NFPA 13 or 13R. The sprinklers must be installed above the ceiling in the combustible concealed space, regardless of whether the referenced standards would require sprinkler heads in those locations.

717.3.3 Other groups. In other groups, draftstopping shall be installed so that horizontal floor areas do not exceed 1,000 square feet (93 m²).

Exception: Draftstopping is not required in buildings equipped throughout with an automatic sprinkler system in accordance with Section 903.3.1.1.

❖ Unless the spaces above and below the ceiling are sprinklered (NFPA 13 system), draftstopping is to be provided in all groups except Groups R-1, R-2, R-3 and R-4 such that the open space does not exceed 1,000 square feet (93 m²) (see Figure 717.3.3).

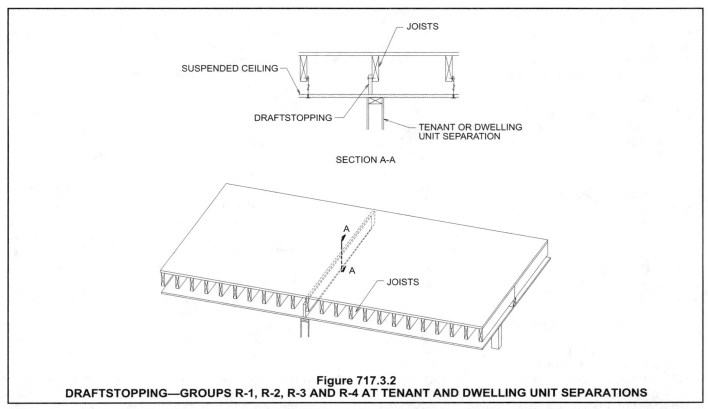

Figure 717.3.2
DRAFTSTOPPING—GROUPS R-1, R-2, R-3 AND R-4 AT TENANT AND DWELLING UNIT SEPARATIONS

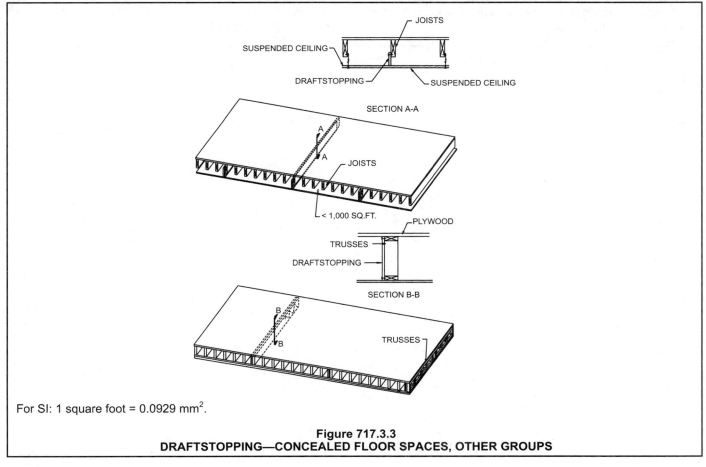

For SI: 1 square foot = 0.0929 mm^2.

Figure 717.3.3
DRAFTSTOPPING—CONCEALED FLOOR SPACES, OTHER GROUPS

717.4 Draftstopping in attics. In combustible construction, draftstopping shall be installed to subdivide attic spaces and concealed roof spaces in the locations prescribed in Sections 717.4.2 and 717.4.3. Ventilation of concealed roof spaces shall be maintained in accordance with Section 1203.2.

❖ Fires that spread to attics that are not properly draftstopped often cause considerable damage. For this reason, draftstopping is required in attic and concealed roof spaces in accordance with Sections 717.4.2 and 717.4.3.

717.4.1 Draftstopping materials. Materials utilized for draftstopping of attic spaces shall comply with Section 717.3.1.

❖ See the commentary to Section 717.3.1.

717.4.1.1 Openings. Openings in the partitions shall be protected by self-closing doors with automatic latches constructed as required for the partitions.

❖ Section 1208.2 requires attic access. The placement of draftstopping in the attic may interfere with the ability of the fire department to gain access to all attic spaces. This section requires that if draftstopping is provided with access openings to adjacent draftstopped portions of the attic, the openings must be constructed of draftstopping materials and be equipped with self-closing mechanisms in order to ensure that the draftstop will perform its intended function.

717.4.2 Groups R-1 and R-2. Draftstopping shall be provided in attics, mansards, overhangs or other concealed roof spaces of Group R-2 buildings with three or more dwelling units and in all Group R-1 buildings. Draftstopping shall be installed above, and in line with, sleeping unit and dwelling unit separation walls that do not extend to the underside of the roof sheathing above.

Exceptions:

1. Where corridor walls provide a sleeping unit or dwelling unit separation, draftstopping shall only be required above one of the corridor walls.

2. Draftstopping is not required in buildings equipped throughout with an automatic sprinkler system in accordance with Section 903.3.1.1.

3. In occupancies in Group R-2 that do not exceed four stories in height, the attic space shall be subdivided by draftstops into areas not exceeding 3,000 square feet (279 m^2) or above every two dwelling units, whichever is smaller.

4. Draftstopping is not required in buildings equipped throughout with an automatic sprinkler system in accordance with Section 903.3.1.2, provided that automatic sprinklers are also installed in the combustible concealed spaces.

❖ To maintain the integrity of tenant and dwelling unit separation walls (see commentary, Section 717.3.2), draftstopping is to be provided above dwelling or sleeping unit separation walls that do not extend to the roof sheathing [see Figures 717.4.2(1) and 717.4.2(2)].

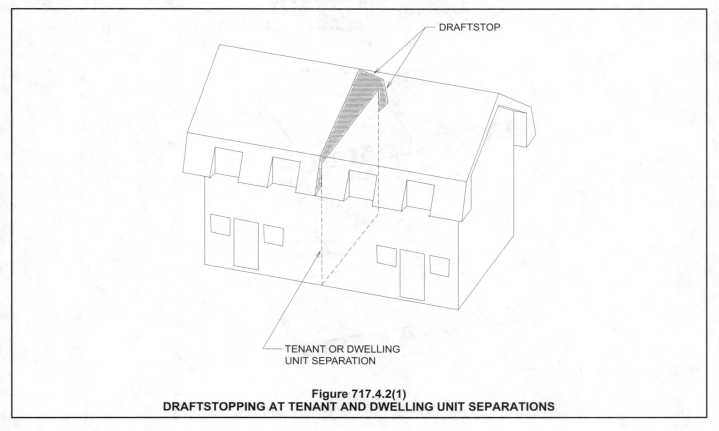

Figure 717.4.2(1)
DRAFTSTOPPING AT TENANT AND DWELLING UNIT SEPARATIONS

Exception 1 clarifies that for corridor walls that serve as dwelling or sleeping unit separations, draftstopping is only required above one of the two corridor walls.

Exceptions 2 and 4 omit the requirement for draftstopping attics that are sprinklered in accordance with Section 903.3.1.1 (see Exception 2) or 903.3.1.2 (see Exception 4). This exception would require the attic to be sprinklered even though it may not be required by the standard. NFPA 13R typically would not require an attic to be sprinklered. This is consistent with the exceptions to Section 717.3.2 for draftstopping of floor spaces.

The draftstopping requirements can be reduced in Group R-2 occupancies of four stories or less by Exception 3. This exception allows the area between draftstopping to be increased to 3,000 square feet (279 m²) or above every two dwelling units, whichever is smaller. This provides the opportunity for a simpler draftstopping installation when the draftstops are not required to be in line with the dwelling or sleeping unit separation walls. Group R-2 buildings are required by Section 903.2.8 to be sprinklered; sprinklers reduce the need for attic draftstopping.

717.4.3 Other groups. Draftstopping shall be installed in attics and concealed roof spaces, such that any horizontal area does not exceed 3,000 square feet (279 m²).

Exception: Draftstopping is not required in buildings equipped throughout with an automatic sprinkler system in accordance with Section 903.3.1.1.

❖ Since nonresidential buildings are not always subdivided into small tenant spaces, concealed roof spaces and attics are to be subdivided into areas not exceeding 3,000 square feet (279 m²). The exception indicates that draftstopping is not required when the spaces above and below the ceiling are provided coverage by an automatic sprinkler system, in accordance with NFPA 13.

717.5 Combustibles in concealed spaces in Type I or II construction. Combustibles shall not be permitted in concealed spaces of buildings of Type I or II construction.

Exceptions:

1. Combustible materials in accordance with Section 603.

2. Combustible materials complying with Section 602 of the *International Mechanical Code.*

3. Class A interior finish materials.

4. Combustible piping within partitions or enclosed shafts installed in accordance with the provisions of this code. Combustible piping shall be permitted within concealed ceiling spaces where installed in accordance with the *International Mechanical Code* and the *International Plumbing Code.*

❖ The use of combustibles in nonstructural applications is limited to the listed exceptions to prevent fire spread within building elements. The severity of a potential problem increases when combustibles are located within concealed spaces that are inaccessible to manual fire fighting. This section regulates the use of combustibles within concealed spaces of noncombustible buildings. This section is referenced from Exception 22 of Section 603.1.

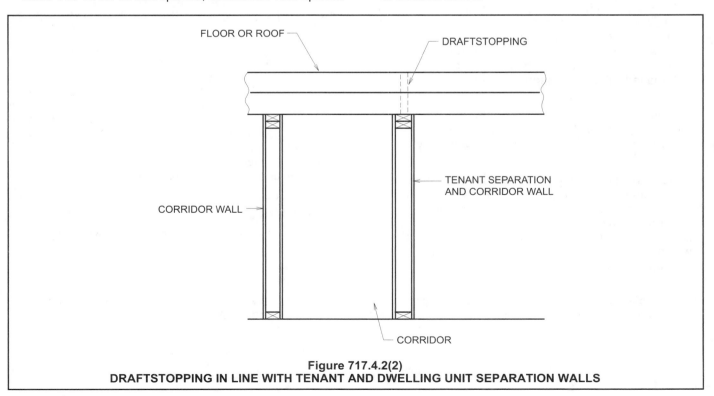

Figure 717.4.2(2)
DRAFTSTOPPING IN LINE WITH TENANT AND DWELLING UNIT SEPARATION WALLS

SECTION 718
FIRE-RESISTANCE REQUIREMENTS
FOR PLASTER

718.1 Thickness of plaster. The minimum thickness of gypsum plaster or portland cement plaster used in a fire-resistance-rated system shall be determined by the prescribed fire tests. The plaster thickness shall be measured from the face of the lath where applied to gypsum lath or metal lath.

❖ The fire-resistance of plaster is dependent on material variations, such as the aggregate mixture (sand content), as well as bonding to the plaster base. The use of wire or wire fabric reinforces the plaster mixes applied to plaster bases and, therefore, assures increased fire resistance. As such, the makeup of the plastered element is required when evaluating the fire-resistance rating of plaster.

Because the type of plaster used affects the fire-resistance rating, it is essential that the tests conducted utilize a plaster mix similar to that which is to be used. Depending on the type of plaster, additional tests may not be required. If the plaster is a type that is judged to be equivalent to a tested type (see Section 718.2), additional testing is not required.

718.2 Plaster equivalents. For fire-resistance purposes, 0.5 inch (12.7 mm) of unsanded gypsum plaster shall be deemed equivalent to 0.75 inch (19.1 mm) of one-to-three gypsum sand plaster or 1 inch (25 mm) of portland cement sand plaster.

❖ This section indicates that prescribed thicknesses of unsanded gypsum plaster, one-to-three sanded gypsum plaster and portland cement sand plaster are considered equivalent in terms of their fire-resistance rating characteristics. Once tests are conducted to determine the fire resistance of an assembly using one of these types, equivalency to the other prescribed types can be determined without subsequent tests.

718.3 Noncombustible furring. In buildings of Type I and II construction, plaster shall be applied directly on concrete or masonry or on approved noncombustible plastering base and furring.

❖ Buildings of Type I and II construction are required to be predominantly of noncombustible construction. This section requires the plastering base and furring to be noncombustible. The intent of this provision is to not permit a combustible cavity behind the finished plaster, which may be an area for a potentially concealed fire buildup.

718.4 Double reinforcement. Plaster protection more than 1 inch (25 mm) in thickness shall be reinforced with an additional layer of approved lath embedded at least 0.75 inch (19.1 mm) from the outer surface and fixed securely in place.

Exception: Solid plaster partitions or where otherwise determined by fire tests.

❖ The additional reinforcement is intended to decrease the likelihood of the plaster loosening when subjected to high temperatures. The plaster relies on the lath to provide a finish that is structurally sound. As the thickness of the plaster increases, additional reinforcement is required to hold the plaster in place.

718.5 Plaster alternatives for concrete. In reinforced concrete construction, gypsum plaster or portland cement plaster is permitted to be substituted for 0.5 inch (12.7 mm) of the required poured concrete protection, except that a minimum thickness of 0.375 inch (9.5 mm) of poured concrete shall be provided in reinforced concrete floors and 1 inch (25 mm) in reinforced concrete columns in addition to the plaster finish. The concrete base shall be prepared in accordance with Section 2510.7.

❖ Concrete elements, such as reinforced slabs and columns, rely in part on the thickness of the element to provide a fire-resistance rating. In order to reduce the concrete thickness, this section permits up to $^1/_2$ inch (12.7 mm) of plaster to be substituted for $^1/_2$ inch (12.7 mm) of concrete without affecting the fire-resistance rating of the element. While plaster can be substituted for concrete for fire-resistance purposes, a minimum amount of concrete is still required to be provided for structural integrity.

SECTION 719
THERMAL- AND SOUND-INSULATING MATERIALS

719.1 General. Insulating materials, including facings such as vapor retarders and vapor-permeable membranes, similar coverings, and all layers of single and multilayer reflective foil insulations, shall comply with the requirements of this section. Where a flame spread index or a smoke-developed index is specified in this section, such index shall be determined in accordance with ASTM E 84. Any material that is subject to an increase in flame spread index or smoke-developed index beyond the limits herein established through the effects of age, moisture, or other atmospheric conditions shall not be permitted.

Exceptions:

1. Fiberboard insulation shall comply with Chapter 23.
2. Foam plastic insulation shall comply with Chapter 26.
3. Duct and pipe insulation and duct and pipe coverings and linings in plenums shall comply with the *International Mechanical Code*.

❖ This section addresses insulating materials installed in building spaces. Other provisions of the code or companion codes regulate insulating materials used for duct and plenum insulation [see the IMC for energy conservation purposes and the *International Energy Conservation Code*® (IECC®)] and foam plastic insulation (see Section 2603). Insulating materials can affect fire development and fire spread and are, therefore, regulated accordingly.

This section addresses the various insulating materials that may be installed in building spaces, including insulating batts, blankets, fills (including vapor barriers and vapor-permeable membranes) or other coverings. Fiberboard insulation is regulated by Chapter 23. Cellulose loose-fill insulation is regulated by Section 719.6. Foam plastic insulation is regulated by Chapter 26. The applicable test method is ASTM E 84 (see commentary, Section 803.1).

719.2 Concealed installation. Insulating materials, where concealed as installed in buildings of any type of construction, shall have a flame spread index of not more than 25 and a smoke-developed index of not more than 450.

> **Exception:** Cellulose loose-fill insulation that is not spray applied, complying with the requirements of Section 719.6, shall only be required to meet the smoke-developed index of not more than 450.

❖ Concealed insulation no longer acts as an interior finish unless the covering material is removed or the fire is in the concealed building space. To limit the contribution of the insulation to a fire condition, the material must have a flame spread index of no more than 25 and a smoke-developed index of no more than 450. This is illustrated in Figure 719.2. Cellulose loose-fill insulation is regulated by the U.S. Consumer Product Safety Commission (CPSC) in accordance with Section 719.6. Spray-applied cellulose is regulated by Section 719.3.1 (see commentary, Section 719.4).

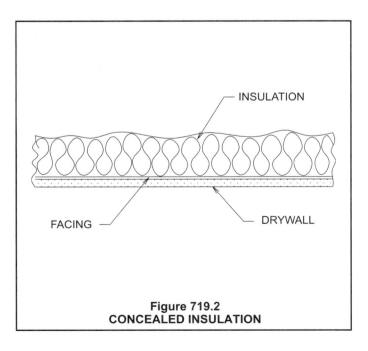

Figure 719.2
CONCEALED INSULATION

719.2.1 Facings. Where such materials are installed in concealed spaces in buildings of Type III, IV or V construction, the flame spread and smoke-developed limitations do not apply to facings, coverings, and layers of reflective foil insulation that are installed behind and in substantial contact with the unexposed surface of the ceiling, wall or floor finish.

❖ In buildings of Type III, IV and V construction, the flame spread indices do not apply to facing material, provided it is installed behind (and in substantial contact with) the unexposed surface of the ceiling, floor or wall finish. For example, when paper-backed insulation is placed directly on top of a ceiling, the paper facing is not required to meet the flame spread index; however, if the same material is applied to the underside of a roof deck and the paper facing is exposed to the attic space, the paper facing must meet the flame spread criteria. The potential for flame spread is greatly diminished when the facings are installed in direct contact with the finish material because of the lack of airspace to support a fire if it were to be exposed to a source of ignition.

The purpose of the vapor retarder being placed on the warm side of the building element is to prevent moisture vapor in the warm interior from reaching the dew-point temperature on a cold surface in the roof, wall or floor system. The generally accepted rating of a vapor retarder is 1 perm [57 mg/(s·m²·Pa)]. Similarly, ASTM C 755 states that for practical purposes, it is assumed that the permeance of an adequate vapor barrier will not exceed 1 perm [57 m8/(s·m²·Pa)] (see commentary, Section 1202.2).

719.3 Exposed installation. Insulating materials, where exposed as installed in buildings of any type of construction, shall have a flame spread index of not more than 25 and a smoke-developed index of not more than 450.

> **Exception:** Cellulose loose-fill insulation that is not spray applied complying with the requirements of Section 719.6 shall only be required to meet the smoke-developed index of not more than 450.

❖ Exposed insulating materials represent the same fire exposure hazard as any other exposed material, such as interior finish. The flame spread index for all applications of insulation materials is limited to 25, compared to 200 for interior finishes installed in rooms or spaces of certain occupancies (see commentary, Chapter 8). As with all interior finishes, the smoke-developed index is limited to 450. Cellulose loose-fill insulation is regulated by the CPSC in accordance with Section 719.6. Spray-applied cellulose is regulated by Section 719.3.1 (see commentary, Section 719.4).

719.3.1 Attic floors. Exposed insulation materials installed on attic floors shall have a critical radiant flux of not less than 0.12 watt per square centimeter when tested in accordance with ASTM E 970.

❖ ASTM E 970 is a test method that was developed by the insulation industry to evaluate the fire hazard of exposed attic insulation, and is referenced in the material standards for insulation. Cellulose loose-fill insulation is required to comply with CPSC 16 CFR, Part 1209 (see Section 719.6), which requires testing by this standard. Spray-applied cellulose insulation which is not subject to the CPSC standard is also subject to the ASTM E 970 testing by virtue of this section.

719.4 Loose-fill insulation. Loose-fill insulation materials that cannot be mounted in the ASTM E 84 apparatus without a screen or artificial supports shall comply with the flame spread and smoke-developed limits of Sections 719.2 and 719.3 when tested in accordance with CAN/ULC S102.2.

> **Exception:** Cellulose loose-fill insulation shall not be required to comply with this test method, provided such insulation complies with the requirements of Section 719.6.

❖ The exception serves to make a distinction between cellulose insulation, which is spray applied using a water-mist applicator, and cellulose loose-fill insulation, which is poured or blown in place. Spray-applied cellulose insulation can be exposed on vertical and horizontal ceiling-type surfaces so it is treated as any other insulating material in regard to testing. Cellulose loose-fill insulation, which is poured or blown in, is regulated by CPSC requirements (see Section 719.6) and is exempt from the test procedure described in this section.

719.5 Roof insulation. The use of combustible roof insulation not complying with Sections 719.2 and 719.3 shall be permitted in any type of construction provided it is covered with approved roof coverings directly applied thereto.

❖ Foam plastic roof insulation is required to comply with Chapter 26. This section allows insulation other than foam plastic insulation to be incorporated in the roof assembly without declassifying the type of construction.

719.6 Cellulose loose-fill insulation. Cellulose loose-fill insulation shall comply with CPSC 16 CFR, Part 1209 and CPSC 16 CFR, Part 1404. Each package of such insulating material shall be clearly labeled in accordance with CPSC 16 CFR, Part 1209 and CPSC 16 CFR, Part 1404.

❖ Cellulose loose-fill insulation is federally regulated by the CPSC. Parts 1209 and 1404 of CPSC 16 CFR contain various requirements that regulate the product to avoid excessive flammability or significant fire hazards. The smoke-developed index for cellulose loose-fill insulation must be determined by the ASTM E 84 test and must be 450 or less. The intent of this section is that the procedure for the smoke-developed index should be done by an ASTM E 84 test and not by the test procedures specified in Section 719.4.

719.7 Insulation and covering on pipe and tubing. Insulation and covering on pipe and tubing shall have a flame spread index

of not more than 25 and a smoke-developed index of not more than 450.

❖ This section maintains the general provision for insulation by requiring a maximum flame spread of 25 and smoke-developed index of 450, as stated in Sections 719.2 and 719.3; however, if exposed in a plenum, the IMC limits the smoke-developed index to 50.

SECTION 720
PRESCRIPTIVE FIRE RESISTANCE

720.1 General. The provisions of this section contain prescriptive details of fire-resistance-rated building elements. The materials of construction listed in Tables 720.1(1), 720.1(2), and 720.1(3) shall be assumed to have the fire-resistance ratings prescribed therein. Where materials that change the capacity for heat dissipation are incorporated into a fire-resistance-rated assembly, fire test results or other substantiating data shall be made available to the building official to show that the required fire-resistance-rating time period is not reduced.

❖ In this section, there are many prescriptive details for fire-resistance-rated construction, particularly those materials and assemblies listed in Table 720.1(1) for structural parts; Table 720.1(2) for walls and partitions and Table 720.1(3) for floor and roof systems. For the most part, the listed items have been tested in accordance with the fire-resistance ratings indicated. In addition, a similar footnote to all of the tables allows the acceptance of generic assemblies that are listed in GA 600. It is important to review all of the applicable footnotes when using a material or assembly from one of the tables.

As stated above, the fire-resistance ratings for the walls and partitions outlined in Table 720.1(2) are based on actual tests. For reinforced concrete walls, it is important to note the type of aggregate. The difference in aggregates is quite significant for a 4-hour fire-resistance-rated wall, as it amounts to a difference in thickness of almost 2 inches (51 mm). For hollow-unit masonry walls, the thickness required for a particular fire-endurance rating is the equivalent thickness as defined in Section 721.3.1 for concrete masonry and Section 721.4.1.1 for clay masonry.

Table 720.1(3) provides fire-resistance ratings for floor/ceiling and roof/ceiling assemblies. Note n, which exempts unusable space from the flooring and ceiling requirements is especially important. This exemption is consistent with Section 711.3.3.

Often, materials such as insulation are added to fire-resistance-rated assemblies. The code requires substantiating fire test data to show that when the materials are added, they do not reduce the required fire-endurance time period. As an example, adding insulation to a floor/ceiling assembly may change its capacity to dissipate heat and, particularly for noncombustible assemblies, the fire-resistance rating may be changed.

Although the primary intent of the provision is to cover those cases where thermal insulation is added, the language is intentionally broad so that it applies to any material that might be added to the assembly.

720.1.1 Thickness of protective coverings. The thickness of fire-resistant materials required for protection of structural members shall be not less than set forth in Table 720.1(1), except as modified in this section. The figures shown shall be the net thickness of the protecting materials and shall not include any hollow space in back of the protection.

❖ In accordance with this section, the required thickness of insulating material used to provide fire-resistance to a structural member cannot be less than the dimension established by Table 720.1(1), except for permitted modifications. An example of the minimum thickness of concrete required for a structural steel column is shown in Figure 720.1.1(1). Figure 720.1.1(2) illustrates the minimum concrete thickness requirements for protecting reinforcing steel in concrete columns, beams, girders and trusses. Refer to Section 714 for additional provisions regarding structural members.

720.1.2 Unit masonry protection. Where required, metal ties shall be embedded in transverse joints of unit masonry for protection of steel columns. Such ties shall be as set forth in Table 720.1(1) or be equivalent thereto.

❖ Items 1-3.1 through 1-3.4 of Table 720.1(1) require horizontal joint reinforcement. This section stipulates that if ties are required to connect the masonry to the steel column, such ties are to be located in the transverse joints and not in the bed joints which is where the horizontal enforcement is to be located.

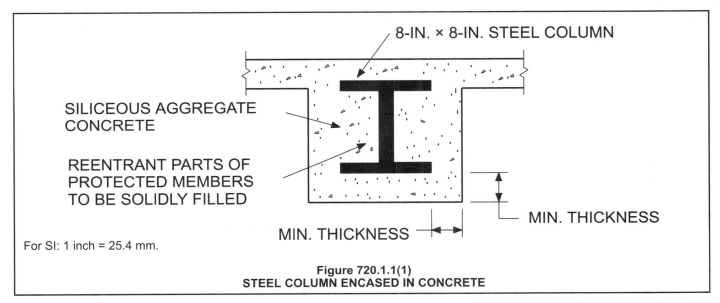

For SI: 1 inch = 25.4 mm.

Figure 720.1.1(1)
STEEL COLUMN ENCASED IN CONCRETE

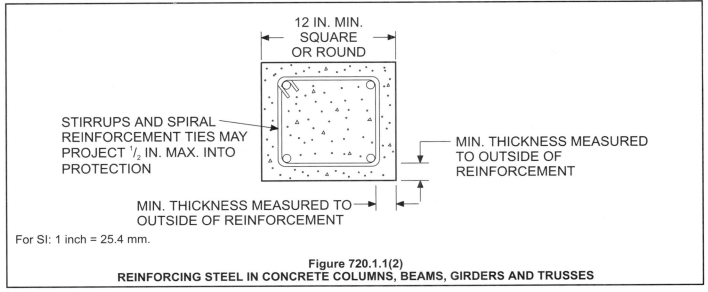

For SI: 1 inch = 25.4 mm.

Figure 720.1.1(2)
REINFORCING STEEL IN CONCRETE COLUMNS, BEAMS, GIRDERS AND TRUSSES

720.1.3 Reinforcement for cast-in-place concrete column protection. Cast-in-place concrete protection for steel columns shall be reinforced at the edges of such members with wire ties of not less than 0.18 inch (4.6 mm) in diameter wound spirally around the columns on a pitch of not more than 8 inches (203 mm) or by equivalent reinforcement.

❖ In order for a concrete-encased steel column to maintain a fire-resistance rating, it is imperative that the concrete remain in place. Similar to a reinforced concrete column with spiral reinforcement (see Section 721.2.4.2), the ties required by this section are necessary for the concrete to remain in place around the column.

720.1.4 Plaster application. The finish coat is not required for plaster protective coatings where they comply with the design mix and thickness requirements of Tables 720.1(1), 720.1(2) and 720.1(3).

❖ The thickness of plaster for prescriptive assemblies, such as Item 1-4.1 in Table 720.1(1), Item 12-1.1 in Table 720.1(2) and Item 11-1.1 in Table 720.1(3), does not include the finish coat. Only the scratch coat and brown coat are needed to achieve the indicated fire-resistance ratings in the tables.

720.1.5 Bonded prestressed concrete tendons. For members having a single tendon or more than one tendon installed with equal concrete cover measured from the nearest surface, the cover shall not be less than that set forth in Table 720.1(1). For members having multiple tendons installed with variable concrete cover, the average tendon cover shall not be less than that set forth in Table 720.1(1), provided:

1. The clearance from each tendon to the nearest exposed surface is used to determine the average cover.

2. In no case can the clear cover for individual tendons be less than one-half of that set forth in Table 720.1(1). A minimum cover of 0.75 inch (19.1 mm) for slabs and 1 inch (25 mm) for beams is required for any aggregate concrete.

3. For the purpose of establishing a fire-resistance rating, tendons having a clear covering less than that set forth in Table 720.1(1) shall not contribute more than 50 percent of the required ultimate moment capacity for members less than 350 square inches (0.226 m²) in cross-sectional area and 65 percent for larger members. For structural design purposes, however, tendons having a reduced cover are assumed to be fully effective.

❖ Figure 720.1.5 depicts the requirements specified in Items 1 and 2 for variable concrete cover for tendons. For all cases of variable concrete cover, the average concrete cover for the tendons must not be less than that specified in Table 720.1(1).

As prestressed concrete members are designed in accordance with their ultimate moment capacity, as well as their performance at service loads, Item 3 provides

two sets of criteria for variable concrete cover for multiple tendons:

1. Those tendons having less concrete cover than specified in Table 720.1(1) are to be considered as furnishing only a reduced portion of the ultimate-moment capacity of the member, depending on the cross-sectional area of the member.

2. No reduction is necessary for those tendons having reduced cover for the design of the member at service loads.

As the ultimate-moment capacity of the member is critical to its behavior under fire conditions, the code requires the reduction for those tendons having cover to be less than that specified by the code. Behavior at service loads, however, is less affected by the heat of a fire; therefore, the code permits those tendons with reduced cover to be fully effective.

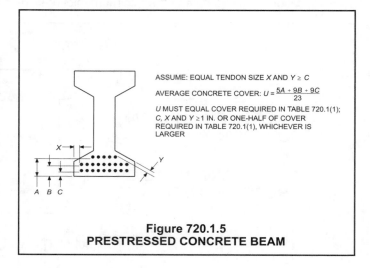

ASSUME: EQUAL TENDON SIZE X AND $Y \geq C$

AVERAGE CONCRETE COVER: $U = \dfrac{5A + 9B + 9C}{23}$

U MUST EQUAL COVER REQUIRED IN TABLE 720.1(1); C, X AND $Y \geq 1$ IN. OR ONE-HALF OF COVER REQUIRED IN TABLE 720.1(1), WHICHEVER IS LARGER

Figure 720.1.5
PRESTRESSED CONCRETE BEAM

TABLE 720.1(1). See page 7-107.

❖ An example of the minimum thickness of concrete required for a structural steel column is shown in Figure 720.1.1(1). As shown, this figure depicts Item 1-1.5 of Table 720.1(1). Figure 720.1.1(2) depicts a reinforced concrete column based on Items 5-1.1 and 5-1.2 (see commentary, Section 720.1).

TABLE 720.1(2). See page 7-112

❖ An example of the minimum thickness of concrete required for a wall is shown in Figure 720.1(2)(1). As shown, this figure depicts Item 4-1.1 of Table 720.1(2). Figure 720.1(2)(2) depicts a noncombustible stud exterior wall based on Item 15-1.4 (see commentary, Section 720.1).

TABLE 720.1(3). See page 7-120

❖ An example of a 1-hour-rated wood floor or roof truss is shown in Figure 720.1(3). As shown, this figure depicts Item 21-1.1 (see commentary, Section 720.1).

TABLE 720.1(1)
MINIMUM PROTECTION OF STRUCTURAL PARTS BASED ON TIME PERIODS
FOR VARIOUS NONCOMBUSTIBLE INSULATING MATERIALS[m]

STRUCTURAL PARTS TO BE PROTECTED	ITEM NUMBER	INSULATING MATERIAL USED	MINIMUM THICKNESS OF INSULATING MATERIAL FOR THE FOLLOWING FIRE-RESISTANCE PERIODS (inches)			
			4 hour	3 hour	2 hour	1 hour
1. Steel columns and all of primary trusses	1-1.1	Carbonate, lightweight and sand-lightweight aggregate concrete, members 6″ × 6″ or greater (not including sandstone, granite and siliceous gravel).[a]	$2^1/_2$	2	$1^1/_2$	1
	1-1.2	Carbonate, lightweight and sand-lightweight aggregate concrete, members 8″ × 8″ or greater (not including sandstone, granite and siliceous gravel).[a]	2	$1^1/_2$	1	1
	1-1.3	Carbonate, lightweight and sand-lightweight aggregate concrete, members 12″ × 12″or greater (not including sandstone, granite and siliceous gravel).[a]	$1^1/_2$	1	1	1
	1-1.4	Siliceous aggregate concrete and concrete excluded in Item 1-1.1, members 6″ × 6″ or greater.[a]	3	2	$1^1/_2$	1
	1-1.5	Siliceous aggregate concrete and concrete excluded in Item 1-1.1, members 8″ × 8″ or greater.[a]	$2^1/_2$	2	1	1
	1-1.6	Siliceous aggregate concrete and concrete excluded in Item 1-1.1, members 12″ × 12″ or greater.[a]	2	1	1	1
	1-.2.1	Clay or shale brick with brick and mortar fill.[a]	$3^3/_4$	—	—	$2^1/_4$
	1-3.1	4″ hollow clay tile in two 2″ layers; $^1/_2$″ mortar between tile and column; $^3/_8$″ metal mesh 0.046″ wire diameter in horizontal joints; tile fill.[a]	4	—	—	—
	1-3.2	2″ hollow clay tile; $^3/_4$″ mortar between tile and column; $^3/_8$″metal mesh 0.046″ wire diameter in horizontal joints; limestone concrete fill;[a] plastered with $^3/_4$″ gypsum plaster.	3	—	—	—
	1-3.3	2″ hollow clay tile with outside wire ties 0.08″ diameter at each course of tile or $^3/_8$″ metal mesh 0.046″ diameter wire in horizontal joints; limestone or trap-rock concrete fill[1] extending 1″ outside column on all sides	—	—	3	—
	1-3.4	2″ hollow clay tile with outside wire ties 0.08″ diameter at each course of tile with or without concrete fill; $^3/_4$″ mortar between tile and column.	—	—	—	2
	1-4.1	Cement plaster over metal lath wire tied to $^3/_4$″ cold-rolled vertical channels with 0.049″ (No. 18 B.W. gage) wire ties spaced 3″ to 6″ on center. Plaster mixed 1:2 $^1/_2$ by volume, cement to sand.	—	—	$2^1/_2$[b]	$^7/_8$
	1-5.1	Vermiculite concrete, 1:4 mix by volume over paperbacked wire fabric lath wrapped directly around column with additional 2″ × 2″ 0.065″/0.065″ (No. 16/16 B.W. gage) wire fabric placed $^3/_4$″ from outer concrete surface. Wire fabric tied with 0.049″ (No. 18 B.W. gage) wire spaced 6″ on center for inner layer and 2″ on center for outer layer.	2	—	—	—
	1-6.1	Perlite or vermiculite gypsum plaster over metal lath wrapped around column and furred $1^1/_4$″ from column flanges. Sheets lapped at ends and tied at 6″ intervals with 0.049″ (No. 18 B.W. gage) tie wire. Plaster pushed through to flanges.	$1^1/_2$	1	—	—
	1-6.2	Perlite or vermiculite gypsum plaster over self-furring metal lath wrapped directly around column, lapped 1″ and tied at 6″ intervals with 0.049″ (No. 18 B.W. gage) wire.	$1^3/_4$	$1^3/_8$	1	—
	1-6.3	Perlite or vermiculite gypsum plaster on metal lath applied to $^3/_4$″ cold-rolled channels spaced 24″ apart vertically and wrapped flatwise around column.	$1^1/_2$	—	—	—
	1-6.4	Perlite or vermiculite gypsum plaster over two layers of $^1/_2$″ plain full-length gypsum lath applied tight to column flanges. Lath wrapped with 1″ hexagonal mesh of No. 20 gage wire and tied with doubled 0.035″ diameter (No. 18 B.W. gage) wire ties spaced 23″ on center. For three-coat work, the plaster mix for the second coat shall not exceed 100 pounds of gypsum to $2^1/_2$ cubic feet of aggregate for the 3-hour system.	$2^1/_2$	2	—	—

(continued)

TABLE 720.1(1)   FIRE-RESISTANCE-RATED CONSTRUCTION

TABLE 720.1(1)—continued
MINIMUM PROTECTION OF STRUCTURAL PARTS BASED ON TIME PERIODS
FOR VARIOUS NONCOMBUSTIBLE INSULATING MATERIALS[m]

STRUCTURAL PARTS TO BE PROTECTED	ITEM NUMBER	INSULATING MATERIAL USED	MINIMUM THICKNESS OF INSULATING MATERIAL FOR THE FOLLOWING FIRE-RESISTANCE PERIODS (inches)			
			4 hour	3 hour	2 hour	1 hour
1. Steel columns and all of primary trusses (continued)	1-6.5	Perlite or vermiculate gypsum plaster over one layer of $1/2''$ plain full-length gypsum lath applied tight to column flanges. Lath tied with doubled 0.049'' (No. 18 B.W. gage) wire ties spaced 23'' on center and scratch coat wrapped with 1'' hexagonal mesh 0.035'' (No. 20 B.W. gage) wire fabric. For three-coat work, the plaster mix for the second coat shall not exceed 100 pounds of gypsum to $2^{1}/_{2}$ cubic feet of aggregate.	—	2	—	—
	1-7.1	Multiple layers of $1/2''$ gypsum wallboard[c] adhesively[d] secured to column flanges and successive layers. Wallboard applied without horizontal joints. Corner edges of each layer staggered. Wallboard layer below outer layer secured to column with doubled 0.049'' (No. 18 B.W. gage) steel wire ties spaced 15'' on center. Exposed corners taped and treated.	—	—	2	1
	1-7.2	Three layers of $5/8''$ Type X gypsum wallboard.[c] First and second layer held in place by $1/8''$ diameter by $1^{3}/_{8}''$ long ring shank nails with $5/16''$ diameter heads spaced 24'' on center at corners. Middle layer also secured with metal straps at mid-height and 18'' from each end, and by metal corner bead at each corner held by the metal straps. Third layer attached to corner bead with 1'' long gypsum wallboard screws spaced 12'' on center.	—	—	$1^{7}/_{8}$	—
	1-7.3	Three layers of $5/8''$ Type X gypsum wallboard,[c] each layer screw attached to $1^{5}/_{8}''$ steel studs 0.018'' thick (No. 25 carbon sheet steel gage) at each corner of column. Middle layer also secured with 0.049'' (No. 18 B.W. gage) double-strand steel wire ties, 24'' on center. Screws are No. 6 by 1'' spaced 24'' on center for inner layer, No. 6 by $1^{5}/_{8}''$ spaced 12'' on center for middle layer and No. 8 by $2^{1}/_{4}''$ spaced 12'' on center for outer layer.	—	$1^{7}/_{8}$	—	—
	1-8.1	Wood-fibered gypsum plaster mixed 1:1 by weight gypsum-to-sand aggregate applied over metal lath. Lath lapped 1'' and tied 6'' on center at all end, edges and spacers with 0.049'' (No. 18 B.W. gage) steel tie wires. Lath applied over $1/2''$ spacers made of $3/4''$ furring channel with 2'' legs bent around each corner. Spacers located 1'' from top and bottom of member and a maximum of 40'' on center and wire tied with a single strand of 0.049'' (No. 18 B.W. gage) steel tie wires. Corner bead tied to the lath at 6'' on center along each corner to provide plaster thickness.	—	—	$1^{5}/_{8}$	—
2. Webs or flanges of steel beams and girders	2-1.1	Carbonate, lightweight and sand-lightweight aggregate concrete (not including sandstone, granite and siliceous gravel) with 3'' or finer metal mesh placed 1'' from the finished surface anchored to the top flange and providing not less than 0.025 square inch of steel area per foot in each direction.	2	$1^{1}/_{2}$	1	1
	2-1.2	Siliceous aggregate concrete and concrete excluded in Item 2-1.1 with 3'' or finer metal mesh placed 1'' from the finished surface anchored to the top flange and providing not less than 0.025 square inch of steel area per foot in each direction.	$2^{1}/_{2}$	2	$1^{1}/_{2}$	1
	2-2.1	Cement plaster on metal lath attached to $3/4''$ cold-rolled channels with 0.049'' (No. 18 B.W. gage) wire ties spaced 3'' to 6'' on center. Plaster mixed 1:2 $^{1}/_{2}$ by volume, cement to sand.	—	—	$2^{1}/_{2}$[b]	$7/8$
	2-3.1	Vermiculite gypsum plaster on a metal lath cage, wire tied to 0.165'' diameter (No. 8 B.W. gage) steel wire hangers wrapped around beam and spaced 16'' on center. Metal lath ties spaced approximately 5'' on center at cage sides and bottom.	—	$7/8$	—	—

(continued)

TABLE 720.1(1)—continued
MINIMUM PROTECTION OF STRUCTURAL PARTS BASED ON TIME PERIODS
FOR VARIOUS NONCOMBUSTIBLE INSULATING MATERIALS[m]

STRUCTURAL PARTS TO BE PROTECTED	ITEM NUMBER	INSULATING MATERIAL USED	MINIMUM THICKNESS OF INSULATING MATERIAL FOR THE FOLLOWING FIRE-RESISTANCE PERIODS (inches)			
			4 hour	3 hour	2 hour	1 hour
2. Webs or flanges of steel beams and girders (continued)	2-4.1	Two layers of $5/8''$ Type X gypsum wallboard[c] are attached to U-shaped brackets spaced 24" on center. 0.018" thick (No. 25 carbon sheet steel gage) $1^5/8''$ deep by 1" galvanized steel runner channels are first installed parallel to and on each side of the top beam flange to provide a $1/2''$ clearance to the flange. The channel runners are attached to steel deck or concrete floor construction with approved fasteners spaced 12" on center. U-shaped brackets are formed from members identical to the channel runners. At the bent portion of the U-shaped bracket, the flanges of the channel are cut out so that $1^5/8''$ deep corner channels can be inserted without attachment parallel to each side of the lower flange. As an alternate, 0.021" thick (No. 24 carbon sheet steel gage) 1" × 2" runner and corner angles may be used in lieu of channels, and the web cutouts in the U-shaped brackets may be omitted. Each angle is attached to the bracket with $1/2''$-long No. 8 self-drilling screws. The vertical legs of the U-shaped bracket are attached to the runners with one $1/2''$ long No. 8 self-drilling screw. The completed steel framing provides a $2^1/8''$ and $1^1/2''$ space between the inner layer of wallboard and the sides and bottom of the steel beam, respectively. The inner layer of wallboard is attached to the top runners and bottom corner channels or corner angles with $1^1/4''$-long No. 6 self-drilling screws spaced 16" on center. The outer layer of wallboard is applied with $1^3/4''$-long No. 6 self-drilling screws spaced 8" on center. The bottom corners are reinforced with metal corner beads.	—	—	$1^1/4$	—
	2-4.2	Three layers of $5/8''$ Type X gypsum wallboard[c] attached to a steel suspension system as described immediately above utilizing the 0.018" thick (No. 25 carbon sheet steel gage) 1" × 2" lower corner angles. The framing is located so that a $2^1/8''$ and 2" space is provided between the inner layer of wallboard and the sides and bottom of the beam, respectively. The first two layers of wallboard are attached as described immediately above. A layer of 0.035" thick (No. 20 B.W. gage) 1" hexagonal galvanized wire mesh is applied under the soffit of the middle layer and up the sides approximately 2". The mesh is held in position with the No. 6 $1^5/8''$-long screws installed in the vertical leg of the bottom corner angles. The outer layer of wallboard is attached with No. 6 $2^1/4''$-long screws spaced 8" on center. One screw is also installed at the mid-depth of the bracket in each layer. Bottom corners are finished as described above.	—	$1^7/8$	—	—
3. Bonded pretensioned reinforcement in prestressed concrete[e]	3-1.1	Carbonate, lightweight, sand-lightweight and siliceous[f] aggregate concrete 　　Beams or girders 　　Solid slabs[h]	4^g 	3^g 2	$2^1/2$ $1^1/2$	$1^1/2$ 1
4. Bonded or unbonded post-tensioned tendons in prestressed concrete[e, i]	4-1.1	Carbonate, lightweight, sand-lightweight and siliceous[f] aggregate concrete Unrestrained members: 　　Solid slabs[h] 　　Beams and girders[j] 　　　8" wide 　　　greater than 12" wide	— 3	2 $4^1/2$ $2^1/2$	$1^1/2$ $2^1/2$ 2	— $1^3/4$ $1^1/2$
	4-1.2	Carbonate, lightweight, sand-lightweight and siliceous aggregate Restrained members:[k] 　　Solid slabs[h] 　　Beams and girders[j] 　　　8" wide 　　　greater than 12" wide	$1^1/4$ $2^1/2$ 2	1 2 $1^3/4$	$3/4$ $1^3/4$ $1^1/2$	— — —

(continued)

TABLE 720.1(1)   FIRE-RESISTANCE-RATED CONSTRUCTION

TABLE 720.1(1)—continued
MINIMUM PROTECTION OF STRUCTURAL PARTS BASED ON TIME PERIODS
FOR VARIOUS NONCOMBUSTIBLE INSULATING MATERIALS[m]

STRUCTURAL PARTS TO BE PROTECTED	ITEM NUMBER	INSULATING MATERIAL USED	MINIMUM THICKNESS OF INSULATING MATERIAL FOR THE FOLLOWING FIRE-RESISTANCE PERIODS (inches)			
			4 hour	3 hour	2 hour	1 hour
5. Reinforcing steel in reinforced concrete columns, beams girders and trusses	5-1.1	Carbonate, lightweight and sand-lightweight aggregate concrete, members 12″ or larger, square or round. (Size limit does not apply to beams and girders monolithic with floors.)	$1^1/_2$	$1^1/_2$	$1^1/_2$	$1^1/_2$
		Siliceous aggregate concrete, members 12″ or larger, square or round. (Size limit does not apply to beams and girders monolithic with floors.)	2	$1^1/_2$	$1^1/_2$	$1^1/_2$
6. Reinforcing steel in reinforced concrete joists[l]	6-1.1	Carbonate, lightweight and sand-lightweight aggregate concrete.	$1^1/_4$	$1^1/_4$	1	$^3/_4$
	6-1.2	Siliceous aggregate concrete.	$1^3/_4$	$1^1/_2$	1	$^3/_4$
7. Reinforcing and tie rods in floor and roof slabs[l]	7-1.1	Carbonate, lightweight and sand-lightweight aggregate concrete.	1	1	$^3/_4$	$^3/_4$
	7-1.2	Siliceous aggregate concrete.	$1^1/_4$	1	1	$^3/_4$

For SI: 1 inch = 25.4 mm, 1 square inch = 645.2 mm^2, 1 cubic foot = 0.0283 m^3.

a. Reentrant parts of protected members to be filled solidly.

b. Two layers of equal thickness with a $^3/_4$-inch airspace between.

c. For all of the construction with gypsum wallboard described in Table 720.1(1), gypsum base for veneer plaster of the same size, thickness and core type shall be permitted to be substituted for gypsum wallboard, provided attachment is identical to that specified for the wallboard and the joints on the face layer are reinforced, and the entire surface is covered with a minimum of $^1/_{16}$-inch gypsum veneer plaster.

d. An approved adhesive qualified under ASTM E 119.

e. Where lightweight or sand-lightweight concrete having an oven-dry weight of 110 pounds per cubic foot or less is used, the tabulated minimum cover shall be permitted to be reduced 25 percent, except that in no case shall the cover be less than $^3/_4$ inch in slabs or $1^1/_2$ inches in beams or girders.

f. For solid slabs of siliceous aggregate concrete, increase tendon cover 20 percent.

g. Adequate provisions against spalling shall be provided by U-shaped or hooped stirrups spaced not to exceed the depth of the member with a clear cover of 1 inch.

h. Prestressed slabs shall have a thickness not less than that required in Table 720.1(3) for the respective fire resistance time period.

i. Fire coverage and end anchorages shall be as follows: Cover to the prestressing steel at the anchor shall be $^1/_2$ inch greater than that required away from the anchor. Minimum cover to steel-bearing plate shall be 1 inch in beams and $^3/_4$ inch in slabs.

j. For beam widths between 8 inches and 12 inches, cover thickness shall be permitted to be determined by interpolation.

k. Interior spans of continuous slabs, beams and girders shall be permitted to be considered restrained.

l. For use with concrete slabs having a comparable fire endurance where members are framed into the structure in such a manner as to provide equivalent performance to that of monolithic concrete construction.

m. Generic fire-resistance ratings (those not designated as PROPRIETARY* in the listing) in GA 600 shall be accepted as if herein listed.

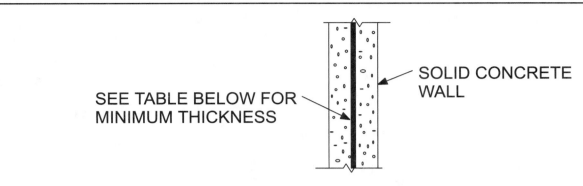

SEE TABLE BELOW FOR
MINIMUM THICKNESS

SOLID CONCRETE
WALL

SOLID CONCRETE WALL

MATERIAL	ITEM NUMBER	CONSTRUCTION	MINIMUM FINISHED THICKNESS FACE-TO-FACE (inches)			
			4 hour	3 hour	2 hour	1 hour
Solid concrete	4-1.1	Siliceous aggregate concrete	7.0	6.2	5.0	3.5
		Carbondate aggregate concrete	6.6	5.7	4.6	3.2
		Sand-lightweight concrete	5.4	4.6	3.8	2.7
		Lightweight concrete	5.1	4.4	3.6	2.5

For SI: 1 inch = 25.4 mm.

Figure 720.1(2)(1)
CONCRETE WALL

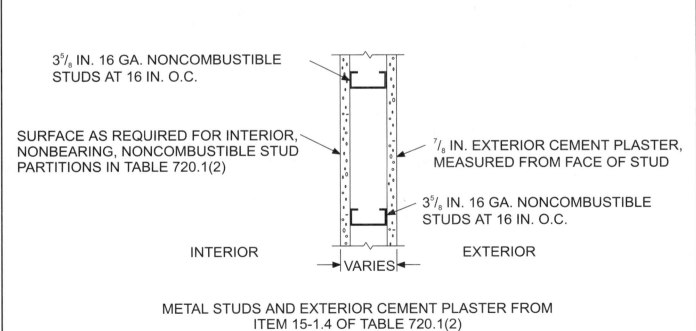

$3^5/_8$ IN. 16 GA. NONCOMBUSTIBLE
STUDS AT 16 IN. O.C.

SURFACE AS REQUIRED FOR INTERIOR,
NONBEARING, NONCOMBUSTIBLE STUD
PARTITIONS IN TABLE 720.1(2)

$^7/_8$ IN. EXTERIOR CEMENT PLASTER,
MEASURED FROM FACE OF STUD

$3^5/_8$ IN. 16 GA. NONCOMBUSTIBLE
STUDS AT 16 IN. O.C.

INTERIOR   EXTERIOR

VARIES

METAL STUDS AND EXTERIOR CEMENT PLASTER FROM
ITEM 15-1.4 OF TABLE 720.1(2)

For SI: 1 inch = 25.4 mm.

Figure 720.1(2)(2)
METAL STUD WALL WITH PLASTER

TABLE 720.1(2)

FIRE-RESISTANCE-RATED CONSTRUCTION

TABLE 720.1(2)
RATED FIRE-RESISTANCE PERIODS FOR VARIOUS WALLS AND PARTITIONS [a,o,p]

MATERIAL	ITEM NUMBER	CONSTRUCTION	MINIMUM FINISHED THICKNESS FACE-TO-FACE [b] (inches)			
			4 hour	3 hour	2 hour	1 hour
1. Brick of clay or shale	1-1.1	Solid brick of clay or shale [c]	6	4.9	3.8	2.7
	1-1.2	Hollow brick, not filled.	5.0	4.3	3.4	2.3
	1-1.3	Hollow brick unit wall, grout or filled with perlite vermiculite or expanded shale aggregate.	6.6	5.5	4.4	3.0
	1-2.1	4″ nominal thick units at least 75 percent solid backed with a hat-shaped metal furring channel $^3/_4$″ thick formed from 0.021″ sheet metal attached to the brick wall on 24″ centers with approved fasteners, and $^1/_2$″ Type X gypsum wallboard attached to the metal furring strips with 1″-long Type S screws spaced 8″ on center.	—	—	5 [d]	—
2. Combination of clay brick and load-bearing hollow clay tile	2-1.1	4″ solid brick and 4″ tile (at least 40 percent solid).	—	8	—	—
	2-1.2	4″ solid brick and 8″ tile (at least 40 percent solid).	12	—	—	—
3. Concrete masonry units	3-1.1 [f, g]	Expanded slag or pumice.	4.7	4.0	3.2	2.1
	3-1.2 [f, g]	Expanded clay, shale or slate.	5.1	4.4	3.6	2.6
	3-1.3 [f]	Limestone, cinders or air-cooled slag.	5.9	5.0	4.0	2.7
	3-1.4 [f, g]	Calcareous or siliceous gravel.	6.2	5.3	4.2	2.8
4. Solid concrete [h, i]	4-1.1	Siliceous aggregate concrete.	7.0	6.2	5.0	3.5
		Carbonate aggregate concrete.	6.6	5.7	4.6	3.2
		Sand-lightweight concrete.	5.4	4.6	3.8	2.7
		Lightweight concrete.	5.1	4.4	3.6	2.5
5. Glazed or unglazed facing tile, nonload-bearing	5-1.1	One 2″ unit cored 15 percent maximum and one 4″ unit cored 25 percent maximum with $^3/_4$″ mortar-filled collar joint. Unit positions reversed in alternate courses.	—	$6^3/_8$	—	—
	5-1.2	One 2″ unit cored 15 percent maximum and one 4″ unit cored 40 percent maximum with $^3/_4$″ mortar-filled collar joint. Unit positions side with $^3/_4$″ gypsum plaster. Two wythes tied together every fourth course with No. 22 gage corrugated metal ties.	—	$6^3/_4$	—	—
	5-1.3	One unit with three cells in wall thickness, cored 29 percent maximum.	—	—	6	—
	5-1.4	One 2″ unit cored 22 percent maximum and one 4″ unit cored 41 percent maximum with $^1/_4$″ mortar-filled collar joint. Two wythes tied together every third course with 0.030″ (No. 22 galvanized sheet steel gage) corrugated metal ties.	—	—	6	—
	5-1.5	One 4″ unit cored 25 percent maximum with $^3/_4$″ gypsum plaster on one side.	—	—	$4^3/_4$	—
	5-1.6	One 4″ unit with two cells in wall thickness, cored 22 percent maximum.	—	—	—	4
	5-1.7	One 4″ unit cored 30 percent maximum with $^3/_4$″ vermiculite gypsum plaster on one side.	—	—	$4^1/_2$	—
	5-1.8	One 4″ unit cored 39 percent maximum with $^3/_4$″ gypsum plaster on one side.	—	—	—	$4^1/_2$

(continued)

TABLE 720.1(2)—continued
RATED FIRE-RESISTANCE PERIODS FOR VARIOUS WALLS AND PARTITIONS [a,o,p]

MATERIAL	ITEM NUMBER	CONSTRUCTION	MINIMUM FINISHED THICKNESS FACE-TO-FACE[b] (inches)			
			4 hour	3 hour	2 hour	1 hour
6. Solid gypsum plaster	6-1.1	$^3/_4$″ by 0.055″ (No. 16 carbon sheet steel gage) vertical cold-rolled channels, 16″ on center with 2.6-pound flat metal lath applied to one face and tied with 0.049″ (No. 18 B.W. Gage) wire at 6″ spacing. Gypsum plaster each side mixed 1:2 by weight, gypsum to sand aggregate.	—	—	—	2^d
	6-1.2	$^3/_4$″ by 0.055″ (No. 16 carbon sheet steel gage) cold-rolled channels 16″ on center with metal lath applied to one face and tied with 0.049″ (No. 18 B.W. gage) wire at 6″ spacing. Perlite or vermiculite gypsum plaster each side. For three-coat work, the plaster mix for the second coat shall not exceed 100 pounds of gypsum to $2^1/_2$ cubic feet of aggregate for the 1-hour system.	—	—	$2^1/_2{}^d$	2^d
	6-1.3	$^3/_4$″ by 0.055″ (No. 16 carbon sheet steel gage) vertical cold-rolled channels, 16″ on center with $^3/_8$″gypsum lath applied to one face and attached with sheet metal clips. Gypsum plaster each side mixed 1:2 by weight, gypsum to sand aggregate.	—	—	—	2^d
	6-2.1	Studless with $^1/_2$″ full-length plain gypsum lath and gypsum plaster each side. Plaster mixed 1:1 for scratch coat and 1:2 for brown coat, by weight, gypsum to sand aggregate.	—	—	—	2^d
	6-2.2	Studless with $^1/_2$″ full-length plain gypsum lath and perlite or vermiculite gypsum plaster each side.	—	—	$2^1/_2{}^d$	2^d
	6-2.3	Studless partition with $^3/_8$″ rib metal lath installed vertically adjacent edges tied 6″ on center with No. 18 gage wire ties, gypsum plaster each side mixed 1:2 by weight, gypsum to sand aggregate.	—	—	—	2^d
7. Solid perlite and portland cement	7-1.1	Perlite mixed in the ratio of 3 cubic feet to 100 pounds of portland cement and machine applied to stud side of $1^1/_2$″ mesh by 0.058-inch (No. 17 B.W. gage) paper-backed woven wire fabric lath wire-tied to 4″-deep steel trussed wire[j] studs 16″ on center. Wire ties of 0.049″ (No. 18 B.W. gage) galvanized steel wire 6″ on center vertically.	—	—	$3^1/_8{}^d$	—
8. Solid neat wood fibered gypsum plaster	8-1.1	$^3/_4$″ by 0.055-inch (No. 16 carbon sheet steel gage) cold-rolled channels, 12″ on center with 2.5-pound flat metal lath applied to one face and tied with 0.049″ (No. 18 B.W. gage) wire at 6″ spacing. Neat gypsum plaster applied each side.	—	—	2^d	—
9. Solid wallboard partition	9-1.1	One full-length layer $^1/_2$″ Type X gypsum wallboard[e] laminated to each side of 1″ full-length V-edge gypsum coreboard with approved laminating compound. Vertical joints of face layer and coreboard staggered at least 3″.	—	—	2^d	—
10. Hollow (studless) gypsum wallboard partition	10-1.1	One full-length layer of $^5/_8$″ Type X gypsum wallboard[e] attached to both sides of wood or metal top and bottom runners laminated to each side of 1″ × 6″ full-length gypsum coreboard ribs spaced 24″ on center with approved laminating compound. Ribs centered at vertical joints of face plies and joints staggered 24″ in opposing faces. Ribs may be recessed 6″ from the top and bottom.	—	—	—	$2^1/_4{}^d$
	10-1.2	1″ regular gypsum V-edge full-length backing board attached to both sides of wood or metal top and bottom runners with nails or $1^5/_8$″ drywall screws at 24″ on center. Minimum width of rumors $1^5/_8$″. Face layer of $^1/_2$″ regular full-length gypsum wallboard laminated to outer faces of backing board with approved laminating compound.	—	—	$4^5/_8{}^d$	—

(continued)

TABLE 720.1(2)

FIRE-RESISTANCE-RATED CONSTRUCTION

TABLE 720.1(2)—continued
RATED FIRE-RESISTANCE PERIODS FOR VARIOUS WALLS AND PARTITIONS [a,o,p]

MATERIAL	ITEM NUMBER	CONSTRUCTION	MINIMUM FINISHED THICKNESS FACE-TO-FACE[b] (inches)			
			4 hour	3 hour	2 hour	1 hour
11. Noncombustible studs—interior partition with plaster each side	11-1.1	$3^1/_4'' \times 0.044''$ (No. 18 carbon sheet steel gage) steel studs spaced 24″ on center. $5/_8''$ gypsum plaster on metal lath each side mixed 1:2 by weight, gypsum to sand aggregate.	—	—	—	$4^3/_4$[d]
	11-1.2	$3^3/_8'' \times 0.055''$ (No. 16 carbon sheet steel gage) approved nailable[k] studs spaced 24" on center. $5/_8''$ neat gypsum wood-fibered plaster each side over $3/_8''$ rib metal lath nailed to studs with 6d common nails, 8″ on center. Nails driven $1^1/_4''$ and bent over.	—	—	$5^5/_8$	—
	11-1.3	$4'' \times 0.044''$ (No. 18 carbon sheet steel gage) channel-shaped steel studs at 16″ on center. On each side approved resilient clips pressed onto stud flange at 16″ vertical spacing, $1/_4''$ pencil rods snapped into or wire tied onto outer loop of clips, metal lath wire-tied to pencil rods at 6″ intervals, 1″ perlite gypsum plaster, each side.	—	$7^5/_8$[d]	—	—
	11-1.4	$2^1/_2'' \times 0.044''$ (No. 18 carbon sheet steel gage) steel studs spaced 16″ on center. Wood fibered gypsum plaster mixed 1:1 by weight gypsum to sand aggregate applied on $3/_4$-pound metal lath wire tied to studs, each side. $3/_4''$ plaster applied over each face, including finish coat.	—	—	$4^1/_4$[d]	—
12. Wood studs interior partition with plaster each side	12-1.1[l, m]	$2'' \times 4''$ wood studs 16″ on center with $5/_8''$ gypsum plaster on metal lath. Lath attached by 4d common nails bent over or No. 14 gage by $1^1/_4''$ by $3/_4''$ crown width staples spaced 6″ on center. Plaster mixed $1:1^1/_2$ for scratch coat and 1:3 for brown coat, by weight, gypsum to sand aggregate.	—	—	—	$5^1/_8$
	12-1.2[l]	$2'' \times 4''$ wood studs 16″ on center with metal lath and $7/_8''$ neat wood-fibered gypsum plaster each side. Lath attached by 6d common nails, 7″ on center. Nails driven $1^1/_4''$ and bent over.	—	—	$5^1/_2$[d]	—
	12-1.3[l]	$2'' \times 4''$ wood studs 16″ on center with $3/_8''$ perforated or plain gypsum lath and $1/_2''$ gypsum plaster each side. Lath nailed with $1^1/_8''$ by No. 13 gage by $19/_{64}''$ head plasterboard blued nails, 4″ on center. Plaster mixed 1:2 by weight, gypsum to sand aggregate.	—	—	—	$5^1/_4$
	12-1.4[l]	$2'' \times 4''$ wood studs 16″ on center with $3/_8''$ Type X gypsum lath and $1/_2''$ gypsum plaster each side. Lath nailed with $1^1/_8''$ by No. 13 gage by $19/_{64}''$ head plasterboard blued nails, 5″ on center. Plaster mixed 1:2 by weight, gypsum to sand aggregate.	—	—	—	$5^1/_4$
13.Noncumbustible studs—interior partition with gypsum wallboard each side	13-1.1	0.018″ (No. 25 carbon sheet steel gage) channel-shaped studs 24″ on center with one full-length layer of $5/_8''$ Type X gypsum wallboard[e] applied vertically attached with 1″ long No. 6 drywall screws to each stud. Screws are 8″ on center around the perimeter and 12″ on center on the intermediate stud. The wallboard may be applied horizontally when attached to $3^5/_8''$ studs and the horizontal joints are staggered with those on the opposite side. Screws for the horizontal application shall be 8″ on center at vertical edges and 12″ on center at intermediate studs.	—	—	—	$2^7/_8$[d]
	13-1.2	0.018″ (No. 25 carbon sheet steel gage) channel-shaped studs 25″ on center with two full-length layers of $1/_2''$ Type X gypsum wallboard[e] applied vertically each side. First layer attached with 1″-long, No. 6 drywall screws, 8″ on center around the perimeter and 12″ on center on the intermediate stud. Second layer applied with vertical joints offset one stud space from first layer using $1^5/_8''$ long, No. 6 drywall screws spaced 9″ on center along vertical joints, 12″ on center at intermediate studs and 24″ on center along top and bottom runners.	—	—	$3^5/_8$[d]	—
	13-1.3	0.055″ (No. 16 carbon sheet steel gage) approved nailable metal studs[e] 24″ on center with full-length $5/_8''$ Type X gypsum wallboard[e] applied vertically and nailed 7″ on center with 6d cement-coated common nails. Approved metal fastener grips used with nails at vertical butt joints along studs.	—	—	—	$4^7/_8$

(continued)

TABLE 720.1(2)—continued
RATED FIRE-RESISTANCE PERIODS FOR VARIOUS WALLS AND PARTITIONS [a,o,p]

MATERIAL	ITEM NUMBER	CONSTRUCTION	MINIMUM FINISHED THICKNESS FACE-TO-FACE[b] (inches)			
			4 hour	3 hour	2 hour	1 hour
14. Wood studs—interior partition with gypsum wallboard each side	14-1.1[h,m]	$2'' \times 4''$ wood studs $16''$ on center with two layers of $3/8''$ regular gypsum wallboard[e] each side, 4d cooler[n] or wallboard[n] nails at $8''$ on center first layer, 5d cooler[n] or wallboard[n] nails at $8''$ on center second layer with laminating compound between layers, joints staggered. First layer applied full length vertically, second layer applied horizontally or vertically	—	—	—	5
	14-1.2[l,m]	$2'' \times 4''$ wood studs $16''$ on center with two layers $1/2''$ regular gypsum wallboard[e] applied vertically or horizontally each side[k], joints staggered. Nail base layer with 5d cooler[n] or wallboard[n] nails at $8''$ on center face layer with 8d cooler[n] or wallboard[n] nails at $8''$ on center.	—	—	—	$5^1/_2$
	14-1.3[l,m]	$2'' \times 4''$ wood studs $24''$ on center with $5/8''$ Type X gypsum wallboard[e] applied vertically or horizontally nailed with 6d cooler[n] or wallboard[n] nails at $7''$ on center with end joints on nailing members. Stagger joints each side.	—	—	—	$4^3/_4$
	14-1.4[l]	$2'' \times 4''$ fire-retardant-treated wood studs spaced $24''$ on center with one layer of $5/8''$ Type X gypsum wallboard[e] applied with face paper grain (long dimension) parallel to studs. Wallboard attached with 6d cooler[n] or wallboard[n] nails at $7''$ on center.	—	—	—	$4^3/_4^d$
	14-1.5[l,m]	$2'' \times 4''$ wood studs $16''$ on center with two layers $5/8''$ Type X gypsum wallboard[e] each side. Base layers applied vertically and nailed with 6d cooler[n] or wallboard[n] nails at $9''$ on center. Face layer applied vertically or horizontally and nailed with 8d cooler[n] or wallboard[n] nails at $7''$ on center. For nail-adhesive application, base layers are nailed $6''$ on center. Face layers applied with coating of approved wallboard adhesive and nailed $12''$ on center.	—	—	6	—
	14-1.6[l]	$2'' \times 3''$ fire-retardant-treated wood studs spaced $24''$ on center with one layer of $5/8''$ Type X gypsum wallboard[e] applied with face paper grain (long dimension) at right angles to studs. Wallboard attached with 6d cement-coated box nails spaced $7''$ on center.	—	—	—	$3^5/_8^d$
15. Exterior or interior walls	15-1.1[l,m]	Exterior surface with $3/4''$ drop siding over $1/2''$ gypsum sheathing on $2'' \times 4''$ wood studs at $16''$ on center, interior surface treatment as required for 1-hour-rated exterior or interior $2'' \times 4''$ wood stud partitions. Gypsum sheathing nailed with $1^3/_4''$ by No. 11 gage by $7/16''$ head galvanized nails at $8''$ on center. Siding nailed with 7d galvanized smooth box nails.	—	—	—	Varies
	15-1.2[l,m]	$2'' \times 4''$ wood studs $16''$ on center with metal lath and $3/4''$ cement plaster on each side. Lath attached with 6d common nails $7''$ on center driven to $1''$ minimum penetration and bent over. Plaster mix 1:4 for scratch coat and 1:5 for brown coat, by volume, cement to sand.	—	—	—	$5^3/_8$
	15-1.3[l,m]	$2'' \times 4''$ wood studs $16''$ on center with $7/8''$ cement plaster (measured from the face of studs) on the exterior surface with interior surface treatment as required for interior wood stud partitions in this table. Plaster mix 1:4 for scratch coat and 1:5 for brown coat, by volume, cement to sand.	—	—	—	Varies
	15-1.4	$3^5/_8''$ No. 16 gage noncombustible studs $16''$ on center with $7/8''$ cement plaster (measured from the face of the studs) on the exterior surface with interior surface treatment as required for interior, nonbearing, noncombustible stud partitions in this table. Plaster mix 1:4 for scratch coat and 1:5 for brown coat, by volume, cement to sand.	—	—	—	Varies[d]

(continued)

TABLE 720.1(2)—continued
RATED FIRE-RESISTANCE PERIODS FOR VARIOUS WALLS AND PARTITIONS [a,o,p]

MATERIAL	ITEM NUMBER	CONSTRUCTION	MINIMUM FINISHED THICKNESS FACE-TO-FACE[b] (inches)			
			4 hour	3 hour	2 hour	1 hour
15. Exterior or interior walls (continued)	15-1.5[m]	$2^1/_4'' \times 3^3/_4''$ clay face brick with cored holes over $^1/_2''$ gypsum sheathing on exterior surface of $2'' \times 4''$ wood studs at $16''$ on center and two layers $^5/_8''$ Type X gypsum wallboard[e] on interior surface. Sheathing placed horizontally or vertically with vertical joints over studs nailed $6''$ on center with $1^3/_4'' \times$ No. 11 gage by $^7/_{16}''$ head galvanized nails. Inner layer of wallboard placed horizontally or vertically and nailed $8''$ on center with 6d cooler[n] or wallboard[n] nails. Outer layer of wallboard placed horizontally or vertically and nailed $8''$ on center with 8d cooler[n] or wallboard[n] nails. All joints staggered with vertical joints over studs. Outer layer joints taped and finished with compound. Nail heads covered with joint compound. 0.035 inch (No. 20 galvanized sheet gage) corrugated galvanized steel wall ties $^3/_4''$ by $6^5/_8''$ attached to each stud with two 8d cooler[n] or wallboard[n] nails every sixth course of bricks.	—	—	10	—
	15-1.6[l, m]	$2'' \times 6''$ fire-retardant-treated wood studs $16''$ on center. Interior face has two layers of $^5/_8''$ Type X gypsum with the base layer placed vertically and attached with 6d box nails $12''$ on center. The face layer is placed horizontally and attached with 8d box nails $8''$ on center at joints and $12''$ on center elsewhere. The exterior face has a base layer of $^5/_8''$ Type X gypsum sheathing placed vertically with 6d box nails $8''$ on center at joints and $12''$ on center elsewhere. An approved building paper is next applied, followed by self-furred exterior lath attached with $2^1/_2''$, No. 12 gage galvanized roofing nails with a $^3/_8''$ diameter head and spaced $6''$ on center along each stud. Cement plaster consisting of a $^1/_2''$ brown coat is then applied. The scratch coat is mixed in the proportion of 1:3 by weight, cement to sand with 10 pounds of hydrated lime and 3 pounds of approved additives or admixtures per sack of cement. The brown coat is mixed in the proportion of 1:4 by weight, cement to sand with the same amounts of hydrated lime and approved additives or admixtures used in the scratch coat.	—	—	$8^1/_4$	—
	15-1.7[l, m]	$2'' \times 6''$ wood studs $16''$ on center. The exterior face has a layer of $^5/_8''$ Type X gypsum sheathing placed vertically with 6d box nails $8''$ on center at joints and $12''$ on center elsewhere. An approved building paper is next applied, followed by $1''$ by No. 18 gage self-furred exterior lath attached with 8d by $2^1/_2''$ long galvanized roofing nails spaced $6''$ on center along each stud. Cement plaster consisting of a $^1/_2''$ scratch coat, a bonding agent and a $^1/_2''$ brown coat and a finish coat is then applied. The scratch coat is mixed in the proportion of 1:3 by weight, cement to sand with 10 pounds of hydrated lime and 3 pounds of approved additives or admixtures per sack of cement. The brown coat is mixed in the proportion of 1:4 by weight, cement to sand with the same amounts of hydrated lime and approved additives or admixtures used in the scratch coat. The interior is covered with $^3/_8''$ gypsum lath with $1''$ hexagonal mesh of 0.035 inch (No. 20 B.W. gage) woven wire lath furred out $^5/_{16}''$ and $1''$ perlite or vermiculite gypsum plaster. Lath nailed with $1^1/_8''$ by No. 13 gage by $^{19}/_{64}''$ head plasterboard glued nails spaced $5''$ on center. Mesh attached by $1^3/_4''$ by No. 12 gage by $^3/_8''$ head nails with $^3/_8''$ furrings, spaced $8''$ on center. The plaster mix shall not exceed 100 pounds of gypsum to $2^1/_2$ cubic feet of aggregate.	—	—	$8^3/_8$	—
	15-1.8[l, m]	$2'' \times 6''$ wood studs $16''$ on center. The exterior face has a layer of $^5/_8''$ Type X gypsum sheathing placed vertically with 6d box nails $8''$ on center at joints and $12''$ on center elsewhere. An approved building paper is next applied, followed by $1^1/_2''$ by No. 17 gage self-furred exterior lath attached with 8d by $2^1/_2''$ long galvanized roofing nails spaced $6''$ on center along each stud. Cement plaster consisting of a $^1/_2''$ scratch coat, and a $^1/_2''$ brown coat is then applied. The plaster may be placed by machine. The scratch coat is mixed in the proportion of 1:4 by weight, plastic cement to sand. The brown coat is mixed in the proportion of 1:5 by weight, plastic cement to sand. The interior is covered with $^3/_8''$ gypsum lath with $1''$ hexagonal mesh of No. 20 gage woven wire lath furred out $^5/_{16}''$ and $1''$ perlite or vermiculite gypsum plaster. Lath nailed with $1^1/_8''$ by No. 13 gage by $^{19}/_{64}''$ head plasterboard glued nails spaced $5''$ on center. Mesh attached by $1^3/_4''$ by No. 12 gage by $^3/_8''$ head nails with $^3/_8''$ furrings, spaced $8''$ on center. The plaster mix shall not exceed 100 pounds of gypsum to $2^1/_2$ cubic feet of aggregate.	—	—	$8^3/_8$	—

(continued)

TABLE 720.1(2)—continued
RATED FIRE-RESISTANCE PERIODS FOR VARIOUS WALLS AND PARTITIONS [a,o,p]

MATERIAL	ITEM NUMBER	CONSTRUCTION	MINIMUM FINISHED THICKNESS FACE-TO-FACE[b] (inches)			
			4 hour	3 hour	2 hour	1 hour
15. Exterior or interior walls (continued)	15-1.9	4″ No. 18 gage, nonload-bearing metal studs, 16″ on center, with 1″ portland cement lime plaster [measured from the back side of the $^3/_4$-pound expanded metal lath] on the exterior surface. Interior surface to be covered with 1″ of gypsum plaster on $^3/_4$-pound expanded metal lath proportioned by weight—1:2 for scratch coat, 1:3 for brown, gypsum to sand. Lath on one side of the partition fastened to $^1/_4$″ diameter pencil rods supported by No. 20 gage metal clips, located 16" on center vertically, on each stud. 3″ thick mineral fiber insulating batts friction fitted between the studs.	—	—	$6^1/_2$[d]	—
	15-1.10	Steel studs 0.060" thick, 4″ deep or 6″ at 16″ or 24″ centers, with $^1/_2$″ Glass Fiber Reinforced Concrete (GFRC) on the exterior surface. GFRC is attached with flex anchors at 24″ on center, with 5″ leg welded to studs with two $^1/_2$″-long flare-bevel welds, and 4″ foot attached to the GFRC skin with $^5/_8$″ thick GFRC bonding pads that extend $2^1/_2$″ beyond the flex anchor foot on both sides. Interior surface to have two layers of $^1/_2$″ Type X gypsum wallboard.[e] The first layer of wallboard to be attached with 1″-long Type S buglehead screws spaced 24″ on center and the second layer is attached with $1^5/_8$″-long Type S screws spaced at 12″ on center. Cavity is to be filled with 5″ of 4 pcf (nominal) mineral fiber batts. GFRC has $1^1/_2$″ returns packed with mineral fiber and caulked on the exterior.	—	—	$6^1/_2$	—
	15-1.11	Steel studs 0.060″ thick, 4″ deep or 6″ at 16″ or 24″ centers, respectively, with $^1/_2$″ Glass Fiber Reinforced Concrete (GFRC) on the exterior surface. GFRC is attached with flex anchors at 24″ on center, with 5″ leg welded to studs with two $^1/_2$″-long flare-bevel welds, and 4″ foot attached to the GFRC skin with $^5/_8$″-thick GFRC bonding pads that extend $2^1/_2$″ beyond the flex anchor foot on both sides. Interior surface to have one layer of $^5/_8$″ Type X gypsum wallboard[e], attached with $1^1/_4$″-long Type S buglehead screws spaced 12″ on center. Cavity is to be filled with 5″ of 4 pcf (nominal) mineral fiber batts. GFRC has $1^1/_2$″ returns packed with mineral fiber and caulked on the exterior.	—	—	—	$6^1/_8$
	15-1.12[q]	2″ × 6″ wood studs at 16″ with double top plates, single bottom plate; interior and exterior sides covered with $^5/_8$″ Type X gypsum wallboard, 4′ wide, applied horizontally or vertically with vertical joints over studs, and fastened with $2^1/_4$″ Type S drywall screws, spaced 12″ on center. Cavity filled with $5^1/_2$″ mineral wool insulation.	—	—	—	$6^3/_4$
	15-1.13[q]	2″ × 6″ wood studs at 16″ with double top plates, single bottom plate; interior and exterior sides covered with $^5/_8$″ Type X gypsum wallboard, 4′ wide, applied horizontally or vertically with vertical joints over studs, and fastened with $2^1/_4$″ Type S drywall screws, spaced 7″ on center. Cavity to be filled with $5^1/_2$″ mineral wool insulation minimum 2.58 pcf (nominal).	—	—	—	$6^3/_4$
	15-1.14[q]	2″ × 4″ wood studs at 16″ with double top plates, single bottom plate; interior and exterior sides covered with $^5/_8$″ Type X gypsum wallboard and sheathing, respectively, 4′ wide, applied horizontally or vertically with vertical joints over studs, and fastened with $2^1/_4$″ Type S drywall screws, spaced 12″ on center. Cavity to be filled with $3^1/_2$″ mineral wool insulation.	—	—	—	$4^3/_4$
	15-1.15[q]	2″ × 4″ wood studs at 16″ with double top plates, single bottom plate; interior sides covered with $^5/_8$″ Type X gypsum wallboard, 4′ wide, applied horizontally unblocked, and fastened with $2^1/_4$″ Type S drywall screws, spaced 12″ on center, wallboard joints covered with paper tape and joint compound, fastener heads covered with joint compound. Exterior covered with $^3/_8$″ wood structural panels (oriented strand board), applied vertically, horizontal joints blocked and fastened with 6d common nails (bright)—12″ on center in the field, 6″ on center panel edges. Cavity to be filled with $3^1/_2$″ mineral wool insulation. Rating established for exposure from interior side only.	—	—	—	$4^1/_2$

(continued)

TABLE 720.1(2)

FIRE-RESISTANCE-RATED CONSTRUCTION

TABLE 720.1(2)—continued
RATED FIRE-RESISTANCE PERIODS FOR VARIOUS WALLS AND PARTITIONS [a,o,p]

MATERIAL	ITEM NUMBER	CONSTRUCTION	MINIMUM FINISHED THICKNESS FACE-TO-FACE[b] (inches)			
			4 hour	3 hour	2 hour	1 hour
15. Exterior or interior walls (continued)	15-1.16[q]	2″ × 6″ (51mm x 152 mm) wood studs at 16″ centers with double top plates, single bottom plate; interior side covered with $^5/_8$″ Type X gypsum wallboard, 4′ wide, applied horizontally or vertically with vertical joints over studs and fastened with $2^1/_4$″ Type S drywall screws, spaced 12″ on center, wallboard joints covered with paper tape and joint compound, fastener heads covered with joint compound, exterior side covered with $^7/_{16}$″; wood structural panels (oriented strand board) fastened with 6d common nails (bright) spaced 12″ on center in the field and 6″ on center along the panel edges. Cavity to be filled with $5^1/_2$″ mineral wool insulation. Rating established from the gypsum-covered side only.	—	—	—	$6^9/_{16}$
	15-1.17[q]	2″ × 6″ wood studs at 24″ centers with double top plates, single bottom plate; interior and exterior side covered with two layers of $^5/_8$″ Type X gypsum wallboard, 4′ wide, applied horizontally with vertical joints over studs. Base layer fastened with $2^1/_4$″ Type S drywall screws, spaced 24″ on center, and face layer fastened with Type S drywall screws, spaced 8″ on center, wallboard joints covered with paper tape and joint compound, fastened heads covered with joint compound. Cavity to be filled with $5^1/_2$″ mineral wool insulation.	—	—	$7^3/_4$	—
16. Exterior walls rated for fire resistance from the inside only in accordance with Section 704.5.	16-1.1[q]	2″ × 4″ wood studs at 16″ centers with double top plates, single bottom plate; interior side covered with $^5/_8$″ Type X gypsum wallboard, 4′ wide, applied horizontally unblocked, and fastened with $2^1/_4$″ Type S drywall screws, spaced 12″ on center, wallboard joints covered with paper tape and joint compound, fastener heads covered with joint compound. Exterior covered with $^3/_8$″ wood structural panels (oriented strand board), applied vertically, horizontal joints blocked and fastened with 6d common nails (bright) — 12″ on center in the field, and 6″ on center panel edges. Cavity to be filled with $3^1/_2$″ mineral wool insulation. Rating established for exposure from interior side only.	—	—	—	$4^1/_2$

For SI: 1 inch = 25.4 mm, 1 square inch = 645.2 mm^2, 1 cubic foot = 0.0283 m^3.

a. Staples with equivalent holding power and penetration shall be permitted to be used as alternate fasteners to nails for attachment to wood framing.

b. Thickness shown for brick and clay tile are nominal thicknesses unless plastered, in which case thicknesses are net. Thickness shown for concrete masonry and clay masonry is equivalent thickness defined in Section 721.3.1 for concrete masonry and Section 721.4.1.1 for clay masonry. Where all cells are solid grouted or filled with silicone-treated perlite loose-fill insulation; vermiculite loose-fill insulation; or expanded clay, shale or slate lightweight aggregate, the equivalent thickness shall be the thickness of the block or brick using specified dimensions as defined in Chapter 21. Equivalent thickness may also include the thickness of applied plaster and lath or gypsum wallboard, where specified.

c. For units in which the net cross-sectional area of cored brick in any plane parallel to the surface containing the cores is at least 75 percent of the gross cross-sectional area measured in the same plane.

d. Shall be used for nonbearing purposes only.

e. For all of the construction with gypsum wallboard described in this table, gypsum base for veneer plaster of the same size, thickness and core type shall be permitted to be substituted for gypsum wallboard, provided attachment is identical to that specified for the wallboard, and the joints on the face layer are reinforced and the entire surface is covered with a minimum of $^1/_{16}$-inch gypsum veneer plaster.

f. The fire-resistance time period for concrete masonry units meeting the equivalent thicknesses required for a 2-hour fire-resistance rating in Item 3, and having a thickness of not less than $7^5/_8$ inches is 4 hours when cores which are not grouted are filled with silicone-treated perlite loose-fill insulation; vermiculite loose-fill insulation; or expanded clay, shale or slate lightweight aggregate, sand or slag having a maximum particle size of $^3/_8$ inch.

g. The fire-resistance rating of concrete masonry units composed of a combination of aggregate types or where plaster is applied directly to the concrete masonry shall be determined in accordance with ACI 216.1/TMS 216. Lightweight aggregates shall have a maximum combined density of 65 pounds per cubic foot.

h. See also Note b. The equivalent thickness shall be permitted to include the thickness of cement plaster or 1.5 times the thickness of gypsum plaster applied in accordance with the requirements of Chapter 25.

i. Concrete walls shall be reinforced with horizontal and vertical temperature reinforcement as required by Chapter 19.

j. Studs are welded truss wire studs with 0.18 inch (No. 7 B.W. gage) flange wire and 0.18 inch (No. 7 B.W. gage) truss wires.

k. Nailable metal studs consist of two channel studs spot welded back to back with a crimped web forming a nailing groove.

l. Wood structural panels shall be permitted to be installed between the fire protection and the wood studs on either the interior or exterior side of the wood frame assemblies in this table, provided the length of the fasteners used to attach the fire protection are increased by an amount at least equal to the thickness of the wood structural panel.

m. The design stress of studs shall be reduced to 78 percent of allowable F'_c with the maximum not greater than 78 percent of the calculated stress with studs having a slenderness ratio l_e/d of 33.

n. For properties of cooler or wallboard nails, see ASTM C 514, ASTM C 547 or ASTM F 1667.

o. Generic fire-resistance ratings (those not designated as PROPRIETARY* in the listing) in the GA 600 shall be accepted as if herein listed.

p. NCMA TEK 5-8, shall be permitted for the design of fire walls.

q. The design stress of studs shall be equal to a maximum of 100 percent of the allowable F'_c calculated in accordance with Section 2306.

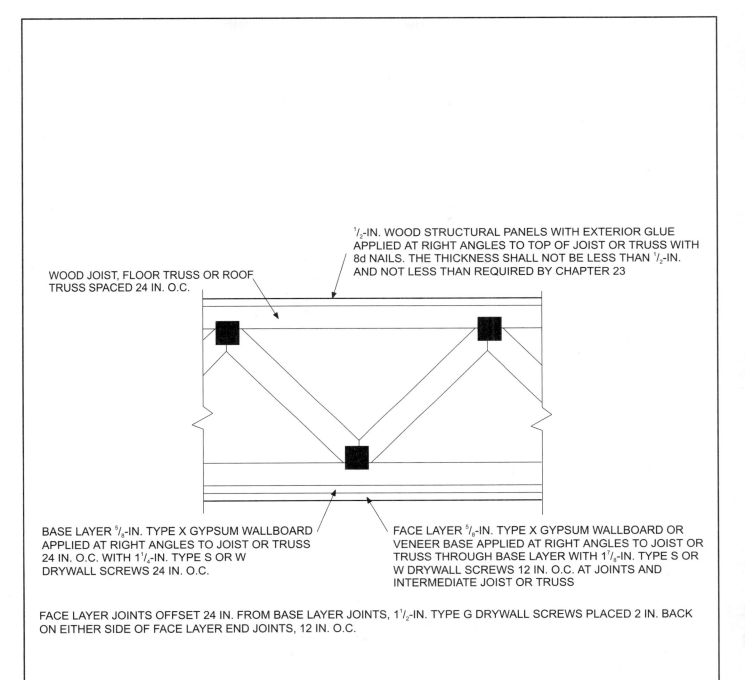

$^1/_2$-IN. WOOD STRUCTURAL PANELS WITH EXTERIOR GLUE APPLIED AT RIGHT ANGLES TO TOP OF JOIST OR TRUSS WITH 8d NAILS. THE THICKNESS SHALL NOT BE LESS THAN $^1/_2$-IN. AND NOT LESS THAN REQUIRED BY CHAPTER 23

WOOD JOIST, FLOOR TRUSS OR ROOF TRUSS SPACED 24 IN. O.C.

BASE LAYER $^5/_8$-IN. TYPE X GYPSUM WALLBOARD APPLIED AT RIGHT ANGLES TO JOIST OR TRUSS 24 IN. O.C. WITH $1^1/_4$-IN. TYPE S OR W DRYWALL SCREWS 24 IN. O.C.

FACE LAYER $^5/_8$-IN. TYPE X GYPSUM WALLBOARD OR VENEER BASE APPLIED AT RIGHT ANGLES TO JOIST OR TRUSS THROUGH BASE LAYER WITH $1^7/_8$-IN. TYPE S OR W DRYWALL SCREWS 12 IN. O.C. AT JOINTS AND INTERMEDIATE JOIST OR TRUSS

FACE LAYER JOINTS OFFSET 24 IN. FROM BASE LAYER JOINTS, $1^1/_2$-IN. TYPE G DRYWALL SCREWS PLACED 2 IN. BACK ON EITHER SIDE OF FACE LAYER END JOINTS, 12 IN. O.C.

For SI: 1 inch = 25.4 mm.

Figure 720.1(3)
WOOD FLOOR TRUSS

TABLE 720.1(3)
MINIMUM PROTECTION FOR FLOOR AND ROOF SYSTEMS[a,q]

FLOOR OR ROOF CONSTRUCTION	ITEM NUMBER	CEILING CONSTRUCTION	THICKNESS OF FLOOR OR ROOF SLAB (inches)				MINIMUM THICKNESS OF CEILING (inches)			
			4 hour	3 hour	2 hour	1 hour	4 hour	3 hour	2 hour	1 hour
1. Siliceous aggregate concrete	1-1.1	Slab (no ceiling required). Minimum cover over nonprestressed reinforcement shall not be less than $3/4$ inch.[b]	7.0	6.2	5.0	3.5	—	—	—	—
2. Carbonate aggregate concrete	2-1.1		6.6	5.7	4.6	3.2	—	—	—	—
3. Sand-lightweight concrete	3-1.1		5.4	4.6	3.8	2.7	—	—	—	—
4. Lightweight concrete	4-1.1		5.1	4.4	3.6	2.5	—	—	—	—
5. Reinforced concrete	5-1.1	Slab with suspended ceiling of vermiculite gypsum plaster over metal lath attached to $3/4''$ cold-rolled channels spaced 12" on center. Ceiling located 6" minimum below joists.	3	2	—	—	1	$3/4$	—	—
	5-2.1	$3/8''$ Type X gypsum wallboard[c] attached to 0.018 inch (No. 25 carbon sheet steel gage) by $7/8''$ deep by $2^5/8''$ hat-shaped galvanized steel channels with 1"-long No. 6 screws. The channels are spaced 24" on center, span 35" and are supported along their length at 35" intervals by 0.033-inch (No. 21 galvanized sheet gage) galvanized steel flat strap hangers having formed edges that engage the lips of the channel. The strap hangers are attached to the side of the concrete joists with $5/32''$ by $1^1/4''$ long power-driven fasteners. The wallboard is installed with the long dimension perpendicular to the channels. All end joints occur on channels and supplementary channels are installed parallel to the main channels, 12" each side, at end joint occurrences. The finished ceiling is located approximately 12" below the soffit of the floor slab.	—	—	$2^1/2$	—	—	—	$5/8$	—
6. Steel joists constructed with a poured reinforced concrete slab on metal lath forms or steel form units[d, e]	6-1.1	Gypsum plaster on metal lath attached to the bottom cord with single No. 16 gage or doubled No. 18 gage wire ties spaced 6" on center. Plaster mixed 1:2 for scratch coat, 1:3 for brown coat, by weight, gypsum-to-sand aggregate for 2-hour system. For 3-hour system plaster is neat.	—	—	$2^1/2$	$2^1/4$	—	—	$3/4$	$5/8$
	6-2.1	Vermiculite gypsum plaster on metal lath attached to the bottom chord with single No.16 gage or doubled 0.049-inch (No. 18 B.W. gage) wire ties 6" on center.	—	2	—	—	—	$5/8$	—	—
	6-3.1	Cement plaster over metal lath attached to the bottom chord of joists with single No. 16 gage or doubled 0.049-inch (No. 18 B.W. gage) wire ties spaced 6" on center. Plaster mixed 1:2 for scratch coat, 1:3 for brown coat for 1-hour system and 1:1 for scratch coat, 1:1 $1/2$ for brown coat for 2-hour system, by weight, cement to sand.	—	—	—	2	—	—	—	$5/8$[f]
	6-4.1	Ceiling of $5/8''$ Type X wallboard[c] attached to $7/8''$ deep by $2^5/8''$ by 0.021 inch (No. 25 carbon sheet steel gage) hat-shaped furring channels 12" on center with 1" long No. 6 wallboard screws at 8" on center. Channels wire tied to bottom chord of joists with doubled 0.049 inch (No. 18 B.W. gage) wire or suspended below joists on wire hangers.[g]	—	—	$2^1/2$	—	—	—	$5/8$	—
	6-5.1	Wood-fibered gypsum plaster mixed 1:1 by weight gypsum to sand aggregate applied over metal lath. Lath tied 6" on center to $3/4''$ channels spaced $13^1/2''$ on center. Channels secured to joists at each intersection with two strands of 0.049 inch (No. 18 B.W. gage) galvanized wire.	—	—	$2^1/2$	—	—	—	$3/4$	—

(continued)

TABLE 720.1(3)—continued
MINIMUM PROTECTION FOR FLOOR AND ROOF SYSTEMS[a,q]

FLOOR OR ROOF CONSTRUCTION	ITEM NUMBER	CEILING CONSTRUCTION	THICKNESS OF FLOOR OR ROOF SLAB (inches)				MINIMUM THICKNESS OF CEILING (inches)			
			4 hour	3 hour	2 hour	1 hour	4 hour	3 hour	2 hour	1 hour
7. Reinforced concrete slabs and joists with hollow clay tile fillers laid end to end in rows $2^1/_2$″ or more apart; reinforcement placed between rows and concrete cast around and over tile.	7-1.1	$^5/_8$″ gypsum plaster on bottom of floor or roof construction.	—	—	8[h]	—	—	—	$^5/_8$	—
	7-1.2	None	—	—	—	$5^1/_2$[i]	—	—	—	—
8. Steel joists constructed with a reinforced concrete slab on top poured on a $^1/_2$″ deep steel deck.[e]	8-1.1	Vermiculite gypsum plaster on metal lath attached to $^3/_4$″ cold-rolled channels with 0.049″ (No. 18 B.W. gage) wire ties spaced 6″ on center.	$2^1/_2$[j]	—	—	—	$^3/_4$	—	—	—
9. 3″ deep cellular steel deck with concrete slab on top. Slab thickness measured to top.	9-1.1	Suspended ceiling of vermiculite gypsum plaster base coat and vermiculite acoustical plaster on metal lath attached at 6″ intervals to $^3/_4$″ cold-rolled channels spaced 12″ on center and secured to $1^1/_2$″ cold-rolled channels spaced 36″ on center with 0.065″ (No. 16 B.W. gage) wire. $1^1/_2$″ channels supported by No. 8 gage wire hangers at 36″ on center. Beams within envelope and with a $2^1/_2$″airspace between beam soffit and lath have a 4-hour rating.	$2^1/_2$	—	—	—	$1^1/_8$[k]	—	—	—
10. $1^1/_2$″-deep steel roof deck on steel framing. Insulation board, 30 pcf density, composed of wood fibers with cement binders of thickness shown bonded to deck with unified asphalt adhesive. Covered with a Class A or B roof covering.	10-1.1	Ceiling of gypsum plaster on metal lath. Lath attached to $^3/_4$″ furring channels with 0.049″ (No. 18 B.W. gage) wire ties spaced 6″ on center. $^3/_4$″ channel saddle tied to 2″ channels with doubled 0.065″ (No. 16 B.W. gage) wire ties. 2″ channels spaced 36″ on center suspended 2″ below steel framing and saddle-tied with 0.165″ (No. 8 B.W. gage) wire. Plaster mixed 1:2 by weight, gypsum-to-sand aggregate.	—	—	$1^7/_8$	1	—	—	$^3/_4$[l]	$^3/_4$[l]
11. $1^1/_2$″-deep steel roof deck on steel-framing wood fiber insulation board, 17.5 pcf density on top applied over a 15-lb asphalt-saturated felt. Class A or B roof covering.	11-1.1	Ceiling of gypsum plaster on metal lath. Lath attached to $^3/_4$″ furring channels with 0.049″ (No. 18 B.W. gage) wire ties spaced 6″ on center. $^3/_4$″ channels saddle tied to 2″ channels with doubled 0.065″ (No. 16 B.W. gage) wire ties. 2″ channels spaced 36″ on center suspended 2″ below steel framing and saddle tied with 0.165″ (No. 8 B.W. gage) wire. Plaster mixed 1:2 for scratch coat and 1:3 for brown coat, by weight, gypsum-to-sand aggregate for 1-hour system. For 2-hour system, plaster mix is 1:2 by weight, gypsum-to-sand aggregate.	—	—	$1^1/_2$	1	—	—	$^7/_8$[g]	$^3/_4$[l]

(continued)

TABLE 720.1(3)   FIRE-RESISTANCE-RATED CONSTRUCTION

TABLE 720.1(3)—continued
MINIMUM PROTECTION FOR FLOOR AND ROOF SYSTEMS[a,q]

FLOOR OR ROOF CONSTRUCTION	ITEM NUMBER	CEILING CONSTRUCTION	THICKNESS OF FLOOR OR ROOF SLAB (inches)				MINIMUM THICKNESS OF CEILING (inches)			
			4 hour	3 hour	2 hour	1 hour	4 hour	3 hour	2 hour	1 hour
12. $1^1/_2$″ deep steel roof deck on steel-framing insulation of rigid board consisting of expanded perlite and fibers impregnated with integral asphalt waterproofing; density 9 to 12 pcf secured to metal roof deck by $1/_2$″ wide ribbons of waterproof, cold-process liquid adhesive spaced 6″ apart. Steel joist or light steel construction with metal roof deck, insulation, and Class A or B built-up roof covering.[e]	12-1.1	Gypsum-vermiculite plaster on metal lath wire tied at 6″ intervals to $3/_4$″ furring channels spaced 12″ on center and wire tied to 2″ runner channels spaced 32″ on center. Runners wire tied to bottom chord of steel joists.	—	—	1	—	—	—	$7/_8$	—
13. Double wood floor over wood joists spaced 16″ on center.[m,n]	13-1.1	Gypsum plaster over $3/_8$″ Type X gypsum lath. Lath initially applied with not less than four $1^1/_8$″ by No. 13 gage by $19/_{64}$″ head plasterboard blued nails per bearing. Continuous stripping over lath along all joist lines. Stripping consists of 3″ wide strips of metal lath attached by $1^1/_2$″ by No. 11 gage by $1/_2$″ head roofing nails spaced 6″ on center. Alternate stripping consists of 3″ wide 0.049″ diameter wire stripping weighing 1 pound per square yard and attached by No.16 gage by $1^1/_2$″ by $3/_4$″ crown width staples, spaced 4″ on center. Where alternate stripping is used, the lath nailing may consist of two nails at each end and one nail at each intermediate bearing. Plaster mixed 1:2 by weight, gypsum-to-sand aggregate.	—	—	—	—	—	—	—	$7/_8$
	13-1.2	Cement or gypsum plaster on metal lath. Lath fastened with $1^1/_2$″ by No. 11 gage by $7/_{16}$″ head barbed shank roofing nails spaced 5″ on center. Plaster mixed 1:2 for scratch coat and 1:3 for brown coat, by weight, cement to sand aggregate.	—	—	—	—	—	—	—	$5/_8$
	13-1.3	Perlite or vermiculite gypsum plaster on metal lath secured to joists with $1^1/_2$″ by No. 11 gage by $7/_{16}$″ head barbed shank roofing nails spaced 5″ on center.	—	—	—	—	—	—	—	$5/_8$
	13-1.4	$1/_2$″ Type X gypsum wallboard[c] nailed to joists with 5d cooler[o] or wallboard[o] nails at 6″ on center. End joints of wallboard centered on joists.	—	—	—	—	—	—	—	$1/_2$
14. Plywood stressed skin panels consisting of $5/_8$″-thick interior C-D (exterior glue) top stressed skin on 2″ × 6″ nominal (minimum) stringers. Adjacent panel edges joined with 8d common wire nails spaced 6″ on center. Stringers spaced 12″ maximum on center.	14-1.1	$1/_2$″-thick wood fiberboard weighing 15 to 18 pounds per cubic foot installed with long dimension parallel to stringers or $3/_8$″ C-D (exterior glue) plywood glued and/or nailed to stringers. Nailing to be with 5d cooler[o] or wallboard[o] nails at 12″ on center. Second layer of $1/_2$″ Type X gypsum wallboard[c] applied with long dimension perpendicular to joists and attached with 8d cooler[o] or wallboard[o] nails at 6″ on center at end joints and 8″ on center elsewhere. Wallboard joints staggered with respect to fiberboard joints.	—	—	—	—	—	—	—	1

(continued)

TABLE 720.1(3)—continued
MINIMUM PROTECTION FOR FLOOR AND ROOF SYSTEMS[a,q]

FLOOR OR ROOF CONSTRUCTION	ITEM NUMBER	CEILING CONSTRUCTION	THICKNESS OF FLOOR OR ROOF SLAB (inches)				MINIMUM THICKNESS OF CEILING (inches)			
			4 hour	3 hour	2 hour	1 hour	4 hour	3 hour	2 hour	1 hour
15. Vermiculite concrete slab proportioned 1:4 (portland cement to vermiculite aggregate) on a $1^1/_2$″-deep steel deck supported on individually protected steel framing. Maximum span of deck 6′-10″ where deck is less than 0.019 inch (No. 26 carbon steel sheet gage) or greater. Slab reinforced with 4″ × 8″ 0.109/0.083″ (No. $^{12}/_{14}$ B.W. gage) welded wire mesh.	15-1.1	None	—	—	—	3[j]	—	—	—	—
16. Perlite concrete slab proportioned 1:6 (portland cement to perlite aggregate) on a $1^1/_4$″-deep steel deck supported on individually protected steel framing. Slab reinforced with 4″ × 8″ 0.109/0.083″ (No. $^{12}/_{14}$ B.W. gage) welded wire mesh.	16-1.1	None	—	—	—	$3^1/_2$[j]	—	—	—	—
17. Perlite concrete slab proportioned 1:6 (portland cement to perlite aggregate) on a $^9/_{16}$″-deep steel deck supported by steel joists 4′ on center. Class A or B roof covering on top.	17-1.1	Perlite gypsum plaster on metal lath wire tied to $^3/_4$″ furring channels attached with 0.065-inch (No. 16 B.W. gage) wire ties to lower chord of joists.	—	2[p]	2[p]	—	—	$^7/_8$	$^3/_4$	—
18. Perlite concrete slab proportioned 1:6 (portland cement to perlite aggregate) on $1^1/_4$″-deep steel deck supported on individually protected steel framing. Maximum span of deck 6′-10″ where deck is less than 0.019″ (No. 26 carbon sheet steel gage) and 8′-0″ where deck is 0.019″ (No. 26 carbon sheet steel gage) or greater. Slab reinforced with 0.042″ (No. 19 B.W. gage) hexagonal wire mesh. Class A or B roof covering on top.	18-1.1	None	—	$2^1/_4$[p]	$2^1/_4$[p]	—	—	—	—	—

(continued)

TABLE 720.1(3) FIRE-RESISTANCE-RATED CONSTRUCTION

TABLE 720.1(3)—continued
MINIMUM PROTECTION FOR FLOOR AND ROOF SYSTEMS[a,q]

FLOOR OR ROOF CONSTRUCTION	ITEM NUMBER	CEILING CONSTRUCTION	THICKNESS OF FLOOR OR ROOF SLAB (inches)				MINIMUM THICKNESS OF CEILING (inches)			
			4 hour	3 hour	2 hour	1 hour	4 hour	3 hour	2 hour	1 hour
19. Floor and beam construction consisting of 3″-deep cellular steel floor unit mounted on steel members with 1:4 (proportion of portland cement to perlite aggregate) perlite-concrete floor slab on top.	19-1.1	Suspended envelope ceiling of perlite gypsum plaster on metal lath attached to $^3/_4$″ cold-rolled channels, secured to $1^1/_2$″ cold-rolled channels spaced 42″ on center supported by 0.203 inch (No. 6 B.W. gage) wire 36″ on center. Beams in envelope with 3″ minimum airspace between beam soffit and lath have a 4-hour rating.	2[p]	—	—	—	1[l]	—	—	—
20. Perlite concrete proportioned 1:6 (portland cement to perlite aggregate) poured to $^1/_8$-inch thickness above top of corrugations of $1^5/_{16}$″-deep galvanized steel deck maximum span 8′-0″ for 0.024-inch (No. 24 galvanized sheet gage) or 6′ 0″ for 0.019-inch (No. 26 galvanized sheet gage) with deck supported by individually protected steel framing. Approved polystyrene foam plastic insulation board having a flame spread not exceeding 75 (1″ to 4″ thickness) with vent holes that approximate 3 percent of the board surface area placed on top of perlite slurry. A 2′ by 4′ insulation board contains six $2^3/_4$″ diameter holes. Board covered with $2^1/_4$″ minimum perlite concrete slab.	20-1.1	None	—	—	Varies	—	—	—	—	—

(continued)

TABLE 720.1(3)—continued
MINIMUM PROTECTION FOR FLOOR AND ROOF SYSTEMS[a,q]

FLOOR OR ROOF CONSTRUCTION	ITEM NUMBER	CEILING CONSTRUCTION	THICKNESS OF FLOOR OR ROOF SLAB (inches)				MINIMUM THICKNESS OF CEILING (inches)			
			4 hour	3 hour	2 hour	1 hour	4 hour	3 hour	2 hour	1 hour
21. Slab reinforced with mesh consisting of 0.042 inch (No. 19 B.W. gage) galvanized steel wire twisted together to form 2″ hexagons with straight 0.065 inch (No. 16 B.W. gage) galvanized steel wire woven into mesh and spaced 3″. Alternate slab reinforcement shall be permitted to consist of 4″ × 8″, 0.109/0.238-inch (No. 12/4 B.W. gage), or 2″ × 2″, 0.083/0.083-inch (No. 14/14 B.W. gage) welded wire fabric. Class A or B roof covering on top.	21-1.1	None	—	—	Varies	—	—	—	—	—
22. Wood joists, floor trusses and flat or pitched roof trusses spaced a maximum 24″ o.c. with $^1/_2$″ wood structural panels with exterior glue applied at right angles to top of joist or top chord of trusses with 8d nails. The wood structural panel thickness shall not be less than nominal $^1/_2$″ less than required by Chapter 23.	22-1.1	Base layer $^5/_8$″ Type X gypsum wallboard applied at right angles to joist or truss 24″ o.c. with $1^1/_4$″ Type S or Type W drywall screws 24″ o.c. Face layer $^5/_8$″ Type X gypsum wallboard or veneer base applied at right angles to joist or truss through base layer with $1^7/_8$″ Type S or Type W drywall screws 12″ o.c. at joints and intermediate joist or truss. Face layer Type G drywall screws placed 2″ back on either side of face layer end joints, 12″ o.c.	—	—	—	Varies	—	—	—	$1^1/_4$
23. Steel joists, floor trusses and flat or pitched roof trusses spaced a maximum 24″ o.c. with $^1/_2$″ wood structural panels with exterior glue applied at right angles to top of joist or top chord of trusses with No. 8 screws. The wood structural panel thickness shall not be less than nominal $^1/_2$″ nor less than required by Chapter 22.	23-1.1	Base layer $^5/_8$″ Type X gypsum board applied at right angles to steel framing 24″ on center with 1″ Type S drywall screws spaced 24″ on center. Face layer $^5/_8$″ Type X gypsum board applied at right angles to steel framing attached through base layer with $1^5/_8$″ Type S drywall screws 12″ on center at end joints and intermediate joints and $1^1/_2$″ Type G drywall screws 12 inches on center placed 2″ back on either side of face layer end joints. Joints of the face layer are offset 24″ from the joints of the base layer.	—	—	—	Varies	—	—	—	$1^1/_4$
24. Wood I-joist (minimum joist depth $9^1/_4$″ with a minimum flange depth of $1^5/_{16}$″ and a minimum flange cross-sectional area of 2.3 square inches) at 24″ o.c. spacing with 1 × 4 (nominal) wood furring strip spacer applied parallel to and covering the bottom of the bottom flange of each member, tacked in place. 2″ mineral fiber insulation, 3.5 pcf (nominal) installed adjacent to the bottom flange of the I-joist and supported by the 1 × 4 furring strip spacer.	24-1.1	$^1/_2$″ deep single leg resilient channel 16″ on center (channels doubled at wallboard end joints), placed perpendicular to the furring strip and joist and attached to each joist by $1^7/_8$″ Type S drywall screws. $^5/_8$″ Type C gypsum wallboard applied perpendicular to the channel with end joints staggered at least 4′ and fastened with $1^1/_8$″ Type S drywall screws spaced 7″ on center. Wallboard joints to be taped and covered with joint compound.	—	—	—	Varies	—	—	—	—

(continued)

Table 720.1(3) Notes.

For SI: 1 inch = 25.4 mm, 1 foot = 304.8 mm, 1 pound = 0.454 kg, 1 cubic foot = 0.0283 m^3,
              1 pound per square inch = 6.895 kPa = 1 pound per lineal foot = 1.4882 kg/m.

a. Staples with equivalent holding power and penetration shall be permitted to be used as alternate fasteners to nails for attachment to wood framing.

b. When the slab is in an unrestrained condition, minimum reinforcement cover shall not be less than $1^5/_8$ inches for 4-hour (siliceous aggregate only); $1^1/_4$ inches for 4- and 3-hour; 1 inch for 2-hour (siliceous aggregate only); and $^3/_4$ inch for all other restrained and unrestrained conditions.

c. For all of the construction with gypsum wallboard described in this table, gypsum base for veneer plaster of the same size, thickness and core type shall be permitted to be substituted for gypsum wallboard, provided attachment is identical to that specified for the wallboard, and the joints on the face layer are reinforced and the entire surface is covered with a minimum of $^1/_{16}$-inch gypsum veneer plaster.

d. Slab thickness over steel joists measured at the joists for metal lath form and at the top of the form for steel form units.

e. (a) The maximum allowable stress level for H-Series joists shall not exceed 22,000 psi.

   (b) The allowable stress for K-Series joists shall not exceed 26,000 psi, the nominal depth of such joist shall not be less than 10 inches and the nominal joist weight shall not be less than 5 pounds per lineal foot.

f. Cement plaster with 15 pounds of hydrated lime and 3 pounds of approved additives or admixtures per bag of cement.

g. Gypsum wallboard ceilings attached to steel framing shall be permitted to be suspended with $1^1/_2$-inch cold-formed carrying channels spaced 48 inches on center, which are suspended with No. 8 SWG galvanized wire hangers spaced 48 inches on center. Cross-furring channels are tied to the carrying channels with No. 18 SWG galvanized wire hangers spaced 48 inches on center. Cross-furring channels are tied to the carrying channels with No. 18 SWG galvanized wire (double strand) and spaced as required for direct attachment to the framing. This alternative is also applicable to those steel framing assemblies recognized under Note q.

h. Six-inch hollow clay tile with 2-inch concrete slab above.

i. Four-inch hollow clay tile with $1^1/_2$-inch concrete slab above.

j. Thickness measured to bottom of steel form units.

k. Five-eighths inch of vermiculite gypsum plaster plus $^1/_2$ inch of approved vermiculite acoustical plastic.

l. Furring channels spaced 12 inches on center.

m. Double wood floor shall be permitted to be either of the following:

   (a) Subfloor of 1-inch nominal boarding, a layer of asbestos paper weighing not less than 14 pounds per 100 square feet and a layer of 1-inch nominal tongue-and-groove finished flooring; or

   (b) Subfloor of 1-inch nominal tongue-and-groove boarding or $^{15}/_{32}$-inch wood structural panels with exterior glue and a layer of 1-inch nominal tongue-and-groove finished flooring or $^{19}/_{32}$-inch wood structural panel finish flooring or a layer of Type I Grade M-1 particleboard not less than $^5/_8$-inch thick.

n. The ceiling shall be permitted to be omitted over unusable space, and flooring shall be permitted to be omitted where unusable space occurs above.

o. For properties of cooler or wallboard nails, see ASTM C 514, ASTM C 547 or ASTM F 1667.

p. Thickness measured on top of steel deck unit.

q. Generic fire-resistance ratings (those not designated as PROPRIETARY* in the listing) in the GA 600 shall be accepted as if herein listed.

SECTION 721
CALCULATED FIRE RESISTANCE

❖ Due to the prescriptive nature of Section 721, this section contains explanatory material, examples and reference materials related to calculating fire-resistance ratings of materials and assemblies. As such, commentary is provided only where necessary to provide the basis for a provision or to illustrate the application of the prescriptive requirement.

721.1 General. The provisions of this section contain procedures by which the fire resistance of specific materials or combinations of materials is established by calculations. These procedures apply only to the information contained in this section and shall not be otherwise used. The calculated fire resistance of concrete, concrete masonry, and clay masonry assemblies shall be permitted in accordance with ACI 216.1/TMS 0216.1. The calculated fire resistance of steel assemblies shall be permitted in accordance with Chapter 5 of ASCE/SFPE 29.

❖ The provisions of this section contain procedures by which the fire resistance of specific materials or combinations of materials is established by calculations. These procedures apply only to the information contained in this section and are not to be otherwise used. The calculated fire resistance of concrete, concrete masonry and clay masonry assemblies are permitted in accordance with

ACI 216.1/TMS 0216.1 (see commentary, Section 703.3).

721.1.1 Definitions. The following words and terms shall, for the purposes of this chapter and as used elsewhere in this code, have the meanings shown herein.

CERAMIC FIBER BLANKET. A mineral wool insulation material made of alumina-silica fibers and weighing 4 to 10 pounds per cubic foot (pcf) (64 to 160 kg/m^3).

❖ This form of insulation is used in conjunction with the provisions of Section 721.2.1.3.1 where joints in precast walls must be insulated (protected) in order to maintain the fire-resistance integrity of the precast wall panel (see Code Figure 721.2.1.3.1).

CONCRETE, CARBONATE AGGREGATE. Concrete made with aggregates consisting mainly of calcium or magnesium carbonate, such as limestone or dolomite, and containing 40 percent or less quartz, chert, or flint.

❖ Concrete made with this type of aggregate is listed in Tables 720.1(1) (Items 1, 2, 3, 4, 5, 6 and 7) and 720.1(2) (Item 4) with prescriptive details regarding concrete fire-resistance ratings. This type of aggregate is also indicated for concrete assemblies that require calculations to determine the fire-resistance rating in accordance with Section 721.2.1 (i.e., Table 721.2.1.1) (see also Section 702.1).

CONCRETE, CELLULAR. A lightweight insulating concrete made by mixing a preformed foam with portland cement slurry and having a dry unit weight of approximately 30 pcf (480 kg/m^3).

❖ Concrete made with this type of aggregate is listed in Table 720.1(3) (Items 9 and 19) with prescriptive details regarding concrete fire-resistance ratings. This type of aggregate is also indicated for concrete assemblies that require calculations to determine the fire-resistance rating in accordance with Section 721.2.1 [i.e., Table 721.2.1.2(1), Note a].

CONCRETE, LIGHTWEIGHT AGGREGATE. Concrete made with aggregates of expanded clay, shale, slag or slate or sintered fly ash or any natural lightweight aggregate meeting ASTM C 330 and possessing equivalent fire-resistance properties and weighing 85 to 115 pcf (1360 to 1840 kg/m^3).

❖ Concrete made with this type of aggregate is listed in Tables 720.1(1) (Items 1, 2, 3, 4, 5, 6 and 7), 720.1(2) (Item 4) and 720.1(3) (Item 4) with prescriptive details regarding concrete fire-resistance ratings. This type of aggregate is also indicated for concrete assemblies that require calculations to determine the fire-resistance rating in accordance with Section 721.2.1 (i.e., Table 721.2.1.1) (see also Section 702.1).

CONCRETE, PERLITE. A lightweight insulating concrete having a dry unit weight of approximately 30 pcf (480 kg/m^3) made with perlite concrete aggregate. Perlite aggregate is produced from a volcanic rock which, when heated, expands to form a glass-like material of cellular structure.

❖ Concrete made with this type of aggregate is listed in Tables 720.1(1) (Item 1), 720.1(2) (Item 1) and 720.1(3) (Items 7, 16, 17, 18, 19 and 20) with prescriptive details regarding concrete fire-resistance ratings. This type of aggregate is also indicated for concrete assemblies that require calculations to determine the fire-resistance rating in accordance with Section 721.2.1 (i.e., Section 721.2.1.1.2).

CONCRETE, SAND-LIGHTWEIGHT. Concrete made with a combination of expanded clay, shale, slag, slate, sintered fly ash, or any natural lightweight aggregate meeting ASTM C 330 and possessing equivalent fire-resistance properties and natural sand. Its unit weight is generally between 105 and 120 pcf (1680 and 1920 kg/m^3).

❖ Concrete made with this type of aggregate is listed in Tables 720.1(1) (Items 1, 2, 3, 4, 5, 6 and 7), 720.1(2) (Item 4) and 720.1(3) (Item 3) with prescriptive details regarding concrete fire-resistance ratings. This type of aggregate is also indicated for concrete assemblies that require calculations to determine the fire-resistance rating in accordance with Section 721.2.1 (i.e., Table 721.2.1.1) (see also Section 702.1).

CONCRETE, SILICEOUS AGGREGATE. Concrete made with normal-weight aggregates consisting mainly of silica or compounds other than calcium or magnesium carbonate, which contains more than 40-percent quartz, chert, or flint.

❖ Concrete made with this type of aggregate is listed in Tables 720.1(1) (Items 1, 2, 4, 5, 6 and 7), 720.1(2) (Items 3 and 4) and 720.1(3) (Item 1) with prescriptive details regarding concrete fire-resistance ratings. This type of aggregate is also indicated for concrete assemblies that require calculations to determine the fire-resistance rating in accordance with Section 721.2.1 (i.e., Table 721.2.1.1) (see also Section 702.1).

CONCRETE, VERMICULITE. A lightweight insulating concrete made with vermiculite concrete aggregate which is laminated micaceous material produced by expanding the ore at high temperatures. When added to a portland cement slurry the resulting concrete has a dry unit weight of approximately 30 pcf (480 kg/m^3).

❖ Concrete made with this type of aggregate is listed in Tables 720.1(1) (Item 1), 720.1(2) (Item 1) and 720.1(3) (Item 15) with prescriptive details regarding concrete fire-resistance ratings. This type of aggregate is also indicated for concrete assemblies that require calculations to determine the fire-resistance rating in accordance with Section 721.2.1 [i.e., Table 721.2.1.2(2), Note a].

GLASS FIBERBOARD. Fibrous glass roof insulation consisting of inorganic glass fibers formed into rigid boards using a binder. The board has a top surface faced with asphalt and kraft reinforced with glass fiber.

❖ Depending on the type and location of insulation in walls, floors and roofs, the insulation may impact the fire-resistance rating (see Section 720). Glass fiber insulation is specifically listed in Table 721.6.2(5).

MINERAL BOARD. A rigid felted thermal insulation board consisting of either felted mineral fiber or cellular beads of expanded aggregate formed into flat rectangular units.

❖ This form of insulation is listed in Table 720.1(1), Items 15-1.9, 15-1.10 and 15-1.11.

721.2 Concrete assemblies. The provisions of this section contain procedures by which the fire-resistance ratings of concrete assemblies are established by calculations.

721.2.1 Concrete walls. Cast-in-place and precast concrete walls shall comply with Section 721.2.1.1. Multiwythe concrete walls shall comply with Section 721.2.1.2. Joints between precast panels shall comply with Section 721.2.1.3. Concrete walls with gypsum wallboard or plaster finish shall comply with Section 721.2.1.4.

721.2.1.1 Cast-in-place or precast walls. The minimum equivalent thicknesses of cast-in-place or precast concrete walls for fire-resistance ratings of 1 hour to 4 hours are shown in Table 721.2.1.1. For solid walls with flat vertical surfaces, the equivalent thickness is the same as the actual thickness. The values in

Table 721.2.1.1 apply to plain, reinforced or prestressed concrete walls.

❖ Although there have been few fire tests of concrete walls (other than concrete masonry), there have been many fire tests of concrete slabs tested as floors or roofs. Fire tests of floors or roofs are considered to be more severe than those of walls because of the loading conditions (tension cracks associated with bending moment). In addition, most ASTM E 119 fire tests of floor assemblies have been conducted while the assembly was supported within restraining frames. As concrete assemblies are heated and tend to expand, the expansion is resisted by the restraining frame. These restraining forces are usually much greater than the superimposed loads supported by load-bearing walls; thus, floor or roof assemblies are subjected to both vertical superimposed loads and horizontal restraining loads during fire tests. By contrast, load-bearing walls are subjected only to superimposed loads.

The fire endurance of masonry or concrete walls is nearly always governed by the ASTM E 119 criteria for temperature rise of the unexposed surface (i.e., the "heat transmission" end point). For flat concrete slabs or panels, the heat transmission fire endurance depends primarily on the aggregate type and thickness and is essentially the same for floors as it is for walls.

TABLE 721.2.1.1
MINIMUM EQUIVALENT THICKNESS OF CAST-IN-PLACE OR PRECAST CONCRETE WALLS, LOAD-BEARING OR NONLOAD-BEARING

CONCRETE TYPE	MINIMUM SLAB THICKNESS (inches) FOR FIRE-RESISTANCE RATING OF				
	1-hour	1¹/₂-hour	2-hour	3-hour	4-hour
Siliceous	3.5	4.3	5.0	6.2	7.0
Carbonate	3.2	4.0	4.6	5.7	6.6
Sand-Lightweight	2.7	3.3	3.8	4.6	5.4
Lightweight	2.5	3.1	3.6	4.4	5.1

For SI: 1 inch = 25.4 mm.

❖ The data in Table 721.2.1.1 was derived from Portland Cement Association (PCA) Bulletin 223, "Fire Endurances of Concrete Slabs as Influenced by Thickness, Aggregate Type, and Moisture," and PCA Publication T-140, "Fire Resistance of Reinforced Concrete Floors."

721.2.1.1.1 Hollow-core precast wall panels. For hollow-core precast concrete wall panels in which the cores are of constant cross section throughout the length, calculation of the equivalent thickness by dividing the net cross-sectional area (the gross cross section minus the area of the cores) of the panel by its width shall be permitted.

❖ The method for determining equivalent thickness of masonry units was developed because the cores in masonry units taper. The method is, of course, also applicable to hollow-core precast concrete panels;

however, because the cores in hollow-core panels do not taper, the equivalent thickness can be calculated by dividing the net cross sectional area of the panel by its width.

721.2.1.1.2 Core spaces filled. Where all of the core spaces of hollow-core wall panels are filled with loose-fill material, such as expanded shale, clay, or slag, or vermiculite or perlite, the fire-resistance rating of the wall is the same as that of a solid wall of the same concrete type and of the same overall thickness.

❖ The PCA report "Tests of Fire Resistance and Strength of Walls of Concrete Masonry Units" shows that filling cores of concrete masonry units with loose lightweight aggregates increases the fire endurance to a duration significantly longer than that of solid masonry units of the same total thickness. It is reasonable to assume that the same relationship exists for walls made of hollow-core panels.

721.2.1.1.3 Tapered cross sections. The thickness of panels with tapered cross sections shall be that determined at a distance $2t$ or 6 inches (152 mm), whichever is less, from the point of minimum thickness, where t is the minimum thickness.

❖ Some precast concrete wall panel sections (e.g., certain single-tee units) have tapered surfaces so the thickness varies, as shown in Code Figure 721.2.2.1.2. In fire tests, it is been customary to monitor the unexposed surface temperature at the location shown in the figure.

721.2.1.1.4 Ribbed or undulating surfaces. The equivalent thickness of panels with ribbed or undulating surfaces shall be determined by one of the following expressions:

for $s \geq 4t$, the thickness to be used shall be t

for $s \leq 2t$, the thickness to be used shall be t_e

for $4t > s > 2t$, the thickness to be used shall be

$$t + \left(\frac{4t}{s} - 1\right)\left(t_e - t\right) \qquad \text{(Equation 7-3)}$$

where:

s = Spacing of ribs or undulations.

t = Minimum thickness.

t_e = Equivalent thickness of the panel calculated as the net cross-sectional area of the panel divided by the width, in which the maximum thickness used in the calculation shall not exceed $2t$.

❖ The portion of a ribbed panel that can be used in calculating equivalent thickness, t_e, is shown in Code Figure 721.2.2.1.3. Note that the procedure outlined gives no credit for stems of double tees or of similar ribbed panels and clearly indicates that the minimum thickness must be used for such sections.

721.2.1.2 Multiwythe walls. For walls that consist of two wythes of different types of concrete, the fire-resistance ratings shall be permitted to be determined from Figure 721.2.1.2.

❖ The graphs in the Code Figure 721.2.1.2 were derived from a report entitled "Fire Endurance of Two-Course Floors and Roofs" in the *ACI Journal*.

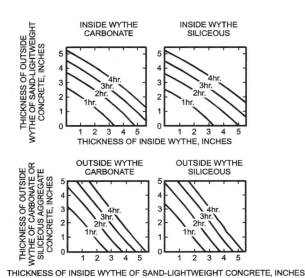

For SI: 1 inch = 25.4 mm.

FIGURE 721.2.1.2
FIRE-RESISTANCE RATINGS OF
TWO-WYTHE CONCRETE WALLS

721.2.1.2.1 Two or more wythes. The fire-resistance rating for wall panels consisting of two or more wythes shall be permitted to be determined by the formula:

$$R = (R_1^{0.59} + R_2^{0.59} + ... + R_n^{0.59})^{1.7} \qquad \textbf{(Equation 7-4)}$$

where:

R = The fire endurance of the assembly, minutes.

R_1, R_2, and R_n = The fire endurances of the individual wythes, minutes. Values of $R_n^{0.59}$ for use in Equation 7-4 are given in Table 721.2.1.2(1). Calculated fire-resistance ratings are shown in Table 721.2.1.2(2).

❖ Equation 7-4 was developed by the U.S. NBS in the early 1940s and first appeared in Appendix B of BMS 92. Verification of the accuracy of the equation is given

in "Fire Endurance of Two-Course Floors and Roofs" in the *ACI Journal*.

TABLE 721.2.1.2(1). See page 7-130.

TABLE 721.2.1.2(2). See page 7-130.

721.2.1.2.2 Foam plastic insulation. The fire-resistance ratings of precast concrete wall panels consisting of a layer of foam plastic insulation sandwiched between two wythes of concrete shall be permitted to be determined by use of Equation 7-4. Foam plastic insulation with a total thickness of less than 1 inch (25 mm) shall be disregarded. The R_n value for thickness of foam plastic insulation of 1 inch (25 mm) or greater, for use in the calculation, is 5 minutes; therefore $R_n^{0.59} = 2.5$.

❖ This value was for foam plastic insulation 1-inch (25 mm) thick and greater, determined from a fire test conducted on a panel that consisted of a 2-inch (51 mm) base slab of carbonate aggregate concrete, a 1-inch (25 mm) layer of cellular polystyrene insulation and a 2-inch (51 mm) face slab of carbonate aggregate concrete. The resulting fire endurance was 2 hours. From Equation 7-4, the contribution of the 1-inch (25 mm) layer of foam polystyrene was calculated to be 5 minutes.

Presumably, a comparable *R*-value for a 1-inch (25 mm) layer of foam polyurethane would be somewhat greater than for a 1-inch (25 mm) layer of foam polystyrene, but test values are not available. The above value for polystyrene is conservative for foam polyurethane. Equation 7-4 can be rewritten as:

$$R^{0.59} = (R1^{0.59} + R2^{0.59} + ... R_{n0.59})$$

This form of the equation is useful in making a quick determination of whether or not an assembly qualifies for a particular classification when used in conjunction with Tables 721.2.1.2(1) and 721.2.1.2(2).

EXAMPLE:

GIVEN: A sandwich wall panel consists of two $2^1/_2$-inch (63 mm) wythes of normal-weight concrete with a 2-inch (51 mm) layer of foam polystyrene between them.

FIND: Does the panel qualify for a 3-hour fire-resistance rating?

SOLUTION: From Table 721.2.1.2(1) in the code, the value of $R_{n0.59}$ for a carbonate aggregate concrete is higher than for siliceous aggregate concrete, but because the type of concrete was not given, the value for siliceous should be used. From Section 721.2.1.2.2, the value of $R_{n0.59}$ for the 2-inch (51 mm) layer of foam polystyrene is 2.5.

$$R^{0.59} = 8.1 + 2.5 + 8.1 = 18.7$$

From Table 721.2.1.2(2), a 3-hour rating is required to have an $R^{0.59}$ value of 21.41.

18.7 < 21.41, thus the wall does not qualify for a 3-hour rating.

TABLE 721.2.1.2(1)
VALUES OF $R_n^{0.59}$ FOR USE IN EQUATION 7-4

TYPE OF MATERIAL	THICKNESS OF MATERIAL (inches)											
	$1^1/_2$	2	$2^1/_2$	3	$3^1/_2$	4	$4^1/_2$	5	$5^1/_2$	6	$6^1/_2$	7
Siliceous aggregate concrete	5.3	6.5	8.1	9.5	11.3	13.0	14.9	16.9	18.8	20.7	22.8	25.1
Carbonate aggregate concrete	5.5	7.1	8.9	10.4	12.0	14.0	16.2	18.1	20.3	21.9	24.7	27.2[c]
Sand-lightweight concrete	6.5	8.2	10.5	12.8	15.5	18.1	20.7	23.3	26.0[c]	Note c	Note c	Note c
Lightweight concrete	6.6	8.8	11.2	13.7	16.5	19.1	21.9	24.7	27.8[c]	Note c	Note c	Note c
Insulating concrete[a]	9.3	13.3	16.6	18.3	23.1	26.5[c]	Note c	Note c	Note c	Note c	Note c	Note c
Airspace[b]	—	—	—	—	—	—	—	—	—	—	—	—

For SI: 1 inch = 25.4 mm, 1 pound per cubic foot = 16.02 kg/m³.
a. Dry unit weight of 35 pcf or less and consisting of cellular, perlite or vermiculite concrete.
b. The $R_n^{0.59}$ value for one $^1/_2$″ to 3 $^1/_2$″ airspace is 3.3. The $R_n^{0.59}$ value for two $^1/_2$″ to 3 $^1/_2$″ airspaces is 6.7.
c. The fire-resistance rating for this thickness exceeds 4 hours.

721.2.1.3 Joints between precast wall panels. Joints between precast concrete wall panels which are not insulated as required by this section shall be considered as openings in walls. Uninsulated joints shall be included in determining the percentage of openings permitted by Table 704.8. Where openings are not permitted or are required by this code to be protected, the provisions of this section shall be used to determine the amount of joint insulation required. Insulated joints shall not be considered openings for purposes of determining compliance with the allowable percentage of openings in Table 704.8.

TABLE 721.2.1.2(2)
FIRE-RESISTANCE RATINGS BASED ON $R^{0.59}$

R^a, MINUTES	$R^{0.59}$
60	11.20
120	16.85
180	21.41
240	25.37

a. Based on Equation 7-4.

721.2.1.3.1 Ceramic fiber joint protection. Figure 721.2.1.3.1 shows thicknesses of ceramic fiber blankets to be used to insulate joints between precast concrete wall panels for various panel thicknesses and for joint widths of $^3/_8$ inch (9.5 mm) and 1 inch (25 mm) for fire-resistance ratings of 1 hour to 4 hours. For joint widths between $^3/_8$ inch (9.5 mm) and 1 inch (25 mm), the thickness of ceramic fiber blanket is allowed to be determined by direct interpolation. Other tested and labeled materials are acceptable in place of ceramic fiber blankets.

❖ Figure 721.2.1.3.1 was derived from data in the report entitled "Fire Tests of Joints Between Precast Concrete Wall Units: Effect of Various Joint Treatments" in the *PCI Journal*.

EXAMPLE:

FIND: Determine the thickness of a ceramic fiber blanket needed for a 2-hour fire-resistance rating for joints between 5-inch-thick (127 mm) precast concrete wall panels made of siliceous aggregate concrete if the maximum joint width is $^7/_8$-inch (22.2 mm).

SOLUTION: Figure 721.2.1.3.1 indicates a minimum 0.7-inch (18 mm) thickness of ceramic fiber blanket for 5-inch (127 mm) panels for a 2-hour rating of a $^3/_4$-inch (19.5 mm) wide joint and 2.1 inches (53 mm) for a 1-inch (25 mm) wide joint. By interpolation, the thickness is computed as follows:

$t = 2.1 - (2.1 - 0.7)(1 - ^7/_8) = 1.93$ inches

Therefore, the required thickness for a 2-hour rating is 1.93 inches (49 mm).

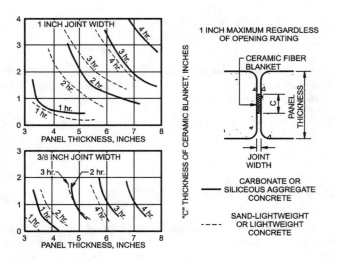

For SI: 1 inch = 25.4 mm.

FIGURE 721.2.1.3.1
CERAMIC FIBER JOINT PROTECTION

721.2.1.4 Walls with gypsum wallboard or plaster finishes. The fire-resistance rating of cast-in-place or precast concrete walls with finishes of gypsum wallboard or plaster applied to one or both sides shall be permitted to be calculated in accordance with the provisions of this section.

❖ The information contained in this section is based on *Fire Endurance Tests on Unit Masonry Walls with Gypsum Wallboard* and *The Supplement to the National Building Code of Canada 1980* (NRCC 17724) by the National Research Council of Canada.

721.2.1.4.1 Nonfire-exposed side. Where the finish of gypsum wallboard or plaster is applied to the side of the wall not exposed to fire, the contribution of the finish to the total fire-resistance rating shall be determined as follows: The thickness of the finish shall first be corrected by multiplying the actual thickness of the finish by the applicable factor determined from Table 721.2.1.4(1) based on the type of aggregate in the concrete. The corrected thickness of finish shall then be added to the actual or equivalent thickness of concrete and fire-resistance rating of the concrete and finish determined from Table 721.2.1.1, Figure 721.2.1.2 or Table 721.2.1.2(1).

❖ The fire resistance of concrete walls is generally determined by temperature rise on the unexposed surface (i.e., the heat transmission end point) (see commentary, Section 721.2.1.1). The time required to reach the heat transmission end point (fire-resistance rating) is primarily dependent upon the thickness of the concrete and the type of aggregate used to make the concrete. When additional finishes are applied to the unexposed side of the wall, the time required to reach the heat transmission end point is delayed and the fire-resistance rating of the wall is thus increased. The increase in the rating contributed by the finish can be determined by considering the finish as adding to the thickness of concrete; however, since the finish material and concrete may have different insulating properties, the actual thickness of the finish may need to be corrected to be compatible with the type of aggregate used in the concrete. The correction is made by multiplying the actual finish thickness by the factor determined from Table 721.2.1.4(1), and then adding the corrected thickness to the thickness of the concrete. This equivalent thickness is used to determine the fire-resistance rating from Table 721.2.1.1, Figure 721.2.1.2 or Table 721.2.1.2(1).

721.2.1.4.2 Fire-exposed side. Where gypsum wallboard or plaster is applied to the fire-exposed side of the wall, the contribution of the finish to the total fire-resistance rating shall be determined as follows: The time assigned to the finish as established by Table 721.2.1.4(2) shall be added to the fire-resistance rating determined from Table 721.2.1.1 or Figure 721.2.1.2, or Table 721.2.1.2(1) for the concrete alone, or to the rating determined in Section 721.2.1.4.1 for the concrete and finish on the nonfire-exposed side.

❖ Where finishes are added to the fire-exposed side of a concrete wall, their contribution to the total fire-resistance rating is based primarily upon the ability of the finish to remain in place, thus affording protection to the concrete wall. Table 721.2.1.4(2) lists the times that have been assigned to finishes on the fire-exposed side of the wall. These time-assigned values are based upon actual fire tests. The time-assigned values are added to the fire-resistance rating of the wall alone or to the rating determined for the wall and any finish on the unexposed side.

TABLE 721.2.1.4(2). See page 7-132.

721.2.1.4.3 Nonsymmetrical assemblies. For a wall having no finish on one side or different types or thicknesses of finish on each side, the calculation procedures of Sections 721.2.1.4.1 and 721.2.1.4.2 shall be performed twice, assuming either side of the wall to be the fire-exposed side. The fire-restance rating of the wall shall not exceed the lower of the two values.

Exception: For an exterior wall with more than 5 feet (1524 mm) of horizontal separation, the fire shall be assumed to occur on the interior side only.

❖ Except for exterior walls having more than 5 feet (1524 mm) of horizontal separation, Section 704.5 requires that walls be rated for exposure to fire from both sides; which therefore, two calculations must be performed, which assumes each side of the wall to be the fire-exposed side. Two calculations are not necessary for exterior walls with more than 5 feet (1524 mm) of horizontal separation or for other walls that are symmetrical (i.e., walls having the same type and thickness of finish on each side). The calculated fire-resistance rating of the wall is the lower of the two ratings determined, assuming that each side is the fire-exposed side.

TABLE 721.2.1.4(1)
MULTIPLYING FACTOR FOR FINISHES ON NONFIRE-EXPOSED SIDE OF WALL

TYPE OF FINISH APPLIED TO MASONRY WALL	TYPE OF AGGREGATE USED IN CONCRETE OR CONCRETE MASONRY			
	Concrete: siliceous or carbonate Masonry: siliceous or calcareous gravel	Concrete: sand lightweight concrete Masonry: limestone, cinders or unexpected slag	Concrete: lightweight concrete Masonry: expanded shale, clay or slate	Concrete: pumice, or expanded slag
Portland cement-sand plaster	1.00	0.75[a]	0.75[a]	0.50[a]
Gypsum-sand plaster or gypsum wallboard	1.25	1.00	1.00	1.00
Gypsum-vermiculite or perlite plaster	1.75	1.50	1.50	1.25

For SI: 1 inch = 25.4 mm.

a. For portland cement-sand plaster $^5/_8$ inch or less in thickness and applied directly to the masonry on the nonfire-exposed side of the wall, the multiplying factor shall be 1.00.

TABLE 721.2.1.4(2)
TIME ASSIGNED TO FINISH MATERIALS ON
FIRE-EXPOSED SIDE OF WALL

FINISH DESCRIPTION	TIME (minute)
Gypsum wallboard	
$^3/_8$ inch	10
$^1/_2$ inch	15
$^5/_8$ inch	20
2 layers of $^3/_8$ inch	25
1 layer $^3/_8$ inch, 1 layer $^1/_2$ inch	35
2 layers $^1/_2$ inch	40
Type X gypsum wallboard	
$^1/_2$ inch	25
$^5/_8$ inch	40
Portland cement-sand plaster applied directly to concrete masonry	See Note a
Portland cement-sand plaster on metal lath	
$^3/_4$ inch	20
$^7/_8$ inch	25
1 inch	30
Gypsum sand plaster on $^3/_8$-inch gypsum lath	
$^1/_2$ inch	35
$^5/_8$ inch	40
$^3/_4$ inch	50
Gypsum sand plaster on metal lath	
$^3/_4$ inch	50
$^7/_8$ inch	60
1 inch	80

For SI: 1 inch = 25.4 mm.

a. The actual thickness of portland cement-sand plaster, provided it is $^5/_8$ inch or less in thickness, shall be permitted to be included in determining the equivalent thickness of the masonry for use in Table 721.3.2.

721.2.1.4.4 Minimum concrete fire-resistance rating. Where finishes applied to one or both sides of a concrete wall contribute to the fire-resistance rating, the concrete alone shall provide not less than one-half of the total required fire-resistance rating. Additionally, the contribution to the fire resistance of the finish on the nonfire-exposed side of a load-bearing wall shall not exceed one-half the contribution of the concrete alone.

❖ Where gypsum wallboard or plaster finishes are applied to a concrete wall, the calculated fire-resistance rating for the concrete alone should not be less than one-half the required fire-resistance rating. This limitation is necessary to ensure that the concrete wall is of sufficient thickness to withstand fire exposure.

EXAMPLE:

GIVEN: An exterior bearing wall of a building of Type IB construction with 4 feet (1216 mm) of horizontal separation is required to have a 2-hour fire-resistance rating. The wall will be cast in place with siliceous aggregate concrete. The interior will be finished with a $^1/_2$-inch (12.7 mm) thickness of gypsum wallboard applied to steel furring members.

FIND: What is the minimum thickness of concrete required?

SOLUTION:

First calculation: Assume the interior to be the fire-exposed side.

1. From Table 721.2.1.4(2), the $^1/_2$-inch (12.7 mm) gypsum wallboard has a time-assigned value of 15 minutes; therefore, the fire-resistance rating that must be developed by the concrete must not be less than $1^3/_4$ hours (2 hours - 15 minutes).

2. Since Table 721.2.1.1 does not include a minimum thickness requirement corresponding to $1^3/_4$ hours, direct interpolation between the values for $1^1/_2$ and 2 hours is acceptable. The interpolation results in a required thickness of 4.65 inches (118 mm) of concrete.

Second calculation: Assume the exterior to be the fire-exposed side.

1. From Table 721.2.1.4(1), the multiplying factor for gypsum wallboard and siliceous aggregate concrete is 1.25; therefore, the corrected thickness for $^1/_2$ inch (12.7 mm) of gypsum wallboard is 0.63 inch (16.02 mm) [1.25 inches (32 mm) × 1/2 inch (12.7 mm)].

2. Table 721.2.1.1 requires 5 inches (127 mm) of siliceous aggregate concrete for a 2-hour fire-resistance rating; therefore, the actual thickness of concrete required is 4.37 inches (111 mm) [5 inches (127 mm) - 0.63 inches (16 mm)].

3. Since the thickness of concrete required when assuming the interior side to be the fire-exposed side is greater, the minimum concrete thickness required to achieve a 2-hour fire-resistance rating is 4.65 inches (118 mm).

4. Section 721.2.1.4.4 requires that the concrete alone provides no less than one-half the total required rating; thus, the concrete must provide at least a 1-hour rating. From Table 721.2.1.1, it can be seen that only 3.5 inches (89 mm) of siliceous aggregate concrete is required for 1 hour, whereas 4.65 inches (118 mm) will be provided.

For a similar example problem, see the commentary to Section 721.3.2.4.

721.2.1.4.5 Concrete finishes. Finishes on concrete walls that are assumed to contribute to the total fire-resistance rating of the wall shall comply with the installation requirements of Section 721.3.2.5.

❖ See the commentary to Section 721.3.2.5.

721.2.2 Concrete floor and roof slabs. Reinforced and prestressed floors and roofs shall comply with Section 721.2.2.1. Multicourse floors and roofs shall comply with Sections 721.2.2.2 and 721.2.2.3, respectively.

❖ The fire test criteria for temperature rise of the unexposed surface and the ability to resist superimposed loads (heat transmission and structural criteria, respectively) must both be considered in determining the fire resistance of floors and roofs. Section 721.2.2 deals with heat transmission and Section 721.2.3 deals with structural criteria.

721.2.2.1 Reinforced and prestressed floors and roofs. The minimum thicknesses of reinforced and prestressed concrete floor or roof slabs for fire-resistance ratings of 1 hour to 4 hours are shown in Table 721.2.2.1.

❖ The criterion limiting the average temperature rise to 250°F (392°C) and the maximum rise at one point to 325°F (527°C) is often referred to as the "heat transmission end point." For solid concrete slabs, the heat transmission end point is primarily a function of slab thickness and aggregate type. Other factors that affect heat transmission to a lesser degree are moisture content of the concrete, aggregate size, mortar content and air content. Factors that have very little effect on heat transmission are cement content and strength; type; amount and location of reinforcement, provided these items are within the normal range of usage. The values in Table 721.2.2.1 apply to concrete slabs reinforced with bars or welded wire fabric as well as to prestressed slabs.

TABLE 721.2.2.1
MINIMUM SLAB THICKNESS (inches)

CONCRETE TYPE	FIRE-RESISTANCE RATING (hour)				
	1	$1^{1}/_{2}$	2	3	4
Siliceous	3.5	4.3	5.0	6.2	7.0
Carbonate	3.2	4.0	4.6	5.7	6.6
Sand-lightweight	2.7	3.3	3.8	4.6	5.4
Lightweight	2.5	3.1	3.6	4.4	5.1

For SI: 1 inch = 25.4 mm.

❖ See the commentary to Section 721.2.2.1.

721.2.2.1.1 Hollow-core prestressed slabs. For hollow-core prestressed concrete slabs in which the cores are of constant cross section throughout the length, the equivalent thickness shall be permitted to be obtained by dividing the net cross-sectional area of the slab including grout in the joints, by its width.

❖ The method for determining equivalent thickness of masonry units was developed because the cores in masonry units taper. The method is, of course, also applicable to hollow-core precast concrete panels; however, because the cores in hollow-core panels do not taper, the equivalent thickness can be calculated by dividing the net cross-sectional area of the panel by its width.

721.2.2.1.2 Slabs with sloping soffits. The thickness of slabs with sloping soffits (see Figure 721.2.2.1.2) shall be determined at a distance 2t or 6 inches (152 mm), whichever is less, from the point of minimum thickness, where *t* is the minimum thickness.

❖ Some precast concrete wall panel sections (e.g., certain single-tee units) have tapered surfaces, so the thickness varies, as shown in Figure 721.2.2.1.2. In fire

tests, it is customary to monitor the unexposed surface temperature at the location shown in the figure.

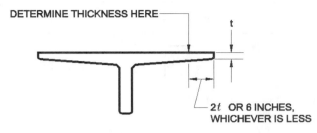

For SI: 1 inch = 25.4 mm.

FIGURE 721.2.2.1.2
DETERMINATION OF SLAB THICKNESS

❖ See the commentary to Section 721.2.2.1.2.

721.2.2.1.3 Slabs with ribbed soffits. The thickness of slabs with ribbed or undulating soffits (see Figure 721.2.2.1.3) shall be determined by one of the following expressions, whichever is applicable:

For $s \geq 4t$, the thickness to be used shall be *t*

For $s \leq 2t$, the thickness to be used shall be t_e

For $4t > s > 2t$, the thickness to be used shall be

$$ t + \left(\frac{4t}{s} - 1\right)\left(t_e - t\right) \qquad \textbf{(Equation 7-5)} $$

where:

s = Spacing of ribs or undulations.

t = Minimum thickness.

t_e = Equivalent thickness of the slab calculated as the net area of the slab divided by the width, in which the maximum thickness used in the calculation shall not exceed $2t$.

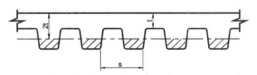

NEGLECT SHADED AREA IN CALCULATION OF EQUIVALENT THICKNESS

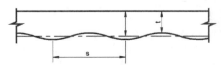

For SI: 1 inch = 25.4 mm.

FIGURE 721.2.2.1.3
SLABS WITH RIBBED OR UNDULATING SOFFITS

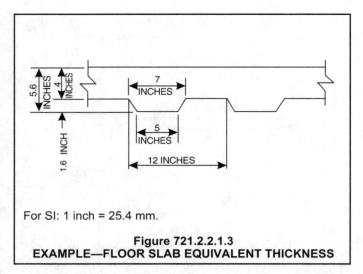

For SI: 1 inch = 25.4 mm.

Figure 721.2.2.1.3
EXAMPLE—FLOOR SLAB EQUIVALENT THICKNESS

❖ The portion of a ribbed slab that can be used in calculating equivalent thickness, t_e, is shown in Figure 721.2.2.1.3. Note that the procedure does not give credit for joists in one-way or two-way joisted floors or in double tees. For such sections, the minimum thickness of the deck slab must be used.

EXAMPLE:

FIND: Determine the fire-resistance rating of the floor section shown in Figure 721.2.2.1.3 if the units were made of siliceous aggregate concrete.

SOLUTION:

s = 12 inches.

t = 4 inches.

Therefore:

$4t > s > 2t$ (16 inches > 12 inches > 8 inches)

$$t_e = \frac{(4")(12") + (5")(1.6") + 2(\tfrac{1}{2})(1")(1.6")}{12"}$$

t_e = 4.8 inches < $2t$.

Therefore the thickness to be used, t_s:

$$t_s = 4" + \left(\frac{(4)(4")}{12} - 1\right)(4.8" - 4")$$

t_s = 4.27 inches.

From Table 721.2.2.1, using interpolation, the fire-resistance rating of the floor section shown is 1.48 hours or 1 hour, 29 minutes.

721.2.2.2 Multicourse floors. The fire-resistance ratings of floors that consist of a base slab of concrete with a topping (overlay) of a different type of concrete shall comply with Figure 721.2.2.2.

❖ The information in this section is based on the report entitled "Fire Endurance of Two-Course Floors and Roofs" in the *ACI Journal*.

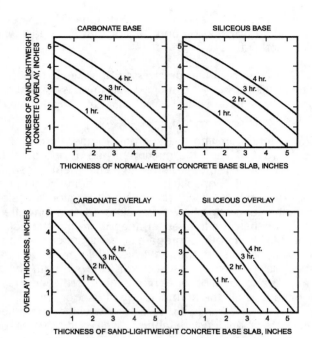

For SI: 1 inch = 25.4 mm.

FIGURE 721.2.2.2
FIRE-RESISTANCE RATINGS FOR TWO-COURSE
CONCRETE FLOORS

721.2.2.3 Multicourse roofs. The fire-resistance ratings of roofs which consist of a base slab of concrete with a topping (overlay) of an insulating concrete or with an insulating board and built-up roofing shall comply with Figures 721.2.2.3(1) and 721.2.2.3(2).

❖ The information in this section is based on the report entitled "Fire Endurance of Two-Course Floors and Roofs" in the *ACI Journal*.

721.2.2.3.1 Heat transfer. For the transfer of heat, three-ply built-up roofing contributes 10 minutes to the fire-resistance rating. The fire-resistance rating for concrete assemblies such as those shown in Figure 721.2.2.3(1) shall be increased by 10 minutes. This increase is not applicable to those shown in Figure 721.2.2.3(2).

721.2.2.4 Joints in precast slabs. Joints between adjacent precast concrete slabs need not be considered in calculating the slab thickness provided that a concrete topping at least 1 inch (25 mm) thick is used. Where no concrete topping is used, joints must be grouted to a depth of at least one-third the slab thickness at the joint, but not less than 1 inch (25 mm), or the joints must be made fire resistant by other approved methods.

❖ Based on data developed by Underwriters Laboratories, where a concrete topping is not used over precast concrete floors, joints must be grouted. If a concrete topping at least 1-inch thick (25 mm) is used, the joints need not be grouted.

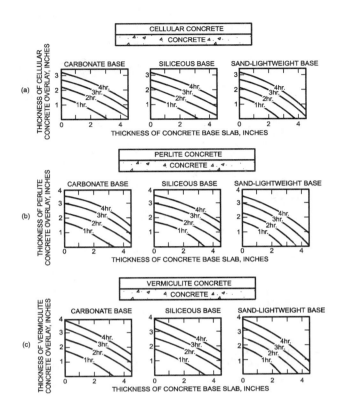

For SI: 1 inch = 25.4 mm.

FIGURE 721.2.2.3(1)
FIRE-RESISTANCE RATINGS FOR CONCRETE
ROOF ASSEMBLIES

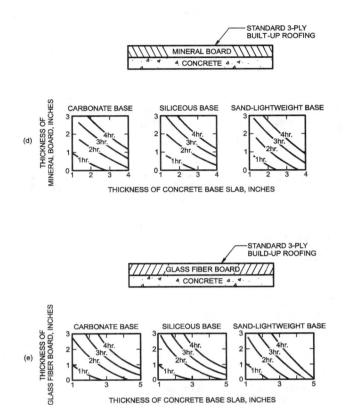

For SI: 1 inch = 25.4 mm.

FIGURE 721.2.2.3(2)
FIRE-RESISTANCE RATINGS FOR CONCRETE
ROOF ASSEMBLIES

721.2.3 Concrete cover over reinforcement. The minimum thickness of concrete cover over reinforcement in concrete slabs, reinforced beams and prestressed beams shall comply with this section.

❖ See the commentary to Sections 721.2.3.1 and 721.2.3.2.

721.2.3.1 Slab cover. The minimum thickness of concrete cover to the positive moment reinforcement shall comply with Table 721.2.3(1) for reinforced concrete and Table 721.2.3(2) for prestressed concrete. These tables are applicable for solid or hollow-core one-way or two-way slabs with flat undersurfaces. These tables are applicable to slabs that are either cast in place or precast. For precast prestressed concrete not covered elsewhere, the procedures contained in PCI MNL 124 shall be acceptable.

❖ The temperature of the tensile reinforcement depends on the thickness of cover and aggregate type. For unrestrained slabs, the values shown in Tables 721.2.3(1) and 721.2.3(2) are the cover thickness needed to keep the tensile reinforcement below 1,100°F (1922°C) and 800°F (1382°C), respectively. For restrained slabs, the

temperature of the tensile reinforcement is not as critical, and thus, a minimum cover of $^3/_4$ inch (19.5 mm) is specified in PCA Bulletin 223, "Fire Endurance of Concrete Slabs as Influenced by Thickness, Aggregate Type, and Moisture."

TABLE 721.2.3(1). See page 7-136.

❖ See the commentary to Section 721.2.3.1.

TABLE 721.2.3(2). See page 7-136.

❖ See the commentary to Section 721.2.3.1.

TABLE 721.2.3(3). See page 7-136.

❖ See the commentary to Section 721.2.3.2.

TABLE 721.2.3(4). See page 7-136.

❖ See the commentary to Section 721.2.3.2.

TABLE 721.2.3(5). See page 7-137.

❖ See the commentary to Section 721.2.3.2.

TABLE 721.2.3(1) – TABLE 721.2.3(4)   FIRE-RESISTANCE-RATED CONSTRUCTION

TABLE 721.2.3(1)
COVER THICKNESS FOR REINFORCED CONCRETE FLOOR OR ROOF SLABS (inches)

CONCRETE AGGREGATE TYPE	FIRE-RESISTANCE RATING (hours)									
	Restrained					Unrestrained				
	1	1½	2	3	4	1	1½	2	3	4
Siliceous	³/₄	³/₄	³/₄	³/₄	³/₄	³/₄	³/₄	1	1¼	1⁵/₈
Carbonate	³/₄	³/₄	³/₄	³/₄	³/₄	³/₄	³/₄	³/₄	1¼	1¼
Sand-lightweight or lightweight	³/₄	³/₄	³/₄	³/₄	³/₄	³/₄	³/₄	³/₄	1¼	1¼

For SI: 1 inch = 25.4 mm.

TABLE 721.2.3(2)
COVER THICKNESS FOR PRESTRESSED CONCRETE FLOOR OR ROOF SLABS (inches)

CONCRETE AGGREGATE TYPE	FIRE-RESISTANCE RATING (hours)									
	Restrained					Unrestrained				
	1	1½	2	3	4	1	1½	2	3	4
Siliceous	³/₄	³/₄	³/₄	³/₄	³/₄	1¹/₈	1¹/₂	1³/₄	2³/₈	2³/₄
Carbonate	³/₄	³/₄	³/₄	³/₄	³/₄	1	1³/₈	1⁵/₈	2¹/₈	2¹/₄
Sand-lightweight or lightweight	³/₄	³/₄	³/₄	³/₄	³/₄	1	1³/₈	1¹/₂	2	2¹/₄

For SI: 1 inch = 25.4 mm.

TABLE 721.2.3(3)
MINIMUM COVER FOR MAIN REINFORCING BARS OF REINFORCED CONCRETE BEAMS[c]
(APPLICABLE TO ALL TYPES OF STRUCTURAL CONCRETE)

RESTRAINED OR UNRESTRAINED[a]	BEAM WIDTH[b] (inches)	FIRE-RESISTANCE RATING (hours)				
		1	1½	2	3	4
Restrained	5	³/₄	³/₄	³/₄	1[a]	1¹/₄[a]
	7	³/₄	³/₄	³/₄	³/₄	³/₄
	≥ 10	³/₄	³/₄	³/₄	³/₄	³/₄
Unrestrained	5	³/₄	1	1¹/₄	—	—
	7	³/₄	³/₄	³/₄	1³/₄	3
	≥ 10	³/₄	³/₄	³/₄	1	1³/₄

For SI: 1 inch = 25.4 mm, 1 foot = 304.8 mm.

a. Tabulated values for restrained assemblies apply to beams spaced more than 4 feet on center. For restrained beams spaced 4 feet or less on center, minimum cover of ³/₄ inch is adequate for ratings of 4 hours or less.

b. For beam widths between the tabulated values, the minimum cover thickness can be determined by direct interpolation.

c. The cover for an individual reinforcing bar is the minimum thickness of concrete between the surface of the bar and the fire-exposed surface of the beam. For beams in which several bars are used, the cover for corner bars used in the calculation shall be reduced to one-half of the actual value. The cover for an individual bar must be not less than one-half of the value given in Table 721.2.3(3) nor less than ³/₄ inch.

TABLE 721.2.3(4)
MINIMUM COVER FOR PRESTRESSED CONCRETE BEAMS 8 INCHES OR GREATER IN WIDTH

RESTRAINED OR UNRESTRAINED[a]	CONCRETE AGGREGATE TYPE	BEAM WIDTH[b] (inches)	FIRE-RESISTANCE RATING (hours)				
			1	1½	2	3	4
Restrained	Carbonate or siliceous	8	1¹/₂	1¹/₂	1¹/₂	1³/₄[a]	2¹/₂[a]
	Carbonate or siliceous	≥ 12	1¹/₂	1¹/₂	1¹/₂	1¹/₂	1⁷/₈[a]
	Sand lightweight	8	1¹/₂	1¹/₂	1¹/₂	1¹/₂	2[a]
	Sand lightweight	≥ 12	1¹/₂	1¹/₂	1¹/₂	1¹/₂	1⁵/₈[a]
Unrestrained	Carbonate or siliceous	8	1¹/₂	1³/₄	2¹/₂	5[c]	—
	Carbonate or siliceous	≥ 12	1¹/₂	1¹/₂	1⁷/₈[a]	2¹/₂	3
	Sand lightweight	8	1¹/₂	1¹/₂	2	3¹/₄	—
	Sand lightweight	≥ 12	1¹/₂	1¹/₂	1⁵/₈	2	2¹/₂

For SI: 1 inch = 25.4 mm, 1 foot = 304.8 mm.

a. Tabulated values for restrained assemblies apply to beams spaced more than 4 feet on center. For restrained beams spaced 4 feet or less on center, minimum cover of ³/₄ inch is adequate for 4-hour ratings or less.

b. For beam widths between 8 inches and 12 inches, minimum cover thickness can be determined by direct interpolation.

c. Not practical for 8-inch-wide beam but shown for purposes of interpolation.

TABLE 721.2.3(5)
MINIMUM COVER FOR PRESTRESSED CONCRETE BEAMS OF ALL WIDTHS

RESTRAINED OR UNRESTRAINED[a]	CONCRETE AGGREGATE TYPE	BEAM AREA[b] A (square inches)	FIRE-RESISTANCE RATING (hours)				
			1	$1^1/_2$	2	3	4
Restrained	All	$40 \leq A \leq 150$	$1^1/_2$	$1^1/_2$	2	$2^1/_2$	—
	Carbonate or	$150 < A \leq 300$	$1^1/_2$	$1^1/_2$	$1^1/_2$	$1^3/_4$	$2^1/_2$
	siliceous	$300 < A$	$1^1/_2$	$1^1/_2$	$1^1/_2$	$1^1/_2$	2
	Sand lightweight	$150 < A$	$1^1/_2$	$1^1/_2$	$1^1/_2$	$1^1/_2$	2
Unrestrained	All	$40 \leq A \leq 150$	2	$2^1/_2$	—	—	—
	Carbonate or	$150 < A \leq 300$	$1^1/_2$	$1^3/_4$	$2^1/_2$	—	—
	siliceous	$300 < A$	$1^1/_2$	$1^1/_2$	2	3^c	4^c
	Sand lightweight	$150 < A$	$1^1/_2$	$1^1/_2$	2	3^c	4^c

For SI:　1 inch = 25.4 mm, 1 foot = 304.8 mm.

a. Tabulated values for restrained assemblies apply to beams spaced more than 4 feet on center. For restrained beams spaced 4 feet or less on center, minimum cover of $^3/_4$ inch is adequate for 4-hour ratings or less.

b. The cross-sectional area of a stem is permitted to include a portion of the area in the flange, provided the width of the flange used in the calculation does not exceed three times the average width of the stem.

c. U-shaped or hooped stirrups spaced not to exceed the depth of the member and having a minimum cover of 1 inch shall be provided.

721.2.3.2 Reinforced beam cover. The minimum thickness of concrete cover to the positive moment reinforcement (bottom steel) for reinforced concrete beams is shown in Table 721.2.3(3) for fire-resistance ratings of 1 hour to 4 hours.

❖ For reinforced concrete beams, the critical steel temperature is 1,100°F (1922°C), so the effect of aggregate type is minimal. For prestressed concrete, the comparable temperature is 800°F (1382°C), so aggregate type must be considered. The data in Tables 721.2.3(3) and 721.2.3(4) was derived from the fire tests of a series of beam specimens that ranged in width between 2 and 24 inches (51 and 607 mm). Other variables in the series included aggregate type and amount of reinforcement. Tests were conducted at the Portland Cement Association and the results are published in "Temperature Distribution in Concrete Beams Subjected to Fire." The results of the fire tests of beams conducted at Underwriters Laboratories Inc., were also analyzed. Charts graphically showing the resulting data are shown in Figure A4 of *PCI Manual 124*, "Design for Fire Resistance of Precast Prestressed Concrete."

721.2.3.3 Prestressed beam cover. The minimum thickness of concrete cover to the positive moment prestressing tendons (bottom steel) for restrained and unrestrained prestressed concrete beams and stemmed units shall comply with the values shown in Tables 721.2.3(4) and 721.2.3(5) for fire-resistance ratings of 1 hour to 4 hours. Values in Table 721.2.3(4) apply to beams 8 inches (203 mm) or greater in width. Values in Table 721.2.3(5) apply to beams or stems of any width, provided the cross-section area is not less than 40 square inches (25 806 mm²). In case of differences between the values determined from Table 721.2.3(4) or 721.2.3(5), it is permitted to use the smaller value. The concrete cover shall be calculated in accordance with Section 721.2.3.3.1. The minimum concrete cover

for nonprestressed reinforcement in prestressed concrete beams shall comply with Section 721.2.3.2.

❖ See the commentary to Section 721.2.3.2.

721.2.3.3.1 Calculating concrete cover. The concrete cover for an individual tendon is the minimum thickness of concrete between the surface of the tendon and the fire-exposed surface of the beam, except that for ungrouped ducts, the assumed cover thickness is the minimum thickness of concrete between the surface of the duct and the fire-exposed surface of the beam. For beams in which two or more tendons are used, the cover is assumed to be the average of the minimum cover of the individual tendons. For corner tendons (tendons equal distance from the bottom and side), the minimum cover used in the calculation shall be one-half the actual value. For stemmed members with two or more prestressing tendons located along the vertical centerline of the stem, the average cover shall be the distance from the bottom of the member to the centroid of the tendons. The actual cover for any individual tendon shall not be less than one-half the smaller value shown in Tables 721.2.3(4) and 721.2.3(5), or 1 inch (25 mm), whichever is greater.

721.2.4 Concrete columns. Concrete columns shall comply with this section.

❖ Most building code provisions for reinforced concrete columns are based on two reports: "Fire Tests of Building Columns" of the Associated Factory Mutual Insurance Companies and "Fire Resistance of Concrete Columns" by the NBS. Sizes of the tested columns were 12 inches (305 mm), 16 inches (406 mm) and 18 inches (457 mm). Nearly all of the columns withstood 4-hour fire tests that were essentially conducted in accordance with ASTM E 119. Most of the tests were stopped after 4 hours, but some were continued to 8 hours. The shortest duration was 3 hours. At the time of the tests few, if

TABLE 721.2.4 – 721.3.2   FIRE-RESISTANCE-RATED CONSTRUCTION

any, concrete columns were smaller than 12 inches (305 mm), but smaller columns have been in use for many years since then.

Fire tests conducted in Europe on smaller columns indicate that fire endurance is greater for larger columns.

TABLE 721.2.4
MINIMUM DIMENSION OF CONCRETE COLUMNS (inches)

TYPES OF CONCRETE	FIRE-RESISTANCE RATING (hours)				
	1	$1^1/_2$	2^a	3^a	4^b
Siliceous	8	9	10	12	14
Carbonate	8	9	10	11	12
Sand-lightweight	8	$8^1/_2$	9	$10^1/_2$	12

For SI: 1 inch = 25 mm.

a. The minimum dimension is permitted to be reduced to 8 inches for rectangular columns with two parallel sides at least 36 inches in length.

b. The minimum dimension is permitted to be reduced to 10 inches for rectangular columns with two parallel sides at least 36 inches in length.

❖ Table 721.2.4 reflects the information cited in the reports listed in the commentary to Section 721.2.4 and in the report "Investigations on Building Fires; Part VI; Fire Resistance of Reinforced Concrete Columns" in the National Building Studies Research Paper No. 18. This table is for columns that may be exposed to fire on all four sides. Notes a and b to Table 721.2.4 permit a smaller minimum dimension for columns than required in the table, provided the longer dimension of the column is at least 36 inches (914 mm).

721.2.4.1 Minimum size. The minimum overall dimensions of reinforced concrete columns for fire-resistance ratings of 1 hour to 4 hours shall comply with Table 721.2.4.

721.2.4.2 Minimum cover for R/C columns. The minimum thickness of concrete cover to the main longitudinal reinforcement in columns, regardless of the type of aggregate used in the concrete, shall not be less than 1 inch (25 mm) times the number of hours of required fire resistance or 2 inches (51 mm), whichever is less.

❖ The minimum concrete cover refers to the main longitudinal reinforcement, not ties or spirals.

721.2.4.3 Columns built into walls. The minimum dimensions of Table 721.2.4 do not apply to a reinforced concrete column that is built into a concrete or masonry wall provided all of the following are met:

1. The fire-resistance rating for the wall is equal to or greater than the required rating of the column;

2. The main longitudinal reinforcing in the column has cover not less than that required by Section 721.2.4.2; and

3. Openings in the wall are protected in accordance with Table 715.4.

Where openings in the wall are not protected as required by Section 715.4, the minimum dimension of columns required to have a fire-resistance rating of 3 hours or less shall be 8 inches

(203 mm), and 10 inches (254 mm) for columns required to have a fire-resistance rating of 4 hours, regardless of the type of aggregate used in the concrete.

❖ Table 721.2.4 does not apply to columns built into concrete or masonry walls that meet all three of the conditions in this section.

721.2.4.4 Precast cover units for steel columns. See Section 721.5.1.4.

721.3 Concrete masonry. The provisions of this section contain procedures by which the fire-resistance ratings of concrete masonry are established by calculations.

721.3.1 Equivalent thickness. The equivalent thickness of concrete masonry construction shall be determined in accordance with the provisions of this section.

721.3.1.1 Concrete masonry unit plus finishes. The equivalent thickness of concrete masonry assemblies, T_{ea}, shall be computed as the sum of the equivalent thickness of the concrete masonry unit, T_e, as determined by Section 721.3.1.2, 721.3.1.3, or 721.3.1.4, plus the equivalent thickness of finishes, T_{ef}, determined in accordance with Section 721.3.2:

$$T_{ea} = T_e + T_{ef} \qquad \text{(Equation 7-6)}$$

$T_e = V_n/LH =$ Equivalent thickness of concrete masonry unit (inch) (mm).

where:

$V_n =$ Net volume of masonry unit (inch^3) (mm^3).

$L =$ Specified length of masonry unit (inch) (mm).

$H =$ Specified height of masonry unit (inch) (mm).

721.3.1.2 Ungrouted or partially grouted construction. T_e shall be the value obtained for the concrete masonry unit determined in accordance with ASTM C 140.

❖ See the commentary to Section 721.2.3.2.

721.3.1.3 Solid grouted construction. The equivalent thickness, T_e, of solid grouted concrete masonry units is the actual thickness of the unit.

721.3.1.4 Airspaces and cells filled with loose-fill material. The equivalent thickness of completely filled hollow concrete masonry is the actual thickness of the unit when loose-fill materials are: sand, pea gravel, crushed stone, or slag that meet ASTM C 33 requirements; pumice, scoria, expanded shale, expanded clay, expanded slate, expanded slag, expanded fly ash, or cinders that comply with ASTM C 331; or perlite or vermiculite meeting the requirements of ASTM C 549 and ASTM C 516, respectively.

721.3.2 Concrete masonry walls. The fire-resistance rating of walls and partitions constructed of concrete masonry units shall be determined from Table 721.3.2. The rating shall be based on the equivalent thickness of the masonry and type of aggregate used.

❖ It has been accepted practice to determine the fire-resistance rating of concrete masonry walls based on the type of aggregate used to manufacture them and the equivalent thickness of solid material in the wall. Equivalent thicknesses shown in Table 721.3.2 have been developed and refined through actual fire testing.

To determine the minimum equivalent thickness, Section 721.3.1.2 references ASTM C 140.

EXAMPLE:

FIND: Determine the equivalent thickness of a standard 8 inch by 8 inch by 16 inch (203 mm by 406 mm) concrete masonry unit as shown in Figure 721.3.2. The unit is normal weight with sand and gravel aggregate.

SOLUTION: The equivalent thickness is actually the average thickness of the solid material in the unit. The equivalent thickness is determined by the following:

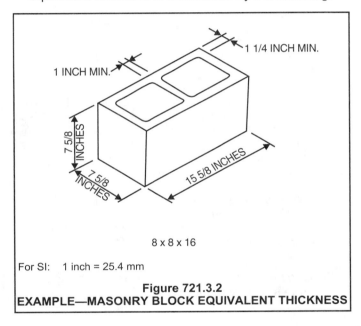

1 1/4 INCH MIN.

1 INCH MIN.

7 5/8 INCHES

7 5/8 INCHES

15 5/8 INCHES

8 x 8 x 16

For SI: 1 inch = 25.4 mm

Figure 721.3.2
EXAMPLE—MASONRY BLOCK EQUIVALENT THICKNESS

$T_e = 1728\, A/(L \times H)$

where:

T_e = Equivalent thickness, in.

A = Net volume of unit, ft^3.

L = Length of unit, in.

H = Height of unit, in.

Net volume, A, is determined by the following equation:

$A = C/D$

where:

C = Dry weight of unit, pounds

D = Density of unit, pounds per cubic foot

From data furnished by the manufacturer, the dry unit weight is 44 pounds (20 kg) and the density is 135 pounds per cubic foot (pcf) (2115 kg/m^2).

$A = C/D = 44\, lbs./135\, pcf$
$= 0.326\, ft^3.$

$$T_e = \frac{(1{,}728\, in.^3/ft.^3)(0.326\, ft.^3)}{(15.625\, in.)(7.625\, in.)}$$

$T_e = 4.73$ inches.

The fire-resistance rating for this type of unit with siliceous gravel aggregate can now be determined from Table 721.3.2. In this case, the fire-resistance rating will fall between 2.25 and 2.5 hours.

TABLE 721.3.2. See below.

❖ See the commentary to Section 721.3.2.

721.3.2.1 Finish on nonfire-exposed side. Where plaster or gypsum wallboard is applied to the side of the wall not exposed to fire, the contribution of the finish to the total fire-resistance rating shall be determined as follows: The thickness of gypsum

TABLE 721.3.2
MINIMUM EQUIVALENT THICKNESS (inches) OF BEARING OR NONBEARING CONCRETE MASONRY WALLS[a,b,c,d]

TYPE OF AGGREGATE	FIRE-RESISTANCE RATING (hours)														
	$^1/_2$	$^3/_4$	1	$1^1/_4$	$1^1/_2$	$1^3/_4$	2	$2^1/_4$	$2^1/_2$	$2^3/_4$	3	$3^1/_4$	$3^1/_2$	$3^3/_4$	4
Pumice or expanded slag	1.5	1.9	2.1	2.5	2.7	3.0	3.2	3.4	3.6	3.8	4.0	4.2	4.4	4.5	4.7
Expanded shale, clay or slate	1.8	2.2	2.6	2.9	3.3	3.4	3.6	3.8	4.0	4.2	4.4	4.6	4.8	4.9	5.1
Limestone, cinders or unexpanded slag	1.9	2.3	2.7	3.1	3.4	3.7	4.0	4.3	4.5	4.8	5.0	5.2	5.5	5.7	5.9
Calcareous or siliceous gravel	2.0	2.4	2.8	3.2	3.6	3.9	4.2	4.5	4.8	5.0	5.3	5.5	5.8	6.0	6.2

For SI: 1 inch = 25.4 mm.

a. Values between those shown in the table can be determined by direct interpolation.

b. Where combustible members are framed into the wall, the thickness of solid material between the end of each member and the opposite face of the wall, or between members set in from opposite sides, shall not be less than 93 percent of the thickness shown in the table.

c. Requirements of ASTM C 55, ASTM C 73 or ASTM C 90 shall apply.

d. Minimum required equivalent thickness corresponding to the hourly fire-resistance rating for units with a combination of aggregate shall be determined by linear interpolation based on the percent by volume of each aggregate used in manufacture.

wallboard or plaster shall be corrected by multiplying the actual thickness of the finish by applicable factor determined from Table 721.2.1.4(1). This corrected thickness of finish shall be added to the equivalent thickness of masonry and the fire-resistance rating of the masonry and finish determined from Table 721.3.2.

❖ The information contained in this section is based on *Fire Endurance Tests on Unit Masonry Walls with Gypsum Wallboard* and *The Supplement to the National Building Code of Canada 1980* by the National Research Council of Canada.

The fire resistance of concrete masonry walls is generally determined by temperature rise on the unexposed surface (i.e., the "heat transmission" end point). The time required to reach the heat transmission end point (fire-resistance rating) is primarily dependent upon the equivalent thickness of the masonry and the type of aggregate used to make the concrete. When additional finishes are applied to the unexposed side of the wall, the time required to reach the heat transmission end point is delayed; thus, the fire-resistance rating of the wall is increased. The increase in the rating contributed by the finish can be determined by considering the finish as an addition to the equivalent thickness of masonry; however, since the finish material and masonry may have different insulating properties, the actual thickness of the finish must be multiplied by a correction factor to be compatible with the type of aggregate used in the concrete. The correction is made by multiplying the actual finish thickness by the factor determined from Table 721.2.1.4(1), then adding the modified thickness to the equivalent thickness of masonry. This combined equivalent thickness is used to determine the fire-resistance rating from Table 721.3.2.

721.3.2.2 Finish on fire-exposed side. Where plaster or gypsum wallboard is applied to the fire-exposed side of the wall, the contribution of the finish to the total fire-resistance rating shall be determined as follows: The time assigned to the finish as established by Table 721.2.1.4(2) shall be added to the fire-resistance rating determined in Section 721.3.2 for the masonry alone, or in Section 721.3.2.1 for the masonry and finish on the nonfire-exposed side.

❖ Where finishes are added to the fire-exposed side of a concrete masonry wall, their contribution to the total fire-resistance rating is based primarily upon their ability to remain in place, thus affording protection to the masonry wall. Table 721.2.1.4(2) lists the times that have been assigned to finishes on the fire-exposed side of the wall. These time-assigned values are based upon actual fire tests. The time-assigned values are added to the fire-resistance rating of the wall alone or to the rating determined for the wall and any finish on the unexposed side.

721.3.2.3 Nonsymmetrical assemblies. For a wall having no finish on one side or having different types or thicknesses of fin-

ish on each side, the calculation procedures of this section shall be performed twice, assuming either side of the wall to be the fire-exposed side. The fire-resistance rating of the wall shall not exceed the lower of the two values calculated.

Exception: For exterior walls with more than 5 feet (1524 mm) of horizontal separation, the fire shall be assumed to occur on the interior side only.

❖ Except for exterior walls having more than 5 feet (1524 mm) of horizontal separation, Section 704.5 requires that walls be rated for exposure to fire from both sides; therefore, two calculations must be performed, which assumes each side to be the fire-exposed side. Two calculations are not necessary for exterior walls with more than 5 feet (1524 mm) of horizontal separation or for other walls that are symmetrical (i.e., walls having the same type and thickness of finish on each side). The calculated fire-resistance rating must not exceed the lower of the two ratings determined, assuming that each side is the fire-exposed side.

721.3.2.4 Minimum concrete masonry fire-resistance rating. Where the finish applied to a concrete masonry wall contributes to its fire-resistance rating, the masonry alone shall provide not less than one-half the total required fire-resistance rating.

❖ Where gypsum wallboard or plaster finishes are applied to a concrete masonry wall, the calculated fire-resistance rating for the concrete alone should be not less than one-half the required fire-resistance rating. This limitation is necessary to ensure that the masonry wall is of sufficient thickness to withstand fire exposure.

EXAMPLE:

GIVEN: A wall required to have a 4-hour fire-resistance rating will be constructed with concrete masonry units of expanded shale aggregate. The wall will be finished on each side with a layer of $^1/_2$-inch (12.7 mm) gypsum wallboard.

FIND: What is the minimum equivalent thickness of concrete masonry required?

SOLUTION:

Since the wall has the same type and thickness of finish on each side, only one calculation is required.

1. The $^1/_2$-inch (12.7 mm) gypsum wallboard on the fire-exposed side has a time-assigned value of 15 minutes in accordance with Table 721.2.1.4(2).

2. Therefore, the fire-resistance required to be provided by the masonry and gypsum wallboard on the unexposed side is 3 hours and 45 minutes (4 hours - 15 minutes).

3. From Table 721.2.1.4(1), the corrected thickness of gypsum wallboard on the unexposed side is 1/2 inch (12.7 mm) (1.00 × $^1/_2$ inch).

4. From Table 721.3.2, the minimum equivalent thickness of masonry, including the corrected thickness of gypsum wallboard, required for a rat-

ing of 3 hours and 45 minutes is 4.9 inches (124 mm).

5. Therefore, the equivalent thickness of masonry required is 4.4 inches (112 mm) [4.9 inches (124 mm) - 1/2 inch (12.7 mm)].

6. From Table 721.3.2, it can be determined that 4.4 inches (112 mm) of expanded shale aggregate concrete masonry will provide a fire-resistance of 3 hours; therefore, the requirement that the masonry alone provide at least one-half of the total required rating is satisfied.

For a similar example problem, see the commentary to Section 721.2.1.4.4.

721.3.2.5 Attachment of finishes. Installation of finishes shall be as follows:

1. Gypsum wallboard and gypsum lath applied to concrete masonry or concrete walls shall be secured to wood or steel furring members spaced not more than 16 inches (406 mm) on center (o.c.).

2. Gypsum wallboard shall be installed with the long dimension parallel to the furring members and shall have all joints finished.

3. Other aspects of the installation of finishes shall comply with the applicable provisions of Chapters 7 and 25.

721.3.3 Multiwythe masonry walls. The fire-resistance rating of wall assemblies constructed of multiple wythes of masonry materials shall be permitted to be based on the fire-resistance rating period of each wythe and the continuous airspace between each wythe in accordance with the following formula:

$$R_A = (R_1^{0.59} + R_2^{0.59} + ... + R_n^{0.59} + A_1 + A_2 + ... + A_n)^{1.7}$$
(Equation 7-7)

where:

R_A = Fire endurance rating of the assembly (hours).

$R_1, R_2, ..., R_n$= Fire endurance rating of wythes for 1, 2, n (hours), respectively.

$A_1, A_2, ..., A_n$ = 0.30, factor for each continuous airspace for 1, 2, ...n, respectively, having a depth of $\frac{1}{2}$ inch (12.7 mm) or more between wythes.

❖ This calculation method was first published in Report BMS 92 of the NBS.

721.3.4 Concrete masonry lintels. Fire-resistance ratings for concrete masonry lintels shall be determined based upon the nominal thickness of the lintel and the minimum thickness of concrete masonry or concrete, or any combination thereof, covering the main reinforcing bars, as determined according to Table 721.3.4, or by approved alternate methods.

TABLE 721.3.4
MINIMUM COVER OF LONGITUDINAL REINFORCEMENT IN FIRE-RESISTANCE-RATED REINFORCED CONCRETE MASONRY LINTELS (inches)

NOMINAL WIDTH OF LINTEL (inches)	FIRE-RESISTANCE RATING (hours)			
	1	2	3	4
6	$1^1/_2$	2	—	—
8	$1^1/_2$	$1^1/_2$	$1^3/_4$	3
10 or greater	$1^1/_2$	$1^1/_2$	$1^1/_2$	$1^3/_4$

For SI: 1 inch = 25.4 mm.

721.3.5 Concrete masonry columns. The fire-resistance rating of concrete masonry columns shall be determined based upon the least plan dimension of the column in accordance with Table 721.3.5 or by approved alternate methods.

TABLE 721.3.5
MINIMUM DIMENSION OF CONCRETE MASONRY COLUMNS (inches)

FIRE-RESISTANCE RATING (hours)			
1	2	3	4
8	10	12	14

For SI: 1 inch = 25.4 mm.

721.4 Clay brick and tile masonry. The provisions of this section contain procedures by which the fire-resistance ratings of clay brick and tile masonry are established by calculations.

721.4.1 Masonry walls. The fire-resistance rating of masonry walls shall be based upon the equivalent thickness as calculated in accordance with this section. The calculation shall take into account finishes applied to the wall and airspaces between wythes in multiwythe construction.

TABLE 721.4.1(1). See page 7-142.

TABLE 721.4.1(2). See page 7-142.

721.4.1.1 Equivalent thickness. The fire-resistance ratings of walls or partitions constructed of solid or hollow clay masonry units shall be determined from Table 721.4.1(1) or 721.4.1(2). The equivalent thickness of the clay masonry unit shall be determined by Equation 7-8 when using Table 721.4.1(1). The fire-resistance rating determined from Table 721.4.1(1) shall be permitted to be used in the calculated fire-resistance rating procedure in Section 721.4.2.

$$T_e = V_n/LH$$
(Equation 7-8)

where:

T_e = The equivalent thickness of the clay masonry unit (inches).

V_n = The net volume of the clay masonry unit (inch^3).

L = The specified length of the clay masonry unit (inches).

H = The specified height of the clay masonry unit (inches).

❖ It has been accepted practice to determine the fire-resistance ratings of clay masonry walls by the ASTM E 119 test method. In fact, Table 721.4.1(1) is based on tests performed by various laboratories in conformance with ASTM E 119. Most notable are the NBS, Ohio State University Experiment Station, and The University of California at Berkeley. New methods, however, have been developed to evaluate the fire-resistance of clay masonry by the equivalent thickness method, similar to concrete masonry. This approach was used in conjunction with extensive testing previously performed on clay masonry to develop Table 721.4.1(1).

721.4.1.1.1 Hollow clay units. The equivalent thickness, T_e, shall be the value obtained for hollow clay units as determined in accordance with ASTM C 67.

721.4.1.1.2 Solid grouted clay units. The equivalent thickness of solid grouted clay masonry units shall be taken as the actual thickness of the units.

TABLE 721.4.1(1)
FIRE-RESISTANCE PERIODS OF CLAY MASONRY WALLS

MATERIAL TYPE	MINIMUM REQUIRED EQUIVALENT THICKNESS FOR FIRE RESISTANCE[a,b,c] (inches)			
	1 hour	2 hour	3 hour	4 hour
Solid brick of clay or shale[d]	2.7	3.8	4.9	6.0
Hollow brick or tile of clay or shale, unfilled	2.3	3.4	4.3	5.0
Hollow brick or tile of clay or shale, grouted or filled with materials specified in Section 721.4.1.1.3	3.0	4.4	5.5	6.6

For SI: 1 inch = 25.4 mm.

a. Equivalent thickness as determined from Section 721.4.1.1.

b. Calculated fire resistance between the hourly increments listed shall be determined by linear interpolation.

c. Where combustible members are framed in the wall, the thickness of solid material between the end of each member and the opposite face of the wall, or between members set in from opposite sides, shall be not less than 93 percent of the thickness shown.

d. For units in which the net cross-sectional area of cored brick in any plane parallel to the surface containing the cores is at least 75 percent of the gross cross-sectional area measured in the same plane.

TABLE 721.4.1(2)
FIRE-RESISTANCE RATINGS FOR BEARING STEEL FRAME
BRICK VENEER WALLS OR PARTITIONS

WALL OR PARTITION ASSEMBLY	PLASTER SIDE EXPOSED (hours)	BRICK FACED SIDE EXPOSED (hours)
Outside facing of steel studs: $1/2''$ wood fiberboard sheathing next to studs, $3/4''$ airspace formed with $3/4'' \times 1\,5/8''$ wood strips placed over the fiberboard and secured to the studs; metal or wire lath nailed to such strips, $3\,3/4''$ brick veneer held in place by filling $3/4''$ airspace between the brick and lath with mortar. Inside facing of studs: $3/4''$ unsanded gypsum plaster on metal or wire lath attached to $5/16''$ wood strips secured to edges of the studs.	1.5	4
Outside facing of steel studs: $1''$ insulation board sheathing attached to studs, $1''$ airspace, and $3\,3/4''$ brick veneer attached to steel frame with metal ties every 5th course. Inside facing of studs: $7/8''$ sanded gypsum plaster (1:2 mix) applied on metal or wire lath attached directly to the studs.	1.5	4
Same as above except use $7/8''$ vermiculite—gypsum plaster or $1''$ sanded gypsum plaster (1:2 mix) applied to metal or wire.	2	4
Outside facing of steel studs: $1/2''$ gypsum sheathing board, attached to studs, and $3\,3/4''$ brick veneer attached to steel frame with metal ties every 5th course. Inside facing of studs: $1/2''$ sanded gypsum plaster (1:2 mix) applied to $1/2''$ perforated gypsum lath securely attached to studs and having strips of metal lath 3 inches wide applied to all horizontal joints of gypsum lath.	2	4

For SI: 1 inch = 25.4 mm.

TABLE 721.4.1(3)
VALUES OF $R_n^{0.59}$

$R_n^{0.59}$	R (hours)
1	1.0
2	1.50
3	1.91
4	2.27

TABLE 721.4.1(4)
COEFFICIENTS FOR PLASTER, pl [a]

THICKNESS OF PLASTER (inch)	ONE SIDE	TWO SIDE
$1/2$	0.3	0.6
$5/8$	0.37	0.75
$3/4$	0.45	0.90

For SI: 1 inch = 25.4 mm.
a. Values listed in table are for 1:3 sanded gypsum plaster.

TABLE 721.4.1(5)
REINFORCED MASONRY LINTELS

NOMINAL LINTEL WIDTH (inches)	MINIMUM LONGITUDINAL REINFORCEMENT COVER FOR FIRE RESISTANCE (inch)			
	1 hour	2 hour	3 hour	4 hour
6	$1^1/_2$	2	NP	NP
8	$1^1/_2$	$1^1/_2$	$1^3/_4$	3
10 or more	$1^1/_2$	$1^1/_2$	$1^1/_2$	$1^3/_4$

For SI: 1 inch = 25.4 mm.
NP = Not permitted.

TABLE 721.4.1(6)
REINFORCED CLAY MASONRY COLUMNS

COLUMN SIZE	FIRE-RESISTANCE RATING (hour)			
	1	2	3	4
Minimum column dimension (inches)	8	10	12	14

For SI: 1 inch = 25.4 mm.

721.4.1.1.3 Units with filled cores. The equivalent thickness of the hollow clay masonry units is the actual thickness of the unit when completely filled with loose-fill materials of: sand, pea gravel, crushed stone, or slag that meet ASTM C 33 requirements; pumice, scoria, expanded shale, expanded clay, expanded slate, expanded slag, expanded fly ash, or cinders in compliance with ASTM C 331; or perlite or vermiculite meeting the requirements of ASTM C 549 and ASTM C 516, respectively.

721.4.1.2 Plaster finishes. Where plaster is applied to the wall, the total fire-resistance rating shall be determined by the formula:

$$R = (R_n^{0.59} + pl)\,1.7 \qquad \text{(Equation 7-9)}$$

where:

R = The fire endurance of the assembly (hours).

R_n = The fire endurance of the individual wall (hours).

pl = Coefficient for thickness of plaster.

Values for $R_n^{0.59}$ for use in Equation 7-9 are given in Table 721.4.1(3). Coefficients for thickness of plaster shall be selected from Table 721.4.1(4) based on the actual thickness of plaster applied to the wall or partition and whether one or two sides of the wall are plastered.

❖ The variables for use in Equation 7-9 for determining the fire-resistance rating of plastered clay masonry walls are based on "Fire Resistance Classifications of Building Construction," BMS 92 of the NBS. These values were derived from fire test results. The average thickness of plaster applied in the series of tests ranged from $1/2$ inch (12.7 mm) to $3/4$ inch (19.1 mm). The thickness for which the coefficients of plaster in Table 721.4.1(4) are given are those most likely to be applied to a clay masonry wall. Ratings for other thicknesses provided can be obtained by substituting the appropriate coefficients in the formula.

A test of four hollow concrete unit walls shows the effect of one coat of plaster on the fire-exposed side to be about the same as for one coat of plaster on the unexposed side. Tests have not been made with plaster on the unexposed side of only clay hollow walls; however, the ratings resulting from Equation 7-9 for clay or tile masonry walls for plaster on one side are believed to have a sufficient margin of safety to be applicable to either the exposed or unexposed side of the wall.

EXAMPLE:

GIVEN: A 4-inch (102 mm) solid brick wall is required to have a 3-hour fire-resistance rating with clay masonry units.

FIND: The thickness of one side of plaster required to attain a 3-hour rating.

SOLUTION:

1. From Table 721.4.1(1), a 2-hour solid brick wall is 3.8 inches (99 mm) of equivalent thickness and a 3-hour wall is 4.9 inches (124 mm). Through interpolation (Note a), a 4-inch (102 mm) wall is approximately a 131 minute (2.18 hours) fire rating.

2. From Table 721.4.1(4), plaster has a coefficient of 0.3 for $1/2$-inch thick (12.7 mm) plaster, 0.37 for $5/8$-inch thick (15.9 mm) plaster or 0.45 for $3/4$-inch (19.1 mm) thick plaster.

3. From Equation 7-9:
$R = [\,R_{n0.59} + pl\,]^{1.7}$
$3 = [\,2.18\,]^{0.59} + pl)^{1.7}$
pl (required) = 0.326.

4. Therefore, one coat of $5/8$-inch-thick (15.9 mm) sanded gypsum plaster on a 4-inch (102 mm) solid brick masonry wall would result in a 3-hour fire-resistance rating.

721.4.1.3 Multiwythe walls with airspace. Where a continuous airspace separates multiple wythes of the wall or partition, the total fire-resistance rating shall be determined by the formula:

$$R = (R_1^{0.59} + R_2^{0.59} + ... + R_n^{0.59} + as)^{1.7} \qquad \textbf{(Equation 7-10)}$$

where:

R = The fire endurance of the assembly (hours).

R_1, R_2 and R_n = The fire endurance of the individual wythes (hours).

as = Coefficient for continuous airspace.

Values for $R_n^{0.59}$ for use in Equation 7-10 are given in Table 721.4.1(3). The coefficient for each continuous airspace of $1/2$ inch to $3^1/_2$ inches (12.7 to 89 mm) separating two individual wythes shall be 0.3.

❖ Tests have shown that a continuous airspace separating two wythes of masonry can also increase the fire-resistance rating of a clay masonry wall. According to BMS 92 of the NBS, the coefficient for continuous airspace (as) between $1/2$ inch (12.7 mm) and $3^1/_2$ inches (89 mm) is 0.3. Equation 7-10 provides the formula for calculating the fire-resistance ratings involving an airspace if the fire-resistance rating of the wythes separated by the airspace are known. The fire-resistance ratings for various wythes can be obtained from Tables 721.4.1(1) and 721.4.1(2). The procedure for using Equation 7-10 is the same as Equation 7-9, except that a coefficient for airspace (as) is substituted for the coefficient of plaster (pl).

721.4.1.4 Nonsymmetrical assemblies. For a wall having no finish on one side or having different types or thicknesses of finish on each side, the calculation procedures of this section shall be performed twice, assuming either side to be the fire-exposed side of the wall. The fire resistance of the wall shall not exceed the lower of the two values determined.

Exception: For exterior walls with more than 5 feet (1524 mm) of horizontal separation, the fire shall be assumed to occur on the interior side only.

❖ Except for exterior walls having more than 5 feet (1524 mm) of horizontal separation, Section 704.5 requires that walls be rated for exposure to fire from both sides; therefore, two calculation procedures must be performed, which assumes each side to be the fire-exposed side. Two calculations are not necessary for exterior walls with more than 5 feet (1524 mm) of horizontal separation or for other walls that are symmetrical (i.e., having the same type and thickness of finish on each side). The calculated fire-resistance rating must not exceed the lower of the two values determined, assuming that each side is the fire-exposed side.

721.4.2 Multiwythe walls. The fire-resistance rating for walls or partitions consisting of two or more dissimilar wythes shall be permitted to be determined by the formula:

$$R = (R_1^{0.59} + R_2^{0.59} + ... + R_n^{0.59})^{1.7} \qquad \textbf{(Equation 7-11)}$$

where:

R = The fire endurance of the assembly (hours).

R_1, R_2 and R_n = The fire endurance of the individual wythes (hours).

Values for $R_n^{0.59}$ for use in Equation 7-11 are given in Table 721.4.1(3).

❖ Typically, the fire-resistance-rating period for clay masonry walls is determined by the temperature rise on the unexposed side of the wall. This criterion is what Equation 7-11 is based on. According to the general theory of heat transmission, if walls of the same material are exposed to a heat source that maintains a constant surface temperature on the exposed side and the unexposed side is protected against heat loss, the time in which a given temperature will be attained on the unexposed side will vary as the square of the wall thickness does.

In the standard ASTM E 119 test, which involves specified conditions of temperature measurement and a fire that increases the temperature at the exposed surface of the wall as the test proceeds, the time required to attain a given temperature rise on the unexposed side will differ when the temperature on the exposed side remains constant at the initial exposure temperature for any period. A degree of correlation between test results and calculations can be obtained by assuming the variation to be according to a lower power of n than the second. The fire resistance of the wall can then be expressed by the formula:

$$R = (CV)^n$$

where

R = Fire-resistance period;

C = Coefficient depending on the material, design of the wall and the units of measurement of R and V;

V = Volume of solid material per unit area of wall surface; and

n = Exponent depending on the rate of increase of temperature at the exposed face of the wall.

For walls of a given material and design, it was found that an increase of 50 percent in volume of solid material per unit area of wall surface resulted in a 100-percent increase in the fire-resistance period. This correlates to a value of 1.7 for n. A value of n less than 2 is expected, since an increasing temperature on the exposed surface would tend to shorten the fire-resistance rating of walls that would qualify for a higher rating.

The fire-resistance rating of a wall may be expressed

in terms of the fire-resistance rating of the conjoined wythes or laminae of the wall as follows:

$$R_1 = (C_1V_1)^n$$
$$R_2 = (C_2V_2)^n$$
$$R_n = (c_nV_n)^n$$

where R_1, R_2, and R_n are the fire resistances of each conjoined wythe.

The fire-resistance period of the composite wall will be:

$$R = \left(\sum_{i=1}^{n} C_iV_i \right)^{n_0}$$

$$R = \left(C_1V_1 + C_2V_2 + \ldots C_nV_n \right)^{n_0}$$

Substituting for C_{1V1}, C_{2V2}, and C_{nVn} from the above relations yields:

$$R = \left(R_1^{1/n_0} + R_2^{1/n_0} + R_n^{1/n_0} \right)^{n_0}$$

Substituting 1.7 for n_0 and 0.59 for $1/n_0$, the general formula becomes:

$$R = \left(R_1^{0.59} + R_2^{0.59} \ldots + R_n^{0.59} \right)^{1.7}$$

Equation 7-11 was developed by the NBS in the early 1940s and first appeared in Appendix B of BMS92 entitled "Fire Resistance Classifications of Building Construction" by the NBS. It is noted that the fire-resistance rating has been expressed in terms of the fire-resistance rating of the component laminae of the wall, which need not be of the same material and design.

721.4.2.1 Multiwythe walls of different material. For walls that consist of two or more wythes of different materials (concrete or concrete masonry units) in combination with clay ma-

sonry units, the fire-resistance rating of the different materials shall be permitted to be determined from Table 721.2.1.1 for concrete; Table 721.3.2 for concrete masonry units or Table 721.4.1(1) or 721.4.1(2) for clay and tile masonry units.

❖ A multiwythe wall (i.e., a wall consisting of two or more dissimilar materials) has a greater fire-resistance rating than a simple summation of the fire endurance rating of the various layers. Equation 7-11 permits a calculated fire-resistance rating if the fire-resistance endurance rating are known for each dissimilar material. This section lists the applicable tables from Section 721 that can be used in conjunction with clay masonry walls to determine the total fire-resistance rating of a multiwythe wall composed of a combination of concrete, concrete masonry or clay masonry units.

721.4.3 Reinforced clay masonry lintels. Fire-resistance ratings for clay masonry lintels shall be determined based on the nominal width of the lintel and the minimum covering for the longitudinal reinforcement in accordance with Table 721.4.1(5).

721.4.4 Reinforced clay masonry columns. The fire-resistance ratings shall be determined based on the last plan dimension of the column in accordance with Table 721.4.1(6). The minimum cover for longitudinal reinforcement shall be 2 inches (51 mm).

721.5 Steel assemblies. The provisions of this section contain procedures by which the fire-resistance ratings of steel assemblies are established by calculations.

721.5.1 Structural steel columns. The fire-resistance ratings of steel columns shall be based on the size of the element and the type of protection provided in accordance with this section.

TABLE 721.5.1(1) through TABLE 721.5.1(10). See page 7-146 to 154.

TABLE 721.5.1(1)   FIRE-RESISTANCE-RATED CONSTRUCTION

TABLE 721.5.1(1)
W/D RATIOS FOR STEEL COLUMNS

STRUCTURAL SHAPE	CONTOUR PROFILE	BOX PROFILE	STRUCTURAL SHAPE	CONTOUR PROFILE	BOX PROFILE
W14 × 233	2.49	3.65	W10 × 112	1.78	2.57
× 211	2.28	3.35	× 100	1.61	2.33
× 193	2.10	3.09	× 88	1.43	2.08
× 176	1.93	2.85	× 77	1.26	1.85
× 159	1.75	2.60	× 68	1.13	1.66
× 145	1.61	2.39	× 60	1.00	1.48
× 132	1.52	2.25	× 54	0.91	1.34
× 120	1.39	2.06	× 49	0.83	1.23
× 109	1.27	1.88	× 45	0.87	1.24
× 99	1.16	1.72	× 39	0.76	1.09
× 90	1.06	1.58	× 33	0.65	0.93
× 82	1.20	1.68			
× 74	1.09	1.53	W8 × 67	1.34	1.94
× 68	1.01	1.41	× 58	1.18	1.71
× 61	0.91	1.28	× 48	0.99S	1.44
× 53	0.89	1.21	× 40	0.83	1.23
× 48	0.81	1.10	× 35	0.73	1.08
× 43	0.73	0.99	× 31	0.65	0.97
			× 28	0.67	0.96
W12 × 190	2.46	3.51	× 24	0.58	0.83
× 170	2.22	3.20	× 21	0.57	0.77
× 152	2.01	2.90	× 18	0.49	0.67
× 136	1.82	2.63			
× 120	1.62	2.36	W6 × 25	0.69	1.00
× 106	1.44	2.11	× 20	0.56	0.82
× 96	1.32	1.93	× 16	0.57	0.78
× 87	1.20	1.76	× 15	0.42	0.63
× 79	1.10	1.61	× 12	0.43	0.60
× 72	1.00	1.48	× 9	0.33	0.46
× 65	0.91	1.35			
× 58	0.91	1.31	W5 × 19	0.64	0.93
× 53	0.84	1.20	× 16	0.54	0.80
× 50	0.89	1.23			
× 45	0.81	1.12	W4 × 13	0.54	0.79
× 40	0.72	1.00			

For SI: 1 pound per linear foot per inch = 0.059 kg/m/mm.

TABLE 721.5.1(2)
PROPERTIES OF CONCRETE

PROPERTY	NORMAL WEIGHT CONCRETE	STRUCTURAL LIGHTWEIGHT CONCRETE
Thermal conductivity (k_c)	0.95 Btu/hr ft °F	0.35 Btu/hr ft °F
Specific heat (c_c)	0.20 Btu/lb °F	0.20 Btu/lb °F
Density (P_c)	145 lb/ft^3	110 lb/ft^3
Equilibrium (free) moisture content (m) by volume	4%	5%

For SI: 1 inch = 25.4 mm, 1 foot = 304.8 mm, 1 lb/ft^3 = 16.0185 kg/m^3, Btu/hr ft °F = 1.731 W/(m · K)

TABLE 721.5.1(3)
THERMAL CONDUCTIVITY OF CONCRETE OR CLAY
MASONRY UNITS

DENSITY (d_m) OF UNITS (lb/ft^3)	THERMAL CONDUCTIVITY (K) OF UNITS (Btu/hr ft °F)
Concrete Masonry Units	
80	0.207
85	0.228
90	0.252
95	0.278
100	0.308
105	0.340
110	0.376
115	0.416
120	0.459
125	0.508
130	0.561
135	0.620
140	0.685
145	0.758
150	0.837
Clay Masonry Units	
120	1.25
130	2.25

For SI: 1 pound per cubic foot = 16.0185 kg/m^3, Btu per hour foot °F = 1.731 W/(m · K).

TABLE 721.5.1(4)   FIRE-RESISTANCE-RATED CONSTRUCTION

TABLE 721.5.1(4)
WEIGHT-TO-HEATED-PERIMETER RATIOS (*W/D*)
FOR TYPICAL WIDE FLANGE BEAM AND GIRDER SHAPES

STRUCTURAL SHAPE	CONTOUR PROFILE	BOX PROFILE	STRUCTURAL SHAPE	CONTOUR PROFILE	BOX PROFILE
W36 × 300	2.47	3.33	× 68	0.92	1.21
× 280	2.31	3.12	× 62	0.92	1.14
× 260	2.16	2.92	× 55	0.82	1.02
× 245	2.04	2.76			
× 230	1.92	2.61	W21 × 147	1.83	2.60
× 210	1.94	2.45	× 132	1.66	2.35
× 194	1.80	2.28	× 122	1.54	2.19
× 182	1.69	2.15	× 111	1.41	2.01
× 170	1.59	2.01	× 101	1.29	1.84
× 160	1.50	1.90	× 93	1.38	1.80
× 150	1.41	1.79	× 83	1.24	1.62
× 135	1.28	1.63	× 73	1.10	1.44
			× 68	1.03	1.35
W33 × 241	2.11	2.86	× 62	0.94	1.23
× 221	1.94	2.64	× 57	0.93	1.17
× 201	1.78	2.42	× 50	0.83	1.04
× 152	1.51	1.94	× 44	0.73	0.92
× 141	1.41	1.80			
× 130	1.31	1.67	W18 × 119	1.69	2.42
× 118	1.19	1.53	× 106	1.52	2.18
			× 97	1.39	2.01
W30 × 211	2.00	2.74	× 86	1.24	1.80
× 191	1.82	2.50	× 76	1.11	1.60
× 173	1.66	2.28	× 71	1.21	1.59
× 132	1.45	1.85	× 65	1.11	1.47
× 124	1.37	1.75	× 60	1.03	1.36
× 116	1.28	1.65	× 55	0.95	1.26
× 108	1.20	1.54	× 50	0.87	1.15
× 99	1.10	1.42	× 46	0.86	1.09
			× 40	0.75	0.96
W27 × 178	1.85	2.55	× 35	0.66	0.85
× 161	1.68	2.33			
× 146	1.53	2.12	W16 × 100	1.56	2.25
× 114	1.36	1.76	× 89	1.40	2.03
× 102	1.23	1.59	× 77	1.22	1.78
× 94	1.13	1.47	× 67	1.07	1.56
× 84	1.02	1.33	× 57	1.07	1.43
			× 50	0.94	1.26
			× 45	0.85	1.15
W24 × 162	1.85	2.57	× 40	0.76	1.03
× 146	1.68	2.34	× 36	0.69	0.93
× 131	1.52	2.12	× 31	0.65	0.83
× 117	1.36	1.91	× 26	0.55	0.70
× 104	1.22	1.71			
× 94	1.26	1.63	W14 × 132	1.83	3.00
× 84	1.13	1.47	× 120	1.67	2.75
× 76	1.03	1.34	× 109	1.53	2.52

(continued)

TABLE 721.5.1(4)—continued
WEIGHT-TO-HEATED-PERIMETER RATIOS (*W/D*)
FOR TYPICAL WIDE FLANGE BEAM AND GIRDER SHAPES

STRUCTURAL SHAPE	CONTOUR PROFILE	BOX PROFILE	STRUCTURAL SHAPE	CONTOUR PROFILE	BOX PROFILE
× 99	1.39	2.31	× 30	0.79	1.12
× 90	1.27	2.11	× 26	0.69	0.98
× 82	1.41	2.12	× 22	0.59	0.84
× 74	1.28	1.93	× 19	0.59	0.78
× 68	1.19	1.78	× 17	0.54	0.70
× 61	1.07	1.61	× 15	0.48	0.63
× 53	1.03	1.48	× 12	0.38	0.51
× 48	0.94	1.35			
× 43	0.85	1.22	W8 × 67	1.61	2.55
× 38	0.79	1.09	× 58	1.41	2.26
× 34	0.71	0.98	× 48	1.18	1.91
× 30	0.63	0.87	× 40	1.00	1.63
× 26	0.61	0.79	× 35	0.88	1.44
× 22	0.52	0.68	× 31	0.79	1.29
			× 28	0.80	1.24
W12 × 87	1.44	2.34	× 24	0.69	1.07
× 79	1.32	2.14	× 21	0.66	0.96
× 72	1.20	1.97	× 18	0.57	0.84
× 65	1.09	1.79	× 15	0.54	0.74
× 58	1.08	1.69	× 13	0.47	0.65
× 53	0.99	1.55	× 10	0.37	0.51
× 50	1.04	1.54			
× 45	0.95	1.40	W6 × 25	0.82	1.33
× 40	0.85	1.25	× 20	0.67	1.09
× 35	0.79	1.11	× 16	0.66	0.96
× 30	0.69	0.96	× 15	0.51	0.83
× 26	0.60	0.84	× 12	0.51	0.75
× 22	0.61	0.77	× 9	0.39	0.57
× 19	0.53	0.67			
× 16	0.45	0.57	W5 × 19	0.76	1.24
× 14	0.40	0.50	× 16	0.65	1.07
W10 × 112	2.14	3.38	W4 × 13	0.65	1.05
× 100	1.93	3.07			
× 88	1.7	2.75			
× 77	1.52	2.45			
× 68	1.35	2.20			
× 60	1.20	1.97			
× 54	1.09	1.79			
× 49	0.99	1.64			
× 45	1.03	1.59			
× 39	0.94	1.40			
× 33	0.77	1.20			

For SI: Pounds per linear foot per inch = 0.059 kg/m/mm.

TABLE 721.5.1(5)   FIRE-RESISTANCE-RATED CONSTRUCTION

TABLE 721.5.1(5)
FIRE RESISTANCE OF CONCRETE MASONRY PROTECTED STEEL COLUMNS

COLUMN SIZE	CONCRETE MASONRY DENSITY POUNDS PER CUBIC FOOT	1-hour	2-hour	3-hour	4-hour	COLUMN SIZE	CONCRETE MASONRY DENSITY POUNDS PER CUBIC FOOT	1-hour	2-hour	3-hour	4-hour
W14 × 82	80	0.74	1.61	2.36	3.04	W10 × 68	80	0.72	1.58	2.33	3.01
	100	0.89	1.85	2.67	3.40		100	0.87	1.83	2.65	3.38
	110	0.96	1.97	2.81	3.57		110	0.94	1.95	2.79	3.55
	120	1.03	2.08	2.95	3.73		120	1.01	2.06	2.94	3.72
W14 × 68	80	0.83	1.70	2.45	3.13	W10 × 54	80	0.88	1.76	2.53	3.21
	100	0.99	1.95	2.76	3.49		100	1.04	2.01	2.83	3.57
	110	1.06	2.06	2.91	3.66		110	1.11	2.12	2.98	3.73
	120	1.14	2.18	3.05	3.82		120	1.19	2.24	3.12	3.90
W14 × 53	80	0.91	1.81	2.58	3.27	W10 × 45	80	0.92	1.83	2.60	3.30
	100	1.07	2.05	2.88	3.62		100	1.08	2.07	2.90	3.64
	110	1.15	2.17	3.02	3.78		110	1.16	2.18	3.04	3.80
	120	1.22	2.28	3.16	3.94		120	1.23	2.29	3.18	3.96
W14 × 43	80	1.01	1.93	2.71	3.41	W10 × 33	80	1.06	2.00	2.79	3.49
	100	1.17	2.17	3.00	3.74		100	1.22	2.23	3.07	3.81
	110	1.25	2.28	3.14	3.90		110	1.30	2.34	3.20	3.96
	120	1.32	2.38	3.27	4.05		120	1.37	2.44	3.33	4.12
W12 × 72	80	0.81	1.66	2.41	3.09	W8 × 40	80	0.94	1.85	2.63	3.33
	100	0.91	1.88	2.70	3.43		100	1.10	2.10	2.93	3.67
	110	0.99	1.99	2.84	3.60		110	1.18	2.21	3.07	3.83
	120	1.06	2.10	2.98	3.76		120	1.25	2.32	3.20	3.99
W12 × 58	80	0.88	1.76	2.52	3.21	W8 × 31	80	1.06	2.00	2.78	3.49
	100	1.04	2.01	2.83	3.56		100	1.22	2.23	3.07	3.81
	110	1.11	2.12	2.97	3.73		110	1.29	2.33	3.20	3.97
	120	1.19	2.23	3.11	3.89		120	1.36	2.44	3.33	4.12
W12 × 50	80	0.91	1.81	2.58	3.27	W8 × 24	80	1.14	2.09	2.89	3.59
	100	1.07	2.05	2.88	3.62		100	1.29	2.31	3.16	3.90
	110	1.15	2.17	3.02	3.78		110	1.36	2.42	3.28	4.05
	120	1.22	2.28	3.16	3.94		120	1.43	2.52	3.41	4.20
W12 × 40	80	1.01	1.94	2.72	3.41	W8 × 18	110	1.22	2.20	3.01	3.72
	100	1.17	2.17	3.01	3.75		100	1.36	2.40	3.25	4.01
	110	1.25	2.28	3.14	3.90		110	1.42	2.50	3.37	4.14
	120	1.32	2.39	3.27	4.06		120	1.48	2.59	3.49	4.28

(continued)

TABLE 721.5.1(5)—continued
FIRE RESISTANCE OF CONCRETE MASONRY PROTECTED STEEL COLUMNS

NOMINAL TUBE SIZE (inches)	CONCRETE MASONRY DENSITY, POUNDS PER CUBIC FOOT	MINIMUM REQUIRED EQUIVALENT THICKNESS FOR FIRE-RESISTANCE RATING OF CONCRETE. MASONRY PROTECTION ASSEMBLY T_e, (inches)				NOMINAL PIPE SIZE (inches)	CONCRETE MASONRY DENSITY, POUNDS PER CUBIC FOOT	MINIMUM REQUIRED EQUIVALENT THICKNESS FOR FIRE-RESISTANCE RATING OF CONCRETE. MASONRY PROTECTION ASSEMBLY T_e, (inches)			
		1-hour	2-hour	3-hour	4-hour			1-hour	2-hour	3-hour	4-hour
4 × 4 × 1/2 wall thickness	80	0.93	1.90	2.71	3.43	4 double extra strong 0.674 wall thickness	80	0.80	1.75	2.56	3.28
	100	1.08	2.13	2.99	3.76		100	0.95	1.99	2.85	3.62
	110	1.16	2.24	3.13	3.91		110	1.02	2.10	2.99	3.78
	120	1.22	2.34	3.26	4.06		120	1.09	2.20	3.12	3.93
4 × 4 × 3/8 wall thickness	80	1.05	2.03	2.84	3.57	4 extra strong 0.337 wall thickness	80	1.12	2.11	2.93	3.65
	100	1.20	2.25	3.11	3.88		100	1.26	2.32	3.19	3.95
	110	1.27	2.35	3.24	4.02		110	1.33	2.42	3.31	4.09
	120	1.34	2.45	3.37	4.17		120	1.40	2.52	3.43	4.23
4 × 4 × 1/4 wall thickness	80	1.21	2.20	3.01	3.73	4 standard 0.237 wall thickness	80	1.26	2.25	3.07	3.79
	100	1.35	2.40	3.26	4.02		100	1.40	2.45	3.31	4.07
	110	1.41	2.50	3.38	4.16		110	1.46	2.55	3.43	4.21
	120	1.48	2.59	3.50	4.30		120	1.53	2.64	3.54	4.34
6 × 6 × 1/2 wall thickness	80	0.82	1.75	2.54	3.25	5 double extra strong 0.750 wall thickness	80	0.70	1.61	2.40	3.12
	100	0.98	1.99	2.84	3.59		100	0.85	1.86	2.71	3.47
	110	1.05	2.10	2.98	3.75		110	0.91	1.97	2.85	3.63
	120	1.12	2.21	3.11	3.91		120	0.98	2.02	2.99	3.79
6 × 6 × 3/8 wall thickness	80	0.96	1.91	2.71	3.42	5 extra strong 0.375 wall thickness	80	1.04	2.01	2.83	3.54
	100	1.12	2.14	3.00	3.75		100	1.19	2.23	3.09	3.85
	110	1.19	2.25	3.13	3.90		110	1.26	2.34	3.22	4.00
	120	1.26	2.35	3.26	4.05		120	1.32	2.44	3.34	4.14
6 × 6 × 1/4 wall thickness	80	1.14	2.11	2.92	3.63	5 standard 0.258 wall thickness	80	1.20	2.19	3.00	3.72
	100	1.29	2.32	3.18	3.93		100	1.34	2.39	3.25	4.00
	110	1.36	2.43	3.30	4.08		110	1.41	2.49	3.37	4.14
	120	1.42	2.52	3.43	4.22		120	1.47	2.58	3.49	4.28
8 × 8 × 1/2 wall thickness	80	0.77	1.66	2.44	3.13	6 double extra strong 0.864 wall thickness	80	0.59	1.46	2.23	2.92
	100	0.92	1.91	2.75	3.49		100	0.73	1.71	2.54	3.29
	110	1.00	2.02	2.89	3.66		110	0.80	1.82	2.69	3.47
	120	1.07	2.14	3.03	3.82		120	0.86	1.93	2.83	3.63
8 × 8 × 3/8 wall thickness	80	0.91	1.84	2.63	3.33	6 extra strong 0.432 wall thickness	80	0.94	1.90	2.70	3.42
	100	1.07	2.08	2.92	3.67		100	1.10	2.13	2.98	3.74
	110	1.14	2.19	3.06	3.83		110	1.17	2.23	3.11	3.89
	120	1.21	2.29	3.19	3.98		120	1.24	2.34	3.24	4.04
8 × 8 × 1/4 wall thickness	80	1.10	2.06	2.86	3.57	6 standard 0.280 wall thickness	80	1.14	2.12	2.93	3.64
	100	1.25	2.28	3.13	3.87		100	1.29	2.33	3.19	3.94
	110	1.32	2.38	3.25	4.02		110	1.36	2.43	3.31	4.08
	120	1.39	2.48	3.38	4.17		120	1.42	2.53	3.43	4.22

For SI: 1 inch = 25.4 mm, 1 pound per cubic feet = 16.02 kg/m³.
Note: Tabulated values assume 1-inch air gap between masonry and steel section.

TABLE 721.5.1(6)   FIRE-RESISTANCE-RATED CONSTRUCTION

TABLE 721.5.1(6)
FIRE RESISTANCE OF CLAY MASONRY PROTECTED STEEL COLUMNS

COLUMN SIZE	CLAY MASONRY DENSITY, POUNDS PER CUBIC FOOT	MINIMUM REQUIRED EQUIVALENT THICKNESS FOR FIRE-RESISTANCE RATING OF CLAY. MASONRY PROTECTION ASSEMBLY T_e, (inches)				COLUMN SIZE	CLAY MASONRY DENSITY, POUNDS PER CUBIC FOOT	MINIMUM REQUIRED EQUIVALENT THICKNESS FOR FIRE-RESISTANCE RATING OF CLAY MASONRY PROTECTION ASSEMBLY T_e, (inches)			
		1-hour	2-hour	3-hour	4-hour			1-hour	2-hour	3-hour	4-hour
W14 × 82	120	1.23	2.42	3.41	4.29	W10 × 68	120	1.27	2.46	3.26	4.35
	130	1.40	2.70	3.78	4.74		130	1.44	2.75	3.83	4.80
W14 × 68	120	1.34	2.54	3.54	4.43	W10 × 54	120	1.40	2.61	3.62	4.51
	130	1.51	2.82	3.91	4.87		130	1.58	2.89	3.98	4.95
W14 × 53	120	1.43	2.65	3.65	4.54	W10 × 45	120	1.44	2.66	3.67	4.57
	130	1.61	2.93	4.02	4.98		130	1.62	2.95	4.04	5.01
W14 × 43	120	1.54	2.76	3.77	4.66	W10 × 33	120	1.59	2.82	3.84	4.73
	130	1.72	3.04	4.13	5.09		130	1.77	3.10	4.20	5.13
W12 × 72	120	1.32	2.52	3.51	4.40	W8 × 40	120	1.47	2.70	3.71	4.61
	130	1.50	2.80	3.88	4.84		130	1.65	2.98	4.08	5.04
W12 × 58	120	1.40	2.61	3.61	4.50	W8 × 31	120	1.59	2.82	3.84	4.73
	130	1.57	2.89	3.98	4.94		130	1.77	3.10	4.20	5.17
W12 × 50	120	1.43	2.65	3.66	4.55	W8 × 24	120	1.66	2.90	3.92	4.82
	130	1.61	2.93	4.02	4.99		130	1.84	3.18	4.28	5.25
W12 × 40	120	1.54	2.77	3.78	4.67	W8 × 18	120	1.75	3.00	4.01	4.91
	130	1.72	3.05	4.14	5.10		130	1.93	3.27	4.37	5.34

Steel tubing						Steel pipe					
Nominal tube size (inches)	Clay masonry density, pounds per cubic foot	Minimum required equivalent thickness for fire-resistance rating of clay. Masonry protection assembly T_e, (inches)				Nominal pipe size (inches)	Clay masonry density, pounds per cubic foot	Minimum required equivalent thickness for fire-resistance rating of clay. Masonry protection assembly T_e, (inches)			
		1-hour	2-hour	3-hour	4-hour			1-hour	2-hour	3-hour	4-hour
4 × 4 × 1/2 wall thickness	120	1.44	2.72	3.76	4.68	4 double extra strong 0.674 wall thickness	120	1.26	2.55	3.60	4.52
	130	1.62	3.00	4.12	5.11		130	1.42	2.82	3.96	4.95
4 × 4 × 3/8 wall thickness	120	1.56	2.84	3.88	4.78	4 extra strong 0.337 wall thickness	120	1.60	2.89	3.92	4.83
	130	1.74	3.12	4.23	5.21		130	1.77	3.16	4.28	5.25
4 × 4 × 1/4 wall thickness	120	1.72	2.99	4.02	4.92	4 standard 0.237 wall thickness	120	1.74	3.02	4.05	4.95
	130	1.89	3.26	4.37	5.34		130	1.92	3.29	4.40	5.37
6 × 6 × 1/2 wall thickness	120	1.33	2.58	3.62	4.52	5 double extra strong 0.750 wall thickness	120	1.17	2.44	3.48	4.40
	130	1.50	2.86	3.98	4.96		130	1.33	2.72	3.84	4.83
6 × 6 × 3/8 wall thickness	120	1.48	2.74	3.76	4.67	5 extra strong 0.375 wall thickness	120	1.55	2.82	3.85	4.76
	130	1.65	3.01	4.13	5.10		130	1.72	3.09	4.21	5.18
6 × 6 × 1/4 wall thickness	120	1.66	2.91	3.94	4.84	5 standard 0.258 wall thickness	120	1.71	2.97	4.00	4.90
	130	1.83	3.19	4.30	5.27		130	1.88	3.24	4.35	5.32
8 × 8 × 1/2 wall thickness	120	1.27	2.50	3.52	4.42	6 double extra strong 0.864 wall thickness	120	1.04	2.28	3.32	4.23
	130	1.44	2.78	3.89	4.86		130	1.19	2.60	3.68	4.67
8 × 8 × 3/8 wall thickness	120	1.43	2.67	3.69	4.59	6 extra strong 0.432 wall thickness	120	1.45	2.71	3.75	4.65
	130	1.60	2.95	4.05	5.02		130	1.62	2.99	4.10	5.08
8 × 8 × 1/4 wall thickness	120	1.62	2.87	3.89	4.78	6 standard 0.280 wall thickness	120	1.65	2.91	3.94	4.84
	130	1.79	3.14	4.24	5.21		130	1.82	3.19	4.30	5.27

TABLE 721.5.1(7)
MINIMUM COVER (inch) FOR STEEL COLUMNS
ENCASED IN NORMAL-WEIGHT CONCRETE[a]
[FIGURE 721.5.1(6)(c)]

STRUCTURAL SHAPE	FIRE-RESISTANCE RATING (hours)				
	1	1½	2	3	4
W14 × 233				1½	2
× 176			1		
× 132		1			2½
× 90	1			2	
× 61			1½		
× 48					3
× 43		1½		2½	
W12 × 152			1		2½
× 96		1		2	
× 65	1				
× 50			1½		3
× 40		1½		2½	
W10 × 88	1			2	
× 49					3
× 45	1	1½	1½		
× 39				2½	3½
× 33			2		
W8 × 67		1			3
× 58			1½		
× 48	1			2½	
× 31		1½			3½
× 21			2		
× 18				3	4
W6 × 25		1½	2		3½
× 20				3	
× 16	1	2			4
× 15					
× 9	1½		2½	3½	

For SI: 1 inch = 25.4 mm.
a. The tabulated thicknesses are based upon the assumed properties of normal-weight concrete given in Table 721.5.1(2).

TABLE 721.5.1(8)
MINIMUM COVER (inch) FOR STEEL COLUMNS
ENCASED IN STRUCTURAL LIGHTWEIGHT CONCRETE[a]
[FIGURE 721.5.1(6)(c)]

STRUCTURAL SHAPE	FIRE-RESISTANCE RATING (HOURS)				
	1	1½	2	3	4
W14 × 233				1	
× 193					1½
× 74	1	1	1	1½	2
× 61					
× 43			1½	2	2½
W12 × 65				1½	2
× 53	1	1	1		
× 40			1½	2	2½
W10 × 112					2
× 88	1		1	1½	
× 60		1			
× 33			1½	2	2½
W8 × 35					2½
× 28	1	1		2	
× 24			1½		3
× 18		1½		2½	

For SI: 1 inch = 25.4 mm.
a. The tabulated thicknesses are based upon the assumed properties of structural lightweight concrete given in Table 721.5.1(2).

TABLE 721.5.1(9) – TABLE 721.5.1(10)　　　　　　　　　　　　　FIRE-RESISTANCE-RATED CONSTRUCTION

TABLE 721.5.1(9)
MINIMUM COVER (inch) FOR STEEL COLUMNS
IN NORMAL-WEIGHT PRECAST COVERS[a]
[FIGURE 721.5.1(6)(a)]

STRUCTURAL SHAPE	FIRE-RESISTANCE RATING (hours)				
	1	1½	2	3	4
W14 × 233			1½	2½	3
× 211					
× 176	1½				3½
× 145		1½	2		
× 109				3	
× 99					
× 61					4
× 43		2	2½	3½	4½
W12 × 190			1½	2½	3½
× 152					
× 120		1½	2		
× 96	1½			3	
× 87					4
× 58					
× 40		2	2½	3½	4½
W10 × 112					3½
× 88		1½	2	3	
× 77	1½				4
× 54		2	2½	3½	
× 33					4½
W8 × 67		1½	2	3	
× 58					4
× 48	1½	2	2½	3½	
× 28					
× 21					4½
× 18		2½	3	4	
W6 × 25		2	2½	3½	
× 20	1½				4½
× 16			3		
× 12	2	2½		4	
× 9					5

For SI:　1 inch = 25.4 mm.

a. The tabulated thicknesses are based upon the assumed properties of normal-weight concrete given in Table 721.5.1(2).

TABLE 721.5.1(10)
MINIMUM COVER (inch) FOR STEEL COLUMNS
IN STRUCTURAL LIGHTWEIGHT PRECAST COVERS[a]
[FIGURE 721.5.1(6)(a)]

STRUCTURAL SHAPE	FIRE-RESISTANCE RATING (hours)				
	1	1½	2	3	4
W14 × 233					2½
× 176				2	
× 145			1½		
× 132	1½	1½			3
× 109					
× 99				2½	
× 68			2		
× 43				3	3½
W12 × 190					2½
× 152				2	
× 136			1½		
× 106					3
× 96	1½	1½		2½	
× 87					
× 65			2		
× 40				3	3½
W10 × 112				2	
× 100			1½		3
× 88					
× 77	1½	1½		2½	
× 60			2		
× 39				3	3½
× 33		2			
W8 × 67			1½	2½	3
× 48		1½			
× 35	1½		2		3½
× 28				3	
× 18		2	2½		4
W6 × 25			2	3	3½
× 15	1½	2			
× 9			2½	3½	4

For SI:　1 inch = 25.4 mm.

a. The tabulated thicknesses are based upon the assumed properties of structural lightweight concrete given in Table 721.5.1(2).

　　　　　　　　　　　　　　　　　　　　　　　　2003 INTERNATIONAL BUILDING CODE® COMMENTARY

721.5.1.1 General. These procedures establish a basis for determining the fire resistance of column assemblies as a function of the thickness of fire-resistant material and, the weight, W, and heated perimeter, D, of steel columns. As used in these sections, W is the average weight of a structural steel column in pounds per linear foot. The heated perimeter, D, is the inside perimeter of the fire-resistant material in inches as illustrated in Figure 721.5.1(1).

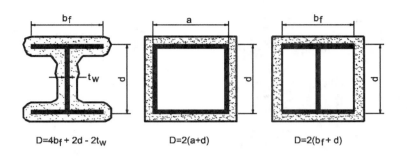

$D = 4b_f + 2d - 2t_w$ $D = 2(a + d)$ $D = 2(b_f + d)$

FIGURE 721.5.1(1)
DETERMINATION OF THE HEATED PERIMETER

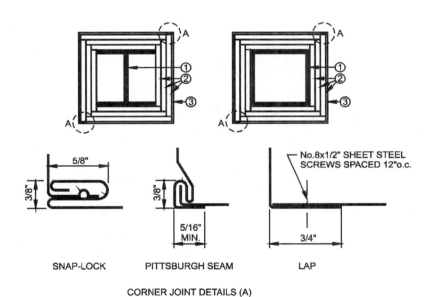

SNAP-LOCK   PITTSBURGH SEAM   LAP

CORNER JOINT DETAILS (A)

For SI: 1 inch = 25.4 mm

FIGURE 721.5.1(2)
GYPSUM WALLBOARD PROTECTED STRUCTURAL STEEL COLUMNS WITH SHEET STEEL COLUMN COVERS

1. Structural steel column, either wide flange or tubular shapes.
2. Type X gypsum wallboard in accordance with ASTM C 36. For single-layer applications, the wallboard shall be applied vertically with no horizontal joints. For multiple-layer applications, horizontal joints are permitted at a minimum spacing of 8 feet, provided that the joints in successive layers are staggered at least 12 inches. The total required thickness of wallboard shall be determined on the basis of the specified fire-resistance rating and the weight-to-heated-perimeter ratio (W/D) of the column. For fire-resistance ratings of 2 hours or less, one of the required layers of gypsum wallboard may be applied to the exterior of the sheet steel column covers with 1-inch long Type S screws spaced 1 inch from the wallboard edge and 8 inches on center. For such installations, 0.0149-inch minimum thickness galvanized steel corner beads with $1^1/_2$-inch legs shall be attached to the wallboard with Type S screws spaced 12 inches on center.

3. For fire-resistance ratings of 3 hours or less, the column covers shall be fabricated from 0.0239-inch minimum thickness galvanized or stainless steel. For 4-hour fire-resistance ratings, the column covers shall be fabricated from 0.0239-inch minimum thickness stainless steel. The column covers shall be erected with the Snap Lock or Pittsburgh joint details.

   For fire-resistance ratings of 2 hours or less, column covers fabricated from 0.0269-inch minimum thickness galvanized or stainless steel shall be permitted to be erected with lap joints. The lap joints shall be permitted to be located anywhere around the perimeter of the column cover. The lap joints shall be secured with $^1/_2$-inch-long No. 8 sheet metal screws spaced 12 inches on center.

   The column covers shall be provided with a minimum expansion clearance of $^1/_8$ inch per linear foot between the ends of the cover and any restraining construction.

FIGURE 721.5.1(3) – FIGURE 721.5.1(6)

FIRE-RESISTANCE-RATED CONSTRUCTION

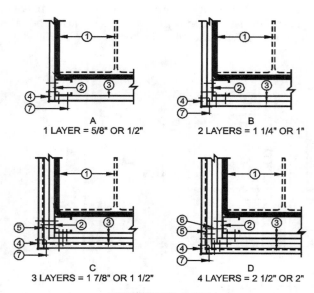

A
1 LAYER = 5/8" OR 1/2"

B
2 LAYERS = 1 1/4" OR 1"

C
3 LAYERS = 1 7/8" OR 1 1/2"

D
4 LAYERS = 2 1/2" OR 2"

FIGURE 721.5.1(3)
GYPSUM WALLBOARD PROTECTED STRUCTURAL STEEL COLUMNS WITH STEEL STUD/SCREW ATTACHMENT SYSTEM

For SI: 1 inch = 25.4 mm.

1. Structural steel column, either wide flange or tubular shapes.

2. $1^5/_8$-inch deep studs fabricated from 0.0179-inch minimum thickness galvanized steel with $1^5/_{16}$ or $1^7/_{16}$-inch legs. The length of the steel studs shall be $^1/_2$ inch less than the height of the assembly.

3. Type X gypsum wallboard in accordance with ASTM C 36. For single-layer applications, the wallboard shall be applied vertically with no horizontal joints. For multiple-layer applications, horizontal joints are permitted at a minimum spacing of 8 feet, provided that the joints in successive layers are staggered at least 12 inches. The total required thickness of wallboard shall be determined on the basis of the specified fire-resistance rating and the weight-to-heated-perimeter ratio (*W/D*) of the column.

4. Galvanized 0.0149-inch minimum thickness steel corner beads with $1^1/_2$-inch legs attached to the wallboard with 1-inch-long Type S screws spaced 12 inches on center.

5. No. 18 SWG steel tie wires spaced 24 inches on center.

6. Sheet metal angles with 2-inch legs fabricated from 0.0221-inch minimum thickness galvanized steel.

7. Type S screws, 1-inch long, shall be used for attaching the first layer of wallboard to the steel studs and the third layer to the sheet metal angles at 24 inches on center. Type S screws $1^3/_4$-inch long shall be used for attaching the second layer of wallboard to the steel studs and the fourth layer to the sheet metal angles at 12 inches on center. Type S screws $2^1/_4$ inches long shall be used for attaching the third layer of wallboard to the steel studs at 12 inches on center.

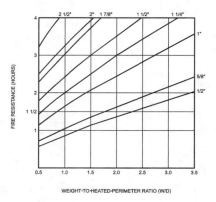

For SI: 1 inch = 25.4 mm, 1 pound per linear foot/ inch = 0.059 kg/m/mm.

a. The W/D ratios for typical wide flange columns are listed in Table 721.5.1(1). For other column shapes, the W/D ratios shall be determined in accordance with Section 720.5.1.1.

FIGURE 720.5.1(4)
FIRE RESISTANCE OF STRUCTURAL STEEL COLUMNS PROTECTED WITH VARIOUS THICKNESSES OF TYPE X GYPSUM WALLBOARD

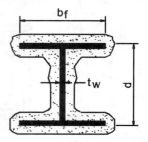

FIGURE 720.5.1(5)
WIDE FLANGE STRUCTURAL STEEL COLUMNS WITH SPRAY-APPLIED FIRE-RESISTANT MATERIALS

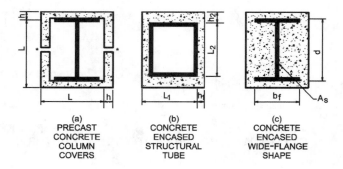

(a)
PRECAST CONCRETE COLUMN COVERS

(b)
CONCRETE ENCASED STRUCTURAL TUBE

(c)
CONCRETE ENCASED WIDE-FLANGE SHAPE

For SI: 1 inch = 25.4 mm.

a. When the inside perimeter of the concrete is not square, L shall be taken as the average of L_1 and L_2. When the thickness of concrete cover is not constant, h shall be taken as the average of h_1 and h_2.

b. Joints shall be protected with a minimum 1-inch thickness of ceramic fiber blanket but in no case less than one-half the thickness of the column cover (see Section 720.2.1.3).

FIGURE 720.5.1(6)
CONCRETE PROTECTED STRUCTURAL STEEL COLUMNS[a]

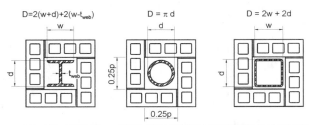

W SHAPE COLUMN STEEL PIPE COLUMN STRUCTURAL TUBE COLUMN

For SI: 1 inch = 25.4 mm.

FIGURE 720.5.1(7)
CONCRETE OR CLAY MASONRY PROTECTED
STRUCTURAL STEEL COLUMNS

d = Depth of a wide flange column, outside diameter of pipe column, or outside dimension of structural tubing column (inches).

t_{web} = Thickness of web of wide flange column (inches).

w = Width of flange of wide flange column (inches).

721.5.1.1.1 Nonload-bearing protection. The application of these procedures shall be limited to column assemblies in which the fire-resistant material is not designed to carry any of the load acting on the column.

721.5.1.1.2 Embedments. In the absence of substantiating fire-endurance test results, ducts, conduit, piping, and similar mechanical, electrical, and plumbing installations shall not be embedded in any required fire-resistant materials.

721.5.1.1.3 Weight-to-perimeter ratio. Table 721.5.1(1) contains weight-to-heated-perimeter ratios (W/D) for both contour and box fire-resistant profiles, for the wide flange shapes most often used as columns. For different fire-resistant protection profiles or column cross sections, the weight-to-heated-perimeter ratios (W/D) shall be determined in accordance with the definitions given in this section.

721.5.1.2 Gypsum wallboard protection. The fire resistance of structural steel columns with weight- to-heated-perimeter ratios (W/D) less than or equal to 3.65 and which are protected with Type X gypsum wallboard shall be permitted to be determined from the following expression:

$$R = 130 \left[\frac{h\left(W'/D\right)}{2} \right]^{0.75}$$ **(Equation 7-12)**

where:

R =   Fire resistance (minutes).

h =   Total thickness of gypsum wallboard (inches).

D =   Heated perimeter of the structural steel column (inches).

W' =   Total weight of the structural steel column and gypsum wallboard protection (pounds per linear foot).

W' = $W + 50hD/144$.

721.5.1.2.1 Attachment. The gypsum wallboard shall be supported as illustrated in either Figure 721.5.1(2) for fire-resistance ratings of 4 hours or less, or Figure 721.5.1(3) for fire-resistance ratings of 3 hours or less.

EXAMPLE:

GIVEN: A W12 × 136 column protected by $^1/_2$-inch (12.7 mm) Type X gypsum wallboard as shown in Figure 721.5.1.2.1.

FIND: The fire-resistance rating of the column.

SOLUTION:

1. The heated perimeter D = 2(12.4 inches + 13.4 - inches) = 51.6 inches (1311 mm).

2. W' = 136 + (50)(0.5)(51.6)/144 = 144.96.

3. W'/Δ = 144.96 / 51.62 = 2.81.

4. R = 130[(0.5)(2.81)/2]$^{0.75}$ = 99.7 minutes (1.66 hours).

5. Alternatively, from Table 721.5.1(1), the W/D ratio for a W12 × 136 is tabulated as 2.63. From Figure 721.5.1(4) using a W/D ratio of 2.63 and $^1/_2$-inch (12.7 mm) gypsum wallboard, an R-value of approximately 1.6 hours is obtained.

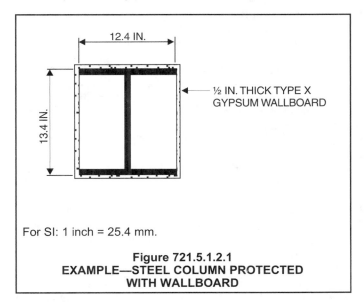

For SI: 1 inch = 25.4 mm.

Figure 721.5.1.2.1
EXAMPLE—STEEL COLUMN PROTECTED
WITH WALLBOARD

721.5.1.2.2 Gypsum wallboard equivalent to concrete. The determination of the fire resistance of structural steel columns from Figure 721.5.1(4) is permitted for various thicknesses of gypsum wallboard as a function of the weight-to-heated-perimeter ratio (W/D) of the column. For structural steel columns with weight-to-heated-perimeter ratios (W/D) greater than 3.65, the thickness of gypsum wallboard required for specified fire-resistance ratings shall be the same as the thickness determined for a W14 x 233 wide flange shape.

721.5.1.3 Spray-applied fire-resistant materials. The fire resistance of wide-flange structural steel columns protected with spray-applied fire-resistant materials, as illustrated in Figure

721.5.1(5), shall be permitted to be determined from the following expression:

$$R = \left[C_1 \left(W / D \right) + C_2 \right] h \qquad \textbf{(Equation 7-13)}$$

where:

R = Fire resistance (minutes).

h = Thickness of spray-applied fire-resistant material (inches).

D = Heated perimeter of the structural steel column (inches).

C_1 and C_2 = Material-dependent constants.

W = Weight of structural steel column (pounds per linear foot).

❖ This section sets forth procedures for determining the fire resistance of structural steel columns protected with spray-applied cementitious or mineral fiber fire protection materials. These procedures are based upon an empirical equation, which includes two material-dependent constants. As a result, in order to apply this equation, the values of these two constants must be determined for specific fire protection materials. The purpose of this section is to provide guidance for the determination of these constants so that the resulting equation will be reasonably accurate, and yet slightly conservative, over the range of column shapes for the test data that is available.

Two different techniques are available for determining the two constants. The first requires a knowledge of thermal conductivity and specific heat of the fire protection material at elevated temperatures. Data of this nature is both difficult and expensive to obtain with any reasonable degree of accuracy. As a result, this technique will probably not be widely used, and accordingly, it will not be described in this section.

The second technique involves the use of Equation 7-13 as a means for interpolating between ASTM E 119 fire endurance test results on different structural steel columns. Since this technique will undoubtedly be the most widely used, it is described in detail in this section. It is, however, important to recognize that a wide variety of both large- and small-scale tests can be used to accurately determine the required constants.

Inherent in Equation 7-13 are two general assumptions. The first is that the ratio of fire endurance time to the thickness of fire protection material (R/h) is essentially constant for a given column of the weight to-heated-perimeter (W/D) ratio. The second assumption is that the ratio of fire endurance time to the thickness of fire protection material (R/h) varies linearly as a function of the W/D ratio of the protected steel column. These concepts are graphically illustrated in the example that follows. It has been found that both of these assumptions are reasonably accurate for lightweight [density less than 50 pcf (801 kg/m³)] spray-applied materials.

If ASTM E 119 fire endurance test results are available for a specific fire protection material on two different structural steel column shapes, the constants C_1 and C_2 can be determined directly. The resulting equation can then be used to determine the thickness of fire protection material required for any specified fire endurance rating when applied to structural steel columns with W/D ratios between the largest and smallest column for the actual test results that are available.

To determine the constants, at least four ASTM E 119 fire endurance tests or a combination of two ASTM E 119 fire endurance tests and six small-scale tests [3-foot-long (914 mm) specimens] are conducted. If the results of the small-scale tests are used, at least two of the test assemblies are necessary to duplicate the ASTM E 119 test assemblies for the purpose of establishing correlation. Regardless of the combination of small and large-scale tests selected, at least two tests are conducted on the largest and two on the smallest columns to establish the limits of applicability to the resulting equation. The constants C_1 and C_2 are to be determined on the basis of the lowest ratios of fire endurance time to fire protection thickness (R/h) for these columns.

In addition, the test data need to be evaluated with respect to the assumption that the ratio of fire endurance to fire protection thickness (R/h) is reasonably constant for a given column shape [(W/D) ratio]. The tests conducted on columns of the same shape are designed so that the resulting fire endurance times are approximately $1^1/_2$ hours and $3^1/_2$ hours. In evaluating the R/h ratios resulting from tests on the same column shape, differences in the range of 10 percent are typical. Differences greater than 20 percent may, however, suggest that Equation 7-13 is not applicable to the specific fire protection material under consideration, and further examination of the test data is warranted.

EXAMPLE:

GIVEN: The fire endurance test data given in Figure 721.5.1.3.

FIND: The thickness of a spray-applied cementitious material required for a 3-hour rating on a W12 × 136.

SOLUTION:

1. Determine C_1 and C_2. From the end points of the graph:
   R/h = 75 for a W/D ratio of 0.6 and
   R/h = 200 for a W/D ratio of 2.5.
   From this, two equations with two unknowns are created:
   75 = C_1 (0.6) + C_2.
   200 = C_1 (2.5) + C_2.
   Solving these results in: C_1 = 65.79; C_2 = 35.53 and Equation 7-13, with "h" in the denominator, becomes:
   R/h = 65.79 (W/D) + 35.53.

2. From Table 721.5.1(1), a W12 × 136 has a W/D ratio of 1.82.

3. The required fire-resistance is 180 minutes. Calculating for h yields:
   180 = [(65.79) (1.82) + 35.53] h
   h = 1.16 inches (29 mm).

Therefore, the required thickness of the spray-applied material is 1.16 inches (29 mm).

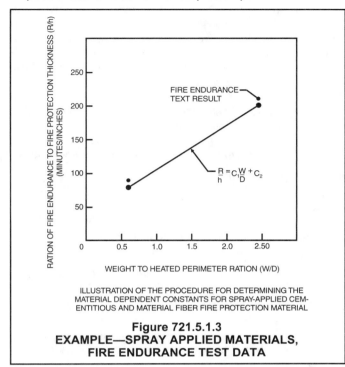

ILLUSTRATION OF THE PROCEDURE FOR DETERMINING THE MATERIAL DEPENDENT CONSTANTS FOR SPRAY-APPLIED CEMENTITIOUS AND MATERIAL FIBER FIRE PROTECTION MATERIAL

Figure 721.5.1.3
EXAMPLE—SPRAY APPLIED MATERIALS, FIRE ENDURANCE TEST DATA

721.5.1.3.1 Material-dependent constants. The material-dependent constants, C1 and C2, shall be determined for specific fire-resistant materials on the basis of standard fire endurance tests in accordance with Section 703.2. Unless evidence is submitted to the building official substantiating a broader application, this expression shall be limited to determining the fire resistance of structural steel columns with weight-to-heated-perimeter ratios (W/D) between the largest and smallest columns for which standard fire-endurance test results are available.

721.5.1.3.2 Spray-applied identification. Spray-applied fire-resistant materials shall be identified by density and thickness required for a given fire-resistance rating.

721.5.1.4 Concrete-protected columns. The fire resistance of structural steel columns protected with concrete, as illustrated in Figure 721.5.1(6) (a) and (b), shall be permitted to be determined from the following expression:

$$R = R_o (1 + 0.03m) \qquad \textbf{(Equation 7-14)}$$

where:

$$R_o = 10 \, (W/D)^{0.7} + 17 \, (h^{1.6}/k_c^{0.2}) \times (1 + 26 \, (H/p_c c_c h \, (L + h))^{0.8})$$

As used in these expressions:

R = Fire endurance at equilibrium moisture conditions (minutes).

R_o = Fire endurance at zero moisture content (minutes).

m = Equilibrium moisture content of the concrete by volume (percent).

W = Average weight of the steel column (pounds per linear foot).

D = Heated perimeter of the steel column (inches).

h = Thickness of the concrete cover (inches).

k_c = Ambient temperature thermal conductivity of the concrete (Btu/hr ft °F).

H = Ambient temperature thermal capacity of the steel column = 0.11W (Btu/ ft °F).

p_c = Concrete density (pounds per cubic foot).

c_c = Ambient temperature specific heat of concrete (Btu/lb °F).

L = Interior dimension of one side of a square concrete box protection (inches).

❖ The values were determined by the procedures indicated in the report "Fire Endurance of Concrete Protected Columns" in the *ACI Journal*.

EXAMPLE:

GIVEN: A W8 × 28 steel column encased in lightweight concrete [density = 110 pcf (1762 kg/m³)] shown in Figure 721.5.1.4, with all reentrant spaces filled with concrete cover 1.25 inches (32 mm) and a moisture content of 5 percent. The web thickness is 0.285².

FIND: The fire-resistance rating.

SOLUTION:

D = 4(6.535) + 2(8.06) - 2(0.285) = 41.69 inches (1059 mm).

W/D = 28 lb/ft/41.69² = 0.67 lb/ft-in.

h = 1.25 inches (32 mm).

k_c = 0.35 Btu/hr ft °F.

c_c = 0.20 Btu/lb °F.

r_c = 110 pcf (1762 kg/m³).

L = (6.535 + 8.06)/2 = 7.30 inches (185 mm).

A_s = 8.25 square inches (.005 m²).

H = .11(28) + [(110)(0.20)/144][(6.535)(8.06) - 8.25] = 9.87

$$R_o = 10(0.67)^{0.7} + 17 \left(\frac{(125)^{1.6}}{(0.35)^{0.2}} \right) \times$$

$$\left(1 + 26 \left[\frac{9.87}{(110)(0.2)(125)(7.3 + 125)} \right]^{0.8} \right)$$

R_o = 99 minutes.

Therefore,

R = 99[1 + 0.33(5)] = 114 minutes.

For comparison purposes, the minimum cover requirement for a W8 × 28 steel column from Table 721.5.1(8) is 1 inch (25 mm) for a 1¹/₂-hour rating. The column does not quite meet the required fire rating for a

2-hour column since 1.5 inches (38 mm) is required; therefore, the fire-resistance of the column is $1^1/_2$ hours.

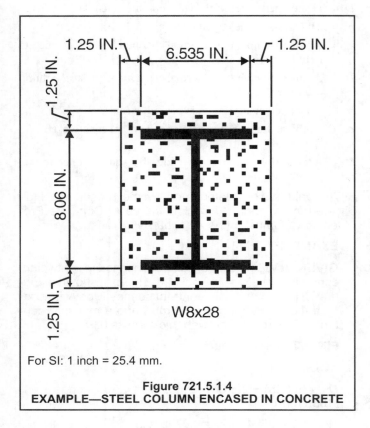

For SI: 1 inch = 25.4 mm.

Figure 721.5.1.4
EXAMPLE—STEEL COLUMN ENCASED IN CONCRETE

721.5.1.4.1 Reentrant space filled. For wide-flange steel columns completely encased in concrete with all reentrant spaces filled [Figure 721.5.1(6)(c)], the thermal capacity of the concrete within the reentrant spaces shall be permitted to be added to the thermal capacity of the steel column, as follows:

$$H = 0.11W + (p_c c_c/144) (b_f d - A_s) \qquad \textbf{(Equation 7-15)}$$

where:

b_f = Flange width of the steel column (inches).

d = Depth of the steel column (inches).

A_s = Cross-sectional area of the steel column (square inches).

721.5.1.4.2 Concrete properties unknown. If specific data on the properties of concrete are not available, the values given in Table 721.5.1(2) are permitted.

721.5.1.4.3 Minimum concrete cover. For structural steel column encased in concrete with all reentrant spaces filled, Figure 721.5.1(6)(c) and Tables 721.5.1(7) and 721.5.1(8) indicate the thickness of concrete cover required for various fire-resistance ratings for typical wide-flange sections. The thicknesses of concrete indicated in these tables also apply to structural steel columns larger than those listed.

721.5.1.4.4 Minimum precast concrete cover. For structural steel columns protected with precast concrete column covers as shown in Figure 721.5.1(6)(a), Tables 721.5.1(9) and 721.5.1(10) indicate the thickness of the column covers required for various fire-resistance ratings for typical wide-flange shapes. The thicknesses of concrete given in these tables also apply to structural steel columns larger than those listed.

721.5.1.4.5 Masonry protection. The fire resistance of structural steel columns protected with concrete masonry units or clay masonry units as illustrated in Figure 721.5.1(7), shall be permitted to be determined from the following expression:

$$R = 0.17 \, (W/D)^{0.7} + [0.285 \, (T_e^{1.6}/K^{0.2}) \,]$$
$$[1.0 + 42.7 \, \{ \, (A_s/d_m \, T_e) \, / \, (0.25p + T_e) \, \}^{\,0.8} \,]$$
$$\textbf{(Equation 7-16)}$$

where:

R = Fire-resistance rating of column assembly (hours).

W = Average weight of steel column (pounds per foot).

D = Heated perimeter of steel column (inches) [see Figure 721.5.1(7)].

T_e = Equivalent thickness of concrete or clay masonry unit (inches) (see Table 721.3.2 Note a or Section 721.4.1).

K = Thermal conductivity of concrete or clay masonry unit (Btu/hr ft °F) [see Table 721.5.1(3)].

A_s = Cross-sectional area of steel column (square inches).

d_m = Density of the concrete or clay masonry unit (pounds per cubic foot).

p = Inner perimeter of concrete or clay masonry protection (inches) [see Figure 721.5.1(7)].

721.5.1.4.6 Equivalent concrete masonry thickness. For structural steel columns protected with concrete masonry, Table 721.5.1(5) gives the equivalent thickness of concrete masonry required for various fire-resistance ratings for typical column shapes. For structural steel columns protected with clay masonry, Table 721.5.1(6) gives the equivalent thickness of concrete masonry required for various fire-resistance ratings for typical column shapes.

721.5.2 Structural steel beams and girders. The fire-resistance ratings of steel beams and girders shall be based upon the size of the element and the type of protection provided in accordance with this section.

721.5.2.1 Determination of fire resistance. These procedures establish a basis for determining resistance of structural steel beams and girders which differ in size from that specified in approved fire-resistance-rated assemblies as a function of the thickness of fire-resistant material and the weight (W) and heated perimeter (D) of the beam or girder. As used in these sections, W is the average weight of a structural steel member in pounds per linear foot (plf). The heated perimeter, D, is the inside perimeter of the fire-resistant material in inches as illustrated in Figure 721.5.2.

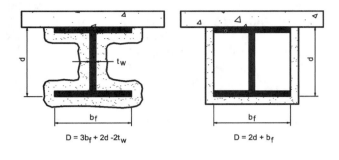

**FIGURE 721.5.2
DETERMINATION OF THE HEATED PERIMETER
OF STRUCTURAL STEEL BEAMS AND GIRDERS**

$D = 3b_f + 2d - 2t_w$

$D = 2d + b_f$

721.5.2.1.1 Weight-to-heated perimeter. The weight-to-heated-perimeter ratios (W/D), for both contour and box fire-resistant protection profiles, for the wide flange shapes most often used as beams or girders are given in Table 721.5.1(4). For different shapes, the weight-to-heated-perimeter ratios (W/D) shall be determined in accordance with the definitions given in this section.

721.5.2.1.2 Beam and girder substitutions. Except as provided for in Section 721.5.2.2, structural steel beams in approved fire-resistance-rated assemblies shall be considered the minimum permissible size. Other beam or girder shapes shall be permitted to be substituted provided that the weight-to-heated-perimeter ratio (*W/D*) of the substitute beam is equal to or greater than that of the beam specified in the approved assembly.

❖ This section defines a general rule for the substitution of different steel beam and girder shapes in all fire-resistant assemblies. In the past, the substitution of larger beams has been permitted based upon the thickness of web and flange elements. Extensive research at Underwriters Laboratories has proven that the heat transfer to a protected steel beam or girder is a direct function of the *W/D ratio*. As a result, beam substitutions should be based upon *W/D* ratios, as defined in this section. The significance of the thickness of web and flange elements is inherently included in the determination of *W/D* ratios. Code Figure 721.5.2 illustrates the procedure for determining heated perimeters (*D*), and Table 721.5.1(4) provides *W/D* ratios for the most commonly used wide flange beam and girder shapes.

721.5.2.2 Spray-applied fire-resistant materials. The provisions in this section apply to unrestrained structural steel beams and girders protected with spray-applied fire-resistance-rated materials. Larger or smaller unrestrained beam and girder shapes shall be permitted to be substituted for beams specified in approved unrestrained or restrained fire-resistance-rated as-

semblies provided that the thickness of the fire-resistant material is adjusted in accordance with the following expression:

$$h_2 = \left[\frac{(W_1 / D_1) + 0.60}{(W_2 / D_2) + 0.60}\right] h_1 \qquad \textbf{(Equation 7-17)}$$

where:

h = Thickness of spray-applied fire-resistant material in inches.

W = Weight of the structural steel beam or girder in pounds per linear foot.

D = Heated perimeter of the structural steel beam or girder in inches.

Subscript 1 refers to the beam and fire-resistant material thickness in the fire-resistance-rated assembly.

Subscript 2 refers to the substitute beam or girder and the required thickness of fire-resistant material.

❖ This section defines an equation for adjusting the thickness of spray-applied cementitious and mineral fiber materials as a function of W/D ratios. This equation was developed by Underwriters Laboratories, and appropriate limitations have been included. The minimum *W/D* ratio of 0.37 in Section 721.5.2.2.1 will prevent the use of this equation for determining the fire-resistance of very small shapes that have not been tested. The $^3/_8$-inch (9.5 mm) minimum thickness of protection is a practical limit based upon the most commonly used spray-applied fire protection materials.

EXAMPLE:

GIVEN: Determine the thickness of spray-applied fire protection required to provide a 2-hour fire-resistance rating for a W12 × 16 beam to be substituted for a W8 × 15 beam requiring 1.44 inches (37 mm) of protection for the same rating.

SOLUTION:

From Table 721.5.1(4):

W_1/D_1 = 0.54 for W8 × 15

W_2/D_2 = 0.45 for W12 × 16

h_1 = 1.44 inches.

$$h_2 = \left[\frac{0.54 + 0.60}{0.45 + 0.60}\right] \times 1.44$$

= 1.56 inches (39 mm).

721.5.2.2.1 Minimum thickness. Equation 7-17 is limited to beams with a weight-to-heated-perimeter ratio (*W/D*) of 0.37 or greater. The minimum thickness of fire-resistant material shall not be less than $^3/_8$ inch (9.5 mm).

❖ See the commentary to Section 721.5.2.2.

721.5.2.3 Structural steel trusses. The fire resistance of structural steel trusses protected with fire-resistant materials spray applied to each of the individual truss elements shall be permitted to be determined in accordance with this section. The thickness of the fire-resistant material shall be determined in accordance with Section 721.5.1.3. The weight-to-heated-perimeter ratio (W/D) of truss elements that can be simultaneously exposed to fire on all sides shall be determined on the same basis as columns, as specified in Section 721.5.1.1. The weight-to-heated-perimeter ratio (W/D) of truss elements that directly support floor or roof construction shall be determined on the same basis as beams and girders, as specified in Section 721.5.2.1.

❖ This section describes the application of spray-applied fire protection to structural steel trusses when each truss element is individually protected with spray-applied materials. The thickness of protection is determined using the column equation specified in Section 721.5.1.3. For trusses, the column equation is more technically correct than the beam equation, since it requires greater thicknesses of protection than the beam equation. Most truss elements can be exposed to fire on all four sides simultaneously. As a result, the heated perimeter of truss elements should be determined in the same manner as for columns. An exception is, however, included for top chord elements that directly support floor or roof construction. The heated perimeter of such elements should be determined in the same manner as for beams and girders.

721.6 Wood assemblies. The provisions of this section contain procedures by which the fire-resistance ratings of wood assemblies are established by calculations.

❖ The information contained in this section is based on Technical Paper No. 222 of the Division of Building Research Council of Canada entitled "Fire Endurance of Light Framed and Miscellaneous Assemblies" and National Bureau of Standards Report BMS 92. The fire-resistance rating is equal to the sum of the time assigned to the membranes [see Table 721.6.2(1)], the time assigned to the framing members [see Table 721.6.2(2)] and the time assigned for additional contribution by other protective measures, such as insulation [see Table 721.6.2(5)]. The membrane on the unexposed side is not included in the calculations. It is assumed that once the structural members fail, the entire assembly fails. The time assigned to the individual membranes is individual contributions to the overall fire-resistance rating of the complete assembly. The assigned time is not to be confused with the fire-resistance rating of the membranes. Fire-resistance rating takes into account the rise in temperature on the unexposed side of the membrane. Times that have been assigned to membranes on the fire-exposed side of the assembly are based on their ability to remain in place during fire tests.

The fire-resistance rating of wood-frame assemblies is equal to the sum of the time assigned to the various components (membranes) on the fire-exposed side and the structural members. Interior walls and partitions; exterior walls within 5 feet (1524 mm) of a lot line or assumed lot line and some shaft walls must be designed

for fire resistance by assuming that either side of the wall is exposed to fire.

Mineral fiber insulation provides additional protection to wood studs by shielding the studs from exposure to the furnace, thus delaying the time of collapse. The use of reinforcement in the membrane exposed to fire also adds to the fire resistance by extending the time to failure. Special care must be taken to insure that all insulation materials used in conjunction with this method satisfy the weight criteria of Table 721.6.2(5).

The following are examples of how to use Section 721.6:

EXAMPLE 1:

GIVEN: The exterior wall assembly shown in Figure 721.6 having a layer of $^{15}/_{32}$-inch (12 mm) wood structural panel bonded with exterior glue covered with a layer of $^{5}/_{8}$-inch (15.8 mm) gypsum wallboard and attached to studs spaced at 16 inches (406 mm) on center (o.c.).

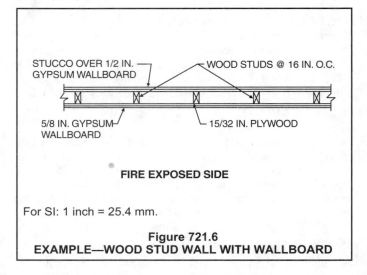

FIRE EXPOSED SIDE

For SI: 1 inch = 25.4 mm.

Figure 721.6
EXAMPLE—WOOD STUD WALL WITH WALLBOARD

FIND: Does the wall assembly qualify as a 1-hour fire-resistant wall assembly?

SOLUTION: From Tables 721.6.2(1) and 721.6.2(2):

Wood structural panel	= 10 minutes.
Gypsum wallboard	= 30 minutes.
Wood studs	= 20 minutes.
Total	= 60 minutes.

Therefore, the wall does qualify as a 1-hour wall. If the wall is an interior wall, both sides would be required to be fire protected with at least 40 minutes of membrane coverings (60 minutes - 20 minutes for wood frame).

It should be noted that Section 721.6.2.3 requires the exterior side to be protected in accordance with Table 721.6.2(3) or any membrane that is assigned a time of at least 15 minutes as listed in Table 721.6.2(1) and as illustrated in the figure.

It should also be noted that if the wall cavities between the studs are filled with mineral fiber batts weighing no less than that specified in Table 721.6.2(5), the $^{15}/_{32}$-inch (12 mm) plywood membrane layer could be eliminated because the insulation adds 15 minutes of fire resistance, as indicated in Section 721.6.2.5 and Table 721.6.2(5). Thus, adding the contribution times for the $^{5}/_{8}$-inch (15.9 mm) gypsum board, the wood framing and the insulation (30 minutes + 20 minutes + 15 minutes), the resultant rating for the wall would be 65 minutes and meet a 1-hour fire-resistance rating.

EXAMPLE 2:

GIVEN: A floor/ceiling assembly using wood joists spaced at 16 inches (406 mm) o.c., protected on the bottom side (ceiling side) with two layers of $^{1}/_{2}$-inch (12.7 mm) Type X gypsum wall board and protected on the upper side (floor side) with a $^{15}/_{32}$-inch (12 mm) plywood subfloor, a $^{3}/_{8}$-inch (15.9 mm) panel-type underlayment and carpet.

FIND: Does the floor/ceiling assembly meet the requirements of a 1-hour fire-resistance-rated assembly?

SOLUTION: Referring to Sections 721.6.2.1 and 721.6.2.4, Table 721.6.2(1) indicates that the time contribution for each layer of $^{1}/_{2}$-inch (12.7 mm) Type X gypsum wallboard is 25 minutes. The time of contribution for wood joists 16 inches (406 mm) o.c. is listed in Table 721.6.2(2) as 10 minutes. By adding the two layers of gypsum board (2 × 25 minutes) to the wood frame (10 minutes), a fire-resistance rating of 60 minutes, or 1 hour, can be obtained. It should be noted that Section 721.6.2.4 requires the upper membrane to be specified as in Table 721.6.2(4) or any membrane that has a time of contribution of at least 15 minutes as listed in Table 721.6.2(1).

If the above example had been a roof/ceiling assembly, the upper membrane would have been treated the same. If the proposed assembly is a ceiling with an attic above, Section 721.6.2.4 notes the exception to Section 711.3.3, which allows the elimination of the upper membrane.

The fastening requirements for assemblies developed by Section 721.6 should be in accordance with Chapter 23 as stated in Section 721.6.2.6.

721.6.1 General. This section contains procedures for calculating the fire-resistance ratings of walls, floor/ceiling and roof/ceiling assemblies based in part on the standard method of testing referenced in Section 703.2.

721.6.1.1 Maximum fire-resistance rating. Fire-resistance ratings calculated using the procedures in this section shall be used only for 1-hour rated assemblies.

721.6.1.2 Dissimilar membranes. Where dissimilar membranes are used on a wall assembly, the calculation shall be made from the least fire-resistant (weaker) side.

❖ All wood assemblies by which the fire-resistance rating is calculated must have a membrane on each side; however, the calculation shall be made from the least fire-resistant side.

721.6.2 Walls, floors and roofs. These procedures apply to both load-bearing and nonload-bearing assemblies.

721.6.2.1 Fire-resistance rating of wood frame assemblies. The fire-resistance rating of a wood frame assembly is equal to the sum of the time assigned to the membrane on the fire-exposed side, the time assigned to the framing members and the time assigned for additional contribution by other protective measures such as insulation. The membrane on the unexposed side shall not be included in determining the fire resistance of the assembly.

TABLE 720.6.2(1)
TIME ASSIGNED TO WALLBOARD MEMBRANES[a,b,c,d]

DESCRIPTION OF FINISH	TIME[e] (minutes)
$^{3}/_{8}$-inch wood structural panel bonded with exterior glue	5
$^{15}/_{32}$-inch wood structural panel bonded with exterior glue	10
$^{19}/_{32}$-inch wood structural panel bonded with exterior glue	15
$^{3}/_{8}$-inch gypsum wallboard	10
$^{1}/_{2}$-inch gypsum wallboard	15
$^{5}/_{8}$-inch gypsum wallboard	30
$^{1}/_{2}$-inch Type X gypsum wallboard	25
$^{5}/_{8}$-inch Type X gypsum wallboard	40
Double $^{3}/_{8}$-inch gypsum wallboard	25
$^{1}/_{2} + ^{3}/_{8}$-inch gypsum wallboard	35
Double $^{1}/_{2}$-inch gypsum wallboard	40

For SI: 1 inch = 25.4 mm.

a. These values apply only when membranes are installed on framing members which are spaced 16 inches o.c.

b. Gypsum wallboard installed over framing or furring shall be installed so that all edges are supported, except $^{5}/_{8}$-inch Type X gypsum wallboard shall be permitted to be installed horizontally with the horizontal joints staggered 24 inches each side and unsupported but finished.

c. On wood-framed floor/ceiling or roof/ceiling assemblies, gypsum board shall be installed with the long dimension perpendicular to framing members and shall have all joints finished.

d. The membrane on the unexposed side shall not be included in determining the fire resistance of the assembly. When dissimilar membranes are used on a wall assembly, the calculation shall be made from the least fire resistant (weaker) side.

e. The time assigned is not a finish rating.

❖ See the commentary to Section 721.6.

721.6.2.2 Time assigned to membranes. Table 721.6.2(1) indicates the time assigned to membranes on the fire-exposed side.

TABLE 721.6.2(2). See below.

TABLE 721.6.2(3). See below.

TABLE 721.6.2(4). See below.

TABLE 721.6.2(5). See below.

TABLE 721.6.2(2)
TIME ASSIGNED FOR CONTRIBUTION OF WOOD FRAME[a,b,c]

DESCRIPTION	TIME ASSIGNED TO FRAME (minutes)
Wood studs 16 inches o.c.	20
Wood floor and roof joists 16 inches o.c.	10

For SI: 1 inch = 25.4 mm.

a. This table does not apply to studs or joists spaced more than 16 inches o.c.

b. All studs shall be nominal 2 × 4 and all joists shall have a nominal thickness of at least 2 inches.

c. Allowable spans for joists shall be determined in accordance with Sections 2308.8, 2308.10.2 and 2308.10.3.

TABLE 721.6.2(3)
MEMBRANE[a] ON EXTERIOR FACE OF WOOD STUD WALLS

SHEATHING	PAPER	EXTERIOR FINISH
$5/8$-inch T & G lumber		Lumber siding
$5/16$-inch exterior glue plywood		Wood shingles and shakes
$1/2$-inch gypsum wallboard	Sheathing paper	$1/4$-inch wood structural panels—exterior type
$5/8$-inch gypsum wallboard		$1/4$-inch hardboard
$1/2$-inch fiberboard		Metal siding
		Stucco on metal lath
		Masonry veneer
None	—	$3/8$-inch exterior-grade wood structural panels

For SI: 1 pound/cubic feet = 16.0185 kg/m^2.

a. Any combination of sheathing, paper, and exterior finish is permitted.

TABLE 721.6.2(4)
FLOORING OR ROOFING OVER WOOD FRAMING[a]

ASSEMBLY	STRUCTURAL MEMBERS	SUBFLOOR OR ROOF DECK	FINISHED FLOORING OR ROOFING
Floor	Wood	$15/32$-inch wood structural panels or $11/16$ inch T & G softwood	Hardwood or softwood flooring on building paper resilient flooring, parquet floor felted-synthetic fiber floor coverings, carpeting, or ceramic tile on $3/8$-inch-thick panel-type underlay Ceramic tile on $1 1/4$-inch mortar bed
Roof	Wood	$15/32$-inch wood structural panels or $11/16$ inch T & G softwood	Finished roofing material with or without insulation

For SI: 1 inch = 25.4 mm.

a. This table applies only to wood joist construction. It is not applicable to wood truss construction.

TABLE 721.6.2(5)
TIME ASSIGNED FOR ADDITIONAL PROTECTION

DESCRIPTION OF ADDITIONAL PROTECTION	FIRE RESISTANCE (minutes)
Add to the fire-resistance rating of wood stud walls if the spaces between the studs are completely filled with glass fiber mineral wool batts weighing not less than 2 pounds per cubic foot (0.6 pound per square foot of wall surface) or rockwool or slag material wool batts weighing not less than 3.3 pounds per cubic foot (1 pound per square foot of wall surface), or cellulose insulation having a nominal density not less than 2.6 pounds per cubic foot.	15

For SI: 1 pound/cubic foot = 16.0185 kg/m^3.

721.6.2.3 Exterior walls. For an exterior wall having more than 5 feet (1524 mm) of horizontal separation, the wall is assigned a rating dependent on the interior membrane and the framing as described in Tables 721.6.2(1) and 721.6.2(2). The membrane on the outside of the nonfire-exposed side of exterior walls having more than 5 feet (1524 mm) of horizontal separation may consist of sheathing, sheathing paper, and siding as described in Table 721.6.2(3).

721.6.2.4 Floors and roofs. In the case of a floor or roof, the standard test provides only for testing for fire exposure from below. Except as noted in Section 703.3, Item 5, floor or roof assemblies of wood framing shall have an upper membrane consisting of a subfloor and finished floor conforming to Table 721.6.2(4) or any other membrane that has a contribution to fire resistance of at least 15 minutes in Table 721.6.2(1).

721.6.2.5 Additional protection. Table 721.6.2(5) indicates the time increments to be added to the fire resistance where glass fiber, rockwool, slag mineral wool, or cellulose insulation is incorporated in the assembly.

721.6.2.6 Fastening. Fastening of wood frame assemblies and the fastening of membranes to the wood framing members shall be done in accordance with Chapter 23.

721.6.3 Design of fire-resistant exposed wood members. The fire-resistance rating, in minutes, of timber beams and columns with a minimum nominal dimension of 6 inches (152 mm) is equal to:

Beams: $2.54Zb (4 -2(b/d))$ for beams which may be exposed to fire on four sides.   **(Equation 7-18)**

   $2.54Zb (4 -(b/d))$ for beams which may be exposed to fire on three sides.   **(Equation 7-19)**

Columns: $2.54Zd (3 -(d/b))$ for columns which may be exposed to fire on four sides   **(Equation 7-20)**

   $2.54Zd (3 -(d/2b))$ for columns which may be exposed to fire on three sides.   **(Equation 7-21)**

where:

b =   The breadth (width) of a beam or larger side of a column before exposure to fire (inches).

d =   The depth of a beam or smaller side of a column before exposure to fire (inches).

Z =   Load factor, based on Figure 720.6.3(1).

721.6.3.1 Equation 7-21. Equation 7-21 applies only where the unexposed face represents the smaller side of the column. If a column is recessed into a wall, its full dimension shall be used for the purpose of these calculations.

721.6.3.2 Allowable loads. Allowable loads on beams and columns are determined using design values given in ANSI/AF&PA NDS.

721.6.3.3 Fastener protection. Where minimum 1-hour fire resistance is required, connectors and fasteners shall be protected

from fire exposure by $1^1/_2$ inches (38 mm) of wood, or other approved covering or coating for a 1-hour rating. Typical details for commonly used fasteners and connectors are shown in AITC Technical Note 7.

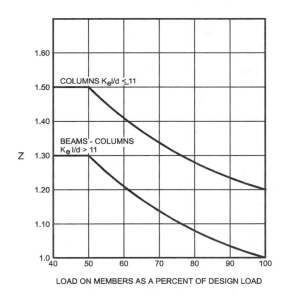

K_e = The effective length factor as noted in Figure 721.6.3(2).
L = The unsupported length of columns (inches).

FIGURE 721.6.3(1)
LOAD FACTOR

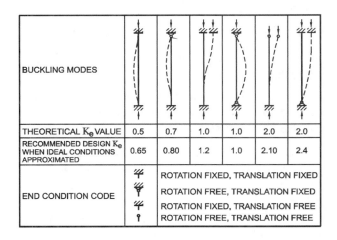

FIGURE 721.6.3(2)
EFFECTIVE LENGTH FACTORS

721.6.3.4 Minimum size. Wood members are limited to dimensions of 6 inches (152 mm) nominal or greater. Glued-laminated timber beams utilize standard laminating combinations except that a core lamination is removed. The tension zone is moved inward and the equivalent of an extra nominal 2-inch-thick (51 mm) outer tension lamination is added.

721.7 Other reference documents. Refer to Section 703.3, Item 1, and NBS BMS 71 and NBSTRBM-44 for fire-resistance ratings of materials and assemblies.

Bibliography

The following resource materials are referenced in this chapter or are relevant to the subject matter addressed in this chapter.

Abrams and Gustaferro. "Fire Endurance of Two Course Floors and Roofs." *Journal of the American Concrete Institute.* American Concrete Institute, February, 1969.

Abrams, M.S. And T.D. Lin. *Temperature Distribution in Concrete Beams Subjected to Fire.*

ACI 216.1/TMS 0216.1-97, *Standard Method for Determining Fire Resistance of Concrete and Masonry Construction Assemblies.* Farmington Hills, MI: American Concrete Institute.

ACI 530/ASCE 5/TMS 402-02, *Building Code Requirements for Masonry Structures.* Detroit, MI: American Concrete Institute; New York: American Society of Civil Engineers; Boulder, CO: The Masonry Council, 2002.

ACI 530.1/ASCE 6/TMS 602-02, *Specifications for Masonry Structures.* Detroit, MI: American Concrete Institute; New York: American Society of Civil Engineers; Boulder, CO: The Masonry Council, 2002.

Allen L.W., M. Galbreath and W.W. Stanzak. *Fire Endurance Tests on Unit Masonry Walls with Gypsum Wallboard* (NRCC 13901). Division of Building Research, National Research Council of Canada.

Analytical Methods of Determining Fire Endurance of Concrete and Masonry Members — Model Code Approved Procedures (SR267.01B). Skokie, IL: Concrete and Masonry Industry Firesafety Committee, 1987.

ASTM C 140-01, *Standard Test Method for Sampling and Testing Concrete Masonry Units.* West Conshohocken, PA: ASTM International, 2001.

ASTM C 755-85, *Practice For Selection of Vapor Retarders for Thermal Insulation.* West Conshohocken, PA: ASTM International,1985.

ASTM E 84-01, *Test Method for Surface Burning Characteristics of Building Materials.* West Conshohocken, PA: ASTM International, 2001.

ASTM E 119-00, *Test Methods for Fire Tests of Building Construction and Materials.* West Conshohocken, PA: ASTM International, 2000.

ASTM E 136-99, *Test Method for Behavior of Materials in a Vertical Tube Furnace at 750° C.* West Conshohocken, PA: ASTM International, 1999.

ASTM E 152-81a, *Methods of Fire Tests of Door Assemblies.* West Conshohocken, PA: ASTM International, 1981.

ASTM E 163-84, *Methods of Fire Tests of Window Assemblies.* West Conshohocken, PA: ASTM International, 1984.

ASTM E 814-00, *Test Method for Fire Tests of Through-Penetration Fire Stops.* West Conshohocken, PA: ASTM International, 2000.

BMS 92, *Fire Resistance Classifications, Building Materials and Structures.* National Bureau of Standards, 1942.

Boring, D.F., J.C. Spence and W.G. Wells. *Fire Protection Through Modern Building Codes.* Washington, DC: American Iron and Steel Institute, 1981.

Building Materials Directory. Northbrook, IL: Underwriters Laboratories Inc., 1996.

Christian, W.J. and J.E. Waterman. "Flame Spread in Corridors: Effects of Location and Area of Wall Finish," *Fire Journal,* July 1971.

CPSC 16 CFR, Part 1209-98, *Interim Safety Standard for Cellulose Insulation.* Washington, DC: Consumer Product Safety Commission, 1998.

CPSC 16 CFR, Part 1404-98, *Cellulose Insulation.* Washington, DC: Consumer Product Safety Commission, 1998.

Designing Fire Protection for Steel Beams. Washington, DC: American Iron and Steel Institute, 1984.

Designing Fire Protection for Steel Columns. Washington, DC: American Iron and Steel Institute, 1980.

Designing Fire Protection for Steel Trusses. Washington, DC: American Iron and Steel Institute, 1981.

"Fire Endurance of Concrete Slabs as Influenced by Thickness, Aggregate Type and Moisture." *PCA Bulletin 223.* Skokie, IL: Portland Cement Association.

Fire Resistance Classifications of Building Construction. National Bureau of Standards, 1942.

Fire Resistance Directory. Northbrook, IL: Underwriters Laboratories Inc., 1996.

Fire Tests of Building Columns. Associated Factory Mutual Insurance Companies, 1921.

"Fire Tests of Joints Between Precast Concrete Wall Units: Effect of Various Joint Treatments." *PCI Journal. September-October, 1975.*

Fisher, F., B. MacCracken and R.B. Williamson. "Room Fire Tests of Textile Wallcoverings." *Service to Industry Report No. 85-4.* University of California Fire Research Laboratory, 1989.

GA 600-00, *Fire Resistance Design Manual.* Washington, DC: Gypsum Association, 2000.

Galbreath, Murdock. "Fire Resistance of Light Framed and Miscellaneous Assemblies." Technical Paper No. 222 of the Division of Building Research, National Research Council of Canada.

Harmanthy, T.Z. And T.T. Lie. "Fire Endurance of Concrete Protected Columns." *ACI Journal,* Proceedings, Vol. 71, No. 2, 1974.

Hull and Inburg. "Fire Resistance of Concrete Columns." Technological Papers of the Bureau of Standards, No. 271, February 24, 1925.

IMC-2003, *International Mechanical Code.* Falls Church, VA: International Code Council, 2003.

Ingberg, S.H. "Tests of Severity of Building Fires," *NFPA Quarterly*, Vol. 22, No. 1, pp. 43-61.

"Investigations on Building Fires, Part VI: Fire Resistance of Reinforced Concrete Columns," National Building Studies Research Paper No. 18. London, England: Her Majesty's Stationary Office, 1953.

IPC-2003, *International Plumbing Code.* Falls Church, VA: International Code Council, 2003.

Menzil, Carl A. *Test of Fire Resistance and Strength of Walls of Concrete Masonry Units.* Skokie, IL: Portland Cement Association, 1934.

National Building Code, 1955. New York: National Board of Fire Underwriters, 1995.

NCMA TEK 6A-91, *Fire Resistance Rating of Concrete Masonry Assembly*. Herndon, VA: National Concrete Masonry Association, 1991.

NCMA TEK 35D-91, *Fire Safety with Concrete Masonry*. Herndon, VA: National Concrete Masonry Association, 1991.

NFPA 13-99, *Installation of Sprinkler Systems.* Quincy, MA: National Fire Protection Association, 1999.

NFPA 13D-99, *Installation of Sprinkler Systems in One- and Two-Family Dwellings and Manufactured Homes.* Quincy, MA: National Fire Protection Association, 1999.

NFPA 13R-99, *Installation of Sprinkler Systems in Residential Occupancies Up to Four Stories in Height.* Quincy, MA: National Fire Protection Association, 1999.

NFPA 72-99, *National Fire Alarm Code.* Quincy, MA: National Fire Protection Association, 1999.

NFPA 80-99, *Fire Doors and Windows.* Quincy, MA: National Fire Protection Association, 1999.

NFPA 252-99, *Methods of Fire Tests of Door Assemblies.* Quincy, MA: National Fire Protection Association, 1999.

NFPA 257-00, *Methods of Fire Tests of Window Assemblies*. Quincy, MA: National Fire Protection Association, 2000.

NRCCC 17724, *Supplement to the National Building Code of Canada 1980.* Associate Committee on the National Building Code. National Research Council of Canada, 1980.

PCA Publication T-140, *Fire Resistance of Referenced Concrete Floors*. Skokie, IL: Portland Cement Association.

PCI MNL 124-89, *Design for Fire Resistance of Precast Prestressed Concrete*. Chicago: Precast/Prestressed Concrete Institute, 1989.

Reinforced Concrete Fire Resistance. Schaumburg, IL: Concrete Reinforcing Steel Institute, 1980.

SFPE Handbook of Fire Protection Engineering, 2nd. Quincy, MA: National Fire Protection Association, 1995.

The Moroney Report, Vol. 5, Issue 2, Second Quarter 1993. Bedford Park, IL: John J. Moroney & Company, 1993.

UL 10A-98, *Standard for Safety Tin-Clad Fire Doors — with Revisions through July 1998*. Northbrook, IL: Underwriters Laboratories Inc., 1998.

UL 14B-98, *Sliding Hardware for Standard, Horizontally Mounted Tin-Clad Fire Doors—with Revisions through October 1996*. Northbrook, IL: Underwriters Laboratories Inc., 1998.

UL 14C-96, *Swinging Hardware for Standard Tin-Clad Fire Doors Mounted Singly and in Pairs—with Revisions through October 1996*. Northbrook, IL: Underwriters Laboratories Inc., 1996.

UL 103-98, *Chimneys, Factory Built, Residential Type and Building Heating Appliance — with Revisions through February 1996*. Northbrook, IL: Underwriters Laboratories Inc., 1998.

UL 127-99, *Factory-Built Fireplaces — with Revisions through November 1994 June 1998*. Northbrook, IL: Underwriters Laboratories Inc., 1999.

UL 555-96, *Fire Dampers*. Northbrook, IL: Underwriters Laboratories Inc., 1996.

UL 555C-96, *Ceiling Dampers*. Northbrook, IL: Underwriters Laboratories Inc., 1996.

UL 555S-99, *Leakage Rated Dampers for Use in Smoke Control Systems*. Northbrook, IL: Underwriters Laboratories Inc. 1999.

Chapter 8:
Interior Finishes

General Comments

Section 801 contains general provisions regarding interior finishes.

Section 802 includes definitions of terms that are primarily used in this chapter.

Section 803 contains regulations for wall and ceiling finishes.

Section 804 includes interior floor finish requirements.

Section 805 contains regulations regarding decorations and trim.

Past fire experience has shown that interior finish and decorative materials are key elements in the development and spread of fire. In some cases, as with Boston's devastating Coconut Grove fire in 1942, interior decorations were the first materials ignited. In many other cases, the interior finish materials became involved in the early stages of fire and contributed to its early growth and spread.

The provisions of Chapter 8 require materials used as interior finishes and decorations to have a flame spread index or to be flame resistant, based on the relative fire hazard associated with the occupancy.

The performance of the material is evaluated based on test standards. The design professional or permit applicant is responsible for determining and providing documentation on the flame spread and flame resistance of a material in order to support the permit application. The building official then evaluates the information supplied to ascertain that compliance with the applicable code provisions has been achieved. It is also critical that a field inspection verifies that the material is installed in accordance with the approved documents and the code.

Purpose

This chapter contains the performance requirements for controlling fire growth within buildings by restricting interior finish materials and decorations.

SECTION 801
GENERAL

801.1 Scope. Provisions of this chapter shall govern the use of materials used as interior finishes, trim and decorative materials.

❖ Chapter 8 contains requirements for materials used as interior finish, trim or decorations. These materials must conform to the flame spread limitations or the flame-resistance criteria established by the chapter.

801.1.1 Interior finishes. These provisions shall limit the allowable flame spread and smoke development based on location and occupancy classification.

Exceptions:

1. Materials having a thickness less than 0.036 inch (0.9 mm) applied directly to the surface of walls or ceilings.

2. Exposed portions of structural members complying with the requirements for buildings of Type IV construction in Section 602.4 shall not be subject to interior finish requirements.

❖ The provisions of Chapter 8 require materials used as interior finishes and decorations to have a flame spread index or to be flame resistant and limit smoke development based on the relative fire hazard associated with the occupancy. The desired performance of the material is evaluated based on test standards.

Exception 1 provides for finish material that does not exceed 0.036 inch (0.9 mm) in thickness and is applied to a noncombustible base or substrate of approved fire-retardant-treated wood (FRTW). For example, standard decorative wallpaper applied to a noncombustible substrate is exempt from the requirements of this section, as is the paper finish on gypsum wallboard; however, gypsum wallboard as a composite material is considered an interior finish. Surface finishes, as limited by this section, will not significantly contribute to fire condition.

Exception 2 recognizes the inherent nature of heavy timber construction relative to fire spread. Exposed structural members in buildings of Type IV construction are exempt from the interior finish requirements. These members are of a size and nature that do not pose a threat to fire spread; therefore, the members can be left exposed.

[F] 801.1.2 Decorative materials and trim. Decorative materials and trim shall be restricted by combustibility and flame resistance in accordance with Section 805.

❖ The intent of this section is to minimize the role of interior finishes in fires by establishing maximum flame spread and smoke-developed indexes for interior finishes as described in Section 801.1.1 and in accordance with Section 805.

801.1.3 Applicability. For buildings in flood hazard areas as established in Section 1612.3, interior finishes, trim and decorative materials below the design flood elevation shall be flood-damage-resistant materials.

❖ All building materials located below the design flood elevation (DFE), including interior finishes, trim and decorative materials, must be resistant to flood damage. Many elevated buildings located in designated flood hazard areas have enclosed lower areas. Flood-damage-resistant materials are defined in Section 1612.2 as "any construction material capable of withstanding direct and prolonged contact with floodwaters without sustaining any damage that requires more than cosmetic repair." The term "prolonged contact" means at least 72 hours, and the term "cosmetic repair" means cleaning the affected surfaces with commercially available cleaners and repainting as necessary. For further guidance, refer to FEMA FIA-TB #2, *Flood Resistant Material Requirements for Buildings Located in Special Flood Hazard Areas* FEMA FIA-TB #7, *Wet Floodproofing Requirements for Structures Located in Special Flood Hazard Areas* and FEMA FIA-TB #8, *Corrosion Protection for Metal Connectors in Coastal Areas for Structures Located in Special Flood Hazard Areas.*

801.2 Application. Combustible materials shall be permitted to be used as finish for walls, ceilings, floors and other interior surfaces of buildings.

❖ The majority of materials that are used as finish materials are combustible in nature; however, all materials are required to have a certain degree of flame spread and smoke-developed indexes.

801.2.1 Windows. Show windows in the first story of buildings shall be permitted to be of wood or of unprotected metal framing.

❖ First-story windows used as display spaces are exempt from finish requirements. These areas are subject to frequent redecoration; however, the enforcement of finish requirements would be impractical. The exemption is limited to the first story, which is considered to provide ready access for the fire department from the exterior for effective manual fire suppression efforts. The materials, however, must still be approved by the building official.

801.2.2 Foam plastics. Foam plastics shall not be used as interior finish or trim except as provided in Section 2603.7 or 2604.

❖ This section establishes that foam-plastic materials are not permitted to be used as interior finish materials unless the foam plastic complies with Section 2603.8 or 2604. At first glance, it would appear that classification

by ASTM E 84 is all that is necessary. Sections 803.1 and 2603.3 establish a maximum smoke-developed index of 450. Section 2604.2.4 sets a maximum flame spread index of 75. When tested, many foam-plastic materials have a flame spread index of less than 25. Chapter 26 also requires that the foam-plastic material be protected from ignition through the use of a thermal barrier (see Section 2603.4).

Sections 2603.4 and 2603.4.1 generally require that foam-plastic materials be covered by an approved thermal barrier, such as 0.5-inch (12.7 mm) gypsum wallboard. When covered in this manner, the foam plastic is no longer the interior finish material. Section 2603.8 provides a means by which foam plastics can be installed without a thermal barrier. When foam-plastic insulation has been tested in accordance with FM Procedure 4880, UL Subject 1040 or UL 1715, a thermal barrier is not required (see commentary, Section 2603.8).

Because compliance with Section 2604 is required by this section, foam plastics can only be used as exposed interior finishes if they pass the rigorous testing requirements of Section 2603.8.

SECTION 802
DEFINITIONS

802.1 General. The following words and terms shall, for the purposes of this chapter and as used elsewhere in this code, have the meanings shown herein.

❖ This section contains definitions of terms that are associated with the subject matter of this chapter. It is important to emphasize that these terms are not exclusively related to this chapter, but are applicable everywhere the term is used in the code.

EXPANDED VINYL WALL COVERING. Wall covering consisting of a woven textile backing, an expanded vinyl base coat layer and a nonexpanded vinyl skin coat. The expanded base coat layer is a homogeneous vinyl layer that contains a blowing agent. During processing, the blowing agent decomposes, causing this layer to expand by forming closed cells. The total thickness of the wall covering is approximately 0.055 inch to 0.070 inch (1.4 mm to 1.78 mm).

❖ Expanded vinyl wall coverings are manufactured with a textile backing, an expanded vinyl base coat layer consisting of closed cells and a vinyl skin coat. Expanded vinyl wall coverings are subject to the requirements of Section 803.7.

FLAME RESISTANCE. That property of materials or combinations of component materials that restricts the spread of flame in accordance with NFPA 701.

❖ All materials or combinations of component materials, single or multilayered, resist the spread of fire to a certain degree because of their inherent properties. The value of this property is determined by two testing methods in accordance with NFPA 701.

FLAME SPREAD. The propagation of flame over a surface.

❖ The rate at which flames travel along the surface of a combustible finish material directly impacts the speed with which a fire spreads within a room or space, and is, therefore, regulated by this chapter.

FLAME SPREAD INDEX. The numerical value assigned to a material tested in accordance with ASTM E 84.

❖ The ASTM E 84 test method renders measurements of surface flame spread (and smoke density) in comparison with test results obtained by using select red oak as a control material. Red oak is used as a control material for furnace calibration because it is a fairly uniform grade of lumber that is readily available nationally, is uniform in thickness and moisture content and generally gives consistent and reproducible results.

INTERIOR FINISH. Interior finish includes interior wall and ceiling finish and interior floor finish.

❖ This is material that is applied to the surfaces of walls, ceilings and floors. Interior finish material is exposed to the interior space enclosed by these building elements.

INTERIOR FLOOR FINISH. The exposed floor surfaces of buildings including coverings applied over a finished floor or stair, including risers.

❖ Materials that are installed above the structural floor element and exposed to the room or space are considered finish materials. Such materials are, therefore, subject to the requirements of Section 805. This section restricts the radiant heat that will be released from the floor finish during a fire condition.

INTERIOR WALL AND CEILING FINISH. The exposed interior surfaces of buildings including, but not limited to: fixed or movable walls and partitions; columns; ceilings; and interior wainscotting, paneling or other finish applied structurally or for decoration, acoustical correction, surface insulation, structural fire resistance or similar purposes, but not including trim.

❖ A material that is applied to ceilings as well as walls, columns, partitions and other vertical interior surfaces whether fixed or movable. The application of this material may be for structural, decorative, acoustical, structural fire resistance and other similar reasons. Trim, such as baseboard, door or window casing, is not considered interior wall and ceiling finish.

SMOKE-DEVELOPED INDEX. The numerical value assigned to a material tested in accordance with ASTM E 84.

❖ The ASTM E 84 test method of measuring the density of smoke emitted from combustible materials determines the smoke-developed index.

TRIM. Picture molds, chair rails, baseboards, handrails, door and window frames and similar decorative or protective materials used in fixed applications.

❖ As interior trim, this material is usually combustible and permanently affixed. Trim is primarily located around door and window openings (frames), around walls at floors (baseboard) and on walls (chair rail). Interior trim material should only constitute 10 percent of the wall or ceiling area.

SECTION 803
WALL AND CEILING FINISHES

803.1 General. Interior wall and ceiling finishes shall be classified in accordance with ASTM E 84. Such interior finish materials shall be grouped in the following classes in accordance with their flame spread and smoke-developed indexes.

Class A: Flame spread 0-25; smoke-developed 0-450.

Class B: Flame spread 26-75; smoke-developed 0-450.

Class C: Flame spread 76-200; smoke-developed 0-450.

Exception: Materials, other than textiles, tested in accordance with Section 803.2.

❖ Interior finish and trim materials are required to have flame spread indexes as prescribed in Sections 803.5 and 803.6, and smoke-developed indexes as prescribed in Section 803.1. Sections 803.5 and 803.6 regulate interior finish of walls, ceilings and interior trim. ASTM E 84 is the required test method for determining the flame spread and smoke-developed indexes that are required by this and other sections of the code.

ASTM E 84 is intended to determine the relative burning behavior of materials on exposed surfaces, such as ceilings and walls, by visually observing the flame spread along the test specimen. Flame spread and smoke density are then reported. The test method may not be appropriate for materials that are not capable of supporting themselves or of being supported in the test tunnel, or that drip, melt or delaminate; therefore, a distinction is made for foam-plastic material and textile materials (see Sections 801.2.2 and 803.6).

ASTM E 84 establishes a flame spread index based on the area under a curve when the actual flame spread distance is plotted as a function of time. The code has divided the acceptable range of flame spread indexes

(0-200) into three classes: Class A (0-25), Class B (26-75) and Class C (76-200). An indication of relative flame spread is as follows: an asbestos-cement board has a flame spread of zero, while a red oak species of wood has a flame spread of 100.

Not to preclude more detailed information resulting from an ASTM E 84 test report, Figure 803.1 identifies the typical flame spread properties of certain building materials. Textile wall coverings and expanded vinyl wall coverings that have been tested in accordance with Section 803.7 are exempt from these requirements because these materials are allowed to be tested for contribution to fire when exposed to radiant conditions.

The exception allows compliance with NFPA 286 in lieu of ASTM E 84. This test is known as a "room corner" fire test and is similar to that referenced for textile wall coverings in Section 803.7 (NFPA 265). In a room corner fire test, a fire source is set in a corner of a small compartment with the material test sample placed on three walls. This test tends to give a more realistic indication of actual field performance. More detail will be provided in Section 803.2.1 regarding the test and the pass/fail criteria.

Material	Flame spread
Glass-fiber sound-absorbing planks	15 to 30
Mineral-fiber sound-absorbing panels	10 to 25
Shredded wood fiberboard (treated)	20 to 25
Sprayed cellulose fibers (treated)	20
Aluminum (with baked enamel finish on one side)	5 to 10
Asbestos-cement board	0
Brick or concrete block	0
Cork	175
Gypsum board (with paper surface on both sides)	10 to 25
Northern pine (treated)	20
Southern pine (untreated)	130 to 190
Plywood paneling (untreated)	75 to 275
Plywood paneling (treated)	100
Carpeting	10 to 600
Concrete	0

Figure 803.1
TYPICAL FLAME SPREAD OF COMMON MATERIALS

803.2 Interior wall or ceiling finishes other than textiles. Interior wall or ceiling finishes, other than textiles, shall be permitted to be tested in accordance with NFPA 286. Finishes tested in accordance with NFPA 286 shall comply with Section 803.2.1.

❖ This section allows the use of NFPA 286, known as a "room corner" fire test, in lieu of ASTM E 84. In this test, a fire source consisting of a wood crib is placed in the corner of a compartment. The materials tested are then

placed on the walls of the compartment (see Figure 803.2). This generally provides a more realistic understanding of the hazards involved with the materials. Section 803.2.1 provides the pass/fail criteria as accepted by the code.

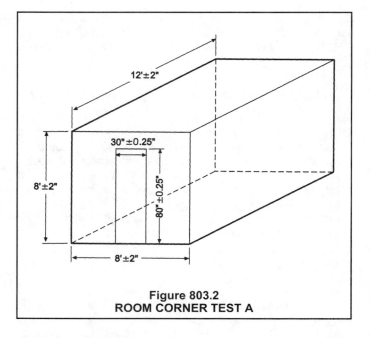

Figure 803.2
ROOM CORNER TEST A

803.2.1 Acceptance criteria. During the 40 kW exposure, the interior finish shall comply with Item 1. During the 160 kW exposure, the interior finish shall comply with Item 2. During the entire test, the interior finish shall comply with Item 3.

1. During the 40kW exposure, flames shall not spread to the ceiling.

2. During the 160 kW exposure, the interior finish shall comply with the following:

    2.1. Flame shall not spread to the outer extremity of the sample on any wall or ceiling.

    2.2. Flashover, as defined in NFPA 286, shall not occur.

3. The total smoke released throughout the NFPA 286 test shall not exceed 1,000 m².

❖ There are two levels of exposure during an NFPA 286 fire test in order to better represent a growing fire: 40kW fire size for 5 minutes and 160 kW for 10 minutes. The 40 kW exposure represents the beginning of a fire where the initial spread is critical; therefore, the stated criteria is that the fire cannot spread to the ceiling. The 160 kW exposure is obviously a more intense fire situation and the criteria relates to preventing flashover (as defined by NFPA 286) and the extent of flame spread throughout the entire test assembly. There is also smoke production criteria of 1,000 m².

It should be noted that the flashover criteria for NFPA 286 and NFPA 265 is as follows:

1. Heat release exceeds 1 MW.

2. Heat flux at the floor exceeds 20 kW/m².

3. Average upper-layer temperature exceeds 600 °C (1112 °F).

4. Flames exit the doorway.

5. Autoignition of paper target on the floor occurs.

803.3 Stability. Interior finish materials regulated by this chapter shall be applied or otherwise fastened in such a manner that such materials will not readily become detached where subjected to room temperatures of 200°F (93°C) for not less than 30 minutes.

❖ Interior finishes are not to become detached under exposure to elevated temperatures [200°F (93°C)] for 30 minutes. There is not a standard test method yet developed to evaluate this requirement. Other sections of the code, however, offer some additional guidance. For example, the performance of the method of attachment of finish materials in a fire-resistance-rated assembly will usually be adequately a ensured by the construction details of the tested assembly. Section 2504 contains requirements for gypsum wallboard that is not part of a fire-resistance-rated assembly. Interior finishes that become detached may add to the spread of fire, as well as create a hazard for fire fighters.

803.4 Application. Where these materials are applied on walls, ceilings or structural elements required to have a fire-resistance rating or to be of noncombustible construction, they shall comply with the provisions of this section.

❖ Where an assembly is required to have a fire-resistance rating or be of noncombustible materials, interior finish materials are required to be applied in accordance with Sections 803.4.1 through 803.4.4.

803.4.1 Direct attachment and furred construction. Where walls and ceilings are required by any provision in this code to be of fire-resistance-rated or noncombustible construction, the interior finish material shall be applied directly against such construction or to furring strips not exceeding 1.75 inches (44 mm) applied directly against such surfaces. The intervening spaces between such furring strips shall be filled with inorganic or Class A material or shall be fireblocked at a maximum of 8 feet (2438 mm) in any direction in accordance with Section 717.

❖ Interior finish materials are required to be applied directly to the exposed surface of structural elements or to the furring attached to such surfaces. All concealed spaces created by furring are to be fireblocked at a maximum of 8-foot (2438 mm) intervals in any direction. By applying the finish directly to the surface, the potential for the back side to contribute to a fire is diminished and, as a result, the effect on the fire-resistance rating and the creation of combustible cavities in noncombustible walls are minimized.

Likewise, fireblocks in furred construction also diminish the likelihood of fire involvement between the finish and the interior wall surface. The use of wood nailing strips is intended to be limited to that which is necessary for the installation of finish and trim materials. As such, the amount of combustible material is restricted so that the performance of the element to which it is attached will not be adversely affected. It should be noted that the type of construction classification of the building is not intended to regulate the combustibility of the interior finish materials. The fire performance characteristics of the interior finish materials are regulated by Sections 803.1 and 803.2.

803.4.2 Set-out construction. Where walls and ceilings are required to be of fire-resistance-rated or noncombustible construction and walls are set out or ceilings are dropped distances greater than specified in Section 803.4.1, Class A finish materials shall be used except where interior finish materials are protected on both sides by an automatic sprinkler system or attached to noncombustible backing or furring strips installed as specified in Section 803.4.1. The hangers and assembly members of such dropped ceilings that are below the main ceiling line shall be of noncombustible materials, except that in Type III and V construction, fire-retardant-treated wood shall be permitted. The construction of each set-out wall shall be of fire-resistance-rated construction as required elsewhere in this code.

❖ Where wide spaces are created, one of the following methods may be used to protect the concealed spaces:

1. Use of Class A finish material;

2. Sprinkler protection within the concealed space and outside;

3. Intimately backing the finish material with noncombustible or FRTW, separating the finish material from the concealed space; or

4. Filling the intervening space between furring strips (see Figure 803.4.2) with inorganic or Class A material or firestopping such spaces a maximum of 8 feet (2438 mm) in any direction.

Class B and C materials may be used on Methods 2, 3 and 4.

803.4.3 Heavy timber construction. Wall and ceiling finishes of all classes as permitted in this chapter that are installed directly against the wood decking or planking of Type IV construction or to wood furring strips applied directly to the wood decking or planking shall be fireblocked as specified in Section 803.4.1.

❖ Attachment directly to heavy timber decking or planking eliminates the concealed spaces (see Section 801.1.1, Exception 2). Furring strips of any depth are allowed when concealed spaces are treated according to Section 803.4.1.

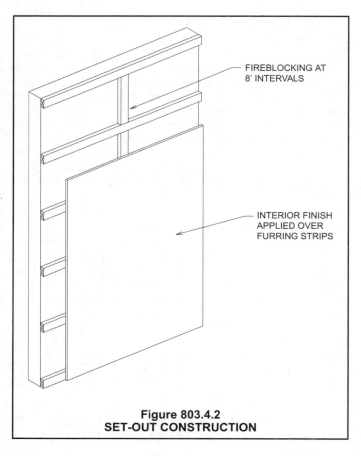

Figure 803.4.2
SET-OUT CONSTRUCTION

FIREBLOCKING AT
8' INTERVALS

INTERIOR FINISH
APPLIED OVER
FURRING STRIPS

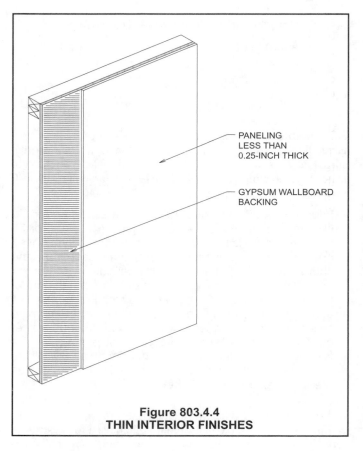

Figure 803.4.4
THIN INTERIOR FINISHES

PANELING
LESS THAN
0.25-INCH THICK

GYPSUM WALLBOARD
BACKING

803.4.4 Materials. An interior wall or ceiling finish that is not more than 0.25 inch (6.4 mm) thick shall be applied directly against a noncombustible backing.

Exceptions:

1. Class A materials.

2. Materials where the qualifying tests were made with the material suspended or furred out from the noncombustible backing.

❖ Combustible materials that have a maximum thickness of 0.25 inch (6.4 mm) when applied directly against a noncombustible backing will not increase significantly the spread of fire. Exception 1 allows Class A material because of its low flame spread index to be furred away from the noncombustible backing. Exception 2 also allows materials that perform similar to directly attached materials when furred or suspended from a noncombustible backing (see Figure 803.4.4).

803.5 Interior finish requirements based on group. Interior wall and ceiling finish shall have a flame spread index not greater than that specified in Table 803.5 for the group and location designated. Interior wall and ceiling finish materials, other than textiles, tested in accordance with NFPA 286 and meeting the acceptance criteria of Section 803.2.1, shall be permitted to

be used where a Class A classification in accordance with ASTM E 84 is required.

❖ The requirements for flame spread indexes for interior finish materials applied to walls and ceilings are contained in Table 803.5 (see Section 804 for floor finish requirements). The referenced test for determining flame spread indexes is ASTM E 84, which establishes a relative measurement of flame spread across the surface of the material. The classifications used in Table 803.5 are defined in Section 803.1. See the commentary to Section 803.1 for additional information on the uses and limitations of the test procedure. This section also allows the use of NFPA 286 as noted in the exception to Sections 803.1 and 803.2. This particular section states that materials that pass this test can be used when a Class A classification is required. In other words, it can be used as a replacement for all catagories, as Class A is the most restrictive (see the commentary to Section 803.2 for more detail on the test).

When evaluating test reports, the method of application and the substrate material are considered. For example, adhesives that soften at moderate temperatures will allow wall or ceiling finishes to drop or peel. This not only increases the susceptibility of the material to ignite but also exposes the substrate material. The substrate material can also affect the interior finish rating of the surface material. It is not uncommon for a thin combustible finish that is applied to a noncombustible material to obtain a lower flame spread index, while the same material applied to a combustible substrate or applied without a substrate could exhibit a significantly higher tendency to spread flame.

TABLE 803.5. See below.

❖ This table prescribes the minimum requirements for interior finishes applied to walls and ceilings; therefore, the use of a Class A material in an area that requires a minimum Class B material is always allowed. Likewise, when the table requires Class C materials, Classes A and B are also permitted.

The requirements are based on the group classification of the space. To determine the applicable criteria, one must determine whether the space is an exit passageway or vertical exit (stairway), a corridor, a room or

an enclosed space. Interior finishes in spaces that are not separated from a corridor (e.g., waiting areas in business or health care facilities) must comply with the requirements for a corridor space. As shown in the table, the code places a higher emphasis in terms of allowable flame spread index for exits than for enclosed rooms. This is because of the critical nature and relative importance that is placed on maintaining the integrity of exits in order to evacuate the building.

There are numerous notes that amend the basic requirements of Table 803.5. Notes a through l apply only where they are specifically referenced in the table.

Note a. A limited amount of combustible wainscotting or other paneling material does not appreciably reduce the level of safety provided in an exit element; therefore, up to 1,000 square feet (93 m²) of Class C wainscotting or paneling is permitted by Note a in a grade-level lobby used as an exit element

TABLE 803.5
INTERIOR WALL AND CEILING FINISH REQUIREMENTS BY OCCUPANCY[k]

GROUP	SPRINKLERED[l]			NONSPRINKLERED		
	Vertical exits and exit passageways[a, b]	Exit access corridors and other exitways	Rooms and enclosed spaces[c]	Vertical exits and exit passageways[a, b]	Exit access corridors and other exitways	Rooms and enclosed spaces[c]
A-1 & A-2	B	B	C	A	A[d]	B[e]
A-3[f], A-4, A-5	B	B	C	A	A[d]	C
B, E, M, R-1, R-4	B	C	C	A	B	C
F	C	C	C	B	C	C
H	B	B	C[g]	A	A	B
I-1	B	C	C	A	B	B
I-2	B	B	B[h, i]	A	A	B
I-3	A	A[j]	C	A	A	B
I-4	B	B	B[h, i]	A	A	B
R-2	C	C	C	B	B	C
R-3	C	C	C	C	C	C
S	C	C	C	B	B	C
U	No restrictions			No restrictions		

For SI: 1 inch = 25.4 mm, 1 square foot = 0.0929 m².

a. Class C interior finish materials shall be permitted for wainscotting or paneling of not more than 1,000 square feet of applied surface area in the grade lobby where applied directly to a noncombustible base or over furring strips applied to a noncombustible base and fireblocked as required by Section 803.4.1.

b. In vertical exits of buildings less than three stories in height of other than Group I-3, Class B interior finish for nonsprinklered buildings and Class C interior finish for sprinklered buildings shall be permitted.

c. Requirements for rooms and enclosed spaces shall be based upon spaces enclosed by partitions. Where a fire-resistance rating is required for structural elements, the enclosing partitions shall extend from the floor to the ceiling. Partitions that do not comply with this shall be considered enclosing spaces and the rooms or spaces on both sides shall be considered one. In determining the applicable requirements for rooms and enclosed spaces, the specific occupancy thereof shall be the governing factor regardless of the group classification of the building or structure.

d. Lobby areas in Group A-1, A-2 and A-3 occupancies shall not be less than Class B materials.

e. Class C interior finish materials shall be permitted in places of assembly with an occupant load of 300 persons or less.

f. For churches and places of worship, wood used for ornamental purposes, trusses, paneling or chancel furnishing shall be permitted.

g. Class B material is required where the building exceeds two stories.

h. Class C interior finish materials shall be permitted in administrative spaces.

i. Class C interior finish materials shall be permitted in rooms with a capacity of four persons or less.

j. Class B materials shall be permitted as wainscotting extending not more than 48 inches above the finished floor in exit access corridors.

k. Finish materials as provided for in other sections of this code.

l. Applies when the vertical exits, exit passageways, exit access corridors or exitways, or rooms and spaces are protected by a sprinkler system installed in accordance with Section 903.3.1.1 or 903.3.1.2.

(see Section 1023, exceptions) when applied in accordance with Section 803.4.1.

Note b. The time required to exit a building that is less than three stories is not as critical as buildings with more than three stories. Thus, the class of material allowed in vertical exits is not as restrictive.

Note c. Since the intended use of certain rooms in buildings is more critical in nature than the group classification for the building, the finish classification must be determined by the group classification for the room.

Note d. Lobby areas in these types of groups provide a primary means of egress for occupancies that by nature have high occupant loads. Thus, the class of materials allowed in such means of egress components is restricted.

Note e. Note e recognizes that with a relatively small number of occupants, egress is accomplished more quickly and the activities tend to be more structured and manageable than with a large number of occupants. As such, where the design occupant load is 300 or less in rooms or spaces of an assembly occupancy, the interior finish materials are permitted to be Class C.

Note f. Churches and places of worship are generally open spaces. The occupants, because of the nature of the activities in such assemblies, are orderly. As such, wood, which is a combustible material, is allowed extensively as a finish material.

Note g. The time required for exiting increases with the number of stories of the structure. Interior finishes can play a major role in fire spread within a structure. Due to the materials found within a high-hazard occupancy, the rate of fire spread can accelerate much more rapidly than in other occupancies. To abate the hazard of rapid fire growth, Note g places a further restriction on interior finishes of rooms and spaces in high-hazard occupancies more than two stories in height.

Note h. Spaces that are used as offices have a low occupant load and the activity in such spaces is considered safe in nature. Since administrative spaces are such, the finish material is less restrictive.

Note i. Rooms with low occupant loads (four or less) do not pose dangerous situations in case of an emergency. Since evacuation of such rooms is accomplished in a short time, finish materials in these rooms are less restrictive.

Note j. This permits interior finish materials applied to a maximum height of 48 inches (1219 mm) above the floor in Class B corridors. This reduction is based on full-scale fire research reported by Christian and Waterman in the NFPA *Fire Journal*, July 1971, in an article entitled "Flame Spread in Corridors: Effects of Location and Area of Wall Finish." The research demonstrated that Class B wall finish used on the

lower 4 feet (1219 mm) of a corridor wall is not likely to spread fire.

Note k. Refer to other sections of the code for material class restrictions for specific groups.

Note l. Sprinklered facilities provide more protection to the occupants than nonsprinklered facilities. As such, in many instances, less-restrictive materials may be used.

803.6 Textiles. Where used as interior wall or ceiling finish materials, textiles, including materials having woven or nonwoven, napped, tufted, looped or similar surface, shall comply with the requirements of this section.

❖ This section is primarily intended to apply to textiles that are carpet and carpet-like coverings, which include materials having woven or nonwoven, napped, tufted, looped or similar surfaces.

803.6.1 Textile wall coverings. Textile wall coverings shall have a Class A flame spread index in accordance with ASTM E 84 and be protected by automatic sprinklers installed in accordance with Section 903.3.1.1 or 903.3.1.2 or the covering shall meet the criteria of Section 803.6.1.1 or 803.6.1.2 when tested in the manner intended for use in accordance with NFPA 265 using the product mounting system, including adhesive.

❖ Materials as described in Section 803.5 increase the fuel load in a room. As such, additional protection is required as a safety factor for rooms finished with this type of material. NFPA 265 is similar to NFPA 286 in that it tests materials with a fire in a corner of a small compartment; this more closely simulates a real fire. The pass/fail criteria is provided for both test methods within Sections 803.6.1.1 and 803.6.1.2. Test B criteria is less restrictive as the test itself is more conservative. The material is applied to three walls instead of two (see Figure 803.6.1).

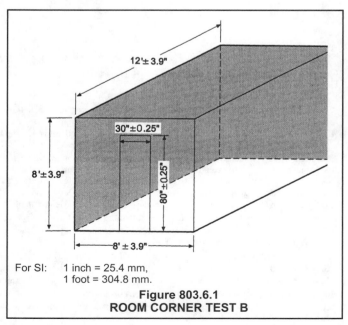

For SI: 1 inch = 25.4 mm,
           1 foot = 304.8 mm.

Figure 803.6.1
ROOM CORNER TEST B

803.6.1.1 Method A test protocol. During the Method A protocol, flame shall not spread to the ceiling during the 40 kW exposure. During the 150 kW exposure, the textile wall covering shall comply with all of the following:

1. Flame shall not spread to the outer extremity of the sample on the 8-foot by 12-foot (203 mm by 305 mm) wall.

2. The specimen shall not burn to the outer extremity of the 2-foot-wide (610 mm) samples mounted in the corner of the room.

3. Burning droplets deemed capable of igniting textile wall coverings or that burn for 30 seconds or more shall not form.

4. Flashover, as defined in NFPA 265, shall not occur.

5. The maximum net instantaneous peak heat release rate, determined by subtracting the burner output from the maximum heat release rate, does not exceed 300 kW.

❖ Full-scale room/corner fire research has shown that flame spread indexes produced by ASTM E 84 may not reliably predict the fire behavior of textile wall coverings. A more reliable test procedure has been developed at the University of California and involves use of a proposed ASTM room/corner method with a gas diffusion burner(s). The research findings are described in a report entitled "Room Fire Experiments of Textile Wall Coverings" by Fred L. Fisher, Bill MacCracken and Robert Brady Williamson. The research was summarized in the July/August 1987 issue of the *Building Official and Code Administrator* magazine published by BOCA.

As an alternative to the identified test procedure and by recognizing the ability of sprinkler systems to control fire growth, Class I textile materials, as determined in accordance with ASTM E 84, may be used in rooms or areas protected with an automatic sprinkler system. It should be noted that only the room or area needs to be suppressed, not the entire building.

The criteria provided for NFPA 265 is somewhat different than that provided for NFPA 286, which is referenced in Section 803.2. First, in NFPA 265, two different tests are offered. Test A is a limited application of the material; therefore, it has more restrictive pass/fail criteria. Test B is more similar to NFPA 200, which places the sample on all three walls. Also similar to NFPA 286, the source fire has two stages in both test methods: 40 kw for 5 minutes and 150 kW for 10 minutes.

Secondly, the criteria in Test A prohibits spread to the ceiling in the 40 kW phase as in NFPA 286, but has more detailed criteria for the second, more intense exposure. In the more intense exposure, the flame cannot spread beyond the 8-inch by 12-inch (203 mm by 305 mm) wall sample or the 2-foot (610 mm) sample mounted in the room. Additionally, burning droplets cannot burn for more than 30 seconds, and flashover (as defined in NFPA 265) cannot occur (see the commentary to Section 803.2.1 for flashover criteria).

The criteria also specifies a maximum instantaneous heat release rate; therefore, the concern is about heat spread and also the intensity of the fire.

803.6.1.2 Method B test protocol. During the Method B protocol, flames shall not spread to the ceiling at any time during the 40 kW exposure. During the 150 kW exposure, the textile wall covering shall comply with the following:

1. Flame shall not spread to the outer extremities of the samples on the 8-foot by 12-foot (203 mm by 305 mm) walls.

2. Flashover, as described in NFPA 265, shall not occur.

❖ As noted, Test B is more conservative as the test sample is placed on three walls (see Figure 803.6.1); therefore there are only two criteria: avoiding flashover and limitations on the flame spread of the sample, just as in NFPA 286.

803.6.2 Textile ceiling finish. Where used as a ceiling finish, carpet and similar textile materials shall have a Class A flame spread index in accordance with ASTM E 84 and be protected by automatic sprinklers.

❖ Materials as described in Section 803.6 increase the fuel load in a room. As such, additional protection is required as a safety factor for rooms finished with this type of material.

803.7 Expanded vinyl wall coverings. Expanded vinyl wall coverings shall comply with the requirements for textile wall and ceiling materials and their use shall comply with Section 803.6.

Exception: Expanded vinyl wall or ceiling coverings complying with Section 803.2 shall not be required to comply with Section 803.1 or 803.6.

❖ Expanded vinyl wall coverings have a woven textile backing so the flame spread indexes produced by ASTM E 84 may not reliably predict their fire behavior (see commentary, Section 803.6). There is a new exception that allows the use of NFPA 286 for testing (see Section 803.2 for further discussion).

803.8 Insulation. Thermal and acoustical insulation shall comply with Section 719.

❖ See the commentary for thermal- and sound-insulating materials in Section 719.

803.9 Acoustical ceiling systems. The quality, design, fabrication and erection of metal suspension systems for acoustical tile and lay-in panel ceilings in buildings or structures shall conform with generally accepted engineering practice, the provisions of this chapter and other applicable requirements of this code.

❖ This section regulates the design and installation of metal suspension systems for acoustical tile and acoustical lay-in panel ceilings. Suspended ceiling systems basically consist of a grid of metal channels or T-bars suspended by steel hangers (wires, rods or flats) from the structure above. Light fixtures, air diffusers and acoustical panels may be placed between the metal framing members.

803.9.1 Materials and installation. Acoustical materials complying with the interior finish requirements of Section 803 shall be installed in accordance with the manufacturer's recommendations and applicable provisions for applying interior finish.

❖ Acoustical materials used in metal suspension systems are regulated for the purpose of limiting flame spread and smoke development in the occupancies shown in Table 803.5.

Metal suspension system components for acoustical ceilings that might be subjected to the severe environmental conditions of high humidity (fog) or salt spray may have coatings ranked according to their ability to protect the components from deterioration. The installation of metal suspension systems in any exterior application must be considered as an installation in a severe environment (see ASTM C 635).

803.9.1.1 Suspended acoustical ceilings. Suspended acoustical ceiling systems shall be installed in accordance with the provisions of ASTM C 635 and ASTM C 636.

❖ ASTM C 635 covers metal ceiling suspension systems used primarily to support acoustical tile or acoustical lay-in panels. Suspension systems have three structural classifications, as described below:

Light-duty systems—Used primarily for residential and light commercial structures where ceiling loads other than acoustical tile or lay-in panels are not anticipated.

Intermediate-duty systems—Used primarily for ordinary commercial structures where some ceiling loads, due to light fixtures and air diffusers, are anticipated.

Heavy-duty systems—Used primarily for commercial structures in which the quantities and weights of ceiling fixtures (lights, air diffusers, etc.) are greater than those for an ordinary commercial structure.

ASTM C 636 covers the installation of suspension systems for acoustical tile and lay-in panels.

While the practices described in this standard have equal application to fire-resistance-rated suspension systems, additional requirements may be imposed to obtain the fire endurance classification of particular floor/ceiling or roof/ceiling assemblies (see Section 803.9.1.2).

803.9.1.2 Fire-resistance-rated construction. Acoustical ceiling systems that are part of fire-resistance-rated construction shall be installed in the same manner used in the assembly tested and shall comply with the provisions of Chapter 7.

❖ A suspended acoustical ceiling system that is by itself does not possess a fire-resistance rating. Acoustical ceiling systems are part of an assembly that has a specified fire- resistance rating. Acoustical lay-in panels must remain in place for the tested assembly to be effective.

SECTION 804
INTERIOR FLOOR FINISH

804.1 General. Interior floor finish and floor covering materials shall comply with this section.

> **Exception:** Floors and floor coverings of a traditional type, such as wood, vinyl, linoleum or terrazzo, and resilient floor covering materials which are not comprised of fibers.

❖ This section regulates the design and installation of floor finish and floor covering materials. Traditional floor coverings such as wood, vinyl, terrazzo and other resilient floor covering material must be exempt from this section.

804.2 Classification. Interior floor finish and floor covering materials required by Section 804.5.1 to be of Class I or II materials shall be classified in accordance with NFPA 253. The classification referred to herein corresponds to the classifications determined by NFPA 253 as follows: Class I, 0.45 watts/cm^2 or greater; Class II, 0.22 watts/cm^2 or greater.

❖ The use of a classification system eliminates the need to state the actual critical radiant flux value for a product to meet the identification requirements of Section 804. Over the years, a classification system has been found to be much easier for the industry to follow and still provides the building official with the information required to verify compliance.

In previous editions of the codes, the exception to Section 804.1 was listed as an exception to Section 804.2. The exception was relocated to follow Section 804.1; otherwise, the exception appears to apply only to the identification and testing section. The original intent of the exception was to exempt traditional floor coverings from the interior finish restrictions because there is no adverse experience to suggest the need for regulation of floor finish materials like wood and vinyl.

804.3 Testing and identification. Floor covering materials shall be tested by an approved agency in accordance with NFPA 253 and identified by a hang tag or other suitable method so as to identify the manufacturer or supplier and style, and shall indicate the interior floor finish or floor covering classification according to Section 804.2. Carpet-type floor coverings shall be tested as proposed for use, including underlayment. Test reports confirming the information provided in the manufacturer's product identification shall be furnished to the building official upon request.

❖ The only method to ascertain that a floor meets the criteria of this section is to request a copy of the test report for the specific material being installed; therefore, it is critical that the carpeting be properly identified in order to verify that acceptable materials are being provided in the appropriate locations. The identification is to be provided on the material itself since a manufacturer's designation is required.

804.4 Application. Combustible materials installed in or on floors of buildings of Type I or II construction shall conform with the requirements of this section.

> **Exception:** Stages and platforms constructed in accordance with Sections 410.3 and 410.4, respectively.

❖ Floors of buildings of Type I and II construction are to be of noncombustible construction except as permitted in this section. Sections 410.3 and 410.4 permit wood flooring in stages, while the floor finish of stairs is regulated by Section 804.5.

804.4.1 Subfloor construction. Floor sleepers, bucks and nailing blocks shall not be constructed of combustible materials, unless the space between the fire-resistance-rated floor construction and the flooring is either solidly filled with approved noncombustible materials or fireblocked in accordance with Section 717, and provided that such open spaces shall not extend under or through permanent partitions or walls.

❖ Sleepers, bucks and nailing blocks are permitted to be of combustible materials only where the void space is filled with a noncombustible material or is fireblocked in accordance with Section 717. Section 717 permits a maximum concealed space area of 100 square feet (9.3 m²). In either case, the open spaces cannot extend under or through permanent partitions or walls. The purpose of the fill or fireblocking is to reduce the impact of a fire in a concealed combustible space in the floor. Likewise, fire spread around partitions or walls through the concealed space in the floor is also intended to be prevented.

804.4.2 Wood finish flooring. Wood finish flooring is permitted to be attached directly to the embedded or fireblocked wood sleepers and shall be permitted where cemented directly to the top surface of approved fire-resistance-rated construction or directly to a wood subfloor attached to sleepers as provided for in Section 804.4.1.

❖ Wood floor finish materials may be applied directly to the wood sleepers, provided that the space is protected in accordance with Section 804.4.1. Wood floor finish materials may also be applied directly to the top surface of a fire-resistance-rated assembly or directly to a wood subfloor that is attached to sleepers in accordance with Section 804.4.2.

Combustible insulating boards not more than a 0.5-inch (12.7 mm) thick in buildings of Type 1 and 2 construction are permitted (see Section 804.4.3) to be attached directly to a noncombustible floor assembly or to wood sleepers protected in accordance with Section 804.4.1. These provisions, similar to the provisions of Section 803.4 for application of interior finishes, permit combustible finishes to be applied to the floor as long as they are not created. If concealed spaces are created, however, they must be properly fireblocked.

804.4.3 Insulating boards. Combustible insulating boards not more than 0.5-inch (12.7 mm) thick and covered with approved finished flooring are permitted, where attached directly to a noncombustible floor assembly or to wood subflooring attached to sleepers as provided for in Section 804.4.1.

❖ The addition of combustible insulating boards with a maximum thickness of 0.5 inch (12.7 mm) will not significantly increase the fuel load when applied as prescribed by this section or attached to sleepers as provided for in Section 804.4.1.

In all occupancies, interior floor finish in vertical exits, exit passageways, exit access corridors and rooms or spaces not separated from exit access corridors by full-height partitions extending from the floor to the underside of the ceiling must withstand a minimum critical radiant flux as specified in Section 804.5.1.

804.5 Interior floor finish requirements. In all occupancies, interior floor finish in vertical exits, exit passageways, exit access corridors and rooms or spaces not separated from exit access corridors by full-height partitions extending from the floor to the underside of the ceiling shall withstand a minimum critical radiant flux as specified in Section 804.5.1.

❖ The finish flooring requirements for all means of egress components such as stairs and corridors, as well as floors to rooms that open to these elements, are regarded as more critical than other building floor surfaces. As such, the minimum critical radiant flux must be as prescribed in Section 804.5.1.

804.5.1 Minimum critical radiant flux. Interior floor finish in vertical exits, exit passageways and exit access corridors shall not be less than Class I in Groups I-2 and I-3 and not less than Class II in Groups A, B, E, H, I- 4, M, R-1, R-2 and S. In all other areas, the interior floor finish shall comply with the DOC FF-1 "pill test" (CPSC 16 CFR, Part 1630).

> **Exception:** Where a building is equipped throughout with an automatic sprinkler system in accordance with Section 903.3.1.1, Class II materials are permitted in any area where Class I materials are required and materials complying with DOC FF-1 "pill test" (CPSC 16 CFR, Part 1630) are permitted in any area where Class II materials are required.

❖ This section prescribes the minimum requirements for interior floor finish materials that are used in vertical exits, exit passageways and exit access corridors. The criteria are based on the occupancy classification and the relationship of the space to the egress system. Similar to Table 803.5, the occupancy classification designation is meant to apply to the actual occupancy of the space and not necessarily the overall building classification.

Classifications I and II as used in this section are defined in Section 804.2, and are based on the results of the ASTM E 648 test procedure. DOC FF-1, also referred to as the "Methenamine Pill Test," was developed as a means of preventing the distribution of highly flammable soft floor coverings. The test essentially evaluates the performance of the floor covering when subject to a cigarette-type ignition by using a small methenamine tablet. All carpeting sold in the United States is required by federal law to pass this test procedure.

Recognizing the ability of automatic sprinkler systems to control a fire and the contribution of interior floor finishes to the early stages of fire growth, the exception allows the required interior floor finish ratings to be reduced when an automatic sprinkler system is provided throughout the building. The reference to Section 903.3.1.1 clarifies that the system is to be installed in accordance with NFPA 13. In cases where Class II materials are required and an automatic sprinkler system is provided, the minimum requirement is that the material meet the DOC FF-1 test criteria.

[F] SECTION 805
DECORATIONS AND TRIM

805.1 General. In occupancies of Groups A, E, I, R-1 and dormitories in Group R-2, curtains, draperies, hangings and other decorative materials suspended from walls or ceilings shall be flame resistant in accordance with Section 805.2 and NFPA 701 or noncombustible.

In Groups I-1 and I-2, combustible decorations shall be flame retardant unless the decorations, such as photographs and paintings, are of such limited quantities that a hazard of fire development or spread is not present. In Group I-3, combustible decorations are prohibited.

❖ These requirements apply to decorative materials installed in occupancies in Groups A, E, I, R-1 and dormitories in Group R-2. The list of groups is based on the number of persons and the condition or capabilities of the occupants with respect to prompt evacuation. Such decorative materials are required to be noncombustible or be maintained flame resistant in accordance with Section 805.2 and NFPA 701.

The requirements for Groups I-1 and I-2 become more stringent because of the nature of the occupants and activities in these types of facilities. Combustible materials in such facilities are required to be flame retardant; however, a limited amount of decorations, such as photographs and paintings, is not required to be flame retardant. No amount of combustible decorations is permitted in Group I-3 facilities.

805.1.1 Noncombustible materials. The permissible amount of noncombustible decorative material shall not be limited.

❖ Where decorative materials are classified as noncombustible, it is presumed that they will contribute little, if any, to the growth and spread of fire; therefore, the quantity of noncombustible decorative materials is not limited.

805.1.2 Flame-resistant materials. The permissible amount of flame-resistant decorative materials shall not exceed 10 percent of the aggregate area of walls and ceilings.

Exception: In auditoriums of Group A, the permissible amount of flame-resistant decorative material shall not exceed 50 percent of the aggregate area of walls and ceilings where the building is equipped throughout with an automatic

sprinkler system in accordance with Section 903.3.1.1 and the material is installed in accordance with Section 803.4.

❖ Flame-resistant materials may contribute to fire growth and spread; therefore, such materials are limited to a maximum of 10 percent of the total wall and ceiling area of the space under consideration. Unlike incidental trim, it is not presumed that such materials are necessarily distributed evenly throughout the room (see Figure 805.1.2).

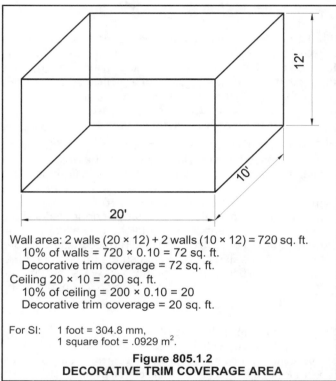

Wall area: 2 walls (20 × 12) + 2 walls (10 × 12) = 720 sq. ft.
   10% of walls = 720 × 0.10 = 72 sq. ft.
   Decorative trim coverage = 72 sq. ft.
Ceiling 20 × 10 = 200 sq. ft.
   10% of ceiling = 200 × 0.10 = 20
   Decorative trim coverage = 20 sq. ft.

For SI: 1 foot = 304.8 mm,
           1 square foot = .0929 m^2.

Figure 805.1.2
DECORATIVE TRIM COVERAGE AREA

805.2 Acceptance criteria and reports. Where required to be flame resistant, decorative materials shall be tested by an approved agency and pass Test 1 or 2, as appropriate, described in NFPA 701 or such materials shall be noncombustible. Reports of test results shall be prepared in accordance with NFPA 701 and furnished to the building official upon request.

❖ The standard test method to be used to evaluate the flame resistance of a material is NFPA 701, which contains two test methods. Test method 1 is used for fabric or composites having a weight of 700 g/m^2 (.14 lb/ft^2). These tests are used to determine whether flame-resistant materials propagate flame beyond the area exposed to the ignition source. The performance tests are not intended to indicate whether the material tested will resist the propagation of flame under severe fire exposure.

Materials that are required to be flame resistant and are treated (instead of being inherently flame resistant) must not generate smoke more dense than that given off by untreated wood or paper burning under comparable conditions. This criterion is in recognition that some treatments may render a material flame resistant and, in so doing, may cause a substantial increase in the amount of generated smoke. This section requires a comparison using NFPA 701 to evaluate the amount of smoke generated; however, NFPA 701 is not intended to evaluate smoke generation. The smoke-developed index provisions of ASTM E 84 may be useful in such an evaluation.

805.3 Foam plastic. Foam plastic used as trim in any occupancy shall comply with Section 2604.2.

❖ This section establishes that foam-plastic materials are permitted to be used as interior finish materials when they comply with Section 2604. At first glance, it would appear that the classification by ASTM E 84 is all that is necessary. When tested, many foam-plastic materials have a flame spread index of less than 25. Chapter 26 also requires that the foam-plastic material be protected from ignition through the use of a thermal barrier.

Because compliance with Section 2604 is required by this section, foam plastics can only be used as exposed interior finishes if they pass the rigorous testing requirements of Section 2603.8.

805.4 Pyroxylin plastic. Imitation leather or other material consisting of or coated with a pyroxylin or similarly hazardous base shall not be used in Group A occupancies.

❖ In occupancies that have normally high occupant loads and because of the hazardous nature of materials consisting of or coated with pyroxylin, this type of finish material is not allowed in Group A occupancies.

805.5 Trim. Material used as interior trim shall have minimum Class C flame spread and smoke-developed indexes. Combustible trim, excluding handrails and guardrails, shall not exceed 10 percent of the aggregate wall or ceiling area in which it is located.

❖ In occupancies of any group, unless otherwise noted in the code, the minimum classification of all trim must be at least Class C. Except for handrails, guardrails, walls and ceilings, combustible trim may be used as long as it does not exceed 10 percent of the area of the walls and ceiling on which it is located. This quantity of combustible material will not significantly increase the fuel load.

Bibliography

The following resource materials are referenced in this chapter or are relevant to the subject matter addressed in this chapter.

ASTM E 84-01, *Test Method for Surface Burning Characteristics of Building Materials*. West Conshohocken, PA: American Society for Testing and Materials, 2001.

ASTM E 648-95, *Test Method for Critical Radiant Flux of Floor Covering Systems Using a Radiant Heat Energy Source*. West Conshohocken, PA: ASTM International, 1995.

Building Materials Directory. Northbrook, IL: Underwriters Laboratories Inc., 1993.

Christian, W.J. and J.E. Waterman. "Flame Spread in Corridors: Effects of Location and Area of Wall Finish." *Fire Journal*, July 1971.

DOC FF-1 (CPSC 16 CFR, Part 1630-70)-98, *Standard for the Surface Flammability of Carpets and Rugs*. Bethesada, MD: Consumer Product Safety Commission, 1970.

FEMA FIA-TB #2, *Flood Resistant Material Requirements for Buildings Located in Special Flood Hazard Areas*, 1993.

FEMA FIA-TB #7, *Wet Floodproofing Requirements for Structures Located in Special Flood Hazard Areas*, 1993.

FEMA FIA-TB #8, *Corrosion Protection for Metal Connectors in Coastal Areas for Structures Located in Special Flood Hazard Areas*, 1996.

Fisher, F.L., B. MacCracken and R.B. Williamson. "Room Fire Tests of Textile Wallcoverings." Service to Industry Report No. 85-4. University of California Fire Research Laboratory, 1986.

GA 600-00, *Fire Resistance Design Manual*. Washington, DC: Gypsum Association, 2000.

NFIP Technical Bulletin Series. Washington, DC: National Flood Insurance Program. [Online]. Available: http://www.fema.gov/MIT/techbul.htm.

NFPA 13-99, *Installation of Sprinkler Systems*. Quincy, MA: National Fire Protection Association, 1999.

NFPA 13R-99, *Installation of Sprinkler Systems in Residential Occupancies Up to Four Stories in Height*. Quincy, MA: National Fire Protection Association, 1999.

NFPA 265-2002, *Standard Methods of Fire Tests for Evaluating Fire Growth Contribution of Textile Coverings on Full Height Panels and Walls*. Quincy, MA: National Fire Protection Association, 2002.

NFPA 286-2000, *Standard Methods of Fire Tests for Evaluating Contribution of Wall and Ceiling Interior Finish to Room Fire Growth*. Quincy, MA: National Fire Protection Association, 2000.

NFPA 701-99, *Standard Methods of Fire Tests for Flame-Resistant Textiles and Film*. Quincy, MA: National Fire Protection Association, 1999.

SFPE *Handbook of Fire Protection Engineering*. Quincy, MA: National Fire Protection Association, 1988.

Chapter 9:
Fire Protection Systems

General Comments

The provisions required by Chapter 9 are just one aspect of the overall fire protection system of a building or structure. All fire protection requirements contained in the code must be considered as a package or overall system. Noncompliance with any part of the overall system may cause other parts of the system to fail. Failure to install the systems in accordance with code provisions may result in an increased loss of life and property due to a reduction in the level of protection provided.

It is important that every effort be made to verify the proper design and installation of a given fire protection system, especially those that result in construction alternatives and other code tradeoffs.

The requirements found in Chapter 9 can be considered active fire safety provisions. They are provisions directed at containing and abating the fire once it has erupted. This chapter is almost a direct copy of Chapter 9 in the *International Fire Code*® (IFC®). The IFC, however, contains specific provisions that are only applicable to existing buildings. The IFC also contains periodic testing criteria that are not duplicated in this chapter. Proper testing, inspection and maintenance of the various systems, however, are critical to establish the reliability of the system. Additionally, Chapter 9 references and adopts numerous National Fire Protection Association (NFPA) standards, including the acceptance testing criteria within the standard. The referenced standards will also contain more specific design and installation criteria than are found in this chapter. As noted in Section 102.4, where differences occur between code provisions and the referenced standard, the code provisions apply.

Purpose

Fire protection systems may serve one or more purposes in providing adequate protection from fire and hazardous material exposure. The purpose of Chapter 9 is to prescribe the minimum requirements for an active system or systems of fire protection to perform the following functions: to detect a fire; to alert the occupants or fire department of a fire emergency; to control smoke and to control or extinguish the fire. Generally, the requirements are based on the occupancy and height and area of the building, as these are the factors that most affect fire-fighting capabilities and the relative hazard of a specific space or area.

SECTION 901
GENERAL

901.1 Scope. The provisions of this chapter shall specify where fire protection systems are required and shall apply to the design, installation and operation of fire protection systems.

❖ Chapter 9 contains requirements for fire protection systems that may be provided in a building. These include automatic suppression systems; standpipe systems; fire alarm and detection systems; smoke control systems; smoke and heat vents and portable fire extinguishers. Besides indicating the conditions under which respective systems are required, this chapter contains the design, installation, maintenance and operational criteria for fire protection systems. While the chapter requires proper maintenance for the reliability of the systems, the actual maintenance provisions (periodic testing, inspections and maintenance) are found in the IFC.

901.2 Fire protection systems. Fire protection systems shall be installed, repaired, operated and maintained in accordance with this code and the *International Fire Code*.

Any fire protection system for which an exception or reduction to the provisions of this code has been granted shall be considered to be a required system.

Exception: Any fire protection system or portion thereof not required by this code shall be permitted to be installed for partial or complete protection provided that such system meets the requirements of this code.

❖ The fire protection system is an integral component of the protection features of the building and is required to be properly installed, repaired, operated and maintained in accordance with the code and the IFC. Improperly installed or maintained systems can negate any anticipated protection and, in fact, create a hazard in itself.

While the code may not require a protection system for a specific building or portion thereof, due to its occupancy, the fire protection system would still be considered a required system if some other code tradeoff, exception or reduction was taken based on the installation of the fire protection system. For example, a typical small office building may not require an automatic sprinkler system solely due to its Group B occupancy classification; however, if an exit access corridor fire-resistance-rating reduction is taken in accordance with Table 1016.1 for buildings equipped throughout with an NFPA 13 sprinkler system, that sprinkler system would be considered a required system.

The exception acknowledges that a building owner or designer may elect to install a fire protection system that is not required by the code. Even though such a system is not required, it must comply with the applicable provisions of this chapter and the IFC. This requirement is predicated on the concept that any fire protection system not installed in accordance with the code is intrinsically lacking because it could give a false impression of properly installed protection.

For example, if a building owner chooses to provide sprinkler protection in a certain area and such protection is not required by the code, the system must be installed in accordance with the applicable NFPA sprinkler standard (13,13R or 13D) and other applicable requirements of the code, such as water supply and supervision. The extent of the protection thus provided would not be regulated.

901.3 Modifications. No person shall remove or modify any fire protection system installed or maintained under the provisions of this code or the *International Fire Code* without approval by the building official.

❖ This section emphasizes the principle that systems installed and maintained in compliance with the codes and standards in effect at the time they were placed in service must remain in an operative condition at all times. Protection must not be diminished in any existing building except for the purpose of conducting tests, maintenance or repairs. The length of service interruptions should be kept to a minimum. The building official must be notified of any service interruptions and should carefully evaluate the continued operation or occupancy of buildings and structures where protection is interrupted.

901.4 Threads. Threads provided for fire department connections to sprinkler systems, standpipes, yard hydrants or any other fire hose connection shall be compatible with the connections used by the local fire department.

❖ Incompatible fire service threads have been a major problem throughout the history of modern fire suppression. Efforts to standardize fire service threads in the United States began as early as 1898. Following the Great Baltimore Fire of 1904, NFPA adopted a national standard for fire hose thread with diameters $2^1/_2$ inches (64 mm) and larger; however, to this day many jurisdictions continue to use their own (historic) thread standards. A number of jurisdictions have pioneered the use of nonthreaded, quarter-turn or "Storz" fire hose couplings in place of the national and local traditional thread standards.

901.5 Acceptance tests. Fire protection systems shall be tested in accordance with the requirements of this code and the *International Fire Code*. When required, the tests shall be conducted in the presence of the building official. Tests required by this code, the *International Fire Code* and the standards listed in this code shall be conducted at the expense of the owner or the owner's representative. It shall be unlawful to occupy portions of a structure until the required fire protection systems within that portion of the structure have been tested and approved.

❖ All fire protection systems are to be subjected to an acceptance test to determine that the system will operate in the manner required by the code. Acceptance tests are usually part of the final inspection procedures required by Section 109.3.10. Specific acceptance test procedures are provided in other sections as well as the referenced standards. In most instances, the acceptance test procedures require 100-percent operation of the testable system components to determine that they are operational and functioning as required. Often, the design professional may require additional testing to verify that the system operates as designed, which may be beyond the code requirements.

The inclusion of the requirement for acceptance tests in the code is not intended to assign responsibility for witnessing the tests. The responsibility to witness the acceptance test is an administrative issue that each municipality must address. Because the acceptance test is critical during design and construction and is a requirement of occupancy, the requirement is located in the code. The section also clarifies that it is the owner's responsibility to conduct the test and the role of the building official to witness—not conduct—the test. Typically, the owner will assign the responsibility to the installing contractor.

Access to all flow test connections, valves and points of fluid discharge required by other sections is to be provided at locations acceptable to the building official. Consideration should be given to the location of sprinkler system inspector test connections and the main drain valves, as well as to the valves necessary to test standpipe and fire pump installations.

Consideration should also be given to the discharge of water from the various test connections for aqueous systems that should be arranged to discharge to the outside or to a drain with sufficient capacity to handle the anticipated water flow. If the test connection is piped to the outside, then protection should be furnished to prevent unnecessary damage to the property by providing splash blocks or similar protection. When the building drainage system is used, the indirect waste provisions of Chapter 7 of the *International Plumbing Code®* (IPC®) are to apply.

Partial occupancy of any structure should not be permitted unless all fire protection systems for the occupied areas have been tested and approved. Even so, the code assumes that full protection for all areas will be provided as expeditiously as possible. The installation of many fire protection systems and the associated code tradeoffs permitted for a given occupancy assume complete building protection, not just in the occupied areas. Final approval of all partial occupancy conditions is subject to the building official.

901.6 Supervisory service. Where required, fire protection systems shall be monitored by an approved supervising station in accordance with NFPA 72.

❖ All required fire protection systems are to be supervised as a means of determining at any time that the system is operational. Acceptable methods of supervision are provided in NFPA 72.

901.6.1 Automatic sprinkler systems. Automatic sprinkler systems shall be monitored by an approved supervising station.

Exceptions:

1. A supervising station is not required for automatic sprinkler systems protecting one- and two-family dwellings.

2. Limited area systems serving fewer than 20 sprinklers.

❖ This section highlights that automatic sprinkler systems must be monitored by an approved supervising station. All water supply control valves and water-flow switches are required to be electrically supervised as indicated in Section 903.4. A central station, remote supervising station or proprietary supervising station are approved services recognized in NFPA 72 (see commentary, Section 903.4.1).

Automatic sprinkler systems in one- and two-family dwellings are typically designed in accordance with NFPA 13D, which does not require electrical supervision (see Exception 1). Limited area sprinkler systems are generally supervised by their connection to the domestic water service (see Exception 2). As such, limited area sprinkler systems with fewer than 20 sprinkler heads are not required to have local alarms or a fire department connection.

901.6.2 Fire alarm systems. Fire alarm systems required by the provisions of Section 907.2 of this code and Section 907.2 of the *International Fire Code* shall be monitored by an approved supervising station in accordance with Section 907.14.

Exceptions:

1. Single- and multiple-station smoke alarms required by Section 907.2.10.

2. Smoke detectors in Group I-3 occupancies.

3. Supervisory service is not required for automatic sprinkler systems in one- and two-family dwellings.

❖ Fire alarm systems are required to automatically transmit an alarm and any trouble signals to the fire department through one of the approved methods in NFPA 72. Exception 1 exempts single- and multiple-station smoke alarms from being supervised due to the potential for unwanted false alarms. Similarly, due to the concern over unwanted alarms, smoke detectors in Group I-3 occupancies need only sound an approved alarm signal, which automatically notifies staff (see Section

907.2.6.2.1). Smoke detectors in Group I-3 occupancies are typically subject to misuse and abuse. Frequent unwanted alarms would negate the effectiveness of the system (see Exception 2).

Exception 3 clarifies that sprinkler systems in one- and two-family dwellings are not part of a dedicated fire alarm system and are typically designed in accordance with NFPA 13D, which does not require electrical supervision.

901.6.3 Group H. Manual fire alarm, automatic fire-extinguishing and emergency alarm systems in Group H occupancies shall be monitored by an approved supervising station.

Exception: When approved by the building official, on-site monitoring at a constantly attended location shall be permitted provided that notifications to the fire department will be equal to those provided by an approved supervising station.

❖ Given the varied nature and quantity of hazardous materials that could be present, all fire protection systems in Group H occupancies are required to be monitored by an approved supervising station. The exception, however, permits the building official to approve an on-site monitoring system. Many companies that routinely deal with hazardous materials employ their own fire brigades and have emergency response procedures in place.

901.7 Fire areas. Where buildings, or portions thereof, are divided into fire areas so as not to exceed the limits established for requiring a fire protection system in accordance with this chapter, such fire areas shall be separated by fire barriers having a fire-resistance rating of not less than that determined in accordance with Section 706.3.7.

❖ This section provides specific guidance on how a building needs to be divided into fire areas in order to avoid the required installation of a fire protection system. A single occupancy group would require fire barriers in order to create multiple fire areas (see Table 706.3.7), each having an area below the threshold for fire protection system installation.

SECTION 902
DEFINITIONS

902.1 Definitions. The following words and terms shall, for the purposes of this chapter, and as used elsewhere in this code, have the meanings shown herein.

❖ Definitions of terms can help in the understanding and application of the code requirements. The purpose for including these definitions within this chapter is to provide more convenient access to them without having to refer back to Chapter 2. For convenience, these terms are also listed in Chapter 2 with a cross reference to this section. The use and application of all defined terms, including those defined herein, are set forth in Section 201.

[F] ALARM NOTIFICATION APPLIANCE. A fire alarm system component such as a bell, horn, speaker, light or text display that provides audible, tactile or visible outputs, or any combination thereof.

❖ The code requires that fire alarm systems be equipped with approved alarm notification appliances so that in an emergency, the fire alarm system will notify the occupants of the need for evacuation or implementation of the fire emergency plan. Alarm notification devices required by the code are of two general types: visible and audible. Except for emergency voice/alarm communication systems, once the fire alarm system has been activated, all visible and audible communication alarms are required to activate. Emergency voice/alarm communication systems are special signaling systems that are activated on a selective basis in response to specific emergency conditions.

[F] ALARM SIGNAL. A signal indicating an emergency requiring immediate action, such as a signal indicative of fire.

❖ This is a general term for all types of supervisory and trouble signals. An example would be a supervisory (tamper) switch on a sprinkler control valve. The activation of the device does not necessarily indicate that there is a fire condition; however, the level of protection may have been compromised (see the definition of "Fire alarm signal").

[F] ALARM VERIFICATION FEATURE. A feature of automatic fire detection and alarm systems to reduce unwanted alarms wherein smoke detectors report alarm conditions for a minimum period of time, or confirm alarm conditions within a given time period, after being automatically reset, in order to be accepted as a valid alarm-initiation signal.

❖ False fire (evacuation) alarms are a nuisance; hence, the code specifies that alarms activated by smoke detectors are not to be sounded until the alarm signal is verified. Verification may be accomplished by cross-zoned detectors in a single protected area or by system features that will retard the alarm until the signal is determined to be valid. Valid alarm initiation signals can be determined by detectors that report alarm conditions for a minimum period of time or which, after being reset, continue to report an alarm condition. The alarm verification feature may not retard signal activation for a period of more than 60 seconds and must not apply to alarm-initiating devices other than smoke detectors (which may be connected to the same circuit). Alarm verification should not be confused with presignal features, which delay an alarm signal for more than 1 minute and are allowed only where specifically permitted by the authority having jurisdiction.

[F] ANNUNCIATOR. A unit containing one or more indicator lamps, alphanumeric displays or other equivalent means in which each indication provides status information about a circuit, condition or location.

❖ This refers to the control panel that provides the status of the monitored fire protection systems and devices.

[F] AUDIBLE ALARM NOTIFICATION APPLIANCE. A notification appliance that alerts by the sense of hearing.

❖ Audible alarms that are part of a fire alarm system must be capable of being heard in every occupied space of a building. Section 907.9.2 prescribes the minimum sound pressure level for all audible alarm notification appliances depending on the occupancy of the building.

[F] AUTOMATIC. As applied to fire protection devices, is a device or system providing an emergency function without the necessity for human intervention and activated as a result of a predetermined temperature rise, rate of temperature rise or combustion products.

❖ This term, when used in conjunction with fire protection systems or devices, means that the system or device will perform its intended function without a person being present or performing any task in its control or operation. Automatic devices and systems operate completely without human presence or intervention.

[F] AUTOMATIC FIRE-EXTINGUISHING SYSTEM. An approved system of devices and equipment which automatically detects a fire and discharges an approved fire-extinguishing agent onto or in the area of a fire.

❖ This term is the generic name for all types of automatic fire-extinguishing systems, including the most common type — the automatic sprinkler system. See Section 904 for requirements for particular alternative automatic fire-extinguishing systems, such as wet-chemical, dry-chemical, foam, carbon dioxide, halon and clean agent systems.

[F] AUTOMATIC SPRINKLER SYSTEM. A sprinkler system, for fire protection purposes, is an integrated system of underground and overhead piping designed in accordance with fire protection engineering standards. The system includes a suitable water supply. The portion of the system above the ground is a network of specially sized or hydraulically designed piping installed in a structure or area, generally overhead, and to which automatic sprinklers are connected in a systematic pattern. The system is usually activated by heat from a fire and discharges water over the fire area.

❖ An automatic sprinkler system is one type of automatic fire-extinguishing system. Automatic sprinkler systems are the most common, and their life safety attributes are widely recognized. The code specifies three types of automatic sprinkler systems: one installed in accordance with NFPA 13, one in accordance with NFPA 13R and the other in accordance with NFPA 13D. To be considered for most code design alternatives, a building's automatic sprinkler system must be installed throughout in accordance with NFPA 13 (see Section 903.3.1.1).

In a fire condition, sprinklers automatically open and discharge water onto a fire in a spray pattern that is designed to cool and extinguish the fire. Originally, automatic sprinkler systems were developed just for the protection of buildings and their contents. Because of the development and improvements in sprinkler head response time and water distribution, however, automatic sprinkler systems are now also considered a life safety system. Proper operation of an automatic sprinkler system requires careful selection of the sprinkler heads so that water in sufficient quantity at adequate pressure and properly distributed will be available to suppress the fire.

[F] AVERAGE AMBIENT SOUND LEVEL. The root mean square, A-weighted sound pressure level measured over a 24-hour period.

❖ The ambient noise that can be expected depends on the occupancy of the building. In order to attract the attention of the occupants, audible alarm devices must be capable of being heard above the ambient noise in the space. As such, the alarm devices must have minimum sound pressure levels above the average ambient sound level. Section 907.9.2 prescribes the minimum sound pressure levels for the audible alarm notification appliances for all occupancy conditions.

[F] CARBON DIOXIDE EXTINGUISHING SYSTEMS. A system supplying carbon dioxide (CO_2) from a pressurized vessel through fixed pipes and nozzles. The system includes a manual- or automatic-actuating mechanism.

❖ Carbon dioxide extinguishing systems are useful in extinguishing fires in specific hazards or equipment in occupancies where an inert electrically nonconductive medium is essential or desirable and where cleanup of other extinguishing agents, such as dry-chemical residue, presents a problem. The system works by displacing the oxygen in an enclosed area by flooding it with carbon dioxide. To effectively flood the enclosure, provisions for automatic door and window closers and dampers for the forced-air ventilation system must be provided. NFPA 12 provides minimum requirements for the design, installation, testing, inspection, approval, operation and maintenance of carbon dioxide extinguishing systems.

[F] CEILING LIMIT. The maximum concentration of an air-borne contaminant to which one may be exposed, as published in DOL 29 CFR Part 1910.1000.

❖ This term represents the maximum level of exposure for employees or occupants to hazardous air contaminants during any part of a normal workday. DOL 29 CFR: 1910.1000 provides acceptable ceiling limits of contamination for various substances.

[F] CLEAN AGENT. Electrically nonconducting, volatile or gaseous fire extinguishant that does not leave a residue upon evaporation.

❖ The two categories of clean agents are halocarbon compounds and inert gas agents. Halocarbon compounds include bromine, carbon, chloride, fluorine, hydrogen and iodine. Halocarbon compounds suppress the fire through a combination of breaking the chemical chain reaction of the fire, reducing the ambient oxygen

supporting the fire and reducing the ambient temperature of the fire origin to reduce the propagation of fire. The clean agents that are inert gas agents contain primary components consisting of helium, neon or argon, or a combination. Inert gases work by reducing the oxygen concentration around the fire origin to a level that does not support combustion (see commentary, Section 904.10).

[F] CONSTANTLY ATTENDED LOCATION. A designated location at a facility staffed by trained personnel on a continuous basis where alarm or supervisory signals are monitored and facilities are provided for notification of the fire department or other emergency services.

❖ These locations are intended to receive trouble, supervisory and alarm signals transmitted by the fire protection equipment installed within a protected facility. It is the intent of this section to have a location and personnel that is acceptable to the authority having jurisdiction, responsible for actions taken when the fire protection system requires attention.

[F] DELUGE SYSTEM. A sprinkler system employing open sprinklers attached to a piping system connected to a water supply through a valve that is opened by the operation of a detection system installed in the same areas as the sprinklers. When this valve opens, water flows into the piping system and discharges from all sprinklers attached thereto.

❖ The purpose of a deluge system is to apply water or foam throughout the protected area by means of a system of open sprinklers. In a fire emergency, a fire detection system activates, making it possible to apply water to a fire more quickly and to cover a larger area than with a conventional automatic sprinkler system, which depends on sprinklers being activated individually as the fire spreads. Deluge systems are particularly beneficial in hazardous areas where the fuel loads (combustible contents) are of such a nature that fire may flash ahead of the operations of conventional automatic sprinklers.

[F] DETECTOR, HEAT. A fire detector that senses heat produced by burning substances. Heat is the energy produced by combustion that causes substances to rise in temperature.

❖ In a fire condition, heat is released that causes the temperature in a room or space to increase. Automatic fire detectors that sense abnormally high temperature or rate-of-temperature rise are known as heat detectors. These include fixed temperature detectors, rate compensation detectors and rate-of-rise detectors. The code requires all automatic fire detectors to be smoke detectors, except that heat detectors tested and approved in accordance with NFPA 72 may be used as an alternative to smoke detectors in rooms and spaces where, during normal operation, products of combustion are present in sufficient quantity to actuate a smoke detector.

[F] DRY-CHEMICAL EXTINGUISHING AGENT. A powder composed of small particles, usually of sodium bicarbonate, potassium bicarbonate, urea-potassium-based bicarbonate, potassium chloride or monoammonium phosphate, with added particulate material supplemented by special treatment to provide resistance to packing, resistance to moisture absorption (caking) and the proper flow capabilities.

❖ A dry-chemical system extinguishes a fire by providing a chemical barrier between the fire and oxygen, which acts to smother a fire. This system is best known for providing protection for commercial ranges, commercial fryers and exhaust hoods. Wet-chemical extinguishing systems, however, are more commonly used for new installations in commercial cooking equipment.

　　The type of dry chemical to be used with the extinguishing system is a function of the hazard to be protected. The type of dry chemical used in a system should not be changed, unless it has been proven changeable by a testing laboratory; is recommended by the manufacturer of the equipment and is acceptable to the building official for the hazard being protected. Additional guidance on the use of various dry-chemical agents can be found in NFPA 17, which provides minimum requirements for the design, installation, testing, inspection, approval, operation and maintenance of dry-chemical extinguishing systems.

[F] EMERGENCY ALARM SYSTEM. A system to provide indication and warning of emergency situations involving hazardous materials.

❖ Due to the potential volatile nature of hazardous materials, an emergency alarm system is required outside of interior building rooms or areas containing hazardous materials in excess of the maximum allowable quantities permitted in Tables 307.7(1) and 307.7(2). The intent of the emergency alarm, upon actuation by an alarm-initiating device such as a pull station, is to alert the occupants of an emergency condition involving hazardous materials.

[F] EMERGENCY VOICE/ALARM COMMUNICATIONS. Dedicated manual or automatic facilities for originating and distributing voice instructions, as well as alert and evacuation signals pertaining to a fire emergency, to the occupants of a building.

❖ An emergency voice/alarm communication system is a special feature of fire alarm systems in buildings with special evacuation considerations, such as a high-rise building. Emergency voice/alarm communication systems automatically communicate a fire emergency message to all occupants of a building on a general or selective basis. Such systems also enable the fire service to transmit manually voice instructions to the building occupants about a fire emergency condition and indicate the action to be taken for evacuation or movement to another area of the building.

[F] EXPLOSION. An effect produced by the sudden violent expansion of gases, that is accompanied by a shock wave or disruption of enclosing materials or structures, or both.

❖ This definition identifies a condition where an expansion of gases is so rapid that the effect is dissipated by a shock wave. Note that the definition for "Explosion" is independent of the origin of the effect. For example, an explosion can be caused by a tank rupture as well as a chemical reaction caused by a material classified as an explosive by the hazardous material regulations of DOTn 49 CFR.

[F] FIRE ALARM BOX, MANUAL. See "Manual Fire Alarm Box."

[F] FIRE ALARM CONTROL UNIT. A system component that receives inputs from automatic and manual fire alarm devices and is capable of supplying power to detection devices and transponder(s) or off-premises transmitter(s). The control unit is capable of providing a transfer of power to the notification appliances and transfer of condition to relays or devices.

❖ The fire alarm control unit (panel) acts as a point where all signals initiated within the protected building are received before the signal is transmitted to a constantly attended location.

[F] FIRE ALARM SIGNAL. A signal initiated by a fire alarm-initiating device such as a manual fire alarm box, automatic fire detector, water flow switch, or other device whose activation is indicative of the presence of a fire or fire signature.

❖ This signal is transmitted to a fire alarm control unit as a warning that requires immediate action. Personnel at the constantly attended location are trained to immediately respond to a fire alarm signal indicating the presence of a fire. A fire alarm signal assumes an actual fire event has been detected (see definition of "Alarm signal").

[F] FIRE ALARM SYSTEM. A system or portion of a combination system consisting of components and circuits arranged to monitor and annunciate the status of fire alarm or supervisory signal-initiating devices and to initiate the appropriate response to those signals.

❖ Fire alarm systems are provided in buildings to limit fire casualties and property losses. This is accomplished by notifying the occupants of the building, the local fire department or both of an emergency condition. The alarm notification appliances associated with fire alarm systems are intended to be evacuation alarms. All fire alarms are required to be designed and installed in accordance with NFPA 72.

[F] FIRE COMMAND CENTER. The principal attended or unattended location where the status of detection, alarm communications and control systems is displayed, and from which the system(s) can be manually controlled.

❖ Fire command centers are communication centers where dedicated manual and automatic facilities are located for the origination, control and transmission of information and instruction pertaining to a fire emergency to the occupants (including fire department personnel) of the building. Fire command centers must provide facilities for the control and display of the status of all fire protection (detection, signaling, etc.) systems. These stations must be located in secure areas as approved by the authority having jurisdiction. Often this is a location near the primary building entrance. Fire command centers also may be combined with other building operations and security facilities when permitted by the authority having jurisdiction; however, operating controls for use by the fire department must be clearly marked.

[F] FIRE DETECTOR, AUTOMATIC. A device designed to detect the presence of a fire signature and to initiate action.

❖ Automatic fire detectors include all approved devices designed to detect the presence of a fire and automatically initiate emergency action. These include smoke-sensing, heat-sensing, flame-sensing, gas-sensing and other fire detectors that operate on other principles as approved by the building official. Automatic fire detectors should be selected based on the type and size of the fire to be detected and the response required. Automatic fire detectors must be approved, installed and tested in accordance with the code and NFPA 72.

[F] FIRE PROTECTION SYSTEM. Approved devices, equipment and systems or combinations of systems used to detect a fire, activate an alarm, extinguish or control a fire, control or manage smoke and products of a fire or any combination thereof.

❖ A fire protection system is any approved device or equipment used singly or in combination, either manually or automatically, which is intended to detect a fire condition, notify the building occupants of a fire condition or suppress a fire. Fire protection systems include fire suppression systems, standpipe systems, fire alarm systems, fire detection systems, smoke control systems and smoke vents. All fire protection systems must be approved by the building official and tested in accordance with the referenced standards and the IFC.

[F] FIRE SAFETY FUNCTIONS. Building and fire control functions that are intended to increase the level of life safety for occupants or to control the spread of harmful effects of fire.

❖ In many cases, automatic fire detectors are installed even in buildings not required to have a fire alarm system. These fire detectors perform specific functions such as elevator recall or air distribution system shutdown (see Section 907.10).

[F] FOAM-EXTINGUISHING SYSTEM. A special system discharging a foam made from concentrates, either mechanically or chemically, over the area to be protected.

❖ Foam-extinguishing systems must be of an approved type and installed and tested in accordance with NFPA 11, 11A and 16. All foams are intended to exclude oxygen from the fire, cool the fire area and insulate adjoining surfaces from heat caused by fires. Foam systems are commonly used to extinguish flammable or combustible liquid fires (see commentary, Section 904).

[F] HALOGENATED EXTINGUISHING SYSTEM. A fire-extinguishing system using one or more atoms of an element from the halogen chemical series: fluorine, chlorine, bromine and iodine.

❖ Halon is a colorless, odorless gas that inhibits the chemical reaction of fire. Halon 1301 extinguishing systems are useful in occupancies such as computer rooms, where an electrically nonconductive medium is essential or desirable and where cleanup of other extinguishing agents presents a problem. The halon extinguishing system is required to be of an approved type and installed and tested in accordance with NFPA 12A.

Halon extinguishing agents have been identified as a source of emissions resulting in the depletion of the stratospheric ozone layer. As such, production of new supplies of halon has been phased out. Alternative extinguishing agents, such as clean agents, have been developed as alternatives to halon.

[F] INITIATING DEVICE. A system component that originates transmission of a change-of-state condition, such as in a smoke detector, manual fire alarm box or supervisory switch.

❖ All fire protection systems consist of devices that upon use or actuation will initiate the intended operation. A manual fire alarm box, for example, upon actuation will activate the building alarm notification appliances and transmit a signal to an approved location.

LISTED. Equipment, materials or services included in a list published by an organization acceptable to the building official and concerned with evaluation of products or services that maintains periodic inspection of production of listed equipment or materials or periodic evaluation of services and whose listing states either that the equipment, material or service meets identified standards or has been tested and found suitable for a specified purpose.

❖ When a product is listed or labeled, it indicates that it has been tested for conformance to an applicable standard and is subject to a third-party inspection quality assurance (QA) program. The QA verifies that the minimum level of quality required by the appropriate standard is maintained. Labeling provides a readily available source of information that is useful for field inspection of installed products. The label identifies the product or material and provides other information that can be further investigated if there is any question as to its suitability for the specific installation. The labeling agency performing the third-party inspection must be approved by the building official, and the basis for this approval may include, but is not limited to, the capacity and capability of the agency to perform the specific testing and inspection.

[F] MANUAL FIRE ALARM BOX. A manually operated device used to initiate an alarm signal.

❖ Manual fire alarm boxes are commonly known as pull stations. Manual fire alarm boxes include all manual devices used to activate a manual fire alarm system and have many configurations depending on the manufacturer. All manual fire alarm devices, however, must be approved and installed in accordance with NFPA 72 for the particular application. Manual fire alarm boxes may be combined in guard tour boxes.

[F] MULTIPLE-STATION ALARM DEVICE. Two or more single-station alarm devices that are capable of interconnection such that actuation of one causes all integral or separate audible alarms to operate. It also can consist of one single-station alarm device having connections to other detectors or to a manual fire alarm box.

❖ This definition refers to a combination of similar or different types of alarm devices that could be interconnected. The actuation of any two devices, whether a smoke detector or manual fire alarm box, will activate the required audible alarms.

[F] MULTIPLE-STATION SMOKE ALARM. Two or more single-station alarm devices that are capable of interconnection such that actuation of one causes all integral or separate audible alarms to operate.

❖ In occupancies with sleeping areas, occupants need to be notified in a fire condition so that they may promptly evacuate the premises. In accordance with the requirements of NFPA 72, multiple-station smoke alarms are self-contained, smoke-activated alarm devices built in accordance with UL 217 or 268 that can be interconnected with other devices such that, when any one device is activated, all integral or separate alarms will operate.

[F] NUISANCE ALARM. An alarm caused by mechanical failure, malfunction, improper installation or lack of proper maintenance, or an alarm activated by a cause that cannot be determined.

❖ A nuisance alarm is essentially any alarm response that occurs due to a condition that does not arise during the normal operation of the equipment.

[F] RECORD DRAWINGS. Drawings ("as builts") that document the location of all devices, appliances, wiring sequences, wiring methods and connections of the components of a fire alarm system as installed.

❖ To verify that the system has been installed in accordance with the code and all applicable referenced standards, complete as-built drawings of the fire alarm system should be available on-site for review.

[F] SINGLE-STATION SMOKE ALARM. An assembly incorporating the detector, the control equipment and the alarm-sounding device in one unit, operated from a power supply either in the unit or obtained at the point of installation.

❖ A single-station smoke alarm is a self-contained alarm device that detects visible or invisible particles of combustion. Its function is to detect a fire condition in the immediate area of the fire incident. Single-station smoke alarms are not interconnected with other devices and are not capable of notifying or controlling any other fire protection equipment or systems. They may be battery powered, directly connected to the building power supply or a combination of both. Single-station smoke alarms must be built in accordance with UL 217 and are to be installed in accordance with NFPA 72.

[F] SMOKE ALARM. A single- or multiple-station alarm responsive to smoke and not connected to a system.

❖ This is a general term that applies to both single or multiple-station smoke alarms that are not part of an automatic fire detection system.

[F] SMOKE DETECTOR. A listed device that senses visible or invisible particles of combustion.

❖ These units are considered early warning devices and have saved many people from smoke inhalation and burns. Smoke detectors have a wide range of uses, from sophisticated fire detection systems for industrial and commercial uses to residential. A smoke detector is a device that activates the fire detection system. This system smoke detector contains only the equipment required to detect the products of combustion and should not be confused with single-station and multiple-station smoke alarms.

Smoke detectors are typically of two types: ionization and photoelectric. An ionization-type detector contains a small amount of radioactive material that ionizes the air in a sensing chamber and causes a current to flow through the air between two charged electrodes. When smoke enters the chamber, the particles cause a reduction in the current. When the level of conductance decreases to a preset level, the detector responds with an alarm.

A photoelectric-type smoke detector consists primarily of a light source, a light beam and a photosensitive device. When smoke particles enter the light beam, they reduce the light intensity in the photosensitive device. When obscuration reaches a preset level, the detector initiates an alarm.

SMOKEPROOF ENCLOSURE. An exit stairway designed and constructed so that the movement of the products of combustion produced by a fire occurring in any part of the building into the enclosure is limited.

❖ A smokeproof enclosure is intended to provide an effective barrier to the entry of smoke into an exit stairway, thereby offering an additional level of protection for occupants of high-rise and underground structures.

[F] STANDPIPE SYSTEM, CLASSES OF. Standpipe classes are as follows:

❖ A standpipe system is typically an arrangement of vertical piping located in exit stairways that allows for fire-fighting personnel to connect hand-carried hoses at each level to manually extinguish fires. Section 905 and NFPA 14 recognize three different classes of standpipe systems.

Class I system. A system providing 2.5-inch (64 mm) hose connections to supply water for use by fire departments and those trained in handling heavy fire streams.

❖ A Class I standpipe system is intended to be used by trained fire service personnel as a readily available water source for manual fire-fighting operations. A Class I standpipe system is equipped with only 2¹/₂-inch (64 mm) hose connections to allow the fire service to attach the appropriate hose and nozzles. A Class I standpipe system is not equipped with hose stations that include a cabinet, hose and nozzle.

Class II system. A system providing 1.5-inch (38 mm) hose stations to supply water for use primarily by the building occupants or by the fire department during initial response.

❖ A Class II standpipe system is intended to be used by building occupants as a device to facilitate manual suppression. The hose stations defined as part of the Class II standpipe system include a cabinet that is equipped with a hose and nozzle readily available to the building occupants.

Class III system. A system providing 1.5-inch (38 mm) hose stations to supply water for use by building occupants and 2.5-inch (64 mm) hose connections to supply a larger volume of water for use by fire departments and those trained in handling heavy fire streams.

❖ A Class III standpipe system is intended for use by building occupants as well as trained fire service personnel. The 1¹/₂-inch (38 mm) hose station is in a cabinet for use by the building occupants for manual fire suppression and the 2¹/₂-inch (64 mm) hose connection is intended for use by fire service personnel.

[F] STANDPIPE, TYPES OF. Standpipe types are as follows:

❖ Section 905 recognizes five types of standpipe systems. The use of each type of system is dependent upon specific occupancy conditions and the presence of an automatic sprinkler system.

Automatic dry. A dry standpipe system, normally filled with pressurized air, that is arranged through the use of a device, such as dry pipe valve, to admit water into the system piping automatically upon the opening of a hose valve. The water supply for an automatic dry standpipe system shall be capable of supplying the system demand.

❖ A typical automatic dry standpipe system has an automatic water supply retained by a dry pipe valve. The dry pipe valve clapper is kept in place by air placed in the standpipe system under pressure. Once a standpipe hose valve is opened, the air is released from the system allowing water to fill the system through the dry pipe valve. This system is traditionally used in areas where the temperature falls below 40°F (4°C), where a wet system would otherwise freeze.

Automatic wet. A wet standpipe system that has a water supply that is capable of supplying the system demand automatically.

❖ An automatic wet standpipe system is used in locations where the entire system is subject to temperatures greater than 40°F (4°C). Since the system is pressurized with water, an immediate release of water occurs when a hose connection valve is opened.

Manual dry. A dry standpipe system that does not have a permanent water supply attached to the system. Manual dry standpipe systems require water from a fire department pumper to be pumped into the system through the fire department connection in order to meet the system demand.

❖ A manual dry standpipe system is filled with water only when the fire service is present. Typically, the fire service connects the discharge from a water source, such as a pumper truck, to the fire department connection of a manual dry standpipe system. When the fire service has suppressed the fire and is preparing to leave, the system is drained of the remaining water. Manual dry standpipe systems are commonly installed in open parking structures.

Manual wet. A wet standpipe system connected to a water supply for the purpose of maintaining water within the system but does not have a water supply capable of delivering the system demand attached to the system. Manual-wet standpipe systems require water from a fire department pumper (or the like) to be pumped into the system in order to meet the system demand.

❖ A manual wet standpipe system is connected to an automatic water supply, but the supply is not capable of providing the system demand. The system demand is met when the fire service provides additional water through the fire department connection from the discharge of a water source, such as a pumper truck.

Semiautomatic dry. A dry standpipe system that is arranged through the use of a device, such as a deluge valve, to admit water into the system piping upon activation of a remote control device located at a hose connection. A remote control activation device shall be provided at each hose connection. The water supply for a semiautomatic dry standpipe system shall be capable of supplying the system demand.

❖ This type of dry standpipe is a special design condition that uses a solenoid-activated valve to retain the automatic water supply. Once the standpipe hose valve is opened, a signal is sent to the deluge valve retaining the automatic water supply to allow water to fill the system. This system is used in areas where the temperature falls below 40°F (4°C), where a wet system would otherwise freeze.

[F] SUPERVISING STATION. A facility that receives signals and at which personnel are in attendance at all times to respond to these signals.

❖ The supervising station is the location where all fire protection system-related signals are sent and where trained personnel are present to respond appropriately to an emergency condition. The supervising station may be an approved central station, remote supervising station, proprietary supervising station or other constantly attended location approved by the building official.

[F] SUPERVISORY SERVICE. The service required to monitor performance of guard tours and the operative condition of fixed suppression systems or other systems for the protection of life and property.

❖ The supervisory service is responsible for maintaining the integrity of the fire protection system by notifying the supervising station of a change in the protection system's status.

[F] SUPERVISORY SIGNAL. A signal indicating the need of action in connection with the supervision of guard tours, the fire suppression systems or equipment or the maintenance features of related systems.

❖ Activation of a supervisory signal-initiating device transmits a signal indicating that a change in the status of the fire protection system has occurred and that the appropriate emergency action must be taken. These signals are the basis for the actions taken by the attendant at the supervising station.

[F] SUPERVISORY SIGNAL-INITIATING DEVICE. An initiation device, such as a valve supervisory switch, water-level indicator or low-air pressure switch on a dry-pipe sprinkler system, whose change of state signals an off-normal condition and its restoration to normal of a fire protection or life safety system, or a need for action in connection with guard tours, fire suppression systems or equipment or maintenance features of related systems.

❖ The supervisory signal-initiating device detects a change in a protection system's status. Examples of a supervisory signal-initiating device include a flow switch to detect movement of water through the system and a tamper switch to detect when someone shuts off a water control valve.

[F] TIRES, BULK STORAGE OF. Storage of tires where the area available for storage exceeds 20,000 cubic feet (566 m³).

❖ This definition provides a threshold beyond which the storage of tires exceeds that which is customary in most typical mercantile and storage occupancies and that poses an extraordinary hazard for fire protection. Buildings utilized for the bulk storage of tires are classified as Group S-1 occupancies in accordance with Section 311.2. All Group S-1 occupancies, regardless of square footage, must be provided with an automatic sprinkler system if utilized for the bulk storage of tires. While Section 903.3.1.1 specifically references NFPA 13, it is intended by secondary reference to require that bulk tire storage buildings be further protected in accordance with the provisions of NFPA 231D.

[F] TROUBLE SIGNAL. A signal initiated by the fire alarm system or device indicative of a fault in a monitored circuit or component.

❖ This type of signal indicates that there has been a change in status of the provided fire detection system or devices and that appropriate measures must be taken.

[F] VISIBLE ALARM NOTIFICATION APPLIANCE. A notification appliance that alerts by the sense of sight.

❖ Visible alarm notification appliances are typically located where occupants may be hearing impaired and in sleeping accommodations of Group I-1 and R-1 occupancies. The alarm devices should be located and oriented so that they will display alarm signals throughout the required space. Visible alarms, when provided, are typically installed in public and common areas of buildings.

[F] WET-CHEMICAL EXTINGUISHING SYSTEM. A solution of water and potassium-carbonate-based chemical, potassium-acetate-based chemical or a combination thereof, forming an extinguishing agent.

❖ This extinguishing agent is a suitable alternative to the use of a dry chemical, especially when protecting commercial kitchen range hoods. There is less cleanup time after system discharge. Wet chemical solutions are considered to be relatively harmless and normally have no lasting effect on the skin or respiratory system. These solutions may produce temporary irritation, which is usually mild and disappears when contact is eliminated. These systems must be preengineered and labeled. NFPA 17A applies to the design, installation, operation, testing and maintenance of wet-chemical extinguishing systems.

[F] WIRELESS PROTECTION SYSTEM. A system or a part of a system that can transmit and receive signals without the aid of wire.

❖ These systems utilize radio-frequency transmitting devices that comply with the special requirements for supervision of low-power wireless systems in accordance with NFPA 72.

[F] ZONE. A defined area within the protected premises. A zone can define an area from which a signal can be received, an area to which a signal can be sent or an area in which a form of control can be executed.

❖ Zoning a system is important to emergency personnel in locating a fire. When an alarm is designated to a specific zone, it allows the fire service to immediately respond to the area where the fire condition is occurring instead of searching the entire building for the origin of an alarm.

SECTION 903
AUTOMATIC SPRINKLER SYSTEMS

[F] 903.1 General. Automatic sprinkler systems shall comply with this section.

❖ This section identifies the conditions requiring an automatic sprinkler system for all occupancies. The need for an automatic sprinkler system may depend on not only the occupancy but also the occupant load, fuel load, height and area of the building as well as fire-fighting capabilities. Section 903.2 addresses all occupancy conditions requiring an automatic sprinkler system. Section 903.3 contains the installation requirements for all sprinkler systems in addition to the requirements of NFPA 13, NFPA 13R and NFPA 13D. The supervision and alarm requirements for sprinkler systems are contained in Section 903.4, whereas Section 903.5 refers to testing and maintenance requirements for sprinkler systems in the IFC.

It is important to note that the area values contained in this section are intended to apply to fire areas, which are comprised of all floor areas within the fire barriers, fire walls or exterior walls. The minimum required fire-resistance rating of fire barrier assemblies that define a fire area is specified in Table 302.3.2. Fire walls are regulated by Section 705 and exterior walls are regulated by Tables 601 and 602 as well as Section 704. Since the areas are defined as fire areas, fire barriers, fire walls or exterior walls are the only acceptable means of subdividing a building into smaller areas in lieu of providing an automatic sprinkler system. Where fire barrier and exterior walls define multiple fire areas within a single building, a fire wall defines separate buildings within one structure. It should also be noted that some of the threshold limitations result in a requirement to provide an automatic sprinkler system throughout the building while others may require only specific fire areas to be sprinklered.

Another important point is that one fire area may be comprised of floor areas in more than one story of a building. See the commentary to the definition of "Fire area" in Section 702.1.

[F] 903.1.1 Alternative protection. Alternative automatic fire-extinguishing systems complying with Section 904 shall be permitted in lieu of automatic sprinkler protection where recognized by the applicable standard and approved by the building official.

❖ This section permits the use of an alternative automatic fire-extinguishing system, when approved by the building official, as a means of compliance with the occupancy requirements of Section 903. While the use of an alternative extinguishing system in accordance with Section 904, such as a carbon dioxide system or clean-agent system, would satisfy the requirements of Section 903.2, it would not be considered an acceptable alternative for the purposes of exceptions, reductions or other code tradeoffs that would be applicable if an automatic sprinkler system were installed.

[F] 903.2 Where required. Approved automatic sprinkler systems in new buildings and structures shall be provided in the locations described in this section.

> **Exception:** Spaces or areas in telecommunications buildings used exclusively for telecommunications equipment, associated electrical power distribution equipment, batteries and standby engines, provided those spaces or areas are equipped throughout with an automatic fire alarm system and are separated from the remainder of the building by a wall with a fire-resistance rating of not less than 1 hour and a floor/ceiling assembly with a fire-resistance rating of not less than 2 hours.

❖ Sections 903.2.1 through 903.2.13 identify the conditions requiring an automatic sprinkler system (see Figure 903.2). The type of sprinkler system must be one that is permitted for the specific occupancy condition. An NFPA 13R sprinkler system, for example, is not permitted to be installed to satisfy the sprinkler threshold requirements for a mercantile occupancy (see Section 903.2.6). As indicated in Section 903.3.1.2, the use of an NFPA 13R sprinkler system is limited to Group R occupancies not exceeding four stories in height.

There is one exception for those spaces or areas used exclusively for telecommunications equipment. The telecommunications industry has continually stressed the need for the continuity of telephone service, and the ability to maintain this service is of prime importance. This service is a vital link between the community and the various life safety services, including fire, police and emergency medical services. The integrity of this communications service can be jeopardized not only by fire, but also by water from whatever the source.

It must be emphasized that the exception applies only to those spaces or areas that are used exclusively for telecommunications equipment. Historically, those spaces have a low incidence of fire events. Fires in telecommunications equipment areas are difficult to start and, if started, grow slowly, thus permitting early detection. Such fires are typically of the smoldering type, do not spread beyond the immediate area and generally self-extinguish.

Note, however, that this exception requires proper fire resistive separation from other portions of the building.

FIGURE 903.2

FIRE PROTECTION SYSTEMS

Occupancy	Threshold	Exception
All occupancies	Buildings with floor level ≥ 55 feet above vehicle access and occupant load ≥ 30.	Airport control towers, open parking structures. F-2, R-3, U
Assembly (A-1, A-3, A-4)	Fire area > 12,000 sq.ft. or fire area occupant load > 300 or fire area above/below level of exit discharge. Multitheater complex (A-1 only)	Participant sport arenas at level of exit discharge. A-3, A-4
Assembly (A-2)	Fire area > 5,000 sq.ft. or fire area occupant load > 300 or fire area above/below level of exit discharge.	None
Assembly (A-5)	Accessory areas > 1,000 sq.ft.	None
Educational (E)	Fire area > 20,000 sq.ft. or level of exit discharge.	Each classroom has exterior door at grade.
Factory (F-1) Mercantile (M) Storage (S-1)	Fire area > 12,000 sq. ft. or Building > 3 stories or Combined fire area > 24,000 sq.ft. Woodworking > 2,500 sq.ft. (F-1 only). Bulk storage of tires > 20,000 cu.ft. (S-1 only).	None
High hazard (H-1, H-2, H-3, H-4, H-5)	Sprinklers required.	None
Institutional (I-1, I-2, I-3, I-4)	Sprinklers required.	None
Residential (R)	Sprinklers required.	None
Repair garage (S-1)	Fire area > 12,000 sq.ft. or ≥ 2 stories (including basement) with fire area > 10,000 sq.ft. or repair garage servicing vehicles in basement.	None
Parking garage (S-2)	Enclosed automobile parking—sprinklers required. Commercial trucks/buses parking area > 5,000 sq.ft.	None
Covered malls (402.8)	Sprinklers required.	Attached open parking structures.
High rises (403.2, 403.3)	> 75 feet above vehicle access.	Airport traffic control towers, open garages. A-5
Unlimited area buildings (507)	A-3, A-4, B, E, F, M, S: 1 story. B, F, M, S: 2 story.	One story F-2 or S-2.

Note: Thresholds located in Section 903.2 unless noted. See also Table 903.2.13 for additional required suppression systems.

For SI: 1 foot = 304.8 mm, 1 square foot = 0.0929 m^2.

Figure 903.2
SUMMARY OF OCCUPANCY-RELATED AUTOMATIC SPRINKLER THRESHOLDS

[F] 903.2.1 Group A. An automatic sprinkler system shall be provided throughout buildings and portions thereof used as Group A occupancies as provided in this section. For Group A-1, A-2, A-3 and A-4 occupancies, the automatic sprinkler system shall be provided throughout the floor area where the Group A-1, A-2, A-3 or A-4 occupancy is located, and in all floors between the Group A occupancy and the level of exit discharge. For Group A-5 occupancies, the automatic sprinkler system shall be provided in the spaces indicated in Section 903.2.1.5.

❖ Occupancies of Group A are characterized by a significant number of people who are not familiar with their surroundings. The requirement for a suppression system reflects the additional time needed for egress. The extent of protection is also intended to extend to the occupants of the assembly group from unobserved fires in other building areas located between the floor level containing the assembly occupancy and the level of exit discharge. The only exception to the coverage is for Group A-5 occupancies, which are open to the atmosphere. Such occupancies require only certain aspects to be sprinklered, such as concession stand (see commentary, Section 903.2.1.5).

[F] 903.2.1.1 Group A-1. An automatic sprinkler system shall be provided for Group A-1 occupancies where one of the following conditions exists:

1. The fire area exceeds 12,000 square feet (1115 m²).

2. The fire area has an occupant load of 300 or more.

3. The fire area is located on a floor other than the level of exit discharge.

4. The fire area contains a multitheater complex.

❖ In accordance with Section 303.1, Group A-1 occupancies are identified as those with fixed seating, such as theaters. In addition to the high occupant load associated with these types of facilities, egress is further complicated by the possibility of low lighting levels customary during performances. The fuel load in such buildings is usually of such a type and quantity to support fairly rapid fire development and sustained duration.

Theaters with stages pose a greater hazard. Sections 410.6 and 410.7 require stages to be equipped with an automatic sprinkler system and a standpipe system, respectively. The proscenium opening is also required to be protected as specified in Section 410.3.5. These features compensate for the additional hazards associated with stages in Group A-1 occupancies.

This section has four conditions that require a suppression system to be installed in a Group A-1 occupancy. Condition 1 requires that, if any one fire area of Group A-1 exceeds 12,000 square feet (1,115 m²), then the automatic fire suppression system is to be installed throughout the entire story or floor level where a Group A-1 occupancy is located, regardless of whether more than one fire area is present in the building.

Condition 2 maintains a threshold of occupants before a suppression system is considered necessary.

The determination of the actual occupant load should be based on Section 1004.

Condition 3 accounts for occupant egress delay when traversing a stairway, requiring a sprinkler system regardless of the size or occupant load.

Condition 4 states that a sprinkler system is required for multitheater complexes to account for the delay associated with the notification of adjacent compartmentalized spaces where the occupants may not be immediately aware of an emergency condition.

[F] 903.2.1.2 Group A-2. An automatic sprinkler system shall be provided for Group A-2 occupancies where one of the following conditions exists:

1. The fire area exceeds 5,000 square feet (464.5 m²).

2. The fire area has an occupant load of 300 or more.

3. The fire area is located on a floor other than the level of exit discharge.

❖ Section 303.1 identifies Group A-2 occupancies as those intended for food or drink consumption, such as banquet halls, nightclubs and restaurants. Occupancies in Group A-2 involve life safety factors such as a high occupant density, flexible fuel loading, movable furnishings and limited lighting and, therefore, must be protected with an automatic sprinkler system under any of the listed conditions.

In the case of an assembly use, the purpose of the automatic sprinkler system is to provide life safety from fire and to preserve property. By requiring fire suppression in areas through which the occupants may egress, including level of exit discharge, the opportunity of unobserved fire development affecting the occupant egress is minimized.

The 5,000-square-foot (465 m²) threshold for the automatic sprinkler system reflects the higher degree of life hazard associated with Group A-2 occupancies. As alluded to earlier, Group A-2 occupancies could have attributes such as low lighting levels, loud music, late hours of operation, dense seating with ill-defined aisles and alcoholic beverage service. These factors in combination could delay fire recognition, confuse the appropriate response and increase egress time. The similar intent of Conditions 2 and 3 is addressed in the commentary to Section 903.2.1.1.

These conditions require sprinklers throughout the fire area containing the Group A-2 occupancies, regardless of the number of fire areas present.

[F] 903.2.1.3 Group A-3. An automatic sprinkler system shall be provided for Group A-3 occupancies where one of the following conditions exists:

1. The fire area exceeds 12,000 square feet (1115 m²).

2. The fire area has an occupant load of 300 or more.

3. The fire area is located on a floor other than the level of exit discharge.

Exception: Areas used exclusively as participant sports areas where the main floor area is located at the same level as the level of exit discharge of the main entrance and exit.

❖ Section 303.1 identifies Group A-3 occupancies as those intended for worship, recreation or amusement and other assembly uses not classified elsewhere in Group A, such as churches, museums and libraries. While Group A-3 occupancies could potentially have a high occupant load, they normally do not have the same potential combination of life safety hazards associated with Group A-2 occupancies. As with most assembly occupancies, occupants are typically not familiar with their surroundings. When any of the three listed conditions is applicable, an automatic sprinkler system is required throughout the fire area containing the A-3 occupancy and in all floors between the Group A occupancy and exit discharge (see commentary, Sections 903.2.1 and 903.2.1.1).

The exception exempts the participant sport area of Group A-3 occupancies from automatic sprinkler system requirements because these areas are typically large open spaces with relatively low fuel loads. The exception includes only the participant sport area, such as an indoor swimming pool or the court area of an indoor tennis court. Note that if the exception is utilized and sprinklers are omitted from the sport area, then the building would not be considered completely sprinklered in accordance with Section 903.3.1.1 for purposes of allowing construction alternatives, such as height and area increases, corridor rating reduction and other code tradeoffs.

[F] 903.2.1.4 Group A-4. An automatic sprinkler system shall be provided for Group A-4 occupancies where one of the following conditions exists:

1. The fire area exceeds 12,000 square feet (1115 m²).

2. The fire area has an occupant load of 300 or more.

3. The fire area is located on a floor other than the level of exit discharge.

Exception: Areas used exclusively as participant sports areas where the main floor area is located at the same level as the level of exit discharge of the main entrance and exit.

❖ Group A-4 occupancies, as identified in Section 303.1, are assembly uses intended for viewing of indoor sporting events and activities such as arenas, skating rinks and swimming pools. The occupant load density may be high depending on the extent and style of seating, such as bleachers or fixed seats, and the potential for standing room viewing.

When any of the three listed conditions is applicable, an automatic sprinkler system is required throughout the fire area containing the Group A-4 occupancy and in all floors between the Group A occupancy and exit dis-

charge (see commentary, Sections 903.2.1 and 903.2.1.1). Similar to Group A-3 occupancies, the participant sport areas on the main floor of Group A-4 occupancies are exempt from the sprinkler system requirement (see commentary, Section 903.2.1.3).

[F] 903.2.1.5 Group A-5. An automatic sprinkler system shall be provided in concession stands, retail areas, press boxes and other accessory use areas in excess of 1,000 square feet (93 m²).

❖ Group A-5 occupancies are assembly uses intended for viewing of outdoor activities. As indicated in Section 303.1, this could include amusement park structures, grandstands and open stadiums. A sprinkler system is not required in the open area of Group A-5 occupancies due to the buildings' inability to accumulate smoke and hot gases. A fire condition in open areas would also be obvious to all spectators. Enclosed areas such as retail areas, press boxes and concession stands would be required to be sprinklered if in excess of 1,000 square feet (93 m²). The 1,000-square-foot (93 m²) accessory use area is not intended to be an aggregate condition.

[F] 903.2.2 Group E. An automatic sprinkler system shall be provided for Group E occupancies as follows:

1. Throughout all Group E fire areas greater than 20,000 square feet (1858 m²) in area.

2. Throughout every portion of educational buildings below the level of exit discharge.

Exception: An automatic sprinkler system is not required in any fire area or area below the level of exit discharge where every classroom throughout the building has at least one exterior exit door at ground level.

❖ Group E occupancies are limited to educational purposes through the twelfth grade and day care centers providing services for children older than 2¹/₂ years of age. The 20,000-square-foot (1,852 m²) fire area threshold for the sprinkler system was established to allow smaller schools and day care centers to be unsprinklered in order to minimize the economic impact on their facility.

The exception would allow the omission of the automatic sprinkler system for the Group E fire area, provided there is a direct exit to the exterior from each classroom at ground level. The students/occupants must be able to egress the classroom directly to the outside without intervening corridors, passageways or exit enclosures.

[F] 903.2.3 Group F-1. An automatic sprinkler system shall be provided throughout all buildings containing a Group F-1 occupancy where one of the following conditions exists:

1. Where a Group F-1 fire area exceeds 12,000 square feet (1115 m²);

2. Where a Group F-1 fire area is located more than three stories above grade; or

3. Where the combined area of all Group F-1 fire areas on all floors, including any mezzanines, exceeds 24,000 square feet (2230 m²).

❖ Because of the difficulty in manually suppressing a fire involving a large area, occupancies of Group F-1 are to be protected throughout with an automatic sprinkler system: if the fire area is in excess of 12,000 square feet (1,115 m²); if the total of all fire areas of Group F-1 in the building is in excess of 24,000 square feet (2,230 m²) or if the Group F-1 fire area is located more than three stories above grade. This is one of the few locations in the code where the total floor area of the building is aggregated for application of a code requirement. The stipulated conditions for when an automatic sprinkler system is required also apply to Group M (see Section 903.2.6) and S-1 (see Section 903.2.8) occupancies.

The following examples illustrate how the criteria should be applied. If a building contains a single fire area of Group F-1 and the fire area is 13,000 square feet (1,208 m²), an automatic sprinkler system is required throughout the entire building. However, if the above fire area is separated into two fire areas and neither is in excess of 12,000 square feet (1,115 m²), an automatic fire sprinkler system is not required. To be considered separate fire areas, the areas must be separated by fire barrier walls or horizontal assemblies having a fire-resistance rating as required in Table 302.3.2.

If a 30,000-square-foot (2,787 m²) Group F-1 building was equally divided into separate fire areas of 10,000 square feet (929 m²), an automatic sprinkler system would still be required throughout the entire building. Since the aggregate area of all fire areas exceeds 24,000 square feet (2,230 m²), additional compartmentation will not eliminate the need for an automatic sprinkler system. Note, however, that the use of a fire wall, which separates the structure into two buildings, thereby reducing the aggregate area of each building to less than 24,000 square feet (2,230 m²) and each fire area to less than 12,000 square feet (1,115 m²), would offset the need for an automatic sprinkler system.

[F] 903.2.3.1 Woodworking operations. An automatic sprinkler system shall be provided throughout all Group F-1 occupancy fire areas that contain woodworking operations in excess of 2,500 square feet (232 m²) in area which generate finely divided combustible waste or use finely divided combustible materials.

❖ Because of the potential amount of combustible dust that could be generated during woodworking operations, an automatic sprinkler system is required throughout a fire area when it contains a woodworking operation that exceeds 2,500 square feet (232 m²) in area. Woodworking operations, such as cabinet making, are considered a Group F-1 occupancy. The intent of the phrase "finely divided combustible waste" is to describe particle concentrations that are in the explosive

range. See the commentary for Chapter 13 of the IFC for discussion of dust-producing operations.

[F] 903.2.4 Group H. Automatic sprinkler systems shall be provided in high-hazard occupancies as required in Sections 903.2.4.1 through 903.2.4.3.

❖ Group H occupancies are those intended for the manufacturing, processing or storage of hazardous materials that constitute a physical or health hazard. To be considered a Group H occupancy, the amount of hazardous materials is assumed to be in excess of the maximum allowable quantities permitted by Tables 307.7(1) and 307.7(2).

[F] 903.2.4.1 General. An automatic sprinkler system shall be installed in Group H occupancies.

❖ This section requires an automatic sprinkler system in all Group H occupancies. While in some instances the hazard associated with the occupancy may be one that is not a fire hazard, an automatic sprinkler system is still required to minimize the potential for fire spreading to the high-hazard use (i.e., protection from fire outside the area of concern). This section does not prohibit the use of an alternative automatic fire-extinguishing system in accordance with Section 904. When the use of a water-based system is not compatible with the hazardous materials involved and thus creates a dangerous condition, an alternative fire-extinguishing system should be utilized. For example, combustible metals, such as magnesium and titanium, have a serious record of involvement with fire and are typically not compatible with water.

[F] 903.2.4.2 Group H-5. An automatic sprinkler system shall be installed throughout buildings containing Group H-5 occupancies. The design of the sprinkler system shall not be less than that required by this code for the occupancy hazard classifications in accordance with Table 903.2.4.2. Where the design area of the sprinkler system consists of a corridor protected by one row of sprinklers, the maximum number of sprinklers required to be calculated is 13.

❖ Group H-5 occupancies are those that are typically used as semiconductor fabrication facilities and comparable research laboratory facilities that utilize hazardous production materials (HPM). Many of the materials used in semiconductor fabrication possess unique hazards. Many of the materials are toxic, and some are corrosive, water reactive or pyrophoric. Fire protection for these facilities is aimed at preventing incidents from escalating and producing secondary threats beyond a fire, such as the release of corrosive or toxic materials. The overall amount of hazardous materials due to the nature of Group H-5 facilities can far exceed the maximum allowable quantities given in Tables 307.7(1) and 307.7(2). While the amount of HPM is restricted in fabrication areas, quantities in storage rooms containing HPM are normally in excess of the maximum allowable quantity per control area in Tables 307.7(1) and 307.7(2). Additional provisions for Group H-5 facilities are located in Section 415.9 and Chapter 18 of the IFC.

This section also specifies the sprinkler design criteria, based on NFPA 13, for various areas in a Group H-5 occupancy (see commentary, Table 903.2.4.2). When the corridor design area sprinkler option is utilized, a maximum of 13 sprinklers must be calculated. This exceeds the requirements of NFPA 13 for typical egress corridors that require a maximum of either five or seven calculated sprinklers depending on the extent of protected openings in the corridor. The increased number of calculated corridor sprinklers is based on the additional hazard associated with the movement of hazardous materials in corridors of Group H-5 facilities.

[F] TABLE 903.2.4.2
GROUP H-5 SPRINKLER DESIGN CRITERIA

LOCATION	OCCUPANCY HAZARD CLASSIFICATION
Fabrication areas	Ordinary Hazard Group 2
Service corridors	Ordinary Hazard Group 2
Storage rooms without dispensing	Ordinary Hazard Group 2
Storage rooms with dispensing	Extra Hazard Group 2
Corridors	Ordinary Hazard Group 2

❖ The purpose of Table 903.2.4.2 is to designate the appropriate occupancy hazard classification for the various areas within a Group H-5 facility. The listed occupancy hazard classifications correspond to specific sprinkler system design criteria in NFPA 13. Ordinary Hazard Group 2 occupancies, for example, require a minimum design density of 0.20 gpm/ft² (8.1 L/min./m²) with a minimum design area of 1,500 square feet (139 m²). An Extra Hazard Group 2 occupancy, in turn, requires a minimum design density of 0.40 gpm/ft² (16.3 L/min./m²) with a minimum operating area of 2,500 square feet (232 m²). The increased overall sprinkler demand for Extra Hazard Group 2 occupancies is based on the potential use and handling of substantial amounts of hazardous materials such as flammable or combustible liquids.

[F] 903.2.4.3 Pyroxylin plastics. An automatic sprinkler system shall be provided in buildings, or portions thereof, where cellulose nitrate film or pyroxylin plastics are manufactured, stored or handled in quantities exceeding 100 pounds (45 kg).

❖ Cellulose nitrate (pyroxylin) plastics and motion picture film pose unusual and substantial fire risks. Pyroxylin plastics are the most dangerous and unstable of all plastic compounds. The chemically bound oxygen in their structure permits them to burn vigorously in the absence of atmospheric oxygen. Although these compounds produce approximately the same amount of energy when they burn as paper, pyroxylin plastics burn at a rate as much as 15 times greater than comparable

common combustibles. When burning, these materials release highly flammable and toxic combustion byproducts. Consequently, cellulose nitrate fires are very difficult to control. While this section specifies a sprinkler threshold quantity of 100 pounds (45 kg), the need for additional fire protection should be considered for pyroxylin plastics in any amount (see also Section 409).

[F] 903.2.5 Group I. An automatic sprinkler system shall be provided throughout buildings with a Group I fire area.

Exception: An automatic sprinkler system installed in accordance with Section 903.3.1.2 or 903.3.1.3 shall be allowed in Group I-1 facilities.

❖ The Group I occupancy is divided into four individual occupancy classifications based on the degree of detention, supervision and physical mobility of the occupants. The evacuation difficulties associated with the occupants creates the need to incorporate a defend-in-place philosophy of fire protection in occupancies of Group I. For this reason, all such occupancies are to be protected with an automatic sprinkler system.

Of particular note, this section encompasses all Group I-3 occupancies where more than five persons are incarcerated. There has been considerable controversy concerning the providing of automatic sprinklers in detention and correctional occupancies. Special design considerations can be taken into account to alleviate the perceived problems with sprinklers in sleeping units. Sprinklers that reduce the likelihood of vandalism as well as the potential to hang oneself are commercially available. Knowledgeable designers can incorporate features to increase reliability and decrease the likelihood of damage to the system.

Group I-4 occupancies would include either adult-only care facilities or occupancies that provide personal care for more than five children 2 ½ years of age or less on a less than 24-hour basis. Since the degree of assistance and the time needed for egress cannot be gauged, an automatic sprinkler system is required. Based on Section 308.5.2, a Group I-4 child care facility located at the level of exit discharge and accommodating 100 children or less, with each child care room having an exit directly to the exterior, would be classified as a Group E occupancy. As such, an automatic sprinkler system would not be required unless dictated by the provisions in Section 903.2.2.

The exception would permit Group I-1 occupancies to be protected throughout with either an NFPA 13R or 13D sprinkler system in lieu of a standard NFPA 13 sprinkler system. The exception recognizes the perceived mobility of the occupants in a Group I-1 facility as well as the basic life safety intent to protect the main occupiable areas. It should be noted, however, that utilization of this section would result in the building not otherwise qualifying as a fully sprinklered building in accordance with NFPA 13 for any applicable code tradeoffs.

[F] 903.2.6 Group M. An automatic sprinkler system shall be provided throughout buildings containing a Group M occupancy where one of the following conditions exists:

1. Where a Group M fire area exceeds 12,000 square feet (1115 m²);

2. Where a Group M fire area is located more than three stories above grade; or

3. Where the combined area of all Group M fire areas on all floors, including any mezzanines, exceeds 24,000 square feet (2230 m²).

❖ The sprinkler threshold requirements for Group M occupancies are identical to those of Group F-1 and S-1 occupancies (see commentary, Section 903.2.3). Automatic sprinkler systems for mercantile occupancies are typically designed for an Ordinary Hazard Group 2 classification in accordance with NFPA 13. If high-piled storage in accordance with Section 903.2.6.1 is anticipated, however, additional levels of fire protection may be required. Also, some merchandise in mercantile occupancies, such as aerosols, rubber tires, paints and certain plastic commodities, even at limited storage heights, are considered beyond the scope of NFPA 13 and may require additional fire protection.

[F] 903.2.6.1 High-piled storage. An automatic sprinkler system shall be provided in accordance with the *International Fire Code* in all buildings of Group M where storage of merchandise is in high-piled or rack storage arrays.

❖ Regardless of the size of the Group M fire area, an automatic sprinkler system may be required depending on the extent of the high-piled storage area. High-piled storage includes piled, palletized, shelf or rack storage of combustibles to a height greater than 12 feet (3,658 mm). All high-piled storage areas must be in accordance with Chapter 23 of the IFC and referenced standards NFPA 231 and NFPA 231C. (Note: The scope of NFPA 13, in the 1999 edition, has been expanded to include protection requirements formerly contained in NFPA 231 and 231C).

[F] 903.2.7 Group R. An automatic sprinkler system installed in accordance with Section 903.3 shall be provided throughout all buildings with a Group R fire area.

❖ It is important to note that the International Code Council® (ICC®) Board, in approving the development of the *International Residential Code®* (IRC®), indicated that it is to be a stand-alone code for the construction of detached one- and two-family dwellings and multiple single-family dwellings (townhouses) not more than three stories in height with a separate means of egress. That is, all of the provisions for new construction that affect those buildings are to be covered exclusively by the IRC and are not to be covered by another *International Code®*. This section is a new construction requirement and, therefore, does not apply to buildings covered by

the IRC. Buildings that do not fall within the scope of the IRC would be classified in Group R and be subject to these provisions.

With respect to life safety, the need for a sprinkler system is dependent on the occupants' proximity to the fire and the ability to respond to a fire emergency. Group R occupancies could contain occupants who may require assistance to evacuate, such as infants and those with a disability or who may simply be asleep. While the presence of a sprinkler system cannot always protect occupants in residential buildings who are aware of the ignition and either do not respond or respond inappropriately, it can prevent fatalities outside of the area of fire origin, regardless of the occupants' response. Section 903.3.2 requires quick-response or residential sprinklers in all Group R occupancies. Full-scale fire tests have demonstrated the ability of quick-response and residential sprinklers to maintain tenability from flaming fires in the room of fire origin.

[F] 903.2.8 Group S-1. An automatic sprinkler system shall be provided throughout all buildings containing a Group S-1 occupancy where one of the following conditions exists:

1. A Group S-1 fire area exceeds 12,000 square feet (1115 m²);

2. A Group S-1 fire area is located more than three stories above grade; or

3. The combined area of all Group S-1 fire areas on all floors, including any mezzanines, exceeds 24,000 square feet (2230 m²).

❖ The sprinkler threshold requirements for Group S-1 ocupancies are identical to those of Groups F-1 and M (see commentary, Sections 903.2.3 and 903.2.6). Group S-1 occupancies such as warehouses and self-storage buildings are assumed to be used for the storage of combustible materials. While high-piled storage does not change the Group S-1 occupancy classification, sprinkler protection, if required, may have to comply with the additional requirements of Chapter 23 of the IFC and NFPA 13. High-piled stock or rack storage in any occupancy, as indicated in Section 413, must comply with the IFC.

[F] 903.2.8.1 Repair garages. An automatic sprinkler system shall be provided throughout all buildings used as repair garages in accordance with Section 406, as shown:

1. Buildings two or more stories in height, including basements, with a fire area containing a repair garage exceeding 10,000 square feet (929 m²).

2. One-story buildings with a fire area containing a repair garage exceeding 12,000 square feet (1115 m²).

3. Buildings with a repair garage servicing vehicles parked in the basement.

❖ Automatic sprinklers are required in repair garages, depending on the quantity of combustibles present, their location and floor area. Repair garages may contain significant quantities of flammable liquids and other combustible materials. These occupancies are typically considered Ordinary Hazard Group 2 occupancies in accordance with NFPA 13. Portions of repair garages used for parts cleaning using flammable or combustible liquids may require automatic sprinkler protection in accordance with the IFC. In the case of Item 1, the area would require sprinklers based upon whether vehicles are actually being repaired there or if they are being parked there.

[F] 903.2.8.2 Bulk storage of tires. Buildings and structures where the area for the storage of tires exceeds 20,000 cubic feet (566 m³) shall be equipped throughout with an automatic sprinkler system in accordance with Section 903.3.1.1.

❖ This section specifies when an automatic sprinkler system is required for the bulk storage of tires based on the volume of the storage area as opposed to a specific number of tires. Even in fully sprinklered buildings, tire fires pose significant problems to local fire departments, as they produce thick smoke and are difficult to extinguish by sprinklers alone. While this section references Section 903.3.1.1 for an NFPA 13 sprinkler system, it is intended, by secondary reference through NFPA 13, that bulk tire storage buildings be further protected in accordance with the provisions of NFPA 231D.

[F] 903.2.9 Group S-2. An automatic sprinkler system shall be provided throughout buildings classified as enclosed parking garages in accordance with Section 406.4 or where located beneath other groups.

Exception: Enclosed parking garages located beneath Group R-3 occupancies as applicable in Section 101.2.

❖ Fire records have shown that fires in parking structures typically fully involve only a single automobile with minor damage to adjacent vehicles. An enclosed parking garage, however, does not allow the dissipation of smoke and hot gases as readily as an open parking structure, which is also considered a Group S-2 occupancy. If the enclosed parking garage is located beneath another occupancy group, the enclosed parking garage must be protected with an automatic sprinkler system. This is based on the potential for a fire to develop undetected, which would impact the occupants of the other occupancy. While this section does not specify a fire area threshold, enclosed parking garages are considered less hazardous than repair garages that are classified as Group S-1 occupancies. It is not intended that enclosed parking garages have a more stringent sprinkler threshold than repair garages.

The exception exempts the enclosed garages of buildings that are classified in Group R-3.

[F] 903.2.9.1 Commercial parking garages. An automatic sprinkler system shall be provided throughout buildings used for storage of commercial trucks or buses where the fire area exceeds 5,000 square feet (464 m²).

❖ Due to the larger-sized vehicles involved in commercial parking structures, such as trucks or buses, a more stringent sprinkler threshold is required. Bus garages may also be located adjacent to passenger terminals (Group A-3) that have a substantial occupant load.

[F] 903.2.10 All occupancies except Groups R-3 and U. An automatic sprinkler system shall be installed in the locations set forth in Sections 903.2.10.1 through 903.2.10.1.3.

Exception: Group R-3 as applicable in Section 101.2 and Group U.

❖ Sections 903.2.10.1 through 903.2.10.2 specify certain conditions when an automatic sprinkler system is required even in otherwise nonsprinklered buildings. As indicated in the exception, the listed conditions in the noted sections are applicable to all occupancies except Groups R-3 and U. The exception for Group R-3 occupancies is consistent with other noted sprinkler exceptions for Group R-3 occupancies, such as Section 903.2.11 for enclosed garages. Most structures that qualify as Group U do not typically have the type of conditions stipulated in Sections 903.2.10.1 through 903.2.10.1.3.

[F] 903.2.10.1 Stories and basements without openings. An automatic sprinkler system shall be installed throughout every story or basement of all buildings where the floor area exceeds 1,500 square feet (139.4 m²) and where there is not provided at least one of the following types of exterior wall openings:

1. Openings below grade that lead directly to ground level by an exterior stairway complying with Section 1009 or an outside ramp complying with Section 1010. Openings shall be located in each 50 linear feet (15 240 mm), or fraction thereof, of exterior wall in the story on at least one side.

2. Openings entirely above the adjoining ground level totaling at least 20 square feet (1.86 m²) in each 50 linear feet (15 240 mm), or fraction thereof, of exterior wall in the story on at least one side.

❖ Because of both the lack of openings in exterior walls for access by the fire department for fire fighting and rescue and the problems associated with venting the products of combustion during fire suppression operations, all stories and basements of buildings that do not have adequate openings as defined in this section are required to be equipped with an automatic sprinkler system. This section applies to all stories and basements without openings that exceed 1,500 square feet (139 m²) in a building that is otherwise not required to be fully sprinklered. The requirement for an automatic sprinkler system in this section only applies to the affected area and does not mandate sprinkler protection throughout the entire building.

Stories without openings, as defined in this section, are stories where there is no less than 20 square feet (186 m²) of opening leading directly to ground level in each 50 lineal feet (15 240 mm) or fraction thereof on at least one side. Since exterior doors will provide openings of 20 square feet (186 m²), or slightly less in some occupancies, exterior stairways and ramps in each 50 lineal feet (15 240 mm) are considered acceptable.

This section intends that the required openings be distributed such that the lineal distance between adjacent openings does not exceed 50 feet (15 240 mm). If the openings in the exterior wall are located without regard to adjacent openings, it is possible that segments of the exterior wall are not provided with the required access to the interior of the building for fire-fighting purposes. Any arrangement of required stairways, ramps or openings that results in a portion of the wall 50 feet (15 240 mm) or more in length with no openings to the exterior does not meet the intent of the code that access be provided in each 50 lineal feet (15 240 mm) (see Figure 903.2.10.1).

[F] 903.2.10.1.1 Opening dimensions and access. Openings shall have a minimum dimension of not less than 30 inches (762 mm). Such openings shall be accessible to the fire department from the exterior and shall not be obstructed in a manner that fire fighting or rescue cannot be accomplished from the exterior.

❖ To qualify, an opening must not be less than 30 inches (762 mm) in length or width and must be accessible to the fire department from the exterior. The purpose of the minimum opening dimension is to permit access to the interior of the story or basement by fire department personnel for fire-fighting and rescue operations and to provide openings that are sufficient in size to vent the products of combustion.

[F] 903.2.10.1.2 Openings on one side only. Where openings in a story are provided on only one side and the opposite wall of such story is more than 75 feet (22 860 mm) from such openings, the story shall be equipped throughout with an approved automatic sprinkler system, or openings as specified above shall be provided on at least two sides of the story.

❖ If openings are only provided on one side, then an automatic sprinkler system would still be required if the opposite wall of the story is more than 75 feet (22 860 mm) from such openings. An alternative to providing the automatic sprinkler system would be to provide openings on at least two sides of the exterior of the building. In basements, if any portion is more than 75 feet (22 860 mm) from the openings, then the entire basement must be provided with an automatic sprinkler system. The purpose of providing openings on more than one wall is so that cross ventilation can be achieved to vent the products of combustion [see Figures 903.2.10.1.2 (1-4)].

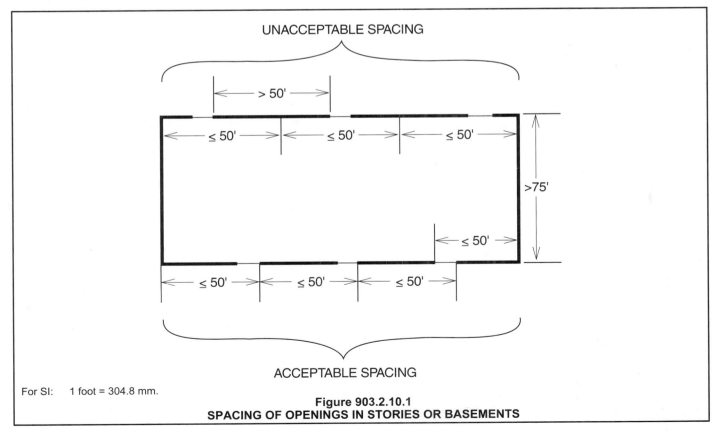

For SI: 1 foot = 304.8 mm.

Figure 903.2.10.1
SPACING OF OPENINGS IN STORIES OR BASEMENTS

FIGURE 903.2.10.1(1) – FIGURE 903.2.10.1(4)

FIRE PROTECTION SYSTEMS

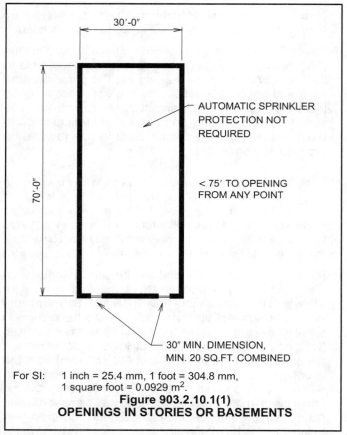

For SI: 1 inch = 25.4 mm, 1 foot = 304.8 mm,
          1 square foot = 0.0929 m².

Figure 903.2.10.1(1)
OPENINGS IN STORIES OR BASEMENTS

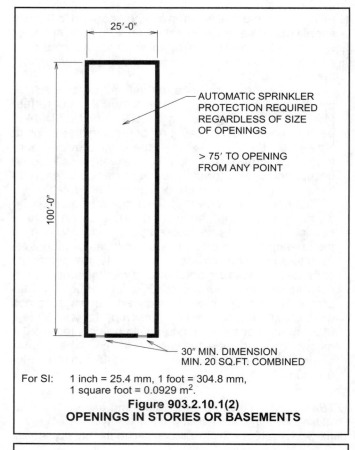

For SI: 1 inch = 25.4 mm, 1 foot = 304.8 mm,
          1 square foot = 0.0929 m².

Figure 903.2.10.1(2)
OPENINGS IN STORIES OR BASEMENTS

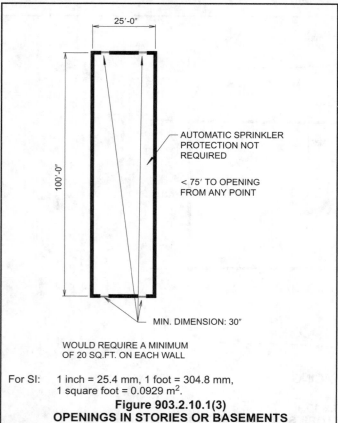

For SI: 1 inch = 25.4 mm, 1 foot = 304.8 mm,
          1 square foot = 0.0929 m².

Figure 903.2.10.1(3)
OPENINGS IN STORIES OR BASEMENTS

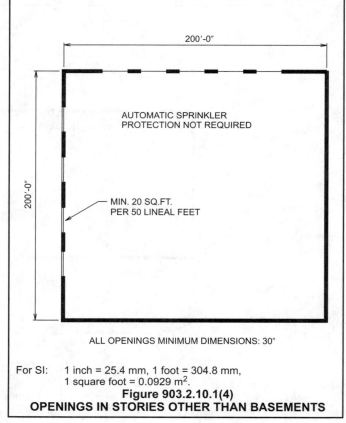

For SI: 1 inch = 25.4 mm, 1 foot = 304.8 mm,
          1 square foot = 0.0929 m².

Figure 903.2.10.1(4)
OPENINGS IN STORIES OTHER THAN BASEMENTS

[F] 903.2.10.1.3 Basements. Where any portion of a basement is located more than 75 feet (22 860 mm) from openings required by Section 903.2.10.1, the basement shall be equipped throughout with an approved automatic sprinkler system.

❖ Where obstructions such as walls or other partitions are present in any given story or basement, the walls and partitions enclosing any room or space must have openings that provide an equivalent degree of fire department access to that which is provided by the openings prescribed in Section 903.2.10.1 for exterior walls. If an equivalent degree of fire department access to all portions of the floor area is not provided, the story or basement would require an automatic sprinkler system.

[F] 903.2.10.2 Rubbish and linen chutes. An automatic sprinkler system shall be installed at the top of rubbish and linen chutes and in their terminal rooms. Chutes extending through three or more floors shall have additional sprinkler heads installed within such chutes at alternate floors. Chute sprinklers shall be accessible for servicing.

❖ This section requires that the chute and termination room associated with a waste or linen system be protected with an automatic sprinkler system. Note that the requirement for suppression is within the chute itself and not within the required shaft that encloses the chute.

[F] 903.2.10.3 Buildings over 55 feet in height. An automatic sprinkler system shall be installed throughout buildings with a floor level having an occupant load of 30 or more that is located 55 feet (16 764 mm) or more above the lowest level of fire department vehicle access.

Exceptions:

1. Airport control towers.
2. Open parking structures.
3. Occupancies in Group F-2.

❖ Because of the difficulties associated with manual suppression of a fire in buildings in excess of 55 feet (16 764 mm) in height, an automatic sprinkler system is required throughout the building regardless of occupancy. It should be noted that buildings that qualify for a sprinkler system by Section 903.2.10.3 are not necessarily high-rise buildings as defined in Section 403.1.

The listed exceptions are occupancies that, based on height only, do not require an automatic sprinkler system. Airport control towers and open parking structures are also exempt from the high-rise provisions of Section 403. While an automatic sprinkler system is not mandated in open parking structures, a system may be required in accordance with Table 406.3.5, depending on the building construction type and the area and number of parking tiers.

[F] 903.2.11 During construction. Automatic sprinkler systems required during construction, alteration and demolition op-

erations shall be provided in accordance with the *International Fire Code.*

❖ Chapter 33 of the code, as well as Chapter 14 of the IFC, address fire safety requirements during construction, alteration or demolition work. Working sprinkler systems should remain in an operative condition at all times unless it is absolutely necessary to shut down the system due to the proposed work. All sprinkler system impairments should be rectified as expeditiously as possible unless specific prior approval has been obtained from the building official. All buildings with a required sprinkler system should not be occupied unless the sprinkler system has been satisfactorily installed and tested.

[F] 903.2.12 Other hazards. Automatic sprinkler protection shall be provided for the hazards indicated in Sections 903.2.12.1 and 903.2.12.2.

❖ This section addresses when an automatic sprinkler system is required for hazardous exhaust ducts and commercial cooking operations. It should be noted that Sections 903.2.12.1 and 903.2.12.2 address conditions where an automatic sprinkler system is the extinguishing system. These sections are not intended to exclude approved alternative extinguishing systems.

[F] 903.2.12.1 Ducts conveying hazardous exhausts. Where required by the *International Mechanical Code,* automatic sprinklers shall be provided in ducts conveying hazardous exhaust, or flammable or combustible materials.

Exception: Ducts in which the largest cross-sectional diameter of the duct is less than 10 inches (254 mm).

❖ To provide protection against the spread of fire within a hazardous exhaust system and to prevent a duct fire from involving the building, an automatic sprinkler system must be installed to protect the exhaust duct system. Where materials conveyed in such ducts are not compatible with water, alternative extinguishing agents should be used. The fire suppression requirement is intended to apply to exhaust systems having an actual fire hazard. An automatic sprinkler system in the duct would be of little value for an exhaust system that conveys only nonflammable or noncombustible materials, fumes, vapors or gases. The exception recognizes the reduced hazard associated with smaller ducts and the impracticality of installing sprinkler protection.

[F] 903.2.12.2 Commercial cooking operations. An automatic sprinkler system shall be installed in commercial kitchen exhaust hood and duct system where an automatic sprinkler system is used to comply with Section 904.

❖ An automatic suppression system is required for all commercial kitchen exhaust hood and duct systems where required by the IFC or the *International Mechanical Code®* (IMC®) to have a Type I hood. Type I hoods are required for all commercial cooking equipment that produces grease-laden vapors or smoke. Section 904.11 recognizes that alternative extinguishing systems other than an automatic sprinkler system may be utilized.

[F] 903.2.13 Other required suppression systems. In addition to the requirements of Section 903.2, the provisions indicated in Table 903.2.13 also require the installation of a suppression system for certain buildings and areas.

❖ In addition to Section 903.2, requirements for automatic fire suppression systems are also found elsewhere in the code as indicated in Table 903.2.13.

[F] TABLE 903.2.13
ADDITIONAL REQUIRED SUPPRESSION SYSTEMS

SECTION	SUBJECT
402.8	Covered malls
403.2, 403.3	High-rise buildings
404.3	Atriums
405.3	Underground structures
407.5	Group I-2
410.6	Stages
411.4	Special amusement buildings
412.2.5, 412.2.6	Aircraft hangars
415.7.2.4	Group H-2
416.4	Flammable finishes
417.4	Drying rooms
507	Unlimited area buildings
IFC	Sprinkler requirements as set forth in Section 903.2.13 of the *International Fire Code*

❖ Table 903.2.13 identifies other sections of the code that require an automatic fire suppression system based on the specific occupancy, process or operation. The table does not identify the various sections of the code that contain design alternatives based on the presence of an automatic fire suppression system, typically an automatic sprinkler system.

[F] 903.3 Installation requirements. Automatic sprinkler systems shall be designed and installed in accordance with Sections 903.3.1 through 903.3.7.

❖ Specific design, installation and testing criteria are provided for automatic sprinkler systems, as well as an indication of the applicability of a nationally recognized standard in the area. The information required to complete a thorough review of an automatic sprinkler system is listed in Figure 903.3.

[F] 903.3.1 Standards. Sprinkler systems shall be designed and installed in accordance with Section 903.3.1.1, 903.3.1.2 or 903.3.1.3.

❖ Automatic sprinkler systems are to be installed in accordance with the code and NFPA 13, 13R or 13D. As provided for in Section 102.4, where differences occur between the code provisions and NFPA 13, 13R or 13D, the code provisions apply. The building official also has the authority to approve the type of sprinkler system to be provided. See Figure 903.3.1 for typical design parameters for each type of sprinkler system.

[F] 903.3.1.1 NFPA 13 sprinkler systems. Where the provisions of this code require that a building or portion thereof be equipped throughout with an automatic sprinkler system in accordance with Section 903.3.1.1, sprinklers shall be installed throughout in accordance with NFPA 13 except as provided in Section 903.3.1.1.1.

❖ NFPA 13 provides the minimum requirements for the design and installation of automatic water sprinkler systems and exposure protection sprinkler systems. The requirements contained in the standard include the character and adequacy of the water supply and the selection of sprinklers, piping, valves and all of the materials and accessories. The standard does not include requirements for the installation of: private fire service mains and their appurtenances; fire pumps or the construction and gravity and pressure tanks and towers (see Figure 903.3.1).

NFPA 13 defines seven classifications or types of water sprinkler systems: wet pipe [see Figure 903.3.1.1(1)]; dry pipe; preaction; deluge; combined dry pipe and preaction; sprinkler systems that are designed for a special purpose and outside sprinklers for exposure protection. While there are numerous variables to be considered in selecting the proper type of sprinkler system, it should be recognized that the wet pipe sprinkler system is the most effective and efficient. The wet pipe system is also the most reliable type of sprinkler system, since water under pressure is available at the sprinkler; therefore, wet pipe sprinkler systems are recommended wherever possible.

Exceptions for the use of NFPA 13R and NFPA 13D systems are addressed throughout the code when exceptions based upon the use of sprinklers are provided. More specifically, if the use of these other standards is appropriate it will be noted within the exception. For a building to be considered "equipped throughout" with an NFPA 13 sprinkler system, complete protection must be provided in accordance with the referenced standard subject to the exempt locations indicated in Section 903.3.1.1.1 [see Figure 903.3.1.1(2)].

1. **Information required on shop drawings includes:**

– Name of owner and occupant
– Location, including street address
– Point of compass
– Ceiling construction
– Full-height cross section
– Location of fire walls
– Location of partitions
– Occupancy of each area or room
– Location and size of blind spaces and closets
– Any questionable small enclosures in which no sprinklers are to be installed
– Size of city main in street, pressure and whether dead end or circulating and, if dead end, direction and distance to nearest circulating main, city main test results
– Other source of water supply, with pressure or elevation
– Make, type and orifice size of sprinkler
– Temperature rating and location of high-temperature sprinklers
– Number of sprinklers on each riser and on each system by floors and total area by each system on each floor
– Make, type, model and size of alarm or dry pipe valve
– Make, type, model and size of preaction or deluge valve
– Type and location of alarm bells
– Total number of sprinklers on each dry pipe system or preaction deluge system
– Approximate capacity in gallons of each dry pipe system
– Cutting lengths of pipe (or center-to-center dimensions)
– Type of fittings, riser nipples and size, and all welds and bends
– Type and location of hangers, inserts and sleeves
– All control valves, checks, drain pipes and test pipes
– Small hand-hose equipment
– Underground pipe size, length, location, weight, material, point of connection to city main; the type of valves, meters and valve pits; and the depth that top of the pipe is laid below grade
– When the equipment is to be installed as an addition to an old group of sprinklers without additional feed from the yard system, enough of the old system shall be indicated on the plans to show the total number of sprinklers to be supplied and to make all connections clear
– Name, address and phone number of contractor and sprinkler designer
– Hydraulic reference points shall be shown by a number and/or letter designation and shall correspond with comparable reference points shown on the hydraulic calculation sheets
– System design criteria showing the minimum rate of water application (density), the design area of water application and the water required for hose streams both inside and outside
– Actual calculated requirements showing the total quantity of water and the pressure required at a common reference point for each system
– Elevation data showing elevations of sprinklers, junction points and supply or reference points

2. **Information required on calculations includes:**

– Location
– Name of owner and occupant
– Building identification
– Description of hazard
– Name and address of contractor and designer
– Name of approving agency

3. **System design requirements include:**

– Design area of water application
– Minimum rate of water application (density)
– Area of sprinkler coverage
– Hazard or commodity classification
– Building height
– Storage height
– Storage method
– Total water requirements, as calculated, including allowance for hose demand water supply information
– Location and elevation of static and residual test gauge with relation to the riser reference point
– Flow location
– Static pressure, psi
– Residual pressure, psi
– Flow, gpm
– Date
– Time
– Test conducted by whom
– Sketch to accompany gridded system calculations to indicate flow quantities and directions for lines with sprinklers operated in the remote area

4. **Additional information necessary for complete review includes:**

– Sprinkler description and discharge constant (K value)
– Hydraulic reference points
– Flow, gpm
– Pipe diameter (actual internal diameter)
– Pipe length
– Equivalent pipe length for fittings and components
– Friction loss in psi per foot of pipe
– Total friction loss between reference points
– Elevation difference between reference points
– Required pressure in psi at each reference point
– Velocity pressures and normal pressure if included in calculations
– Notes to indicate starting points, reference to other sheets or classification of data

5. **Included with the submittal must be a graph sheet showing water supply curves and system requirements including:**

– Hose demand plotted on semi-logarithmic graph paper so as to present a graphic summary of the complete hydraulic calculations

Figure 903.3
SAMPLE SPRINKLER SYSTEM SHOP DRAWING SUBMITTALS

	NFPA 13	NFPA 13R	NFPA 13D
Extent of protection	Equip throughout (Section 903.3.1.1)	Occupied spaces (Section 903.3.1.2)	Occupied spaces (Section 903.3.1.3)
Scope	All occupancies	Low-rise residential	One- and two-family dwellings
Sprinkler design	Density/area concept	4-head design	2-head design
Sprinklers	All types	Residential only	Residential only
Duration	30 minutes (minimum)	30 minutes	10 minutes
Advantages	Property and life protection	Life safety/tenability	Life safety/tenability

Figure 903.3.1
NFPA 13, NFPA 13R, NFPA 13D SYSTEMS

[F] 903.3.1.1.1 Exempt locations. Automatic sprinklers shall not be required in the following rooms or areas where such rooms or areas are protected with an approved automatic fire detection system in accordance with Section 907.2 that will respond to visible or invisible particles of combustion. Sprinklers shall not be omitted from any room merely because it is damp, of fire-resistance-rated construction or contains electrical equipment.

1. Any room where the application of water, or flame and water, constitutes a serious life or fire hazard.

2. Any room or space where sprinklers are considered undesirable because of the nature of the contents, when approved by the building official.

3. Generator and transformer rooms separated from the remainder of the building by walls and floor/ceiling or roof/ceiling assemblies having a fire-resistance rating of not less than 2 hours.

4. In rooms or areas that are of noncombustible construction with wholly noncombustible contents.

❖ This section allows the omission of sprinkler protection in certain conditions, provided that an approved automatic fire detection system is installed. Buildings in compliance with one of the four listed conditions would still be considered fully sprinklered throughout in accordance with the code and NFPA 13 and thus are eligible for all applicable code tradeoffs, exceptions or reductions.

Condition 1 addresses restrictions where the application of water could create a hazardous condition. For example, sprinkler protection should be avoided where it is not compatible with certain stored material (i.e., some water-reactive hazardous materials). Combustible metals, such as magnesium and aluminum, may burn so intensely that the use of water to attempt fire control will only intensify the reaction.

The elimination of the sprinkler system in a sensitive area is subject to the approval of the building official. It is not the intent of Condition 2 to omit sprinklers solely due to a potential for water damage. It should also be noted that a desire not to sprinkler a certain area (such as a computer room or operating room) does not fall within the limitations of the exception unless there is something unique about the space that would result in water being incompatible.

Condition 3 recognizes the low fuel load and low occupancy hazards associated with generator and transformer rooms and thereby allows the omission of sprinkler protection provided the rooms are separated from adjacent areas by 2-hour fire-resistance-rated construction. This condition assumes the room is not utilized for combustible storage.

Condition 4 requires the construction of the room or area, as well as the contents, to be noncombustible. An example would be an area in an unprotected steel frame building (Type IIB construction) used for steel or concrete block storage. Neither involves any significant combustible packaging or sources of ignition, and few combustibles are present (see Figure 903.3.1).

[F] 903.3.1.2 NFPA 13R sprinkler systems. Where allowed in buildings of Group R, up to and including four stories in height, automatic sprinkler systems shall be installed throughout in accordance with NFPA 13R.

❖ NFPA 13R provides design and installation requirements for a sprinkler system to aid in the detection and control of fires in low-rise (four stories or less) residential occupancies. Sprinkler systems designed in accordance with NFPA 13R are intended to prevent flashover (total involvement) in the room of fire origin and to improve the chance for occupants to escape or be evacuated. The design criteria in NFPA 13R are similar to those in NFPA 13 except that sprinklers may be omitted from areas in which fatal fires in residential occupancies do not typically originate (bathrooms, closets, attics, porches, garages and concealed spaces) (see Figure 903.3.1).

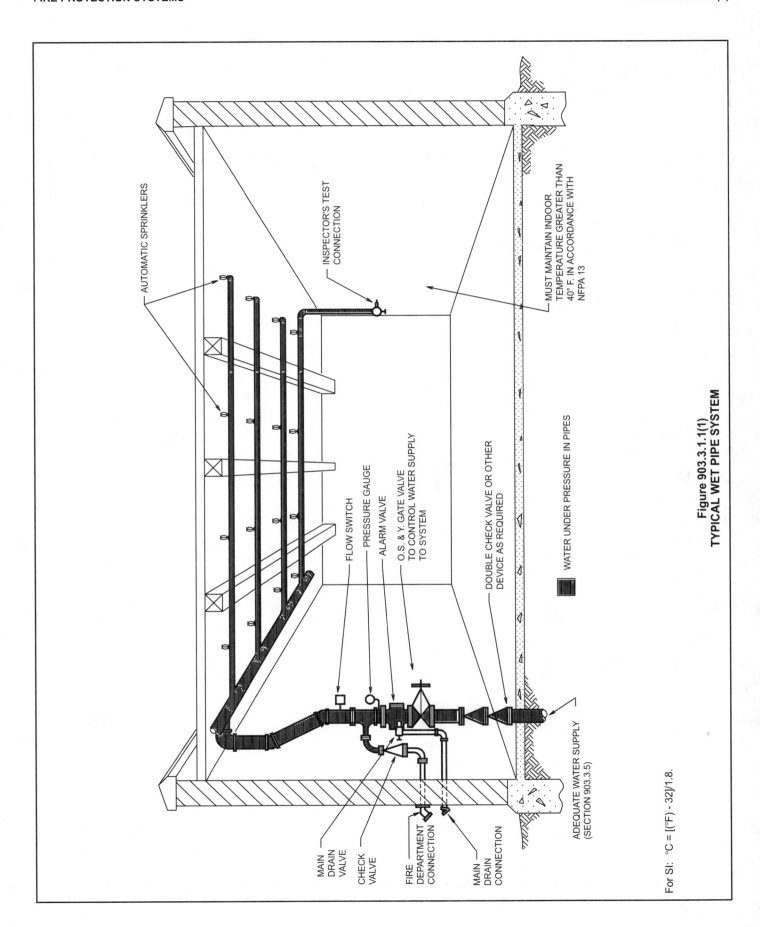

AUTOMATIC SPRINKLERS

INSPECTOR'S TEST CONNECTION

MUST MAINTAIN INDOOR TEMPERATURE GREATER THAN 40° F. IN ACCORDANCE WITH NFPA 13

FLOW SWITCH

PRESSURE GAUGE

ALARM VALVE

O.S. & Y. GATE VALVE TO CONTROL WATER SUPPLY TO SYSTEM

DOUBLE CHECK VALVE OR OTHER DEVICE AS REQUIRED

WATER UNDER PRESSURE IN PIPES

MAIN DRAIN VALVE

CHECK VALVE

FIRE DEPARTMENT CONNECTION

MAIN DRAIN CONNECTION

ADEQUATE WATER SUPPLY (SECTION 903.3.5)

Figure 903.3.1.1(1)
TYPICAL WET PIPE SYSTEM

For SI: °C = [(°F) - 32]/1.8.

CODE SECTION	MODIFICATION	NFPA 13	NFPA 13R	NFPA 13D
Increases				
504.2	Height increase	yes	yes	no
506.3	Area increase	yes	no	no
Table 1005.1	Egress width	yes	yes	no
Table 1015.1	Travel distance	yes	yes	no
Rating Reductions				
302.3.2	Separated uses	yes	no	no
Table 601	Type VA construction	yes	no	no
708.3	Fire partitions (dwelling units, guestrooms)	yes	no	no
Table 1016.1	Corridor walls	yes	yes	no
Miscellaneous				
Tables 307.7(1), 307.7(2)	Hazardous material increase	yes	no	no
403.3	High-rise modification	yes	no	no
404.2	Atriums	yes	no	no
507.2, 507.3	Unlimited area buildings	yes	no	no
704.8.1	Allowable area of opening	yes	no	no
704.9	Vertical separation of openings	yes	yes	no
717.4.2	Residential attics	yes	yes	no
Table 803.5	Interior finish	yes	yes	no
804.5.1	Floor finish	yes	no	no
907.2	Fire alarm system	yes (B, F, M)	yes (R-1, R-2)	no
1007.2.1	Accessible egress	yes	yes	no
1023.1	Exit discharge	yes	yes	no
1025.1	Emergency escape openings	yes	yes	no
1406.3	Balconies	yes	yes*	yes*

* Sprinkler protection must be extended to the affected areas.

Figure 903.3.1.1(2)
EXAMPLES OF REQUIREMENTS MODIFIED THROUGH USE OF AUTOMATIC SPRINKLER SYSTEMS

[F] 903.3.1.2.1 Balconies. Sprinkler protection shall be provided for exterior balconies and ground-floor patios of dwelling units where the building is of Type V construction. Sidewall sprinklers that are used to protect such areas shall be permitted to be located such that their deflectors are within 1 inch (25 mm) to 6 inches (152 mm) below the structural members, and a maximum distance of 14 inches (356 mm) below the deck of the exterior balconies that are constructed of open wood joist construction.

❖ This section requires additional sprinklers on balconies and patios when Type V construction is used for Group R occupancies. This is in addition to the requirements of NFPA 13R. The intent is to address hazards such as grilling and similar activities. Since NFPA 13R does not require such coverage, there is the potential that a fire on a balcony could grow too large for the system within the building to handle.

[F] 903.3.1.3 NFPA 13D sprinkler systems. Where allowed, automatic sprinkler systems in one- and two-family dwellings shall be installed throughout in accordance with NFPA 13D.

❖ NFPA 13D provides design and installation requirements for a sprinkler system to aid in the detection and control of fires in one- and two-family dwellings and mobile homes. Similar to NFPA 13R, sprinkler systems designed in accordance with NFPA 13D are intended to prevent flashover (total involvement) in the room of fire origin and to improve the chance for occupants to escape or be evacuated. Whereas NFPA 13D has similar provisions to NFPA 13R regarding the allowable omission of sprinklers in certain areas of the dwelling unit, the water supply requirements are less restrictive. NFPA 13D utilizes a two-head sprinkler design with a 10-minute duration requirement, while NFPA 13R utilizes a four-head sprinkler design with a 30-minute duration requirement. The decreased water supply requirement emphasizes the main intent of NFPA 13D—to control the fire and maintain tenability during evacuation of the residence (see Figure 903.3.1).

Since the building official has the authority to approve the type of sprinkler system, this section may be used to prevent the use of a specific type of sprinkler system that may be inappropriate for the type of occupancy.

[F] 903.3.2 Quick-response and residential sprinklers. Where automatic sprinkler systems are required by this code, quick-response or residential automatic sprinklers shall be installed in the following areas in accordance with Section 903.3.1 and their listings:

1. Throughout all spaces within a smoke compartment containing patient sleeping units in Group I-2 in accordance with this code.

2. Dwelling units, and sleeping units in Group R and I-1 occupancies.

3. Light-hazard occupancies as defined in NFPA 13.

❖ This section requires the use of either listed quick-response or residential automatic sprinklers depending upon the type of sprinkler system required to facilitate

faster and more effective suppression in certain areas. Residential sprinklers are required in all types of residential buildings that would permit the use of an NFPA 13R or 13D sprinkler system.

Condition 1 reiterates the requirements of Section 407.5 for the use of approved quick-response or residential sprinklers in smoke compartments containing patient sleeping units in Group I-2 occupancies (see commentary, Section 407.6).

Because of the kind of occupants sleeping in Group R and I-1 occupancies, as indicated in Condition 2, a fast response-type sprinkler is desirable.

Condition 3 recognizes light-hazard occupancies in accordance with NFPA 13. These could include restaurants, schools, office buildings, churches and similar occupancies where the fire load and potential heat release of combustible contents are low.

[F] 903.3.3 Obstructed locations. Automatic sprinklers shall be installed with due regard to obstructions that will delay activation or obstruct the water distribution pattern. Automatic sprinklers shall be installed in or under covered kiosks, displays, booths, concession stands, or equipment that exceeds 4 feet (1219 mm) in width. Not less than a 3-foot (914 mm) clearance shall be maintained between automatic sprinklers and the top of piles of combustible fibers.

Exception: Kitchen equipment under exhaust hoods protected with a fire-extinguishing system in accordance with Section 904.

❖ In order to provide adequate sprinkler coverage, sprinkler protection should be extended under any obstruction that exceeds 4 feet (1,219 mm) in width. Large air ducts are another common obstruction where sprinklers are routinely extended beneath the duct. The 3-foot (914 mm) storage clearance requirement for combustible fibers is due to their potential high heat release. Most storage conditions only require a minimum 18-inch (457 mm) storage clearance to combustibles dependent upon the type of sprinklers utilized and their actual storage conditions.

The exception recognizes that an alternative extinguishing system is permitted for commercial cooking systems in lieu of sprinkler protection for exhaust hoods that may exceed 4 feet (1,219 mm) in width.

[F] 903.3.4 Actuation. Automatic sprinkler systems shall be automatically actuated unless specifically provided for in this code.

❖ The intent of this section is to eliminate the need for occupant intervention during a fire situation.

Wet pipe and dry pipe sprinkler systems, for example, are essentially fail-safe systems in the sense that, if the system is in proper operating condition, it will operate once a sprinkler fuses. Dry systems have an inherent "time lag" for water to reach the sprinkler; therefore, the response is not the same as for a wet pipe system. Other types of sprinkler systems, such as preaction and deluge, rely on the actuation of a detection system to operate the sprinkler valve.

[F] 903.3.5 Water supplies. Water supplies for automatic sprinkler systems shall comply with this section and the standards referenced in Section 903.3.1. The potable water supply shall be protected against backflow in accordance with the requirements of this section and the *International Plumbing Code.*

❖ In order to be effective, all sprinkler systems must have an adequate supply of water. The criteria for an acceptable water supply are contained in the standards referenced in Section 903.3.1. For example, NFPA 13 contains criteria for different types of water supplies as well as the methods to determine the pressure, flow capabilities and capacity necessary to achieve the intended performance of a sprinkler system. An acceptable water supply could consist of a reliable municipal supply, a gravity tank or a fire pump with a pressure tank or a combination thereof.

This section also establishes the requirements for protecting the potable water system against a nonpotable source, such as stagnant water retained within the sprinkler piping. In accordance with Section 608.16.4 of the IPC, an approved double check valve device or reduced pressure principle backflow preventer is required.

[F] 903.3.5.1 Domestic services. Where the domestic service provides the water supply for the automatic sprinkler system, the supply shall be in accordance with this section.

❖ This section establishes the scope of domestic services for limited area sprinkler systems and residential combination services.

[F] 903.3.5.1.1 Limited area sprinkler systems. Limited area sprinkler systems serving fewer than 20 sprinklers on any single connection are permitted to be connected to the domestic service where a wet automatic standpipe is not available. Limited area sprinkler systems connected to domestic water supplies shall comply with each of the following requirements:

1. Valves shall not be installed between the domestic water riser control valve and the sprinklers.

   Exception: An approved indicating control valve supervised in the open position in accordance with Section 903.4.

2. The domestic service shall be capable of supplying the simultaneous domestic demand and the sprinkler demand required to be hydraulically calculated by NFPA 13, NFPA 13R or NFPA 13D.

❖ Limited area sprinkler systems are primarily limited to fire areas or other areas where the number of sprinklers does not exceed 20.

The use of limited area sprinkler systems is restricted to cases in which the code requires a limited number of sprinklers and not a complete automatic sprinkler system. For example, limited area sprinkler systems may be used to protect stages; storage and workshop areas; stories and basements without openings; painting rooms; trash rooms and chutes; furnace rooms; kitchens and hazardous exhaust systems and incidental use areas as defined in Section 302.1.1. When a wet automatic standpipe is not available, limited area sprinkler systems may be connected to the domestic water supply.

The water supply to the sprinkler system is to be controlled only by the same valve that controls the domestic water supply to the building and no shutoff valves are permitted in the sprinkler system piping. These provisions increase the likelihood that the sprinkler system will be operational should a fire occur. Likewise, if the sprinkler system needs restoration after having operated in response to a fire or needs repairs requiring that the water supply be shut off, this section increases the probability that the system will be restored in a timely manner, since having the domestic water supply to the entire building shut off is an inconvenience to occupants that will not be tolerated long.

The exception recognizes the value of standard sprinkler system valve supervision in accordance with Section 903.4 as providing the level of system reliability contemplated by this section.

Documentation, usually in the form of hydraulic calculations, must be submitted demonstrating that the domestic water system is adequate to supply the sprinkler demand in addition to the peak domestic demand. The domestic demand would normally be determined in accordance with the IPC.

[F] 903.3.5.1.2 Residential combination services. A single combination water supply shall be permitted provided that the domestic demand is added to the sprinkler demand as required by NFPA 13R.

❖ NFPA 13R permits a common supply main to a building to serve both the sprinkler system and domestic services provided the domestic demand is added to the sprinkler demand. NFPA 13R systems do not provide the same level of property protection as NFPA 13 systems.

[F] 903.3.5.2 Secondary water supply. A secondary on-site water supply equal to the hydraulically calculated sprinkler demand, including the hose stream requirement, shall be provided for high-rise buildings in Seismic Design Category C, D, E or F as determined by this code. The secondary water supply shall have a duration not less than 30 minutes as determined by the occupancy hazard classification in accordance with NFPA 13.

Exception: Existing buildings.

❖ In order to increase the reliability of the sprinkler system should an earthquake disable the primary water supply, a secondary on-site water supply is required for high-rise buildings in Seismic Design Category C, D, B or F (see commentary, Section 1616.3). The amount of water required is equal to the hydraulically calculated sprinkler demand plus hose stream demand for a 30-minute period, related to the appropriate hazard classification in NFPA 13.

The exception recognizes the infeasibility of requiring a secondary water supply in existing high-rise buildings.

[F] 903.3.6 Hose threads. Fire hose threads used in connection with automatic sprinkler systems shall be approved and compatible with fire department hose threads.

❖ The threads on connections in which the fire department may connect a hose must be compatible with the fire department's threads. The majority of fire departments in the United States use the "American National Fire Hose Connection Screw Thread" (NH), commonly known as NST and NS. NFPA 1963 provides the screw thread dimensions and the size thread of threaded connections, with nominal sizes ranging from $3/4$ inch (19 mm) to 6 inches (152 mm) for the NH thread. Although efforts to standardize fire hose threads were initiated after the Boston conflagration in 1872, there are still many different screw threads—some of which give the appearance of compatibility with the NH thread. While NFPA 1963 may be used as a guide, the code does not require any particular standard to be adhered to. Design documents should specify the type of thread to be provided in order to be compatible with the fire department's equipment. The criteria typically apply to fire department connections for sprinkler and standpipe systems, standpipe hose connections, yard hydrants and wall hydrants.

[F] 903.3.7 Fire department connections. The location of fire department connections shall be approved by the building official.

❖ Fire department connections are required as part of a water-based suppression system as the auxiliary water supply. These connections provide the fire department with the capability of supplying the necessary water to the automatic sprinkler or standpipe system at a sufficient pressure without pressurizing the underground supply. The fire department connection also serves as an alternative source of water should a valve in the primary water supply be closed. A fire department connection does not, however, constitute an automatic water source.

Section 912 of the IFC provides additional guidance with respect to the location and accessibility of fire department connections, which should be readily visible from the street and unobstructed. See Figure 903.3.7 for typical fire department connection arrangements.

[F] 903.3.7.1 Locking fire department connection (FDC) caps. The fire code official is authorized to require locking FDC caps on fire department connections for water-based fire protection systems where the responding fire department carries appropriate key wrenches for removal.

❖ This section allows for the fire department connection (FDC) caps to be locked as long as the fire departments that respond to that building or facility have the appropriate key wrenches. This avoids unnecessary vandalism and ensures a more functional FDC when needed.

[F] 903.4 Sprinkler system monitoring and alarms. All valves controlling the water supply for automatic sprinkler systems, pumps, tanks, water levels and temperatures, critical air pressures and water-flow switches on all sprinkler systems shall be electrically supervised.

Exceptions:

1. Automatic sprinkler systems protecting one- and two-family dwellings.

2. Limited area systems serving fewer than 20 sprinklers.

3. Automatic sprinkler systems installed in accordance with NFPA 13R where a common supply main is used to supply both domestic water and the automatic sprinkler systems and a separate shutoff valve for the automatic sprinkler system is not provided.

4. Jockey pump control valves that are sealed or locked in the open position.

5. Control valves to commercial kitchen hoods, paint spray booths or dip tanks that are sealed or locked in the open position.

6. Valves controlling the fuel supply to fire pump engines that are sealed or locked in the open position.

7. Trim valves to pressure switches in dry, preaction and deluge sprinkler systems that are sealed or locked in the open position.

❖ The reliability data on automatic sprinkler systems clearly indicate that a closed valve is the leading cause of sprinkler system failure. There are also a number of other critical elements that contribute to successful sprinkler system operation including, but not limited to, pumps, water tanks and air pressure maintenance devices; therefore, this section requires that the various critical elements that contribute to an available water supply and to the function of the sprinkler system be electrically supervised.

Automatic sprinkler systems in one- and two-family dwellings are typically designed to comply with NFPA 13D, which does not require electrical supervision (see Exception 1).

Limited-area sprinkler systems are generally supervised by their connection to the domestic water service (see Exception 2). Electrical supervision is required only if a control valve is installed between the riser control valve and the sprinkler system piping. Similar to limited-area sprinkler systems, electrical supervision is not required for NFPA 13R residential combination services when a shutoff valve is not installed (see Exception 3). NFPA 13R sprinkler systems are supervised in that the only way to shut off the sprinkler system is to also shut off the domestic water supply. The valves discussed in Exceptions 4 through 7 can be sealed or locked in the open position because they do not control the sprinkler system water supply.

FIGURE 903.3.7

FIRE PROTECTION SYSTEMS

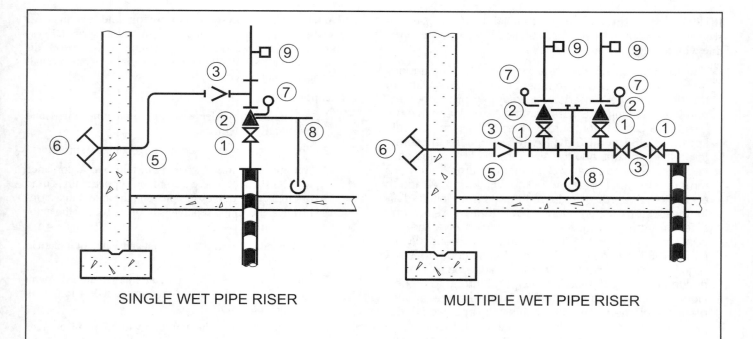

SINGLE WET PIPE RISER

MULTIPLE WET PIPE RISER

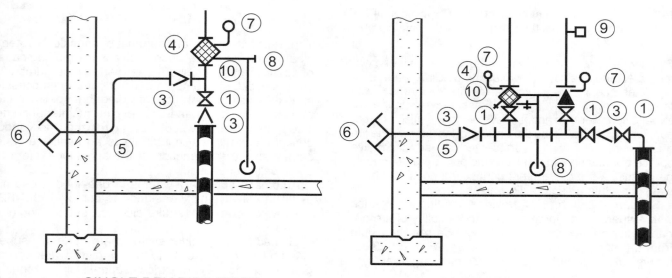

SINGLE DRY PIPE RISER

COMBINATION RISER

LEGEND:

1. INDICATING CONTROL VALVE
2. ALARM CHECK VALVE
3. SWING CHECK VALVE
4. DRY PIPE VALVE
5. 1/2" BALL DRIP

6. FIRE DEPARTMENT CONNECTION
7. PRESSURE GAUGE
8. MAIN DRAIN
9. WATER-FLOW SWITCH
10. PRESSURE FLOW SWITCH

For SI: 1 inch = 25.4 mm.

Figure 903.3.7
FIRE DEPARTMENT CONNECTION DETAILS

[F] 903.4.1 Signals. Alarm, supervisory and trouble signals shall be distinctly different and automatically transmitted to an approved central station, remote supervising station or proprietary supervising station as defined in NFPA 72 or, when approved by the building official, shall sound an audible signal at a constantly attended location.

Exceptions:

1. Underground key or hub valves in roadway boxes provided by the municipality or public utility are not required to be monitored.

2. Backflow prevention device test valves, located in limited area sprinkler system supply piping, shall be locked in the open position. In occupancies required to be equipped with a fire alarm system, the backflow preventer valves shall be electrically supervised by a tamper switch installed in accordance with NFPA 72 and separately annunciated.

❖ Automatic sprinkler systems are required to be supervised as a means of determining that the system is operational. Supervision is usually accomplished by a valve supervisory switch operating as a normally open or normally closed switch. NFPA 72 does not permit valve supervisory switches to be located on the same zone circuit as the water-flow switch unless specifically arranged to actuate a distinctive signal from the circuit trouble condition.

Required sprinkler systems are to be monitored by an approved supervising service in accordance with NFPA 72. Types of supervising stations recognized in NFPA 72 indicate central station, remote supervising station or proprietary supervising station.

A central station is an independent off-site facility operated and maintained by personnel whose primary business is to furnish, maintain, record and supervise a signaling system. A proprietary system is similar to a central station; however, a proprietary system is typically an on-site facility monitoring a number of buildings on the same site for the same owner. A remote station system has an alarm signal that is transmitted to a remote location acceptable to the jurisdiction that is attended 24 hours a day. The receiving equipment is usually located at a fire station, police station or telephone answering service. Alternatively to the three previous supervising methods, an audible signal can be transmitted to a constantly attended location approved by the building official.

Exception 1 recognizes that underground key or hub valves in roadway boxes are not normally supervised or required to be supervised by this section or NFPA 13.

Exception 2 acknowledges that local water utilities and environmental authorities in many instances require, by local ordinances, that backflow prevention devices be installed in limited-area sprinkler system piping. In order to facilitate the testing and maintenance of backflow prevention devices, test valves are installed on each side of the device. Those valves are typically indicating-type valves and can function as shutoff valves

for the sprinkler system, thus requiring some level of supervision.

In consideration of the fact that these infrequently used valves may be the only feature of protection requiring supervision in occupancies not otherwise required to be equipped with a fire alarm system, Exception 2 permits these valves to be locked in the open position. However, if the occupancy is protected by a fire alarm system, then these valves must be equipped with approved valve supervisory devices connected to the fire alarm control panel on a separate (supervisory) zone so that the supervisory signal is transmitted to the appropriate receiving station. Backflow preventers in sprinkler systems and the testing of such devices are regulated in Sections 312.9 (testing) and 608.16.4 (devices) of the IPC.

[F] 903.4.2 Alarms. Approved audible devices shall be connected to every automatic sprinkler system. Such sprinkler water-flow alarm devices shall be activated by water flow equivalent to the flow of a single sprinkler of the smallest orifice size installed in the system. Alarm devices shall be provided on the exterior of the building in an approved location. Where a fire alarm system is installed, actuation of the automatic sprinkler system shall actuate the building fire alarm system.

❖ The audible alarm, sometimes referred to as the "outside ringer" or "water-motor gong," sounds when the sprinkler system has activated. The alarm device may be electrically operated or it may be a true water-motor gong operated by a paddle-wheel-type attachment to the sprinkler system riser that responds to the flow of water in the piping. Though no longer the alarm device of choice, water-motor gongs do have the advantage of not being subject to power failures within or outside the protected building (see Sections 3-10 and 5-15 of NFPA 13 for further information on these devices). The alarm must be installed on the exterior of the building in a location approved by the building official. This location is often in close proximity to the FDC and also serves a collateral function of helping the responding fire apparatus engineer to more promptly locate the FDC. The alarm is not intended to be an evacuation alarm; however, when a fire alarm system is installed, the sprinkler system must be interconnected with the fire alarm system so that when the sprinkler system actuates, it sounds the evacuation alarms required for the fire alarm system.

[F] 903.4.3 Floor control valves. Approved supervised indicating control valves shall be provided at the point of connection to the riser on each floor in high-rise buildings.

❖ In high-rise buildings, sprinkler control valves with supervisory initiating devices must be provided at the point of connection to the riser on each floor. Sprinkler control valves on each floor are intended to permit servicing activated systems without impairing the water supply to large portions of the building.

[F] 903.5 Testing and maintenance. Sprinkler systems shall be tested and maintained in accordance with the *International Fire Code*.

❖ Section 901 of the IFC contains requirements for the testing and maintenance of sprinkler systems. Acceptance tests are necessary to verify that the system performs as intended by design and by the code. Periodic testing and maintenance are essential to verify that the level of protection designed in the building will be provided whenever a fire emergency occurs. All water-based extinguishing systems are required to be tested and maintained in accordance with NFPA 25.

SECTION 904
ALTERNATIVE AUTOMATIC
FIRE-EXTINGUISHING SYSTEMS

[F] 904.1 General. Automatic fire-extinguishing systems, other than automatic sprinkler systems, shall be designed, installed, inspected, tested and maintained in accordance with the provisions of this section and the applicable referenced standards.

❖ The provisions of Section 904 are for alternative fire-extinguishing systems that utilize extinguishing agents other than water. Alternative automatic fire-extinguishing systems include wet-chemical, dry-chemical, foam, carbon dioxide, halon and clean-agent suppression systems. In addition to the provisions of Section 904, the indicated referenced standards provide specific installation, maintenance and testing requirements for all systems.

[F] 904.2 Where required. Automatic fire-extinguishing systems installed as an alternative to the required automatic sprinkler systems of Section 903 shall be approved by the building official. Automatic fire-extinguishing systems shall not be considered alternatives for the purposes of exceptions or reductions permitted by other requirements of this code.

❖ One of the main considerations in selecting an extinguishing agent should be the compatibility of the agent with the hazard. The building official has the responsibility to approve an alternative extinguishing agent. The approval should be based on the compatibility of the agent with the hazard and the potential effectiveness of the agent to suppress a fire involving the hazards present. It is important to note that the code places limitations on alternative systems in that they will not be credited toward a building being "equipped throughout" with an automatic sprinkler system where the sprinkler system is an alternative or tradeoff to a code requirement.

[F] 904.2.1 Hood system suppression. Each required commercial kitchen exhaust hood and duct system required by the *International Fire Code* or the *International Mechanical Code* to have a Type I hood shall be protected with an approved auto-

matic fire-extinguishing system installed in accordance with this code.

❖ This section requires an effective suppression system to combat fire on the cooking surfaces of grease-producing appliances and within the hood and exhaust system of a commercial kitchen installation. All Type I hoods, including the duct system, are required to be suppressed. Type I hoods are utilized for handling grease-laden vapors or smoke whenever Type II hoods handle fumes, steam, heat and odors. Type I hoods are typically required for commercial food heat-processing equipment, such as deep fryers, griddles, charbroilers, broilers and open burner stoves and ranges.

[F] 904.3 Installation. Automatic fire-extinguishing systems shall be installed in accordance with this section.

❖ The installation of all automatic fire-extinguishing systems is required to comply with the provisions of Sections 904.3.1 through 904.3.5 in addition to the installation criteria contained in the referenced standard for the proposed type of alternative extinguishing system.

[F] 904.3.1 Electrical wiring. Electrical wiring shall be in accordance with the ICC *Electrical Code*.

❖ The ICC *Electrical Code*® (ICC EC™) in turn references NFPA 70, *National Electrical Code*, for the design and installation of all electrical systems and equipment. All electrical work should also be in compliance with any specific electrical classifications and conditions contained in the referenced standards for each type of system.

[F] 904.3.2 Actuation. Automatic fire-extinguishing systems shall be automatically actuated and provided with a manual means of actuation in accordance with Section 904.11.1.

❖ To increase the reliability of the system and to provide the opportunity to initiate the system as a preventive measure, a manual means to activate the system is required (see commentary, Section 904.11.1).

[F] 904.3.3 System interlocking. Automatic equipment interlocks with fuel shutoffs, ventilation controls, door closers, window shutters, conveyor openings, smoke and heat vents and other features necessary for proper operation of the fire-extinguishing system shall be provided as required by the design and installation standard utilized for the hazard.

❖ Shutting off fuel supplies will eliminate potential ignition sources in the protected area. Automatic door and window closers and dampers for forced-air ventilation systems are intended to maintain the desired concentration level of the extinguishing agent in the protected area.

[F] 904.3.4 Alarms and warning signs. Where alarms are required to indicate the operation of automatic fire-extinguishing systems, distinctive audible and visible alarms and warning signs shall be provided to warn of pending agent discharge. Where exposure to automatic-extinguishing agents poses a hazard to persons and a delay is required to ensure the evacuation of occupants before agent discharge, a separate warning signal shall be provided to alert occupants once agent discharge has begun. Audible signals shall be in accordance with Section 907.9.2.

❖ Steps and safeguards are necessary in order to prevent injury or death to personnel in areas where the atmosphere will be made hazardous by the pending agent discharge. Provisions are to be made for alarms to operate on fire detection and agent discharge. A continuous alarm that will sound until the room atmosphere is restored to normal should be provided at entrances to such areas. Warning and instructional signs are to be provided at the entrances to and within the protected area.

[F] 904.3.5 Monitoring. Where a building fire alarm system is installed, automatic fire-extinguishing systems shall be monitored by the building fire alarm system in accordance with NFPA 72.

❖ Automatic fire-extinguishing systems are not required to be electrically supervised unless the building is provided with a fire alarm system. This section recognizes the fact that not all buildings are required to be equipped with a fire alarm system; however, when a fire alarm system is provided with corresponding evacuation alarms, this section also recognizes the increased level of hazard notification available to the occupants that can be obtained by interconnecting to the present building fire alarm system.

[F] 904.4 Inspection and testing. Automatic fire-extinguishing systems shall be inspected and tested in accordance with the provisions of this section prior to acceptance.

❖ The completed installation is required to be tested and inspected to determine that the system has been installed in compliance with the code and will function as required. Full-scale acceptance tests should be conducted as required in the applicable referenced standard.

[F] 904.4.1 Inspection. Prior to conducting final acceptance tests, the following items shall be inspected:

1. Hazard specification for consistency with design hazard.
2. Type, location and spacing of automatic- and manual-initiating devices.
3. Size, placement and position of nozzles or discharge orifices.
4. Location and identification of audible and visible alarm devices.
5. Identification of devices with proper designations.
6. Operating instructions.

❖ This section identifies those items that need to be verified or visually inspected prior to conducting the final ac-

ceptance tests. All equipment should be listed and approved and installed in accordance with the manufacturer's recommendations.

[F] 904.4.2 Alarm testing. Notification appliances, connections to fire alarm systems and connections to approved supervising stations shall be tested in accordance with this section and Section 907 to verify proper operation.

❖ All components of fire-extinguishing systems related to alarm devices and supervision thereof should be tested prior to approval of the system. All alarm devices should be tested in accordance with the applicable provisions of NFPA 72.

[F] 904.4.2.1 Audible and visible signals. The audibility and visibility of notification appliances signaling agent discharge or system operation, where required, shall be verified.

❖ This section requires the audibility and visibility of all notification appliances to be verified upon installation. For example, Section 907.9.2 prescribes minimum sound pressure levels for audible alarm notification appliances depending upon the occupancy of the building. All audible and visible alarm notification devices are also required to comply with the applicable requirements of NFPA 72.

[F] 904.4.3 Monitor testing. Connections to protected premises and supervising station fire alarm systems shall be tested to verify proper identification and retransmission of alarms from automatic fire-extinguishing systems.

❖ Where fire-extinguishing systems are required to be monitored, such as by Section 904.3.5, all connections related to the supervision of the system must be tested to verify they are in proper working order.

[F] 904.5 Wet-chemical systems. Wet-chemical extinguishing systems shall be installed, maintained, periodically inspected and tested in accordance with NFPA 17A and their listing.

❖ NFPA 17A provides minimum requirements for the design, installation, operation, testing and maintenance of wet-chemical preengineered extinguishing systems. Equipment that is typically protected with wet-chemical extinguishing systems includes restaurant, commercial and institutional hoods, plenums, ducts and associated cooking equipment. Strict compliance with the manufacturer's installation instructions is vital for a viable installation.

Wet-chemical solutions used in extinguishing systems are relatively harmless and there is usually no lasting significant effect on a person's skin, respiratory system or clothing. These solutions may produce a mild, temporary irritation but the symptoms will usually disappear when contact is eliminated.

[F] 904.6 Dry-chemical systems. Dry-chemical extinguishing systems shall be installed, maintained, periodically inspected and tested in accordance with NFPA 17 and their listing.

❖ NFPA 17 provides the minimum requirements for the design, installation, testing, inspection, approval, operation and maintenance of dry-chemical extinguishing systems.

The building official has the authority to approve the type of dry-chemical extinguishing system to be provided. NFPA 17 identifies three types of dry-chemical extinguishing systems: total flooding, local application and hand hose-line systems. Only total flooding and local application systems are considered automatic extinguishing systems.

The types of hazards and equipment that can be protected with dry-chemical extinguishing systems include: flammable and combustible liquids and combustible gases; combustible solids, which melt when involved in a fire; electrical hazards, such as transformers or oil circuit breakers; textile operations subject to flash surface fires; ordinary combustibles, such as wood, paper or cloth and restaurant and commercial hoods, ducts and associated cooking appliance hazards, such as deep fat fryers and some plastics, depending on the type of material and configuration.

Total flooding dry-chemical extinguishing systems are used only where there is a permanent enclosure about the hazard that is adequate to enable the required concentration to be built up. The total area of unenclosable openings must not exceed 15 percent of the total area of the sides, top and bottom of the enclosure. Consideration must be given to eliminating the probable sources of reignition within the enclosure because the extinguishing action of dry-chemical systems is transient.

Local application of dry-chemical extinguishing systems is to be used for extinguishing fires where the hazard is not enclosed or where the enclosure does not conform to the requirements for total flooding systems. Local application systems have successfully protected hazards involving flammable or combustible liquids, gases and shallow solids, such as paint deposits.

NFPA 17 also addresses preengineered dry-chemical systems consisting of components designed to be installed in accordance with pretested limitations as tested and labeled by a testing agency. Preengineered systems must be installed within the limitations that have been established by the testing agency and may include total flooding, local application or a combination of both types of systems.

The type of dry chemical to be used with the extinguishing system is a function of the hazard to be protected. The type of dry chemical used in a system should not be changed, unless it has been proven changeable by a testing laboratory, is recommended by the manufacturer of the equipment and is acceptable to the building official for the hazard being protected. Additional guidance on the use of various dry-chemical agents can be found in NFPA 17.

[F] 904.7 Foam systems. Foam-extinguishing systems shall be installed, maintained, periodically inspected and tested in accordance with NFPA 11 and NFPA 16 and their listing.

❖ NFPA 11 covers the characteristics of foam-producing materials used for fire protection and the requirements for design, installation, operation, testing and maintenance of equipment and systems, including those used in combination with other fire-extinguishing agents. The minimum requirements are covered for flammable and combustible liquid hazards in local areas within buildings, storage tanks and indoor and outdoor processing areas.

Low-expansion foam is defined as an aggregation of air-filled bubbles resulting from the mechanical expansion of a foam solution by air with a foam-to-solution volume ratio of less than 20:1. It is principally used to protect flammable and combustible liquid hazards. Also, low-expansion foam may be used for heat radiation protection. Combined agent systems involve the application of low-expansion foam to a hazard simultaneously or sequentially with dry-chemical powder.

NFPA 11A provides minimum requirements for the installation, design, operation, testing and maintenance of medium- and high-expansion foam systems. Medium-expansion foam is defined as an aggregation of air-filled bubbles resulting from the mechanical expansion of a foam solution by air or other gases with a foam-to-solution volume ratio of 20:1 to 200:1. High-expansion foam has a foam-to-solution volume ratio of 200:1 to approximately 1,000:1.

Medium-expansion foam may be used on solid fuel and liquid fuel fires where some degree of in-depth coverage is necessary (e.g., for the total flooding of small, enclosed or partially enclosed volumes, such as engine test cells, transformer rooms, etc.). High-expansion foam is most suitable for large volume spaces in which fires could exist at various levels. For example, high-expansion foam can be used effectively against high-rack storage fires in enclosures, such as in underground passages, where it may be dangerous to send personnel to control fires involving liquefied natural gas (LNG) and liquefied petroleum gas (LPG), and to provide vapor dispersion control for LNG and ammonia spills. High-expansion foam is particularly suited for indoor fires in confined spaces, since it is highly susceptible to wind and lack of confinement effects.

NFPA 16 provides the minimum requirements for open-head deluge-type foam-water sprinkler systems and foam-water spray systems. The systems are especially applicable to the protection of most flammable liquid hazards and have been successfully used to protect aircraft hangars and truck loading racks.

[F] 904.8 Carbon dioxide systems. Carbon dioxide extinguishing systems shall be installed, maintained, periodically inspected and tested in accordance with NFPA 12 and their listing.

❖ NFPA 12 provides minimum requirements for the design, installation, testing, inspection, approval, operation and maintenance of carbon dioxide extinguishing systems.

Carbon dioxide extinguishing systems are useful in extinguishing fires in specific hazards or equipment in occupancies where an inert electrically nonconductive medium is essential or desirable and where cleanup of other extinguishing agents, such as dry-chemical residue, presents a problem. Carbon dioxide systems have satisfactorily protected the following: flammable liquids; electrical hazards, such as transformers, oil switches, rotating and electronic equipment; engines using gasoline and other flammable liquid fuels; ordinary combustibles, such as paper, wood and textiles and hazardous solids.

The building official has the authority to approve the type of carbon dioxide system to be installed. NFPA 12 defines four types of carbon dioxide systems: total flooding; local application; hand hose lines and standpipe and mobile supply systems. Only total flooding and local application systems are automatic suppression systems.

Total flooding systems may be used where there is a permanent enclosure surrounding the hazard that is adequate to enable the required concentration to be built up and maintained for the required period of time, which varies for different hazards. Examples of hazards that have been successfully protected by total flooding systems include rooms, vaults, enclosed machines, ducts, ovens, containers and the contents thereof.

Local application systems may be used for extinguishing surface fires in flammable liquids, gases and shallow solids where the hazard is not enclosed or the enclosure does not conform to the requirements for a total flooding system. Examples of hazards that have been successfully protected by local application systems include dip tanks, quench tanks, spray booths, oil-filled electric transformers and vapor vents.

[F] 904.9 Halon systems. Halogenated extinguishing systems shall be installed, maintained, periodically inspected and tested in accordance with NFPA 12A and their listing.

❖ NFPA 12A provides minimum requirements for the design, installation, testing, inspection, approval, operation and maintenance of Halon 1301 extinguishing systems. Halon 1301 fire-extinguishing systems are useful in specific hazards, in equipment or occupancies where an electrically nonconductive medium is essential or desirable and where cleanup of other extinguishing agents presents a problem.

Halon 1301 systems have satisfactorily protected the following: gaseous and liquid flammable materials; electrical hazards, such as transformers, oil switches and rotating equipment; engines using gasoline and other flammable fuels; ordinary combustibles, such as paper, wood and textiles and hazardous solids. Halon 1301 systems have also satisfactorily protected electronic computers, data processing equipment and control rooms.

The building official has the authority to approve the type of halogenated extinguishing system to be installed. NFPA 12A defines two types of halogenated extinguishing systems: total flooding and local application. Total flooding systems may be used where there is a fixed enclosure about the hazard that is adequate to enable the required Halon concentration to be built up and maintained for the required period of time to enable effective extinguishing of the fire. Total flooding systems may provide fire protection for rooms, vaults, enclosed machines, ovens, containers, storage tanks and bins.

Local application systems are used where there is not a fixed enclosure surrounding the hazard or where the fixed enclosure is not adequate to enable an extinguishing concentration to be built up and maintained in the space. Hazards that may be successfully protected by local application systems include dip tanks, quench tanks, spray booths, oil-filled electric transformers and vapor vents.

Two other considerations in selecting the proper extinguishing system are ambient temperature and the personnel hazards associated with the agent. The ambient temperature of the enclosure for a total flooding system must be above 70°F (21°C) for Halon 1301 systems. Special consideration must also be given to the use of Halon systems when the temperatures are in excess of 900°F (482°C), because Halon will readily decompose at such temperatures and the products of decomposition can be extremely irritating if inhaled, even in small amounts.

Halon 1301 total flooding systems must not be used in concentrations greater than 10 percent in normally occupied areas. Where egress cannot be accomplished within 1 minute, Halon 1301 total flooding systems must not be used in normally occupied areas with concentrations greater than 7 percent. Halon 1301 total flooding systems may be used with concentrations up to 15 percent if the area is not normally occupied, and egress can be accomplished from the area within 30 seconds.

The use of halogenated extinguishing systems has become a concern with respect to the potential environmental effects of Halon. Halongenated fire-extinguishing agents have been identified as a source of emissions resulting in the depletion of the stratospheric ozone layer and, in accordance with the Montreal protocol, the production of Halon ceased in January 1994; therefore, the supply of Halon is limited and new supplies of halogenated extinguishing agents will not be available in the future. Existing supplies of Halon can, however, continue to be used in existing, undischarged systems or to recharge discharged systems. This newfound need for Halon supplies has given rise to new industries geared to the banking, recycling and reclamation of existing Halon supplies. Alternative extinguishing agents to replace halogenated agents have been developed (see Section 904.10).

[F] 904.10 Clean-agent systems. Clean-agent fire-extinguishing systems shall be installed, maintained, periodically inspected and tested in accordance with NFPA 2001 and their listing.

❖ NFPA 2001 provides minimum requirements for the design, installation, testing, inspection and operation of clean-agent fire-extinguishing systems. A clean agent is an electrically nonconducting suppression agent that is volatile or gaseous at discharge and does not leave a residue on evaporation. Clean-agent fire-extinguishing systems are installed in locations that are enclosed and have the ability to seal all openings in the protected area on activation of the alarm to provide effective clean-agent concentrations. The clean-agent fire-extinguishing system should not be placed in locations that cannot be sealed unless tests are performed that show adequate concentrations can be developed and maintained. The two categories of clean agents are halocarbon compounds and inert gas agents. Halocarbon compounds include bromine, carbon, chlorine, fluorine, hydrogen and iodine. Halocarbon compounds suppress fire by a combination of breaking the chemical chain reaction of the fire, reducing the ambient oxygen supporting the fire and reducing the ambient temperature of the fire origin to reduce the propagation of the fire. Inert gas agents contain primary components consisting of helium, neon or argon, or a combination of both. Inert gases work by reducing the oxygen concentration around the fire origin to a level that does not support combustion.

The development of clean-agent fire-extinguishing systems was in response to the demise of Halon as an acceptable fire-extinguishing agent because of its deleterious effect on the environment. While the original hope for a Halon substitute was that these new clean agents could be directly and proportionally substituted for Halon agents in existing systems ("drop in" replacements), research has shown that clean agents are less efficient in extinguishing fires than are the Halons they were intended to replace. They require approximately 60 percent more agent by weight and volume in storage to do the same job. Additionally, the physical and chemical characteristics of clean agents differ sufficiently from Halon so as to require different nozzles in addition to the need for larger storage vessels. Existing piping systems should only be salvaged for use with clean agents if they are carefully evaluated and determined to be hydraulically compatible with the flow characteristics of the new agent.

This section also relies on strict adherence to the system manufacturer's design and installation instructions for code compliance. As with many of the alternative fire suppression systems covered in this chapter, clean agent systems are, for the most part, subjected by their manufacturers to a testing and listing program conducted by an approved testing agency. In such testing and listing programs, the clean agent is listed for use with specific equipment and equipment is listed for use with specific clean agents. The resultant listings include reference to the manufacturer's installation manuals, thereby providing the building official with another valuable resource for reviewing and approving clean-agent systems.

While clean agents have found a limited market for lo-

cal application uses, such as a replacement for Halon 1211 in portable fire extinguishers, their primary application is in total flooding systems and they are available in both engineered and preengineered configurations. Engineered clean-agent systems are specifically designed for protection of a particular hazard, whereas preengineered systems are designed to operate within predetermined limitations up to the noted maximums, thus allowing broader applicability to a variety of hazard applications. Total flooding systems are used where there is a fixed enclosure surrounding the hazard that is adequate to enable the required clean-agent concentration to build up and be maintained within the space in order to effect extinguishment. Such applications can include vaults, ovens, containers, tanks, computer rooms, paint lockers or enclosed machinery. In selecting the clean agent to be used in a given application, careful consideration must be given to whether the protected area is a normally occupied space, since different agents have different levels of concentration at which they may be a health hazard to occupants of the area.

The building official has the authority to approve the type of clean-agent system to be installed and should become familiar with the unique characteristics and hazards of clean-agent extinguishing systems using all available resources on the subject.

[F] 904.11 Commercial cooking systems. The automatic fire-extinguishing system for commercial cooking systems shall be of a type recognized for protection of commercial cooking equipment and exhaust systems of the type and arrangement protected. Preengineered automatic dry- and wet-chemical extinguishing systems shall be tested in accordance with UL 300 and listed and labeled for the intended application. Other types of automatic fire-extinguishing systems shall be listed and labeled for specific use as protection for commercial cooking operations. The system shall be installed in accordance with this code, its listing and the manufacturer's installation instructions. Automatic fire-extinguishing systems of the following types shall be installed in accordance with the referenced standard indicated, as shown:

1. Carbon dioxide extinguishing systems, NFPA 12.

2. Automatic sprinkler systems, NFPA 13.

3. Foam-water sprinkler system or foam-water spray systems, NFPA 16.

4. Dry-chemical extinguishing systems, NFPA 17.

5. Wet-chemical extinguishing systems, NFPA 17A.

> **Exception:** Factory-built commercial cooking recirculating systems that are tested in accordance with UL 197, and listed, labeled and installed in accordance with Section 304.1 of the *International Mechanical Code.*

❖ The history of commercial kitchen exhaust systems shows that the mixture of flammable grease and effluents carried by such systems and the potential for the cooking equipment to act as an ignition source contribute to a higher level of hazard for kitchen exhaust systems than is normally found in many other exhaust systems. Furthermore, fire in a grease exhaust duct can produce temperatures of 2,000°F (1093°C) or greater and heat radiating from the duct can ignite nearby combustibles. As a result, the code requires exhaust systems serving grease-producing equipment to include provisions for fire suppression to protect the cooking surfaces, hood, filters and the exhaust duct in order to confine a fire to the hood and duct system, thus reducing the likelihood of it spreading to the structure.

In addition to the general provisions of this section, five industry standards are referenced for the installation of fire-extinguishing systems protecting commercial food heat-processing equipment and kitchen exhaust systems. Design professionals should specify and design fire-extinguishing systems in accordance with these referenced standards. It should be noted that only the installation of fire-extinguishing systems is regulated by these references. Where pre-engineered automatic dry- and wet-chemical extinguishing systems are installed, they must be listed and labeled for the specific cooking operation and tested in accordance with UL 300. Design and construction requirements for the specific types of fire-extinguishing systems are found in the respective sections of the referenced standards.

Regulatory requirements for the approval and installation of fire-extinguishing systems are no different than the approval required for all mechanical equipment and appliances. As such, this section requires all extinguishing systems to be listed and labeled by an approved agency and installed in accordance with their listing and the manufacturers' installation instructions.

The exception allows factory-built commercial cooking recirculating systems to be installed if they have been tested and listed in accordance with UL 197. It is important that they be installed in accordance with the manufacturer's installation instructions to ensure the listing requirements are met. An improper installation could result in hazardous vapors being discharged back into the kitchen.

Commercial cooking recirculating systems consist of an electric cooking appliance and an integral or matched packaged hood assembly. The hood assembly consists of a fan, collection hood, grease filter, fire damper, fire-extinguishing system and air filter, such as an electrostatic precipitator. These systems are tested for fire safety and emissions. The grease vapor (condensible particulate matter) in the effluent at the system discharge is not allowed to exceed a concentration of 5.0 mg/m^3. Recirculating systems are not used with fuel-fired appliances because the filtering systems do not remove combustion products. Kitchens require ventilation in accordance with Chapter 4 of the IMC.

[F] 904.11.1 Manual system operation. A manual actuation device shall be located at or near a means of egress from the cooking area, a minimum of 10 feet (3048 mm) and a maximum of 20 feet (6096 mm) from the kitchen exhaust system. The manual actuation device shall be located a minimum of 4 feet (1219 mm) and a maximum of 5 feet (1524 mm) above the floor. The manual actuation shall require a maximum force of 40 pounds (178 N) and a maximum movement of 14 inches (356 mm) to actuate the fire suppression system.

Exception: Automatic sprinkler systems shall not be required to be equipped with manual actuation means.

❖ The manual device, usually a pull station, mechanically activates the suppression system. The typical system employs a mechanical circuit of cables under tension to hold the system in the armed (cocked) mode. Melting of a fusible link or actuation of a manual pull station causes the cable to lose tension that, in turn, actuates the discharge of the suppression agent. The manual actuation device must be readily and easily usable by the building occupants; therefore, the device must not require excessive force or range of movement to cause actuation.

Manual actuation is not required for automatic sprinkler systems because the typical system design will employ closed heads and wet system piping. A manual actuation valve would serve no purpose since sprinkler heads are already supplied with pressurized water and will only discharge water when the individual fusible elements open the heads.

[F] 904.11.2 System interconnection. The actuation of the fire suppression system shall automatically shut down the fuel or electrical power supply to the cooking equipment. The fuel and electrical supply reset shall be manual.

❖ The actuation of any fire suppression system must automatically shut off all sources of fuel or power to all cooking equipment located beneath the exhaust hood and protected by the suppression system. This requirement is intended to shut off all heat sources that could reignite or intensify a fire. Shutting off a fuel and power supply to cooking appliances will eliminate an ignition source and allow the cooking surfaces to cool down. This shutdown is accomplished with mechanical or electrical interconnections between the suppression system and a shutoff valve or switch located on the fuel or electrical supply. Common fuel shutoff valves include mechanical-type gas valves and electrical solenoid-type gas valves. Contractor-type switches or shunt-trip circuit breakers can be used for electrically heated appliances. The fuel or electric source must not be automatically restored after the suppression system has been actuated. Chemical-type fire-extinguishing systems have a limited duration for discharge and can discharge only once before recharge and reset; therefore, precautions must be taken to prevent a fire from reigniting. After a fire event, the fuel and power supply will be locked out, thereby preventing the operation of the appliances until all systems are again ready for operation. Fuel and power supply shutoff must be manually restored by resetting a mechanical linkage or holding (latching)-type circuit.

[F] 904.11.3 Carbon dioxide systems. When carbon dioxide systems are used, there shall be a nozzle at the top of the ventilating duct. Additional nozzles that are symmetrically arranged to give uniform distribution shall be installed within vertical ducts exceeding 20 feet (6096 mm) and horizontal ducts exceeding 50 feet (15 240 mm). Dampers shall be installed at either the top or the bottom of the duct and shall be arranged to operate automatically upon activation of the fire-extinguishing system. Where the damper is installed at the top of the duct, the top nozzle shall be immediately below the damper. Automatic carbon dioxide fire-extinguishing systems shall be sufficiently sized to protect against all hazards venting through a common duct simultaneously.

❖ This section requires specific design requirements for nozzle locations, dampers and ducts for carbon dioxide extinguishing systems that may be used to protect commercial cooking systems. These provisions are intended to supercede the more general provisions in NFPA 12 (see Section 102.4).

[F] 904.11.3.1 Ventilation system. Commercial-type cooking equipment protected by an automatic carbon dioxide-extinguishing system shall be arranged to shut off the ventilation system upon activation.

❖ The purpose of shutting down the ventilation system upon activation of the carbon dioxide extinguishing system is to maintain the desired concentration of carbon dioxide in order to suppress the fire. Leakage of gas from the protected area should be kept to a minimum. Where leakage is anticipated, additional quantities of carbon dioxide should be provided to compensate for any losses.

[F] 904.11.4 Special provisions for automatic sprinkler systems. Automatic sprinkler systems protecting commercial-type cooking equipment shall be supplied from a separate, readily accessible, indicating-type control valve that is identified.

❖ This section requires a separate control valve in the water line to the sprinklers protecting the cooking and ventilating system. The additional valve allows the flexibility to shut off the system due to repairs or for clean ups after sprinkler discharge without taking the entire system out of service.

[F] 904.11.4.1 Listed sprinklers. Sprinklers used for the protection of fryers shall be listed for that application and installed in accordance with their listing.

❖ Sprinklers specifically listed for such use must be used when protecting deep-fat fryers. These specially listed sprinklers utilize finer water droplets than standard spray sprinklers. The water spray lowers the temperature below a point where the fire can sustain itself and reduces the possibility of expanding the fire.

SECTION 905
STANDPIPE SYSTEMS

[F] 905.1 General. Standpipe systems shall be provided in new buildings and structures in accordance with this section. Fire hose threads used in connection with standpipe systems shall be approved and shall be compatible with fire department hose threads. The location of fire department hose connections shall be approved. In buildings used for high-piled combustible stor-

age, fire protection shall be in accordance with the *International Fire Code*.

❖ Standpipe systems are required in buildings to provide a quick, convenient water source for fire department use where hose lines would otherwise be impractical, such as in high-rise buildings. Standpipe systems can also be utilized prior to deployment of hose lines from fire department apparatus. The requirements for standpipes are based on practical requirements of typical fire-fighting operations and the nationally recognized standard, NFPA 14.

The threads on connections in which the fire department may connect a hose must be compatible with the fire department's hose threads (see commentary, Section 903.3.6). Chapter 23 of the IFC requires a Class I standpipe system in exit passageways of buildings utilized for high-piled storage. High-piled storage involves the solid piled, palletized or rack storage of combustible materials over 12 feet (3658 mm) in height or that contain high-hazard commodities over 6 feet (1829 mm) in height.

[F] 905.2 Installation standards. Standpipe systems shall be installed in accordance with this section and NFPA 14.

❖ This section requires the installation of all standpipe systems to be in accordance with the applicable provisions of NFPA 14 in addition to Section 905. NFPA 14 provides the minimum requirements for the installation of standpipe and hose systems for buildings and structures. The standard addresses additional requirements not in the code, such as pressure limitations, minimum flow rates, piping specifications, hose connection details, valves, fittings, hangers and the testing and inspection of standpipes. The periodic inspection, testing and maintenance of standpipe systems are required to be in accordance with NFPA 25.

Section 905 and NFPA 14 recognize three classes of standpipe systems: Class I, II or III. The type of system required depends on building height, building area, type of occupancy and the extent of automatic sprinkler protection. Section 905 also recognizes five types of standpipe systems: automatic dry, automatic wet, manual dry, manual wet and semiautomatic dry. The use of each type of system is limited to the building conditions and locations identified in Section 905.3. The classes and types of standpipe systems are defined in Section 902.1.

[F] 905.3 Required installations. Standpipe systems shall be installed where required by Sections 905.3.1 through 905.3.6 and in the locations indicated in Sections 905.4, 905.5 and 905.6. Standpipe systems are permitted to be combined with automatic sprinkler systems.

Exception: Standpipe systems are not required in Group R-3 occupancies as applicable in Section 101.2.

❖ Standpipe systems are required in buildings based on the occupancy, fire department accessibility and special provisions that may mandate manual fire suppression capability exceeding the capacity of a fire extinguisher. Standpipe systems are most commonly required for buildings that exceed the height threshold requirement in Section 905.3.1 or the area threshold requirement in Section 905.3.2. Specific occupancies such as covered mall buildings, stages and underground buildings, due to the nature of their use or occupancy, also require a standpipe system.

This section also clarifies that the standpipe system does not have to be separate from the sprinkler system, when provided. It is common practice in multistory buildings for the standpipe system risers to serve as risers for the automatic sprinkler systems.

In these instances, precautions need to be taken so that the operation of one system will not adversely affect the operation of the other system; therefore, control valves for the sprinkler system are to be installed where the sprinklers are connected to the standpipe riser at each floor level. This enables the standpipe system to remain operational, even if the sprinkler system is shut off at the floor control valve.

The exception recognizes that standpipe systems in Group R-3 occupancies would be of minimal value to the fire department and would send the wrong message to the occupants of a dwelling unit. In the case of multiple single-family dwellings, each dwelling unit in such buildings has a separate entrance and is separated by 1-hour fire-resistance-rated construction. These conditions permit ready access to fires and also provide for a degree of fire containment through compartmentation, which is not always present in other occupancies.

[F] 905.3.1 Building height. Class III standpipe systems shall be installed throughout buildings where the floor level of the highest story is located more than 30 feet (9144 mm) above the lowest level of fire department vehicle access, or where the floor level of the lowest story is located more than 30 feet (9144 mm) below the highest level of fire department vehicle access.

Exceptions:

1. Class I standpipes are allowed in buildings equipped throughout with an automatic sprinkler system in accordance with Section 903.3.1.1 or 903.3.1.2.

2. Class I manual standpipes are allowed in open parking garages where the highest floor is located not more than 150 feet (45 720 mm) above the lowest level of fire department vehicle access.

3. Class I manual dry standpipes are allowed in open parking garages that are subject to freezing temperatures, provided that the hose connections are located as required for Class II standpipes in accordance with Section 905.5.

4. Class I standpipes are allowed in basements equipped throughout with an automatic sprinkler system.

❖ Given the available manpower on the fire department vehicle, standard fire-fighting operations and standard hose sizes, a 30-foot (9,144 mm) vertical distance is generally considered the maximum height at which a typical fire department engine company can practically

and readily extend its hose lines. Thus, the maximum vertical travel (height) threshold is based on the time it would take a typical fire department engine (pumper) company to manually suppress a fire. The standpipe connection reduces the time needed for the fire department to extend hose lines up or down stairways to advance and apply water to the fire. As such, a minimum Class III standpipe system is required.

With respect to the height of the building, the threshold is measured from the level at which the fire department can gain access to the building directly from its vehicle. In the case of floor levels above grade, the measurement is made from the lowest level of fire department vehicle access to the highest floor level above [see Figure 905.3.1(1)]. If a building contains floor levels below the level of fire department vehicle access, the measurement is made from the highest level of fire department vehicle access to the lowest floor level. In cases where a building has more than one level of fire department vehicle access, the most restrictive measurement is utilized since it is not known at which level the fire department will access the building. In other words, the vertical distance is to be measured from the more restrictive level of fire department vehicle access to the level of the highest (or lowest, if below) floor [see Figure 905.3.1(2)].

It should be noted that the threshold based on the height of the building is independent of the occupancy of the building, the area of the building or the presence of an automatic sprinkler system. This is based on the universal need to be able to provide a water supply for fire suppression in any building and on the limitations of the physical effort necessary to extend hose lines vertically.

[F] 905.3.2 Group A. Class I automatic wet standpipes shall be provided in nonsprinklered Group A buildings having an occupant load exceeding 1,000 persons.

Exceptions:

1. Open-air-seating spaces without enclosed spaces.

2. Class I automatic dry and semiautomatic dry standpipes or manual wet standpipes are allowed in buildings where the highest floor surface used for human occupancy is 75 feet (22 860 mm) or less above the lowest level of fire department vehicle access.

❖ The main concern in assembly occupancies with a high occupant load is evacuation. Many occupants may not be familiar with their surroundings and the egress arrangement in the building. This section also assumes the building is not sprinklered; therefore, control and suppression of the fire is left to the fire department.

Exception 1 exempts open-air seating without enclosed spaces, such as grandstands and bleachers. In such occupancies, a buildup of smoke and hot gases is not possible because these structures are open to the atmosphere.

Exception 2 states that in lieu of a Class I automatic wet standpipe, automatic-dry and semiautomatic dry Class I standpipes are permitted in buildings that are not considered to be a high rise. Class III standpipes may be installed as an alternate to the required Class I standpipes where occupant-use stations are desired.

[F] 905.3.3 Covered mall buildings. A covered mall building shall be equipped throughout with a standpipe system where required by Section 905.3. Covered mall buildings not required to be equipped with a standpipe system by Section 905.3 shall be equipped with Class I hose connections connected to a system sized to deliver 250 gallons per minute (946.4 L/min.) at the most hydraulically remote outlet. Hose connections shall be provided at each of the following locations:

1. Within the mall at the entrance to each exit passageway or corridor.

2. At each floor-level landing within enclosed stairways opening directly on the mall.

3. At exterior public entrances to the mall.

❖ Covered mall buildings are only required to have a standpipe system if Section 905.3.1 requires such features. If standpipes are not required, Class I hose connections are still required at key locations, such as en-

trances to exit passageways (see commentary, Section 402.8).

[F] 905.3.4 Stages. Stages greater than 1,000 square feet in area (93 m²) shall be equipped with a Class III wet standpipe system with 1.5-inch and 2.5-inch (38 mm and 64 mm) hose connections on each side of the stage.

Exception: Where the building or area is equipped throughout with an automatic sprinkler system, the hose connections are allowed to be supplied from the automatic sprinkler system and shall have a flow rate of not less than that required by NFPA 14 for Class III standpipes.

❖ Because of the potentially large fuel load and three-dimensional aspect of the fire hazard associated with stages greater than 1,000 square feet (93 m²) in area, Class III standpipes are required on each side of such stages. The standpipes are required to be equipped with 1¹/₂-inch (38 mm) and 2¹/₂-inch (64 mm) hose connections. The 1¹/₂-inch (38 mm) connection is for the hose requirement in Section 905.3.4.1. The 2¹/₂-inch (64 mm) connection is to provide greater flexibility to the fire department during its fire-fighting operations.

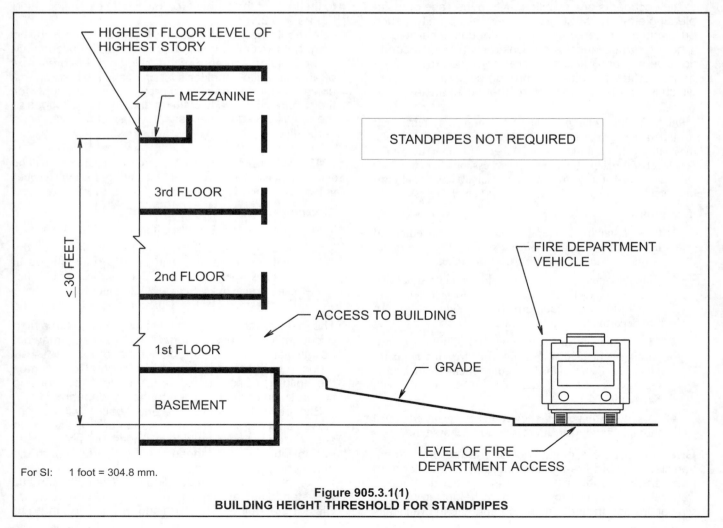

For SI: 1 foot = 304.8 mm.

Figure 905.3.1(1)
BUILDING HEIGHT THRESHOLD FOR STANDPIPES

The exception recognizes the benefit of the building or area being sprinklered. As noted, the area that is sprinklered may only be the area containing the stage as required by Section 410.6.

[F] 905.3.4.1 Hose and cabinet. The 1.5-inch (38 mm) hose connections shall be equipped with sufficient lengths of 1.5-inch (38 mm) hose to provide fire protection for the stage area. Hose connections shall be equipped with an approved adjustable fog nozzle and be mounted in a cabinet or on a rack.

❖ The $1^1/_2$-inch (38 mm) standpipe hose is required to be provided for stages greater than 1,000 square feet (93 m²) in area for stage personnel who have been trained to use it. The length of hose required to be provided is a function of the size and configuration of the stage. This includes by definition the entire performance area and adjacent backstage and support areas not fire separated from the performance area. The effective reach of the fire stream from the fog nozzle is a function of the available water supply, in particular, the pressure. Fog nozzles typically require 100 pounds per square inch (psi) (689 kPa) for optimum performance.

[F] 905.3.5 Underground buildings. Underground buildings shall be equipped throughout with a Class I automatic wet or manual wet standpipe system.

❖ Underground buildings present unique hazards to life safety due to their isolation and inaccessibility. Additional fire protection measures are required for safe egress for the occupants, due to the lack of exterior fire suppression and rescue operations (see commentary, Section 405).

[F] 905.3.6 Helistops and heliports. Buildings with a helistop or heliport that are equipped with a standpipe shall extend the standpipe to the roof level on which the helistop or heliport is located in accordance with Section 1107.5 of the *International Fire Code*.

❖ If a building already has a standpipe (required or not) and also contains a helistop or heliport, the standpipe is required to extend to the roof level. Essentially, this is to make use of the fact that a protection feature is already available. Section 1107.5 of the IFC requires a $2^1/_2$-inch (63 mm) standpipe outlet to be within 150 feet (45 720 mm) of all portions of the heliport or helistop and be either Class I or III.

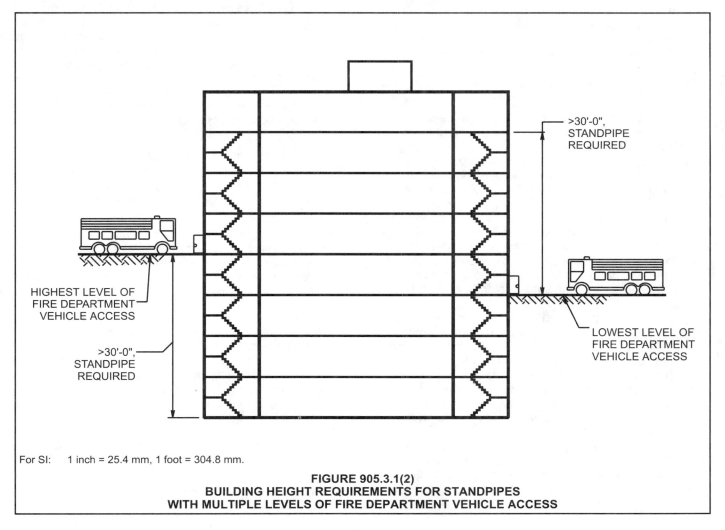

For SI: 1 inch = 25.4 mm, 1 foot = 304.8 mm.

FIGURE 905.3.1(2)
BUILDING HEIGHT REQUIREMENTS FOR STANDPIPES
WITH MULTIPLE LEVELS OF FIRE DEPARTMENT VEHICLE ACCESS

[F] 905.4 Location of Class I standpipe hose connections.
Class I standpipe hose connections shall be provided in all of the following locations:

1. In every required stairway, a hose connection shall be provided for each floor level above or below grade. Hose connections shall be located at an intermediate floor level landing between floors, unless otherwise approved by the building official.

2. On each side of the wall adjacent to the exit opening of a horizontal exit.

3. In every exit passageway at the entrance from the exit passageway to other areas of a building.

4. In covered mall buildings, adjacent to each exterior public entrance to the mall and adjacent to each entrance from an exit passageway or exit corridor to the mall.

5. Where the roof has a slope less than four units vertical in 12 units horizontal (33.3-percent slope), each standpipe shall be provided with a hose connection located either on the roof or at the highest landing of stairways with stair access to the roof. An additional hose connection shall be provided at the top of the most hydraulically remote standpipe for testing purposes.

6. Where the most remote portion of a nonsprinklered floor or story is more than 150 feet (45 720 mm) from a hose connection or the most remote portion of a sprinklered floor or story is more than 200 feet (60 960 mm) from a hose connection, the building official is authorized to require that additional hose connections be provided in approved locations.

❖ Hose connections are required for the fire department to make use of the standpipe system. Since the fire department will typically access the building using the stairways, and most fire departments do not permit entry to the fire floor without an operating hose line, a hose connection is to be provided for each floor level of each enclosed stairway.

Item 1 specifies that the hose connections are to be located at intermediate landings between floors. This reduces congestion at the stairway door and may reduce the hose lay distance. Hose connections, however, are still permitted at each floor level of the exit stair instead of at the intermediate landing if approved by the building official.

Since horizontal exits are also primary entrances to the fire floor, hose connections must also be provided at each horizontal exit (see Item 2). The construction of the fire separation assembly used as the horizontal exit will provide protection to the fire fighters while connecting to the standpipe system. The hose connections are to be located on each side of the horizontal exit to enable fire fighters to be in a protected area, regardless of the location of the fire.

An exit passageway, in a building required to have a standpipe system, is typically utilized as an extension of a required exit stairway. As such, the exit passageway is used for fire-fighting staging operations the same as an exit stair (see Item 3).

In covered mall buildings, hose connections are required at the entrance to exits and the exterior entrances to allow fire personnel to have a support line as soon as they enter the building (see Item 4).

Depending on the slope of the roof, Item 5 requires a hose connection to be available to aid in the suppression of roof fires either due to the nature of the construction of the roof or the equipment on the roof and for exposure protection.

Hose connections located in each exit stairway are based on the travel distances permitted in Table 1015.1. It is recognized that most fire departments carry standpipe hose packs with 150 feet (45 720 mm) of hose or possibly 100 feet (30 480 mm) of hose with an additional 50-foot (15 240 mm) section that could be easily connected. With the typical travel distance permitted in nonsprinklered buildings of 200 feet (60 960 mm), reasonable coverage is provided when the effective reach of a fire stream is considered. Depending on the arrangement of the floor, however, all areas may not be effectively protected. While this situation could easily be corrected by locating additional hose connections on the floor, it is recognized that such connections may be rarely used because of the difficulty in identifying their location during a fire and the fact that most fire departments require an operational hose line prior to entering the fire floor. Since travel distances may be greater in sprinklered buildings, the problem is increased but the need for prompt manual suppression is reduced by the presence of the sprinkler system. As such, Item 6 provides the building official the authority to require additional hose connections if needed.

[F] 905.4.1 Protection. Risers and laterals of Class I standpipe systems not located within an enclosed stairway or pressurized enclosure shall be protected by a degree of fire resistance equal to that required for vertical enclosures in the building in which they are located.

Exception: In buildings equipped throughout with an approved automatic sprinkler system, laterals that are not located within an enclosed stairway or pressurized enclosure are not required to be enclosed within fire-resistance-rated construction.

❖ To minimize the potential for damage to the standpipe systems from a fire, the risers and laterals must be located in an enclosure having the same fire-resistance rating as required for a vertical or shaft enclosure within the building. The required fire-resistance rating for the enclosure can be determined as detailed in Section 707.4. The enclosure is not required if the building is equipped throughout with an approved automatic sprinkler system. The potential for damage to the standpipe system is minimized by the protection provided by the sprinkler system. The automatic sprinkler system may be either an NFPA 13 or 13R system depending on what was permitted for the building occupancy.

[F] 905.4.2 Interconnection. In buildings where more than one standpipe is provided, the standpipes shall be interconnected in accordance with NFPA 14.

❖ In cases where there are multiple Class I standpipe risers, such risers must be supplied from and interconnected to a common supply line. The use of the required fire department connection should be such that it will serve all of the sprinklers or standpipes in the building.

[F] 905.5 Location of Class II standpipe hose connections. Class II standpipe hose connections shall be accessible and located so that all portions of the building are within 30 feet (9144 mm) of a nozzle attached to 100 feet (30 480 mm) of hose.

❖ Sections 905.5.1 through 905.5.3 specify the requirements for Class II standpipe hose connections. Class II standpipe systems are primarily intended for use by the building occupants.

[F] 905.5.1 Groups A-1 and A-2. In Group A-1 and A-2 occupancies with occupant loads of more than 1,000, hose connections shall be located on each side of any stage, on each side of the rear of the auditorium, on each side of the balcony and on each tier of dressing rooms.

❖ Due to the high occupant load density in Group A-1 and A-2 occupancies, providing additional means for controlling fires in their initial stage is important to enable prompt evacuation of the building. This section is independent of the Class I standpipe requirement for stages based on square footage as indicated in Section 905.3.4.

[F] 905.5.2 Protection. Fire-resistance-rated protection of risers and laterals of Class II standpipe systems is not required.

❖ Class II standpipe systems are normally not located in exit stairways. As such, standpipe hose connections are located near the protected area to allow quick access.

[F] 905.5.3 Class II system 1-inch hose. A minimum 1-inch (25 mm) hose shall be permitted to be used for hose stations in light-hazard occupancies where investigated and listed for this service and where approved by the building official.

❖ This section permits the use of a 1-inch (25 mm) listed hose as an alternative to a 1 ½-inch (38 mm) hose, subject to the approval of the building official. This alternative is limited to light-hazard occupancies, such as office buildings and certain assembly occupancies that tend to have lower fuel loads.

[F] 905.6 Location of Class III standpipe hose connections. Class III standpipe systems shall have hose connections located as required for Class I standpipes in Section 905.4 and shall have Class II hose connections as required in Section 905.5.

❖ Class III standpipe systems that have both a 2 ½-inch (64 mm) and a 1½-inch (38 mm) hose connection must comply with the applicable requirements of Sections 905.4, 905.5 and 905.6.

[F] 905.6.1 Protection. Risers and laterals of Class III standpipe systems shall be protected as required for Class I systems in accordance with Section 905.4.1.

❖ Since Class III standpipe systems are intended for use by fire-suppression personnel, they must be located in construction equivalent to the vertical or shaft enclosure requirements of the building with respect to fire resistance (see commentary, Section 905.4.1).

[F] 905.6.2 Interconnection. In buildings where more than one Class III standpipe is provided, the standpipes shall be interconnected at the bottom.

❖ As indicated in Section 905.4.2 for Class I standpipe systems, multiple standpipe risers should be interconnected with a common supply line. An indicating valve is typically installed at the base of each riser to permit individual risers to be taken out of service without affecting the water supply or the operation of other standpipe risers.

[F] 905.7 Cabinets. Cabinets containing fire-fighting equipment such as standpipes, fire hoses, fire extinguishers or fire department valves shall not be blocked from use or obscured from view.

❖ Cabinets must be readily visible and accessible at all times. Sections 905.7.1 and 905.7.2 contain additional criteria for the construction and identification of the cabinets. Where cabinets are located in fire-resistance-rated assemblies, the integrity of the assembly must be maintained.

[F] 905.7.1 Cabinet equipment identification. Cabinets shall be identified in an approved manner by a permanently attached sign with letters not less than 2 inches (51 mm) high in a color that contrasts with the background color, indicating the equipment contained therein.

Exceptions:

1. Doors not large enough to accommodate a written sign shall be marked with a permanently attached pictogram of the equipment contained therein.

2. Doors that have either an approved visual identification clear glass panel or a complete glass door panel are not required to be marked.

❖ This section specifies the minimum criteria to make the signs readily visible. Different color combinations may be approved by the building official provided the color contrast between the letters and the background is vivid enough to make the sign visible at an approved distance. The exceptions address alternatives to letter signage, provided the cabinet is still conspicuously identified or the contents readily visible.

[F] 905.7.2 Locking cabinet doors. Cabinets shall be unlocked.

Exceptions:

1. Visual identification panels of glass or other approved transparent frangible material that is easily broken and allows access.

2. Approved locking arrangements.

3. Group I-3.

❖ Ready access to all fire-fighting equipment in the cabinet is essential. The exceptions, however, recognize the need to lock cabinets for security reasons and to prevent theft or vandalism.

[F] 905.8 Dry standpipes. Dry standpipes shall not be installed.

Exception: Where subject to freezing and in accordance with NFPA 14.

❖ Wet standpipe systems are preferred, since they tend to be the most reliable type of standpipe system; therefore, dry standpipes are prohibited unless subject to freezing. For example, Class I manual standpipe systems, which do not have a permanent water supply, are permitted in open parking structures. This recognizes that open parking structures are not heated and that most fires are limited to the vehicle of origin. The use of any dry standpipe system in lieu of a wet standpipe should take into consideration the inherent response time with respect to the occupancy characteristics of the building.

[F] 905.9 Valve supervision. Valves controlling water supplies shall be supervised in the open position so that a change in the normal position of the valve will generate a supervisory signal at the supervising station required by Section 903.4. Where a fire alarm system is provided, a signal shall also be transmitted to the control unit.

Exceptions:

1. Valves to underground key or hub valves in roadway boxes provided by the municipality or public utility do not require supervision.

2. Valves locked in the normal position and inspected as provided in this code in buildings not equipped with a fire alarm system.

❖ Similar to sprinkler systems, all water control valves for standpipe systems are required to be electrically supervised as a means of determining that the system is operational (see commentary, Section 903.4).

Exception 1 recognizes that underground key or hub valves in roadway boxes are not normally supervised or required to be supervised whether the building contains a standpipe system or an automatic sprinkler system.

Exception 2 does not require the control valves for the standpipes to be electrically monitored provided they are locked in the normal position and a fire alarm system is not provided in the building. When a fire alarm system is installed, control valves for the standpipes should be electrically monitored and tied into the supervision required for the fire alarm system.

[F] 905.10 During construction. Standpipe systems required during construction and demolition operations shall be provided in accordance with Section 3311.

❖ In accordance with Section 3311.1, not less than one standpipe is required during construction of buildings four stories or more in height. Standpipe systems must be operable during construction and demolition operations to assist in any potential fire condition (see commentary, Section 3311).

SECTION 906
PORTABLE FIRE EXTINGUISHERS

[F] 906.1 General. Portable fire extinguishers shall be provided in occupancies and locations as required by the *International Fire Code.*

❖ Portable fire extinguishers need to be provided where they can contribute to the protection of the occupants due to evacuation difficulties of the occupancy or the specific hazard within the area. Portable fire extinguishers provide occupants with an opportunity to suppress a fire in its incipient stage. In order to be effective, personnel must be properly trained in the use of portable fire extinguishers.

Section 906.1 requires portable fire extinguishers to be provided in occupancies and locations as required by the IFC. As such, Section 906 of the IFC specifies occupancy conditions that require portable fire extinguishers. In most cases, the need for portable fire extinguishers is dependent on the capability of the occupants to evacuate a building and the hazardous nature of the building contents. The location, size and hazard compatibility of portable fire extinguishers must be in accordance with the applicable provisions in the IFC and NFPA 10. For additional requirements, see Section 906 of the IFC.

SECTION 907
FIRE ALARM AND DETECTION SYSTEMS

[F] 907.1 General. This section covers the application, installation, performance and maintenance of fire alarm systems and their components.

❖ Fire alarm systems, which typically include manual fire alarm systems and automatic fire detection systems, are required to be installed in accordance with the provisions of Section 907 and NFPA 72.

Manual fire alarm systems are provided in buildings to limit fire casualties and property losses. Fire alarm systems do this by promptly notifying the occupants of the building of an emergency condition, which increases the time available for evacuation. Similarly, when fire alarm systems are required to be supervised, the fire department will be promptly notified and its response time relative to the onset of the fire incident will be reduced.

Automatic fire detection systems are required under certain conditions to increase the likelihood that fire is detected and occupants are alerted in a timely manner. The detection system is a system of devices and associated hardware that activate the alarm system. The automatic detecting devices are to be smoke detectors, unless a condition exists mandating the use of a different type of detector.

[F] 907.1.1 Construction documents. Construction documents for fire alarm systems shall be submitted for review and approval prior to system installation. Construction documents shall include, but not be limited to, all of the following:

1. A floor plan which indicates the use of all rooms.
2. Locations of alarm-initiating and notification appliances.
3. Alarm control and trouble signaling equipment.
4. Annunciation.
5. Power connection.
6. Battery calculations.
7. Conductor type and sizes.
8. Voltage drop calculations.
9. Manufacturers, model numbers and listing information for equipment, devices and materials.
10. Details of ceiling height and construction.
11. The interface of fire safety control functions

❖ Construction documents for all fire alarm systems must be submitted for review to determine compliance with the code and NFPA 72. First, the floor plan indicating the use of each space is required. In terms of the actual alarm system, construction documents are to show the location and number of all alarm-initiating devices and alarm-notification appliances, and also provide a description of all equipment to be used, proposed zoning, location of the control panel(s) and annunciator(s) and a complete sequence of operation for the system. All of the information required by this section may not be available during the design stage and initial permit process. Submission of more detailed shop drawings may be required. In order to facilitate the review process, the building official may elect to develop a checklist beyond what is listed in Section 907.1.1 stating the information required to complete a thorough review.

[F] 907.1.2 Equipment. Systems and their components shall be listed and approved for the purpose for which they are installed.

❖ The components of the fire alarm system must be approved for use in such a system. NFPA 72 requires all devices, combinations of devices, appliances and equipment to be labeled for the purposes for which they are used. The testing agency will test the components for use in various types of systems and stipulate the use of the component on the label. Evidence of labeling of the system components should be submitted with the shop construction documents.

In some instances, the entire system may be labeled.

At least one major testing agency, Underwriters Laboratories Inc. (UL) has a program whereby alarm installation and service companies are issued a certificate and become listed by UL as being qualified to design, install and maintain local, auxiliary, remote station or proprietary fire alarm systems. Such listed companies may then issue a certificate that the system is in compliance with NFPA 72. Terms of the company's certification by UL include the company being responsible for keeping accurate system documentation, including as-built record drawings, acceptance test records and complete maintenance records on a given system. The company is also responsible for the required periodic inspection and testing of the system under contract with the owner. A similar program has been available for many years for central station alarm service, whereas this program is relatively new to the industry. While this company and system listing program is not required by the code or NFPA 72, it can be a valuable tool for the building official in determining compliance with the referenced standard if it is provided on a given system.

Another issue that must be considered is the compatibility of the system components as required by NFPA 72. The labeling of system components discussed above should include any compatibility restrictions for components. Compatibility is primarily an issue of the ability of smoke detectors and fire alarm control panels (FACPs) to function properly when interconnected and affects the two-wire type of smoke detectors, which obtain their operating power over the same pair of wires over which signals are transmitted to the FACP (i.e., the control unit initiating device circuits). Laboratories will test for such compatibility either by actual testing or by reviewing the circuit parameters of both the detector and the FACP. It is generally true that if both the two-wire detector and the FACP are of the same brand, there should not be a compatibility problem. Nevertheless, the building official must be satisfied that the components are listed as being compatible. Failure to comply with the compatibility requirements of NFPA 72 can lead to system malfunction or failure when it may be needed the most.

[F] 907.2 Where required. An approved manual, automatic or manual and automatic fire alarm system shall be provided in accordance with Sections 907.2.1 through 907.2.23. Where automatic sprinkler protection, installed in accordance with Section 903.3.1.1 or 903.3.1.2, is provided and connected to the building fire alarm system, automatic heat detection required by this section shall not be required. An approved automatic fire detection system shall be installed in accordance with the provisions of this code and NFPA 72. Devices, combinations of devices, appliances and equipment shall comply with Section 907.1.2. The automatic fire detectors shall be smoke detectors, except that an approved alternative type of detector shall be installed in spaces such as boiler rooms where, during normal operation, products of combustion are present in sufficient quantity to actuate a smoke detector.

❖ This section specifies the occupancies or conditions that require some form of fire alarm system, either a manual fire alarm system (manual fire alarm boxes) or an automatic fire detection system (smoke and heat detectors).

Manual fire alarm systems are required to be installed in certain occupancies depending on the number of occupants, capabilities of the occupants and height of the building. An automatic fire detection system is required to be installed in those occupancies and conditions in which the need to detect the fire is essential to evacuation or protection of the occupants. The requirements for automatic smoke detection are generally based on the evacuation difficulty of the occupants and whether the occupancy provides sleeping accommodations.

This section also states that automatic heat detection is not required when buildings are fully sprinklered in accordance with NFPA 13 or 13R. It should be made clear that the presence of a sprinkler system exempts areas where a heat detector is permitted to be installed in lieu of a smoke detector, such as in storage or furnace rooms. The sprinkler head in this case essentially acts as a heat detection device.

Automatic fire detection systems are to be installed in accordance with the code and NFPA 72. NFPA 72 provides minimum requirements for the performance of automatic fire detectors to provide timely warning for purposes of life safety and property protection. The standard covers minimum performance, location, mounting, testing and maintenance requirements of automatic fire detectors. Smoke detectors must be utilized, except when such use would result in unwanted alarms or when a smoke detector may not provide the desired detection. The manufacturer's literature will identify the limitations on the use of smoke detectors, including environmental conditions such as humidity, temperature and airflow.

Only certain occupancies are required to have either a manual fire alarm or automatic fire detection system installed (see Figure 907.2). The necessity of either system is determined by the number of occupants and the height of the building. Figure 907.2 contains the conditions that require either system to be installed in a building. Note the requirements of Section 907.2 do not specifically state that the systems are required to be installed and maintained throughout buildings, only in the area that contains the occupancies.

In summary, the fire alarm/detection system contained in mixed-occupancy buildings is limited to the fire area that requires the system. Buildings that have more than one occupancy and do not contain more than one fire area are identified as nonseparated mixed-use buildings. A nonseparated mixed-use building that contains one fire area that requires a fire alarm/detection system is required to have the system extended throughout the building in accordance with Section 302.3.1.

Figure 907.2 summarizes the threshold requirements for when a manual fire alarm system or automatic fire detection system is required based on the occupancy group. Sections 907.2.11 through 907.2.23 contain additional requirements for fire alarm systems depending on special occupancy conditions such as atriums, high-rise buildings or covered mall buildings.

[F] 907.2.1 Group A. A manual fire alarm system shall be installed in accordance with NFPA 72 in Group A occupancies having an occupant load of 300 or more. Portions of Group E occupancies occupied for assembly purposes shall be provided with a fire alarm system as required for the Group E occupancy.

> **Exception:** Manual fire alarm boxes are not required where the building is equipped throughout with an automatic sprinkler system and the notification appliances will activate upon sprinkler water flow.

❖ Group A occupancies are typically occupied by a significant number of people who are not familiar with their surroundings. As such, a manual fire alarm system is required in all Group A occupancies with an occupant load of 300 or more to aid in the prompt evacuation of the occupants, especially in nonsprinklered buildings.

The exception allows the omission of manual fire alarm boxes in buildings equipped throughout with an automatic sprinkler system, provided activation of the sprinkler system will activate the building evacuation alarms associated with the manual fire alarm system.

FIGURE 907.2. See page 9-47.

❖ This section also permits assembly-type areas in Group E occupancies to comply with the provisions of Section 907.2.3 in lieu of the requirements of this section. A typical high school contains many areas used for assembly purposes, such as a gymnasium, cafeteria, auditorium or library; however, they all exist to serve the main function as an educational facility (see Section 303.1).

[F] 907.2.1.1 System initiation in Group A occupancies with an occupant load of 1,000 or more. Activation of the fire alarm in Group A occupancies with an occupant load of 1,000 or more shall initiate a signal using an emergency voice/alarm communications system in accordance with NFPA 72.

> **Exception:** Where approved, the prerecorded announcement is allowed to be manually deactivated for a period of time, not to exceed 3 minutes, for the sole purpose of allowing a live voice announcement from an approved, constantly attended location.

❖ Due to the unfamiliarity of occupants in large assembly buildings with respect to their surroundings and egress arrangements, the required manual fire alarm system, upon activation, must initiate a signal from an emergency voice/alarm communication system (see commentary, Section 907.2.12.2).

The exception allows the automatic alarm signal to be overridden, provided the live voice instructions do not exceed 3 minutes. While this section does not specifically require a fire command center, the location from which the live voice announcement originates should be constantly attended and approved by the building official.

Manual Fire Alarms	
Occupancy	**Threshold**
Assembly (A-1, A-2, A-3, A-4, A-5)	All with occupant load $\geq$ 300
Business (B)	Total occupant load $\geq$ 500 or occupant load, or $>$ 100 above/ below level of exit discharge
Educational (E)	$>$ 50 occupants (several exceptions for manual fire alarm box placement)
Factory (F-1, F-2)	$\geq$ 2 stories with occupant load $\geq$ 500 above /below level of exit discharge
High hazard (H-1, H-2, H-3, H-4, H-5)	H-5 occupancies and occupancies used for manufacture of organic coatings
Institutional (I-1, I-2, I-3, I-4)	All
Mercantile (M)	Total occupant load $\geq$ 500 or occupant load or $>$ 100 above/ below level of exit discharge
Hotels (R-1)	All unsprinklered buildings except $\leq$ 2 stories with sleeping units having exit directly to exterior sleeping units
Apartments (R-2)	Unsprinklered 3 stories or 1 story below level of exit discharge or $>$ 16 units without exits directly to exterior
Automatic Fire Detection	
Occupancy	**Threshold**
High hazard (H-1, H-2, H-3, H-4, H-5)	Highly toxic gases, organic peroxides and oxidizers in accordance with IFC
Institutional (I-1, I-2, I-3, I-4)	All
Hotels (R-1)	All except building without interior corridors and guestrooms having exit directly to exterior sleeping units

FIGURE 907.2
SUMMARY OF MANUAL FIRE ALARM AND AUTOMATIC FIRE DETECTION THRESHOLDS

[F] 907.2.1.2 Emergency power. Emergency voice/alarm communications systems shall be provided with an approved emergency power source.

❖ Due to the critical nature of the emergency voice/alarm communication system to aid in evacuating the building, the system must be connected to an approved emergency power source in accordance with Chapter 27 and NFPA 70.

[F] 907.2.2 Group B. A manual fire alarm system shall be installed in Group B occupancies having an occupant load of 500 or more persons or more than 100 persons above or below the lowest level of exit discharge.

Exception: Manual fire alarm boxes are not required where the building is equipped throughout with an automatic sprinkler system and the alarm notification appliances will activate upon sprinkler water flow.

❖ Group B occupancies generally involve individuals or groups of people in separate office areas. As a result, the occupants are not necessarily aware of what is going on in other parts of the building. Group B buildings with large occupant loads, even in single-story buildings

or where a substantial number of occupants are above or below the level of exit discharge, increase the difficulty in the ability to alert the occupants of a fire condition. This is especially true in nonsprinklered buildings with given occupant load thresholds.

The exception does not omit the fire alarm system, but rather permits it to be initiated automatically by the sprinkler water flow switch(es) in lieu of the manual fire alarm boxes.

[F] 907.2.3 Group E. A manual fire alarm system shall be installed in Group E occupancies. When automatic sprinkler systems or smoke detectors are installed, such systems or detectors shall be connected to the building fire alarm system.

Exceptions:

1. Group E occupancies with an occupant load of less than 50.

2. Manual fire alarm boxes are not required in Group E occupancies where all the following apply:

   2.1. Interior corridors are protected by smoke detectors with alarm verification.

2.2. Auditoriums, cafeterias, gymnasiums and the like are protected by heat detectors or other approved detection devices.

2.3. Shops and laboratories involving dusts or vapors are protected by heat detectors or other approved detection devices.

2.4. Off-premises monitoring is provided.

2.5. The capability to activate the evacuation signal from a central point is provided.

2.6. In buildings where normally occupied spaces are provided with a two-way communication system between such spaces and a constantly attended receiving station from where a general evacuation alarm can be sounded, except in locations specifically designated by the building official.

❖ Group E occupancies involve groups of people distributed throughout a number of classrooms or small rooms. Occupants in one area of the building would not necessarily be aware of an emergency in another part of the building unless a system is provided to alert them. Due to the age and maturity of the occupants, more time may be needed to safely evacuate the building. The requirement for a manual fire alarm system in Group E occupancies is not dependent on the location of the level of exit discharge.

Exception 1 exempts Group E occupancies from requiring a fire alarm system with an occupant load of less than 50. This would exempt small day care centers that serve children older than 2 $^1/_2$ years of age or a small Sunday school classroom at a church (see Section 305.2).

Exception 2 exempts manual fire alarm boxes in interior corridors, laboratories, auditoriums, cafeterias, gymnasiums and similar spaces depending on the use of heat/smoke detectors and the extent of supervision. The applicability of Exception 2 is independent of whether an automatic sprinkler system is provided. In accordance with Section 903.2.2, an automatic sprinkler system is required throughout all Group E fire areas in excess of 20,000 square feet (1858 m²). If an automatic sprinkler system or smoke detectors are installed, however, they must be connected to the building fire alarm system.

[F] 907.2.4 Group F. A manual fire alarm system shall be installed in Group F occupancies that are two or more stories in height and have an occupant load of 500 or more above or below the lowest level of exit discharge.

Exception: Manual fire alarm boxes are not required if the building is equipped throughout with an automatic sprinkler system and the notification appliances will activate upon sprinkler water flow.

❖ This section is only intended to apply to large multistory manufacturing facilities. As such, a manual fire alarm system would only be required if the building were at least two stories in height and had 500 or more occupants above or below the level of exit discharge. An unlimited area two-story Group F occupancy complying with Section 507.3 would be indicative of an occupancy requiring a manual fire alarm system.

Buildings in compliance with Section 507.3 and large manufacturing facilities in general, however, are required to be fully sprinklered and would thus be eligible for the exception. The exception does not omit the fire alarm system but rather permits it to be initiated automatically by the sprinkler system water flow switch(es) in lieu of the manual fire alarm boxes.

[F] 907.2.5 Group H. A manual fire alarm system shall be installed in Group H-5 occupancies and in occupancies used for the manufacture of organic coatings. An automatic smoke detection system shall be installed for highly toxic gases, organic peroxides and oxidizers in accordance with Chapters 37, 39 and 40, respectively, of the *International Fire Code*.

❖ Due to the nature and potential quantity of hazardous materials in Group H-5 occupancies, a manual means of activating an evacuation alarm is essential for the safety of the occupants. In accordance with Section 415.9.8, the activation of the alarm system is to initiate a local alarm and transmit a signal to the emergency control station. The manual fire alarm system requirement for the building is in addition to the emergency alarm requirements in Section 415.9.4.6 (see also Section 414.7).

Occupancies involved in the manufacture of organic coatings present special hazardous conditions due to the unstable character of the materials, such as nitrocellulose. Good housekeeping and control of ignition sources is critical. Chapter 20 of the IFC contains additional requirements for organic coating manufacturing processes.

This section also requires an automatic smoke detection system in certain occupancy conditions involving either highly toxic gases, organic peroxides or oxidizers. The need for the automatic smoke detection system may depend upon the class of materials and additional levels of fire protection provided in accordance with the IFC. This requirement also assumes the quantity of materials is in excess of the maximum allowable permitted by Tables 307.7(1) and 307.7(2).

[F] 907.2.6 Group I. A manual fire alarm system and an automatic fire detection system shall be installed in Group I occupancies. An electrically supervised, automatic smoke detection system shall be provided in waiting areas that are open to corridors.

Exception: Manual fire alarm boxes in patient sleeping areas of Group I-1 and I-2 occupancies shall not be required at exits if located at all nurses' control stations or other constantly attended staff locations, provided such stations are visible and continuously accessible and that travel distances required in Section 907.3.1 are not exceeded.

❖ Since the protection and possible evacuation of the occupants in Group I occupancies are most often dependent on the response by staff, all occupancies in Group I are required to be protected with a manual fire alarm system and an automatic fire detection system. Occupancies in Group I also tend to be compartmentalized into small rooms so that a fire in one area of the building would not easily be noticed by occupants in another part of the building.

To reduce the likelihood that a fire within such a waiting area could develop beyond the incipient stage, thereby jeopardizing the integrity of the corridor, the area is to be equipped with an automatic smoke detection system. Whereas the areas also often serve as designated smoking areas, arrangements should be made to minimize the potential for unwanted alarms caused by activating the smoke detection system. The detectors are to be located so as to provide the appropriate coverage to the waiting area space.

In order to reduce the potential for unwanted alarms, manual fire alarm boxes are permitted to be located at the nurses' control stations or other constantly attended location. The exception assumes the approved location is always accessible by staff and within 200 feet (60 960 mm) of travel distance.

[F] 907.2.6.1 Group I-2. Corridors in nursing homes (both intermediate-care and skilled nursing facilities), detoxification facilities and spaces open to the corridors shall be equipped with an automatic fire detection system.

Exceptions:

1. Corridor smoke detection is not required in smoke compartments that contain patient sleeping rooms where patient sleeping units are provided with smoke detectors that comply with UL 268. Such detectors shall provide a visual display on the corridor side of each patient sleeping unit and an audible and visual alarm at the nursing station attending each unit.

2. Corridor smoke detection is not required in smoke compartments that contain patient sleeping rooms where patient sleeping unit doors are equipped with automatic door-closing devices with integral smoke detectors on the unit sides installed in accordance with their listing, provided that the integral detectors perform the required alerting function.

❖ Automatic fire detection is required in areas open to corridors in occupancies classified as Group I-2 and corridors in nursing homes and detoxification facilities. In recognition of quick-response sprinkler technology, and the fact that the sprinkler system is electronically supervised and because the doors to patient sleeping units are supervised by staff on a continual basis when in the open position, it is now believed that smoke detectors are not required for adequate fire safety in patient sleeping units.

In nursing homes and detoxification facilities, however, some redundance is appropriate because such facilities typically have less control over furnishings and personal items and thereby result in a less predictable and usually higher fire hazard load than other Group I-2 occupancies. Also, there is generally less staff supervision in these facilities than in other health care facilities and thus less control over patient smoking and other fire

causes. Therefore, to provide additional protection against fires spreading from the room of origin, automatic fire detection is required in corridors of nursing homes and detoxification facilities.

It should be noted that fire detection is not required in corridors of other Group I-2 occupancies except where otherwise specifically required in the code (see Section 907.2.6). Similarly, since areas open to the corridor very often are the room of fire origin and because such areas are no longer required by the code to be under visual supervision by staff, some redundance to protection by the sprinkler system is necessary. Accordingly, all areas open to corridors must be protected by an automatic fire detection system. This requirement provides an additional level of protection against sprinkler system failures or lapses in staff supervision.

There are two exceptions to the requirement for an automatic fire detection system in corridors of nursing homes and detoxification facilities. In both exceptions, the alternative methods of protection specifically provide an equivalent level of safety to what is required in patient sleeping units.

Exception 1 requires smoke detectors to be located in all of the patients' sleeping units, which activate both a visual display on the corridor side of the patients, sleeping units and a visual and audible alarm at the nurses' station serving or attending the sleeping unit. Detectors complying with UL 268 are intended for open area protection and for connection to a normal power supply or as part of a fire alarm system.

This exception, however, is specifically designed so as not to require the detectors to activate the building's fire alarm system where approved patient sleeping unit smoke detectors are installed and where visual and audible alarms are provided. This is in response to the concern over unwanted alarms. It should be noted that the required alarm signals will not necessarily indicate to staff that a fire emergency exists. The nursing call system may typically be used to identify numerous conditions within the room.

Exception 2 addresses the situation where smoke detectors are incorporated within automatic door-closing devices. The units are acceptable as long as the required alarm functions are still provided. Such units are usually listed as combination door closer and hold-open devices.

[F] 907.2.6.2 Group I-3. Group I-3 occupancies shall be equipped with a manual and automatic fire alarm system installed for alerting staff.

❖ Because of the evacuation difficulties associated with Group I-3 occupancies and the dependence on adequate staff response, a manual fire alarm system and automatic fire detection system is required subject to the special occupancy conditions in Sections 907.2.6.2.1 through 907.2.6.2.3. This section recognizes that the evacuation of Group I-3 occupancies depends on an effective staff response. Section 408.7 of the IFC sets forth the requirements for an emergency plan, including employee training, staff availability, the need for occupants to notify staff and the need for the proper keys for unlocking doors for staff in Group I-3 occupancies.

[F] 907.2.6.2.1 System initiation. Actuation of an automatic fire-extinguishing system, a manual fire alarm box or a fire detector shall initiate an approved fire alarm signal which automatically notifies staff. Presignal systems shall not be used.

❖ This section specifies the systems that upon actuation must initiate the required alarm signal to the staff. So that staff will respond in a timely manner, a presignal system is not permitted (see Section 1-5.4.10 of NFPA 72 for further information on the presignal feature).

[F] 907.2.6.2.2 Manual fire alarm boxes. Manual fire alarm boxes are not required to be located in accordance with Section 907.3 where the fire alarm boxes are provided at staff-attended locations having direct supervision over areas where manual fire alarm boxes have been omitted. Manual fire alarm boxes shall be permitted to be locked in areas occupied by detainees, provided that staff members are present within the subject area and have keys readily available to operate the manual fire alarm boxes.

❖ Because of the potential for intentional false alarms and the resulting disruption to the facility, manual fire alarm boxes in Group I-3 occupancies may be either locked or made inaccessible to the occupants. The locking of manual fire alarm boxes is only permitted in areas where staff members are present and keys are readily available to them to unlock the boxes, or where the alarm boxes are located in a manned staff location which has direct supervision of the Group I-3 area.

[F] 907.2.6.2.3 Smoke detectors. An approved automatic smoke detection system shall be installed throughout resident housing areas, including sleeping areas and contiguous day rooms, group activity spaces and other common spaces normally accessible to residents.

Exceptions:

1. Other approved smoke detection arrangements providing equivalent protection including, but not limited to, placing detectors in exhaust ducts from cells or behind protective guards listed for the purpose are allowed when necessary to prevent damage or tampering.

2. Sleeping units in Use Conditions 2 and 3.

3. Smoke detectors are not required in sleeping units with four or fewer occupants in smoke compartments that are equipped throughout with an approved automatic sprinkler system.

❖ Evacuation of Group I-3 facilities is impractical because of the need to maintain security. As such, an automatic smoke detection system is required to provide early warning of a fire situation.

As indicated in Exception 1, the installation of automatic smoke detectors must take into account the need to protect the detector from vandalism by residents. As a result, detectors may have to be located in return air

ducts or protected by a substantial physical barrier.

Occupants in Use Condition II or III are not locked in their sleeping units, which reduces the need for smoke detection (Exception 2).

Exception 3 allows smoke detectors to be omitted in sleeping units with four or less occupants on the basis that in a building that is protected throughout with an approved automatic sprinkler system, the system will provide both detection and suppression functions. It should be noted that all Group I facilities are assumed to be fully sprinklered throughout in accordance with NFPA 13 as required by Section 903.2.5. The limitation of four or less occupants reduces the potential fuel load (mattresses, clothes, etc.) and likelihood of involvement over an extended area.

[F] 907.2.7 Group M. A manual fire alarm system shall be installed in Group M occupancies, other than covered mall buildings complying with Section 402, having an occupant load of 500 or more persons or more than 100 persons above or below the lowest level of exit discharge.

Exception: Manual fire alarm boxes are not required if the building is equipped throughout with an automatic sprinkler system and the alarm notification appliances will activate upon sprinkler water flow.

❖ Group M occupancies have the potential for a large number of occupants who may not be familiar with their surroundings. The installation of a fire alarm system increases the ability to alert the occupants of a fire condition. It should be noted that the occupant thresholds are independent. If the total occupant load is 500 or more persons or more than 100 persons are above or below the level of exit discharge, a manual fire alarm system is required.

The exception does not omit the fire alarm system, but rather permits it to be initiated automatically by sprinkler system water flow switch(es) in lieu of manual fire alarm boxes. All buildings with a fire area containing a Group M occupancy in excess of 12,000 square feet (1115 m²) must be provided with an automatic sprinkler system in accordance with Section 903.2.6.

[F] 907.2.7.1 Occupant notification. During times that the building is occupied, in lieu of the automatic activation of alarm notification appliances, the manual fire alarm system shall be allowed to activate an alarm signal at a constantly attended location from which evacuation instructions shall be initiated over an emergency voice/alarm communication system installed in accordance with Section 907.2.12.2. The emergency voice/alarm communication system shall be allowed to be used for other announcements, provided the manual fire alarm use takes precedence over any other use.

❖ Occupants in a mercantile occupancy may assume the alarm is a false alarm or act inappropriately and thus delay evacuation of the building. In order to prevent a panic situation, the manual fire alarm system is permitted to be part of an emergency voice/alarm communication system. The signal is to be sent to a constantly attended location on site from which evacuation instructions can be given.

[F] 907.2.8 Group R-1. Fire alarm systems shall be installed in Group R-1 occupancies as required in Sections 907.2.8.1 through 907.2.8.3.

❖ Since residents of Group R-1 occupancies may be asleep and are usually of a transient nature and unfamiliar with the building, and since such buildings contain numerous small rooms so that the occupants may not notice a fire in another part of the building, occupancies in Group R-1 are required to be provided throughout with a manual fire alarm system and an automatic fire detection system. Requirements for single- or multiple-station smoke alarms in sleeping units are contained in Section 907.2.10.1.1.

[F] 907.2.8.1 Manual fire alarm system. A manual fire alarm system shall be installed in Group R-1 occupancies.

Exceptions:

1. A manual fire alarm system is not required in buildings not over two stories in height where all individual guestrooms and contiguous attic and crawl spaces are separated from each other and public or common areas by at least 1-hour fire partitions and each individual guestroom has an exit directly to a public way, exit court or yard.

2. Manual fire alarm boxes are not required throughout the building when the following conditions are met:

   2.1. The building is equipped throughout with an automatic sprinkler system installed in accordance with Section 903.3.1.1 or 903.3.1.2.

   2.2. The notification appliances will activate upon sprinkler water flow, and

   2.3. At least one manual fire alarm box is installed at an approved location.

❖ This section is specific to manual fire alarm systems and requires such systems to be included in all Group R-1 occupancies, with two exceptions.

Exception 1 eliminates the requirement for a manual fire-alarm system, if the sleeping units have an exit discharging directly to a public way, exit court or yard. Even though the building may be two stories in height, the sleeping units on each floor must have access directly to an approved exit at grade level. The use of an exterior exit access balcony with exterior stairs serving the second floor does not constitute an exit directly at grade. The minimum 1-hour fire-resistance rating required for adequate separation of the sleeping units must be maintained.

Exception 2 does not omit the fire alarm system, but rather permits it to be initiated automatically by sprinkler system water flow switch(es) in lieu of manual fire alarm boxes. The sprinkler system is to be equipped with local audible alarms that can be heard throughout the building and at least one manual fire alarm box installed at an approved location.

[F] 907.2.8.2 Automatic fire alarm system. An automatic fire alarm system shall be installed throughout all interior corridors serving guestrooms.

Exception: An automatic fire detection system is not required in buildings that do not have interior corridors serving guestrooms and each guestroom has a means of egress door opening directly to an exterior exit access that leads directly to an exit.

❖ This section requires an automatic fire alarm system within interior corridors. Such systems primarily make use of smoke detectors for alarm initiation in accordance with Section 907.2, with one exception.

Automatic fire detectors are not required in motels and hotels that do not have interior corridors and in which sleeping units have direct access to an exterior exit. The intent of the exception is that the exit access from the sleeping unit door be exterior and not require reentering the building prior to entering the exit. Since the exit access is outside, the need for detectors other than in sleeping units is greatly reduced. Unlike Exception 1 to Section 907.2.8.1, exit balconies and exterior stairs are allowed.

[F] 907.2.8.3 Smoke alarms. Smoke alarms shall be installed as required by Section 907.2.10. In buildings that are not equipped throughout with an automatic sprinkler system installed in accordance with Section 903.3.1.1 or 903.3.1.2, the smoke alarms in guestrooms shall be connected to an emergency electrical system and shall be annunciated by guestroom at a constantly attended location from which the fire alarm system is capable of being manually activated

❖ The actual requirements for single- and multiple-station smoke detectors are located in Section 907.2.10. This section requires that, in unsprinklered buildings, single- and multiple-station smoke alarms within sleeping units be connected to the emergency electrical system and that they be annunciated near guest rooms at a location from which the fire alarm can be manually activated. Automatic activation is avoided to reduce unnecessary alarms within such buildings. There will be some cases where the above conditions exist, but Exception 1 of Section 907.2.8.1 would apply and a fire alarm system would not be required.

[F] 907.2.9 Group R-2. A manual fire alarm system shall be installed in Group R-2 occupancies where:

1. Any dwelling unit or sleeping unit is located three or more stories above the lowest level of exit discharge;

2. Any dwelling unit or sleeping unit is located more than one story below the highest level of exit discharge of exits serving the dwelling unit or sleeping unit; or

3. The building contains more than 16 dwelling units or sleeping units.

Exceptions:

1. A fire alarm system is not required in buildings not over two stories in height where all dwelling units or sleeping units and contiguous attic and crawl spaces

are separated from each other and public or common areas by at least 1-hour fire partitions and each dwelling unit or sleeping unit has an exit directly to a public way, exit court or yard.

2. Manual fire alarm boxes are not required throughout the building when the following conditions are met:

   2.1. The building is equipped throughout with an automatic sprinkler system in accordance with Section 903.3.1.1 or 903.3.1.2.

   2.2. The notification appliances will activate upon sprinkler flow, and

   2.3. At least one manual fire alarm box is installed at an approved location.

3. A fire alarm system is not required in buildings that do not have interior corridors serving dwelling units and are protected by an approved automatic sprinkler system installed in accordance with Section 903.3.1.1 or 903.3.1.2, provided that dwelling units either have a means of egress door opening directly to an exterior exit access that leads directly to the exits or are served by open-ended corridors designed in accordance with Section 1022.6, Exception 4.

❖ The occupants of Group R-2 occupancies are not considered to be as transient as those of Group R-1; thus, this increases the probability that residents can more readily notify each other of a fire emergency. Therefore, whereas all Group R-1 occupancies are required to have a manual fire alarm system subject to the exceptions in Section 907.2.8.1, Group R-2 occupancies are only required to have a manual fire alarm system as stipulated in one of the three listed conditions. The threshold conditions are meant to be applied independently of each other.

Exceptions 1 and 2 are essentially identical to Exceptions 1 and 2 in Section 907.2.8.1 (see commentary, Section 907.2.8.1).

Exception 3 allows the omission of a fire alarm system in fully sprinklered buildings (NFPA 13 or 13R) with no interior corridors and that exit directly to an exterior exit access or have open-ended corridors. The important thing to note is that the sprinkler system is not required to activate notification devices since fire alarm systems would not be required.

[F] 907.2.10 Single- and multiple-station smoke alarms. Listed single- and multiple-station smoke alarms shall be installed in accordance with the provisions of this code and the household fire-warning equipment provisions of NFPA 72.

❖ Single- and multiple-station smoke alarms have evolved as one of the most important fire safety features in residential and similar occupancies having sleeping occupants. The value of early fire warning in these occupancies has been repeatedly demonstrated in fire incidents involving both successful and unsuccessful smoke alarm performance.

In order to provide an opportunity for successful smoke alarm operation and performance, single- and multiple-station smoke alarms are to be installed in ac-

cordance with the code and NFPA 72, which provides the minimum requirements for the selection, installation, operation and maintenance of fire warning equipment for use within family living units. They are termed "smoke alarms" versus "smoke detectors" because they are independent of a fire alarm system and include an integral alarm notification device.

[F] 907.2.10.1 Where required. Single- or multiple-station smoke alarms shall be installed in the locations described in Sections 907.2.10.1.1 through 907.2.10.1.3.

❖ Section 907.2.10.1 establishes the conditions where single- or multiple-station smoke alarms are required. Single- and multiple-station smoke alarms are typically required where occupants may be sleeping, and therefore are unaware of a fire in the room or a fire that may affect their egress or escape path, such as in hallways that are adjacent to the bedrooms. Smoke alarms are intended to alert the occupants within the dwelling unit or sleeping unit—not all of the occupants within the building. It is presumed that after the occupant in the unit of origin has safely evacuated the unit, he or she will notify other occupants of the building by either word of mouth or initiation of the manual fire alarm system, if provided.

[F] 907.2.10.1.1 Group R-1. Single- or multiple-station smoke alarms shall be installed in all of the following locations in Group R-1:

1. In sleeping areas.

2. In every room in the path of the means of egress from the sleeping area to the door leading from the sleeping unit.

3. In each story within the sleeping unit, including basements. For sleeping units with split levels and without an intervening door between the adjacent levels, a smoke alarm installed on the upper level shall suffice for the adjacent lower level provided that the lower level is less than one full story below the upper level.

❖ Whereas the occupant(s) of a sleeping unit or suite may be asleep and unaware of a fire developing in the room or egress path, single- or multiple-station smoke alarms must be provided in the sleeping unit and any intervening room between the sleeping unit and the exit access door from the room. If the sleeping unit or suite involves more than one level, a smoke alarm must be provided on every level.

Smoke alarms are required in split-level arrangements, except those that meet the conditions described in Item 3. In accordance with Section 907.2.10.3, all smoke alarms within a sleeping unit must be interconnected so that actuation of one alarm will actuate all smoke alarms within the sleeping unit or suite.

Section 907.2.8.3 contains requirements for single- or multiple-station smoke alarms in the sleeping unit to be connected to an emergency electrical system and be annunciated at a constantly attended location from which manual activation of the fire alarm system can be accomplished (see commentary, Section 907.2.8.3).

[F] 907.2.10.1.2 Groups R-2, R-3, R-4 and I-1. Single- or multiple-station smoke alarms shall be installed and maintained in Groups R-2, R-3, R-4 and I-1, regardless of occupant load at all of the following locations:

1. On the ceiling or wall outside of each separate sleeping area in the immediate vicinity of bedrooms.

2. In each room used for sleeping purposes.

3. In each story within a dwelling unit, including basements but not including crawl spaces and uninhabitable attics. In dwellings or dwelling units with split levels and without an intervening door between the adjacent levels, a smoke alarm installed on the upper level shall suffice for the adjacent lower level provided that the lower level is less than one full story below the upper level.

❖ Whereas the occupants of a dwelling unit may be asleep and unaware of a fire developing in the room or in an area within the dwelling unit that will affect their ability to escape, single- or multiple-station smoke alarms must be provided within every bedroom, in the vicinity of all bedrooms (e.g., hallways leading to the bedrooms) and on each story of the dwelling unit (see Figure 907.2.10.1.2). This section also requires smoke alarms in all sleeping areas of Group I-1 occupancies, subject to the automatic fire detection system modification in Section 907.2.10.1.3.

It should be noted that if a sprinkler system was installed throughout the building in accordance with either NFPA 13, 13R or 13D, if applicable, smoke alarms would still be required in the bedrooms even if residential sprinklers were utilized.

Smoke alarms are required in split-level arrangements. In accordance with Section 907.2.10.3, all smoke alarms within a dwelling unit must be interconnected so that actuation of one alarm will actuate the alarms in all detectors within the dwelling unit.

[F] 907.2.10.1.3 Group I-1. Single- or multiple-station smoke alarms shall be installed and maintained in sleeping areas in occupancies in Group I-1. Single- or multiple-station smoke alarms shall not be required where the building is equipped throughout with an automatic fire detection system in accordance with Section 907.2.6.

❖ Whereas the occupants of a sleeping unit in Group I-1 occupancy may be asleep, they are still considered capable of self-preservation. Regardless, smoke alarms are required in sleeping units. Single- or multiple-station smoke alarms need not be provided within the room if an automatic fire detection system that includes room system smoke detectors is provided in accordance with Section 907.2.6.

[F] 907.2.10.2 Power source. In new construction, required smoke alarms shall receive their primary power from the building wiring where such wiring is served from a commercial source and shall be equipped with a battery backup. Smoke alarms shall emit a signal when the batteries are low. Wiring

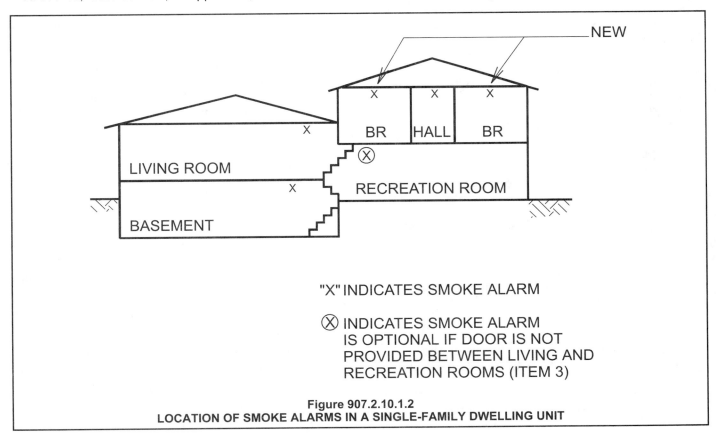

"X" INDICATES SMOKE ALARM

⊗ INDICATES SMOKE ALARM
IS OPTIONAL IF DOOR IS NOT
PROVIDED BETWEEN LIVING AND
RECREATION ROOMS (ITEM 3)

Figure 907.2.10.1.2
LOCATION OF SMOKE ALARMS IN A SINGLE-FAMILY DWELLING UNIT

shall be permanent and without a disconnecting switch other than as required for overcurrent protection.

Exception: Smoke alarms are not required to be equipped with battery backup in Group R-1 where they are connected to an emergency electrical system.

❖ Smoke alarms are required to utilize AC as a primary power source and battery power as a secondary source in order to enhance their reliability. For example, during a power outage, the probability of fire is increased because of the use of candles or lanterns for temporary light. Required backup battery power is intended to provide for continued performance of the smoke alarms. Smoke alarms are commonly designed to emit a recurring signal when batteries are low and need to be replaced.

Certain Group R-1 occupancies may already have an emergency electrical system provided in the building to monitor other building system conditions. The emergency electrical system provides a level of reliability equivalent to battery backup.

[F] 907.2.10.3 Interconnection. Where more than one smoke alarm is required to be installed within an individual dwelling unit in Group R-2, R-3 or R-4, or within an individual dwelling unit or sleeping unit in Group R-1, the smoke alarms shall be interconnected in such a manner that the activation of one alarm will activate all of the alarms in the individual unit. The alarm shall be clearly audible in all bedrooms over background noise levels with all intervening doors closed.

❖ The installation of smoke alarms in areas remote from the sleeping area will be of minimal value if the alarm is not heard by the occupants. Interconnection of multiple smoke alarms within an individual dwelling unit or sleeping unit is required in order to alert a sleeping occupant of a remote fire within the unit before the combustion products reach the smoke alarm in the sleeping area, thus providing additional time for evacuation.

It should be noted that the term "interconnection" is intended to allow the use of not only hard-wired systems, but also those that use radio signals (wireless systems) (see Section 907.5). Underwriters Laboratories Inc. (UL) has listed smoke detectors that use this technology. It is presumed that on safely evacuating the unit or room of fire origin, an occupant will notify other occupants by actuating the manual fire alarm system or use other means available.

[F] 907.2.10.4 Acceptance testing. When the installation of the alarm devices is complete, each detector and interconnecting wiring for multiple-station alarm devices shall be tested in accordance with the household fire warning equipment provisions of NFPA 72.

❖ In order to determine that smoke alarms have been properly installed and are ready to perform as intended, all smoke alarms are required to be actuated during an acceptance test. The test also confirms that interconnected detectors will operate simultaneously as required. The responsibility for conducting the accep-

tance tests rests with the owner or the owner's representative in accordance with Section 901.5.

[F] 907.2.11 Special amusement buildings. An approved automatic smoke detection system shall be provided in special amusement buildings in accordance with this section.

Exception: In areas where ambient conditions will cause a smoke detection system to alarm, an approved alternative type of automatic detector shall be installed.

❖ Special amusement buildings are buildings in which the means of egress is not readily apparent, is intentionally confounded or is not readily available. Special amusement buildings must also comply with the provisions of Section 411.

The approved automatic smoke detection system is required to provide early warning of a fire condition. The detection system is required regardless of the presence of staff in the building. The exception recognizes that the ambient conditions in some special amusement buildings may preclude the use of automatic smoke detectors. In such instances, an alternative detection device must be used that can provide an early detection of a fire.

[F] 907.2.11.1 Alarm. Activation of any single smoke detector, the automatic sprinkler system or any other automatic fire detection device shall immediately sound an alarm at the building at a constantly attended location from which emergency action can be initiated, including the capability of manual initiation of requirements in Section 907.2.11.2.

❖ Upon activation of either a smoke detector or other automatic fire detection device or the automatic sprinkler system, an alarm is required to be sounded at a constantly attended location. The staff at the location is expected to be capable of then providing the required egress illumination, stopping the conflicting or confusing sounds and distractions and activating the exit marking required by Section 907.2.11.2. It is also anticipated that the staff would be capable of preventing additional people from entering the building.

[F] 907.2.11.2 System response. The activation of two or more smoke detectors, a single smoke detector with alarm verification, the automatic sprinkler system or other approved fire detection device shall automatically:

1. Cause illumination of the means of egress with light of not less than 1 foot-candle (11 lux) at the walking surface level;

2. Stop any conflicting or confusing sounds and visual distractions; and

3. Activate an approved directional exit marking that will become apparent in an emergency. Such system response shall also include activation of a prerecorded message, clearly audible throughout the special amusement building, instructing patrons to proceed to the nearest exit. Alarm signals used in conjunction with the prerecorded message shall produce a sound which is distinctive from other sounds used during normal operation. The wiring to the auxiliary devices and equipment used to accomplish

the above fire safety functions shall be monitored for integrity in accordance with NFPA 72.

❖ Once a fire has been detected, measures must be taken to stop the confusion or distractions. Additionally, the egress path is required to be illuminated and marked. These measures must occur automatically upon detection of the fire or sprinkler water flow. A prerecorded message is to be automatically activated that can be heard throughout the building instructing the occupants to proceed to the nearest exit. The message and alarm signals should be such so as not to create a panic condition. The prerecorded message capability is in addition to the emergency voice/alarm communication system requirement of Section 907.2.11.3. The wiring of all devices must be in accordance with NFPA 72.

[F] 907.2.11.3 Emergency voice/alarm communication system. An emergency voice/alarm communication system, which is also allowed to serve as a public address system, shall be installed in accordance with NFPA 72, and shall be audible throughout the entire special amusement building.

❖ Due to the problem associated with evacuating special amusement buildings, an emergency voice/alarm communication system is required (see Section 907.2.12.2). This section permits the system to also serve as a public address system in order to have the capability to alert the occupants of a fire emergency and provide them with the proper instructions.

[F] 907.2.12 High-rise buildings. Buildings having floors used for human occupancy located more than 75 feet (22 860 mm) above the lowest level of fire department vehicle access shall be provided with an automatic fire alarm system and an emergency voice/alarm communication system in accordance with Section 907.2.12.2.

Exceptions:

1. Airport traffic control towers in accordance with Sections 412 and 907.2.22.

2. Open parking garages in accordance with Section 406.3.

3. Buildings with an occupancy in Group A-5.

4. Low-hazard special occupancies in accordance with Section 503.1.2.

5. Buildings with an occupancy in Group H-1, H-2 or H-3 in accordance with Section 415.

❖ High-rise buildings require additional fire protection systems because of the difficulties with smoke movement, egress time and fire department access. As such, this section requires both an automatic fire alarm system and an emergency voice/alarm communication system (see commentary, Section 907.2.12.2).

The listed exceptions are the same as those indicated in Section 403.1 regarding the applicability of the high-rise provisions; therefore, additional fire protection systems are not required for the occupancy conditions in the listed exceptions (see commentary, Section 403.1).

[F] 907.2.12.1 Automatic fire detection. Smoke detectors shall be provided in accordance with this section. Smoke detectors shall be connected to an automatic fire alarm system. The activation of any detector required by this section shall operate the emergency voice/alarm communication system. Smoke detectors shall be located as follows:

1. In each mechanical equipment, electrical, transformer, telephone equipment or similar room which is not provided with sprinkler protection, elevator machine rooms and in elevator lobbies.

2. In the main return air and exhaust air plenum of each air-conditioning system having a capacity greater than 2,000 cubic feet per minute (cfm) (0.94 m^3/s). Such detectors shall be located in a serviceable area downstream of the last duct inlet.

3. At each connection to a vertical duct or riser serving two or more stories from a return air duct or plenum of an air-conditioning system. In Group R-1 and R-2 occupancies a listed smoke detector is allowed to be used in each return air riser carrying not more than 5,000 cfm (2.4 m^3/s) and serving not more than 10 air inlet openings.

❖ Automatic smoke detectors are required in all high-rise buildings in certain locations so that a fire will be detected in its early stages of development. Smoke detectors must be installed in rooms that are not typically occupied. Spaces specifically identified as requiring automatic fire detection are: mechanical equipment; electrical; transformer; telephone equipment and elevator machine rooms. The detectors must be connected to the automatic fire alarm system and be capable of initiating operation of the emergency voice/alarm communication system.

Smoke detectors are required to be installed in the main return air and exhaust air plenum of each air-conditioning system having a design capacity exceeding 2,000 cubic feet per minute (cfm) (0.94 m^3/s). Systems with design capacities equal to or less than 2,000 cfm (0.94 m^3/s) are exempt from this requirement because their small size inherently limits their capacity for spreading smoke to parts of the building not already involved with fire.

The area that could be served by a 2,000-cfm (0.94 m^3/s) system (approximately 5 tons of cooling capacity) is comparatively small; therefore, the distribution of smoke in a system of such size would be minimal. Smoke detectors must be located so that they monitor the total airflow within the system. Where a single detector would be unable to sample the total airflow at all times, multiple detectors would be required. The smoke detectors must be made accessible for maintenance and inspection purposes. Many failures and false alarms are caused by a lack of maintenance and cleaning of the smoke detectors.

Consistent with Section 606.2.3 of the IMC, return-air risers serving two or more stories are required to have smoke detectors installed at each story. Item 3 permits the use of a single listed smoke detector in each return-air riser in a Group R-1 or R-2 occupancy, provided the capacity of each riser did not exceed 5,000 cfm (2.4

m³/s) and did not serve more than 10 air-inlet openings. This alternative recognizes that it is not as necessary in buildings dedicated to residential occupancies only to monitor the return air from each story prior to intermixing it in the common riser.

[F] 907.2.12.2 Emergency voice/alarm communication system. The operation of any automatic fire detector, sprinkler water-flow device or manual fire alarm box shall automatically sound an alert tone followed by voice instructions giving approved information and directions on a general or selective basis to the following terminal areas on a minimum of the alarming floor, the floor above and the floor below in accordance with the *International Fire Code*.

1. Elevator lobbies.

2. Corridors.

3. Rooms and tenant spaces exceeding 1,000 square feet (93 m²) in area.

4. Dwelling units or sleeping units in Group R-2 occupancies.

5. Sleeping units in Group R-1 occupancies.

6. Areas of refuge as defined in Section 1002.

Exception: In Group I-1 and I-2 occupancies, the alarm shall sound in a constantly attended area and a general occupant notification shall be broadcast over the overhead page.

❖ The section identifies the areas of coverage when an emergency voice/alarm communication system is required. The system may sound a general alarm or be a selective system in which only selected areas of the building receive the alarm indication. This section requires a minimum area of notification that must include the alarming floor and the floors above and below it.

The requirement for alarm notification within dwelling units in Group R-2 occupancies and within sleeping units in Group R-1 occupancies recognizes the need for a sound pressure level that will awaken sleeping occupants. Although Section 907.10.2 requires a minimum sound pressure level of 70 decibels (dBA) in Group R occupancies, a sound pressure level of 75 dBA at the head of the bed is generally considered the minimum to alert sleeping individuals.

This section also indicates that the emergency voice/alarm system is to be initiated as all other fire alarm systems are initiated. The functional operation of the system begins with an alert tone (usually 3 to 10 seconds in duration) followed by the evacuation signal (message).

The exception recognizes the supervised environment typical of institutional uses and the reliance placed on staff to act appropriately in an emergency.

[F] 907.2.12.2.1 Manual override. A manual override for emergency voice communication shall be provided for all paging zones.

❖ The intent of this section is to provide the ability to transmit live voice instructions over any previously initiated signals or messages for all zones.

[F] 907.2.12.2.2 Live voice messages. The emergency voice/alarm communication system shall also have the capability to broadcast live voice messages through speakers located in elevators, exit stairways and throughout a selected floor or floors.

❖ An adequate number of speakers should be provided such that the emergency voice/alarm communication system has the capability of broadcasting live voice messages to all desired areas, including elevators, exit stairs and selected floors. Speakers utilized for background music should not be used unless specifically listed for fire alarm system use. NFPA 72 provides additional requirements for the placement, location and audibility of speakers used as part of an emergency voice/alarm communication system.

[F] 907.2.12.2.3 Standard. The emergency voice/alarm communication system shall be designed and installed in accordance with NFPA 72.

❖ NFPA 72 contains the minimum requirements for the design, installation, maintenance and acceptance testing for emergency voice/alarm communication systems. The primary purpose of the emergency voice/alarm communication system is to provide dedicated manual and automatic facilities for the origination, control and transmission of information and instructions pertaining to a fire alarm emergency to the occupants (including fire department personnel) of the building.

[F] 907.2.12.3 Fire department communication system. An approved two-way, fire department communication system designed and installed in accordance with NFPA 72 shall be provided for fire department use. It shall operate between a fire command center complying with Section 911 and elevators, elevator lobbies, emergency and standby power rooms, fire pump rooms, areas of refuge and inside enclosed exit stairways. The fire department communication device shall be provided at each floor level within the enclosed stairway.

Exception: Fire department radio systems where approved by the fire department.

❖ High-rise buildings have also posed a challenge to the traditional communication systems used by the fire service for fire-to-ground communications; therefore, a fire department communication system must be provided to assist fire ground officers in communicating with the fire fighters working in various areas of the building. The system must be capable of operating between the fire command center and every elevator, elevator lobby, emergency/standby power room, fire pump room, area of refuge and exit stairway.

The exception permits the use of hand-held radios where approved by the fire department for that specific building. The use of radio systems may not be effective in buildings with a significant amount of structural steel.

[F] 907.2.13 Atriums connecting more than two stories. A fire alarm system shall be installed in occupancies with an atrium that connects more than two stories. The system shall be activated in accordance with Section 907.6. Such occupancies in Group A, E or M shall be provided with an emergency voice/alarm communication system complying with the requirements of Section 907.2.12.2.

❖ In order that occupants can be notified of a fire emergency and begin to evacuate, all buildings containing an atrium that connects more than two stories are to be equipped with a fire alarm system. The alarm system must be initiated in accordance with Section 907.7, which requires that in buildings containing an atrium, the alarm system is to be initiated by the sprinkler system and any automatic or manual fire alarm initiating devices found in the atrium as well as elsewhere in the building. It does not intend to require certain features to be installed within the atrium but, rather, is simply requiring that any such features present initiate the fire alarm notification system. It would not necessarily be appropriate to also initiate the smoke control system upon activation of the alarm system within a building containing an atrium (see Section 909.12.2).

Groups A, E and M are to be provided with an emergency voice/alarm communication system that complies with the requirements of Section 907.2.12.2. This is in recognition of the number of persons to be evacuated and the lack of familiarity with the location of exits that is typical of occupants in Groups A and M. The alarm system is intended to warn occupants entering the atrium since smoke is being drawn to the atrium.

[F] 907.2.14 High-piled combustible storage areas. An automatic fire detection system shall be installed throughout high-piled combustible storage areas where required by the *International Fire Code*.

❖ Section 2306.5 of the IFC requires an automatic fire detection system in high-piled combustible storage areas depending on the commodity class, size of the high-piled storage area and the presence of an automatic sprinkler system. High-piled storage is considered to be the storage of combustible materials in piles, on pallets or in racks greater than 12 feet (3658 mm) in height. Chapter 23 of the IFC contains fire protection requirements for all high-piled storage conditions in addition to the requirements of NFPA 13.

[F] 907.2.15 Delayed egress locks. Where delayed egress locks are installed on means of egress doors in accordance with Section 1008.1.8.6, an automatic smoke or heat detection system shall be installed as required by that section.

❖ This section alerts the code user to additional provisions in Section 1008.1.8.6 in which the operation of egress doors may be tied into the activation of an automatic fire detection system. A smoke or heat detection system is required to unlock delayed egress locks upon activation. The heat detection system can be the sprinkler system. For example, Section 1008.1.3.3 requires horizontal sliding doors utilized as a component of the means of egress, where required to be rated, to be self-closing or automatic-closing by smoke detection. Also, access-controlled egress doors in occupancies as

required by Section 1008.1.3.4 must be capable of being automatically unlocked by activation of an automatic fire detection system, if provided.

[F] 907.2.16 Aerosol storage uses. Aerosol storage rooms and general-purpose warehouses containing aerosols shall be provided with an approved manual fire alarm system where required by the *International Fire Code*.

❖ Chapter 28 of the IFC and NFPA 30B provide additional guidance on storage and fire protection requirements for aerosol products. The requirements for storing the various levels of aerosol products are dependent on the level of sprinkler protection, type of storage condition and quantity of aerosol products. While it is recognized that aerosol product fires generally involve property loss as opposed to loss of life, a manual fire alarm system, if provided, could aid in the prompt evacuation of the occupants. Fires involving aerosol products can spread rapidly through a building unless properly protected and controlled.

[F] 907.2.17 Lumber, plywood and veneer mills. Lumber, plywood and veneer mills shall be provided with a manual fire alarm system.

❖ Any facility using mechanical methods to process wood into finished products such as waferboard, oriented strandboard, composite wood panels or plywood produces debris and the potential for combustible dust. As such, good housekeeping and control of ignition sources are essential. To aid in the quick evacuation of occupants in an emergency condition, Chapter 19 of the IFC requires a manual fire alarm system in lumber, plywood and veneer mills that contain product dryers, due to their potential as a source of ignition. A manual fire alarm system is not required, however, if the dryers and all other potential sources of ignition are protected by a supervised automatic sprinkler system.

[F] 907.2.18 Underground buildings with smoke exhaust system. Where a smoke exhaust system is installed in an underground building in accordance with this code, automatic fire detectors shall be provided in accordance with this section.

❖ As indicated in Section 405.5.2, each compartment of an underground building is to be provided with a smoke control/exhaust system and be capable of being both automatically and manually activated. Compartmentation must be provided when there are floor levels more than 60 feet (18 288 mm) below the lowest level of exit discharge. Compartmentation is a key element in the egress and fire access plan for floor areas in an underground building. The smoke exhaust system is required to not only facilitate egress during a fire, but also improve fire department access to the fire source by maintaining visibility that is otherwise impossible given the inability of the fire service to ventilate manually the underground portion of the building. In order to reduce potential involvement of other compartments, the smoke exhaust system for each compartment must be independent (see also Section 909.21).

[F] 907.2.18.1 Smoke detectors. A minimum of one smoke detector listed for the intended purpose shall be installed in the following areas:

1. Mechanical equipment, electrical, transformer, telephone equipment, elevator machine or similar rooms.

2. Elevator lobbies.

3. The main return and exhaust air plenum of each air-conditioning system serving more than one story and located in a serviceable area downstream of the last duct inlet.

4. Each connection to a vertical duct or riser serving two or more floors from return air ducts or plenums of heating, ventilating and air-conditioning systems, except that in Group R occupancies, a listed smoke detector is allowed to be used in each return air riser carrying not more than 5,000 cfm (2.4 m³/s) and serving not more than 10 air inlet openings.

❖ Automatic smoke detectors are required in all underground buildings in certain locations so that a fire will be detected in its early stages of development. Underground buildings are similar to high-rise buildings in that they present an unusual hazard since they are virtually inaccessible to exterior fire department suppression and rescue operations with the increased potential to trap occupants inside the structure. As such, the smoke detector location requirements for underground buildings are similar to those in Section 907.2.12.1 for high-rise buildings (see commentary, Section 907.2.12.1). The requirement for a smoke detector in the main return and exhaust air plenum of an air-conditioning system in an underground building, however, differs from that of a high-rise building in that it is not a function of capacity [2,000 cfm (0.94 m³/s)] but rather a function of whether the system serves more than one floor level. There is more concern over the threat of smoke movement from floor to floor due to the inability to manually vent the products of combustion directly to the atmosphere.

[F] 907.2.18.2 Alarm required. Activation of the smoke exhaust system shall activate an audible alarm at a constantly attended location.

❖ The audible alarm requirement is so that qualified personnel are notified immediately that the smoke exhaust system has activated and appropriate emergency procedures are promptly put into action.

[F] 907.2.19 Underground buildings. Where the lowest level of a structure is more than 60 feet (18 288 mm) below the lowest level of exit discharge, the structure shall be equipped throughout with a manual fire alarm system, including an emergency voice/alarm communication system installed in accordance with Section 907.2.12.2.

❖ The ability to communicate and offer warning of a fire scenario can increase the time available for egress from

the building. Underground structures located greater than 60 feet (18 288 mm) below the level of exit discharge are, therefore, required to be provided with a manual fire alarm system. A voice/alarm communication system is also required as part of this system (see commentary, Section 907.2.12.2).

[F] 907.2.19.1 Public address system. Where a fire alarm system is not required by Section 907.2, a public address system shall be provided that shall be capable of transmitting voice communications to the highest level of exit discharge serving the underground portions of the structure and all levels below.

❖ In underground structures where a fire alarm system is not required, a public address system must be provided. This communication between the highest level of discharge and the underground levels can be a vital communication link between emergency personnel and building occupants. This is especially important in an environment in which visibility is likely to be impaired in order to determine the status of emergency evacuation from a fire scene below grade.

[F] 907.2.20 Covered mall buildings. Covered mall buildings exceeding 50,000 square feet (4645 m²) in total floor area shall be provided with an emergency voice/alarm communication system. An emergency voice/alarm communication system serving a mall, required or otherwise, shall be accessible to the fire department. The system shall be provided in accordance with Section 907.2.12.2.

❖ Due to the potential number of occupants and their unfamiliarity with their surroundings, an emergency voice/alarm communication system, accessible by the fire department, is required in order to aid in evacuation of the facility (see commentary, Section 402.13).

[F] 907.2.21 Residential aircraft hangars. A minimum of one listed smoke alarm shall be installed within a residential aircraft hangar as defined in Section 412.3.1 and shall be interconnected into the residential smoke alarm or other sounding device to provide an alarm that will be audible in all sleeping areas of the dwelling.

❖ Residential aircraft hangars are assumed to be on the same property as a one- or two-family dwelling. Section 412.3 provides additional requirements for the construction of residential aircraft hangars. As noted in Section 412.3, the hangar could be located immediately adjacent to the dwelling unit, provided it is separated by 1-hour fire-resistance-rated construction. As such, due to the potentially close proximity of the aircraft and its flammability and fuel source, a minimum of one smoke alarm is required in the hangar that is interconnected to the residential smoke alarms. It should be noted, however, that the requirement for a smoke alarm is also applicable to residential aircraft hangars that are detached from the dwelling unit. Since a minimum separation distance is not specified, a fire in the hangar could still present a serious conflagration hazard to the dwelling unit.

[F] 907.2.22 Airport traffic control towers. An automatic fire detection system shall be provided in airport traffic control towers.

❖ Airport traffic control towers are required to be designed in accordance with the provisions of Section 412. These structures are unique in that they can be built to excessive heights, depending upon construction type, are permitted to have one exit stairway and are typically nonsprinklered. Section 903.2.10.3 specifically exempts airport control towers from the requirements of an automatic sprinkler system. Due to these factors, an automatic fire detection system is required to provide early warning notification of an emergency condition for the occupants.

[F] 907.2.23 Battery rooms. An approved automatic smoke detection system shall be installed in areas containing stationary lead-acid battery systems having a liquid capacity of more than 50 gallons (189.3 L). The detection system shall be supervised by an approved central, proprietary or remote station service or a local alarm that will sound an audible signal at a constantly attended location.

❖ Stationary lead-acid battery systems are commonly used for providing standby power, emergency power or uninterrupted power supplies. The release of hydrogen gas during the operation of the battery system is usually minimal. Adequate ventilation will disperse the small amounts of liberated hydrogen. Since standby power and emergency power systems control many important building emergency systems and functions, a supervised automatic smoke detection system is required for early warning notification of a hazardous condition. Sections 608 and 609 of the IFC contain additional requirements, including the need for safety venting; room enclosure requirements; spill control and neutralization provisions; ventilation criteria; signage and seismic protection.

[F] 907.3 Manual fire alarm boxes. Manual fire alarm boxes shall be installed in accordance with Sections 907.3.1 through 907.3.5.

❖ This section specifies the requirements for manual fire alarm boxes that are part of a manual fire alarm system.

[F] 907.3.1 Location. Manual fire alarm boxes shall be located not more than 5 feet (1524 mm) from the entrance to each exit. Additional manual fire alarm boxes shall be located so that travel distance to the nearest box does not exceed 200 feet (60 960 mm).

  Exception: Manual fire alarm boxes shall not be required in Group E occupancies where the building is equipped

throughout with an approved automatic sprinkler system, the notification appliances will activate on sprinkler water flow and manual activation is provided from a normally occupied location.

❖ Manual fire alarm boxes are to be located in the path of egress and are to be readily available to the occupants.
   They are required to be located within 5 feet (1,524 mm) of the entrance to each exit on every story of the building. This would include the need to locate manual fire alarm boxes near each horizontal exit, as well as entrances to stairs and exit doors to the exterior.
   The location of manual fire alarm boxes near exits is so that an adequate number of devices are provided so an alarm can be transmitted in a timely manner. Such locations also encourage the actuation of a manual fire alarm box on the fire floor prior to entering the stair, resulting in the alarm being received from the actual fire floor and not another floor along the path of egress.
   The location also presumes that individuals will be evacuating the area where the fire originated. When evacuation of the fire area is unlikely, consideration could be given to putting manual fire alarm boxes in more convenient places. Examples of such instances would be officer stations in Group I-3 occupancies and nurses' stations in Group I-2 occupancies.
   The 200-foot (60 960 mm) travel distance limitation is consistent with the exit access travel distance permitted for most nonsprinklered occupancies. If the 200-foot (60 960 mm) travel distance to a manual fire alarm box is exceeded, even in a fully sprinklered building, additional manual fire alarm boxes would be required.
   The exception allows the omission of the manual fire alarm boxes in Group E occupancies equipped throughout with an automatic sprinkler system, provided the actuation of the sprinkler system will activate the building evacuation alarms associated with the manual fire alarm system.

[F] 907.3.2 Height. The height of the manual fire alarm boxes shall be a minimum of 42 inches (1067 mm) and a maximum of 48 inches (1219 mm), measured vertically, from the floor level to the activating handle or lever of the box.

❖ Manual fire alarm boxes must be reachable by the occupants of the building. They must also be mounted at a certain height to reduce the likelihood of damage or false alarms from something accidentally striking the device. Therefore, manual fire alarm boxes must be mounted a minimum of 42 inches (1067 mm) and a maximum of 48 inches (1219 mm) above the floor level. The 48-inch (1219 mm) measurement corresponds to the maximum unobstructed side reach height by a person in a wheelchair.

[F] 907.3.3 Color. Manual fire alarm boxes shall be red in color.

❖ Manual fire alarm boxes are to be painted a distinctive red to assist the building occupants in identifying the device.

[F] 907.3.4 Signs. Where fire alarm systems are not monitored by a supervising station, an approved permanent sign that reads: WHEN ALARM SOUNDS—CALL FIRE DEPARTMENT shall be installed adjacent to each manual fire alarm box.

Exception: Where the manufacturer has permanently provided this information on the manual fire alarm box.

❖ This section has limited application because, as indicated in Section 901.6.2, fire alarm systems generally are required to be monitored by an approved supervising station. A nonmonitored system, such as possibly a nonrequired fire alarm system, must be provided with adequate signage indicating the appropriate response to be taken. Most building occupants assume that when an alarm device is activated, the fire department will be automatically notified as well. The sign should be conspicuously located adjacent to the manual fire alarm box unless such signage is provided on the manual fire alarm box itself by the manufacturer.

[F] 907.3.5 Protective covers. The building official is authorized to require the installation of listed manual fire alarm box protective covers to prevent malicious false alarms or provide the manual fire alarm box with protection from physical damage. The protective cover shall be transparent or red in color with a transparent face to permit visibility of the manual fire alarm box. Each cover shall include proper operating instructions. A protective cover that emits a local alarm signal shall not be installed unless approved.

❖ While manual fire alarm boxes should be readily available to all occupants in buildings required to have a manual fire alarm system, this section permits the use of protective covers subject to the approval of the building official. Protective covers are commonly used to reduce either the potential for intentional false alarms or vandalism.

[F] 907.4 Power supply. The primary and secondary power supplies for the fire alarm system shall be provided in accordance with NFPA 72.

❖ The operation of fire alarm systems is essential to life safety in buildings and must be reliable in the event the normal power supply fails. To insure proper operation of fire alarm systems, this section requires that the primary and secondary power supplies are to be provided in accordance with NFPA 72. Note that Section 907.4 mandates the primary and secondary power supplies to be provided in accordance with NFPA 72 for all fire alarm systems. This is an additional requirement to the general requirements for electrical installations set forth in Chapter 27.

NFPA 72 offers three alternatives for secondary supply: a 24-hour storage battery; storage batteries with a 4-hour capacity and a generator or multiple generators.

Within the provisions of NFPA 72, it should be noted that the primary and secondary power supplies for remotely located control equipment essential to the system operation must also conform to the requirements for primary and secondary power supplies set forth in NFPA 72. Also, NFPA 72 contains requirements to monitor the integrity of primary power supplies and requires a backup power supply.

[F] 907.5 Wiring. Wiring shall comply with the requirements of the ICC *Electrical Code* and NFPA 72. Wireless protection systems utilizing radio-frequency transmitting devices shall comply with the special requirements for supervision of low-power wireless systems in NFPA 72.

❖ Wiring for fire alarm systems must be installed so that it is secure and will function reliably in an emergency. The code requires that the performance of wiring for fire alarm systems meet the requirements of NFPA 72. This requirement is in addition to the general requirements for electrical installations set forth in Chapter 27 and the ICC EC. For reliability, systems that use radio-frequency transmitting devices for signal transmission are required to have appropriate supervised transmitting and receiving equipment. Such supervisory equipment and its installation must conform to the special requirements contained in NFPA 72. This requirement is in addition to the general requirements for supervision set forth in Section 907.14.

[F] 907.6 Activation. Where an alarm notification system is required by another section of this code, it shall be activated by:

1. A required automatic fire alarm system.

2. Sprinkler water-flow devices.

3. Required manual fire alarm boxes.

❖ It is not the intent of this section to require that the various initiating devices contained in the list be identified. Rather, the section indicates that when such systems or devices are provided, they must serve as alarm-initiating devices. This section assumes that alarm notification appliances are required due to another section of the code that may require either a manual fire alarm or an automatic fire detection system.

[F] 907.7 Presignal system. Presignal systems shall not be installed unless approved by the building official and the fire department. Where a presignal system is installed, 24-hour personnel supervision shall be provided at a location approved by the fire department, in order that the alarm signal can be actuated in the event of fire or other emergency.

❖ Presignal fire alarm systems have been a contributing factor in several multiple-death fire incidents. In most instances, the staff failed to activate the general alarm in a timely manner and the occupants of the building were unaware of the fire emergency; therefore, the use of presignal systems is discouraged by the code. Presignal systems may only be used if approved by the building official and the fire department.

[F] 907.8 Zones. Each floor shall be zoned separately and a zone shall not exceed 22,500 square feet (2090 m²). The length of any zone shall not exceed 300 feet (91 440 mm) in any direction.

> **Exception:** Automatic sprinkler system zones shall not exceed the area permitted by NFPA 13.

❖ Since the fire alarm system also aids emergency personnel in locating the fire, the system must be zoned to enable a quick response to the fire area. Zoning is also critical if the fire alarm system initiates certain other fire protection systems or control features, such as smoke control systems.

At a minimum, each floor of a building must constitute one zone of the system. If the floor area exceeds 22,500 square feet (2,090 m²), additional zones are required. The maximum length of a zone is 300 feet (91 440 mm).

The exception clarifies that NFPA 13 provides maximum areas permitted to be protected by one sprinkler system and that the sprinkler system need not be designed to meet the 22,500-square-foot area (2,090 m²) limitations for a fire alarm system zone. For example, NFPA 13 permits a sprinkler system riser in a light-hazard occupancy to protect an area of 52,000 square feet (4,831 m²) per floor. In accordance with the exception, a single water-flow switch, and consequently a single fire alarm system zone, could be provided. If other alarm-initiating devices are present on the floor, they would need to be zoned separately to meet the 22,500-square-foot (2,090 m²) limitation.

[F] 907.8.1 Zoning indicator panel. A zoning indicator panel and the associated controls shall be provided in an approved location. The visual zone indication shall lock in until the system is reset and shall not be canceled by the operation of an audible alarm-silencing switch.

❖ The zoning indicator panel must be provided in a location approved by the building official. The panel should be located to permit ready access by emergency responders. Once an alarm-initiating device within a zone has been activated, the annunciation of the zone must lock in until the system is reset.

[F] 907.8.2 High-rise buildings. In buildings used for human occupancy that have floors located more than 75 feet (22 860 mm) above the lowest level of fire department vehicle access, a separate zone by floor shall be provided for all of the following types of alarm-initiating devices where provided:

1. Smoke detectors.
2. Sprinkler water-flow devices.
3. Manual fire alarm boxes.
4. Other approved types of automatic fire detection devices or suppression systems.

❖ In addition to at least one zone per floor, a separate zone for each indicated type of alarm-initiating device is to be provided in high-rise buildings. While such a feature may be desirable in all buildings, it is recognized that the incremental cost difference is substantially higher in low-rise buildings in which basic fire alarm systems are installed. State-of-the-art fire alarm systems provided in high-rise buildings enable such distinctive zoning to occur at a minimal cost difference.

[F] 907.9 Alarm notification appliances. Alarm notification appliances shall be provided and shall be listed for their purpose.

❖ The code requires that fire alarm systems be equipped with approved alarm notification appliances so that in an emergency, the fire alarm system will notify the occupants of the need for evacuation or implementation of the fire emergency plan. Alarm notification devices required by the code are of two general types: visible and audible. Except for voice/alarm signaling systems, once the system has been activated, all visible and audible alarms are required to activate. Voice/alarm signaling systems are special signaling systems that are activated on a selective basis in response to specific emergency conditions.

[F] 907.9.1 Visible alarms. Visible alarm notification appliances shall be provided in accordance with Sections 907.9.1.1 through 907.9.1.3.

Exceptions:

1. Visible alarm notification appliances are not required in alterations, except where an existing fire alarm system is upgraded or replaced, or a new fire alarm system is installed.

2. Visible alarm notification appliances shall not be required in exits as defined in Section 1002.1.

❖ The purpose of this section is to provide alarm system requirements for occupants who are hearing impaired. Visible alarm notification appliances should be located and oriented so that they will display alarm signals throughout a space.

Exception 1 clarifies that visible alarm devices are not required to be provided in previously approved existing fire alarm systems or as part of minor alterations to existing fire alarm systems. Extensive modifications to an existing fire alarm system such as an upgrade or replacement would require the installation of visible alarm devices even if the previous existing system neither had them nor required them.

In Exception 2, visible alarm devices are not required in exit elements due to the potential distraction during egress. Exits as defined in Section 1002.1 could include exit enclosures or exit passageways, but not exit access corridors.

[F] 907.9.1.1 Public and common areas. Visible alarm notification appliances shall be provided in public areas and common areas.

❖ Visible alarm notification appliances must provide coverage in all areas open to the public as well as all shared or common areas (e.g., corridors, public restrooms, shared offices, classrooms, etc.) Areas where visible alarm notification appliances are not required include private offices, mechanical rooms or similar spaces.

[F] 907.9.1.2 Employee work areas. Where employee work areas have audible alarm coverage, the wiring systems shall be designed so that visible alarm notification appliances can be integrated into the alarm system.

❖ This allows for those with hearing impairments to be accomodated as necessary. This provision also reduces the initial construction costs, as such alarms may not be necessary in every situation.

[F] 907.9.1.3 Groups I-1 and R-1. Group I-1 and R-1 sleeping units in accordance with Table 907.9.1.3 shall be provided with a visible alarm notification appliance, activated by both the in-room smoke alarm and the building fire alarm system.

❖ Fire alarm systems in Group I-1 and R-1 sleeping accommodations are required to be equipped with visible alarms in accordance with Table 907.9.1.3.

 The visible alarm notification devices in these rooms are to be activated by both the required in-room smoke alarm and the building fire alarm system. All visible alarm notification appliances in a building, however, need not be activated by individual room detectors. It is not a requirement that the accessible sleeping units be provided with visible alarm notification appliances even though some elderly patients or residents may be both mobility and hearing impaired.

[F] TABLE 907.9.1.3
VISIBLE AND AUDIBLE ALARMS

NUMBER OF SLEEPING UNITS	SLEEPING UNITS WITH VISIBLE AND AUDIBLE ALARMS
6 to 25	2
26 to 50	4
51 to 75	7
76 to 100	9
101 to 150	12
151 to 200	14
201 to 300	17
301 to 400	20
401 to 500	22
501 to 1,000	5% of total
1,001 and over	50 plus 3 for each 100 over 1,000

❖ This table specifies the minimum number of sleeping units that are to be equipped with visible and audible alarms. The numbers are based on the total number of sleeping accommodations provided. The requirements in this table are intended to be in concert with the *Americans with Disabilities Act Accessibility Guidelines for Buildings and Facilities* (ADAAG).

[F] 907.9.1.4 Group R-2. In Group R-2 occupancies required by Section 907 to have a fire alarm system, all dwelling units and sleeping units shall be provided with the capability to sup-

port visible alarm notification appliances in accordance with ICC A117.1.

❖ Group R-2 occupancies with a fire alarm system are required to have all dwelling units wired to support visible alarm notification appliances. This includes all dwelling and sleeping units, not just those classified as either Type A or B. By reference to Sections 1004.2 through 1004.4.4 of *ICC A117.1*, the building alarm system wiring must be extended to the unit smoke detectors so that audible/visible alarm notification appliances may be connected to the building fire alarm system to notify residents with hearing impairments of an emergency situation. Chapter 11 of this code contains additional information on the classification criteria and requirements for accessible dwelling units.

[F] 907.9.2 Audible alarms. Audible alarm notification appliances shall be provided and shall sound a distinctive sound that is not to be used for any purpose other than that of a fire alarm. The audible alarm notification appliances shall provide a sound pressure level of 15 decibels (dBA) above the average ambient sound level or 5 dBA above the maximum sound level having a duration of at least 60 seconds, whichever is greater, in every occupied space within the building. The minimum sound pressure levels shall be: 70 dBA in occupancies in Groups R and I-1; 90 dBA in mechanical equipment rooms and 60 dBA in other occupancies. The maximum sound pressure level for audible alarm notification appliances shall be 120 dBA at the minimum hearing distance from the audible appliance. Where the average ambient noise is greater than 105 dBA, visible alarm notification appliances shall be provided in accordance with NFPA 72 and audible alarm notification appliances shall not be required.

 Exception: Visible alarm notification appliances shall be allowed in lieu of audible alarm notification appliances in critical-care areas of Group I-2 occupancies.

❖ In order to attract the attention of building occupants, audible alarms must be distinctive, utilizing a sound that is unique to the signaling system and capable of being heard above the ambient noise in the space. In no case may the sound pressure exceed 120 dBA at the minimum hearing distance from the audible appliance. Sound pressures above 120 dBA can cause pain or even permanent hearing loss.

 It should be noted that in certain work areas, the Occupational Safety and Health Administration (OSHA) requires employees to wear hearing protection, possibly preventing them from hearing an audible alarm. Additionally, the noise factor in these areas is such that an audible alarm is not discernible. The primary method of indicating a fire can be by a visible signal. Employees must be capable of identifying such a signal as indicating a fire condition. The code user is advised that audible alarm-notification appliances are not required when visible alarm notification appliances are provided.

 The exception recognizes the incapacitated nature of the occupants in critical care areas of Group I-2 occupancies. The audible alarms may have the effect of unnecessarily disrupting the patients who are most likely not capable of self-preservation. Critical care areas are also assumed to be adequately staffed at all times and ready to respond appropriately upon activation of a visible alarm device.

[F] 907.10 Fire safety functions. Automatic fire detectors utilized for the purpose of performing fire safety functions shall be connected to the building's fire alarm control panel where a fire alarm system is required by Section 907.2. Detectors shall, upon actuation, perform the intended function and activate the alarm notification appliances or a visible and audible supervisory signal at a constantly attended location. In buildings not required to be equipped with a fire alarm system, the automatic fire detector shall be powered by normal electrical service and, upon actuation, perform the intended function. The detectors shall be located in accordance with NFPA 72.

❖ When the code requires the installation of automatic fire detectors in order to perform a specific function—such as elevator recall (see Section 3003.2), smokeproof enclosure ventilation (see Section 909.20) or when detectors are installed to comply with a permitted alternative, such as door-closing devices (see Section 715.3.7.3)—these detectors must be connected to the automatic fire alarm system if the building is required by the code to have such a system.

In addition to performing its intended function (e.g., closing a door), if a detector is activated, it must also activate either the building's alarm devices or a supervisory signal at a constantly attended location. This requirement recognizes that these detectors and the devices they control are part of the building's fire protection system and are expected to perform as designed. By requiring them to be connected to the automatic fire alarm system, they will have the supervision necessary to provide increased operational reliability.

An exception is provided for fire safety function detectors in buildings not required to have a fire alarm system. Fire safety function detectors must be powered by the building's electrical system and be located as required by NFPA 72. Without this exception, these detectors could not be expected to perform as intended, since the characteristics of a fire alarm system (e.g., power supply) would not be provided.

[F] 907.11 Duct smoke detectors. Duct smoke detectors shall be connected to the building's fire alarm control panel when a fire alarm system is provided. Activation of a duct smoke detector shall initiate a visible and audible supervisory signal at a constantly attended location. Duct smoke detectors shall not be used as a substitute for required open-area detection.

Exceptions:

1. The supervisory signal at a constantly attended location is not required where duct smoke detectors activate the building's alarm notification appliances.

2. In occupancies not required to be equipped with a fire alarm system, actuation of a smoke detector shall activate a visible and audible signal in an approved location. Smoke detector trouble conditions shall activate a

visible or audible signal in an approved location and shall be identified as air duct detector trouble.

❖ It is not the intent of this section to send a signal to the fire department or to activate the alarm notification devices within a building. Instead, this section requires a supervisory signal to be sent to a constantly attended location. Smoke detectors are required to be connected to a fire alarm system where such systems are required to be installed. Connection to the fire alarm system will activate a visible and audible supervisory signal at a constantly attended location that will alert building supervisory personnel that a smoke alarm has activated, and will also provide electronic supervision of the duct detectors, thereby monitoring any problems that may develop in the detector system circuitry or power supply. The signal must be sent to a location that is constantly attended by supervisory personnel while the building is occupied.

Exception 1 allows activation of the building's alarm notification appliances in lieu of a supervisory signal. Causing the building's fire alarm system to sound and indicate an alarm would alert the occupants of the building that an alarm condition existed within the air distribution system, thereby performing the same function as a supervisory signal sent to a constantly attended location.

Exception 2 recognizes the fact that not all buildings are required to be equipped with a fire alarm system. A visible and audible signal must be activated at an approved location that will alert building supervisory personnel to take appropriate action. Additionally, the duct smoke detectors must be electronically supervised to indicate trouble (system fault) in the detector system circuitry or power supply. A trouble condition must activate a distinct visible or audible signal at a location that will alert the responsible personnel.

[F] 907.12 Access. Access shall be provided to each detector for periodic inspection, maintenance and testing.

❖ Automatic fire detectors, especially smoke detectors, require periodic cleaning to reduce the likelihood of malfunction. As such, they are required by Section 907.20 of the IFC and NFPA 72 to be inspected and tested at regular intervals. Access to perform the required inspections, necessary maintenance and testing must be provided. This is a particularly important consideration for those detectors that are installed within a concealed space, such as an air duct.

[F] 907.13 Fire-extinguishing systems. Automatic fire-extinguishing systems shall be connected to the building fire alarm system where a fire alarm system is required by another section of this code or is otherwise installed.

❖ This section requires alternative automatic fire-extinguishing systems, such as a wet-chemical system for a commercial kitchen exhaust hood and duct system, to be monitored by the building fire alarm system when provided. Again, it should be noted that this section does not require electrical supervision of all fire-extinguishing systems but only those systems in buildings that contain a fire alarm system.

[F] 907.14 Monitoring. Where required by this chapter or the *International Fire Code*, an approved supervising station in accordance with NFPA 72 shall monitor fire alarm systems.

Exception: Supervisory service is not required for:

1. Single- and multiple-station smoke alarms required by Section 907.2.10.

2. Smoke detectors in Group I-3 occupancies.

3. Automatic sprinkler systems in one- and two-family dwellings.

❖ Fire alarm systems required by Section 907 are to be electrically supervised in accordance with NFPA 72. Exception 1 exempts single- and multiple-station smoke alarms from being supervised due to the potential for unwanted false alarms. Similarly, due to the concern over unwanted alarms, smoke detectors in Group I-3 occupancies need only sound an approved alarm signal that automatically notifies staff (see Section 907.2.6.2.1). Exception 2 recognizes that smoke detectors in Group I-3 occupancies are typically subject to misuse and abuse. Frequent unwanted alarms would negate the effectiveness of the system. Exception 3 clarifies that sprinkler systems in one- and two-family dwellings are not part of a dedicated fire alarm system and are typically designed in accordance with NFPA 13D, which does not require electrical supervision.

[F] 907.15 Automatic telephone-dialing devices. Automatic telephone-dialing devices used to transmit an emergency alarm shall not be connected to any fire department telephone number unless approved by the fire chief.

❖ Upon initiation of an alarm, supervisory or trouble signal, an automatic telephone-dialing device takes control of the telephone line for the reliability for the transmission of all signals. The device, however, should not be connected to the fire department telephone number so as to not disrupt any potential emergency (911) calls. NFPA 72 provides additional guidance on such devices including digital alarm communicator systems.

[F] 907.16 Acceptance tests. Upon completion of the installation of the fire alarm system, alarm notification appliances and circuits, alarm-initiating devices and circuits, supervisory-signal initiating devices and circuits, signaling line circuits, and primary and secondary power supplies shall be tested in accordance with NFPA 72.

❖ A complete performance test of the fire alarm system must be conducted to determine that the system is operating as required by the code. The acceptance test must include a test of each circuit, alarm-initiating device, alarm notification appliance and any supplementary functions, such as activation of closers and dampers. The operation of the primary and secondary

(emergency) power supplies must also be tested, as well as the supervisory function of the control panel. In accordance with Section 901.5, the responsibility for conducting the acceptance tests rests with the owner or the owner's representative.

NFPA 72 provides specific acceptance test procedures. Additional guidance on periodic testing and inspection can be also obtained from Section 907.20 of the IFC.

[F] 907.17 Record of completion. A record of completion in accordance with NFPA 72 verifying that the system has been installed in accordance with the approved plans and specifications shall be provided.

❖ In accordance with NFPA 72, this section requires a written statement from the installing contractor that the fire alarm system has been tested and installed in accordance with the approved plans and the manufacturer's specifications.

[F] 907.18 Instructions. Operating, testing and maintenance instructions, and record drawings ("as builts") and equipment specifications shall be provided at an approved location.

❖ To permit adequate testing, maintenance and troubleshooting for the installed fire alarm system, an owner's manual with complete installation instructions should be kept on site or at another approved location. The instructions should include a description of the system, operating procedures and testing and maintenance provisions.

[F] 907.19 Inspection, testing and maintenance. The maintenance and testing schedules and procedures for fire alarm and fire detection systems shall be in accordance with the *International Fire Code*.

❖ All fire alarms and fire detection systems are to be inspected, tested and maintained in accordance with Section 907.20 of the IFC and the applicable requirements in NFPA 72. It is the building owner's responsibility to maintain these systems in an operable condition at all times.

SECTION 908
EMERGENCY ALARM SYSTEMS

[F] 908.1 Group H occupancies. Emergency alarms for the detection and notification of an emergency condition in Group H occupancies shall be provided in accordance with Section 414.7.

❖ Emergency alarm systems provide indication and warning of emergency situations involving hazardous materials. An emergency alarm system is required in all Group H occupancies as indicated in Section 414.7 as well as Group H-5 HPM facilities. The Group H occupancy classification assumes the storage or use of hazardous materials exceeds the maximum allowable quantities specified in Tables 307.7(1) and 307.7(2).

An emergency alarm system should include an emergency alarm-initiating device outside each interior door of all hazardous material storage areas, a local alarm device and adequate supervision.

While Sections 908.4 (ozone gas-generator rooms), 908.5 (repair garages) and 908.6 (refrigerant detector) are not typically classified as Group H occupancies, the potential hazards associated with the noted occupancy conditions are such that additional means of early warning detection are required.

[F] 908.2 Group H-5 occupancy. Emergency alarms for notification of an emergency condition in an HPM facility shall be provided as required in Section 415.9.4.6. A continuous gas-detection system shall be provided for HPM gases in accordance with Section 415.9.7.

❖ In addition to hazardous material storage areas as regulated by Section 414.7, Section 415.9.4.6 also requires emergency alarms for all service corridors, exit access corridors and exit enclosures due to the potential transport of hazardous materials through these areas. Section 415.9.7 requires a continuous gas detection system in areas where HPM gas is used as a means of early detection of leaks. All gas detection systems are required to initiate a local alarm and transmit a signal to the emergency control station upon detection (see commentary, Sections 415.9.4.6 and 415.9.7).

[F] 908.3 Highly toxic and toxic materials. A gas detection system shall be provided for indoor storage and use of highly toxic and toxic gases to detect the presence of gas at or below the permissible exposure limit (PEL) or ceiling limit of the gas for which detection is provided. The system shall be capable of monitoring the discharge from the treatment system at or below one-half the IDLH limit.

Exception: A gas detection system is not required for toxic gases when the physiological warning properties are at a level below the accepted PEL for the gas.

❖ A gas detection system in the room or area utilized for indoor storage or the use of highly toxic or toxic gases provides early notification of a leak that is occurring before the escaping gas reaches hazardous exposure concentration levels. The exception recognizes that certain toxic compressed gases do not pose a severe exposure hazard. Those toxic gases whose properties under standard conditions are still below the 8-hour weighted average concentration for the permitted exposure limit (PEL) are exempt from the requirement for a gas detection system.

This section also specifies the discharge requirements for treatment system performance to establish a maximum allowable concentration of highly toxic or toxic gases at the point of discharge to the atmosphere. The concentration level of one-half immediately dangerous to life and health (IDLH) represents a minimum acceptable level of dilution at the point of discharge where the location of discharge is away from the general public. Where the treatment system processes more than one type of compressed gas, the maximum allowable concentration must be based on the release rate, quantity and IDLH for the gas that poses the worst-case release scenario.

[F] 908.3.1 Alarms. The gas detection system shall initiate a local alarm and transmit a signal to a constantly attended control station when a short-term hazard condition is detected. The alarm shall be both visible and audible and shall provide warning both inside and outside the area where gas is detected. The audible alarm shall be distinct from all other alarms.

Exception: Signal transmission to a constantly attended control station is not required when not more than one cylinder of highly toxic or toxic gas is stored.

❖ The required local alarm is intended to alert occupants to a hazardous condition in the vicinity of the inside storage room or area. The alarm is not intended to be an evacuation alarm; however, it is required to be monitored to hasten emergency personnel response. The exception allows the omission of supervision for the gas detection system where a single cylinder of highly toxic or toxic gas is stored. It should be noted that this section is only intended to apply when the maximum allowable quantity of a specific gas per control area is exceeded. A single cylinder would not require a gas detection system if it contained less than the maximum allowable quantities indicated in Tables 307.7(1) and 307.7(2).

[F] 908.3.2 Shutoff of gas supply. The gas detection system shall automatically close the shutoff valve at the source on gas supply piping and tubing related to the system being monitored for whichever gas is detected.

Exception: Automatic shutdown is not required for reactors utilized for the production of highly toxic or toxic compressed gases where such reactors are:

1. Operated at pressures less than 15 pounds per square inch gauge (psig) (103.4 kPa).

2. Constantly attended.

3. Provided with readily accessible emergency shutoff valves.

❖ Actuation of the gas detection system is required to close automatically all valves controlling highly toxic or toxic gases. This degree of protection is deemed necessary to provide a life safety measure in areas where highly toxic and toxic gases are being utilized for filling, dispensing or other operations.

In most situations, dispensing and use operations are assumed to be constantly attended because of the potential hazard and need to monitor operations involving highly toxic and toxic compressed gases.

The exception recognizes the decreased level of hazard for constantly attended operations involving reactors in manufacturing processes with low operating pressures. With low operating pressures, the rate at which the volume of the affected space would be filled by the released gas is minimized. Readily accessible manual shutoff valves would thus be permitted in lieu of an automatic-closing shutoff valve.

[F] 908.3.3 Valve closure. The automatic closure of shutoff valves shall be in accordance with the following:

1. When the gas-detection sampling point initiating the gas detection system alarm is within a gas cabinet or exhausted enclosure, the shutoff valve in the gas cabinet or exhausted enclosure for the specific gas detected shall automatically close.

2. Where the gas-detection sampling point initiating the gas detection system alarm is within a gas room and compressed gas containers are not in gas cabinets or exhausted enclosures, the shutoff valves on all gas lines for the specific gas detected shall automatically close.

3. Where the gas-detection sampling point initiating the gas detection system alarm is within a piping distribution manifold enclosure, the shutoff valve for the compressed container of specific gas detected supplying the manifold shall automatically close.

Exception: When the gas-detection sampling point initiating the gas-detection system alarm is at a use location or within a gas valve enclosure of a branch line downstream of a piping distribution manifold, the shutoff valve in the gas valve enclosure for the branch line located in the piping distribution manifold enclosure shall automatically close.

❖ This section specifies the conditions where the shutoff valve for the gas supply is required to automatically close. The requirement depends on the location of the gas detection sampling point, which would initiate the actuation of the gas detection system, and the use of gas cabinets, exhaust enclosures or a piping distribution manifold enclosure.

[F] 908.4 Ozone gas-generator rooms. Ozone gas-generator rooms shall be equipped with a continuous gas-detection system that will shut off the generator and sound a local alarm when concentrations above the PEL occur.

❖ In order to monitor the potential buildup of dangerous levels of ozone gas, a gas detection system is required that upon actuation will shut off the generator and sound a local alarm. Ozone gas generators are commonly used in water treatment applications. The ozone gas-generator room should not be a normally occupied area or be used for the storage of combustibles or other hazardous materials. Section 3705 of the IFC provides additional requirements for ozone gas generators.

[F] 908.5 Repair garages. A flammable-gas detection system shall be provided in repair garages for vehicles fueled by nonodorized gases in accordance with Section 406.6.6.

❖ As indicated in Section 406.6.6, an approved flammable-gas detection system is required for garages used for repair of vehicles fueled by nonodorized gases, such as hydrogen and nonodorized LNG. In order to prevent a hazardous potential buildup of flammable gas due to normal leakage and use conditions, the flammable-gas detection system is required to activate when the level of flammable gas exceeds 25 percent of the lower explosive limit (LEL) (see commentary, Section 406.6.6).

[F] 908.6 Refrigerant detector. Machinery rooms shall contain a refrigerant detector with an audible and visual alarm. The detector, or a sampling tube that draws air to the detector, shall be located in an area where refrigerant from a leak will concentrate. The alarm shall be actuated at a value not greater than the corresponding TLV-TWA values for the refrigerant classification indicated in the *International Mechanical Code*. Detectors and alarms shall be placed in approved locations.

Exception: Detectors are not required in ammonia system machinery rooms equipped with a vapor detector in accordance with the *International Mechanical Code*.

❖ A refrigerant-specific detector is required for the purpose of leak detection, early warning and actuation of emergency exhaust systems. Depending on the density of the refrigerants, leakage may collect near the floor, near the ceiling or disperse equally throughout. Refrigerant detector locations must be carefully considered. Most refrigerants are heavier than air; thus, floor depressions and pits are natural areas for accumulation. The code is silent on the location of sensors due to the endless variety of equipment room designs. The key to properly locating a detector in the machinery room is to remember that occupant safety is the primary objective, and the danger is in breathing refrigerant. Placing the sensor below the common breathing height of 5 feet (1525 mm) provides an additional safety margin because all commonly used halocarbon refrigerants are three to five times heavier than air. When undistributed by airflow, such escaping refrigerant will fall to the floor, seeking the lowest levels, and fill the room from the bottom up. Since pits, stairwells or trenches are likely to fill with refrigerant first, detectors should be also be placed in any of these areas that may be occupied. The alarm actuation threshold is dictated by the TLV-TWA values for the refrigerant classification in accordance with Table 1103.1 of the IMC.

Manufacturer's instructions for detectors provide installation guidance for the location and required number of detectors for any given room size.

The exception recognizes that ammonia refrigerant is "self-alarming" because of its strong odor and that ammonia machinery rooms are required to be continuously ventilated in accordance with the IMC.

Most general machinery rooms are unoccupied for long periods of time. Because of this, a refrigeration leak may go undetected, allowing a buildup of refrigerant that can pose a threat to the building occupants and the maintenance personnel who will be required to enter the machinery room. Also, the refrigerants may or may not be detectable by the sense of smell, depending upon the chemical nature and concentration in air of the refrigerant. This can be especially critical when a toxic refrigerant is used in the refrigeration system.

There are three levels of refrigerant exposure that are defined by the American Conference of Government Industrial Hygienists (ACGIH). Level 1 is the allowable exposure limit (AEL), which is the level at which a person can be exposed for 8 hours per day for 40 hours per week without having an adverse effect on health. At this level, a person exposed to the refrigerant should not suffer any adverse health effects. Level 2 is the short-term exposure limit (STEL), which is defined as three times the AEL. At this level, a person should not be exposed for more than 30 minutes at a time. Persons working in a machinery room having this concentration of refrigerant should be equipped with respiratory protection. Level 3 is the emergency exposure limit (EEL). At this level, persons should not be in the room without a self-contained breathing apparatus.

Early detection of leaking refrigerant is dependent upon the location of the refrigerant detectors. If improperly located, a refrigerant leak could go undetected for an undesirable period of time, thus allowing a significant amount of refrigerant to escape. It is therefore advantageous to locate the detectors in positions that will provide early detection and minimize the amount of refrigerant leakage.

Factors to be considered when choosing locations for detectors are the airflow patterns of the room, the particular refrigerant density and the fact that the primary hazard to the occupants is through inhalation. Specifically, halocarbon refrigerants are oxygen displacers; that is, they are heavier than air and will occupy the lowest areas of the room first. This suggests that the location of detectors should be below the breathing zone height. Additionally, detectors should be located so as to prevent the normal ventilation system from interfering with detection. Placing detectors between the refrigeration system and exhaust fan inlets should help to ensure that the presence of refrigerant will be detected. Depending on the size of the machinery room and the number and type of refrigeration systems, more than one detector may be necessary. Manufacturer's installation instructions for the refrigeration detection system should be followed when choosing the location and number of sensors for a particular machinery room application.

SECTION 909
SMOKE CONTROL SYSTEMS

❖ The intent of this section, as well as Sections 909 of the IFC and 513 of the IMC, is to provide a tenable environment for building occupants to or through that building occupants can evacuate or relocate in case of a fire. These provisions are not intended to protect contents, enable timely restoration of operations or facilitate fire suppression or overhaul activities. It is important to note that these provisions only apply when smoke control is required by other sections of the code. The only time the code mandates smoke control, in accordance with these methods, is for atriums. A covered mall would only require smoke control when it contains an atrium. Underground buildings require smoke control in accordance with Section 909.21.

In the last several years, smoke control provisions have become more complex. The reason is related to the fact that smoke is a complex problem, while a generic solution of six air changes has repeatedly and sci-

entifically been shown to be inadequate. Six air changes per hour does not take into account factors such as buoyancy; expansion of gases; wind; the geometry of the space and of communicating spaces; the dynamics of the fire; the production and distribution of smoke and the interaction of the building systems.

Smoke control systems can be either passive or active. Active systems are sometimes referred to as mechanical. Passive smoke control systems take advantage of smoke barriers surrounding the zone in which the fire event occurs or high bay areas that act as reservoirs to control the movement of smoke to other areas of the building. Active systems utilize pressure differences to contain smoke within the event zone or exhaust flow rates sufficient to retard the descent of the upper-level smoke accumulation to some predetermined position above necessary exit paths through the event zone. On rare occasions, there is also a possibility of controlling the movement of smoke horizontally by opposed airflow, but this method requires a specific architectural geometry to function properly that does not create an even greater hazard.

Essentially, there are three methods of mechanical or active smoke control that can be used separately or in combination within a design: pressurization, exhaust and, in rare and very special circumstances, opposed airflow.

Of course, all of these active approaches can be used in combination with the passive method.

Typically, the mechanical pressurization method is used in high-rise buildings when pressurizing stairways and for zoned smoke control. Pressurization is not practical in large open spaces such as atriums or malls, since it is difficult to develop the required pressure differences due to the large volume of the space.

The exhaust method is typically used in large open spaces such as atriums and malls. As noted, the pressurization method would not be practical within large spaces. The opposed airflow method, which basically uses a velocity of air horizontally to slow the movement of smoke, is typically applied in combination with either a pressurization method or exhaust method within hallways or openings into atriums and malls.

The application of each of these methods will be dependent on the specifics of the building design, which is why the provisions are contained primarily within the code. Smoke control within a building is fundamentally an architecturally driven problem. Different architectural geometries first dictate the need or lack thereof for smoke control, and then define the bounds of available solutions to the problem.

909.1 Scope and purpose. This section applies to mechanical or passive smoke control systems when they are required by other provisions of this code. The purpose of this section is to establish minimum requirements for the design, installation and acceptance testing of smoke control systems that are intended to provide a tenable environment for the evacuation or relocation of occupants. These provisions are not intended for the preservation of contents, the timely restoration of operations or for assistance in fire suppression or overhaul activities. Smoke control systems regulated by this section serve a different purpose than the smoke- and heat-venting provisions found in Section 910. Mechanical smoke control systems shall not be considered exhaust systems under Chapter 5 of the *International Mechanical Code*.

❖ This section is clarifying the intent of smoke control provisions, which is to provide a tenable environment to occupants during evacuation and relocation and not to protect the contents, enable timely restoration of operations or facilitate fire suppression and overhaul activities. Another element addressed in Section 909.1 is that smoke control systems serve a different purpose than smoke and heat vents. This eliminates any confusion that smoke and heat vents can be used as a substitution for smoke control. Additionally, a clarification is provided to note that smoke control systems are not considered an exhaust system in accordance with Chapter 5 of the IMC. This is due to the fact that such systems are unique in their operation and are not necessarily designed to exhaust smoke. It should be noted that smoke control provisions are also located in Chapter 5 of the IMC.

909.2 General design requirements. Buildings, structures or parts thereof required by this code to have a smoke control system or systems shall have such systems designed in accordance with the applicable requirements of Section 909 and the generally accepted and well-established principles of engineering relevant to the design. The construction documents shall include sufficient information and detail to adequately describe the elements of the design necessary for the proper implementation of the smoke control systems. These documents shall be accompanied by sufficient information and analysis to demonstrate compliance with these provisions.

❖ This section simply states that when smoke control systems are required by the code, the design is required to be in accordance with the provisions of this section. It is also stressed that such designs need to follow "generally accepted and well-established principles of engineering relevant to the design," essentially requiring a certain level of qualifications in the area of engineering to prepare such designs. It should be noted that each state in the U.S. typically requires minimum qualifications to undertake engineering design. An important resource when designing smoke control systems is the American Society of Heating, Refrigerating and Air-Conditioning Engineers' (ASHRAE's) *Design of Smoke Management Systems*.

A key element covered in this section is the need for detailed and clear construction documents so that the system is installed correctly. In most complex designs, the key to success is appropriate communication to the contractors as to what needs to be installed. The more complex a design becomes, the more likely there is to be construction errors. Most smoke control systems are complex, which is why special inspections in accordance with Section 909.3 and Chapter 17 of the code are critical for smoke control systems. Additionally, in order for the design to be accepted, analyses and justifi-

cations need to be provided in enough detail to evaluate for compliance.

909.3 Special inspection and test requirements. In addition to the ordinary inspection and test requirements which buildings, structures and parts thereof are required to undergo, smoke control systems subject to the provisions of Section 909 shall undergo special inspections and tests sufficient to verify the proper commissioning of the smoke control design in its final installed condition. The design submission accompanying the construction documents shall clearly detail procedures and methods to be used and the items subject to such inspections and tests. Such commissioning shall be in accordance with generally accepted engineering practice and, where possible, based on published standards for the particular testing involved. The special inspections and tests required by this section shall be conducted under the same terms in Section 1704.

❖ Due to the complexity and uniqueness of each design, special inspection and testing must be conducted. The designer needs to provide specific recommendations for special inspection and testing within his or her documentation. In fact, the code specifies in Chapter 17 that special inspection agencies for smoke control have expertise in fire protection engineering, mechanical engineering and certification as air balancers. Since the designs are unique to each building, there probably will not be a generic approach available to inspect and test such systems. The designer can and should, however, use any available published standards or guides when developing the special inspection and testing requirements for that particular design. ASHRAE Guideline 5 is a good starting place but only as a general outline. Each system will require a unique commissioning plan that can be developed only after careful and thoughtful examination of the final design and all of its components and interrelationships. Generally, these provisions may be included in design standards or engineering guides.

909.4 Analysis. A rational analysis supporting the types of smoke control systems to be employed, their methods of operation, the systems supporting them and the methods of construction to be utilized shall accompany the submitted construction documents and shall include, but not be limited to, the items indicated in Sections 909.4.1 through 909.4.6.

❖ This section indicates that simply determining airflow, exhaust rates and pressures to maintain tenable conditions is not adequate. There are many factors that could alter the effectiveness of a smoke control system, including stack effect, temperature effect of fire, wind effect, HVAC system interaction and climate. Each of these factors is addressed in the sections that follow. Additionally, the duration of operation of any smoke control system is mandated at 20 minutes or greater from the time the fire is detected. It should be noted that a proper engineering analysis, which takes into account people movement, may determine that a duration more than 20 minutes is necessary. There are also occasions when the movement of people can be demonstrated to require significantly less time than that required for the smoke to present a credible risk to their safety and, in such situations, rational analysis can serve as the technical justification for an alternate means of protection. The code cannot reasonably anticipate every conceivable building arrangement.

909.4.1 Stack effect. The system shall be designed such that the maximum probable normal or reverse stack effect will not adversely interfere with the system's capabilities. In determining the maximum probable stack effect, altitude, elevation, weather history and interior temperatures shall be used.

❖ Stack effect is the tendency for air to rise within a heated building when the temperature is colder on the exterior of the building. Reverse stack effect is the tendency for air to flow downward within a building when the interior is cooler than the exterior of the building. This air movement can affect the intended operation of a smoke control system. If stack effect is great enough, it may overcome the pressures determined during the design analyses and allow smoke to enter areas outside the zone of origin.

909.4.2 Temperature effect of fire. Buoyancy and expansion caused by the design fire in accordance with Section 909.9 shall be analyzed. The system shall be designed such that these effects do not adversely interfere with the system's capabilities.

❖ This section requires that the design account for the effect temperature may have on the success of the system. When air or any gases are heated they will expand. This expansion makes the gases lighter and therefore more buoyant. The buoyancy of hot gases is important when the design is to exhaust such gases from a location in or close to the ceiling; therefore, if sprinklers are part of the design, as required by Section 909, the gases may be significantly cooler than an unsprinklered fire, making it more difficult to remove the smoke and alter the plume dynamics. The fact that air expands when heated needs to be accounted for in the design.

When using the pressurization method, the expansion of hot gases needs to be accounted for, since it will take a larger volume of air to create the necessary pressure differences to maintain the area of fire origin in negative pressure. The expansion of the gases has the effect of pushing the hot gases out of the area of fire origin. Since sprinklers will tend to cool the gases, the effect of expansion is lower. The pressure differences required in Section 909.6.1 are specifically based on a sprinklered building. If the building is nonsprinklered, higher pressure differences may be required. The minimum pressure difference for certain unsprinklered ceiling height buildings is as follows:

Ceiling height (feet)	Minimum pressure difference (inch water gage)
9	.10
15	.14
21	.18

This is a very complex issue that needs to be part of the design analysis. It needs to address the type and reaction of the fire protection systems, ceiling heights and the size of the design fire.

909.4.3 Wind effect. The design shall consider the adverse effects of wind. Such consideration shall be consistent with the wind-loading provisions of Chapter 16.

❖ The effect of wind on a smoke control system within a building is very complex. It is generally known that wind exerts a load upon a building. The loads are looked at as windward (positive pressure) and leeward (negative pressure). The velocity of winds will vary based on the terrain and the height above grade; therefore, the height of the building and surrounding obstructions will have an effect on these velocities. These pressures alter the operation of fans, especially propeller fans, thus altering the pressure differences and airflow direction in the building. There is not an easy solution to dealing with these effects. In fact, little research has been done in this area.

It should be noted that in larger buildings a wind study is normally undertaken for the structural design. The data from those studies can be used in the analysis of the effects on the pressures and airflow within the building with regard to the performance of the smoke control system.

909.4.4 HVAC systems. The design shall consider the effects of the heating, ventilating and air-conditioning (HVAC) systems on both smoke and fire transport. The analysis shall include all permutations of systems status. The design shall consider the effects of the fire on the HVAC systems.

❖ If not properly configured to shut down or included as part of the design, the HVAC system can alter the smoke control design. More specifically, if dampers are not provided between smoke zones within the HVAC system ducts, smoke could be transported from one zone to another. Additionally, if the HVAC system places more supply air than assumed for the smoke control system design, the velocity of the air may adversely affect the fire plume or a positive pressure may be created. Generally, an analysis of the smoke control design and the HVAC system in all potential modes must occur and be noted within the design documentation.

909.4.5 Climate. The design shall consider the effects of low temperatures on systems, property and occupants. Air inlets and exhausts shall be located so as to prevent snow or ice blockage.

❖ This section is focused on properly protecting equipment from weather conditions that may affect the reliability of the design. For instance, extremely cold or hot air may damage critical equipment within the system when pulled directly from the outside. Some listings of duct smoke detectors are for specific temperature ranges; therefore, placing such detectors within areas exposed to extreme temperatures may void the listing. Also, the equipment and air inlets and outlets should be designed and located so as to not collect snow and ice that could block air from entering or exiting the building.

909.4.6 Duration of operation. All portions of active or passive smoke control systems shall be capable of continued operation after detection of the fire event for not less than 20 minutes.

❖ The intent of the smoke control provisions is to provide a tenable environment for occupants to either evacuate or relocate to a safe place. Evacuation and relocation activities include notifying occupants, possible investigation time for the occupants, decision time and the actual travel time. In order to achieve this goal, the code has established 20 minutes as a minimum time for evacuation or relocation. A proper engineering analysis of people movement in relationship to smoke development may result in a longer duration of operations; therefore, systems must be designed to keep running or be effective for at least 20 minutes. It is stressed that the 20-minute duration begins after the detection of the fire event, since occupants need to be alerted before evacuation can occur.

909.5 Smoke barrier construction. Smoke barriers shall comply with Section 709, and shall be constructed and sealed to limit leakage areas exclusive of protected openings. The maximum allowable leakage area shall be the aggregate area calculated using the following leakage area ratios:

1. Walls: $A/A_w = 0.00100$
2. Exit enclosures: $A/A_w = 0.00035$
3. All other shafts: $A/A_w = 0.00150$
4. Floors and roofs: $A/A_F = 0.00050$

where:

A = Total leakage area, square feet (m^2).

A_F = Unit floor or roof area of barrier, square feet (m^2).

A_w = Unit wall area of barrier, square feet (m^2).

The leakage area ratios shown do not include openings due to doors, operable windows or similar gaps. These shall be included in calculating the total leakage area.

❖ Part of the strategy of smoke control systems is the use of smoke barriers to divide a building into separate smoke zones. This strategy is used in both passive and mechanical systems. It should be noted that not all walls, ceilings or floors would be considered smoke barriers. Only walls that designate separate smoke zones within a building need to be constructed as smoke barriers. This section is simply providing requirements for walls, floors and ceilings that are used as smoke barriers. It should be noted that it is possible that a smoke control system utilizing the exhaust method may not need to utilize a smoke barrier to divide the building into separate smoke zones; therefore, the evaluation of barrier construction and leakage area may not be necessary.

In order for smoke to not travel from one smoke zone to another, specific construction requirements are necessary in accordance with the code. It should be noted

that openings such as doors and windows are dealt with separately within Section 909.5.2 from openings such as cracks or penetrations.

909.5.1 Leakage area. The total leakage area of the barrier is the product of the smoke barrier gross area monitored by the allowable leakage area ratio, plus the area of other openings such as gaps and operable windows. Compliance shall be determined by achieving the minimum air pressure difference across the barrier with the system in the smoke control mode for mechanical smoke control systems. Passive smoke control systems tested using other approved means such as door fan testing shall be as approved by the building official.

❖ It is impossible for walls and floors to be constructed that are completely free from openings that may allow the migration of smoke; therefore, leakage needs to be compensated for within the design by calculating the leakage area of walls, ceilings and floors. The factors provided in Section 909.5, which originate from ASHRAE's provisions on leaky buildings, are used to calculate the total leakage area. The total leakage area is then used in the design process to determine the proper amount of air to create the required pressure differences across these surfaces that form smoke zones. These pressure differences then need to be verified when the system is in smoke control mode.

Additionally, Section 909.5 provides ratios to determine the maximum allowable leakage in walls, exit enclosures, shafts, floors and roofs. These leakage areas are critical in determining whether the proper pressure differences are provided when utilizing the pressurization method of smoke control. Pressure differences will decrease as the openings get larger.

909.5.2 Opening protection. Openings in smoke barriers shall be protected by automatic-closing devices actuated by the required controls for the mechanical smoke control system. Door openings shall be protected by fire door assemblies complying with Section 715.3.3.

Exceptions:

1. Passive smoke control systems with automatic-closing devices actuated by spot-type smoke detectors listed for releasing service installed in accordance with Section 907.10.

2. Fixed openings between smoke zones which are protected utilizing the airflow method.

3. In Group I-2, where such doors are installed across corridors, a pair of opposite-swinging doors without a center mullion shall be installed having vision panels with approved fire-rated glazing materials in approved fire-rated frames, the area of which shall not exceed that tested. The doors shall be close fitting within operational tolerances and shall not have undercuts, louvers or grilles. The doors shall have head and jamb stops, astragals or rabbets at meeting edges, and automatic-closing devices. Positive-latching devices are not required.

4. Group I-3.

5. Openings between smoke zones with clear ceiling heights of 14 feet (4267 mm) or greater and bank-down capacity of greater than 20 minutes as determined by the design fire size.

❖ Similar to concerns of smoke leakage between smoke zones, openings may compromise the necessary pressure differences between smoke zones. Openings in smoke barriers, such as doors and windows, must be either constantly or automatically closed when the smoke control system is operating. This section requires that doors be automatically closed through the activation of an automatic closing device linked to the smoke control system. Essentially, when the smoke control system is activated, all openings are automatically closed. This most likely would mean that the mechanism that activates the smoke control system would also automatically close all openings. More than likely, the smoke control system will be activated by a specifically zoned smoke detection or sprinkler system as required by Sections 909.12.2 and 909.12.3.

In terms of actual opening protection, Section 909.5.2 is simply referring the user to Section 715.3.3 for specific construction requirements for doors located in smoke barriers. Note that smoke barriers are different from a fire-resistance-rated barrier, since the intended measure of performance is different. One is focused on fire spread from the perspective of heat, the other from the perspective of smoke passage.

There are several exceptions to this particular section. Exception 1 is specifically for passive systems. Passive systems, as noted, are systems in which there is no use of mechanical systems. Instead, the system operates primarily upon the configuration of barriers and layout of the building to provide smoke control. Passive systems can use spot-type detectors to close doors that constitute portions of a smoke barrier. Essentially, this means a full fire alarm system would not be required. Instead, single station detectors would be allowed to close the doors. Such doors would need to fail in the closed position if power is lost. The specifics as to approved devices would be found in NFPA 72.

Exception 2 is based on the fact that some systems take advantage of the opposed airflow method such that smoke is prevented from migrating past the doors. Therefore, since the design already accounts for potential smoke migration at these openings through the use of air movement, it is unnecessary to require the barrier to be closed.

Exception 3 is specifically related to the unique requirements for Group I-2 occupancies. Essentially, a very specific alternative, which meets the functional needs of Group I-2 occupancies, is provided. One aspect of the alternative approach is that doors have vision panels with approved fire-rated glazing in fire-rated frames of a size that does not exceed the type tested.

Exception 4 allows an exemption from the automatic-closing requirements for all Group I-3 occupancies. This is related to the fact that facilities that have occupants under restraint or with specific security restrictions have unique requirements in accordance with Section 408 of the code. These requirements accomplish the intent of providing reliable barriers between each smoke zone since, for the most part, such facilities will have a majority of doors closed and in a locked position due to the nature of the facility. The staff very closely controls these types of facilities.

Exception 5 relates to the behavior of smoke. The assumption is that smoke rises due to the buoyancy of hot gases and if the ceiling is sufficiently high, the smoke layer will be contained for a longer period of time before it begins to move into the next smoke zone. Therefore, it is not as critical that the doors automatically close. This allowance is dependent on the specific design fire for a building. Different size design fires create different amounts of smoke that, depending on the layout of the building, may migrate in different ways throughout the building. This section mandates that smoke cannot begin to migrate into the next smoke zone for at least 20 minutes. This is consistent with the 20-minute duration of operation of smoke control systems in Section 909.4.6. It should be noted that a minimum of 14-foot (4267 mm) ceilings are required to take advantage of this exception.

909.5.2.1 Ducts and air transfer openings. Ducts and air transfer openings are required to be protected with a minimum Class II, 250°F (121°C) smoke damper complying with Section 716.

❖ Another factor that adds to the reliability of smoke barriers is the protection of ducts and air transfer openings within smoke barriers. Left open, these openings may allow the transfer of smoke between smoke zones. These ducts and air transfer openings most often are part of the HVAC system. Damper operation and the reaction with the smoke control system will be evaluated during acceptance testing. It should be noted that there are duct systems used within a smoke control design that are controlled by the smoke control system and should not automatically close upon detection of smoke via a smoke damper.

It should be noted that a smoke damper works differently than a fire damper. Fire dampers react to heat via a fusible link, while smoke dampers activate upon the detection of smoke. The dampers used should be rated as Class II, 250°F (121°C). The class of the smoke damper refers to its level of performance relative to leakage. The temperature rating is related to its ability to withstand the heat of smoke resulting from a fire. It should be noted that although smoke barriers are only required to utilize smoke dampers, there may be many instances where a fire damper is also required. For instance, the smoke barrier may also be used as a fire barrier. Also, Section 716.5.3.1 would require penetration of shafts to contain both a smoke and fire damper. Therefore, in some cases both a smoke damper and fire damper would be required. There are listings specific to combination smoke and fire dampers.

More specific requirements about dampers can be found in Chapter 7 of the code and Chapter 6 of the IMC.

909.6 Pressurization method. The primary mechanical means of controlling smoke shall be by pressure differences across smoke barriers. Maintenance of a tenable environment is not required in the smoke control zone of fire origin.

❖ There are several methods or strategies that may be used to control smoke. One of these methods is pressurization, wherein the system primarily utilizes pressure differences across smoke barriers to control the movement of smoke. Basically, if the area of fire origin maintains a negative pressure, then the smoke will be contained to that smoke zone. A typical approach used to obtain a negative pressure is to exhaust the fire floor. This is a fairly common practice in high-rise buildings. Stairway enclosures also utilize the concept of pressurization. The pressurization method in large open spaces, such as malls and atria, is impractical since it would take a large quantity of supply air to create the necessary pressure differences. It should be noted that pressurization is mandated as the primary method for mechanical smoke control design. Airflow and exhaust methods are only allowed when appropriate.

This type of smoke control does not require that tenable conditions be maintained in the smoke zone where the fire originates. Maintaining this area tenable would be impossible, based on the fact that pressures from the surrounding smoke zones would be placing a negative pressure within the zone of origin to keep the smoke from migrating.

Pressurization is used often with exit stair enclosures. This method provides a positive pressure within the stair enclosure to resist the passage of smoke. Stair pressurization is one method of compliance for stairways in high-rise or underground buildings where the floor surface is located more than 75 feet (22 860 mm) above the lowest level of fire fighter access or more than 30 feet (9144 mm) below the level of exit discharge. It should be noted that there are two methods found in the code that address smoke movement— smokeproof enclosures or pressurized stairs. A smokeproof enclosure requires a certain fire-resistance rating along with access through a ventilated vestibule or an exterior balcony. The vestibule can be ventilated in two ways: using natural ventilation or mechanical ventilation as outlined in Sections 909.20.3 and 909.20.4. The pressurization method requires a sprinklered building and a minimum pressure difference of .15 inch (37 Pa) of water and a maximum of .35 inches (87 Pa) of water. These pressure differences are to be available with all doors closed under maximum stack pressures (see Sections 909.20 and 1019.1.8 for more details).

As noted, the pressurization method utilizes pressure differences across smoke barriers to achieve control of smoke. Sections 909.6.1 and 909.6.2 provide the criteria for smoke control design in terms of minimum and maximum pressure differences.

In summary, the pressurization method is used in two ways. The first is through the use of smoke zones where the zone of origin is exhausted, creating a negative

pressure. The second is stair pressurization that creates a positive pressure within the stair to avoid the penetration of smoke. Note that the code allows the use of a smokeproof enclosure instead of pressurization.

909.6.1 Minimum pressure difference. The minimum pressure difference across a smoke barrier shall be 0.05-inch water gage (0.0124 kPa) in fully sprinklered buildings. In buildings permitted to be other than fully sprinklered, the smoke control system shall be designed to achieve pressure differences at least two times the maximum calculated pressure difference produced by the design fire.

❖ The minimum pressure difference is established as .05-inch (12 Pa) water gauge in fully sprinklered buildings. This particular criterion is related to the pressures needed to overcome buoyancy and the pressures generated by the fire, which include expansion. This particular criterion is based upon a sprinklered building. The pressure differences would need to be higher in a building that is not sprinklered. Additionally, these pressure differences need to be provided based upon the possible stack and wind effects present.

909.6.2 Maximum pressure difference. The maximum air pressure difference across a smoke barrier shall be determined by required door-opening or closing forces. The actual force required to open exit doors when the system is in the smoke control mode shall be in accordance with Section 1008.1.2. Opening and closing forces for other doors shall be determined by standard engineering methods for the resolution of forces and reactions. The calculated force to set a side-hinged, swinging door in motion shall be determined by:

$$F = F_{dc} + K(WA\Delta P)/2(W\text{-} d) \qquad \textbf{(Equation 9-1)}$$

where:

A = Door area, square feet (m²).

d = Distance from door handle to latch edge of door, feet (m).

F = Total door opening force, pounds (N).

F_{dc} = Force required to overcome closing device, pounds (N).

K = Coefficient 5.2 (1.0).

W = Door width, feet (m).

ΔP = Design pressure difference, inches of water (Pa).

❖ The maximum pressure difference is based primarily upon the force needed to open and close doors. The code establishes maximum opening forces for doors. This maximum opening force cannot be exceeded, taking into account the pressure differences across a doorway in a pressurized environment. Essentially, based on the opening force requirements of Section 1008.1.2, the maximum pressure difference can be calculated in accordance with Equation 9-1. In accordance with

Chapter 10, the maximum opening force of a door has three components, including:

Door latch release:
Maximum of 15 pounds (67 N)
Set door in motion:
Maximum of 30 pounds (134 N)
Swing to full open position:
Maximum of 15 pounds (67 N)

Equation 9-1 is used to calculate the total force to set the door into motion when in the smoke control mode; therefore, the limiting criteria would be 30 pounds (134 N). Accessibility requirements would further limit these maximum opening forces.

909.7 Airflow design method. When approved by the building official, smoke migration through openings fixed in a permanently open position, which are located between smoke control zones by the use of the airflow method, shall be permitted. The design airflow shall be in accordance with this section. Airflow shall be directed to limit smoke migration from the fire zone. The geometry of openings shall be considered to prevent flow reversal from turbulent effects.

❖ This method is only allowed when approved by the building official. As the title states, this method utilizes airflow to avoid the migration of smoke across smoke barriers. This has been referred to as opposed airflow. Specifically, this method is suited for the protection of smoke migration through doors and related openings fixed in a permanently open position. This method consists of providing a particular velocity of air based upon the temperature of the smoke and the height of the opening. The temperature of the smoke will depend on the design fire that is established for the particular building. The higher the temperature of the smoke and the larger the opening, the higher the velocity necessary to maintain the smoke from migrating into the smoke zone. It should be noted that the airflow method seldom works for large openings, since the velocity to oppose the smoke becomes too high. This method tends to work better for smaller openings, such as pass-through windows. Equation 9-2 provides the method to calculate the necessary velocity.

909.7.1 Velocity. The minimum average velocity through a fixed opening shall not be less than:

$$v = 217.2 \left[h \left(T_f - T_o \right) / \left(T_f + 460 \right) \right]^{1/2} \qquad \textbf{(Equation 9-2)}$$

For SI: $v = 119.9 \left[h \left(T_f - T_o \right) / T_f \right]^{1/2}$

where:

h = Height of opening, feet (m).

T_f = Temperature of smoke, °F (°K).

T_o = Temperature of ambient air, °F (°K).

v = Air velocity, feet per minute (m/minute).

❖ This section provides the formula for the minimum average velocity through a fixed opening. The minimum ve-locity is based on the velocity needed to prevent the smoke from migrating into the smoke zone. See the commentary to Section 909.7 for further discussion.

909.7.2 Prohibited conditions. This method shall not be employed where either the quantity of air or the velocity of the airflow will adversely affect other portions of the smoke control system, unduly intensify the fire, disrupt plume dynamics or interfere with exiting. In no case shall airflow toward the fire exceed 200 feet per minute (1.02 m/s). Where the formula in Section 909.7.1 requires airflow to exceed this limit, the airflow method shall not be used.

❖ The airflow method has a limitation on maximum velocity. This limitation is based upon the fact that air may distort the flame and cause additional entrainment and turbulence; therefore, having a high velocity of air entering the zone of fire origin has the potential of increasing the amount of smoke produced. The velocity may also interact with other portions of the smoke control design. For instance, the pressure differences in other areas of the building may be altered, which may exceed the limitations of Sections 909.6.1 and 909.6.2. This section requires that when a velocity of over 200 feet per minute (1.02 m/sec) is calculated, the airflow method is not allowed. The solution may result in requiring a barrier such as a wall or door.

If the airflow design method is chosen to protect areas communicating with an atrium, the air added to the smoke layer needs to be accounted for in the exhaust rate.

909.8 Exhaust method. When approved by the building official, mechanical smoke control for large enclosed volumes, such as in atriums or malls, shall be permitted to utilize the exhaust method. The design exhaust volumes shall be in accordance with this section.

❖ This method is only allowed when approved by the building official. The primary application of the exhaust method is in large spaces, such as atriums and malls. The strategy of this method is to pull the smoke out of the space based on the understanding of the amount of smoke being produced by a particular size fire. Essentially, fires produce different amounts and patterns of smoke based on the material being burned and the placement of the fire; therefore, several equations representing different fire plume configurations are presented to determine the mass loss rate that is then converted to an exhaust rate. The plume calculation that yields the highest exhaust rate will determine the exhaust rate for the space; however, only realistic plume configurations should be considered. The three plume configurations that will each be analyzed in this section are: axisymmetric plumes, balcony spill plumes and window plumes.

Additionally, when a fire plume is in contact with a wall, the mass flow rate can be adjusted in accordance with Equation 9-7 in Section 909.8.5.

909.8.1 Exhaust rate. The height of the lowest horizontal surface of the accumulating smoke layer shall be maintained at least 10 feet (3048 mm) above any walking surface which forms a portion of a required egress system within the smoke zone. The required exhaust rate for the zone shall be the largest of the calculated plume mass flow rates for the possible plume configurations. Provisions shall be made for natural or mechanical supply of air from outside or adjacent smoke zones to make up for the air exhausted. Makeup airflow rates, when measured at the potential fire location, shall not exceed 200 feet per minute (60 960 mm per minute) toward the fire. The temperature of the makeup air shall be such that it does not expose temperature-sensitive fire protection systems beyond their limits.

❖ The design criteria of this method is to keep smoke at least 10 feet (3048 mm) above any walking surface that is considered part of the required egress within the particular smoke zone, such as an atrium, for at least 20 minutes. Chapter 10 of the code considers the majority of occupiable space as part of the means of egress system. Also keep in mind that the criteria of 10 feet (3048 mm) does not apply just to the floor surface of the mall or atrium but to any level where occupants may be exposed. Again, the required exhaust volume is based upon the largest mass loss rate from each of the plume configurations in Sections 909.8.2 through 909.8.5; however, only realistic plume configurations should be considered.

In order to exhaust the air from the space, make-up air is required. There are a few considerations related to make-up air. First, similar to the airflow method, a 200-foot-per-minute (1.02 m/sec) limitation is placed upon the speed of the make-up air, since it may disturb the plume dynamics and potentially increase smoke production. The second consideration is the temperature of the incoming air, primarily when the air could be very low in temperature. Fire protection equipment may be temperature sensitive and could fail as a result of exposure to extreme temperatures.

There are several considerations when automatic doors are depended upon for the make-up air in a smoke control system. First, depending on the climate, the temperature of the air needs to be accounted for just like any another make-up air entering the building. Additionally, the climate and location of the opening may have varying air velocities due to the effects of winds. Finally, the owners and users of the building must fully understand the implication of a door that is designed to remain open during a fire with regard to security concerns. If the owner and user cannot tolerate doors being open, then other forms of intake must be explored.

909.8.2 Axisymmetric plumes. The plume mass flow rate (m_p), in pounds per second (kg/s), shall be determined by placing the design fire center on the axis of the space being analyzed. The limiting flame height shall be determined by:

$$z_l = 0.533 Q_c^{2/5} \qquad \text{(Equation 9-3)}$$

For SI: $z_l = 0.166 Q_c^{2/5}$

where:

m_p = Plume mass flow rate, pounds per second (kg/s).

Q = Total heat output.

Q_c = Convective heat output, British thermal units per second (kW). (The value of Q_c shall not be taken as less than $0.70Q$).

z = Height from top of fuel surface to bottom of smoke layer, feet (m).

z_l = Limiting flame height, feet (m). The z_l value must be greater than the fuel equivalent diameter (see Section 909.9).

for $z > z_l$

$$m_p = 0.022 Q_c^{1/3} z^{5/3} + 0.0042 Q_c$$

For SI: $m_p = 0.071 \, Q_c^{1/3} z^{5/3} + 0.0018 Q_c$

for $z = z_l$

$$m_p = 0.011 \, Q_c$$

For SI: $m_p = 0.035 Q_c$

for $z < z_l$

$$m_p = 0.0208 Q_c^{3/5} z$$

For SI: $m_p = 0.032 Q_c^{3/5} z$

To convert m_p from pounds per second of mass flow to a volumetric rate, the following equation shall be used:

$$V = 60 \, m_p / \rho \qquad \text{(Equation 9-4)}$$

where:

V = Volumetric flow rate, cubic feet per minute (m³/s).

ρ = Density of air at the temperature of the smoke layer, pounds per cubic feet (T: in °F) [kg/m³ (T: in °C)].

❖ This section provides the methodology to determine the required exhaust rate based upon a fire burning away from walls, windows and other factors that may affect it working in a symmetric manner. This type of fire is able to entrain air into the plume from all sides until it reaches the bottom of the smoke layer. See Figure 909.8.2(1) for an illustration of an axisymmetric plume in an atria.

The user must first determine the limiting flame height through Equation 9-3. This height is then used as the criteria to determine the mass flow rate equation. Figure 909.8.2(1) illustrates the limiting flame height, Z_l. Different behaviors are observed at different flame height ranges. This is based upon the fact that at different heights, air is entrained differently and ultimately affects the output of the fire. When the smoke layer is at the same height or lower than the flame height, less air is being entrained into the fire and the hot gases released from the fire are reradiating heat back onto the fire; therefore, less smoke can be produced, since less air is being entrained into the plume when the height of the smoke layer is equal to or less than the flame height.

The height, z, noted in the equation will be based upon the design criteria of 10 feet (3048 mm) above the occupants for 20 minutes. For instance, Figure 909.8.2(2) indicates an atrium 80 feet (24 384 mm) tall with a balcony with a walking surface at 56 feet (17 069 mm) would require that the smoke layer be at least 66 feet (20 117 mm) above the floor; therefore, $z = 66$ feet (20 117 mm). The design will then be based upon keeping the smoke layer at least 66 feet (20 117 mm) (56 feet + 10 feet) above the fuel source. It is assumed that the fuel source is close to the floor.

The actual exhaust volume is determined based on the mass flow rate. The mass flow rate expresses how much mass is being released based upon the rate of combustion and air entrainment. As noted above, the air entrainment is reduced when the smoke layer height is less than or equal to the flame height and increases as the height differential is increased. When the height differential is positive (i.e., z is greater than the limiting flame height), the mass flow rate increases to the 5/3 power. The ability to remove the products of combustion is based upon the calculated mass flow rate. An exhaust volume that can pull that mass out of the space is necessary. The exhaust volume relates to the density of the air at the fire temperature and mass flow rate, as Density = Mass/Volume. Equation 9-4 uses this relationship to determine the necessary exhaust volume.

909.8.3 Balcony spill plumes. The plume mass flow rate (m_p) for spill plumes shall be determined using the geometrically probable width based on architectural elements and projections in the following equation:

$$m_p = 0.124(QW^2)^{1/3}(z_b + 0.25H) \qquad \textbf{(Equation 9-5)}$$

For SI: $m_p = 0.36(QW^2)^{1/3}(z_b + 0.25H)$

where:

H = Height above fire to underside of balcony, feet (m).

m_p = Plume mass flow rate, pounds per second (kg/s).

Q = Total heat output.

W = Plume width at point of spill, feet (m).

z_b = Height from balcony, feet (m).

❖ The next type of plume for which the mass flow rate is to be determined is a balcony spill plume. This plume occurs when a fire occurs under a balcony and the smoke travels upward and around the balcony. Equation 9-5 specifically provides a method to calculate the mass flow rate for this particular configuration. The mass flow rate resulting from this calculation is then taken and applied to Equation 9-4 to determine the necessary exhaust volume. This particular calculation is somewhat difficult since it is difficult to determine the plume width

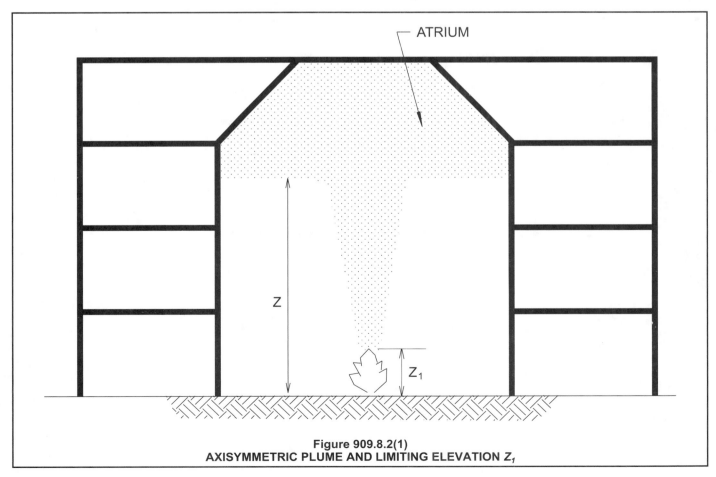

Figure 909.8.2(1)
AXISYMMETRIC PLUME AND LIMITING ELEVATION Z_1

at the point of spill. Additionally, the larger the width used in Equation 9-7, the larger the mass flow rate. Since this width is difficult to determine and because a smaller width yields a lower exhaust rate, architectural features such as glass or plexiglass draft curtains are installed or can be used to determine a particular plume width. The mass flow rate also depends on the height above the fire to the underside of the balcony.

909.8.4 Window plumes. The plume mass flow rate (m_p) shall be determined from:

$$m_p = 0.077(A_wH_w^{1/2})^{1/3}(z_w+a)^{5/3} + 0.18A_wH_w^{1/2}$$

(Equation 9-6)

For SI: $m_p = 0.68(A_wH_w^{1/2})^{1/3}(z_w + a)^{5/3} + 1.5A_wH_w^{1/2}$

where:

A_w = Area of the opening, square feet (m²).

H_w = Height of the opening, feet (m).

m_p = plume mass flow rate, pounds per second (kg/s).

z_w = Height from the top of the window or opening to the bottom of the smoke layer, feet (m).

a $= 2.4A_w^{2/5}H_w^{1/5} - 2.1H_w.$

❖ Another scenario in terms of possible fire plumes and smoke generation is the window plume. The type of fire this equation is addressing could be a fire in a hotel room that is located within an atrium. The fire may break through the window and allow smoke to be released into the atrium. The key data necessary for this equation beyond the heat release rate (fire size) are the area and the height of the opening. These two elements affect the behavior of the fire and how the smoke is generated. Also, the variable z_w, which is the height from the top of the window or opening to the bottom of the smoke layer, must be determined. Again, the criteria z_w will depend on the design criteria of 10 feet (3048 mm). This height is important since it helps determine where the plume stops and the smoke layer begins. It is at this point that the plume is limited in the mass loss rate it can produce. This calculation is only to be used with one opening. Again, as with the axisymmetric and balcony spill plumes, the mass loss rate in pounds per second needs to be converted to a volumetric flow rate in accordance with Equation 9-4.

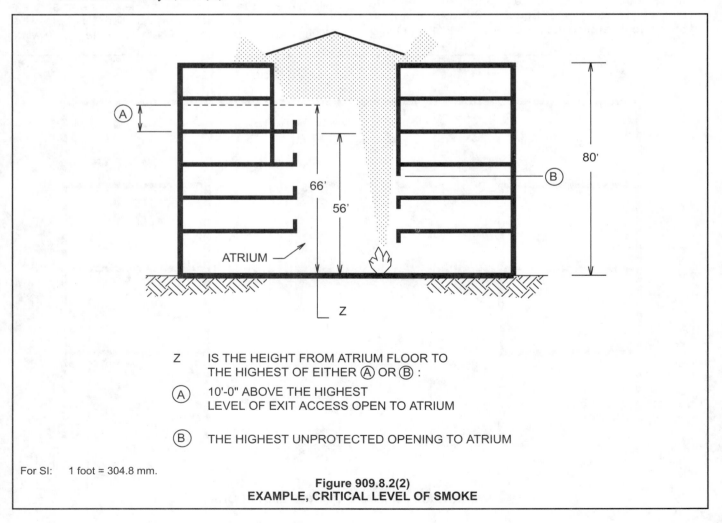

Z   IS THE HEIGHT FROM ATRIUM FLOOR TO
      THE HIGHEST OF EITHER (A) OR (B) :

(A)   10'-0" ABOVE THE HIGHEST
      LEVEL OF EXIT ACCESS OPEN TO ATRIUM

(B)   THE HIGHEST UNPROTECTED OPENING TO ATRIUM

For SI: 1 foot = 304.8 mm.

Figure 909.8.2(2)
EXAMPLE, CRITICAL LEVEL OF SMOKE

909.8.5 Plume contact with walls. When a plume contacts one or more of the surrounding walls, the mass flow rate shall be adjusted for the reduced entrainment resulting from the contact provided that the contact remains constant. Use of this provision requires calculation of the plume diameter, that shall be calculated by:

$$d = 0.48 \left[(T_c + 460)/(T_a + 460) \right]^{1/2} z \qquad \textbf{(Equation 9-7)}$$

For SI: $d = 0.48 \left(T_c/T_a \right)^{1/2} z$

where:

d = Plume diameter, feet (m).

T_a = Ambient air temperature, °F (°K).

T_c = Plume centerline temperature, °F (°K).

= $0.60 (T_a + 460) Q_c^{2/3} z^{-5/3} + T_a$

z = Height at which T_c is determined, feet (m).

For SI: $T_c = 0.08 T_a Q_c^{2/3} z^{-5/3} + T_a$

❖ When a fire plume is situated either against the wall or in a corner, the plume dynamics are altered and less air is being entrained into the plume. This decrease is 50 percent and 75 percent for walls and corners, respectively. The result is a hotter fire that tends to have longer flames. The lack of air entrainment reduces the mass loss rate. If a plume is large enough in a space such as an atrium, the mass loss rate may be limited by the size of the atrium. When the width of the plume becomes large enough, contact is made with the walls and air can no longer be entrained or is limited in the areas where it can be entrained, thus limiting the mass loss rate and production of smoke. To determine whether the plume makes contact with the wall, the diameter of the plume needs to be calculated. The calculation is based upon the centerline temperature of the plume at height z. When calculating the width, it may be found that the width obtained with the design criteria height z is larger than the dimensions of the atrium. That is an indication that the plume makes contact with the walls before the design criteria height z is obtained; therefore, several different heights may need to be used to determine when the plume makes contact with the wall. Once determined, that information can be used to recalculate the mass loss rate for the various plume calculations as appropriate, which would then result in a lower exhaust volume requirement. Generally, it should be noted that it is more conservative to base the necessary exhaust volume on the initial design criteria height z than a reduced height. It should also be noted that if the plume contacts the walls prior to the criteria height z, then consideration of the smoke exposure to the communicating spaces needs to be evaluated.

909.9 Design fire. The design fire shall be based on a Q of not less than 5,000 Btu/s (5275 kW) unless a rational analysis is per-formed by the registered design professional and approved by the building official. The design fire shall be based on the analysis in accordance with Section 909.4 and this section.

❖ The design fire is the critical element in the smoke control system design. The fire is what produces the smoke to be controlled by the system; therefore, it is considered important to establish a minimum fire size of 5,000 Btu per second (5275 kw) for the application of this section. This particular fire in many cases is fairly conservative. Generally, heat release data is fairly difficult to find for individual combustibles found within buildings. Fire size is sometimes estimated through heat release per square foot with figures such as 44 Btu/ft²/sec for mercantile occupancies and 20 Btu/ft²/sec for office buildings (ASHRAE). The 5,000-Btu-per-second (5275 kw) fire would equate to the following fire area using the above figures:

Mercantile   (5,000 Btu/sec) / (44 Btu/ft²/sec) =
114 ft² (10.7 ft × 10.7 ft)

Office   (5,000 Btu/sec) / (20 Btu/ft²/sec) =
250 ft² (16 ft × 16 ft)

This does not mean that all fires will be 5,000 Btu per second (5275 kw). It is simply stating that this is the minimum fire size that should be considered when conducting an analysis to determine the necessary exhaust rate to remove the smoke. The higher the heat release rate, the faster the products of combustion will be released (i.e., the mass flow rate). This section does provide the option that after a rational analysis is conducted, the fire size could be established at less than 5,000 Btu per second. This analysis would be based upon the specific hazards associated with a particular facility. For instance, a building may have very sparse furnishings and after reviewing fire data and other background information, such as sprinkler activation, it is determined that the fire size in that particular space would only be 3,500 Btu per second (3693 kw). For example, after a detailed analysis, the fire size may be reduced based upon sprinkler activation. It is recommended that the analysis to alter the fire size be conducted by a registered design professional with experience in the area of fire dynamics and possibly in the area of smoke control. Additionally, future use of the building needs to address reasonably foreseeable changes to the content for example, holiday displays. In addition to the allowance of a reduction in fire size, the designer should also conduct an analysis to determine if 5,000 Btu per second (5275 kw) is adequate.

It should also be noted that the fire size provided in Section 909.9 is a steady state fire versus an unsteady fire. A steady fire assumes a constant heat release rate over a period of time, where unsteady does not. An unsteady fire includes the growth and decay phases of the fire. An unsteady fire will hit a peak heat release rate when burning in the open, like an axisymmetric fire. To provide an understanding of fire sizes obtained from various combustibles, the following data from fire tests is provided. The following heat release rates, found in Section 3, Chapter 3-1 of the 2nd edition of the *SFPE Handbook of Fire Protection Engineering*, are peak heat release rates:

Plastic trash bags/paper trash:
        114-332 Btu/sec (120-350 kw)

Latex foam pillow:
        114 Btu/sec approximately (120 kw)

Dry Christmas tree:
        475-618 Btu/sec (500-650 kw)

Sofa:
        2,852 Btu/sec approximately (3,000 kw)

Plywood wardrobe:
        2,947-6,084 Btu/sec (3,100-6,400 kw)

   Section 909.9 assumes a constant or steady heat release rate of 5,000 Btu per second (5275 kw) without consideration of the growth or decay phase. Section 909.4.6 requires that this particular heat release rate fire be countered by a smoke control system for 20 minutes; therefore, the designer needs to assume a fire of 5,000 Btu per second (5275 kw) is burning for a period of 20 minutes. This assumption is fairly conservative in most cases but accounts for other uncertainties such as ineffectiveness of sprinklers due to ceiling height. Also, assuming a steady heat release simplifies the design process.

909.9.1 Factors considered. The engineering analysis shall include the characteristics of the fuel, fuel load, effects included by the fire and whether the fire is likely to be steady or unsteady.

❖ The design fire is required to be a minimum of 5,000 Btu per second (5275 kw), but it must also be further determined if this is adequate. To determine the appropriate fire size, an engineering analysis is necessary that takes into account the following elements: fuel (potential burning rates), fuel load (how much), effects included by the fire (smoke particulate size and density), steady or unsteady (burn steadily or simply peak and dissipate) and likelihood of sprinkler activation (based on height and distance from the fire).

909.9.2 Separation distance. Determination of the design fire shall include consideration of the type of fuel, fuel spacing and configuration. The ratio of the separation distance to the fuel equivalent radius shall not be less than 4. The fuel equivalent radius shall be the radius of a circle of equal area to floor area of the fuel package. The design fire shall be increased if other combustibles are within the separation distance as determined by:

$$R = [Q/(12\pi q'')]^{1/2} \qquad \textbf{(Equation 9-8)}$$

where:

q'' = Incident radiant heat flux required for nonpiloted ignition, Btu/ft$^2 \cdot$ s (W/m^2).

Q = Heat release from fire, Btu/s (kW).

R = Separation distance from target to center of fuel package, feet (m).

❖ The design fire size may also be affected by surrounding combustibles, which may have the effect of increasing the fire size. More specifically, there is concern that if sufficient separation is not maintained between combustibles, then a larger design fire is likely. Equation 9-8 calculates the critical separation distance R. This distance is where a nonpiloted ignition will occur adjacent to the combustibles. Nonpiloted ignition means that the radiated heat without direct flame contact or impingement will ignite the adjacent combustibles. This equation is based on the fire size, which in most cases is about 5,000 Btu per second (5275 kw) and the incident radiant heat flux required to ignite the specific combustible item in question, such as a polyurethane sofa. This particular data would need to be obtained from sources such as the *SFPE Handbook of Fire Protection Engineering*. The units of the radiant heat flux variable are in Btu/ft^{2}sec (kw/m^2). If the separation distance is less than R as determined in Equation 9-8, then the fire size needs to be reevaluated to include additional combustibles. It is suggested that the heat release of the combustibles simply be added to the fire size.

   Equation 9-8 calculates the distance R from the center of the fuel package. It is very possible that due to the area covered by the fuel package itself, the design fire could be much closer to the surrounding combustibles than the distance R; therefore, a minimum distance is set by the following ratio:

$$\text{Minimum distance} = \frac{R(\text{Equation } 9-8)}{4r^3}$$

Where r = equivalent fuel radius.

   The equivalent fuel radius is determined by first determining the area of the fuel package, then assuming that area as a circle. For example, the fuel package for a 4-foot by 3-foot (1219 mm by 914 mm) kiosk in an atrium is calculated as:

Area = 4 ft x 3 ft = 12 ft^2

Area of circle, $A = \pi r^2$

$r = \sqrt{A/\pi}$

$r = \sqrt{12/\pi} = 1.96$ feet

   Therefore, the equivalent fuel radius for this scenario is 1.96 feet ≥ 2 feet.

   Looking at the ratio established in Section 909.9.2, R would be limited as follows in the above example:

$\dfrac{R}{r} \geq 4$

$R \geq 4r$

   Therefore, R should be greater than or equal to $R \geq 4 \times r \geq 4 \times 2 \geq 8$ feet.

909.9.3 Heat-release assumptions. The analysis shall make use of best available data from approved sources and shall not be based on excessively stringent limitations of combustible material.

❖ This section is merely stressing the fact that data obtained for use in a rational analysis needs to come from relevant and appropriate sources. Data can be obtained from groups such as the National Institute for Standards and Technology (NIST). Data from fire tests is available and is a good resource for such analysis. As noted earlier, such data is not prevalent (see also Section 3, Chapter 3-1 of the *SFPE Handbook of Fire Protection Engineering*).

909.9.4 Sprinkler effectiveness assumptions. A documented engineering analysis shall be provided for conditions that assume fire growth is halted at the time of sprinkler activation.

❖ This section raises a question concerning an assumption that sprinklers will immediately control the fire as soon as they are activated. This assumption may be true in some cases, but for high ceilings the sprinkler may not activate or may be ineffective. Sprinklers may be ineffective in high spaces, since by the time they are activated the fire is too large to control. Essentially, the fire plume may push away and evaporate the water before it actually reaches the seat of the fire. This is a common problem with high-piled storage. Also, if the fire becomes too large before the sprinklers are activated, the available water supply and pressure for the system may be compromised. Additionally, based on the layout of the room and the movement of the fire effluents, the wrong sprinklers could be activated, which leads to a larger fire size and depletion of the available water supply and pressure. Therefore, each scenario needs to be looked at individually to determine whether sprinklers would be effective in halting the growth of fire. More specifically, the evaluation should include droplet size, density and area of coverage and should also be based on actual test results.

909.10 Equipment. Equipment such as, but not limited to, fans, ducts, automatic dampers and balance dampers, shall be suitable for its intended use, suitable for the probable exposure temperatures that the rational analysis indicates, and as approved by the building official.

❖ Section 909.10 and subsequent sections are primarily related to the reliability of the system components to provide a smoke control system that works according to the design. One of the largest concerns when using smoke control provisions is the overall reliability of the system. Such systems have many different components, such as smoke and fire dampers; fans; ducts and controls associated with such components. The more components a system has, the less reliable it becomes. In fact, one approach in providing a higher level of reliability is utilizing the normal building systems such as the HVAC to provide the smoke control system. Basically, systems used every day are more likely to be working appropriately, since they are essentially being tested daily; however, there are many components that are specific to the smoke control system, such as exhaust fans in an atrium or the smoke control panel.

Also, there is not a generic prescriptive set of requirements as to how all smoke control system elements should operate, since each design may be fairly unique. The specifics on operation of such a system need to be included within the design and construction documents. Most components used in smoke control systems are elements used in many other applications such as HVAC systems; therefore, the basic mechanisms of a fan used in a smoke control system may not be different, although they may be applied differently.

909.10.1 Exhaust fans. Components of exhaust fans shall be rated and certified by the manufacturer for the probable temperature rise to which the components will be exposed. This temperature rise shall be computed by:

$$T_s = (Q_c/mc) + (T_a) \qquad \textbf{(Equation 9-9)}$$

where:

c = Specific heat of smoke at smoke layer temperature, Btu/lb°F (kJ/kg · K).

m = Exhaust rate, pounds per second (kg/s).

Q_c = Convective heat output of fire, Btu/s (kW).

T_a = Ambient temperature, °F (°K).

T_s = Smoke temperature, °F (°K).

Exception: Reduced T_s as calculated based on the assurance of adequate dilution air.

❖ Fans used for smoke control systems must be able to tolerate the possible elevated temperatures to which they will be exposed. Again, like many other factors this depends upon the specifics of the design fire. Essentially, Equation 9-9 requires the calculation of the potential temperature rise. The exhaust fans must be specifically rated and certified by the manufacturer to be able to handle these rises in temperature. There is an exception that allows reduction of the temperature if it can be shown that adequate temperature reduction will occur. In many cases if the exhaust fans are near the ceiling, the smoke will be much cooler than the value resulting from Equation 9-9 since the smoke may cool considerably by the time it reaches the ceiling. Also, sprinkler activation will assist in cooling the smoke further.

909.10.2 Ducts. Duct materials and joints shall be capable of withstanding the probable temperatures and pressures to which they are exposed as determined in accordance with Section 909.10.1. Ducts shall be constructed and supported in accordance with the *International Mechanical Code*. Ducts shall be leak tested to 1.5 times the maximum design pressure in accordance with nationally accepted practices. Measured leakage shall not exceed 5 percent of design flow. Results of such testing shall be a part of the documentation procedure. Ducts shall be supported directly from fire-resistance-rated structural elements of the building by substantial, noncombustible supports.

Exception: Flexible connections (for the purpose of vibration isolation) complying with the *International Mechanical Code*, that are constructed of approved fire-resistance-rated materials.

❖ The next essential component of a smoke control system is the integrity of the ducts to transport supply and exhaust air. The integrity of ducts is also important for an HVAC system, but is more critical in this case since it is not simply a comfort issue but one of life safety. The key concern with ducts in smoke control systems is that they can withstand elevated temperatures and that there will be minimal leakage. The concern with leakage is the potential of leaking smoke into another smoke zone or not providing the proper amount of supply air to support the system.

More specifically, all ducts need to be leak tested to 1.5 times the maximum static design pressure. The leakage resulting should be no more than 5 percent of the design flow. For example, a duct that has a design flow of 300 cubic feet per minute (cfm) (.141 m³/s) would be allowed 15 cfm (0.007 m³/s) of leakage when exposed to a pressure equal to 1.5a times the design pressure for that duct. The tests should be in accordance with nationally accepted practices. This criterion will often limit ductwork for smoke control systems to lined, systems, since the amount of leakage in such systems is much less.

As part of the concern for possible exposure to fire and fire products, the ducts are required to be supported by way of substantial noncombustible supports connected to the fire-resistance-rated structural elements of the building. As noted, the system needs to able to run for 20 minutes starting from the detection of the fire. The supports are allowed to be other than noncombustible when they are flexible connections provided to mitigate the effects of vibration, perhaps as part of a building exposed to seismic loads. The flexible connections still need to be constructed of approved fire-resistance-rated materials.

909.10.3 Equipment, inlets and outlets. Equipment shall be located so as to not expose uninvolved portions of the building to an additional fire hazard. Outside air inlets shall be located so as to minimize the potential for introducing smoke or flame into the building. Exhaust outlets shall be so located as to minimize reintroduction of smoke into the building and to limit exposure of the building or adjacent buildings to an additional fire hazard.

❖ The intent of this section is to minimize the likelihood of smoke being reintroduced into the building due to poorly placed outdoor air inlets and exhaust air outlets; therefore, placing one right next to another on the exterior of the building would be inappropriate. In addition, wind and other adverse conditions should be considered when choosing locations for these inlets and outlets. Particular attention should be paid to introducing exhausted smoke into another smoke zone. Also,

smoke should be exhausted in a direction that will not introduce it into surrounding buildings or facilities. Within the building itself, the supply air and exhaust outlets should also be strategically located. The exhaust inlets and supply air should be evenly distributed to reduce the likelihood of a high velocity of air which may disrupt the fire plume and also push smoke back into occupied areas.

909.10.4 Automatic dampers. Automatic dampers, regardless of the purpose for which they are installed within the smoke control system, shall be listed and conform to the requirements of approved, recognized standards.

❖ This section addresses the reliability of any dampers used within a smoke control system. This particular provision requires that the dampers be listed and conform to the appropriate recognized standards. More specifically, Section 716 contains more detailed information on the specific requirements for smoke and fire dampers. Smoke and fire dampers should be listed in accordance with UL 555S and 555, respectively. Also, remember that each smoke control design is unique and the sequence and methods used to activate the dampers may vary from design to design. This information needs to be addressed in the construction documents.

Another factor to take into account, with regard to timing of the system, is the fact that some dampers react more quickly than others, simply due to the particular smoke damper characteristics. Additionally, during the commissioning of the system, the damper is going to be exposed to many repetitions. These repetitions need to be accounted for in the overall reliability of the system.

909.10.5 Fans. In addition to other requirements, belt-driven fans shall have 1.5 times the number of belts required for the design duty, with the minimum number of belts being two. Fans shall be selected for stable performance based on normal temperature and, where applicable, elevated temperature. Calculations and manufacturer's fan curves shall be part of the documentation procedures. Fans shall be supported and restrained by noncombustible devices in accordance with the requirements of Chapter 16. Motors driving fans shall not be operated beyond their nameplate horsepower (kilowatts), as determined from measurement of actual current draw, and shall have a minimum service factor of 1.15.

❖ Part of the overall reliability requires that fans used to provide supply air and exhaust capacity will be functioning when necessary; therefore, a safety factor of 1.5 is placed upon the required belts for fans. All fans used as part of a smoke control system must provide 1.5 times the number of required belts with a minimum of two belts for all fans.

This section also points out that the fan chosen should fit the specific application. It should be able to withstand the temperature rise as calculated in Section 909.10.1 and generally be able to handle typical exposure conditions, such as location and wind. For instance, propeller fans are highly sensitive to the effects of wind. When located on the windward side of a building, wall-mounted, nonhooded propeller fans are not able to compensate for wind effects. Additionally, even hooded propeller fans located on the leeward side of the building may not adequately compensate for the decrease in pressure caused by wind effects. In general, when designing a system, it should be remembered that field conditions might vary from the calculations; therefore, flexibility should be built into the design that would account for things such as variations in wind conditions.

Finally, this section stresses that fan motors not be operated beyond their rated horsepower.

909.11 Power systems. The smoke control system shall be supplied with two sources of power. Primary power shall be the normal building power systems. Secondary power shall be from an approved standby source complying with the ICC *Electrical Code*. The standby power source and its transfer switches shall be in a separate room from the normal power transformers and switch gear and shall be enclosed in a room constructed of not less than 1-hour fire-resistance-rated fire barriers ventilated directly to and from the exterior. Power distribution from the two sources shall be by independent routes. Transfer to full standby power shall be automatic and within 60 seconds of failure of the primary power. The systems shall comply with the ICC *Electrical Code*.

❖ As with any life safety system, a level of redundancy with regard to power supply is required to enable the functioning of the system during a fire. The primary source is the building's normal power system. The secondary power system is by means of standby power. One of the key elements is that standby power systems are intended to operate within 60 seconds of loss of primary power. It should be noted that the primary difference between standby power and emergency power is that emergency power must operate within 10 seconds of loss of primary power versus 60 seconds. This section also requires isolation from normal building power systems via a 1-hour fire barrier. This increases the reliability and reduces the likelihood that a single event could remove both power supplies.

909.11.1 Power sources and power surges. Elements of the smoke management system relying on volatile memories or the like shall be supplied with uninterruptable power sources of sufficient duration to span a 15-minute primary power interruption. Elements of the smoke management system susceptible to power surges shall be suitably protected by conditioners, suppressors or other approved means.

❖ Smoke management systems have many components, sometimes highly sensitive electronics, that are adversely affected by any interruption in or sudden surges of power. Therefore, Section 909.11.1 requires that any components of a smoke control system, such as volatile memories, be supplied with an uninterruptible power system for the first 15 minutes of loss of primary power. Volatile memory components will lose memory upon any loss of power no matter how short the time period. Once the 15 minutes elapses, these elements can be

transitioned to the already operating standby power supply.

With regard to components sensitive to power surges, they need to be provided with surge protection in the form of conditioners, suppressors or other approved means.

909.12 Detection and control systems. Fire detection systems providing control input or output signals to mechanical smoke control systems or elements thereof shall comply with the requirements of Section 907. Such systems shall be equipped with a control unit complying with UL 864 and listed as smoke control equipment.

Control systems for mechanical smoke control systems shall include provisions for verification. Verification shall include positive confirmation of actuation, testing, manual override, the presence of power downstream of all disconnects and, through a preprogrammed weekly test sequence report, abnormal conditions audibly, visually and by printed report.

❖ This section requires that fire detection elements used in mechanical smoke control systems comply with Chapter 9 of the code and NFPA 72, which is the fire alarm standard. Specific to smoke control systems is the requirement for a control unit complying with UL 864 that is listed as smoke control equipment to be used. UL 864 has a subcategory specific to fire alarm control panels (UUKL).

Additionally, the supervisory system for smoke control systems needs to verify certain activities of the system via audible, visual and a printed report. As noted in Section 909.12, the following verifications would be included: positive confirmation of actuation, testing, manual override, presence of power downstream of all disconnects and abnormal conditions obtained by a weekly preprogrammed test sequence.

909.12.1 Wiring. In addition to meeting requirements of the ICC *Electrical Code*, all wiring, regardless of voltage, shall be fully enclosed within continuous raceways.

❖ Wiring is required to be placed within continuous raceways. This provides an additional level of reliability for the system. It should be noted that such wiring would not necessarily need to be rated for plenums unless it is located within a plenum.

[F] 909.12.2 Activation. Smoke control systems shall be activated in accordance with this section.

❖ The activation of a smoke control system is dependent on when such a system is required. Mechanical smoke control systems, which could include pressurization, airflow or exhaust methods, require an automatic activation mechanism. When using a passive system, which depends upon compartmentation, spot-type detectors are acceptable for the release of door closers and similar openings.

[F] 909.12.2.1 Pressurization, airflow or exhaust method. Mechanical smoke control systems using the pressurization, airflow or exhaust method shall have completely automatic control.

❖ See Sections 909.6 for the pressurization method, 909.7 for the airflow design method and 909.8 for the exhaust method.

[F] 909.12.2.2 Passive method. Passive smoke control systems actuated by approved spot-type detectors listed for releasing service shall be permitted.

❖ This section recognizes that a passive system does not require a fire alarm system and would allow single station detectors to close openings where required by the design.

[F] 909.12.3 Automatic control. Where completely automatic control is required or used, the automatic-control sequences shall be initiated from an appropriately zoned automatic sprinkler system complying with Section 903.3.1.1, manual controls that are readily accessible to the fire department and any smoke detectors required by engineering analysis.

❖ When automatic activation is required, it must be accomplished by a properly zoned automatic sprinkler system and, if the engineering analysis requires them, smoke detectors. Manual control for the fire department needs to be provided. An important point with this particular requirement is that smoke control systems are engineered systems and a prescribed smoke detection system may not fit the needs of the specific design. Other types of detectors, such as beam detectors (within an atrium), may be used and could be more practical from a maintenance standpoint. The fire alarm system within the building also should not generally activate such systems, as it may alter the effectiveness of the system by pulling smoke through the building versus removing or containing the smoke.

909.13 Control air tubing. Control air tubing shall be of sufficient size to meet the required response times. Tubing shall be flushed clean and dry prior to final connections and shall be adequately supported and protected from damage. Tubing passing through concrete or masonry shall be sleeved and protected from abrasion and electrolytic action.

❖ Control tubing is a method that uses pneumatics to operate components such as the opening and closing of dampers. Due to the sophistication of electronic systems today, control tubing is becoming less common.

These particular requirements provide the criteria for properly designing and installing control tubing. Essentially, it is up to the design professional to determine the size requirements and to properly design appropriate supports. This information needs to be detailed within the construction documents. Additionally, due to the effect of moisture and other contaminants on control tubing, it must be flushed clean then dried before installation.

909.13.1 Materials. Control air tubing shall be hard drawn copper, Type L, ACR in accordance with ASTM B 42, ASTM B 43,

ASTM B 68, ASTM B 88, ASTM B 251 and ASTM B 280. Fittings shall be wrought copper or brass, solder type, in accordance with ASME B 16.18 or ASME B 16.22. Changes in direction shall be made with appropriate tool bends. Brass compression-type fittings shall be used at final connection to devices; other joints shall be brazed using a BCuP5 brazing alloy with solidus above 1,100°F (593°C) and liquids below 1,500°F (816°C). Brazing flux shall be used on copper-to-brass joints only.

Exception: Nonmetallic tubing used within control panels and at the final connection to devices, provided that all of the following conditions are met:

1. Tubing shall be listed by an approved agency for flame and smoke characteristics.

2. Tubing and connected devices shall be completely enclosed within galvanized or paint-grade steel enclosure of not less than 0.030 inch (0.76 mm) (No. 22 galvanized sheet gage) thickness. Entry to the enclosure shall be by copper tubing with a protective grommet of neoprene or teflon or by suitable brass compression to male-barbed adapter.

3. Tubing shall be identified by appropriately documented coding.

4. Tubing shall be neatly tied and supported within enclosure. Tubing bridging cabinet and door or moveable device shall be of sufficient length to avoid tension and excessive stress. Tubing shall be protected against abrasion. Tubing serving devices on doors shall be fastened along hinges.

❖ This section addresses the materials allowed for control air tubing along with approved methods of connection. All of this information needs to be documented, as it will be subject to review by the special inspector.

909.13.2 Isolation from other functions. Control tubing serving other than smoke control functions shall be isolated by automatic isolation valves or shall be an independent system.

❖ This section requires separation of control tubing used for other functions through the use of isolation valves or a completely separate system. This is due to the difference in requirements for control tubing used in a smoke control system versus other building systems. The isolation of the control air tubing for a smoke control system needs to be specifically noted on the construction documents.

909.13.3 Testing. Control air tubing shall be tested at three times the operating pressure for not less than 30 minutes without any noticeable loss in gauge pressure prior to final connection to devices.

❖ As part of the acceptance testing of the smoke control system, the control air tubing will be pressure tested three times the operating pressure for 30 minutes or more. The performance criteria as to whether the control tubing is considered a failure is when there is any noticeable loss in gauge pressure prior to final connection of devices during the 30-minute duration test.

909.14 Marking and identification. The detection and control systems shall be clearly marked at all junctions, accesses and terminations.

❖ This section requires that all portions of the fire detection system that activate the smoke control system be marked and identified appropriately. This includes all applicable fire alarm-initiating devices, the respective junction boxes, all data-gathering panels and fire alarm control panels. Additionally, all components of the smoke control system, which are not considered a fire detection system, are required to be properly identified and marked. This would include all applicable junction boxes, control tubing, temperature control modules, relays, damper sensors, automatic door sensors and air movement sensors.

[F] 909.15 Control diagrams. Identical control diagrams showing all devices in the system and identifying their location and function shall be maintained current and kept on file with the building official, the fire department and in the fire command center in format and manner approved by the fire chief.

❖ The purpose of control diagrams is to provide consistent information on the system in several key locations, including the building department, the fire department and the fire command center. This information is intended to assist in the use and operation of the smoke control system. The format of the control diagram is as approved by the fire chief. This is necessary since the fire department is the agency that will be using such a system during a fire and when the system is tested in the future. The more clearly the information is communicated, the more effective the smoke control system will be.

It should be noted that the fire department may want all smoke control systems within a jurisdiction to follow a particular protocol for control diagrams. Generally, the control diagrams should indicate the required reaction of the system in all scenarios. The status or position of every fan and damper in every scenario must be clearly identified.

[F] 909.16 Fire-fighter's smoke control panel. A fire-fighter's smoke control panel for fire department emergency response purposes only shall be provided and shall include manual control or override of automatic control for mechanical smoke control systems. The panel shall be located in a fire command center complying with Section 911, and shall comply with Sections 909.16.1 through 909.16.3.

❖ One of the elements that makes a smoke control system effective is that its activity is successfully communicated to the fire department and the fire department is able to manually operate the system. The following sections provide requirements for a control panel specifically for smoke control systems. This panel is required to be located within the fire command center. There are two components that include the requirements for the display and for the controls. This control panel will provide an ability to override any other controls whether manual or automatic within the building as they relate to the smoke control system.

[F] 909.16.1 Smoke control systems. Fans within the building shall be shown on the fire-fighter's control panel. A clear indication of the direction of airflow and the relationship of components shall be displayed. Status indicators shall be provided for all smoke control equipment, annunciated by fan and zone, and by pilot-lamp-type indicators as follows:

1. Fans, dampers and other operating equipment in their normal status—WHITE.

2. Fans, dampers and other operating equipment in their off or closed status—RED.

3. Fans, dampers and other operating equipment in their on or open status—GREEN.

4. Fans, dampers and other operating equipment in a fault status—YELLOW/AMBER.

❖ This section denotes what should be displayed on the control panel. The display is required to include all fans, an indication of the direction of airflow and the relationship of the components. Also, status lights are required, and this section sets out specific standardized colors to indicate normal status, closed status, open status and fault status. A standardized approach increases the likelihood that the fire department will be able to quickly become familiar with a system. Since the fire department has the ability to override the automatic functions of the system, this information is critical.

[F] 909.16.2 Smoke control panel. The fire-fighter's control panel shall provide control capability over the complete smoke-control system equipment within the building as follows:

1. ON-AUTO-OFF control over each individual piece of operating smoke control equipment that can also be controlled from other sources within the building. This includes stairway pressurization fans; smoke exhaust fans; supply, return and exhaust fans; elevator shaft fans and other operating equipment used or intended for smoke control purposes.

2. OPEN-AUTO-CLOSE control over individual dampers relating to smoke control and that are also controlled from other sources within the building.

3. ON-OFF or OPEN-CLOSE control over smoke control and other critical equipment associated with a fire or smoke emergency and that can only be controlled from the fire-fighter's control panel.

Exceptions:

1. Complex systems, where approved, where the controls and indicators are combined to control and indicate all elements of a single smoke zone as a unit.

2. Complex systems, where approved, where the control is accomplished by computer interface using approved, plain English commands.

❖ This section sets the requirements as to which controls need to be provided for the fire department on the control panel.

There are two aspects to the controls. The controls will include on-auto-off and open-auto-close settings or will be strictly on-off or open-close. If the system or component can be set on automatic (auto), it can be controlled from other locations beyond the fire command center. This would include an automatic smoke detection system or by manual activation. If a control only contains on-off or open-close settings, the only way the system component can be controlled is in the fire command center.

It should be noted that components such as fans are usually associated with on-off-type controls, whereas components such as dampers are associated with open-close-type controls.

[F] 909.16.3 Control action and priorities. The fire-fighter's control panel actions shall be as follows:

1. ON-OFF, OPEN-CLOSE control actions shall have the highest priority of any control point within the building. Once issued from the fire-fighter's control panel, no automatic or manual control from any other control point within the building shall contradict the control action. Where automatic means are provided to interrupt normal, nonemergency equipment operation or produce a specific result to safeguard the building or equipment (i.e., duct freezestats, duct smoke detectors, high-temperature cutouts, temperature-actuated linkage and similar devices), such means shall be capable of being overridden by the fire-fighter's control panel. The last control action as indicated by each fire-fighter's control panel switch position shall prevail. In no case shall control actions require the smoke control system to assume more than one configuration at any one time.

   Exception: Power disconnects required by the ICC *Electrical Code.*

2. Only the AUTO position of each three-position fire-fighter's control panel switch shall allow automatic or manual control action from other control points within the building. The AUTO position shall be the NORMAL, nonemergency, building control position. Where a fire-fighter's control panel is in the AUTO position, the actual status of the device (on, off, open, closed) shall continue to be indicated by the status indicator described above. When directed by an automatic signal to assume an emergency condition, the NORMAL position shall become the emergency condition for that device or group of devices within the zone. In no case shall control actions require the smoke control system to assume more than one configuration at any one time.

❖ This section clarifies that when a component of the system is placed in an on-off or open-close configuration, no other control point in the building, whether automatic or manual, can override the action established in the fire command center. If a system component is configured in auto mode, it can be controlled from locations within the building beyond the fire command center. Some controls are specifically designed to only allow an action from the fire command center.

[F] 909.17 System response time. Smoke-control system activation shall be initiated immediately after receipt of an appropriate automatic or manual activation command. Smoke control systems shall activate individual components (such as dampers and fans) in the sequence necessary to prevent physical damage to the fans, dampers, ducts and other equipment. For purposes of smoke control, the fire-fighter's control panel response time shall be the same for automatic or manual smoke control action initiated from any other building control point. The total response time, including that necessary for detection, shutdown of operating equipment and smoke control system startup, shall allow for full operational mode to be achieved before the conditions in the space exceed the design smoke condition. The system response time for each component and their sequential relationships shall be detailed in the required rational analysis and verification of their installed condition reported in the required final report.

❖ This particular section provides the criteria as to when the smoke control system is required to begin operation. Whether or not the activation is manual or automatic, this criteria clarifies that the system be initiated immediately. Also, it requires that components activate in a sequence that will not potentially damage the fans, dampers, ducts and other equipment. Unrealistic timing of the system has the potential of creating an unsuccessful system. Delays in the system can be seen in slow dampers, fans that ramp up or down, systems that poll slowly and intentional built-in delays. These factors can add significantly to the reaction time of the system and may hamper achieving the design goals.

The key element is that the system be fully operational before the smoke conditions exceed the design parameters. The sequence of events need to be justified within the design analysis and described clearly in the construction documents.

[F] 909.18 Acceptance testing. Devices, equipment, components and sequences shall be individually tested. These tests, in addition to those required by other provisions of this code, shall consist of determination of function, sequence and, where applicable, capacity of their installed condition.

❖ In order to achieve a certain level of performance, the smoke control system needs to be thoroughly tested. Section 909.18 requires that all devices, equipment components and sequences be individually tested.

[F] 909.18.1 Detection devices. Smoke or fire detectors that are a part of a smoke control system shall be tested in accordance with Chapter 9 in their installed condition. When applicable, this testing shall include verification of airflow in both minimum and maximum conditions.

❖ Detection devices are required to be tested in accordance with the fire protection requirements found in Chapter 9. Also, since such detectors may be subject to higher air velocities than typical detectors, their operation needs to be verified in the minimum and maximum anticipated airflow conditions.

[F] 909.18.2 Ducts. Ducts that are part of a smoke control system shall be traversed using generally accepted practices to determine actual air quantities.

❖ This section requires ducts that are part of the smoke control system to be tested to show that the proper amount of air is flowing. It should be noted that Section 909.10.2 requires that the ducts be leak tested to 1.5 times the maximum design pressure. Such leakage is not allowed to exceed 5 percent of the design flow.

[F] 909.18.3 Dampers. Dampers shall be tested for function in their installed condition.

❖ This section notes that all dampers need to be inspected to meet the function for which they are installed. For instance, a damper that is to be open when the system is in smoke control mode should be verified to be open when testing the system. Also, a damper may have a specific timing associated with its operation that would need to be verified though testing.

[F] 909.18.4 Inlets and outlets. Inlets and outlets shall be read using generally accepted practices to determine air quantities.

❖ Similar to ducts, the appropriate amount of air that is entering or exiting the inlets and outlets, respectively, must be checked.

[F] 909.18.5 Fans. Fans shall be examined for correct rotation. Measurements of voltage, amperage, revolutions per minute (rpm) and belt tension shall be made.

❖ This section requires the testing of fans for the following: correct rotation, voltage, amperage, revolutions per minute and belt tension.

A common problem with fans is that they are often installed in the reverse direction. Also, to verify the reliability of the fans, elements such as the appropriate voltage and belt tension need to be tested.

[F] 909.18.6 Smoke barriers. Measurements using inclined manometers or other approved calibrated measuring devices shall be made of the pressure differences across smoke barriers. Such measurements shall be conducted for each possible smoke control condition.

❖ As discussed in Section 909.5.1, the testing of pressure differences across smoke barriers needs to be measured in the smoke control mode. As noted in Section 909.18.6, such testing is to be performed for every possible smoke control condition, and the measurements will be taken using an inclined manometer or other approved methods. Electronic devices are also available. Qualified individuals must calibrate these types of devices. Additionally, before using an alternate method of testing, the building official needs to approve it.

[F] 909.18.7 Controls. Each smoke zone, equipped with an automatic-initiation device, shall be put into operation by the actuation of one such device. Each additional device within the zone shall be verified to cause the same sequence without requiring the operation of fan motors in order to prevent damage. Control sequences shall be verified throughout the system, including verification of override from the fire-fighter's control panel and simulation of standby power conditions.

❖ This section requires the overall testing of the system. More specifically, each zone needs to individually initiate the smoke control system by the activation of an automatic initiation device. Once that has occurred, all other devices within each zone need to be verified that they will activate the system, but to avoid damage, the fans do not need to be activated.

In addition to determining that all the appropriate devices initiate the system, it must also be verified that all of the controls on the fire-fighter control panel initiate the appropriate aspects of the smoke control system, including the override capability.

Finally, the initiation and availability of the standby power system need to be verified.

[F] 909.18.8 Special inspections for smoke control. Smoke control systems shall be tested by a special inspector.

❖ Smoke control systems require special inspection since they are life safety systems.

[F] 909.18.8.1 Scope of testing. Special inspections shall be conducted in accordance with the following:

1. During erection of ductwork and prior to concealment for the purposes of leakage testing and recording of device location.

2. Prior to occupancy and after sufficient completion for the purposes of pressure-difference testing, flow measurements, and detection and control verification.

❖ Special inspections need to occur at two different stages during construction to facilitate the necessary inspections. The first round of special inspections occurs before concealment of the ductwork or fire protection elements. The special inspector needs to verify the leakage as noted in Section 909.10.2. Additionally, the location of all fire protection devices needs to be verified and documented at this time.

The second round of special inspections occurs just prior to occupancy. The inspections include the verification of pressure differences across smoke barriers as required in Sections 909.18.6 and 909.5.1, the verification of appropriate volumes of airflow as noted in the design and finally the verification of the appropriate operation of the detection and control mechanisms as required in Sections 909.18.1 and 909.18.7. These tests need to occur just prior to occupancy, since the test result will more clearly represent actual conditions.

[F] 909.18.8.2 Qualifications. Special inspection agencies for smoke control shall have expertise in fire protection engineering, mechanical engineering and certification as air balancers.

❖ As noted in Section 909.3, special inspections are required for smoke control systems. This means a certain level of qualification that would include the need for expertise in fire protection engineering, mechanical engineering and certification as air balancers.

[F] 909.18.8.3 Reports. A complete report of testing shall be prepared by the special inspector or special inspection agency. The report shall include identification of all devices by manufacturer, nameplate data, design values, measured values and identification tag or mark. The report shall be reviewed by the responsible registered design professional and, when satisfied that the design intent has been achieved, the responsible registered design professional shall seal, sign and date the report.

❖ Once the special inspections are complete, documentation of the activity is required. This documentation is to be prepared in the form of a report that identifies all devices by manufacturer, nameplate data, design values, measured values and identification or mark.

[F] 909.18.8.3.1 Report filing. A copy of the final report shall be filed with the building official and an identical copy shall be maintained in an approved location at the building.

❖ The report needs to be reviewed, approved and then signed, sealed and dated. This report is to be provided to the building official and a copy is also to remain in the building in an approved location. The fire command center is probably the most appropriate location.

[F] 909.18.9 Identification and documentation. Charts, drawings and other documents identifying and locating each component of the smoke control system, and describing its proper function and maintenance requirements, shall be maintained on file at the building as an attachment to the report required by Section 909.18.8.3. Devices shall have an approved identifying tag or mark on them consistent with the other required documentation and shall be dated indicating the last time they were successfully tested and by whom.

❖ Additional documentation that needs to be maintained includes charts, drawings and other related documentation that assists in the identification of each aspect of the smoke control system. This documentation is where information, such as the last time a device or component was successfully tested and by whom, is recorded. This will serve as the main documentation for the system. Again, the fire command center is most likely the most appropriate location for such information.

[F] 909.19 System acceptance. Buildings, or portions thereof, required by this code to comply with this section shall not be issued a certificate of occupancy until such time that the building official determines that the provisions of this section have been fully complied with, and that the fire department has received satisfactory instruction on the operation, both automatic and manual, of the system.

Exception: In buildings of phased construction, a temporary certificate of occupancy, as approved by the building official, shall be permitted provided that those portions of the building to be occupied meet the requirements of this section and that the remainder does not pose a significant hazard to the safety of the proposed occupants or adjacent buildings.

❖ This section stipulates that the certificate of occupancy cannot be issued unless the smoke control system has been accepted. It is essential that the system be inspected and approved since it is a life safety system. There is an exception for buildings that are constructed in phases where a temporary certificate of occupancy is allowed.

909.20 Smokeproof enclosures. Where required by Section 1019.1.8, a smokeproof enclosure shall be constructed in accordance with this section. A smokeproof enclosure shall consist of an enclosed interior exit stairway that conforms to Section 1019.1 and an outside balcony or ventilated vestibule meeting the requirements of this section. Where access to the roof is required by the *International Fire Code*, such access shall be from the smokeproof enclosure where a smokeproof enclosure is required.

❖ In a building that serves stories where the floor surface is located more than 75 feet (22 860 mm) above the level of fire department access or more than 30 feet (9144 mm) below the level of exit discharge stairways, either a smokeproof enclosure or a pressurized stairway is required.

A smokeproof enclosure essentially takes advantage of fire-resistance-rated construction surrounding the stair shaft and a buffer zone created by an outside balcony or ventilated vestibule to avoid the accumulation of smoke within the exit stairway. The premise is that if access to the smokeproof enclosure provides a sufficient buffer to the effects of smoke via ventilation or simply being located outside, the smoke will not migrate into the enclosure.

The vestibule as noted is required to be ventilated. There are two alternatives provided in Section 909.20. They include natural ventilation, covered in Section 909.20.3, and mechanical ventilation, covered in Section 909.20.4. It should be noted that a smokeproof enclosure is not considered a pressurized stairway. The method of pressurizing stairways is recognized as an alternative to the smokeproof enclosures in Section 909.20.5, but is not required.

909.20.1 Access. Access to the stair shall be by way of a vestibule or an open exterior balcony. The minimum dimension of the vestibule shall not be less than the required width of the corridor leading to the vestibule but shall not have a width of less than 44 inches (1118 mm) and shall not have a length of less than 72 inches (1829 mm) in the direction of egress travel.

❖ As noted, access to the stair in a smokeproof enclosure is via an exterior balcony or a ventilated vestibule. This section provides the minimum dimensions for access to the smokeproof enclosure through a vestibule and then travel space to the entrance into the stairway. If a vestibule is chosen, the minimum width is established as the larger of 44 inches (1118 mm) or the width of the corridor. Additionally, when entering a vestibule there needs to be at least 72 inches (1829 mm) in the direction of travel (see Figure 909.20.1).

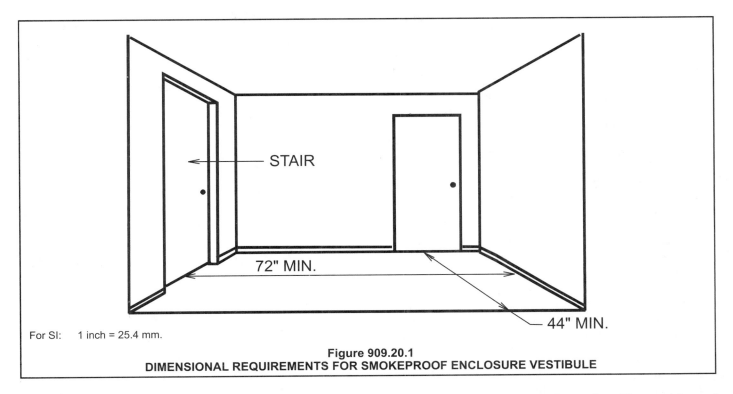

For SI: 1 inch = 25.4 mm.

Figure 909.20.1
DIMENSIONAL REQUIREMENTS FOR SMOKEPROOF ENCLOSURE VESTIBULE

909.20.2 Construction. The smokeproof enclosure shall be separated from the remainder of the building by not less than a 2-hour fire-resistance-rated fire barrier without openings other than the required means of egress doors. The vestibule shall be separated from the stairway by not less than a 2-hour fire-resistance-rated fire barrier. The open exterior balcony shall be constructed in accordance with the fire-resistance-rating requirements for floor construction.

❖ This section sets the basic construction requirements for the stairway and the associated vestibule or exterior balcony. Essentially, a 2-hour fire-resistance-rated barrier is required between the enclosure and the rest of the building.

Additionally, vestibules need to be separated from the stairway by a 2-hour fire-resistance-rated barrier. The exterior balcony, if used, would only be required to comply with the fire-rating requirements for the floor construction.

909.20.2.1 Door closers. Doors in a smokeproof enclosure shall be self-closing or automatic-closing by actuation of a smoke detector installed at the floor-side entrance to the smokeproof enclosure in accordance with Section 715.3.7. The actuation of the smoke detector on any door shall activate the closing devices on all doors in the smokeproof enclosure at all levels. Smoke detectors shall be installed in accordance with Section 907.10.

❖ In order to maintain the separation between the smokeproof enclosure and the rest of the building, the doors need to be self-closing or automatic-closing through the actuation of a smoke detector. Also, when a smoke detector is activated on any floor, all doors are required to close for that particular smokeproof enclosure.

909.20.3 Natural ventilation alternative. The provisions of Sections 909.20.3.1 through 909.20.3.3 shall apply to ventilation of smokeproof enclosures by natural means.

❖ Smoke and hot gases generated by fire may enter smokeproof enclosures through the doorways. This section provides for the diffusion of smoke and gases to the outside of the building by means of unenclosed openings to the outside.

909.20.3.1 Balcony doors. Where access to the stairway is by way of an open exterior balcony, the door assembly into the enclosure shall be a fire door in accordance with Section 715.3.

❖ Where an open exterior balcony or a smokeproof enclosure assembly is used for entry into the exit stairway, the doorways to the stairway enclosure are required to have a 1^1/$_2$-hour fire protection rating as required for 2-hour fire-resistance-rated wall assemblies in accordance with the requirements of Section 715.3.

909.20.3.2 Vestibule doors. Where access to the stairway is by way of a vestibule, the door assembly into the vestibule shall be a fire door complying with Section 715.3. The door assembly from the vestibule to the stairway shall have not less than a 20-minute fire protection rating complying with Section 715.3.

❖ When a smokeproof enclosure is of the type that employs a vestibule as the means of access to an enclosed exit stairway, the entry into the vestibule from the exit access must consist of a fire door assembly having a 1^1/$_2$-hour fire protection rating. This requirement is commensurate with the 2-hour fire-resistance-rated smokeproof enclosure, as specified in Section 715.3. Essentially, the fire door assemblies and the enclosure walls serve as the primary exit stairway protection from fires occuring in adjacent spaces.

The entry from a vestibule into the stairway requires a 20-minute fire-protection-rated door assembly. These doorways are not normally subject to direct fire exposure and, therefore, only a nominal fire protection rating is required as a standard of construction that would minimize air leakage and thus reduce the possibilities of smoke and hot gases penetrating the stairway enclosure in amounts that could become a threat to life safety.

Door assemblies must comply with the requirements of Section 715.3 [see Figures 909.20.3.2(1) and 909.20.3.2(2)].

909.20.3.3 Vestibule ventilation. Each vestibule shall have a minimum net area of 16 square feet (1.5 m²) of opening in a wall facing an outer court, yard or public way that is at least 20 feet (6096 mm) in width.

❖ Natural ventilation is the first option for the ventilation of vestibules. This alternative would only be allowed for exterior balconies or for vestibules that have a minimum net opening of 16 square feet (1.49 m²) and open into an outer court, yard or public way that is at least 20 feet (6096 mm) in width. Basically, this option is intending to take advantage of the fact that smoke may disperse into the open area before entering the vestibule.

909.20.4 Mechanical ventilation alternative. The provisions of Sections 909.20.4.1 through 909.20.4.4 shall apply to ventilation of smokeproof enclosures by mechanical means.

❖ When an exterior balcony is not possible or when the clear open area cannot be achieved, then the mechani-

cal ventilation alternative is necessary. This option will be necessary for entrances into smokeproof enclosures located within buildings.

909.20.4.1 Vestibule doors. The door assembly from the building into the vestibule shall be a fire door complying with Section 715.3. The door assembly from the vestibule to the stairway shall have not less than a 20-minute fire protection rating in accordance with Section 715.3. The door from the building into the vestibule shall be provided with gaskets or other provisions to minimize air leakage.

❖ The first of the requirements is that the doors have a 20-minute fire-resistance rating and also contain gaskets or another method to minimize air leakage.

909.20.4.2 Vestibule ventilation. The vestibule shall be supplied with not less than one air change per minute and the exhaust shall not be less than 150 percent of supply. Supply air shall enter and exhaust air shall discharge from the vestibule through separate, tightly constructed ducts used only for that purpose. Supply air shall enter the vestibule within 6 inches (152 mm) of the floor level. The top of the exhaust register shall be located at the top of the smoke trap but not more than 6 inches (152 mm) down from the top of the trap, and shall be entirely within the smoke trap area. Doors in the open position shall not obstruct duct openings. Duct openings with controlling dampers are permitted where necessary to meet the design requirements, but dampers are not otherwise required.

❖ This section provides the basic criteria for the mechanical ventilation of the vestibule during a fire situation. Such ventilation is required to be activated via a smoke detector in accordance with Section 909.20.6. The basic requirements are for one air change per minute

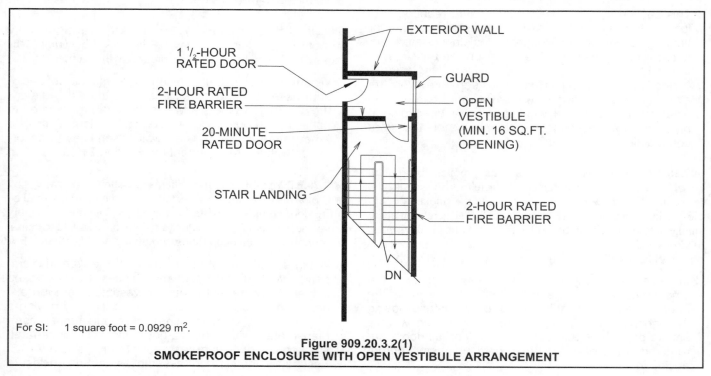

For SI: 1 square foot = 0.0929 m².

Figure 909.20.3.2(1)
SMOKEPROOF ENCLOSURE WITH OPEN VESTIBULE ARRANGEMENT

within the vestibule. Also the exhaust is required to be 150 percent of the supply. The exhaust is required to exceed the supply so that air will not be pushed into the stairway due to a positive pressure being created within the vestibule. It should be noted that there is not a minimum pressure difference requirement. Instead, the code simply prescribes a particular air change rate.

This section also prescribes criteria as to where supply air and exhaust air are to enter and exit, respectively, and requires that those paths are clear (see Figure 909.20.4.2).

909.20.4.2.1 Engineered ventilation system. Where a specially engineered system is used, the system shall exhaust a quantity of air equal to not less than 90 air changes per hour from any vestibule in the emergency operation mode and shall be sized to handle three vestibules simultaneously. Smoke detectors shall be located at the floor-side entrance to each vestibule and shall activate the system for the affected vestibule. Smoke detectors shall be installed in accordance with Section 907.10.

❖ If the method prescribed in Section 909.20.4.2 for mechanical ventilation of vestibules is not chosen, then minimum criteria is provided for engineered systems. Essentially, the designer can create a design that meets the flexibility needs of the building and at the same time meets basic criteria. Once designed and installed, the vestibules must be ventilated at 90 air changes per hour, which is 30 more air changes per hour than the prescriptive solution would require. Additionally, this system needs to be sized to handle at least three vestibules simultaneously. Finally, the engineered systems

are required to operate in accordance with the initiation of a smoke detector located on the floor-side entrance of the affected vestibule; therefore, a smoke detector would be required on each floor at the entrance to each vestibule.

909.20.4.3 Smoke trap. The vestibule ceiling shall be at least 20 inches (508 mm) higher than the door opening into the vestibule to serve as a smoke and heat trap and to provide an upward-moving air column. The height shall not be decreased unless approved and justified by design and test.

❖ This section requires a 20-inch (508 mm) difference in height between the door openings into the vestibule and the ceiling. This is intended to allow an additional safeguard to assist in the containment of smoke inside the vestibule and outside the stair shaft. Essentially, as the title states, it provides a smoke trap. This arrangement also has the effect of keeping the air movement upward.

909.20.4.4 Stair shaft air movement system. The stair shaft shall be provided with a dampered relief opening and supplied with sufficient air to maintain a minimum positive pressure of 0.10 inch of water (25 Pa) in the shaft relative to the vestibule with all doors closed.

❖ In order for smoke to not migrate into the stairway from the vestibule, a positive pressure of .10 inch (25 Pa) of water in the stair shaft relative to the vestibule is required. This pressure difference is to be available when all doors are closed. This pressure difference is lower than that required by the stair pressurization alternative. This would not be considered a pressurized stair.

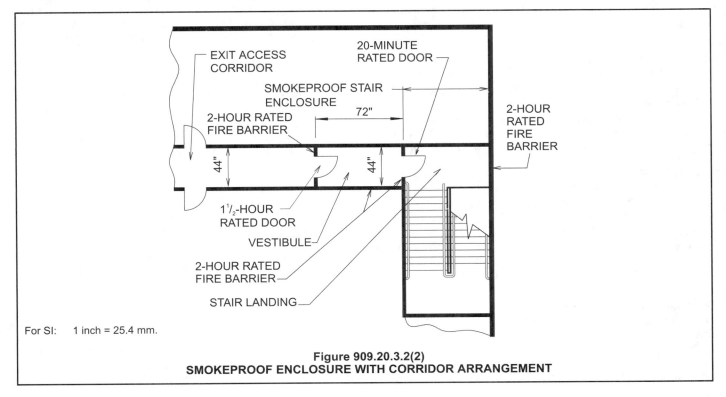

For SI: 1 inch = 25.4 mm.

Figure 909.20.3.2(2)
SMOKEPROOF ENCLOSURE WITH CORRIDOR ARRANGEMENT

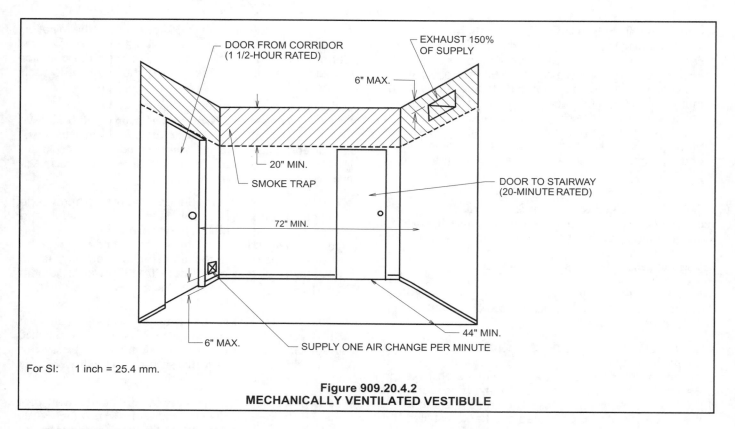

For SI: 1 inch = 25.4 mm.

Figure 909.20.4.2
MECHANICALLY VENTILATED VESTIBULE

909.20.5 Stair pressurization alternative. Where the building is equipped throughout with an automatic sprinkler system in accordance with Section 903.3.1.1, the vestibule is not required, provided that interior exit stairways are pressurized to a minimum of 0.15 inch of water (37 Pa) and a maximum of 0.35 inch of water (87 Pa) in the shaft relative to the building measured with all stairway doors closed under maximum anticipated stack pressures.

❖ This method is allowed only when the building is fully sprinklered. It should be noted that smokeproof enclosures are not required in fully sprinklered buildings, but the areas where smokeproof enclosures are required are often sprinklered buildings. This alternative would not require vestibules or an exterior exit balcony. The criteria for smoke control design is provided in terms of minimum and maximum pressure differences of .15 inch (37 Pa) of water and .35 inch (87 Pa) of water, respectively, between the shaft and the building. This pressure difference is to be achieved when all doors are closed and maximum stack effect has been taken into account. It should be noted that additional limitations may be placed on the maximum pressure differences for pressurized stairs due to the lower opening forces required to comply with accessibility requirements.

909.20.6 Ventilating equipment. The activation of ventilating equipment required by the alternatives in Sections 909.20.4 and 909.20.5 shall be by smoke detectors installed at each floor level at an approved location at the entrance to the smokeproof enclosure. When the closing device for the stair shaft and vestibule doors is activated by smoke detection or power failure, the me-

chanical equipment shall activate and operate at the required performance levels. Smoke detectors shall be installed in accordance with Section 907.10.

❖ This section clarifies that the activation mechanism for both mechanical means of smoke management in Sections 909.20.4 and 909.20.5 should be via a smoke detector located at each level outside the door leading into the vestibule and stairway, respectively. For systems that use automatic-closing devices on the doors, whether for vestibules in smokeproof enclosures or for pressurized stairs, if the door closes, the system must activate. This includes normal activation of the smoke detector or in the event of a power failure. Essentially, if there is a power failure it will operate in a fail-safe manner even if there is not a fire.

909.20.6.1 Ventilation systems. Smokeproof enclosure ventilation systems shall be independent of other building ventilation systems. The equipment and ductwork shall comply with one of the following:

1. Equipment and ductwork shall be located exterior to the building and directly connected to the smokeproof enclosure or connected to the smokeproof enclosure by ductwork enclosed by 2-hour fire-resistance-rated fire barriers.

2. Equipment and ductwork shall be located within the smokeproof enclosure with intake or exhaust directly from and to the outside or through ductwork enclosed by 2-hour fire-resistance-rated fire barriers.

3. Equipment and ductwork shall be located within the building if separated from the remainder of the building, including other mechanical equipment, by 2-hour fire-resistance-rated fire barriers.

❖ Smokeproof enclosures and pressurized stair shaft ventilation systems must be independent of other building ventilating systems. This section provides three options for the location of the ductwork associated with smokeproof enclosures and stair pressurization mechanical equipment. The three options include the following:

- Located on the exterior of the building and directly connected to the smokeproof enclosure,
- Within the smokeproof enclosure or
- Within the building but separated from the remainder of the building by 2-hour fire-resistance-rated barriers.

909.20.6.2 Standby power. Mechanical vestibule and stair shaft ventilation systems and automatic fire detection systems shall be powered by an approved standby power system conforming to Section 403.10.1 and Chapter 27.

❖ This section requires standby power for mechanical systems for pressurized stair shafts, mechanically ventilated vestibules, stair shafts for smokeproof enclosures and any automatic fire detection systems. Note that Section 909.20.6.2 requires ventilation and fire alarm systems to be available in case of power loss.

909.20.6.3 Acceptance and testing. Before the mechanical equipment is approved, the system shall be tested in the presence of the building official to confirm that the system is operating in compliance with these requirements.

❖ This section requires verification of successful testing of the system to the building official. It should be noted that although this does not specifically state "fire" official, it would be in the best interest to also notify the fire department of such testing and acceptance. It should be noted that identical provisions are located within the IFC. The IFC refers to the building official, which would be designated by the local jurisdiction. This will depend on the local authority having jurisdiction.

909.21 Underground building smoke exhaust system. Where required in accordance with Section 405.5 for underground buildings, a smoke exhaust system shall be provided in accordance with this section.

❖ Buildings that qualify in accordance with Section 405 as underground buildings require smoke control. It should be noted that this is a prescriptive smoke control approach. The intent of the smoke control provisions is to restrict movement of smoke to the general area of fire origin.

909.21.1 Exhaust capability. Where compartmentation is required, each compartment shall have an independent, automatically activated smoke exhaust system capable of manual operation. The system shall have an air supply and smoke exhaust capability that will provide a minimum of six air changes per hour.

❖ Section 405.4.1 requires compartmentation of an underground building when the building has a floor level more than 60 feet (18 288 mm) below the lowest level of exit discharge. If compartmentation is required, each compartment must have the ability to exhaust the smoke independently of the other compartments. The exhaust system for each compartment is required to provide six air changes per hour.

[F] 909.21.2 Operation. The smoke exhaust system shall be operated in the compartment of origin by the following, independently of each other:

1. Two cross-zoned smoke detectors within a single protected area of a single smoke detector monitored by an alarm verification zone or an approved equivalent method.

2. The automatic sprinkler system.

3. Manual controls that are readily accessible to the fire department.

❖ The smoke exhaust system is required to activate in the compartment of fire origin in three ways: two cross-zoned smoke detectors within a single protected area, the automatic sprinkler system and manual control. It is stressed that the exhaust systems only operate within the compartment of fire origin.

[F] 909.21.3 Alarm required. Activation of the smoke exhaust system shall activate an audible alarm at a constantly attended location.

❖ Regardless of how the system is activated, an audible alarm is required to notify someone in a constantly attended location such as a security booth, fire command center or central station of the activation of the smoke exhaust system. This will provide notification of a fire or of a malfunction related to the system.

SECTION 910
SMOKE AND HEAT VENTS

[F] 910.1 General. Where required by this code or otherwise installed, smoke and heat vents or mechanical smoke exhaust systems and draft curtains shall conform to the requirements of this section.

Exception: Frozen-food warehouses used solely for storage of Class I and II commodities where protected by an approved automatic sprinkler system.

❖ Smoke and heat vents must be provided in buildings, structures or portions thereof where required by Section 910.2 or as an alternative to another protection feature (see commentary, Section 1015.2). It should be noted that Chapter 23 of the IFC would also be applicable (see commentary, Section 910.2.3). The systems must be designed, installed, maintained and operated in accordance with the provisions of this section.

The purpose of smoke and heat vents has historically been related to the needs of fire fighters. More specifically, smoke and heat vents, when activated, have the potential effect of lifting the height of the smoke layer and providing more tenable conditions to undertake fire-fighting activities. Other potential benefits tend to be a decrease in property damage and creating more tenable conditions for occupants. The purpose of draft curtains, as addressed in Section 910.3.4, is both to contain the smoke in certain areas and potentially increase the speed in the activation of the smoke and heat vents.

The exception recognizes the "building-within-a-building" nature of typical frozen food warehouses. As such, smoke from a fire within a freezer would be contained within the freezer, thus negating the usefulness of smoke and heat vents at the roof level.

[F] 910.2 Where required. Approved smoke and heat vents shall be installed in the roofs of one-story buildings or portions thereof occupied for the uses set forth in Sections 910.2.1 through 910.2.4.

❖ Smoke and heat vents are only required in single-story buildings and then only as required by the provisions of Sections 910.2.1 through 910.2.3 or as a tradeoff for increased exit access travel distances in buildings or portions thereof classified in Group F-1 or S-1.

[F] 910.2.1 Groups F-1 and S-1. Buildings and portions thereof used as a Group F-1 or S-1 occupancy having more than 50,000 square feet (4645 m²) in undivided area.

Exception: Group S-1 aircraft repair hangars.

❖ Large-area buildings with moderate to heavy fire loads present special challenges to the fire department in disposing of the smoke generated in a fire. In order to provide the fire department with the capability to rapidly and efficiently dispose of smoke in buildings classified in Occupancy Groups F-1 and S-1 and exceeding 50,000 square feet (4,65 m²) in area without the exposure of personnel to the danger associated with cutting ventilation holes in the roof, smoke and heat vents or, alternatively, mechanical smoke removal facilities must be provided.

[F] 910.2.2 Group H. Buildings and portions thereof used as a Group H occupancy as shown:

1. In occupancies classified as Group H-2 or H-3, any of which are over 15,000 square feet (1394 m²) in single floor area.

Exception: Buildings of noncombustible construction containing only noncombustible materials.

2. In areas of buildings in Group H used for storing Class 2, 3, and 4 liquid and solid oxidizers, Class 1 and unclassified detonable organic peroxides, Class 3 and 4 unstable (reactive) materials, or Class 2 or 3 water-reactive materi-

als as required for a high-hazard commodity classification.

Exception: Buildings of noncombustible construction containing only noncombustible materials.

❖ Due to the explosion, deflagration, flammability or combustibility hazard of the contents found in Group H-1, H-2 and H-3 occupancies, smoke and heat vents are required to enhance confinement of the fire and to assist the fire department in ventilation. See the commentary to Section 415.6 for further information on these occupancy groups.

[F] 910.2.3 High-piled combustible storage. Buildings and portions thereof containing high-piled combustible stock or rack storage in any occupancy group in accordance with Section 413 and the *International Fire Code*.

❖ This section alerts the code user to the specific high-piled combustible storage requirements contained in Chapter 23 of the IFC. High-piled storage, whether solid piled, palletized or in racks, in excess of 12 feet (3658 m²) in height requires specific consideration, including fire protection design features and smoke and heat vents in order to be adequately protected. Not all high-piled storage will require the use of smoke and heat vents and draft curtains. In fact, if the high-piled storage is properly sprinklered (in accordance with Chapter 23 of the IFC and NFPA 13), draft curtains are not required (see commentary to Chapter 23 and Table 2306.2 of the IFC).

[F] 910.2.4 Exit access travel distance increase. Buildings and portions thereof used as a Group F-1 or S-1 occupancy where the maximum exit access travel distance is increased in accordance with Section 1015.2.

❖ This section applies to buildings or portions thereof in Occupancy Groups F-1 and S-1 (moderate-hazard industrial/factory and moderate-hazard storage, respectively) when the exit access travel distance is increased beyond that permitted in Table 1015.1. Table 1015.1 already permits an exit access travel distance of 400 feet (121.9 m) in sprinklered Group F-2 and S-2 occupancies without smoke and heat vents. Note that this section does not require the installation of smoke and heat venting but rather allows them to be utilized as a tradeoff to offset the inherent hazards of longer-than-normal travel distances.

The venting system provides a mechanism for relieving smoke and products of combustion from the building, thus increasing the amount of time required for smoke to obscure visibility in the path of egress travel and providing better visibility and additional time to reach the exits.

[F] 910.3 Design and installation. The design and installation of smoke and heat vents and draft curtains shall be as specified in this section and Table 910.3.

❖ Careful design and installation of smoke and heat vents is vital to their efficient operation in case of fire. The design criteria for these fire protection tools are organized for convenience and ready reference in Table 910.3, which is referenced by this section.

TABLE 910.3. See below.

❖ Table 910.3 identifies the required vent area in terms of ratio of vent area to floor area. The table is essentially divided into two parts. The first part is for Group F-1 occupancies, while the second portion is for Group S-1 and high-piled combustible storage. The way in which the table is currently written makes it appear as if only high-piled storage in Group S-1 occupancies requires smoke and heat venting. If Chapter 23 of the IFC referenced, smoke and heat venting requirements are independent of occupancy and simply focus upon the commoditity classification and area of storage. Additionally, Chapter 23 of the IFC only requires smoke and heat venting and draft curtains for larger storage areas. The required vent areas vary based upon the commodity classification (I-IV or high hazard) and height of storage. The higher the storage, the higher the potential for a larger fire. Two options are given for high hazard and classifications I-IV. The only significance to these options is that one allows a lower vent-to-floor area ratio if a deeper draft curtain is chosen, simply providing some credit for the fact that the deeper draft curtains are more likely to contain more smoke than a smaller draft curtain; thus, the area contained by the draft curtains also varies. When in a situation where smoke and heat vents are required and draft curtains are not, the second option would be appropriate.

[F] 910.3.1 Vent operation. Smoke and heat vents shall be approved and labeled and shall be capable of being operated by approved automatic and manual means. Automatic operation of smoke and heat vents shall conform to the provisions of this section.

❖ For the purpose of establishing the dependability of vents, this section requires that they be approved by the building official and labeled. While there is no standard referenced in the code for vent labeling, standards such as UL 33, UL 521, UL 793 and FM 4430 may be used.

Since vents are used as a component of an active venting system, the releasing device is required to be automatic, such as a fusible link. If large volumes of smoke could be generated prior to the vent being operated by heat-sensitive devices, then approved automatic smoke detectors should be installed in accordance with Section 907.10. Special care should be taken when using a method such as smoke detection actuation, since opening the vents prior to the sprinkler activation may interrupt the effectiveness of the sprinkler system. In addition to automatic operation of the vents, a manual means of opening them by the fire department during fire suppression operations must also be provided. It should be remembered that one of the main reasons smoke and heat vents were initially introduced was to keep fire fighters from having to get on a roof of a burning building; therefore, the mechanisms for release and the needs of the fire departement should be carefully considered.

[F] 910.3.1.1 Gravity-operated drop-out vents. Automatic smoke and heat vents containing heat-sensitive glazing designed to shrink and drop out of the vent opening when exposed to fire shall fully open within 5 minutes after the vent cavity is exposed to a simulated fire, represented by a time-temperature

TABLE 910.3
REQUIREMENTS FOR DRAFT CURTAINS AND SMOKE AND HEAT VENTS[a]

OCCUPANCY GROUP AND COMMODITY CLASSIFICATION	DESIGNATED STORAGE HEIGHT (feet)	MINIMUM DRAFT CURTAIN DEPTH (feet)	MAXIMUM AREA FORMED BY DRAFT CURTAINS (square feet)	VENT AREA TO FLOOR AREA RATIO	MAXIMUM SPACING OF VENT CENTERS (feet)	MAXIMUM DISTANCE TO VENTS FROM WALL OR DRAFT CURTAINS[b] (feet)
Group F-1	—	$0.2 \times H$ but ≥ 4	50,000	1:100	120	60
Group S-1 I-IV (Option 1)	≤ 20	6	10,000	1:100	100	60
	> 20 ≤ 40	6	8,000	1:75	100	55
Group S-1 I-IV (Option 2)	≤ 20	4	3,000	1:75	100	55
	> 20 ≤ 40	4	3,000	1:50	100	50
Group S-1 High hazard (Option 1)	≤ 20	6	6,000	1:50	100	50
	> 20 ≤ 30	6	6,000	1:40	90	45
Group S-1 High hazard (Option 2)	≤ 20	4	4,000	1:50	100	50
	> 20 ≤ 30	4	2,000	1:30	75	40

For SI: 1 foot = 304.8 mm, 1 square foot = 0.0929 m^2.

a. Requirements for rack storage heights in excess of those indicated shall be in accordance with Chapter 23 of the *International Fire Code*. For solid-piled storage heights in excess of those indicated, an approved engineered design shall be used.

b. The distance specified is the maximum distance from any vent in a particular draft curtained area to walls or draft curtains which form the perimeter of the draft curtained area.

gradient that reaches an air temperature of 500°F (260°C) within 5 minutes.

❖ This section establishes minimum performance criteria for drop-out vents, which include a nonmetallic, clear or opaque glazing element designed to shrink from its frame and fall away when exposed to heat from a fire. Such vent design must be capable of completely opening the roof vent within 5 minutes of exposure to a simulated fire represented by a time-temperature gradient that reaches an air temperature of 500° F (260°C) within 5 minutes. Drop-out vents tested in accordance with UL 793 must begin to operate at a maximum temperature of 286° F (141°C) in order to be labeled.

[F] 910.3.1.2 Sprinklered buildings. Where installed in buildings provided with an approved automatic sprinkler system, smoke and heat vents shall be designed to operate automatically.

❖ Where smoke and heat vents are installed in sprinklered buildings, their operation must be automatic and coordinated with the operation of the sprinkler system. Caution should be exercised in the design of smoke and heat vents and the required draft curtain so that the curtain boards do not interfere with the operation of the automatic sprinklers since locating a draft curtain too close to a sprinkler head could prevent proper water distribution over the fire. In addition, draft curtains will contain smoke and hot gases that can direct them away from the area where the fire is actually burning, thus activating sprinklers in the wrong area. This has the potential of overwhelming the sprinkler system. This is especially an issue with specially designed systems such as Early Suppression Fast Response (ESFR) sprinklers.

Coordination of fusible link operating temperatures with sprinkler head operating temperatures is also critical to the timely and effective operation of both the vents and the sprinklers. For example, the premature operation of a vent-operating mechanism could retard the operation of higher temperature-rated sprinkler heads by dissipating the level of heat needed to make the fusible link of the sprinkler(s) operate.

Delaying the operation of sprinklers can have the negative effect of causing an excessive number of sprinklers to operate, including some located outside the immediate area of fire danger. Concern over this issue has increased with the introduction of new sprinkler technology, such as the use of ESFR sprinklers, which are designed to act quickly to apply larger volumes of water to extinguish rather than simply control the fire.

[F] 910.3.1.3 Nonsprinklered buildings. Where installed in buildings not provided with an approved automatic sprinkler system, smoke and heat vents shall operate automatically by ac-

tuation of a heat-responsive device rated at between 100°F (38°C) and 220°F (104°C) above ambient.

Exception: Gravity-operated drop-out vents complying with Section 910.3.1.1

❖ Where smoke and heat vents are installed in buildings that are not equipped with an automatic sprinkler system, their operation must be automatic, with their operating elements set at between 100 and 220° F (38 and 104°C). The exception indicates that gravity-operated drop-out vents are not subject to this requirement.

[F] 910.3.2 Vent dimensions. The effective venting area shall not be less than 16 square feet (1.5 m²) with no dimension less than 4 feet (1219 mm), excluding ribs or gutters having a total width not exceeding 6 inches (152 mm).

❖ This section prescribes the minimum clear area required for each individual smoke and heat vent, exclusive of any obstructions. The design of the aggregate vent area actually needed is based, in part, on the area defined by the draft curtains and the depth of the curtained area, the objective of the design being to prevent either smoke from spilling out of the curtained area or the smoke interface from interfering with egress visibility. It has also been argued that draft curtains are intended to speed the operation of the smoke and heat vents by keeping the smoke in a smaller area.

[F] 910.3.3 Vent locations. Smoke and heat vents shall be located 20 feet (6096 mm) or more from adjacent lot lines and fire walls and 10 feet (3048 mm) or more from fire barrier walls. Vents shall be uniformly located within the roof area above high-piled storage areas, with consideration given to roof pitch, draft curtain location, sprinkler location and structural members.

❖ This section intends to minimize the effects of the discharge from operating smoke and heat vents on adjacent properties, especially if the smoke being vented contains products of combustion from burning hazardous materials. Since fire walls define separate buildings and in order not to compromise their integrity by placing a nonfire-resistance-rated roof opening too close to the wall, the clearance between vents and fire walls is the same as for property lines. The integrity of fire barriers is similarly protected; however, being a less restrictive assembly than a fire wall, the vent clearance requirements are reduced. This section also lists the considerations that must be taken into account when designing roof vent locations for high-piled storage areas.

[F] 910.3.4 Draft curtains. Where required, draft curtains shall be provided in accordance with this section.

Exception: Where areas of buildings are equipped with early suppression fast-response (ESFR) sprinklers, draft curtains shall not be provided within these areas. Draft curtains shall only be provided at the separation between the ESFR sprinklers and the conventional sprinklers.

❖ Draft curtains, sometimes termed "curtain boards," are required to be installed in conjunction with smoke and heat vents in accordance with Table 910.3 and as required by Section 413 and Chapter 23 of the IFC. They are installed within and at the perimeter of a protected area to restrict smoke and heat movement beyond the area of fire origin or the protected area and enhance smoke and heat removal through the roof vents. Table 2306.2 does not require draft curtains in sprinklered buildings. Instead, only smoke and heat vents are required in certain cases (larger areas of high-piled storage). The extent of the protection is addressed in Chapter 23 of the IFC (how much of a building would need smoke and heat vents and draft curtains)

[F] 910.3.4.1 Construction. Draft curtains shall be constructed of sheet metal, lath and plaster, gypsum board or other approved materials which provide equivalent performance to resist the passage of smoke. Joints and connections shall be smoke tight.

❖ In order not to contribute to the fire load of a building and to increase the likelihood that draft curtains will remain intact under fire conditions, they must be constructed of noncombustible materials or an approved equivalent (see the commentary to Section 703.4 for further information on noncombustibility), but are not required to possess a fire-resistance rating. Draft curtains need only be capable of resisting the passage of smoke.

[F] 910.3.4.2 Location and depth. The location and minimum depth of draft curtains shall be in accordance with Table 910.3.

❖ The requirements for depth and location of draft curtains are provided in Table 910.3 based on the occupancy group (in the case of Group F-1, S-1 and H occupancies) of the building, the commodity classification of the stored materials and the height of the storage.

[F] 910.4 Mechanical smoke exhaust. Where approved by the building official, engineered mechanical smoke exhaust shall be an acceptable alternate to smoke and heat vents.

❖ This section recognizes that providing a mechanical smoke exhaust system may, under certain circumstances, be more desirable, practical or efficient than installing automatic smoke and heat roof vents. The intent of Sections 910.4.1 through 910.4.6 is to create a mechanical system that performs at least as efficiently as smoke and heat vents designed in accordance with Section 910.3. Installation of an alternative mechanical smoke exhaust system is subject to the specific approval of the building official so that the design can be reviewed and the operational sequence and control information can be shared with the fire department. This smoke exhaust system is different from that required in Section 909.

[F] 910.4.1 Location. Exhaust fans shall be uniformly spaced within each draft-curtained area and the maximum distance between fans shall not be greater than 100 feet (30 480 mm).

❖ One or more smoke exhaust fans must be provided in each area defined by draft curtains and when more than one fan is provided in a curtained area, the fans must be spaced uniformly within that area, no more that 100 feet (30 480 mm) apart. Locating fans in this manner will enhance the uniform removal of smoke from curtained areas and reduce the likelihood of smoke spillage under the draft curtains.

[F] 910.4.2 Size. Fans shall have a maximum individual capacity of 30,000 cfm (14.2 m³/s). The aggregate capacity of smoke exhaust fans shall be determined by the equation:

$$C = A \times 300 \hspace{2cm} \textbf{(Equation 9-10)}$$

where:

C = Capacity of mechanical ventilation required, in cubic feet per minute (m³/s).

A = Area of roof vents provided in square feet (m²) in accordance with Table 910.3.

❖ The intent of the sizing requirements of this section is to provide a smoke exhaust rate at least equivalent to the venting capacity provided by roof vents. The exhaust rate required by this section, based on Equation 9-10, is equivalent to 300 cubic feet per minute per square foot (153 m³/s·m²) of the roof vent area required by Table 910.3, with no single fan exceeding a 30,000 cubic feet per minute (14.2 m³/s) rate. For example, a Group F-1 factory with maximum-sized curtained areas of 50,000 square feet (4545 m²) would be required to have a total vent area of 500 square feet (46 m²) in each curtained area in accordance with Table 910.3. The mechanical exhaust rate required based on Equation 9-10 would then be 500 x 300 = 150,000 cfm (70.8 m³/s), which could be supplied by five 30,000 cfm (14.2 m³/s) fans spaced in accordance with Section 910.4.1 in each curtained area.

[F] 910.4.3 Operation. Mechanical smoke exhaust fans shall be automatically activated by the automatic sprinkler system or by heat detectors having operating characteristics equivalent to those described in Section 910.3.1. Individual manual controls of each fan unit shall also be provided.

❖ The activation of the mechanical smoke exhaust system must be capable of being accomplished by actuation of the automatic sprinkler system or, in nonsprinklered buildings, by heat detectors with a temperature rating of between 100 and 220° F (38 and 104°C) as required for smoke vents in Section 910.3.1 and manual controls. The manual control is for fire department use to increase the reliability of the system and to allow the fire department to activate the exhaust system to assist in the removal of smoke during or after a fire. Since manual control of the system is primarily for fire department use, the location of the controls should be subject to approval by the fire department. While not specifically stated in this section, a manual fire alarm system is more prone to false activations; therefore, activation of the smoke exhaust system should not be permitted by this means.

[F] 910.4.4 Wiring and control. Wiring for operation and control of smoke exhaust fans shall be connected ahead of the main disconnect and protected against exposure to temperatures in excess of 1,000°F (538°C) for a period of not less than 15 minutes. Controls shall be located so as to be immediately accessible to the fire service from the exterior of the building and protected against interior fire exposure by fire barriers having a fire-resistance rating not less than 1 hour.

❖ Unless the mechanical smoke exhaust system also functions as a component of a smoke control system, standby power is not specifically required (see commentary, Sections 909.11 and 2702). In order to provide an enhanced level of operational reliability, this section requires that the power supply to smoke exhaust fans must be provided from a circuit connected ahead of the building's main electrical service disconnecting means. Note that this is one of the sources of standby power recognized by the ICC EC via the reference to NFPA 70, Section 701.11(E).

This section also requires that the wiring for smoke exhaust fans be thermally protected in a manner approved by the building official that will protect the wiring from heat damage in the event of an interior fire. This protection could be provided by an approved wiring material listed for the temperature application, by physical protection with approved materials or assemblies or by installation outside of the building.

Since smoke exhaust systems are a vital fire-fighting tool, their operating controls are also required to be protected from interior fire exposure by 1-hour fire-resistance-rated construction that complies with Section 706. Exterior access to the controls allows fire department personnel to promptly operate the system from a protected area without entering the building. Controls should also be clearly identified in an approved, permanent manner.

[F] 910.4.5 Supply air. Supply air for exhaust fans shall be provided at or near the floor level and shall be sized to provide a minimum of 50 percent of required exhaust. Openings for supply air shall be uniformly distributed around the periphery of the area served.

❖ The introduction of makeup air is critical to the proper operation of all exhaust systems. Too little makeup air will cause a negative pressure to develop in the area being exhausted, thereby reducing the exhaust flow.

This section requires that makeup air be introduced to the area equipped with a mechanical smoke exhaust system in order to maintain the required exhaust flow. Since the system can only exhaust as much air as is introduced into the area, and this section allows mechanical or gravity makeup air openings to provide only 50 percent of the required makeup air, this section allows

the designer to rely upon infiltration air to provide up to the remaining 50 percent of the design makeup air required to allow the system to perform. Although not specifically stated in this section, where a mechanical makeup air source is utilized, it should be electrically interlocked and controlled by a single start switch, such that makeup air is always being supplied when the smoke exhaust system is in operation.

Even distribution of makeup air is important because, if too much air is coming from one particular direction, it has the potential to vary the dynamics of the fire and the ability of the system to address the smoke.

[F] 910.4.6 Interlocks. In combination comfort air-handling/smoke removal systems or independent comfort air-handling systems, fans shall be controlled to shut down in accordance with the approved smoke control sequence.

❖ For HVAC systems and combination HVAC/smoke removal systems, this section requires that any fan shutdown requirements, such as those required by the IMC, defer to the approved smoke control sequence established in accordance with Section 909.

SECTION 911
FIRE COMMAND CENTER

[F] 911.1 Features. Where required by other sections of this code, a fire command center for fire department operations shall be provided. The location and accessibility of the fire command center shall be separated from the remainder of the building by not less than a 1-hour fire-resistance-rated fire barrier. The room shall be a minimum of 96 square feet (9 m²) with a minimum dimension of 8 feet (2438 mm). A layout of the fire command center and all features required by the section to be contained therein shall be submitted for approval prior to installation. The fire command center shall comply with NFPA 72 and shall contain the following features.

1. The emergency voice/alarm communication system unit.

2. The fire department communications unit.

3. Fire detection and alarm system annunciator unit.

4. Annunciator unit visually indicating the location of the elevators and whether they are operational.

5. Status indicators and controls for air-handling systems.

6. The fire-fighter's control panel required by Section 909.16 for smoke control systems installed in the building.

7. Controls for unlocking stairway doors simultaneously.

8. Sprinkler valve and water-flow detector display panels.

9. Emergency and standby power status indicators.

10. A telephone for fire department use with controlled access to the public telephone system.

11. Fire pump status indicators.

12. Schematic building plans indicating the typical floor plan and detailing the building core, means of egress, fire protection systems, fire-fighting equipment and fire department access.

13. Worktable.

14. Generator supervision devices, manual start and transfer features.

15. Public address system, where specifically required by other sections of this code.

❖ Fire command centers are typically found in high-rise buildings (see Section 403.8). Fire ground operations usually involve establishing an incident command post where the fire fighter can observe what is happening, control arriving personnel and equipment and direct resources and fire-fighting operations effectively. Because of the difficulties in controlling a fire in a high-rise building, a separate room within the building must be established to assist the fire fighter. The room must be provided at a location that is acceptable to the fire department, usually along the front of the building or near the main entrance.

For the survivability of the fire command center, it must be separated from the remainder of the building by a minimum 1-hour fire-resistance-rated fire barrier. The minimum specified room dimensions are required so that adequate clearance is provided for the fire command center control equipment. The room must contain equipment necessary to monitor or control fire protection and other building service systems as specified in Section 911.1. The fire command center must be in compliance with Section 911 and the applicable provisions in NFPA 72.

Bibliography

The following resource materials are referenced in this chapter or are relevant to the subject matter addressed in this chapter.

Americans with Disabilities Act Accessibility Guidelines for Buildings and Facilities (ADAAG). Washington, DC: U.S. Architectural and Transportation Barriers Compliance Board, 1998.

Automatic Sprinkler Systems Handbook, 7th edition. Quincy, MA: National Fire Protection Association, 1996.

Bryan, John L. Automatic Sprinkler and Standpipe Systems, 3rd edition. Quincy, MA: National Fire Protection Association, 1997.

Budnick, E.K. Estimating Effectiveness of State-of-the-Art Detectors and Automatic Sprinklers on Life Safety in Residential Occupancies. Washington, DC National Bureau of Standards, NBS IR 84-2819.

Bukowski, R.W. And R.J. O' Laughlin. Fire Alarm Signaling Systems. Quincy, MA: National Fire Protection Association, 1994.

Design of Smoke Management Systems. Atlanta, GA: American Society of Heating, Refrigerating and Air-Conditioning Engineers, Inc., 1992.

DOJ 28 CFR, Part 36-91, Americans With Disabilities Act. Washington, DC: U.S. Department of Justice, 1991.

DOJ 28 CFR, Part 36 (Appendix A)-91, ADA Guidelines for Buildings and Facilities. Washington, DC: U.S. Department of Justice, 1991.

DOTn 49 CFR, Parts 100-178 & 179-199-95, Specification for Transportation of Explosive and Other Dangerous Articles, Shipping Containers. Washington, DC: U.S. Department of Transportation, 1995.

Evans, D. and J. Klote. Smoke Control Provisions of the 2000 IBC an Applications Guide. Country Club Hills, IL: International Code Council, 2002.

Fire Protection Equipment Directory. Northbrook, IL, Underwriters Laboratories Inc., 2002.

Fire Protection Handbook, 18th edition. Quincy, MA: National Fire Protection Association, 1997.

Fire Pump Handbook. Quincy, MA: National Fire Protection Association, 1998.

FM 4430-88, Approved Standard for Heat and Smoke Vents. Norwood, MA: Factory Mutual, 1988.

Grant, Casey Cavanaugh. "Halon and Beyond: Developing New Alternatives." FPA Journal, November/December 1994.

Harrington, Jeff L. "The Halon Phaseout Speeds Up." NFPA Journal, March/April 1993.

Health Care Facilities. Falls Church, VA: CABO Board for the Coordination of the Model Codes Report, 1985.

ICC A117.1-98, Accessible and Usable Buildings and Facilities. Falls Church, VA: International Code Council, 1998.

Klote, J. and J. Milke. Principles of Smoke Management. ASHRAE, 2002.

Milke, James A. "Smoke Management for Covered Malls and Atria." Fire Technology, August 1990.

National Fire Alarm Code Handbook. Quincy, MA: National Fire Protection Association, 1999.

NFPA 10-98, Portable Fire Extinguishers. Quincy, MA: National Fire Protection Association, 1998.

NFPA 11-98, Low Expansion Foam and Combined Agent Systems. Quincy, MA: National Fire Protection Association, 1998.

NFPA 11A-99, Medium and High Expansion Foam Systems. Quincy, MA: National Fire Protection Association, 1999.

NFPA 12-00, Carbon Dioxide Extinguishing Systems. Quincy, MA: National Fire Protection Association, 2000.

NFPA 12A-97, *Halon 1301 Fire Extinguishing Systems.* Quincy, MA: National Fire Protection Association, 1997.

NFPA 13-99, *Installation of Sprinkler Systems.* Quincy, MA: National Fire Protection Association, 1999.

NFPA 13D-99, *Installation of Sprinkler Systems in One- and Two-Family Dwellings and Manufactured Homes.* Quincy, MA: National Fire Protection Association, 1999.

NFPA 13R-99, *Installation of Sprinkler Systems in Residential Occupancies Up to and Including Four Stories in Height.* Quincy, MA: National Fire Protection Association, 1999.

NFPA 14-00, *Standpipe and Hose Systems.* Quincy, MA: National Fire Protection Association, 2000.

NFPA 16-99, *Installation of Deluge Foam-Water Sprinkler and Foam-Water Spray Systems.* Quincy, MA: National Fire Protection Association, 1999.

NFPA 17-98, *Dry Chemical Extinguishing Systems.* Quincy, MA: National Fire Protection Association, 1998.

NFPA 17A-98, *Wet Chemical Extinguishing Systems.* Quincy, MA: National Fire Protection Association, 1998.

NFPA 20-99, *Installation of Centrifugal Fire Pumps.* Quincy, MA: National Fire Protection Association, 1999.

NFPA 24-95, *Installation of Private Fire Service Mains.* Quincy, MA: National Fire Protection Association, 1995.

NFPA 25-98, *Inspection, Testing and Maintenance of Water Based Fire Protection Systems.* Quincy, MA: National Fire Protection Association, 1998.

NFPA 30-00, *Flammable and Combustible Liquids Code.* Quincy, MA: National Fire Protection Association, 2000.

NFPA 30B-00, *Manufacture and Storage of Aerosol Products.* Quincy, MA: National Fire Protection Association, 2000.

NFPA 68-02, *Guide for Venting of Deflagrations.* Quincy, MA: National Fire Protection Association, 2002.

NFPA 69-97, *Explosion Prevention Systems.* Quincy, MA: National Fire Protection Association, 1997.

NFPA 72-99, *National Fire Alarm Code.* Quincy, MA: National Fire Protection Association, 1999.

NFPA 92A-00, *Smoke Control Systems.* Quincy, MA: National Fire Protection Association, 2000.

NFPA 92B-00, *Smoke Management Systems in Malls, Atria and Large Areas.* Quincy, MA: National Fire Protection Association, 2000.

NFPA 204-00, *Standard for Smoke and Heat Venting.* Quincy, MA: National Fire Protection Association, 2000.

NFPA 231-98, *General Storage.* Quincy, MA: National Fire Protection Association, 1998.

NFPA 231C-98, *Rack Storage of Materials.* Quincy, MA: National Fire Protection Association, 1998.

NFPA 231D-98, *Storage of Rubber Tires.* Quincy, MA: National Fire Protection Association, 1998.

NFPA 495-96, *Explosive Materials Code.* Quincy, MA: National Fire Protection Association, 1996.

NFPA 1963-03, *Standard for Screw Threads and Gaskets for Fire Hose Connections.* Quincy, MA: National Fire Protection Association, 2003.

NFPA 2001-00, *Clean Agent Fire Extinguishing Systems.* Quincy, MA: National Fire Protection Association, 2000.

Smoke Control in Fire Safety Design. London: E. & F.N., Spon Ltd., 1979.

The SFPE Handbook of Fire Protection Engineering, 3rd ed. Quincy, MA: National Fire Protection Association, 2002

UL 33-03, *Standard for Heat-Responsive Links.* Northbrook, IL: Underwriters Laboratories Inc., 2003.

UL 217-97, *Single and Multiple Station Smoke Detectors.* Northbrook, IL: Underwriters Laboratories Inc., 1997.

UL 268-96, *Smoke Detectors for Fire Protective Signaling Systems.* Northbrook, IL: Underwriters Laboratories Inc., 1996.

UL 300-96, *Standard for Fire Testing of Fire Extinguishing Systems for Protection of Restaurant Cooking Areas with Revisions through December 1998.* Northbrook: IL, Underwriters Laboratories Inc., 1996.

UL 521-99, *Standard for Heat Detectors for Fire Protective Signaling Systems.* Northbrook, IL: Underwriters Laboratories Inc., 1999.

UL 793-03, *Standard for Automatically Operated Roof Vents for Smoke and Heat.* Northbrook, IL: Underwriters Laboratories Inc., 2003.

UL 1058-95, *Standard for Halogenated Agent Extinguishing System Units with Revisions through April 1998.* Northbrook, IL: Underwriters Laboratories Inc., 1995.

Chapter 10
Means of Egress

General Comments

Chapter 10 provides the minimum requirements for means of egress in all buildings and structures. The requirements detail the size, arrangement, number and protection of means of egress components. Functional and operational characteristics also are specified for the components that will permit their safe use without special knowledge or effort.

Chapter 10 of the coded has been reorganized from the 2000 edition as a result of an approved code change proposal. This resulted in a renumbering of the chapter from nine sections to 25. The presentation of text predominantly follows that of the 2000 edition; however, the section numbers have been revised. A comprehensive 2000/2003 Chapter 10 section number cross index is posted on the ICC website at www.iccsafge.org.

Section 1001 includes the administrative provisions. Section 1002 shows the definitions of terms that are primarily associated with Chapter 10. Sections 1003 through 1012 include general provisions that apply to all three components of a means of egress system: exit access, exit and exit discharge. The exit access requirements are in Sections 1013 through 1016, the exit requirements are in Sections 1017 through 1022 and the exit discharge requirements are in Section 1023. Section 1024 includes those means of egress requirements that are unique to an assembly occupancy. Emergency escape and rescue requirements are in Section 1025.

The evolution of means of egress requirements has been influenced by lessons learned from real fire incidents. While contemporary fires may reinforce some of these lessons, one must view each incident as an opportunity to assess critically the safety and reasonability of current regulations.

Cooperation among the developers of model codes and standards has resulted in agreement on many basic terms and concepts. The text of the code, including this chapter, is consistent with these national uniformity efforts.

National uniformity in an area such as means of egress has many benefits for the code official and other code us-ers. At the top of the list are the lessons to be learned from experiences throughout the nation and the world, which can be reported in commonly used terminology and conditions that we can all relate to and clearly understand.

Purpose

A primary purpose of codes in general and building codes in particular is to safeguard life in the presence of a fire. Integral to this purpose is the path of egress travel for occupants to escape and avoid a fire. Means of egress can be considered the lifeline of a building. The principles on which means of egress are based and that form the fundamental criteria for requirements are to provide a system:

1. That will give occupants alternative paths of travel to a place of safety to avoid fire.

2. That will shelter occupants from fire and the products of combustion.

3. That will accommodate all occupants of a structure.

4. That is clear, unobstructed, well marked and illuminated and in which all components are under control of the user without requiring any tools, keys or special knowledge or effort.

History is marked with the loss of life from fire. Early as well as contemporary multiple fire fatalities can be traced to a compromise of one or more of the above principles.

Life safety from fire is a matter of successfully evacuating or relocating the occupants of a building to a place of safety. As a result, life safety is a function of time: time for detection, time for notification and time for safe egress. The fire growth rate over a period of time is also a critical factor in addressing life safety. Other sections of the code, such as protection of vertical openings (Chapter 7), interior finish (Chapter 8), fire suppression and detection systems (Chapter 9) and numerous others, also have an impact on life safety. This chapter addresses the issues related to the means available to relocate or evacuate building occupants.

SECTION 1001
ADMINISTRATION

1001.1 General. Buildings or portions thereof shall be provided with a means of egress system as required by this chapter. The provisions of this chapter shall control the design, construction and arrangement of means of egress components required to provide an approved means of egress from structures and portions thereof.

❖ The minimum requirements for means of egress are to be incorporated in all new structures as specified in this chapter. The system shall include exit access, exit and exit discharge. Such application would be effective on the date the code is adopted and placed into effect.

1001.2 Minimum requirements. It shall be unlawful to alter a building or structure in a manner that will reduce the number of exits or the capacity of the means of egress to less than required by this code.

❖ A fundamental concept in life safety design is that the means of egress system is to be constantly available throughout the life of a building. Any change in the building or its contents, either by physical reconstruction, alteration or by a change of occupancy, is cause to review the resulting egress system. As a minimum, a building's means of egress is to be continued as initially approved. If a building or portion thereof has a change of occupancy, the complete egress system is to be evaluated and approved for compliance with the current code requirements for new occupancies (see Chapter 34).

The means of egress in an existing building that experiences a change of occupancy, such as from Group S-2 (storage) to A-3 (assembly), would require reevaluation for code compliance based on the new occupancy. Similarly, the means of egress in an existing Group A-3 occupancy in which additional seating is to be provided, thereby increasing the occupant load, would require re-evaluation for code compliance based on the increased occupant load.

The temptation is to temporarily remove egress components or other fire protection features from service during an alteration, repair to or temporary occupancy of a building. During such times, a building is frequently more vulnerable to fire and the rapid spread of products of combustion. Either the occupants should not occupy those spaces where the means of egress has been compromised by the construction or compensating fire safety features should be considered that will provide equivalent safety for the occupants. Occupants in adjacent areas may also require access to the egress facilities in the area under construction.

[F] 1001.3 Maintenance. Means of egress shall be maintained in accordance with the *International Fire Code.*

❖ This section provides a cross reference to the code requirements that address the maintenance of the means of egress in an existing building. The means of egress must be maintained so that occupants are not pre-vented from exiting the building quickly in case of an emergency condition.

Sections 1002 through 1025 in the code are repeated in the *International Fire Code*® (IFC®). These sections are maintained by the building code committees so that there will be consistency between the two documents. Note the [B] in front of the main section headings in the IFC. Note the [F] in front of this section. This means that this section is maintained by the fire code committee. Additionally, the IFC includes Sections 1026 and 1027, which apply to existing buildings.

SECTION 1002
DEFINITIONS

1002.1 Definitions. The following words and terms shall, for the purposes of this chapter and as used elsewhere in this code, have the meanings shown herein.

❖ Definitions of terms can help in the understanding and application of the code requirements. The purpose for including these definitions in this chapter is to provide more convenient access to them without having to refer back to Chapter 2.

For convenience, these terms are also listed in Chapter 2 with a cross reference to this section.

The use and application of all defined terms, including those defined herein, are set forth in Section 201.

ACCESSIBLE MEANS OF EGRESS. A continuous and unobstructed way of egress travel from any point in a building or facility that provides an accessible route to an area of refuge, a horizontal exit or a public way.

❖ Accessible means of egress requirements are needed to provide those persons with physical disabilities a means of egress to a safe area in the building or to exit the building.

AISLE ACCESSWAY. That portion of an exit access that leads to an aisle.

❖ As illustrated in Figure 1002.1(1), an aisle accessway is intended for one-way travel or limited two-way travel. The space between tables, seats, displays or other furniture (i.e., aisle accessway) utilized for means of egress will lead to a main aisle.

ALTERNATING TREAD DEVICE. A device that has a series of steps between 50 and 70 degrees (0.87 and 1.22 rad) from horizontal, usually attached to a center support rail in an alternating manner so that the user does not have both feet on the same level at the same time.

❖ An alternating tread device is commonly used in areas that would otherwise be provided with a ladder. The device is used extensively in industrial facilities for worker access to platforms or equipment.

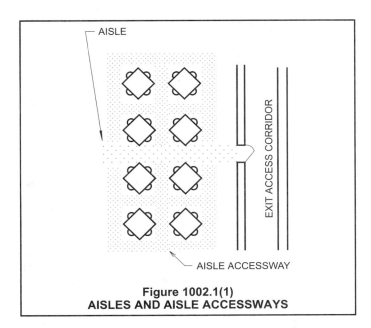

**Figure 1002.1(1)
AISLES AND AISLE ACCESSWAYS**

AREA OF REFUGE. An area where persons unable to use stairways can remain temporarily to await instructions or assistance during emergency evacuation.

❖ The area of refuge provides a safe area in the building during an emergency condition for temporary use by persons who are unable to exit the building using the stairways.

BLEACHERS. Tiered seating facilities.

❖ Bleachers, folding and telescopic seating and grandstands are essentially unique forms of tiered seating. All types are addressed in ICC 300, the new safety standard for these types of seating arrangements. Bleachers typically do not have backrests. The travel path across the bleachers is not restricted to designated rows, aisles and aisle accessways. Without backrests, occupants can traverse from row to row without traveling to the designated egress aisles (see Section 1024.1.1).

COMMON PATH OF EGRESS TRAVEL. That portion of exit access which the occupants are required to traverse before two separate and distinct paths of egress travel to two exits are available. Paths that merge are common paths of travel. Common paths of egress travel shall be included within the permitted travel distance.

❖ The common path of egress travel is a concept used to refine travel distance criteria. Similar to dead ends in corridors and passageways, a common path of travel is the course an occupant will travel and, regardless of which route is chosen, he or she must pass through the same location. The length of common path of egress travel is limited so that the route to a remote means of egress begins to diverge before the occupant has traveled an excessive distance. This reduces the possibility that, although the exits are remote from one another, a single fire condition will render both paths unavailable.

CORRIDOR. An enclosed exit access component that defines and provides a path of egress travel to an exit.

❖ Corridors are regulated in the code because they serve as principal elements of travel in the means of egress systems within buildings. Corridors have walls that extend from the floor to at least the ceiling. They need not extend above the ceiling or have doors in their openings unless a fire-resistance rating is required (see Section 1016). The enclosed character of the corridor restricts the sensory perception of the user. A fire located on the other side of the corridor wall, for example, may not be as readily seen, heard or smelled by the occupants traveling through the egress corridor. An egress path bounded by partial-height walls, such as work-station partitions in an office, are not corridors by definition since they are not enclosed by full-height walls.

DOOR, BALANCED. A door equipped with double-pivoted hardware so designed as to cause a semicounterbalanced swing action when opening.

❖ Balanced doors are commonly used to decrease the force necessary to open the door or to reduce the length of the door swing.

EGRESS COURT. A court or yard which provides access to a public way for one or more exits.

❖ The egress court requirements address situations where the exit discharge portion of the means of egress passes through confined areas near the building and therefore faces a hazard not normally found in the exit discharge.

EMERGENCY ESCAPE AND RESCUE OPENING. An operable window, door or other similar device that provides for a means of escape and access for rescue in the event of an emergency.

❖ These are commonly windows that are big enough and located such that they can be used to exit a building directly from a basement or bedroom during an emergency condition. The openings are also used by emergency personnel to rescue the occupants in a building (see Section 1025).

EXIT. That portion of a means of egress system which is separated from other interior spaces of a building or structure by fire-resistance-rated construction and opening protectives as required to provide a protected path of egress travel between the exit access and the exit discharge. Exits include exterior exit doors at ground level, exit enclosures, exit passageways, exterior exit stairs, exterior exit ramps and horizontal exits.

❖ Exits are the critical element of the means of egress system that the building occupants travel through to reach the exterior grade level. Exit stairways from upper and lower stories, as well as horizontal exits, must be separated from adjacent areas with fire-resistance-rated construction. The fire-resistance-rated construction serves as a barrier between the fire and the means of egress and protects the occupants while they travel through the exit. Separation

FIGURE 1002.1(2) – FIGURE 1002.1(5)　　　　　　　　　　　　　　　　　　　　　　MEANS OF EGRESS

by fire-resistance-rated construction is not required, however, where the exit leads directly to the exterior at the level of exit discharge (e.g., exterior door at grade).

Figure 1002.1(2) illustrates three different types of exits: interior exit stairway, exterior exit stairway and exterior exit door.

EXIT ACCESS. That portion of a means of egress system that leads from any occupied portion of a building or structure to an exit.

❖ The exit access portion of the means of egress consists of all floor areas that lead from usable spaces within the building to the exit or exit(s) serving that floor area. Crawl spaces and concealed attic and roof spaces are not considered to be part of the exit access. As shown in Figure 1002.1(3), the exit access begins at the furthest points within each room or space and ends at the entrance to the exit.

EXIT DISCHARGE. That portion of a means of egress system between the termination of an exit and a public way.

❖ The exit discharge will typically begin when the building occupants reach the exterior at grade level. It provides oc-

cupants with a path of travel away from the building. All components between the building and the public way are considered to be the exit discharge, regardless of the distance. The exit discharge is part of the means of egress and, therefore, its components are subject to the requirements of the code [see Figures 1002.1(2) and 1002.1(4)].

EXIT DISCHARGE, LEVEL OF. The horizontal plane located at the point at which an exit terminates and an exit discharge begins.

❖ The term is intended to describe the level at which the occupants, during egress, leave an exit and are no longer required to ascend or descend to be at the level at which exit discharge begins. At this level, the occupant need only move in a substantially horizontal path to reach the start of exit discharge. Hence, an exit may be composed of both interior and exterior exit stairs which together terminate at the level of exit discharge [see Figure 1002.1(5)].

EXIT ENCLOSURE. An exit component that is separated from other interior spaces of a building or structure by fire-resistance-rated construction and opening protectives, and provides

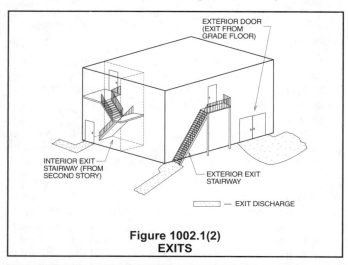

Figure 1002.1(2)
EXITS

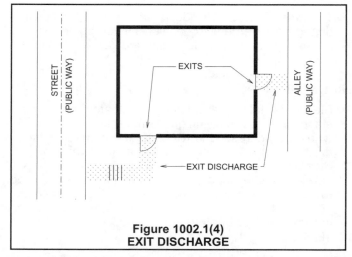

Figure 1002.1(4)
EXIT DISCHARGE

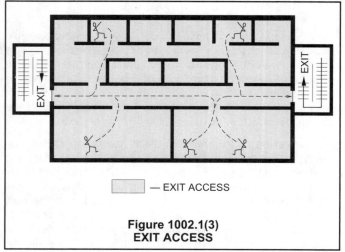

Figure 1002.1(3)
EXIT ACCESS

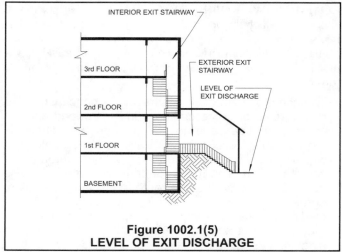

Figure 1002.1(5)
LEVEL OF EXIT DISCHARGE

for a protected path of egress travel in a vertical or horizontal direction to the exit discharge or the public way.

❖ This term is used to describe an exit that is within a fire-resistance-rated enclosure for a generally vertical path of travel (e.g., a stair) or a generally horizontal path of travel (e.g., a ramp).

EXIT, HORIZONTAL. A path of egress travel from one building to an area in another building on approximately the same level, or a path of egress travel through or around a wall or partition to an area on approximately the same level in the same building, which affords safety from fire and smoke from the area of incidence and areas communicating therewith.

❖ This term refers to fire-resistance-rated wall that subdivides a building or buildings into multiple compartments and provides an effective barrier to protect occupants from a fire condition within one of the compartments. After occupants pass through a horizontal exit, they must be provided with sufficient space to gather and must also be provided with another exit, such as an exterior door or exit stairway, through which they can exit the building. Figure 1002.1(6) depicts the exits serving a single building that is subdivided with a fire-resistance-rated wall.

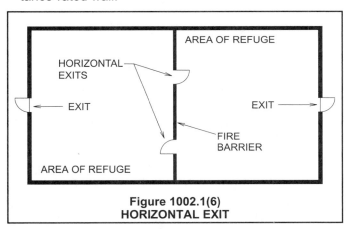

**Figure 1002.1(6)
HORIZONTAL EXIT**

EXIT PASSAGEWAY. An exit component that is separated from all other interior spaces of a building or structure by fire-resistance-rated construction and opening protectives, and provides for a protected path of egress travel in a horizontal direction to the exit discharge or the public way.

❖ This term refers to the portion of the means of egress that serves as a horizontal exit element and leads to the exit discharge. Since an exit passageway is considered an exit element, it must be protected and separated as required by the code for exits. An exit passageway may be located before or after a vertical exit enclosure. Exit passageways that lead to an exterior exit door are commonly used in malls to satisfy the travel distance in buildings having a large floor area. Exit passageways between a vertical exit enclosure and an exterior exit door are typically found on the level of exit discharge to provide a protected path from a centrally located exit stairway to the exit discharge.

FIRE EXIT HARDWARE. Panic hardware that is listed for use on fire door assemblies.

❖ Where a door that is required to be fire-resistance-rated construction has panic hardware, the hardware is required to be listed for use on the door. Thus, fire door hardware has been demonstrated to function properly when exposed to the effects of a fire.

FLOOR AREA, GROSS. The floor area within the inside perimeter of the exterior walls of the building under consideration, exclusive of vent shafts and courts, without deduction for corridors, stairways, closets, the thickness of interior walls, columns or other features. The floor area of a building, or portion thereof, not provided with surrounding exterior walls shall be the usable area under the horizontal projection of the roof or floor above. The gross floor area shall not include shafts with no openings or interior courts.

❖ Gross floor area is that area measured within the perimeter formed by the inside surface of the exterior walls. The area of all occupiable and nonoccupiable spaces, including mechanical and elevator shafts, toilets, closets, mechanical equipment rooms, etc., is included in the gross floor area. This area could also include any covered porches or other exterior space intended to be used as part of the building's occupiable space. This dimension is primarily used for the determination of occupant load.

FLOOR AREA, NET. The actual occupied area not including unoccupied accessory areas such as corridors, stairways, toilet rooms, mechanical rooms and closets.

❖ This area is intended to be only the room areas that are used for specific occupancy purposes and does not include circulation areas such as corridors or stairways and service and utility spaces, such as toilet rooms and mechanical and electrical equipment rooms.

FOLDING AND TELESCOPIC SEATING. Tiered seating facilities having an overall shape and size that are capable of being reduced for purposes of moving or storing.

❖ Bleachers, folding and telescopic seating and grandstands are essentially unique forms of tiered seating. All types are addressed in ICC 300, the new ICC safety standard for these types of seating arrangements. Folding and telescopic seating are commonly used in gymnasiums and sports arenas where the seating can be configured in a variety of ways for various types of events (see Section 1024.1.1).

GRANDSTAND. Tiered seating facilities.

❖ Bleachers, folding and telescopic seating and grandstands are essentially unique forms of tiered seating. All types are addressed in the new ICC safety standard for these types of seating arrangements. Grandstands can be found at a county fairground, along a parade route or within indoor facilities. Examples are sports arenas and public auditoriums, as well as churches and gallery-type lecture halls (see Section 1024.1.1).

GUARD. A building component or a system of building components located at or near the open sides of elevated walking surfaces that minimizes the possibility of a fall from the walking surface to a lower level.

❖ Guards are sometimes mistakenly referred to as "guardrails." In actuality, the guard consists of the entire vertical portion of the barrier, not just the top rail (see Section 1012).

HANDRAIL. A horizontal or sloping rail intended for grasping by the hand for guidance or support.

❖ Handrails are provided along walking surfaces that lead from one elevation to another, such as ramps, stairways and landings and are generally circular in shape. Noncircular shapes could also be acceptable provided that they can be gripped by hand for support and guidance and for checking possible falls on the adjacent walking surface. In addition to being necessary in normal day-to-day use, handrails are especially needed in times of emergency when the pace of egress travel is hurried and the probability for occupant instability while traveling along the sloped or stepped walking surface increases. Handrails are not intended to be used in place of guards to prevent people from falling over the edge.

MEANS OF EGRESS. A continuous and unobstructed path of vertical and horizontal egress travel from any occupied portion of a building or structure to a public way. A means of egress consists of three separate and distinct parts: the exit access, the exit and the exit discharge.

❖ The means of egress is the path traveled by building occupants to leave the building and the site on which it is located. It includes all interior and exterior elements that the occupants must utilize as they make their way from every room and usable space within the building to a public way such as a street or alley. The elements that make up the means of egress create the lifeline that occupants utilize to travel out of the structure and to a safe distance from the structure. The means of egress provisions of this chapter strive to provide a reasonable level of life safety in every structure. The means of egress provisions are subdivided into three distinct portions (see the definitions of "Exit access," "Exit" and "Exit discharge").

NOSING. The leading edge of treads of stairs and of landings at the top of stairway flights.

❖ Limiting the extent of the tread nosings results in a stair that is easy to use. If too large, they are a tripping hazard when walking up a stairway, and reduce the effective tread depth when walking down the stairway [see Figures 1009.3(1) and 1009.3(2)].

OCCUPANT LOAD. The number of persons for which the means of egress of a building or portion thereof is designed.

❖ In addition to the limitation on the maximum occupant load for a space, the code also requires the determina-

tion of the occupant load that is to be utilized for the design of the means of egress system. This occupant load is also utilized to determine the required number of plumbing fixtures (see Chapter 29).

PANIC HARDWARE. A door-latching assembly incorporating a device that releases the latch upon the application of a force in the direction of egress travel.

❖ Panic hardware is commonly used in educational and assembly-type spaces where the number of occupants who would use a doorway during a short time frame in an emergency is high in relation to an occupancy with a less dense occupant load, such as an office building. The hardware is required so that the door can be easily opened during an emergency.

PUBLIC WAY. A street, alley or other parcel of land open to the outside air leading to a street, that has been deeded, dedicated or otherwise permanently appropriated to the public for public use and which has a clear width and height of not less than 10 feet (3048 mm).

❖ The public way marks the termination of the exit discharge portion of the means of egress system. It is the final destination for occupants, and is presumed to be safe from the emergency in the structure.

RAMP. A walking surface that has a running slope steeper than one unit vertical in 20 units horizontal (5-percent slope).

❖ This definition is needed to determine the threshold at which the ramp requirements apply to a walking surface. Walking surfaces steeper than specified in the definition are subject to the ramp requirements.

SCISSOR STAIR. Two interlocking stairways providing two separate paths of egress located within one stairwell enclosure.

❖ A scissor or interlocking stair is sometimes used in high-rise buildings or to increase exit capacity of a stairway enclosure. In this configuration, two independent stairways are located within the same exit enclosure and open to one another. If interlocking stairways are separated from each other with appropriate fire barrier assemblies, they are not considered scissor stairways that may serve as only one exit (see Section 1014.2.1).

SMOKE-PROTECTED ASSEMBLY SEATING. Seating served by means of egress that is not subject to smoke accumulation within or under a structure.

❖ An example of smoke-protected assembly seating is an open outdoor grandstand or an indoor arena with a smoke control system. The code has less stringent requirements for certain aspects of smoke-protected assembly seating than for seating that is not smoke protected, since occupants are subject to less hazard during a fire event. For example, an assembly dead-end aisle is permitted to be longer for a smoke-protected assembly area.

STAIR. A change in elevation, consisting of one or more risers.

❖ All steps, even a single step, are defined as a stair. This makes the stair requirements applicable to all steps unless specifically exempt in the code.

STAIRWAY. One or more flights of stairs, either exterior or interior, with the necessary landings and platforms connecting them, to form a continuous and uninterrupted passage from one level to another.

❖ It is important to note that this definition characterizes a stairway as connecting one level to another. The term "level" is not to be confused with "story." Steps that connect two floor levels, one or both of which are not a "story" of the structure, would be considered a stairway. For example, a set of steps between the basement level in an areaway and the outside ground level would be considered a stairway. A series of steps between the floor of a story and a mezzanine within that story would also be considered a stairway.

STAIRWAY, EXTERIOR. A stairway that is open on at least one side, except for required structural columns, beams, handrails and guards. The adjoining open areas shall be either yards, courts or public ways. The other sides of the exterior stairway need not be open.

❖ This definition is needed since the code requirements for an exterior stairway are different than for an interior stairway. For specific openness requirements, see Section 1022.

STAIRWAY, INTERIOR. A stairway not meeting the definition of an exterior stairway.

❖ This definition is needed since the requirements for an interior stairway are more stringent than those for an exterior stairway (see the definition for "Stairway, exterior").

STAIRWAY, SPIRAL. A stairway having a closed circular form in its plan view with uniform section-shaped treads attached to and radiating from a minimum-diameter supporting column.

❖ Spiral stairways are commonly used where a small number of occupants use the stairway and the floor space for the stair is very limited. Spiral stairways are typically supported by a center pole.

WINDER. A tread with nonparallel edges.

❖ Winders are used as components of stairs that change direction, just as fliers (straight treads) are components in straight stairs. A winder performs the same function as a tread, but its shape allows the additional function of a gradual turning of the stairway direction. The tread depth of a winder at the walk line and the minimum tread depth at the narrow end can control the turn made by each winder.

SECTION 1003
GENERAL MEANS OF EGRESS

1003.1 Applicability. The general requirements specified in Sections 1003 through 1012 shall apply to all three elements of the means of egress system, in addition to those specific requirements for the exit access, the exit and the exit discharge detailed elsewhere in this chapter.

❖ The text of Chapter 10 is subdivided into 25 sections. The requirements in the chapter address the three parts of a means of egress system: the exit access, the exit and the exit discharge. This section specifies that the requirements of Sections 1003 through 1012 apply to the components of all three parts of the system. For example, the stair tread and riser dimensions in Section 1009 apply to exit access stairs such as those leading from a small mezzanine and also apply to enclosed exit stairs according to the vertical exit enclosure requirements in Section 1019.

1003.2 Ceiling height. The means of egress shall have a ceiling height of not less than 7 feet (2134 mm).

Exceptions:

1. Sloped ceilings in accordance with Section 1208.2.
2. Ceilings of dwelling units and sleeping units within residential occupancies in accordance with Section 1208.2.
3. Allowable projections in accordance with Section 1003.3.
4. Stair headroom in accordance with Section 1009.2.
5. Door height in accordance with Section 1008.1.1.

❖ Generally, the specified ceiling height is the minimum allowed in any part of the egress path. The exceptions are intended to address conditions where the code allows the ceiling height to be lower than specified in this section. This is also consistent with the headroom requirements at ramps in Section 1010.5.2.

The ceiling height for other areas is specified in Section 1208.

1003.3 Protruding objects. Protruding objects shall comply with the requirements of Sections 1003.3.1 through 1003.3.4.

❖ This section identifies the sections that apply to protruding objects and helps to improve awareness of these safety and accessibility-related provisions.

1003.3.1 Headroom. Protruding objects are permitted to extend below the minimum ceiling height required by Section 1003.2 provided a minimum headroom of 80 inches (2032 mm) shall be provided for any walking surface, including walks, corridors, aisles and passageways. Not more than 50 percent of the ceiling area of a means of egress shall be reduced in height by protruding objects.

Exception: Door closers and stops shall not reduce headroom to less than 78 inches (1981 mm).

A barrier shall be provided where the vertical clearance is less than 80 inches (2032 mm) high. The leading edge of such a barrier shall be located 27 inches (686 mm) maximum above the floor.

❖ This provision is applicable to all components of the means of egress. Specifically, the limitations in this section and those in Sections 1003.3.2 and 1003.3.3 provide a reasonable level of safety for those who are preoccupied or not paying attention while walking, as well as for people with impaired vision.

Minimum dimensions for headroom clearance also are specified in this section. The minimum headroom clearance over all walking surfaces is required to be maintained at 80 inches (2032 mm). This minimum headroom clearance is consistent with the requirements in Section 1009.2 for stairs and Section 1010.5.2 for ramps. Allowance must be made for door closers and stops, since their design and function necessitates placement within the door opening. The minimum headroom clearance for door closers and stops is allowed to be 78 inches (1981 mm) (see Figure 1003.3.1). The 2-inch (51 mm) projection into the doorway height is reasonable, since these devices are normally mounted away from the center of the door opening, thus minimizing the potential for contact with a person moving through the opening.

1003.3.2 Free-standing objects. A free-standing object mounted on a post or pylon shall not overhang that post or pylon more than 12 inches (305 mm) where the lowest point of the

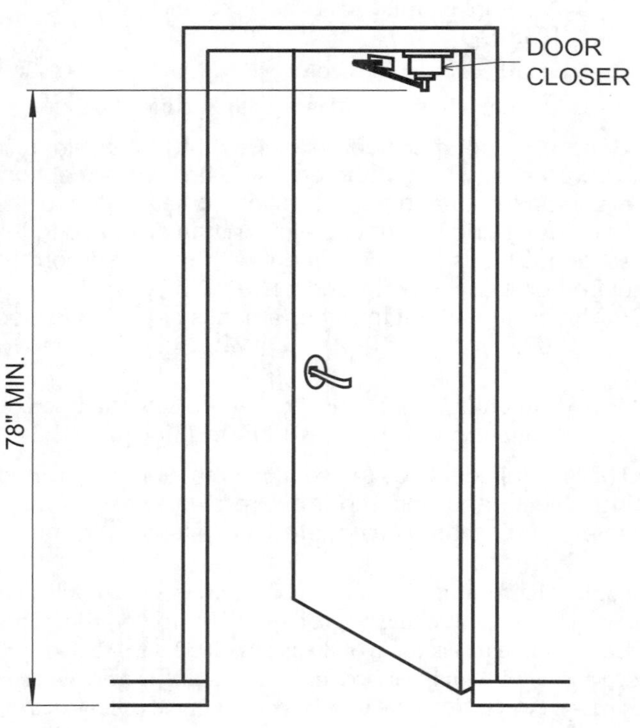

For SI: 1 inch = 25.4 mm.

Figure 1003.3.1
DOOR CLOSER HEADROOM PROTRUSIONS FOR
WALKING SURFACE

leading edge is more than 27 inches (686 mm) and less than 80 inches (2032 mm) above the walking surface. Where a sign or other obstruction is mounted between posts or pylons and the clear distance between the posts or pylons is greater than 12 inches (305 mm), the lowest edge of such sign or obstruction shall be 27 inches (685 mm) maximum or 80 inches (2030 mm) minimum above the finish floor or ground.

Exception: This requirement shall not apply to sloping portions of handrails serving stairs and ramps.

❖ Free-standing objects, such as signs mounted on posts, are not permitted to overhang more than 12 inches (305 mm) over the edges of the post where located higher than 27 inches (686 mm) above the walking surface (see Figure 1003.3.2). Since the minimum required height of doorways, stairways and ramps in the means of egress is 80 inches (2032 mm), protruding objects located higher than 80 inches (2032 mm) above the walking surface are not regulated. Protrusions that are located lower than 27 inches (686 mm) above the walking surface are also permitted, since they are more readily detected by a walking cane. The projection of objects located lower than 27 inches (686 mm) is not limited, provided that the minimum required width of the egress element is maintained.

When signs are provided on multiple posts, the posts must be located closer than 12 inches (305 mm) apart, and the bottom edge of the sign must be lower than 27 inches (686 mm) so it is within detectable cane range or above 80 inches (2032 mm) so that it is above headroom clearances.

The exception is intended for handrails that are located along a stairway or ramp. The extensions at the top and bottom of stairways and ramps must meet the requirements for protruding objects.

1003.3.3 Horizontal projections. Structural elements, fixtures or furnishings shall not project horizontally from either side more than 4 inches (102 mm) over any walking surface between the heights of 27 inches (686 mm) and 80 inches (2032 mm) above the walking surface.

Exception: Handrails serving stairs and ramps are permitted to protrude 4.5 inches (114 mm) from the wall.

❖ Protruding objects could slow down the egress flow through a passageway and injure someone hurriedly passing by or someone with a visual impairment. Persons with a visual impairment, who use a long cane for guidance, must have sufficient warning of a protruding object. Where protrusions are located higher than 27 inches (686 mm) above the walking surface, the cane will most likely not encounter the protrusion before the person collides with the object.

Additionally, people with poor visual acuity or poor depth perception may have difficulty identifying protruding objects higher than 27 inches (686 mm). Therefore, objects such as lights, signs and door hardware, located between 27 inches (686 mm) and 80 inches (2032 mm) above the walking surface, are not permitted to extend more than 4 inches (102 mm) from each wall (see Figure 1003.3.3).

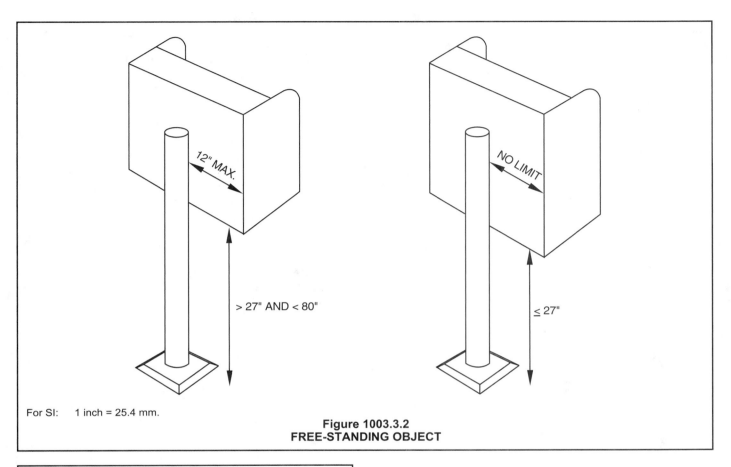

For SI: 1 inch = 25.4 mm.

Figure 1003.3.2
FREE-STANDING OBJECT

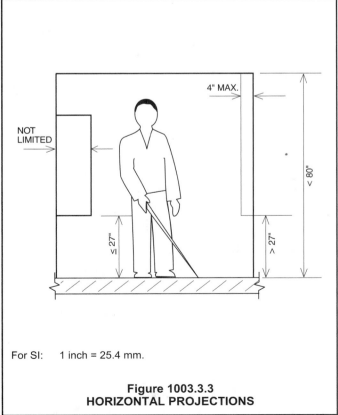

For SI: 1 inch = 25.4 mm.

Figure 1003.3.3
HORIZONTAL PROJECTIONS

1003.3.4 Clear width. Protruding objects shall not reduce the minimum clear width of accessible routes as required in Section 1104.

❖ The intent of this section is to limit the projections into an accessible route to those specified in Sections 1003.3 through 1003.3.3. The accessible route requirements are in Section 1104.

1003.4 Floor surface. Walking surfaces of the means of egress shall have a slip-resistant surface and be securely attached.

❖ As the pace of exit travel becomes hurried during emergency situations, the probability of slipping on smooth or slick floor surfaces increases. To minimize the hazard, all floor surfaces in the means of egress are required to be slip resistant. The use of hard floor materials with highly polished, glazed, glossy or finely finished surfaces should be avoided.

   Field testing and uniform enforcement of the concept of slip resistance are not practical. One method used to establish slip resistance is that the static coefficient of friction between leather [Type 1 (Vegetable Tanned) of Federal Specification KK-L-165C] and the floor surface is greater than 0.5. Laboratory test procedures such as ASTM D 2047 can determine the static coefficient of resistance. Bulletin No. 4 entitled "Surfaces" issued by the U.S. Architectural and Transportation Barriers Compliance Board (ATBCB) contains further information regarding slip resistance.

1003.5 Elevation change. Where changes in elevation of less than 12 inches (305 mm) exist in the means of egress, sloped surfaces shall be used. Where the slope is greater than one unit vertical in 20 units horizontal (5-percent slope), ramps complying with Section 1010 shall be used. Where the difference in elevation is 6 inches (152 mm) or less, the ramp shall be equipped with either handrails or floor finish materials that contrast with adjacent floor finish materials.

Exceptions:

1. A single step with a maximum riser height of 7 inches (178 mm) is permitted for buildings with occupancies in Groups F, H, R-2 and R-3 as applicable in Section 101.2, and Groups S and U at exterior doors not required to be accessible by Chapter 11.

2. A stair with a single riser or with two risers and a tread is permitted at locations not required to be accessible by Chapter 11, provided that the risers and treads comply with Section 1009.3, the minimum depth of the tread is 13 inches (330 mm) and at least one handrail complying with Section 1009.11 is provided within 30 inches (762 mm) of the centerline of the normal path of egress travel on the stair.

3. An aisle serving seating that has a difference in elevation less than 12 inches (305 mm) is permitted at locations not required to be accessible by Chapter 11, provided that the risers and treads comply with Section 1024.11 and the aisle is provided with a handrail complying with Section 1024.13.

Any change in elevation in a corridor serving nonambulatory persons in a Group I-2 occupancy shall be by means of a ramp or sloped walkway.

❖ Minor changes in elevation, such as a single step that is located in any portion of the means of egress (i.e., exit access, exit or exit discharge), may not be readily apparent during normal use or emergency egress and are considered to present a potential tripping hazard. Where the elevation change is less than 12 inches (305 mm), a ramp is specified to make the transition from higher to lower levels. The ramp is intended to reduce accidental falls associated with tripping hazards and must be constructed in accordance with Section 1010.1. The presence of the ramp must be readily apparent from the directions from which it is approached. Handrails are one method of identifying the change in elevation. In lieu of handrails, the surface of the ramp must be finished with materials that contrast with the surrounding floor surfaces. The walking surface of the ramp should contrast both visually and physically.

The first exception allows up to a 7-inch (178 mm) step at exterior doors to avoid blocking the outward swing of the door by a buildup of snow or ice in locations that are not used by the public on a regular basis (see Figure 1003.5). This exception supersedes the general provisions of Section 1008.1.4 and is only applicable in occupancies that have relatively low occupant densities, such as factory and industrial structures. This exception is not applicable to exterior doors that are re-

quired to serve as an accessible entrance or that are part of a required accessible route.

The second exception allows the transition from higher to lower elevations to be accomplished through the construction of stairs with one or two risers. The pitch of the stairway, however, must be shallower than that required for typical stairways (see Section 1009.3). Since the total elevation change is limited to 12 inches (305 mm), each riser must be approximately 6 inches (152 mm) in height. The presence of the elevation change must be readily apparent from the directions from which it is approached. At least one handrail is required, constructed in accordance with Section 1009.11 and located so as to provide a graspable surface from the normal walking path.

The third exception is basically a cross reference to the assembly provisions in Section 1024.

None of the exceptions are permitted in a Group I-2 occupancy (e.g., nursing home, hospital) in areas where nonambulatory persons may need access. The mobility impairments of these individuals require additional consideration.

1003.6 Means of egress continuity. The path of egress travel along a means of egress shall not be interrupted by any building element other than a means of egress component as specified in this chapter. Obstructions shall not be placed in the required width of a means of egress except projections permitted by this

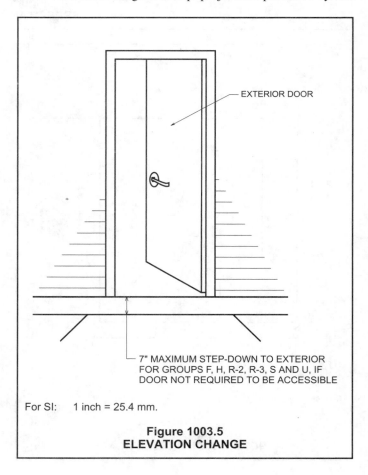

For SI: 1 inch = 25.4 mm.

Figure 1003.5
ELEVATION CHANGE

chapter. The required capacity of a means of egress system shall not be diminished along the path of egress travel.

❖ The purpose of this section requires that the entire means of egress path is clear of obstructions that could reduce the egress capacity at any point. The egress path is also not allowed to be reduced in width such that the design occupant load would not be served. Note, however, that the egress path could be reduced in width in situations where it is wider than required by the code based on the occupant load. For example, if the required width of a corridor were 52 inches (1321 mm) based on the number of occupants using the corridor and the corridor provided was 96 inches (2438 mm) in width, the corridor would be allowed to be reduced to the required width of 52 inches (1321 mm) since that width would still serve the number of occupants required by the code.

1003.7 Elevators, escalators and moving walks. Elevators, escalators and moving walks shall not be used as a component of a required means of egress from any other part of the building.

> **Exception:** Elevators used as an accessible means of egress in accordance with Section 1007.4.

❖ Generally the code does not allow elevators, escalators and moving sidewalks to be used as a required means of egress. Elevators are allowed to be part of an accessible means of egress, provided they comply with the requirements of Section 1007.4. The concern is that escalators and moving sidewalks may not provide a safe and reliable means of egress that is available for use at all times.

SECTION 1004
OCCUPANT LOAD

1004.1 Design occupant load. In determining means of egress requirements, the number of occupants for whom means of egress facilities shall be provided shall be established by the largest number computed in accordance with Sections 1004.1.1 through 1004.1.3.

❖ The design occupant load is the number of people that are intended to occupy a building or portion thereof at any one time; consequently, the number for which the means of egress is to be designed. It is the largest number derived by the application of Sections 1004.1 through 1004.1.3. There is a limit to the density of occupants permitted in an area to enable a reasonable amount of freedom of movement (see Section 1004.2). The design occupant load is also utilized to determine the required plumbing fixture count (see commentary, Chapter 29).

1004.1.1 Actual number. The actual number of occupants for whom each occupied space, floor or building is designed.

❖ The number of occupants that will occupy a space is the actual number and is only limited by Section 1004.2. If the construction documents indicate that the actual occupant load of a space exceeds that determined by

Sections 1004.1.2 and 1004.1.3, then the actual number is to be used as the design occupant load of that space. Where the actual number is less than the occupant load determined in accordance with Section 1004.1.2 or 1004.1.3, the largest number must be used in the egress design. For example, if a proposed conference room has a calculated occupant load—using 15 net square feet (1.39 m²) per person, for assembly without fixed seats, unconcentrated tables and chairs (see Table 1004.1.2)—of 55, but the owner indicates that the actual number of occupants will not exceed 25, the design occupant load of the room is 55. Therefore, in accordance with Table 1014.1, at least two means of egress must be provided from the conference room. Conversely, if the actual occupant load planned for is 65, the design occupant load is then 65.

1004.1.2 Number by Table 1004.1.2. The number of occupants computed at the rate of one occupant per unit of area as prescribed in Table 1004.1.2.

❖ This number reflects common and traditional occupant density based on empirical data for the density of similar spaces. The number determined using the occupant load rates in Table 1004.1.2 generally establishes the minimum occupant load for which the egress facilities of the rooms, spaces and building must be designed.

It is difficult to predict the many conditions by which a space within a building will be occupied over time. An assembly banquet room in a hotel, for example, could be arranged with rows of chairs to host a business seminar one day and with mixed tables and chairs to host a dinner reception the next day. In some instances, the room will be arranged with no tables and very few chairs to accommodate primarily standing occupants. In such a situation, the egress facilities must safely accommodate the maximum number of persons permitted to occupy the space. When determining the occupant load of this type of occupancy, the various arrangements (e.g., tables and chairs, chairs only, standing space) should be recognized. The worse case scenario should be utilized to determine the requirements for the means of egress elements.

While some of the values in the table utilize the net floor area, most utilize the gross floor area. See the commentary to Table 1004.1.2 for additional discussion and examples.

The occupant load determined in accordance with this section is the minimum occupant load upon which means of egress requirements are to be based. This limitation is true regardless of any indication by the owner that the space will be occupied by fewer people (see commentary, Section 1004.1.1).

Some occupancies may not typically contain an occupant load totally consistent with the occupant load density factors of Table 1004.1.2. Any special considerations for such unique uses must be documented and justified. Additionally, the owner must be aware that such special considerations will impact the future use of the building with respect to the means of egress and other protection features.

TABLE 1004.1.2   MEANS OF EGRESS

TABLE 1004.1.2
MAXIMUM FLOOR AREA ALLOWANCES PER OCCUPANT

OCCUPANCY	FLOOR AREA IN SQ. FT. PER OCCUPANT
Agricultural building	300 gross
Aircraft hangars	500 gross

Airport terminal
    Baggage claim
    Baggage handling
    Concourse
    Waiting areas
	
	20 gross
	300 gross
	100 gross
	15 gross

Assembly Gaming floors (keno, slots, etc.)	11 gross
Assembly with fixed seats	See Section 1004.7

Assembly without fixed seats
    Concentrated (chairs only—not fixed)
    Standing space
    Unconcentrated (tables and chairs)
	
	7 net
	5 net
	15 net

Bowling centers, allow 5 persons for each lane including 15 feet of runway, and for additional areas	7 net
Business areas	100 gross
Courtrooms—other than fixed seating areas	40 net
Dormitories	50 gross

Educational
    Classroom area
    Shops and other vocational room areas
	
	20 net
	50 net

Exercise rooms	50 gross
H-5 Fabrication and manufacturing areas	200 gross
Industrial areas	100 gross

Institutional areas
    Inpatient treatment areas
    Outpatient areas
    Sleeping areas
	
	240 gross
	100 gross
	120 gross

Kitchens, commercial	200 gross

Library
    Reading rooms
    Stack area
	
	50 net
	100 gross

Locker rooms	50 gross

Mercantile
    Areas on other floors
    Basement and grade floor areas
    Storage, stock, shipping areas
	
	60 gross
	30 gross
	300 gross

Parking garages	200 gross
Residential	200 gross

Skating rinks, swimming pools
    Rink and pool
    Decks
	
	50 gross
	15 gross

Stages and platforms	15 net
Accessory storage areas, mechanical equipment room	300 gross
Warehouses	500 gross

For SI: 1 square foot = 0.0929 m^2.

❖ The table presents the maximum floor area allowance per occupant based on studies and counts of the number of occupants in typical buildings. The use of this table, then, results in the minimum occupant load for which rooms, spaces and the building must be designed. While an assumed normal occupancy may be viewed as somewhat less than that determined by the use of the table factors, such a normal occupant load is not necessarily an appropriate design criterion. The greatest hazard to the occupants occurs when an unusually large crowd is present. The code does not limit the occupant load density of an area, except as provided for in Section 1004.2, but once the occupant load is established, the means of egress must be designed for at least that capacity. If it is intended that the occupant load will exceed that calculated in accordance with the table, then the occupant load is to be based on the estimated actual number of people in accordance with Section 1004.1.1. Table 1004.1.2 establishes minimum occupant densities based on the occupancy (not group classification) of the space. Therefore, the occupant load of the office or business areas in a storage warehouse or nightclub is to be determined using the occupant load factor most appropriate to that space—one person for each 100 square feet (9 m^2) of gross floor area.

The use of net and gross floor areas as defined in Section 1002.1 is intended to provide a refinement in the occupant load determination. The gross floor area technique applied to a building only allows the deduction of the plan area of the exterior walls, vent shafts and interior courts from the plan area of the building.

The net floor area permits the exclusion of certain spaces which would be included in the gross floor area. The net floor area is intended to apply to the actual occupied floor areas. The area used for permanent building components, such as shafts, fixed equipment, thicknesses of walls, corridors, stairways, toilet rooms, mechanical rooms and closets, is not included in net floor area. For example, consider a restaurant dining area with dimensions measured from the inside of the enclosing walls of 80 feet by 60 feet (24 384 mm by 18 288 mm) (see Figure 1004.1.2). Within the restaurant area is a 6-inch (152 mm) privacy wall running the length of the room [80 feet by 0.5 feet = 40 square feet (3.7 m^2)], a fireplace [40 square feet (3.7 m^2)] and a cloak room [60 square feet (5.6 m^2)]. Each of these areas is deducted from the restaurant area, resulting in a net floor area of 4,660 square feet (433 m^2). Since the restaurant intends to have unconcentrated seating that involves loose tables and chairs, the resulting occupant load is 311 persons (4,660 divided by 15). As the definition of "Floor area, net" indicates, certain spaces are to be excluded from the gross floor area to derive the net floor area. The key point in this definition is that the net floor area is to include the actual occupied area and does not include spaces uncharacteristic of that occupancy.

In determining the occupant load of a building with mixed groups, each floor area of a single occupancy must be separately analyzed, such as required by Section 1004.9. The occupant load of the business portion of an office/warehouse building is determined at a rate of one person for each 100 square feet (9 m²) of office space, whereas the occupant load of the warehouse portion is determined at the rate of one person for each 300 square feet (28 m²).

If a specific type of facility is not found in the table, the occupancy it most closely resembles should be utilized. For example, a training room in a business office may utilize the 20 square feet (1.86 m²) net established for educational classroom areas.

Table 1004.1.2, in accordance with Section 1004.1.2, presents a method of determining the absolute base minimum occupant load of a space that the means of egress is to accommodate.

In addition to the table, Section 402 contains the basis for calculating the occupant load of a covered mall building; however, Table 1004.1.2 should be used for determining the occupant load of each anchor store.

1004.1.3 Number by combination. Where occupants from accessory spaces egress through a primary area, the calculated occupant load for the primary space shall include the total occupant load of the primary space plus the number of occupants egressing through it from the accessory space.

❖ This section provides a method by which the occupant load of adjacent areas of a building is calculated. The resulting occupant load is what must be considered. For example, the means of egress from a lobby must be sized for the cumulative occupant load of the adjacent office spaces if the occupants must travel through the lobby to reach an exit. Likewise, if an adjacent room has an egress route independent of the lobby, the occupant load of that room would not be combined with the occupant loads of the other rooms that pass through that lobby. If a portion of the adjacent room's occupant load is to travel through the lobby, only that portion would be combined with the lobby occupant load for the design of the means of egress from the lobby (see Figure 1004.1.3). This is particularly important in determining the capacity and the number of means of egress.

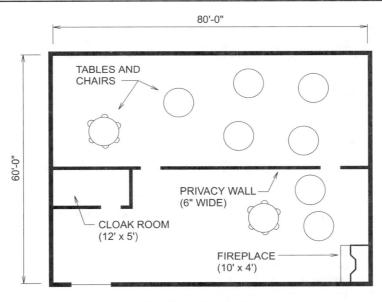

80' x 60' = 4,800 SQ.FT.

PRIVACY WALL:	40 SQ.FT.
FIREPLACE:	40 SQ.FT.
CLOAK ROOM:	60 SQ.FT.
TOTAL:	140 SQ.FT.

(TOTAL AREA WITHIN WALLS) - (EXCLUDED ITEMS) = (NET FLOOR AREA)
4,800 SQ.FT. - 140 SQ.FT. = 4,660 SQ.FT.

(NET FLOOR AREA)/[TABLE 1004.1.2 VALUE] = (OCCUPANT LOAD)
4,660 SQ.FT./15 SQ.FT. PER OCCUPANT = 311 OCCUPANTS

For SI: 1 inch = 25.4 mm, 1 foot = 304.8 mm,
1 square foot = 0.0929 m².

**Figure 1004.1.2
TYPICAL NET FLOOR AREA OCCUPANT LOAD CALCULATION**

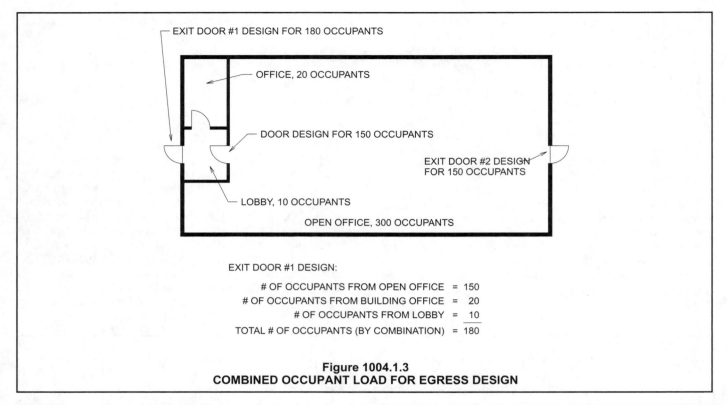

EXIT DOOR #1 DESIGN FOR 180 OCCUPANTS

OFFICE, 20 OCCUPANTS

DOOR DESIGN FOR 150 OCCUPANTS

EXIT DOOR #2 DESIGN FOR 150 OCCUPANTS

LOBBY, 10 OCCUPANTS

OPEN OFFICE, 300 OCCUPANTS

EXIT DOOR #1 DESIGN:

# OF OCCUPANTS FROM OPEN OFFICE	=	150
# OF OCCUPANTS FROM BUILDING OFFICE	=	20
# OF OCCUPANTS FROM LOBBY	=	10
TOTAL # OF OCCUPANTS (BY COMBINATION)	=	180

Figure 1004.1.3
COMBINED OCCUPANT LOAD FOR EGRESS DESIGN

1004.2 Increased occupant load. The occupant load permitted in any building or portion thereof is permitted to be increased from that number established for the occupancies in Table 1004.1.2 provided that all other requirements of the code are also met based on such modified number and the occupant load shall not exceed one occupant per 5 square feet (0.47 m²) of occupiable floor space. Where required by the building official, an approved aisle, seating or fixed equipment diagram substantiating any increase in occupant load shall be submitted. Where required by the building official, such diagram shall be posted.

❖ An increased occupant load is permitted above that developed by using Table 1004.1.2; for example, utilizing the actual occupant load alternative in Section 1004.1.1. However, if the occupant load exceeds that which is determined in accordance with Section 1004.1.2, the code official has the authority to require aisle, seating and equipment diagrams to confirm that: all occupants have access to an exit, the exits provide sufficient capacity for all occupants and compliance with this section is attained.

1004.3 Posting of occupant load. Every room or space that is an assembly occupancy shall have the occupant load of the room or space posted in a conspicuous place, near the main exit or exit access doorway from the room or space. Posted signs shall be of an approved legible permanent design and shall be maintained by the owner or authorized agent.

❖ Each room or space used for an assembly occupancy is required to display the approved occupant load. The placard must be posted in a visible location (near the main entrance) (see Figure 1004.3 for an example of an occupant load limit sign).

NOTICE

FOR YOUR SAFETY

OCCUPANCY

IS LIMITED TO:

428

PERSONS

BY ORDER OF
THE CODE OFFICIAL
Keep Posted Under Penalty Of Law

Figure 1004.3
EXAMPLE OF OCCUPANT LOAD LIMIT SIGN

The posting is required to provide a means by which to determine that the maximum approved occupant load is not exceeded. This permanent and readily visible sign provides a constant reminder to building personnel and is a reference for code officials during periodic inspections.

While the composition and organization of information in the sign are not specified, information must be recorded in a permanent manner. This means that a sign with changeable numbers would not be acceptable.

1004.4 Exiting from multiple levels. Where exits serve more than one floor, only the occupant load of each floor considered individually shall be used in computing the required capacity of the exits at that floor, provided that the exit capacity shall not decrease in the direction of egress travel.

❖ The sum total capacity of the exits that serve a floor is not to be less than the occupant load of the floor as determined by Section 1004.1. If an exit such as a stairway also serves a second floor, and the required capacity of the exit serving the occupants of the second floor is greater than the first floor, the greater capacity would govern the egress components that the occupants of the floors share. For example, if an exit stairway serves two floors, with occupant loads of 300 on the lower floor and 500 on the upper floor, assuming that two stairways serve each floor, the two stairways would be designed for a capacity of 250 people each, using the upper-floor occupant load of 500 as the basis of determination. Note that the doors to the stairways on the lower floor would be designed for a capacity of 150 and the doors to the stairways on the upper floor would be designed for a capacity of 250. Reversing these two floors would result in the portion of the stairways that serves the upper floor to be designed for a capacity of 150 and the stairways that serve the lower floor to be designed for 250. Requiring the egress component to be designed for the largest tributary occupant load accommodates the worst-case situation.

Also note that the capacity of the exits is based on the occupant load of one floor. The occupant loads are not combined with other floors for the exit design. It is assumed that the peak demand or flow of occupants from more than one floor level at a common point in the means of egress will not occur simultaneously, except as provided for in Sections 1004.5 (Egress convergence) and 1004.6 (Mezzanine levels).

1004.5 Egress convergence. Where means of egress from floors above and below converge at an intermediate level, the capacity of the means of egress from the point of convergence shall not be less than the sum of the two floors.

❖ Convergence of occupants can occur whenever the occupants of one floor travel down and occupants of a lower floor travel up and meet at a common, intermediate egress component. The intermediate component may or may not be another occupiable floor and, most often, is an exit discharge door [see Figures 1004.5(1) and 1004.5(2)].

The entire premise of egress convergence is based on the assumption of simultaneous notification (i.e., all occupants of all floors begin moving toward the exits at the same time). As illustrated in Figure 1004.5(3), the occupants of the first floor will have exited the building by the time the occupants of the second floor have reached the exit discharge door. However, as illustrated in Figure 1004.5(1), the occupants of a basement will reach the discharge door simultaneously with the second-floor occupants, thereby creating the need for sizing the components for a larger combined occupant load.

An egress convergence situation can also be created when an intermediate floor level is not present, as illustrated in Figure 1004.5(2). Again, under the assumption of simultaneous notification, occupants of both floors would reach the exit discharge door at approximately the same time, invoking the requirements for a larger egress capacity.

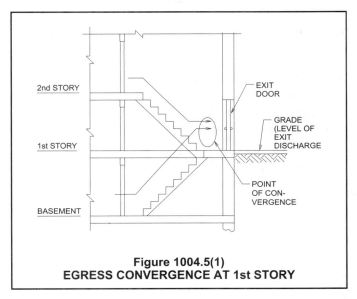

Figure 1004.5(1)
EGRESS CONVERGENCE AT 1st STORY

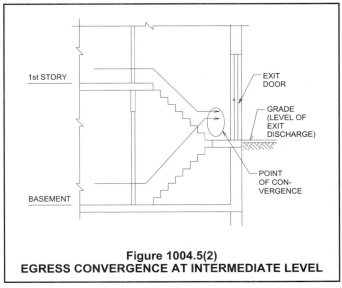

Figure 1004.5(2)
EGRESS CONVERGENCE AT INTERMEDIATE LEVEL

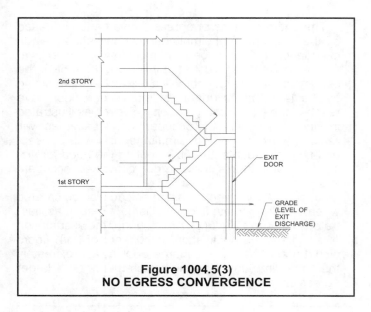

Figure 1004.5(3)
NO EGRESS CONVERGENCE

1004.6 Mezzanine levels. The occupant load of a mezzanine level with egress onto a room or area below shall be added to that room or area's occupant load, and the capacity of the exits shall be designed for the total occupant load thus established.

❖ The egress requirements for mezzanines are handled similar to those addressed in Section 1004.1.3 versus the requirements for exiting from multiple levels in Section 1004.4. That is, that portion of the mezzanine occupant load that discharges to the floor below is to be added to the occupant load of the space on the floor below. The sizing and number of the egress components must reflect this combined occupant load. This does not apply to the means of egress from a mezzanine that does not require travel through another level (i.e., an exit stairway serving the mezzanine). Section 505 contains additional criteria for the means of egress from mezzanines.

1004.7 Fixed seating. For areas having fixed seats and aisles, the occupant load shall be determined by the number of fixed seats installed therein.

For areas having fixed seating without dividing arms, the occupant load shall not be less than the number of seats based on one person for each 18 inches (457 mm) of seating length.

The occupant load of seating booths shall be based on one person for each 24 inches (610 mm) of booth seat length measured at the backrest of the seating booth.

❖ The occupant load in an area with fixed seats is readily determined. In spaces with a combination of fixed and loose seating, the occupant load is determined by a combination of the occupant density number from Table 1004.1.2 and a count of the fixed seats.

For bleachers, booths and other seating facilities without dividing arms, the occupant load is simply based on the number of people that can be accommodated in the length of the seat. Measured at the hips, an average person occupies about 18 inches (457 mm) on a bench. In a booth, additional space is necessary for "elbow room" while eating. In a circular or curved booth or bench, the measurement should be taken just a few inches from the back of the seat, which is where a person's hips would be located (see Figure 1004.7).

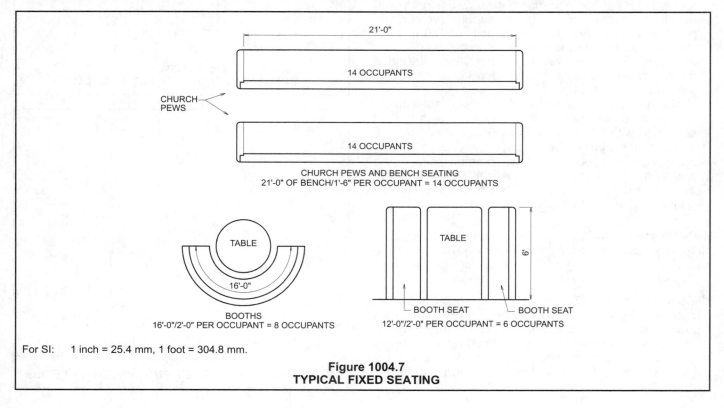

For SI: 1 inch = 25.4 mm, 1 foot = 304.8 mm.

Figure 1004.7
TYPICAL FIXED SEATING

1004.8 Outdoor areas. Yards, patios, courts and similar outdoor areas accessible to and usable by the building occupants shall be provided with means of egress as required by this chapter. The occupant load of such outdoor areas shall be assigned by the building official in accordance with the anticipated use. Where outdoor areas are to be used by persons in addition to the occupants of the building, and the path of egress travel from the outdoor areas passes through the building, means of egress requirements for the building shall be based on the sum of the occupant loads of the building plus the outdoor areas.

Exceptions:

1. Outdoor areas used exclusively for service of the building need only have one means of egress.

2. Both outdoor areas associated with Group R-3 and individual dwelling units of Group R-2, as applicable in Section 101.2.

❖ This section addresses the means of egress of outdoor areas such as yards, patios and courts. The primary concern is for outdoor areas that are used for functions that would include occupants other than the building occupants and the egress from the outdoor area is back through the building to reach the exit discharge. An example is an interior court of an office building where assembly functions are held during normal business hours for persons other than the building occupants. Where the occupants have to egress from the interior court back through the building, the building's egress system is to be designed for the building occupants plus the assembly occupants from the interior court.

The occupant load is to be assigned by the building official based on use. It is suggested that the design occupant load be determined in accordance with Section 1004.1.2.

The exceptions describe conditions where the combination of both occupant loads is not a concern.

1004.9 Multiple occupancies. Where a building contains two or more occupancies, the means of egress requirements shall apply to each portion of the building based on the occupancy of that space. Where two or more occupancies utilize portions of the same means of egress system, those egress components shall meet the more stringent requirements of all occupancies that are served.

❖ Since the means of egress systems are designed for the specific occupancy of a space, the provisions of this chapter are to be applied based on the actual occupancy conditions of the space served.

For example, a hospital is classified as Group I-2 and normally includes the associated administrative or business functions found in the same building. Chapter 3 would permit the entire building to be constructed to the more restrictive provisions for Group I-2; however, each area of the building need only have the means of egress designed in accordance with the actual occupancy conditions, such as Groups I-2 and B. If the corridor serves only the occupants in the business use (i.e., administrative staff), and is not intended to serve as a required means of egress for patients, the corridor need only be 36 or 44 inches (914 or 1118 mm) in width, depending on the occupant load.

Where the corridor is used by both I-2 and B occupancies, it must meet the most stringent requirement. For example, if a corridor in the business area is also used for the movement of beds (i.e., exit access from a patient care area), it would need to be a minimum of 96 inches (2438 mm) in clear width.

SECTION 1005
EGRESS WIDTH

1005.1 Minimum required egress width. The means of egress width shall not be less than required by this section. The total width of means of egress in inches (mm) shall not be less than the total occupant load served by the means of egress multiplied by the factors in Table 1005.1 and not less than specified elsewhere in this code. Multiple means of egress shall be sized such that the loss of any one means of egress shall not reduce the available capacity to less than 50 percent of the required capacity. The maximum capacity required from any story of a building shall be maintained to the termination of the means of egress.

Exception: Means of egress complying with Section 1024.

❖ The sum of the capacities of the individual means of egress components that serve each space must equal or exceed the occupant load of that space. For example, the two exit access doorways from a room with an occupant load of 300 would each have a required capacity of not less than 150. Likewise, the two exits from a story of a building with a total occupant load of 450 would each have a required capacity of not less than 225. The code does require that when multiple means of egress are required, the loss of any one path would not reduce the available capacity to less than 50 percent. This requirement does not, however, require that the capacities be equally distributed when more than two means of egress are provided. An egress design with a dramatic imbalance of egress component capacities relative to occupant load distribution should be reviewed closely to avoid a needless delay in egressing a floor or area. The balancing of the means of egress components, in accordance with the distribution of the occupant load, is reasonable and, in some cases, necessary for facilities having mixed occupancies with dramatically different occupant loads.

The code requires the utilization of two methods to determine the minimum width of egress components. While this section provides a methodology for determining required widths based on the design occupant load, calculated in accordance with Section 1004.1, other sections provide minimum widths of various components. The actual width that is provided is to be the larger of the two widths. Also note that the width of stairways that are part of the accessible means of egress is further regulated by Section 1007.3. Obviously, an egress door opening with a clear width of 12 inches (305 mm) would be an impossible egress component, al-

TABLE 1005.1   MEANS OF EGRESS

though in accordance with Table 1005.1, the capacity of such an opening can be calculated to be 60 persons (60 by 0.2 = 12 inches). Section 1008.1.1, however, specifies that the minimum required clear width of each door opening is 32 inches (813 mm) with certain exceptions.

The 50-percent minimum of the required capacity results in a fairly uniform distribution of egress paths. For example, the doors for large occupancy rooms are to be located to meet this requirement.

The requirement that the maximum capacity from any floor is to be provided results in an egress width that is adequate to the exit discharge.

The purpose of the exception is to require assembly seating to comply with Section 1024.

TABLE 1005.1
EGRESS WIDTH PER OCCUPANT SERVED

OCCUPANCY	WITHOUT SPRINKLER SYSTEM		WITH SPRINKLER SYSTEM[a]	
	Stairways (inches per occupant)	Other egress components (inches per occupant)	Stairways (inches per occupant)	Other egress components (inches per occupant)
Occupancies other than those listed below	0.3	0.2	0.2	0.15
Hazardous: H-1, H-2, H-3 and H-4	0.7	0.4	0.3	0.2
Institutional: I-2	NA	NA	0.3	0.2

For SI: 1 inch = 25.4 mm. NA = Not applicable.

a. Buildings equipped throughout with an automatic sprinkler system in accordance with Section 903.3.1.1 or 903.3.1.2.

❖ This table establishes the necessary width of each egress component on a "per occupant" basis. When the required occupant capacity of an egress component is determined, multiplication by the appropriate factor from Table 1005.1 results in the required clear width of the component in inches, based on capacity. Similarly, if the clear width of a component is known, division by the appropriate factor from Table 1005.1 results in the permitted capacity of that component.

The following typical calculations illustrate the various uses of Table 1005.1:

1. Determine the minimum required width of a stairway to accommodate 175 occupants in a two-story office building (Business Group B):

   a. Without an automatic sprinkler system: 175 by 0.3 (from Table 1005.1) = 52$^1/_2$ inches (1334 mm), minimum; or

   b. With an automatic sprinkler system: 175 by 0.2 (from Table 1005.1) = 35 inches (889 mm), minimum.

   Section 1009.1, however, prescribes that the width of an interior stairway cannot be less than 44 inches (1118 mm); therefore, 44 inches (1118 mm) is the governing dimension when an automatic sprinkler system is installed. When a sprinkler system is not installed, the capacity criteria are more restrictive and, therefore, the minimum required width is 52$^1/_2$ inches (1334 mm).

2. Determine the minimum width of an exit access doorway required to accommodate 240 occupants in a new six-story hotel (Group R-1) (Assume the occupant load on the floor is larger and that the capacity of this egress path is being designed to serve 240 of the occupants):

   a. Since Section 903.2.7 requires this building to be equipped throughout with an automatic fire sprinkler system, the proper width per occupant is 0.15 (see Table 1005.1). The required width is 240 by 0.15 = 36 inches (914 mm), minimum. Therefore, the egress doorway must be at least 36 inches (914 mm) in clear width [not merely 32 inches (813 mm), as required by Section 1008.1.1].

   b. It is important to remember that the total width of egress from an area may be distributed among the available means of egress. If Example 2 had a total story occupant load of 240 and two exits available, then two doors that met the 32-inch (813 mm) requirement would be acceptable. [240 occupants by 0.15 = 36 inches (914 mm) of total width]. Since the loss of any one path may not reduce the available capacity below 50 percent, each doorway would need to provide a minimum of 18 inches of that width. However, Section 1008.1.1 would then require each door to be sized to provide 32 inches (813 mm) of clear width.

3. What is the egress capacity of a pair of entrance/exit doors that provide a clear width of 64 inches (1626 mm) in an existing retail store?

   a. If an unsprinklered building or only partially sprinklered: egress capacity = 64 inches by 0.2 = 320 occupants.

   b. If a building is sprinklered throughout: egress capacity = 64 inches by 0.15 = 426.67 occupants. (Note that where the clear width provides capacity for a fraction of an occupant, the capacity should be rounded down to the nearest whole number. In this case, the doors can accommodate 426 occupants.)

The examples above demonstrate the use of Table 1005.1 in conjunction with other applicable sections of the code that require minimum widths of various components of the means of egress.

The traditional unit of measurement of egress capacity was based on a "unit exit width," which was to simulate the body ellipse with a basic dimensional width of 22 inches (559 mm)—approximately the shoulder width of an average adult male. This unit exit width was combined with assumed egress movement (such as single file or staggered file) to result in an egress capacity per unit exit width for various occupancies, as shown in Table 1005.1. This assumption simplifies the dynamic egress process, since contemporary studies have indicated that people do not egress in such precise and predictable movements. As traditionally used in the codes, the method of determining capacity per unit of clear width implies a higher level of accuracy than can realistically be achieved. The resulting factors in Table 1005.1 preserve the features of the past practices that can be documented, while providing a more straightforward method of determining egress capacity.

The considerations embodied in Table 1005.1 include the nature and conditions of the occupants, the fire hazard of the occupancy, the efficiency of the column of occupants to utilize the component, the physical dimensioning of the components and the presence of an automatic sprinkler system. The presence of an automatic sprinkler system is deemed to control or limit the fire threat to the occupants, thus permitting greater time for the occupants to egress as well as reducing the need for total and immediate evacuation. The footnote reference to Section 903.3.1.1 or 903.3.1.2 indicates that the sprinkler system is to be designed and installed in accordance with either NFPA 13 or NFPA 13R.

It should be recognized that although an automatic sprinkler system is required for all Group H fire areas, egress width per occupant is shown for such occupancies without an automatic sprinkler system. These factors should be applied in special cases where the use of a sprinkler system is detrimental or of no value to the protection of the hazardous condition and can also be used to establish a recommended minimum safety guideline for existing nonsuppressed Group H occupancies or where the sprinkler system is allowed to be omitted through the appeals process.

The increased factors for buildings of Group I-2 recognize the decreased mobility of occupants in these buildings. A number is not given for Group I-2 buildings without sprinklers since all Group I-2 facilities are required to be sprinklered. A similar increase is not required in buildings of Group I-3 since the population, while not capable of self preservation, is typically a mobile group and once locks are released, are as capable as the general population with respect to using the egress components.

1005.2 Door encroachment. Doors opening into the path of egress travel shall not reduce the required width to less than one-half during the course of the swing. When fully open, the door shall not project more than 7 inches (178 mm) into the required width.

Exception: The restrictions on a door swing shall not apply to doors within individual dwelling units and sleeping units of Group R-2 and dwelling units of Group R-3.

❖ Projections or restrictions in the required width can impede and restrict occupant travel, causing egress to occur less efficiently than contemplated. The swing of a door, such as from a room into a corridor, is a permitted projection.

However, the arc created by the door's outside edge cannot project into more than one-half of the required corridor width. When opened to its fullest extent, the door cannot project more than 7 inches (178 mm) into the required width, which is approximately the sum dimension of a door leaf thickness and protrusion of operational hardware on each side of the leaf as shown in Figure 1005.2. These projections are permitted because they are considered to be temporary and do not significantly impede the flow. Occupants will compensate for the projection by a reduction in the natural cushion they retain between themselves and a boundary, known as the edge effect.

The door swing restrictions do not apply within dwelling units since the occupant load is very low.

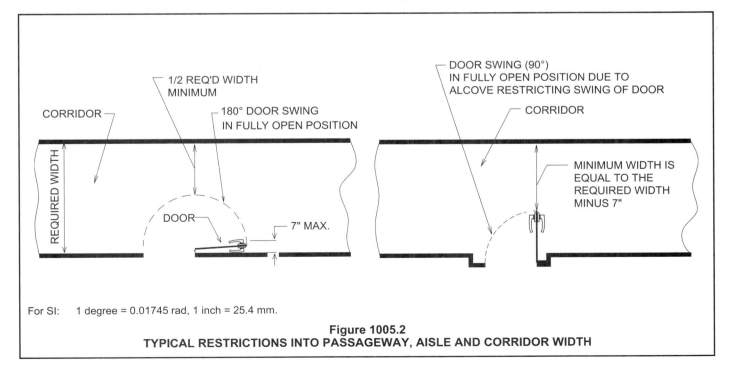

For SI: 1 degree = 0.01745 rad, 1 inch = 25.4 mm.

Figure 1005.2
TYPICAL RESTRICTIONS INTO PASSAGEWAY, AISLE AND CORRIDOR WIDTH

SECTION 1006
MEANS OF EGRESS ILLUMINATION

1006.1 Illumination required. The means of egress, including the exit discharge, shall be illuminated at all times the building space served by the means of egress is occupied.

Exceptions:

1. Occupancies in Group U.

2. Aisle accessways in Group A.

3. Dwelling units and sleeping units in Groups R-1, R-2 and R-3.

4. Sleeping units of Group I occupancies.

❖ All means of egress must be illuminated by artificial lighting during the entire time a building is occupied, so that the paths of exit travel are always visible and available for evacuation of the occupants during emergencies. The code makes a special point of noting that the exit discharge must also be provided with adequate illumination so that occupants can safely find the public way should the emergency occur at night.

The exceptions are for occupancies where the constant illumination of the means of egress would interfere with the use of space such as sleeping areas or theater aisles during a performance.

Bear in mind that means of egress lighting is not emergency lighting. For emergency lighting requirements, see Sections 1006.3 and 1006.4.

1006.2 Illumination level. The means of egress illumination level shall not be less than 1 foot-candle (11 lux) at the floor level.

Exception: For auditoriums, theaters, concert or opera halls and similar assembly occupancies, the illumination at the floor level is permitted to be reduced during performances to not less than 0.2 foot-candle (2.15 lux) provided that the required illumination is automatically restored upon activation of a premise's fire alarm system where such system is provided.

❖ The intensity of floor lighting illuminating the entire means of egress, including open plan spaces, aisles, corridors and exit access passageways, exit stairways, exit doors and places of exit discharge must not be less than 1 foot-candle (11 lux). It has been found that this low level of lighting renders enough visibility for the occupants to evacuate the building safely.

It is important to note that this lighting level is measured at the floor in order to make the floor surface visible. Levels of illumination above the floor may be higher or lower, thus allowing lights on the steps to be used rather than general area lights.

The exception addresses occupancies where low light level is needed for the function of the space. The level of intensity of aisle lighting in such spaces may be reduced to 0.2 foot-candle (2.15 lux), but only during the time of a performance. This intensity of illumination is sufficient to distinguish the aisles and stairs leading to the egress doors and is not a source of distraction dur-

ing a performance. It is not the intent of the exception to require a fire alarm system but to require a connection to the egress lighting where a fire alarm system is provided.

1006.3 Illumination emergency power. The power supply for means of egress illumination shall normally be provided by the premise's electrical supply.

In the event of power supply failure, an emergency electrical system shall automatically illuminate the following areas:

1. Exit access corridors, passageways and aisles in rooms and spaces which require two or more means of egress.

2. Exit access corridors and exit stairways located in buildings required to have two or more exits.

3. Exterior egress components at other than the level of exit discharge until exit discharge is accomplished for buildings required to have two or more exits.

4. Interior exit discharge elements, as permitted in Section 1023.1, in buildings required to have two or more exits.

5. The portion of the exterior exit discharge immediately adjacent to exit discharge doorways in buildings required to have two or more exits.

The emergency power system shall provide power for a duration of not less than 90 minutes and shall consist of storage batteries, unit equipment or an on-site generator. The installation of the emergency power system shall be in accordance with Section 2702.

❖ The means of egress must be illuminated, especially in times of emergency when the occupants must have a lighted path of exit travel in order to evacuate the building safely. The code is very specific in the description of the areas that are required to be illuminated by the emergency power system.

The locations include:

1. Dedicated egress routes in larger rooms or spaces. For example, an aisle in an open office plan but not within individual offices that egress through the open area.

2. Public or common egress elements. Essential portions of the interior egress system, such as stairways and corridors in larger buildings, must be illuminated.

3. Means of egress systems normally used in exit balconies, which would provide emergency exit illumination to these components until egress at grade is "reached" or "achieved."

4. Interior exit discharge elements such as fully sprinklered lobbies and vestibules where stairways discharge into these exit elements.

5. Exterior portions of the exit discharge. Note that only the portion of the exterior discharge that is immediately adjacent to the building exit discharge door is required to have emergency illumi-

nation and not the entire exterior discharge path to the public way.

So that there will be a continuing source of electrical energy for maintaining the illumination of the means of egress when there is a loss of the main power supply, the means of egress lighting system must be connected to an emergency electrical system that consists of storage batteries, unit equipment or an on-site generator. This emergency power-generating facility must be capable of supplying electricity for at least 90 minutes, thereby giving the occupants sufficient time to leave the premises. In most cases, where the loss of the main electrical supply is attributed to a malfunction in the distribution system of the electric power company, experience has shown that such power outages do not usually last as long as 90 minutes.

1006.4 Performance of system. Emergency lighting facilities shall be arranged to provide initial illumination that is at least an average of 1 foot-candle (11 lux) and a minimum at any point of 0.1 foot-candle (1 lux) measured along the path of egress at floor level. Illumination levels shall be permitted to decline to 0.6 foot-candle (6 lux) average and a minimum at any point of 0.06 foot-candle (0.6 lux) at the end of the emergency lighting time duration. A maximum-to-minimum illumination uniformity ratio of 40 to 1 shall not be exceeded.

❖ This section provides the criteria for the illumination levels of the emergency lighting system. The initial average level is the same as for the means of egress illumination in Section 1006.2. The reduction of illumination recognizes the performance characteristics over time of some types of power supplies, such as batteries. The minimum levels are sufficient for the occupants to egress from the building.

The maximum illumination uniformity ratio of 40 means that the variation in the illumination levels is not to exceed that number. For example, a minimum of 0.06 foot-candle (0.6 lux) would establish a maximum illumination of 2.4 foot-candle (24 lux) in an adjacent area. This is to establish a variation limit such that the means of egress can be seen as a person walks from bright to darker areas along the egress path.

SECTION 1007
ACCESSIBLE MEANS OF EGRESS

1007.1 Accessible means of egress required. Accessible means of egress shall comply with this section. Accessible spaces shall be provided with not less than one accessible means of egress. Where more than one means of egress is required by Section 1014.1 or 1018.1 from any accessible space, each accessible portion of the space shall be served by not less than two accessible means of egress.

Exceptions:

1. Accessible means of egress are not required in alterations to existing buildings.

2. One accessible means of egress is required from an accessible mezzanine level in accordance with Section 1007.3 or 1007.4.

3. In assembly spaces with sloped floors, one accessible means of egress is required from a space where the common path of travel of the accessible route for access to the wheelchair spaces meets the requirements in Section 1024.8.

❖ This section establishes the minimum requirements for means of egress facilities serving all spaces that are required to be accessible to people with physical disabilities. Previously, attention had been focused on response to the civil-rights-based issue of providing adequate access for people with physical disabilities into and throughout buildings. Concerns about life safety and evacuation of people with mobility impairments were frequently cited as reasons for not embracing widespread building accessibility, in the best interest of the disabled community.

The provisions for accessible means of egress are predominantly, though not exclusively, intended to address the safety of persons with a mobility impairment. These requirements reflect the balanced philosophy that accessible means of egress are to be provided for occupants who have gained access into the building but are incapable of independently utilizing the typical means of egress facilities, such as the exit stairways. By making such provisions, the code now addresses means of egress for all building occupants, with and without physical disabilities.

Any space that is not required by the code to be accessible in accordance with Chapter 11 is not required to be provided with accessible means of egress. This may include an entire story, a portion of a story or an individual room.

In new construction and additions, accessible means of egress are required in the same number as the general means of egress, up to a maximum of two. For example, in buildings, stories or spaces required by Section 1014.1 or 1018 to have three or more exits or exit access doors, a minimum of two accessible means of egress is required. The number of exits or exit access doors is based on occupant load, therefore, no matter how large the total occupant load of the space, two fully complying accessible means of egress are considered to provide sufficient capacity for those building occupants with a mobility impairment.

An accessible means of egress is required to provide a continuous path of travel to a public way. This principle is consistent with the general requirements for all means of egress, as reflected in Section 1003.1 and in the definition of "Means of egress" in Section 1002. This section also emphasizes the intent that accessible means of egress must be usable by a person with a mobility impairment, such as a person in a wheelchair.

The exceptions address special situations where accessible means of egress requirements need special consideration. Note that these are exceptions for accessible means of egress; not an exception for accessible entrance requirements (see Section 1105).

Exception 1 indicates that existing buildings that are undergoing alterations are not required to be provided with accessible means of egress as part of that alteration. In many cases, meeting the requirements for accessible means of egress, especially the 48-inch (1219 mm) clear stair width required in nonsprinklered buildings, would be considered technically infeasible.

Exception 2 is a special consideration for mezzanines. As an example, consider an open mezzanine that does not meet Section 1014.1, therefore, two means of egress are required. In accordance with Section 505.3, open mezzanines are permitted to utilize two open exit access stairways leading to the main level to meet the general means of egress requirements. Stairways are not typically navigable by persons with mobility impairments without assistance. Additionally, if a mezzanine is large enough to be required to be accessible (see Section 1104.4), two accessible means of egress are required. The accessible means of egress do not permit exit access stairways. An accessible route must be available to two exits. In Exception 2, the open mezzanine is permitted to have only one accessible means of egress provided by either an enclosed exit stairway or an elevator with standby power. If the building is not sprinklered throughout, an area of refuge must also be provided.

Exception 3 is in consideration of the practical difficulties of providing accessible routes in assembly areas with sloped floors. Rooms with more than 50 persons are required to have two means of egress; therefore, each accessible seating location is required to have access to two accessible means of egress. Depending on the slope of the seating arrangement, this can be difficult to achieve, especially in small theaters. A maximum travel distance of 30 feet (9144 mm) for ambulatory persons moving from the last seat in dead-end aisles or from box-type seating arrangements to where they have access to a choice of means of egress routes has been established in Section 1024.8. In accordance with Exception 3, persons using wheelchair seating spaces have the same maximum 30-foot (9144 mm) travel distance from the accessible seating locations to a cross aisle or out of the room to an adjacent corridor or space where two choices for accessible means of egress are provided. Note that there are increases in travel distance for smoke-protected seating and small spaces, such as boxes, galleries or balconies. For additional information, see Section 1024.8.

1007.2 Continuity and components. Each required accessible means of egress shall be continuous to a public way and shall consist of one or more of the following components:

1. Accessible routes complying with Section 1104.
2. Stairways within exit enclosures complying with Sections 1007.3 and 1019.1.
3. Elevators complying with Section 1007.4.
4. Platform lifts complying with Section 1007.5.
5. Horizontal exits.

6. Smoke barriers.

Exceptions:

1. Where the exit discharge is not accessible, an exterior area for assisted rescue must be provided in accordance with Section 1007.8.
2. Where the exit stairway is open to the exterior, the accessible means of egress shall include either an area of refuge in accordance with Section 1007.6 or an exterior area for assisted rescue in accordance with Section 1007.8.

❖ This section identifies the various building features that can serve as elements of an accessible means of egress. Accessible routes are readily recognizable as to how they can provide accessible means of egress, however, some nontraditional principles have been established for the total concept of accessible means of egress. This is evident in that stairways and elevators are also identified as elements that can comprise part of an accessible means of egress. For example, elevators are generally not available for egress during a fire, while stairways are not independently usable by a person in a wheelchair. The concept of accessible means of egress includes the idea that evacuation of people with a mobility impairment may require the assistance of others. In buildings not equipped with an automatic sprinkler system, provisions are also included for the creation of an area of refuge, wherein people can safely await either further instructions or evacuation assistance. Refuge areas can also be established by utilizing either smoke barriers or horizontal exits. All of these elements can be arranged in the manner prescribed in this section to provide accessible means of egress.

Typically, the accessible way into a single-story facility is also one of the accessible means of egress. The exceptions are intended to address problems that most typically arise for the second accessible means of egress. Site constraints or configurations may result in difficulty providing an accessible route for exit discharge. The alternative for exterior areas for assisted rescue is offered for these types of situations. A change in elevation across a site may result in exit doors leading to exterior steps. These steps are considered exit stairways. For these types of situations, providing either an area of refuge inside the facility or an exterior area of rescue assistance is viable.

1007.2.1 Buildings with four or more stories. In buildings where a required accessible floor is four or more stories above or below a level of exit discharge, at least one required accessible means of egress shall be an elevator complying with Section 1007.4.

Exceptions:

1. In buildings equipped throughout with an automatic sprinkler system installed in accordance with Section 903.3.1.1 or 903.3.1.2, the elevator shall not be required on floors provided with a horizontal exit and located at or above the level of exit discharge.

2. In buildings equipped throughout with an automatic sprinkler system installed in accordance with Section 903.3.1.1 or 903.3.1.2, the elevator shall not be required on floors provided with a ramp conforming to the provisions of Section 1010.

❖ Elevators are the most common and convenient means of providing access in multistory buildings. As such, elevators represent a prime candidate for accessible means of egress from such buildings, especially in light of the difficulties involved in carrying a person in a wheelchair up or down a stairway. The primary consideration for elevators as an accessible means of egress is that the elevator most likely will be available and protected during a fire event.

This section addresses the use of an elevator as part of an accessible means of egress by requiring a standby source of power for the elevator. The standby power requirement establishes a higher degree of reliability that the elevator will be available and usable by reducing the likelihood of power loss caused by fire or other conditions of power failure.

In buildings having four or more stories above or below the level of exit discharge, it is unreasonable to rely solely on exit stairways for all of the required accessible means of egress. This is the point at which complete reliance on assisted evacuation will not be effective or adequate because of the limited availability of either experienced personnel who are trained to handle wheelchair evacuation or special devices, such as self-braking stairway descent equipment or evacuation chairs. In this case, the code requires that at least one elevator, serving all floors of the building, is to serve as one of the required accessible means of egress. This should not represent a hardship, since elevators are typically provided in such buildings for the convenience of the occupants.

The presence of an automatic sprinkler system significantly reduces the potential fire hazard and provides for increased evacuation time. Exception 1 establishes that accessible egress elevator service to floor levels at or above the level of exit discharge is not necessary under specified conditions. The conditions are that the building is equipped throughout with an automatic sprinkler system in accordance with NFPA 13 or NFPA 13R (see Section 903.3.1.1 or 903.3.1.2) and the floors not serviced by an accessible egress elevator are provided with a horizontal exit. The combination of automatic sprinklers and a horizontal exit provides adequate protection for the occupants despite their distance to the level of exit discharge. This exception does not apply to floor levels below the level of exit discharge, since such levels are typically below grade and do not have the added advantage of exterior openings that are available for fire-fighting or rescue purposes. Exception 2 specifies that a building sprinklered throughout in accordance with NFPA 13 or NFPA 13R (see Section 903.3.1.1 or 903.3.1.2), with ramp access to each level, such as in a sports stadium, is not required to also have an elevator for accessible means of egress. The reasoning behind this is that the issue of carrying people down stairways does not occur because the ramps may be utilized instead.

1007.3 Enclosed exit stairways. An enclosed exit stairway, to be considered part of an accessible means of egress, shall have a clear width of 48 inches (1219 mm) minimum between handrails and shall either incorporate an area of refuge within an enlarged floor-level landing or shall be accessed from either an area of refuge complying with Section 1007.6 or a horizontal exit.

Exceptions:

1. Open exit stairways as permitted by Section 1019.1 are permitted to be considered part of an accessible means of egress.

2. The area of refuge is not required at open stairways that are permitted by Section 1019.1 in buildings or facilities that are equipped throughout with an automatic sprinkler system installed in accordance with Section 903.3.1.1.

3. The clear width of 48 inches (1219 mm) between handrails and the area of refuge is not required at exit stairways in buildings or facilities equipped throughout with an automatic sprinkler system installed in accordance with Section 903.3.1.1 or 903.3.1.2.

4. The clear width of 48 inches (1219 mm) between handrails is not required for enclosed exit stairways accessed from a horizontal exit.

5. Areas of refuge are not required at exit stairways serving open parking garages.

❖ The fundamental condition is under which a stairway can function as part of an accessible means of egress are that the stairway is sufficiently wide to enable the assisted evacuation.

Sufficient width is accomplished by requiring the stairway to have a minimum clear width between handrails of 48 inches (1219 mm). This dimension is sufficient to enable two or three persons to carry the person in a wheelchair down (or up) to the level of exit discharge from which access to a public way is afforded.

The exit stairway, in combination with an area of refuge, can provide for safety from fire in one of two ways. One approach is for the fire-resistance-rated stairway enclosure to afford the necessary safety. To accomplish this, the landing within the stairway enclosure must be able to contain the wheelchair. The concept is that the person in the wheelchair will remain on the stairway landing for a period of time awaiting further instructions or evacuation assistance; therefore, the stairway landing must be able to accommodate the wheelchair without obstructing the use of the stairway by other egressing occupants. An enlarged, story-level landing is required within the stairway enclosure and must be of sufficient size to accommodate the number of wheelchairs required by Section 1007.6.1 [see Figure 1007.3(1)].

The other approach is to utilize a stairway that is accessed from an area of refuge complying with Section 1007.6. Under this approach, the stairway is made safe by virtue of its access being in an area that is separated and protected from the point of fire origin. An area of refuge can be created by constructing a vestibule adjacent and with direct access to the stair enclosure in accordance with Section 1007.6 [see Figure 1007.3(2)]. This is similar in theory to the approach of an enlarged landing within the stairway enclosure.

Note that the exceptions are for areas of refuge or stairway width requirements. They are not intended to indicate that an accessible means of egress is not required.

In the case of a horizontal exit [see Figure 1007.3(3)], each floor area on either side of the exit is each considered a refuge area (see commentary, Section 1021.1) by virtue of the construction and separation requirements for horizontal exits. The discharge area is always assumed to be the nonfire side and, therefore, is protected from fire. Any stairway located within the horizontal exit discharge area constitutes an accessible means of egress. Exception 4 relieves the requirement for a minimum 48-inch (1219 mm) clear width for stairways that are accessed from a horizontal exit. This exception considers that an area of refuge created by a horizontal exit will be of considerable size so that movement of occupants from the area of refuge down the stairway can be more deliberate.

Exception 5 for open parking structures is in recognition of the natural ventilation of the products of combustion that will be afforded by the exterior openings required of such structures (see Section 406.3). The most immediate hazard for occupants in a fire incident is exposure to smoke and fumes. Floor areas in open parking structures communicate sufficiently with the outdoors such that the need for protection from smoke is not necessary; therefore, open parking garages are exempted from the requirements for an area of refuge. It is due to this level of natural ventilation that parking garage exit stairways are not required to be enclosed (see Section 1019.1, Exception 5).

Exception 1 is in recognition of the open exit stairways permitted by Section 1019.1. Therefore, the same exit stairway that serves the ambulatory occupants in the building can also serve as part of the accessible means of egress. Note that this is for exit stairways only, not exit access stairways. If the building is not sprinklered, an area of refuge and 48-inch (1219 mm) clear width on the stairway would still be required.

Exception 2 states that for an exit stairway permitted to be unenclosed by any exception in Section 1019.1 and in a building sprinklered in accordance with NFPA 13, an area of refuge is not required, but the 48-inch (1210 mm) clear width for the stairway is still required. This exception was not added with the attempt to override any options permitted by a combination of Exceptions 1 and 3.

Exception 3 exempts the 48-inch-width (1219 mm) requirement and the area of refuge or a horizontal exit for any occupancy that is equipped throughout with an automatic sprinkler system in accordance with NFPA 13 or NFPA 13R (see Section 903.3.1.1 or 903.3.1.2). With

this exception and Exception 2 to Section 1007.4, the intent of the code is that areas of refuge are not required in fully sprinklered buildings. This exception is in recognition of the increased level of safety and evacuation time that is afforded in a sprinklered occupancy. The expectation is that a supervised system will reduce the threat of fire by reliably controlling and confining the fire to the immediate area of origin. There is also the additional safety of the sprinkler system requirements for automatic notification when the system is activated. This has been substantiated by a study of accessible means of egress conducted for the General Services Administration (GSA). A report issued by the National Institute for Standards and Technology (NIST), NIST IR 4770, concluded that the operation of a properly designed NFPA 13 sprinkler system eliminates the life threat to all building occupants, regardless of their individual physical abilities and is a superior form of protection as compared to areas of refuge. This logic is extended to an NFPA 13R system. One intent of an NFPA 13R system is to provide a minimum level of protection to allow for evacuation of the occupants.

1007.4 Elevators. An elevator to be considered part of an accessible means of egress shall comply with the emergency operation and signaling device requirements of Section 2.27 of ASME A17.1. Standby power shall be provided in accordance with Sections 2702 and 3003. The elevator shall be accessed from either an area of refuge complying with Section 1007.6 or a horizontal exit.

Exceptions:

1. Elevators are not required to be accessed from an area of refuge or horizontal exit in open parking garages.

2. Elevators are not required to be accessed from an area of refuge or horizontal exit in buildings and facilities equipped throughout with an automatic sprinkler system installed in accordance with Section 903.3.1.1 or 903.3.1.2.

❖ Elevators are the most common and convenient means of providing access in multistory buildings. As such, elevators represent a prime candidate for accessible means of egress from such buildings, especially in light of the difficulties involved in carrying a person in a wheelchair up or down a stairway. The primary consideration for elevators as an accessible means of egress is that the elevator most likely will be available and protected during a fire event.

This section addresses the use of an elevator as part of an accessible means of egress by requiring both a backup source of power for the elevator and access to the elevator from an area of refuge or a horizontal exit. For situations where elevators are required to be part of one of the accessible means of egress, see Section 1007.2.1. The backup power requirement establishes a higher degree of reliability that the elevator will be available and usable by reducing the likelihood of power loss caused by fire or other conditions. Requiring access from an area of refuge or a horizontal exit affords the same degree of fire safety as described for stairways (see commentary, Section 1007.3). Additionally, the reference to Sections 2702 and 3003 clarifies that the elevator will comply with the emergency operation features that relate to elevator operation under fire conditions (see commentary, Sections 2702.2.18 and 3003). Elevators on an accessible route are also required to meet the accessibility provisions of ICC A117.1 (see commentary, Sections 1109.6 and 3001.3).

Note that the exceptions are for areas of refuge requirements only. They are not intended to indicate that an accessible means of egress is not required.

If a level in an open parking garage is accessible, that level is required to have an accessible means of egress. Exception 1, for open parking structures, is in recognition of the natural ventilation of the products of combustion that will be afforded by the exterior openings required of such structures (see Section 406.3). The most immediate hazard for occupants in a fire incident is exposure to smoke and fumes. Floor areas in open parking structures communicate sufficiently with the out-

doors, thus the need for protection from smoke is not necessary. Therefore, open parking garages are exempt from the requirements for an area of refuge or horizontal exit to access and elevator that is utilized as part of the accessible means of egress. This is consistent with the exception for stairways in Section 1007.3.

As previously discussed, the presence of an automatic sprinkler system significantly reduces the potential fire hazard and provides for increased evacuation time. Exception 2 indicates that in an occupancy equipped throughout with an automatic sprinkler system in accordance with NFPA 13 or NFPA 13R (see Section 903.3.1.1 or 903.3.1.2), the accessible egress elevator is not required to be accessed from an area of refuge or a horizontal exit. This is consistent with the exception for stairways in Section 1007.3 and recognizes the value of a sprinkler system in limiting fire growth and spread. Note that an elevator lobby that is off a fire-resistance-rated corridor must also comply with Section 707.14.1.

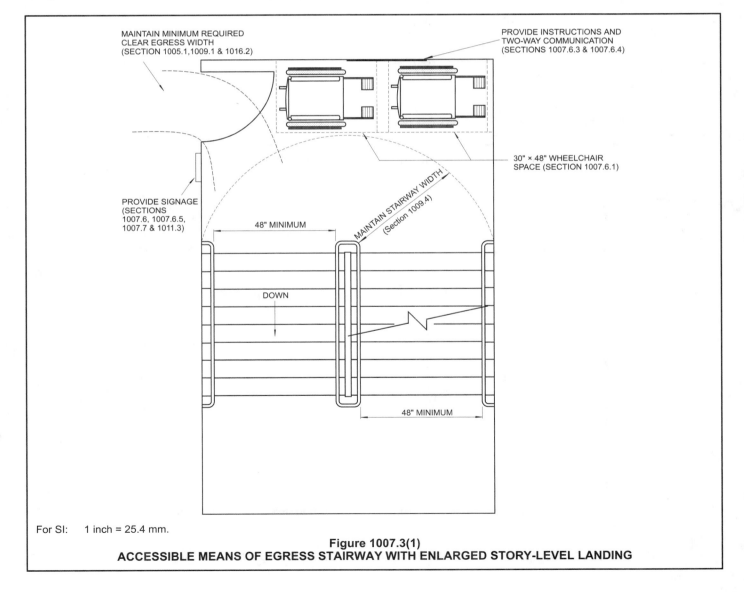

For SI: 1 inch = 25.4 mm.

Figure 1007.3(1)
ACCESSIBLE MEANS OF EGRESS STAIRWAY WITH ENLARGED STORY-LEVEL LANDING

FIGURE 1007.3(2) – FIGURE 1007.3(3)

MEANS OF EGRESS

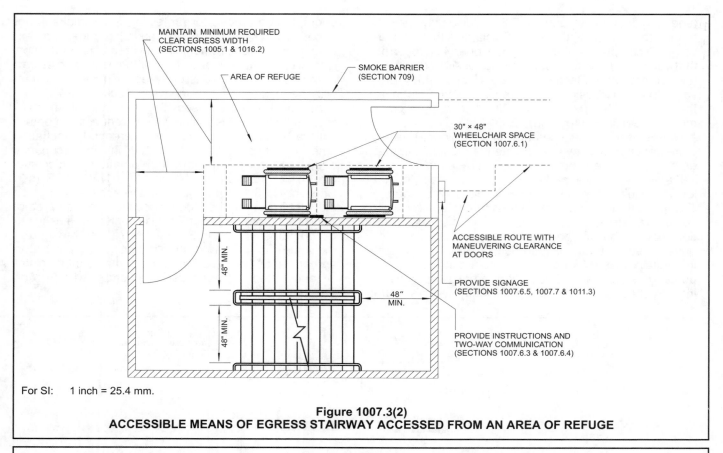

For SI: 1 inch = 25.4 mm.

Figure 1007.3(2)
ACCESSIBLE MEANS OF EGRESS STAIRWAY ACCESSED FROM AN AREA OF REFUGE

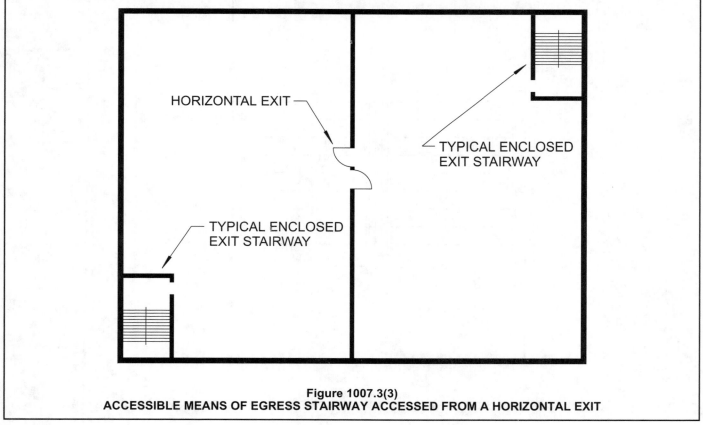

Figure 1007.3(3)
ACCESSIBLE MEANS OF EGRESS STAIRWAY ACCESSED FROM A HORIZONTAL EXIT

1007.5 Platform lifts. Platform (wheelchair) lifts shall not serve as part of an accessible means of egress, except where allowed as part of a required accessible route in Section 1109.7. Platform lifts in accordance with Section 2702 shall be installed in accordance with ASME A18.1. Standby power shall be provided for platform lifts permitted to serve as part of a means of egress.

❖ Previously, there have been concerns about whether a platform lift will be reliably available at all times. However, ASME A181, the new standard for platform lifts, no longer requires key operation. It is important to note that platform lifts are not prohibited by the code. They simply cannot be counted as a required accessible means of egress in other than locations where they are allowed as part of the accessible route into a space (see commentary, Section 1109.7). When platform lifts are utilized as part of an accessible means of egress, they must come equipped with standby power.

In existing buildings undergoing alterations, platform lifts are allowed as part of an accessible route (see commentary, Section 3409.7.3). Note that accessible means of egress are not required in existing buildings undergoing an alteration (see Section 1007.1).

1007.6 Areas of refuge. Every required area of refuge shall be accessible from the space it serves by an accessible means of egress. The maximum travel distance from any accessible space to an area of refuge shall not exceed the travel distance permitted for the occupancy in accordance with Section 1015.1. Every required area of refuge shall have direct access to an enclosed stairway complying with Sections 1007.3 and 1019.1 or an elevator complying with Section 1007.4. Where an elevator lobby is used as an area of refuge, the shaft and lobby shall comply with Section 1019.1.8 for smokeproof enclosures except where the elevators are in an area of refuge formed by a horizontal exit or smoke barrier.

❖ An area of refuge is of no value as part of an accessible means of egress if it is not accessible. The code states an obvious but essential requirement: the path that leads to an area of refuge must qualify as an accessible means of egress. This provision is required so that there will be an accessible route leading from every accessible space to each required area of refuge. For consistency in principle with the general means of egress design concepts, the code also limits the travel distance to the area of refuge. The limitation is the same distance as specified in Section 1015.1 for maximum exit access travel distance. This equates the maximum travel distance required to reach an exit with the maximum distance required to reach an area of refuge. It should be noted that an area of refuge is not necessarily an exit in the classic sense. For example, when the area of refuge is an enlarged, story-level landing within an exit stairway, the area of refuge is within the exit and the maximum travel distance for both the conventional exit and the accessible area of refuge is measured to the same point (the entrance to the exit stairway). If the area of refuge is a vestibule immediately adjacent to an enclosed exit stairway, however, the maximum travel distance for purposes of the required accessible means of egress is measured to the entrance to the area of refuge and, for purposes of the travel distance to the exit, to the entrance to the exit stairway [see Figure 1007.6]. In the case of accessible means of egress with an elevator, the maximum travel distance may end up being mea-

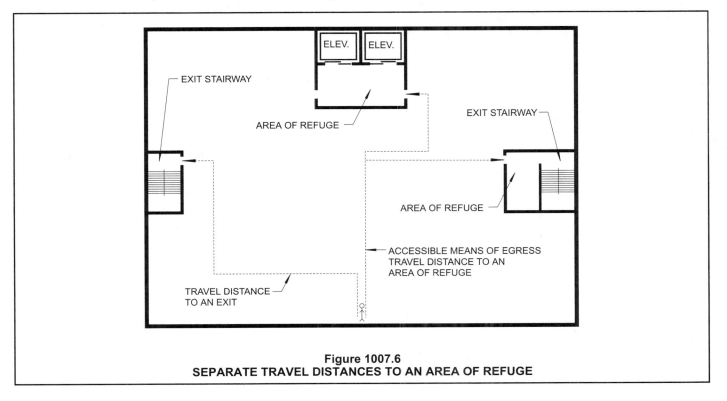

**Figure 1007.6
SEPARATE TRAVEL DISTANCES TO AN AREA OF REFUGE**

sured along two different paths, with the only consistency between the conventional means of egress and the accessible means of egress being the maximum travel distance [see Figure 1007.6]. The travel distance within an area of refuge is not directly regulated, but it will be limited by the general provisions for maximum exit access travel distance, which are always applicable.

In summary, the code takes a reasonably consistent approach for both conventional and accessible means of egress by limiting the distance one must travel to reach a safe area from which further egress to a public way is available.

So there is continuity in an accessible means of egress, the code requires that every area of refuge have direct access to either an exit stairway (see Sections 1007.3 and 1019.1) or an elevator (see Section 1007.4). This, again, may be viewed as stating the obvious, but it is necessary so that the egress layout does not involve entering an area of refuge and then having to leave that protected area before gaining access to a stairway or elevator. Once an occupant reaches the safety of an area of refuge, that level of protection must be continuous until the vertical transportation element (the stairway or elevator) is reached.

If one chooses to comply with the accessible means of egress requirements by providing an accessible elevator and an area of refuge in the form of an elevator lobby, the elevator shaft and the lobby are required to be constructed as a smokeproof enclosure in accordance with Section 1019.1.8. Elevator hoistway door assemblies are, by nature, not substantially air tight. Significant quantities of air can move throughout the shaft and adjacent lobby because of stack effect, especially in tall buildings. The requirement for smokeproof enclosure construction provides additional assurance that the elevator will not be rendered unavailable due to smoke movement into the elevator shaft. If the elevator is in an area of refuge that is formed by the use of a horizontal exit or smoke barrier, which essentially means there is no need for an additional elevator lobby, it is presumed that the area of refuge is free from fire involvement; therefore, smokeproof enclosure construction is not necessary

1007.6.1 Size. Each area of refuge shall be sized to accommodate one wheelchair space of 30 inches by 48 inches (762 mm by 1219 mm) for each 200 occupants or portion thereof, based on the occupant load of the area of refuge and areas served by the area of refuge. Such wheelchair spaces shall not reduce the required means of egress width. Access to any of the required wheelchair spaces in an area of refuge shall not be obstructed by more than one adjoining wheelchair space.

❖ The number of wheelchair spaces that is required to be provided in an area of refuge is intended to represent broadly the expected population of the average building. As one point of measurement, a 1977 survey conducted by the National Center for Health indicated that one in 333 civilian, noninstitutionalized persons uses a wheelchair. Originally, a ratio of one space for each 100

occupants was considered, based on the area served by the area of refuge. The Americans with Disabilities Act (ADA) currently utilizes the criterion of one space for each 200 occupants, based on the area served by the area of refuge. Given the variations and difficulties involved in accurately predicting a representative ratio for application to all occupancies, it was concluded that a requirement for one space for each 200 occupants based on the area of refuge itself, plus the areas served by the area of refuge, represents a reasonable criterion.

Arrangement of the required wheelchair spaces is critical so as not to interfere with the means of egress for ambulatory occupants. Since the design concept is that wheelchair occupants will move to the area of refuge and await further instructions or evacuation assistance, the spaces must be located so as not to reduce the required means of egress width of the stairway, door, corridor or other egress path through the area of refuge.

In order to provide for orderly maneuvering of wheelchairs, this section states that access to any of the required wheelchair spaces cannot be obstructed by more than one adjoining wheelchair space. For example, this precludes an arrangement in which three or more wheelchairs could be stacked down a dead-end corridor. This also effectively limits the difficulty any given wheelchair occupant would have in reaching or leaving a given wheelchair space, as well as providing easier access to all wheelchair spaces by persons providing evacuation assistance.

1007.6.2 Separation. Each area of refuge shall be separated from the remainder of the story by a smoke barrier complying with Section 709. Each area of refuge shall be designed to minimize the intrusion of smoke.

Exceptions:

1. Areas of refuge located within a stairway enclosure.

2. Areas of refuge where the area of refuge and areas served by the area of refuge are equipped throughout with an automatic sprinkler system installed in accordance with Section 903.3.1.1 or 903.3.1.2.

❖ The minimum standard for construction of an area of refuge is a smoke barrier, in accordance with Section 709. This establishes a minimum degree of performance by means of a 1-hour fire-resistance rating, including opening protectives and a minimum degree of performance against the intrusion of smoke into an enclosed area of refuge, as specified in Sections 709.4 and 709.5. This section does not require an area of refuge within an exit stairway to be designed to prevent the intrusion of smoke. This was based on a study of areas of refuge conducted by the National Institute for Standards and Technology (NIST) for the General Services Administration (GSA), which concluded that a story-level landing within a fire-resistance-rated exit stairway would provide a satisfactory staging area for evacuation assistance.

This section also exempts the requirement that the area of refuge be designed to prevent intrusion of smoke when the area of refuge and all areas served by it are sprinklered, in accordance with an NFPA 13 or NFPA 13R system (see Section 903.3.1.1 or 903.3.1.2.2). This is based on the value of sprinkler protection. This differs from the exceptions for enclosed stairways and elevators (see Sections 1007.3 and 1007.4), where areas of refuge are not required in buildings equipped throughout with an automatic sprinkler system. Instead, this section covers partially sprinklered buildings or buildings that choose to provide areas of refuge in addition to full suppression.

1007.6.3 Two-way communication. Areas of refuge shall be provided with a two-way communication system between the area of refuge and a central control point. If the central control point is not constantly attended, the area of refuge shall also have controlled access to a public telephone system. Location of the central control point shall be approved by the fire department. The two-way communication system shall include both audible and visible signals.

❖ Use of an elevator, stair enclosure or other area of refuge as part of an accessible means of egress requires a disabled person to wait for evacuation assistance or relevant instructions. The two-way communication system allows this person to inform emergency personnel of his or her location and to receive additional instructions or assistance as needed. It is important that both visual and audible signals be provided as part of the communication system, so that a person with any disability may utilize the system. The central control point can be where emergency management of the elevators is located, the main desk or lobby or other location where it is likely that trained or experienced personnel will be available to communicate with the occupants of the area of refuge. A suitable central control point is often not available in low-rise buildings or in a high-rise building the central control point may not be manned on a 24-hour basis. In situations such as these, a public telephone where a caller may reach an appropriate emergency location, such as the fire department, must also be provided in the area of refuge. The fire department must approve the configuration of the system.

1007.6.4 Instructions. In areas of refuge that have a two-way emergency communications system, instructions on the use of the area under emergency conditions shall be posted adjoining the communications system. The instructions shall include all of the following:

1. Directions to find other means of egress.

2. Persons able to use the exit stairway do so as soon as possible, unless they are assisting others.

3. Information on planned availability of assistance in the use of stairs or supervised operation of elevators and how to summon such assistance.

4. Directions for use of the emergency communications system.

❖ The required instructions on the proper use of the area of refuge and the communication system provide a higher degree of assurance that the communication

system will accomplish the intended function and occupants will behave as expected. A two-way communication system will not be of much value if a person in that area does not know how to operate it. Also, since the area of refuge is required by Section 1007.6.5 to be identified as such, ambulatory occupants may mistakenly conclude that they should remain in that area. The instructions remind ambulatory occupants that they should continue to egress as soon as possible.

1007.6.5 Identification. Each door providing access to an area of refuge from an adjacent floor area shall be identified by a sign complying with ICC A117.1, stating: AREA OF REFUGE, and including the International Symbol of Accessibility. Where exit sign illumination is required by Section 1011.2, the area of refuge sign shall be illuminated. Additionally, tactile signage complying with ICC A117.1 shall be located at each door to an area of refuge.

❖ Signage enables an occupant to become aware of an area of refuge. The approach that the code takes for identification of the area of refuge is comparable to the general provisions for identification of exits, including the requirement for lighted signage. Tactile signage is also required for the benefit of persons with a visual disability.

1007.7 Signage. At exits and elevators serving a required accessible space but not providing an approved accessible means of egress, signage shall be installed indicating the location of accessible means of egress.

❖ The additional signage required by this section is intended to advise persons of the location of the accessible means of egress. Since not all of the exits will necessarily be accessible means of egress, it is appropriate to provide this information at exit stairways and particularly at elevators that are not the approved accessible means of egress. This section is not intended to require that signage be provided at an accessible exit giving information as to the location of any other accessible means of egress.

1007.8 Exterior area for assisted rescue. The exterior area for assisted rescue must be open to the outside air and meet the requirements of Section 1007.6.1. Separation walls shall comply with the requirements of Section 704 for exterior walls. Where walls or openings are between the area for assisted rescue and the interior of the building, the building exterior walls within 10 feet (3048 mm) horizontally of a nonrated wall or unprotected opening shall be constructed as required for a minimum 1-hour fire-resistance rating with $^3/_4$-hour opening protectives. This construction shall extend vertically from the ground to a point 10 feet (3048 mm) above the floor level of the area for assisted rescue or to the roof line, whichever is lower.

❖ The exterior area of assisted rescue is an alternative for situations where the exit is via an unenclosed exterior stairway or either the exit discharge is not accessible. The protection required would be equivalent to that required for an area of refuge. Note that there is no exception for the exterior area of assisted rescue for buildings that contain sprinkler systems. The separation requirements are similar to exterior exit stairways (see Section 1022.6). The exceptions for exterior exit stairway protection would not be applicable where they would also include an exterior area for assisted rescue. Providing a location that was 10 feet (3048 mm) away from the exterior wall would not serve as a viable alternative for having a rated exterior wall. Persons waiting for assistance to move away from the building must have a minimum level of protection from a fire in the building.

An example where the exterior area for assisted rescue would be a good alternative is along the rear of a strip mall. The front of the facility may be accessible, but if two accessible means of egress are required, than an accessible route must be provided out the rear of each of the tenant spaces. Often in these types of malls, the rear entrance doubles as the loading dock or service entrance. Due to the associated change in elevation, the ramps required to allow accessible exit discharge could be extensive. In addition, the ramps could become damaged over time by trucks maneuvering into the loading dock areas. In this situation, interior areas of refuge are not always a positive alternative. Tenants may tend to use them as convenient storage areas. If persons with a mobility impairment wait for assisted rescue outside of the building, they are already protected from smoke and fumes—the deadliest of the fire hazards. Being visible should also result in a shorter period

of time before assisted rescue is affected [see Figures 1007.8(1) and 1007.8(2)].

1007.8.1 Openness. The exterior area for assisted rescue shall be at least 50 percent open, and the open area above the guards shall be so distributed as to minimize the accumulation of smoke or toxic gases.

❖ The openness criteria for exterior areas of assisted rescue are similar to the requirements for exterior balconies. The purpose is to ensure that a person at an exterior area of rescue assistance is not in danger from smoke and fumes. The criteria are to address the situation where the area is open to outside air, but a combination of roof overhangs and perimeter walls or guards could still trap enough smoke that the safety of the occupants would be jeopardized.

1007.8.2 Exterior exit stairway. Exterior exit stairways that are part of the means of egress for the exterior area for assisted rescue shall provide a clear width of 48 inches (1219 mm) between handrails.

❖ Any steps that lead from an exterior area for assisted rescue to grade must have a clear width of 48 inches (1219 mm) between handrails. The additional width is to permit adequate room to assist a mobility impaired person down the steps and to a safe location.

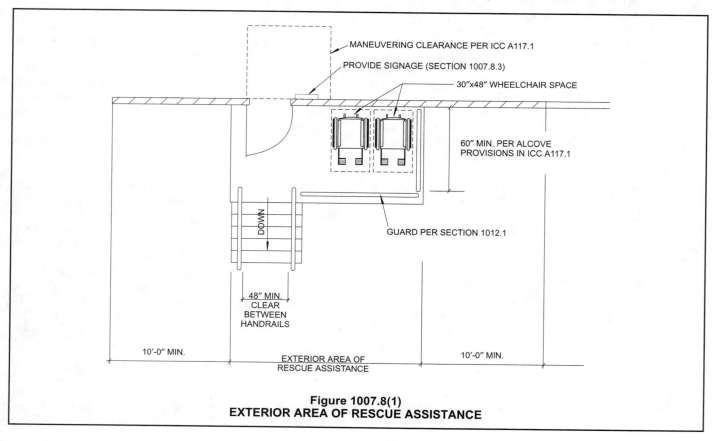

Figure 1007.8(1)
EXTERIOR AREA OF RESCUE ASSISTANCE

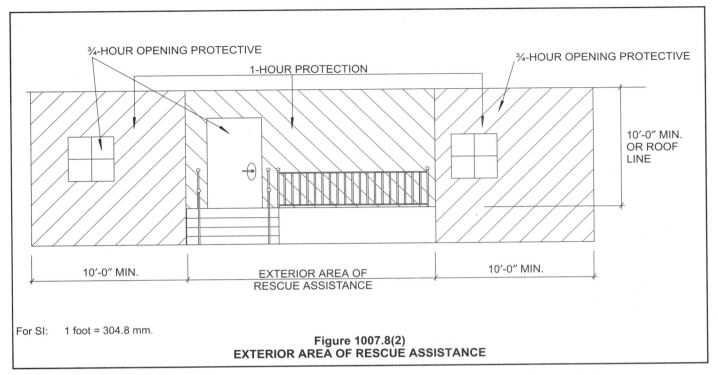

¾-HOUR OPENING PROTECTIVE

1-HOUR PROTECTION

¾-HOUR OPENING PROTECTIVE

10'-0" MIN. OR ROOF LINE

10'-0" MIN.

EXTERIOR AREA OF RESCUE ASSISTANCE

10'-0" MIN.

For SI: 1 foot = 304.8 mm.

Figure 1007.8(2)
EXTERIOR AREA OF RESCUE ASSISTANCE

1007.8.3 Identification. Exterior areas for assisted rescue shall have identification as required for area of refuge that complies with Section 1007.6.5.

❖ The exterior area for assisted rescue must have both visual and tactile signage that states "EXTERIOR AREA FOR ASSISTED RESCUE" and includes the International Symbol of Accessibility. This signage should be illuminated similar to the "Exit" signage. Due to immediate visibility of a person at an exterior area of rescue assistance, two-way communication and instructions are not required.

SECTION 1008
DOORS, GATES AND TURNSTILES

1008.1 Doors. Means of egress doors shall meet the requirements of this section. Doors serving a means of egress system shall meet the requirements of this section and Section 1017.2. Doors provided for egress purposes in numbers greater than required by this code shall meet the requirements of this section.

Means of egress doors shall be readily distinguishable from the adjacent construction and finishes such that the doors are easily recognizable as doors. Mirrors or similar reflecting materials shall not be used on means of egress doors. Means of egress doors shall not be concealed by curtains, drapes, decorations or similar materials.

❖ The general requirements for doors are in this section and the following subsections. Doors need to be easily recognizable for immediate use in an emergency condition. Thus, the code specifies that doors are not to be

hidden in such a manner that a person would have trouble seeing where to egress.

1008.1.1 Size of doors. The minimum width of each door opening shall be sufficient for the occupant load thereof and shall provide a clear width of not less than 32 inches (813 mm). Clear openings of doorways with swinging doors shall be measured between the face of the door and the stop, with the door open 90 degrees (1.57 rad). Where this section requires a minimum clear width of 32 inches (813 mm) and a door opening includes two door leaves without a mullion, one leaf shall provide a clear opening width of 32 inches (813 mm). The maximum width of a swinging door leaf shall be 48 inches (1219 mm) nominal. Means of egress doors in an occupancy in Group I-2 used for the movement of beds shall provide a clear width not less than $41^{1}/_{2}$ inches (1054 mm). The height of doors shall not be less than 80 inches (2032 mm).

Exceptions:

1. The minimum and maximum width shall not apply to door openings that are not part of the required means of egress in occupancies in Groups R-2 and R-3 as applicable in Section 101.2.

2. Door openings to resident sleeping units in occupancies in Group I-3 shall have a clear width of not less than 28 inches (711 mm).

3. Door openings to storage closets less than 10 square feet (0.93 m²) in area shall not be limited by the minimum width.

4. Width of door leafs in revolving doors that comply with Section 1008.1.3.1 shall not be limited.

5. Door openings within a dwelling unit or sleeping unit shall not be less than 78 inches (1981 mm) in height.

6. Exterior door openings in dwelling units and sleeping units, other than the required exit door, shall not be less than 76 inches (1930 mm) in height.

7. Interior egress doors within a dwelling unit or sleeping unit which is not required to be adaptable or accessible.

8. Door openings required to be accessible within Type B dwelling units shall have a minimum clear width of $31^3/_4$ inches (806 mm).

❖ The size of a door opening determines its capacity as a component of egress and its ability to fulfill its function in normal use. A door opening must meet certain minimum criteria as to its width and height in order to be used safely and to provide accessibility to people with physical disabilities. Doorways that are not in the means of egress are not limited in size by this section. However, doors that are used for egress purposes, including additional doors over and above the number of means of egress required by the code, are required to meet the requirements of this section unless one of the exceptions applies.

The minimum clear width of an egress doorway for occupant capacity is based on the portion of the occupant load (see Section 1004.1) intended to utilize the doorway for egress purposes multiplied by the egress width per occupant from Table 1005.1. The clear width of a swinging door opening is the horizontal dimension measured between the face of the door and the door stops when the door is in the 90-degree position [see Figure 1008.1.1(1)].

Using the face of the door as the measurement point is consistent with the provisions of ICC A117.1 and the ADAAG Review Advisory Committee. Further, this measurement is not intended to prohibit other projections into the required clear width, such as latching or panic hardware. See the commentary to Section 1008.1.1.1 and Figure 1008.1.1(2) for further discussion on the specific projections allowed in the required clear width. For nonswinging means of egress doors, such as a sliding door, the clear width is to be measured from the face of the door jambs.

The minimum clear width in a doorway of 32 inches (813 mm) is to allow passage of a wheelchair as well as persons utilizing walking devices or other support apparatus. Similarly, because of the difficulties that a person with physical disabilities would have in opening two doors simultaneously, the 32-inch (813 mm) minimum must be provided by a single door leaf.

Note that in some cases, with standard door construction and hardware, a 36-inch-wide (914 mm) door is the narrowest door that can be used while still providing the minimum clear width of 32 inches (813 mm). A standard 34-inch-wide (864 mm) door has less than a 32-inch (813 mm) clear opening depending on the thickness of the opposing doorstop and where the door hinges are located. The building designer must verify that the swinging door specified will in fact provide the required clear width. A minimum clear width of $41^1/_2$ inches (1054 mm) is required for doors in any portion of Group I-2 where bed movement is needed to evacuate

patients from the area in the event of a fire.

The maximum width for a means of egress door leaf in a swinging door is 48 inches (1219 mm) because larger doors are difficult to handle and are of sizes that typically are not fire tested. The maximum width only applies to swinging doors and not to horizontal sliding doors.

Minimum door heights are required to provide clear headroom for the users. A minimum height of 80 inches (2032 mm) has been empirically derived as sufficient for most users. Note that although the clear height of a doorway is not specified, typical door frame dimensions will render an opening very close to 80 inches (2032 mm) in clear height.

Exception 1 is very limited in scope and is primarily intended to permit decorative-type doors, e.g., café doors, in dwelling units. This exception addresses spaces that are provided with two or more means of egress when only one is required. These nonrequired egress elements are exempted from the minimum and maximum dimensions.

Exception 2 permits the continued use of doors to resident sleeping rooms (cells) in occupancies in Group I-3 according to current practices.

Exception 3 permits doors to storage closets in any occupancy classification less than 10 square feet (0.9 m²) in area to be less than 32 inches (813 mm). This provision is intended to include those closets that can be reached in an arm's length and thus do not require full passage into the closet to be functional.

Exception 4 permits the door leaves in a revolving door assembly to be of any width when the revolving door passage width complies with Section 1008.1.3.1, which provides for adequate egress width when collapsed into a book-fold position.

Exception 5 permits the doorway within a dwelling or sleeping unit to be a minimum of 78 inches (1981 mm) in clear height. This is deemed acceptable because of the familiarity persons in a dwelling or sleeping unit usually have with the egress system and the lack of adverse injury statistics relating to such doors. Note that this exception does not apply to exterior doors of the dwelling unit.

Exception 6 permits exterior doorways to a dwelling or sleeping unit, except for the required exit door, to be a minimum of 76 inches (1930 mm) in clear height. Accordingly, the required exterior exit door to a dwelling or sleeping unit must be 80 inches (2032 mm) in height (exterior doors are not within the scope of Exception 5), but other exterior doors are allowed to be a height of only 76 inches (1930 mm). This provision allows for the continued use of 76-inch-high (1930 mm) sliding patio doors and swinging doors sized to replace such doors.

Exception 7 allows interior doors in a means of egress in dwelling or sleeping units not required to be accessible to have a clear width less than 32 inches (813 mm). Since the doors specified do not serve a dwelling or sleeping unit that is required to be Accessible, Type A or B units, smaller doors are allowed.

Exception 8 addresses the doors in a Type B dwelling

or sleeping unit that are required to be accessible. Refer to Chapter 11 for additional information related to Type B dwelling and sleeping units.

1008.1.1.1 Projections into clear width. There shall not be projections into the required clear width lower than 34 inches (864 mm) above the floor or ground. Projections into the clear opening width between 34 inches (864 mm) and 80 inches (2032 mm) above the floor or ground shall not exceed 4 inches (102 mm).

❖ This section of the code provides specific allowances for projection into the required clear widths of means of egress doors. These allowances directly correspond with the method of measuring the required clear width of the door as specified in Section 1008.1.1. A reasonable range of projections for door hardware and trim has been established by these requirements. The use of the means of egress door by a wheelchair occupant will not be significantly impacted by small projections located in inconspicuous areas. The key to these allowances is their location. Projections are allowed at a height between 34 inches (864 mm) and 80 inches (2032 mm). Below the 34-inch (864 mm) height, the code does not permit any projections since they would decrease the available width for wheelchair operation. The full 32-inch (813 mm) width must be provided at this location. At 34 inches (864 mm) and higher, projections of up to and including 4 inches (102 mm) are permitted. The 4-inch (102 mm) projection is consistent with the allowances of Section 1003.3.3. This section permits door hardware such as panic hardware to extend into the clear width yet maintain accessibility for persons

with physical disabilities [see Figure 1008.1.1(2)].

The protrusion permitted for door closers and stops in Section 1003.3.1 is more specific than this 80-inch (2032 mm) headroom requirement; therefore, a door closer or stop is not limited to a 4-inch (102 mm) protrusion (see Figure 1003.3.1).

1008.1.2 Door swing. Egress doors shall be side-hinged swinging.

Exceptions:

1. Private garages, office areas, factory and storage areas with an occupant load of 10 or less.

2. Group I-3 occupancies used as a place of detention.

3. Doors within or serving a single dwelling unit in Groups R-2 and R-3 as applicable in Section 101.2.

4. In other than Group H occupancies, revolving doors complying with Section 1008.1.3.1.

5. In other than Group H occupancies, horizontal sliding doors complying with Section 1008.1.3.3 are permitted in a means of egress.

6. Power-operated doors in accordance with Section 1008.1.3.2.

Doors shall swing in the direction of egress travel where serving an occupant load of 50 or more persons or a Group H occupancy.

The opening force for interior side-swinging doors without closers shall not exceed a 5-pound (22 N) force. For other side-swinging, sliding and folding doors, the door latch shall release when subjected to a 15-pound (67 N) force. The door shall

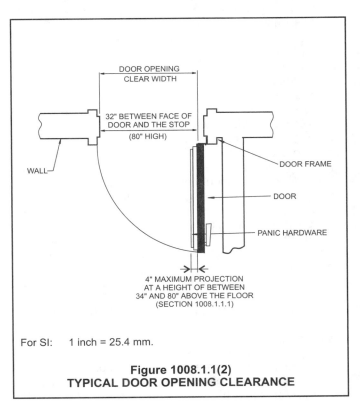

For SI: 1 inch = 25.4 mm.

Figure 1008.1.1(1)
TYPICAL DOOR OPENING CLEARANCE

For SI: 1 inch = 25.4 mm.

Figure 1008.1.1(2)
TYPICAL DOOR OPENING CLEARANCE

be set in motion when subjected to a 30-pound (133 N) force. The door shall swing to a full-open position when subjected to a 15-pound (67 N) force. Forces shall be applied to the latch side.

❖ Generally, egress doors are required to be side-hinged swinging type. Side-hinged swinging doors are familiar to all occupants in the method of operation. Some contemporary door designs have pivots at the top and bottom, which are intended to be permitted by this section, since the door action itself has little difference between the side-hinged-type door.

The code has several conditions where it allows doors that are not side-hinged swinging type.

Examples of the doors discussed in Exception 1 are an overhead garage door and sliding door. Exception 1 allows doors other than the swinging type for the listed occupancies where the number of occupants is very low.

Exception 2 allows for sliding-type doors that are commonly used in prisons and jails.

Exception 3 allows for sliding-type doors within or serving a residential occupancy. The use of sliding doors on the interior of dwelling units is permitted by HUD accessibility requirements and such doors often are needed because of the code requirements that large doors comply with the accessibility provisions.

Exception 4 allows for revolving doors that meet the requirements of Section 1008.1.3.1.

Exception 5 allows for horizontal sliding doors that meet the requirements of Section 1008.1.3.3. This exception is intended to allow wide span openings to be used in a means of egress serving areas of refuge formed by horizontal exits, smoke barriers or elevator lobbies. This is to enhance the movement of people with mobility impairments to areas of safety without obstructions, since the horizontal sliding doors specified in Section 1008.1.3.3 afford simple operation by persons with disabilities.

Exception 6 provides for power-operated doors that comply with the requirements of Section 1008.1.3.2.

A side-hinged door must swing in the direction of egress travel where the required occupant capacity of the room is 50 or more. As such, a room with two doors and an occupant load of 99 would require both doors to swing in the direction of egress travel, even though each door has a calculated occupant usage of less than 50. At this level of occupant load, the possibility exists that, in an emergency situation, a compact line of people could form at a closed door that swings in a direction opposite the egress flow. This could delay or eliminate the first person's ability to open the door inward, with the rest of the queue behind the person.

In a Group H occupancy, the threat of rapid fire buildup or worse is such that any delay in egress caused by door swing may jeopardize the opportunity for all occupants to evacuate the premises. For this reason, all egress doors in Group H occupancies are to swing in the direction of egress.

The ability of all potential users to be physically capable of opening an egress door is a function of the forces required to open the door. The 5-pound (22 N) maximum force for pushing and pulling interior side-swinging doors without closers is based on that which has been deemed appropriate for people with a physical disability that makes it difficult to operate a door. The operating force for other doors is permitted to be higher. This recognizes that doors with closers, particularly fire doors, require greater operating forces in order to close fully in an emergency where combustion gases may be exerting pressure on the door assembly. Similarly, exterior doors are exempted because air pressure differentials and strong winds may prevent doors from being automatically closed. A maximum force of 15 pounds (7 kg) is required for operating the latching mechanism. Once unlatched, a maximum force of 30 pounds (14 kg) is applied to the latch side of the leaf to start the door in motion by overcoming its stationary inertia. Once in motion, it must not take more than 15 pounds (7 kg) of force to keep the door in motion until it reaches its full open position and the required clear width is available. To conform to this requirement on a continuing basis, door closers must be adjusted periodically and door fits must also be checked and adjusted when necessary.

1008.1.3 Special doors. Special doors and security grilles shall comply with the requirements of Sections 1008.1.3.1 through 1008.1.3.5.

❖ This section simply defines the scope of the code requirements for special doors and security grilles.

1008.1.3.1 Revolving doors. Revolving doors shall comply with the following:

1. Each revolving door shall be capable of collapsing into a bookfold position with parallel egress paths providing an aggregate width of 36 inches (914 mm).

2. A revolving door shall not be located within 10 feet (3048 mm) of the foot of or top of stairs or escalators. A dispersal area shall be provided between the stairs or escalators and the revolving doors.

3. The revolutions per minute (rpm) for a revolving door shall not exceed those shown in Table 1008.1.3.1.

4. Each revolving door shall have a side-hinged swinging door which complies with Section 1008.1 in the same wall and within 10 feet (3048 mm) of the revolving door.

❖ Revolving doors must comply with all four provisions.

Item 1: One of the causes contributing to the loss of lives in the 1942 Coconut Grove fire in Boston was that the revolving doors at the club's entrance could not collapse for emergency egress and there was not an alternative means of egress adjacent to the revolving doors. Thus, in the panic of the fire, the door became jammed and the club's occupants were trapped.

As a result of this fire experience, all revolving doors, including those for air structures, now are required to be equipped with a collapse feature. A book-fold operation is where all leafs collapse parallel to each other and to the direction of egress [see Figure 1008.1.3.1(1)]. A book-fold operation creates two openings of approximately equal width. The sum of the widths is not to be less than 36 inches (914 mm) so that a stream of pedestrians may use each side of the opening.

Item 2: If a stairway or escalator delivers users to a landing in front of a revolving door at a greater rate than the capacity of the door, a compact line of people will develop. Lines of people formed on a stairway or escalator create an unsafe situation, since stairways and escalators are not intended to be used as standing space for persons who may be waiting to use the revolving doors. Therefore, to avoid congestion at a revolving door, which under normal operation has a maximum delivery capacity of users, a dispersal area is required between the stairways or escalators and the revolving doors to allow for the queuing of people as they enter the door. Accordingly, to create a dispersal area for users of a revolving door, the door is not to be placed closer than 10 feet (3048 mm) from the foot or top of a stairway or escalator.

Item 3: Door speeds also directly relate to the capacity of a revolving door, which is calculated by multiplying the number of leafs (wings) by the revolutions per minute (rpm). For example, if you have a four-leaf door moving at 10 rpm, the door will allow 40 people to move in either direction in 1 minute.

Item 4: In case a revolving door malfunctions or becomes obstructed, the adjacent area is to be equipped with a conventional side-hinged door to provide users with an immediate alternative way to exit a building. The side-hinged door is intended to be used as a relief device for people lined up to use the revolving door or who desire to avoid it because of a physical disability or other reason. It also can be used when the revolving door is obstructed or out of service. The swinging door is to be immediately adjacent to the revolving door so that its availability is obvious [see Figure 1008.1.3.1(2)]. A single swinging door can be located between two revolving doors and comply with this provision.

TABLE 1008.1.3.1
REVOLVING DOOR SPEEDS

INSIDE DIAMETER (feet-inches)	POWER-DRIVEN-TYPE SPEED CONTROL (rpm)	MANUAL-TYPE SPEED CONTROL (rpm)
6-6	11	12
7-0	10	11
7-6	9	11
8-0	9	10
8-6	8	9
9-0	8	9
9-6	7	8
10-0	7	8

For SI: 1 inch = 25.4 mm, 1 foot = 304.8 mm.

❖ Door speeds also directly relate to the capacity of a revolving door, which is calculated by multiplying the number of leafs (wings) by the revolutions per minute (rpm). For example, if you have a four-leaf door moving at 10 rpm, the door will allow 40 people to move in either direction in 1 minute.

1008.1.3.1.1 Egress component. A revolving door used as a component of a means of egress shall comply with Section 1008.1.3.1 and the following three conditions:

1. Revolving doors shall not be given credit for more than 50 percent of the required egress capacity.

2. Each revolving door shall be credited with no more than a 50-person capacity.

3. Each revolving door shall be capable of being collapsed when a force of not more than 130 pounds (578 N) is applied within 3 inches (76 mm) of the outer edge of a wing.

❖ A revolving door can be incorporated to a very limited extent in a means of egress. Compliance with these three additional conditions is required.

Condition 1 limits the exit capacity that revolving doors can provide in a building. This is so that 50 percent of the capacity has conventional egress components and is not dependent on mechanical devices or fail-safe mechanisms.

Condition 2 limits the capacity of any one revolving door for the same reasons as stated in Condition 1. Each revolving door is therefore limited to a 50-person capacity.

Condition 3 limits the collapse force to 130 pounds (59 kg), as opposed to the 180-pound (82 kg) value listed in Section 1008.1.3.1.2. Revolving doors used as means of egress are not permitted to have the collapse force exceed 130 pounds (59 kg) under any circumstances.

PARALLEL PATHS

A + B = 36"
MINIMUM COMBINED WIDTH

For SI: 1 inch = 25.4 mm.

Figure 1008.1.3.1(1)
REVOLVING DOORS IN BOOK-FOLD POSITION

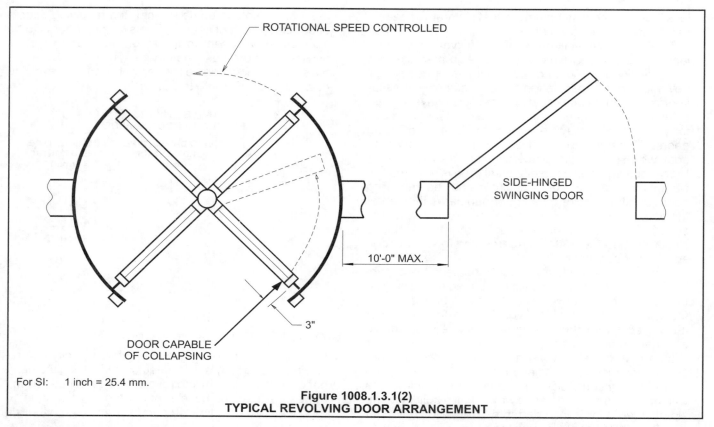

ROTATIONAL SPEED CONTROLLED

SIDE-HINGED
SWINGING DOOR

10'-0" MAX.

3"

DOOR CAPABLE
OF COLLAPSING

For SI: 1 inch = 25.4 mm.

Figure 1008.1.3.1(2)
TYPICAL REVOLVING DOOR ARRANGEMENT

1008.1.3.1.2 Other than egress component. A revolving door used as other than a component of a means of egress shall comply with Section 1008.1.3.1. The collapsing force of a revolving door not used as a component of a means of egress shall not be more than 180 pounds (801 N).

Exception: A collapsing force in excess of 180 pounds (801 N) is permitted if the collapsing force is reduced to not more than 130 pounds (578 N) when at least one of the following conditions is satisfied:

1. There is a power failure or power is removed to the device holding the door wings in position.

2. There is an actuation of the automatic sprinkler system where such system is provided.

3. There is an actuation of a smoke detection system which is installed in accordance with Section 907 to provide coverage in areas within the building which are within 75 feet (22 860 mm) of the revolving doors.

4. There is an actuation of a manual control switch, in an approved location and clearly defined, which reduces the holding force to below the 130-pound (578 N) force level.

❖ This section addresses revolving doors that are not used to serve any portion of the occupant egress capacity. For example, where adjacent side-hinged doors have more than the required egress capacity, the revolving door would not be part of the required means of egress.

The maximum collapse force of 180 pounds (82 kg), applied within 3 inches (76 mm) of the outer edge of a wing, is based on industry standards to accommodate normal use conditions and other forces that may act on the leafs, such as that caused by wind or air pressure. An exception for revolving doors that are not a component of a required means of egress allows the collapse force to exceed 180 pounds (82 kg) in normal operating conditions provided that a force of not more than 130 pounds (59 kg) is required whenever any one of the listed conditions is satisfied.

1008.1.3.2 Power-operated doors. Where means of egress doors are operated by power, such as doors with a photoelectric-actuated mechanism to open the door upon the approach of a person, or doors with power-assisted manual operation, the design shall be such that in the event of power failure, the door is capable of being opened manually to permit means of egress travel or closed where necessary to safeguard means of egress. The forces required to open these doors manually shall not exceed those specified in Section 1008.1.2, except that the force to set the door in motion shall not exceed 50 pounds (220 N). The door shall be capable of swinging from any position to the full width of the opening in which such door is installed when a force is applied to the door on the side from which egress is made. Full-power-operated doors shall comply with BHMA A156.10. Power-assisted and low-energy doors shall comply with BHMA A156.19.

Exceptions:

1. Occupancies in Group I-3.

2. Horizontal sliding doors complying with Section 1008.1.3.3.

3. For a biparting door in the emergency breakout mode, a door leaf located within a multiple-leaf opening shall be exempt from the minimum 32-inch (813 mm) single-leaf requirement of Section 1008.1.1, provided a minimum 32-inch (813 mm) clear opening is provided when the two biparting leaves meeting in the center are broken out.

❖ For convenience purposes, power-operated doors are intended to facilitate the normal nonemergency flow of persons through a doorway. Where a power-operated or assisted door is also required to be an egress door, the door must conform to the requirements of this section. The essential characteristic is that the door is to be manually openable from any position to its full open position at any time, with or without a power failure or a failure of a door mechanism. Hence, both swinging and horizontal sliding doors, complying with this section, may be used, provided the door can be operated manually from any position as a swinging door and that the minimum required clear width for egress capacity is not less than 32 inches (813 mm). Note that the opening forces of Section 1008.1.2 are applicable, except that the 30-pound (14 kg) force needed to set the door in motion is increased to 50 pounds (23 kg) as an operational tolerance in the design of the power-operated door.

In accordance with Exception 1, power-operated doors in detention and correctional occupancies (Group I-3) are not required to be manually operable by the occupants (inmates) for security reasons, but otherwise are required to conform to Section 408. Section 1008.1.3.2 does not apply to horizontal sliding doors that do not meet the provisions of this section (i.e., breakout panels for horizontal power-operated doors) but do comply with Section 1008.1.3.3.

Exception 2 states that power-operated doors that meet the requirements of this section are not required to meet the requirements of Section 1008.1.3.3 for horizontal sliding doors that are not capable of swing operation in the event of power failure.

Exception 3 allows an individual leaf of a four-panel biparting door to be less than 32 inches (813 mm) wide provided 32 inches (813 mm) of clear space is available when the two center biparting leaves are broken out.

1008.1.3.3 Horizontal sliding doors. In other than Group H occupancies, horizontal sliding doors permitted to be a component of a means of egress in accordance with Exception 5 to Section 1008.1.2 shall comply with all of the following criteria:

1. The doors shall be power operated and shall be capable of being operated manually in the event of power failure.

2. The doors shall be openable by a simple method from both sides without special knowledge or effort.

3. The force required to operate the door shall not exceed 30 pounds (133 N) to set the door in motion and 15 pounds (67 N) to close the door or open it to the minimum required width.

4. The door shall be openable with a force not to exceed 15 pounds (67 N) when a force of 250 pounds (1100 N) is applied perpendicular to the door adjacent to the operating device.

5. The door assembly shall comply with the applicable fire protection rating and, where rated, shall be self-closing or automatic-closing by smoke detection, shall be installed in accordance with NFPA 80 and shall comply with Section 715.

6. The door assembly shall have an integrated standby power supply.

7. The door assembly power supply shall be electrically supervised.

8. The door shall open to the minimum required width within 10 seconds after activation of the operating device.

❖ Horizontal sliding doors are permitted in the means of egress, in other than rooms or areas of Group H, under the conditions set forth in this section. Horizontal sliding doors are not permitted to be used in Group H occupancies because of the potential for delaying or impeding egress from those areas and the additional risk to occupants in hazardous occupancies. Note that this section regulates egress doors that do not meet all of the requirements of Section 1008.1.3.2 (e.g., a power-operated horizontal sliding door that does not have "breakout" capabilities to allow the door panels to swing if power is lost).

Such doors permitted in other groups are typically in the open position and close either because of a fire or to provide some degree of separation, often for security purposes to restrict movement to certain parts of a building. When in the closed position, the doors must be easily operated with forces similar to the limitations for swinging doors.

All eight of the criteria listed in this section must be met for a horizontal sliding door since there is a concern that it must be able to be easily opened under all conditions.

Additionally, the door must be openable even if a force of 250 pounds (114 kg) is being applied perpendicular to it, as may occur if a group of people were pushing on it.

Since the doors are manually operable, they need not automatically open or close during a loss of power. However, a standby power supply must be provided. The primary power supply must be supervised so that an alarm is received at a constantly attended location (such as a security desk) on loss of the primary power. If the doors are also serving as fire doors, they must be automatic- or self-closing in accordance with Section 715.

Since the maximum swinging door leaf width limitations of Section 1008.1.1 do not apply, a maximum opening time of 10 seconds is permitted. It should be noted, however, that the door need not open fully within the 10 seconds; rather, it must open to the required width. For example, if the door is protecting an opening that is 10 feet (3048 mm) wide, but the minimum required width of the opening is 32 inches (813 mm) (as determined by Section 1008.1.1), the door need only open 32 inches (813 mm) within the 10-second criterion. In fact, the door may have controls such that the automatic opening feature only opens the door to a width of 32 inches (813 mm). If additional width is required, it can be accomplished by manual means and, possibly, by an additional activation of the operating device.

1008.1.3.4 Access-controlled egress doors. The entrance doors in a means of egress in buildings with an occupancy in Group A, B, E, M, R-1 or R-2 and entrance doors to tenant spaces in occupancies in Groups A, B, E, M, R-1 and R-2 are permitted to be equipped with an approved entrance and egress access control system which shall be installed in accordance with all of the following criteria:

1. A sensor shall be provided on the egress side arranged to detect an occupant approaching the doors. The doors shall be arranged to unlock by a signal from or loss of power to the sensor.

2. Loss of power to that part of the access control system which locks the doors shall automatically unlock the doors.

3. The doors shall be arranged to unlock from a manual unlocking device located 40 inches to 48 inches (1016 mm to 1219 mm) vertically above the floor and within 5 feet (1524 mm) of the secured doors. Ready access shall be provided to the manual unlocking device and the device shall be clearly identified by a sign that reads "PUSH TO EXIT." When operated, the manual unlocking device shall result in direct interruption of power to the lock—independent of the access control system electronics—and the doors shall remain unlocked for a minimum of 30 seconds.

4. Activation of the building fire alarm system, if provided, shall automatically unlock the doors, and the doors shall remain unlocked until the fire alarm system has been reset.

5. Activation of the building automatic sprinkler or fire detection system, if provided, shall automatically unlock the doors. The doors shall remain unlocked until the fire alarm system has been reset.

6. Entrance doors in buildings with an occupancy in Group A, B, E or M shall not be secured from the egress side during periods that the building is open to the general public.

❖ Security in buildings is a major concern to owners and occupants from the perspective of property preservation and personal physical safety. Since many occupancies are partially occupied around the clock, after normal business hours or on weekends, it is necessary that an adequate level of security be provided without jeopardizing the egress capabilities of the occupants.

This section permits the building entrance doors in a means of egress and entrance doors to tenant spaces in occupancies of Groups A, B, E, M, R-1 and R-2 to be secured while maintaining them as a means of egress. Items 1 through 6 provide additional life safety measures to permit easier egress during normal and emergency situations. Occupancies in Groups F, S and H are not included here because of their increased hazard due to an increase in fuel load and other potentially life-threatening activities. This potential increase in life-threatening circumstances requires an immediate egress capacity without the necessary "waiting period" afforded by the access control of this section. Item 1 requires that such doors be provided with an automatic exit sensor typically operating on an infrared, microwave or sonic principle. This sensor is required to release automatically the lock upon an occupant approaching the door. Item 2 requires that if there is a loss of power to this device or to the access control system itself, the doors must unlock. Item 3 requires that there be a manual exit device, such as a push button, within 5 feet (1524 mm) of the door, mounted 40 to 48 inches (1016 to 1219 mm) above the floor, unobstructed and with a clearly identifiable sign that says "PUSH TO EXIT." When operated, the manual exit device is to interrupt (independent of the access control electronics) the power to the lock directly and cause the doors to remain unlocked for a minimum of 30 seconds. Items 4 and 5 require the building fire alarm system, automatic fire detection system or sprinkler system if provided be interfaced with the access control system to unlock automatically the doors on activation. The doors are to remain unlocked until the fire alarm system is reset. This is so that the building entrance doors with controlled access will remain unlocked and open until fire fighters responding to the alarm have entered.

Item 6 requires that during the hours the building is open to the general public, doors equipped with an access control system will not be secured from the egress side in Group A, B, E and M occupancies. Thus, the building entrance doors in these occupancies are allowed to be secured only during off hours when the building occupant load will generally be reduced. In Groups R-1 and R-2, there is no restriction on the time period that access-controlled egress doors may be used, because the residents of the building are expected to be familiar with the building entrance and locking systems. Hence, with a building access control system, the number of individuals who are not familiar with the building is limited and they are usually accompanied by a resident. Note that access-controlled tenant entrance doors in Group A, B, E and M occupancies may be secured at any time.

1008.1.3.5 Security grilles. In Groups B, F, M and S, horizontal sliding or vertical security grilles are permitted at the main exit and shall be openable from the inside without the use of a key or special knowledge or effort during periods that the space is occupied. The grilles shall remain secured in the full-open position during the period of occupancy by the general public. Where two or more means of egress are required, not more than one-half of the exits or exit access doorways shall be equipped with horizontal sliding or vertical security grilles.

❖ This section really functions as an exception to several sections, including Sections 1008.1.2 (Door swing) and 1008.1.8.3 (Locks and latches) and permits the use of these security grilles under conditions that are similar to those found in Section 402 for covered mall buildings. These security grilles will be open when the space is occupied and will therefore not obstruct any egress path.

1008.1.4 Floor elevation. There shall be a floor or landing on each side of a door. Such floor or landing shall be at the same elevation on each side of the door. Landings shall be level except for exterior landings, which are permitted to have a slope not to exceed 0.25 unit vertical in 12 units horizontal (2-percent slope).

Exceptions:

1. Doors serving individual dwelling units in Groups R-2 and R-3 as applicable in Section 101.2 where the following apply:

    1.1. A door is permitted to open at the top step of an interior flight of stairs, provided the door does not swing over the top step.

    1.2. Screen doors and storm doors are permitted to swing over stairs or landings.

2. Exterior doors as provided for in Section 1003.5, Exception 1, and Section 1017.2, which are not on an accessible route.

3. In Group R-3 occupancies, the landing at an exterior doorway shall not be more than $7^3/_4$ inches (197 mm) below the top of the threshold, provided the door, other than an exterior storm or screen door, does not swing over the landing.

4. Variations in elevation due to differences in finish materials, but not more than 0.5 inch (12.7 mm).

5. Exterior decks, patios or balconies that are part of Type B dwelling units and have impervious surfaces, and that are not more than 4 inches (102 mm) below the finished floor level of the adjacent interior space of the dwelling unit.

❖ Changes in floor surface elevation at a door, however small, often are slip or trip hazards. This is because persons passing through a door, including those who are physically disabled, usually do not expect changes in floor surface elevation or are not able to recognize them because of the intervening door leaf. Under emergency conditions, a fall in a doorway could result not only in injury to the falling occupant but also interruption of orderly egress by other occupants.

The first exception that applies to residential occupancies recognizes that occupants are familiar with the stair and landing arrangements. Note that an interior or exterior door (other than screen or storm doors) is not allowed to swing over a stair.

Except for exterior doors that meet Exceptions 2 and 3, the floor surface elevation of the area, described by the doorway width and length measured from the face of the door on both sides of the threshold at a distance equal to the door width, is to be at the same elevation plus or minus ½ inch (12.7 mm) (see Figure 1008.1.4).

Certain exterior doors of Type B dwelling or sleeping units are also exempt from the floor surface requirements of Section 1008.1.4. Please note that this exception is not applicable for the primary entrance door (see Section 1105.1.6). Exterior doors that open out onto an exterior deck, patio or balcony are allowed a 4-inch (102 mm) step. Type B units are established by Chapter 11 for residential occupancies containing four or more dwelling or sleeping units. The exterior decks, patios or balconies must be of solid and impervious construction, such as concrete or wood. A 4-inch (102 mm) step from inside the unit down to the exterior surfaces is allowed for weather purposes. This allowance is consistent with the provisions of ICC A117.1 and the *Federal Fair Housing Accessibility Guidelines*.

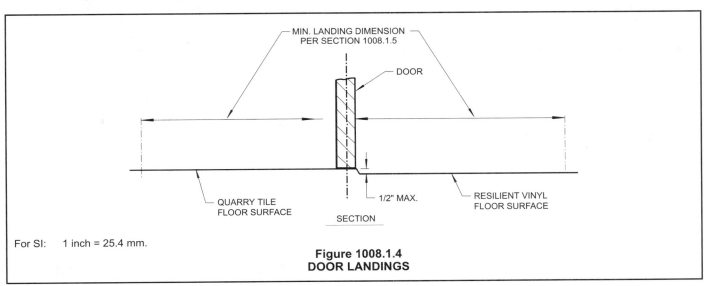

For SI: 1 inch = 25.4 mm.

**Figure 1008.1.4
DOOR LANDINGS**

1008.1.5 Landings at doors. Landings shall have a width not less than the width of the stairway or the door, whichever is the greater. Doors in the fully open position shall not reduce a required dimension by more than 7 inches (178 mm). When a landing serves an occupant load of 50 or more, doors in any position shall not reduce the landing to less than one-half its required width. Landings shall have a length measured in the direction of travel of not less than 44 inches (1118 mm).

Exception: Landing length in the direction of travel in Group R-3 as applicable in Section 101.2 and Group U and within individual units of Group R-2 as applicable in Section 101.2, need not exceed 36 inches (914 mm).

❖ This section is intended to address landings at doors (also see Section 1009.4 for stairway landings). The width of a landing at a door in a stairway is to be not less than the width of the stairway or the door, whichever is greater [see Figure 1009.4(4) for an example of these provisions]. Door landings are to have the floor elevation requirements of Section 1008.1.4 extending at least 44 inches (1118 mm) in the direction of egress travel.
The reduction in landing length for certain residential occupancies is a result of their low occupant load.

1008.1.6 Thresholds. Thresholds at doorways shall not exceed 0.75 inch (19.1 mm) in height for sliding doors serving dwelling units or 0.5 inch (12.7 mm) for other doors. Raised thresholds and floor level changes greater than 0.25 inch (6.4 mm) at doorways shall be beveled with a slope not greater than one unit vertical in two units horizontal (50-percent slope).

Exception: The threshold height shall be limited to 7 $^3/_4$ inches (197 mm) where the occupancy is Group R-2 or R-3 as applicable in Section 101.2, the door is an exterior door that is not a component of the required means of egress and the doorway is not on an accessible route.

❖ A threshold is a potential tripping hazard and a barrier to accessibility by people with mobility impairments. For these reasons, thresholds for all doorways, except exterior sliding doors serving dwelling units, are to be a maximum of ½-inch (12.7 mm) high. Exterior sliding doors serving dwelling units, however, are permitted to be $^3/_4$ inch (19.1 mm) high because of practical design considerations, concern for deterioration of the doorway due to snow and ice buildup and lack of adequate drainage in severe climates. Raised threshold and floor level changes at doorways without edge treatment [see Figure 1008.1.6(1)] are permitted to be $^1/_4$ inch (6.3 mm) high vertically.
Raised threshold and floor level changes with edges beveled with a slope not greater than one unit vertical in two units horizontal (1:2) [see Figure 1008.1.6(2)] are permitted to be ½-inch (12.7 mm) high, with its parts no more than $^1/_4$-inch (6.3 mm) high and the remainder with beveled edges no more than 1:2 [see Figure 1008.1.6(3)]. This kind of threshold treatment provides for minimum obstructions for wheelchair users and limits the trip hazard for those with other mobility disabilities.
The exception permits a step-down at exterior doors

for dwelling or sleeping units not required to be Accessible, Type A or B units.

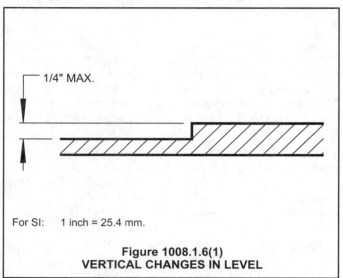

For SI: 1 inch = 25.4 mm.

Figure 1008.1.6(1)
VERTICAL CHANGES IN LEVEL

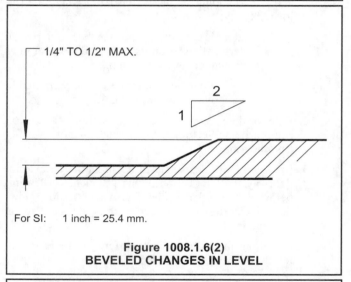

For SI: 1 inch = 25.4 mm.

Figure 1008.1.6(2)
BEVELED CHANGES IN LEVEL

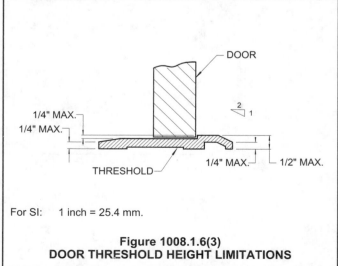

For SI: 1 inch = 25.4 mm.

Figure 1008.1.6(3)
DOOR THRESHOLD HEIGHT LIMITATIONS

1008.1.7 Door arrangement. Space between two doors in series shall be 48 inches (1219 mm) minimum plus the width of a door swinging into the space. Doors in series shall swing either in the same direction or away from the space between doors.

Exceptions:

1. The minimum distance between horizontal sliding power-operated doors in a series shall be 48 inches (1219 mm).

2. Storm and screen doors serving individual dwelling units in Groups R-2 and R-3 as applicable in Section 101.2 need not be spaced 48 inches (1219 mm) from the other door.

3. Doors within individual dwelling units in Groups R-2 and R-3 as applicable in Section 101.2 other than within Type A dwelling units.

❖ Door arrangement is required to be such that an occupant's use of a means of egress doorway is not hampered by the operation of a preceding door located in the same line of travel so that the occupant flow can be smooth through the openings. Successive doors in a single egress path (i.e., in series) can cause such interference. The 4-foot (1219 mm) clear distance between doors when the first door is open allows an occupant, including one in a wheelchair, to move past one door and its swing before beginning the operation of the next door [see Figure 1008.1.7(1)]. Note that where doors in series are not arranged in a straight line, the intent of the code is to provide sufficient space to enable occupants to negotiate the second door without being encumbered by the first door's swing arc. To facilitate accessibility, the space between doors should provide sufficient clear space for a wheelchair [30 inches by 48 inches (762 mm by 1219 mm)] beyond the arc of the door swing [see Figure 1008.1.7.(2)]. Additionally, the approach and ac-

cess provisions of ICC A117.1 should be considered.

The exception is to permit horizontal sliding power-operated doors (see Section 1008.1.3.2) to be designed with a lesser distance between them in a series arrangement because they are customarily designed to open simultaneously or in sequence such that movement through them is unhampered. Storm and screen doors on residential dwelling units need not be spaced at 48 inches (1219 mm) since it would be impractical. Doors within dwelling units of Group R-2 or R-3 that are not Type A dwelling units (see Chapter 11) are also permitted to have a lesser distance between doors, because the accessibility provisions do not apply.

1008.1.8 Door operations. Except as specifically permitted by this section egress doors shall be readily openable from the egress side without the use of a key or special knowledge or effort.

❖ When installed for security purposes, locks and latches can intentionally prohibit the use of an egress door and thus interfere with or prevent the egress of occupants at the time of a fire. While the security of property is important for many, the life safety of occupants is essential for everyone. Where security and life safety objectives conflict, alternative measures, such as those permitted by each of the exceptions in Section 1008.1.8.3, may be applicable.

Egress doors are permitted to be locked, but must be capable of being unlocked and readily openable from the side from which egress is to be made. The outside of a door can be key locked as long as the inside—the side from which egress is to be made—can be unlocked without the use of tools, keys or special knowledge or effort. For example, an unlocking operation that is inte-

4' MIN.

For SI: 1 foot = 304.8 mm.

Figure 1008.1.7(1)
SPACING OF DOORS IN SERIES

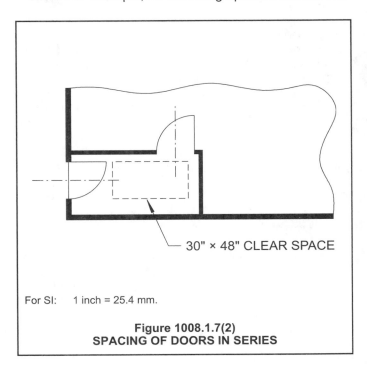

30" × 48" CLEAR SPACE

For SI: 1 inch = 25.4 mm.

Figure 1008.1.7(2)
SPACING OF DOORS IN SERIES

gral with an unlatching operation is acceptable.

Examples of special knowledge would be a combination lock or an unlocking device in an unknown, unexpected or hidden location. Special effort would dictate the need for unusual and unexpected physical ability to unlock or make the door fully available for egress.

Where a pair of egress door leafs is installed, with or without a center mullion, the general requirement is that each leaf must be provided with its own releasing or unlatching device so as to be readily openable. Door arrangements or devices that depend on the release of one door before the other can be opened are not to be used, except as permitted by Section 1008.1.8.4.

1008.1.8.1 Hardware. Door handles, pulls, latches, locks and other operating devices on doors required to be accessible by Chapter 11 shall not require tight grasping, tight pinching or twisting of the wrist to operate.

❖ Any doors that are located along an accessible route for ingress or egress must have door hardware that is easy to operate by a person with limited mobility. This would include all elements of the door hardware used in typical door operation, such as door levers, locks, security changes, etc. This requirement is also an advantage for persons with arthritis in their hands. Items such as small, full-twist thumb turns or smooth circular knobs are examples of hardware that is not acceptable.

1008.1.8.2 Hardware height. Door handles, pulls, latches, locks and other operating devices shall be installed 34 inches (864 mm) minimum and 48 inches (1219 mm) maximum above the finished floor. Locks used only for security purposes and not used for normal operation are permitted at any height.

❖ The requirements in this section place the door hardware at a level that is useable by most people, including a person in a wheelchair. The exception allows security locks to be placed at any height. An example would be an unframed glass door at the front door of a tenant space in a mall that has the lock near the floor level. The lock is only used when the store is not open for business. Such locks are not required for the normal operation of the door.

1008.1.8.3 Locks and latches. Locks and latches shall be permitted to prevent operation of doors where any of the following exists:

1. Places of detention or restraint.

2. In buildings in occupancy Group A having an occupant load of 300 or less, Groups B, F, M and S, and in churches, the main exterior door or doors are permitted to be equipped with key-operated locking devices from the egress side provided:

   2.1. The locking device is readily distinguishable as locked,

   2.2. A readily visible durable sign is posted on the egress side on or adjacent to the door stating: THIS DOOR TO REMAIN UNLOCKED WHEN BUILDING IS OCCUPIED. The sign shall be in letters 1 inch (25 mm) high on a contrasting background,

   2.3. The use of the key-operated locking device is revokable by the building official for due cause.

3. Where egress doors are used in pairs, approved automatic flush bolts shall be permitted to be used, provided that the door leaf having the automatic flush bolts has no doorknob or surface-mounted hardware.

4. Doors from individual dwelling or sleeping units of Group R occupancies having an occupant load of 10 or less are permitted to be equipped with a night latch, dead bolt or security chain, provided such devices are openable from the inside without the use of a key or tool.

❖ Where security and life safety objectives conflict, alternative measures, such as those permitted by each of the exceptions, may be applicable.

Exception 1 is needed for jails and prisons.

Exception 2 permits a locking device, such as a double-cylinder dead bolt, on the main entrance door. Such locking devices must have an integral indicator that automatically reflects the "locked" or "unlocked" status of the device. In addition to being an indicating lock, a sign must be provided that clearly states that the door is to be unlocked when the building is occupied. The sign on the door not only reminds employees to unlock the door, but also advises the public that an unacceptable arrangement exists if one finds the door locked. Ideally, the individual who encounters the locked door will notify management and possibly the building official. Note that the use of the key-locking device is revokable by the building official. The locking arrangement is not permitted on any door other than the main exit and, therefore, the employees, security and cleaning crews will have access to other exits without requiring the use of a key. This allowance is not limited just to multiple-exit buildings but also to small buildings with one exit.

In Exception 3, an automatic flush bolt device is one that is internal to the inactive leaf of a pair of doors. The device has a small "knuckle" that extends from the inactive leaf into the opening of the active leaf. When the active leaf is opened, the bolt is automatically retracted. When the active leaf is closed, the knuckle is pressed into the inactive leaf by the active leaf, extending the flush bolt(s) in the head or sill of the inactive leaf (see Figure 1008.1.8.3).

Automatic flush bolts on one leaf of a pair of egress doors are acceptable, provided the leaf with the automatic flush bolts is not equipped with a door knob or other hardware that would imply to the user that the door leaf is unlatched independently of the companion leaf.

Exception 4 addresses the need for security in residential units. The occupants are familiar with the operation of the indicated devices, which are intended to be relatively simple to operate without the use of a key or tool. Note that this exception only applies to the door from the dwelling unit.

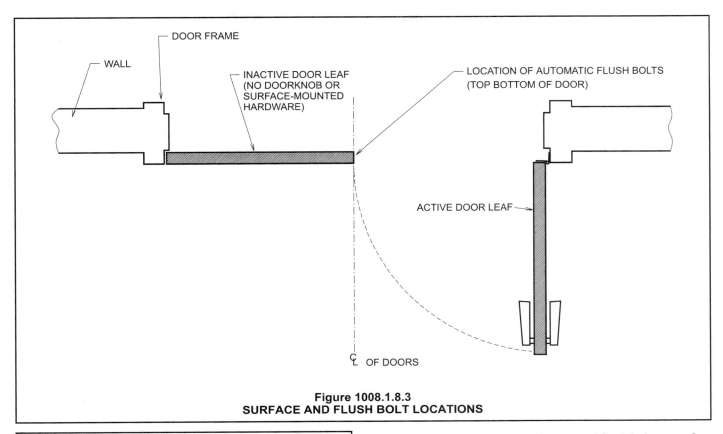

Figure 1008.1.8.3
SURFACE AND FLUSH BOLT LOCATIONS

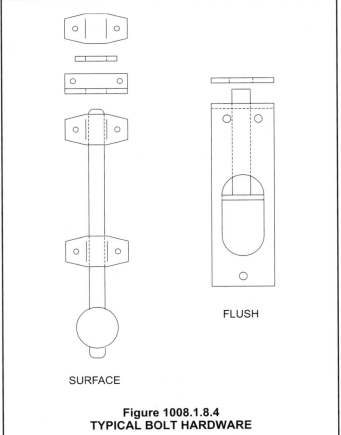

Figure 1008.1.8.4
TYPICAL BOLT HARDWARE

1008.1.8.4 Bolt locks. Manually operated flush bolts or surface bolts are not permitted.

Exceptions:

1. On doors not required for egress in individual dwelling units or sleeping units.

2. Where a pair of doors serves a storage or equipment room, manually operated edge- or surface-mounted bolts are permitted on the inactive leaf.

❖ This section is applicable to doors that are intended and required to be for means of egress purposes or are identified as a means of egress, such as by an "Exit" sign or other device. Doors, as well as a second leaf in a doorway that is provided for a purpose other than means of egress, such as for convenience or building operations, should be arranged or identified so as not to be mistaken as a means of egress.

This section prohibits installation of manually operated flush and surface bolts except in an individual dwelling or sleeping unit. Even then, such bolts may only be used on doors not required for egress (see Section 1008.1.8.3, Exception 4 for security of doors from individual dwelling and sleeping units). Flush and surface bolts represent locking devices that are difficult to operate because of their location and operation (see Figure 1008.1.8.4).

Exception 1 provides for edge-mounted or surface-mounted bolts on the inactive leaf of a pair of door(s) from these limited use areas.

1008.1.8.5 Unlatching. The unlatching of any leaf shall not require more than one operation.

> **Exception:** More than one operation is permitted for unlatching doors in the following locations:
>
> 1. Places of detention or restraint.
> 2. Where manually operated bolt locks are permitted by Section 1008.1.8.4.
> 3. Doors with automatic flush bolts as permitted by Section 1008.1.8.3, Exception 3.
> 4. Doors from individual dwelling units and guestrooms of Group R occupancies as permitted by Section 1008.1.8.3, Exception 4.

❖ The code prohibits the use of multiple locks or latching devices on a door, which would be a safety hazard in an emergency situation. The exceptions address locations where multiple locks or latching devices are accessible.

1008.1.8.6 Delayed egress locks. Approved, listed, delayed egress locks shall be permitted to be installed on doors serving any occupancy except Group A, E and H occupancies in buildings that are equipped throughout with an automatic sprinkler system in accordance with Section 903.3.1.1 or an approved automatic smoke or heat detection system installed in accordance with Section 907, provided that the doors unlock in accordance with Items 1 through 6 below. A building occupant shall not be required to pass through more than one door equipped with a delayed egress lock before entering an exit.

1. The doors unlock upon actuation of the automatic sprinkler system or automatic fire detection system.

2. The doors unlock upon loss of power controlling the lock or lock mechanism.

3. The door locks shall have the capability of being unlocked by a signal from the fire command center.

4. The initiation of an irreversible process which will release the latch in not more than 15 seconds when a force of not more than 15 pounds (67 N) is applied for 1 second to the release device. Initiation of the irreversible process shall activate an audible signal in the vicinity of the door. Once the door lock has been released by the application of force to the releasing device, relocking shall be by manual means only.

   > **Exception:** Where approved, a delay of not more than 30 seconds is permitted.

5. A sign shall be provided on the door located above and within 12 inches (305 mm) of the release device reading: PUSH UNTIL ALARM SOUNDS. DOOR CAN BE OPENED IN 15 [30] SECONDS.

6. Emergency lighting shall be provided at the door.

❖ For security reasons, special locking arrangements are permitted for doors in a means of egress serving occupancies other than those in Groups A, E and H. The arrangements are not permitted in assembly or educational occupancies because the resulting delay in egress is not acceptable given the greater possibility of mass panic. Such a delay would also be inconsistent with Section 1008.1.9, which requires the installation of panic hardware on doors in such uses. Also, the delay from Group H would be unreasonable given the potential for rapid fire buildup in such areas.

Because of the possible delay caused by the controlled locking device in the available use of the egress door, the building must be provided throughout with compensating fire protection features to promptly warn occupants of a fire condition. All of the listed conditions must be met in order to permit use of such a locking device.

Item 1 interconnects the lock with an automatic sprinkler system in accordance with NFPA 13 or, alternatively, an automatic fire detection system in accordance with Section 907 which is required to be installed throughout the building. Such systems are to provide occupants with an early warning of a fire event and thus additional time for egress. Note that the provision for an automatic fire detection system does not include the use of single- or multiple-station detectors. Also note that actuation of the automatic sprinkler or fire detection system is to unlock the control device so as to permit the egress door to be readily and immediately openable.

Item 2 is so that the control device is "fail safe." Since the operation of the device is dependent on electrical power, in the event of electrical power loss to the lock or locking mechanism, the egress doors must be readily and immediately openable from the side from which egress is to be made.

Item 3 specifies that the door must be capable of being manually unlocked by a signal sent from a fire command center. Personnel at that location are intended to be the first alerted to an emergency event and are expected to take appropriate action to unlock all egress doors equipped with special locking arrangements. This will permit the locks to be deactivated in some cases prior to the sprinklers or smoke detection system detecting the problem or in case of other nonfire emergencies such as an earthquake.

Item 4 specifies the operational characteristics of the locking control device, which is similar to a panic device. A user must apply a minimum 15-pound (7 kg) force to the release device for at least 1 second, at which time an audible alarm will sound and the device will automatically start to unlock the door. The 1-second duration is to prevent initiation of the unlocking process because of an inadvertent bump or accidental contact against the device. The unlocking cycle is irreversible; once it is started, it does not stop. Once the cycle starts, the door is required to be unlocked in no more than 15 seconds. When the door is unlocked at the end of the 15-second delay, it stays unlocked until someone comes to the door and manually relocks it. A method of automatically relocking the door from a remote location such as a central control station or security office is not permitted. Therefore, the first users to the door may face a delay, but after that other users would be able to exit immediately.

The exception will permit the building official to allow the time delay prior to opening to be increased beyond the basic 15 seconds but never to the point where the delay would exceed 30 seconds.

If a user continues to exert the 15-pound (7 kg) force for more than 1 second, the door is required to be openable after 15 seconds from the start of the force application.

The sign required by Item 5 informs the user of the type of unlocking device and that the door will become available for egress. An undesirable consequence of the door not being immediately unlocked is if the user assumes it will never be available and proceeds to another exit door.

Item 6 provides emergency lighting at the door so that the user can read the sign required by Item 5.

1008.1.8.7 Stairway doors. Interior stairway means of egress doors shall be openable from both sides without the use of a key or special knowledge or effort.

Exceptions:

1. Stairway discharge doors shall be openable from the egress side and shall only be locked from the opposite side.

2. This section shall not apply to doors arranged in accordance with Section 403.12.

3. In stairways serving not more than four stories, doors are permitted to be locked from the side opposite the egress side, provided they are openable from the egress side.

❖ Based on adverse fire experience where occupants have become trapped in smoke-filled stairway enclosures, generally stairway doors must be arranged to permit reentry into the building without the use of any tools, keys or special knowledge or effort. For security reasons, this restriction does not apply to the discharge door from the stairway enclosure—which is often to the outside. Section 403 for high-rise buildings permits the locking of the doors from the stairway side, provided the doors are capable of being unlocked from a fire command station and there is a communication system within the stairway enclosure that allows contact with the fire command station. It would be reasonable to permit this arrangement in buildings other than high-rise buildings.

Exception 3 addresses the need for security. The exception is limited to four-story buildings to provide a short travel distance to the stairway discharge door.

1008.1.9 Panic and fire exit hardware. Where panic and fire exit hardware is installed, it shall comply with the following:

1. The actuating portion of the releasing device shall extend at least one-half of the door leaf width.

2. A maximum unlatching force of 15 pounds (67 N).

Each door in a means of egress from an occupancy of Group A or E having an occupant load of 100 or more and any occupancy of Group H-1, H-2, H-3 or H-5 shall not be provided with a latch or lock unless it is panic hardware or fire exit hardware.

If balanced doors are used and panic hardware is required, the panic hardware shall be the push-pad type and the pad shall not extend more than one-half the width of the door measured from the latch side.

❖ As its name implies, panic hardware is special unlatching and unlocking hardware that is intended to simplify the unlatching and unlocking operation to a single 15-pound (7 kg) force applied in the direction of egress (see Figure 1008.1.9). In a panic situation with a rush of persons trying to utilize a door, the conventional devices, such as doorknobs or thumb turns, may cause sufficient delay so as to create a crush at the door and prevent or slow the opening operation.

In addition to hazardous occupancies, panic hardware is required on all doors that provide means of egress to rooms and spaces of assembly and educational (Groups A and E) occupancies with an occupant load of 100 or more. These uses are characterized by higher occupant load densities. Whereas doors from an assembly or educational room with an occupant load of less than 100 do not require panic hardware, a door that provides means of egress for two such rooms could require panic hardware when the combination of spaces has a total occupant load of 100 or more.

The locational specifications for the activating panel or bar are based on ready availability and access to the unlatching device. Note that the section requires the activating portion to extend at least one-half the width of the door leaf.

Where a fire door, such as to an exit stairway, is required to be equipped with panic hardware, the hardware must accomplish the dual objectives of panic hardware and continuity of the enclosure in which it is located. In this case, fire exit hardware is to be provided that meets both objectives and requirements, since panic hardware is not tested for use on fire doors.

There are standard test procedures designed to evaluate the performance of panic and fire exit hardware from the panic standpoint as well as from a fire protection standpoint.

The provisions for balanced doors ensure that the occupants push on the latch side of the door, which will swing in the direction of egress travel.

1008.2 Gates. Gates serving the means of egress system shall comply with the requirements of this section. Gates used as a component in a means of egress shall conform to the applicable requirements for doors.

Exception: Horizontal sliding or swinging gates exceeding the 4-foot (1219 mm) maximum leaf width limitation are permitted in fences and walls surrounding a stadium.

❖ This section specifies that all of the requirements for doors also apply to gates, except gates that surround a stadium are allowed to exceed 4 feet (1219 mm) in width. Usually a large gate is required to adequately serve a stadium crowd for egress purposes.

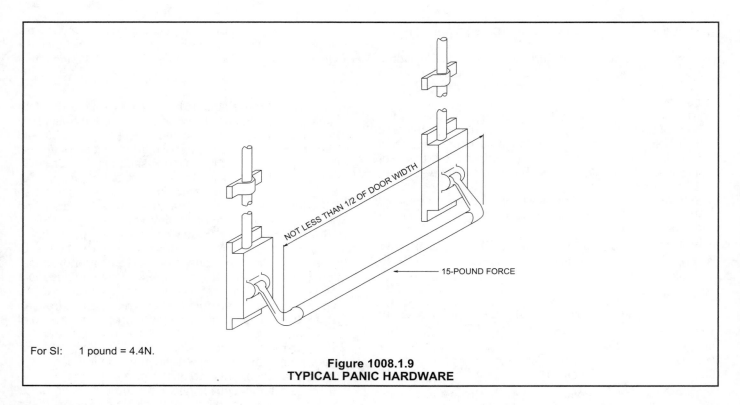

For SI: 1 pound = 4.4N.

Figure 1008.1.9
TYPICAL PANIC HARDWARE

1008.2.1 Stadiums. Panic hardware is not required on gates surrounding stadiums where such gates are under constant immediate supervision while the public is present, and further provided that safe dispersal areas based on 3 square feet (0.28 m²) per occupant are located between the fence and enclosed space. Such required safe dispersal areas shall not be located less than 50 feet (15 240 mm) from the enclosed space. See Section 1017 for means of egress from safe dispersal areas.

❖ Panic hardware is impractical for large gates that surround stadiums. Normally these gates are opened and closed by the stadium's ground crew, which is constantly in attendance during their use. The safe dispersal area requirement provides for the safety of the crowd if for some reason the gate is not open. The safe dispersal area is to be between the stadium enclosure and the surrounding fence and the area to be occupied is not to be closer than 50 feet (15 240 mm) to the stadium enclosure.

1008.3 Turnstiles. Turnstiles or similar devices that restrict travel to one direction shall not be placed so as to obstruct any required means of egress.

   Exception: Each turnstile or similar device shall be credited with no more than a 50-person capacity where all of the following provisions are met:

   1. Each device shall turn free in the direction of egress travel when primary power is lost, and upon the manual release by an employee in the area.

   2. Such devices are not given credit for more than 50 percent of the required egress capacity.

   3. Each device is not more than 39 inches (991 mm) high.

   4. Each device has at least 16.5 inches (419 mm) clear width at and below a height of 39 inches (991 mm) and at least 22 inches (559 mm) clear width at heights above 39 inches (991 mm).

   Where located as part of an accessible route, turnstiles shall have at least 36 inches (914 mm) clear at and below a height of 34 inches (864 mm), at least 32 inches (813 mm) clear width between 34 inches (864 mm) and 80 inches (2032 mm) and shall consist of a mechanism other than a revolving device.

❖ This section provides for a limited use of turnstiles to serve as a means of egress component. The exception to this section limits each turnstile to a maximum of 50 persons of egress capacity. The turnstile must comply with all four listed items to be considered serving any part of the occupant load means of egress. The turnstiles must rotate freely both when there is a loss of power and when they are manually released. Note that the 50-person limit applies to each individual turnstile.

1008.3.1 High turnstile. Turnstiles more than 39 inches (991 mm) high shall meet the requirements for revolving doors.

❖ Where a turnstile is higher than 39 inches (991 mm), the restriction to egress is much like a revolving door. Thus, the egress limitations for revolving doors in Section 1008.1.3.1 apply to this type of turnstile. If a high turnstile does not meet the revolving door requirements for doors that are an egress component, it is not to be included as serving a portion of the means of egress. It would be necessary to provide doors in these areas for egress.

1008.3.2 Additional door. Where serving an occupant load greater than 300, each turnstile that is not portable shall have a side-hinged swinging door which conforms to Section 1008.1 within 50 feet (15 240 mm).

❖ This section addresses a common egress condition for sports areas where a number of turnstiles are installed for ticket taking. Portable turnstiles are moved from the egress path for proper exiting capacity. Permanent turnstiles are not considered as providing any of the required egress capacity when serving an occupant load greater than 300, no matter how many turnstiles are installed. Doors are required to provide occupants with a path of egress other than through the turnstiles. The doors are to be located within 50 feet (15 240 mm) of the turnstiles.

SECTION 1009
STAIRWAYS AND HANDRAILS

1009.1 Stairway width. The width of stairways shall be determined as specified in Section 1005.1, but such width shall not be less than 44 inches (1118 mm). See Section 1007.3 for accessible means of egress stairways.

Exceptions:

1. Stairways serving an occupant load of 50 or less shall have a width of not less than 36 inches (914 mm).

2. Spiral stairways as provided for in Section 1009.9.

3. Aisle stairs complying with Section 1024.

4. Where a stairway lift is installed on stairways serving occupancies in Group R-3, or within dwelling units in occupancies in Group R-2, both as applicable in Section 101.2, a clear passage width not less than 20 inches (508 mm) shall be provided. If the seat and platform can be folded when not in use, the distance shall be measured from the folded position.

❖ To provide adequate space for occupants traveling in opposite directions and to permit the intended full egress capacity to be developed, minimum dimensions are dictated for means of egress stairways. A minimum width of 44 inches (1118 mm) is required for stairway construction to permit two columns of users to travel in the same or opposite directions. The reference to Section 1005.1 is for the determination of stairway width based on occupant load. The larger of the two widths is to be used.

Exception 1 recognizes the relatively small occupant loads of 50 or less that permit a staggered file of users when traveling in the same direction. When traveling in opposite directions, one column of users must stop their ascent (or descent) to permit the opposite column to continue. Again, considering the relatively small occupant loads, any disruption of orderly flow will be infrequent.

Exception 2 permits a spiral stairway to have a minimum width of 26 inches (660 mm) when it conforms to

Section 1009.9, on the basis that the configuration of a spiral stairway is such that anything other than single-file travel is unlikely.

Exception 3 provides for the aisle and aisle stair widths that are specified in Section 1024.

Exception 4 addresses the use of stairway lifts for individual dwelling units. Stairway lifts are typically used by mobility-impaired people in their homes. The code and ASME A 18.1 call for a minimum 20 inches (508 mm) of clear passageway to be maintained on a stairway where a stairway lift is located. If a portion of the stairway lift, such as a seat, can be folded, the minimum clear dimension is to be measured from the folded position. If the lift cannot be folded, then the 20 inches (508 mm) is measured from the fixed position. The track for these lifts typically extends 9 to 12 inches (229 to 305 mm) from the wall, making the 20-inch (508 mm) clear measurement actually 24 to 27 inches (610 to 686 mm) from the edge of the track.

1009.2 Headroom. Stairways shall have a minimum headroom clearance of 80 inches (2032 mm) measured vertically from a line connecting the edge of the nosings. Such headroom shall be continuous above the stairway to the point where the line intersects the landing below, one tread depth beyond the bottom riser. The minimum clearance shall be maintained the full width of the stairway and landing.

Exception: Spiral stairways complying with Section 1009.9 are permitted a 78-inch (1981 mm) headroom clearance.

❖ This requirement is necessary to avoid an obstruction to orderly flow and to provide visibility to the users so that the desired path of travel can be planned and negotiated. Height is a vertical measurement above every point along the stairway stepping and walking surfaces, with minimum height measured vertically from the tread nosing or from the surface of a landing or platform up to the ceiling (see Figure 1009.2). The exception for a clear headroom of 6 feet 6 inches (1981 mm) for spiral stairs correlates with the provisions of Section 1009.9.

1009.3 Stair treads and risers. Stair riser heights shall be 7 inches (178 mm) maximum and 4 inches (102 mm) minimum. Stair tread depths shall be 11 inches (279 mm) minimum. The riser height shall be measured vertically between the leading edges of adjacent treads. The greatest riser height within any flight of stairs shall not exceed the smallest by more than 0.375 inch (9.5 mm). The tread depth shall be measured horizontally between the vertical planes of the foremost projection of adjacent treads and at right angle to the tread's leading edge. The greatest tread depth within any flight of stairs shall not exceed the smallest by more than 0.375 inch (9.5 mm). Winder treads shall have a minimum tread depth of 11 inches (279 mm) measured at a right angle to the tread's leading edge at a point 12 inches (305 mm) from the side where the treads are narrower and a minimum tread depth of 10 inches (254 mm). The greatest winder tread depth at the 12-inch (305 mm) walk line within any

FIGURE 1009.2   **MEANS OF EGRESS**

flight of stairs shall not exceed the smallest by more than 0.375 inch (9.5 mm).

Exceptions:

1. Circular stairways in accordance with Section 1009.7.

2. Winders in accordance with Section 1009.8.

3. Spiral stairways in accordance with Section 1009.9.

4. Aisle stairs in assembly seating areas where the stair pitch or slope is set, for sightline reasons, by the slope of the adjacent seating area in accordance with Section 1024.11.2.

5. In occupancies in Group R-3, as applicable in Section 101.2, within dwelling units in occupancies in Group R-2, as applicable in Section 101.2, and in occupancies in Group U, which are accessory to an occupancy in Group R-3, as applicable in Section 101.2, the maximum riser height shall be 7.75 inches (197 mm) and the minimum tread depth shall be 10 inches (254 mm), the minimum winder tread depth at the walk line shall be 10 inches (254 mm), and the minimum winder tread depth shall be 6 inches (152 mm). A nosing not less than 0.75 inch (19.1 mm) but not more than 1.25 inches (32 mm) shall be provided on stairways with solid risers where the tread depth is less than 11 inches (279 mm).

6. See the *International Existing Building Code* for the replacement of existing stairways.

❖ The provisions for treads and risers not only determine the slope of a stairway, but also contribute to the efficient use of the stairway by facilitating smooth and orderly travel.

The riser height—the vertical dimension from tread surface to tread surface or tread surface to landing surface—is typically limited to no more than 7 inches (178 mm) nor less than 4 inches (102 mm). The minimum tread depth—the horizontal distance from the leading edge (nosing) of one tread to the leading edge (nosing) of the next adjacent tread—is typically limited to no less than 11 inches (279 mm) [see Figure 1009.3(1)]. The minimum tread depth of 11 inches (279 mm) is intended to accommodate the largest shoe size found in 95 percent of the adult population, allowing for an appropriate overhang of the foot beyond the tread nosing while descending a stairway. Tread depths under 11 inches (279 mm) could cause an abnormal overhang (depending on the size of the foot) and could force users to descend a stairway with their feet pointing sideways in a crab-like manner. Such a condition does not promote orderly stairway travel. Based on research of geometrical possibilities of adequate foot placement, the rate of misstep with various step sizes and consideration for user's comfort and energy expenditure, it was found that the 11-inch (279 mm) minimum tread depth and maximum 7-inch (178 mm) riser height resulted in the best proportions for stairway construction.

The size for a winder tread is also considered for proper foot placement along the walking line [see Figure 1009.3(3)]. The dimensional requirements are consistent with the straight tread.

A minimum riser height of 4 inches (102 mm) is necessary to enable the user to identify visually the presence of the riser in ascent or descent.

The precise location of tread depth and riser measurements is to be perpendicular to the tread's or riser's leading edge or nosing. This is to duplicate the user's anticipated foot placement in traveling the stairway [see Figure 1009.3(2)].

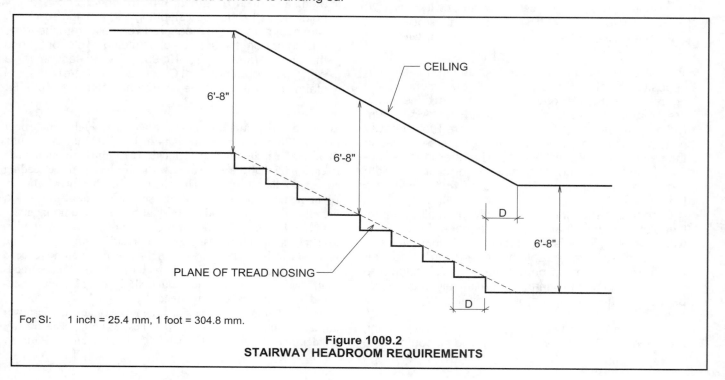

For SI: 1 inch = 25.4 mm, 1 foot = 304.8 mm.

Figure 1009.2
STAIRWAY HEADROOM REQUIREMENTS

The maximum variance between the minimum and maximum riser height along a run of a stairway is $^3/_8$ inch (9.5 mm). A cadence is established as a person moves up or down a stairway. A change of more than $^3/_8$ inch (9.5 mm) along a run would become a tripping hazard. A $^3/_8$-inch (9.5 mm) maximum difference between the tread depth of a winder-type stairway is also based on an expected foot placement once a cadence is established.

The exceptions apply only to the extent of the text of each exception. For example, the entire text of Section 1009.3 is set aside for spiral stairways conforming to Section 1009.9 (see Exception 3). However, Exception 5 allows a different maximum riser and minimum tread under limited conditions, but retains the minimum riser height and measurement method of Section 1009.3.

At this time, the requirements for dimensional uniformity are found in Sections 1009.3 and 1009.3.1.

Exceptions 1, 2 and 3 are discussed in Sections 1009.7, 1009.8 and 1009.9, respectively.

Exception 4 provides a practical exception where assembly facilities are designed for viewing. See Sections 1024.11 through 1024.11.3 for assembly aisle stair limiting dimensions.

Exception 5 allows revisions to the standard 7-11 riser/tread requirements for residential occupancies of Group R-3, within dwelling units of Group R-2 and U

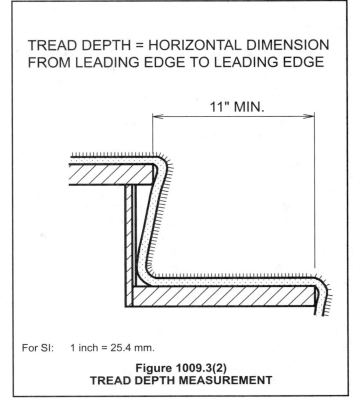

TREAD DEPTH = HORIZONTAL DIMENSION FROM LEADING EDGE TO LEADING EDGE

11" MIN.

For SI: 1 inch = 25.4 mm.

Figure 1009.3(2)
TREAD DEPTH MEASUREMENT

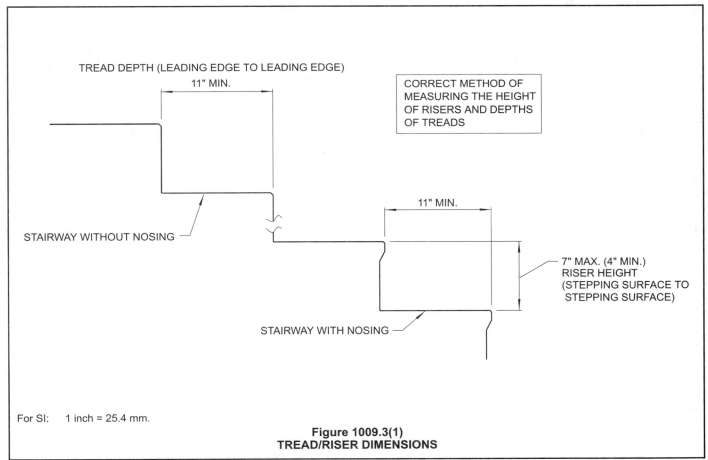

TREAD DEPTH (LEADING EDGE TO LEADING EDGE)
11" MIN.

STAIRWAY WITHOUT NOSING

CORRECT METHOD OF MEASURING THE HEIGHT OF RISERS AND DEPTHS OF TREADS

11" MIN.

7" MAX. (4" MIN.) RISER HEIGHT (STEPPING SURFACE TO STEPPING SURFACE)

STAIRWAY WITH NOSING

For SI: 1 inch = 25.4 mm.

Figure 1009.3(1)
TREAD/RISER DIMENSIONS

buildings where such structures are accessory to a Group R-3 occupancy (such as barns or detached garages). This increase is allowed because of the low occupant load and the high degree of occupant familiarity with the stairways. When this exception is taken for stairways that have solid risers, each tread is required to have a nosing with a minimum dimension of $^3/_4$ inch (19.1 mm) and maximum dimension of $1^1/_4$ inches (32 mm), where the tread depth is less than 11 inches (279 mm). Nosings are not required for residential stairs with open risers and 10-inch (254 mm) treads. The nosing provides a greater stepping surface for those ascending the stairway [see Figure 1009.3(1)].

Exception 6 allows for the replacement of an existing stair. Where a change of occupancy would require compliance with current standards, this exception allows a stairway that may be steeper than that permitted, provided it does not constitute a hazard [see the International Existing Building Code® (IEBC®)].

1009.3.1 Dimensional uniformity. Stair treads and risers shall be of uniform size and shape. The tolerance between the largest and smallest riser or between the largest and smallest tread shall not exceed 0.375 inch (9.5 mm) in any flight of stairs.

Exceptions:

1. Nonuniform riser dimensions of aisle stairs complying with Section 1024.11.2.

2. Consistently shaped winders, complying with Section 1009.8, differing from rectangular treads in the same stairway flight.

Where the bottom or top riser adjoins a sloping public way, walkway or driveway having an established grade and serving as a landing, the bottom or top riser is permitted to be reduced along the slope to less than 4 inches (102 mm) in height with the variation in height of the bottom or top riser not to exceed one unit vertical in 12 units horizontal (8-percent slope) of stairway

width. The nosings or leading edges of treads at such nonuniform height risers shall have a distinctive marking stripe, different from any other nosing marking provided on the stair flight. The distinctive marking stripe shall be visible in descent of the stair and shall have a slip-resistant surface. Marking stripes shall have a width of at least 1 inch (25 mm) but not more than 2 inches (51 mm).

❖ Dimensional uniformity in the design and construction of interior stairways contributes to safe stairway use. When ascending or especially descending a stair, a user sets a natural cadence or rhythmic movement based on the unconscious expectation or "feel" that each step taken will be at the same height and will land in approximately the same balanced position on the tread as the previous steps in the pattern. Any substantial change in tread or riser dimensions in a flight of stairways can break the rhythm and cause a misstep, stumbling or undue physical strain that may result in a fall or serious injury. In emergency situations, building occupants tend to use stairways at a faster pace than under normal conditions, increasing the risk to the user. Therefore, this section limits the dimensional variations to a tolerance of $^3/_8$ inch (9.5 mm) between the largest and smallest riser or tread dimension in a flight of stairs. A "flight" of stairs is commonly defined as the stairs between landings.

For special conditions of construction and as a practical matter, this section allows some greater variations in stairway tread and riser dimensions than the general limitations specified above. Exception 1 addresses conditions where the seating in assembly facilities is on a sloping gradient (for sightline purposes) and the aisle stairs become an integral part of the arrangement. Exception 2 addresses winder treads, which must be consistent along the walking line (see Figure 1009.3.1). At this time, the requirements for dimensional uniformity are found in Sections 1009.3 and 1009.3.1. This section

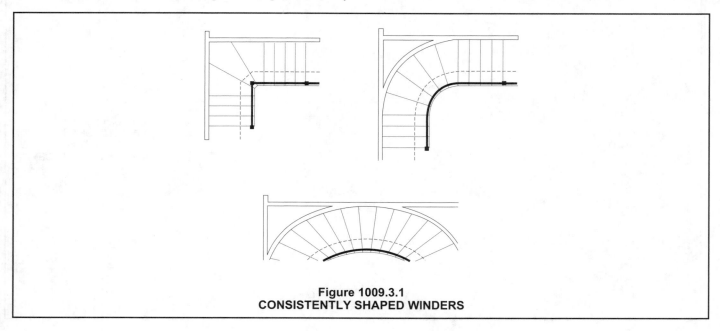

Figure 1009.3.1
CONSISTENTLY SHAPED WINDERS

also addresses the situation where the bottom riser of a flight of stairways meets a sloped landing such as a public way, walk or driveway.

1009.3.2 Profile. The radius of curvature at the leading edge of the tread shall be not greater than 0.5 inch (12.7 mm). Beveling of nosings shall not exceed 0.5 inch (12.7 mm). Risers shall be solid and vertical or sloped from the underside of the leading edge of the tread above at an angle not more than 30 degrees (0.52 rad) from the vertical. The leading edge (nosings) of treads shall project not more than 1.25 inches (32 mm) beyond the tread below and all projections of the leading edges shall be of uniform size, including the leading edge of the floor at the top of a flight.

Exceptions:

1. Solid risers are not required for stairways that are not required to comply with Section 1007.3, provided that the opening between treads does not permit the passage of a sphere with a diameter of 4 inches (102 mm).

2. Solid risers are not required for occupancies in Group I-3.

❖ The profiles of treads and risers contribute to stairway safety. The maximum radius of curvature at the leading edge of the tread is intended to allow descending foot placement on a surface that does not pitch the foot forward or allow the ball of the foot to slide off the treads. If a stairway design uses a beveled nosing configuration, the bevel is limited to a depth of ½ inch (12.7 mm). Solid risers with either no nosing or a slope at the underside of the nosing are required so that the user's foot does not catch while ascending the stairway [see Figure 1009.3.2(1)].

Exception 1 establishes that open risers on stairs are allowed where the accessible means of egress stairway provisions of Section 1007.3 do not apply. The maximum radius for the leading edge, however, is still required. Since the opening limitation in guards is 4 inches (102 mm), even where the riser is allowed to be open, the opening is limited to be consistent with the requirements for guards [see Figure 1009.3.2(2)].

Open risers are not permitted where accessibility is required based primarily on the difficulty open risers pose to people with sight or mobility impairments. In fact, ICC A117.1 specifically prohibits open risers.

Exception 2 recognizes that open risers are commonly used for stairs in occupancies such as detention facilities for practical reasons. Open risers provide a greater degree of security and supervision due to the fact that people cannot effectively conceal themselves behind the stair. The opening limitations of Exception 1 are not applicable to these stairs.

1009.4 Stairway landings. There shall be a floor or landing at the top and bottom of each stairway. The width of landings shall not be less than the width of stairways they serve. Every landing shall have a minimum dimension measured in the direction of travel equal to the width of the stairway. Such dimension need not exceed 48 inches (1219 mm) where the stairway has a straight run.

Exceptions:

1. Aisle stairs complying with Section 1024.

2. Doors opening onto a landing shall not reduce the landing to less than one-half the required width. When fully open, the door shall not project more than 7 inches (178 mm) into a landing.

❖ A level portion of a stairway provides users with a place to rest in their ascent or descent, to enter a stairway and to adjust their gait before continuing. Landings also

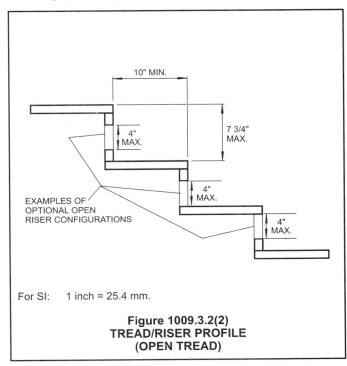

1/2" BEVEL MAX.

SOLID RISER

30° MAX.

1/2" RADIUS MAX.

1 1/4" MAX.
(3/4" MIN. FOR
RESIDENTIAL)
SEE EXCEPTION 5
TO SECTION 1009.3

30° MAX.

1 1/4" MAX.

For SI: 1 inch = 25.4 mm, 1 degree = 0.01745 rad.

**Figure 1009.3.2(1)
TREAD/RISER PROFILE
(SOLID TREAD)**

10" MIN.

4" MAX.

7 3/4" MAX.

4" MAX.

EXAMPLES OF
OPTIONAL OPEN
RISER CONFIGURATIONS

4" MAX.

For SI: 1 inch = 25.4 mm.

**Figure 1009.3.2(2)
TREAD/RISER PROFILE
(OPEN TREAD)**

FIGURE 1009.4(1) – FIGURE 1009.4(2)

MEANS OF EGRESS

break up the run of a stairway that continues for a considerable distance to arrest falls that may occur (see Section 1009.6).

The minimum size (width and depth) of all landings in a stairway is determined by the actual width of the stairway. If Section 1009.1 requires a stairway to have a width of at least 44 inches (1118 mm) and the stairway is constructed with that minimum width, then all landings serving that stairway must be at least 44 inches (1118 mm) wide and 44 inches (1118 mm) deep [see Figure 1009.4(1) and 1009.4(3)]. If a stairway is constructed wider than required, landings must increase accordingly

so as to not create a bottleneck situation in the egress travel. However, when a stairway is configured so that it has a straight run, the minimum dimension of the landing between flights in the direction of egress travel is not required to exceed 48 inches (1219 mm), even though the actual width of the stair may exceed 48 inches (1219 mm) [see Figure 1009.4(2)].

It is not the intent of this section to require that a stairway landing be shaped as a square or rectangle. A landing shaped as a half-circle would be permitted, as long as the landing provides an area described by an arc with a radius equal to the actual stairway width [see Figure

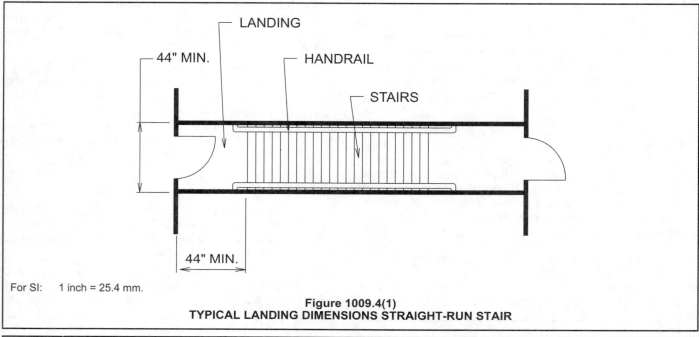

For SI: 1 inch = 25.4 mm.

Figure 1009.4(1)
TYPICAL LANDING DIMENSIONS STRAIGHT-RUN STAIR

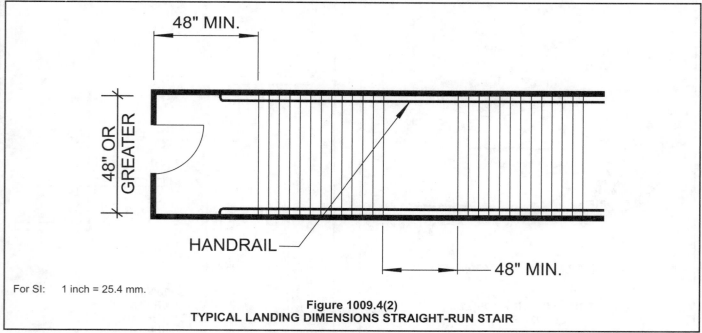

For SI: 1 inch = 25.4 mm.

Figure 1009.4(2)
TYPICAL LANDING DIMENSIONS STRAIGHT-RUN STAIR

1009.4(3)]. In this case, the space necessary for means of egress will be available.

Exception 1 provides for aisle stairs where the requirements are in Section 1024 and this section does not apply. Exception 2 limits the arc of the door swing on a landing so that the effect on the means of egress is minimized [see Figure 1009.4(4)].

1009.5 Stairway construction. All stairways shall be built of materials consistent with the types permitted for the type of construction of the building, except that wood handrails shall be permitted for all types of construction.

❖ In keeping with the different levels of fire protection provided by each of the five basic types of construction

designated in Chapter 6, the materials used for stairway construction must meet the appropriate combustibility/noncombustibility requirements indicated in Section 602 for the particular type of construction of the building in which the stairway is located. This is required whether or not the stair is part of the required means of egress.

If desired, wood handrails may be used on the basis that the fuel load contributed by this combustible component of stairway construction is insignificant and will not pose a fire hazard.

1009.5.1 Stairway walking surface. The walking surface of treads and landings of a stairway shall not be sloped steeper than one unit vertical in 48 units horizontal (2-percent slope) in any

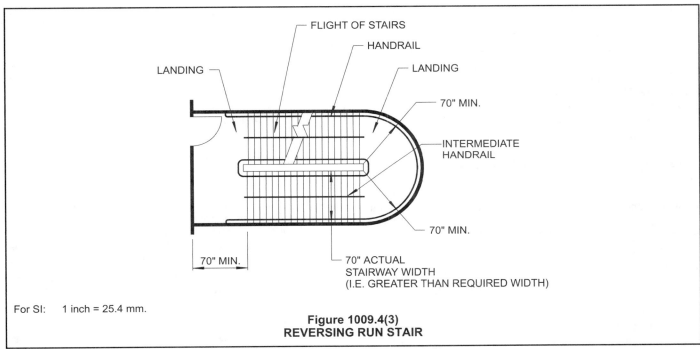

For SI: 1 inch = 25.4 mm.

Figure 1009.4(3)
REVERSING RUN STAIR

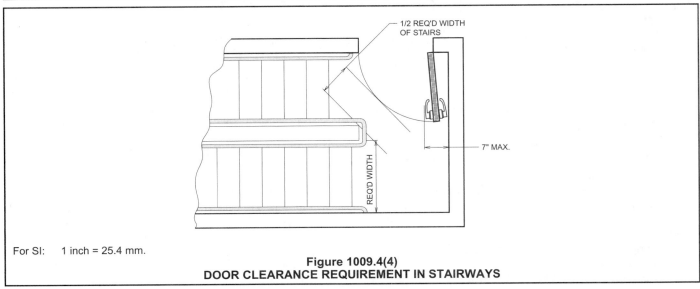

For SI: 1 inch = 25.4 mm.

Figure 1009.4(4)
DOOR CLEARANCE REQUIREMENT IN STAIRWAYS

direction. Stairway treads and landings shall have a solid surface. Finish floor surfaces shall be securely attached.

Exception: In Group F, H and S occupancies, other than areas of parking structures accessible to the public, openings in treads and landings shall not be prohibited provided a sphere with a diameter of $1^1/_8$ inches (29 mm) cannot pass through the opening.

❖ It is the intent of this section that both landing and stair treads be solid.

The exception permits the use of open grate-type material for stairway treads and landings in factory, industrial, storage and high-hazard occupancies. This provision is intended to apply primarily to stairs that provide access to areas not required to be accessible, such as pits, catwalks, tanks, equipment platforms, roofs or mezzanines. Walking surfaces with limited-size openings are typically used because open grate-type material is less susceptible to accumulation of dirt, debris or moisture as well as being more resistant to corrosion. Most commercially available grate material is manufactured with a maximum nominal 1-inch (25 mm) opening, therefore, the limitation that the opening not allow the passage of a sphere of $1^1/_8$-inch (29 mm) diameter allows the use of most material as well as accounting for manufacturing tolerances.

1009.5.2 Outdoor conditions. Outdoor stairways and outdoor approaches to stairways shall be designed so that water will not accumulate on walking surfaces. In other than occupancies in Group R-3, and occupancies in Group U that are accessory to an occupancy in Group R-3, treads, platforms and landings that are part of exterior stairways in climates subject to snow or ice shall be protected to prevent the accumulation of same.

❖ Outdoor stairways and approaches to stairways are to be constructed with a slope that complies with Section 1009.5.1 or are required to be protected such that walking surfaces do not accumulate water.

Where exterior stairways are used in moderate or severe climates, they must be protected from accumulations of snow and ice to provide a safe path of egress travel at all times, including winter. Typical methods for protecting these egress elements include roof overhangs or canopies, heated slab and, when approved by the building official, a reliable snow removal maintenance program.

1009.6 Vertical rise. A flight of stairs shall not have a vertical rise greater than 12 feet (3658 mm) between floor levels or landings.

Exception: Aisle stairs complying with Section 1024.

❖ Between landings and platforms, the vertical rise is to be measured from one landing walking surface to another (see Figure 1009.6). The limited height provides a reasonable interval for users with physical limitations to rest on a level surface and also serves to alleviate potential negative psychological effects of long and uninterrupted stairway flights.

The exception provides for aisle stairs in assembly

occupancies that are regulated by Section 1024 and not by this section.

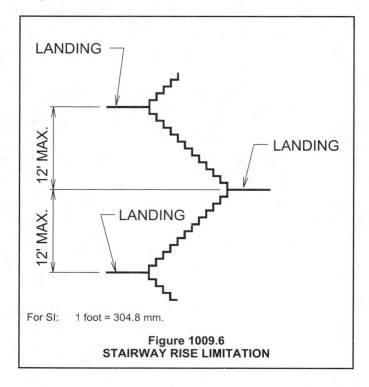

For SI: 1 foot = 304.8 mm.

Figure 1009.6
STAIRWAY RISE LIMITATION

1009.7 Circular stairways. Circular stairways shall have a minimum tread depth and a maximum riser height in accordance with Section 1009.3 and the smaller radius shall not be less than twice the width of the stairway. The minimum tread depth measured 12 inches (305 mm) from the narrower end of the tread shall not be less than 11 inches (279 mm). The minimum tread depth at the narrow end shall not be less than 10 inches (254 mm).

Exception: For occupancies in Group R-3, and within individual dwelling units in occupancies in Group R-2, both as applicable in Section 101.2.

❖ Circular stairway construction consists of a series of tapered treads that form a stairway with a circular configuration. The commentary to Section 1009.8 regarding the eccentricity of movement on stairways with winders also applies to circular stairways. This type of stairway is only allowed to be used as a component of a means of egress when tread and riser dimensions meet the requirements of Section 1009.3. This means that tapered treads must have a minimum depth of 11 inches (279 mm) measured 12 inches (305 mm) in from the side of the stairway having the shorter radius and that risers are not to exceed 7 inches (178 mm) in height. This section also requires that the shorter radius must be equal to or greater than twice the actual width of the stairway (see Figure 1009.7). If these minimum dimensions are not met, the stairway would be considered a stairway with winders and be subject to the requirements of Section 1009.8.

This exception provides for less stringent requirements for residential units where the occupants are familiar with the circular stair arrangement.

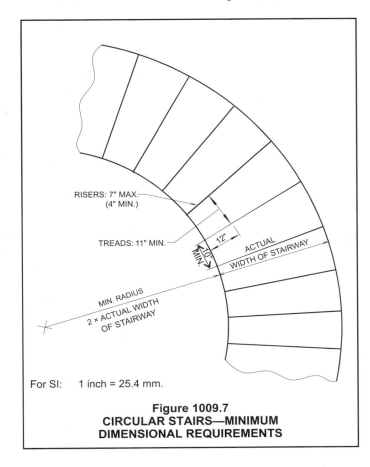

For SI: 1 inch = 25.4 mm.

**Figure 1009.7
CIRCULAR STAIRS—MINIMUM
DIMENSIONAL REQUIREMENTS**

1009.8 Winders. Winders are not permitted in means of egress stairways except within a dwelling unit.

❖ This section specifies the minimum dimensional requirements for the construction of stairway winders. Winders are used to form a bend in a flight of stairways to change the direction of the run.

The risk of injury in the use of stairways constructed with winders is greater than for stairways constructed as straight runs. This is particularly true in emergency situations where the rate of travel up or down a stairway is increased from the pace set under normal conditions of stairway use.

The employment of winders in stairway construction creates a special hazard because of the tapered configuration of the treads. For example, a person descending a straight flight of stairs will set up a natural cadence or rhythmic movement. However, in a stairway constructed with winders, the rhythmic movement of descent is suddenly disturbed when the section of stairway with the winders is reached. Because of the tapered treads, the horizontal distance traveled by each of the footsteps nearest the radial center of the winding section is necessarily shorter than the distance that

must be traveled by each if the footsteps are nearest the periphery or outer edge of the stairway (see Figure 1009.8). This condition sets up an eccentric movement. The hazard is further amplified because the inner footsteps (nearest to the radial center of the turn) must land on those portions of the tapered treads that are smaller in depth than the portions receiving the outer footsteps.

Because of the inherent dangers of stairways with winders, this section prohibits winders except for stairways serving a single dwelling unit. This section does not prohibit winders from being used in stairways that are not a required means of egress.

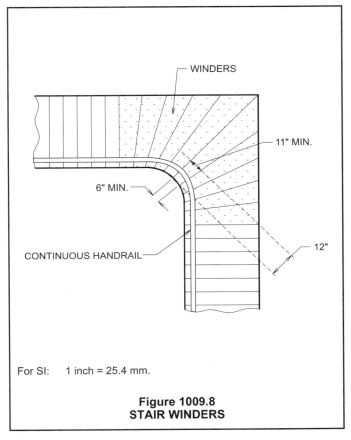

For SI: 1 inch = 25.4 mm.

**Figure 1009.8
STAIR WINDERS**

1009.9 Spiral stairways. Spiral stairways are permitted to be used as a component in the means of egress only within dwelling units or from a space not more than 250 square feet (23 m²) in area and serving not more than five occupants, or from galleries, catwalks and gridirons in accordance with Section 1014.6.

A spiral stairway shall have a 7.5-inch (191 mm) minimum clear tread depth at a point 12 inches (305 mm) from the narrow edge. The risers shall be sufficient to provide a headroom of 78 inches (1981 mm) minimum, but riser height shall not be more than 9.5 inches (241 mm). The minimum stairway width shall be 26 inches (660 mm).

❖ Spiral stairways are generally constructed with a fixed center pole which serves as either the primary or the only means of support from which pie-shaped treads radiate to form a winding stairway.

The commentary to Section 1009.8 regarding the eccentricity of movement on stairways with winders also applies to spiral stairways. The nature of stairway construction is such that it does not serve well when used in emergencies that require immediate evacuation, nor does a spiral stairway configuration permit the handling of a large occupant load in an efficient and safe manner. Therefore, this section allows only very limited use of spiral stairways. Like stairways with winders, spiral stairways may be used in any occupancy as long as such stairways are not a component of a required means of egress.

Spiral stairways are required to have dimensional uniformity. Treads must be at least 26 inches (660 mm) wide and the depth of the treads must not be less than $7^1/_2$ inches (191 mm) measured at a point that is 12 inches (305 mm) out from the narrow edge (see Figure 1009.9). Riser heights are required to be the same throughout the stairway, but are not to exceed $9^1/_2$ inches (241 mm). A minimum headroom of 6 feet 6 inches (1981 mm) is required.

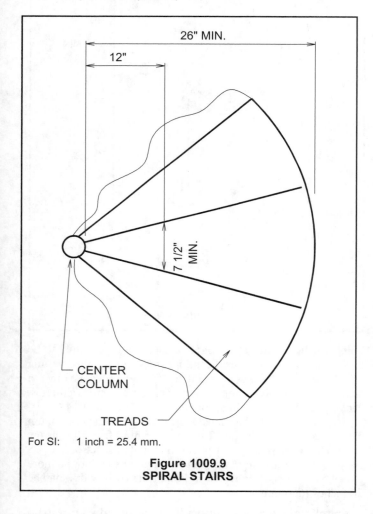

26" MIN.

12"

7 1/2" MIN.

CENTER COLUMN

TREADS

For SI: 1 inch = 25.4 mm.

Figure 1009.9
SPIRAL STAIRS

1009.10 Alternating tread devices. Alternating tread devices are limited to an element of a means of egress in buildings of Groups F, H and S from a mezzanine not more than 250 square feet (23 m²) in area and which serves not more than five occu-

pants; in buildings of Group I-3 from a guard tower, observation station or control room not more than 250 square feet (23 m²) in area and for access to unoccupied roofs.

❖ This type of device is constructed in such a way that each tread alternates with each adjacent tread so that the device consists of a system of right-footed and left-footed treads (see Figure 1009.10).

The use of center stringer construction, half-treads and an incline that is considerably steeper than allowed for ordinary stairway construction makes the alternating tread device unique. However, because of its structural features, only single-file use of the device (between handrails) is possible, thus preventing the occupants from passing one another. The pace of occupant travel is set by the slowest user, a condition that could become critical in an emergency situation. Furthermore, it is impossible for fire service personnel to use an alternating tread device at the same time and in a direction opposite that being used by occupants to exit the premises, possibly causing a serious delay in fire-fighting operations. For these reasons, this section greatly restricts the use of alternating tread devices as a means of egress.

Alternating tread devices are considered a modest improvement to ladder construction and, therefore, can be used as an unoccupied roof access in accordance with the requirements of Section 1009.12.

1009.10.1 Handrails of alternating tread devices. Handrails shall be provided on both sides of alternating tread devices and shall conform to Section 1009.11.

❖ For the safety of occupants, this section references the dimensional requirements for handrail locations to be used in conjunction with the special construction features of alternating tread devices provided in Section 1009.10. Due to the steepness of these devices, additional clearances are required so that handrail movement will not be encumbered by obstructions.

1009.10.2 Treads of alternating tread devices. Alternating tread devices shall have a minimum projected tread of 5 inches (127 mm), a minimum tread depth of 8.5 inches (216 mm), a minimum tread width of 7 inches (178 mm) and a maximum riser height of 9.5 inches (241 mm). The initial tread of the device shall begin at the same elevation as the platform, landing or floor surface.

Exception: Alternating tread devices used as an element of a means of egress in buildings from a mezzanine area not more than 250 square feet (23 m²) in area which serves not more than five occupants shall have a minimum projected tread of 8.5 inches (216 mm) with a minimum tread depth of 10.5 inches (267 mm). The rise to the next alternating tread surface should not be more than 8 inches (203 mm).

❖ Alternating tread stairways (see Section 1009.10) are required to have tread depths of at least $8^1/_2$ inches (216 mm) and widths of 7 inches (178 mm) or more.

The risers are to be not more than $9^1/_2$ inches (241 mm) when measured from tread to alternating tread (next adjacent surface). Tread projections are not to be less than 5 inches (127 mm) when measured (in a horizontal plane) from tread nosing to the next tread nosing (see Figure 1009.10). Applying the limiting dimensions stated above would result in a device with a very steep incline (approximately 4:1).

For alternating tread devices used as a means of egress from small-area mezzanines as prescribed in the exception, the treads must project at least $8^1/_2$ inches (216 mm) as compared to the 5 inches (127 mm) stated above; treads are to be at least $10^1/_2$ inches (267 mm) in depth [compared to $8^1/_2$ inches (216 mm)] and risers are not to exceed 8 inches (203 mm) in height [compared to $9^1/_2$ inches (341 mm)]. Applying these latter limiting dimensions would result in a device of lesser incline (approximate slope of 2:1) and a more comfortable and safer device to use for egress travel.

1009.11 Handrails. Stairways shall have handrails on each side. Handrails shall be adequate in strength and attachment in accordance with Section 1607.7. Handrails for ramps, where required by Section 1010.8, shall comply with this section.

Exceptions:

1. Aisle stairs complying with Section 1024 provided with a center handrail need not have additional handrails.

2. Stairways within dwelling units, spiral stairways and aisle stairs serving seating only on one side are permitted to have a handrail on one side only.

3. Decks, patios and walkways that have a single change in elevation where the landing depth on each side of the change of elevation is greater than what is required for a landing do not require handrails.

4. In Group R-3 occupancies, a change in elevation consisting of a single riser at an entrance or egress door does not require handrails.

5. Changes in room elevations of only one riser within dwelling units and sleeping units in Group R-2 and R-3 occupancies do not require handrails.

❖ Falls are the leading cause of nonfatal injuries in the United States, exceeding even motor vehicle injuries. To protect the user from falls to surfaces below and to aid in the use of the stairway, guards and handrails are to be provided. In cases of fire where vision might be obscured by smoke, handrails serve as guides directing the user along the path of egress travel.

This section requires that handrails be continuous along, and be placed on both sides of, a stairway so that a mobility-impaired person can support his or her "strong side" in both ascent and descent [see Figures 1009.4(1) and 1009.4(2)].

This section is referenced when handrails are required on ramps. Handrails on ramps must comply with Sections 1009.11 through 1009.11.7.

The exceptions state conditions where handrails are only required on one side or are not needed at all.

1009.11.1 Height. Handrail height, measured above stair tread nosings, or finish surface of ramp slope, shall be uniform, not less than 34 inches (864 mm) and not more than 38 inches (965 mm).

❖ It has been demonstrated that for safe use, the height of handrails must not be less than 34 inches (864 mm) nor more than 38 inches (965 mm) above the leading edge

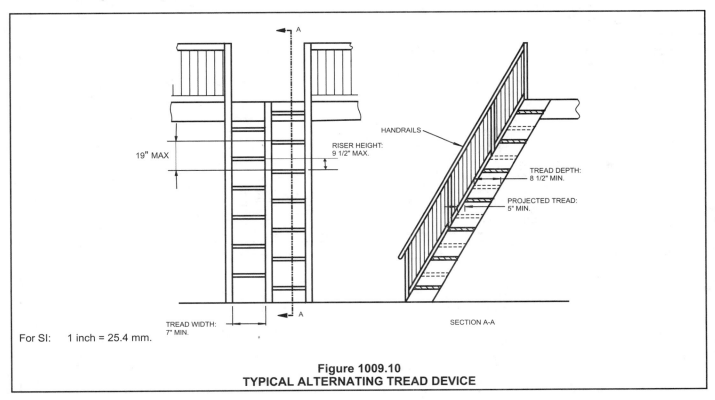

For SI: 1 inch = 25.4 mm.

Figure 1009.10
TYPICAL ALTERNATING TREAD DEVICE

of stairway treads, landings or other walking surfaces (see Figure 1009.11.1). This requirement is applicable for all uses, including handrails within a dwelling unit.

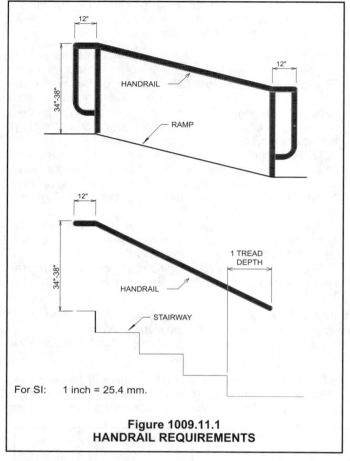

For SI: 1 inch = 25.4 mm.

Figure 1009.11.1
HANDRAIL REQUIREMENTS

1009.11.2 Intermediate handrails. Intermediate handrails are required so that all portions of the stairway width required for egress capacity are within 30 inches (762 mm) of a handrail. On monumental stairs, handrails shall be located along the most direct path of egress travel.

❖ In order to always be available to the user of the stairway, the maximum distance to a handrail from within the required width is to be not more than 30 inches (762 mm). People tend to walk adjacent to handrails, and if intermediate handrails are not provided for very wide stairways, the center portion of such stairways will normally receive limited use. More importantly, in emergencies, the center portions of wide stairways with handrails would more aptly be used to speed up egress travel rather than delay it by overcrowding at the sides with the handrails. This would especially be true under panic conditions. Without the requirement for intermediate handrails, the use of wide interior stairways could become particularly hazardous.

The distance to the handrail applies to the "required width" of the stairway. If a stairway is greater than 60 inches (1524 mm) in width, but only 60 inches (1524 mm) are required based on occupant load (see Section 1005.1), intermediate handrails are not required. Adequate safety is provided since the occupants can use the 30 inches (762 mm) within the handrails provided on each side.

The requirement for monumental stairways deals with the very wide stairway in relation to the required width. While handrails on both sides of the stairway may be sufficient to accommodate the required width, the handrails may not be near the stream of traffic or even apparent to the user. In this case, the handrails are to be placed in a location more reflective of the egress path

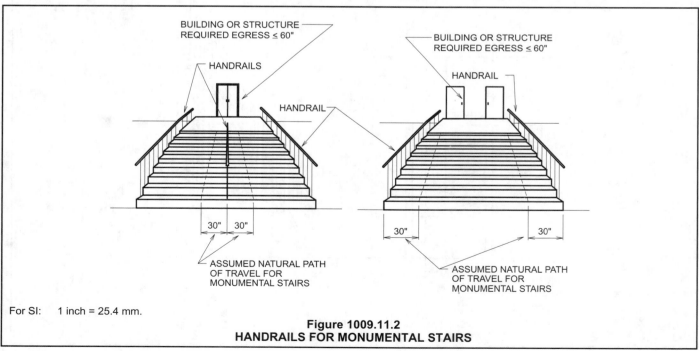

For SI: 1 inch = 25.4 mm.

Figure 1009.11.2
HANDRAILS FOR MONUMENTAL STAIRS

(see Figure 1009.11.2 for handrail locations for monumental stairs).

1009.11.3 Handrail graspability. Handrails with a circular cross section shall have an outside diameter of at least 1.25 inches (32 mm) and not greater than 2 inches (51 mm) or shall provide equivalent graspability. If the handrail is not circular, it shall have a perimeter dimension of at least 4 inches (102 mm) and not greater than 6.25 inches (160 mm) with a maximum cross-section dimension of 2.25 inches (57 mm). Edges shall have a minimum radius of 0.01 inch (0.25 mm).

❖ The ability of grasping a handrail firmly and sliding the hand along the rail without meeting obstructions are important factors in the safe use of stairways and ramps. This section requires that handrails have a circular cross section with an outside diameter of at least 1.25 inches (32 mm) but not greater than 2 inches (51 mm). A handrail with either a very narrow or a large cross section is not graspable in a power grip by all able-bodied users and certainly not by those with hand-strength or flexibility deficiencies. Noncircular cross sections can be approved by way of the alternative noncircular criteria in this section, as long as they provide a suitable gripping surface. Edges must be rounded so that they are not sharp. An example is shown in Figure 1009.11.3.

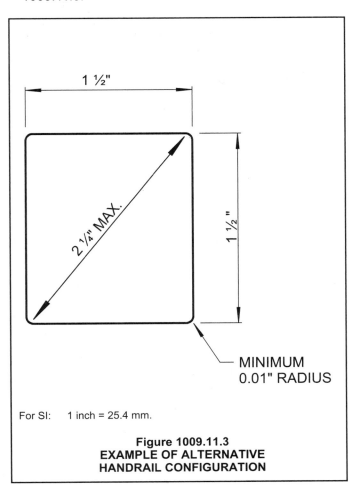

For SI: 1 inch = 25.4 mm.

Figure 1009.11.3
EXAMPLE OF ALTERNATIVE
HANDRAIL CONFIGURATION

1009.11.4 Continuity. Handrail-gripping surfaces shall be continuous, without interruption by newel posts or other obstructions.

Exceptions:

1. Handrails within dwelling units are permitted to be interrupted by a newel post at a stair landing.

2. Within a dwelling unit, the use of a volute, turnout or starting easing is allowed on the lowest tread.

3. Handrail brackets or balusters attached to the bottom surface of the handrail that do not project horizontally beyond the sides of the handrail within 1.5 inches (38 mm) of the bottom of the handrail shall not be considered to be obstructions and provided further that for each 0.5 inch (13 mm) of additional handrail perimeter dimension above 4 inches (102 mm), the vertical clearance dimension of 1.5 inches (38 mm) shall be permitted to be reduced by 0.125 inch (3 mm).

❖ The degree of occupant safety as it relates to handrail use is a function of the features of handrail construction.

Handrails must be usable for their entire length without requiring the users to release their grasp. Typically, in traveling the means of egress, an individual's fingers will trail along the rail. If handrails are to be of service to the occupants, they must be uninterrupted and continuous. Oversize newels or changes in the guard system can cause interruption of the handrail, requiring the occupants to release their grip [see Figure 1009.11.4(1)]. Exceptions 1 and 2 provide for handrail details that have been used for years in dwelling units. Exception 3 provides for conventional methods of handrail support while providing the user with an uninterrupted gripping surface. The larger handrail size permits shorter brackets since geometrically the finger clearance is still maintained. For example, a handrail with a perimeter of 5 inches (127 mm) would be permitted to have a 1.25-inch (32 mm) clearance to the bottom bracket [see Figures 1009.11.4(2)].

1009.11.5 Handrail extensions. Handrails shall return to a wall, guard or the walking surface or shall be continuous to the handrail of an adjacent stair flight. Where handrails are not continuous between flights, the handrails shall extend horizontally at least 12 inches (305 mm) beyond the top riser and continue to slope for the depth of one tread beyond the bottom riser.

Exceptions:

1. Handrails within a dwelling unit that is not required to be accessible need extend only from the top riser to the bottom riser.

2. Aisle handrails in Group A occupancies in accordance with Section 1024.13.

❖ The purpose of the handrail return requirements is to prevent a person from being injured by falling onto the end of a handrail or catching an article of loose clothing on it.

The length that a handrail extends beyond the top and bottom of a stairway ramp or other location where handrails are otherwise not continuous is an important

FIGURE 1009.11.4(1) – FIGURE 1009.11.4(2)

MEANS OF EGRESS

factor for the safety of the users. An occupant must be able to grasp securely a handrail beyond the last riser of a stairway or the last sloped segment of a ramp.

For stairways, handrails must be extended 12 inches (305 mm) horizontally beyond the top riser and sloped a distance of one tread depth beyond the bottom riser. For ramps, handrails must be extended 12 inches (305 mm) horizontally beyond the last sloped ramp segment at both the top and bottom locations. These handrail ex-

tensions are not only required at the top and bottom of stairways and ramps, but also at other places where handrails are not continuous, such as landings and platforms. These changes to previous code requirements are intended to reflect the current provisions of ICC A117.1 (see Figure 1009.11.1). Note that if the handrail extension is at a location that could be considered a protruding object, the handrail must return to the post at a height of less than 27 inches (686 mm) above the floor.

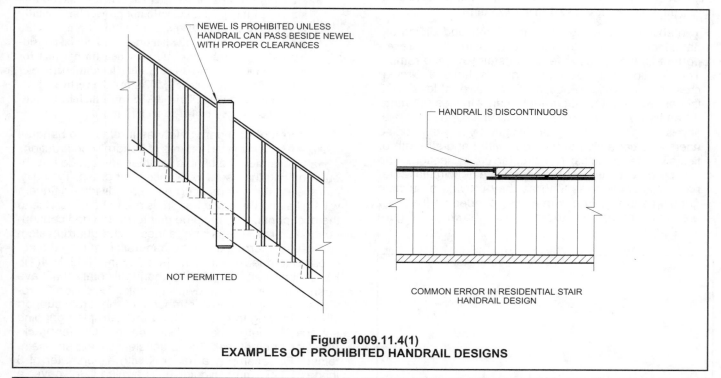

Figure 1009.11.4(1)
EXAMPLES OF PROHIBITED HANDRAIL DESIGNS

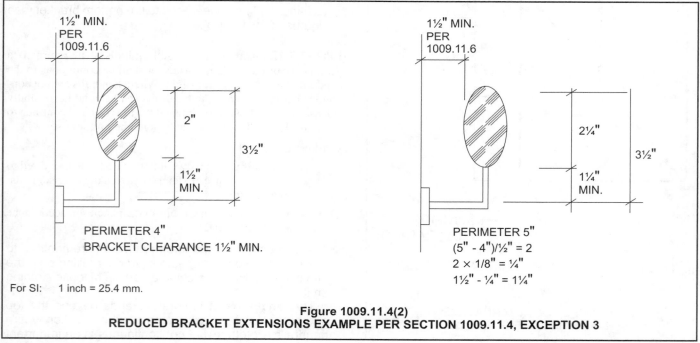

For SI: 1 inch = 25.4 mm.

Figure 1009.11.4(2)
REDUCED BRACKET EXTENSIONS EXAMPLE PER SECTION 1009.11.4, EXCEPTION 3

In accordance with Exception 1, handrail extensions are not required where a dwelling unit is not required to meet any level of accessibility (i.e. accessible unit, Type A unit or Type B unit). Handrail extensions are not essential for these circumstances. Exception 2 provides for handrails along aisles in assembly seating areas. It is necessary to have discontinuous handrails for assembly installations to provide for circulation of the occupants from the aisle to the seating areas.

1009.11.6 Clearance. Clear space between a handrail and a wall or other surface shall be a minimum of 1.5 inches (38 mm). A handrail and a wall or other surface adjacent to the handrail shall be free of any sharp or abrasive elements.

❖ See Figures 1009.11.4(2) and 1009.11.7(2) for an illustration of handrail clearance. A clear space is needed between a handrail and the wall or other surface to allow the user to slide his or her hand along the rail with fingers in the gripping position without contacting the wall surface, which could have an abrasive texture. In climates where persons may be expected to be wearing heavy gloves during the winter, a larger clearance would be desirable at an exterior stairway, or a stairway directly inside the entrance to a building.

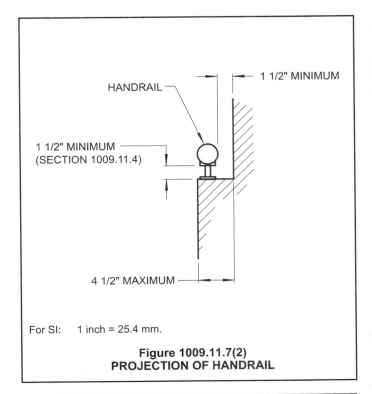

For SI: 1 inch = 25.4 mm.

Figure 1009.11.7(2)
PROJECTION OF HANDRAIL

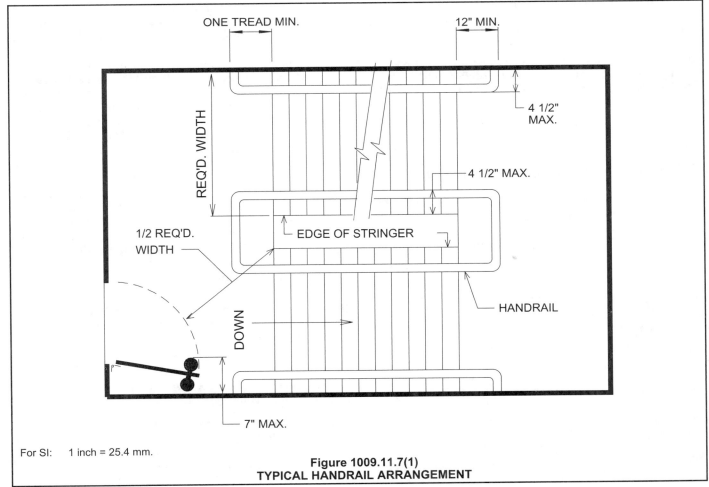

For SI: 1 inch = 25.4 mm.

Figure 1009.11.7(1)
TYPICAL HANDRAIL ARRANGEMENT

1009.11.7 Stairway projections. Projections into the required width at each handrail shall not exceed 4.5 inches (114 mm) at or below the handrail height. Projections into the required width shall not be limited above the minimum headroom height required in Section 1009.2.

❖ Handrails may not project more than 4¹/₂ inches (114 mm) into the required width of a stairway, so that the clear width of the passage will not be seriously reduced [see Figure 1009.11.7(1)]. This projection may exist below the handrail height as well [see Figure 1009.11.7(2)].

1009.12 Stairway to roof. In buildings four or more stories in height above grade, one stairway shall extend to the roof surface, unless the roof has a slope steeper than four units vertical in 12 units horizontal (33-percent slope). In buildings without an occupied roof, access to the roof from the top story shall be permitted to be by an alternating tread device.

❖ Because of safety considerations, roofs used for habitable purposes such as roof gardens, observation decks, sporting facilities (including jogging or walking tracks and tennis courts) or other similar occupancies must be provided with conventional stairways that will serve as required means of egress. Access by ladders or an alternating tread device for such uses is not permitted.

In buildings four or more stories high, roofs that are not used for habitable purposes must be made accessible by conventional stairways or by an alternating tread device (see Section 1009.10). Two reasons for this are access for roof or equipment repair and fire department access during a fire event. Sloping roofs with a rise greater than 4 inches (102 mm) for every 12 inches (305 mm) in horizontal measurement (4:12) are exempt from the requirements of this section because of the steepness of the construction and the inherent dangers to life safety.

While it is not specifically required that roof access be through an exit stairway enclosure, since part of the intent is for fire department access to the roof, it is advisable. Section 1019.1.7 requires signage at the level of exit discharge indicating whether the stairway has roof access.

1009.12.1 Roof access. Where a stairway is provided to a roof, access to the roof shall be provided through a penthouse complying with Section 1509.2.

Exception: In buildings without an occupied roof, access to the roof shall be permitted to be a roof hatch or trap door not less than 16 square feet (1.5 m²) in area and having a minimum dimension of 2 feet (610 mm).

❖ The purpose of the penthouse or stairway bulkhead requirement in this section is to protect the walking surface of the stairway to the roof. The exception provides for situations where roof access is only needed for service or maintenance purposes.

SECTION 1010
RAMPS

1010.1 Scope. The provisions of this section shall apply to ramps used as a component of a means of egress.

Exceptions:

1. Other than ramps that are part of the accessible routes providing access in accordance with Sections 1108.2.2 through 1108.2.4.1, ramped aisles within assembly rooms or spaces shall conform with the provisions in Section 1024.11.

2. Curb ramps shall comply with ICC A117.1.

3. Vehicle ramps in parking garages for pedestrian exit access shall not be required to comply with Sections 1010.3 through 1010.9 when they are not an accessible route serving accessible parking spaces, other required accessible elements or part of an accessible means of egress.

❖ Ramps provide an alternative method of vertical means of access to or egress from a building. Ramps are required for access to building areas for mobility-impaired persons (see Chapter 11) and for small changes in floor elevations, which are a safety hazard in themselves. All ramps, whether required or otherwise provided, must comply with the requirements of this section. The code considers any walking surface that has a slope steeper than one unit vertical in 20 units horizontal (5-percent slope) to be a ramp.

Exception 1 clarifies that in assembly rooms with sloped floors, ramps other than the routes utilized to and from accessible wheelchair seating locations (see Section 1108.2) are permitted to be designed in accordance with the sloped aisle provisions in Section 1024.

Exception 2 references specific curb cut requirements found in ICC A117.1.

Exception 3 addresses parking garages. An accessible route is required to and from any accessible parking space, and all ramp provisions must be followed. However, ramps that provide access to and from nonaccessible spaces in the remainder of the parking garage need only comply with the provisions for slope and guard requirements. Ramps that are strictly for vehicles, such as jump ramps, are not required to meet any of the ramp provisions.

1010.2 Slope. Ramps used as part of a means of egress shall have a running slope not steeper than one unit vertical in 12 units horizontal (8-percent slope). The slope of other ramps shall not be steeper than one unit vertical in eight units horizontal (12.5-percent slope).

Exception: Aisle ramp slope in occupancies of Group A shall comply with Section 1024.11.

❖ Maximum slope is limited to facilitate the ease of ascent and to control the descent of persons with or without a mobility impairment. The maximum slope of a ramp in the direction of travel is limited to 1 unit vertical in 12 units horizontal (1:12) (see Figure 1010.2). Ramps in

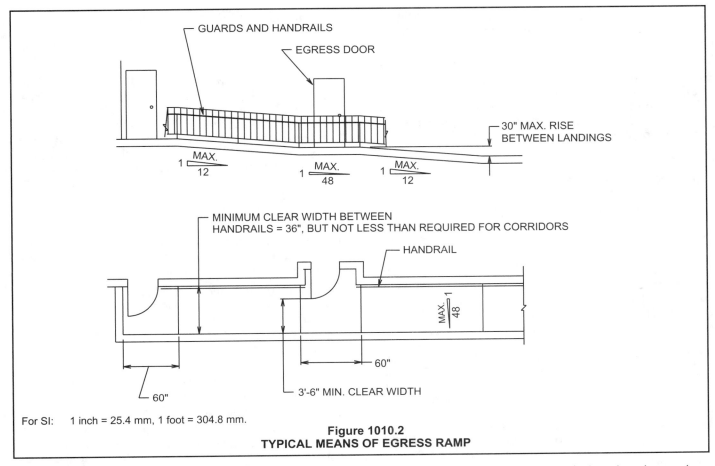

For SI: 1 inch = 25.4 mm, 1 foot = 304.8 mm.

Figure 1010.2
TYPICAL MEANS OF EGRESS RAMP

existing buildings may be permitted to have a steeper slope at small changes in elevation (see Section 3409.7.5 or IEBC Section 506.1.4). An example of a ramp that is not part of a means of egress and, therefore, allowed to be a maximum slope of 1:8, is an industrial access ramp that provides access to a raised floor level around a piece of equipment.

The exception is a reference to the assembly requirements. However, the ramps that are part of an accessible route for ingress or egress are still required to meet the 1:12 maximum slope. Only the ramps used elsewhere in the assembly space may utilize the 1:8 maximum slope permitted in Section 1024.11.

1010.3 Cross slope. The slope measured perpendicular to the direction of travel of a ramp shall not be steeper than one unit vertical in 48 units horizontal (2-percent slope).

❖ The limitation of 1 unit vertical in 48 units horizontal (1:48) on the slope across the direction of travel is to prevent a severe cross slope that would pitch a user to one side (see Figure 1010.2).

1010.4 Vertical rise. The rise for any ramp run shall be 30 inches (762 mm) maximum.

❖ Because pushing a wheelchair up a ramp requires great energy, landings must be situated so that a person can

rest after each 30-inch (762 mm) elevation change (see Figure 1010.2).

1010.5 Minimum dimensions. The minimum dimensions of means of egress ramps shall comply with Sections 1010.5.1 through 1010.5.3.

❖ These minimum dimension requirements allow the ramp to function as a means of egress.

1010.5.1 Width. The minimum width of a means of egress ramp shall not be less than that required for corridors by Section 1016.2. The clear width of a ramp and the clear width between handrails, if provided, shall be 36 inches (914 mm) minimum.

❖ The requirements for the width of a means of egress ramp are 36 inches (914 mm) minimum, similar to that established by Section 1016.2 for corridors. Note that the clear width of 36 inches (914 mm) is required between the handrails for proper clearance for a person in a wheelchair. This is different from stairways where handrails are permitted to project into the required width. The 36-inch (914 mm) minimum clear width between handrails is consistent with ICC A117.1 and the recommendations of the ADAAG Review Advisory Committee.

1010.5.2 Headroom. The minimum headroom in all parts of the means of egress ramp shall not be less than 80 inches (2032 mm).

❖ The requirement for headroom on any part of an egress ramp is identical to the requirement of a conventional (nonspiral) stairway (see Section 1009.2). The ceiling heights of Section 1003.2 may also be applicable depending on the occupancy of the space.

1010.5.3 Restrictions. Means of egress ramps shall not reduce in width in the direction of egress travel. Projections into the required ramp and landing width are prohibited. Doors opening onto a landing shall not reduce the clear width to less than 42 inches (1067 mm).

❖ The purpose of not allowing ramps to reduce in width in the direction of egress travel is to prevent a restriction that would interfere with the flow of occupants out of a facility. This would include ramp landings per Section 1010.6.2.

Doors that open onto a ramp landing must not reduce the clear width to less than 42 inches (1067 mm). This is a more restrictive provision than for stairways, which would permit the reduction to one-half the required width. Since one of the purposes of a ramp is to accommodate persons with physical disabilities, it must provide the additional clear width for access by those confined to wheelchairs without the interference or potential blockage caused by the swing of a door (see Figures 1010.2 and 1010.5.3).

1010.6 Landings. Ramps shall have landings at the bottom and top of each ramp, points of turning, entrance, exits and at doors. Landings shall comply with Sections 1010.6.1 through 1010.6.5.

❖ Landings must be provided to allow users of a ramp to rest on a level floor surface and to adjust to the change in floor surface pitch.

Landings are required at the top and bottom of each ramp run. In addition, Section 1010.4 requires a landing every 30 inches (762 mm) of vertical rise of the ramp. The requirements for landings allow those occupants of the structure the ability to negotiate all changes in direction, prepare themselves to either ascend or descend the ramp and to rest.

1010.6.1 Slope. Landings shall have a slope not steeper than one unit vertical in 48 units horizontal (2-percent slope) in any direction. Changes in level are not permitted.

❖ Landings must be almost flat. This allows persons confined to a wheelchair to come to a complete stop without having to activate the brake or hold themselves stationary at the landing. The maximum slope or cross slope of the landing in any direction is 1:48 (see Figure 1010.2).

1010.6.2 Width. The landing shall be at least as wide as the widest ramp run adjoining the landing.

❖ The width of all landings must be consistently as wide as the widths of the ramp runs leading to them. Means of egress ramps cannot be reduced in width in the direction of egress travel. This is also applicable to the landings connecting the ramp runs.

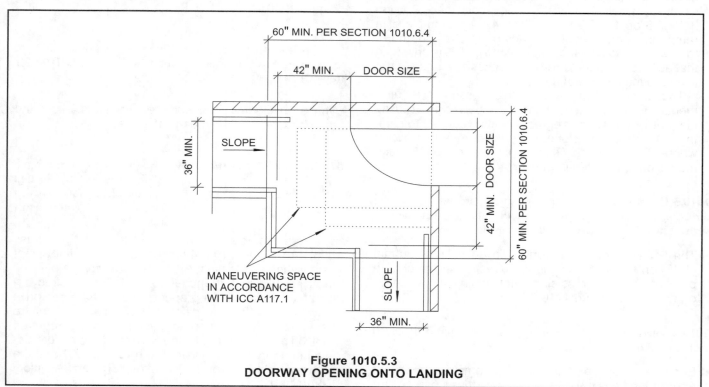

Figure 1010.5.3
DOORWAY OPENING ONTO LANDING

1010.6.3 Length. The landing length shall be 60 inches (1525 mm) minimum.

> **Exception:** Landings in nonaccessible Group R-2 and R-3 individual dwelling units, as applicable in Section 101.2, are permitted to be 36 inches (914 mm) minimum.

❖ The landings for ramps must be at least 60 inches (1524 mm) long. This allows persons confined to wheelchairs a sufficient distance to stop and rest along with any persons who may be assisting them. This requirement is directly applicable to straight-run ramps that may require an intermediate landing at every 30 inches (762 mm) of vertical rise (see Figure 1010.2). If the landing is also to be used to negotiate a change in the ramp's direction, Section 1010.6.4 is applicable. If a door overlaps the landing, Section 1010.5.3 is applicable.

The exception provides for smaller landings in dwelling units that are not required to be accessible.

1010.6.4 Change in direction. Where changes in direction of travel occur at landings provided between ramp runs, the landing shall be 60 inches by 60 inches (1524 mm by 1524 mm) minimum.

> **Exception:** Landings in nonaccessible Group R-2 and R-3 individual dwelling units, as applicable in Section 101.2, are permitted to be 36 inches by 36 inches (914 mm by 914 mm) minimum.

❖ When a change in direction is made in the ramp at a landing, the landing must be at least 60 square inches (0.039 m²). This allows the person confined to a wheelchair enough room to negotiate the turn with minimal effort. The length of the landing may need to exceed 60 inches (1524 mm) to match the widths of the two ramp runs. In any case, the landing would still need to be 60 inches (1524 mm) wide (see Figure 1010.5.3). If a door overlaps the landing, Section 1010.5.3 is applicable.

The exception provides for smaller landings in dwelling units that are not required to be accessible.

1010.6.5 Doorways. Where doorways are located adjacent to a ramp landing, maneuvering clearances required by ICC A117.1 are permitted to overlap the required landing area.

❖ This section specifies that the area required for maneuvering to open the door and the area of the landing are allowed to overlap. It is not necessary to provide the sum of the two area requirements (see Figure 1010.5.3).

1010.7 Ramp construction. All ramps shall be built of materials consistent with the types permitted for the type of construction of the building; except that wood handrails shall be permitted for all types of construction. Ramps used as an exit shall conform to the applicable requirements of Sections 1019.1 and 1019.1.1 through 1019.1.3 for vertical exit enclosures.

❖ Material requirements for the type of construction as required by Section 602 for floors are also the material requirements for ramp construction. The ramp, if used as

an exit, must be enclosed and protected similar to an exit stairway within a vertical exit enclosure.

1010.7.1 Ramp surface. The surface of ramps shall be of slip-resistant materials that are securely attached.

❖ As the pace of exit travel becomes hurried during emergency situations, the probability of slipping on smooth or slick floor surfaces increases. To minimize the hazard, all floor surfaces in the means of egress are required to be slip resistant. The use of hard floor materials with highly polished, glazed, glossy or finely finished surfaces should be avoided.

Field testing and uniform enforcement of the concept of slip resistance is not practical. One method used to establish slip resistance is that the static coefficient of friction between leather [Type 1 (Vegetable Tanned) of Federal Specification KK-L-165C] and the floor surface is greater than 0.5. Laboratory test procedures such as ASTM D 2047 can determine the static coefficient of resistance. Bulletin No. 4 entitled "Surfaces" issued by the U.S. Architectural and Transportation Barriers Compliance Board (ATBCB) contains further information regarding slip resistance.

1010.7.2 Outdoor conditions. Outdoor ramps and outdoor approaches to ramps shall be designed so that water will not accumulate on walking surfaces. In other than occupancies in Group R-3, and occupancies in Group U that are accessory to an occupancy in Group R-3, surfaces and landings which are part of exterior ramps in climates subject to snow or ice shall be designed to minimize the accumulation of same.

❖ Outdoor ramps must be kept free of water, snow and ice to provide a safe path of egress travel at all times, including winter. Sheltering the ramp can serve to minimize snow and ice accumulation but, in some areas, may not completely eliminate the need to remove the snow or ice by other means. Typical methods for protecting these egress elements include roof overhangs or canopies, heated slab and, when approved by the building official, a reliable snow removal maintenance program. This section does not apply to Group R-3 occupancies or their associated utility structures.

1010.8 Handrails. Ramps with a rise greater than 6 inches (152 mm) shall have handrails on both sides complying with Section 1009.11.

❖ To aid in the use of a ramp, handrails are to be provided. Handrails are intended to provide the user with a graspable surface for guidance and support. All ramps with a vertical rise greater than 6 inches (152 mm) between landings are to be provided with handrails on both sides (see Figures 1010.8 and 1009.11.1). Note that if the handrail extension is at a location that could be considered a protruding object, the handrail must return to the post at a height of less that 27 inches (686 mm) above the floor.

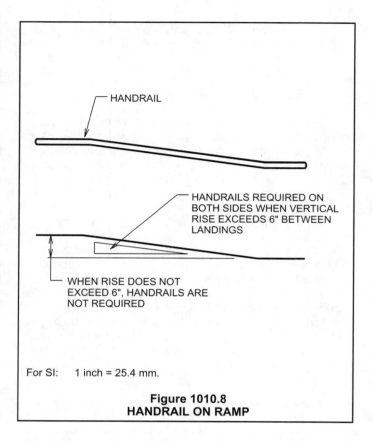

HANDRAIL

HANDRAILS REQUIRED ON
BOTH SIDES WHEN VERTICAL
RISE EXCEEDS 6" BETWEEN
LANDINGS

WHEN RISE DOES NOT
EXCEED 6", HANDRAILS ARE
NOT REQUIRED

For SI: 1 inch = 25.4 mm.

**Figure 1010.8
HANDRAIL ON RAMP**

1010.9 Edge protection. Edge protection complying with Section 1010.9.1 or 1010.9.2 shall be provided on each side of ramp runs and at each side of ramp landings.

Exceptions:

1. Edge protection is not required on ramps not required to have handrails, provided they have flared sides that comply with the ICC A117.1 curb ramp provisions.

2. Edge protection is not required on the sides of ramp landings serving an adjoining ramp run or stairway.

3. Edge protection is not required on the sides of ramp landings having a vertical dropoff of not more than 0.5 inch (13 mm) within 10 inches (254 mm) horizontally of the required landing area.

❖ This section of the code now addresses the comprehensive requirements for edge protection for all ramps. It must be noted that edge protection is not the same as the requirements for guards. The presence of a guard does not necessarily provide adequate edge protection and the presence of adequate edge protection does not satisfy the requirements for a guard. Edge protection is necessary to prevent the wheels of a wheelchair from leaving the ramp surface or becoming lodged between the edge of the ramp and any adjacent construction. For example, a ramp may be located relatively adjacent to the exterior wall of a building. However, between the ramp edge and the exterior wall there is a strip of earth

for landscape purposes. Without adequate edge protection, persons confined to wheelchairs could possibly have their wheels run off the side of the ramp into the landscape causing them to tip. These requirements are consistent with ICC A117.1 and those from the ADAAG Review Advisory Committee.

The first exception allows a ramp to have minimal edge protection as long as its vertical use is 6 inches (152 mm) or less. The exception is predicated on the ramp not needing any handrails, which is established by the provisions of Section 1010.8. Such a ramp would only need flared sides or returned curbs. For specific details of these types of edge protection, the provisions of ICC A117.1 for curb ramps must be followed.

The second exception reiterates that edge protection is not literally required around each side of a ramp landing. Obviously, edge protection is not required along that portion of the landing that directly adjoins the next ramp run; it is only required along the unprotected sides of ramp landings.

The third exception states that edge protection is not required for those sides of a ramp landing directly adjacent to the ground surface that gently slopes away from the edge of the landing. If the grade adjacent to the ramp landing slopes no more than 1/2:10 (which equals 1:20) away from the landing, additional edge protection is not required. Such a gradual slope would not be detrimental to persons confined to wheelchairs as they negotiate the ramp landing.

1010.9.1 Railings. A rail shall be mounted below the handrail 17 inches to 19 inches (432 mm to 483 mm) above the ramp or landing surface.

❖ The purpose of the railing in this section is to prevent a wheelchair from leaving the ramp surface or becoming lodged between the edge of the ramp and any adjacent construction (see Figure 1010.9.1).

1010.9.2 Curb or barrier. A curb or barrier shall be provided that prevents the passage of a 4-inch-diameter (102 mm) sphere, where any portion of the sphere is within 4 inches (102 mm) of the floor or ground surface.

❖ Edge protection for ramps and ramp landings may be achieved with a built-up curb or other effective barrier. The curb or barrier must be located near the surface of the ramp and landing such that a 4-inch-diameter (102 mm) sphere cannot pass through any openings. An example of an effective barrier would be the bottom rail of a guard system. If the bottom rail is located less than 4 inches (102 mm) above the ramp and landing surface, edge protection has been provided. The curb or barrier prevents the wheel of a wheelchair from running off the edge of the surface and provides people with visual disabilities a toe stop at the edge of the walking surface (see Figure 1010.9.1).

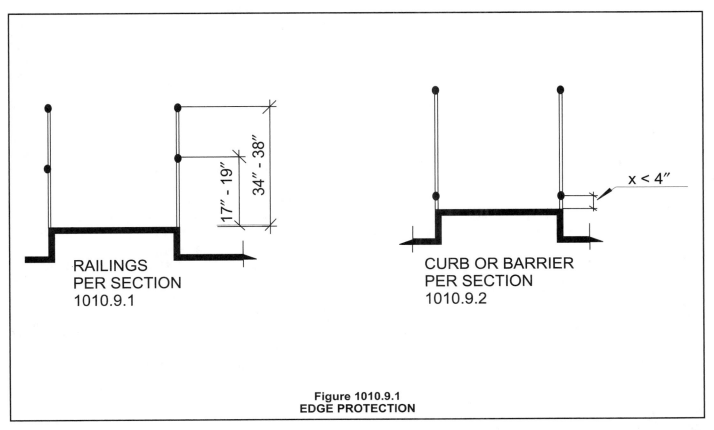

Figure 1010.9.1
EDGE PROTECTION

1010.10 Guards. Guards shall be provided where required by Section 1012 and shall be constructed in accordance with Section 1012.

❖ To protect the user from falls to surfaces below, guards are to be provided.

Guards are to be provided where the sides of a ramp or landing are more than 30 inches (762 mm) above the adjacent grade. Guards are to be constructed in accordance with Section 1012, including the minimum height of 42 inches (1067 mm)(see Figure 1010.10).

SECTION 1011
EXIT SIGNS

1011.1 Where required. Exits and exit access doors shall be marked by an approved exit sign readily visible from any direction of egress travel. Access to exits shall be marked by readily visible exit signs in cases where the exit or the path of egress travel is not immediately visible to the occupants. Exit sign placement shall be such that no point in an exit access corridor is more than 100 feet (30 480 mm) or the listed viewing distance for the sign, whichever is less, from the nearest visible exit sign.

Exceptions:

1. Exit signs are not required in rooms or areas which require only one exit or exit access.

2. Main exterior exit doors or gates which obviously and clearly are identifiable as exits need not have exit signs where approved by the building official.

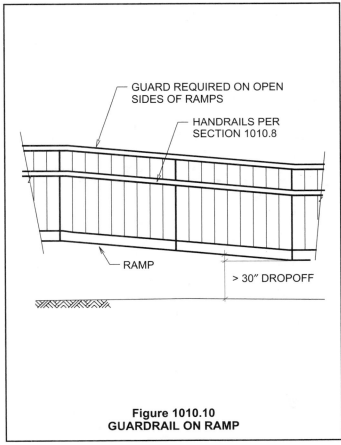

Figure 1010.10
GUARDRAIL ON RAMP

3. Exit signs are not required in occupancies in Group U and individual sleeping units or dwelling units in Group R-1, R-2 or R-3.

4. Exit signs are not required in sleeping areas in occupancies in Group I-3.

5. In occupancies in Groups A-4 and A-5, exit signs are not required on the seating side of vomitories or openings into seating areas where exit signs are provided in the concourse that are readily apparent from the vomitories. Egress lighting is provided to identify each vomitory or opening within the seating area in an emergency.

❖ Where an occupancy has two or more required exits or exit accesses, the means of egress must be provided with illuminated signs that readily identify the location of the exits and indicate the path of travel to the exits. The signs must be illuminated with letters reading "Exit." The illumination may be internal or external to the sign. The signs should be visible from all directions in the exit access route. In cases where the signs are not visible to the occupants because of turns in the corridor or for other reasons, additional illuminated signs must be provided indicating the direction of egress to an exit. "Exit" signs must be located so that, where required, the nearest one is within 100 feet (30 480 mm) of the sign's listed viewing distance. While not a referenced standard, UL 924 permits exit signs to be listed with a viewing distance of less than 100 feet (30 480 mm). When a sign is listed for a viewing distance of less than 100 feet, the label on the sign will indicate the appropriate viewing distance. If such a sign is used, the spacing of the signs should be based on the listed viewing distance.

The exceptions identify conditions where exit signs are not necessary since they would not increase the safety of the egress path.

For Exceptions 1 and 3, the assumption is that the occupants are familiar enough with the space to know the way out, which in most cases is also the way they went in.

In accordance with Exception 2, when the exit is identifiable in itself and is the main exterior door through which the occupants would enter the building, "Exit" signs are not required. For example, a two-story Group B building has a main employee/customer entrance. The entrance consists of a storefront arrangement with glass doors and sidelights. The entrance is centrally located within the building. These main exterior exit doors can be quickly observed as being an exit and would not need to be marked with an "Exit" sign.

In accordance with Exception 4, "Exit" signs are not required in sleeping room areas of Group I-3 buildings. In cases of emergency, occupants in Group I-3 are escorted by staff to the exits and to safety. The "Exit" signs also represent potential weapons when they are accessible to the residents.

In the Group A-4 and A-5 occupancies described in Exception 5, the egress path is obvious and thus exit signs are not needed. Additionally, due to the configuration of the vomitories, the exit signs are not readily visible to the persons immediately adjacent to or above the vomitory.

1011.2 Illumination. Exit signs shall be internally or externally illuminated.

Exception: Tactile signs required by Section 1011.3 need not be provided with illumination.

❖ This section simply provides the scope for illumination of regulated exit signs. Special requirements are established for tactile exit signs in Section 1011.3.

1011.3 Tactile exit signs. A tactile sign stating EXIT and complying with ICC A117.1 shall be provided adjacent to each door to an egress stairway, an exit passageway and the exit discharge.

❖ This signage is needed to provide directions to the exits for those persons with visual impairments. Signs are needed on the required exits in the building, including at doors leading to exit stairway enclosures and passageways, within the exit enclosures leading to the outside and any exit doors that lead directly to the outside. While not specifically stated, this provision is intended to be applicable to ramps within exit enclosures.

Tactile signage in accordance with ICC A117.1 includes both raised lettering and braille. While not required to be illuminated, illumination of this sign would be advantageous for a person with partial sight.

1011.4 Internally illuminated exit signs. Internally illuminated exit signs shall be listed and labeled and shall be installed in accordance with the manufacturer's instructions and Section 2702. Exit signs shall be illuminated at all times.

❖ While not required to be listed, exit signage may comply with UL 924. Listed internally illuminated exit signs are required by UL 924 to meet the graphics, illumination and power sources defined in Sections 1011.5.1 through 1011.5.3. Exit signs must be illuminated at all times, including when the building may not be fully occupied. If a fire occurs late at night, there may be cleaning crews or persons working overtime in the building who will need to be able to find the exits.

1011.5 Externally illuminated exit signs. Externally illuminated exit signs shall comply with Sections 1011.5.1 through 1011.5.3.

❖ Externally illuminated exit signage must meet the graphic, illumination and emergency power requirements in the referenced sections. The requirements are the same as for internally illuminated signage.

1011.5.1 Graphics. Every exit sign and directional exit sign shall have plainly legible letters not less than 6 inches (152 mm) high with the principal strokes of the letters not less than 0.75 inch (19.1 mm) wide. The word "EXIT" shall have letters having a width not less than 2 inches (51 mm) wide except the letter "I," and the minimum spacing between letters shall not be less than 0.375 inch (9.5 mm). Signs larger than the minimum established in this section shall have letter widths, strokes and spacing in proportion to their height.

The word "EXIT" shall be in high contrast with the background and shall be clearly discernible when the exit sign illumination means is or is not energized. If an arrow is provided as part of the exit sign, the construction shall be such that the arrow direction cannot be readily changed.

❖ Every "Exit" sign and directional sign located in the exit access route is required to have a color contrast vivid enough to make the signs readily visible, even when not illuminated. Letters must be at least 6 inches (152 mm) high and their stroke not less than $^3/_4$-inch (19.1 mm) wide (see Figure 1011.5.1). The sizing of the letters is predicated on the readability of the wording from a distance of 100 feet (30 480 mm).

While red letters are common for exit signs, sometimes green on black is used in auditorium areas with low lighting levels, such as theaters, because that color combination tends not to distract the audience's attention. It is more important that the "Exit" sign be readily visible with respect to the background.

"Exit" signs may be larger than the minimum size specified. However, the standardized proportion of the letters must be maintained. Externally illuminated signage that was smaller could use the requirements in UL 924 for guidance; however, sign spacing would need to be adjusted, and alternative approval would be through the building official having jurisdiction.

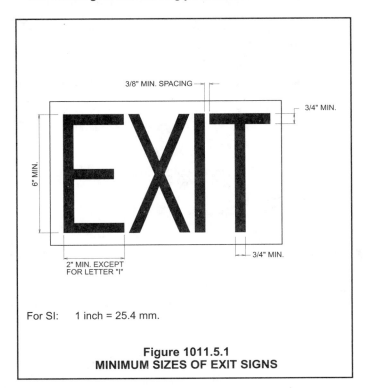

For SI: 1 inch = 25.4 mm.

Figure 1011.5.1
MINIMUM SIZES OF EXIT SIGNS

1011.5.2 Exit sign illumination. The face of an exit sign illuminated from an external source shall have an intensity of not less than 5 foot-candles (54 lux).

❖ Every "Exit" sign and directional sign must be continuously illuminated to provide a light intensity at the illumi-

nated surface of at least 5 foot-candles (54 lux). It is not a requirement that the "Exit" signs be internally illuminated. An external illumination source with the power capabilities specified by Section 1011.5.3 is acceptable.

1011.5.3 Power source. Exit signs shall be illuminated at all times. To ensure continued illumination for a duration of not less than 90 minutes in case of primary power loss, the sign illumination means shall be connected to an emergency power system provided from storage batteries, unit equipment or an on-site generator. The installation of the emergency power system shall be in accordance with Section 2702.

Exception: Approved exit sign illumination means that provide continuous illumination independent of external power sources for a duration of not less than 90 minutes, in case of primary power loss, are not required to be connected to an emergency electrical system.

❖ "Exit" signs must be illuminated on a continuous basis so that when a fire emergency occurs, occupants will be able to identify the locations of the exits. The reliability of the power sources supplying the electrical energy required for maintaining the illumination of exit signs is important. When power interruptions occur, "Exit" sign illumination must be obtained from an emergency power system. This does not imply that the sign must be internally illuminated. Whatever illumination system is used, whether internal or external, it must be connected to a system designed to pick up the power load required by the "Exit" signs after loss of the normal power supply.

Where self-luminous signs are used, connection to the emergency electrical supply system is not required.

SECTION 1012
GUARDS

1012.1 Where required. Guards shall be located along open-sided walking surfaces, mezzanines, industrial equipment platforms, stairways, ramps and landings which are located more than 30 inches (762 mm) above the floor or grade below. Guards shall be adequate in strength and attachment in accordance with Section 1607.7. Guards shall also be located along glazed sides of stairways, ramps and landings that are located more than 30 inches (762 mm) above the floor or grade below where the glazing provided does not meet the strength and attachment requirements in Section 1607.7.

Exception: Guards are not required for the following locations:

1. On the loading side of loading docks or piers.

2. On the audience side of stages and raised platforms, including steps leading up to the stage and raised platforms.

3. On raised stage and platform floor areas such as runways, ramps and side stages used for entertainment or presentations.

4. At vertical openings in the performance area of stages and platforms.

5. At elevated walking surfaces appurtenant to stages and platforms for access to and utilization of special lighting or equipment.

6. Along vehicle service pits not accessible to the public.

7. In assembly seating where guards in accordance with Section 1024.14 are permitted and provided.

❖ Where one or more sides of a walking surface is open to the floor level or grade below, a guard system must be provided to minimize the possibility of occupants accidentally falling to the surface below (see Figure 1012.1). A guard is required only where the difference in elevation between the higher walking surface and the surface below is greater than 30 inches (762 mm). The loads for guard design are addressed in Section 1607. Additionally, this section regulates where glazing is installed in a guard on the side of a stairway, ramp or landing when the glazing has not been designed to resist the forces from a fall.

Most of the exceptions identify typical situations where guards are not practical, such as along loading docks, stages and their approaches and vehicle service pits. Exception 7 references the lower guards permitted at locations where a line of sight for assembly spaces is part of the considerations.

1012.2 Height. Guards shall form a protective barrier not less than 42 inches (1067 mm) high, measured vertically above the leading edge of the tread, adjacent walking surface or adjacent seatboard.

Exceptions:

1. For occupancies in Group R-3, and within individual dwelling units in occupancies in Group R-2, both as

applicable in Section 101.2, guards whose top rail also serves as a handrail shall have a height not less than 34 inches (864 mm) and not more than 38 inches (965 mm) measured vertically from the leading edge of the stair tread nosing.

2. The height in assembly seating areas shall be in accordance with Section 1024.14.

❖ Guards must not be less than 42 inches (1067 mm) in height as measured vertically from the top of the guard down to the leading edge of the tread or to an adjacent walking surface (see Figures 1012.1 and 1012.2). Experience has shown that 42 inches (1067 mm) or more provides adequate height for protection purposes. This puts the top of the guard above the center of gravity of the average adult. Note that with this height requirement, at locations where both a guard and handrail are required, the handrail cannot be at the top of the guard except as permitted in Exception 1.

Due to safety concerns, the designer may want to install a guard where there is a drop-off of less than 30 inches (762 mm). Decorative guards may be utilized to support handrails or serve as part of the edge protection along a ramp. When nonrequired guards are provided, the 42-inch (1067 mm) minimum height is not required.

Exception 1 is for certain residential occupancies and allows for the handrail to be the top of a lower guard. The reduced allowable guard height is consistent with current construction practice. Exception 2 references the lower guards permitted at locations where a line of sight for assembly spaces is part of the consideration.

1012.3 Opening limitations. Open guards shall have balusters or ornamental patterns such that a 4-inch-diameter (102 mm)

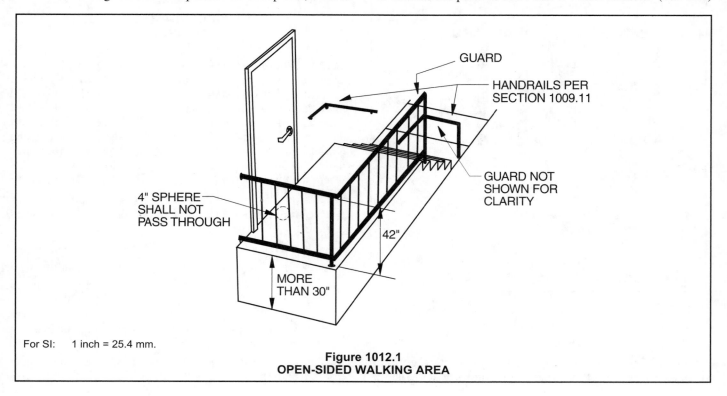

For SI: 1 inch = 25.4 mm.

Figure 1012.1
OPEN-SIDED WALKING AREA

sphere cannot pass through any opening up to a height of 34 inches (864 mm). From a height of 34 inches (864 mm) to 42 inches (1067 mm) above the adjacent walking surfaces, a sphere 8 inches (203 mm) in diameter shall not pass.

Exceptions:

1. The triangular openings formed by the riser, tread and bottom rail at the open side of a stairway shall be of a maximum size such that a sphere of 6 inches (152 mm) in diameter cannot pass through the opening.

2. At elevated walking surfaces for access to and use of electrical, mechanical or plumbing systems or equipment, guards shall have balusters or be of solid materials such that a sphere with a diameter of 21 inches (533 mm) cannot pass through any opening.

3. In areas which are not open to the public within occupancies in Group I-3, F, H or S, balusters, horizontal intermediate rails or other construction shall not permit a sphere with a diameter of 21 inches (533 mm) to pass through any opening.

4. In assembly seating areas, guards at the end of aisles where they terminate at a fascia of boxes, balconies and galleries shall have balusters or ornamental patterns such that a 4-inch-diameter (102 mm) sphere cannot pass through any opening up to a height of 26 inches (660 mm). From a height of 26 inches (660 mm) to 42 inches (1067 mm) above the adjacent walking surfaces, a sphere 8 inches (203 mm) in diameter shall not pass.

❖ The basis for limiting openings in a guard to a 4-inch (102 mm) sphere is in research that indicates that a 4-inch (102 mm) opening will prevent nearly all children 1 year in age or older from falling through the guard. The allowable opening increases to an 8-inch (203 mm) sphere at heights where falling through the guard is not an issue.

An exception to the 4-inch (102 mm) spacing require-

ment is that a 6-inch (152 mm) opening is allowed for openings formed by the riser, tread and bottom rail of guards at the open side of a stairway. This is because the geometry of the openings is such that the entire body cannot pass through the triangular opening. In the case of a standard stair, limiting such openings to a 4-inch (102 mm) sphere is impractical to achieve with a sloped bottom member in the guard (see Figure 1012.2).

Exceptions 2 and 3 address areas where the presence of small children is unlikely and often prohibited. Guards along walkways leading to electrical, mechani-

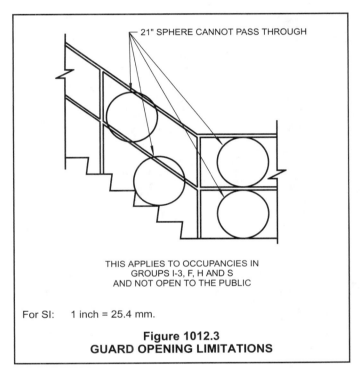

21" SPHERE CANNOT PASS THROUGH

THIS APPLIES TO OCCUPANCIES IN
GROUPS I-3, F, H AND S
AND NOT OPEN TO THE PUBLIC

For SI: 1 inch = 25.4 mm.

Figure 1012.3
GUARD OPENING LIMITATIONS

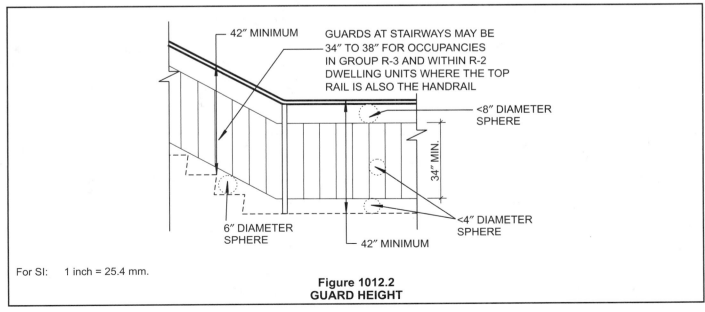

42" MINIMUM

GUARDS AT STAIRWAYS MAY BE
34" TO 38" FOR OCCUPANCIES
IN GROUP R-3 AND WITHIN R-2
DWELLING UNITS WHERE THE TOP
RAIL IS ALSO THE HANDRAIL

<8" DIAMETER SPHERE

34" MIN.

6" DIAMETER SPHERE

<4" DIAMETER SPHERE

42" MINIMUM

For SI: 1 inch = 25.4 mm.

Figure 1012.2
GUARD HEIGHT

cal and plumbing systems or equipment and in occupancies in Groups I-3, F, H and S may be constructed in such a way that a sphere 21 inches (533 mm) in diameter will not pass through any of the openings (see Figure 1012.3). This requirement allows the use of horizontal intermediate members.

The exception for the guard infill near the top of the guard in assembly seating areas is to reduce sightline problems.

1012.4 Screen porches. Porches and decks which are enclosed with insect screening shall be provided with guards where the walking surface is located more than 30 inches (762 mm) above the floor or grade below.

❖ Insect screening located on the open sides of porches and decks does not provide an adequate barrier to reasonably protect an occupant from falling to the surface below. Guards are required on the open sides of porches and decks where the floor is located more than 30 inches (762 mm) above the surface below. The guards must comply with the provisions of Section 1012.

1012.5 Mechanical equipment. Guards shall be provided where appliances, equipment, fans or other components that require service are located within 10 feet (3048 mm) of a roof edge or open side of a walking surface and such edge or open side is located more than 30 inches (762 mm) above the floor, roof or grade below. The guard shall be constructed so as to prevent the passage of a 21-inch-diameter (533 mm) sphere.

❖ The purpose of this requirement is to protect workers from falls off of roofs or from open-sided walking surfaces when doing maintenance work on equipment. The guard opening is allowed to be up to 21 inches (533 mm) since children are not likely to be in such areas. This requirement allows the use of horizontal intermediate members.

SECTION 1013
EXIT ACCESS

1013.1 General. The exit access arrangement shall comply with Sections 1013 through 1016 and the applicable provisions of Sections 1003 through 1012.

❖ Sections 1013 through 1016 include the design requirements for exit access and exit access components. The general requirements that also apply to the exit access are in Sections 1003 through 1012.

1013.2 Egress through intervening spaces. Egress from a room or space shall not pass through adjoining or intervening rooms or areas, except where such adjoining rooms or areas are accessory to the area served; are not a high-hazard occupancy and provide a discernible path of egress travel to an exit. Egress shall not pass through kitchens, storage rooms, closets or spaces used for similar purposes. An exit access shall not pass through a room that can be locked to prevent egress. Means of egress from dwelling units or sleeping areas shall not lead through other sleeping areas, toilet rooms or bathrooms.

Exceptions:

1. Means of egress are not prohibited through a kitchen area serving adjoining rooms constituting part of the same dwelling unit or sleeping unit.

2. Means of egress are not prohibited through adjoining or intervening rooms or spaces in a Group H occupancy when the adjoining or intervening rooms or spaces are the same or a lesser hazard occupancy group.

❖ This section allows adjoining spaces to be considered a part of the room or space from which egress originates, provided that there are reasonable assurances that the continuous egress path will always be available. For example, such egress paths must remain unobstructed and must not pass through an extraordinary fire hazard, such as an area of high-hazard use (Group H). Requiring occupants to egress from an area and pass through an adjoining area that can be characterized by rapid fire buildup or worse, places them in an unreasonable risk situation [see Figure 1013.2(1)]. An occupant should be provided with an equivalent or increased level of safety as he or she approaches the exit. The code does not limit the number of intervening or adjoining rooms through which egress can be made, provided that all other code requirements (i.e., travel distance, number of doorways, etc.) are met. An exit access route, for example, may be laid out such that an occupant leaves a room or space, passes through an adjoining space, enters an exit access corridor, passes through another room and, finally, into an exit [see Figure 1013.2(2)], as long as all other code requirements are satisfied.

Relying on an egress path through an adjacent dwelling unit to be available at all times is not a reasonable expectation. Egress through an adjacent business tenant space can be unreasonable given the security and privacy measures the adjacent tenant may take to secure such a space. However, egress through a reception area that serves a suite of offices of the same tenant is clearly accessory and is permitted.

A common code enforcement problem is a locked door in the egress path. Twenty-five workers perished in September 1991 when they were trapped inside the Imperial Food Processing Plant in Hamlet, North Carolina, due, in part, to locked exit doors. As long as the egress door is readily openable in the direction of egress travel without the use of keys, special knowledge or effort (see Section 1008.1.8.5), the occupants can move unimpeded away from a fire emergency.

Exception 1 allows egress through a kitchen provided it is part of a dwelling unit or guestroom. Egress through a restaurant kitchen is not permitted.

Exception 2 allows egress through high-hazard rooms, provided the adjoining rooms are also high-hazard occupancies.

1013.2.1 Multiple tenants. Where more than one tenant occupies any one floor of a building or structure, each tenant space, dwelling unit and sleeping unit shall be provided with access to the required exits without passing through adjacent tenant spaces, dwelling units and sleeping units.

❖ Where a floor is occupied by multiple tenants, each tenant must be provided with full and direct access to the required exits serving that floor without passing through another tenant space. Tenants frequently lock the doors to their spaces for privacy and security. Should an egress door that is shared by both tenants be locked, occupants in one of the spaces could be trapped and unable to reach an exit. Therefore, an egress layout where occupants from one tenant space travel through another tenant space to gain access to one of the required exits from that floor is prohibited.

This limitation is so that occupants from all tenant spaces will have unrestricted access to the required egress elements while maintaining the security and privacy of the individual tenants. This limitation is based on one of the fundamental principles of egress: to provide a means of egress where all components are capable of being used by the occupants without keys, tools, special knowledge or effort (see Section 1008.1.8.5).

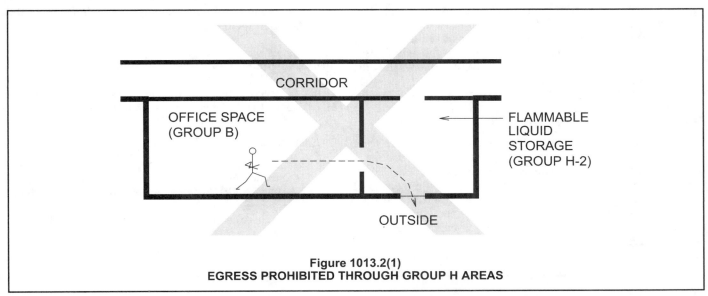

Figure 1013.2(1)
EGRESS PROHIBITED THROUGH GROUP H AREAS

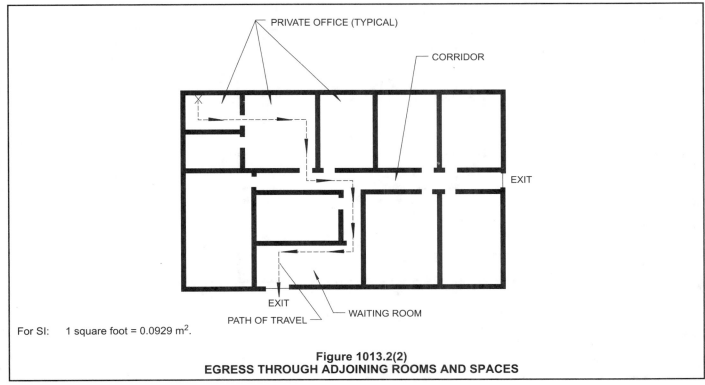

For SI: 1 square foot = 0.0929 m².

Figure 1013.2(2)
EGRESS THROUGH ADJOINING ROOMS AND SPACES

1013.2.2 Group I-2. Habitable rooms or suites in Group I-2 occupancies shall have an exit access door leading directly to an exit access corridor.

Exceptions:

1. Rooms with exit doors opening directly to the outside at ground level.

2. Patient sleeping rooms are permitted to have one intervening room if the intervening room is not used as an exit access for more than eight patient beds.

3. Special nursing suites are permitted to have one intervening room where the arrangement allows for direct and constant visual supervision by nursing personnel.

4. For rooms other than patient sleeping rooms, suites of rooms are permitted to have one intervening room if the travel distance within the suite to the exit access door is not greater than 100 feet (30 480 mm) and are permitted to have two intervening rooms where the travel distance within the suite to the exit access door is not greater than 50 feet (15 240 mm).

Suites of sleeping rooms shall not exceed 5,000 square feet (465 m²). Suites of rooms, other than patient sleeping rooms, shall not exceed 10,000 square feet (929 m²). Any patient sleeping room, or any suite that includes patient sleeping rooms, of more than 1,000 square feet (93 m²) shall have at least two exit access doors remotely located from each other. Any room or suite of rooms, other than patient sleeping rooms, of more than 2,500 square feet (232 m²) shall have at least two access doors remotely located from each other. The travel distance between

any point in a Group I-2 occupancy and an exit access door in the room shall not exceed 50 feet (15 240 mm). The travel distance between any point in a suite of sleeping rooms and an exit access door of that suite shall not exceed 100 feet (30 480 mm).

❖ The purpose of this section is to establish the means of egress requirements that are unique to Group I-2 occupancies. Patient rooms are permitted to be arranged as suites (see Figure 1013.2.2 for an example plan of a patient suite). The figure illustrates a patient suite of more than 1,000 square feet (93 m²) where two remote egress doors are required. The criteria in this section recognizes the low patient-to-staff ratio of these facilities where the staff is directly responsible for the safety of the patients in the event of a fire.

1013.3 Common path of egress travel. In occupancies other than Groups H-1, H-2 and H-3, the common path of egress travel shall not exceed 75 feet (22 860 mm). In occupancies in Groups H-1, H-2, and H-3, the common path of egress travel shall not exceed 25 feet (7620 mm).

Exceptions:

1. The length of a common path of egress travel in an occupancy in Groups B, F and S shall not be more than 100 feet (30 480 mm), provided that the building is equipped throughout with an automatic sprinkler system installed in accordance with Section 903.3.1.1.

2. Where a tenant space in an occupancy in Groups B, S and U has an occupant load of not more than 30, the

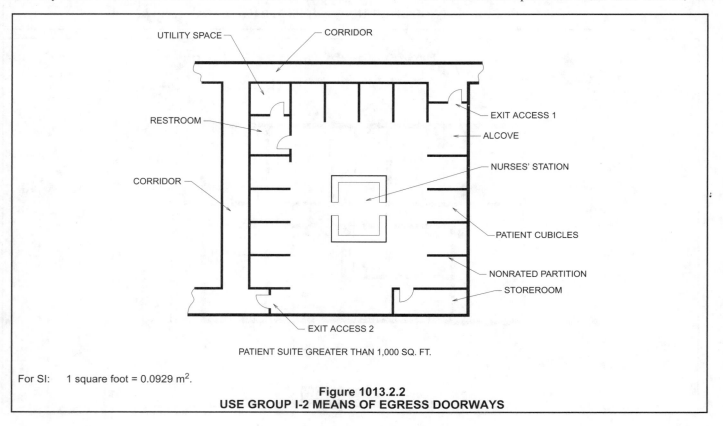

PATIENT SUITE GREATER THAN 1,000 SQ. FT.

For SI: 1 square foot = 0.0929 m².

Figure 1013.2.2
USE GROUP I-2 MEANS OF EGRESS DOORWAYS

length of a common path of egress travel shall not be more than 100 feet (30 480 mm).

3. The length of a common path of egress travel in occupancies in Group I-3 shall not be more than 100 feet (30 480 mm).

❖ The common path of travel is the distance measured from the most remote point in a space to the point in the exit path where the occupant has access to two required exits in separate directions. The definition of "Common path of egress travel" is found in Section 1002. The distance limitations are applicable to all paths of travel that lead out of a space or building. An illustration of this distance is found in Figure 1013.3. The illustration reflects two examples of a common path of travel where the occupants at points A and B are able to travel in only one direction before they reach a point at which they have a choice of two paths of travel to the required exits from the building. Note that from point A, the occupants have two available paths, but these merge to form a single path out of the space. This is also considered a common path of travel.

The exceptions increase the allowable length of the common path of travel based on the installation of a sprinkler system, a low occupant load or for an I-3 occupancy.

1013.4 Aisles. Aisles serving as a portion of the exit access in the means of egress system shall comply with the requirements of this section. Aisles shall be provided from all occupied portions of the exit access which contain seats, tables, furnishings, displays and similar fixtures or equipment. Aisles serving as-

sembly areas, other than seating at tables, shall comply with Section 1024. Aisles serving reviewing stands, grandstands and bleachers shall also comply with Section 1024.

The required width of aisles shall be unobstructed.

Exception: Doors, when fully opened, and handrails shall not reduce the required width by more than 7 inches (178 mm). Doors in any position shall not reduce the required width by more than one-half. Other nonstructural projections such as trim and similar decorative features are permitted to project into the required width 1.5 inches (38 mm) from each side.

❖ This section addresses aisles for other than assembly areas which are covered in Section 1024. Aisle accessways for tables and seating are covered by Section 1013.4.2.

The term "Aisle accessway" is defined in Section 1002 as "that portion of exit access that leads to an aisle." While the term "aisle" is not defined in the code, the dictionary indicates an aisle to be "a walkway between or along sections of seats, shelves, counters, etc., as in a theater, church or department store." Given the many possible configurations of fixtures and furniture, both permanent and moveable, the determination of where aisle accessways stop and aisles begin is often subject to interpretation.

Since the aisle serves as a path for means of egress similar to a corridor, the requirements for doors obstructing the aisle are similar. The exception also includes provisions for the protrusion of handrails or trim along an aisle.

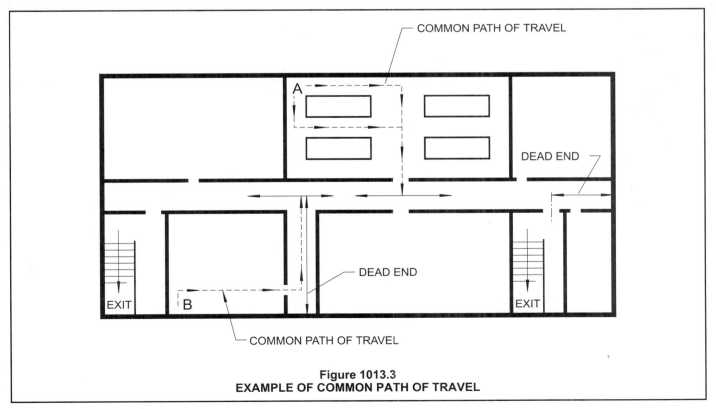

Figure 1013.3
EXAMPLE OF COMMON PATH OF TRAVEL

1013.4.1 Groups B and M. In Group B and M occupancies, the minimum clear aisle width shall be determined by Section 1005.1 for the occupant load served, but shall not be less than 36 inches (914 mm).

> **Exception:** Nonpublic aisles serving less than 50 people, and not required to be accessible by Chapter 11 need not exceed 28 inches (711 mm) in width.

❖ This requirement establishes aisle width criteria for Group B and M occupancies based on the occupant load served by the aisle. The reference to Section 1005.1 would trigger a requirement for aisles wider than 36 inches (914 mm) when the anticipated occupant load that the aisle served was larger than 180 in nonsprinklered buildings and larger than 240 in sprinklered buildings. The exception addresses aisles that may be found in an archival file room or stock storage racks.

At this time, the code does not include criteria for aisle accessways in these types of facilities. If fixtures are permanent, such as in a typical grocery store, the aisle provisions would be applicable throughout. In a situation where there were groups of displays separated by aisles, the area within the displays may be considered aisle accessways.

Note that Section 105.2, Item 13, exempts movable cases, counters and partitions not over 5 feet, 9 inches (1753 mm) in height from requiring a building permit. It is not practicable to require a means of egress inspection every time a store changes a display or a business moves someone into a new office. At least a minimal amount of space will be provided throughout to allow access to the displays, furniture or equipment. Concerns that should be addressed in the aisle accessways would be unobstructed access to an aisle and common path of travel constraints in Section 1013.3. While common path of travel is limited to 30 feet (9144 mm) (see Sections 1013.4.2.3 and 1024.8) in dining and assembly seating, the path of travel for these arrangements is typically limited to one direction. Very few Group B and M occupancies have this type of arrangement of fixtures and furniture.

1013.4.2 Seating at tables. Where seating is located at a table or counter and is adjacent to an aisle or aisle accessway, the measurement of required clear width of the aisle or aisle accessway shall be made to a line 19 inches (483 mm) away from and parallel to the edge of the table or counter. The 19-inch (483 mm) distance shall be measured perpendicular to the side of the table or counter. In the case of other side boundaries for aisle or aisle accessways, the clear width shall be measured to walls, edges of seating and tread edges, except that handrail projections are permitted.

> **Exception:** Where tables or counters are served by fixed seats, the width of the aisle accessway shall be measured from the back of the seat.

❖ Seating is often provided adjacent to aisles and aisle accessways. In measuring the width of an aisle or aisle accessway for movable seating, the measurement is taken at a distance of 19 inches (483 mm) perpendicular to the side of the table or counter. This 19-inch (483 mm) space from the edge of the table or counter to the line where the aisle or aisle accessway measurement begins is intended to represent the space occupied by a typical seated occupant. This dimension is also considered to be adequate to accommodate seats with armrests that are too high to fit under the table where fixed seats are used. The aisle width is permitted to be measured from the back of the seat based upon the exception. As indicated in Figure 1013.4.2, where seating abuts an aisle or aisle accessway, 19 inches (483 mm) must be added to the required aisle or aisle accessway width for seating on only one side and 38 inches (965 mm) for seating on both sides. When seating will not be adjacent to the aisles or aisle passageways, as is the case when tables are at an angle to the aisle or aisle accessway, the measurement may be taken to the edge of the seating, table, counter or tread (see Figure 1013.4.2).

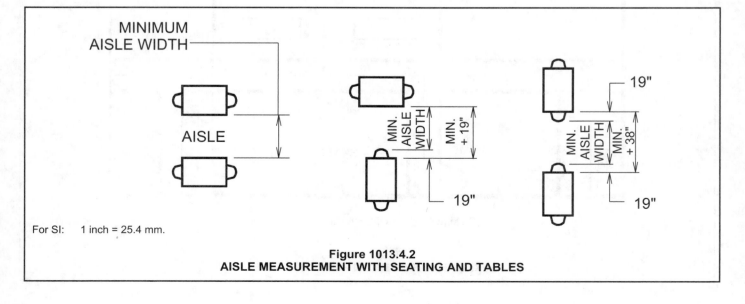

For SI: 1 inch = 25.4 mm.

Figure 1013.4.2
AISLE MEASUREMENT WITH SEATING AND TABLES

1013.4.2.1 Aisle accessway for tables and seating. Aisle accessways serving arrangements of seating at tables or counters shall have sufficient clear width to conform to the capacity requirements of Section 1005.1 but shall not have less than the appropriate minimum clear width specified in Section 1013.4.1.

❖ This section specifies two criteria for the determination of the required width of aisle accessways: the requirements of Sections 1005.1 and 1013.4.1. The aisle accessway width is to be the wider of the two requirements.

The requirements of Sections 1013.4.2.1, 1013.4.2.2 and 1013.4.2.3 are illustrated in Figure 1013.4.2.1.

1013.4.2.2 Table and seating accessway width. Aisle accessways shall provide a minimum of 12 inches (305 mm) of width plus 0.5 inch (12.7 mm) of width for each additional 1 foot (305 mm), or fraction thereof, beyond 12 feet (3658 mm) of aisle accessway length measured from the center of the seat farthest from an aisle.

Exception: Portions of an aisle accessway having a length not exceeding 6 feet (1829 mm) and used by a total of not more than four persons.

❖ The relationship of tables and seating results in a situation in which it is difficult to determine which chairs are served by which aisle accessway. Therefore, the width

of the aisle accessway is a function of the distance from the aisle. The same minimum 12 inches (305 mm) is used and is increased 0.5 inch (12.7 mm) for each additional foot of travel beyond 12 feet (3658 mm).

Recognizing that the normal use of table and chair seating will require some clearance for access and service, the exception eliminates the minimum width criteria if the distance to the aisle [or an aisle accessway of at least 12 inches (305 mm)] is less than 6 feet (1829 mm) and the number of people served is not more than four. Therefore, the first 6 feet (1829 mm) are not required to meet any minimum width criteria. After the first 6 feet (1829 mm), the requirements for an aisle accessway will apply. The length of the aisle accessway is then restricted by Section 1013.4.2.3. When the maximum length of the aisle accessway is reached, either an aisle, corridor or exit access door must be provided (see Figure 1013.4.2.1).

1013.4.2.3 Table and seating aisle accessway length. The length of travel along the aisle accessway shall not exceed 30 feet (9144 mm) from any seat to the point where a person has a choice of two or more paths of egress travel to separate exits.

❖ At some point in the exit access travel, it is necessary to reach an aisle complying with the minimum widths of Section 1013.4. An aisle accessway travel distance is

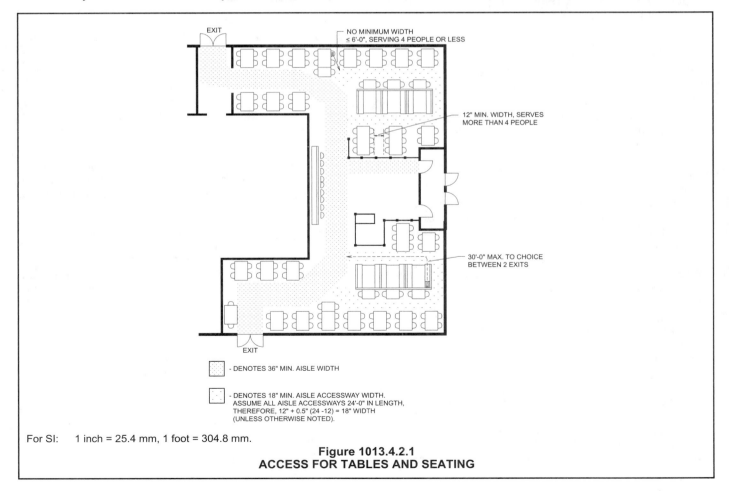

EXIT

NO MINIMUM WIDTH
≤ 6'-0", SERVING 4 PEOPLE OR LESS

12" MIN. WIDTH, SERVES
MORE THAN 4 PEOPLE

30'-0" MAX. TO CHOICE
BETWEEN 2 EXITS

EXIT

▦ - DENOTES 36" MIN. AISLE WIDTH

▢ - DENOTES 18" MIN. AISLE ACCESSWAY WIDTH.
ASSUME ALL AISLE ACCESSWAYS 24'-0" IN LENGTH,
THEREFORE, 12" + 0.5" (24 -12) = 18" WIDTH
(UNLESS OTHERWISE NOTED).

For SI: 1 inch = 25.4 mm, 1 foot = 304.8 mm.

Figure 1013.4.2.1
ACCESS FOR TABLES AND SEATING

not to exceed 30 feet (9144 mm), which may represent a dead-end condition (see Figure 1013.4.2.1).

1013.5 Egress balconies. Balconies used for egress purposes shall conform to the same requirements as corridors for width, headroom, dead ends and projections. Exterior balconies shall be designed to minimize accumulation of snow or ice that impedes the means of egress.

Exception: Exterior balconies and concourses in outdoor stadiums shall be exempt from the design requirement to protect against the accumulation of snow or ice.

❖ This section regulates balconies that are used as an exit access and requires that they meet the same requirements as exit access corridors, except for the enclosure. Exterior exit access balconies must be kept free of snow and ice. Sheltering the exterior balcony can serve to minimize snow and ice accumulation but, in some areas, may not completely eliminate the need to remove the snow or ice by other means. If the exterior side of the balcony is also enclosed to prevent snow and ice accumulation, the balcony essentially creates the same conditions as an interior corridor and must be so protected.

The exception for protection of snow and ice accumulation for outdoor stadiums is based on the assumption that an adequate snow removal or maintenance policy exists for these facilities.

1013.5.1 Wall separation. Exterior egress balconies shall be separated from the interior of the building by walls and opening protectives as required for corridors.

Exception: Separation is not required where the exterior egress balcony is served by at least two stairs and a dead-end travel condition does not require travel past an unprotected opening to reach a stair.

❖ An exterior exit access balcony has a valuable attribute in that the products of combustion may be freely vented to the open air. In the event of a fire in an adjacent space, the products of combustion would not be expected to build up in the balcony area as would commonly occur in an interior corridor. However, there is still a concern for the egress of occupants who must use the balcony for exit access, and consequently, may have to pass the room or space where the fire is located. Therefore, an exterior exit access balcony is required to be separated from interior spaces by fire partitions, as is required for interior corridors. The other provisions of Section 1016 relative to dead ends and opening protectives also apply.

If there are no dead-end conditions that require travel past an unprotected opening and the balcony is provided with at least two stairways, then the wall separating the balcony from the interior spaces need not have a fire-resistance rating (see Figure 1013.5.1). Such an arrangement reduces the probability that occupants will need to pass the area with the fire to gain access to an exit.

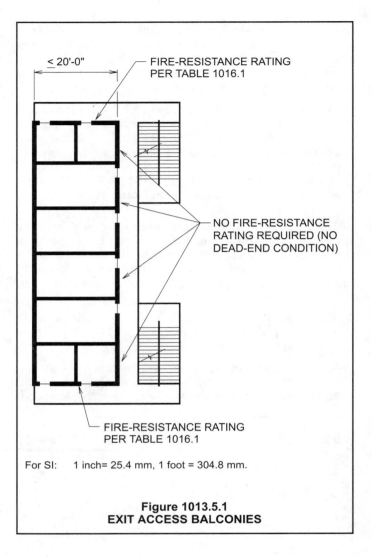

For SI: 1 inch= 25.4 mm, 1 foot = 304.8 mm.

Figure 1013.5.1
EXIT ACCESS BALCONIES

1013.5.2 Openness. The long side of an egress balcony shall be at least 50 percent open, and the open area above the guards shall be so distributed as to minimize the accumulation of smoke or toxic gases.

❖ This section provides an opening requirement that is intended to preclude the rapid buildup of smoke and toxic gases. A minimum of one side of the exterior balcony is required to have a minimum open exterior area of 50 percent of the side area of the balcony. The side openings are to be fairly uniformly distributed along the length of the balcony.

SECTION 1014
EXIT AND EXIT ACCESS DOORWAYS

1014.1 Exit or exit access doorways required. Two exits or exit access doorways from any space shall be provided where one of the following conditions exists:

1. The occupant load of the space exceeds the values in Table 1014.1.

2. The common path of egress travel exceeds the limitations of Section 1013.3.

3. Where required by Sections 1014.3, 1014.4 and 1014.5.

Exception: Group I-2 occupancies shall comply with Section 1013.2.2.

❖ This section dictates the minimum number of paths of travel an occupant is to have available to avoid a fire incident in the occupied room or space. While providing multiple egress doorways from every room is unrealistic, a point does exist where alternative egress paths must be provided based on the number of occupants at risk, the distance any one occupant must travel to reach a doorway and the relative hazards associated with the occupancy of the space. Generally, the number of egress doorways required from any room or space coincides with the occupant load threshold criteria set forth for the minimum number of exits required in a building (see Section 1018.1). The limiting criteria in Table 1014.1 for rooms or spaces permitted to have a single exit access doorway are based on an empirical judgement of the associated risks.

If the occupants of a room are required to egress through another room, as permitted in Section 1013.2, the rooms are to be combined to determine if multiple doorways are required from the combined rooms. For example, if a suite of offices share a common reception area, the entire suite with the reception area must meet both the occupant load and the travel distance criteria.

It should be noted that where two doorways are required, the remoteness requirement of Section 1014.2 is applicable.

Item 2 sets the limits for a single means of egress based on travel distance. Where the common path of travel exceeds the limits in Section 1013.3, two egress paths are required for safe egress from the space.

Item 3 addresses when two means of egress may be required in boiler, incinerator and furnace rooms; refrigerator machinery rooms or refrigerated rooms and spaces.

Group I-2 occupancies are not addressed in Table 1014.1. The exception refers to Section 1013.2.2 for Group I-2 means of egress requirements.

TABLE 1014.1
SPACES WITH ONE MEANS OF EGRESS

OCCUPANCY	MAXIMUM OCCUPANT LOAD
A, B, E, F, M, U	50
H-1, H-2, H-3	3
H-4, H-5, I-1, I-3, I-4, R	10
S	30

❖ The table represents an empirical judgement of the risks associated with a single means of egress from a room or space based on the occupant load in the room, the travel distance to the exit access door and the inherent risks associated with the occupancy (such as occupant mobility, occupant familiarity with the building, occupant response and the fire growth rate).

Since the occupants of Groups I and R may be sleeping and, therefore, may not be able to detect a fire in its early stages without staff supervision or room detectors, the number of occupants in a single egress room or space is limited to 10.

Because of the potential for rapidly developing hazardous conditions, the single egress condition in Groups H-1, H-2 and H-3 is limited to a maximum of three persons. Because the materials contained in Groups H-4 and H-5 do not represent the same fire hazard potential as those found in Groups H-1, H-2 and H-3, the occupant load for spaces with one means of egress is increased.

Because of the reduced occupant density in Group S and the occupants' normal familiarity with the building, the single egress condition is permitted with an occupant load of 30.

1014.1.1 Three or more exits. Access to three or more exits shall be provided from a floor area where required by Section 1018.1.

❖ This section provides a reference to Section 1018.1 for conditions where three or more exits are required. The reference is provided in this section so that the requirements of Section 1018.1 will be obvious.

1014.2 Exit or exit access doorway arrangement. Required exits shall be located in a manner that makes their availability obvious. Exits shall be unobstructed at all times. Exit and exit access doorways shall be arranged in accordance with Sections 1014.2.1 and 1014.2.2.

❖ Exits need to be unobstructed and obvious at all times for the safety of occupants to evacuate the building in an emergency situation. This is consistent with the requirements in Section 1008.1 for exit or exit access doors to not be concealed by curtains, drapes, decorations or mirrors.

1014.2.1 Two exits or exit access doorways. Where two exits or exit access doorways are required from any portion of the exit access, the exit doors or exit access doorways shall be placed a distance apart equal to not less than one-half of the length of the maximum overall diagonal dimension of the building or area to be served measured in a straight line between exit doors or exit access doorways. Interlocking or scissor stairs shall be counted as one exit stairway.

Exceptions:

1. Where exit enclosures are provided as a portion of the required exit and are interconnected by a 1-hour fire-resistance-rated corridor conforming to the requirements of Section 1016, the required exit separation shall be measured along the shortest direct line of travel within the corridor.

2. Where a building is equipped throughout with an automatic sprinkler system in accordance with Section 903.3.1.1 or 903.3.1.2, the separation distance of the exit doors or exit access doorways shall not be less than

FIGURE 1014.2.1(1)

MEANS OF EGRESS

one-third of the length of the maximum overall diagonal dimension of the area served.

❖ This section provides a method to determine quantitatively remoteness between exits and exit access doors, based on the dimensional characteristics of the space served. This measure has been common practice for some years with significant success. Very simply, the method involves determining the maximum dimension between any two points in a floor or a room (e.g., a diagonal between opposite corners in a rectangular room or building or the diameter in a circular room or building). If two doors or exits are required from the room or building (see Sections 1014.1 and 1018.1), the straight-line distance between the center of the thresholds of the doors must be at least one-half of the maximum dimension [see Figure 1014.2.1(1)].

While technical proof is not available to substantiate this method of determining remoteness, it has been found to be realistic and practical for building designs except for the common building with exits in a center core and office spaces around the perimeter.

If a scissors stairway is utilized, regardless of the separation of the two entrances, the scissors stair may only be counted as one exit. Two stairways within the same enclosure could result in both stairways being unnavigable in an emergency if smoke penetrated the single enclosure. (see the definition for "Scissor stairways" in Section 1002).

In Exception 1, a method of permitting the distance between exits to be measured along a complying corridor connecting the exits has served to mitigate the disruption to this design concept [see Figure 1014.2.1(2)].

As reflected in Exception 2, the protection provided by an automatic sprinkler system can reduce the threat of fire buildup so that the reduction in remoteness to one-third of the diagonal dimension is not unreasonable, based on the presumption that it provides the occupants with an acceptable level of safety from fire [see Figure 1014.2.1(3)]. The automatic sprinkler system must be installed throughout the building in accordance with Section 903.3.1.1 or 903.3.1.2 (NFPA 13 or 13R). This reduced separation (one-third diagonal) may also be used when applying the requirements of Exception 1.

In applying the provisions of this section, it is important to recognize any convergence of egress paths that may exist. Figure 1014.2.1(4) illustrates that although the assembly room has remotely located exit access doors, the doors from the entire space do not meet the criteria of this section.

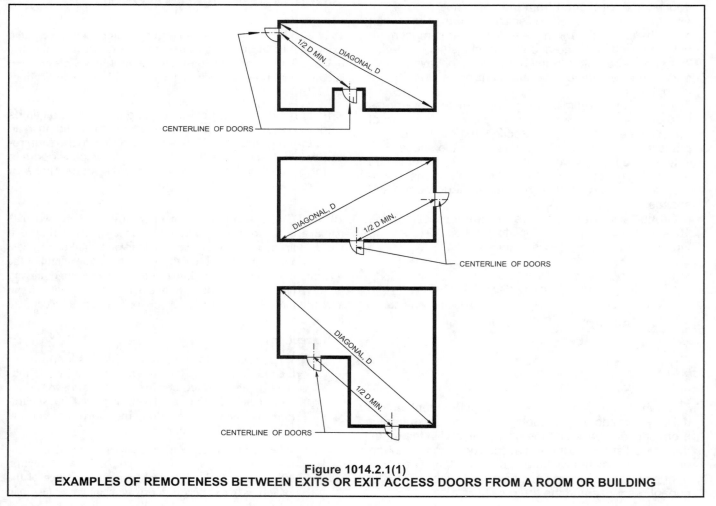

Figure 1014.2.1(1)
EXAMPLES OF REMOTENESS BETWEEN EXITS OR EXIT ACCESS DOORS FROM A ROOM OR BUILDING

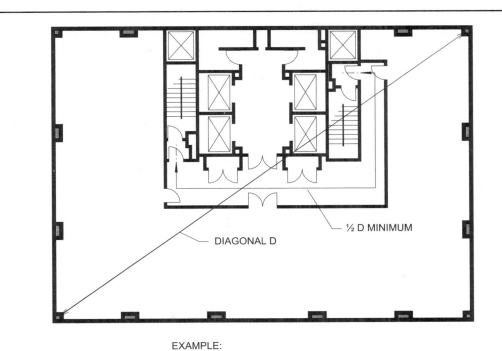

½ D MINIMUM

DIAGONAL D

EXAMPLE:
DIAGONAL DIMENSION = 134'-0"
MIN. SEPARATION OF EXITS = 134/2 = 67'-0"

For SI: 1 inch = 25.4 mm, 1 foot = 304.8 mm.

Figure 1014.2.1(2)
REMOTENESS OF EXITS INTERCONNECTED BY A 1-HOUR FIRE-RESISTANT-RATED CORRIDOR

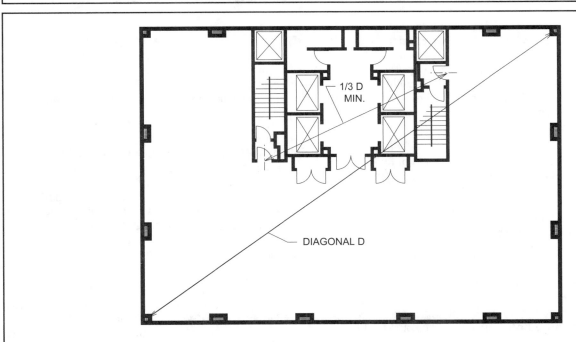

1/3 D MIN.

DIAGONAL D

EXAMPLE:
DIAGONAL DIMENSION = 134'-0"
MIN. SEPARATION OF EXITS = 134/3 = 44'-8"

For SI: 1 inch = 25.4 mm, 1 foot = 304.8 mm.

Figure 1014.2.1(3)
REMOTENESS OF EXITS IN A BUILDING WITH AN AUTOMATIC SPRINKLER SYSTEM

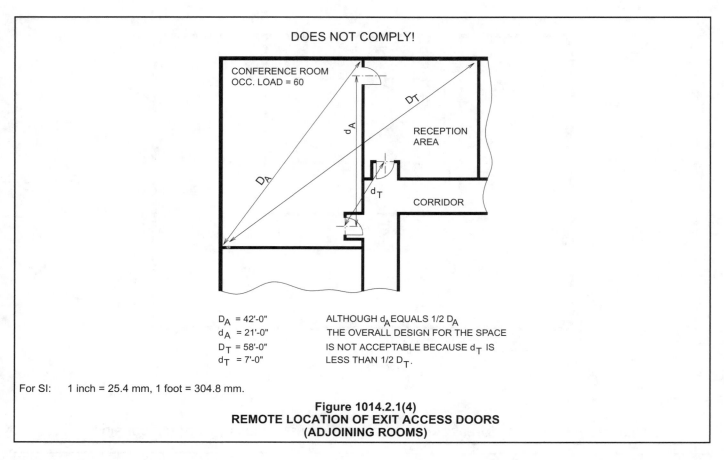

DOES NOT COMPLY!

CONFERENCE ROOM
OCC. LOAD = 60

RECEPTION
AREA

CORRIDOR

D_A = 42'-0"   ALTHOUGH d_A EQUALS 1/2 D_A
d_A = 21'-0"   THE OVERALL DESIGN FOR THE SPACE
D_T = 58'-0"   IS NOT ACCEPTABLE BECAUSE d_T IS
d_T = 7'-0"   LESS THAN 1/2 D_T.

For SI: 1 inch = 25.4 mm, 1 foot = 304.8 mm.

Figure 1014.2.1(4)
REMOTE LOCATION OF EXIT ACCESS DOORS
(ADJOINING ROOMS)

1014.2.2 Three or more exits or exit access doorways. Where access to three or more exits is required, at least two exit doors or exit access doorways shall be placed a distance apart equal to not less than one-half of the length of the maximum overall diagonal dimension of the area served measured in a straight line between such exit doors or exit access doorways. Additional exits or exit access doorways shall be arranged a reasonable distance apart so that if one becomes blocked, the others will be available.

Exception: Where a building is equipped throughout with an automatic sprinkler system in accordance with Section 903.3.1.1 or 903.3.1.2, the separation distance of at least two of the exit doors or exit access doorways shall not be less than one-third of the length of the maximum overall diagonal dimension of the area served.

❖ When there are three or more required exits from a building or exit access doors from a room, they are to be analyzed identically to the method described in Section 1014.2.1. Two of the exits or exit access doors must meet the remoteness test and any additional exits or doors can be located anywhere within the floor plan that meets the code requirements, including independence, accessibility, capacity and continuity.

The exception is for consistency with Exception 2 in Section 1014.2.1 for sprinklered buildings.

1014.3 Boiler, incinerator and furnace rooms. Two exit access doorways are required in boiler, incinerator and furnace

rooms where the area is over 500 square feet (46 m²) and any fuel-fired equipment exceeds 400,000 British thermal units (Btu) (422 000 KJ) input capacity. Where two exit access doorways are required, one is permitted to be a fixed ladder or an alternating tread device. Exit access doorways shall be separated by a horizontal distance equal to one-half the maximum horizontal dimension of the room.

❖ This section requires two exit access doorways for the specified mechanical equipment spaces because of the level of hazards in this type of occupancy. A fixed ladder or an alternating tread device is permitted for the occupants to egress where two doorways are required. The remoteness of the exit access doorways specified in this section is to give the occupants two paths of travel to exit the room so that if one doorway is not available, the alternate path can be used.

1014.4 Refrigeration machinery rooms. Machinery rooms larger than 1,000 square feet (93 m²) shall have not less than two exits or exit access doors. Where two exit access doorways are required, one such doorway is permitted to be served by a fixed ladder or an alternating tread device. Exit access doorways shall be separated by a horizontal distance equal to one-half the maximum horizontal dimension of room.

All portions of machinery rooms shall be within 150 feet (45 720 mm) of an exit or exit access doorway. An increase in travel distance is permitted in accordance with Section 1015.1.

Doors shall swing in the direction of egress travel, regardless of the occupant load served. Doors shall be tight fitting and self-closing.

❖ The reasons for these requirements are the same as for Section 1014.3. Travel distance is to be limited in accordance with Section 1015.1. The travel distance is to be increased in accordance with Section 1015.1. For example, the travel distance limit for a large refrigeration machinery room classified as Group F-1 that has a sprinkler system throughout the entire building in accordance with Section 903.3.1.1 would be 250 feet (76 200 mm) based on Table 1015.1. The 150-foot (45 720 mm) maximum distance to an exit or exit access doorway that is specified in this section would not apply in this example. The 150-foot (45 720 mm) travel distance is intended to be applied where a sprinkler system is not installed and to shorten the time that occupants would be exposed to the hazards within the machinery room.

1014.5 Refrigerated rooms or spaces. Rooms or spaces having a floor area of 1,000 square feet (93 m²) or more, containing a refrigerant evaporator and maintained at a temperature below 68°F (20°C), shall have access to not less than two exits or exit access doors.

Travel distance shall be determined as specified in Section 1015.1, but all portions of a refrigerated room or space shall be within 150 feet (45 720 mm) of an exit or exit access door where such rooms are not protected by an approved automatic sprinkler system. Egress is allowed through adjoining refrigerated rooms or spaces.

Exception: Where using refrigerants in quantities limited to the amounts based on the volume set forth in the *International Mechanical Code.*

❖ The commentary to Section 1014.4 also applies to this section. The exception is intended to apply if Chapter 11 of the *International Mechanical Code®* (IMC®) does not require a refrigeration machinery room due to the small amount of refrigerant used (see the commentary for Section 1104 of the IMC for further explanation of the machinery room requirements).

1014.6 Stage means of egress. Where two means of egress are required, based on the stage size or occupant load, one means of egress shall be provided on each side of the stage.

❖ Two means of egress are required from stages in accordance with Section 410.5.4. The stage means of egress paths are to be separate so the two means of egress are independent.

1014.6.1 Gallery, gridiron and catwalk means of egress. The means of egress from lighting and access catwalks, galleries and gridirons shall meet the requirements for occupancies in Group F-2.

Exceptions:
1. A minimum width of 22 inches (559 mm) is permitted for lighting and access catwalks.

2. Spiral stairs are permitted in the means of egress.

3. Stairways required by this subsection need not be enclosed.

4. Stairways with a minimum width of 22 inches (559 mm), ladders, or spiral stairs are permitted in the means of egress.

5. A second means of egress is not required from these areas where a means of escape to a floor or to a roof is provided. Ladders, alternating tread devices or spiral stairs are permitted in the means of escape.

6. Ladders are permitted in the means of egress.

❖ The purpose of this section is to specify the various options that are allowed for means of egress from theater lighting and access catwalks, galleries and gridirons. The requirements are consistent with the use being limited to service personnel.

SECTION 1015
EXIT ACCESS TRAVEL DISTANCE

1015.1 Travel distance limitations. Exits shall be so located on each story such that the maximum length of exit access travel, measured from the most remote point within a story to the entrance to an exit along the natural and unobstructed path of egress travel, shall not exceed the distances given in Table 1015.1.

Where the path of exit access includes unenclosed stairways or ramps within the exit access or includes unenclosed exit ramps or stairways as permitted in Section 1019.1, the distance of travel on such means of egress components shall also be included in the travel distance measurement. The measurement along stairways shall be made on a plane parallel and tangent to the stair tread nosings in the center of the stairway.

Exceptions:
1. Travel distance in open parking garages is permitted to be measured to the closest riser of open stairs.

2. In outdoor facilities with open exit access components and open exterior stairs or ramps, travel distance is permitted to be measured to the closest riser of a stair or the closest slope of the ramp.

3. Where an exit stair is permitted to be unenclosed in accordance with Exception 8 or 9 of Section 1019.1, the travel distance shall be measured from the most remote point within a building to an exit discharge.

❖ The length of travel, as measured from the most remote point within a structure to an exit, is limited to restrict the amount of time that the occupant is exposed to a potential fire condition [see Figure 1015.1(1)]. The route must be assumed to be the natural path of travel without obstruction. This commonly results in a rectilinear path similar to what can be experienced in most occupancies, such as a schoolroom or an office with rows of desks [see Figure 1015.1(2)]. The "arc" method, using

FIGURE 1015.1(1) – FIGURE 1015.1(2)   MEANS OF EGRESS

an "as the crow flies" linear measurement, must be used with caution, as it seldom represents typical floor design and layout and, in most cases, would not be deemed to be the natural, unobstructed path.

The travel distance is measured from each and every occupiable point on a floor to the closest exit. While each occupant may be required to have access to a second or third exit, the travel distance limitation is only applicable to the distance to the nearest exit. In effect, this means that the distance an occupant must travel to the second or third exit is not regulated.

Travel distance is measured along the exit access path. Exit access travel distance may include travel on a exit stairway if it is not constructed to meet the definition of an exit (i.e., enclosure, discharge, etc.). An example of this would be an unenclosed exit access stairway from a mezzanine level or steps along the path of travel in a split floor-level situation. When Section 1019.1 permits an exit stairway to be unenclosed, the travel distance would also include travel down the open exit stairway and to a vertical exit enclosure, a horizontal exit or an exit door to the outside. An example of this would be an open exit stairway within an individual dwelling unit. See Section 1019.1, Exception 3 or an open exit stairway from a small space not open to the public (see Section 1019.1, Exception 1).

Exceptions 1 and 2 provide for a travel distance terminating at the top of an open exit stair in an open parking

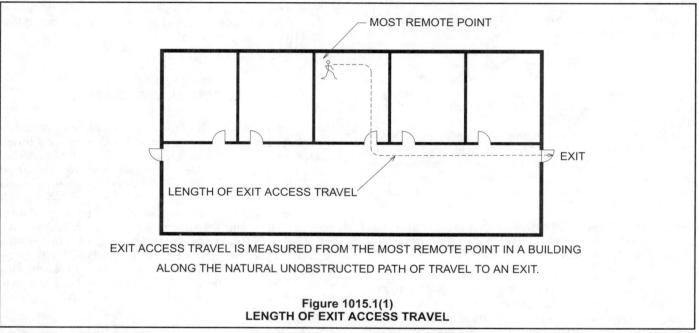

EXIT ACCESS TRAVEL IS MEASURED FROM THE MOST REMOTE POINT IN A BUILDING ALONG THE NATURAL UNOBSTRUCTED PATH OF TRAVEL TO AN EXIT.

Figure 1015.1(1)
LENGTH OF EXIT ACCESS TRAVEL

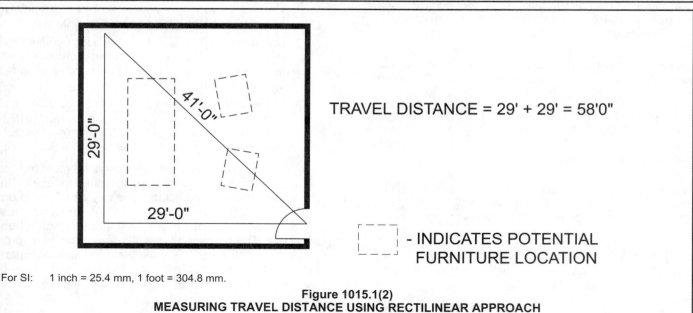

For SI: 1 inch = 25.4 mm, 1 foot = 304.8 mm.

Figure 1015.1(2)
MEASURING TRAVEL DISTANCE USING RECTILINEAR APPROACH

structure or an open exit stair or ramp in outdoor facilities (e.g., stadiums, exterior stairways from open balconies, observation decks and amusement structures see Section 1019.1 Exceptions 2 and 5). This is appropriate in view of the low hazard in these facilities.

Exception 3 addresses the special concerns for an open exit stairway as permitted in Section 1019.1, Exceptions 8 and 9. The measurement for the travel distance must be from the most remote point, down the exit stairway and out of the building to the exit discharge. Therefore, when applying Exception 8 for open exit stairways between upper levels this exception literally would require the total travel distance measurement to include any travel distance that was inside an exit enclosure as well as any exterior exit stairways or ramps until the occupants reached grade level. It may be a reasonable interpretation to measure the travel distance to an enclosed exit, horizontal exit or exterior exit stairway.

The distance of travel within an exit encloser and in the exit discharge portion of the means of egress is also not regulated.

Section 1018.2 permits certain buildings to be provided with a single exit. In instances where there is a single exit, travel distances less than those permitted in Table 1015.1 apply (see Table 1018.2).

TABLE 1015.1
EXIT ACCESS TRAVEL DISTANCE[a]

OCCUPANCY	WITHOUT SPRINKLER SYSTEM (feet)	WITH SPRINKLER SYSTEM (feet)
A, E, F-1, I-1, M, R, S-1	200	250[b]
B	200	300[c]
F-2, S-2, U	300	400[b]
H-1	Not Permitted	75[c]
H-2	Not Permitted	100[c]
H-3	Not Permitted	150[c]
H-4	Not Permitted	175[c]
H-5	Not Permitted	200[c]
I-2, I-3, I-4	150	200[c]

For SI: 1 foot = 304.8 mm.

a. See the following sections for modifications to exit access travel distance requirements:
Section 402: For the distance limitation in malls.
Section 404: For the distance limitation through an atrium space.
Section 1015.2: For increased limitation in Groups F-1 and S-1.
Section 1024.7: For increased limitation in assembly seating.
Section 1024.7: For increased limitation for assembly open-air seating.
Section 1018.2: For buildings with one exit.
Chapter 31: For the limitation in temporary structures.

b. Buildings equipped throughout with an automatic sprinkler system in accordance with Section 903.3.1.1 or 903.3.1.2. See Section 903 for occupancies where sprinkler systems according to Section 903.3.1.2 are permitted.

c. Buildings equipped throughout with an automatic sprinkler system in accordance with Section 903.3.1.1.

❖ This table reflects the maximum distance a person is allowed to travel from any point in a building floor area to the nearest exit along a natural and unobstructed path. While quantitative determinations or formulas are not available to substantiate the tabular distances, empirical factors are utilized to make relative judgements as to reasonable limitations. Such considerations include the nature and fitness of the occupants; the typical configurations and physical conditions of each group; the level of fire hazard with respect to the specific uses of the facilities, including fire spread and the potential intensity of a fire. The inclusion of an automatic sprinkler system throughout the building can serve to control, confine or possibly eliminate the fire threat to the occupants so an increased travel distance is permitted. Increased travel distances are permitted when an automatic sprinkler system is installed in accordance with NFPA 13 or 13R (see Section 903.3.1.1 or 903.3.1.2).

When measuring travel distance, it is important to consider the natural path of travel [see Figure 1015.1(1)]. In many cases, the actual layout of furnishings and equipment is not known or is not identified on the plans submitted with the permit application. In such instances, it may be necessary to measure travel distance using the legs of a triangle instead of the hypotenuse [see Figure 1015.1(2)]. Since most people tend to migrate to more open spaces while egressing, measurement of the natural path of travel frequently excludes areas of the building within approximately 1 foot (305 mm) of walls, corners, columns and other permanent construction. Where the travel path includes passage through a doorway, the natural route is generally measured through the centerline of the door openings.

1015.2 Roof vent increase. In buildings which are one story in height, equipped with automatic heat and smoke roof vents complying with Section 910 and equipped throughout with an automatic sprinkler system in accordance with Section 903.3.1.1, the maximum exit access travel distance shall be 400 feet (122 m) for occupancies in Group F-1 or S.

❖ This section permits an increase in travel distances when roof vents are installed because of the increased visibility provided in a fire event if the roof-venting system is properly designed and installed in accordance with Section 910. While smoke/heat vents and automatic sprinklers serve different life safety functions, they are both active systems of fire protection and can work well together. However, due care must be taken in the design and installation of venting systems so that when they are used in conjunction with automatic sprinklers, they will not cause cooling air drafts to occur at the sprinklers so as to delay activation of the sprinkler system or even render it inoperative. The travel distance increase is limited to single-story buildings of factory (Group F-1) and storage (Group S-1) occupancies because of the characteristic open areas, fire department accessibility and limited occupant densities. Note that the building must be sprinklered throughout by an automatic sprinkler system designed and installed in accordance with NFPA 13 (see Section 903.3.1.1.)

1015.3 Exterior egress balcony increase. Travel distances specified in Section 1015.1 shall be increased up to an additional 100 feet (30 480 mm) provided the last portion of the exit access leading to the exit occurs on an exterior egress balcony constructed in accordance with Section 1013.5. The length of such balcony shall not be less than the amount of the increase taken.

❖ This section allows an additional travel distance on exterior egress balconies since the accumulation of smoke is much less on a balcony. Note that the length of the increase is not to be more than the length of the exterior balcony. For example, if the length of the balcony is 75 feet (22 860 mm), the additional travel distance is limited to 75 feet (22 860). In order for the increase to apply, the exterior balcony must be located at the end of the path of egress travel and not in some other portion of the egress path.

SECTION 1016
CORRIDORS

1016.1 Construction. Corridors shall be fire-resistance rated in accordance with Table 1016.1. The corridor walls required to be fire-resistance rated shall comply with Section 708 for fire partitions.

Exceptions:

1. A fire-resistance rating is not required for corridors in an occupancy in Group E where each room that is used for instruction has at least one door directly to the exterior and rooms for assembly purposes have at least one-half of the required means of egress doors opening directly to the exterior. Exterior doors specified in this exception are required to be at ground level.

2. A fire-resistance rating is not required for corridors contained within a dwelling or sleeping unit in an occupancy in Group R.

3. A fire-resistance rating is not required for corridors in open parking garages.

4. A fire-resistance rating is not required for corridors in an occupancy in Group B which is a space requiring only a single means of egress complying with Section 1014.1.

❖ The purpose of corridor enclosures is to provide fire protection to occupants as they travel the confined path, perhaps unaware of a fire buildup in an adjacent floor area. The base protection is a fire partition having a 1-hour fire-resistance rating (see Table 1016.1). The table allows a reduction or elimination of the fire-resistance rating depending on the occupant load and the presence of an NFPA 13 or 13R automatic sprinkler system throughout the building.

Section 708 addresses the continuity of fire partitions serving as corridor walls. In addition to allowing the fire partitions to terminate at the underside of a fire-resistance-rated floor/ceiling or roof/ceiling assembly, the supporting construction need not have the same fire-resistance rating in buildings of Type IIB, IIIB and VB construction as specified in Section 708. If such walls were required to be supported by fire-resistance-rated construction, the use of these construction types would be

severely restricted when the corridors are required to have a fire-resistance rating. Section 407.3 requires that corridor walls in Group I-2 occupancies that are required to have a fire-resistance rating must be continuous to the underside of the floor or roof deck above or at a smoke-limiting ceiling membrane. Continuity is required because of the defend-in-place protection strategy utilized in such buildings. Requirements for corridor construction within Group I-3 occupancies are found in Section 408.7. For additional requirements for an elevator lobby that is adjacent to or part of a corridor, see the commentary to Section 707.14.1.

Exception 1 indicates a fire-resistance rating is not required for corridors in Group E when any room adjacent to the corridor that is used for instruction or assembly purposes has a door directly to the outside. Because these rooms are provided with an alternate egress path due to the requirement for exterior exits, the need for a fire-resistance-rated corridor is eliminated.

In accordance with Exception 2, a fire-resistance rating for a corridor contained within a single dwelling unit (e.g., apartment, townhouse) or sleeping unit (e.g., hotel guestroom, assistive living suite) is not required for practical reasons. It is unreasonable to expect fire doors and the associated hardware in homes.

Given the relatively smoke-free environment of open parking structures, Exception 3 does not require rated corridors in these types of facilities.

If an office suite is small enough that only one means of egress is required from the suite, Exception 4 indicates that a rated corridor would not be required in that area. The main corridor that connected these suites to the exits would be rated in accordance with Table 1016.1.

TABLE 1016.1. See page 10-87.

❖ The required fire-resistance ratings of corridors serving adjacent spaces are provided in Table 1016.1. The fire-resistance rating is based on the group classification (considering characteristics such as occupant mobility, density and knowledge of the building as well as the fire hazard associated with the classification), the total occupant load served by the corridor and the presence of an automatic sprinkler system.

Where the corridor serves a limited number of people (see the second column in Table 1016.1), the fire-resistance rating is eliminated because of the limited size of the facility and the likelihood that the occupants would become aware of a fire buildup in sufficient time to exit the structure safely. The total occupant load that the corridor serves is used to determine the requirement for a rated corridor enclosure. Corridors serving a total occupant load equal to or less than that indicated in the second column of Table 1016.1 are not required to be enclosed with fire-resistance-rated construction. For example, a corridor serving an occupant load of 30 or less in an unsprinklered Group B occupancy is not required to be enclosed with fire-resistance-rated construction. This example is illustrated in Figure 1016.1.

The purpose of corridor enclosures is to provide fire

protection to occupants as they travel the confined path, perhaps unaware of a fire buildup in an adjacent floor area. The base protection is a fire partition having a 1-hour fire-resistance rating. The table allows a reduction or elimination of the fire-resistance rating depending on the occupant load and the presence of an NFPA 13 or 13R automatic sprinkler system throughout the building.

A common mistake is assuming a building is sprinklered throughout and utilizing the corridor rating reductions, when in fact certain requirements in NFPA 13 would not consider the building sprinklered throughout. For example, a health club installs a sprinkler system, but chooses to eliminate the sprinklers over the pool in accordance with the exception in Section 903.2.1.3. Any corridors within the building that served greater than 30 occupants would still have to be rated because the building would not be considered sprinklered throughout in accordance with NFPA 13 requirements.

Note that because of the hazardous nature of occu-pancies in Groups H-1, H-2 and H-3, fire-resis-tance-rated corridors are required under all conditions. Regardless of the presence of a fire sprinkler system, a 1-hour-rated corridor enclosure is typically required in high-hazard occupancies. The only exception is for Group H-4 and H-5 occupancies. Occupancies that contain materials or operations constituting a health hazard do not pose the same relative fire or explosion hazard as Group H-1, H-2 or H-3 materials. As such, in Group H-4 or H-5, where the corridor serves a total oc-cupant load of 30 or less, a fire-resistance-rated enclo-sure is not required. The "not permitted" in the third col-umn is in coordination with Section 903.2.4 which requires all Group H buildings to be fully sprinklered.

The code acknowledges that an automatic sprinkler system can serve to control or eliminate fire development that could threaten the exit access corridor. Most occu-pancies where sleeping rooms are not present (Groups A, B, E, F, M, S and U) are permitted to have nonfire-resis-tance-rated corridors if a sprinkler system is installed throughout the building in accordance with NFPA 13.

TABLE 1016.1
CORRIDOR FIRE-RESISTANCE RATING

OCCUPANCY	OCCUPANT LOAD SERVED BY CORRIDOR	REQUIRED FIRE-RESISTANCE RATING (hours)	
		Without sprinkler system	With sprinkler system[c]
H-1, H-2, H-3	All	Not Permitted	1
H-4, H-5	Greater than 30	Not Permitted	1
A, B, E, F, M, S, U	Greater than 30	1	0
R	Greater than 10	1	0.5
I-2[a], I-4	All	Not Permitted	0
I-1, I-3	All	Not Permitted	1[b]

a. For requirements for occupancies in Group I-2, see Section 407.3.
b. For a reduction in the fire-resistance rating for occupancies in Group I-3, see Section 408.7.
c. Buildings equipped throughout with an automatic sprinkler system in accordance with Section 903.3.1.1 or 903.3.1.2 where allowed.

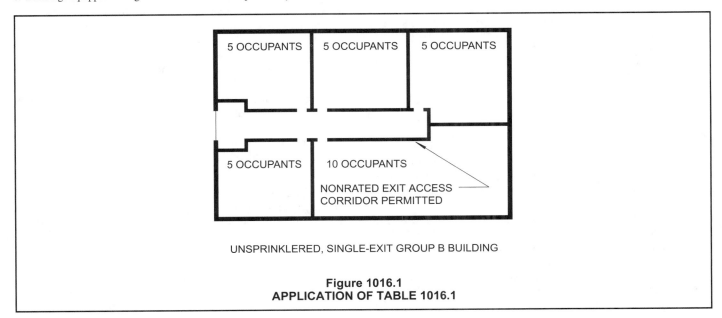

UNSPRINKLERED, SINGLE-EXIT GROUP B BUILDING

Figure 1016.1
APPLICATION OF TABLE 1016.1

In residential facilities, the response time to a fire may be delayed because the residents may be sleeping. With this additional safety concern, the requirements for corridors are more restrictive than nonresidential occupancies. If the corridor serves more than 10 occupants, it is required to be rated. If the building is sprinklered throughout with either an NFPA 13 or 13R system, then the rating on the corridor may be reduced to $1/2$ hour. Note the exception for rated corridors within an individual dwelling or sleeping unit in Section 1016.1. Also note that the reduction in the rating of the corridor walls is not permitted when an NFPA 13D sprinkler system is provided.

While all Group I facilities are supervised environments, the level of supervision in Group I-2 and I-4 occupancies would permit assisted evacuation by staff in an emergency; therefore, corridors are not required to be rated. Corridors in Groups I-2 are also regulated by Section 407.3. Due to the lower level of staff/resident ratio in Group I-1 and the limitation on free egress in Group I-3, corridors must have a 1-hour fire-resistance rating (see Section 408.7 for a reduction in the corridors in Group I-3). The "not permitted" in the third column is in coordination with Section 903.2.5, which requires all Group I buildings to be fully sprinklered.

1016.2 Corridor width. The minimum corridor width shall be as determined in Section 1005.1, but not less than 44 inches (1118 mm).

Exceptions:

1. Twenty-four inches (610 mm)—For access to and utilization of electrical, mechanical or plumbing systems or equipment.

2. Thirty-six inches (914 mm)—With a required occupant capacity of 50 or less.

3. Thirty-six inches (914 mm)—Within a dwelling unit.

4. Seventy-two inches (1829 mm)—In Group E with a corridor having a required capacity of 100 or more.

5. Seventy-two inches (1829 mm)—In corridors serving surgical Group I, health care centers for ambulatory patients receiving outpatient medical care, which causes the patient to be not capable of self-preservation.

6. Ninety-six inches (2438 mm)—In Group I-2 in areas where required for bed movement.

❖ The corridor widths specified in this section are minimums (see Section 1005.1 to determine the corridor width based on the number of occupants served by the corridor). Remember, if a corridor goes in two directions, this number can be divided in half. The wider of the two requirements must be utilized for corridor design.

The width of passageways, aisles and corridors is a functional element of building construction that allows the occupants to circulate freely and comfortably throughout the floor area under nonemergency conditions. Under emergency situations, the egress passageways must provide the needed width to accommodate the number of occupants that must utilize the corridor for egress.

When the occupant load of the space exceeds 50, the minimum width of the passageway, aisle or corridor serving that space is required to be 44 inches (1118 mm) to permit two unimpeded parallel columns of users to travel in opposite directions. When the occupant load served is 50 or less, a minimum width of 36 inches (914 mm) is permitted and the users are expected to encounter some intermittent travel interference from fellow users, but the lower occupant load makes those occasions infrequent and tolerable. The 36-inch (914 mm) minimum width is also required within a dwelling unit.

Passageways that lead to building equipment and systems must be at least 24 inches (610 mm) in width to provide a means to access and service the equipment when needed. Due to the frequency of the servicing intervals and the limited number of occupants in these areas, a reduced width is warranted. This minimum width criteria applies to many common situations, such as stage lighting and special effects catwalks; catwalks leading to heating and cooling equipment as well as passageways providing access to boilers, furnaces, transformers, pumps and other equipment.

Except for small buildings, Group E occupancies are required to have minimum 72-inch-wide (1829 mm) corridors where they corridor serve educational areas. This width is needed not only for proper functional use, but because of the edge effect caused by student lockers and other boundary attractions and defects. Service and other corridors outside of educational areas, such as an administrative area, would be regulated by the remaining criteria.

In Group I-2 occupancies, where the corridor is utilized during a fire emergency for moving patients confined to beds, it is required to be at least 96 inches (2438 mm) in clear width. This width requirement is only applicable to Group I-2, since occupants in other occupancy classifications are assumed to be ambulatory. This minimum width allows two beds to pass in a corridor and permits the movement of a bed into the corridor through a room door. In Group I-2 areas, where the movement of beds is not anticipated, such as administrative and some outpatient areas of a hospital, the corridor would not be required to be 96 inches (2438 mm) wide. The minimum width would be determined by one of the remaining applicable criteria. For outpatient medical care, where the patient may not be capable of self-preservation, such as some outpatient surgery areas or dialysis treatment areas, the 72-inch-wide (1829 mm) corridor is required based on Exception 5.

1016.3 Dead ends. Where more than one exit or exit access doorway is required, the exit access shall be arranged such that there are no dead ends in corridors more than 20 feet (6096 mm) in length.

Exceptions:

1. In occupancies in Group I-3 of Occupancy Condition 2, 3 or 4 (see Section 308.4), the dead end in a corridor shall not exceed 50 feet (15 240 mm).

2. In occupancies in Groups B and F where the building is equipped throughout with an automatic sprinkler system in accordance with Section 903.3.1.1, the length of dead-end corridors shall not exceed 50 feet (15 240 mm).

3. A dead-end corridor shall not be limited in length where the length of the dead-end corridor is less than 2.5 times the least width of the dead-end corridor.

❖ The requirements of this section apply where a space is required to have more than one means of egress according to Section 1014.1

Dead ends in corridors and passageways can seriously increase the time needed for an occupant to locate the exits. More importantly, dead ends will allow a single fire event to eliminate access to all of the exits by trapping the occupants in the dead-end area. A dead end exists if the occupant of the corridor or passageway has only one direction to travel to reach any of the building exits [see Figure 1016.3(1)]. While a preferred building layout would be one without dead ends, a maximum dead-end length of 20 feet (6096 mm) is permitted and is to be measured from the extreme point in the dead end to the point where the occupants have a choice of two directions to the exits. Having to go back only 20 feet (6096 mm) after coming to a dead end is not such a significant distance as to cause a serious delay in reaching an exit during an emergency situation.

A dead end results whether or not egress elements open into it. A dead end is a hazard for occupants who enter the area from adjacent spaces, travel past an exit into a dead end or enter a dead end with the mistaken assumption that an exit is directly accessible from the dead end.

Note that Section 402.4.5 deals with dead-end distances in a covered mall and assumes that, with a sufficiently wide mall in relation to its length, alternative paths of travel will be available in the mall itself to reach an exit (i.e., the mall is not to be confused as being a corridor).

Under special conditions, exceptions to the 20-foot (6096 mm) dead-end limitation apply.

Exception 1 is permitted based on the considerations of the functional needs of Group I-3 Occupancy Conditions 2, 3 or 4, the requirements for smoke compartmentalization in Section 408.6 and the requirement for automatic sprinkler protection of the facility in Section 903.2.5.

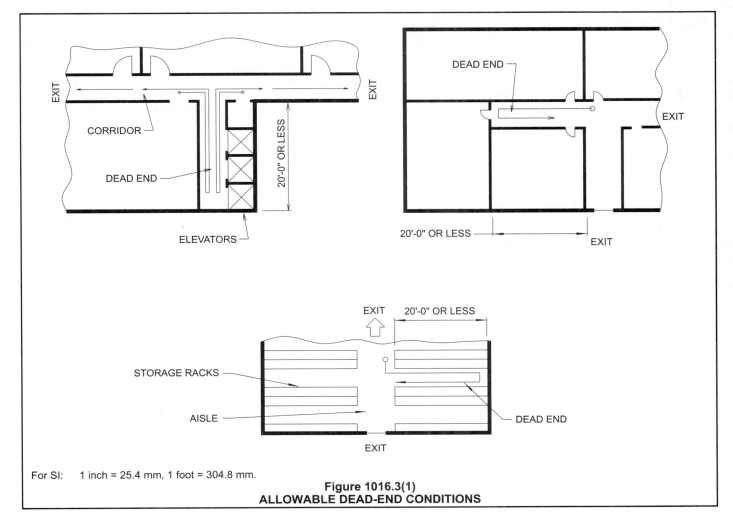

For SI: 1 inch = 25.4 mm, 1 foot = 304.8 mm.

Figure 1016.3(1)
ALLOWABLE DEAD-END CONDITIONS

Exception 2 recognizes the fire protection benefits and performance history of automatic fire sprinkler systems. While the degree of hazard in Group B occupancies does not generally require an automatic fire suppression system, the length of a dead-end corridor or passageway is permitted to be extended to 50 feet (15 240 mm) where an automatic fire sprinkler system in accordance with NFPA 13 (see Section 903.3.1.1) is provided throughout the building. This exception is also permitted in Group F occupancies.

Exception 3 addresses the condition presented by "cul-de-sac" elevator lobbies directly accessible from exit access corridors. In such an elevator lobby, lengths of 20 to 30 feet (6096 to 9144 mm) are common for three- or four-car elevator banks. Typically, the width of this elevator lobby is such that the possibility of confusion with a path of egress is minimized. Below the 2.5:1 ratio, the dead end becomes so wide that it is less likely to be perceived as a corridor leading to an exit. For example, based on the 2.5:1 ratio limitation, a 25-foot-long (7620 mm) dead end, over 10 feet (3048 mm) in width, would not be considered a dead-end corridor [see Figure 1016.3(2)]. For additional elevator lobby requirements, see the commentary to Section 707.14.1.

1016.4 Air movement in corridors. Exit access corridors shall not serve as supply, return, exhaust, relief or ventilation air ducts.

Exceptions:

1. Use of a corridor as a source of makeup air for exhaust systems in rooms that open directly onto such corridors, including toilet rooms, bathrooms, dressing rooms, smoking lounges and janitor closets, shall be permitted provided that each such corridor is directly supplied with outdoor air at a rate greater than the rate of makeup air taken from the corridor.

2. Where located within a dwelling unit, the use of corridors for conveying return air shall not be prohibited.

3. Where located within tenant spaces of 1,000 square feet (93 m²) or less in area, utilization of corridors for conveying return air is permitted.

❖ Two of the most critical elements of the means of egress are the required exit stairways and exit access corridors. Exit stairways serve as protected areas in the building that provide occupants with safe passage to the level of exit discharge. Corridors that provide access to the required exits frequently limit the direction of egress travel (e.g., travel forward or backward only). Since required exits and exit access corridors are critical elements in the means of egress, the potential spread of smoke and fire into these spaces must be minimized. The scope of this section is exit access corridors. For requirements for the exits, see Section 1019.

The use of these corridors as part of the air distribution system could render those egress elements unusable. Therefore, any air movement condition that could introduce smoke into these vital egress elements is prohibited. It is not the intent of this section to prohibit the air movement necessary for ventilation and space conditioning of exit access corridors, but rather to prevent those spaces from serving as conduits for the distribution of air to or the collection of air from, adjacent

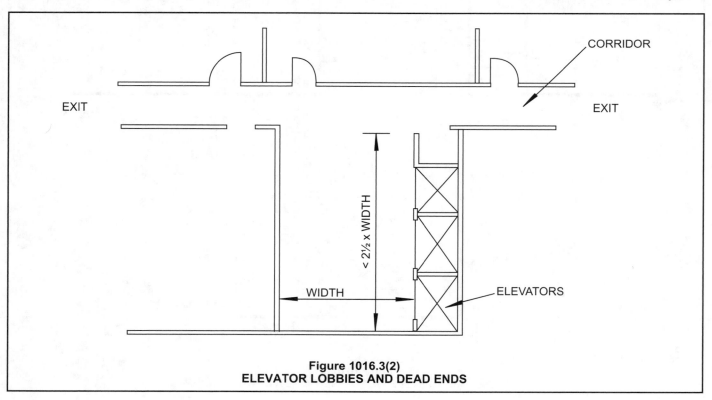

Figure 1016.3(2)
ELEVATOR LOBBIES AND DEAD ENDS

spaces. This restriction also extends to door transoms and door grilles that would allow the spread of smoke into an exit access corridor. This limitation is not, however, intended to restrict slight pressure differences across corridor doors, such as a negative pressure differential maintained in kitchens to prevent odor migration into dining rooms.

The three exceptions to this section identify conditions where an exit access corridor can be utilized as part of the air distribution system. The exceptions do not apply to exits.

Exception 1 addresses the common practice of using air from the corridor as makeup air for small exhaust fans in adjacent rooms. Where the corridor is supplied directly with outdoor air at a rate equal to or greater than the makeup air rate, negative pressure will not be created in the corridor with respect to the adjoining rooms and smoke would generally not be drawn into the corridor.

Regarding Exception 2, it is common practice to locate return air openings in the corridors of dwelling units and draw return air from adjoining spaces through the corridor. Such use of dwelling unit exit access corridors for conveying return air is not considered to be a significant hazard and is permitted. Dwelling units are permitted to have unprotected openings between floors. Corridors in dwelling units serve small occupant loads, are short in length and are not required to be fire-resistance rated. For these reasons, the use of the corridor or the space above a corridor ceiling for conveying return air does not constitute an unacceptable hazard.

Exception 3 permits exit access corridors located in small tenant spaces to be used for conveying return air based on the relatively low occupant load and the relatively short length of the corridor. These conditions do not pose a significant hazard. In the event of an emergency, the occupants of the space would tend to simply retrace their steps to the entrance.

1016.4.1 Corridor ceiling. Use of the space between the corridor ceiling and the floor or roof structure above as a return air plenum is permitted for one or more of the following conditions:

1. The corridor is not required to be of fire-resistance-rated construction;

2. The corridor is separated from the plenum by fire-resistance-rated construction;

3. The air-handling system serving the corridor is shut down upon activation of the air-handling unit smoke detectors required by the *International Mechanical Code*.

4. The air-handling system serving the corridor is shut down upon detection of sprinkler waterflow where the building is equipped throughout with an automatic sprinkler system; or

5. The space between the corridor ceiling and the floor or roof structure above the corridor is used as a component of an approved engineered smoke control system.

❖ This section identifies five different conditions where the space above the corridor ceiling is permitted to serve as

a return air plenum only. Since a return air plenum operates at a negative pressure with respect to the corridor, any smoke and gases within the plenum should be contained within that space. Conversely, a supply plenum operates at a positive pressure with respect to the corridor, thus increasing the likelihood that smoke and gases will infiltrate the corridor enclosure. Where any one of the five conditions is present, the use of the corridor ceiling space as a return air plenum is permitted.

Where the corridor is permitted to be constructed without a fire-resistance rating (see Section 1016.1), Item 1 permits the space above the ceiling to be utilized as a return air plenum without requiring it to be separated from the corridor with fire-resistance-rated construction.

Item 2 is only applicable to corridors that are required to be enclosed with fire-resistance-rated construction. Compliance with this item requires the plenum to be separated from the corridor by fire-resistance-rated construction equivalent to the rating of the corridor enclosure itself. Therefore, the ceiling membrane itself must provide the fire-resistance rating required of the corridor enclosure. Section 708.4, Exception 3, is an example of this method of construction.

Items 3 and 4 recognize that the hazard associated with smoke spread through a plenum is minimized if the air movement is stopped.

It is not uncommon for an above-ceiling plenum to be utilized as part of the smoke removal system. This practice is permitted by Item 5. Due to the way these systems are designed, the higher equipment ratings and the power supply provisions, this is considered acceptable.

1016.5 Corridor continuity. Fire-resistance-rated corridors shall be continuous from the point of entry to an exit, and shall not be interrupted by intervening rooms.

Exception: Foyers, lobbies or reception rooms constructed as required for corridors shall not be construed as intervening rooms.

❖ This section requires the fire protection offered by a corridor to be continuous from the point of entry into the corridor and to an exit. This is to protect occupants from the accumulation of smoke or fire exposure and to allow for sufficient time to evacuate the building. For requirements for elevator lobbies that are adjacent to rated corridors, see Section 707.14.1. For requirements for exit enclosures, see Section 1019. Note that when vertical exit enclosures are not required around an exit stairway or ramp, the "point of entry to the exit" would be the top of the stairway or ramp.

The exception allows a "reception room" to be located on the path of egress from a corridor, provided the room has the same fire-resistance-rated walls and doors as required for the corridor. While the term "reception room" is not defined, use of the term should be viewed as limiting the types of uses that may occur within the protected corridor. Occupied spaces within the corridor should have very limited uses and hazards. Foyers and lobbies are included in this exception based on the low fire hazard of the contents in such rooms.

SECTION 1017
EXITS

1017.1 General. Exits shall comply with Sections 1017 through 1022 and the applicable requirements of Sections 1003 through 1012. An exit shall not be used for any purpose that interferes with its function as a means of egress. Once a given level of exit protection is achieved, such level of protection shall not be reduced until arrival at the exit discharge.

❖ The use of required exterior exit doors, exit stairways, exit passageways and horizontal exits for any purpose other than exiting is prohibited, because it might interfere with use as an exit. For example, the use of an exit stairway landing for storage, vending machines, copy machines or any purpose other than for exiting is not permitted. Such a situation could not only lead to obstruction of the path of exit travel, thereby creating a hazard to life safety, but if the contents consist of combustible materials, then the use of the stairway as a means of egress could be jeopardized by a fire in the exit enclosure. The restriction does not apply solely to combustible contents, because of the potential for obstruction and difficulties associated in limiting the materials to noncombustible.

It is recognized that standpipe risers are provided within the stair enclosure and that vertical electrical conduit may be necessary for power or lighting. However, such risers must be located so as not to interfere with the required clear width of the exit. For example, a standpipe riser located in the corner of a stairway will not reduce the required clear width of the landing. Electrical conduit and mechanical equipment are permitted when required for stairway equipment.

Section 1017 through 1022 applies to all exits but does not apply to elements of the means of egress that are not exits, such as exit access corridors and passageways or elements of the exit discharge. For exit discharge options for the enclosure, other than a door leading directly to the outside, see the requirements for exit passageways in Section 1020 or the options permitting usage of a lobby or vestibule in the exceptions to Section 1023.1.

1017.2 Exterior exit doors. Buildings or structures used for human occupancy shall have at least one exterior door that meets the requirements of Section 1008.1.1.

❖ The purpose of this section is to specify that at least one exterior exit door is required to meet the door size requirements in Section 1008.1.1. It is not the intent of this section to specify the number of exit doors required, which is addressed in Section 1018.1.

1017.2.1 Detailed requirements. Exterior exit doors shall comply with the applicable requirements of Section 1008.1.

❖ The purpose of this section is simply to provide a cross reference from the exit section to all of the detailed requirements for doors that are included in Section 1008.1 and all of its subsections. For example, the requirements for door operation on exterior exit doors are controlled by Section 1008.1.8.

1017.2.2 Arrangement. Exterior exit doors shall lead directly to the exit discharge or the public way.

❖ The exterior exit door is to be the entry point of the exit discharge or lead directly to the public way. When a person reaches the exterior exit door, he or she is directly outside, where smoke and toxic gases are not a health hazard. Additionally, this section will keep exterior doors at other locations, such as to an exit balcony, from being viewed as an exit.

SECTION 1018
NUMBER OF EXITS AND CONTINUITY

1018.1 Minimum number of exits. All rooms and spaces within each story shall be provided with and have access to the minimum number of approved independent exits as required by Table 1018.1 based on the occupant load, except as modified in Section 1014.1 or 1018.2. For the purposes of this chapter, occupied roofs shall be provided with exits as required for stories. The required number of exits from any story, basement or individual space shall be maintained until arrival at grade or the public way.

❖ This section requires every floor of a building to be served by at least two exits (see Figure 1018.1). Similarly, every portion of a floor must also be provided with access to at least two exits. Where more than 500 occupants are located on a single floor, additional exits must be provided. For example, if a floor has an occupant load of 750, each occupant of that floor must have access to not less than three exits (see Table 1018.1).

The need for exits to be independent of each other cannot be overstated. Each occupant of each floor must be provided with the required number of exits without having to pass through one exit to gain access to another. Each exit is required to be independent of other exits to prohibit such areas from merging downstream and becoming, in effect, one exit.

As discussed in conjunction with Section 1014.2, the location of two exits is to accomplish certain remoteness, and any additional exits are to be accessible from any point on the floor.

This section also addresses the need for exits from roofs that are occupied by the public, such as roof-top restaurants. It is important that the exit enclosures serve the roof level in addition to the other floor levels.

The text references Sections 1018.2 and 1014.1 (see Section 1018.2 for a discussion of buildings that are allowed to have only one exit). Where a building requires more than one exit, that number, determined in accordance with Table 1018.1, is required from each floor level. This is true even if the floor level qualifies as a space with one means of egress in accordance with Table 1014.1. Table 1014.1 is intended to be applicable to rooms and spaces, but not entire floor levels. One of the main concerns has been that vertical travel takes longer than horizontal travel in emergency exiting situations. This is consistent with Section 1018.2, Item 3.

While not specifically stated in this section, there are two situations where the single means of egress can be used from a multilevel space. Section 505 permits mezzanines to be considered part of the floor below for purposes of means of egress. When a mezzanine meets the occupant load in Table 1014.1 and the common path of travel distance (see Section 1013.3) measured from the most remote point to the bottom of the stairway, it can be considered a space with one means of egress. The second situation is more interpretive, but is based on multiple years of practice within an individual dwelling or sleeping unit. In buildings with multistory dwelling or sleeping units, the means of egress from a dwelling or sleeping unit is typically permitted to be from one level only. If the unit has an occupant load of 10 or less (see Table 1014.1), and the common path of travel from the most remote point on any level to the exit door from the unit itself is 75 feet (22 860 mm) maximum (see Section 1013.3), that unit may have only one means of egress. However, once the occupants exit the unit itself, the floor level must have access to two or more means of egress for all tenants, depending on the number required for the building as a whole. This would be consistent with the residential exception found in Section 1018.2, Item 2.

TABLE 1018.1
MINIMUM NUMBER OF EXITS FOR OCCUPANT LOAD

OCCUPANT LOAD	MINIMUM NUMBER OF EXITS
1-500	2
501-1,000	3
More than 1,000	4

❖ Table 1018.1 specifies that the minimum number of exits available to each occupant of a floor is based on the total occupant load of that floor. This is so that at least one exit will be available in case of a fire emergency and to provide increased assurance that a larger number of occupants can be accommodated by the remaining exits, when one exit is not available. While an equal distribution of exit capacity among all of the exits is not required, a proper design would not only balance capacity with the occupant load distribution, but also consider a reasoned distribution of capacity to avoid a severe dependence on one exit.

1018.1.1 Open parking structures. Parking structures shall not have less than two exits from each parking tier, except that only one exit is required where vehicles are mechanically parked.

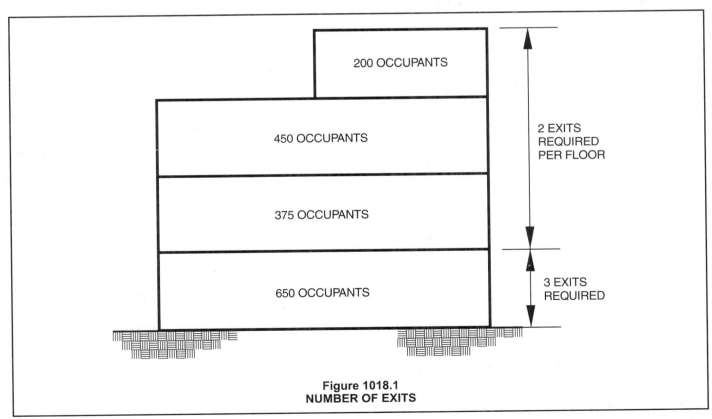

Figure 1018.1
NUMBER OF EXITS

Unenclosed vehicle ramps shall not be considered as required exits unless pedestrian facilities are provided.

❖ At least two exits from each parking tier are required in open parking structures, except where the resulting occupant load is minimal, such as where the vehicles are mechanically parked. A vehicle ramp may be considered as one of the required exits if pedestrian walkways are provided along the ramp. According to Section 1019.1, Exception 5, the exit stairways are not required to be enclosed, since an open parking structure is designed to permit the ready ventilation of the products of combustion to the outside by exterior wall openings (see Section 406.3). Also, parking structures are characterized by open floor areas that allow the occupants to observe a fire condition and choose a travel path that would avoid the fire threat.

1018.1.2 Helistops. The means of egress from helistops shall comply with the provisions of this chapter, provided that landing areas located on buildings or structures shall have two or more exits. For landing platforms or roof areas less than 60 feet (18 288 mm) long, or less than 2,000 square feet (186 m²) in area, the second means of egress is permitted to be a fire escape or ladder leading to the floor below.

❖ This section provides occupants of helistops with adequate exit facilities. The reduction in the exit requirements for small-area helistops is based on the low number of occupants that are associated with these facilities.

1018.2 Buildings with one exit. Only one exit shall be required in buildings as described below:

1. Buildings described in Table 1018.2, provided that the building has not more than one level below the first story above grade plane.

2. Buildings of Group R-3 occupancy.

3. Single-level buildings with the occupied space at the level of exit discharge provided that the story or space complies with Section 1014.1 as a space with one means of egress.

❖ Buildings with one exit are permitted where the configuration and occupancy meet certain characteristics so as not to present an unacceptable fire risk to the occupants. Buildings that are relatively small in size have a shorter travel distance and fewer occupants; thus, having access to a single exit does not significantly compromise the safety of the occupants since they will also be alerted to and get away from the fire more quickly. It is important to note that the provisions in Section 1018.2 apply to entire buildings only, not individual stories or fire areas.

Occupants of a story of limited size and configuration may have access to a single exit, provided that the building does not have more than one level below the first story above the grade plane (see Table 1018.2 for a list of groups and building characteristics where the occupants are permitted to have access to a single exit). The limitation on the number of levels below the first story is intended to limit the vertical travel an occupant

must accomplish to reach the exit discharge in a single-exit building. Taken to its extreme, the code would otherwise allow a single exit in a building with one or two stories above grade and an unlimited number of stories below grade. This would be clearly inadequate for those stories below grade.

Group R-3 building occupancies are permitted to have a single exit since they do not have more than two dwelling units.

Item 3 is a correlation with the requirements of Section 1014.1, which permits a single egress path for certain spaces having a low occupant load. When a single egress doorway is permitted from a given space, it is consistent to permit a single exit to serve a one-story building of the same size as that space.

While the term "building" limits the area addressed to that bordered by exterior walls or fire walls, a common application of Item 3 is on a tenant-by-tenant basis. For example, a strip mall may not meet the provisions for a building with one means of egress. However, assume a tenant meets the provisions for a space with one means of egress in accordance with Section 1014.1. This tenant could exist as either a stand-alone single-exit building or a single-exit tenant space that exited into an interior corridor. Is it not as safe to permit this tenant to exist as part of a larger building with the door exiting directly to the exterior? Final approval of this alternative approach would be subject to the building official having jurisdiction.

TABLE 1018.2
BUILDINGS WITH ONE EXIT

OCCUPANCY	MAXIMUM HEIGHT OF BUILDING ABOVE GRADE PLANE	MAXIMUM OCCUPANTS (OR DWELLING UNITS) PER FLOOR AND TRAVEL DISTANCE
A, B[d], E, F, M, U	1 Story	50 occupants and 75 feet travel distance
H-2, H-3	1 Story	3 occupants and 25 feet travel distance
H-4, H-5, I, R	1 Story	10 occupants and 75 feet travel distance
S[a]	1 Story	30 occupants and 100 feet travel distance
B[b], F, M, S[a]	2 Stories	30 occupants and 75 feet travel distance
R-2	2 Stories[c]	4 dwelling units and 50 feet travel distance

For SI: 1 foot = 304.8 mm.

a. For the required number of exits for open parking structures, see Section 1018.1.1.

b. For the required number of exits for air traffic control towers, see Section 412.1.

c. Buildings classified as Group R-2 equipped throughout with an automatic sprinkler system in accordance with Section 903.3.1.1 or 903.3.1.2 and provided with emergency escape and rescue openings in accordance with Section 1025 shall have a maximum height of three stories above grade.

d. Buildings equipped throughout with an automatic sprinkler system in accordance with Section 903.3.1.1 with an occupancy in Group B shall have a maximum travel distance of 100 feet.

❖ Table 1018.2 lists the characteristics a building must have to be of single-exit construction, including occupancy, maximum height of building above grade plane, maximum occupants or dwelling units per floor and exit access travel distance per floor. The occupant load of each floor is determined in accordance with the provisions of Section 1004.1. The exit access travel distance is measured along the natural and unobstructed path to the exit, as described in Section 1015.1. If the occupant load is exceeded, as indicated in Table 1018.2, two exits are required from each floor in the building. Likewise, if the travel distance or number of dwelling units is exceeded, as indicated in Table 1018.2, two exits are required from each floor of the building.

The exit enclosure required in a two-story, single-exit building is identical to any other complying exit (i.e., interior stairs, exterior stairs, etc.). Similarly, the fire-resistance rating required for opening protectives is identical to that required by Section 714.

1018.3 Exit continuity. Exits shall be continuous from the point of entry into the exit to the exit discharge.

❖ The intent of this section is to provide safety in all portions of the exit by requiring continuity of the fire protection characteristics of the exit enclosure. This would include, but not be limited to, the fire-resistance rating of the exit enclosure walls and the opening protection rating of the doors. The code provides no exception for this requirement, which also reinforces the protection issues addressed in Section 1017.1.

1018.4 Exit door arrangement. Exit door arrangement shall meet the requirements of Sections 1014.2 through 1014.2.2.

❖ The intent of this section is to provide a cross reference to the exit door requirements.

SECTION 1019
VERTICAL EXIT ENCLOSURES

1019.1 Enclosures required. Interior exit stairways and interior exit ramps shall be enclosed with fire barriers. Exit enclosures shall have a fire-resistance rating of not less than 2 hours where connecting four stories or more and not less than 1 hour where connecting less than four stories. The number of stories connected by the shaft enclosure shall include any basements but not any mezzanines. An exit enclosure shall not be used for any purpose other than means of egress. Enclosures shall be constructed as fire barriers in accordance with Section 706.

Exceptions:

1. In other than Group H and I occupancies, a stairway serving an occupant load of less than 10 not more than one story above the level of exit discharge is not required to be enclosed.

2. Exits in buildings of Group A-5 where all portions of the means of egress are essentially open to the outside need not be enclosed.

3. Stairways serving and contained within a single residential dwelling unit or sleeping unit in occupancies in

Group R-2 or R-3 and sleeping units in occupancies in Group R-1 are not required to be enclosed.

4. Stairways that are not a required means of egress element are not required to be enclosed where such stairways comply with Section 707.2.

5. Stairways in open parking structures which serve only the parking structure are not required to be enclosed.

6. Stairways in occupancies in Group I-3 as provided for in Section 408.3.6 are not required to be enclosed.

7. Means of egress stairways as required by Section 410.5.4 are not required to be enclosed.

8. In other than occupancy Groups H and I, a maximum of 50 percent of egress stairways serving one adjacent floor are not required to be enclosed, provided at least two means of egress are provided from both floors served by the unenclosed stairways. Any two such interconnected floors shall not be open to other floors.

9. In other than occupancy Groups H and I, interior egress stairways serving only the first and second stories of a building equipped throughout with an automatic sprinkler system in accordance with Section 903.3.1.1 are not required to be enclosed, provided at least two means of egress are provided from both floors served by the unenclosed stairways. Such interconnected stories shall not be open to other stories.

❖ This section requires that all interior exit stairways or ramps are to be enclosed with fire barriers having a fire-resistance rating of at least 1 hour. The fire-resistance rating of the enclosure must be increased to 2 hours if the stairway or ramp connects four or more stories. Note that the criteria are based on the number of stories connected by the stairway or ramp and not the height of the building. Therefore, a building that has three stories located entirely above the grade plane and a basement would require an enclosure with a 2-hour fire-resistance rating if the stairway or ramp connects all four stories.

The enclosure is needed because an exit stairway or ramp penetrates the floor/ceiling assemblies between the levels, thus creating a vertical opening or shaft. In cases of fire, a vertical opening may act as a chimney, causing smoke, hot gases and light-burning products to flow upward (buoyant force). If an opening is unprotected, these products of combustion will be forced by positive pressure differentials to spread horizontally into the building spaces.

The enclosure of interior stairways or ramps with construction having a fire-resistance rating is intended to prevent the spread of fire from floor to floor. Another important purpose is to provide a safe path of travel for the building occupants and to serve as a protected means of access to the fire floor by fire department personnel. For this reason, this section and Section 1019.1.1 limit the penetrations and openings permitted in a stairway enclosure.

It is important that an exit stairway not be used for any purpose other than as a means of egress. For example, there is a tendency to use stairway landings for storage purposes. Such a situation obstructs the path of exit travel and if the stored contents consist of combustible materials, the use of the stairway as a means of egress may be jeopardized, creating a hazard to life safety. However, the restriction on the use of an exit stairway is not limited to situations when the contents are combustible.

For travel distance measurements at the exit stairways in the exceptions, see the commentary to Section 1015.1. The exceptions are only for the vertical exit enclosure. Any exit stairway or ramp must still comply with exit discharge requirements of Section 1023 or go into an exit passageway per Section 1020.

This section allows exceptions to the enclosure requirements stated above:

Exception 1 allows an open exit stairway from the first to the second floor where the stairway does not serve more than 10 occupants. An example of the application of this exception is a retail space that has a small office located on the second story. This exception allows the stair serving the office to be unenclosed based on the small number of occupants served.

In Exception 2, stairways in occupancies in Group A-5 in which the means of egress is essentially open to the outside need not be enclosed because of the ability to vent the fire to the outside. The criteria specified in Section 1022.1 should be used to determine if the stairway is "essentially open."

In Exception 3, the exit stairways in the listed Group R occupancies are not required to have enclosures because of the small occupancy load. This exception is limited to stairways located within the individual dwelling or sleeping units.

In Exception 4, stairways that do not serve as a means of egress are not subject to the enclosure requirements (see commentary, Section 707.2). Such stairways may include ornamental stairways within an atrium or any stairways provided in excess of the minimum number required for egress and not designed as an exit. While not specifically stated in this exception, Section 707.2 would also permit a ramp that was not part of a required means of egress from enclosure requirements.

In Exception 5, stairways located in open parking structures are exempt from the enclosure requirements because of the ease of accessibility by the fire services, the natural ventilation of such structures, the low level of fire hazard, the small number of people using the structure at any one time and the excellent fire record of such structures (see commentary, Section 1018.1.1).

In Exception 6, because of security needs in detention facilities, one of the exit stairs is permitted to be glazed in a manner similar to atrium enclosures. Specific limitations and requirements are discussed in Section 408.3.6.

Regarding Exception 7, due to the nature of stages and platforms, stage exit stairways from the side of a stage, spaces under the stage, fly galleries and gridirons are not required to be enclosed.

Exception 8 allows certain portions of a required exit stairway to be unenclosed. This exception may be used in either sprinklered or nonsprinklered buildings; however, it may not be used in Group H or I occupancies. Since this exception is limited to 50 percent of the egress stairways, it may not be used in a single-exit building.

There are two instances where this exception has been used. The first is for two-story buildings where one of the required stairways is required to be enclosed but the other is not. The second case is for a multistory building where one of the required exit stairways between levels may be open between two adjacent levels. However, in this situation, after moving down one level, the required number of enclosed exits must be available. Once the enclosed exit is entered, that level of protection for the occupants must be maintained until they reach the exit discharge. The most typical example of the second case would be to provide a second exit for a tenant who did not have access to two enclosed exits on the upper floor because the second stair was only available through another tenant space. Access to exits may not be through another tenant space in accordance with Sections 1013.2 and 1013.2.1.

It should be noted that a stair shaft could not be discontinued at some point in the height of a multistory building and then continued one floor level lower, such that a person would have to come out of the stair enclosure and then back into it at the next level lower. This arrangement would not be allowed according to Sections 1017.1 and 1018.3, which require the level of protection of an exit to be continuous from the point of entry into the exit to the exit discharge.

Exception 9 is limited to buildings sprinklered throughout by an NFPA 13 system (see Section 903.3.1.1). This exception is not available for Group H or I occupancies. Literally, this exception would permit the second floor of any building to use two, three or four unenclosed stairways to serve as the required exits for that level. However, in most cases in a multistory building, the second level would utilize the enclosed exit stairways moving down from the upper floors. Sections 1017.1 and 1018.3 require the level of protection of an exit to be continuous from the point of entry, to the exit and to the exit discharge; therefore, the exit stairways coming down from the upper level could not be open between the second and first level. Therefore, this particular exception will most typically be utilized in two-story buildings.

1019.1.1 Openings and penetrations. Exit enclosure opening protectives shall be in accordance with the requirements of Section 715.

Except as permitted in Section 402.4.6, openings in exit enclosures other than unexposed exterior openings shall be limited to those necessary for exit access to the enclosure from normally occupied spaces and for egress from the enclosure.

Where interior exit enclosures are extended to the exterior of a building by an exit passageway, the door assembly from the exit enclosure to the exit passageway shall be protected by a fire door conforming to the requirements in Section 715.3. Fire door assemblies in exit enclosures shall comply with Section 715.3.4.

❖ The only openings that are permitted in fire-resistance-rated exit stairways or ramp enclosures are doors that lead either from normally occupied spaces or from the enclosure. This restriction on openings essentially prohibits the use of windows in an exit enclosure except for those exterior windows that are not exposed to any hazards. This requirement is not intended to prohibit windows or other openings in the exterior walls of the exit enclosure. The verbiage "unexposed exterior openings" includes windows or doors not required to be protected by either Section 704.8 or 1019.1.4. The only exception would be window assemblies that have been tested as wall assemblies in accordance with ASTM E 119. The objective of this provision is to minimize the possibility of fire spreading into an exit enclosure and endangering the occupants or even preventing the use of the exit at a time when it is most needed. The limitation on openings applies regardless of the fire protection rating of the opening protective. The limitation on openings from normally occupied areas is intended to reduce the probability of a fire occurring in an unoccupied area, such as a storage closet, which has an opening into the stairway, thereby possibly resulting in fire spread into the stairway. Other spaces that are not normally occupied include, but are not limited to, toilet rooms, electrical/mechanical equipment rooms and janitorial closets.

The reference to the specific criteria for covered malls in Section 402.4.6 is not applicable to exit enclosures between levels. Its application is limited to exit passageways, which are addressed in Section 1020.

The last paragraph addresses when the exit enclosure could move from the vertical enclosure of a stairway or ramp to the horizontal enclosure of an exit passageway. While the exit passageway is an extension of the protection offered by the vertical exit, there must still be an opening protective at the bottom of the stairway or ramp. This is to prevent any smoke that may migrate into the exit passageway from also moving up the exit stairway or ramp. A door is also required at the bottom of the vertical enclosure for the exit discharge options for lobbies and vestibules (see Section 1023.1).

These opening requirements are very similar to those required for an exit passageway (see Section 1020.4).

1019.1.2 Penetrations. Penetrations into and openings through an exit enclosure are prohibited except for required exit doors, equipment and ductwork necessary for independent pressurization, sprinkler piping, standpipes, electrical raceway for fire department communication and electrical raceway serving the exit enclosure and terminating at a steel box not exceeding 16 square inches (0.010 m²). Such penetrations shall be protected in accordance with Section 712. There shall be no penetrations or communication openings, whether protected or not, between adjacent exit enclosures.

❖ This section specifically lists the items that are allowed to penetrate a vertical exit enclosure. This is consistent for all types of exit enclosures, including stair or ramp vertical exit enclosures and exit passageways (see Section 1020.5). In general, only portions of the building service systems that serve the exit enclosure are allowed to penetrate the exit enclosure. As indicated in the commentary to Section 1017.1, standpipe systems are commonly located in the exit stair enclosures. If two exit enclosures are adjacent to one another there must be no penetrations between them to limit the chances of smoke being in both stairwells. This requirement is not intended to prohibit windows on exterior walls of stairways that are not required to be rated.

1019.1.3 Ventilation. Equipment and ductwork for exit enclosure ventilation shall comply with one of the following items:

1. Such equipment and ductwork shall be located exterior to the building and shall be directly connected to the exit enclosure by ductwork enclosed in construction as required for shafts.

2. Where such equipment and ductwork is located within the exit enclosure, the intake air shall be taken directly from the outdoors and the exhaust air shall be discharged directly to the outdoors, or such air shall be conveyed through ducts enclosed in construction as required for shafts.

3. Where located within the building, such equipment and ductwork shall be separated from the remainder of the building, including other mechanical equipment, with construction as required for shafts.

In each case, openings into the fire-resistance-rated construction shall be limited to those needed for maintenance and operation and shall be protected by self-closing fire-resistance-rated devices in accordance with Chapter 7 for enclosure wall opening protectives.

Exit enclosure ventilation systems shall be independent of other building ventilation systems.

❖ The purpose of the requirements for the ventilation system equipment and ductwork is to maintain the fire resistance of the exit enclosure. The exit enclosure ventilation system is to be independent of other building systems to prevent smoke in the exit enclosure from traveling to other areas of the building.

1019.1.4 Vertical enclosure exterior walls. Exterior walls of a vertical exit enclosure shall comply with the requirements of Section 704 for exterior walls. Where nonrated walls or unprotected openings enclose the exterior of the stairway and the walls or openings are exposed by other parts of the building at an angle of less than 180 degrees (3.14 rad), the building exterior walls within 10 feet (3048 mm) horizontally of a nonrated wall or unprotected opening shall be constructed as required for a minimum 1-hour fire-resistance rating with $^3/_4$-hour opening protectives. This construction shall extend vertically from the ground to a point 10 feet (3048 mm) above the topmost landing of the stairway or to the roof line, whichever is lower.

❖ This section does not require exterior walls of a stairway enclosure to have the same fire-resistance rating as interior walls; however, to minimize the potential fire spread into the stairway from the exterior, the issue of the exterior wall of the stairway and adjacent exterior walls of the building is addressed. Essentially, there are two alternatives where an exposure hazard exists: either provide protection to the stairway by having a fire-resistance rating on its exterior wall or provide a fire-resistance rating to the walls adjacent to the stairway for a distance of 10 feet (3048 mm) measured horizontally and vertically from the stairway enclosure where those walls are at an angle of less than 180 degrees (3.14 rad) from the enclosure's exterior wall [see Figure 1019.1.4(1)]. When the adjacent exterior wall is protected in lieu of the stairway enclosure wall, the protection is to extend from the ground to a level of 10 feet (3048 mm) above the highest landing of the stairway. However, the protection is not required to extend beyond the normal roof line of the building.

The 180-degree (3.14 rad) angle criteria is based on the scenario where the exterior wall of the stair enclosure is in the same plane and flush with the exterior wall of the building. In this scenario, a fire would need to travel 180 degrees (3.14 rad) around in order to impinge on the stair. Based on studies of existing buildings, this spread of fire 180 degrees (3.14 rad) does not appear to be a problem. This criteria is only applicable when the angle between the walls is 180 degrees (3.14 rad) or less.

As the fire exposure on the exterior is different than can be expected on the interior, the fire-resistance rating of the exterior wall is not required to exceed 1 hour, regardless of whether it is the stairway enclosure wall or the adjacent exterior wall, unless the exterior wall is required by other sections of the code to have a higher fire-resistance rating. The fire protection rating on any openings in the exterior wall of a stairway enclosure or adjacent exterior wall within 180 degrees (3.14 rad) is to be a minimum of $^3/_4$ hour [see Figure 1019.1.4(2)].

1019.1.5 Enclosures under stairways. The walls and soffits within enclosed usable spaces under enclosed and unenclosed stairways shall be protected by 1-hour fire-resistance-rated construction, or the fire-resistance rating of the stairway enclosure, whichever is greater. Access to the enclosed usable space shall not be directly from within the stair enclosure.

Exception: Spaces under stairways serving and contained within a single residential dwelling unit in Group R-2 or R-3 as applicable in Section 101.2.

There shall be no enclosed usable space under exterior exit stairways unless the space is completely enclosed in 1-hour fire-resistance-rated construction. The open space under exterior stairways shall not be used for any purpose.

❖ This section addresses the fire hazard of storage under a stairway. The stairway must be protected from a storage area under it, even if the stair is not required to be enclosed. The section also requires that the storage area not open into a stair enclosure. This is so that a fire that starts in the storage area would not be mistakenly left open to the stair enclosure. The exception does not require stair protection from storage areas under the stair for the indicated residential occupancies.

1019.1.6 Discharge identification. A stairway in an exit enclosure shall not continue below the level of exit discharge unless an approved barrier is provided at the level of exit discharge to prevent persons from unintentionally continuing into levels below. Directional exit signs shall be provided as specified in Section 1011.

❖ So that building occupants using an exit stairway during an emergency situation will be prevented from going past the level of exit discharge, the run of the stairway is to be interrupted by partitions, doors, gates or other approved means. These devices help the users of the stairway to recognize when they have reached the point that is the level of exit discharge. Exit signs, including tactile, are to be provided for occupant guidance. Furthermore, signs are to be placed at each floor landing in all interior exit stairways connecting more than three floor levels that designate the level or story of the landings in accordance with Section 1019.1.7

The code does not specify the type of material or construction of the barrier used to identify the level of exit discharge. The key issues to be considered in the selection and approval of the type of barrier to be used are: (1) will the barrier provide a visible and physical means of alerting occupants who are exiting under emergency conditions that they have reached the level of exit discharge and (2) is the barrier constructed of materials that are permitted by the construction type of the building? In an emergency situation, some occupants are likely to come in contact with the barrier during exiting before realizing that they are at the level of exit discharge. Therefore, the barrier should be constructed in a manner that is substantial enough to withstand the anticipated physical contact, such as pushing or shoving. It would be reasonable, as a minimum, to design the barrier to withstand the structural load requirements of Section 1607.5 for interior walls and partitions. The barrier could be opaque (such as gypsum wallboard and stud framing) or not (such as a wire grid-type material).

The use of signage only or relatively insubstantial barriers, such as ropes or chains strung across the opening, is typically not sufficient to prevent occupants from attempting to continue past the level of exit discharge during an emergency.

Figure 1019.1.6 is an example of one method of discharge identification.

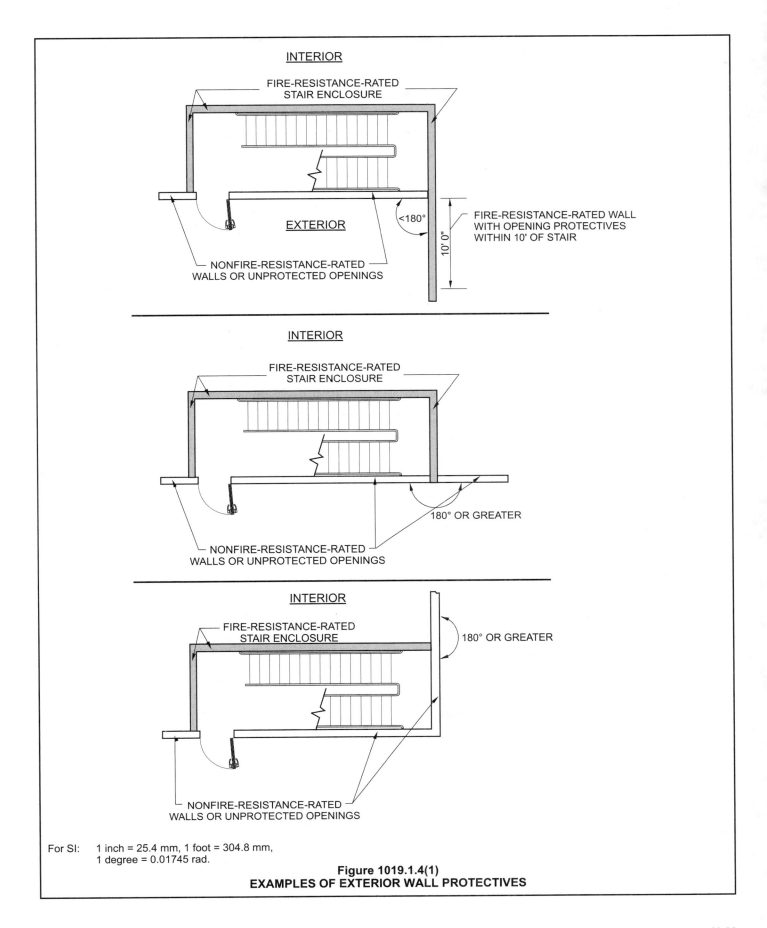

For SI:　　1 inch = 25.4 mm, 1 foot = 304.8 mm,
　　　　　　1 degree = 0.01745 rad.

Figure 1019.1.4(1)
EXAMPLES OF EXTERIOR WALL PROTECTIVES

FIGURE 1019.1.4(2)

MEANS OF EGRESS

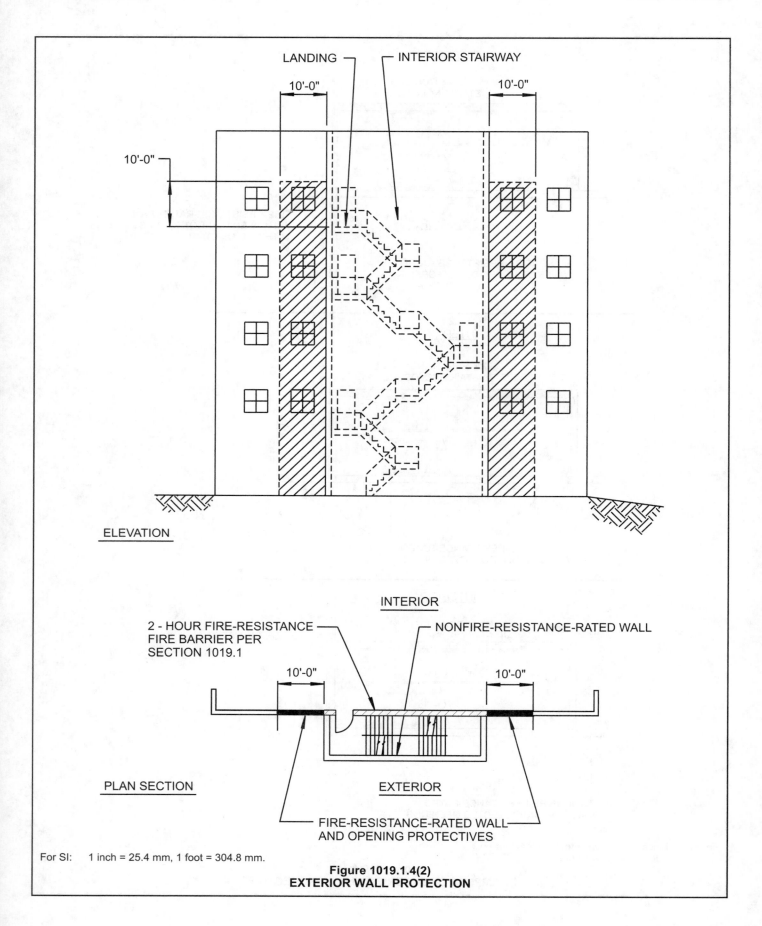

For SI: 1 inch = 25.4 mm, 1 foot = 304.8 mm.

Figure 1019.1.4(2)
EXTERIOR WALL PROTECTION

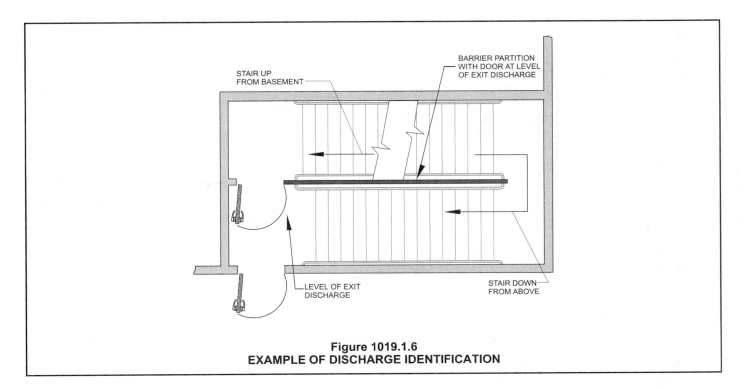

Figure 1019.1.6
EXAMPLE OF DISCHARGE IDENTIFICATION

1019.1.7 Stairway floor number signs. A sign shall be provided at each floor landing in interior vertical exit enclosures connecting more than three stories designating the floor level, the terminus of the top and bottom of the stair enclosure and the identification of the stair. The signage shall also state the story of, and the direction to the exit discharge and the availability of roof access from the stairway for the fire department. The sign shall be located 5 feet (1524 mm) above the floor landing in a position which is readily visible when the doors are in the open and closed positions.

❖ Signs are to be placed at each floor landing in all exit stairways connecting more than three stories. The signs are to designate the level or story of the landings above or below the level of exit discharge. The purpose is to inform the occupants of their location with respect to the level of exit discharge as they use the stairway to leave the building. More importantly, it allows the fire service to locate and gain quick access to the fire floor. At each level, the direction to the exit discharge is required to be indicated. The identification of the level that is the exit discharge also is to be indicated at each level. The identification of the roof access availability is for the fire department. Roof access is required by Section 1009.12. For visibility, the signs are required to be located approximately 5 feet (1524 mm) above the floor surface and be visible when the stairway door is open. The need to designate levels remaining to reach the level of exit discharge may mean that the numbering is other than that designation used by building management. For example, a designation of P1, P2, P3, etc., would not be acceptable for stairways in the basement parking garage, since in themselves they do not designate the floor level below the level of exit discharge.

1019.1.8 Smokeproof enclosures. In buildings required to comply with Section 403 or 405, each of the exits of a building that serves stories where the floor surface is located more than 75 feet (22 860 mm) above the lowest level of fire department vehicle access or more than 30 feet (9144 mm) below the level of exit discharge serving such floor levels shall be a smokeproof enclosure or pressurized stairway in accordance with Section 909.20.

❖ While smokeproof enclosures for exit stairways can, at the designer's option, be used in buildings of any occupancy, height and area, this section specifically requires smokeproof enclosures to be provided when either of two conditions occur.

The first condition requires all exit stairways in serving buildings with floor levels higher than 75 feet (22 860 mm) above the level of exit discharge to be smokeproof enclosures or pressurized stairways. The reason for this provision is that in very tall buildings, often during fire emergencies, total and immediate evacuation of the occupants cannot be readily accomplished. In such situations, exit stairways become places of safety for the occupants and must be adequately protected with smokeproof enclosures to provide a safe egress environment. In order to provide this safe environment, the smokeproof enclosure must be constructed to resist the migration of smoke caused by the "stack effect." Stack effect occurs in tall enclosures such as chimneys, when a fluid such as smoke, which is less dense than the ambient air, is introduced into the enclosure. The smoke will rise due to the effect of buoyancy and will induce additional flow into the enclosure through openings at the lower levels.

The second condition applies when an occupiable floor level is located more than 30 feet (9144 mm) below the level of exit discharge serving such floor levels. Stairways serving those levels are also required to be protected by smokeproof enclosures because underground portions of a building present unique problems in providing not only for life safety but also access for fire-fighting purposes. The choice of a 30-foot (9144 mm) threshold for this requirement is intended to provide a reasonable limitation on vertical travel distance before the requirement for smokeproof enclosures applies.

Detailed system requirements for a smokeproof enclosure are in Section 909.20.

1019.1.8.1 Enclosure exit. A smokeproof enclosure or pressurized stairway shall exit into a public way or into an exit passageway, yard or open space having direct access to a public way. The exit passageway shall be without other openings and shall be separated from the remainder of the building by 2-hour fire-resistance-rated construction.

Exceptions:

1. Openings in the exit passageway serving a smokeproof enclosure are permitted where the exit passageway is protected and pressurized in the same manner as the smokeproof enclosure, and openings are protected as required for access from other floors.

2. Openings in the exit passageway serving a pressurized stairway are permitted where the exit passageway is protected and pressurized in the same manner as the pressurized stairway.

❖ The walls that comprise the smokeproof enclosure or pressurized stairway, which includes the stairway shaft and the vestibules, must be fire barriers having a fire-resistance rating of at least 2 hours. This level of fire endurance is specified because exit stairways in high-rise buildings serve as principal components of the egress system and as the source of fire service access to the fire floor. This supersedes any allowed reduction of enclosure rating, even if the stair from the level that is more than 30 feet (9144 mm) below exit discharge connects three stories or less. Smokeproof enclosure requirements in this section are more specific; therefore, the exceptions in Section 1023.1 would not apply.

The exceptions apply to openings in the exit passageway, which are permitted provided the exit passageway is protected and pressurized.

1019.1.8.2 Enclosure access. Access to the stairway within a smokeproof enclosure shall be by way of a vestibule or an open exterior balcony.

Exception: Access is not required by way of a vestibule or exterior balcony for stairways using the pressurization alternative complying with Section 909.20.5.

❖ See Figures 1019.1.8.2(1) and 1019.1.8.2(2) for illustrations of access to the smokeproof stairway by way of a vestibule or an exterior balcony. The purpose of this requirement is to keep the enclosure clear of smoke. Where a pressurized stairway is used, these elements are not necessary.

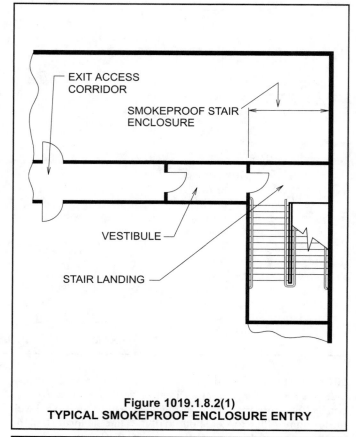

Figure 1019.1.8.2(1)
TYPICAL SMOKEPROOF ENCLOSURE ENTRY

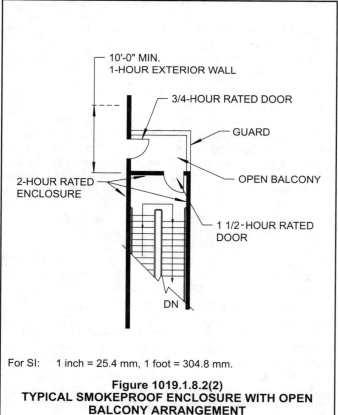

For SI: 1 inch = 25.4 mm, 1 foot = 304.8 mm.

Figure 1019.1.8.2(2)
TYPICAL SMOKEPROOF ENCLOSURE WITH OPEN BALCONY ARRANGEMENT

SECTION 1020
EXIT PASSAGEWAYS

1020.1 Exit passageway. Exit passageways serving as an exit component in a means of egress system shall comply with the requirements of this section. An exit passageway shall not be used for any purpose other than as a means of egress.

❖ See Figure 1020.1 for an illustration of an exit passageway arrangement. In the case of office buildings or similar types of structures, the exit stairways are often located at the central core or in line with the centrally located exit access corridors. Exit passageways may be used to connect the exit stair to the exterior exit door. Such an arrangement provides great flexibility in the design use of the building. Without the passageway at the grade floor or the level of exit discharge, the occupants of the upper floors or basement levels would have to leave the safety of the exit stairway to travel to the exterior doors. Such a reduction of protection is not acceptable. This section provides acceptable methods of continuing the protected path of travel for building occupants. The building designer/owner is given these different options for achieving this protected path of travel. Exit passageways may also be used on their own in locations not connected with a stair enclosure. Sometimes on large floor plans, an exit passageway may be used to extend an exit into areas that would not otherwise be able to meet the travel distance requirements.

1020.2 Width. The width of exit passageways shall be determined as specified in Section 1005.1 but such width shall not be less than 44 inches (1118 mm), except that exit passageways serving an occupant load of less than 50 shall not be less than 36 inches (914 mm) in width.

The required width of exit passageways shall be unobstructed.

Exception: Doors, when fully opened, and handrails, shall not reduce the required width by more than 7 inches (178 mm). Doors in any position shall not reduce the required width by more than one-half. Other nonstructural projections such as trim and similar decorative features are permitted to project into the required width 1.5 inches (38 mm) on each side.

❖ The width of an exit passageway is to be determined in accordance with Section 1005.1, based on the number of occupants served in the same manner as for corridors. The greater of the minimum width or the width determined based on occupancy is to be used. In situations where the exit passageway also serves as an exit access corridor for the first floor, the corridor width must comply with the stricter requirement.

The limit on the restriction from a door that opens into the exit passageway is the same as for general means of egress requirements for door encroachment (see Section 1005.2) and stairway landings (see Section 1009.4).

1020.3 Construction. Exit passageway enclosures shall have walls, floors and ceilings of not less than 1-hour fire-resistance rating, and not less than that required for any connecting exit enclosure. Exit passageways shall be constructed as fire barriers in accordance with Section 706.

❖ The entire exit passageway enclosure is to be fire-resistance rated as specified. The floors and ceilings are required to be rated in addition to the walls. When used separately, a minimum 1-hour fire-resistance rating is required. Where extending an exit enclosure, the rating

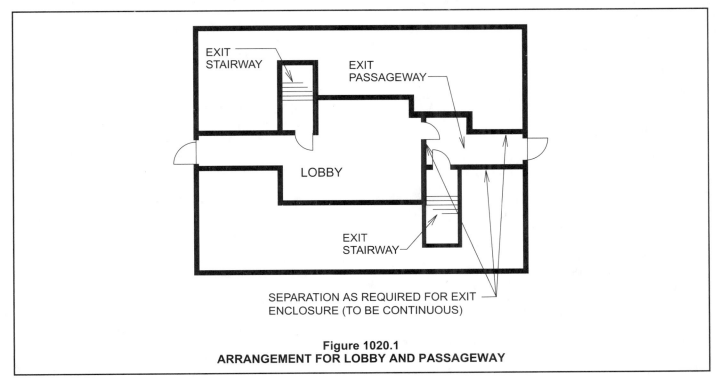

Figure 1020.1
ARRANGEMENT FOR LOBBY AND PASSAGEWAY

must not be less than the exit enclosure so that the degree of protection is kept at the same level. Remember that if the exit passageway extends over a basement, in accordance with the continuity requirements for fire barriers and horizontal assemblies, all supporting construction is to have the same fire-resistance rating as the elements supported (see Sections 706.4 and 711.4). The continuity requirements would also be a concern for the rated ceiling of the exit passageway. An alternative for the ceiling of the exit passageway could be a top shaft enclosure (see Section 707.12).

1020.4 Openings and penetrations. Exit passageway opening protectives shall be in accordance with the requirements of Section 715.

Except as permitted in Section 402.4.6, openings in exit passageways other than unexposed exterior openings shall be limited to those necessary for exit access to the exit passageway from normally occupied spaces and for egress from the exit passageway.

Where interior exit enclosures are extended to the exterior of a building by an exit passageway, the door assembly from the exit enclosure to the exit passageway shall be protected by a fire door conforming to the requirements in Section 715.3. Fire door assemblies in exit passageways shall comply with Section 715.3.4.

Elevators shall not open into an exit passageway.

❖ The requirements for exit passageways are very similar to those required for vertical exit enclosures (see Section 1019.1.1). The only openings that are permitted in fire-resistance-rated exit passageways are doors that lead either from normally occupied spaces or from the vertical exit enclosure. This restriction on openings essentially prohibits the use of windows in an exit passageway except for those exterior windows that are not exposed to any hazards by fire separation distance (see Section 704.8) or adjacent exterior walls similar to what is specified for vertical exit enclosures (see Section 1019.1.4). The only exception would be window assemblies that have been tested as wall assemblies in accordance with ASTM E 119. The objective of this provision is to minimize the possibility of fire spreading into an exit passageway and endangering the occupants or even preventing the use of the exit at a time when it is most needed. The limitation on openings applies regardless of the fire protection rating of the opening protective. The limitation on openings from normally occupied areas is intended to reduce the probability of a fire occurring in an unoccupied area, such as a storage closet, which has an opening into the stairway, thereby resulting in fire spread into the stairway. Other spaces that are not normally occupied include, but are not limited to, toilet rooms, electrical/mechanical equipment rooms and janitorial closets. Note that exit passageways have an additional requirement that prohibits elevators from opening directly into the passageway. While not specifically stated in Section 1019.1.1, the prohibition for openings from spaces that were not normally occupied could be interpreted as prohibiting elevators from opening into stairway enclosures as well. There are some exceptions for these unoccupied spaces in exit passage-

ways in covered malls (see Section 402.4.6).

The third paragraph addresses when the exit enclosure could move from the vertical enclosure of a stairway or ramp to the horizontal enclosure of an exit passageway. While the exit passageway is an extension of the protection offered by the vertical exit, there must still be an opening protective at the bottom of the stairway or ramp. This is to prevent any smoke that may migrate into the exit passageway from also moving up the exit stairway or ramp.

1020.5 Penetrations. Penetrations into and openings through an exit passageway are prohibited except for required exit doors, equipment and ductwork necessary for independent pressurization, sprinkler piping, standpipes, electrical raceway for fire department communication and electrical raceway serving the exit passageway and terminating at a steel box not exceeding 16 square inches (0.010 m²). Such penetrations shall be protected in accordance with Section 712. There shall be no penetrations or communicating openings, whether protected or not, between adjacent exit passageways.

❖ This section specifically lists the items that are allowed to penetrate an exit passageway. This is consistent for all types of exit enclosures, including stair or ramp vertical exit enclosures (see Section 1019.1.2) and exit passageways. In general, only portions of the building service systems that serve the exit enclosure are allowed to penetrate the exit enclosure.

SECTION 1021
HORIZONTAL EXITS

1021.1 Horizontal exits. Horizontal exits serving as an exit in a means of egress system shall comply with the requirements of this section. A horizontal exit shall not serve as the only exit from a portion of a building, and where two or more exits are required, not more than one-half of the total number of exits or total exit width shall be horizontal exits.

Exceptions:

1. Horizontal exits are permitted to comprise two-thirds of the required exits from any building or floor area for occupancies in Group I-2.

2. Horizontal exits are permitted to comprise 100 percent of the exits required for occupancies in Group I-3. At least 6 square feet (0.6 m²) of accessible space per occupant shall be provided on each side of the horizontal exit for the total number of people in adjoining compartments.

Every fire compartment for which credit is allowed in connection with a horizontal exit shall not be required to have a stairway or door leading directly outside, provided the adjoining fire compartments have stairways or doors leading directly outside and are so arranged that egress shall not require the occupants to return through the compartment from which egress originates.

The area into which a horizontal exit leads shall be provided with exits adequate to meet the occupant requirements of this chapter, but not including the added occupant capacity imposed by persons entering it through horizontal exits from another area. At least one of its exits shall lead directly to the exterior or to an exit enclosure.

❖ Horizontal exits can be up to 50 percent of the exits from an area of a building. The percentage is higher for Group I-2 and I-3 occupancies where the evacuation strategy is defend in place rather than direct egress. (see Figure 1021.1 for a typical horizontal exit arrangement).

A horizontal exit may be an element of a means of egress when in compliance with the requirements of this section. The actual horizontal exit is the protected door opening in a wall, open-air balcony or bridge that separates two areas of a building. A horizontal exit is often used in hospitals and in prisons where it is not feasible or desirable that the occupants exit the facility. See Section 1002 for the definition of a "Horizontal exit."

Note that the capacity of the exit (such as an exit stairway) from the area that the horizontal exit leads is required to be sufficient for the number of occupants in the area, not including those who come into the space from

other areas using the horizontal exit. This is because the adjacent area of refuge is of sufficient safety to house occupants during a fire or until the egress system is available. The occupant capacity of the refuge area is addressed in Section 1021.4.

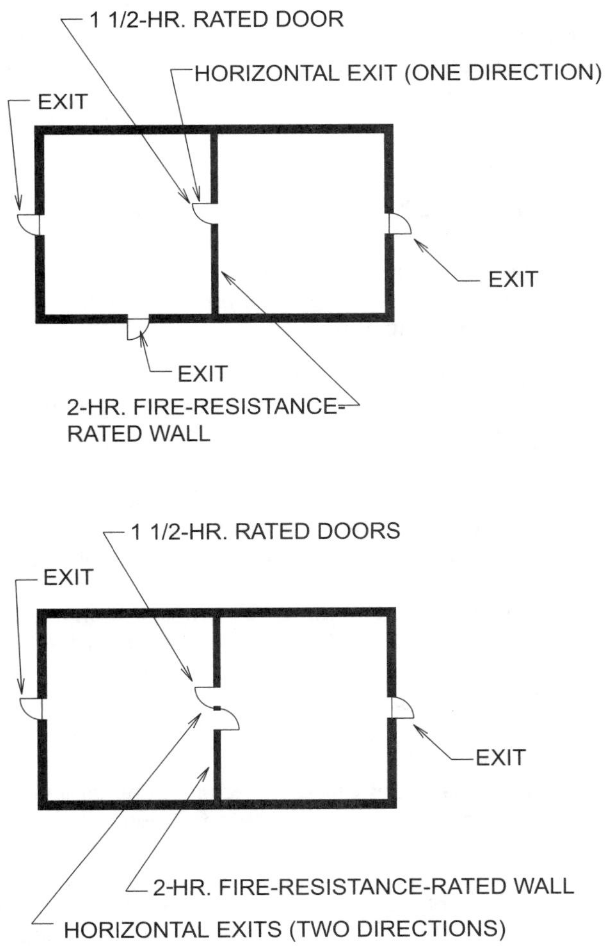

1021.2 Separation. The separation between buildings or areas of refuge connected by a horizontal exit shall be provided by a fire wall complying with Section 705 or a fire barrier complying with Section 706 and having a fire-resistance rating of not less than 2 hours. Opening protectives in horizontal exit walls shall also comply with Section 715. The horizontal exit separation shall extend vertically through all levels of the building unless floor assemblies are of 2-hour fire resistance with no unprotected openings.

> **Exception:** A fire-resistance rating is not required at horizontal exits between a building area and an above-grade pedestrian walkway constructed in accordance with Section 3104, provided that the distance between connected buildings is more than 20 feet (6096 mm).

Horizontal exit walls constructed as fire barriers shall be continuous from exterior wall to exterior wall so as to divide completely the floor served by the horizontal exit.

❖ The basic concept of a horizontal exit is that during a fire emergency, the occupants of a floor will transfer from one side of a fire wall or fire barrier to the other. Separation between areas of a building can be accomplished by either a fire wall (see Section 705) or a fire barrier (see Section 706), with a fire-resistance rating not less than 2 hours. Any fire shutters or fire door must have an opening protective of not less than 1 ½ hours (see Table 715.3).

In buildings of Groups I-2 and I-3, it may also be desirable (while not mandatory) for the horizontal exit to serve as a smoke barrier. In such cases, the wall containing the horizontal exit must also comply with the requirements for a smoke barrier (see Section 709).

In order to decrease the amount of smoke able to migrate around the edges of a horizontal exit, the horizontal exit must extend from at least the floor to the deck above (i.e., fire barrier), as well as across the floor level from one side of the building to another. Moving up from floor to floor, there are two choices. One option is that the horizontal exit can extend vertically through all levels of the building (i.e., fire wall or fire barriers). The second option is to utilize fire barriers that are not lined up vertically, but then the floor must have a 2-hour fire-resistance rating and no unprotected openings are permitted between any two "refuge areas."

The exception is permitting a pedestrian walkway or skybridge to act as a horizontal exit when buildings are at least 20 feet (6096 mm) apart.

1021.3 Opening protectives. Fire doors in horizontal exits shall be self-closing or automatic-closing when activated by a smoke detector installed in accordance with Section 907.10. Opening protectives in horizontal exits shall be consistent with the fire-resistance rating of the wall. Such doors where located in a

Figure 1021.1
TYPICAL HORIZONTAL EXITS

cross-corridor condition shall be automatic-closing by activation of a smoke detector installed in accordance with Section 907.10.

❖ For the safety of occupants using a horizontal exit, it is important that the doors be fire doors that are self-closing or automatic-closing by activation of a smoke detector. Smoke detectors that initiate automatic closing should be located at both sides of the doors (see the commentary to Section 907.10 for an additional explanation of the installation requirements). Any openings in the fire barriers or fire walls used as horizontal exits must be protected in coordination with the rating of the wall. There is a reference to Section 715 for opening protectives in Section 1021.2.

1021.4 Capacity of refuge area. The refuge area of a horizontal exit shall be spaces occupied by the same tenant or public areas and each such area of refuge shall be adequate to house the original occupant load of the refuge space plus the occupant load anticipated from the adjoining compartment. The anticipated occupant load from the adjoining compartment shall be based on the capacity of the horizontal exit doors entering the area of refuge. The capacity of areas of refuge shall be computed on a net floor area allowance of 3 square feet (0.2787 m²) for each occupant to be accommodated therein, not including areas of stairways, elevators and other shafts or courts.

> **Exception:** The net floor area allowable per occupant shall be as follows for the indicated occupancies:
>
> 1. Six square feet (0.6 m²) per occupant for occupancies in Group I-3.
>
> 2. Fifteen square feet (1.4 m²) per occupant for ambulatory occupancies in Group I-2.
>
> 3. Thirty square feet (2.8 m²) per occupant for nonambulatory occupancies in Group I-2.

❖ The building area on the discharge side of a horizontal exit must serve as a refuge area for the occupants of both sides of the floor areas connected by the horizontal exit. Therefore, adequate space must be available on each side of the wall to hold the full occupant load of that side, plus the number of occupants from the other side who may be required to use the horizontal exit. These refuge areas are meant to hold temporarily the occupants in a safe place until they can evacuate the premises in an orderly manner or, in the case of hospitals and like facilities, to hold bedridden patients and other nonambulatory occupants in a protected area until the fire emergency has ended. The size of the refuge area is based on the nature of the expected occupants. In the case of Group I-3, the area will be used to hold the occupants until deliberate egress can be accomplished with staff assistance or supervision. In other cases, it is assumed that the occupants simply wait in line to egress through the required exit facilities provided on the discharge side. Although similar language is used in describing the "area of refuge" for an accessible means of egress, Section 1007.6 specifies area requirements that may be insufficient for use as a refuge area for a

horizontal exit. Care must be taken when applying both principles.

The 3 square feet (0.28 m²) per occupant is based on the maximum permitted occupant density at which orderly movement to the exits is reasonable. The 30 square feet (2.8 m²) per hospital or nursing home patient is based on the space necessary for a bed or litter. It should be noted that 30 square feet (2.8 m²) is not based on the total occupant load, as would be determined in accordance with Section 1004.1, but rather on the number of nonambulatory patients. The 15-square-feet (1.4 m²) requirement for occupancies in Group I-2 facilities is based on each ambulatory patient having a staff attendant.

In a single-tenant facility, any of the spaces that are constantly available (e.g., not lockable) can be used as places of refuge. However, in spaces housing more than one tenant, public areas of refuge such as corridors or passageways must be provided and be accessible at all times. This requirement is necessary because if a horizontal exit connected two areas occupied by different tenants, the tenants could (for privacy and security purposes) render the necessary free access through the horizontal exit ineffective. When the horizontal exit discharges into a public or common space, such as a corridor leading to an exit, each tenant can obtain the desired security.

SECTION 1022
EXTERIOR EXIT RAMPS AND STAIRWAYS

1022.1 Exterior exit ramps and stairways. Exterior exit ramps and stairways serving as an element of a required means of egress shall comply with this section.

> **Exception:** Exterior exit ramps and stairways for outdoor stadiums complying with Section 1019.1, Exception 2.

❖ Exterior exit ramps and stairways are an important element of the means of egress system and must be designed and constructed so that they will serve as a safe path of travel. The general requirements in Section 1009 also apply to exterior stairways (for ramp provisions, see Section 1010).

The exception references Exception 2 in Section 1019.1, which allows exterior ramps and stairways that serve outdoor assembly facilities to be open. The openness criteria in Section 1022.3 would be an appropriate guideline for what is intended.

1022.2 Use in a means of egress. Exterior exit ramps and stairways shall not be used as an element of a required means of egress for occupancies in Group I-2. For occupancies in other than Group I-2, exterior exit ramps and stairways shall be permitted as an element of a required means of egress for buildings not exceeding six stories or 75 feet (22 860 mm) in height.

❖ This section specifies the conditions where an exterior ramp or stairway can be used as a required exit. Exterior exit stairways are not permitted for Group I-2 since quick evacuation of nonambulatory patients from buildings is impractical. Some of the patients may not be capable of self-preservation and, therefore, may require assistance from the staff. The period of evacuation of nonambulatory patients could become lengthy.

With the exception of outdoor stadiums (see Section 1022.1), exterior ramps or stairways are not allowed to be a required exit in buildings that exceed 6 stories or 75 feet (22 860 mm) in height due to the hazard of using such a ramp or stairway in poor weather. Some persons may not use such a stair due to vertigo. When confronted with a view from a great height, vertigo sufferers can become confused, disoriented and dizzy. They could injure themselves, become disabled or refuse to move (freeze). In a fire situation, they could become an obstruction in the path of travel, possibly causing panic and injuries to other users of the exit.

1022.3 Open side. Exterior exit ramps and stairways serving as an element of a required means of egress shall be open on at least one side. An open side shall have a minimum of 35 square feet (3.3 m²) of aggregate open area adjacent to each floor level and the level of each intermediate landing. The required open area shall be located not less than 42 inches (1067 mm) above the adjacent floor or landing level.

❖ An important factor in considering exterior exit ramps or stairways is that natural ventilation is assumed to occur. This is so that smoke will not be trapped above the ramp or stairway walking surfaces and obscure safe egress.

The exterior ramp or stairway must have at least one of its sides directly facing an outer court, yard or public way. This will allow the products of combustion escaping from the interior of the building to quickly vent to the outdoor atmosphere and let the building occupants egress down the exterior ramp or stairway. Since exterior ramps or stairways are occasionally partially enclosed within the building construction, minimum amounts of exterior openings are specified by the code.

The openings on each and every floor level and landing must total 35 square feet (3.3 m²) or greater. The opening is to occur higher than 42 inches (1067 mm) from the floor and intermediate landing levels. The high openings dissipate the smoke buildup from the exterior ramp or stairway (See Figure 1022.3). The bottom edge of the opening is consistent with the height requirements for guards (see Section 1012.2).

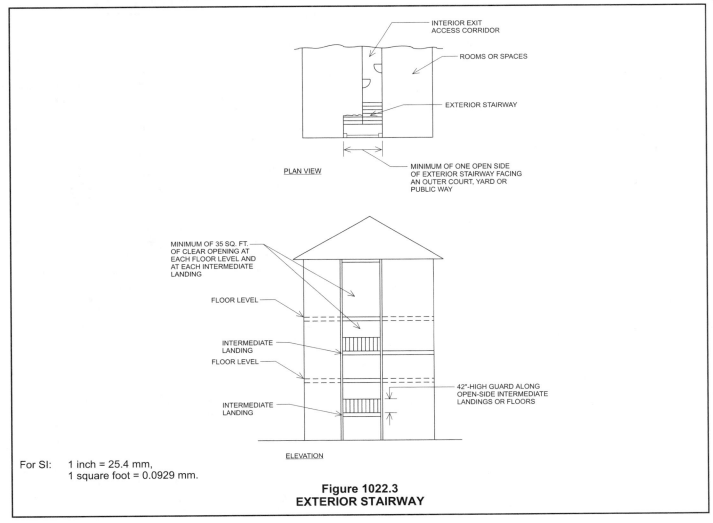

For SI: 1 inch = 25.4 mm,
           1 square foot = 0.0929 mm.

Figure 1022.3
EXTERIOR STAIRWAY

1022.4 Side yards. The open areas adjoining exterior exit ramps or stairways shall be either yards, courts or public ways; the remaining sides are permitted to be enclosed by the exterior walls of the building.

❖ See Section 1022.3 for a discussion of the opening requirements. This section simply specifies the type of areas that the exterior opening of the exterior ramp or stair is to adjoin. These open spaces will enable the smoke to dissipate from the exterior ramp or stairway so it will be useable as a required exit.

1022.5 Location. Exterior exit ramps and stairways shall be located in accordance with Section 1023.3.

❖ The location requirements of this section protect the users of the exterior exit ramp or stairway from the effects of a fire in another building on the same lot or an adjacent lot. The separation distance reduces the exposure to heat and smoke. If the stairway is closer than specified, then adjacent buildings' exterior walls and openings are to be protected in accordance with Section 704 such that the users of the exterior exit are protected.

1022.6 Exterior ramps and stairway protection. Exterior exit ramps and stairways shall be separated from the interior of the building as required in Section 1019.1. Openings shall be limited to those necessary for egress from normally occupied spaces.

Exceptions:

1. Separation from the interior of the building is not required for occupancies, other than those in Group R-1 or R-2, in buildings that are no more than two stories above grade where the level of exit discharge is the first story above grade.

2. Separation from the interior of the building is not required where the exterior ramp or stairway is served by an exterior ramp and/or balcony that connects two remote exterior stairways or other approved exits, with a perimeter that is not less than 50 percent open. To be considered open, the opening shall be a minimum of 50 percent of the height of the enclosing wall, with the top of the openings no less than 7 feet (2134 mm) above the top of the balcony.

3. Separation from the interior of the building is not required for an exterior ramp or stairway located in a building or structure that is permitted to have unenclosed interior stairways in accordance with Section 1019.1.

4. Separation from the interior of the building is not required for exterior ramps or stairways connected to open-ended corridors, provided that Items 4.1 through 4.4 are met:

    4.1. The building, including corridors and ramps and/or stairs, shall be equipped throughout with an automatic sprinkler system in accordance with Section 903.3.1.1 or 903.3.1.2.

    4.2. The open-ended corridors comply with Section 1016.

    4.3. The open-ended corridors are connected on each end to an exterior exit ramp or stairway complying with Section 1022.

    4.4. At any location in an open-ended corridor where a change of direction exceeding 45 degrees (0.79 rad) occurs, a clear opening of not less than 35 square feet (3.3 m²) or an exterior ramp or stairway shall be provided. Where clear openings are provided, they shall be located so as to minimize the accumulation of smoke or toxic gases.

❖ Exterior exit ramps or stairways must be protected from interior fires that may project through windows or other openings adjacent to the ramp or stairway, possibly endangering the occupants using this means of egress to reach grade. The protection of an exterior ramp or stairway is to be obtained by separating the exterior exit from the interior of the building using walls having a fire-resistance rating of at least 1 hour with opening protectives. Consistent with the protection required in Sections 1019.1 and 1019.1.4 for interior exit stairways, the fire-resistance rating must be provided for a distance of 10 feet (3048 mm) horizontally and vertically from the ramp or stairway edges, and from the ground to a level of 10 feet (3048 mm) above the highest landing.

All window and door openings falling inside the 10-foot (3048 mm) horizontal separation distance as well as all window and door openings 10 feet (3048 mm) above the topmost landing and below the stairway must be protected with minimum $^3/_4$-hour fire-resistance-rated opening protectives [see Figure 1022.6(1)].

Openings within the width of the stairway must only to be from normally occupied spaces. This is consistent with the requirements for vertical exit enclosures (see Sections 1019.1.1 and 1019.1.2).

Exception 1 indicates that opening protectives are not required for occupancies other than Groups R-1 and R-2, which are two stories or less above grade when the level of exit discharge is at the lower story. The reason for this exception is that in cases of fire in low buildings, the occupants are usually able to evacuate the premises before the fire can emerge through exterior wall openings and endanger the exit ramp or stairways. In hotels and apartments, however, the occupants' response to a fire emergency could be significantly reduced because they may be unfamiliar with the surroundings or sleeping.

Exception 2 allows the opening protectives to be omitted when an exterior exit access balcony is served by two exits and when they are remote from each other. Remoteness is regulated by Section 1014.2. This exception is applicable to all groups. In such instances, it is unlikely that the users of the exterior ramp or stairway will become trapped by fire, since they have the option of using the balcony to gain access to either of the two available exits, and the products of combustion will be vented directly to the outside (see Section 1013.5 regarding exterior balconies). At least one-half of the total perimeter of the exterior balcony must be permanently open to the outside. The requirement for at least one-half the height of that level to be open allows for columns, solid guards and decorative elements, such as arches. With the top of the opening at least 7 feet (2134 mm) above the walking surface, products of combustion can vent and allow passage below the smoke layer [see Figure 1022.6(2)].

Exception 3 exempts exterior ramps or stairways from protection when Section 1019.1 permits unenclosed interior exit stairways.

Exception 4 deletes the requirement for a separation between the interior of the building and the exterior wall area where an open-ended corridor (breezeway) interfaces with an exterior ramp or stairway. The separation is not needed as a result of the sprinkler system in all areas of the building, including the open-ended corridor

and the exterior exit. The other characteristics of the open-ended corridor described in this exception are needed so that it is safe to be used in the event of a fire. The requirements for an exterior ramp or stairway at each end and the opening or an exterior ramp or stairway where the open-ended corridor has a change of direction of greater than 45 degrees (0.79 erad) are for adequate ventilation of the open-ended corridor.

Similar language is used in describing the exterior wall requirements for an exterior area for assisted rescue (see Section 1007.8). If this option is utilized at an exterior stairway or nonaccessible ramp, the exceptions in this section would not be permitted since occupants of the exterior area for assisted rescue could not immediately move away from the building.

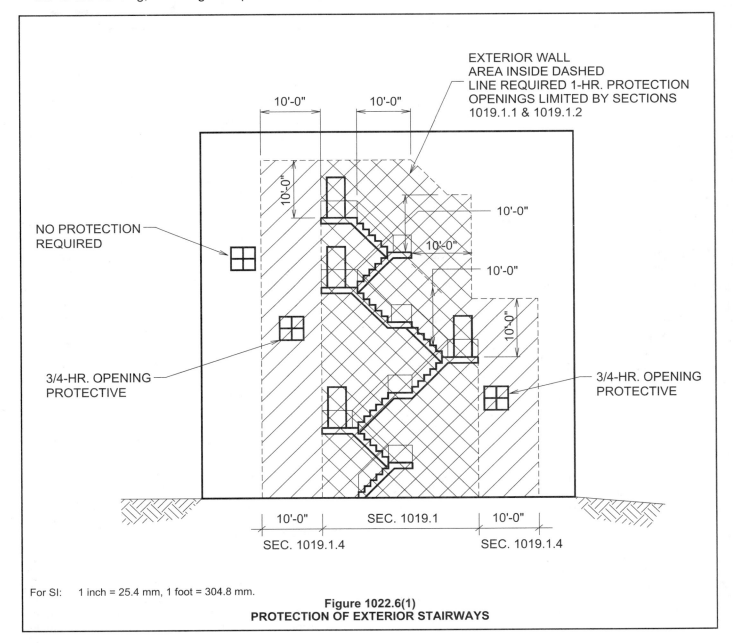

For SI: 1 inch = 25.4 mm, 1 foot = 304.8 mm.

Figure 1022.6(1)
PROTECTION OF EXTERIOR STAIRWAYS

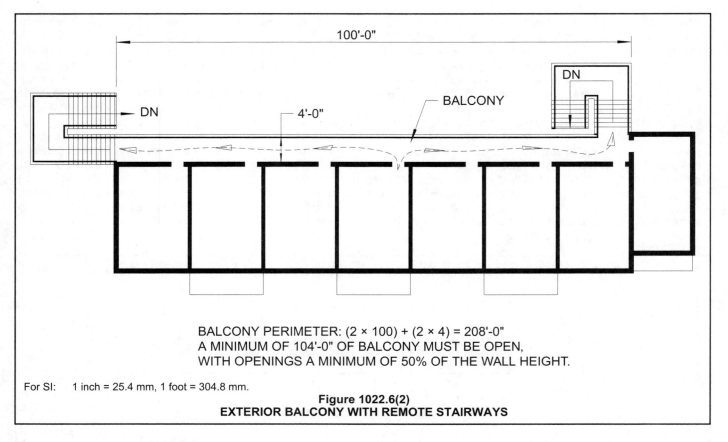

100'-0"

DN

4'-0"

BALCONY

DN

BALCONY PERIMETER: (2 × 100) + (2 × 4) = 208'-0"
A MINIMUM OF 104'-0" OF BALCONY MUST BE OPEN,
WITH OPENINGS A MINIMUM OF 50% OF THE WALL HEIGHT.

For SI: 1 inch = 25.4 mm, 1 foot = 304.8 mm.

Figure 1022.6(2)
EXTERIOR BALCONY WITH REMOTE STAIRWAYS

SECTION 1023
EXIT DISCHARGE

1023.1 General. Exits shall discharge directly to the exterior of the building. The exit discharge shall be at grade or shall provide direct access to grade. The exit discharge shall not reenter a building.

Exceptions:

1. A maximum of 50 percent of the number and capacity of the exit enclosures is permitted to egress through areas on the level of discharge provided all of the following are met:

    1.1. Such exit enclosures egress to a free and unobstructed way to the exterior of the building, which way is readily visible and identifiable from the point of termination of the exit enclosure.

    1.2. The entire area of the level of discharge is separated from areas below by construction conforming to the fire-resistance rating for the exit enclosure.

    1.3. The egress path from the exit enclosure on the level of discharge is protected throughout by an approved automatic sprinkler system. All portions of the level of discharge with access to the egress path shall either be protected throughout with an automatic sprinkler system installed in accordance with Section 903.3.1.1 or

903.3.1.2, or separated from the egress path in accordance with the requirements for the enclosure of exits.

2. A maximum of 50 percent of the number and capacity of the exit enclosures is permitted to egress through a vestibule provided all of the following are met:

    2.1. The entire area of the vestibule is separated from areas below by construction conforming to the fire-resistance rating for the exit enclosure.

    2.2. The depth from the exterior of the building is not greater than 10 feet (3048 mm) and the length is not greater than 30 feet (9144 mm).

    2.3. The area is separated from the remainder of the level of exit discharge by construction providing protection at least the equivalent of approved wired glass in steel frames.

    2.4. The area is used only for means of egress and exits directly to the outside.

3. Stairways in open parking garages complying with Section 1019.1, Exception 5, are permitted to egress through the open parking garage at the level of exit discharge.

❖ The exit discharge is the third piece of the means of egress system, which includes exit access, exit and exit discharge. The general provisions for means of egress in Sections 1003 through 1012 are applicable to the exit discharge. The basic provision is that exits must discharge directly to the outside of the building. The exit discharge is the path from the termination of the exit to the public way. When this is not practical, there are three alternatives: an exit passageway (see Section 1020), an exit discharge lobby (see Section 1023.1, Exception 1) or an exit discharge passageway (see Section 1023.1, Exception 2). Open parking garages are a special case, and are addressed in Exception 3.

While Exceptions 1 and 2 could be applicable for exit passageways and exit ramps, most of the real life application of the exceptions is for exit stairways. This commentary will be limited to enclosed exit stairways. Up to 50 percent of the exit stairways in a building may use Exception 1 or 2; therefore, neither exception is viable for a single-exit building.

An interior exit discharge lobby is permitted to receive the discharge from an exit stairway in lieu of the stairway discharging directly to the exterior. A fire door must be provided at the point where the exit stairway discharges into the lobby. Without an opening protective between the stairway and a lobby, it would be possible for the stairway to be directly exposed to smoke movement from a fire in the lobby. The opening protective provides for full continuity of the vertical component of the exit arrangement. Additionally, in buildings where stair towers must be pressurized, pressurization would not be possible without a door at the lobby level.

An exit discharge lobby is the sole location recognized in the code where an exit element can be used for purposes other than pedestrian travel for means of egress. The lobby may contain furniture or decoration and nonoccupiable spaces may open directly into the lobby. The lobby, and all other areas on the same level

that are not separated from the lobby by fire barriers consistent with the rating of the stair enclosure, must be sprinklered in accordance with an NFPA 13 or NFPA 13R system [see Figure 1023.1(1)]. If the entire level is sprinklered, no separation is required. In this case, the automatic sprinkler system is anticipated to control and (perhaps) eliminate the fire threat so as not to jeopardize the path of egress of the occupants. The lobby floor and any supporting construction must be rated the same as the stairway enclosure. If the lobby is slab on grade, this requirement is not applicable. This is consistent with the fundamental concept that an exit enclosure provides the necessary level of protection from adjacent areas. A path of travel through the lobby must be continually clear and available. The exit door leading out of the building must be visible and identifiable immediately when a person leaves the exit. This does not mean the exterior exit door must be directly in front of the door at the bottom of the stairway, but the intent is that it should be within the general range of vision. It should not be required that a person must turn completely around or go around a corner to be able to see the way out.

An exit is also allowed to discharge through a vestibule provided it complies with the specified requirements of Exception 2. Note that a vestibule is not allowed to have any vending machines within the space or any storage rooms opening into the vestibule. The vestibule floor and any supporting construction must be rated the same as the stairway enclosure. If the vestibule is slab on grade, this requirement is not applicable. The walls of the vestibule itself must have something at least equivalent with wired glass in steel frames. This is typically an opening protective with a fire protection rat-

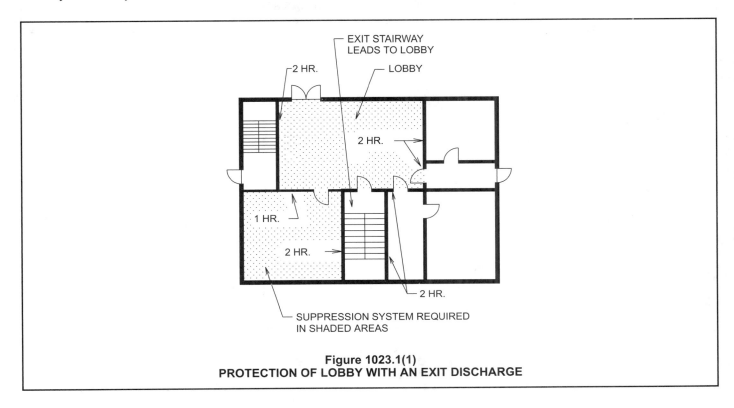

Figure 1023.1(1)
PROTECTION OF LOBBY WITH AN EXIT DISCHARGE

FIGURE 1023.1(2) – FIGURE 1023.1(3)

MEANS OF EGRESS

ing of at least 20 minutes. The size of the vestibule is limited so that it cannot be used for other activities, and the travel distance from the exit stairway to the exterior exit doorway is limited [see Figures 1023.1(2) and 1023.1(3)].

The reason that stairs in open parking garages are allowed to be unenclosed is due to the low hazard and free venting aspects. When an occupant leaves the stair in an open parking garage, typically he or she can move in several directions to reach the perimeter of the build-

ing. This, coupled with the open nature of the parking garage structure, allows occupants a safe path of travel from the bottom of the stairway and across the parking level to the exterior and public way without the additional protection of an exit passageway, exit lobby or exit vestibule. This path must meet the general means of egress requirements and may not be down a driveway unless a pedestrian path is also provided (see Section 1010.1).

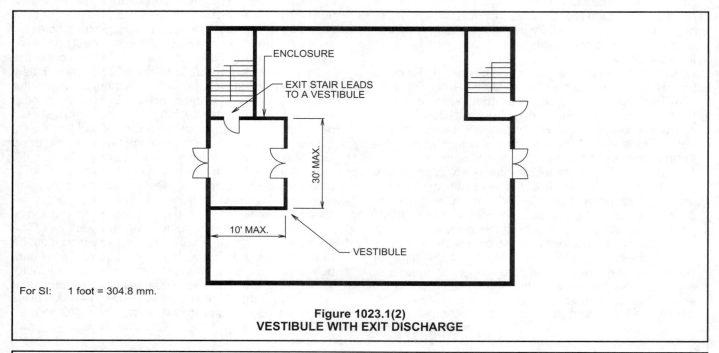

For SI: 1 foot = 304.8 mm.

Figure 1023.1(2)
VESTIBULE WITH EXIT DISCHARGE

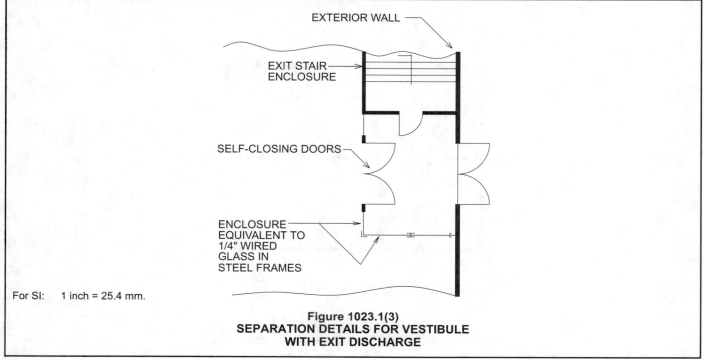

For SI: 1 inch = 25.4 mm.

Figure 1023.1(3)
SEPARATION DETAILS FOR VESTIBULE
WITH EXIT DISCHARGE

1023.2 Exit discharge capacity. The capacity of the exit discharge shall be not less than the required discharge capacity of the exits being served.

❖ This section specifies the exit discharge capacity. The exit discharge is required to be designed for the required capacity of all the exits it serves. If the exit discharge serves two exits, it is to be designed for the sum of the occupants served by both exits. Note that the capacity of the exit discharge is not required to be the capacity of both exits, which is higher than the sum of the occupants served by both exits.

1023.3 Exit discharge location. Exterior balconies, stairways and ramps shall be located at least 10 feet (3048 mm) from adjacent lot lines and from other buildings on the same lot unless the adjacent building exterior walls and openings are protected in accordance with Section 704 based on fire separation distance.

❖ The purpose of this section is to protect the users of exterior balconies, stairways and ramps from fire in an adjacent building by requiring specific separation distance from that adjacent building. The reason for the required distance to a lot line is to provide for a future building that could be built on the adjacent lot.

1023.4 Exit discharge components. Exit discharge components shall be sufficiently open to the exterior so as to minimize the accumulation of smoke and toxic gases.

❖ This section includes exit discharge components that currently only include egress courts. It should be noted that the general requirements in Sections 1003 through 1012 will still apply when those components are used in the exit discharge.

1023.5 Egress courts. Egress courts serving as a portion of the exit discharge in the means of egress system shall comply with the requirements of Section 1023.

❖ See Figure 1023.5 for an illustration of an exit discharge, including an egress court.

This section and the following subsections address the detailed requirements for egress courts. It is essential that exterior egress courts that serve occupants from an exit to a public way be sufficiently open to prevent the accumulation of smoke and toxic gases in the event of a fire.

1023.5.1 Width. The width of egress courts shall be determined as specified in Section 1005.1, but such width shall not be less than 44 inches (1118 mm), except as specified herein. Egress courts serving occupancies in Group R-3 applicable in Section 101.2 and Group U shall not be less than 36 inches (914 mm) in width.

The required width of egress courts shall be unobstructed to a height of 7 feet (2134 mm).

Exception: Doors, when fully opened, and handrails shall not reduce the required width by more than 7 inches (178 mm). Doors in any position shall not reduce the required width by more than one-half. Other nonstructural projections such as trim and similar decorative features are permitted to project into the required width 1.5 inches (38 mm) from each side.

Where an egress court exceeds the minimum required width and the width of such egress court is then reduced along the path of exit travel, the reduction in width shall be gradual. The transi-

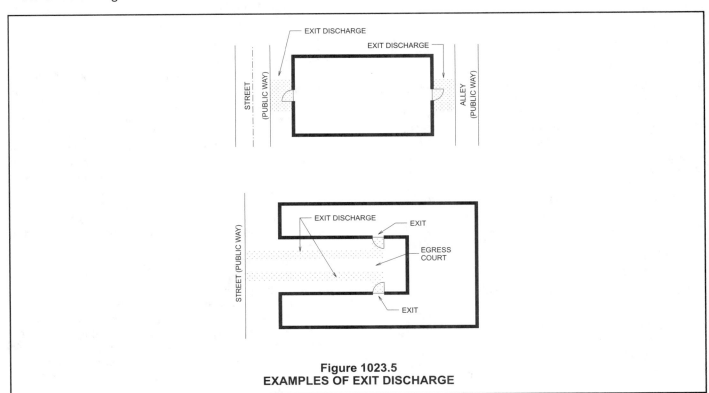

Figure 1023.5
EXAMPLES OF EXIT DISCHARGE

tion in width shall be affected by a guard not less than 36 inches (914 mm) in height and shall not create an angle of more than 30 degrees (0.52 rad) with respect to the axis of the egress court along the path of egress travel. In no case shall the width of the egress court be less than the required minimum.

❖ The width of an exterior court is to be determined in the same fashion as for an interior corridor. The width is not to be less than required to serve the number of occupants from the exit or exits and not less than the minimum specified in the section. The protrusion limitations are consistent with the general requirements for door encroachment, aisles and exit discharge (see Sections 1005.2, 1013.4 and 1020.3).

Many egress courts are significantly larger than required. Thus, the code allows such an egress court to decrease in width along the path of travel to the public way. The gradual transition requirement is so the flow of the occupants will be uniform without pockets of congestion. The transition requirements should be applied to egress courts where a reduction results in a width that is near the minimum based on the number of occupants served. It is this condition where the uniform flow of occupants is essential.

1023.5.2 Construction and openings. Where an egress court serving a building or portion thereof is less than 10 feet (3048 mm) in width, the egress court walls shall be not less than 1-hour fire-resistance-rated exterior walls complying with Section 704 for a distance of 10 feet (3048 mm) above the floor of the court, and openings therein shall be equipped with fixed or self-closing, $^3/_4$-hour opening protective assemblies.

Exceptions:

1. Egress courts serving an occupant load of less than 10.

2. Egress courts serving Group R-3 as applicable in Section 101.2.

❖ The purpose of this section is to protect the occupants served by the egress court from the building that they are exiting from. If occupants must walk closely by the exterior walls of the exit court, the walls are required to have the specified fire-resistance rating and the openings are required to be protected as specified. This requirement is only for the first 10 feet (3048 mm) above the level of the egress court since the exposure hazard from walls and openings above this level is reduced.

The two exceptions provide for egress courts that serve a very low number of occupants and the specified residential occupancy where the protection requirement would be excessive.

1023.6 Access to a public way. The exit discharge shall provide a direct and unobstructed access to a public way.

Exception: Where access to a public way cannot be provided, a safe dispersal area shall be provided where all of the following are met:

1. The area shall be of a size to accommodate at least 5 square feet (0.28 m²) for each person.

2. The area shall be located on the same property at least 50 feet (15 240 mm) away from the building requiring egress.

3. The area shall be permanently maintained and identified as a safe dispersal area.

4. The area shall be provided with a safe and unobstructed path of travel from the building.

❖ There are instances where the path of travel to the public way is not safe or not achievable due to site constraints or security concerns. The provisions in this section specify what would constitute a safe area to allow occupants of a building to assemble in an emergency. The 5 square feet (0.28 m²) would allow adequate space for standing persons as well as some space for persons in wheelchairs or on stretchers.

SECTION 1024
ASSEMBLY

1024.1 General. Occupancies in Group A which contain seats, tables, displays, equipment or other material shall comply with this section.

❖ Assembly spaces that contain elements that would affect the path of travel for the means of egress must comply with this section. Assembly spaces require special consideration due to the large occupant loads and possible low lighting (e.g., nightclubs, theaters).

1024.1.1 Bleachers. Bleachers, grandstands, and folding and telescopic seating shall comply with ICC 300.

❖ On February 24, 1999, the Bleacher Safety Act of 1999 was introduced in the House of Representatives. The bill, which cites the International Code Council® (ICC®) and the *International Building Code®* (IBC®), authorizes the Consumer Product Safety Commission (CPSC) to issue a standard for bleacher safety. This was in response to concerns relative to accidents on bleacher-type structures. As a result, the CPSC developed and revised the *Guidelines for Retrofitting Bleachers*. The ICC Board of Directors decided that a comprehensive standard dealing with all aspects of both new and existing bleachers was warranted and authorized the formation of the ICC Consensus Committee on Bleacher Safety. The committee is comprised of 12 members, including the requisite balance of general, user interest and producer interest.

The *ICC Standard on Bleachers, Folding and Telescopic Seating and Grandstands* was completed in December 2001, and submitted to ANSI on January 1, 2002. While the term "bleachers" is generic, the standard addresses all aspects of tiered seating associated with bleachers, grandstands and folding and telescopic seating. The bleacher standard references Chapter 11 of the code and ICC A117.1 for accessibility requirements.

1024.2 Assembly main exit. Group A occupancies that have an occupant load of greater than 300 shall be provided with a main exit. The main exit shall be of sufficient width to accommodate not less than one-half of the occupant load, but such width shall not be less than the total required width of all means of egress leading to the exit. Where the building is classified as a Group A occupancy, the main exit shall front on at least one street or an unoccupied space of not less than 10 feet (3048 mm) in width that adjoins a street or public way.

> **Exception:** In assembly occupancies where there is no well-defined main exit or where multiple main exits are provided, exits shall be permitted to be distributed around the perimeter of the building provided that the total width of egress is not less than 100 percent of the required width.

❖ Assembly buildings present an unusual life safety problem which includes frequent high occupant densities and therefore large occupant loads and the opportunity for irrational mass response to a perceived emergency, i.e., panic. For this reason, the code requires a specific arrangement of the exits. Studies have indicated that in any emergency, occupants will tend to egress via the same path of travel used to enter the room and building. Therefore, a main entrance to the building must also be designed as the main exit to accommodate this behavior, even if the required exit capacity might be more easily accommodated elsewhere. The main entrance (and exit) must be sized to accommodate at least 50 percent of the total occupant load of the structure and must front on a large, open space, such as a street, for rapid dispersal of the occupants outside the building. The remaining exits must also accommodate at least 50 percent of the total occupant load from each level (see Figure 1024.2). The total occupant load includes those within the theater seating area, the foyer and of any other space (e.g., ticket booth, concession stand, offices, storage and the like).

The required width of the means of egress in places of assembly is more often determined by the occupant load than in most other occupancies. In other occupancies, the minimum required widths and the travel distances will often determine the required widths of the exits.

This section only requires the main exit to accommodate 50 percent of the occupant load when there is a single main entrance. Therefore, a large stadium or civic center, in which there are numerous entrances (and exits), need not comply with the main entrance criteria. This condition is addressed in the exception.

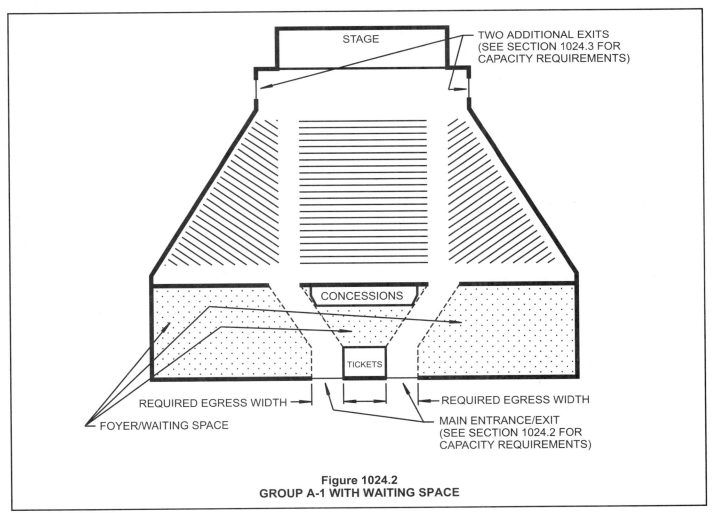

Figure 1024.2
GROUP A-1 WITH WAITING SPACE

1024.3 Assembly other exits. In addition to having access to a main exit, each level of an occupancy in Group A having an occupant load of greater than 300 shall be provided with additional exits that shall provide an egress capacity for at least one-half of the total occupant load served by that level and comply with Section 1014.2.

> **Exception:** In assembly occupancies where there is no well-defined main exit or where multiple main exits are provided, exits shall be permitted to be distributed around the perimeter of the building provided that the total width of egress is not less than 100 percent of the required width.

❖ This section provides for the egress of one-half of the total occupant load by way of exits other than the main exit that is described in Section 1024.2. The exception addresses a large stadium or civic center, in which there are numerous entrances (and exits), none of which are a main entrance or main exit.

1024.4 Foyers and lobbies. In Group A-1 occupancies, where persons are admitted to the building at times when seats are not available and are allowed to wait in a lobby or similar space, such use of lobby or similar space shall not encroach upon the required clear width of the means of egress. Such waiting areas shall be separated from the required means of egress by substantial permanent partitions or by fixed rigid railings not less than 42 inches (1067 mm) high. Such foyer, if not directly connected to a public street by all the main entrances or exits, shall have a straight and unobstructed corridor or path of travel to every such main entrance or exit.

❖ In every case, the main entrance (exit) and all other exits are to be constantly available for the entire building occupant load.

For example, because of the queuing of large crowds, particularly in theaters where a performance may be in progress and people must wait to attend the next one, standing space is often provided. For reasons of safety, such spaces cannot be located in or interfere with established paths of egress from the assembly areas. It is required to designate these areas using partitions or railings so that the means of egress remains clear (see Figure 1024.2).

1024.5 Interior balcony and gallery means of egress. For balconies or galleries having a seating capacity of over 50 located in Group A occupancies, at least two means of egress shall be provided, one from each side of every balcony or gallery, with at least one leading directly to an exit.

❖ This section states the threshold where two means of egress are required based on the occupant load of the interior balcony. Note that one of the means of egress must lead directly to an exit. It is not allowed to go through any other spaces such as an adjoining room prior to reaching an exit such as an exterior exit doorway. However, Section 1024.5.1 does not require the stairways serving the balcony to be enclosed in an exit enclosure. These requirements will assure that at least one path of travel is always available and occupants face a minimum number of hazards.

For balconies with 50 or fewer occupants, see Section 1024.8.

1024.5.1 Enclosure of balcony openings. Interior stairways and other vertical openings shall be enclosed in a vertical exit enclosure as provided in Section 1019.1, except that stairways are permitted to be open between the balcony and the main assembly floor in occupancies such as theaters, churches and auditoriums. At least one accessible means of egress is required from a balcony or gallery level containing accessible seating locations in accordance with Section 1007.3 or 1007.4.

❖ This section allows the stairways that lead from interior balconies to be unenclosed from the balcony to the main floor where the interior balconies are within theaters, churches and auditoriums. Thus, vertical exit enclosures are not required for interior balconies for these facilities. When balconies or galleries contain accessible wheelchair spaces (see Section 1108.2), at least one means of egress must be accessible. While the section references only indicate exit stairways or elevators (see Sections 1007.3 and 1007.4), there are special allowances for the use of platform lifts (see Sections 1007.5 and 1109.7, Item 2) in assembly spaces to allow for dispersion of wheelchair spaces to a variety of locations. This is especially important in assembly spaces with sloped or tiered seating. Section 1007.5 states that if a platform lift is permitted as part of an accessible route, it should also be permitted as part of the accessible means of egress.

1024.6 Width of means of egress for assembly. The clear width of aisles and other means of egress shall comply with Section 1024.6.1 where smoke-protected seating is not provided and with Section 1024.6.2 or 1024.6.3 where smoke-protected seating is provided. The clear width shall be measured to walls, edges of seating and tread edges except for permitted projections.

❖ The means of egress width for assembly occupancy is to be in accordance with this section and the referenced sections instead of the criteria specified in Section 1005.1. Different means of egress width criteria is also specified for assembly seating where smoke protection is provided versus areas it is not provided. The egress width for smoke-protected seating is allowed to be less than for areas where smoke protection is not provided, since the smoke level is required to be maintained at least 6 feet (1829 mm) above the floor of the means of egress, according to Section 1024.6.2.1.

1024.6.1 Without smoke protection. The clear width of the means of egress shall provide sufficient capacity in accordance with all of the following, as applicable:

1. At least 0.3 inch (7.6 mm) of width for each occupant served shall be provided on stairs having riser heights 7 inches (178 mm) or less and tread depths 11 inches (279 mm) or greater, measured horizontally between tread nosing.

2. At least 0.005 inch (0.127 mm) of additional stair width for each occupant shall be provided for each 0.10 inch (2.5 mm) of riser height above 7 inches (178 mm).

3. Where egress requires stair descent, at least 0.075 inch (1.9 mm) of additional width for each occupant shall be provided on those portions of stair width having no handrail within a horizontal distance of 30 inches (762 mm).

4. Ramped means of egress, where slopes are steeper than one unit vertical in 12 units horizontal (8-percent slope), shall have at least 0.22 inch (5.6 mm) of clear width for each occupant served. Level or ramped means of egress, where slopes are not steeper than one unit vertical in 12 units horizontal (8-percent slope), shall have at least 0.20 inch (5.1 mm) of clear width for each occupant served.

❖ This section prescribes the criteria needed to calculate the clear widths of aisles and aisle accessways in order to provide sufficient capacity to handle the occupant loads established by the "catchment areas" described in Section 1024.9.2. Clear width is to be measured to walls, edges of seating and tread edges.

The criteria for determining the required widths are based on analytical studies and field tests that used people to model egress situations [see Figures 1024.6.1(1) and 1024.6.1(2)].

Criterion 1 addresses the method for determining the required egress width for aisles and aisle accessways that are stepped. This method corresponds with the requirements of Table 1005.1 for egress width per occupant of stairways in an unsprinklered building.

Criterion 2 addresses the method for determining the additional stair width required for aisle and aisle accessway stairs with risers greater than 7 inches (178 mm).

Criterion 3 addresses the method for determining the additional stair width where a handrail is not located within 30 inches (762 mm).

Criterion 4 addresses the method for determining the required widths for level or ramped means of egress.

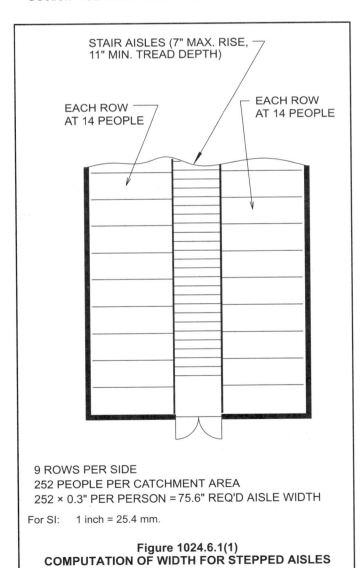

9 ROWS PER SIDE
252 PEOPLE PER CATCHMENT AREA
252 × 0.3" PER PERSON = 75.6" REQ'D AISLE WIDTH

For SI: 1 inch = 25.4 mm.

Figure 1024.6.1(1)
COMPUTATION OF WIDTH FOR STEPPED AISLES

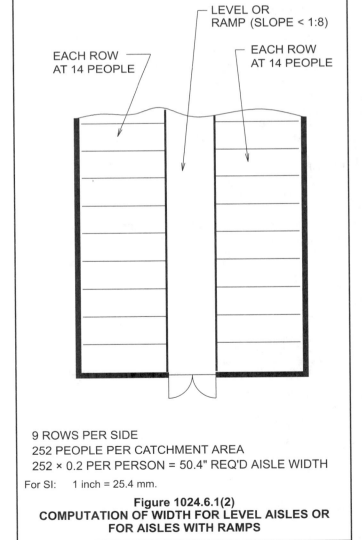

9 ROWS PER SIDE
252 PEOPLE PER CATCHMENT AREA
252 × 0.2 PER PERSON = 50.4" REQ'D AISLE WIDTH

For SI: 1 inch = 25.4 mm.

Figure 1024.6.1(2)
COMPUTATION OF WIDTH FOR LEVEL AISLES OR FOR AISLES WITH RAMPS

1024.6.2 Smoke-protected seating. The clear width of the means of egress for smoke-protected assembly seating shall be not less than the occupant load served by the egress element multiplied by the appropriate factor in Table 1024.6.2. The total number of seats specified shall be those within a single assembly space and exposed to the same smoke-protected environment. Interpolation is permitted between the specific values shown. A life safety evaluation, complying with NFPA 101, shall be done for a facility utilizing the reduced width requirements of Table 1024.6.2 for smoke-protected assembly seating.

> **Exception:** For an outdoor smoke-protected assembly with an occupant load not greater than 18,000, the clear width shall be determined using the factors in Section 1024.6.3.

❖ Special consideration is given to facilities with features that will prevent the means of egress from being blocked by smoke. Facilities to be considered smoke protected by Sections 1024.6.2.1 through 1024.6.2.3 are permitted increases in travel distance, egress capacity, longer dead-end aisles and increased row lengths. All of these result in an increase of allowable egress time. Typically, model codes based on research by Dr. John Fruin and others recognize the need for occupants exposed to the fire environment to evacuate to a safe area within 90 seconds of notification and to reach an area of refuge within 5 minutes. With the increases permitted for smoke-protected facilities, these times are effectively doubled, since the time available for safe egress also increases.

The exception is a pointer to the specific criteria for outdoor seating areas. For outdoor stadiums with 18,000 seats or greater, use Table 1024.6.2.

TABLE 1024.6.2. See below.

❖ This section requires the egress component to be of adequate size to accommodate the occupant load. The egress width per occupant for nonsmoke-protected seating is to be based on Section 1024.6.1 and is similar to the provisions in Table 1005.1. For smoke-protected

seating, the egress width per occupant is based on Table 1024.6.2.

1024.6.2.1 Smoke control. Means of egress serving a smoke-protected assembly seating area shall be provided with a smoke control system complying with Section 909 or natural ventilation designed to maintain the smoke level at least 6 feet (1829 mm) above the floor of the means of egress.

❖ The means of egress and the assembly seating area are required to have some type of smoke control system that will prevent smoke buildup from encroaching on the egress path. This may be a mechanical smoke control system, designed in accordance with Section 909, or a natural ventilation system.

In either type of system, the major consideration is that a smoke-free environment be maintained at least 6 feet (1829 mm) above the floor of the means of egress for a period of at least 20 minutes.

1024.6.2.2 Roof height. A smoke-protected assembly seating area with a roof shall have the lowest portion of the roof deck not less than 15 feet (4572 mm) above the highest aisle or aisle accessway.

> **Exception:** A roof canopy in an outdoor stadium shall be permitted to be less than 15 feet (4572 mm) above the highest aisle or aisle accessway provided that there are no objects less than 80 inches (2032 mm) above the highest aisle or aisle accessway.

❖ One element of a smoke-protected assembly seating facility is that the lowest portion of the roof is required to be at least 15 feet (4572 mm) above the highest aisle or aisle accessway. The objective of this provision is to have a minimum 6-foot (1829 mm) smoke-free height to accommodate safe egress through the area. The additional 9 feet (2743 mm) of height is to provide a volume of space that will act to dissipate smoke. The measurement of the height is shown in Figures 1024.6.2.2(1) and 1024.6.2.2(2).

TABLE 1024.6.2
WIDTH OF AISLES FOR SMOKE-PROTECTED ASSEMBLY

TOTAL NUMBER OF SEATS IN THE SMOKE-PROTECTED ASSEMBLY OCCUPANCY	INCHES OF CLEAR WIDTH PER SEAT SERVED			
	Stairs and aisle steps with handrails within 30 inches	Stairs and aisle steps without handrails within 30 inches	Passageways, doorways and ramps not steeper than 1 in 10 in slope	Ramps steeper than 1 in 10 in slope
Equal to or less than 5,000	0.200	0.250	0.150	0.165
10,000	0.130	0.163	0.100	0.110
15,000	0.096	0.120	0.070	0.077
20,000	0.076	0.095	0.056	0.062
Equal to or greater than 25,000	0.060	0.075	0.044	0.048

For SI: 1 inch = 25.4 mm.

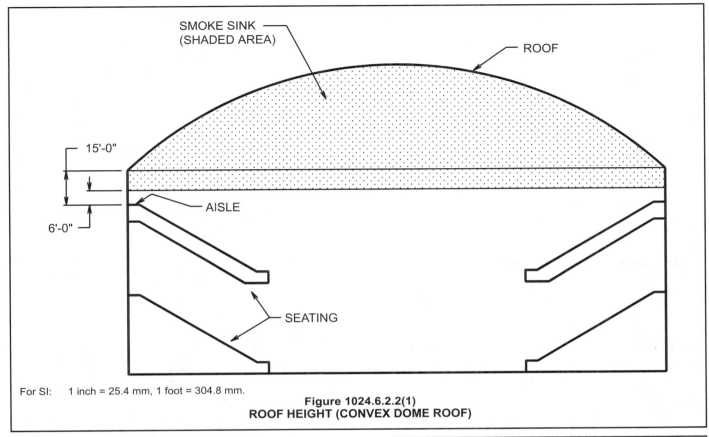

For SI: 1 inch = 25.4 mm, 1 foot = 304.8 mm.

Figure 1024.6.2.2(1)
ROOF HEIGHT (CONVEX DOME ROOF)

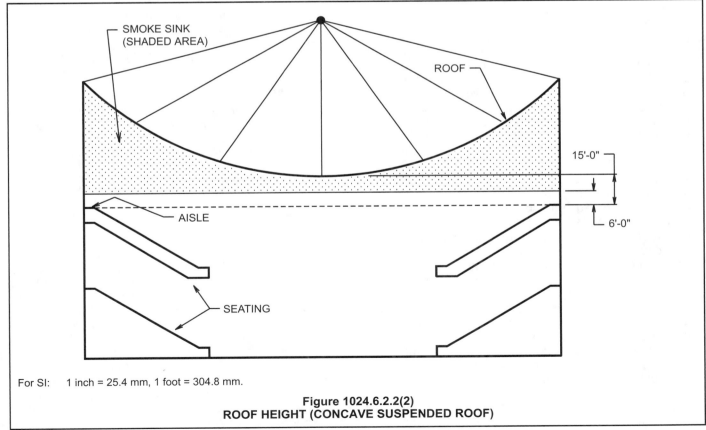

For SI: 1 inch = 25.4 mm, 1 foot = 304.8 mm.

Figure 1024.6.2.2(2)
ROOF HEIGHT (CONCAVE SUSPENDED ROOF)

1024.6.2.3 Automatic sprinklers. Enclosed areas with walls and ceilings in buildings or structures containing smoke-protected assembly seating shall be protected with an approved automatic sprinkler system in accordance with Section 903.3.1.1.

Exceptions:

1. The floor area used for contests, performances or entertainment provided the roof construction is more than 50 feet (15 240 mm) above the floor level and the use is restricted to low fire hazard uses.

2. Press boxes and storage facilities less than 1,000 square feet (93 m²) in area.

3. Outdoor seating facilities where seating and the means of egress in the seating area are essentially open to the outside.

❖ If there are areas in the smoke-protected assembly seating structure enclosed by walls and ceilings, the entire structure is to be provided with an automatic sprinkler designed to meet the requirements of NFPA 13. NFPA 13R systems are not acceptable for this use.

Exception 1 indicates that the area over the playing field or performance area is not required to be sprinklered if the use of the floor area is restricted. If the facility is used for conventions, trade shows, displays or similar purposes, sprinklers would be required throughout, since the occupancy would no longer be a low-fire-hazard use. A characteristic of a low-fire-hazard occupancy is that the fuel load due to combustibles is approximately 2 pounds per square foot (9.8 kg/m²) or less.

In order for the contest, performance or entertainment area to be unsprinklered, the roof over that area must be at least 50 feet (15 240 mm) above the floor in addition to the floor area meeting the low-fire-hazard criteria. The 50-foot (15 240 mm) criterion was selected because the response time for sprinklers at this height is extremely slow. It is estimated that the response time for standard sprinklers [50 feet (15 240 mm) above a floor with a fire having a heat release rate of 5 British thermal units (Btu) per square foot per second] exceeds 15 minutes. Therefore, it is not reasonable to install sprinklers at that height with little expectation of timely activation [see Figure 1024.6.2.3(1)]. Note that if this exception is utilized, the trade-offs for a fully sprinklered building, such as increased height and area limitations or decreased corridor ratings, are no longer permitted.

Exception 2 indicates that automatic sprinklers are not required in small spaces in buildings. Sprinklers are required in press box and storage areas of outdoor facilities when the aggregate area exceeds 1,000 square feet (93 m²). The primary reasons for sprinklers in these areas is that both are anticipated to have a relatively large combustible load when compared to the main seating and participant areas. Additionally, in the case of storage areas, there is an increased potential for an undetected fire condition to occur [see Figure 1024.6.2.3(2)].

Exception 3 provides for outdoor seating facilities where smoke entrapment is not a safety concern.

1024.6.3 Width of means of egress for outdoor smoke-protected assembly. The clear width in inches (mm) of aisles and other means of egress shall be not less than the total occupant load served by the egress element multiplied by 0.08 (2.0 mm) where egress is by aisles and stairs and multiplied by 0.06 (1.52 mm) where egress is by ramps, corridors, tunnels or vomitories.

Exception: The clear width in inches (mm) of aisles and other means of egress shall be permitted to comply with Section 1024.6.2 for the number of seats in the outdoor smoke-protected assembly where Section 1024.6.2 permits less width.

❖ This section has the coefficients for the determination of the width of egress required for outdoor smoke-protected assembly areas. Note that the coefficients are very low compared to the values in Section 1024.6.1 for assembly areas without smoke protection. The coefficients are also lower than those for smoke-protected assembly seating in Table 1024.6.2 except for very large assembly areas. The exception in this section would apply where the coefficients in Table 1024.6.2 are less than those in this section.

Low coefficients are a result of the very low hazard of outdoor smoke-protected assembly areas.

Generally, an outdoor assembly area meets the smoke control requirements of Section 1024.6.1 by natural ventilation and does not require an automatic sprinkler system according to Section 1024.6.3, Exception 3.

1024.7 Travel distance. Exits and aisles shall be so located that the travel distance to an exit door shall be not greater than 200 feet (60 960 mm) measured along the line of travel in nonsprinklered buildings. Travel distance shall not be more than 250 feet (76 200 mm) in sprinklered buildings. Where aisles are provided for seating, the distance shall be measured along the aisles and aisle accessway without travel over or on the seats.

Exceptions:

1. Smoke-protected assembly seating: The travel distance from each seat to the nearest entrance to a vomitory or concourse shall not exceed 200 feet (60 960 mm). The travel distance from the entrance to the vomitory or concourse to a stair, ramp or walk on the exterior of the building shall not exceed 200 feet (60 960 mm).

2. Open-air seating: The travel distance from each seat to the building exterior shall not exceed 400 feet (122 m). The travel distance shall not be limited in facilities of Type I or II construction.

❖ This section includes the travel distance limits for an assembly occupancy, which are the same as those in Table 1015.1 for an assembly occupancy. The travel distance is to be measured in the same path as the occupants would normally take to exit the facility.

Exception 1 provides an extended travel distance for smoke-protected assembly seating that meets the requirements of Sections 1024.6.1 through 1024.6.3. Exception 2 applies to outdoor open-air seating areas where the smoke and fire hazard is very low. The Type I and II construction referred to in this exception is described in Section 602.

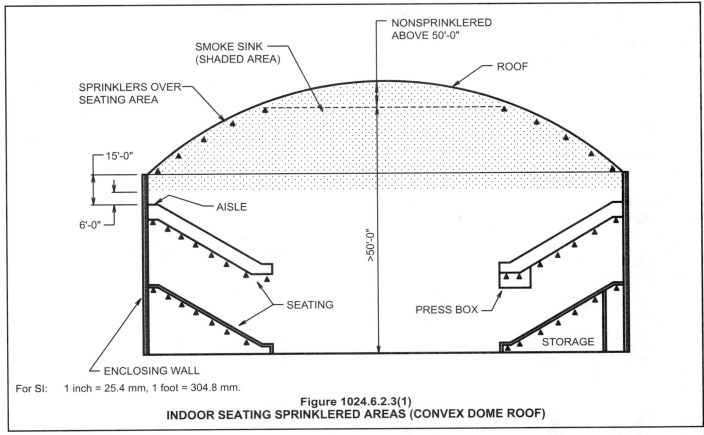

For SI: 1 inch = 25.4 mm, 1 foot = 304.8 mm.

Figure 1024.6.2.3(1)
INDOOR SEATING SPRINKLERED AREAS (CONVEX DOME ROOF)

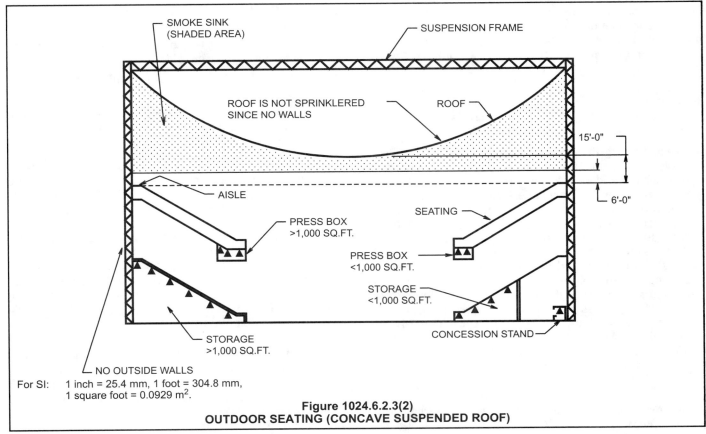

For SI: 1 inch = 25.4 mm, 1 foot = 304.8 mm,
1 square foot = 0.0929 m².

Figure 1024.6.2.3(2)
OUTDOOR SEATING (CONCAVE SUSPENDED ROOF)

1024.8 Common path of travel. The common path of travel shall not exceed 30 feet (9144 mm) from any seat to a point where a person has a choice of two paths of egress travel to two exits.

Exceptions:

1. For areas serving not more than 50 occupants, the common path of travel shall not exceed 75 feet (22 860 mm).

2. For smoke-protected assembly seating, the common path of travel shall not exceed 50 feet (15 240 mm).

❖ The maximum travel distance down a single access row of seating to a location where a patron would have two choices for a way out of the space is 30 feet (9144 mm). In smoke-protected seating, the common path of travel can be up to 50 feet (15 240 mm).

   If the room or space (e.g., box, gallery or balcony) has 50 or fewer occupants, the travel distance can be increased to 75 feet (22 860 mm). For example, this allows for a path of travel from a box seat, out of the box and to a main aisle or even a corridor located outside the assembly room itself. When this section is referenced for accessible means of egress (see Section 1007.1, Exception 3), the utilization of Exception 1 would include the entire occupant load of the box, gallery or balcony, not just the number of wheelchair spaces and/or companion seats.

1024.8.1 Path through adjacent row. Where one of the two paths of travel is across the aisle through a row of seats to another aisle, there shall be not more than 24 seats between the two aisles, and the minimum clear width between rows for the row between the two aisles shall be 12 inches (305 mm) plus 0.6 inch (15.2 mm) for each additional seat above seven in the row between aisles.

Exception: For smoke-protected assembly seating there shall not be more than 40 seats between the two aisles and the minimum clear width shall be 12 inches (305 mm) plus 0.3 inch (7.6 mm) for each additional seat.

❖ In establishing the point where the occupants of a row served by a single access aisle have two distinct paths of travel, the code allows one of those paths to be through the rows of an adjacent seating area or section. This requirement increases the row widths for the single-access seating section and the adjacent dual-access seating section. This allows the occupants to either travel down the single access aisle or readily traverse the oversized row widths to gain access to a second means of egress. This exception allows a greater number of seats spaced with a minimum clearance of 12 inches (305 mm) for smoke-protected assembly seating that complies with Sections 1024.6.2.1 through 1024.6.2.3 or Section 1024.6.3. For the base width requirements for single- and dual-access rows, see the commentary to Sections 1024.10 through 1024.10.2.

1024.9 Assembly aisles are required. Every occupied portion of any occupancy in Group A that contains seats, tables, dis-

plays, similar fixtures or equipment shall be provided with aisles leading to exits or exit access doorways in accordance with this section. Aisle accessways for tables and seating shall comply with Section 1013.4.2.

❖ This section requires that each assembly area have designated aisles. For aisle accessway requirements, see Section 1024.10. Assembly area aisle accessways between tables and chairs are to comply with the width requirements in Section 1013.4.2.

1024.9.1 Minimum aisle width. The minimum clear width of aisles shall be as shown:

1. Forty-eight inches (1219 mm) for aisle stairs having seating on each side.

   Exception: Thirty-six inches (914 mm) where the aisle does not serve more than 50 seats.

2. Thirty-six inches (914 mm) for aisle stairs having seating on only one side.

3. Twenty-three inches (584 mm) between an aisle stair handrail or guard and seating where the aisle is subdivided by a handrail.

4. Forty-two inches (1067 mm) for level or ramped aisles having seating on both sides.

   Exceptions:

   1. Thirty-six inches (914 mm) where the aisle does not serve more than 50 seats.

   2. Thirty inches (762 mm) where the aisle does not serve more than 14 seats.

5. Thirty-six inches (914 mm) for level or ramped aisles having seating on only one side.

   Exception: Thirty inches (762 mm) where the aisle does not serve more than 14 seats.

6. Twenty-three inches (584 mm) between an aisle stair handrail and seating where an aisle does not serve more than five rows on one side.

❖ The clear widths of aisles and other means of egress established by the formulas given in Section 1024.6 must not be less than the minimum width requirements of this section. The development of minimum width requirements is based on the association of aisle capacity with the path of exit travel as influenced by the different features of aisle construction. The purpose is to create an aisle system that would provide an even flow of occupant egress. The minimum width of the aisles is also based on an anticipated movement of people in two directions. The exceptions are only intended to be applicable to the item directly above.

1024.9.2 Aisle width. The aisle width shall provide sufficient egress capacity for the number of persons accommodated by the catchment area served by the aisle. The catchment area served by an aisle is that portion of the total space that is served by that section of the aisle. In establishing catchment areas, the assumption shall be made that there is a balanced use of all means of egress, with the number of persons in proportion to egress capacity.

❖ The determination of required aisle and aisle accessway width is a function of the occupant load. In calculating the required widths, the assumption is that in a system or network of aisles and aisle accessways serving an occupied area, people will normally exit the area in a way that will distribute the occupant load throughout the system in proportion to the egress capacity of the aisles and aisle accessways. Each aisle and aisle accessway would take its tributary share (catchment area) of the total occupant load (see Figure 1024.9.2).

In addition to the provisions in this section, the requirement for the capacity of the main exit and other exits must also be considered (see Section 1024.2). While this section assumes an equal distribution, Section 1024.2 requires that where the facility has a main exit, the main exit and the access thereto must be capable of handling 50 percent of the occupant load.

1024.9.3 Converging aisles. Where aisles converge to form a single path of egress travel, the required egress capacity of that path shall not be less than the combined required capacity of the converging aisles.

❖ Where one or more aisles or aisle accessways meet to form a single path of egress travel, that path must be sized to handle the combined occupant capacity of the converging aisles and aisle accessways (see Figure 1024.9.3). The reason for this requirement is to maintain the natural pace of travel all the way to the exits and to minimize the queuing of occupants.

This section requires combining the required occupant capacity of converging aisles and aisle accessways, but not necessarily the required widths. For example, if two 48-inch (1219 mm) aisles converge, the result need not be a 96-inch (2438 mm) aisle unless the 48-inch (1219 mm) width of the aisles is required based on the requirements of Section 1024.6 for the actual occupant load served. However, if the 48-inch (1219 mm) width is not based on the occupant load but is required to comply with the minimum aisle width requirements of Section 1024.9.1, the resulting aisle width must be capable of serving the total occupant load served by the converging aisles, as determined by Section 1024.6, but not less than the minimum widths of Section 1024.9.1.

1024.9.4 Uniform width. Those portions of aisles, where egress is possible in either of two directions, shall be uniform in required width.

❖ Aisles that connect or lead to opposite exits must be of uniform width throughout their entire length to allow for exit travel in two directions without creating a traffic bottleneck (see Figure 1024.9.4).

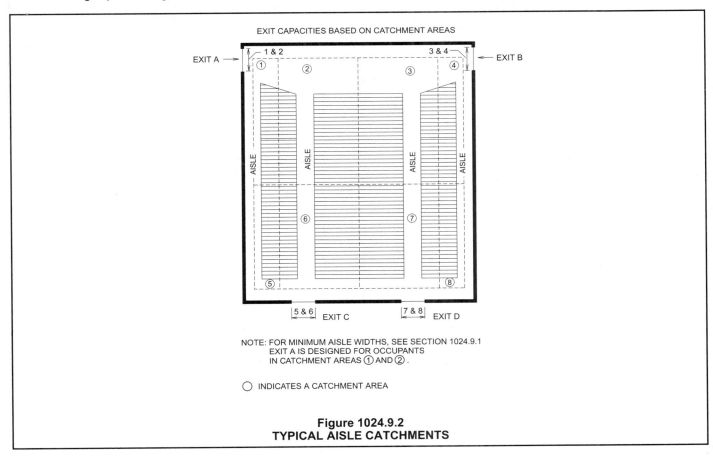

EXIT CAPACITIES BASED ON CATCHMENT AREAS

NOTE: FOR MINIMUM AISLE WIDTHS, SEE SECTION 1024.9.1.
EXIT A IS DESIGNED FOR OCCUPANTS
IN CATCHMENT AREAS ① AND ② .

◯ INDICATES A CATCHMENT AREA

**Figure 1024.9.2
TYPICAL AISLE CATCHMENTS**

FIGURE 1024.9.3 – FIGURE 1024.9.4

MEANS OF EGRESS

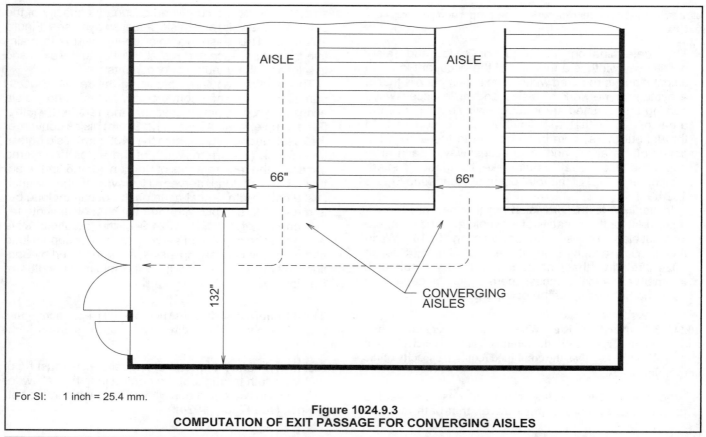

For SI: 1 inch = 25.4 mm.

Figure 1024.9.3
COMPUTATION OF EXIT PASSAGE FOR CONVERGING AISLES

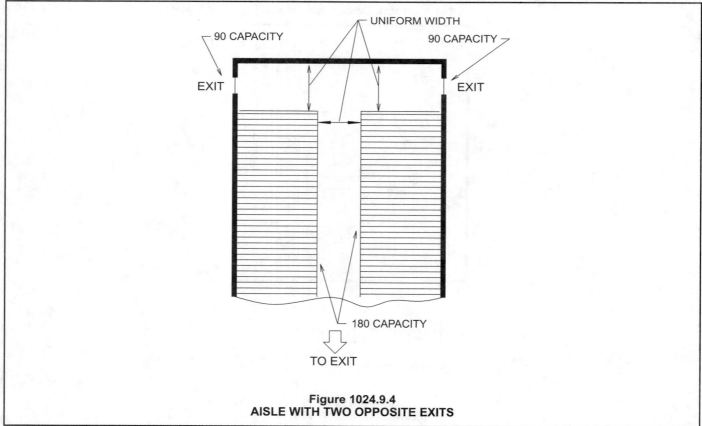

Figure 1024.9.4
AISLE WITH TWO OPPOSITE EXITS

1024.9.5 Assembly aisle termination. Each end of an aisle shall terminate at cross aisle, foyer, doorway, vomitory or concourse having access to an exit.

Exceptions:

1. Dead-end aisles shall not be greater than 20 feet (6096 mm) in length.

2. Dead-end aisles longer than 20 feet (6096 mm) are permitted where seats beyond the 20-foot (6096 mm) dead-end aisle are no more than 24 seats from another aisle, measured along a row of seats having a minimum clear width of 12 inches (305 mm) plus 0.6 inch (15.2 mm) for each additional seat above seven in the row.

3. For smoke-protected assembly seating, the dead-end aisle length of vertical aisles shall not exceed a distance of 21 rows.

4. For smoke-protected assembly seating, a longer dead-end aisle is permitted where seats beyond the 21-row dead-end aisle are not more than 40 seats from another aisle, measured along a row of seats having an aisle accessway with a minimum clear width of 12 inches (305 mm) plus 0.3 inch (7.6 mm) for each additional seat above seven in the row.

❖ Both ends of a cross aisle must terminate at either an intersecting aisle, a foyer, a doorway or a vomitory (lane) that gives access to an exit(s). Dead-end aisles (similar to corridors and passageways) that terminate at one end of a cross aisle or at a foyer, doorway or vomitory must not be more than 20 feet (6096 mm) in length (see Figure 1024.9.5). The intent of the row width requirements in the exceptions is to provide sufficient clear width between rows of seating to allow the occupants in times of emergency to pass quickly from a dead-end aisle to the aisle at the opposite end. In Exception 2, the 0.6-inch (15 mm) increase beyond seven seats is consistent with the minimum width determined in accordance with Section 1024.10.2 for single access rows. The code recognizes that one dead-end aisle may not be usable, thus creating a single access row condition. Exceptions 3 and 4 allow longer dead-end aisles for smoke-protected assembly seating that complies with Sections 1024.6.2.1 through 1024.6.2.3 or Section 1024.6.3.

The overall purpose of this section is to provide aisle/seating arrangements that would allow the occupants to seek safe and rapid passage to exits in case of fire or other emergency.

1024.9.6 Assembly aisle obstructions. There shall be no obstructions in the required width of aisles except for handrails as provided in Section 1024.13.

❖ Except for handrails, aisles are required to be clear of any obstructions so that the full width is available for

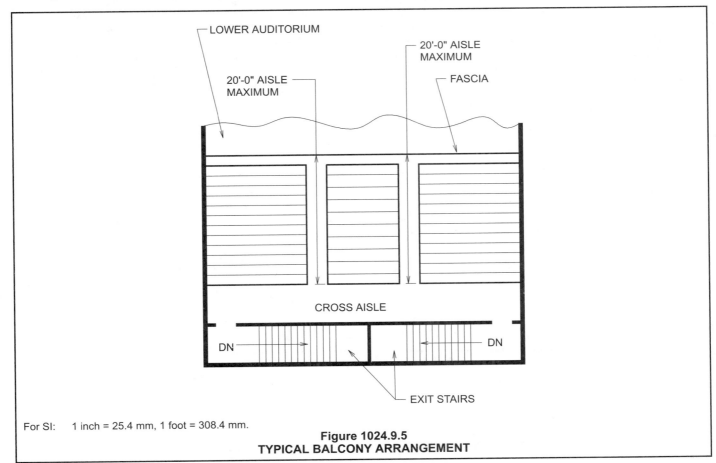

For SI: 1 inch = 25.4 mm, 1 foot = 308.4 mm.

Figure 1024.9.5
TYPICAL BALCONY ARRANGEMENT

egress purposes. Handrails are allowed to project into the required aisle width in the same manner as handrail projections in stairways.

1024.10 Clear width of aisle accessways serving seating. Where seating rows have 14 or fewer seats, the minimum clear aisle accessway width shall not be less than 12 inches (305 mm) measured as the clear horizontal distance from the back of the row ahead and the nearest projection of the row behind. Where chairs have automatic or self-rising seats, the measurement shall be made with seats in the raised position. Where any chair in the row does not have an automatic or self-rising seat, the measurements shall be made with the seat in the down position. For seats with folding tablet arms, row spacing shall be determined with the tablet arm down.

❖ The requirements of this section are applicable to theater-type seating arrangements. This includes both "continental" and "traditional" seating arrangements. Theater-type seating is characterized by a number of seats arranged side by side and in rows. In this type of seating arrangement, the potential exists for a large number of occupants to be present in a confined environment where the ability of the occupants to move is limited. In order to egress, people are required to move within a row before reaching an aisle or aisle accessway, and the aisle or aisle accessway also limits movement toward an exit. To provide adequate passage between rows of seats, this section requires that the clear width between the back of a row to the nearest projection of the seating immediately behind must be at least 12 inches (305 mm) (see Figure 1024.10). Where chairs are manufactured with automatic or self-lifting seats, the minimum width requirement may be measured with the seats in a raised position. When tablet arm chairs are used, the required width is to be deter-

mined with the tablet arm in its usable position. It is not acceptable to eliminate the tablet arm from the measurement, even though it is retractable.

With respect to self-rising seats. ASTM F 851 provides one method of determining acceptability.

1024.10.1 Dual access. For rows of seating served by aisles or doorways at both ends, there shall not be more than 100 seats per row. The minimum clear width of 12 inches (305 mm) between rows shall be increased by 0.3 inch (7.6 mm) for every additional seat beyond 14 seats, but the minimum clear width is not required to exceed 22 inches (559 mm).

Exception: For smoke-protected assembly seating, the row length limits for a 12-inch-wide (305 mm) aisle accessway, beyond which the aisle accessway minimum clear width shall be increased, are in Table 1024.10.1.

❖ Where rows of seating are served by aisles or doorways located at both ends of the path of row travel, the number of seats that may be used in a row may be up to, but not more than, 100 (continental seating) and the minimum required clear width aisle accessway of 12 inches (305 mm) between rows of seats must be increased by 0.3 inch (8 mm) for every additional seat beyond 14, but not more than a total of 22 inches (559 mm) (see Figure 1024.10.1). For example, in a row of 24 seats, the minimum clear width would compute to 15 inches (381 mm) [12 + (0.3 by 10)]. For a row of 34 seats, a clear width of 18 inches (457 mm) would be required. Increases in the clear width between rows of seats would occur up to a row of 46 seats. From 47 to 100 seats, a maximum clear width between rows of 22 inches (559 mm) would apply.

Since the row is to provide access to an aisle in both directions, the minimum width applies to the entire length of the row aisle accessway.

The exception allows more seats in a row with the

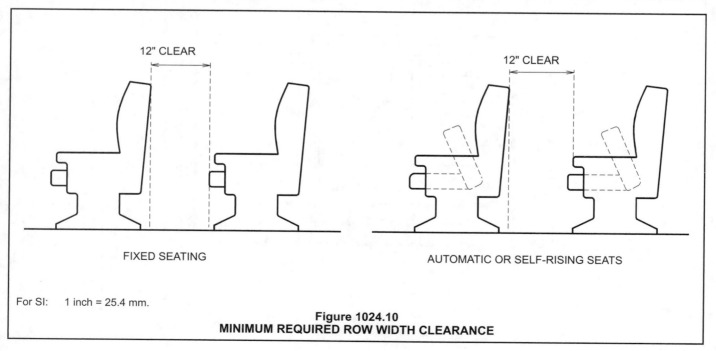

12" CLEAR

12" CLEAR

FIXED SEATING

AUTOMATIC OR SELF-RISING SEATS

For SI: 1 inch = 25.4 mm.

Figure 1024.10
MINIMUM REQUIRED ROW WIDTH CLEARANCE

minimum 12-inch (305 mm) seat spacing since safe egress time is extended for this condition.

For additional aisle accessway width requirements when one of the means of egress at the end of the single access row is through a dual access row, see Section 1024.8.1.

TABLE 1024.10.1
SMOKE-PROTECTED
ASSEMBLY AISLE ACCESSWAYS

TOTAL NUMBER OF SEATS IN THE SMOKE-PROTECTED ASSEMBLY OCCUPANCY	MAXIMUM NUMBER OF SEATS PER ROW PERMITTED TO HAVE A MINIMUM 12-INCH CLEAR WIDTH AISLE ACCESSWAY	
	Aisle or doorway at both ends of row	Aisle or doorway at one end of row only
Less than 4,000	14	7
4,000	15	7
7,000	16	8
10,000	17	8
13,000	18	9
16,000	19	9
19,000	20	10
22,000 and greater	21	11

For SI: 1 inch = 25.4 mm.

❖ Table 1024.10.1 recognizes the increased egress time available in smoke-protected assembly seating areas. Therefore, the table permits greater lengths of rows that have the minimum 12 inches (305 mm) of clear width. When a row exceeds the lengths identified in the table, the row width is to be increased in accordance with Section 1024.10.1 [0.3 inch (8 mm) per additional seat] for

dual access rows and Section 1024.10.2 [0.6 inch (15 mm) per additional seat] for single access rows. The requirements of this table are based on the total number of seats contained within the assembly space.

1024.10.2 Single access. For rows of seating served by an aisle or doorway at only one end of the row, the minimum clear width of 12 inches (305 mm) between rows shall be increased by 0.6 inch (15.2 mm) for every additional seat beyond seven seats, but the minimum clear width is not required to exceed 22 inches (559 mm).

Exception: For smoke-protected assembly seating, the row length limits for a 12-inch-wide (305 mm) aisle accessway, beyond which the aisle accessway minimum clear width shall be increased, are in Table 1024.10.1.

❖ Where rows of seating are served by an aisle or doorway at only one end of a row, the minimum clear width of 12 inches (305 mm) between rows of seats must be increased by 0.6 inch (15 mm) for every additional seat beyond seven, but not more than a total of 22 inches (559 mm) (see Figure 1024.10.2). While this section does not specify the maximum number of seats permitted in a row, the 30-foot (9144 mm) common path of travel limitation (see Section 1024.8) essentially restricts the single access row to approximately 20 seats, based on an 18-inch (457 mm) width per seat. A row of 12 seats would compute to a required minimum width of 15 inches [12 + (0.5 by 5)]. Similarly, a row of 17 seats would require a clear width of 18 inches (457 mm) and so on. Since dual access is not provided, incremental increases would be permitted in the aisle accessway width as shown in Figure 1024.10.2. Incremental increases in the required width would occur up to the maximum number of seats, which is determined by the 30-foot (9144 mm) dead-end limitation.

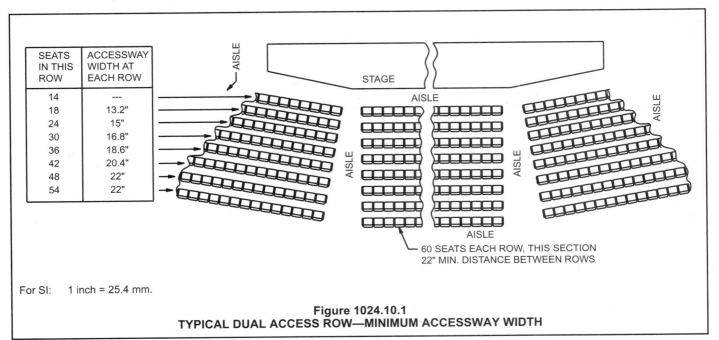

SEATS IN THIS ROW	ACCESSWAY WIDTH AT EACH ROW
14	---
18	13.2"
24	15"
30	16.8"
36	18.6"
42	20.4"
48	22"
54	22"

60 SEATS EACH ROW, THIS SECTION
22" MIN. DISTANCE BETWEEN ROWS

For SI: 1 inch = 25.4 mm.

Figure 1024.10.1
TYPICAL DUAL ACCESS ROW—MINIMUM ACCESSWAY WIDTH

The reason for increasing the row accessway widths incrementally with increases in the number of seats per row is to provide more efficient passage for the occupants who are using the aisle accessway. As a practical matter, where dual-access (see Section 1024.10.1) and single-access seating arrangements are used together, the largest computed clear width dimension would normally be applied by the designer to both arrangements so that the rows of seats will be in alignment. For additional aisle accessway width requirements when one of the means of egress at the end of the single access row is through a dual access row, see Section 1024.8.1.

1024.11 Assembly aisle walking surfaces. Aisles with a slope not exceeding one unit vertical in eight units horizontal (12.5-percent slope) shall consist of a ramp having a slip-resistant walking surface. Aisles with a slope exceeding one unit vertical in eight units horizontal (12.5-percent slope) shall consist of a series of risers and treads that extends across the full width of aisles and complies with Sections 1024.11.1 through 1024.11.3.

❖ Assembly facilities such as theaters and auditoriums often require sloping or stepped floors to provide seated occupants with preferred sightlines for viewing presentations. Aisles must, therefore, be designed to accommodate the changing elevations of the floor in such a manner that the path of travel will allow occupants to leave the area at a rapid pace with minimal possibilities for stumbling or falling during times of emergency.

This section requires that aisles with a gradient of one in eight (1:8) or less must consist of a ramp with a slip-resistant surface. Aisles with a gradient exceeding one in eight (1:8) must consist of a series of treads and risers that comply with the requirements of Sections 1024.11.1 through 1024.11.3. Note that ramps that

serve as part of an accessible route to and from accessible wheelchair spaces must comply with the more restrictive requirements for ramps in Section 1010 (see Section 1010.1, Exception 1).

While not specifically indicated for stepped aisles, such floor surfaces must also be slip resistant in accordance with Section 1003.4. Field testing and uniform enforcement of the concept of slip resistance is not practical. One method used to establish slip resistance is that the static coefficient of friction between leather [Type 1 (Vegetable Tanned) of Federal Specification KK-L-165C] and the floor surface is greater than 0.5. Laboratory test procedures can determine the static coefficient of resistance.

What must be recognized here is that stepped aisles are part of the floor construction and are intended to provide horizontal egress. Tread and riser construction for this purpose should not be compared to the requirements for treads and risers in conventional stairways that serve as means of vertical egress. Sometimes, because of design considerations, the gradient of an aisle is required to change from a level floor, to a ramp and then to steps. In cases where there is no uniformity in the path of travel, occupants tend to be considerably more cautious, particularly in the use of stepped aisles, than they would normally be in the use of conventional stairways.

1024.11.1 Treads. Tread depths shall be a minimum of 11 inches (279 mm) and shall have dimensional uniformity.

Exception: The tolerance between adjacent treads shall not exceed 0.188 inch (4.8 mm).

❖ Depths of treads are not to be less than 11 inches (279 mm) and uniform throughout each flight, except that a

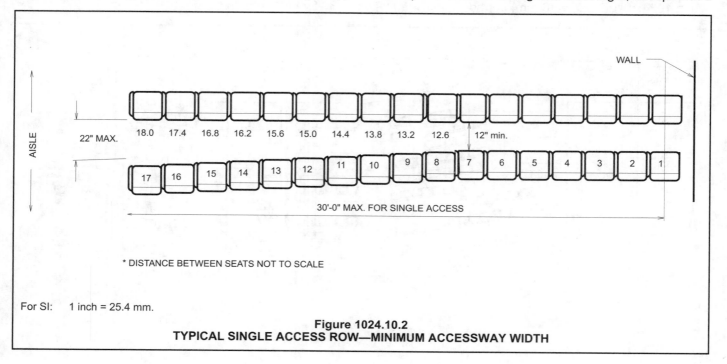

AISLE

22" MAX.

18.0 17.4 16.8 16.2 15.6 15.0 14.4 13.8 13.2 12.6 | 12" min.

WALL

| 17 | 16 | 15 | 14 | 13 | 12 | 11 | 10 | 9 | 8 | 7 | 6 | 5 | 4 | 3 | 2 | 1 |

30'-0" MAX. FOR SINGLE ACCESS

* DISTANCE BETWEEN SEATS NOT TO SCALE

For SI: 1 inch = 25.4 mm.

Figure 1024.10.2
TYPICAL SINGLE ACCESS ROW—MINIMUM ACCESSWAY WIDTH

variance of not more than 0.188 inch (4.8 mm) is permitted between adjacent treads to accommodate variations in construction. While this provision is the same as the limiting dimension for treads in interior stairways (see Section 1009.3), it rarely applies in the construction of stepped aisles. A more common form of stepped aisle construction is to provide a tread depth equal to the back-to-back distance between rows of seats. This way the treads can be extended across the full length of the row and serve as a supporting platform for the seats. Other arrangements might require two treads between rows of seats.

In theaters, for example, the back-to-back distance between rows of fixed seats usually ranges somewhere between 3 and 4 feet (914 and 1219 mm), depending on seat style and seat dimensions as well as the ease of passage between the rows (see Figure 1024.11.1). The selection of single-tread or two-tread construction between rows of seats depends on the gradient and suitable riser height (see Section 1024.11.2), as needed for sightlines.

In comparing this section with Section 1024.11.2, it is significant to note the emphasis placed on the tread dimension. While not desirable, the code permits riser heights to deviate; however, tread dimensions must not vary beyond the 0.188-inch (4.8 mm) tolerance.

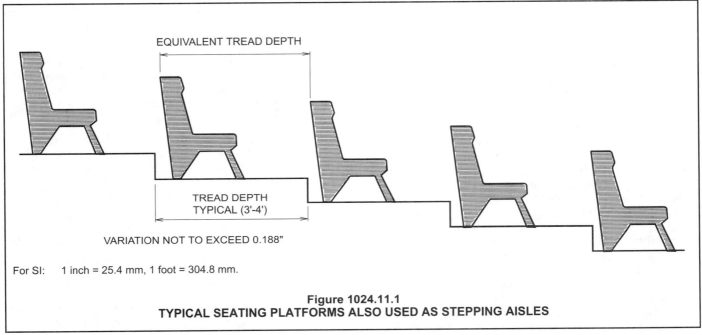

For SI: 1 inch = 25.4 mm, 1 foot = 304.8 mm.

Figure 1024.11.1
TYPICAL SEATING PLATFORMS ALSO USED AS STEPPING AISLES

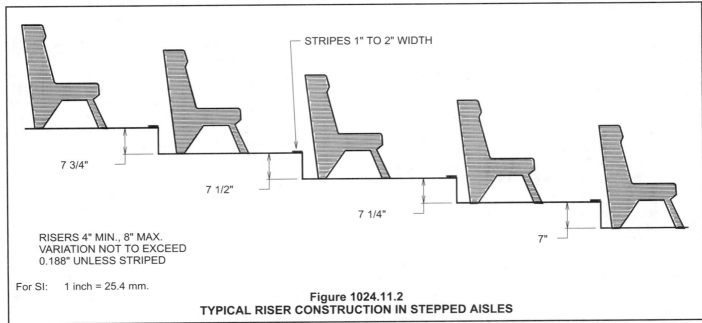

For SI: 1 inch = 25.4 mm.

Figure 1024.11.2
TYPICAL RISER CONSTRUCTION IN STEPPED AISLES

1024.11.2 Risers. Where the gradient of aisle stairs is to be the same as the gradient of adjoining seating areas, the riser height shall not be less than 4 inches (102 mm) nor more than 8 inches (203 mm) and shall be uniform within each flight.

Exceptions:

1. Riser height nonuniformity shall be limited to the extent necessitated by changes in the gradient of the adjoining seating area to maintain adequate sightlines. Where nonuniformities exceed 0.188 inch (4.8 mm) between adjacent risers, the exact location of such nonuniformities shall be indicated with a distinctive marking stripe on each tread at the nosing or leading edge adjacent to the nonuniform risers. Such stripe shall be a minimum of 1 inch (25 mm), and a maximum of 2 inches (51 mm), wide. The edge marking stripe shall be distinctively different from the contrasting marking stripe.

2. Riser heights not exceeding 9 inches (229 mm) shall be permitted where they are necessitated by the slope of the adjacent seating areas to maintain sightlines.

❖ In stepped aisles where the gradient of the aisle is the same as the gradient of the adjoining seating area, riser heights are not to be less than 4 inches (102 mm) nor more than 8 inches (203 mm) (see Figure 1024.11.2). For the safety of the occupants, risers should have uniform heights, where possible, throughout each flight of steps. However, nonuniformity of riser heights is permitted in cases where changes to the gradient in the adjoining seating area are required because of sightlines and other seating layout considerations.

Where variations in height exceed 0.188 inch (4.8 mm) between adjacent risers, a distinctive marking stripe between 1 and 2 inches (25 and 51 mm) wide is to be located on the nosings of each tread where the variations occur as a visual warning to the occupants to be cautious. Frequently, this is done with "runway" lights. Note that this stripe must be different from the tread contrast marking stripes required in Section 1024.11.3.

In comparing this section with Section 1024.11.1, it is significant to note the emphasis placed on the tread dimension. While not desirable, the code permits riser heights to deviate; however, Section 1024.11.1 does not permit tread dimensions to vary beyond the 0.188-inch (4.8 mm) tolerance.

1024.11.3 Tread contrasting marking stripe. A contrasting marking stripe shall be provided on each tread at the nosing or leading edge such that the location of each tread is readily apparent when viewed in descent. Such stripe shall be a minimum of 1 inch (25 mm), and a maximum of 2 inches (51 mm), wide.

Exception: The contrasting marking stripe is permitted to be omitted where tread surfaces are such that the location of each tread is readily apparent when viewed in descent.

❖ The exception provides for the omission of the contrasting marking stripe where the tread is readily apparent such as when aisle stair treads are provided with a roughened metal nosing strip or where lighted nosings occur. In this situation, the user is aware of the treads

without the marking stripe. This stripe must be different from the marking stripe required for nonuniform risers in Section 1024.11.2, Exception 1.

1024.12 Seat stability. In places of assembly, the seats shall be securely fastened to the floor.

Exceptions:

1. In places of assembly or portions thereof without ramped or tiered floors for seating and with 200 or fewer seats, the seats shall not be required to be fastened to the floor.

2. In places of assembly or portions thereof with seating at tables and without ramped or tiered floors for seating, the seats shall not be required to be fastened to the floor.

3. In places of assembly or portions thereof without ramped or tiered floors for seating and with greater than 200 seats, the seats shall be fastened together in groups of not less than three or the seats shall be securely fastened to the floor.

4. In places of assembly where flexibility of the seating arrangement is an integral part of the design and function of the space and seating is on tiered levels, a maximum of 200 seats shall not be required to be fastened to the floor. Plans showing seating, tiers and aisles shall be submitted for approval.

5. Groups of seats within a place of assembly separated from other seating by railings, guards, partial height walls or similar barriers with level floors and having no more than 14 seats per group shall not be required to be fastened to the floor.

6. Seats intended for musicians or other performers and separated by railings, guards, partial height walls or similar barriers shall not be required to be fastened to the floor.

❖ The purpose of this section is to require that assembly seating be fastened to the floor where it would be a significant hazard if loose and subject to tipping over. The exceptions allow loose assembly seating for situations where the hazard is lower, such as floors where ramped or tiered seating is not used, where no more than 200 seats are used and for box seating arrangements where a limited number of seats are within railings, guards or partial height walls.

1024.13 Handrails. Ramped aisles having a slope exceeding one unit vertical in 15 units horizontal (6.7-percent slope) and aisle stairs shall be provided with handrails located either at the side or within the aisle width.

Exceptions:

1. Handrails are not required for ramped aisles having a gradient no greater than one unit vertical in eight units horizontal (12.5-percent slope) and seating on both sides.

2. Handrails are not required if, at the side of the aisle, there is a guard that complies with the graspability requirements of handrails.

❖ For the safety of occupants, handrails must be provided in aisles where ramps exceed a gradient of one in 15 (1:15) (see Figure 1024.13).

Exception 1 omits the handrail requirements where ramped aisles are not steep and seats are on both sides to reduce the fall hazard.

Exception 2 allows handrails to be omitted where there is a guard at the side of the aisle with a top rail that complies with the requirements for handrail graspability (see Section 1009.11.3). Note that the guard must meet the height and opening requirements specified in Section 1012 or 1024.14 as applicable.

1024.13.1 Discontinuous handrails. Where there is seating on both sides of the aisle, the handrails shall be discontinuous with gaps or breaks at intervals not exceeding five rows to facilitate access to seating and to permit crossing from one side of the aisle to the other. These gaps or breaks shall have a clear width of at least 22 inches (559 mm) and not greater than 36 inches (914 mm), measured horizontally, and the handrail shall have rounded terminations or bends.

❖ Where aisles have seating on both sides, handrails may be located at the sides of the aisles, but are typically located in the center of the aisle. The width of each section of the subdivided aisle between the handrail and the edge of seating is not to be less than 23 inches (584 mm) (see Section 1024.9.1, Item 3).

For reasons of life safety in fire situations and also as a practical matter in the efficient use of the facility, a handrail down the middle of an aisle should not be continuous along its entire length. Crossovers must be provided by means of gaps or breaks in the handrail installation. Such openings must not be less than 22 inches (559 mm) nor more than 36 inches (914 mm) wide, and

must be provided at intervals not exceeding the distance of five rows of seats (see Figure 1024.13.1). All handrail terminations should be designed to have rounded ends or bends to avoid possible injury to the occupants (see Figure 1024.13).

1024.13.2 Intermediate handrails. Where handrails are provided in the middle of aisle stairs, there shall be an additional intermediate handrail located approximately 12 inches (305 mm) below the main handrail.

❖ Handrail installations down the middle of an aisle must be constructed with intermediate rails located 12 inches (305 mm) below and parallel to main handrails. This is to provide handholds for children, to prevent people from using the handrail like a gym apparatus and possibly injuring themselves and from ducking to get under the rail (see Figure 1024.13).

1024.14 Assembly guards. Assembly guards shall comply with Sections 1024.14.1 through 1024.14.3.

❖ This section establishes the scope of the assembly guard provisions.

1024.14.1 Cross aisles. Cross aisles located more than 30 inches (762 mm) above the floor or grade below shall have guards in accordance with Section 1012.

Where an elevation change of 30 inches (762 mm) or less occurs between a cross aisle and the adjacent floor or grade below, guards not less than 26 inches (660 mm) above the aisle floor shall be provided.

Exception: Where the backs of seats on the front of the cross aisle project 24 inches (610 mm) or more above the adjacent floor of the aisle, a guard need not be provided.

❖ The purpose of this section is to provide for occupant safety with guards along elevated cross aisles. The min-

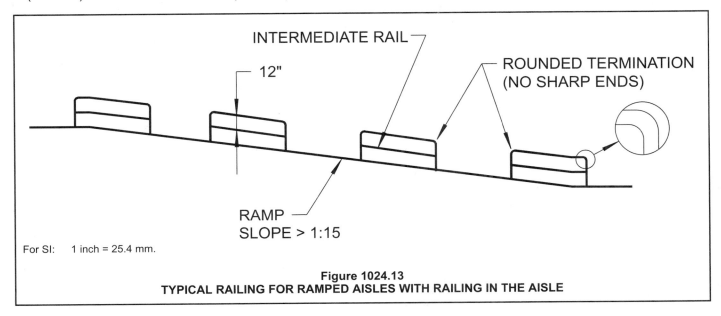

For SI: 1 inch = 25.4 mm.

Figure 1024.13
TYPICAL RAILING FOR RAMPED AISLES WITH RAILING IN THE AISLE

imum height of the guard is a function of the cross-aisle elevation above the adjacent floor or grade below [i.e., 42 inches (1067 mm) high with more than a 30-inch (762 mm) drop-off and 26 inches (660 mm) high with a 30-inch (762 mm) or less drop-off]. When the back of the seats adjacent to the cross aisle are a minimum of 24 inches (610 mm) above the floor level of the cross aisle, they will serve as the guard (see Figure 1024.14.2 for an illustration of the requirements in this section).

1024.14.2 Sightline-constrained guard heights. Unless subject to the requirements of Section 1024.14.3, a fascia or railing system in accordance with the guard requirements of Section 1012 and having a minimum height of 26 inches (660 mm) shall be provided where the floor or footboard elevation is more than 30 inches (762 mm) above the floor or grade below and the fascia or railing would otherwise interfere with the sightlines of immediately adjacent seating. At bleachers, a guard must be provided where the floor or footboard elevation is more than 24 inches (610 mm) above the floor or grade below and the fascia or

railing would otherwise interfere with the sightlines of the immediately adjacent seating.

❖ This section specifies a guard height of 26 inches (660 mm) for guards within assembly seating areas other than at the end of aisles where a vertical 36-inch (914 mm) height is required. This is to provide a reasonable degree of safety while providing for sightlines within the viewing area. The guard opening configuration must comply with Section 1012.3. (see Figure 1024.14.2 for an illustration of the requirements in this section).

At bleachers, the maximum drop-off before a guard is required is 24 inches (610 mm), rather than 30 inches (762 mm). This is consistent with the requirements for guards found in the bleacher standard referenced in Section 1024.1.1. The guard heights and opening requirement would be the same. With the lower drop-off in bleacher configurations, a guard could be required at the locations where cut-ins were made in the lower rows to accommodate wheelchair seating spaces.

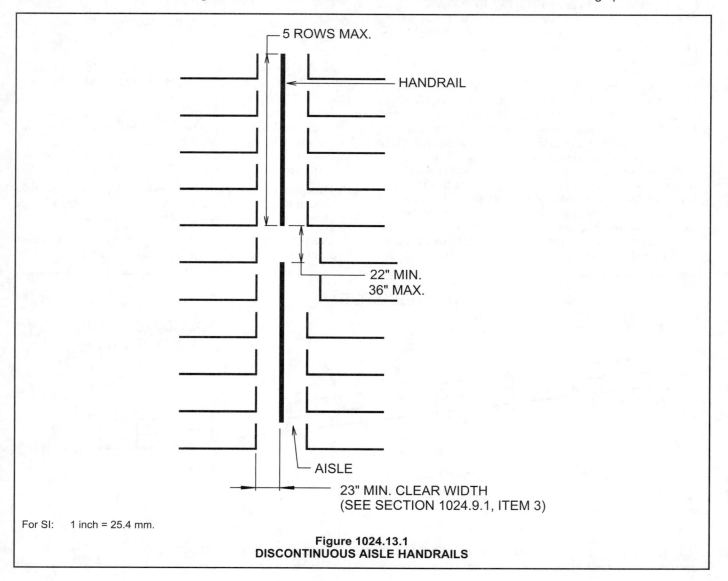

For SI: 1 inch = 25.4 mm.

Figure 1024.13.1
DISCONTINUOUS AISLE HANDRAILS

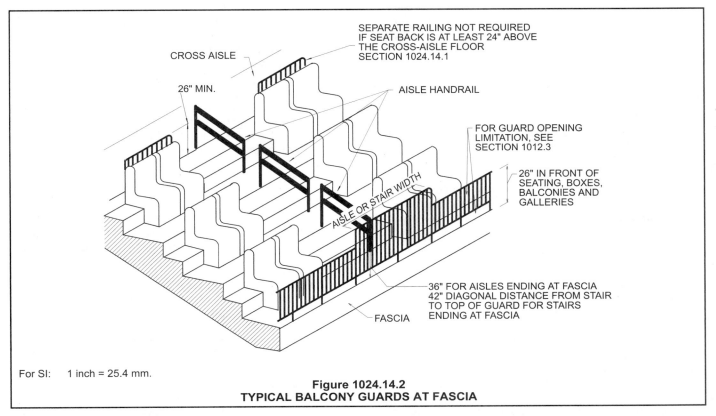

CROSS AISLE

SEPARATE RAILING NOT REQUIRED
IF SEAT BACK IS AT LEAST 24" ABOVE
THE CROSS-AISLE FLOOR
SECTION 1024.14.1

26" MIN.

AISLE HANDRAIL

FOR GUARD OPENING
LIMITATION, SEE
SECTION 1012.3

26" IN FRONT OF
SEATING, BOXES,
BALCONIES AND
GALLERIES

AISLE OR STAIR WIDTH

36" FOR AISLES ENDING AT FASCIA
42" DIAGONAL DISTANCE FROM STAIR
TO TOP OF GUARD FOR STAIRS
ENDING AT FASCIA

FASCIA

For SI: 1 inch = 25.4 mm.

Figure 1024.14.2
TYPICAL BALCONY GUARDS AT FASCIA

1024.14.3 Guards at the end of aisles. A fascia or railing system complying with the guard requirements of Section 1012 shall be provided for the full width of the aisle where the foot of the aisle is more than 30 inches (762 mm) above the floor or grade below. The fascia or railing shall be a minimum of 36 inches (914 mm) high and shall provide a minimum 42 inches (1067 mm) measured diagonally between the top of the rail and the nosing of the nearest tread.

❖ This section applies only at the end of aisles where the foot (the lower end) of the aisle is greater than 30 inches (762 mm) above the adjacent floor or grade below. The guard must satisfy both of the specified height requirements to provide safety for persons at the end of the aisle. The 36-inch (914 mm) minimum height is measured from the floor vertically to the top of the guard. The minimum 42-inch (1067 mm) diagonal dimension from the nosing of the nearest stair tread to the top of the fascia or guard is to provide sufficient height for a fall from the nearest stair tread. (see Figure 1024.14.2 for an illustration of the requirements in this section).

1024.15 Bench seating. Where bench seating is used, the number of persons shall be based on one person for each 18 inches (457 mm) of length of the bench.

❖ The purpose of this section is to specify the length of bench for each occupant for bench and bleacher seating. This is commonly used to calculate the occupant load of bench or bleacher seating for egress purposes and is not intended to limit any individual to an 18-inch

(457 mm) area. This is consistent with the fixed seating occupant loads indicated in Section 1004.7.

SECTION 1025
EMERGENCY ESCAPE AND RESCUE

1025.1 General. In addition to the means of egress required by this chapter, provisions shall be made for emergency escape and rescue in Group R as applicable in Section 101.2 and Group I-1 occupancies. Basements and sleeping rooms below the fourth story above grade plane shall have at least one exterior emergency escape and rescue opening in accordance with this section. Where basements contain one or more sleeping rooms, emergency egress and rescue openings shall be required in each sleeping room, but shall not be required in adjoining areas of the basement. Such opening shall open directly into a public street, public alley, yard or court.

Exceptions:

1. In other than Group R-3 occupancies as applicable in Section 101.2, buildings equipped throughout with an approved automatic sprinkler system in accordance with Section 903.3.1.1 or 903.3.1.2.

2. In other than Group R-3 occupancies as applicable in Section 101.2, sleeping rooms provided with a door to a fire-resistance-rated corridor having access to two remote exits in opposite directions.

3. The emergency escape and rescue opening is permitted to open onto a balcony within an atrium in accordance with the requirements of Section 404, provided the bal-

cony provides access to an exit and the dwelling unit or sleeping unit has a means of egress that is not open to the atrium.

4. Basements with a ceiling height of less than 80 inches (2032 mm) shall not be required to have emergency escape and rescue windows.

5. High-rise buildings in accordance with Section 403.

6. Emergency escape and rescue openings are not required from basements or sleeping rooms which have an exit door or exit access door that opens directly into a public street, public alley, yard, egress court or to an exterior exit balcony that opens to a public street, public alley, yard or egress court.

7. Basements without habitable spaces and having no more than 200 square feet (18.6 square meters) in floor area shall not be required to have emergency escape windows.

❖ This section requires emergency escape and rescue provisions in groups where occupants may be sleeping during a potential fire buildup, but are capable of self-preservation (Groups R and I-1). A basement and each sleeping room are to be provided with an exterior window or door that meets the minimum size requirements and is operable for emergency escape by methods that are obvious and clearly understood by all users. Sleeping rooms four stories or more above grade are not required to be so equipped, since fire service access at that height, as well as escape through such an opening, may not be practical or reliable. In accordance with Chapter 9, such buildings will also be equipped throughout with an automatic fire suppression system. The provision for basements is in recognition that such types of spaces typically only have a single path of egress and often have no alternate routes available as other levels do.

It is important to note that this window is an element of escape and does not comprise any part of the means of egress unless it is a door with appropriate egress component characteristics.

Exception 1 assumes that the automatic sprinkler system can control fire buildup and reduce, if not eliminate, the need for an occupant to use an emergency escape window. The exception applies to buildings equipped throughout with an NFPA 13 or 13R sprinkler system (see Sections 903.3.1.1 and 903.3.1.2).

Exception 2 allows another acceptable means of escape; that is, a door directly from the sleeping room to a corridor with exits in opposite directions, to substitute for the escape window.

Exception 3 provides for dwelling and sleeping units that have egress windows to a balcony that is within an atrium. The exception specifies that the dwelling or sleeping unit is to have another means of egress that does not pass through the atrium so that an independent route of egress is provided.

Exceptions 4 and 7 are intended to exempt basements that would not be likely to have sleeping rooms in them from the requirement to have emergency escape and rescue openings.

Exception 5 is in correlation with the exception for emergency escape windows in high-rise buildings addressed in Section 403.4.

The intent of Exception 6 is to permit sleeping rooms with a direct access to an exterior-type environment, such as a street or exit balcony, to not have an emergency escape window. The open atmosphere of the escape route would increase the likelihood that the means of egress be available even with the delayed response time for sleeping residents.

1025.2 Minimum size. Emergency escape and rescue openings shall have a minimum net clear opening of 5.7 square feet (0.53 m²).

> **Exception:** The minimum net clear opening for emergency escape and rescue grade-floor openings shall be 5 square feet (0.46 m²).

❖ The dimensional criteria of the opening are intended to permit fire service personnel (in full protective clothing with a breathing apparatus) to enter from a ladder, as well as permit occupants to escape. The net clear opening area and minimum dimensions are intended to provide a clear opening through which an occupant can pass to escape the building or a fire fighter can pass to enter the building for rescue or fire suppression activities. Since the emergency escape windows must be usable to all occupants, including children and guests, the required opening dimensions must be achieved by the normal operation of the window from the inside (e.g., sliding, swinging or lifting the sash). It is impractical to assume that all occupants can operate a window that requires a special sequence of operations to achieve the required opening size. While most occupants are familiar with the normal operation by which to open the window, children and guests are frequently unfamiliar with special procedures necessary to remove the sashes. The time spent in comprehending the special operation unnecessarily delays egress from the bedroom and could lead to panic and further confusion. Thus, windows that achieve the required opening dimensions only through operations such as the removal of sashes or mullions are not permitted. It should be noted that the minimum area cannot be achieved by using both the minimum height and minimum width specified in Section 1025.2.1 (see Figure 1025.2).

1025.2.1 Minimum dimensions. The minimum net clear opening height dimension shall be 24 inches (610 mm). The minimum net clear opening width dimension shall be 20 inches (508 mm). The net clear opening dimensions shall be the result of normal operation of the opening.

❖ Note that the minimum dimensions in this section and the minimum area requirements in Section 1025.2 both apply. Thus, a grade floor window that is only 24 inches (610 mm) in height must be 30 inches (762 mm) wide to meet the 5-square-foot (0.46 m²) area requirement of Section 1025.2 for grade-floor window (see Figure 1025.2).

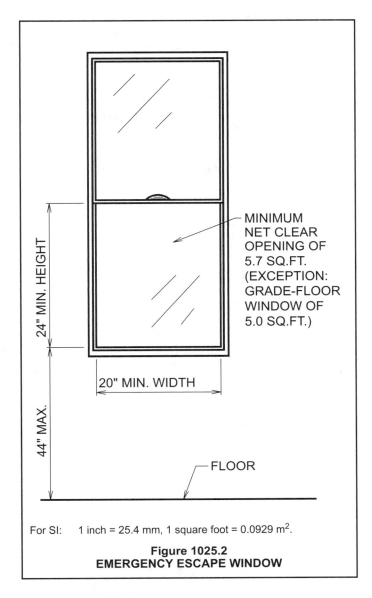

For SI: 1 inch = 25.4 mm, 1 square foot = 0.0929 m².

Figure 1025.2
EMERGENCY ESCAPE WINDOW

(In figure: MINIMUM NET CLEAR OPENING OF 5.7 SQ.FT. (EXCEPTION: GRADE-FLOOR WINDOW OF 5.0 SQ.FT.); 24" MIN. HEIGHT; 20" MIN. WIDTH; 44" MAX.; FLOOR)

1025.3 Maximum height from floor. Emergency escape and rescue openings shall have the bottom of the clear opening not greater than 44 inches (1118 mm) measured from the floor.

❖ This section limits the height of the bottom of the clear opening to 44 inches (1118 mm) or less such that it can be used effectively as an emergency escape (see Figure 1025.2).

1025.4 Operational constraints. Emergency escape and rescue openings shall be operational from the inside of the room without the use of keys or tools. Bars, grilles, grates or similar devices are permitted to be placed over emergency escape and rescue openings provided the minimum net clear opening size complies with Section 1025.2 and such devices shall be releasable or removable from the inside without the use of a key, tool or force greater than that which is required for normal operation of the escape and rescue opening. Where such bars, grilles, grates or similar devices are installed in existing buildings,

smoke alarms shall be installed in accordance with Section 907.2.10 regardless of the valuation of the alteration.

❖ If security grilles, decorations or similar devices are installed on escape windows, such items must be readily removable to permit occupant escape without the use of any tools, keys or a force greater than that required for the normal operation of the window.

Where bars, grilles or grates are placed over the emergency escape and rescue opening, it is important that they are easily removable. Thus, the requirements for ease of operation are the same as required for windows.

The smoke alarms that are required for existing buildings where such items are installed provides advance warning of a fire for safety purposes.

1025.5 Window wells. An emergency escape and rescue opening with a finished sill height below the adjacent ground level shall be provided with a window well in accordance with Sections 1025.5.1 and 1025.5.2.

❖ Emergency escape and rescue openings that are partially or completely below grade need to have window wells so that they can be used effectively (see Figure 1025.5).

1025.5.1 Minimum size. The minimum horizontal area of the window well shall be 9 square feet (0.84 m²), with a minimum dimension of 36 inches (914 mm). The area of the window well shall allow the emergency escape and rescue opening to be fully opened.

❖ This section specifies the size of the window well that is needed for a rescue person in full protective clothing and breathing apparatus to use the rescue opening. The required 9 square feet (0.84 m²) is the horizontal cross-sectional area of the window well. Thus, if the window well projects away from the plane of the window 3 feet (914 mm), the required dimension in the plane of the window along the wall is also 3 feet (914 mm) (see Figure 1025.5).

1025.5.2 Ladders or steps. Window wells with a vertical depth of more than 44 inches (1118 mm) shall be equipped with an approved permanently affixed ladder or steps. Ladders or rungs shall have an inside width of at least 12 inches (305 mm), shall project at least 3 inches (76 mm) from the wall and shall be spaced not more than 18 inches (457 mm) on center (o.c.) vertically for the full height of the window well. The ladder or steps shall not encroach into the required dimensions of the window well by more than 6 inches (152 mm). The ladder or steps shall not be obstructed by the emergency escape and rescue opening. Ladders or steps required by this section are exempt from the stairway requirements of Section 1009.

❖ This section specifies that a ladder or steps be provided for ease of getting into and out of window wells that are more than 44 inches (1118 mm) deep.

Usually ladder rungs are embedded in the wall of the window well. The 44-inch (1118 mm) dimension is the depth of the window well, not the distance from the bottom of the window well to grade. Thus, if the floor of a

FIGURE 1025.5

MEANS OF EGRESS

window well is 40 inches (1016 mm) below grade, but the wall of the window well projects above grade by 6 inches (152 mm), steps or a ladder are required since the vertical depth is 46 inches (1168 mm).

It is important that the ladder not obstruct the operation of the emergency escape window (see Figure 1025.5).

Bibliography

The following resource materials are referenced in this chapter or are relevant to the subject matter addressed in this chapter.

ASCE 7-02, *Minimum Design Loads for Buildings and Other Structures.* American Society of Civil Engineers, 2002.

ASME A17.1-00, *Safety Code for Elevators and Escalator.* New York: American Society of Mechanical Engineers, 2000.

ASME A18.1-99, *Safety Standard for Platform Lifts and Stairway Chairlifts - with Addenda A18.1a-2001.* New York: American Society of Mechanical Engineers, 2001.

ASTM C 1028-96, *Test Method for Determining the Static Coefficient of Friction of Ceramic Tile and Other Like Surfaces by the Horizontal Dynamometer Pull Meter Method.* West Conshohocken, PA: ASTM International, 1996.

ASTM D 2047-99, *Test Method for Static Coefficient of Friction Polish-Coated Floor Surfaces as Measured by the James Machine.* West Conshohocken, PA: ASTM International, 1999.

ASTM D 2394-83 (Reapproved 1999), *Methods for Simulated Service Testing of Wood and Wood-Base Finish Flooring.* West Conshohocken, PA: ASTM International, 1999.

ASTM E 119-00, *Test Method for Fire Tests of Building Construction and Materials.* West Conshohocken, PA: ASTM International, 2000.

ASTM F 851-87 (Reapproved 2000), *Test Method for Self-Rising Seat Mechanisms.* West Conshohocken, PA: ASTM International, 2000.

Cole, Ron, ed. *Life Safety Code Handbook,* 8th ed. Quincy, MA: National Fire Protection Association, 2000.

DOJ 28 CFR, Part 36 (Appendix A)-91, ADA *Accessibility Guidelines for Buildings and Facilities.* Washington, DC: U.S. Department of Justice, 1991.

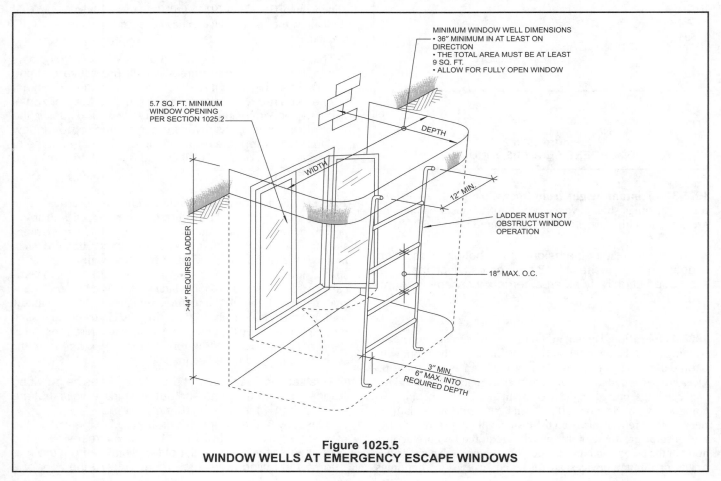

Figure 1025.5
WINDOW WELLS AT EMERGENCY ESCAPE WINDOWS

DOJ 28 CFR, Part 36-91, *Americans with Disabilities Act (ADA)*. Washington, DC: U.S. Department of Justice, 1991.

DOL 29 CFR, Part 1910-92, *Occupational Safety and Heath Standards.* Washington, DC: U.S. Department of Labor; Occupational Safety and Health Administration, 1992.

Fire Protection Handbook, 19th ed. Quincy, MA: National Fire Protection Association, 2003.

Fruin, John J. *Pedestrian Planning and Design.* Mobile, AL: Elevator World, Inc., 1987.

ICC 300-02, *ICC Standards* on *Bleachers, Folding and Telescopic Seating and Grandstands.*

ICC A117.1-98, *Accessible and Usable Buildings and Facilities.* Falls Church, VA: International Code Council, 1998.

ICC EC-03, *ICC Electrical Code.* Falls Church, VA: International Code Council, 2003.

IMC-03, *International Mechanical Code.* Falls Church, VA: International Code Council, 2003.

IPMC-03, *International Property Maintenance Code.* Falls Church, VA: International Code Council, 2003.

IRC-03, *International Residential Code.* Falls Church, VA: International Code Council, 2003.

"Means of Egress." CABO Board of the Coordination of the Model Codes Report, 1985.

NFPA 13-99, *Installation of Sprinkler Systems.* Quincy, MA: National Fire Protection Association, 1999.

NFPA 13D-99, *Installation of Sprinkler Systems in One- and Two-Family Dwellings and Manufactured Homes.* Quincy, MA: National Protection Association, 1999.

NFPA 13R-99, *Installation of Sprinkler Systems in Residential Occupancies Up to Four Stories in Height.* Quincy, MA: National Fire Protection Association, 1999.

NFPA 92A-00, *Recommended Practice for Smoke Control Systems.* Quincy, MA: National Fire Protection Association, 2000.

NFPA 92B-00, *Guide for Smoke Management Systems in Malls, Atria and Large Areas.* Quincy, MA: National Fire Protection Association, 2000.

NFPA 204-02, *Standard for Smoke and Heat Venting.* Quincy, MA: National Fire Protection Association, 2002.

NFPA 252-99, *Methods of Fire Tests of Door Assemblies.* Quincy, MA: National Fire Protection Association, 1999.

NIST IR 4770-92, *Report on Staging Areas for Persons with Mobility Limitations.* Washington, DC: National Institute of Standards and Technology, 1992.

SFPE *Fire Protection Engineering Handbook, 3rd ed.* Quincy, MA: National Fire Protection Association, 2002.

UL 199-97, *Automatic Sprinklers for Fire-Protection Service.* Northbrook, IL: Underwriters Laboratories Inc., 1997.

UL 1626-01, *Residential Sprinklers for Fire-Protection Service - with Revisions thru March 1997.* Northbrook, IL: Underwriters Laboratories Inc., 2001.

UL 1784-95, *Air Leakage Tests of Door Assemblies.* Northbrook, IL: Underwriters Laboratories Inc., 1995.

Chapter 11:
Accessibility

General Comments

Chapter 11 contains provisions that set forth requirements for accessibility of buildings and their associated sites and facilities for people with physical disabilities. Existing building criteria is addressed in Section 3409. Appendix E was added to address accessibility for items in the new *Americans with Disabilities Act Accessibility Guidelines* (ADAAG) that were not typically enforceable through the standard traditional building code enforcement approach system (e.g., beds, room signage).

The *International Residential Code*® (IRC®) references Chapter 11 for accessibility provisions, therefore, this chapter may be applicable to housing covered under the IRC (see commentary, Sections 1107.6 and 1107.6.3). Structures referenced to Chapter 11 from the IRC would be considered Group R-3.

Section 1101 contains the broad scope statement of the chapter and identifies the baseline criteria for accessibility as being in compliance with this chapter and ICC A117.1. ICC A117.1 is the consensus national standard that sets forth the details, dimensions and construction specifications for accessibility.

Section 1102 contains definitions of terms that are associated with accessibility.

Section 1103 describes the applicability of the provisions of this chapter. Accessibility is broadly required in all buildings, structures, sites and facilities. Those specific circumstances in which accessibility is not required are set forth as exceptions.

Section 1104 contains the requirements for interior and exterior accessible routes. An accessible route is a key component of the built environment that provides a person with a disability access to spaces, elements, facilities and buildings. Note that ramps are addressed in Section 1010.

Section 1105 contains requirements for accessible entrances to buildings and structures. Note that requirements for accessible means of egress are addressed in Section 1007.

Section 1106 sets forth the requirements for accessible parking facilities and passenger loading zones.

Section 1107 contains various accessibility requirements that are unique to occupancies that contain dwelling and sleeping units and are applicable in addition to other general requirements of this chapter. Specific provisions unique to Group I and R occupancies are included. Requirements in this section are coordinated with the requirements found in the *Fair Housing Accessibility Guidelines* (FHAG).

Section 1108 contains various accessibility requirements that are unique to specific occupancies other than Groups I and R and are applicable in addition to other general requirements of this chapter. Specific provisions unique to assembly seating, performance areas, self-storage and judicial facilities, such as courtrooms and jail facilities, are included.

Section 1109 contains various requirements that are applicable to features and facilities that are not occupancy related, including requirements for toilet and bathing facilities; sinks; kitchens; drinking fountains; elevators; lifts; storage facilities; detectable warnings; seating at tables, counters and work surfaces; service facilities; controls and operating mechanisms and recreational facilities.

Section 1110 sets forth requirements for signage identifying certain required accessible elements.

Access to buildings and structures for people with physical disabilities has been a subject that the building codes have regulated since the early 1970s. They have consistently relied on a consensus national standard, CABO/ANSI A117.1, as the technical basis for accessibility. The title of CABO/ANSI A117.1 has been revised to ICC/ANSI A117.1 to reflect that the International Code Council® (ICC®) is the secretariat for this standard. Accessibility is not a new subject to the construction regulatory community. There has been a great deal of emphasis and awareness recently placed on the subject of accessibility through the passage of two federal laws. The Americans with Disabilities Act (ADA) and the Fair Housing Amendment Act (FHA) are federal regulations that affect building construction as it relates to accessibility.

ADA, signed into law in July 1990, is a very broad civil rights law designed to protect persons with disabilities. There are five sections of the ADA that address different aspects of civil rights for people with disabilities. Title I deals with employment and generally prohibits discrimination against disabled people in employment. Title II deals with public services, including access to all services and facilities receiving federal funding, and access to public transportation, including buses, rail lines, and passenger transit facilities. Title III requires access to a large category of buildings and structures in a manner that is equal or comparable to that available to the general public. (Note that FHA is a companion document to ADA. FHA deals with residential and institutional living arrangements that, for the most part, are not covered under ADA.) Title IV deals with accessible telecommunication facilities. Title V further defines miscellaneous issues dealing with ADA compliance.

Titles II and III of the ADA overlap requirements customarily regulated by building codes. Most Title II facilities may opt to follow either the *Uniform Federal Accessibility Standards* (UFAS) or ADAAG for compliance with accessibility requirements. This group includes anyone receiving federal funding, i.e., colleges, schools, park districts, local and state governments, hospitals, etc. Implementation of Title III of the ADA is based on the requirements of ADAAG. Both UFAS and ADAAG set forth scoping and technical requirements for building design and construction. The federal government is interested in bringing more uniformity to federal accessibility standards and will be looking at ways in which harmonization of all federal and private sector standards can be accomplished. In many ways, UFAS and ADAAG are similar to a building code and cover many of the same matters that are dealt with in such codes. It is important to emphasize, however, that the ADA is written as civil rights legislation and not as a building code. The enforcement mechanism for civil rights laws is significantly different than that for traditional building regulation.

The ADA does not preempt the adoption or enforcement of accessibility-related codes by state and local governments. All buildings constructed within a jurisdiction must comply with locally adopted building codes and any applicable state codes, as well as ADAAG; however, as a federal law, state and local governments have neither the authority nor the responsibility to enforce ADA. Enforcement of ADA is the responsibility of the U.S. Department of Justice (DOJ). As previously described, ADA covers a large range of issues dealing with disabilities. Under Titles II and III, the scoping and technical requirements of UFAS or ADAAG deal directly with accessibility of buildings. Typically, enforcement of these regulations can only take place after the construction process. Any differences between the regulations must be ultimately reconciled by the building owner or designer. This creates a difficult situation for the building owners and designers who are faced with multiple sets of regulations. In an effort to alleviate this problem, some states have adopted ADAAG as the referenced standard for their accessibility laws. Currently, this creates difficulties in enforcement, since ADAAG is not written in language that is customarily used in building codes and relied on for enforcement by state and local enforcement officials. Fortunately, there are many similarities between the requirements of ADAAG and the code. In addition, compliance with the code typically takes place during the plan review and construction process. This is a less costly time to make required changes to a facility that are necessary for compliance with applicable requirements. Again, the building official is expected to enforce the code adopted by the state or local jurisdiction, and any applicable state laws. The building official is neither required to nor responsible for interpretation or enforcement of ADA unless the state or local jurisdiction has adopted ADAAG in its accessibility laws. No attempt should be made to represent that ADAAG is being interpreted or enforced by the building official or the jurisdiction when ADAAG is not specifically adopted by the jurisdiction.

The concept of two completely separate and independent, broad-based sets of regulations that affect building design and construction may sound onerous, but the situation may not be quite as bad as it appears. The reason is that the code has a high level of consistency with ADAAG.

Since the early 1990s, code change activity has incorporated the recommendations of the Council of American Building Officials' (CABO) Board for the Coordination of the Model Codes (BCMC) [succeeded in 1995 by the ICC Board for the Development of the Model Codes (BDMC)]. BDMC undertook the effort to review comprehensively all facets of the accessibility issue. This effort included a comparison of ADAAG with the provisions of the model codes. Based on the results of its studies, BMDC then developed recommended code language for consideration by all three of the model code groups: Building Officials and Code Administrators International® (BOCA®), International Conference of Building Officials® (ICBO®) and Southern Building Code Congress internatinal® (SBCCI®). BDMC's recommendations have also been adopted and are reflected in the code. There are still differences between the code and ADAAG, but the differences have been minimized. Substantial activity to revise the federal standards is also taking place. The United States Architectural and Transportation Barriers Compliance Board (Access Board) convened an advisory committee to develop recommendations to comprehensively revise ADAAG. A primary objective of that effort was to integrate the federal standards with the BDMC provisions and ICC A117.1. The federal rule-making process to revise ADAAG is not yet complete, but many of the advisory committee's recommendations will be included in the new ADAAG. The new ADAAG started the public comment cycle in November 1999. The code already includes the majority of those same provisions and, therefore, could have a high level of consistency with the revised federal standards. That, of course, is dependent on the outcome of the future federal rule-making process. Building owners and designers can have a high degree of confidence that when a building complies with the code, it will also be in substantial compliance with ADAAG. It is essential that building owners and designers recognize that there are differences in some of the provisions that are covered in both the code and ADAAG. Additionally, there are some matters that are covered by ADAAG that are not traditionally included in the scope of the code. Such subjects as public telephones, automatic teller machines (ATMs), signage for identification of rooms and spaces (i.e., room numbers, functions, etc.), among others, are matters that are not effectively enforceable through traditional building code techniques. For example, room names and numbers are routinely changed without involving building code enforcement. Another example is that building codes traditionally do not regulate the type, placement or location of telephones or ATMs. All differences must be addressed and reconciled by the building owner or designer in order to provide building design and construction that comply with all applicable laws. The ADAAG provisions in those "noncode" subjects have been incorporated into Appendix E as a convenience for those who are required to comply with federal law and for the benefit of any jurisdictions that may wish to adopt these provisions and, therefore, go beyond the traditional scope of building code enforcement.

The FHA is federal legislation promulgated by the

United States Department of Housing and Urban Development (HUD) that extends fair housing protection against discrimination on the basis of family status or persons with disabilities. The standard for FHA is the FHAG, which set forth accessibility scoping and technical requirements for a broad category of residential construction. The scope of ADAAG generally excludes occupancies that are covered by FHAG; however, FHA is being interpreted to overlap the scope of the coverage of ADA in some respects (e.g., dorms, nursing homes). Residential buildings that are covered by FHAG must also comply with the locally adopted building code. The same circumstances with respect to differences in the requirements and responsibility for enforcement described for ADA also exist with FHA.

The ICC recognizes the value of consistency between federal laws and the codes. Through efforts to incorporate the work of the ADAAG Review Federal Advisory Committee and BCMC, the ICC has worked toward, and will continue to strive for, accessibility regulations that reflect the highest possible degree of consistency with federal regulations and, more importantly, reasonable and appropriate provisions to meet the needs of people with disabilities.

Again, as a federal law, building officials have neither the responsibility nor the authority to enforce the requirements of FHAG. Inquiries regarding those laws should be referred to the Department of Justice (DOJ) or HUD. Until such time that their laws are coordinated with and consider traditional code enforcement mechanisms and methods, building officials are advised not to attempt to interpret or enforce these laws. Permit applicants should be advised that the work they propose has not been reviewed for compliance with FHAG.

Due to concerns about residential construction complying with all applicable codes and laws, the National Home Builders Association (NHBA), along with others, approached the model code groups about incorporating the FHAG requirements into the model code requirements. Since the FHA is a civil rights law rather than a building code, careful study was needed to interpret them into enforceable language that could be utilized in model codes. Through the efforts of the BDMC, recommendations to the model codes were proposed in 1994. The "adaptable" dwelling unit requirements were carried forward as a Type A unit, with the FHA requirements reflected in the Type B units.

In 2000, HUD reviewed the 2000 edition of the *International Building Code®* (IBC®) and the referenced accessibility standard ICC/ANSI A117.1-1998 for compliance with FHAG. As is typical in such evaluations, there was good news and not so good news. The good news in the report was that for apartments and townhouses, the 2000 IBC basically had the scoping requirements correct. The not so good news was that there were other types of facilities that the FHA considered places or residences, and, therefore, should also be covered. These types of residences, classified by the code as Groups I and R, include boarding houses, congregate residences, institutional facilities, assisted living facilities, vacation timeshares, etc.

Based on HUD's report, a series of modifications were proposed as part of the 2000 code change cycle. The proposed modifications were accepted by the voting members and were incorporated into the 2001 *Supplement to the IBC*. As a result of these changes, HUD issued a press release that stated the 2000 edition of the IBC with the 2001 Supplement and ICC/ANSI A117.1-1998 could be considered "safe harbor" for anyone wanting to comply with FHAG scoping and technical requirements. In addition to the press release, the preamble of HUD's *Fair Housing Act Design Manual* states that this document is safe harbor. This is viewed as a positive step forward for the housing industry.

With the code being a safe-harbor document, architects and developers can design and be reviewed for compliance with the IBC through the typical building code review process. At the same time, they would also be in compliance with the building requirements for the FHAG. At the time this commentary was written, HUD had not given the 2003 edition of the IBC official safe-harbor status. Since the IBC is developed through a code change/public comment process, there have been changes to items in Chapter 11. This is typically a good way for new concerns, concepts or procedures to be incorporated into the requirements. It is the ICC's opinion that none of these changes has resulted in jeopardizing the safe-harbor status between FHAA and IBC requirements. The ICC hopes to have official safe-harbor status for the 2003 edition of the IBC from HUD very soon.

In addition, the ICC has produced a document entitled *Code Requirements for Housing Accessibility* (CRHA), which, if complied with, is intended to provide safe harbor for compliance with FHAG. This document includes both the scoping requirements from the 2000 edition of the IBC with the 2001 Supplement as well as the technical requirements from ICC A117.1 for residential and institutional facilities covered under the FHAG.

Efforts for coordination with the federal accessibility requirements are ongoing. Representatives from interested accessibility groups, HUD and the Access Board have been attending and participating in the code change process for the IBC and ICC A117.1. In addition, the ICC has participated in the public comment process on the development of federal regulations.

The fundamental philosophy of the code on the subject of accessibility is that everything is required to be accessible. This is reflected in the basic applicability requirement (see Section 1103.1). The code's scoping requirements then address the conditions under which accessibility is not required in terms of exceptions to this general mandate. In the early 1990s, building codes tended to describe where accessibility was required in each occupancy, and any circumstance not specifically identified was excluded. The more recent codes represent a fundamental change in approach. Now one must think of accessibility in terms of "if it is not specifically exempted, it must be accessible."

Another important concept is that of "mainstreaming."

There are many accessibility issues that not only benefit people with disabilities, but also provide a tangible benefit to people without disabilities. This type of requirement can be set forth in the code as generally applicable without necessarily identifying it specifically as an accessibility-related issue. Such a requirement would then be considered as having been mainstreamed. For example, the limitation on objects protruding into corridors, aisles and passageways (see Section 1003.3) is intended to aid people with a vision impairment by reducing the potential for unintended contact that may cause injury. Clearly, this has an additional benefit to people without a vision impairment. A protruding object can be encountered by a sighted person who is simply not paying close attention. The concept of mainstreaming is responsive to the desire of people with disabilities to not be singled out and categorized separately from the remainder of society. It is therefore important to recognize that, while the provisions of Chapter 11 are specifically grouped and identified as accessibility related, they do not represent all of the issues that must be taken into consideration when evaluating accessibility in buildings.

There are many items that have some basis in accessibility requirements, but actually result in more "user-friendly" buildings for all of us. Physical disabilities can be permanent or temporary; can affect all age groups; can involve all levels of abilities and can range from persons with minor visual, hearing or mobility impairments to persons who are blind, deaf or confined to a wheelchair. Everyone will benefit from accessible features in buildings, either directly or indirectly through those they care for. With some foresight in designing our built environment, access into and throughout buildings can be better for everyone throughout their lifetimes.

Purpose

The purpose of this chapter is to set forth requirements for accessibility applicable to those elements of the built environment that are included within the scope of the code. There are two categories or types of requirements that must be addressed in order to accomplish accessibility: scoping and technical. Scoping requirements describe what and where accessibility is required, or how many accessible features or elements must be provided. For example, a requirement that at least one of the first 25 sleeping units in a hotel must be accessible is referred to as a scoping requirement because it describes what and how many accessible features are required. Another example of scoping is the exception for detached one- and two-family dwellings from accessibility requirements because it defines what is not required to be accessible. A technical requirement is intended to refer to a statement indicating how accessibility is to be accomplished. For example, a requirement indicating that in order for a parking space to be accessible, it must have a width of 96 inches (2438 mm) with an adjacent 60-inch (1524 mm) access aisle is considered a technical requirement. The provisions of Chapter 11 are scoping requirements in that they all address what and where accessibility is required or how many accessible elements are required. ICC A117.1 indicates how to make something accessible and is the consensus national standard that is referenced for establishing technical requirements. In addition to the provisions of Chapter 11, there are accessibility-related technical requirements elsewhere in the code. These are subject-matter-specific provisions that have been mainstreamed into the chapter or section of the code that deals with those subjects. As such, the code and its references regulate accessibility comprehensively by setting forth the scoping and technical requirements that establish the minimum level of accessibility required in the built environment.

SECTION 1101
GENERAL

1101.1 Scope. The provisions of this chapter shall control the design and construction of facilities for accessibility to physically disabled persons.

❖ This section establishes the scope of the chapter as providing for design and construction of facilities for accessibility to disabled persons. The scope is broadly inclusive of all aspects of construction that affect the ability of disabled people to approach, enter and utilize a facility. The term "facility," as defined in Section 1102.1, includes not only buildings and structures, but also the site on which they are located. Features of a site, such as parking areas and paths of travel from a public way to a structure, affect accessibility and are, therefore, within

the scope of Chapter 11. Chapter 11, in conjunction with mainstreamed provisions throughout the code, sets forth scoping requirements.

1101.2 Design. Buildings and facilities shall be designed and constructed to be accessible in accordance with this code and ICC A117.1.

❖ This section establishes the primary and fundamental relationship of ICC A117.1 to the code. The code text is intended to "scope" or provide thresholds for application of required accessibility features. The referenced standard contains technical provisions indicating how compliance with the code is achieved. In short, Chapter 11 specifies what, when and how many accessible features are required; the referenced standard indicates how to make that feature accessible. Compliance with both the code and the standard is required.

SECTION 1102
DEFINITIONS

1102.1 Definitions. The following words and terms shall, for the purposes of this chapter and as used elsewhere in the code, have the following meanings:

❖ This section contains definitions of terms that are associated with the subject matter of this chapter. It is important to emphasize that these terms are not exclusively related to this chapter but are applicable everywhere the term is used in the code.

Definitions of terms can help in the understanding and application of the code requirements. The purpose for including these definitions within this chapter is to provide more convenient access to them without having to refer back to Chapter 2. For convenience, these terms are also listed in Chapter 2 with a cross reference to this section. Terms that are italicized provide a visual identification throughout the code that a definition exists for that term. The use and application of all defined terms, including those defined herein, are set forth in Section 201.

ACCESSIBLE. A site, building, facility or portion thereof, that complies with this chapter.

❖ This definition identifies the fundamental concept of Chapter 11. Accessibility is deemed to be accomplished if a building, site or facility complies with the applicable provisions of the code and ICC A117.1. It is not the intent of the code to accommodate fully every type and range of disability. It would not be feasible to do so. The extent to which the code requires accessible features in the various occupancies covered by Chapter 11 (scoping) and the characteristics those features are required to meet through reference to ICC A117.1 (technical requirements) establish that which the code considers fully accessible.

ACCESSIBLE ROUTE. A continuous, unobstructed path that complies with this chapter.

❖ An accessible route enables a disabled person to approach and utilize a facility's accessible fixtures and features. The design and construction of an accessible route is governed predominantly by provisions necessary for accessibility to a person in a wheelchair. There are typically more physical barriers in the built environment to people with a mobility impairment than in any other category of disability. An accessible route must also be safe and usable by people with other disabilities and, therefore, requirements are set forth in consideration of those needs. For example, there are restrictions on objects that protrude into an accessible route in consideration of a person with a visual impairment.

ACCESSIBLE UNIT. A dwelling unit or sleeping unit that complies with this code and Chapters 1 through 9 of ICC A117.1.

❖ The intent of this definition is to clarify that the technical criteria for an Accessible dwelling or sleeping unit is found in ICC A117.1, Chapters 1 through 9. Accessible units are required to be constructed fully accessible. The adaptable features for dwelling and sleeping units indicated in Chapter 10 of ICC A117.1 are limited to Type A and Type B units; Accessible units are considered more accessible than Type A and Type B units (see the commentary for the definitions of "Dwelling unit" and "Sleeping Unit" and to Section 1107.2).

CIRCULATION PATH. An exterior or interior way of passage from one place to another for pedestrians.

❖ Examples of circulation paths include walks, hallways, courtyards, ramps, stairways and stair landings.

COMMON USE. Interior or exterior circulation paths, rooms, spaces or elements that are not for public use and are made available for the shared use of two or more people.

❖ Some buildings include areas that are restricted to employees only or where public access is limited. Common use spaces may be part of employee work areas. Common use spaces do not include public use spaces. Any space that is shared by two or more persons, such as copy areas, break rooms, toilet rooms or circulation paths, are common use areas. A grade school classroom would be another example of a common use space (see also the commentary for the definition of "Public-use areas" and "Employee work area").

DETECTABLE WARNING. A standardized surface feature built in or applied to walking surfaces or other elements to warn visually impaired persons of hazards on a circulation path.

❖ A detectable warning is a change in texture that is detectable by a person with a vision impairment. Specifications for a detectable warning surface are set forth in ICC A117.1 only for transit platform edges so that users can be confident of the warning that is intended to be communicated and not be confused by multiple, different surfaces intended to convey the same warning (see commentary, Section 1109.9).

DWELLING UNIT OR SLEEPING UNIT, MULTISTORY. A dwelling unit or sleeping unit with habitable space located on more than one story.

❖ A multistory dwelling or sleeping unit has living, sleeping, eating, cooking or bathroom space on more than one floor level within the unit (see Section 1107.7.2). A residence with only a garage underneath or an unfinished basement would not be a multistory unit.

DWELLING UNIT OR SLEEPING UNIT, TYPE A. A dwelling unit or sleeping unit designed and constructed for accessibility in accordance with ICC A117.1.

❖ A Type A unit has some elements that are constructed for accessibility [e.g., 32-inch (813 mm) clear width doors with maneuvering clearances] and some elements that are constructed adaptable (e.g., blocking for future installation of grab bars). A Type A dwelling unit is designed and constructed to provide accessibility for wheelchair users throughout the unit, and as such, is considered more accessible than a Type B dwelling unit. The technical requirements for the interior of Type A units are in Sections 1002 and 1004 of ICC A117.1 (see the commentary for the definitions of "Dwelling unit" and "Sleeping unit" and to Section 1107.2).

DWELLING UNIT OR SLEEPING UNIT, TYPE B. A dwelling unit or sleeping unit designed and constructed for accessibility in accordance with ICC A117.1, consistent with the design and construction requirements of the federal Fair Housing Act.

❖ A Type B dwelling or sleeping unit is designed and constructed to provide a minimal level of accessibility, and as such, is considered less accessible than either an Accessible unit or a Type A unit. The requirements for Type B units are intended to be consistent with the Fair Housing Amendments Act (FHA). The technical requirements for the interior of Type B units are in Sections 1003 and 1004 of ICC A117.1 (see the commentary for the definitions of "Dwelling unit" and "Sleeping unit" and to Section 1107.2).

EMPLOYEE WORK AREA. All or any portion of a space used only by employees and only for work. Corridors, toilet rooms, kitchenettes and break rooms are not employee work areas.

❖ An employee work area is different in an office versus on a factory line. An employee work area may expand past the station or desk where an employee performs his or her job. An employee work area could include common use spaces, but not public use spaces. Depending on the duties of the employee, it may also include copy areas, stock rooms, filing areas, an assembly line, etc. (see also the commentary for the definitions of "Common use" and "Public-use areas").

FACILITY. All or any portion of buildings, structures, site improvements, elements and pedestrian or vehicular routes located on a site.

❖ This term is intentionally broad and includes all portions within a site and all aspects of that site that contain features required to be accessible. This includes parking areas, exterior walkways leading to accessible features, recreational facilities, such as playgrounds and picnic areas, as well as any structures on the site (see also the commentary to the definition of "Site").

INTENDED TO BE OCCUPIED AS A RESIDENCE. This refers to a dwelling unit or sleeping unit that can or will be used all or part of the time as the occupant's place of abode.

❖ A unit that is a person's home, rather than a unit used for a more transient nature, is a place of abode. Fair Housing regulations do not include a 30-day criteria for transient/nontransient, similar to what has been traditionally used by the building codes (see commentary, Section 1107); therefore, beach homes, timeshares, extended stay hotels, etc., may be included.

MULTILEVEL ASSEMBLY SEATING. Seating that is arranged in distinct levels where each level is comprised of either multiple rows, or a single row of box seats accessed from a separate level.

❖ Assembly rooms may include a sloped seating arrangement (i.e., either ramped or stepped) to improve the viewing of the event for the occupants. These spaces can be single- or multiple-level arrangements. For example, for an auditorium with a sloped floor, the entire main floor is a single level. A level can be a balcony or a separate section of seating in an arena or stadium, such as sky boxes. The upper-seating bowl in the coliseum is a separate level, as is the loge in the theater, however, it is not the intent of this provision that each row of seats be considered a separate level.

PUBLIC ENTRANCE. An entrance that is not a service entrance or a restricted entrance.

❖ A public entrance is one that provides access for the general public or employees, other than the service entrance or a restricted entrance (see the commentary for the definition of "Service entrance" and "Restricted entrance" and to Section 1105.1).

PUBLIC-USE AREAS. Interior or exterior rooms or spaces that are made available to the general public.

❖ This term is utilized to describe all interior and exterior spaces or rooms that may be occupied by the general public for any amount of time. Spaces that are utilized by the general public may be located in facilities that are publicly or privately owned. Examples include the lobby in an office building, a high-school basketball gymnasium with bleachers or a multipurpose room, a doctor's exam room, etc. (see also the commentary for the definitions of "Common use" and "Employee work area").

RESTRICTED ENTRANCE. An entrance that is made available for common use on a controlled basis, but not public use, and that is not a service entrance.

❖ The key to this provision is that the entrance has a controlled access or some type of limiting basis. This may be an entrance for juror's only at a courthouse, visitors only at a jail or employees only at a factory. A sports facility where there is control at the entrance for ticket holders only is not typically considered a restricted entrance (see the commentary for the definitions of "Service entrance" and "Public entrance" and to Section 1105.1.3).

SELF-SERVICE STORAGE FACILITY. Real property designed and used for the purpose of renting or leasing individual storage spaces to customers for the purpose of storing and removing personal property on a self-service basis.

❖ A portion or space within these facilities can be rented by persons to store personal property. Movement of items into and out of the space is handled by the individual.

SERVICE ENTRANCE. An entrance intended primarily for delivery of goods or services.

❖ This entrance is utilized primarily for accepting or sending deliveries of goods and services. Often this entrance is directly associated with a loading dock, and is not considered a public or restricted entrance (see the commentary for the definitions of "Service entrance" and "Public entrance" and to Section 1105.1.5).

SITE. A parcel of land bounded by a property line or a designated portion of a public right-of-way.

❖ A site, for purposes of accessibility requirements, is the same as that which is considered in the application of other code requirements. The property within the boundaries of the site is under the control of the owner. The owner can be held responsible for code compliance of the site and all facilities on it.

WHEELCHAIR SPACE. A space for a single wheelchair and its occupant.

❖ A wheelchair space is a designated space for a person to be stationary in his or her wheelchair as part of a fixed assembly seating configuration. The wheelchair space must be sized in accordance with ICC A117.1 (see Section 1108.2 for wheelchair space dispersion requirements).

SECTION 1103
SCOPING REQUIREMENTS

1103.1 Where required. Buildings and structures, temporary or permanent, including their associated sites and facilities, shall be accessible to persons with physical disabilities.

❖ This section establishes the broad principle that all buildings, structures and their associated sites and facilities are required to be accessible to persons with physical disabilities. This would include anyone who utilizes a space, including occupants, employees, students, spectators, participants and visitors. The approach taken by the code on the subject of accessibility is to require all construction to be accessible and then provide for the acceptable level of inaccessibility that is reasonable and logical. In codes created before the early 1990s, the approach was to list the conditions and occupancies to which the accessibility requirements applied; however, this is no longer practical, since the exceptions are far fewer than the circumstances to which

accessibility applies. The 14 exceptions to this section, Sections 1103.2.1 to 1103.2.14, reflect the extent to which accessibility in construction is either exempt or reduced in scope.

1103.2 General exceptions. Sites, buildings, facilities and elements shall be exempt from this chapter to the extent specified in this section.

❖ Accessibility is generally applicable to all building sites and facilities except as specifically exempted in the subsections to Section 1103.2.

1103.2.1 Specific requirements. Accessibility is not required in buildings and facilities, or portions thereof, to the extent permitted by Sections 1104 through 1110.

❖ This section provides a correlative reference to the various sections in Chapter 11 that identify when the intended number of accessible elements in various occupancies is less than 100 percent. The number of accessible fixtures and elements required by the referenced sections are deemed to provide adequate accessibility for those circumstances. For example, Section 1106 does not require all parking spaces in a parking facility to be accessible, Section 1107.6.1 does not require all sleeping units in Group R-1 to be fully accessible and Section 1109.5 does not require all drinking fountains to be accessible.

1103.2.2 Existing buildings. Existing buildings shall comply with Section 3409.

❖ The second section is a reference to the existing building criteria in Section 3409 for accessibility concerns in existing buildings. In accordance with Section 101.2, Exception 2, the *International Existing Building Code®* (IEBC™) could also be used (see commentary, Section 3409).

1103.2.3 Employee work areas. Spaces and elements within employee work areas shall only be required to comply with Sections 907.9.1.2, 1007 and 1104.3.1 and shall be designed and constructed so that individuals with disabilities can approach, enter and exit the work area. Work areas, or portions of work areas, that are less than 150 square feet (14 m²) in area and elevated 7 inches (178 mm) or more above the ground or finish floor where the elevation is essential to the function of the space shall be exempt from all requirements.

❖ This section states that elements within individual work stations are not required to be accessible, with the exception of visible alarms (see Section 907.9.1.2), accessible means of egress (see Section 1107) and circulation paths (see Section 1104.3.1). The assumption is that the employment nondiscrimination requirements of Americans with Disabilities Act (ADA) will provide for "reasonable accommodations" to the disability of the employee at that station. In other words, employers will modify individual work stations for the specific requirements of the individual utilizing the space. An accessible route will be required to each work station. An example of this is an individual work station in a laboratory. Installing sinks and built-in counters at accessible levels (see commentary, Sections 1109.3 and 1109.11) could make the station impractical for use by a person without a disability. When a station is required to be adapted for an individual, it would be revised based on the individual's needs and abilities. An accessible route to each work station in the laboratory would be required so that access to and from that station would be available. Note that the 36-inch (914 mm) clear width for the accessible route is the same as the minimum required width of an exit access aisle.

There is an additional exception for work areas that need to be raised 7 inches (178 mm) or more above the floor and have an area of less than 150 square feet (14 m²). Examples would include a raised area around a metal stamping machine, a safety manager's observation station on a production line or the pulpit area in a church (see commentary, Section 1107.4).

1103.2.4 Detached dwellings. Detached one- and two-family dwellings and accessory structures, and their associated sites and facilities as applicable in Section 101.2, are not required to be accessible.

❖ This section exempts detached one- and two-family dwellings from accessibility requirements. The key word here is detached. For example, a structure containing four or more dwelling units, even if the dwelling units are separated by fire walls, would still be required to be accessible as indicated in Section 1107.

Although one- and two-family detached dwellings are typically regulated by the IRC, this exception in the code is still necessary. Single-family dwellings or duplexes that are four stories or higher would be designed and constructed under the IBC Group R-3 requirements. The IRC references Chapter 11 of the code for accessibility requirements in Section R322. Those multiple-family structures with four or more dwelling or sleeping units are required to comply with the requirements for Group R-3 in Section 1107.6.3.

1103.2.5 Utility buildings. Occupancies in Group U are exempt from the requirements of this chapter other than the following:

1. In agricultural buildings, access is required to paved work areas and areas open to the general public.

2. Private garages or carports that contain required accessible parking.

❖ This section exempts Group U from accessibility requirements except as indicated, on the basis that such structures are a low priority when considering the need for accessibility. Areas in utility buildings that are required to be accessible would be paved work areas or areas open to the general public, such as if a farmer included a farmstand within his barn or buildings in which accessible parking spaces are located.

1103.2.6 Construction sites. Structures, sites and equipment directly associated with the actual processes of construction including, but not limited to, scaffolding, bridging, materials hoists, materials storage or construction trailers are not required to be accessible.

❖ This section exempts structures directly associated with the construction process because the need for accessibility on a continuous or regular basis in those circumstances is unlikely to arise. Note that all structures that may be involved during a construction project are not exempt — only those specifically involved in the actual process of construction. For example, if mobile units are brought into house classrooms during a school addition project, those mobile units would have to comply with accessibility provisions.

1103.2.7 Raised areas. Raised areas used primarily for purposes of security, life safety or fire safety including, but not limited to, observation galleries, prison guard towers, fire towers or lifeguard stands are not required to be accessible or to be served by an accessible route.

❖ If there is a reason to elevate an area for concerns about security or safety, these areas are not required to be accessible.

1103.2.8 Limited access spaces. Nonoccupiable spaces accessed only by ladders, catwalks, crawl spaces, freight elevators or very narrow passageways are not required to be accessible.

❖ Nonoccupiable, limited access spaces are considered areas where work could not reasonably be performed by a person in a wheelchair. These areas are not required to be accessible.

1103.2.9 Equipment spaces. Spaces frequented only by personnel for maintenance, repair or monitoring of equipment are not required to be accessible. Such spaces include, but are not limited to, elevator pits, elevator penthouses, mechanical, electrical or communications equipment rooms, piping or equipment catwalks, water or sewage treatment pump rooms and stations, electric substations and transformer vaults, and highway and tunnel utility facilities.

❖ Spaces that only contain heating, ventilation and air conditioning (HVAC), electrical, elevator, communication and similar types of equipment are considered areas where work could not reasonably be performed by a person utilizing a wheelchair. These areas are not required to be accessible.

1103.2.10 Single-occupant structures. Single-occupant structures accessed only by passageways below grade or elevated above grade including, but not limited to, toll booths that are accessed only by underground tunnels, are not required to be accessible.

❖ This section addresses facilities such as toll booths, which are raised to facilitate access to higher vehicles as well as to provide protection from being struck by vehicles. Access to these spaces are often lanes by tunnel or bridge. Some of these types of facilities could also be covered by Section 1103.2.3.

1103.2.11 Residential Group R-1. Buildings of Group R-1 containing not more than five sleeping units for rent or hire that are also occupied as the residence of the proprietor are not required to be accessible.

❖ This section exempts small bed-and-breakfast or transient boarding house facilities that are also the home of the owner.

1103.2.12 Day care facilities. Where a day care facility (Groups A-3, E, I-4 and R-3) is part of a dwelling unit, only the portion of the structure utilized for the day care facility is required to be accessible.

❖ When an adult day care or child day care facility is part of a person's home, the accessibility requirements are only applicable to the portion of the home that constitutes the day care facility, not the home itself.

1103.2.13 Detention and correctional facilities. In detention and correctional facilities, common use areas that are used only by inmates or detainees and security personnel, and that do not serve holding cells or housing cells required to be accessible, are not required to be accessible or to be served by an accessible route.

❖ Section 1107.5.5 addresses when sleeping units, special holding or housing cells and medical care units in detention and correctional facilities are required to be accessible. If the purpose of any common or shared space is to serve only the associated cells that are not required to be accessible (e.g., shared bathrooms or living space serving a specific group of cells), then those common or shared spaces are not required to be accessible.

1103.2.14 Fuel-dispensing systems. The operable parts on fuel-dispensing devices shall comply with ICC A117.1, Section 308.2.1 or 308.3.1.

❖ This section exempts gas and diesel fuel pumps at service stations from all accessibility provisions, except for the unobstructed reach range provisions in ICC A117.1. Basically, all the operable parts (e.g., pump controls, credit card readers) must be located between 15 inches and 48 inches (381 mm and 1219 mm) above the parking surface.

SECTION 1104
ACCESSIBLE ROUTE

1104.1 Site arrival points. Accessible routes within the site shall be provided from public transportation stops, accessible parking and accessible passenger loading zones and public streets or sidewalks to the accessible building entrance served.

Exception: An accessible route shall not be required between site arrival points and the building or facility entrance if the only means of access between them is a vehicular way not providing for pedestrian access.

❖ The intent of this section is to require an accessible route from the point at which one enters the site to any buildings or facilities that are required to be accessible on that site. It is presumed that people with disabilities are capable of gaining access to the site from such locations as accessible parking, public transportation stops, loading zones, public streets or sidewalks.

  If a vehicular route is the only route provided between an arrival point and an accessible entrance, an accessible route for pedestrian access is not required. For example, if there is a bus stop at the front of an industrial complex, but the only route to the building entrance is via a long driveway, an accessible pedestrian route to that entrance from the bus stop is not required. For special considerations in residential developments with recreational facilities, see the commentary to Sections 1107.3 and 1109.14.

1104.2 Within a site. At least one accessible route shall connect accessible buildings, accessible facilities, accessible elements and accessible spaces that are on the same site.

Exception: An accessible route is not required between accessible buildings, accessible facilities, accessible elements and accessible spaces that have, as the only means of access between them, a vehicular way not providing for pedestrian access.

❖ Developments may include several buildings on the same site. The intent of this section is to require an accessible route to all facilities offered on a site. Often sites are designed such that the only way to reach a building or facility is by automobile. If there are multiple, separated parking areas serving one or more buildings on a site, an accessible route is required between all such parking facilities and the buildings they serve. If there is an exterior feature, such as a swimming pool, located on a site containing multiple buildings, an accessible route is required from each building to the swimming pool. The exception clarifies that an accessible route is not required where no pedestrian access is otherwise intended or provided to a particular building or feature on the site. For special considerations in residential developments with recreational facilities, see the commentary to Sections 1107.3 and 1109.14.

1104.3 Connected spaces. When a building, or portion of a building, is required to be accessible, an accessible route shall be provided to each portion of the building, to accessible building entrances connecting accessible pedestrian walkways and the public way. Where only one accessible route is provided, the accessible route shall not pass through kitchens, storage rooms, restrooms, closets or similar spaces.

Exceptions:

1. In assembly areas with fixed seating required to be accessible, an accessible route shall not be required to

serve fixed seating where wheelchair spaces or designated aisle seats required to be on an accessible route are not provided.

2. Accessible routes shall not be required to mezzanines provided that the building or facility has no more than one story, or where multiple stories are not connected by an accessible route as permitted by Section 1104.4.

3. A single accessible route is permitted to pass through a kitchen or storage room in an accessible dwelling unit.

❖ This section requires that there be at least one route from an accessible entrance to all required accessible features within a building. If an area is addressed with a specific exception, then an accessible route is not required. For example, mechanical penthouses are exempt under Section 1103.2.9; therefore, an accessible route would not be required to this area. When only one accessible route is provided, this section puts limits on the types of spaces through which that accessible route may pass so that it is readily available. Spaces such as storage rooms, restrooms, closets and kitchens can be subject to locking or their availability may be restricted to authorized personnel; therefore, they would not serve as a reliable accessible routes for all building occupants. Exception 3 allows a single accessible route to pass through a kitchen or storage room within a dwelling unit that is required to be accessible (i.e., Accessible, Type A or Type B dwelling or sleeping unit). Given the space limitations and typical arrangements of dwelling units, it is not unreasonable to allow access, for example, through a kitchen, to reach an eating area or an exterior patio as the only means of access.

Exception 1 reemphasizes that areas or levels in assembly seating that do not contain wheelchair seating locations or designated aisle seats are not required to be accessed by an accessible route; however, the types of services available in the facility must be considered when determining what services must be available to accessible seats. A route is required to services from accessible seats in accordance with Section 1108.2.1.

Exception 2 references Section 1104.4. Mezzanines in a single-story building are required to be accessed by an accessible route unless they meet one of the exceptions in Section 1104.4. For example, a mezzanine in a storage warehouse is not required to be accessible if the area is less than 3,000 square feet (279 m²). The intent is that the same limitations for nonaccessible levels be applied in a single-story building with a mezzanine as applicable in a multistory building.

1104.3.1 Employee work areas. Common use circulation paths within employee work areas shall be accessible routes.

Exceptions:

1. Common use circulation paths, located within employee work areas that are less than 300 square feet (27.9 m²) in size and defined by permanently installed partitions, counters, casework or furnishings, shall not be required to be accessible routes.

2. Common use circulation paths, located within employee work areas, that are an integral component of equipment, shall not be required to be accessible routes.

3. Common use circulation paths, located within exterior employee work areas that are fully exposed to the weather, shall not be required to be accessible routes.

❖ This requirement for common use circulation paths within employee work areas is consistent with the exception in Section 1103.2.3. An accessible route is required to each employee work area. When employees share work areas, an accessible route must be available throughout that area. Note that the accessible route minimum width of 36 inches (914 mm) clear is consistent with the minimum means of egress pathways.

Exception 1 addresses the accessible route within small employee work areas. Shared work areas that are less than 300 square feet (28 m²) and confined by walls, partitions, permanently installed equipment, cabinets and counters are not required to have an accessible route through that particular area. The intent was to allow such areas as the last two workstations down an aisle, two or three workstations in a small office, etc. An accessible route is required to this area, but not necessarily throughout the area. Again, modifications would be performed at a later date based on employee needs.

Exception 2 permits nonaccessible areas around and through pieces of equipment. An example would be the shared work areas around a piece of assembly equipment in a factory where one or more persons are required to monitor and operate the machine and incoming or outgoing product.

Exception 3 is an exception for outdoor work areas. This would be applicable to landscapers, sewer workers, gravel pit crews, etc.

1104.3.2 Press boxes. Press boxes in assembly areas shall be on an accessible route.

Exceptions:

1. An accessible route shall not be required to press boxes in bleachers that have points of entry at only one level, provided that the aggregate area of all press boxes is 500 square feet (46 m²) maximum.

2. An accessible route shall not be required to free-standing press boxes that are elevated above grade 12 feet (3660 mm) minimum provided that the aggregate area of all press boxes is 500 square feet (46 m²) maximum.

❖ Press boxes are required to be served by an accessible route. If the occupant load is five or less, this could be provided by a platform lift (see Section 1109.7, Item 3).

Exception 1 is mainly applicable to press boxes located at the back of the bleacher seating in an outdoor sports facility. The "point of entry at only one level" refers to access to bleacher seating.

Exception 2 is mainly applicable to the free-standing

press box. If that press box is either less than 12 feet (3677 mm) above the ground or more than 500 square feet (46 m²) in area, it must be served by an accessible route.

For both exceptions, if one press box is provided, the total area of the press box must be less than 500 square feet (46 m²). If more than one press box overlooks the same playing field, the aggregate area of both press boxes must be less than 500 square feet (46 m²). The aggregate area is not intended to be applicable to press boxes that happen to be on the same multiple-facility site. For example, If a high school has a football field with two press boxes, the area of those two press boxes must be added together; however, if the same high school also has a press box for the baseball field, the area of the press box in the baseball field would not be included with the area of the press boxes in the football field.

1104.4 Multilevel buildings and facilities. At least one accessible route shall connect each accessible level, including mezzanines, in multilevel buildings and facilities.

Exceptions:

1. An accessible route is not required to stories and mezzanines above and below accessible levels that have an aggregate area of not more than 3,000 square feet (278.7 m²). This exception shall not apply to:

    1.1. Multiple tenant facilities of Group M occupancies containing five or more tenant spaces;

    1.2. Levels containing offices of health care providers (Group B or I); or

    1.3. Passenger transportation facilities and airports (Group A-3 or B).

2. In Group A, I, R and S occupancies, levels that do not contain accessible elements or other spaces required by Section 1107 or 1108 are not required to be served by an accessible route from an accessible level.

3. In air traffic control towers, an accessible route is not required to serve the cab and the floor immediately below the cab.

4. Where a two-story building or facility has one story with an occupant load of five or fewer persons that does not contain public use space, that story shall not be required to be connected by an accessible route to the story above or below.

❖ At least one accessible route is required between levels in a facility. This requirement does not mandate an elevator. The accessible route between levels can be via ramps, platform lifts (where permitted), limited use/limited access (LULA) elevators, passenger elevators, etc. The intent of the exceptions is to allow limited areas to be inaccessible without restricting access to services available to the general public.

Exception 1 addresses conditions under which it may not be practical or economical to provide an accessible route in multilevel buildings. The primary economic consideration is that, in the vast majority of circumstances, the means of providing an accessible route to floor levels above or below the entrance level of the building will be by an elevator. This exception applies to levels above and below the entry level that have an aggregate area of 3,000 square feet (279 m²) or less. For example, if a building had a floor area of 2,000 square feet (186 m²) above the entrance level and a floor area of 2,000 square feet (186 m²) below the entrance level, since the aggregate area is 4,000 square feet (372 m²), at least one of the two floor areas would be required to be connected to the entrance level by an accessible route. There are certain facilities that, despite being relatively small in size, are not exempt from the requirement for an accessible route. Due to the critical nature of the services provided, offices of health care providers, passenger transportation facilities and airports are not included in this exception. Multi-tenant mercantile occupancies (i.e., five or more tenants in the facility), which include facilities such as shopping malls, are not included in the exception because individual retail stores have limited types of commodities (i.e., shoes, eyeglasses, sporting goods, etc.), and the exception would mean that people with disabilities would not have access to the same range of goods that are available to the general public. In these cases, access must be provided such that people with disabilities may patronize all the establishments within the facility. If the mercantile facility has less than five tenants, they can use the 3,000-square feet (279 m²) exception.

Exception 2 is a direct reference to the specific items addressed in Sections 1107 and 1108. It is not a general exception for Groups A, I, R and S. The purpose of Exception 2 is to address areas within specific occupancies where a single element would be repeated multiple times on a level. Where Sections 1107 and 1108 would allow construction of a floor level with no accessible features, this section is not intended to trigger an accessible route to that otherwise inaccessible level. For example, an accessible route would not be required to upper levels in a hotel with all public spaces and all required accessible sleeping units located on the ground floor. If an elevator is to be provided in a building that qualifies for this exception, the elevator will be required to conform fully to the requirements for an accessible elevator (see Sections 1109.6 and 3001.3). If an elevator is not part of an accessible route, signage must be provided in accordance with Section 1110.2.

Exception 3 is specific to the unique situations found in air traffic control towers.

Exception 4 permits small nonpublic second floors to not require access. An example would be the second floor in a doctor's office that is used only for storage. The occupant load table in Section 1004.1.2 would limit this storage area to 1,500 square feet (139 m²) [i.e., 300 square feet (28 m²) per occupant x 5 occupants maximum = 1,500 square feet (28 m²)]. If the second floor also contained a mechanical room, that area would be exempt under Section 1103.2.9; however, if the doctor chose later to expand exam rooms into this second floor, he or she would have to provide an accessible route. Another example would be a second level in an airport that included only operational offices. In this case, the area would be limited by the occupant load table in Section 1004.1.2 to 500 square feet (46 m²) [i.e., 100 square feet (9 m²) per occupant x 5 occupants maximum = 500 square feet (46 m²)]. While the exception is specifically stated as applicable only to second-story buildings, it could be interpreted to apply to a one-story building with a basement level or a one-story building with a mezzanine.

1104.5 Location. Accessible routes shall coincide with or be located in the same area as a general circulation path. Where the circulation path is interior, the accessible route shall also be interior.

> **Exception:** Accessible routes from parking garages contained within and serving Type B dwelling units are not required to be interior.

❖ One of the objectives of accessibility requirements is to normalize, to the extent possible, the facilities provided for disabled persons and those comparable facilities that exist for people without disabilities. In addition, the intent of this section is to avoid the circumstance where an interior path between facilities is provided but the only accessible route between those same facilities is an exterior path. The exception is to address individual dwelling units that include their own garages or residential facilities with shared garage levels. The exception is intended to allow a person to exit the garage and enter through the front door, instead of requiring access from the garage into the unit directly. Section 1106.2 requires that when parking is provided within a building, accessible parking must also be located within the building. If this exception is utilized in accordance with Sections 1104.1 and 1107.3, an accessible route must be provided from the accessible parking spaces to the front door.

1104.6 Security barriers. Security barriers including, but not limited to, security bollards and security check points shall not obstruct a required accessible route or accessible means of egress.

> **Exception:** Where security barriers incorporate elements that cannot comply with these requirements, such as certain metal detectors, fluoroscopes or other similar devices, the accessible route shall be permitted to be provided adjacent to security screening devices. The accessible route shall permit persons with disabilities passing around security barriers to maintain visual contact with their personal items to the same extent provided others passing through the security barrier.

❖ This requirement provides guidance for when a security feature is required along an accessible route. An example would be the security checkpoints in airports. The intent is that the route for a person with mobility impairments should move through security checkpoints as close to the typical route as possible. It is recognized that some people in wheelchairs or with braces could not move through standard security barriers, such as metal detectors.

SECTION 1105
ACCESSIBLE ENTRANCES

1105.1 Public entrances. In addition to accessible entrances required by Sections 1105.1.1 through 1105.1.6, at least 50 percent of all public entrances shall be accessible.

Exceptions:

1. An accessible entrance is not required to areas not required to be accessible.

2. Loading and service entrances that are not the only entrance to a tenant space.

❖ A facility is not accessible if the entrances into it are inaccessible. This section establishes a reasonable criteria for providing accessible entrances. A facility is not required to have all of its entrances accessible in order to provide reasonable accommodation to disabled persons. If a facility has multiple public entrances, as a minimum, it is not considered unreasonable to require at least 50 percent of the entrances to be accessible. In addition to the 50-percent accessible public entrances, entrances that have a specific function or provide access to only certain portions of the facility must be addressed (see commentary, Sections 1105.1.1 through 1105.1.6). Sections 1110.1 and 1110.2 require signage at all entrances if all entrances are not accessible.

Exception 1 is self-evident in that if a facility or portion of a facility is not required to be accessible, then the entrances to such facilities or spaces are not required to be accessible. An example would be an exterior entrance to the sprinkler room (see Section 1103.2.9).

Exception 2 exempts loading and service entrances from the accessibility requirement on the basis that such entrances are unlikely to be used on a regular basis by people with disabilities. Loading and service entrances are required to be accessible if they are the only means of access into a facility or tenant space. This is consistent with Section 1105.1.5.

The intent of the reference to Sections 1105.1.1 through 1105.1.6 is to provide reasonable and convenient availability of an accessible entrance from the accessible facilities provided on the site. All entrances that serve distinct arrival points on the site are required to be accessible. For example, a public entrance from a public parking structure must be accessible (see Section 1105.1.1), as well as the employees-only entrance adjacent to separate employee parking (see Section 1105.1.3). This should deter the formerly common practice of designating an entrance as "the handicap entrance" (which is inappropriate) without regard to its location relative to such arrival points as accessible parking facilities, transportation facilities, passenger loading zones, taxi stands, public streets or sidewalks and tunnels or elevated walkways. This section also is intended to provide reasonable and convenient availability of an accessible entrance to the building's accessible vertical access, such as the elevators. This would allow a person with disabilities to move readily throughout all the levels of the building.

1105.1.1 Parking garage entrances. Where provided, direct access for pedestrians from parking structures to buildings or facility entrances shall be accessible.

❖ Occasionally, a parking garage is attached to an office building or mall. An accessible route must be provided from the accessible parking spaces in that parking garage to an accessible entrance into the office or mall portion of the building.

1105.1.2 Entrances from tunnels or elevated walkways. Where direct access is provided for pedestrians from a pedestrian tunnel or elevated walkway to a building or facility, at least one entrance to the building or facility from each tunnel or walkway shall be accessible.

❖ Elevated walkways are sometimes provided between adjacent buildings for easy access or protection from the weather for people who must commonly move between certain buildings. Tunnels may serve the same purpose. Washington, DC, has tunnels between many of its congressional buildings. Chicago has a "pedway" system connecting the commuter train stations with the basement level of many buildings in the downtown area. When these types of walkways or tunnels are provided, at least one of the building entrances off of these walkways or tunnels must be an accessible entrance.

1105.1.3 Restricted entrances. Where restricted entrances are provided to a building or facility, at least one restricted entrance to the building or facility shall be accessible.

❖ Where access into a specific entrance is restricted, such as an employees-only entrance, then at least one of that type of entrance must be accessible.

1105.1.4 Entrances for inmates or detainees. Where entrances used only by inmates or detainees and security personnel are provided at judicial facilities, detention facilities or correctional facilities, at least one such entrance shall be accessible.

❖ Where access to a specific entrance is limited for security reasons to inmates or detainees and the staff supervising them, that entrance must be accessible.

1105.1.5 Service entrances. If a service entrance is the only entrance to a building or a tenant space in a facility, that entrance shall be accessible.

❖ Loading and service entrances are required to be accessible if they are the only means of access into a facility or tenant space. Typically, loading and service entrances are exempted from the accessibility requirement on the basis that such entrances are unlikely to be used on a regular

basis by people with disabilities. This is consistent with Section 1105.1, Exception 2.

1105.1.6 Tenant spaces, dwelling units and sleeping units. At least one accessible entrance shall be provided to each tenant, dwelling unit and sleeping unit in a facility.

Exceptions:

1. An accessible entrance is not required to tenants that are not required to be accessible.

2. An accessible entrance is not required to dwelling units and sleeping units that are not required to be Accessible units, Type A units or Type B units.

❖ Each tenant space must have at least one accessible entrance. Dwelling and sleeping units that are Accessible, Type A or Type B units must have at least one accessible entrance. If a building is a single-tenant building, the 50-percent entrance requirement in Section 1005.1 is also applicable. If a space, whether a tenant, dwelling unit or sleeping unit, does not have accessibility requirements, then an accessible entrance is not required.

SECTION 1106
PARKING AND PASSENGER LOADING FACILITIES

1106.1 Required. Where parking is provided, accessible parking spaces shall be provided in compliance with Table 1106.1, except as required by Sections 1106.2 through 1106.4. The number of accessible parking spaces shall be determined based on the total number of parking spaces provided for the facility.

Exception: This section does not apply to parking spaces used exclusively for buses, trucks, other delivery vehicles, law enforcement vehicles or vehicular impound and motor pools where lots accessed by the public are provided with an accessible passenger loading zone.

❖ Accessible parking facilities are an important component of building and site accessibility. This section addresses the number and type of parking spaces required and the location of such parking spaces in relation to accessible building entrances. Parking facility requirements are primarily intended to facilitate accessibility for people with a mobility impairment.

This section is not intended to require that parking facilities be provided; rather, when parking is provided, the minimum number of accessible parking spaces specified herein must be provided. ICC A117.1 contains detailed requirements on the size and configuration of accessible parking spaces, including the required access aisle adjacent to the space, as well as the dimensional requirements for van-accessible parking spaces required by Section 1106.5.

Multitenant facilities, such as a mall, or large facilities, such as hospitals, may have multiple parking lots or parking garages. If multiple parking lots, or a combination of parking garages and parking lots are provided for a facility, the total number of spaces in all of the parking areas would be utilized to determine the number of accessible spaces required. Accessible parking spaces do not have to be determined on a lot-by-lot or tenant-by-tenant basis.

In certain types of lots it is not logical to require accessible parking spaces. The exception helps to clarify these situations. Parking lots that contain only parking for buses, trucks and delivery vehicles would not need accessible parking spaces. Lots where the parked cars are limited to vehicles used by law enforcement, impound lots or motor pools are also not required to provide accessible spaces. An accessible passenger drop-off is required to be provided on the site, so that when a person with a mobility impairment comes to that site , he or she has a place to exit the arrival vehicle safely and enter a vehicle coming from the lot.

TABLE 1106.1
ACCESSIBLE PARKING SPACES

TOTAL PARKING SPACES PROVIDED	REQUIRED MINIMUM NUMBER OF ACCESSIBLE SPACES
1 to 25	1
26 to 50	2
51 to 75	3
76 to 100	4
101 to 150	5
151 to 200	6
201 to 300	7
301 to 400	8
401 to 500	9
501 to 1,000	2% of total
More than 1,000	20, plus one for each 100 over 1,000

❖ The ratio of required number of accessible parking spaces is based on the accessible parking requirements of the Uniform Federal Accessibility Standards (UFAS). It does not reflect the demographic statistics on wheelchair usage that were used to scope other requirements in Chapter 11, because the majority of disabled-parking permit and license-plate holders in most states are ambulatory, mobility-impaired persons. The required ratios are intended to be responsive to the anticipated demand for all facilities, except as provided for in Sections 1106.2 through 1106.4, such that accessible parking spaces will be reasonably available on demand. Section 1110.1 states that signage is not required on the one required accessible parking space when the total number of parking spaces provided is four or less. This could be burdensome for the building tenant in that the accessible parking space, which is restricted for use only by authorized vehicles, could constitute anywhere from 25 to 100 percent of the available parking. This may unduly restrict the availability of parking for all other vehicles and patrons of the facility.

1106.2 Groups R-2 and R-3. Two percent, but not less than one, of each type of parking space provided for occupancies in Groups R-2 and R-3, which are required to have Accessible, Type A or Type B dwelling or sleeping units, shall be accessible. Where parking is provided within or beneath a building, accessible parking spaces shall also be provided within or beneath the building.

❖ When Accessible, Type A or Type B units are required in a Group R-2 or R-3 facility, 2 percent of parking spaces provided for all the occupants shall be accessible. Where parking is provided within or beneath a building, accessible parking spaces also are to be provided within or beneath the building. If a combination of surface and covered parking is provided, accessible parking may be provided in both locations. This is intended to establish consistency in the type and location of parking spaces available to all people. Note that parking provided for visitors should comply with the numbers indicated in Table 1106.1.

This section provides a separate criterion for the required number of accessible parking spaces for occupancies in Groups R-2 and R-3 that include Accessible, Type A or Type B units. The requirements are based on HUD's Fair Housing Accessibility Guidelines (FHAG).

1106.3 Hospital outpatient facilities. Ten percent of patient and visitor parking spaces provided to serve hospital outpatient facilities shall be accessible.

❖ Certain facilities can be expected to have a higher demand for accessible parking spaces than that reflected in Table 1106.1. Medical outpatient facilities are one such facility. This section requires that 10 percent of the parking provided for patients and visitors of these types of facilities be accessible. Parking provided for the employees of these facilities should comply with Table 1106.1. If a hospital contains multiple clinics or types of facilities, the amount of parking dedicated to the outpatient facility is interpretive. Options for determining the number of accessible spaces could be based on the area percentage of the outpatient facilities to other facilities or the anticipated number of patients and visitors to these facilities in relation to other facilities.

1106.4 Rehabilitation facilities and outpatient physical therapy facilities. Twenty percent, but not less than one, of the portion of patient and visitor parking spaces serving rehabilitation facilities and outpatient physical therapy facilities shall be accessible.

❖ Medical facilities that specialize in treatment or other services for people with mobility impairments can be expected to have a higher demand for accessible parking spaces. In these cases, 20 percent of the parking spaces provided for visitors and patients are required to be accessible. Parking provided for the employees of these facilities should comply with Table 1106.1. If a hospital contains multiple types of facilities, the amount of parking dedicated to the outpatient facility is interpretive. Options for determining the number of accessible spaces could be based on the area percentage of the outpatient facilities to other facilities or the anticipated number of patients and visitors to these facilities in relation to other facilities.

1106.5 Van spaces. For every six or fraction of six accessible arking spaces, at least one shall be a van-accessible parking space.

❖ Vans that are specially equipped to accommodate a wheelchair user require larger parking spaces than the typical accessible parking space. This is because they are usually equipped with a mechanized loading/unloading platform that extends outward from the passenger side of the vehicle. ICC A117.1 contains the dimensional requirements for van-accessible spaces. In order to accommodate the growing usage of these specially equipped vans, this section requires that one in every six accessible parking spaces, or a fraction thereof, be of a size to accommodate these vans. The requirement for van-accessible spaces is not intended to require a greater number of accessible spaces than required by Table 1106.1. For example, if Table 1106.1 requires six accessible parking spaces, one of those six must be van accessible. If seven accessible spaces are required by Table 1106.1, two of those seven are required to be van accessible.

1106.6 Location. Accessible parking spaces shall be located on the shortest accessible route of travel from adjacent parking to an accessible building entrance. Accessible parking spaces shall be dispersed among the various types of parking facilities provided. In parking facilities that do not serve a particular building, accessible parking spaces shall be located on the shortest route to an accessible pedestrian entrance to the parking facility. Where buildings have multiple accessible entrances with adjacent parking, accessible parking spaces shall be dispersed and located near the accessible entrances.

> **Exception:** In multilevel parking structures, van-accessible parking spaces are permitted on one level.

❖ As previously stated, the majority of disabled-parking permits and license-plate holders in most states are ambulatory, mobility-impaired persons. Travel distance, as well as severe weather conditions encountered when traversing from the parking lot to the building entrance, are more difficult for mobility-impaired persons to deal with than the general population. The intent of this section is to locate the parking so that the people utilizing the accessible parking spaces have to travel a minimum distance to an accessible entrance. This requirement is stated in performance terms and requires a degree of subjective judgement on the part of both the designer and the building official in determining the appropriate location for accessible spaces that meets the intent of this section. If a facility has multiple accessible entrances, accessible parking spaces are to be dispersed consistent with the location of these entrances. The exception is intended to acknowledge a practical difficulty associated with multilevel parking structures. In many cases, a multilevel parking structure will serve accessible building entrances on more than one or all of its

parking levels. Specially equipped vans for disabled persons are often modified by raising the roof of the vehicle in order to provide greater interior headroom. Consequently, such vans require greater vertical clearances. Typical parking structure design may not easily accommodate the necessary vertical clearance for accessible vans due to their low floor-to-ceiling heights. It would be impractical and economically unjustified to require parking structures to be designed solely for the purpose of enabling van-accessible spaces to be located on upper levels. Accordingly, the exception allows the required van-accessible parking spaces to be located on only one level of a multilevel parking structure. The route to and from the space, as well as at the van space and associated access aisle must meet the minimum clear height of 98 inches (2489 mm) as specified in ICC A117.1 Section 502.5. This will usually be the entry level of the parking facility.

1106.7 Passenger loading zones. Passenger loading zones shall be designed and constructed in accordance with ICC A117.1.

❖ This section does not require passenger loading zones to be provided, however, where provided, they must be accessible in accordance with ICC A117.1. For locations where passenger loading zones are required, see Sections 1106.7.1 through 1106.7.3.

1106.7.1 Continuous loading zones. Where passenger loading zones are provided, one passenger loading zone in every continuous 100 linear feet (30.4 m) maximum of loading zone space shall be accessible.

❖ The intent of this section is to address the continuous loading zones found at most larger airports as well as some other types of transportation facilities, such as commuter train "kiss-n-ride" drop-off areas. At least one 20-foot (6096 mm) section for every 100 feet (30,480 mm) of passenger loading zone must meet the accessibility provisions in ICC A117.1.

1106.7.2 Medical facilities. A passenger loading zone shall be provided at an accessible entrance to licensed medical and long-term care facilities where people receive physical or medical treatment or care and where the period of stay exceeds 24 hours.

❖ The requirement for accessible passenger loading zones is applicable to both medical and long-term care facilities. The most common examples are hospitals and nursing homes; however, this could also be applicable to some rehabilitation and assisted living facilities (typically Group I-2). Medical and long-term care facilities typically have a higher percentage of people with mobility impairments living in the facility or visiting on a regular basis. Due to their mobility impairment, such people will be frequently driven by a friend or family member, and it is desirable and convenient to drop them off at the entrance rather than parking and requiring them to travel from the parking lot to the facility's entrance. As such, this section requires that a passenger loading zone be provided at no less than one accessible entrance to the facility.

1106.7.3 Valet parking. A passenger loading zone shall be provided at valet parking services.

❖ Valet parking is typically the public access point to a facility. When valet parking is provided, the passenger loading zone must be accessible so that it may be utilized by persons requiring wheelchairs. This would include the access aisle and an accessible route from the access aisle to the front entrance. Note that providing valet parking services does not eliminate the need for accessible parking spaces. Some vehicles that are modified for persons with disabilities may not be drivable by a valet attendant. In addition, valet parking may only be provided during certain periods of time, such as during dinner hours at a restaurant.

SECTION 1107
DWELLING UNITS AND SLEEPING UNITS

❖ The development of this section was part of the coordination effort with the Fair Housing Amendments Act (FHA) and the Fair Housing Accessibility Guidelines (FHAG). For further explanation see the general comments for this chapter.

One of the phrases that was developed as part of this coordination was "intended to be occupied as a residence." This defines a type of dwelling or sleeping unit that is a person's home or place of abode. The intent of this language is to clarify that, consistent with the FHA, all dwelling units or sleeping units that can or will be used as a place of abode—even for a short period of stay—must meet the requirements for Type B dwellings. There is a presumption in the code that if a building is nontransient (i.e., occupants stay more than 30 days), it is covered by the requirements in this section. In addition, in coordination with the FHA, occupancies of fewer than 30 days may be required to meet Type B design and construction requirements.

Accordingly, this commentary will provide guidance for determining when short-term occupancies that might not typically be viewed as housing are covered by this section. Such short-term Group R occupancies may include residential hotels and motels; corporate housing; seasonal vacation units; timeshares; boarding houses; dormitories and migrant-farm worker housing. Also included are some occupancies in Group I, such as nursing homes, assisted living facilities, hospices and homeless shelters.

The key factor in determining whether any of these occupancies are subject to Type B requirements is whether the occupant will have the right to return to the property and whether he or she would have anywhere else in which to return to. If it is intended that an occupant will have a right to return to the property and will not have anywhere else to return, the unit is "intended to be occupied as a residence" and must meet the Type B requirements regardless of the length of stay. Thus, for example, homeless shelters where occupants have a right to return nightly must comply with Type B criteria even if the occupants stay only a few nights. In addition, nursing homes in which a resident moves after vacating his or her primary residence are subject to Type B requirements.

If the occupants have a right to return to the property but also have another place in which to return to, the unit may still be subject to Type B criteria. Additional factors must be considered to determine whether the property is a short-term dwelling or sleeping unit that must meet Type B criteria or a transient property that is not required to comply with Type B criteria. These factors must be considered by owners, builders, developers, architects and other designers to determine whether or not a building must be designed and constructed in accordance with Type B unit requirements.

For additional discussion on the elements related to hotels, motels, corporate housing and seasonal vacation units and timeshares, see the commentary to Section 1107.6.1 (Group R-1). With respect to most other occupancies, the following factors are also relevant to determine whether the units are subject to Type B requirements:

1. Whether the property is to be marketed as short-term housing;

2. Whether the terms and length of occupancy will be established through a written agreement;

3. How payment will be calculated (e.g., on a daily, weekly, monthly or yearly basis) and

4. What types of amenities and services are offered with the occupancy.

For example, an assisted living facility that provides sleeping units and medical services to its occupants, and bills them on a monthly basis is subject to Type B design and construction requirements. In addition, housing for migrant farm workers that is provided in conjunction with the worker's employment (whether or not rent is paid) and contains amenities for cooking and sleeping would be subject to Type B criteria.

1107.1 General. In addition to the other requirements of this chapter, occupancies having dwelling units or sleeping units shall be provided with accessible features in accordance with this section.

❖ There are two basic types of facilities that this section covers: dwelling units and sleeping units. A dwelling unit is defined in Section 310.2 as a single unit that contains permanent provisions for "living, sleeping, eating, cooking and sanitation." A sleeping unit is defined in Section 202 as a room in which people sleep, which can include some of the provisions found in a dwelling unit but not all. Occupancy of dwelling units or sleeping units can be transient or nontransient. Dwelling units are typically apartments, condominiums, detached homes or townhouses. Dwelling units can be located in hotels that offer cabins, suites or rooms with kitchen facilities. Bedrooms within dwelling units are not considered sleeping units.

A sleeping unit could be a typical hotel guestroom; a bedroom in a congregate residence, such as a dorm, sorority house, fraternity house, convent, monastery or boarding house; a nursing home room or a jail cell.

1107.2 Design. Dwelling units and sleeping units which are required to be Accessible units shall comply with this code and the applicable portions of Chapters 1 through 9 of ICC A117.1. Type A and Type B units shall comply with the applicable portions of Chapter 10 of ICC A117.1. Units required to be Type A units are permitted to be designed and constructed as Accessible units. Units required to be Type B units are permitted to be designed and constructed as Accessible units or as Type A units.

❖ There are three levels of accessibility that can be required in a dwelling or sleeping unit: Accessible units, Type A units and Type B units.

An Accessible unit is constructed for full accessibility in accordance with the requirements in Chapters 3 through 9 of ICC A117.1. For example, grab bars are in place in the bathrooms, a clear floor space is provided for front approach at the kitchen sink and bathroom lavatories, 32-inch (813 mm) clear width doors with maneuvering clearances and lever hardware are provided, etc. None of the elements in the unit are constructed for adaptability. The requirements for an Accessible unit are more restrictive than either a Type A or a Type B unit.

A Type A unit has some elements that are constructed accessible [e.g., 32-inch (813 mm) clear width doors with maneuvering clearances and lever hardware] and some elements designed to be altered when needed (e.g., blocking in the walls of the bathroom for future installation of grab bars). This type of unit is less accessible than an Accessible unit and more accessible than a Type B unit.

The scoping or technical requirements for Type B units are consistent with the requirements for units required by the FHAG. A Type B unit is constructed to a lower level of accessibility than either an Accessible unit or Type A unit. While a person who uses a wheelchair could maneuver in a Type B unit, the technical requirements are geared more towards persons with mobility impairments. Areas of a Type B unit may be totally nonaccessible (e.g., sunken living room, extra bedrooms on a mezzanine level). Side approach is permitted to sinks in the kitchen and lavatories in the bathroom rather than planning for a front approach. Some elements are constructed with a minimal level of accessibility [e.g., doors within the unit are 31³/₄-inch (806 mm) clear width but do not require maneuvering clearances], while some elements are designed to be altered when needed (e.g., blocking in the walls of the bathroom for future installation of grab bars).

This section also takes into consideration the fact that Accessible unit requirements are more stringent than Type A requirements, and Type A requirements are more stringent than Type B requirements. Units are permitted to be constructed to a higher level of accessibility than required.

1107.3 Accessible spaces. Rooms and spaces available to the general public or available for use by residents and serving Accessible units, Type A units or Type B units shall be accessible. Accessible spaces shall include toilet and bathing rooms, kitchen, living and dining areas and any exterior spaces, including patios, terraces and balconies.

Exception: Recreational facilities in accordance with Section 1109.14.

❖ All spaces available to the general public or for the use of residents of the Accessible units, Type A units or Type B units are required to be accessible under the general provisions for this chapter. For example, common areas utilized for laundry facilities or storage are required to be fully accessible. This also includes lobby areas and corridors that provide access to Accessible, Type A or Type B dwelling units or sleeping units.

If shared facilities are intended for use by persons utilizing the Accessible, Type A or Type B sleeping units, those facilities must also be accessible. Examples would be lounges, common bathrooms and shared kitchens in a congregate residence. The net result is that a disabled person, either occupant or visitor, can approach and enter all Accessible, Type A or Type B units or any associated facilities and public areas on the site.

1107.4 Accessible route. At least one accessible route shall connect accessible building or facility entrances with the primary entrance of each Accessible unit, Type A unit and Type B unit within the building or facility and with those exterior and interior spaces and facilities that serve the units.

Exceptions:

1. If the slope of the finished ground level between accessible facilities and buildings exceeds one unit vertical in 12 units horizontal (1:12), or where physical barriers prevent the installation of an accessible route, a vehicular route with parking that complies with Section 1106 at each public or common use facility or building is permitted in place of the accessible route.

2. Exterior decks, patios or balconies that are part of Type B units and have impervious surfaces, and that are not more than 4 inches (102 mm) below the finished floor level of the adjacent interior space of the unit.

❖ The intent of this section is that there will be at least one accessible route that connects all accessible building or facility entrances with the entrance of all Accessible, Type A or Type B units. If a dwelling or sleeping unit is accessible, it is of little value if there is no accessible route provided from the entrance of the building to that dwelling or sleeping unit. This section is also intended to confirm that an accessible route is available connecting accessible building or facility entrances with all interior and exterior spaces and facilities that serve the accessible dwelling or sleeping unit. For example, a recreational facility, such as a common swimming pool area, within a multiple-family dwelling complex that is required to be accessible in accordance with Section 1109.14 will be of little value to a disabled person if there is not an accessible route from that person's building leading to the swimming pool area.

Exception 1 is intended to recognize that there may be site features or constraints that make it impractical to provide an accessible route, such as a paved walkway or path. In such cases, it is reasonable to provide access by means of a vehicular route to a parking area serving the accessible facility or building, from which access can then be gained. For example, a community building that is available to occupants of a multiple-family complex may be located on a hilly site such that the slope of the ground level between a building containing an accessible dwelling unit and the community building exceeds 1:12. The exception allows access to be provided by means of a vehicular route leading from the parking areas serving the dwelling unit and the community building. Accessible parking must be provided in each parking area. Accessibility in this circumstance may be considered less convenient; however, the impracticality of providing a conventional accessible route due to the site constraints is the governing factor.

Exception 2 is concerning decks, patios or balconies associated with a Type B dwelling or sleeping unit. Due to weather concerns (i.e., rain or snow penetration), a 4-inch (102 mm) step-down to an impervious surface would be permitted at those locations.

1107.5 Group I. Occupancies in Group I shall be provided with accessible features in accordance with Sections 1107.5.1 through 1107.5.5.

❖ This section introduces five subsections that set forth the threshold for accessibility in institutional occupancies. Note that in Accessible units the associated bathing and toilet facilities in each case must comply with the requirements in ICC A117.1, Chapter 6. Only Type B units can utilize dwelling unit criteria for bathing facilities as specified in ICC A117.1, Chapter 10.

1107.5.1 Group I-1. Group I-1 occupancies shall be provided with accessible features in accordance with Sections 1107.5.1.1 and 1107.5.1.2.

❖ Group I-1 is comparable in many respects to Group R-2 in that it is a residential setting. The difference is that Group I-1 is a supervised environment because of age, mental disability or other such characteristics of the occupants (see Section 308.2).

1107.5.1.1 Accessible units. At least 4 percent, but not less than one, of the dwelling units and sleeping units shall be Accessible units.

❖ The threshold for accessibility in apartments (Group R-2), in accordance with Section 1107.6.2.1.1, is that 2 percent of the units are required to be Type A units. It is anticipated that Group I-1 will experience a greater demand for accessible facilities because of the nature of the occupants and the frequency of occupant turnover. As such, this section requires that 4 percent of the

sleeping units and their associated facilities (e.g., bathing and toilet facilities) must meet the criteria for Accessible units.

1107.5.1.2 Type B units. In structures with four or more dwelling or sleeping units intended to be occupied as a residence, every dwelling and sleeping unit intended to be occupied as a residence shall be a Type B unit.

Exception: The number of Type B units is permitted to be reduced in accordance with Section 1107.7.

❖ As discussed in the general comments to this section, most Group I-1 assisted living facilities are intended to serve as a person's place of residence. Since a Group I-1 starts at an occupant load of 16 residents, the Type B criteria for any units that are not Accessible units are almost always required. The exception is a general reference to Section 1107.7 which addresses situations where it is logical to back off on the requirements for Type B units within a structure.

1107.5.2 Group I-2 Nursing homes. Nursing homes of Group I-2 shall be provided with accessible features in accordance with Sections 1107.5.2.1 and 1107.5.2.2.

❖ The requirements for accessibility in Group I-2 are based on the anticipated frequency of usage by disabled persons. There are different thresholds established for different types of health care occupancies, based on the nature of their activity. A nursing home is anticipated to have a high percentage of residents who use wheelchairs or some type of mobility aid.

1107.5.2.1 Accessible units. At least 50 percent, but not less than one, of the dwelling units and sleeping units shall be Accessible units.

❖ Nursing homes, which generally provide long-term care, have a higher degree of likelihood that the patients will be physically disabled. As such, 50 percent of the patient sleeping rooms and their associated facilities (e.g., bathing and toilet facilities) must meet the criteria for Accessible units.

1107.5.2.2 Type B units. In structures with four or more dwelling or sleeping units intended to be occupied as a residence, every dwelling and sleeping unit intended to be occupied as a residence shall be a Type B unit.

Exception: The number of Type B units is permitted to be reduced in accordance with Section 1107.7.

❖ As discussed in the general comments to this section, most Group I-2 nursing homes are intended to serve as a person's place of residence. The Type B criteria for any units that are not Accessible units are almost always required. The exception is a general reference to Section 1107.7 which addresses situations where it is logical to back off on the requirements for Type B units within a structure.

1107.5.3 Group I-2 Hospitals. General-purpose hospitals, psychiatric facilities, detoxification facilities and residential care/assisted living facilities of Group I-2 shall be provided with accessible features in accordance with Sections 1107.5.3.1 and 1107.5.3.2.

❖ The requirements for accessibility in Group I-2 are based on the anticipated frequency of usage by disabled persons. There are different thresholds established for different types of health care occupancies based on the nature of their activity. The anticipated need in hospitals is partially based on patients needing assistance due to temporary disabilities resulting from illness or operations.

1107.5.3.1 Accessible units. At least 10 percent, but not less than one, of the dwelling units and sleeping units shall be Accessible units.

❖ While the anticipated percentage of patients in general-purpose hospitals is typically higher than 10 percent, there is also an anticipation that many of the patients will be routinely assisted by staff. Section 1109.2, Exception 6 exempts the bathrooms associated with critical-care or intensive-care Accessible sleeping units from the requirements in ICC A117.1. In the case of psychiatric facilities, detoxification facilities or residential care/assisted facilities there is a lesser likelihood that the typical patient will be physically disabled.

Housing the elderly should not automatically be classified as a "nursing home." Some residential care/assisted living facilities provide quarters for residents that more closely represent a residential setting (i.e., sleeping, eating, cooking and sanitation within the individual living accommodations) and the residents are more ambulatory than in a nursing home. Based on the anticipated frequency of usage by disabled persons, these facilities might be more appropriately classified with the 10-percent requirement for Accessible units.

1107.5.3.2 Type B units. In structures with four or more dwelling or sleeping units intended to be occupied as a residence, every dwelling and sleeping unit intended to be occupied as a residence shall be a Type B unit.

Exception: The number of Type B units is permitted to be reduced in accordance with Section 1107.7.

❖ As discussed in the general comments to this section, most Group I-2 hospitals are not intended to serve as a person's place of residence; however, residential care/assisted living facilities are typically a person's place of residence, and the Type B criteria for any units that are not Accessible units are required. Whether or not a psychiatric or detoxification facility is a person's place of abode is subject for interpretation. Please see the general discussion to this section for additional information. The exception is a general reference to Section 1107.7, which addresses situations where it is logical to back off on the requirements for Type B units within a structure.

1107.5.4 Group I-2 Rehabilitation facilities. In hospitals and rehabilitation facilities of Group I-2 which specialize in treating conditions that affect mobility, or units within either which specialize in treating conditions that affect mobility, 100 percent of the dwelling units and sleeping units shall be Accessible units.

❖ The requirements for accessibility in Group I-2 are based on the anticipated frequency of usage by disabled persons. There are different thresholds established for different types of health care occupancies based on the nature of their activity. In rehabilitation facilities that specialize in treating conditions that affect mobility, the anticipated need is very high compared to other Group I-2 facilities. In such cases, it is realistic to presume that the majority, if not all, of the patients will have some degree of mobility impairment and will, thus, require accessible features. Accordingly, all patient rooms and their associated bathing and toilet facilities are required to be Accessible units. While this facility may also be intended to be occupied as a residence, Accessible unit requirements are more restrictive than Type B requirements, so those needs are already addressed.

1107.5.5 Group I-3. Buildings, facilities or portions thereof with Group I-3 occupancies shall comply with Sections 1107.5.5.1 through 1107.5.5.3.

❖ The requirement for Accessibility in Group I-3 facilities was taken from the draft for the new ADAAG. In consideration of the need for accessibility in Group I-3, the CABO Board for the Coordination of Model Codes (BCMC) found a lack of statistical data on which to base a scoping requirement. As such, consistency with ADAAG was deemed to be appropriate. Section 1107.5.5.1 contains requirements for general housing cells. Sections 1107.5.5.2 and 1107.5.5.3 are additional requirements for special types of cells or holding areas. While jail cells may be considered a detainee's residence, since the FHA does not cover detention and correctional facilities, Type B units are not required in this occupancy.

1107.5.5.1 Group I-3 sleeping units. In occupancies in Group I-3, at least 2 percent, but not less than one, of the dwelling units and sleeping units shall be Accessible units.

❖ The requirement for Accessible sleeping units or cells in Group I-3 is for 2 percent of the general housing cells and their associated facilities to be accessible. The current trend in jail design is to group housing cells into "pods." Since the administration of the correctional facility will determine where detainees will reside, there are no requirements for distribution of the Accessible units. A correctional facility may decide to locate all its Accessible cells in one pod due to suicide prevention and other safety concerns. Sections 1103.2.13 and 1109.11 have exceptions for spaces shared by detainees for areas within the jail where Accessible units are not provided.

1107.5.5.2 Special holding cells and special housing cells or rooms. In addition to the units required to be accessible by Section 1107.5.5.1, where special holding cells or special housing cells or rooms are provided, at least one serving each purpose shall be accessible. Cells or rooms subject to this requirement include, but are not limited to, those used for purposes of orientation, protective custody, administrative or disciplinary detention or segregation, detoxification and medical isolation.

Exception: Cells or rooms specially designed without protrusions and that are used solely for purposes of suicide prevention shall not be required to include grab bars.

❖ If special types of holding or housing cells are provided within a facility (e.g., orientation, protective custody, disciplinary detention, segregation, detoxification, medical isolation), at least one of each type must be an Accessible unit. A cell that is designed specifically for a suicide watch is not required to be an Accessible cell.

1107.5.5.3 Medical care facilities. Patient sleeping units or cells required to be accessible in medical care facilities shall be provided in addition to any medical isolation cells required to comply with Section 1107.5.5.2.

❖ When medical care facilities are provided within a detention or correctional facility, 10 percent of the units are required to be Accessible units. The intent is to reference the hospital requirements found in Section 1107.5.3.1.

1107.6 Group R. Occupancies in Group R shall be provided with accessible features in accordance with Sections 1107.6.1 through 1107.6.4.

❖ This section introduces four subsections that address accessibility requirements for occupancies in Group R. Note that in Accessible units, the associated bathing and toilet facilities must comply with requirements in ICC A117.1, Chapter 6. Only Type A units and Type B units can utilize the dwelling unit criteria for bathing facilities as specified in ICC A117.1, Chapter 10.

The IRC references this chapter for accessibility provisions; therefore, this chapter may be applicable to housing covered under the IRC (see commentary, Section 1107.6.3). Structures referenced to Chapter 11 from the IRC would be considered Group R-3.

1107.6.1 Group R-1. Group R-1 occupancies shall be provided with accessible features in accordance with Sections 1107.6.1.1 and 1107.6.1.2.

❖ The terms "dwelling unit" and "sleeping unit" are utilized instead of "guestrooms" so that the amenities in the units themselves are reflected. For an exception for small bed-and-breakfast style hotels, see Section 1103.2.11.

To be able to accommodate persons with disabilities as they travel or when they may need short-term housing, hotels, motels and boarding houses are required to provide Accessible units. If these facilities are also intended to be occupied as a residence, Type B unit criteria would be applicable. For a complete discussion,

please see the general comments to this chapter and this section.

The following factors should be considered in determining whether hotels, motels, corporate housing and seasonal vacation units where occupants can stay more or less than 30 days are subject to Type B unit criteria:

1. What amenities will be included inside the units, including kitchen facilities;

2. Whether the property is to be marketed to the public as short-term housing;

3. Whether the terms and length of the occupancy will be established through a lease or other written agreement and

4. How payment will be calculated (e.g., on a daily, weekly, monthly or yearly basis).

If the amenities and operation of the units are closer to those of apartments than of hotels, they are subject to Type B requirements. For example, if a hotel is marketed as short-term housing, payment is made monthly, and if the units contain kitchens, the hotel would be subject to Type B unit criteria.

1107.6.1.1 Accessible units. In occupancies in Group R-1, Accessible dwelling units and sleeping units shall be provided in accordance with Table 1107.6.1.1. All facilities on a site shall be considered to determine the total number of Accessible units. Accessible units shall be dispersed among the various classes of units. Roll-in showers provided in Accessible units shall include a permanently mounted folding shower seat.

❖ The required number of Accessible dwelling or sleeping units in Group R-1 is indicated in Table 1107.6.1.1. The table requires a certain number of Accessible units to include roll-in-type showers. These showers are intended to serve the dual purpose of roll-in and transfer; therefore, they must include folding shower seats mounted on the shower wall.

For sites where the hotel rooms are provided in multiple buildings, all the units on the site should be considered to determine the number of Accessible units. For example, if a hotel consists of several small buildings on a site, the same number of Accessible units would be required as if the hotel was constructed as a single structure.

Accessible units must be dispersed among the different types of units. This is not an automatic requirement to disperse the Accessible units to different floor levels. When different classes of dwelling or sleeping units are available, special amenities must also be made available in the Accessible units, including suites or larger rooms, kitchenettes, executive levels, etc. For example, if double, queen and king sleeping units are provided and two Accessible units are required, the Accessible units must be provided in two of the three options. If no other types of amenities are provided on other levels, such as executive levels, the hotel could provide the Accessible units on the main level. Some hotels chose to provide all Accessible units on the main level due to concerns for disabled guests' ease of escape from the building during an emergency, such as a fire.

Table 1107.6.1.1. See below.

❖ The number of Accessible dwelling and sleeping units required is coordinated with the Americans with Disabilities Act Accessibility Guidelines (ADAAG) requirements. The ADAAG Review Federal Advisory Committee reviewed all the information and data provided by the industry in support of a lower number of Accessible units being required. The committee determined that accessible features are useful to persons with disabilities beyond those who use wheelchairs, and that good design would lessen the "institutional" impression that may make Accessible units less attractive to nondisabled guests. To accommodate wheelchair users in relatively large facilities, this table requires a minimum number of units to be equipped with roll-in-type showers.

1107.6.1.2 Type B units. In structures with four or more dwelling or sleeping units intended to be occupied as a residence, every dwelling and sleeping unit intended to be occupied as a residence shall be a Type B unit.

Exception: The number of Type B units is permitted to be reduced in accordance with Section 1107.7.

❖ Most hotels and motels are not intended to be occupied as a residence; therefore, the criteria for Type B units would not be applicable. However, in certain situations, extended-stay hotels or similar corporate housing arrangements are required to meet Type B unit criteria. For additional information on how to determine if this section is applicable, see the comments to this chapter, Section 1107 and Section 1107.6.1. The exception is a general reference to Section 1107.7, which addresses

situations where it is logical to back off on the requirements for Type B units within a structure.

1107.6.2 Group R-2. Accessible units, Type A units and Type B units shall be provided in occupancies in Group R-2 in accordance with Sections 1107.6.2.1 and 1107.6.2.2.

❖ The terms "dwelling unit" and "sleeping unit" are utilized instead of "apartment," "dorm room," "sleeping accommodations," "suites," etc., so that the amenities in the units themselves are reflected and for consistent application within all Group R-2 facilities. Since these types of facilities are intended to be occupied as a residence, Type B unit criteria would be applicable. For a complete discussion, please see the comments to this chapter and this section.

Timeshare properties require a different analysis. Timeshare owners have an interest in the property, yet they typically stay less than 30 days. For a timeshare property to be subject to Type B requirements, the owners must have an ownership interest in the property itself, rather than in just coming to the area for a vacation without ties to a particular property. The following additional factors must be considered to determine whether timeshare units must meet Type B design and construction requirements:

1. Whether traditional rights of ownership are to be unrestricted (e.g., whether the timeshare owner has the right to occupy, alter or exercise control over a particular unit over a period of time),

2. The nature of the ownership interest conveyed (e.g., fee simple) and

3. The extent to which operations resemble those of a hotel, motel or inn (e.g., reservation, central registration, meals, laundry service).

If an owner's rights regarding the units are subject to few restrictions and the operation of the units is closer

TABLE 1107.6.1.1
ACCESSIBLE DWELLING AND SLEEPING UNITS

TOTAL NUMBER OF UNITS PROVIDED	MINIMUM REQUIRED NUMBER OF ACCESSIBLE UNITS ASSOCIATED WITH ROLL-IN SHOWERS	TOTAL NUMBER OF REQUIRED ACCESSIBLE UNITS
1 to 25	0	1
26 to 50	0	2
51 to 75	1	4
76 to 100	1	5
101 to 150	2	7
151 to 200	2	8
201 to 300	3	10
301 to 400	4	12
401 to 500	4	13
501 to 1,000	1% of total	3% of total
Over 1,000	10, plus 1 for each 100 over 1,000	30, plus 2 for each 100 over 1,000

to that of condominiums/apartments than hotels, the units are subject to Type B requirements.

1107.6.2.1 Apartment houses, monasteries and convents. Type A and Type B units shall be provided in apartment houses, monasteries and convents in accordance with Sections 1107.6.2.1.1 and 1107.6.2.1.2.

❖ Unless they receive some type of federal funding, apartments and condominiums are not covered by ADAAG. Convents and monasteries are associated with religious organizations, so they also are not covered by ADAAG. Since all these facilities may serve as a person's residence, they are covered by FHAG. A convent or monastery cannot be sued under FHA for limiting its residents; however, it must comply with the FHAG construction requirements for the residential units that it provides. The scope for Type B units is consistent with FHAG requirements.

1107.6.2.1.1 Type A units. In occupancies in Group R-2 containing more than 20 dwelling units or sleeping units, at least 2 percent, but not less than one, of the units shall be a Type A unit. All units on a site shall be considered to determine the total number of units and the required number of Type A units. Type A units shall be dispersed among the various classes of units.

Exceptions:

1. The number of Type A units is permitted to be reduced in accordance with Section 1107.7.

2. Existing structures on a site shall not contribute to the total number of units on a site.

❖ To be able to better accommodate a person who uses a wheelchair in his or her search for housing, apartments and condominiums are required to provide Type A units. Congregate residences, such as convents and monasteries, must also provide Type A units. The housing industry has been concerned that dwelling units equipped with the full range of features to accomplish accessibility will not be marketable to people without disabilities. The Type A requirements are intended to accomplish a middle ground that will satisfy both needs. Allowances made during the construction process for future conversion for accessibility, such as reinforcement in bathroom walls for future installation of grab bars, will allow a unit to be altered later at a considerably lesser cost.

Type A units are required when the site contains more than 20 dwelling units or sleeping units. So that there is a consistent number of Type A units in multibuilding sites based on the size of a development as a whole, all the buildings are added together to determine the number required. For example, a 300-unit building would require six Type A units, and a development with 150 buildings with four units per building would also require six Type A units. Exception 2 is in recognition that a development may be built in stages. Only the units that are being constructed as part of that development phase are considered in determining the number of Type A units.

The Type A units must be dispersed among the classes of units provided. For example, if one-, two- and three-bedroom units are available with the development and two Type A units are required, it is the designer's choice as to which two of the options to provide as Type A units. The designer, however, cannot choose to only provide the one-bedroom option with both Type A units. This is not intended to require Type A units to be provided in different buildings in a multibuilding site. Many times in multibuilding developments, there are shared facilities such as clubhouses or pools. The designer may choose to locate the Type A units in the building closest to those amenities for ease of access for the residents. This is acceptable as long as the dispersion requirement is met.

Exception 1 allows the number of Type A units to be reduced in accordance with Section 1107.7.5 when the building's first-floor elevation is required to be raised due to flood-plain regulations.

1107.6.2.1.2 Type B units. Where there are four or more dwelling units or sleeping units intended to be occupied as a residence in a single structure, every dwelling unit and sleeping unit intended to be occupied as a residence shall be a Type B unit.

Exception: The number of Type B units is permitted to be reduced in accordance with Section 1107.7.

❖ When four or more dwelling or sleeping units are provided in a single structure, those units must meet Type B criteria. Note the use of the term "structure" instead of "building." The criteria is applicable if four dwelling units are built together, regardless of fire walls [see Figures 1107.6.2.1.2(1) and 1107.6.2.1.2(2)]. The exception is a general reference to Section 1107.7, which addresses situations where it is logical to back off on the requirements for Type B units within a structure.

1107.6.2.2 Boarding houses, dormitories, fraternity houses and sorority houses. Accessible units and Type B dwelling units shall be provided in boarding houses, dormitories, fraternity houses and sorority houses in accordance with Sections 1107.6.2.2.1 and 1107.6.2.2.2.

❖ Dormitories are typically located in universities. Since universities typically receive some type of federal funding, their dormitories are covered by ADAAG. Boarding houses, fraternity houses and sorority houses have similar types of living arrangements; therefore, it seems logical that the same anticipated need should result in the same level of required Accessible units. Since all these facilities typically serve as a person's residence, they are covered by FHAG. The scope for Accessible units is consistent with ADAAG requirements for dormitories. The scope for Type B units is consistent with FHAG requirements.

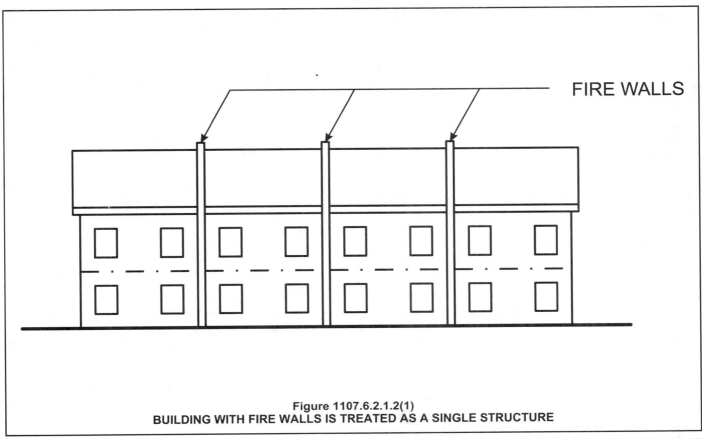

Figure 1107.6.2.1.2(1)
BUILDING WITH FIRE WALLS IS TREATED AS A SINGLE STRUCTURE

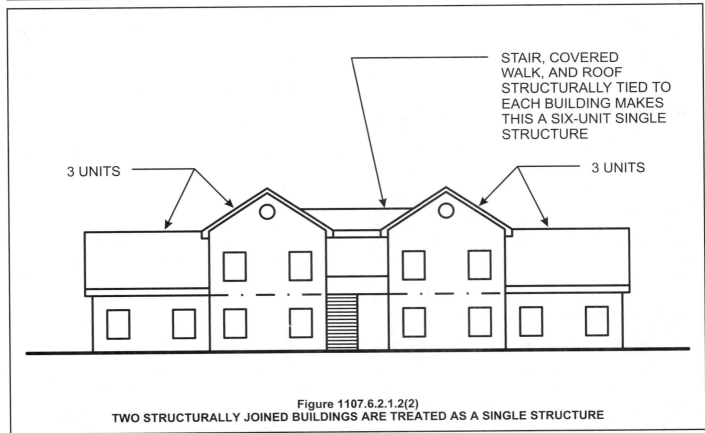

Figure 1107.6.2.1.2(2)
TWO STRUCTURALLY JOINED BUILDINGS ARE TREATED AS A SINGLE STRUCTURE

1107.6.2.2.1 Accessible units. Accessible dwelling units and sleeping units shall be provided in accordance with Table 1107.6.1.1.

❖ The number of Accessible units required in boarding houses, dormitories, fraternity houses and sorority houses is the same as that required in hotels and motels. All associated and shared areas (e.g., bathrooms, kitchens, study rooms) must also be accessible in accordance with ICC A117.1, Chapters 3 through 9. A certain number of the Accessible units must have associated bathrooms equipped with a roll-in shower.

1107.6.2.2.2 Type B units. Where there are four or more dwelling units or sleeping units intended to be occupied as a residence in a single structure, every dwelling unit and every sleeping unit intended to be occupied as a residence shall be a Type B unit.

Exception: The number of Type B units is permitted to be reduced in accordance with Section 1107.7.

❖ When four or more sleeping units are provided in a single structure, those units must meet Type B criteria. The exception is a general reference to Section 1107.7, which addresses situations where it is logical to back off on the requirements for Type B units within a structure.

1107.6.3 Group R-3. In occupancies in Group R-3 where there are four or more dwelling units or sleeping units intended to be occupied as a residence in a single structure, every dwelling and sleeping unit intended to be occupied as a residence shall be a Type B unit.

Exception: The number of Type B units is permitted to be reduced in accordance with Section 1107.7.

❖ When four or more dwelling or sleeping units are provided in a single structure, those units must meet Type B criteria. Note the use of the term "structure" instead of "building." The criteria is applicable if four dwelling units are built together, regardless of fire walls. The exception is a general reference to Section 1107.7, which addresses situations where it is logical to back off on the requirements for Type B units within a structure.

Section 101.2, Exception 1 states that detached one- and two-family townhouses and multiple-family dwellings that are both three stories or less and have an independent means of egress must comply with the *International Residential Code*® (IRC®). In the IRC, townhouses are further defined as extending from foundation to roof and open on at least two sides; therefore, the typical side-by-side townhouse is constructed using the IRC. Any configuration of townhouses that does not meet all four criteria has to be constructed under the IBC as Group R-2 or possibly Group R-3. If the structure is divided into one- or two-dwelling unit buildings with fire walls, the structure is a Group R-3 (see commentary, Section 310). Institutional facilities, other than Group I-3, with five or fewer residents and Group R-4 facilities have the option of complying with the IBC or the IRC, since they often operate similar to a single-family home.

IRC Section 322.1 refers any structure with four or more sleeping units or dwelling units back to the IBC as

Group R-3 buildings for accessibility requirements. Since these types of facilities typically serve as a person's permanent residence, the units must meet Type B criteria. The exception is a general reference to Section 1107.7, which addresses situations where it is logical to back off on the requirements for Type B units within a structure.

1107.6.4 Group R-4. Group R-4 occupancies shall be provided with accessible features in accordance with Sections 1107.6.4.1 and 1107.6.4.2.

❖ Group R-4 facilities are limited to between six and 16 residents. A Group R-4 is basically a small Group I-1 facility. The Group R-4 occupancy was originally developed in response to some lawsuits filed under FHA concerning homes for mentally disabled adults. Typically, a structure where persons live in a supervised environment is a Group I occupancy. In neighborhoods zoned residential only, a variance was required if a group of mentally disabled adults wanted to move into a single-family style dwelling in these neighborhoods. In addition, Group I facilities are typically required to be sprinklered throughout with an NFPA 13 system. This type of sprinkler system is not what is typically used in a single-family home. The 16-resident criteria is based on two things: In the last census, 98 percent of the households in the United States that identified themselves as single family had 16 or fewer residents. The number 16 also happens to be the limit of residents permitted in a building where an NFPA 13D sprinkler system can be utilized. Establishing these types of facilities as part of Group R eliminated potential conflict with the zoning issue. The limit on the number of residents and allowing the alternative sprinkler system addressed the discrimination in housing concerns and provided a reasonable level of sprinkler protection for the residents.

1107.6.4.1 Accessible units. At least one of the dwelling or sleeping units shall be an accessible unit.

❖ The requirement for one sleeping unit and its associated facilities (e.g., bathing room) to meet Accessible unit criteria is consistent with Group I-1 requirements. All common rooms are required to be fully accessible. ICC A117.1, Chapters 3 through 9, provide the technical criteria for Accessible sleeping units.

1107.6.4.2 Type B units. In structures with four or more dwelling or sleeping units intended to be occupied as a residence, every dwelling and sleeping unit intended to be occupied as a residence shall be a Type B unit.

Exception: The number of Type B units is permitted to be reduced in accordance with Section 1107.7.

❖ Since these types of facilities typically serve as a person's permanent residence, the sleeping unit must meet Type B criteria. The exception is a general reference to Section 1107.7, which addresses situations where it is logical to back off on the requirements for Type B units within a structure. Since this a congregate residence, this section would be applicable when four or more sleeping units are provided.

1107.7 General exceptions. Where specifically permitted by Section 1107.5 or 1107.6, the required number of Type A and Type B units is permitted to be reduced in accordance with Sections 1107.7.1 through 1107.7.5.

❖ Section 1107.5 or 1107.6 establish when Accessible, Type A or Type B units are expected to be provided. Section 1107.7 covers the general exceptions where it is reasonable and logical to not provide accessibility for some of the units. This would be consistent with the general provisions of Section 1103. Code users should start out with the assumption that everything, in this case Group I and R dwelling and sleeping units, is required to be accessible, and then back off from that level of accessibility when specific exceptions are indicated. Note that there are no exceptions for Accessible units. While Type A units are mentioned in Section 1107.7.1, the only exception for Type A units is found in Section 1107.7.5.

Sections 1107.7.1 through 1107.7.4 primarily deal with nonelevator buildings. If elevators are provided, except as addressed in Section 1107.7.3, all units in the building are required to meet Type B criteria or better. Providing access to the upper floors in these cases would require either an elevator or a ramp system (or multiple elevators or ramps), both of which would be unreasonable. These exceptions are intended to be consistent with the scope of FHAG. It should be noted that these are only exceptions to the requirements for Type B dwelling units.

1107.7.1 Buildings without elevator service. Where no elevator service is provided in a building, only the dwelling and sleeping units that are located on stories indicated in Sections 1107.7.1.1 and 1107.7.1.2 are required to be Type A and Type B units. The number of Type A units shall be determined in accordance with Section 1107.6.2.1.1.

❖ Only the units located on the floor levels defined in Sections 1107.7.1.1 and 1107.7.1.2 are required to contain Type A or Type B units when no elevator service is provided in the building. Floor levels that do not meet the criteria are not required to have an accessible route to that level, nor do the dwelling units or sleeping units on those levels have to meet Type A or Type B unit criteria. The building must still have the minimum number of Type A units required on the first floor.

1107.7.1.1 One story with Type B units required. At least one story containing dwelling units or sleeping units intended to be occupied as a residence shall be provided with an accessible entrance from the exterior of the building and all units intended to be occupied as a residence on that story shall be Type B units.

❖ This section basically states that Type B dwelling units or sleeping units must be provided on at least one level of a building without elevator service. For example, on a flat site, a two-story structure with Type B dwelling units or sleeping units on the first floor would not require an accessible route to or Type B units on the second floor (see Figure 1107.7.1.1).

1107.7.1.2 Additional stories with Type B units. On all other stories that have a building entrance in proximity to arrival points intended to serve units on that story, as indicated in Items 1 and 2, all dwelling units and sleeping units intended to be occupied as a residence served by that entrance on that story shall be Type B units.

1. Where the slopes of the undisturbed site measured between the planned entrance and all vehicular or pedestrian

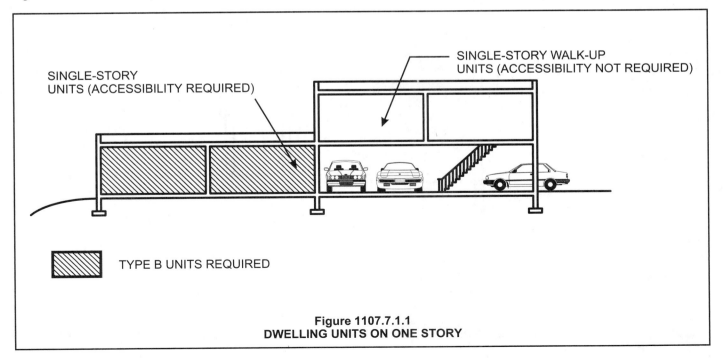

**Figure 1107.7.1.1
DWELLING UNITS ON ONE STORY**

FIGURE 1107.7.1.2(1) – 1107.7.2   ACCESSIBILITY

arrival points within 50 feet of the planned entrance are 10 percent or less, and

2. Where the slopes of the planned finished grade measured between the entrance and all vehicular or pedestrian arrival points within 50 feet of the planned entrance are 10 percent or less.

Where no such arrival points are within 50 feet (15 240 mm) of the entrance, the closest arrival point shall be used unless that arrival point serves the story required by Section 1107.7.1.1.

❖ This section addresses the idea that a building could have two levels that would be provided with accessible routes. This could be a structure built into a hill or a building with the first level a few feet down and the second level a few feet up from grade level [see Figure 1107.7.1.2(1)].

The basic test for multiple accessible levels is to check for the slope between the building entrance and any arrival points within a 50-foot (15 240 mm) arc. If no arrival points (e.g., sidewalk, parking) are within that 50 feet (15 240 mm), the closest arrival point should be used as a reference. If the slope between these two points is 10 percent or less both before and after grading of the site, than an accessible route is required to that level [see Figures 1107.7.1.2(2) and 1107.7.1.2(3)].

In the case of sidewalks, the closest point to the entrance will be where a public sidewalk entering the site intersects with the sidewalk to the entrance. In the case of resident parking areas, the closest point to the planned entrance will be measured from the entry point to the parking area that is located closest to the planned entrance. The measurement for elevation should be taken from the center of the entrance door to the top of the pavement at the arrival point.

1107.7.2 Multistory units. A multistory dwelling or sleeping unit which is not provided with elevator service is not required to be a Type B unit. Where a multistory unit is provided with external elevator service to only one floor, the floor provided with elevator service shall be the primary entry to the unit, shall comply with the requirements for a Type B unit and a toilet facility shall be provided on that floor.

❖ Section 1107.7.2 addresses multistory dwelling units (see the definitions in Section 1102.1). For example, the typical townhouse scenario, where you have a series of two-story units adjacent to each other in a single structure, would not require Type B units. If multistory units are provided in a building containing elevators and access is provided at only one floor level, that level must be accessible and contain a toilet facility [see Figures 1107.7.2(1) and 1107.7.2(2)].

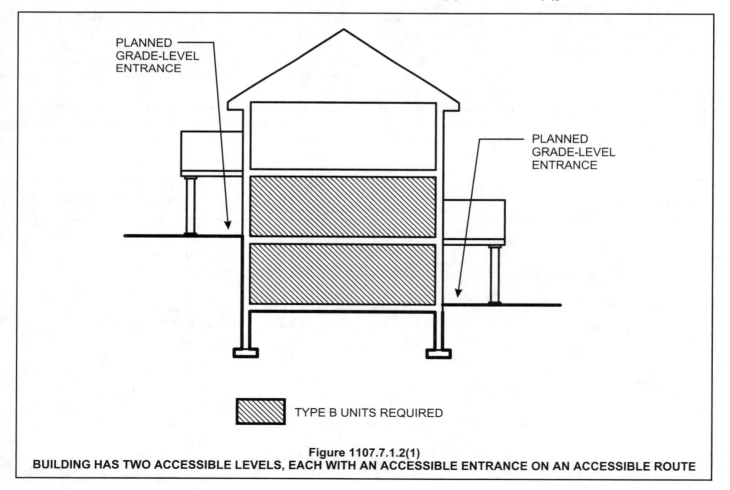

PLANNED GRADE-LEVEL ENTRANCE

PLANNED GRADE-LEVEL ENTRANCE

[hatched] TYPE B UNITS REQUIRED

Figure 1107.7.1.2(1)
BUILDING HAS TWO ACCESSIBLE LEVELS, EACH WITH AN ACCESSIBLE ENTRANCE ON AN ACCESSIBLE ROUTE

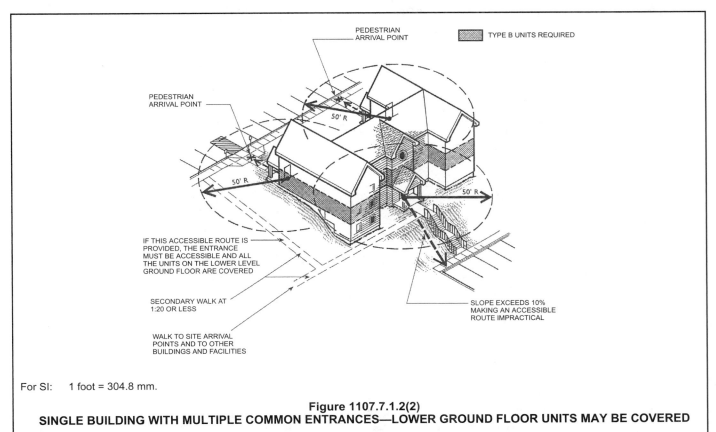

For SI: 1 foot = 304.8 mm.

Figure 1107.7.1.2(2)
SINGLE BUILDING WITH MULTIPLE COMMON ENTRANCES—LOWER GROUND FLOOR UNITS MAY BE COVERED

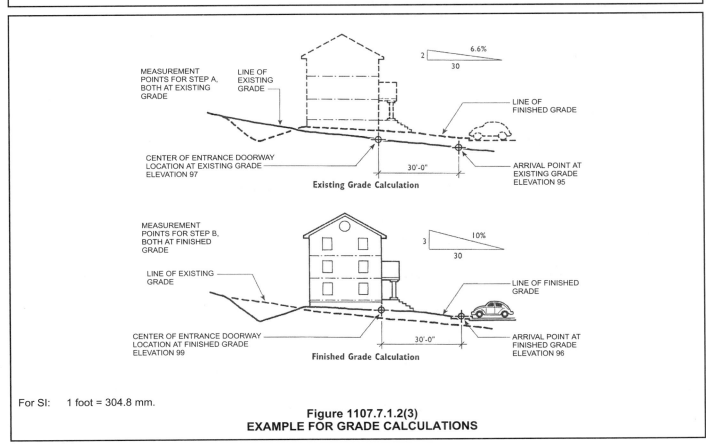

For SI: 1 foot = 304.8 mm.

Figure 1107.7.1.2(3)
EXAMPLE FOR GRADE CALCULATIONS

FIGURE 1107.7.2(1) – FIGURE 1107.7.2(2)

ACCESSIBILITY

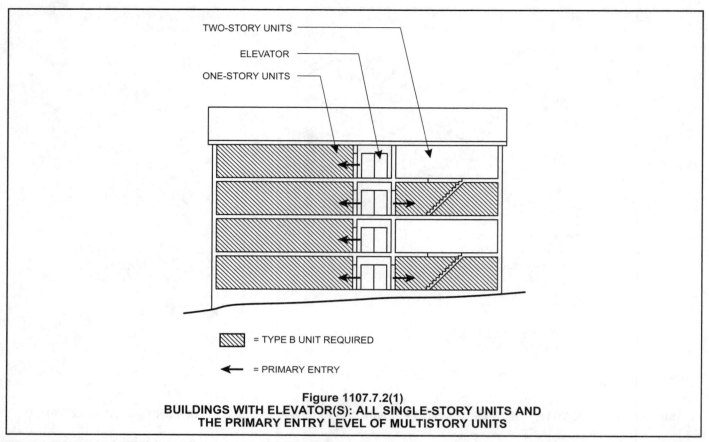

Figure 1107.7.2(1)
BUILDINGS WITH ELEVATOR(S): ALL SINGLE-STORY UNITS AND
THE PRIMARY ENTRY LEVEL OF MULTISTORY UNITS

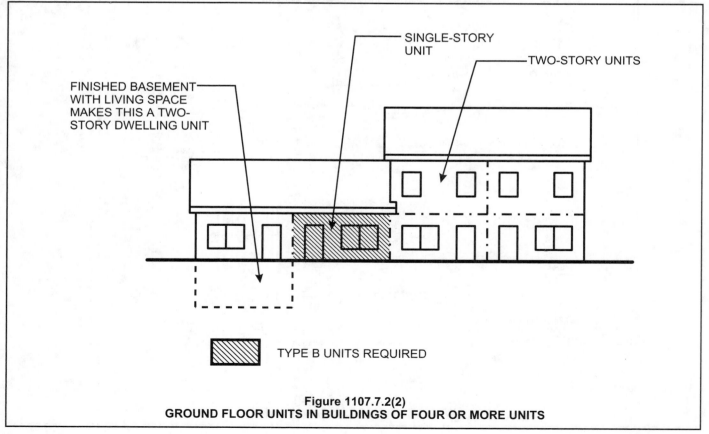

Figure 1107.7.2(2)
GROUND FLOOR UNITS IN BUILDINGS OF FOUR OR MORE UNITS

1107.7.3 Elevator service to the lowest story with units. Where elevator service in the building provides an accessible route only to the lowest story containing dwelling or sleeping units intended to be occupied as a residence, only the units on that story which are intended to be occupied as a residence are required to be Type B units.

❖ Section 1107.7.3 exempts dwelling units located on upper floors as long as the dwelling units on the first floor containing dwelling units are at least Type B units. For example, a three-story structure has a business on the first floor with apartments on the second and third levels. An accessible route is required to the second level, and all dwelling units on that level are required to be Type B units; however, an accessible route and Type B units are not required to the third level. (Note: If an elevator is utilized to provide access to only the lowest level containing dwelling or sleeping units, this building would not be considered an "elevator" building) [see Figures 1107.7.3(1) and 1107.7.3(2)].

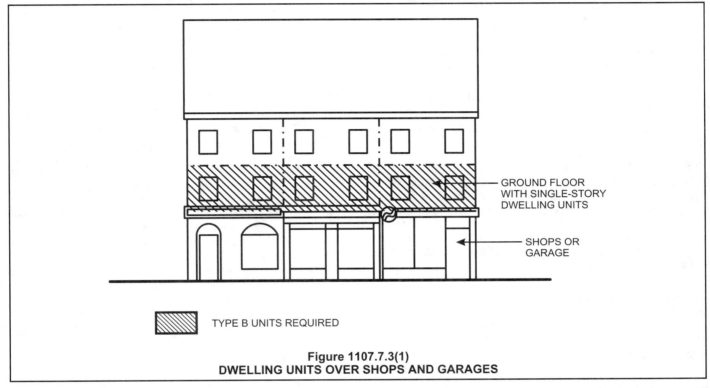

TYPE B UNITS REQUIRED

Figure 1107.7.3(1)
DWELLING UNITS OVER SHOPS AND GARAGES

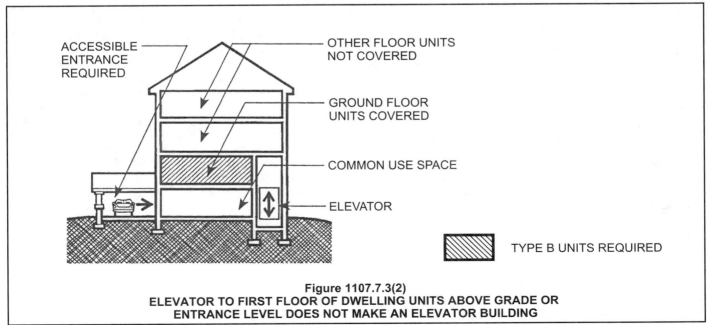

TYPE B UNITS REQUIRED

Figure 1107.7.3(2)
ELEVATOR TO FIRST FLOOR OF DWELLING UNITS ABOVE GRADE OR
ENTRANCE LEVEL DOES NOT MAKE AN ELEVATOR BUILDING

1107.7.4 Site impracticality. On a site with multiple nonelevator buildings, the number of units required by Section 1107.7.1 to be Type B units is permitted to be reduced to a percentage which is equal to the percentage of the entire site having grades, prior to development, which are less than 10 percent, provided that all of the following conditions are met:

1. Not less than 20 percent of the units required by Section 1107.7.1 on the site are Type B units;

2. Units required by Section 1107.7.1, where the slope between the building entrance serving the units on that story and a pedestrian or vehicular arrival point is no greater than 8.33 percent, are Type B units;

3. Units required by Section 1107.7.1, where an elevated walkway is planned between a building entrance serving the units on that story and a pedestrian or vehicular arrival point and the slope between them is 10 percent or less are Type B units; and

4. Units served by an elevator in accordance with Section 1107.7.3 are Type B units.

❖ Section 1107.7.4 addresses multiple buildings on a sloping site. For example, if an apartment complex was built on a steep or hilly site, the number of Type B dwelling units required on the ground floor could be reduced due to the difficulty of providing accessible routes. A minimum of 20 percent of the total ground floor units on the site must be Type B units, regardless of site complications (see Figure 1107.7.4).

1107.7.5 Design flood elevation. The required number of Type A and Type B units shall not apply to a site where the lowest floor or the lowest structural building members of nonelevator buildings are required to be at or above the design flood elevation resulting in:

1. A difference in elevation between the minimum required floor elevation at the primary entrances and vehicular and pedestrian arrival points within 50 feet (15 240 mm) exceeding 30 inches (762 mm), and

2. A slope exceeding 10 percent between the minimum required floor elevation at the primary entrances and vehicular and pedestrian arrival points within 50 feet (15 240 mm).

Where no such arrival points are within 50 feet (15 240 mm) of the primary entrances, the closest arrival point shall be used.

❖ Section 1107.7.5 is intended to address residential construction located in areas prone to flooding. These structures typically must be raised above the design flood elevation. If the criteria in either Item 1 or 2 are met, it is considered that an accessible route to the units is not feasible. This exception includes both Type A and Type B dwelling units [see Figures 1107.7.5(1) and 1107.7.5(2)].

SECTION 1108
SPECIAL OCCUPANCIES

1108.1 General. In addition to the other requirements of this chapter, the requirements of Sections 1108.2 through 1108.4 shall apply to specific occupancies.

❖ The criteria provided herein are occupancy specific, and are intended to result in a reasonable level of accessibility in areas with assembly seating, self-service storage facilities and judicial facilities.

1108.2 Assembly area seating. Assembly areas with fixed seating shall comply with Sections 1108.2.1 through 1108.2.8. Dining areas shall comply with Section 1108.2.9.

❖ Sections 1108.2.1 through 1108.2.8 specifically address facilities with fixed seating utilized for purposes of viewing an event, typically facilities and spaces of occupancies of Groups A-1, A-3, A-4 and A-5. These criteria would also be applicable in assembly-type spaces with fixed seats that are located in buildings of other occupancies (see Section 1109.10). The requirement for access to services; number and dispersion of wheelchair spaces; requirements for companion seats; designated aisle seats; assistive listening systems and performance areas are addressed. The requirements for seating in areas for eating or drinking, typically Group A-2 spaces or facilities, and the dispersion of this seating are addressed in Section 1108.2.9.

1108.2.1 Services. Services and facilities provided in areas not required to be accessible shall be provided on an accessible level and shall be accessible.

❖ This section establishes an important concept. The intent is that all of the services provided by a facility must be accessible. For example, a stadium may have a particular section of seating that is not accessible (assuming that the required number and dispersion of spaces is in compliance with the code). This section establishes that if a souvenir stand is located in or about that section, one of three circumstances would apply: an additional souvenir stand must be provided in the facility on an accessible level, the souvenir stand would have to be relocated to an accessible level or the section where the souvenir stand is located must be made accessible. This section also emphasizes that not only must that service or facility be located on an accessible level, but the service itself must also be accessible. Obviously, it would be inappropriate to locate a souvenir stand on an accessible level if the approach or entrance to that area is inaccessible.

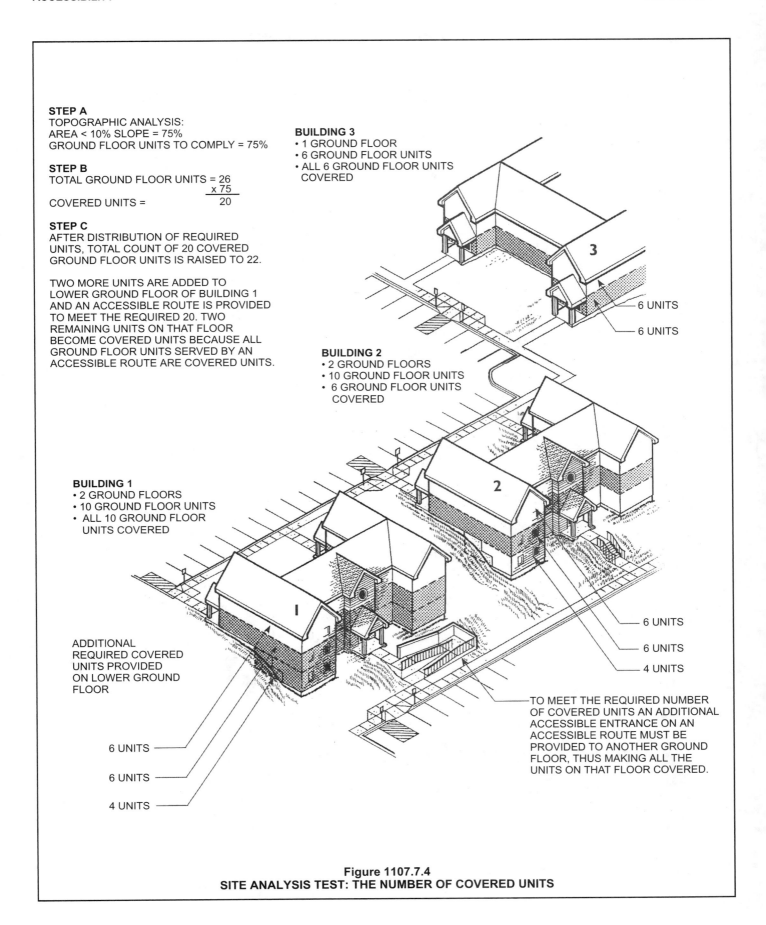

STEP A
TOPOGRAPHIC ANALYSIS:
AREA < 10% SLOPE = 75%
GROUND FLOOR UNITS TO COMPLY = 75%

STEP B
TOTAL GROUND FLOOR UNITS = 26
$\times$ 75
COVERED UNITS = 20

STEP C
AFTER DISTRIBUTION OF REQUIRED
UNITS, TOTAL COUNT OF 20 COVERED
GROUND FLOOR UNITS IS RAISED TO 22.

TWO MORE UNITS ARE ADDED TO
LOWER GROUND FLOOR OF BUILDING 1
AND AN ACCESSIBLE ROUTE IS PROVIDED
TO MEET THE REQUIRED 20. TWO
REMAINING UNITS ON THAT FLOOR
BECOME COVERED UNITS BECAUSE ALL
GROUND FLOOR UNITS SERVED BY AN
ACCESSIBLE ROUTE ARE COVERED UNITS.

BUILDING 3
• 1 GROUND FLOOR
• 6 GROUND FLOOR UNITS
• ALL 6 GROUND FLOOR UNITS
  COVERED

6 UNITS

6 UNITS

BUILDING 2
• 2 GROUND FLOORS
• 10 GROUND FLOOR UNITS
• 6 GROUND FLOOR UNITS
  COVERED

BUILDING 1
• 2 GROUND FLOORS
• 10 GROUND FLOOR UNITS
• ALL 10 GROUND FLOOR
  UNITS COVERED

ADDITIONAL
REQUIRED COVERED
UNITS PROVIDED
ON LOWER GROUND
FLOOR

6 UNITS

6 UNITS

4 UNITS

6 UNITS

6 UNITS

4 UNITS

TO MEET THE REQUIRED NUMBER
OF COVERED UNITS AN ADDITIONAL
ACCESSIBLE ENTRANCE ON AN
ACCESSIBLE ROUTE MUST BE
PROVIDED TO ANOTHER GROUND
FLOOR, THUS MAKING ALL THE
UNITS ON THAT FLOOR COVERED.

Figure 1107.7.4
SITE ANALYSIS TEST: THE NUMBER OF COVERED UNITS

FIGURE 1107.7.5(1) – FIGURE 1107.7.5(2)

ACCESSIBILITY

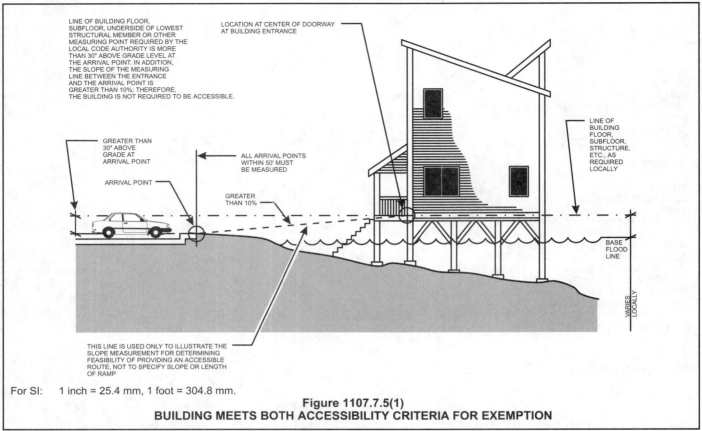

For SI: 1 inch = 25.4 mm, 1 foot = 304.8 mm.

Figure 1107.7.5(1)
BUILDING MEETS BOTH ACCESSIBILITY CRITERIA FOR EXEMPTION

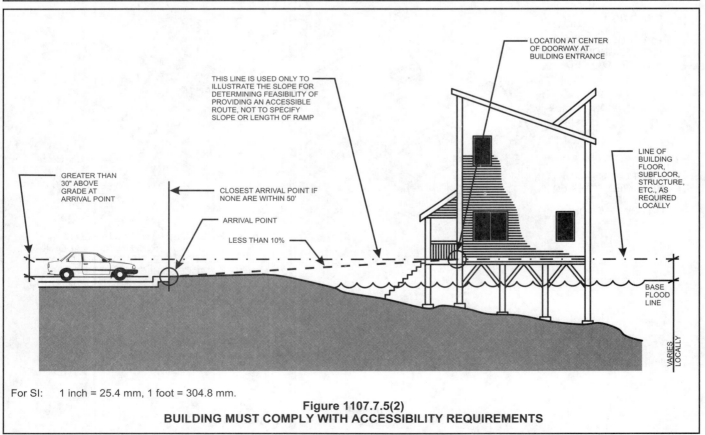

For SI: 1 inch = 25.4 mm, 1 foot = 304.8 mm.

Figure 1107.7.5(2)
BUILDING MUST COMPLY WITH ACCESSIBILITY REQUIREMENTS

1108.2.2 Wheelchair spaces. In theaters, bleachers, grandstands, stadiums, arenas and other fixed seating assembly areas, accessible wheelchair spaces complying with ICC A117.1 shall be provided in accordance with Sections 1108.2.2.1 through 1108.2.2.5.

❖ The intent of this section is to provide a reasonable number of spaces in an assembly occupancy with fixed seating to accommodate persons who use wheelchairs. Demographic statistics from the National Center for Health Statistics on the number of noninstitutionalized Americans who use wheelchairs indicate that these requirements are realistic. These required wheelchair spaces consist of an open, available floor space in which the wheelchair takes the place of the fixed seat that would otherwise occupy that space. Companion seating is required adjacent to each wheelchair space so that friends and family can enjoy the activity presented with the person utilizing the wheelchair space (see Section 1108.2.5).

1108.2.2.1 General seating. Wheelchair spaces shall be provided in accordance with Table 1108.2.2.1

❖ The number of required wheelchair spaces is indicated in Table 1108.2.2.1. and is based on the total number of fixed seats set up to view the same event, with the exception of box seats covered in Sections 1108.2.2.2 and 1108.2.2.3. For example, the fixed seating on all levels and all types provided in a sports stadium, excluding box seats, is used to calculate the number of required wheelchair spaces for the general seating. The requirements for box- or suite-type seating are calculated separately. Seating provided in different rooms, such as in a series of lecture halls in a university, must be calculated separately. Note that the percentage of wheelchair spaces is less for facilities with a capacity of more than 5,000. The industry has been able to provide statistics to show that the higher percentage of wheelchair spaces is not typically utilized in large facilities.

TABLE 1108.2.2.1
ACCESSIBLE WHEELCHAIR SPACES

CAPACITY OF SEATING IN ASSEMBLY AREAS	MINIMUM REQUIRED NUMBER OF WHEELCHAIR SPACES
4 to 25	1
26 to 50	2
51 to 100	4
101 to 300	5
301 to 500	6
501 to 5,000	6, plus 1 for each 150, or fraction thereof, between 501 through 5,000
5,001 and over	36 plus 1 for each 200, or fraction thereof, over 5,000

❖ This table sets forth the required number of wheelchair spaces based on the capacity of seating in the space containing the fixed seating. Any fixed seating with less than four seats is not required to provide wheelchair

spaces. There is special criteria for wheelchair spaces in facilities with a seating capacity of over 500 and over 5,000. For example, six wheelchair spaces would be required for an assembly space with a seating capacity of 500. Seven wheelchair spaces would be required for a seating capacity of 501 through 650. Both of the last two rows use the term "or fraction thereof" to designate the next step up.

1108.2.2.2 Luxury boxes, club boxes and suites. In each luxury box, club box, and suite within arenas, stadiums and grandstands, wheelchair spaces shall be provided in accordance with Table 1108.2.2.1.

❖ When luxury boxes, club boxes or suites are provided, each luxury box, club box or suite must have an accessible route to that box and at least one wheelchair space and associated companion space. If the luxury box, club box or suite has more than 25 seats, the number of required wheelchair spaces is increased in accordance with Table 1108.2.2.1. For example, if a stadium has three luxury boxes with 51 seats in each, at least four wheelchair spaces and their associated companion seats must be provided in each box. While the total of all three luxury boxes (i.e., 153 occupants) would only require five total wheelchair spaces, these specific types of boxes must be calculated individually.

1108.2.2.3 Other boxes. In boxes other than those required to comply with Section 1108.2.2.2, the total number of wheelchair spaces provided shall be determined in accordance with Table 1108.2.2.1. Wheelchair spaces shall be located in not less than 20 percent of all boxes provided.

❖ When boxes other than luxury boxes, club boxes or suites are provided, at least 20 percent of the boxes must have an accessible route to that box and at least one wheelchair space and associated companion space in that box. The total number of box seats is utilized to calculate the total number of required wheelchair spaces in accordance with Table 1108.2.2.1. For example, if a stadium has five boxes with 51 seats in each, the total of all five boxes (i.e., 255 occupants) would require five total wheelchair spaces. The wheelchair spaces could be located in one box (i.e., 20 percent minimum) or dispersed to all five boxes. Note that there is additional dispersion criteria for assembly seating based on lines of sight in ICC A117.1, Section 802.8 as well as the integration of wheelchair spaces indicated in Section 1108.2.3 and dispersion by level in Section 1108.2.4.

1108.2.3 Integration. Wheelchair spaces shall be an integral part of the seating plan.

❖ Wheelchair spaces must be integrated into the general seating plan as much as possible. This requirement is interpretive since the issue for line of sight over standing spectators may require raising the area where the wheelchair seating is provided. The 2003 edition of ICC

A117.1 addresses integration and dispersion require-ments to a much greater extent than the 1998 edition.

1108.2.4 Dispersion of wheelchair spaces. Dispersion of wheelchair spaces shall be based on the availability of accessible routes to various seating areas including seating at various levels in multilevel facilities.

❖ The availability of an accessible route to seating areas must be considered in the decisions made for dispersion of wheelchair spaces. It is not logical to provide wheelchair spaces on a level that is not accessible (see commentary, Sections 1108.2.4.1 and 1109.7). Dispersion for a comparable line of sight is a requirement in the 1998 edition of the ICC A117.1. The 2003 edition of ICC A117.1 also addresses criteria for dispersion based on lines of sight; dispersion by locations, groups or spaces; distance away from an event; dispersion across the seating from side to side; and by seating types or amenities.

1108.2.4.1 Multilevel assembly seating areas. In multilevel assembly seating areas, wheelchair spaces shall be provided on the main floor level and on one of each two additional floor or mezzanine levels. Wheelchair spaces shall be provided in each luxury box, club box and suite within assembly facilities.

Exceptions:

1. In multilevel assembly spaces utilized for worship services, where the second floor or mezzanine level contains 25 percent or less of the total seating capacity, wheelchair spaces shall be permitted to all be located on the main level.

2. In multilevel assembly seating where the second floor or mezzanine level provides 25 percent or less of the total seating capacity and 300 or fewer seats, wheelchair spaces shall be permitted to all be located on the main level.

❖ When facilities provide seating on multiple levels, wheelchair spaces must also be provided on multiple levels. For example, if a theater has a main level and a balcony level, wheelchair spaces must be provided on both. If a theater has a main level and two balcony levels, wheelchair spaces must be provided on the main level and at least one of the two balcony levels. While the term "level" is difficult to define in some assembly arrangements, the intent is to provide the wheelchair user with a choice of seating similar to what is available to all persons in the space. A "level" is not each time that there is a step-down to another row of seating. Subjective judgement on the part of both the designer and the building official is required for each individual facility.

There are two specific exceptions to this multilevel dispersion requirement. Exception 1 is applicable to assembly spaces utilized for worship services. There is typically not a cost or line-of-sight issue in this type of facility, as there is in other assembly seating areas. The balcony is not required to contain accessible seating or have an accessible route, as long as accessible seating is provided on the main level and the balcony does not contain more than 25 percent of the seating. Exception 2 is applicable to balconies in assembly facilities with a limited number of seats in the upper level. If accessible seating is provided on the main level and the balcony contains less than 25 percent of the seating (and less than 300 seats), then the balcony is not required to contain accessible seating or have an accessible route.

1108.2.5 Companion seats. At least one companion seat complying with ICC A117.1 shall be provided for each wheelchair space required by Section 1108.2.2.

❖ The purpose of a companion seat is to allow someone to accompany a person using a wheelchair to an event and to share in the experience the same as others attending the event. ICC A117.1 permits two wheelchair spaces to be located next to each other; however, each wheelchair space must also be adjacent to a companion seat.

1108.2.6 Designated aisle seats. At least five percent, but not less than one, of the total number of aisle seats provided shall be designated aisle seats.

❖ The purpose for a designated aisle seat is for persons with mobility impairments to be able to easily access their seat. Due to their impairment, it may be difficult or impossible for them to move further into the row. These seats may be used by persons using wheelchairs who wish to transfer to a seat for the event. The 2003 edition of ICC A117.1 requires designated aisle seats to have folding or retractable armrests when they are provided for adjacent seating. Note that while 5 percent of the aisle seats are required to be designated aisle seats, there is not a requirement for them to be located on an accessible route. It would be good design practice to locate the designated aisle seats as close to an accessible route as possible. A person using a walker may want to use a designated aisle seat since he or she would have almost as much difficulty with steps as a person in a wheelchair.

1108.2.7 Assistive listening systems. Each assembly area where audible communications are integral to the use of the space shall have an assistive listening system.

Exception: Other than in courtrooms, an assistive listening system is not required where there is no audio amplification system.

❖ This section is intended to accommodate people with a hearing impairment. Assembly occupancies, such as stadiums, theaters, auditoriums, lecture halls and similar spaces, are required to make provisions for the use of assistive listening devices in accordance with Section 1108.2.7.1 and Table 1108.2.7.1. In these assembly areas, audible communication is often integral to the use and full enjoyment of the space. This requirement offers the possibility for individuals with hearing impairments to attend functions in these facilities without having to give advance notice and without disrupting the event in order to have a portable assistive listening system set up and made ready for use.

There are three primary types of listening systems available: induction loop, AM/FM and infrared. Each type of system has certain advantages and disadvantages that the designer should take into consideration when choosing the type of system that is most appropriate for the intended application. Signage notifying the general public of the availability of these systems must be provided in accordance with Section 1110.3.

If an audio amplification system is not provided within a facility, an assistive listening system is not required. Given the essential nature of the proceedings, this exception is not applicable to courtrooms.

1108.2.7.1 Receivers. Receivers shall be provided for assistive listening systems in accordance with Table 1108.2.7.1.

> **Exception:** Where a building contains more than one assembly area, the total number of required receivers shall be permitted to be calculated according to the total number of seats in the assembly areas in the building provided that all receivers are usable with all systems, and if assembly areas required to provide assistive listening are under one management.

❖ Table 1108.2.7.1 specifies the number of required receivers in an assembly occupancy. If a facility has more than one assembly area with an audio amplification system, such as a multiplex theater, then the total seating for all the spaces may be used to determine the number of receivers required.

Table 1108.2.7.1. See below.

❖ Table 1108.2.7.1 specifies the number of required receivers in an assembly occupancy. At least 25 percent of the receivers provided (but not less than two receivers) must be hearing-aid compatible, as persons with hearing aids typically cannot use earpieces or headphone-equipped receivers. When determining the general number of receivers required, the table is to be applied as though it read "2 plus 1 for each additional 25 over 50, or a fraction thereof." For example, where 51 to 75 total seats are provided, three assistive listening devices are required. Where 76 to 100 total seats are provided, four assistive listening devices are required.

The number of required receivers is based on the capacity of the assembly areas. Note that as the size of the assembly area increases, the number of required receivers also increases; however, the percentage of receivers to seats decreases. This is based on the actual usage in large assembly areas.

1108.2.7.2 Public address systems. Where stadiums, arenas and grandstands provide audible public announcements, they shall also provide equivalent text information regarding events and facilities in compliance with Sections 1108.2.7.2.1 and 1108.2.7.2.2.

❖ If stadiums, arenas or grandstands provide public announcements, the same information should be displayed on some type of electronic signage. Most stadiums, arenas and grandstands have electronic scoreboards that are capable of display text messages. If electronic signage is not provided, compliance with Sections 1108.2.7.2.1 and 1108.2.7.2.2 is not required.

1108.2.7.2.1 Prerecorded text messages. Where electronic signs are provided and have the capability to display prerecorded text messages containing information that is the same, or substantially equivalent, to information that is provided audibly, signs shall display text that is equivalent to audible announcements.

> **Exception:** Announcements that cannot be prerecorded in advance of the event shall not be required to be displayed.

❖ If prerecorded messages are part of the event, the same information should be displayed in a text format. Text display is not required for announcements that are not prerecorded. Note that text information is only required when electronic signage is provided.

1108.2.7.2.2 Real-time messages. Where electronic signs are provided and have the capability to display real-time messages containing information that is the same, or substantially equivalent, to information that is provided audibly, signs shall display text that is equivalent to audible announcements.

❖ If the electronic signage in the facility is capable of displaying real-time messages, then the same information being provided to the general audience through audible means should also be displayed in text. Note that text

TABLE 1108.2.7.1
RECEIVERS FOR ASSISTIVE LISTENING SYSTEMS

CAPACITY OF SEATING IN ASSEMBLY AREAS	MINIMUM REQUIRED NUMBER OF RECEIVERS	MINIMUM NUMBER OF RECEIVERS TO BE HEARING-AID COMPATIBLE
50 or less	2	2
51 to 200	2, plus 1 per 25 seats over 50 seats*	2
201 to 500	2, plus 1 per 25 seats over 50 seats.*	1 per 4 receivers*
501 to 1,000	20, plus 1 per 33 seats over 500 seats*	1 per 4 receivers*
1,001 to 2,000	35, plus 1 per 50 seats over 1,000 seats*	1 per 4 receivers*
Over 2,000	55, plus 1 per 100 seats over 2,000 seats*	1 per 4 receivers*

NOTE: * = or fraction thereof

information is only required when electronic signage is provided.

1108.2.8 Performance areas. An accessible route shall directly connect the performance area to the assembly seating area where a circulation path directly connects a performance area to an assembly seating area. An accessible route shall be provided from performance areas to ancillary areas or facilities used by performers.

❖ Performance areas, such as stages, orchestra pits, band platforms, choir lofts and similar spaces, must be accessible. If there is a direct route from the seating to the performance area, there must also be an accessible route. For example, if steps are provided from the assembly seating area to the stage within the theater, then an accessible route (e.g., ramp or platform lift) to the stage must also be provided within the theater. An accessible route must also be provided to any ancillary areas, such as greenrooms or practice/warm-up rooms. The intent is that a person with mobility impairments could participate in the event. This could include high school graduation with students coming from the audience up onto the stage to receive their diplomas; participating with the community band; playing in the orchestra for a performance; acting in a production or giving a speech.

1108.2.9 Dining areas. In dining areas, the total floor area allotted for seating and tables shall be accessible.

Exceptions:

1. In buildings or facilities not required to provide an accessible route between levels, an accessible route to a mezzanine seating area is not required, provided that the mezzanine contains less than 25 percent of the total area and the same services are provided in the accessible area.

2. In sports facilities, tiered dining areas providing seating required to be accessible shall be required to have accessible routes serving at least 25 percent of the dining area, provided that accessible routes serve accessible seating and where each tier is provided with the same services.

❖ Dining areas most frequently occur in Group A-2 (restaurants, cafeterias, portions of nightclubs, dinner theaters, etc.). The provisions of this section are intended to govern such areas, rather than the criteria specified in Section 1108.2. This section requires the total floor area allotted for seating and tables to be accessible, with two exceptions. Exception 1 is intended to acknowledge a practical and reasonable limitation in buildings with a mezzanine level that is not required to be served by an accessible route in accordance with Section 1104.4 [i.e., greater than 3,000 square feet (279 m²)]. This exception exempts mezzanine areas that represent a relatively small portion of the total dining area of the facility on the condition that all services provided by the facility are available in an accessible area. Any mezzanine condition that does not meet the criteria of this exception, for example, a mezzanine area that contains 25 percent or more of the total dining area or any mezzanine in a building that is required to be served by an accessible route, does not qualify for the exception and must be accessible.

Exception 2 is for sports facilities. Some sports facilities also have accommodations for dining or picnicking while watching the event. For line-of-sight issues, the dining terraces are tiered. If the same services are available on the accessible level as any other level, only 25 percent of the dining area is required to be on an accessible route. Section 1109.7 allows platform lifts to provide access to these levels when the sporting event and dining facilities are outdoors, such as at a baseball park.

1108.2.9.1 Dining surfaces. Where dining surfaces for the consumption of food or drink are provided, at least 5 percent, but not less than one, of the seating and standing spaces at the dining surfaces shall be accessible and be distributed throughout the facility.

❖ Section 1108.2.9.1 establishes the criteria for the percentage of tables, booths, bars and counters that will be used for eating or drinking that must be accessible. This criteria is consistent with Section 1109.11 for the required percentage of accessible built-in surfaces in all other occupancies. The required accessible surfaces are also required to be distributed throughout the facility such that a comparable choice of locations and types (i.e., tables, booths, counters, etc.) is available. This requirement, in conjunction with Section 1108.2.9, provides a reasonable and appropriate degree of accessibility throughout dining areas. The result is that a person with a mobility impairment will be able to approach, enter and move about in virtually all portions of a dining area. The entire dining or drinking area must be accessible. In addition, 5 percent of the total surfaces provided must be accessible. The issue of whether a portion of the bar in a restaurant is required to be accessible is subjective. The assumption is that if other types of seating are provided adjacent to the counter, then services provided at the counter will also be available at the adjacent seating; therefore, if adequate accessible seating is available adjacent to the bar area, the bar is not required to be lowered. If the bar is the only eating or dining surface, however, then a portion of the bar must be made accessible.

1108.3 Self-service storage facilities. Self-service storage facilities shall provide accessible individual self-storage spaces in accordance with Table 1108.3.

❖ This section addresses facilities that provide self-storage units or spaces. The intent is to provide access for persons with disabilities to this service without requiring the entire facility to be accessible.

TABLE 1108.3
ACCESSIBLE SELF-SERVICE STORAGE FACILITIES

TOTAL SPACES IN FACILITY	MINIMUM NUMBER OF REQUIRED ACCESSIBLE SPACES
1 to 200	5%, but not less than 1
Over 200	10, plus 2% of total number of units over 200

❖ The minimum number of accessible spaces is based on the total number of self-storage spaces available in a facility.

1108.3.1 Dispersion. Accessible individual self-service storage spaces shall be dispersed throughout the various classes of spaces provided. Where more classes of spaces are provided than the number of required accessible spaces, the number of accessible spaces shall not be required to exceed that required by Table 1108.3. Accessible spaces are permitted to be dispersed in a single building of a multibuilding facility.

❖ Self-storage facilities may offer a variety of spaces, such as heated/nonheated, different sizes, etc. If the variety offered is greater than the number of accessible spaces required, the accessible spaces should be dispersed as much as possible. For example, if a facility offers the choice of a 10-foot by 10-foot (3048 mm by 3048 mm) unit and a 10-foot by 20-foot (3048 mm by 6096 mm) unit, but only one accessible space is required, only one unit is required to be accessible. The choice of which unit is made accessible is up to the building owner or designer. When a facility has multiple buildings, accessible spaces are permitted to be located in a single building.

1108.4 Judicial facilities. Judicial facilities shall comply with Sections 1108.4.1 through 1108.4.3.

❖ Accessibility in judicial facilities (e.g., courthouses) includes access to all public areas as well as special requirements for the courtrooms, holding cells and visitation areas. This section was added for coordination with new requirements for judicial facilities from the DOJ.

1108.4.1 Courtrooms. Each courtroom shall be accessible.

❖ Every courtroom must be fully accessible, including any raised areas, such as the judge's bench, clerk's station, jury boxes, witness stands and any audience seating (see Sections 1108.2.2, 1109.7 and 1109.10). There are also requirements for assistive listening systems (see Section 1108.2.7).

1108.4.2 Holding cells. Where provided, central holding cells and court-floor holding cells shall comply with Sections 1108.4.2.1 and 1108.4.2.2.

❖ If holding cells are provided, Sections 1108.4.2.1 and 1108.4.2.2 indicate which cells shall be accessible.

1108.4.2.1 Central holding cells. Where separate central holding cells are provided for adult males, juvenile males, adult females or juvenile females, one of each type shall be accessible. Where central holding cells are provided and are not separated by age or sex, at least one accessible cell shall be provided.

❖ Some courthouse facilities have a central holding area for detainees that are in the courthouse for judicial proceedings. At least one central holding cell is required to be accessible. If separate holding cells are provided for males and females or to separate juvenile offenders from the adult offenders, at least one of each type is required to be accessible.

1108.4.2.2 Court-floor holding cells. Where separate court-floor holding cells are provided for adult males, juvenile males, adult females or juvenile females, each courtroom shall be served by one accessible cell of each type. Where court-floor holding cells are provided and are not separated by age or sex, courtrooms shall be served by at least one accessible cell. Accessible cells shall be permitted to serve more than one courtroom.

❖ Some courthouse facilities have a holding cell located in the area of the courtroom for detainees who are in the courthouse for judicial proceedings. For security reasons, these holding cells may be provided on each floor of the facility in order to have the detainees as close to the courtrooms they are to appear in as possible. At least one holding cell per floor is required to be accessible. One accessible holding cell could serve all the courtrooms on one floor. If separate holding cells are provided for males and females or to separate juvenile offenders from the adult offenders, at least one of each type is required to be accessible.

1108.4.3 Visiting areas. Visiting areas shall comply with Sections 1108.4.3.1 and 1108.4.3.2.

❖ Persons appearing in court proceedings may need to talk with their lawyer, social worker, parole officer, etc. Typically, these visitation areas include cubicles or counters in supervised areas.

1108.4.3.1 Cubicles and counters. At least 5 percent, but no fewer than one, of cubicles shall be accessible on both the visitor and detainee sides. Where counters are provided, at least one shall be accessible on both the visitor and detainee sides.

Exception: This requirement shall not apply to the detainee side of cubicles or counters at noncontact visiting areas not serving holding cells.

❖ If detainee/visitor counters or cubicles are provided, at least 5 percent must be accessible. Since it is not known if the person who will need to use the accessible counter is on the visitor side or the detainee side, at least one counter location shall be accessible on both sides. In accordance with the exception, visiting areas that do not serve accessible holding cells would still require 5 percent of the counters or cubicles to be accessible on the visitor's side. This is consistent with Section 1103.2.13.

1108.4.3.2 Partitions. Where solid partitions or security glazing separate visitors from detainees, at least one of each type of cubicle or counter partition shall be accessible.

❖ Section 1108.4.3.1 requires 5 percent of cubicles or counters in courthouse visiting areas to be accessible in accordance with the built-in counter/work surface requirements. This section deals with the visiting situation where solid partitions of security glazing separate visitors from detainees. To facilitate communication, the system may include some type of handset, headphones or microphone. The system must consider in its design that either user may use a wheelchair or have difficulty bending or stooping. Any type of telephone or headphone system must have a volume control.

SECTION 1109
OTHER FEATURES AND FACILITIES

1109.1 General. Accessible building features and facilities shall be provided in accordance with Sections 1109.2 through 1109.15.

Exception: Type A and Type B dwelling and sleeping units shall comply with ICC A117.1.

❖ This section clarifies that the provisions of Section 1109 are applicable in addition to all other requirements of Chapter 11. For example, although Section 1109.2 sets forth accessibility requirements for toilet rooms and bathing facilities within a store, it is not intended to mean that these are the only accessibility requirements that are applicable to stores. The building features and facilities covered by this section are required to be accessible to the extent set forth herein. The exception is a reminder that Type A and Type B dwelling units should comply with ICC A117.1 within the individual dwelling units. Common areas in Group R-2 and R-3 occupancies should comply with Section 1109.

1109.2 Toilet and bathing facilities. Toilet rooms and bathing facilities shall be accessible. Where a floor level is not required to be connected by an accessible route, the only toilet rooms or bathing facilities provided within the facility shall not be located on the inaccessible floor. At least one of each type of fixture, element, control or dispenser in each accessible toilet room and bathing facility shall be accessible.

Exceptions:

1. In toilet rooms or bathing facilities accessed only through a private office, not for common or public use,

and intended for use by a single occupant, any of the following alternatives are allowed:

  1.1. Doors are permitted to swing into the clear floor space provided the door swing can be reversed to meet the requirements in ICC A117.1,

  1.2. The height requirements for the water closet in ICC A117.1 are not applicable,

  1.3. Grab bars are not required to be installed in a toilet room, provided that reinforcement has been installed in the walls and located so as to permit the installation of such grab bars, and

  1.4. The requirement for height, knee and toe clearance shall not apply to a lavatory.

2. This section is not applicable to toilet and bathing facilities that serve dwelling units or sleeping units that are not required to be accessible by Section 1107.

3. Where multiple single-user toilet rooms or bathing facilities are clustered at a single location and contain fixtures in excess of the minimum required number of plumbing fixtures, at least 5 percent, but not less than one room for each use at each cluster, shall be accessible.

4. Toilet room fixtures that are in excess of those required by the *International Plumbing Code* and that are designated for use by children in day care and primary school occupancies.

5. Where no more than one urinal is provided in a toilet room or bathing facility, the urinal is not required to be accessible.

6. Toilet rooms that are part of critical-care or intensive-care patient sleeping rooms are not required to be accessible.

❖ This section generally requires toilet rooms and bathing facilities to be accessible. A person using a wheelchair must be able to approach and enter the room. Within the toilet room and bathing facility, a minimum of one of each element or fixture is required to be accessible. Elements and fixtures include such things as water closets, lavatories, mirrors, towel dispensers, hand dryers and any other device that is installed and intended for use by the occupants of the room. Requirements for the total number of bathing and toilet facilities are in Chapter 29 of the code [which is duplicated from Section 403 of the *International Plumbing Code®* (IPC®)]. Large assembly and mercantile occupancies must also include a unisex accessible bathroom in addition to their other toilet facilities (see commentary, Section 1109.2.1).

ICC A117.1 is the document referenced for all accessible toilet room requirements. The technical requirements in ICC A117.1 are based on allowing a person in a wheelchair to perform a side transfer [see Figure 1109.2(1)]. Maintaining a clear floor space at each fixture is important [see Figures 1109.2(2) and 1109.2(3) for possible configurations].

Exception 1 addresses a condition in which a toilet room or bathing facility is permitted to be adaptable rather than fully accessible. The intent is that if a toilet room is part of an individual office and serves only the occupant of that office, the adaptable toilet room can be readily modified to be fully accessible based on that individual's needs. Preplanning during construction and design, such as installing blocking for grab bars and arranging plumbing fixtures to have adequate clear floor space, will facilitate future alterations.

Exception 2 is intended to correlate with the provisions of Section 1107, which establish the minimum number of facilities required to be Accessible, Type A or Type B units. Without this exception, the code would literally be requiring accessible fixtures in an inaccessible toilet room or bathing facility. It is important to note that the bathrooms associated with required Accessible dwelling and sleeping units must comply with the requirements in ICC A117.1, Chapter 6. Type A and Type B unit toilet room technical requirements are located in ICC A117.1, Chapter 10.

Exception 3 specifies that all toilet rooms in excess of those required by the IPC and clustered together need not be accessible. In such configurations, typically found in a doctor's office or drug test center, the requirement is reduced to a 5-percent minimum. If these toilet rooms are clustered in separate locations, such as in a multiclinic facility, the 5-percent minimum would be applied to each cluster.

Exception 4 specifies that all toilet rooms in excess of those required by the IPC and provided specifically for children in day care and primary school classrooms are not required to be accessible. Child care facilities commonly provide toilet facilities in the classrooms for the use of the children. The desired mounting height for the fixtures in these rooms is based on the size of small children. These lower dimensions are to facilitate use by the children with minimum assistance and maximum safety. When this is done, the fixtures will not meet the requirements for accessibility in ICC A117.1-1998, which are based on adult sizes. Technical criteria for child sizes are provided in the 2003 edition of ICC A117.1.

The IPC permits urinals to be substituted for water closets to a maximum of 67 percent in each toilet room (see IPC Section 419.2). Exception 5 states that if only one urinal is provided within a toilet room, that urinal is not required to be accessible. This situation would typically only occur in bathrooms with one water closet compartment.

While Exception 2 would exempt all the non-accessible patient rooms in a hospital from accessible bathroom requirements, the intent of Exception 6 is also to exempt the Accessible units that may be provided within the critical-care or intensive-care units from requiring accessible bathrooms. In critical-care or intensive-care units the patients are often too ill to use the bathroom without assistance; therefore, assistance is offered and expected in these areas to all patients. In addition, critical-care and intensive-care rooms often must be designed to maximize free space for equipment and personnel in case of emergency care situations.

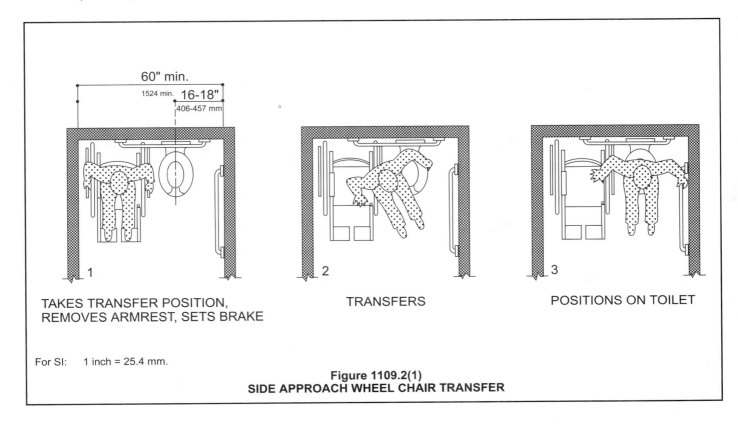

For SI: 1 inch = 25.4 mm.

Figure 1109.2(1)
SIDE APPROACH WHEEL CHAIR TRANSFER

FIGURE 1109.2(2) – FIGURE 1109.2(3)

ACCESSIBILITY

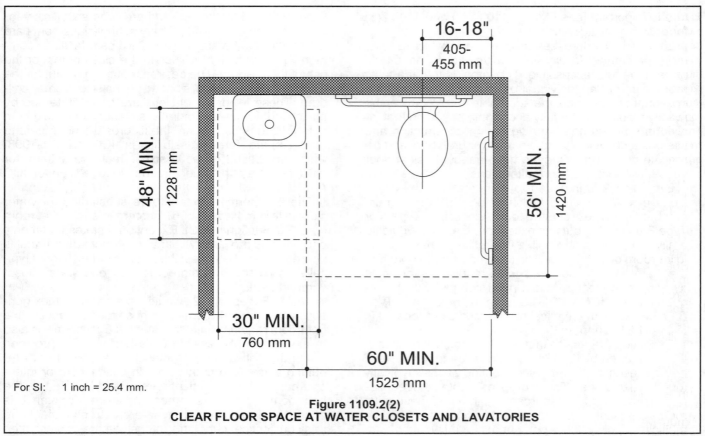

For SI: 1 inch = 25.4 mm.

Figure 1109.2(2)
CLEAR FLOOR SPACE AT WATER CLOSETS AND LAVATORIES

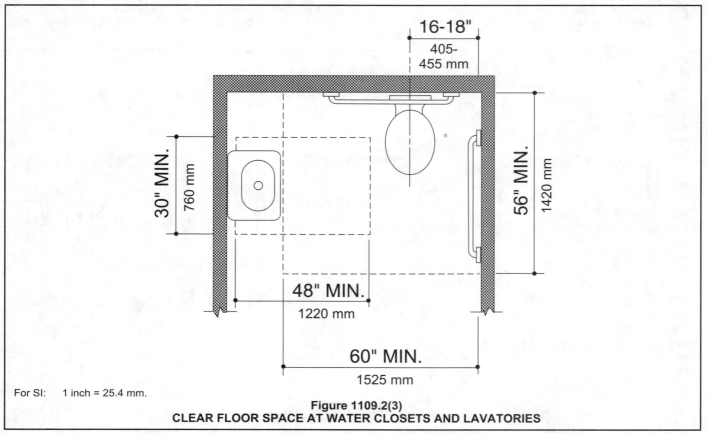

For SI: 1 inch = 25.4 mm.

Figure 1109.2(3)
CLEAR FLOOR SPACE AT WATER CLOSETS AND LAVATORIES

1109.2.1 Unisex toilet and bathing rooms. In assembly and mercantile occupancies, an accessible unisex toilet room shall be provided where an aggregate of six or more male and female water closets is required. In buildings of mixed occupancy, only those water closets required for the assembly or mercantile occupancy shall be used to determine the unisex toilet room requirement. In recreational facilities where separate-sex bathing rooms are provided, an accessible unisex bathing room shall be provided. Fixtures located within unisex toilet and bathing rooms shall be included in determining the number of fixtures provided in an occupancy.

> **Exception:** Where each separate-sex bathing room has only one shower or bathtub fixture, a unisex bathing room is not required.

❖ The requirements for unisex toilet and bathing rooms were recommended for adoption into the codes by the Board for the Coordination of the Model Codes (BCMC), in conjunction with representatives of the United States Architectural and Transportation Barriers Compliance Board (Access Board). The primary issue relative to unisex toilet/bathing facilities is that some people with disabilities require assistance to utilize them. If that attendant is of the opposite sex, a toilet or bathing facility that can accommodate both persons is required. It is important to note that these provisions will typically not result in a substantial cost burden to the building owners, since the fixtures provided also count towards the minimum number required (see Section 2902.1.1 or IPC Section 403.1.1).

The unisex requirements, although beneficial to all occupancies and building sizes, have been limited to those structures that typically have large transient occupant loads, namely assembly and mercantile. The section also identifies how mixed-occupancy buildings are to be addressed relative to the calculation. The fixtures provided in unisex toilet and bathing rooms count towards the number of required fixtures for the occupancy. The number of fixtures to be located in such rooms is limited, based on the premise that these rooms are securable (see commentary, Section 1109.2.1.2).

The provision for bathing facilities is primarily geared towards recreational facilities. It is only required where the designer has chosen to provide separate-sex facilities for bathing and has provided more than one tub or shower for each sex. The exception acknowledges that smaller bathing rooms such as those described can be utilized by a disabled person with assistance without much hardship to a nondisabled person.

These types of facilities have also been identified as "family bathrooms." For example, a person requiring assistance can also be a small child or an elderly adult. A parent shopping or attending an event with a child of the opposite sex can utilize this facility as well as an adult who may need assistance from a spouse. Many facilities have been installing these family bathrooms as part of their standard bathroom layouts and viewing them as "customer friendly."

Note that the requirement for unisex accessible toilet rooms does not exempt separate-sex bathrooms from providing accessible or ambulatory stalls. For a discussion of unisex accessible bathrooms in existing buildings, see Section 3409.7.9.

1109.2.1.1 Standard. Unisex toilet and bathing rooms shall comply with Sections 1109.2.1.2 through 1109.2.1.7 and ICC A117.1.

❖ The facilities are required to comply with the provisions of ICC A117.1 as well as the additional criteria established in the following subsections.

1109.2.1.2 Unisex toilet rooms. Unisex toilet rooms shall include only one water closet and only one lavatory. A unisex bathing room in accordance with Section 1109.2.1.3 shall be considered a unisex toilet room.

> **Exception:** A urinal is permitted to be provided in addition to the water closet in a unisex toilet room.

❖ A unisex toilet room must include a lavatory and a water closet. The exception permits a urinal to also be installed in the unisex toilet room if desired. If a unisex bathing room is provided within a facility, it can serve as the required unisex toilet room.

1109.2.1.3 Unisex bathing rooms. Unisex bathing rooms shall include only one shower or bathtub fixture. Unisex bathing rooms shall also include one water closet and one lavatory. Where storage facilities are provided for separate-sex bathing rooms, accessible storage facilities shall be provided for unisex bathing rooms.

❖ A unisex bathing facility is required to have one shower or tub, one water closet and one lavatory. The shower can be a transfer type, roll-in type or a combination of the two. Accessible storage facilities, such as lockers, are also required if storage facilities are provided in the separate-sex bathing facilities. A unisex bathing room can also serve a dual purpose (bathing and toilet) as the required unisex toilet room (see commentary, Section 1109.2.1.2).

1109.2.1.4 Location. Unisex toilet and bathing rooms shall be located on an accessible route. Unisex toilet rooms shall be located not more than one story above or below separate-sex toilet rooms. The accessible route from any separate-sex toilet room to a unisex toilet room shall not exceed 500 feet (152 m).

❖ A one-story, 500-foot (1524 mm) limitation for access to customer toilet facilities is currently in Section 2902.6 (which is duplicated from Section 403.6 of the IPC). The distance to the customer toilet facilities is measured from the main entrance of a store or space to the toilet facility. The distance in this section is measured from the separate-sex facility to the unisex facility. The general travel distance requirement, in combination with this section, could result in a total travel distance for a unisex accessible toilet of two stories or 1,000 feet (3048 mm) maximum. An accessible route is required to provide access to the unisex accessible facilities. Signage is required at both the separate-sex and unisex facilities in accordance with Sections 1110.1 and 1110.2.

1109.2.1.5 Prohibited location. In passenger transportation facilities and airports, the accessible route from separate-sex toilet rooms to a unisex toilet room shall not pass through security checkpoints.

❖ Security checkpoints in airports and similar facilities represent a potential delay, which may cause missed flights or connections.

1109.2.1.6 Clear floor space. Where doors swing into a unisex toilet or bathing room, a clear floor space not less than 30 inches by 48 inches (762 mm by 1219 mm) shall be provided, within the room, beyond the area of the door swing.

❖ The clear floor space provisions are intended to provide a room that is large enough to allow a person in a wheelchair to enter and close the door before utilizing the fixtures. This requirement is also in ICC A117.1.

1109.2.1.7 Privacy. Doors to unisex toilet and bathing rooms shall be securable from within the room.

❖ Since privacy while utilizing bathing and toilet facilities is an issue, the door to the facility must be securable from the inside. The securing mechanism must be within reach and not require any tight pinching, grasping or sharp turning of the wrist to operate.

1109.2.2 Water closet compartment. Where water closet compartments are provided in a toilet room or bathing facility, at least one wheelchair-accessible compartment shall be provided. Where the combined total water closet compartments and urinals provided in a toilet room or bathing facility is six or more, at least one ambulatory-accessible water closet compartment shall be provided in addition to the wheelchair-accessible compartment. Wheelchair-accessible and ambulatory-accessible compartments shall comply with ICC A117.1.

❖ There are different configurations of water closet compartments that facilitate different degrees of physical disability. The provisions of ICC A117.1 establish the configuration and dimensional requirements for various types of water closet compartments. A wheelchair-accessible compartment is one in which sufficient space is provided for the wheelchair to enter completely the water closet compartment [see Figures 1109.2.2(2) and 1109.2.2(3)]. The wheelchair user then transfers from the wheelchair to the water closet in order to utilize the fixture. It is important that the required clear floor space be maintained. It is not intended for a lavatory to be provided within the minimum-sized stall. An ambulatory-accessible compartment is intended to facilitate use by a person with a mobility impairment that necessitates the use of a walking aid, such as a cane or walker [see Figure 1109.2.2(1)]. An ambulatory-accessible water closet compartment is not intended to be utilized by a person in a wheelchair. The 36-inch (914 mm) width is intended to allow standing persons to support themselves utilizing the grab bars on both sides. A wider stall would not allow adequate bearing. Since these provisions are intended to address, within reason, the needs of both ranges of mobility impairment, this section requires a wheelchair-accessible compartment in all cases. In larger toilet rooms (i.e., those with a total of six or more water closet compartments and urinals), one ambulatory-accessible compartment is required in addition to the wheelchair-accessible compartment.

This section is not intended to increase the required number of fixtures beyond that required by the IPC. For example, if a toilet room contains 10 water closets, eight water closet compartments may be of conventional design, one water closet compartment must be wheelchair accessible and one must be ambulatory accessible.

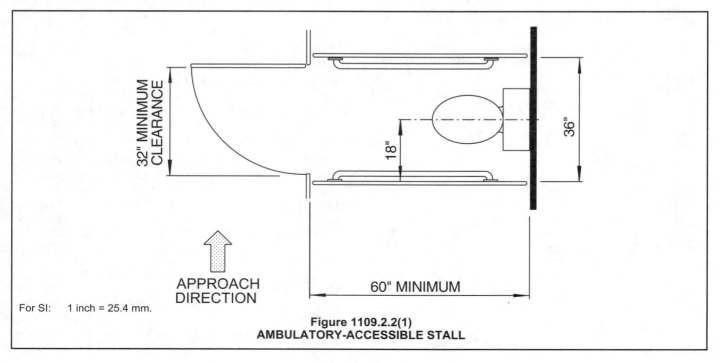

For SI: 1 inch = 25.4 mm.

Figure 1109.2.2(1)
AMBULATORY-ACCESSIBLE STALL

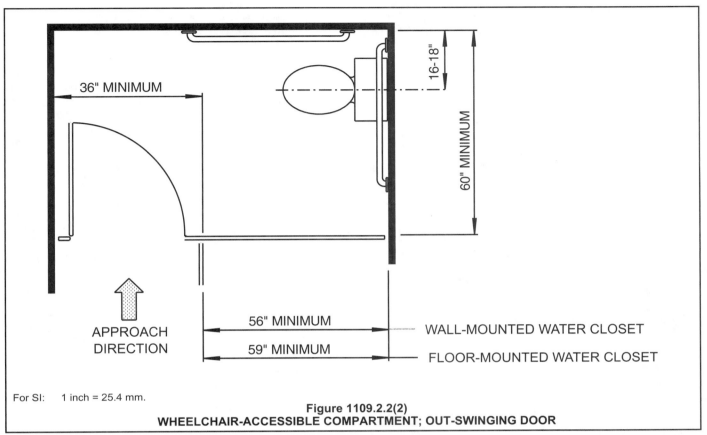

36" MINIMUM

16-18"

60" MINIMUM

56" MINIMUM — WALL-MOUNTED WATER CLOSET

APPROACH DIRECTION

59" MINIMUM — FLOOR-MOUNTED WATER CLOSET

For SI: 1 inch = 25.4 mm.

Figure 1109.2.2(2)
WHEELCHAIR-ACCESSIBLE COMPARTMENT; OUT-SWINGING DOOR

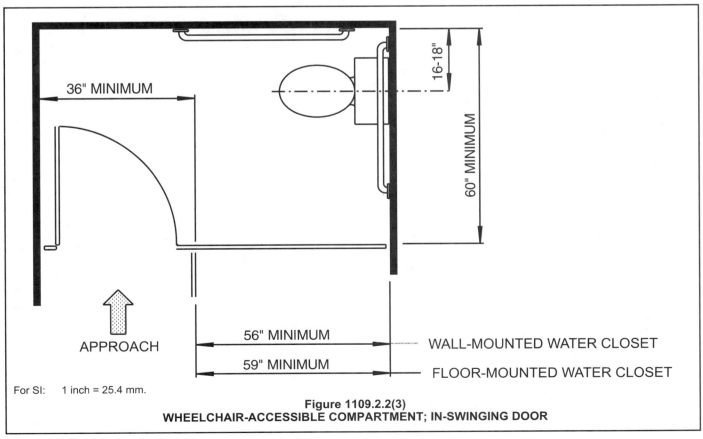

36" MINIMUM

16-18"

60" MINIMUM

56" MINIMUM — WALL-MOUNTED WATER CLOSET

APPROACH

59" MINIMUM — FLOOR-MOUNTED WATER CLOSET

For SI: 1 inch = 25.4 mm.

Figure 1109.2.2(3)
WHEELCHAIR-ACCESSIBLE COMPARTMENT; IN-SWINGING DOOR

1109.3 Sinks. Where sinks are provided, at least 5 percent, but not less than one, provided in accessible spaces shall comply with ICC A117.1.

Exceptions:

1. Mop or service sinks are not required to be accessible.

2. Sinks designated for use by children in day care and primary school occupancies.

❖ This section is intended to provide scoping requirements for sinks located in spaces not specifically exempted by Section 1109.2. If a sink is provided in toilet and bathing rooms specifically exempted by Section 1109.2, it is not required to be accessible. When sinks are provided in other areas, a minimum of 5 percent, but not less than one, is required to be accessible. There are specific exceptions for mop or service sinks, and sinks intended for use by children in day care or primary school occupancies. In addition, if a sink is part of an individual work station, that sink is not required to be accessible in accordance with Section 1103.2.3.

Front access to sinks, similar to what is required within a toilet room, is not always required (see commentary, Section 1109.4).

1109.4 Kitchens and kitchenettes. Where kitchens and kitchenettes are provided in accessible spaces or rooms, they shall be accessible in accordance with ICC A117.1.

❖ Kitchens, kitchenettes and wet bars in accessible spaces, other than within Type A and Type B dwelling units, must be accessible. When a cooktop or conventional range is not provided, side approach to the sink is acceptable.

1109.5 Drinking fountains. On floors where drinking fountains are provided, at least 50 percent, but not less than one fountain, shall be accessible.

❖ This section establishes a reasonable threshold for the required number of accessible drinking fountains. It should be noted that this section does not require the installation of drinking fountains where none are required or provided; however, when provided, at least 50 percent, but not less than one, is required to be accessible. Chapter 29 (which is duplicated from Section 403 of the IPC) contains criteria indicating the number of drinking fountains that are required based on occupancy.

1109.6 Elevators. Passenger elevators on an accessible route shall be accessible and comply with Section 3001.3.

❖ This section requires all passenger elevators on an accessible route to be accessible. The reference to Section 3001.3 is somewhat redundant, in that it requires all passenger elevators to conform to ICC A117.1. Compliance with ICC A117.1 is also required by Section 1101.2. It is not the intent of this section to prohibit limited access/limited use elevators (LULA) elevators or platform lifts (as permitted by Section 1109.7).

It should also be noted that unlike plumbing fixtures, this section does not establish that a certain number of passenger elevators are required to be accessible while other passenger elevators may be of a design that is not fully accessible. All passenger elevators that are on an accessible route are required to be accessible. If a bank of passenger elevators is provided as part of an accessible route, all elevators in that bank must be accessible.

The reference to an accessible route is important because some areas in certain buildings may not be required to be accessible (see the exceptions to Section 1104.4). In addition, certain types of elevators (e.g., service elevator) may not be part of an accessible route.

1109.7 Lifts. Platform (wheelchair) lifts are permitted to be a part of a required accessible route in new construction where indicated in Items 1 through 7. Platform (wheelchair) lifts shall be installed in accordance with ASME A18.1.

1. An accessible route to a performing area and speakers' platforms in occupancies in Group A.

2. An accessible route to wheelchair spaces required to comply with the wheelchair space dispersion requirements of Section 1108.2.2 through 1108.2.4.

3. An accessible route to spaces that are not open to the general public with an occupant load of not more than five.

4. An accessible route within a dwelling or sleeping unit.

5. An accessible route to wheelchair seating spaces located in outdoor dining terraces in A-5 occupancies where the means of egress from the dining terraces to a public way are open to the outdoors.

6. An accessible route to raised judges' benches, clerks' stations, jury boxes, witness stands and other raised or depressed areas in a court.

7. An accessible route where existing exterior site constraints make use of a ramp or elevator infeasible.

❖ This section indicates that platform (wheelchair) lifts are only permitted to be part of a required accessible route in new construction in limited situations. If a platform lift is permitted as part of the accessible route into a space, it can also serve as part of the accessible route used for means of egress out of the space if it has standby power (see Section 1007.5). Platform lifts may be used more extensively in existing construction (see Section 3409.7.3). This is in recognition that circumstances in existing buildings may make it impractical to accomplish accessibility by use of an elevator, in which case a platform lift is a reasonable alternative. Accessible means of egress is not required in existing buildings undergoing alterations (see Section 1007.1, Exception 1 or Section 3409.5, Exception 2); therefore, a platform lift as part of the accessible means of egress in these situations is not an issue.

The previous technical standard (i.e., ASME A17.1-1996) required key operation to platform lifts, which necessarily inhibits independent access by persons with physical disabilities. The requirements for platform lifts have been removed from the elevator standard and now have their own standard, ASME A18.1, *Safety Standard for Platform Lifts and Stairway Chairlifts*. This new standard allows push-button operation of platform lifts, thus making independent access much easier.

The listed items indicate when a platform can be utilized as part of a required accessible route in new construction. Items 1 and 2 allow platform lifts to be utilized for access to performing and viewing areas in assembly occupancies. Item 3 specifies that a platform lift may be used to provide access to a nonpublic area with five or less occupants, such as a projection booth. Item 4 allows for a platform lift within an individual dwelling or sleeping unit. Item 5 is intended to address an outdoor dining area in Group A-5, such as a picnic terrace in a baseball park. Item 6 permits platform lifts to provide access to the raised areas typically found in a courtroom setting. Item 7 recognizes that existing site constraints may make installation of a ramp or elevator infeasible. An example would be the situation of dealing with existing public sidewalks, easements and public ways in downtown urban areas.

Note that a platform lift that was not part of the required accessible route could be used to facilitate access to any space. The governing factor in those situations would be the limitations of the product itself for capacity and travel distance. Item 4 permits platform lifts to be utilized within individual dwelling or sleeping units as part of a required accessible route. Single-family homes are not required to be accessible; however, persons with mobility impairments may choose to install stairway chairlifts in their private homes. Section 1009.1, Exception 4 specifies a minimum stairway width for platform lifts or chair lifts provided within dwelling units.

A platform lift is an electrically operated, mechanical device designed to transport a person who cannot use stairs over a short vertical distance. Platform lifts must be sized to accommodate a wheelchair user. Platform lifts can be used by wheelchair users and persons with limited mobility and are sometimes also equipped with folding seats. A fold-down seat that moves up the stairway is not a platform lift. A stairway chair lift cannot serve as part of a required accessible route. A platform lift is most suitable for changes of elevation of one story or less where the installation of a ramp is not feasible (see Figures 1109.7(1) and 1109.7(2)].

There are two kinds of platform lifts: vertical lifts and inclined lifts. Vertical lifts are similar to elevators in that they travel only up and down in a fixed vertical space. Inclined platform lifts are usually installed in conjunction with a stairway and travel along the slope of the stairway. Inclined lifts are a design consideration for long flights of stairs where a vertical platform lift is not practical, where headroom is limited or where ceilings are low.

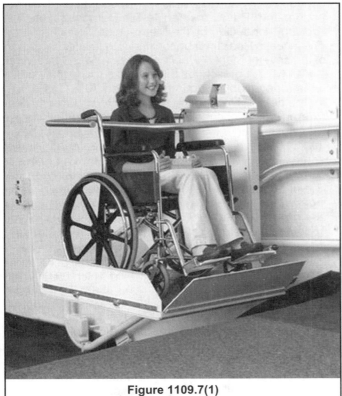

Figure 1109.7(1)
PLATFORM (WHEELCHAIR) LIFT
(Photo courtesy of Wheel-evator)

Figure 1109.7(2)
STAIRWAY CHAIRLIFT
(Figure courtesy of Wheel-evator)

1109.8 Storage. Where fixed or built-in storage elements such as cabinets, shelves, medicine cabinets, closets and drawers are provided in required accessible spaces, at least one of each type shall contain storage space complying with ICC A117.1.

❖ Fixed or built-in storage elements are common in many occupancies, including dwelling and sleeping units and many nonresidential occupancies such as office buildings, recreational facilities, etc. Examples would include mailboxes, coat closets, storage closets, lockers, etc. The code does not require storage facilities to be provided, but when such facilities are provided, this section requires at least one of each type to contain storage space in accordance with ICC A117.1.

1109.8.1 Lockers. Where lockers are provided in accessible spaces, at least five percent, but not less than one, of each type shall be accessible.

❖ Lockers are often provided in schools, health clubs, etc. At least 5 percent of the lockers in each location must have adequate clear floor space for wheelchair approach with exterior locks and latches and interior coat hooks and shelves within reach ranges. In accordance with Section 803.4 of ICC A117.1, an accessible bench must be provided in all locker rooms. These benches should be located adjacent to the accessible lockers.

1109.8.2 Shelving and display units. Self-service shelves and display units shall be located on an accessible route. Such shelving and display units shall not be required to comply with reach-range provisions.

❖ The intent of this section is that persons who utilize wheelchairs can see items on shelves in stores and libraries, but someone may be required to assist them in reaching the items.

1109.8.3 Coat hooks and folding shelves. Where coat hooks and folding shelves are provided in toilet rooms, toilet compartments, or in dressing, fitting or locker rooms, at least one of each type shall be accessible and shall be provided in accessible toilet rooms without toilet compartments, accessible toilet compartments and accessible dressing, fitting and locker rooms.

❖ If amenities such as coat hooks or shelves are provided for use by the general public (in nonaccessible spaces), then those same amenities must be provided in an accessible space.

1109.9 Detectable warnings. Passenger transit platform edges bordering a drop-off and not protected by platform screens or guards shall have a detectable warning.

Exception: Detectable warnings are not required at bus stops.

❖ A detectable warning is a standardized feature built in or applied to walking surfaces to warn a visually impaired person of a hazard on or near his or her path of travel that may otherwise go unnoticed and could result in injury to that person. A typical example of the need for a detectable warning is at any walking surface where a

significant drop-off occurs that, if unnoticed, has the potential to cause injury should a person fall.

This section requires a detectable warning at the edges of passenger transit platforms where they border a drop-off. For example, loading platforms in both light- and heavy-rail transit stations where passengers await the arrival of trains are a serious potential hazard to a person with a vision impairment when there is no train at the station. There have been incidents where people with severe vision impairments have fallen off the transit platform and been killed or seriously injured. The presence of a detectable warning at the platform edge where it borders the drop-off would be encountered and recognized by a person with a vision impairment either through detection by a long cane or by foot contact with the detectable warning surface.

The exception for bus stops is established so as not to require a detectable warning at the curb between a sidewalk and the street. Detectable warnings at curbs are contradictory to the envisioned application of detectable warnings in the built environment. The curb itself is a recognizable and well-known cue to people with vision impairments to proceed with caution.

There is a great deal of controversy and general disagreement as to the benefit or advisability of detectable warning surfaces throughout the built environment. The use of such warnings at transit stations is, to date, the only well-documented use of detectable warnings, as discussed in studies such as "Tactile Warnings to Promote Safety in the Vicinity of Transit Platform Edges," conducted by the Urban Mass Transportation Administration, and "Pathfinder Tactile Tile Demonstration Test Project," conducted by the Metro-Dade Transit Agency. The Access Board will be looking into this issue as part of the development of *Public Rights-of-Way Guidelines*.

ICC A117.1 prescribes the type of surface that constitutes a detectable warning in Section 705. One option for the detectable warning surface that is currently considered suitable consists of raised, truncated domes with a diameter of 0.9 inch (23 mm), a height of approximately 0.2 inch (5 mm) and center-to-center spacing of approximately 2.35 inches (60 mm). However, there is current and ongoing research into the use and application of truncated domes as a suitable detectable-warning surface for various applications, including interior and exterior locations. ICC A117.1 requires that the type of detectable warning surface utilized must be standard throughout a building, facility, site or complex of buildings. If different types of detectable warning surfaces are utilized, their usefulness is diminished. The different messages they would convey to a person with a vision impairment would not be consistent and may be confusing and easily misinterpreted.

It is anticipated that future applications of detectable warnings will be considered when additional research and documentation of their usefulness and suitability is available. One issue, among many, is the durability of a detectable warning surface in an exterior application and the potential difficulty that truncated domes may present to a person in a wheelchair, as well as to nondisabled persons who may have to negotiate the surface.

1109.10 Assembly area seating. Assembly areas with fixed seating in every occupancy shall comply with Section 1108.2 for accessible seating and assistive listening devices.

❖ This is a cross reference to the assembly seating requirements in Section 1108.2. The cross reference is placed here to clarify that wheelchair seating is required in any occupancy that includes fixed seating, not just Group A structures.

1109.11 Seating at tables, counters and work surfaces. Where seating or standing space at fixed or built-in tables, counters or work surfaces is provided in accessible spaces, at least 5 percent of the seating and standing spaces, but not less than one, shall be accessible. In Group I-3 occupancy visiting areas at least 5 percent, but not less than one, cubicle or counter shall be accessible on both the visitor and detainee sides.

Exceptions:

1. Check-writing surfaces at check-out aisles not required to comply with Section 1109.12.2 are not required to be accessible.

2. In Group I-3 occupancies, the counter or cubicle on the detainee side is not required to be accessible at noncontact visiting areas or in areas not serving accessible holding cells or sleeping units.

❖ Many occupancies such as libraries, classrooms, restaurants and other public spaces are equipped with fixed seating, tables or both. Correctional facilities often have cubicles or counters that are set up for visiting sessions between detainees and their lawyers or visitors. This section requires that at least 5 percent, but not less than one, be accessible. The threshold of 5 percent is consistent with the requirement for accessible seating in dining areas as specified in Section 1108.2.9.1. Fixed or built-in seating that is part of an individual work station is not required to be accessible (see Section 1103.2.3). These accessible table and work surfaces are expected to include adequate knee and toe clearances as specified in ICC A117.1. For requirements for points of sales or service that do not include a work station, see Section 1109.12.3. For example, the check-writing station in the center of the bank lobby is considered a workstation; however, the teller window is typically considered a service counter. The accessible check-writing station would require knee and toe clearances, while the accessible teller window would not.

Note that the requirement addresses built-in counters, table or workstations for standing or seating. Built-in surfaces where persons are typically standing are most often installed at a height of 36 to 48 inches (914 to 1219 mm). Depending on the overall configuration of the counters in the space, a portion of the counter for standing persons may need to be lowered to an accessible height of 28 to 34 inches (711 to 864 mm). The intent of Exception 1 is that if a check-writing sur-

face is provided at a nonaccessible checkout aisle, this surface is not required to meet the counter and work surface requirements in ICC A117.1.

Exception 2 is consistent with the Group I-3 exception found in Section 1103.2.13. If Accessible units are not required by Section 1107.5.5, accessible elements are not required on the detainee side of the cubicles. Since the detainees may have a disabled person visiting them on the public side, at least 5 percent of the stations must be accessible.

1109.11.1 Dispersion. Accessible fixed or built-in seating at tables, counters or work surfaces shall be distributed throughout the space or facility containing such elements.

❖ An appropriate distribution of seating and tables is required and is intended to be consistent with the distribution of all such seating and tables in the facility. For example, a library may contain several separate areas in which fixed seating or tables are installed. This provision would prohibit the location of all accessible seating and tables in a single specific area of the facility. However, this should not be construed to mean that every separate area in which fixed seating or tables are provided is required to contain an accessible seat or table. For example, if there are four areas in which fixed seating or tables are located, and two such seats or tables are required by this section to be accessible, the intent of this section is to require that one accessible seat or table be located in two of the four areas.

1109.12 Service facilities. Service facilities shall provide for accessible features in accordance with Sections 1109.12.1 through 1109.12.5.

❖ This section introduces five subsections that address accessibility of dressing rooms; service counters and windows; check-out aisles; food service; queues and waiting lines, etc. These types of elements are broadly termed "service facilities" and are intended to permit usage by physically disabled persons.

1109.12.1 Dressing, fitting and locker rooms. Where dressing rooms, fitting rooms or locker rooms are provided, at least 5 percent, but not less than one, of each type of use in each cluster provided shall be accessible.

❖ This section establishes a reasonable minimum threshold for accessible dressing, fitting and locker rooms. This section does not require that dressing, fitting and locker rooms be provided, but when such facilities are provided, at least 5 percent are required to be accessible. This section also requires an appropriate distribution of accessible facilities based on distinct and different functions of each group of facilities provided. For example, a department store may have one group of fitting rooms serving the menswear department and a separate group of fitting rooms serving the women's wear department. This section would not permit the location of all accessible fitting rooms in one department and no accessible fitting rooms in another department. This may result in a number of accessible fitting rooms greater than 5 percent of the total number of all fitting rooms in the facility. For example, if there are two dressing rooms in each of three different departments in a retail store, a total of three accessible fitting rooms would be required: one in the three areas. Where amenities such as coat hooks, lockers or shelves are provided in inaccessible dressing, fitting or locker rooms, they must also be provided in the accessible rooms (see Section 1109.8). In accordance with Section 803.4 of ICC A117.1, an accessible bench must be provided in all accessible dressing, fitting and locker rooms.

1109.12.2 Check-out aisles. Where check-out aisles are provided, accessible check-out aisles shall be provided in accordance with Table 1109.12.2. Where check-out aisles serve different functions, at least one accessible check-out aisle shall be provided for each function. Where checkout aisles serve different functions, accessible check-out aisles shall be provided in accordance with Table 1109.12.2 for each function. Where check-out aisles are dispersed throughout the building or facility, accessible check-out aisles shall also be dispersed. Traffic control devices, security devices and turnstiles located in accessible check-out aisles or lanes shall be accessible.

> **Exception:** Where the area of the selling space is less than 5,000 square feet (465 m²), only one check-out aisle is required to be accessible.

❖ This section establishes the required number of accessible checkout aisles, such as are typically found in supermarkets, drugstores, discount retail stores, etc. If checkout aisles serve different functions, at least one checkout aisle that serves each function shall be accessible. If a facility offers checkouts at a variety of locations, the accessible locations must also be dispersed. Where security devices such as turnstiles and automatically activated gates are provided, they are required to be accessible. The required number of check-out aisles for each function set forth in Table 1109.12.2. If the selling spaces less than 5,000 square feet (465 m²), only one accessible checkout aisle is required. Signage is required in accordance with Section 1110.1.

TABLE 1109.12.2
ACCESSIBLE CHECK-OUT AISLES

TOTAL CHECK-OUT AISLES OF EACH FUNCTION	MINIMUM NUMBER OF ACCESSIBLE CHECK-OUT AISLES OF EACH FUNCTION
1 to 4	1
5 to 8	2
9 to 15	3
Over 15	3, plus 20% of additional aisles

❖ This table establishes the minimum required number of accessible check-out aisles based on the total number provided. Where more than 15 total check-out aisles are provided, the intent of Table 1109.12.2 is to be applied as though it reads "3 plus 1 for each additional 5 over 15, or a fraction thereof." For example, where 15 check-out aisles are provided, three accessible check-out aisles are required. Where 16 total check-out aisles are provided, four accessible check-out aisles are required.

1109.12.3 Point of sale and service counters. Where counters are provided for sales or distribution of goods or services, at least one of each type provided shall be accessible. Where such counters are dispersed throughout the building or facility, accessible counters shall also be dispersed.

❖ Wherever sales or service counters or windows are provided, at least one window or a portion of a counter is required to be accessible. For example, most hotel registration and check-out functions occur at a counter. The intent of this section is to require that one portion of the counter area be accessible. Accessibility is accomplished by providing a lower counter height to accommodate a person using a wheelchair and locating that counter on an accessible route. This is not necessarily intended to require that multiple lower counter heights be provided based on the different types of services that are offered. For example, this is not intended to require two separate, lower counter heights: one for hotel registration and one for check-out. One lower counter height can be provided at which both functions are accomplished. In the above described situation, dispersion of the accessible counters would be required if the counters themselves were dispersed in separate locations.

The intent of the accessible service window is to permit the customer to interact with the service representative the same way they interact with the general public. An extra piece of counter stuck on the front of the service counter [typically 42 to 48 inches (1067 to 1219 mm) high] may be appropriate as a temporary fix for barrier removal but is not acceptable in new construction. ICC A117.1 permits a maximum counter height of 36 inches (914 mm) at a service window (see commentary, Section 1109.11 and Figure 1109.12.3).

1109.12.4 Food service lines. Food service lines shall be accessible. Where self-service shelves are provided, at least 50 percent, but not less than one, of each type provided shall be accessible.

❖ All lines that provide access to food service must meet the accessible route provisions in ICC A117.1. Where self-service shelves are utilized, such as in a cafeteria, at least half of each type must be on an accessible route and within reach ranges.

1109.12.5 Queue and waiting lines. Queue and waiting lines servicing accessible counters or check-out aisles shall be accessible.

❖ If patrons must form a line for service, then the waiting line must meet the accessible route provisions of ICC A117.1. When dealing with permanent crowd control barriers, special attention must be paid to turns around obstructions or dividers. Accessible waiting lines are required at locations where accessible counters, windows or check-out aisles are provided, such as banks, ticket counters, fast-food establishments and similar facilities. A separate waiting line for the disabled is not a viable alternative.

Figure 1109.12.3
SERVICE COUNTERS

1109.13 Controls, operating mechanisms and hardware. Controls, operating mechanisms and hardware intended for operation by the occupant, including switches that control lighting and ventilation, and electrical convenience outlets, in accessible spaces, along accessible routes or as parts of accessible elements shall be accessible.

Exceptions:

1. Operable parts that are intended for use only by service or maintenance personnel shall not be required to be accessible.

2. Electrical or communication receptacles serving a dedicated use shall not be required to be accessible.

3. Where two or more outlets are provided in a kitchen above a length of counter top that is uninterrupted by a sink or appliance, one outlet shall not be required to be accessible.

4. Floor electrical receptacles shall not be required to be accessible.

5. HVAC diffusers shall not be required to be accessible.

6. Except for light switches, where redundant controls are provided for a single element, one control in each space shall not be required to be accessible.

❖ This section requires that any controls intended to be utilized by the occupants of the space be accessible when such controls are located in an accessible space, along accessible routes or as part of an accessible element. For example, the switch controlling the general lighting of a room that is required to be accessible also must be accessible. This is also true for most of the electrical outlets that are located in the room. Thermostats that are intended to be adjustable by the occupants must also be accessible. Another example would be the button or switch that activates an automatic hand dryer in a toilet room. If the fixture is mounted such that the outlet for the air is at an accessible height, the purpose is defeated if the switch to activate the device is inaccessible.

The exceptions listed are similar to the exceptions already located in ICC A117.1 for Type A and Type B dwelling units. Since the same problems exist in nonresidential facilities, the exceptions are appropriate. Exception 1 deals with items that are not intended for use by the general occupants of the space. This would include controls such as thermostats that are intended to be adjustable only by authorized personnel. Another example would be the controls restricted to use by authorized personnel for a public address system in a meeting room. Exception 2 deals with connections for a dedicated use. For example, the outlet behind the refrigerator that is to plug in the refrigerator is not required to be accessible. Exception 3 is for kitchens and kitchenettes. If more than one electrical outlet is provided over a portion of the countertop, only one is required to be accessible. Exception 4 permits floor receptacles not to be accessible, since by being in the floor they are out of the reach range. Exception 5 is for HVAC diffusers. The typical locations for these elements are most often in the floor, high up on the walls or on the ceiling. Exception 6 allows for redundant controls for everything but lighting. For example, a ceiling fan could be operated by a wall switch as well as by the chain on the fan itself.

1109.13.1 Operable windows. Where operable windows are provided in rooms that are required to be accessible in accordance with Sections 1107.5.1.1, 1107.5.2.1, 1107.5.3.1, 1107.5.4, 1107.6.1.1, 1107.6.2.2.1 and 1107.6.4.1, at least one window in each room shall be accessible and each required operable window shall be accessible.

> **Exception:** Accessible windows are not required in bathrooms or kitchens.

❖ This section specifies when operable windows must be accessible, limiting the requirements to Accessible dwelling and sleeping units in Groups I-1, I-2, R-1, R-4 and some R-2 occupancies. All operable windows in all groups are not required to be accessible. Operable windows in Type A must comply with the provisions in Section 1002.13 of ICC A117.1. There are no requirements for access to operable windows in Type B units. The concern is that such a requirement would force designers to opt for nonoperable windows when permitted by the code. If an operable window is required for natural ventilation or an emergency escape window within an Accessible unit, then that window must be accessible. Due to their typical locations within the space (e.g., over the sink in a kitchen, raised in a bathroom for privacy), operable windows in kitchens or bathrooms are exempted.

1109.14 Recreational facilities. Recreational facilities shall be provided with accessible features in accordance with Sections 1109.14.1 through 1109.14.3.

❖ Recreational facilities such as tennis courts, swimming pools, baseball fields, etc., may be part of development on a site such as a school or apartment complex. Group R-2 and R-3 facilities are addressed in Sections 1109.14.1 and 1109.14.2 so that equitable opportunity for use by all occupants of a multibuilding site would be evaluated. All other occupancies should comply with Section 1109.14.3.

1109.14.1 Facilities serving a single building. In Group R-2 and R-3 occupancies where recreational facilities are provided serving a single building containing Type A or Type B units, 25 percent, but not less than one, of each type of recreational facility shall be accessible. Every recreational facility of each type on a site shall be considered to determine the total number of each type that is required to be accessible.

❖ Many multiple-family developments include common recreational facilities, such as a swimming pool, community meeting room, tennis court, playground, etc., which are available for use by residents of the complex. In some cases, there may be more than one type of recreational facility located at different points within the development. This section establishes the criterion that 25 percent, but not less than one, of each type of recreational facility be accessible. For example, if two separate tennis court areas are provided, and one contains two tennis courts and another contains seven tennis courts, then a total of three tennis courts must be accessible. In a single building site, the three accessible ten-

nis courts could all be in one area or divided between the two areas. This satisfies the requirement that at least 25 percent of the total number of courts be made accessible.

1109.14.2 Facilities serving multiple buildings. In Group R-2 and R-3 occupancies on a single site where multiple buildings containing Type A or Type B units are served by recreational facilities, 25 percent, but not less than one, of each type of recreational facility serving each building shall be accessible. The total number of each type of recreational facility that is required to be accessible shall be determined by considering every recreational facility of each type serving each building on the site.

❖ In a multibuilding site, the same criteria applies as for single building sites; however, in addition, if certain facilities only serve certain buildings, at least 25 percent of them in each area must be accessible. For example, if two separate tennis court areas are provided, each serving a specific building, and one contains two tennis courts and another contains seven tennis courts, then 25 percent at each location, for a total of three tennis courts, must be accessible. While the number is the same for the single building, the distribution requirement is different. In this case, one tennis court in the two-court area must be made accessible and two of the tennis courts in the seven-court area must be made accessible. This satisfies the requirement that at least 25 percent of the courts for each building are made accessible. If all tennis courts serve all the buildings, then the distribution could be the same as discussed in Section 1109.14.1.

1109.14.3 Other occupancies. All recreational facilities not falling within the purview of Section 1109.14.1 or 1109.14.2 shall be accessible.

❖ When tennis courts, pools, baseball diamonds, playgrounds or other recreational facilities are provided in occupancies other than Group R-2 or R-3, all facilities must be accessible. The Department of Justice (DOJ) released guidelines for play areas in 2000 and recreational facilities in 2002. Requirements for outdoor developed areas are still being developed. These documents would be a good resource when addressing accessibility issues for different types of recreational facilities.

1109.15 Stairways. Stairways located along accessible routes connecting floor levels that are not connected by an elevator shall be designed and constructed to comply with ICC A117.1 and Chapter 10.

❖ Stairways that are located adjacent to a ramp must comply with Chapter 10 as well as ICC A117.1. The intent of this provision is to provide allowances for mobility-impaired persons who may be utilizing the stairway at that location.

SECTION 1110
SIGNAGE

1110.1 Signs. Required accessible elements shall be identified by the International Symbol of Accessibility at the following locations:

1. Accessible parking spaces required by Section 1106.1 except where the total number of parking spaces provided is four or less.

2. Accessible passenger loading zones.

3. Accessible areas of refuge required by Section 1007.6.

4. Accessible rooms where multiple single-user toilet or bathing rooms are clustered at a single location.

5. Accessible entrances where not all entrances are accessible.

6. Accessible check-out aisles where not all aisles are accessible. The sign, where provided, shall be above the check-out aisle in the same location as the check-out aisle number or type of check-out identification.

7. Unisex toilet and bathing rooms.

8. Accessible dressing, fitting and locker rooms where not all such rooms are accessible.

❖ Identification of accessible elements can be accomplished by use of an international symbol of accessibility (see Figure 1110.1. These figures are international in that they are recognized throughout the world as that which identifies accessibility.

There are eight specific circumstances in which required accessible elements are to be identified, as indicated in Items 1 through 8 of this section. Generally, these are locations in which not all of the facilities provided are accessible and, therefore, it is necessary to identify those that are accessible so that they can be readily recognized by the intended user. For example, Section 1106.1 specifies the required number of accessible parking spaces. If these are not identified by signage, it would be difficult and unnecessarily inconvenient for one to identify their location. (Note that the requirement for parking signage is limited to when more than four parking spaces are provided. To require an accessible space to be reserved when it involves 25 to 100 percent of the parking provided is an undue hardship to the building tenant.) One of the concepts embodied in the code is to mainstream accessibility in recognition that many of the features that make facilities and elements accessible are also useful and of benefit to people without disabilities. Part of this principle includes the idea that if an element is universally usable by people both with and without disabilities, there is no need for signage specifically identifying the element as being accessible. Hence, the signage requirements generally address circumstances in which not all of the elements will be accessible.

1110.2 Directional signage. Directional signage indicating the route to the nearest like accessible element shall be provided at

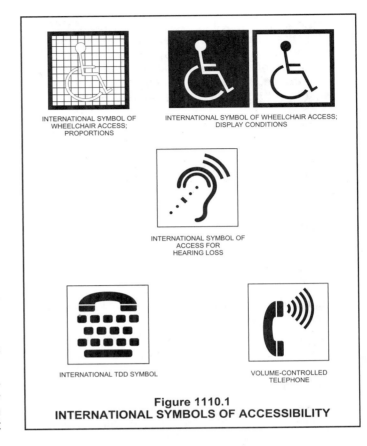

Figure 1110.1
INTERNATIONAL SYMBOLS OF ACCESSIBILITY

the following locations. These directional signs shall include the International Symbol of Accessibility:

1. Inaccessible building entrances.

2. Inaccessible public toilets and bathing facilities.

3. Elevators not serving an accessible route.

4. At each separate-sex toilet and bathing room indicating the location of the nearest unisex toilet or bathing room where provided in accordance with Section 1109.2.1.

5. At exits and elevators serving a required accessible space, but not providing an approved accessible means of egress, signage shall be provided in accordance with Section 1007.7.

❖ There are circumstances in which it is useful and necessary to locate directional signage at certain inaccessible elements to indicate the route to the nearest like accessible element. For example, not all building entrances are required to be accessible (see Section 1105). Should a person in a wheelchair happen to approach an inaccessible building entrance at an unfamiliar facility, it is appropriate to provide direction indicating where the nearest accessible entrance is located. The same circumstance presents itself at inaccessible public toilet and bathing facilities, at elevators that do not serve an accessible route and in assembly and mercantile occupancies where unisex accessible toilet and bathing facilities are required.

This requirement for directional signage works in conjunction with the fact that everything is initially assumed to be accessible; therefore, items are not required to be identified by the international symbols of accessibility in the interest of mainstreaming. See the general comments to this chapter and the commentary to Section 1110.1 for a further discussion of mainstreaming.

1110.3 Other signs. Signage indicating special accessibility provisions shall be provided as shown:

1. Each assembly area required to comply with Section 1108.2.7 shall provide a sign notifying patrons of the availability of assistive listening systems.

   Exception: Where ticket offices or windows are provided, signs are not required at each assembly area provided that signs are displayed at each ticket office or window informing patrons of the availability of assistive listening systems.

2. At each door to an egress stairway, exit passageway and exit discharge, signage shall be provided in accordance with Section 1011.3.

3. At areas of refuge, signage shall be provided in accordance with Sections 1007.6.3 through 1007.6.5.

4. At areas for assisted rescue, signage shall be provided in accordance with Section 1007.8.3.

❖ It is considered desirable that all the requirements for accessible signage be in one location. Item 1 requires signage indicating that assistive listening systems are provided when they are required by Section 1108.2.7. Items 2 through 4 are references to signage criteria in specific sections that have been mainstreamed in Chapter 10.

Bibliography

The following resource materials are referenced in this chapter or are relevant to the subject matter addressed in this chapter.

24 CFR, *Fair Housing Accessibility Guidelines* (FHAG). Washington, DC: Department of Housing and Urban Development, 1991.

36 CFR Parts 1190 and 1191 Draft, *The Americans with Disabilities Act Accessibility Guidelines* (ADAAG), Washington, DC: Architectural and Transportation Barriers Compliance Board, April 2, 2002.

42 USC 3601-88, *Fair Housing Amendments Act* (FHA). Washington, DC: United States Code, 1988.

"Accessibility and Egress for People with Physical Disabilities." CABO Board for the Coordination of the Model Codes Report, 1993.

"Code Requirements for Housing Accessibility." (CRHA) Falls Church, VA: International Code Council, 2000.

DOJ 28 CFR, Part 36-91, *Americans with Disabilities Act* (ADA). Washington, DC: Department of Justice, 1991.

DOJ 28 CFR, Part 36-91 (Appendix A), *ADA Accessibility Guidelines for Building and Facilities* (ADAAG). Washington, DC: Department of Justice, 1991.

Fair Housing Act Design Manual. U.S. Department of Housing and Urban Development, Revised 1998.

FED-STD-795-88, *Uniform Federal Accessibility Standards.* Washington, DC: General Services Administration; Department of Defense; Department of Housing and Urban Development; U.S. Postal Service, 1988.

"Final Report, Recommendations for a New ADAAG." ADAAG Review Federal Advisory Committee, September 30, 1996.

HUD 24 CFR, Part 100, *Fair Housing Accessibility Guidelines.* Washington, DC: U.S. Department of Housing and Urban Development.

ICC/A117.1-98, *Accessible and Usable Buildings and Facilities.* Falls Church, VA: International Code Council, 1998.

IPC-2003, *International Plumbing Code.* Falls Church, VA: International Code Council, 2000.

"Pathfinder Tactile Tile Demonstration Test Project" Metro-Dade Transit Agency, 1988.

"Tactile Warnings to Promote Safety in the Vicinity of Transit Platform Edges." Urban Mass Transportation Administration, 1987.

Vital Health Statistics Series 10, No. 135, Tables 1 & 2. 1977 National Health Interview Survey, National Center for Health Statistics.

Chapter 12:
Interior Environment

General Comments

Chapter 12 contains provisions governing the interior environment requirements of all buildings and structures intended for human occupancy.

Section 1201 identifies the scope of the chapter.

Section 1202 provides definitions applicable to terms used in this chapter.

Section 1203 identifies special enclosed spaces where accumulated moisture must be removed by ventilation and establishes criteria (minimum openable area) for the only method of measuring compliance with the natural ventilation requirements for occupied spaces.

Section 1204 identifies the minimum space-heating requirement for interior spaces intended for human occupancy.

Section 1205 requires light for every room or space intended for human occupancy. The method of compliance is the choice of the designer, who may elect to provide artificial instead of natural light. Prescriptive requirements for stairway lighting in dwelling units are also included.

Section 1206 specifies the minimum requirements for courts and yards, including area width, accessibility for cleaning and the location of air intakes when natural light or natural ventilation is the chosen design option.

Section 1207 establishes the sound transmission control requirements for air-borne and structure-borne sound in residential buildings.

Section 1208 addresses the minimum ceiling height for all habitable and occupiable spaces along with other spaces as specified (i.e., hallways, toilet rooms and bathrooms). The minimum floor area for rooms in dwelling units is also specified.

Section 1209 provides minimal opening requirements for access to crawl spaces and attics.

Section 1210 contains requirements for toilet room surfaces and fixture surrounds.

The environmental and physiological justification for the code requirements relevant to light and ventilation of occupiable spaces is based on knowledge, technology and practices developed over centuries of building structures in which humans live and work. These design practices have been further validated by studies during the nineteenth and twentieth centuries.

The greatest impact on these provisions, and those that have resulted in changes during the past 30 years, has been from the interest in energy conservation through construction practices, including the minimization of exterior wall openings.

The requirements related to interior sound transmission and control are considered an important aspect of human comfort in residential occupancies.

Purpose

The purpose of Chapter 12 is to establish minimum conditions for the interior environment of a building. The size of spaces, light, ventilation and noise intrusion are all addressed in order to define the minimum acceptable conditions to which any occupant may be exposed. Design options of natural and mechanical systems are introduced and the criteria for performance are specified.

Even though it was not completely understood, the need for "fresh air" had been recognized for centuries. Designers of centuries-old adobe buildings in the Southwest, hide-covered Indian tepees of the plains and frame houses of early settlers in the East all relied upon the buoyancy of warm air, enabling it to rise and cooler air to flow in to replace it. Whether the design relied on solar energy, thermal mass or even wind velocity to cause the movement, it still reflected a natural movement and, as a result, has been termed "natural ventilation." Only recently have we begun to recognize the reasons for ventilation and the implications of failing to provide an adequate quantity and acceptable quality of air for all occupants. The expression "sick building syndrome" has crept into our vocabulary and reflects the increased understanding of the relationship between interior environment requirements and the physiological well-being of the occupants.

The other purpose of regulating the interior environment is psychological. Merely providing adequate conditions is not sufficient if the occupant does not perceive them as adequate. Minimum space requirements (floor area, yard dimensions or ceiling height) address the need to perceive adequate light, ventilation and space to promote psychological well-being. Regulation of sound transmission also bears directly on the psychological and long-term physical well-being of the occupant.

Finally, adequate lighting from natural sources also meets the physical and psychological needs of the occupants and contributes directly to their overall safety. Safe use of any building under ordinary and emergency conditions depends greatly on proper illumination of the space. This chapter references the *International Mechanical Code*® (IMC®) as the performance standard to which ventilation must be compared and the installation standard for mechanical systems used in buildings regulated by the code.

SECTION 1201
GENERAL

1201.1 Scope. The provisions of this chapter shall govern ventilation, temperature control, lighting, yards and courts, sound transmission, room dimensions, surrounding materials and rodent proofing associated with the interior spaces of buildings.

❖ This section identifies the scope of Chapter 12. The requirements of this chapter are intended to govern and regulate the need for light, ventilation, sound transmission control, interior space dimensions and materials surrounding plumbing fixtures in all buildings. It is the intent of the code that the user must comply with these regulations for all newly constructed buildings and structures and for all buildings and structures, or portions thereof, when there is to be a change of occupancy.

SECTION 1202
DEFINITIONS

1202.1 General. The following words and terms shall, for the purposes of this chapter and as used elsewhere in this code, have the meanings shown herein.

❖ Definitions of terms can help in the understanding and application of the code requirements. The purpose for including these definitions within this chapter is to provide more convenient access to them without having to refer back to Chapter 2. For convenience, these terms are also listed in Chapter 2 with a cross reference to this section. The use and application of all defined terms, including those defined herein, are set forth in Section 201.

SUNROOM ADDITION. A one-story addition added to an existing building with a glazing area in excess of 40 percent of the gross area of the structure's exterior walls and roof.

❖ This terminology is added in order to deal with separate requirements for sunroom additions with regard to ventilation of adjoining spaces (see Section 1203.4.1.1).

THERMAL ISOLATION. A separation of conditioned spaces, between a sunroom addition and a dwelling unit, consisting of existing or new wall(s), doors and/or windows.

❖ This terminology is required for the same reason provided in the definition of "Sunroom addition" (see above).

SECTION 1203
VENTILATION

1203.1 General. Buildings shall be provided with natural ventilation in accordance with Section 1203.4, or mechanical ventilation in accordance with the *International Mechanical Code.*

❖ Every room or space must be provided with ventilation. The selection of natural versus mechanical ventilation on a room-by-room or space-by-space basis is the designer's prerogative. Certain conditions require mechanical exhaust even though natural ventilation systems have been selected by the designer. Section 1203.4.2 of the code and Sections 401.7 and 403 of the IMC direct the treatment of those special situations. Existence of these conditions, however, does not compel mechanical ventilation other than to address the condition. Other rooms or spaces not affected by those conditions may be served by natural systems.

1203.2 Attic spaces. Enclosed attics and enclosed rafter spaces formed where ceilings are applied directly to the underside of roof framing members shall have cross ventilation for each separate space by ventilating openings protected against the entrance of rain and snow. Blocking and bridging shall be arranged so as not to interfere with the movement of air. A minimum of 1 inch (25 mm) of airspace shall be provided between the insulation and the roof sheathing. The net free ventilating area shall not be less than 1/150 of the area of the space ventilated, with 50 percent of the required ventilating area provided by ventilators located in the upper portion of the space to be ventilated at least 3 feet (914 mm) above eave or cornice vents with the balance of the required ventilation provided by eave or cornice vents.

> **Exception:** The minimum required net free ventilating area shall be $1/300$ of the area of the space ventilated, provided a vapor retarder having a transmission rate not exceeding 1 perm in accordance with ASTM E 96 is installed on the warm side of the attic insulation and provided 50 percent of the required ventilating area provided by ventilators located in the upper portion of the space to be ventilated at least 3 feet (914 mm) above eave or cornice vents, with the balance of the required ventilation provided by eave or cornice vents.

❖ All attic spaces and each separate space formed between solid roof rafters are required to be cross ventilated where the ceiling is applied directly to the underside of the roof rafters. Care must be taken, however, to provide cross ventilation in a manner that does not introduce moisture to the attic area.

Snow infiltration can occur when the attic ventilation openings are not sufficiently protected against the entrance of snow or rain, or when more than 50 percent of the ventilation openings are located along the ridge or gable wall of the roof rather than at the eave. When the wind blows perpendicular to a roof ridge vent, a negative pressure builds up across the ridge that draws air out of the attic space through the attic vents. Cross flow of air through the attic can be achieved when outside air is drawn into the attic through the eave or cornice and exits through the ridge or gable vents. In order for this to occur, eave or cornice vents must be greater than or equal to the area of the ridge or gable vents.

Vents that permit snow or rain to infiltrate the attic are not permitted. While there is no specific test standard for this performance aspect of vents, snow infiltration through roof vents can be addressed by what is referred to as "balanced" venting. Balanced venting is providing at least 50 percent of the required ventilating area in the upper third of the space being ventilated (e.g., through ridge or gable vents). The balance of the required ventilation is provided by eave or cornice vents that are greater than or equal to the ventilation area provided by the ridge or gable vents. Most ridge vent manufacturers require slightly more ventilation area in the eaves than provided in the ridge vent to help prevent snow infiltration.

If insufficient eave or cornice ventilation is provided, air will be drawn through the ridge or gable vents. Snow-laden air can enter the attic space through a ridge or gable vent; therefore, it is preferable to have both eave and ridge vents, with the eave vent area being greater than or equal to the ridge vent area (i.e., "balanced" venting).

If an adequate amount of ventilation area in the upper portion of the space is not provided and the ventilation area is provided mainly at the eave or cornice vents, then air will enter and leave the attic space at the eave, and very little cross flow of air will occur.

A test method that has been devised for ridge vent manufacturers involves the use of a 13-foot, 6-inch-diameter (4115 mm) propeller of a 2,650 horsepower aircraft engine wind generator. Snow is simulated with fine, soft wood sawdust, added to the airstream at about 5 pounds (2 kg) per minute. Using this method, the wind speed is varied, because it is not known at what speed the most snow infiltration will occur or the factors that will determine that each wind speed was sustained for a period of 5 minutes. The entire roof system, with all vents, must be installed in the test set-up in order to get a true measure of the potential (or lack thereof) for snow infiltration.

Attic ventilation openings cannot be placed in roof areas subject to snow drifts. These roof areas are subject to greater concentrations of snow, which could increase the chances of snow entering the attic through the ventilation openings. In addition, the ventilation openings are required to be covered with corrosion-resistant mesh or similar material in accordance with Section 1203.2.1.

If roof spaces are not created (e.g., solid concrete roof sections), ventilation is not required, as there is no concealed space for condensation, etc., to accumulate. The amount of area needed for ventilating a roof space is also established in this section. The minimum required ventilating area is reduced in accordance with the exception where the upper ventilation area is provided in the upper portion of the attic space [at least 3 feet (914 mm) above eave or cornice vents] to maximize cross ventilation from the eave or soffit vents. Also required for the reduction of the ventilating area is an approved vapor retarder installed to keep interior moisture from reaching the insulation [see Figure 1203.2(1) for the location of a vapor retarder].

The following example illustrates the calculation of required ventilation areas for an attic space (with only soffit vents) [see Figure 1203.2(2)]:

Note: The area of the attic must include the area of the eave or soffit.

Area of attic	=	1,100 sq. ft.
Eave or soffit vents only	=	$1/150$ of area
Required ventilating area	=	1,100/150
	=	7.33 sq. ft.
	=	1,056 sq. in.
Minimum required ventilation in upper portion of attic	=	.5 x 1,056 sq. in.
	=	528 sq. in. (Balance to be provided in soffit vents.)

Common methods used to provide soffit ventilation include manufactured units, strips or soffit panels and holes or slots (with screening) that meet the criteria and are approved.

Note: When an approved vapor retarder is provided as noted in Figure 1203.2(1), the ventilation requirements can be reduced to $1/300$.

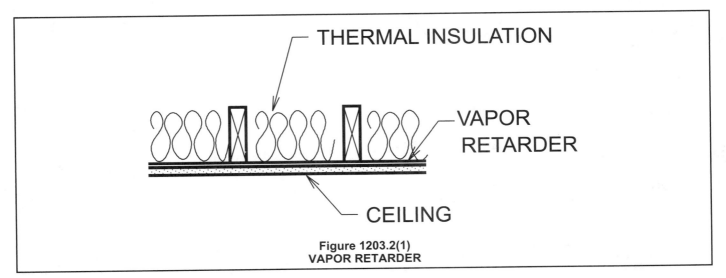

THERMAL INSULATION

VAPOR RETARDER

CEILING

Figure 1203.2(1)
VAPOR RETARDER

The following example illustrates the calculation of required ventilation areas where both ridge and soffit vents are provided [see Figure 1203.2(3)]:

Eave or soffit and ridge or roof vents	$= \frac{1}{300}$ of area
Required ventilating area	$= 1,100/300$
	$= 3.67$ sq. ft.
	$= 528$ sq. in.
Provide 50% by	$= 528 \times 0.5$ ridge or roof vent $= 264$ sq. in.
Provide soffit ventilation	$= 264$ sq.in. total or $264/2(50) = 2.64$ sq. in./ft.

It is important to note that the distribution of ventilation openings should be uniform along the length of the soffits and ridge.

Common methods used to provide roof ridge ventilation include manufactured roof units, ridge vents and gable louvers.

Common methods used to provide soffit ventilation include manufactured units, strips or soffit panels and holes or slots (with screening) that meet the criteria and are approved.

Note that the net-free ventilation area of equipment used for soffit, roof and gable ventilation must be determined. One cannot simply calculate the ventilation area based on the opening created in the roof, wall or soffit. Products vary by manufacturer, and a review of the listing and specifications is necessary to verify actual or net-free ventilation areas.

Where an attic space is not created, but the ceiling membrane is applied directly to the bottom of the solid roof rafters, each rafter space is to be ventilated separately. In this type of installation, it is particularly important that cross ventilation is developed between each rafter space by providing vents at the ridge and eave [see Figure 1203.2(4)].

1203.2.1 Openings into attic. Exterior openings into the attic space of any building intended for human occupancy shall be covered with corrosion-resistant wire cloth screening, hardware cloth, perforated vinyl or similar material that will prevent the entry of birds, squirrels, rodents, snakes and other similar creatures. The openings therein shall be a minimum of $\frac{1}{8}$ inch (3.2 mm) and shall not exceed $\frac{1}{4}$ inch (6.4 mm). Where combustion air is obtained from an attic area, it shall be in accordance with Chapter 7 of the *International Mechanical Code*.

❖ Ventilation openings that would permit the entrance of small animals into the structure must be protected in accordance with this section. Hardware cloth is a particular kind of metal wire cloth screening, and perforated vinyl is a plastic screening or grid with openings of similar dimensions. Metal or vinyl are specified because of their resistance to deterioration over time; therefore whatever material is used, it must be nondeteriorating in addition to having the minimum and maximum opening dimensions of $\frac{1}{8}$ and $\frac{1}{4}$ inches.

Combustion air is air supplied to the room where a fuel-burning appliance is located, so that combustion of the fuel can take place in a safe and complete manner. Chapter 7 of the IMC permits combustion air to be taken from an attic space that is ventilated by openings to the exterior, under certain conditions. Those conditions are based on the configuration of the attic space and the ventilation openings themselves (see Chapter 7 of the IMC and the IMC commentary for more information).

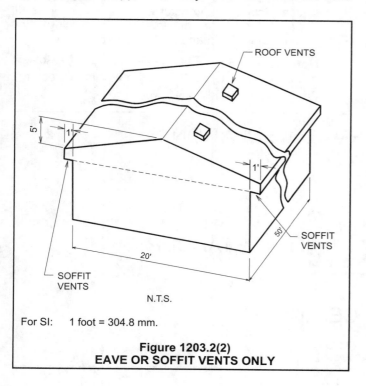

For SI: 1 foot = 304.8 mm.

Figure 1203.2(2)
EAVE OR SOFFIT VENTS ONLY

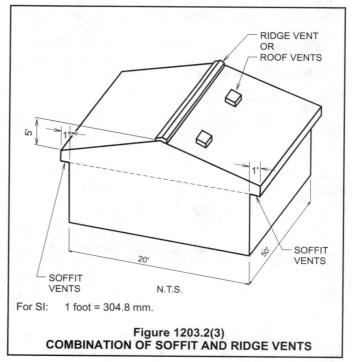

For SI: 1 foot = 304.8 mm.

Figure 1203.2(3)
COMBINATION OF SOFFIT AND RIDGE VENTS

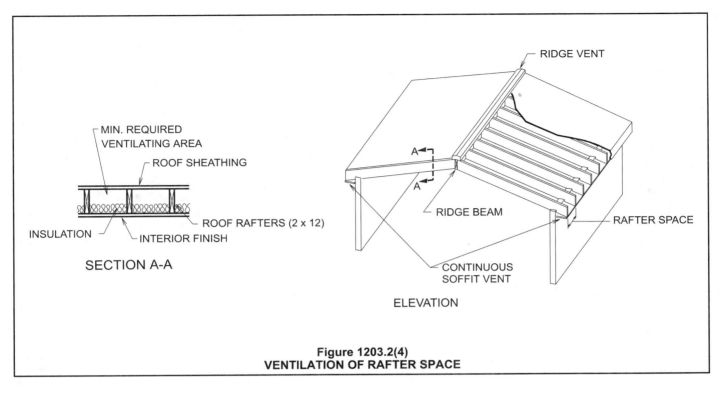

Figure 1203.2(4)
VENTILATION OF RAFTER SPACE

1203.3 Under-floor ventilation. The space between the bottom of the floor joists and the earth under any building except spaces occupied by a basement or cellar shall be provided with ventilation openings through foundation walls or exterior walls. Such openings shall be placed so as to provide cross ventilation of the under-floor space.

❖ The intent of this section is to create an adequate flow of air through crawl spaces to achieve the ventilation goals of controlling temperature, humidity and accumulation of gases. The entire space must be properly ventilated by openings that are distributed to affect cross flow and include corner areas. Although the code does not specify the exact location of openings, an equal distribution of openings on at least three sides of a building, with at least one opening near each corner of the building, is typically sufficient.

Mechanical ventilating devices also can be installed to force air movement and ventilate the space, in which case the location and number of ventilation openings are less critical (see Exception 3 in Section 1203.3.2). The amount of ventilation openings required can be drastically reduced if a vapor retarder is used on the ground surface in the crawl space, in accordance with Exception 2 in Section 1203.3.2. Also, in accordance with Exception 3, when a vapor retarder is used, the installation of operable louvers (to close the openings in the coldest times of the year) is permitted.

1203.3.1 Openings for under-floor ventilation. The minimum net area of ventilation openings shall not be less than 1 square foot for each 150 square feet (0.67 m² for each 100 m²) of crawl-space area. Ventilation openings shall be covered for their height and width with any of the following materials, provided that the least dimension of the covering shall not exceed $^{1}/_{4}$ inch (6 mm):

1. Perforated sheet metal plates not less than 0.070 inch (1.8 mm) thick.

2. Expanded sheet metal plates not less than 0.047 inch (1.2 mm) thick.

3. Cast-iron grills or gratings.

4. Extruded load-bearing vents.

5. Hardware cloth of 0.035 inch (0.89 mm) wire or heavier.

6. Corrosion-resistant wire mesh, with the least dimension not exceeding $^{1}/_{8}$ inch (3.2 mm).

❖ The following is an example of the area calculation: A rectangular building that is 60 feet (18 288 mm) long and 20 feet (6096 mm) wide has a plan area of 1,200 square feet (111.5 m²). The amount of ventilation opening required is 1,200/150 = 8 square feet (0.74 m²) by 144 square inches per square foot = 1,152 square inches (0.74 m²). This is the total (aggregate) amount of ventilation opening that must be distributed among all the openings.

This required amount of openings may be reduced by a factor of 10 if a vapor retarder is used on the ground surface in accordance with Exception 2 of Section 1203.3.2.

The requirement for covering the openings with perforated plates, corrosion-resistant wire mesh or other covering is to keep small animals out. Six alternatives are given for this covering, and they all must have openings that have no dimension exceeding $^1/_4$ inch (6.4 mm).

1203.3.2 Exceptions. The following are exceptions to Sections 1203.3 and 1203.3.1:

1. Where warranted by climatic conditions, ventilation openings to the outdoors are not required if ventilation openings to the interior are provided.

2. The total area of ventilation openings is permitted to be reduced to $^1/_{1,500}$ of the under-floor area where the ground surface is treated with an approved vapor retarder material and the required openings are placed so as to provide cross ventilation of the space. The installation of operable louvers shall not be prohibited.

3. Ventilation openings are not required where continuously operated mechanical ventilation is provided at a rate of 1.0 cubic foot per minute (cfm) for each 50 square feet (1.02 L/s for each 10 m^2) of crawl-space floor area and the ground surface is covered with an approved vapor retarder.

4. Ventilation openings are not required when the ground surface is covered with an approved vapor retarder, the perimeter walls are insulated and the space is conditioned in accordance with the *International Energy Conservation Code.*

5. For buildings in flood hazard areas as established in Section 1612.3, the openings for under-floor ventilation shall be deemed as meeting the flood opening requirements of ASCE 24 provided that the ventilation openings are designed and installed in accordance with ASCE 24.

❖ This section lists the locations and conditions where ventilation openings can be omitted entirely or the area of required openings can be reduced. Exception 1 could be used in extremely cold climates, where ventilation openings are a serious breach of the structure in terms of energy usage. It provides for ventilating the crawl space to the interior conditioned space of the building, which is heated and can accept moisture from the underground space without detrimental effects on the building structure.

The use of a vapor retarder material on the ground surface inhibits the flow of moisture from the ground surface into the crawl space and thus reduces, if not virtually eliminates, the need for ventilation; therefore, Exception 2 provides for a drastic reduction in the amount of ventilation openings required. While the vapor retarder may significantly reduce the moisture accumulation, ventilation openings are still required but may be equipped with manual dampers to permit them to be closed during the coldest weeks of the year in northern climates.

Exception 3 provides for the use of mechanical ventilation, such as an exhaust fan similar to a bathroom exhaust fan, to keep air moving through the crawl space. A vapor retarder on the ground surface is also required when using Exception 3.

When a crawl space is provided with a vapor retarder on the ground surface and is mechanically conditioned

and insulated, it becomes like any other space in the conditioned structure and ventilation openings are not required in accordance with Exception 4. Requirements for insulating structures are found in the *International Energy Conservation Code* (IECC®).

Section 1612.4 of the code requires buildings located in flood hazard areas [as identified by the Federal Emergency Management Agency (FEMA), see Section 1612.3] to be constructed in accordance with ASCE 24. Exception 5 is necessary to coordinate the requirements for openings in ASCE 24 with this section. In most cases, the ventilation requirements of Section 1203.3 are not satisfied by installation of the flood openings that meet the requirements for enclosed areas beneath elevated buildings and structures located in designated flood hazard areas. This is because the flood opening requirements found in ASCE 24-98 and Section 1612.5 specify that flood openings are to be no more than 1 foot (305 mm) above the adjacent grade, while most airflow vents that meet the requirements of Section 1203.3 are installed immediately below the elevated floor. If there is a sufficient number of airflow vents that are located within 1 foot (305 mm) of the adjacent grade, they may satisfy both the flood opening requirements of ASCE 24-98 and Section 1612.5 and the airflow ventilation requirements of Section 1203.3. In either case, both airflow ventilation and flood opening requirements must be met for buildings and structures located in designated flood hazard areas. For further guidance, refer to FEMA publication FIA-TB 1, *Openings in Foundation Walls for Buildings Located in Special Flood Hazard Areas.*

1203.4 Natural ventilation. Natural ventilation of an occupied space shall be through windows, doors, louvers or other openings to the outdoors. The operating mechanism for such openings shall be provided with ready access so that the openings are readily controllable by the building occupants.

❖ This section provides the standard of natural ventilation for all occupied spaces. Openings to the outdoor air, such as doors, windows, louvers, etc., provide natural ventilation. The section does not, however, state or intend that the doors, windows or openings actually be constantly open. The intent is that they be maintained in an operable condition so that they are available for use at the discretion of the occupant.

1203.4.1 Ventilation area required. The minimum openable area to the outdoors shall be 4 percent of the floor area being ventilated.

❖ This section specifies the ratio of openable doors, windows or openings to the floor space being ventilated but does not address the distribution around the space or location of these openings. It is the designer's prerogative to distribute openings in such a manner as to accomplish the natural ventilation of the space. When inadequate natural ventilation is provided, mechanical ventilation can supplement any inadequacy (see Chapter 4 of the IMC). The plan reviewer can determine compliance with this section. For example, in Figure 1203.4.1, the combined openable area (the net-free area of a door, window, louver, vent or skylight, etc., when fully open) of double-hung windows B and C is equal to 4 percent of the floor area [300 × 0.04 = 12 square feet (1 m²)]. The openable area of window A is not required and need not open onto a court or yard complying with Section 1206.

1203.4.1.1 Adjoining spaces. Where rooms and spaces without openings to the outdoors are ventilated through an adjoining room, the opening to the adjoining room shall be unobstructed

> GLAZED AREA (TYP) 4' x 3'
>
> A   B   C
>
> WIDTH
> LENGTH 20'
>
> 15'
>
> FLOOR AREA = (WIDTH) x (LENGTH)
> 15 x 20 = 300 sq.ft.
>
> For SI: 1 foot = 304.8 mm, 1 square foot = 0.0929 m².
>
> **Figure 1203.4.1**
> **NATURAL LIGHT AND VENTILATION WINDOWS**

and shall have an area of not less than 8 percent of the floor area of the interior room or space, but not less than 25 square feet (2.3 m²). The minimum openable area to the outdoors shall be based on the total floor area being ventilated.

Exception: Exterior openings required for ventilation shall be permitted to open into a thermally isolated sunroom addition or patio cover provided that the openable area between the sunroom addition or patio cover and the interior room shall have an area of not less than 8 percent of the floor area of the interior room or space, but not less than 20 square feet (1.86 m²). The minimum openable area to the outdoors shall be based on the total floor area being ventilated.

❖ Adjacent spaces with large connecting openings may share sources of light and ventilation. This section deals with the natural ventilation of connecting interior spaces, and it is the designer's obligation to place openings between rooms with exterior openings and connecting spaces without exterior openings in such a manner as to accomplish natural ventilation of the connected space. For purposes of ventilation, this section establishes a minimum openness requirement for the common wall between a room with openings to the exterior and an interior room without openings to the exterior. The minimum amount of openness required in that common wall is 8 percent of the floor area of the interior room or 25 square feet (2.33 m²), whichever is greater. The openable area of the exterior openings in the "outer" room is required to be equal to or greater than 4 percent (in accordance with Section 1203.4.1) of the total combined floor areas served.

Figure 1203.4.1.1 shows a cut-away of an interior room (Room A) adjacent to a room with openings to the exterior (room B). The openable area of exterior openings in space B is required to be equal to or greater than 0.04 times the area of the entire space (floor area of

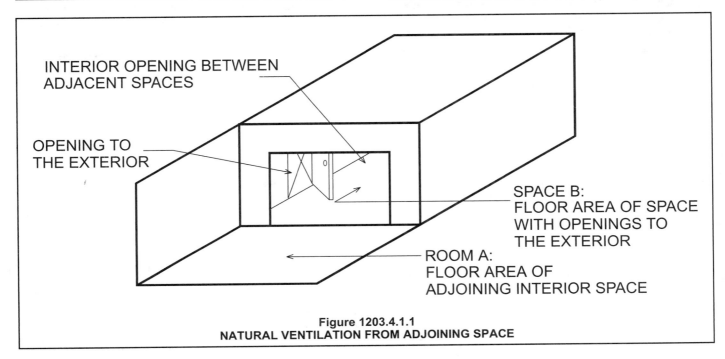

INTERIOR OPENING BETWEEN
ADJACENT SPACES

OPENING TO
THE EXTERIOR

SPACE B:
FLOOR AREA OF SPACE
WITH OPENINGS TO
THE EXTERIOR

ROOM A:
FLOOR AREA OF
ADJOINING INTERIOR SPACE

Figure 1203.4.1.1
NATURAL VENTILATION FROM ADJOINING SPACE

space A plus floor area of interior space B). The opening in the wall between adjacent spaces must be a minimum of 25 square feet (2.33 m²) but not less than .08 times the floor area of interior space A. Since the opening between the adjacent spaces must be unobstructed in accordance with this section, a door cannot be installed in the opening.

The exception deals with a very common circumstance, especially in residential construction. As long as the sunroom addition is large enough and is thermally isolated, the building owner need not move ventilation openings when installing an addition that falls within the definition of "Sunroom addition."

1203.4.1.2 Openings below grade. Where openings below grade provide required natural ventilation, the outside horizontal clear space measured perpendicular to the opening shall be one and one-half times the depth of the opening. The depth of the opening shall be measured from the average adjoining ground level to the bottom of the opening.

❖ This section is applicable whenever occupied spaces below grade are dependent upon natural ventilation through structures like window wells. In order to provide adequate ventilation, this section sets the minimum horizontal clear space adjacent to the opening used for natural ventilation. Without this minimum horizontal area, there will be inadequate air movement through the opening.

As illustrated in Figure 1203.4.1.2, the opening area required for the story below grade intended for human occupancy is:

$$A = 0.04 (L \times W)$$

The area of the window in the vertical plane (w × h) must equal or exceed the required opening area. Additionally, the horizontal dimension from the window to the well wall must equal one and a half times the depth of the openable portion of the window at the lowest

point. If the story below grade is not intended for human occupancy, ventilation is required to be provided in accordance with Section 1203.3 for under-floor spaces.

1203.4.2 Contaminants exhausted. Contaminant sources in naturally ventilated spaces shall be removed in accordance with the *International Mechanical Code* and the *International Fire Code*.

❖ Contaminants in the air are to be collected and exhausted by special means. Chapters 4 and 5 of the IMC specify areas or conditions that must be separately addressed. For example, there are many operations listed in Chapter 5 of the IMC that produce contaminants that cannot be properly or safely treated by natural means. Natural ventilation only anticipates normal occupancy by people and not the extra heat loads, dust, vapors and other contaminants generated by some activities.

1203.4.2.1 Bathrooms. Rooms containing bathtubs, showers, spas and similar bathing fixtures shall be mechanically ventilated in accordance with the *International Mechanical Code*.

❖ Chapter 4 of the IMC contains provisions for bathroom ventilation, requiring mechanical exhaust without recirculation of air at specific rates that depend on the occupancy group.

1203.4.3 Openings on yards or courts. Where natural ventilation is to be provided by openings onto yards or courts, such yards or courts shall comply with Section 1206.

❖ In order that adequate air movement will be provided through openings to naturally ventilated rooms, the openings must directly connect to yards or courts with the minimum dimensions specified in Section 1206.

1203.5 Other ventilation and exhaust systems. Ventilation and exhaust systems for occupancies and operations involving flammable or combustible hazards or other contaminant sources

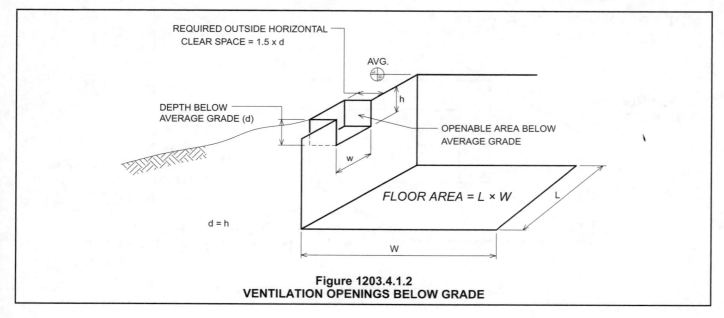

Figure 1203.4.1.2
VENTILATION OPENINGS BELOW GRADE

as covered in the *International Mechanical Code* or the *International Fire Code* shall be provided as required by both codes.

❖ Chapter 5 of the IMC contains specific provisions for hazardous exhaust systems in occupancies such as vehicle repair garages, aircraft fueling stations, dry cleaning plants, spray painting operations and hazardous production materials (HPM) facilities. Many of these provisions are duplicated from the *International Fire Code®* (IFC®), which also contains provisions for the handling and storage of hazardous materials.

SECTION 1204
TEMPERATURE CONTROL

1204.1 Equipment and systems. Interior spaces intended for human occupancy shall be provided with active or passive space-heating systems capable of maintaining a minimum indoor temperature of 68°F (20°C) at a point 3 feet (914 mm) above the floor on the design heating day.

Exception: Interior spaces where the primary purpose is not associated with human comfort.

❖ Heating facilities are required for comfort in all new construction. The systems may be either active (such as a forced-air furnace) or passive (such as solar systems), as long as the specified performance is achieved.

The outdoor design temperatures are taken from the ASHRAE *Handbook of Fundamentals* and are listed in Appendix D of the *International Plumbing Code®* (IPC®). Outdoor design temperatures provide a baseline from which heat load calculations are made. Heating system capacity is dependent upon the predicted outdoor temperatures during the heating season. As the outdoor temperature falls, the heat input to a building must increase to offset the increasing heat losses through the building envelope. Heating systems are designed to have the capacity to maintain the desired indoor temperature when the outdoor temperature is at or above the outdoor design temperature. When the outdoor temperatures are below the outdoor design temperature, the heating system will not be able to maintain a desired indoor temperature. It would be impractical, for example, to design a heating system based on the assumption that someday it might be -20°F (-29°C) outdoors if the outdoor temperature in that region rarely, if ever, dropped that low. In such a case, the heating system would be oversized and, thereby, less efficient and economical.

The winter outdoor design temperature is defined as follows: For 97.5-percent of the total hours in the northern hemisphere heating season, from December through February, the predicted outdoor temperatures will be at or above the values given in Appendix D of the IPC. It would be unreasonable to expect any heating system to maintain a desired indoor temperature when the outdoor temperature is below the design temperature. When the 97.5-percent column in Appendix D of

the IPC is used, it can be assumed that the actual outdoor temperature will be at or below the design temperature for roughly 54 hours of the total of 2,160 hours in the months of December through February (2,160 hours by 2.5% = 54).

SECTION 1205
LIGHTING

1205.1 General. Every space intended for human occupancy shall be provided with natural light by means of exterior glazed openings in accordance with Section 1205.2 or shall be provided with artificial light in accordance with Section 1205.3. Exterior glazed openings shall open directly onto a public way or onto a yard or court in accordance with Section 1206.

❖ This section establishes that an option can be exercised on a room-by-room or space-by-space basis. The option allows the designer to provide either natural light in accordance with this chapter or equivalent levels of artificial lighting.

1205.2 Natural light. The minimum net glazed area shall not be less than 8 percent of the floor area of the room served.

❖ This section establishes the minimum glazed area required based on the floor area served by the window. This is required only for spaces that are not provided with artificial light in accordance with Section 1205.3. It is the intent of the code to establish this ratio as the minimum glazed opening onto yards or courts, in accordance with Section 1205.1.

Early codes set this standard at 10 percent of the floor area served. This ratio was derived from certain architectural styles that yielded adequate light and ventilation; however, this is a more than adequate amount and has been reduced to the current levels because of energy conservation issues. Openings in excess of that minimum area are permitted to open onto areas other than a complying court or yard. In Figure 1203.4.1, the room dimensions are 15 feet times 20 feet (4572 mm times 6096 mm), or 300 square feet (27.9 m²) of area. If windows B and C are double hung, with a combined glazed area of 24 square feet (2.23 m²), they provide the minimum area required of 8 percent of the floor area (24/300 = .08). In this example, glazing unit A is not required for natural light; therefore, it need not face onto a required yard or court.

1205.2.1 Adjoining spaces. For the purpose of natural lighting, any room is permitted to be considered as a portion of an adjoining room where one-half of the area of the common wall is open and unobstructed and provides an opening of not less than one-tenth of the floor area of the interior room or 25 square feet (2.32 m²), whichever is greater.

Exception: Openings required for natural light shall be permitted to open into a thermally isolated sunroom addition or patio cover where the common wall provides a glazed area of not less than one-tenth of the floor area of the interior room or 20 square feet (1.86 m²), whichever is greater.

❖ In a case where a space (or room) has no glazed area open to the required courts or yards but is adjacent to one that does, it may "borrow" natural lighting from the adjacent space if (1) the wall between the adjoining spaces is at least one-half open and unobstructed; (2) the opening equals at least 10 percent of the floor area of the interior space and (3) the opening is not less than 25 square feet (2.33 m²). The required glazed area facing the required court or yard must not be less than 8 percent of the total floor area of all rooms served.

For example, in Figure 1205.2.1, the glazed area in space B is required to be equal to or greater than 0.08 (floor area of space A + floor area of space B).

In the figure, the opening between the adjacent spaces must meet all three criteria: the wall must be at least half open and unobstructed it must be a minimum of 25 square feet (2.33 m²) and it must be not less than one tenth of the floor area of space A.

The exception deals with a very common circumstance, especially in residential construction. As long as the sunroom addition is large enough and is thermally isolated, the building owner need not move openings for lighting when installing an addition that falls within the definition of "Sunroom addition."

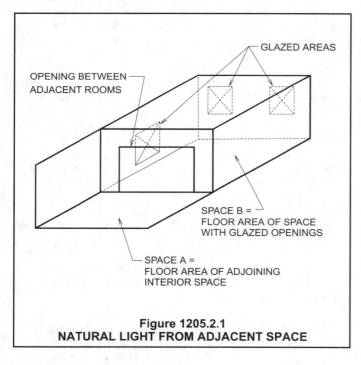

Figure 1205.2.1
NATURAL LIGHT FROM ADJACENT SPACE

1205.2.2 Exterior openings. Exterior openings required by Section 1205.2 for natural light shall open directly onto a public way, yard or court, as set forth in Section 1206.

Exceptions:

1. Required exterior openings are permitted to open into a roofed porch where the porch:

   1.1. Abuts a public way, yard or court.

   1.2. Has a ceiling height of not less than 7 feet (2134 mm).

   1.3. Has a longer side at least 65 percent open and unobstructed.

2. Skylights are not required to open directly onto a public way, yard or court.

❖ In order that enough light will be provided through openings to naturally lit rooms, the openings must open onto yards or courts with the minimum dimensions specified in Section 1206. Skylights admit light directly from above and, therefore, are not required to face a court or yard in accordance with Exception 2. Exception 1 gives the criteria by which a roofed porch may be located directly outside required openings without significantly obstructing the entrance of light to the space.

1205.3 Artificial light. Artificial light shall be provided that is adequate to provide an average illumination of 10 foot-candles (107 lux) over the area of the room at a height of 30 inches (762 mm) above the floor level.

❖ The section establishes the minimum required illumination for rooms without the minimum required natural light (see Figure 1205.3).

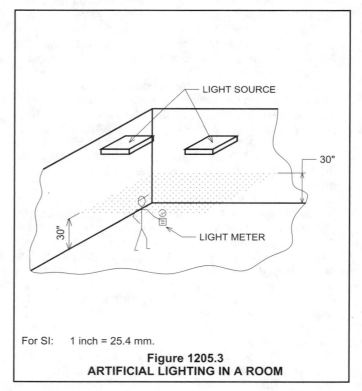

For SI: 1 inch = 25.4 mm.

Figure 1205.3
ARTIFICIAL LIGHTING IN A ROOM

1205.4 Stairway illumination. Stairways within dwelling units and exterior stairways serving a dwelling unit shall have an illumination level on tread runs of not less than 1 foot-candle (11 lux). Stairs in other occupancies shall be governed by Chapter 10.

❖ Stairway illumination is essential to the safe use of stairs during normal use as well as during egress in an emergency. The lighting must be operable by switches in the vicinity of the stair, located as required by the ICC *Electrical Code®* (ICC EC™) and the *National Electrical Code*. Emergency egress lighting, also referred to as "means of egress illumination," is required in occupancies other than dwelling units at a lower rate of illumination (see commentary, Sections 1006 and 1205.5).

1205.4.1 Controls. The control for activation of the required stairway lighting shall be in accordance with the ICC *Electrical Code*.

❖ The *National Electrical Code*, referenced by the ICC EC, provides for controls at the top and bottom of stairways within dwelling units, allowing an occupant to illuminate the stairways before traversing any stairs, regardless of the direction of travel. Illuminated switches, where required, allow an occupant to quickly find the switches when the stairs are dark.

Illumination controls for exterior stairs that are operable from the inside of a dwelling unit allow an occupant to safely egress by activating exterior stair illumination prior to leaving the building. Exterior stairs must be provided with the minimum illumination level specified in Section 1205.4.

1205.5 Emergency egress lighting. The means of egress shall be illuminated in accordance with Section 1006.1.

❖ Means of egress illumination is required in all buildings to allow occupants enough light to negotiate the exit access (such as corridors) and exits (such as enclosed stairs) at all times the building is occupied (see commentary, Section 1006.1).

SECTION 1206
YARDS OR COURTS

1206.1 General. This section shall apply to yards and courts adjacent to exterior openings that provide natural light or ventilation. Such yards and courts shall be on the same property as the building.

❖ These provisions are intended to regulate those exterior areas of a building or structure that are supposed to supply required natural light or ventilation to interior spaces. These requirements are intended to increase the likelihood that the exterior walls are provided with enough adjacent open space to allow the required light and ventilating air to freely enter the exterior wall openings. These exterior areas are defined as courts and yards. Courts and yards must be open, uncovered and on the same lot as the building. They may be either partly or wholly enclosed by the building. Requirements are provided in Section 1206, which regulates the minimum width, area, air intake and drainage of courts and yards. The requirements of Sections 1206.2 through 1206.3.3 do not apply if artificial ventilation and lighting is provided for the spaces opening onto the court or yard in accordance with Section 1203.1 or 1205.1.

1206.2 Yards. Yards shall not be less than 3 feet (914 mm) in width for one- and two-story buildings. For buildings more than two stories in height, the minimum width of the yard shall be increased at the rate of 1 foot (305 mm) for each additional story. For buildings exceeding 14 stories in height, the required width of the yard shall be computed on the basis of 14 stories.

❖ A yard is distinguished from a court by the definitions in Chapter 2 (see commentary, Chapter 2). A court is bounded on at least three sides, whereas a yard is open at both ends.

The required width of a yard is measured perpendicular from the face of the wall to the opposing wall on the other side of the yard. A five-story building would be required to have a yard at least 6 feet (1829 mm) in width [3 feet (914 mm) plus 1 foot (305 mm) each for stories three through five]. A 20-story building is required to have a yard at least 15 feet (4572 mm) [3 feet (914 mm) plus 1 foot (305 mm) each for stories three through 14]. The last sentence of the section simply requires a minimum yard width of 15 feet (4572 mm) for all buildings over 14 stories in height. If the building is adjacent to a court rather than a yard, the requirements of Section 1206.3 apply. Neither Section 1206.2 nor 1206.3 is applicable if artificial lighting and ventilation is provided for spaces facing the yard or court in accordance with Sections 1203.1 and 1205.1.

1206.3 Courts. Courts shall not be less than 3 feet (914 mm) in width. Courts having windows opening on opposite sides shall not be less than 6 feet (1829 mm) in width. Courts shall not be less than 10 feet (3048 mm) in length unless bounded on one end by a public way or yard. For buildings more than two stories in height, the court shall be increased 1 foot (305 mm) in width and 2 feet (310 mm) in length for each additional story. For buildings exceeding 14 stories in height, the required dimensions shall be computed on the basis of 14 stories.

❖ Courts are defined in Chapter 2 as being bounded on no less than three sides by walls or other enclosing construction. A court adjacent to a five-story building would be required to have a width measured perpendicular from the wall facing the court of at least 6 feet (1829 mm) when required openings are on one wall and 9 feet (2743 mm) when required openings are on opposing walls [3 feet (914 mm) plus 1 foot (305 mm) each for stories three through five, or 6 feet (1829 mm) plus 1 foot (305 mm) each for stories three through five]. If the same court is bounded on all sides, the required minimum length would be 16 feet (4877 mm) [10 feet (3048 mm) plus 2 feet (610 mm) each for stories three through five]. The last sentence simply requires all buildings higher than 14 stories to have a minimum court width of 15 feet (4572 mm) without opposing required openings, a minimum width of 18 feet (5486 mm) where required openings oppose each other and a minimum length of 34 feet (10 363 mm) if bounded on all sides [width equals 3 feet (914 mm) or 6 feet (1829 mm) plus 1 foot (305 mm) each for stories three through 14, and length equals 10 feet (3048 mm) plus 2 feet (610 mm) each for stories three through 14]. The requirements of this section are not applicable if artificial lighting and ventilation are provided for spaces facing the court in accordance with Sections 1203.1 and 1205.1.

1206.3.1 Court access. Access shall be provided to the bottom of courts for cleaning purposes.

❖ Courts must be conveniently accessed for maintenance. Clearly, a court intended to be a source of ventilation air must be maintained in a manner conducive to its purpose.

1206.3.2 Air intake. Courts more than two stories in height shall be provided with a horizontal air intake at the bottom not less than 10 square feet (0.93 m²) in area and leading to the exterior of the building unless abutting a yard or public way.

❖ This section is applicable only to courts that are bounded on all four sides by walls or other construction. A fully bounded court takes on characteristics similar to a chimney during summer weather when the building mass is heated from the daytime sun. In order for a fully bounded court to function as an efficient source of natural ventilation, the bottom of the court must have a source of fresh air (similar to a chimney). This source of fresh air is supplied through the required opening of 10 square feet (0.93 m²) connected directly to a street or yard. The requirements of this section are not applicable if artificial lighting and ventilation are provided for spaces facing the court in accordance with Sections 1203.1 and 1206.1.

1206.3.3 Court drainage. The bottom of every court shall be properly graded and drained to a public sewer or other approved disposal system complying with the *International Plumbing Code.*

❖ A court is an inherent water trap. A court that is not both graded and drained will accumulate water and remain in a saturated condition, which will promote an insanitary condition, including odors. Based on the design and nature of the soil after construction, paving the court may be the best solution to eliminate a problem.

SECTION 1207
SOUND TRANSMISSION

1207.1 Scope. This section shall apply to common interior walls, partitions and floor/ceiling assemblies between adjacent dwelling units or between dwelling units and adjacent public areas such as halls, corridors, stairs or service areas.

❖ Since noise transmission can be quantified and affects the quality of life, the code incorporates regulations that address noise transmission in multiple-family residential construction, wherein the occupants may have no control over noise. The regulated components of con-

struction are those through which noise is primarily transmitted.

1207.2 Air-borne sound. Walls, partitions and floor/ceiling assemblies separating dwelling units from each other or from public or service areas shall have a sound transmission class (STC) of not less than 50 (45 if field tested) for air-borne noise when tested in accordance with ASTM E 90. Penetrations or openings in construction assemblies for piping; electrical devices; recessed cabinets; bathtubs; soffits; or heating, ventilating or exhaust ducts shall be sealed, lined, insulated or otherwise treated to maintain the required ratings. This requirement shall not apply to dwelling unit entrance doors; however, such doors shall be tight fitting to the frame and sill.

❖ The code requires common walls between dwelling units and between dwelling units and public areas to have a minimum sound transmission class (STC) of 50. The STC is a measure of an assembly's ability to resist sound transmission. The higher the number (rating), the higher the resistance (less sound transmission). Standard architectural wall construction assemblies have been tested for sound transmission ratings, and reference to the construction specifications will yield such information. Air-borne noise originates in the air, such as voice or music. For structure-borne sound, see the commentary to Section 1207.3.

As a rule, vertical assemblies meeting the requirements of this section consist of double walls or walls containing insulation similar to exterior walls.

1207.3 Structure-borne sound. Floor/ceiling assemblies between dwelling units or between a dwelling unit and a public or service area within the structure shall have an impact insulation class (IIC) rating of not less than 50 (45 if field tested) when tested in accordance with ASTM E 492.

❖ The impact insulation class (IIC) is a measure of an assembly's ability to resist sound transmission. The higher the number (rating), the higher the resistance (less sound transmission). Floors between dwelling units and those between dwelling units and public areas are required to have a minimum IIC rating of 50.

Usually, floor assemblies that are carpeted meet the minimum requirement for an IIC rating of 50. Other areas with hard-surfaced finishes may require additional treatment or insulation to comply with these requirements.

There are various resource documents containing STC ratings and IIC ratings, including the following: GA 600, *Fire Resistance Design Manual*, by the Gypsum Association; NCMA TEK 69A, *New Data on Sound Reduction with Concrete Masonry Walls*, by the National Concrete Masonry Association and BIA TN 5A, *Sound Insulation—Clay Masonry Walls*, by the Brick Institute of America. These or any other similar sources can be submitted as a basis for approval once it is demonstrated to the building official that the data are based on ASTM E 90 and E 492.

SECTION 1208
INTERIOR SPACE DIMENSIONS

1208.1 Minimum room widths. Habitable spaces, other than a kitchen, shall not be less than 7 feet (2134 mm) in any plan dimension. Kitchens shall have a clear passageway of not less than 3 feet (914 mm) between counter fronts and appliances or counter fronts and walls.

❖ This room dimension provision specifies the minimum horizontal dimensions required for all habitable rooms. Any room that functions as a living room, bedroom, dining room or any other similar habitable room must be sized such that a cylinder with a diameter of 7 feet (2134 mm) may be placed in it. Only kitchens are exempt from this requirement. This code provision allows the circulation of ventilation air through the space while maintaining reasonably sized living quarters for the occupants.

1208.2 Minimum ceiling heights. Occupiable spaces, habitable spaces and corridors shall have a ceiling height of not less than 7 feet 6 inches (2286 mm). Bathrooms, toilet rooms, kitchens, storage rooms and laundry rooms shall be permitted to have a ceiling height of not less than 7 feet (2134 mm).

Exceptions:

1. In one- and two-family dwellings, beams or girders spaced not less than 4 feet (1219 mm) on center and projecting not more than 6 inches (152 mm) below the required ceiling height.

2. If any room in a building has a sloped ceiling, the prescribed ceiling height for the room is required in one-half the area thereof. Any portion of the room measuring less than 5 feet (1524 mm) from the finished floor to the ceiling shall not be included in any computation of the minimum area thereof.

3. Mezzanines constructed in accordance with Section 505.1.

❖ Occupiable spaces or rooms (including habitable spaces) are required to have a specific minimum ceiling height. Bathrooms, toilet rooms, kitchens, storage rooms and laundry rooms are permitted to have a lower minimum ceiling height in accordance with this section. Ceiling height is one of the variables that affects the circulation of air in a space. Additionally, there is a psychological need for spaciousness in a living space or in one of the accessory spaces.

Figure 1208.2(1) illustrates the application of Exception 1 for beams and girders spaced no more than 4 feet (1219 mm) on center in one- and two-family dwellings.

Figure 1208.2(2) illustrates the application of Exception 2. Rooms with sloped ceilings are required to meet two distinct conditions. First, the area of the room having a floor-to-ceiling clearance of less than 5 feet (1524 mm) does not contribute to the minimum floor area required by Section 1208.3. Second, at least one-half of the actual total area of the room must meet the minimum ceiling height requirements of Section 1208.2 [see Figure 1208.2(2)].

Finally, Exception 3 is consistent with Section 505.1,

which establishes the minimum ceiling height for mezzanines at 7 feet (2134 mm). In accordance with Section 505.2, mezzanines cannot exceed one-third of the area of the room in which they are located.

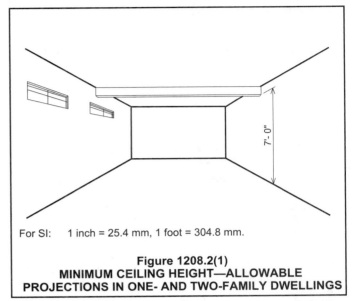

For SI: 1 inch = 25.4 mm, 1 foot = 304.8 mm.

Figure 1208.2(1)
**MINIMUM CEILING HEIGHT—ALLOWABLE
PROJECTIONS IN ONE- AND TWO-FAMILY DWELLINGS**

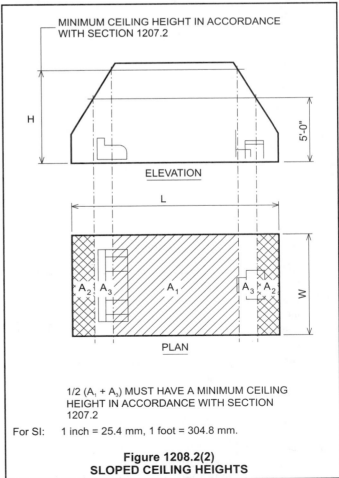

$1/2 (A_1 + A_3)$ MUST HAVE A MINIMUM CEILING HEIGHT IN ACCORDANCE WITH SECTION 1207.2

For SI: 1 inch = 25.4 mm, 1 foot = 304.8 mm.

Figure 1208.2(2)
SLOPED CEILING HEIGHTS

1208.2.1 Furred ceiling. Any room with a furred ceiling shall be required to have the minimum ceiling height in two-thirds of the area thereof, but in no case shall the height of the furred ceiling be less than 7 feet (2134 mm).

❖ This section only applies to rooms required to have a ceiling height of no less than 7 feet, 6 inches (2286 mm). In Figure 1208.2.1, floor area A_1 must be greater than or equal to two-thirds (length times width). Note that only those ceiling heights furred to a height of less than 7 feet, 6 inches (2286 mm) affect area A_1.

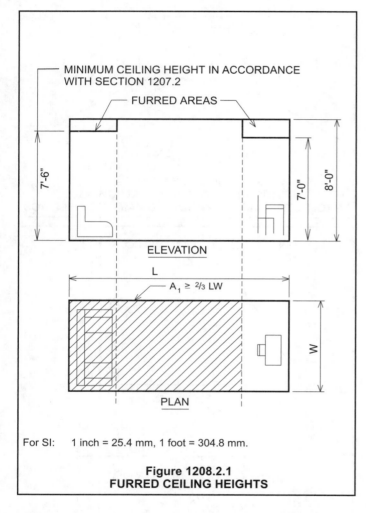

MINIMUM CEILING HEIGHT IN ACCORDANCE WITH SECTION 1207.2

FURRED AREAS

7'-6"

7'-0"

8'-0"

ELEVATION

L

$A_1 \geq \frac{2}{3} LW$

W

PLAN

For SI: 1 inch = 25.4 mm, 1 foot = 304.8 mm.

Figure 1208.2.1
FURRED CEILING HEIGHTS

1208.3 Room area. Every dwelling unit shall have at least one room that shall have not less than 120 square feet (13.9 m²) of net floor area. Other habitable rooms shall have a net floor area of not less than 70 square feet (6.5 m²).

Exception: Every kitchen in a one- and two-family dwelling shall have not less than 50 square feet (4.64 m²) of gross floor area.

❖ This section applies only to dwelling units. A minimum area of 120 square feet (13.9 m²) for at least one room of each dwelling unit and a minimum of 70 square feet (6.5 m²) for all other habitable rooms, except kitchens, are the minimum standards for habitable rooms. These

minimums reflect the physiological requirements of light and ventilation and also preserve the individual's perception of space and the elements necessary for a psychological sense of well-being. The code does not regulate the minimum area of rooms and spaces in other than residential occupancies.

The exception is consistent with minimum requirements for kitchens given in the *International Residential Code®* (IRC®). Kitchens in one- and two-family dwelling units are required to be 50 square feet (4.65 m²) minimum.

1208.4 Efficiency dwelling units. An efficiency living unit shall conform to the requirements of the code except as modified herein:

1. The unit shall have a living room of not less than 220 square feet (20.4 m) of floor area. An additional 100 square feet (9.3 m) of floor area shall be provided for each occupant of such unit in excess of two.

2. The unit shall be provided with a separate closet.

3. The unit shall be provided with a kitchen sink, cooking appliance and refrigeration facilities, each having a clear working space of not less than 30 inches (762 mm) in front. Light and ventilation conforming to this code shall be provided.

4. The unit shall be provided with a separate bathroom containing a water closet, lavatory and bathtub or shower.

❖ Efficiency units are very small apartments consisting of one or two rooms and a bathroom. Efficiency units that comply with this section are not required to comply with the minimum area requirements in Section 1208.3; however, the total allowable number of occupants in the dwelling is limited, depending on the area of the unit in accordance with Item 1. The purpose of both efficiency units and this section is to provide for combined use of spaces in an economical or "efficient" manner without jeopardizing health or comfort. This is possible because of the occupant limitation.

In addition to the living room, the apartment must have a separate closet and bathroom. There are no minimum area requirements for these spaces; however, the fixture clearances in the bathroom must be as required in the IPC. There is not a requirement for a separate kitchen, but if the required sink and appliances are located in the living room, they cannot encroach on the required minimum floor space.

SECTION 1209
ACCESS TO UNOCCUPIED SPACES

1209.1 Crawl spaces. Crawl spaces shall be provided with a minimum of one access opening not less than 18 inches by 24 inches (457 mm by 610 mm).

❖ The requirements of this section establish 18 inches by 24 inches (457 mm by 610 mm) as the minimum size opening for crawl spaces. If access is through a wall, the 18-inch (457 mm) minimum would be the height and the

24-inch (610 mm) minimum would be the width.

Items such as plumbing and wiring installations pass through crawl spaces at times. Required initial and periodic inspections, and repairs cannot be carried out without access to such crawl spaces.

1209.2 Attic spaces. An opening not less than 20 inches by 30 inches (559 mm by 762 mm) shall be provided to any attic area having a clear height of over 30 inches (762 mm). A 30-inch (762 mm) minimum clear headroom in the attic space shall be provided at or above the access opening.

❖ Access to the attic provides a convenient and nondestructive means for fire department personnel to visually check for an attic fire and, if need be, gain entry to the concealed spaces and suppress a fire. Access to attic spaces can be provided through the ceiling within each compartment that is created by draftstops or through openings within the draftstops themselves. Openings located within the draftstop are required to be self-closing and the opening protective must provide structural fire integrity (the ability to remain in place) similar to the draftstop. Access is required when the attic space has a clear height greater than 30 inches (762 mm) measured from the top of the ceiling joists (or top of the floor sheathing, if present) to the underside of the roof rafters.

1209.3 Mechanical appliances. Access to mechanical appliances installed in under-floor areas, in attic spaces and on roofs or elevated structures shall be in accordance with the *International Mechanical Code.*

❖ Access to mechanical appliances is needed to maintain and service the equipment. See Section 306 in the IMC for detailed requirements.

SECTION 1210
SURROUNDING MATERIALS

1210.1 Floors. In other than dwelling units, toilet and bathing room floors shall have a smooth, hard, nonabsorbent surface that extends upward onto the walls at least 6 inches (152 mm).

❖ The purpose of this requirement is to provide nonabsorbent surfaces that can be maintained in a sanitary condition. The 6-inch (152 mm) extension of the surface up the surrounding walls is so that the wall will not absorb moisture during cleaning and, thus, will be left in a clean condition.

1210.2 Walls. Walls within 2 feet (610 mm) of urinals and water closets shall have a smooth, hard, nonabsorbent surface, to a height of 4 feet (1219 mm) above the floor, and except for structural elements, the materials used in such walls shall be of a type that is not adversely affected by moisture.

Exceptions:

1. Dwelling units and sleeping units.

2. Toilet rooms that are not accessible to the public and which have not more than one water closet.

Accessories such as grab bars, towel bars, paper dispensers and soap dishes, provided on or within walls, shall be installed and sealed to protect structural elements from moisture.

❖ The walls near urinals and water closets need to have the surface specified in this section since they are subject to moisture. Exception 1 recognizes that water closet facilities in dwelling units and sleeping units are not exposed to as much use as those that serve the public and, thus, are easier to maintain. Exception 2 acknowledges that toilet fixtures that do not serve the public are also subject to less use.

This section requires protection of the structural supports for accessories so that they will maintain their strength.

1210.3 Showers. Shower compartments and walls above bathtubs with installed shower heads shall be finished with a smooth, nonabsorbent surface to a height not less than 70 inches (1778 mm) above the drain inlet.

❖ The 70-inch (1778 mm) requirement in this section is based on the height of the shower compartment and walls that are exposed to significant moisture that would cause the surface to become insanitary over a long period of time.

1210.4 Waterproof joints. Built-in tubs with showers shall have waterproof joints between the tub and adjacent wall.

❖ The joint between the tub and wall must be sealed to prevent moisture from getting into the supporting floor and framing. Waterproof joints are also needed to keep the concealed area of the wall in a sanitary condition.

1210.5 Toilet rooms. Toilet rooms shall not open directly into a room used for the preparation of food for service to the public.

❖ The requirement that toilet rooms not open directly into rooms where food is prepared for the public is necessary to keep the food preparation area in a sanitary condition.

BIBLIOGRAPHY

The following resource materials are referenced in this chapter or are relevant to the subject matter addressed in this chapter.

ASCE 24-98, *Flood Resistance Design and Construction Standard.* Reston, VA: American Society of Civil Engineers, 1998.

ASHRAE-2001, *ASHRAE Fundamentals Handbook.* Atlanta: American Society of Heating, Refrigerating and Air-Conditioning Engineers, Inc., 2001.

BIA TN 5A-83, *Sound Insulation—Clay Masonry Walls.* Reston, VA: Brick Institute of America, 1983.

FEMA FIA-TB 1, *Openings in Foundation Walls for Buildings Located in Special Flood Hazard Areas.* Washington, DC: Federal Emergency Management Agency, 1997.

GA 600-00, *Fire Resistance Design Manual*. Evanston, IL: Gypsum Association, 1997.

NCMA TEK 69A-92, *New Data on Sound Reduction with Concrete Masonry Walls*. Herndon, VA: National Concrete Masonry Association, 1992.

Chapter 13:
Energy Efficiency

General Comments

Chapter 13 provides for the design and construction of energy-efficient buildings and structures or portions thereof intended primarily for human occupancy by direct reference to the *International Energy Conservation Code*® (IECC®).

The need for energy conservation can be traced to the increased demand for energy in this country coupled with the decline of domestic energy resource development. The vulnerability of a one-dimensional, "national" outlook was illustrated to our nation by the Arab States' oil embargo of 1973. This event highlighted the United States' dependency on foreign sources of energy and awakened the nation to the crippling effects that might occur should offshore supply lines be interrupted. In 1975, the American Society of Heating, Refrigerating and Air-Conditioning Engineers, Inc. (ASHRAE) published Standard 90. This standard was a culmination of efforts that began in 1972, when the National Conference of States on Building Codes and Standards, Inc. (NCSBCS) voted to request continued National Bureau of Standards (NBS) activities investigating the feasibility of a building-related standard for energy conservation.

In August 1973, NBS agreed to develop design and evaluation criteria for energy conservation in new buildings. In February 1974, ASHRAE accepted the responsibility to develop a national voluntary consensus standard based on the NBS document. After two public reviews, ASHRAE Standard 90-75 (Standard 90) was approved for publication in August 1975. This standard was subsequently revised in 1980, and the first nine sections were published as ASHRAE Standard 90A (Standard 90A), with the remainder being published as ASHRAE Standards 90B and 90C.

Over the next several years, all 50 states eventually enacted regulations or developed their own regional-based, energy-related codes based on the 1975 edition of Standard 90, or one of several regional-based energy codes that also used the standard for technical support.

First published in 1983, the Council of American Building Officials (CABO) *Model Energy Code* (MEC) was also based on the Standard 90 series, specifically the 1980 edition of Standard 90A. The CABO MEC was developed jointly by the Building Officials and Code Administrators International® (BOCA®), the International Conference of Building Officials® (ICBO®), NCSBCS and Southern Building Code Congress International (SBCCI), under a contract funded by the United States Department of Energy (DOE).

During that same year, the results of an extensive research program initiated by ASHRAE and DOE into energy conservation in building design demonstrated that significant cost-effective improvements could be made to the existing Standard 90 series. The ASHRAE Standing Standards Project Review Committee (SSPRC) 90R, thereby became responsible for maintaining the provisions the Standard 90 series applicable to buildings "other than low-rise residential buildings" (i.e., buildings three stories or less in height). It took six years to finalize revisions to Standards 90A and 90B as Standard 90.1. After three public reviews and two appeals, these revisions were published in 1989 as ASHRAE/IES Standard 90.1 (Standard 90.1). Since 1989, numerous addenda to this standard have been developed. Some have been published and others are undergoing continuous development and review. After 10 years and several exhaustive public review periods, the successor to the 1989 standard, ASHRAE/IESNA Standard 90.1-1999, received final approval in June 1999.

With the 1990 Iraqi invasion of Kuwait, lawmakers in Washington again saw the need to lessen the nation's precarious dependence on foreign sources of oil. The federal government issued a mandate to regulate energy usage for the United States. This federal mandate began as two congressional bills: the National Energy Efficiency Act of 1991 for the House of Representatives and the National Energy Security Act of 1991 for the Senate. After months of debate, the bills were combined into one document that was renamed the Energy Policy Act of 1992 (EPAct). On October 24, 1992, then President George Herbert Walker Bush signed EPAct into law (Public Law 102-486).

EPAct established the 1992 CABO MEC (applicable to detached one- and two-family dwellings and low-rise residential buildings three stories or less in height) and ASHRAE Standard 90.1 (applicable to all other buildings) as the acceptance criteria for several building energy-related requirements. By October 24, 1994, each state had to certify to the secretary of energy that it had reviewed the provisions of its residential building code regarding energy efficiency and made a determination as to whether a revision of that code was needed to meet or exceed the 1992 CABO MEC. The states were not required to update their residential building energy codes; only to review the code and determine whether it was appropriate to update. If, for whatever reasons, a state determined that it was not feasible to revise its residential energy code, the state was required to submit to the secretary of energy, in writing, the reasons for such determination.

EPAct also mandates that whenever a new edition of the CABO MEC is published, the secretary of energy has one year to make a determination as to whether the new edition "would improve energy efficiency in residential buildings." States then have two years from this determination to repeat the review process previously described. This analysis is also intended to assist the Department of Housing and Urban Development (HUD) in determining whether the latest edition of the CABO MEC meets EPAct's criteria for justifying its adoption for HUD loan programs.

Since the signing of EPAct in October 1992, two subsequent editions of the CABO MEC have been published and made available for adoption by the states: the 1993 and 1995 editions. While the CABO MEC is revised every three years, it was published again in 1993 because of a greater-than-average number of changes introduced during the 1992 CABO Code Development Cycle.

The 1993 edition introduced more stringent ceiling and wall insulation requirements for single-family and low-rise multiple-family buildings in warmer (southern) locations; had new requirements for heating, ventilating and air-conditioning equipment efficiencies, which are consistent with the National Appliance Energy Conservation Act of 1987 (Public Law 100-12); had less stringent requirements for duct insulation and adopted by reference Standard 90.1 for commercial buildings and high-rise residential buildings.

The 1995 edition added a reference to the National Fenestration Rating Council (NFRC) standard for glazing U-factors and provided default U-factors for a variety of fenestration types; added criteria to specifically account for metal frame wall construction in building thermal envelope calculations; strengthened the duct-sealing provisions and applied them to all supply and return ducts; adopted by reference the 1993 ASHRAE *Handbook of Fundamentals* in place of the 1989 edition, thereby directing users to assume a higher fraction of wall area as framing and adopted the 1993 edition of ASHRAE 90.1—the codified counterpart to Standard 90.1 by reference in place of the then-current reference to Standard 90.1.

On July 15, 1994, the DOE issued a Federal Register notice declaring that the 1993 CABO MEC "would achieve greater energy efficiency in residential buildings" than the 1992 CABO MEC (94 FR 17259). On December 6, 1996, the DOE issued a similar finding that the 1995 CABO MEC "would achieve greater energy efficiency in low-rise residential buildings" than the 1993 CABO MEC (96 FR 31065). In tandem, it was implied in these determinations that the states must certify whether it was appropriate to revise their residential building energy code to meet or exceed the 1995 CABO MEC by December 6, 1998. Furthermore, these determinations justified the adoption of the 1995 CABO MEC for HUD-assisted housing.

Shortly thereafter, effective December 4, 1995, CABO assigned all rights and responsibilities related to the MEC to the International Code Council® (ICC®). Through its efforts to develop a complete set of international construction codes without regional limitations and to provide proper interface with the *International Codes*®, the ICC subsequently introduced the first edition of the IECC in February 1998. In its first edition, the IECC incorporated the provisions of the 1995 MEC promulgated by CABO and included the technical content of the MEC as modified by the approved changes from the remaining 1995, 1996 and 1997 CABO Code Development Cycles. Note that until the publishing of the 1998 IECC, code development activities during 1995, 1996 and 1997 were carried out under the CABO code change procedures.

Significant changes that were incorporated in the 1998 IECC include: an organizational restructuring of the code's chapters to accommodate differences in format between what was a CABO code and is now an ICC *International Code*; the addition of maximum solar heat gain coefficient (SHGC) criteria for glazing in cooling-dominated climates; revisions to the default U-factor tables for fenestration products; heat traps for storage hot water tanks in noncirculating systems and the addition of an entirely new simplified compliance approach for commercial buildings not over three stories in height having "reasonable" glass areas and "simple" mechanical systems.

In its 2000 edition, the IECC expanded in scope to include energy-related provisions for all commercial buildings, with the exception of those having more than 50-percent glazing in the gross area of above-grade walls. Comprehensive language was also added to introduce useable provisions specific to complex mechanical systems. A simplified and enforceable "total building performance" alternative to the estimated energy cost budget provisions of Chapter 7 of the IECC was another improvement. Likewise, the interior lighting power requirements were revised and are now fully autonomous from related requirements found by reference in Chapter 7 of the IECC.

A completely new stand-alone chapter for one- and two-family dwellings, townhomes and low-rise buildings containing multiple dwelling units was also added to the 2000 edition of the IECC. The new chapter, entitled "Simplified Prescriptive Requirements for Residential Buildings," creates a platform for technical coordination of related energy conservation provisions in the *International Residential Code*® (IRC®).

The updating and progression of the IECC is still an ongoing process. The code currently references the 2003 edition of the IECC, which has continued the tradition of providing a document that is easier to use and enforce and improves the energy efficiency of buildings. This process of continually updating the code ensures that the document stays current and takes advantage of the latest knowledge and technical improvements that the industry has to offer.

The majority of the work involved in determining code compliance can be done during the plan review stage, thereby establishing a focal point for energy-related code enforcement.

Purpose

The purpose of Chapter 13 is to provide minimum design requirements that will promote efficient utilization of energy in building construction. The requirements are directed toward the design of building envelopes with adequate thermal resistance and low air leakage and toward the design and selection of mechanical, service water-heating, electrical and illumination systems that will promote the effective use of depletable energy resources and encourage the use of nondepletable energy resources.

SECTION 1301
GENERAL

1301.1 Scope. This chapter governs the design and construction of buildings for energy efficiency.

❖ The scope of Chapter 13 is applicable to all buildings and structures, as well as their components and systems, that are regulated by the IECC. The IECC, thereby addresses the design of energy-efficient building envelopes and the selection and installation of energy-efficient mechanical, service water heating, electrical distribution and illumination systems and equipment for the effective use of energy in both residential and commercial buildings [see Figures 1301.1(1) and 1301.1(2), respectively].

1301.1.1 Criteria. Buildings shall be designed and constructed in accordance with the *International Energy Conservation Code*.

❖ The energy conservation requirements of this chapter rely exclusively on the technical provisions of the IECC. As indicated in Figure 1301.1.1(1), the IECC contains six alternative building thermal envelope design procedures for detached one- and two-family dwellings and low-rise residential buildings, three stories or less in height. For commercial and high-rise residential buildings, the technical provisions of either the IECC or ASHRAE 90.1 (by way of secondary reference through

the IECC) are to be used to establish compliance with the energy conservation requirements of this chapter [see Figure 1301.1.1(2)].

Compliance with the IECC is the sole means of demonstrating compliance with the technical provisions of Chapter 13. The following is an overview of the design methods found therein:

• *International Energy Conservation Code—Residential*

Building design by systems analysis and design of buildings utilizing renewable energy sources: Chapter 4 provides an alternative way to meet the code's goal for the "effective use of energy" by showing that the predicted annual energy use of the proposed design (the building intended to be built) is less than or equal to that of the same home if it had been built to exactly meet the prescriptive criteria in Chapter 5. Chapter 4 does not prescribe a single set of requirements; rather, it provides a process to reach the energy-efficiency goal based on establishing equivalence with the intent of the code. Because of the level of detail required of the analysis, this method of design is not often used for residential buildings, if ever.

Note that there are two fundamental requirements for using Chapter 4. First, Chapter 4 compliance is based on total estimated annual energy usage across all building energy-using systems: envelope, lighting, mechanical and service water

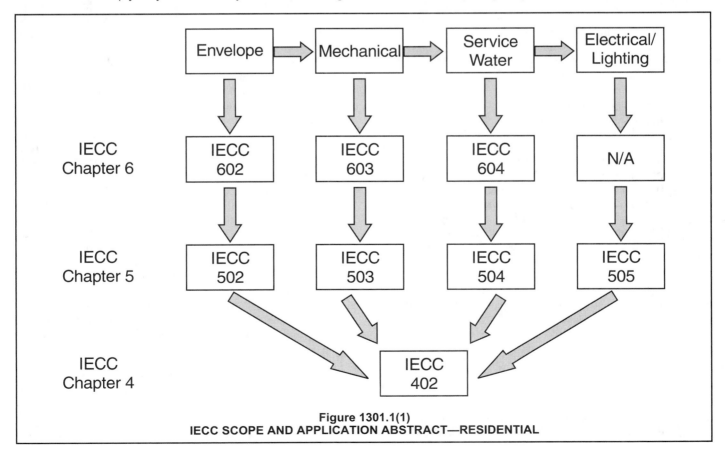

Figure 1301.1(1)
IECC SCOPE AND APPLICATION ABSTRACT—RESIDENTIAL

FIGURE 1301.1(2)   **ENERGY EFFICIENCY**

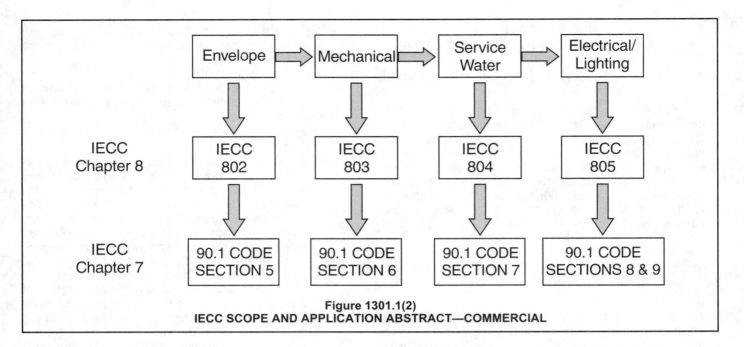

Figure 1301.1(2)
IECC SCOPE AND APPLICATION ABSTRACT—COMMERCIAL

heating. Second, Chapter 4 compares the energy use of the "proposed design" and the "standard design." The standard design is the same building design as that proposed (intended to be built), except that the energy features required by the code (e.g., insulation, windows, HVAC, infiltration) are modified to exactly meet the minimum prescriptive requirements in Chapter 5. The standard design is only used as a benchmark for comparison and is never actually built.

Chapter 4 sets both general principles and specific guidelines, or rules, for use in computing the estimated annual energy use of the proposed and standard designs. These guidelines constitute a large portion of Chapter 4, but are necessary to maintain fairness and consistency between the proposed and standard designs. While the systems analysis method to compliance is the most complex, it gives the design professional the flexibility to introduce exterior walls, roof/ceiling components, etc., which do not meet the requirements of the component performance approach of Chapter 5, but are acceptable where the annual energy use of the proposed building is equal to or less than that of the standard-design building. Envelope features that lower energy consumption (window orientation, passive-solar features and light-colored roofs in cooling-dominated climates) and mechanical, electrical and service water-heating systems that are more efficient than those required by the minimum prescriptive requirements in Chapter 5 could be used to offset the potentially high thermal transmittance of an innovative exterior envelope design in this instance.

The systems analysis method also allows energy supplied by renewable energy sources to be discounted from the total energy consumption of the proposed-design building. Because renewable energy comes from nondepletable sources (i.e., solar radiation, wind, plant byproducts and geothermal sources), its use is not counted as part of the proposed building's energy use.

Building design by component performance approach: Most of Chapter 5 focuses on building envelope requirements (e.g., insulation and windows). The building envelope is defined by those elements of a building that enclose conditioned spaces through which heat is transferred to the outdoors or unconditioned spaces. Unconditioned garage, attic and crawl spaces are considered outside the building envelope. The insulation requirement for each building envelope component is based on the jurisdiction's annual Fahrenheit heating degree days (HDD) and the type of envelope component (i.e., walls, roof/ceiling assemblies, floors over unconditioned spaces, basement walls, etc.). Building envelope compliance for residential buildings must be demonstrated utilizing any one of the four methods provided in Chapter 5 or one of the other two options the code provides.

Compliance by performance on an individual component basis: This envelope compliance method provides specific insulation requirements for each building envelope component based on building type and annual Fahrenheit HDD for the jurisdiction.

Compliance by total building envelope performance: Alternatively, a proposed design's component U_o, U-factors or R-values may vary from the required values when demonstrating compliance by total building envelope performance. This is often referred to as the "tradeoff approach."

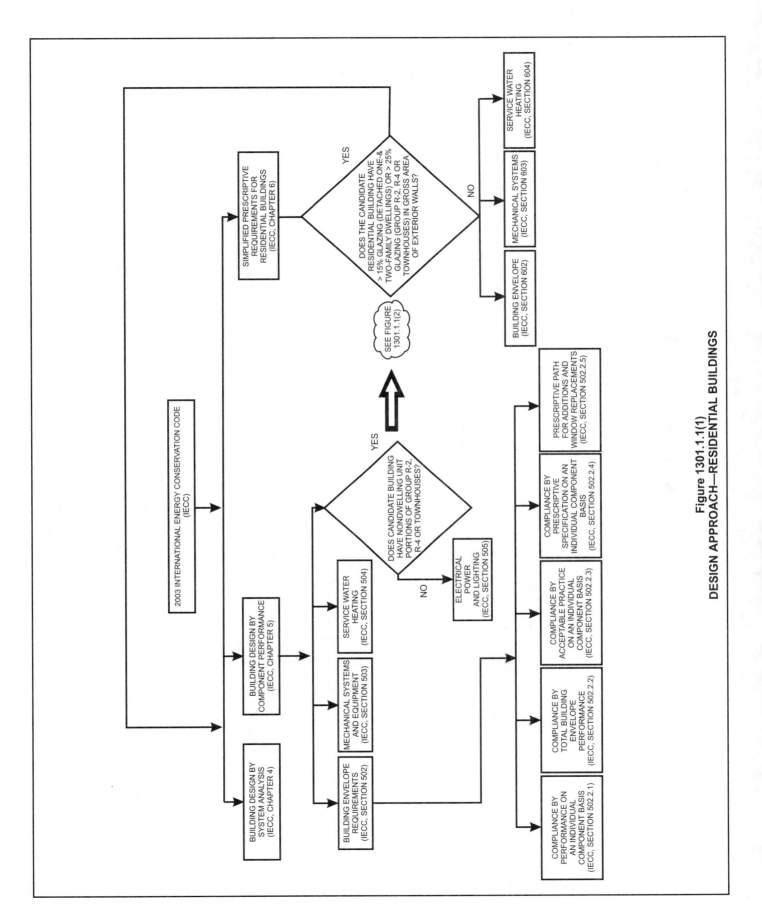

Figure 1301.1.1(1)
DESIGN APPROACH—RESIDENTIAL BUILDINGS

FIGURE 1301.1.1(2)

ENERGY EFFICIENCY

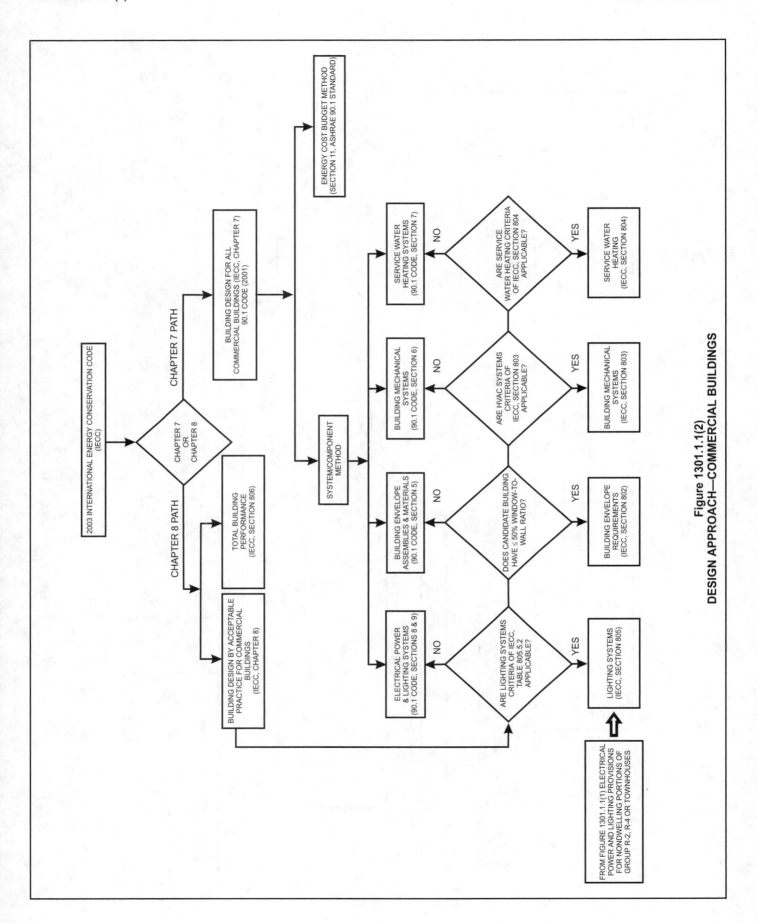

Figure 1301.1.1(2)
DESIGN APPROACH—COMMERCIAL BUILDINGS

Using the tradeoff approach, the insulating value of an envelope component, where the thermal performance level may fall short of meeting the individual component requirement, can be "traded off" for another envelope component in the same building having a thermal performance level better than the code's individual component requirement. Tradeoff calculations are based on the UA for the whole building. The UA is the sum of the U_o times the area (A) for each building envelope component. Accordingly, the building design can be shown to comply with the IECC where the resultant total UA of the proposed design is no greater than the total UA of the standard design of the same building insulated to meet the individual component criteria of Chapter 5.

Compliance by acceptable practice on an individual component basis: The methods used to establish compliance by acceptable practice on an individual component basis are founded on the principles described earlier for envelope compliance by performance on an individual component basis. This compliance approach, however, is further limited by the select number of construction details offered in the code to meet building thermal envelope requirements.

Compliance by prescriptive specification on an individual component basis: This envelope compliance method is by far the simplest of the six building thermal envelope compliance options. It requires one to determine the amount of window area in the gross area of exterior walls. Based upon that information and the annual Fahrenheit HDD for the jurisdiction, the building's thermal envelope criteria is determined.

Simplified prescriptive requirements for residential buildings: Chapter 6 was created to facilitate technical coordination of related provisions in Chapter 11 of the IRC; thus, the provisions of Chapter 6 are built upon elements of Chapter 5 while removing the perceived complexity (from the point of view of builders, designers, plan reviewers and field contractors) when using and enforcing the IECC. It is worth noting that Chapter 6 consists of four pages of code text in its entirety.

- *International Energy Conservation Code—Commercial*
  Commercial buildings include all buildings except one- and two-family dwellings and low-rise, multiple-family residential buildings. Chapter 8 of the IECC is designed to provide a simplified and enforceable alternative to ASHRAE 90.1 for all commercial buildings, with the exception of those that have more than 50-percent glazing in the gross area of above-grade walls. Envelope compliance for commercial buildings having glazing no greater than 50 percent of the above-grade wall area is demonstrated by acceptable practice using the prescriptive methods and tables of Chapter 8. The tables prescribe the minimum level of thermal performance required for a variety of roof, floor and wall configurations typical of commercial construction practice based upon building location and percentage of glass-to-wall area. For buildings having glazing greater than 50 percent of the above-grade wall area, envelope compliance must be demonstrated in accordance with Chapter 7 of the IECC (i.e., ASHRAE 90.1). Mechanical system energy code compliance for commercial HVAC equipment is addressed by Section 803 of the IECC. Buildings served by mechanical systems and equipment not otherwise covered by Chapter 8 must be evaluated for energy code compliance using Chapter 7 of the IECC. Buildings with simple lighting systems, such as standard lighting controls (e.g., manual switches, occupancy sensors and dimmers), are to be evaluated using the lighting criteria for the entire building or each specific building area type as provided in Chapter 8.

  Using Chapter 8, each major system (envelope, mechanical, lighting, service water heating) must comply on its own. Simply put, when a candidate commercial building (or building system) does not meet the criteria set forth in Chapter 8, the candidate building (or building system) must demonstrate compliance under the applicable provisions of Chapter 7 (i.e., ASHRAE 90.1).

  Innovative commercial and high-rise residential building designs or technically sophisticated building envelope, mechanical and lighting systems serving these building types must meet the applicable provisions of the codified counterpart to Standard 90.1, ASHRAE 90.1 as referenced by Chapter 7 of the IECC.

  Buildings evaluated for compliance under ASHRAE 90.1 must be designed for the code's basic requirements and either the system/component or the energy cost budget (ECB) methods to compliance. These design methods are comprehensive and somewhat complex. In fact, the requisite calculations are usually done by design professionals who have extensive experience with sophisticated energy conservation design methods. Using the ECB method, proposed designs may utilize varying amounts of different types of energy, with the total annual energy cost (versus annual energy usage) being the limiting design parameter. Using the system/component method, either prescriptive or system performance criteria may be used for lighting, exterior envelope compliance or both, whereas mechanical system compliance is evaluated using only prescriptive criteria.

  The user of the IECC should note that Section 806 now provides a simplified alternative to meeting the ECB methods prescribed by ASHRAE 90.1, adopted by reference in Chapter 7. In addition to being an alternative building performance approach to that contained in Section 13 of ASHRE 90.1, which is adopted by secondary reference through Section 102 of ASHRAE 90.1, Section 806 now provides a simple prescriptive plan that, if followed, will result in a building design that complies with ASHRAE 90.1. Using the provisions of Section 806, it is possible to establish the annual energy use and cost for a candidate proposed design, assuming it just satisfied the minimum requirements in Chapter 8. This information can form the basis for evaluating proposed designs based on total performance. This concept is consistent with the provisions in Chapter 1 of other *International Codes* as well as Section 13 of ASHRE 90.1.

Bibliography

The following resource materials are referenced in this chapter or are relevant to the subject matter addressed in this chapter.

ASHRAE 90.1-01, *Energy Efficient Design of New Buildings Except New Low-Rise Residential Buildings*. Atlanta: American Society of Heating, Refrigerating and Air-Conditioning Engineers, Inc., 2001.

ASHRAE 90A-80, *Energy Conservation in New Building Design - with Addendum 90A-a-87*. Atlanta: American Society of Heating, Refrigerating and Air-Conditioning Engineers, Inc., 1987.

ASHRAE 90B-75, *Energy Conservation in New Building Design*. Atlanta: American Society of Heating, Refrigerating and Air-Conditioning Engineers, Inc., 1975.

ASHRAE 90C-77, *Energy Conservation in New Building Design*. Atlanta: American Society of Heating, Refrigerating and Air-Conditioning Engineers, Inc., 1977.

ASHRAE-2001, *Handbook of Fundamentals*. Atlanta: American Society of Heating, Refrigerating and Air-Conditioning Engineers, Inc., 2001.

ASHRAE/IESNA-93, *Energy Code for Commercial and High-Rise Residential Buildings—Based on ASHRAE/IES 90.1-1989*. Atlanta, GA: American Society of Heating, Refrigerating and Air-Conditioning Engineers, Inc., 1993.

CABO-92, *Model Energy Code*. Falls Church, VA: Council of American Building Officials, 1992.

CABO-93, *Model Energy Code*. Falls Church, VA: Council of American Building Officials, 1993.

CABO-95, *Model Energy Code*. Falls Church, VA: Council of American Building Officials, 1995.

CABO-95, *Model Energy Code Commentary*. Falls Church, VA: Council of American Building Officials, 1998.

IECC-98, *International Energy Conservation Code*. Falls Church, VA: International Code Council, 1998.

IECC-2000, *International Energy Conservation Code*. Falls Church, VA: International Code Council, 2000.

IECC-2003, *International Energy Conservation Code*. Falls Church, VA: International Code Council, 2003.

IRC-2000, *International Residential Code*. Falls Church, VA: International Code Council, 2000.

IRC-2003, *International Residential Code*. Falls Church, VA: International Code Council, 2003.

PNNL-97, *Assessment of the 1995 Model Energy Code for Adoption*. Richland, WA: Prepared for U.S. Department of Housing and Urban Development Office of Policy Development and Research by Pacific Northwest National Laboratory, 1997.

U.S. DOE-97, *Technical Support Document for ComCheck-EZ, V1.0*. Washington, DC: U.S. Department of Energy, 1997.

Chapter 14:
Exterior Walls

General Comments

Chapter 14 provides requirements for the materials and construction creating finished exterior surfaces of a building or structure. The chapter provides requirements resulting in minimum permitted weather resistance and fire performance.

In the past, there were relatively few materials available for application to exterior walls for protection against weather and exposure to other outside elements. With the development of new methods and materials over the years, architects and builders began to use a variety of materials for different appearances, improved insulating quality, sound transmission control and fire resistance.

The code has developed prescriptive and performance regulations to control these aspects and the types and thickness of exterior wall coverings.

Purpose

The purpose of Chapter 14 is to provide the minimum requirements for exterior wall coverings. This chapter also includes the minimum regulations for materials (and the minimum thicknesses and installation requirements for exterior weather coverings) and various wall veneers. Limitations on the use of combustible exterior wall finishes are also included.

SECTION 1401
GENERAL

1401.1 Scope. The provisions of this chapter shall establish the minimum requirements for exterior walls, exterior wall coverings, exterior wall openings, exterior windows and doors, architectural trim, balconies and bay windows.

❖ The requirements for exterior wall construction, including components such as openings, doors, windows, trim and balconies, are specified herein. Also specified are provisions intended to prevent damage to a building from the effects of weather and from moisture intrusion into or through the exterior walls.

SECTION 1402
DEFINITIONS

1402.1 General. The following words and terms shall, for the purposes of this chapter and as used elsewhere in this code, have the meanings shown herein.

❖ Definitions of terms can help in the understanding and application of the code requirements. The purpose for including these definitions within this chapter is to provide more convenient access to them without having to refer back to Chapter 2.

For convenience, these terms are also listed in Chapter 2 with a cross reference to this section. Terms that are italicized provide a visual identification throughout the code that a definition exists for that term.

The use and application of all defined terms, including those defined herein, are set forth in Section 201.

ADHERED MASONRY VENEER. Veneer secured and supported through the adhesion of an approved bonding material applied to an approved backing.

❖ This type of masonry veneer relies on the backing surface for both vertical and lateral load resistance. The components of adhered masonry veneer construction generally include the masonry veneer, the adhering material and the backing to which the veneer is attached. It should be noted that the term "approve" means components are subject to approval by the building official or the authority having jurisdiction, in accordance with Section 104.

ANCHORED MASONRY VENEER. Veneer secured with approved mechanical fasteners to an approved backing.

❖ This type of masonry veneer is generally supported from below and anchored to the sheathing, studs or other structural portion of the wall. Veneers provide little, if any, strength to the wall and are, therefore, considered to be nonstructural. Anchored masonry veneer is unique in that it is usually supported from below by a footing, lintels or shelf angles. Anchored masonry veneer must not, however, support loads other than its own dead loads, wind loads and seismic loads resulting from the dead load of the wall.

BACKING. The wall or surface to which the veneer is secured.

❖ The backing is the portion of a structure that provides support for the exterior veneer. The backing also typically resists lateral and transverse loads imposed by the veneer and may be load bearing or nonload bearing.

EXTERIOR WALL. A wall, bearing or nonbearing, that is used as an enclosing wall for a building, other than a fire wall, and that has a slope of 60 degrees (1.05 rad) or greater with the horizontal plane.

❖ A wall is defined as an exterior element that encloses a structure and that has a slope equal to or greater than 60 degrees (1.05 rad) from the horizontal plane. Exterior enclosing elements with slopes less than this are generally subjected to more severe weather exposure than vertical surfaces and, thus, may experience a greater amount of water intrusion. These sloped surfaces, which may include elements such as inset windowsills, sloped parapets and other architectural elements, should be designed to resist water penetration in a manner similar to a roof.

EXTERIOR WALL COVERING. A material or assembly of materials applied on the exterior side of exterior walls for the purpose of providing a weather-resisting barrier, insulation or for aesthetics, including but not limited to, veneers, siding, exterior insulation and finish systems, architectural trim and embellishments such as cornices, soffits, facias, gutters and leaders.

❖ Materials such as wood, masonry, metal, concrete, structural glass and plastics are used for exterior wall coverings. It should be noted that exterior wall coverings that are combustible are to meet the combustible materials requirements of Section 1406.2.

EXTERIOR WALL ENVELOPE. A system or assembly of exterior wall components, including exterior wall finish materials, that provides protection of the building structural members, including framing and sheathing materials, and conditioned interior space, from the detrimental effects of the exterior environment.

❖ This definition is needed for the understanding and proper application of the weather protection requirements in Section 1403.2.

FIBER CEMENT SIDING. A manufactured, fiber-reinforcing product made with an inorganic hydraulic or calcium silicate binder formed by chemical reaction and reinforced with organic or inorganic nonasbestos fibers, or both. Additives that enhance manufacturing or product performance are permitted. Fiber cement siding products have either smooth or textured faces and are intended for exterior wall and related applications.

❖ Fiber cement siding is also a material used for weather-resistant siding. It is manufactured from fiber-reinforced cement, and the code permits its use as either panel siding or horizontal lap siding.

METAL COMPOSITE MATERIAL (MCM). A factory-manufactured panel consisting of metal skins bonded to both faces of a plastic core.

❖ These types of panels are sandwich constructions composed of thin metal (usually aluminum or steel) sheets covering a solid plastic core. The panels typically have

thicknesses that do not exceed 0.25 inches (6.4 mm). Metal composite material (MCM) panels are different from other types of sandwich panels in that they do not contain foam plastic cores and are not typically intended to provide thermal insulation.

METAL COMPOSITE MATERIAL (MCM) SYSTEM. An exterior wall finish system fabricated using MCM in a specific assembly including joints, seams, attachments, substrate, framing and other details as appropriate to a particular design.

❖ This section provides a definition of the components that are typically part of an MCM installation. The components of the exterior wall covering system consist of framing members for the attachment and support of the MCM panels; the types of joints and seams used to maintain the weather resistance of the system and the means for attaching the entire system to the building substrate or structural frame. Figures 1402.1(1) through (3) illustrate the details of some types of MCM systems.

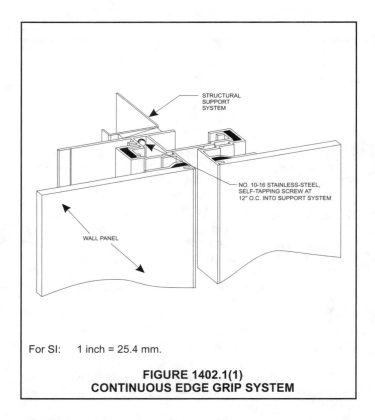

STRUCTURAL SUPPORT SYSTEM

NO. 10-16 STAINLESS-STEEL, SELF-TAPPING SCREW AT 12" O.C. INTO SUPPORT SYSTEM

WALL PANEL

For SI: 1 inch = 25.4 mm.

FIGURE 1402.1(1)
CONTINUOUS EDGE GRIP SYSTEM

VENEER. A facing attached to a wall for the purpose of providing ornamentation, protection or insulation, but not counted as adding strength to the wall.

❖ Veneers are wall facings or claddings of various materials that are used for environmental protection or ornamentation on the exterior of walls. Veneers are nonstructural in that they do not carry any load other than their own weight.

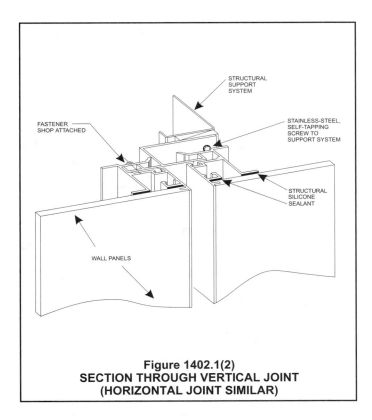

Figure 1402.1(2)
SECTION THROUGH VERTICAL JOINT
(HORIZONTAL JOINT SIMILAR)

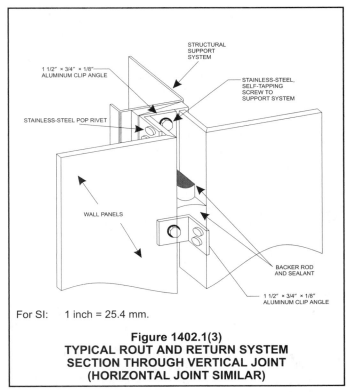

For SI: 1 inch = 25.4 mm.

Figure 1402.1(3)
TYPICAL ROUT AND RETURN SYSTEM
SECTION THROUGH VERTICAL JOINT
(HORIZONTAL JOINT SIMILAR)

SECTION 1403
PERFORMANCE REQUIREMENTS

1403.1 General. The provisions of this section shall apply to exterior walls, wall coverings and components thereof.

❖ The exterior walls of buildings provide support for the floor and roof systems; act as barriers against the outside environment; reduce heat loss from the interior and, in some instances, provide the sole resistance to such destructive forces as earthquakes and floods. This section provides references to other applicable sections of the code that are intended to detail specific requirements to accomplish these objectives. Additionally, this chapter contains detailed requirements related to the performance of exterior wall coverings and the exterior wall envelope.

1403.2 Weather protection. Exterior walls shall provide the building with a weather-resistant exterior wall envelope. The exterior wall envelope shall include flashing, as described in Section 1405.3. The exterior wall envelope shall be designed and constructed in such a manner as to prevent the accumulation of water within the wall assembly by providing a water-resistive barrier behind the exterior veneer, as described in Section 1404.2 and a means for draining water that enters the assembly to the exterior of the veneer, unless it is determined that penetration of water behind the veneer shall not be detrimental to the building performance. Protection against condensation in the

exterior wall assembly shall be provided in accordance with the *International Energy Conservation Code.*

Exceptions:

1. A weather-resistant exterior wall envelope shall not be required over concrete or masonry walls designed in accordance with Chapters 19 and 21, respectively.

2. Compliance with the requirements for a means of drainage, and the requirements of Sections 1405.2 and 1405.3, shall not be required for an exterior wall envelope that has been demonstrated through testing to resist wind-driven rain, including joints, penetrations and intersections with dissimilar materials, in accordance with ASTM E 331 under the following conditions:

    2.1. Exterior wall envelope test assemblies shall include at least one opening, one control joint, one wall/eave interface and one wall sill. All tested openings and penetrations shall be representative of the intended end-use configuration.

    2.2. Exterior wall envelope test assemblies shall be at least 4 feet by 8 feet (1219 mm by 2438 mm) in size.

    2.3. Exterior wall envelope assemblies shall be tested at a minimum differential pressure of 6.24 pounds per square foot (psf) (0.297 kN/m^2).

FIGURE 1403.2   **EXTERIOR WALLS**

2.4. Exterior wall envelope assemblies shall be subjected to a minimum test exposure duration of 2 hours.

The exterior wall envelope design shall be considered to resist wind-driven rain where the results of testing indicate that water did not penetrate control joints in the exterior wall envelope, joints at the perimeter of openings or intersections of terminations with dissimilar materials.

❖ All exterior walls of buildings must be protected against damage caused by precipitation, wind and other weather conditions.

The main text of this section prescribes three basic components of a weather-resistive exterior wall assembly: a water-resistive barrier installed over the building substrate; flashings at penetrations and terminations of the exterior wall finish and a means of draining moisture that may penetrate behind the finish back to the exterior.

Section 1404.2 is referenced for the requirements of the water-resistive barrier and Section 1405.3 is referenced for requirements for the flashings (see commentary, Sections 1404.2 and 1405.3). This section does not, however, contain a prescriptive requirement for the means of drainage to be provided. The method to provide the means of drainage is a performance criterion and must be evaluated based upon the ability to allow moisture that may penetrate behind the exterior wall covering to drain back to the exterior. This may be as complicated as a rain-screen pressure-equalized type of exterior assembly or as simple as providing discontinuities or gaps between the surface of the substrate and the back side of the finish, such as through the use of noncorrodible furring.

For common types of construction, such as vinyl siding or brick veneer, the typical practice of installing building paper and weeps will comply with the intent of this section. Discontinuities between the exterior covering and substrate must be such that they encourage the flow of moisture via gravity or capillary action to a location where the water may exit, such as at flashings and weeps. The absence of a means of drainage may result in the accumulation of moisture that becomes trapped between the finish and the substrate. Over time, extended exposure to moisture may contribute to the degradation of the finish, building substrate or even the structural elements of the exterior wall.

Exception 1 states that where the exterior wall envelope is designed and constructed of concrete or masonry materials in accordance with the requirements of Chapters 19 and 21, respectively, the water-resistive barrier and means of drainage may be omitted. This is because the penetration of moisture behind the exterior wall finish is not detrimental to concrete and masonry substrates.

Exception 2 permits the use of exterior wall finishes that do not meet the prescriptive requirements of Section 1403.2, provided that the system, with penetration details, is tested for wind-driven rain resistance. The test specimen(s) must incorporate those penetration and termination details intended for use, and the system

will be limited to use with those details that successfully pass the test. The minimum panel size specified represents that which is commonly used in testing to ASTM E 331; however, this does not preclude the testing of larger panels if desired. The modifications to the test pressure differential and test duration are intended to represent more closely conditions that will be encountered in service. The pass/fail criteria is based upon the visual observation of moisture on the rear side of the wall assembly. In cavity-type assemblies, such as stud walls, this requires the observation of locations such as the rear face of the exterior wall sheathing and wall framing members for the presence of moisture. The test method is intended to assess the performance of the method(s) intended for use in sealing the interface between the termination of the exterior wall finish and the penetration items or abutting construction. The test is not necessarily intended to test the performance of the penetrating item.

Walls designed and constructed in accordance with this chapter must also comply with the requirements of the *International Energy Conservation Code*® (IECC®). This requires that wall cavities be protected from moisture infiltration from the building interior through the use of a vapor retarder (see Figure 1403.2) or by the ventilation of the wall cavity.

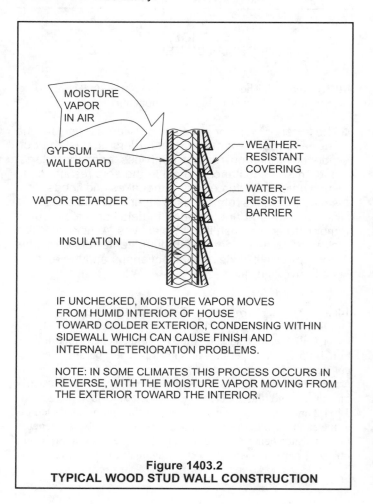

IF UNCHECKED, MOISTURE VAPOR MOVES FROM HUMID INTERIOR OF HOUSE TOWARD COLDER EXTERIOR, CONDENSING WITHIN SIDEWALL WHICH CAN CAUSE FINISH AND INTERNAL DETERIORATION PROBLEMS.

NOTE: IN SOME CLIMATES THIS PROCESS OCCURS IN REVERSE, WITH THE MOISTURE VAPOR MOVING FROM THE EXTERIOR TOWARD THE INTERIOR.

Figure 1403.2
TYPICAL WOOD STUD WALL CONSTRUCTION

1403.3 Vapor retarder. An approved vapor retarder shall be provided.

Exceptions:

1. Where other approved means to avoid condensation and leakage of moisture are provided.

2. Plain and reinforced concrete or masonry exterior walls designed and constructed in accordance with Chapter 19 or 21, respectively.

❖ Vapor retarders are intended to prevent the movement of moisture-laden air from the warm side of the wall to the cold side. The IECC requires that a vapor retarder have a maximum rating of 1.0 perm [5 mg/(s · m² · Pa)] when tested in accordance with Procedure A [Desiccant Method at 73.4°F (23°C)] of ASTM E 96. The material used for the vapor retarder must be resistant to degradation from moisture in order for it to be effective over the expected life of the building.

Water vapor flow through interior finish materials into a wall cavity can result from the lack of or improper installation of a vapor retarder on the warm-in-winter side of the insulation. The water vapor can then condense when it comes in contact with a colder surface, resulting in an accumulation of moisture in the wall cavity. The prolonged presence of moisture can eventually lead to the degradation of both the structural and thermal resistance properties of the wall assembly. It must be noted that in some climatic areas, the dynamic of vapor drive is reversed, such as in hot, humid locations. In areas where the wet-bulb temperature is 67°F (19.4°C) or higher for 3,000 or more hours during the warmest six consecutive months of the year or 73°F (22.8°C) or higher for 1,500 or more hours during the warmest six consecutive months of the year, the vapor retarder should not be placed on the interior side of the exterior wall (refer to IECC Sections 502.1.1 and 802.1.2). Under these climatic conditions, the main source of moisture-laden air is the exterior environment, which is driven toward the cooler, drier air in the interior of the building.

1403.4 Structural. Exterior walls, and the associated openings, shall be designed and constructed to resist safely the superimposed loads required by Chapter 16.

❖ Exterior walls and their associated openings are required to resist all structural loads in accordance with the provisions of Chapter 16. This section is a correlative cross reference to emphasize the applicability of Chapter 16.

1403.5 Fire resistance. Exterior walls shall be fire-resistance rated as required by other sections of this code with opening protection as required by Chapter 7.

❖ The required fire-resistance rating of exterior walls is set forth in Tables 601 and 602. Table 602 is applicable to both load-bearing and nonload-bearing walls, since it addresses the prevention of fire spread from one building to an adjacent building. Load-bearing walls must comply with the greater of the requirements in Tables 601 and 602, based on the type of construction of the building. The size of openings in exterior walls is limited to prevent the spread of fire to other buildings. The commentary to Section 704 should also be reviewed when designing exterior walls. Trim on exterior walls is regulated by Section 1406.

The allowable size of openings in exterior walls is listed in Table 704.8. Section 715.3 specifies the required fire protection rating for fire door and fire shutter opening protectives, and Section 715.4 specifies the required rating for windows when a rating is required by Table 704.8. Where fire-resistance rating is not required for the wall and unprotected openings are allowed, the glazing and the sash or frame may be of any material permitted by the code.

1403.6 Flood resistance. For buildings in flood hazard areas as established in Section 1612.3, exterior walls extending below the design flood elevation shall be resistant to water damage. Wood shall be pressure-preservative treated in accordance with AWPA C1, C2, C3, C4, C9, C15, C18, C22, C23, C24, C28, P1, P2 and P3, or decay-resistant heartwood of redwood, black locust or cedar.

❖ Flood-resistant construction of exterior walls requires special consideration. Construction materials used in exterior walls of buildings that are located in flood hazard areas must be resistant to the effects of flooding. Some of the properties these materials must posses include resistance to prolonged contact with water, the ability to be cleaned and disinfected after the water recedes and negligible loss of physical properties after exposure to water. Additionally, systems must possess structural strength to resist the hydrodynamic and impact forces related to flooding. Wood construction materials in locations below the base flood elevation are required to be resistant to decay and exposure to water (see the commentary to Section 2303.1.8 for more discussion on preservative treatment of wood). Appendix G contains requirements specific to construction within areas that are prone to flooding.

1403.7 Flood resistance for high-velocity wave action areas. For buildings in flood hazard areas subject to high-velocity wave action as established in Section 1612.3, electrical, mechanical and plumbing system components shall not be mounted on or penetrate through exterior walls that are designed to break away under flood loads.

❖ ASCE 24 and Section 1612.3 provide for the option to design exterior walls to break away under flood loads. This section prohibits the installation of penetrations for electrical, plumbing or mechanical systems through such exterior walls. The breakaway provisions are a method of protecting the remaining building frame and building system. Installation of penetrations through these walls could serve in part to defeat that purpose or to reduce the ability of the breakaway system to perform as designed.

SECTION 1404
MATERIALS

1404.1 General. Materials used for the construction of exterior walls shall comply with the provisions of this section. Materials not prescribed herein shall be permitted, provided that any such alternative has been approved.

❖ Section 1404 contains performance requirements and detailed installation specifications for a number of common materials used in or on exterior walls of buildings. Reference is made to the applicable chapters that contain additional requirements. These chapters should be reviewed by the reader.

It is not the intent of the code to prevent the use of any material that is equivalent in performance to those specified in Section 1404.

1404.2 Water-resistive barrier. A minimum of one layer of No. 15 asphalt felt, complying with ASTM D 226 for Type 1 felt, shall be attached to the sheathing, with flashing as described in Section 1405.3, in such a manner as to provide a continuous water-resistive barrier behind the exterior wall veneer.

❖ Many exterior veneers provide weather resistance but may allow either penetration of water through joints or seams or the development of condensation to occur behind the veneer. To increase the weather resistance of the wall, a layer of asphalt felt is required to be installed over the wall backing. The felt layer and the flashing provide a water-resistive barrier behind the exterior veneer. Section 1403.2 allows for the omission of felt where it can be shown that the presence of moisture will not be detrimental to the exterior wall envelope (see commentary, Section 1403.2).

1404.3 Wood. Exterior walls of wood construction shall be designed and constructed in accordance with Chapter 23.

❖ Chapter 23 contains general requirements regulating the use of wood in buildings; however, this section contains specific provisions for the minimum thicknesses of wood siding, wood backing surfaces and hardboard siding, as well as installation and nailing requirements for wood used as weatherboarding or a wall covering.

1404.3.1 Basic hardboard. Basic hardboard shall conform to the requirements of AHA A135.4.

❖ Basic hardboard is required to comply with AHA A135.4. This material is not intended for use as exposed siding or as a weather covering, but it is often a component of veneered wall construction.

1404.3.2 Hardboard siding. Hardboard siding shall conform to the requirements of AHA A135.6 and, where used structurally, shall be so identified by the label of an approved agency.

❖ Hardboard conforming to AHA A135.6 is manufactured in panels, usually 4 feet by 8 feet (1219 mm by 2438 mm) in size or in sizes simulating drop siding where the material is to be applied horizontally over structural

sheathing. When used as a panel, hardboard siding is applied directly to the wood framing members as a backing material. Hardboard panels are not adequate to serve as a nailing base, spanning between studs, for coverings; however, they can be used to provide some lateral resistance to the wall [see Table 2308.9.3(6)]. Coverings are fastened directly at framing locations, through the panels.

1404.4 Masonry. Exterior walls of masonry construction shall be designed and constructed in accordance with this section and Chapter 21. Masonry units, mortar and metal accessories used in anchored and adhered veneer shall meet the physical requirements of Chapter 21. The backing of anchored and adhered veneer shall be of concrete, masonry, steel framing or wood framing.

❖ Chapter 21 contains the general requirements for the use of masonry in buildings. The material classification of masonry in this section includes materials such as concrete masonry; bricks of clay, shale or calcium silicate; clay-tile units; glazed masonry units; terra cotta; natural or cast stone and ceramic tile. Chapter 21 and this section contain provisions for the minimum thickness of masonry units, limitations on loading, heights of veneers, backing and attachment.

1404.5 Metal. Exterior walls of formed steel construction, structural steel or lightweight metal alloys shall be designed in accordance with Chapters 22 and 20, respectively.

❖ Chapters 20 and 22 contain the general requirements for use of metal in buildings. Chapters 20 and 22 as well as this section contain provisions for the minimum thickness of metal weather coverings and requirements for protection from corrosion, installation and grounding.

1404.5.1 Aluminum siding. Aluminum siding shall conform to the requirements of AAMA 1402.

❖ Although aluminum siding is addressed in Chapter 20, it is more specifically addressed here as an exterior wall covering. Aluminum siding is required to conform to the referenced standard. The thickness, installation and protection provisions of this section are also applicable (see commentary, Sections 1405.10, 1405.10.1, 1405.10.2, 1405.10.3 and 1405.10.4).

1404.6 Concrete. Exterior walls of concrete construction shall be designed and constructed in accordance with Chapter 19.

❖ Chapter 19 contains the general requirements for use of concrete in buildings, including provisions for the minimum thickness of concrete units used as veneers and wall coverings and their installation and protection.

1404.7 Glass-unit masonry. Exterior walls of glass-unit masonry shall be designed and constructed in accordance with Chapter 21.

❖ Structural glass block is generally regulated by the provisions of Sections 2103 and 2104; however, structural glass veneers used as wall coverings must comply with the thickness requirement of Table 1405.2. Criteria for the support of glass unit masonry on wood-frame construction are also contained in Exception 4 of Section 2304.12.

1404.8 Plastics. Plastic panel, apron or spandrel walls as defined in this code shall not be limited in thickness, provided that such plastics and their assemblies conform to the requirements of Chapter 26 and are constructed of approved weather-resistant materials of adequate strength to resist the wind loads for cladding specified in Chapter 16.

❖ This section provides for the use of plastic panels for exterior walls. The code uses the term "light-transmitting plastic" to apply to plastics used for exterior wall panels and roof panels, as well as plastics used for glazing and skylights. The panels must meet the requirements of Chapter 26, as well as certain requirements of Chapter 14 for weather resistance and Chapter 16 for structural capabilities.

1404.9 Vinyl siding. Vinyl siding shall conform to the requirements of ASTM D 3679.

❖ Plastics are addressed in Chapter 26; however, polyvinyl chloride (PVC) siding is specifically addressed here as an exterior wall covering. PVC siding is required to conform to the provisions of ASTM D 3679. Installation of vinyl siding is prescribed by Section 1405.13 (see the commentary to Section 1405.13 for additional discussion on the installation of vinyl siding).

1404.10 Fiber cement siding. Fiber cement siding shall conform to the requirements of ASTM C 1186 and shall be so identified on labeling listing an approved quality control agency.

❖ Fiber cement siding is also a material used for weather-resistant siding. It is manufactured from fiber-reinforced cement, and the code permits its use as either panel siding or horizontal lap siding. The material standard is ASTM C 1186. Note that the product is required to bear a label, which means that the manufacturer must have regular inspections by a third- party quality control agency.

SECTION 1405
INSTALLATION OF WALL COVERINGS

1405.1 General. Exterior wall coverings shall be designed and constructed in accordance with the applicable provisions of this section.

❖ Exterior wall coverings used to provide weather protection to the structure are required to comply with all the applicable provisions contained in Section 1405.

1405.2 Weather protection. Exterior walls shall provide weather protection for the building. The materials of the minimum nominal thickness specified in Table 1405.2 shall be acceptable as approved weather coverings.

❖ The exterior walls of a structure must be designed and constructed to provide for the health and safety of the occupants and to protect the structure from the detrimental effects of weather exposure.

This section introduces Table 1405.2, which is a tabulation of the minimum acceptable nominal thicknesses for a number of common weather coverings.

TABLE 1405.2
MINIMUM THICKNESS OF WEATHER COVERINGS

COVERING TYPE	MINIMUM THICKNESS (inches)
Adhered masonry veneer	0.25
Anchored masonry veneer	2.625
Aluminum siding	0.019
Asbestos-cement boards	0.125
Asbestos shingles	0.156
Cold-rolled copper[d]	0.0216 nominal
Copper shingles[d]	0.0162 nominal
Exterior plywood (with sheathing)	0.313
Exterior plywood (without sheathing)	See Section 2304.6
Fiberboard siding	0.5
Fiber cement lap siding	0.25[c]
Fiber cement panel siding	0.25[c]
Glass-fiber reinforced concrete panels	0.375
Hardboard siding[c]	0.25
High-yield copper[d]	0.0162 nominal
Lead-coated copper[d]	0.0216 nominal
Lead-coated high-yield copper	0.0162 nominal
Marble slabs	1
Particleboard (with sheathing)	See Section 2304.6
Particleboard (without sheathing)	See Section 2304.6
Precast stone facing	0.625
Steel (approved corrosion resistant)	0.0149
Stone (cast artificial)	1.5
Stone (natural)	2
Structural glass	0.344
Stucco or exterior portland cement plaster	
Three-coat work over:	
Metal plaster base	0.875[b]
Unit masonry	0.625[b]
Cast-in-place or precast concrete	0.625[b]
Two-coat work over:	
Unit masonry	0.5[b]
Cast-in-place or precast concrete	0.375[b]

(continued)

**TABLE 1405.2—continued
MINIMUM THICKNESS OF WEATHER COVERINGS**

Terra cotta (anchored)	1
Terra cotta (adhered)	0.25
Vinyl siding	0.035
Wood shingles	0.375
Wood siding (without sheathing)[a]	0.5

For SI: 1 inch = 25.4 mm.

a. Wood siding of thicknesses less than 0.5 inch shall be placed over sheathing that conforms to Section 2304.6.

b. Exclusive of texture.

c. As measured at the bottom of decorative grooves.

d. 16 ounces per square foot for cold-rolled copper and lead-coated copper, 12 ounces per square foot for copper shingles, high-yield copper and lead-coated high-yield copper.

❖ This table should be used in addition to all other applicable chapters and requirements of the code for the specific material. Testing and experience have determined that the tabulated minimum thicknesses will be durable and protect the building against the elements for relatively long periods of time when attached and maintained as required.

Note a of Table 1405.2 allows material less than 0.5-inch-thick (12.7 mm) wood siding to be used, provided that it is installed over an approved sheathing conforming to the requirements of Section 2304.6.1 (see commentary, Section 2304.6.1).

Note b of Table 1405.2 establishes that the minimum permitted nominal thickness is based on the narrowest solid thickness of material, exclusive of texture.

Section 2304.6.1 allows fiberboard and particleboard for sheathing. Although permitted as sheathing, fiberboard and particleboard materials as well as other similar materials do not provide a nailing base for weatherboarding. When used over such materials, 0.5-inch (12.7 mm) wood siding is to be fastened to the wood framing members or otherwise adequately secured. Wood siding having a thickness of less than 0.5 inch (12.7 mm) is not permitted to be fastened to foam plastic and other sheathing materials not having adequate strength as a nailing base spanning between studs or framing members.

1405.3 Flashing. Flashing shall be installed in such a manner so as to prevent moisture from entering the wall or to redirect it to the exterior. Flashing shall be installed at the perimeters of exterior door and window assemblies, penetrations and terminations of exterior wall assemblies, exterior wall intersections with roofs, chimneys, porches, decks, balconies and similar projections and at built-in gutters and similar locations where moisture could enter the wall. Flashing with projecting flanges shall be installed on both sides and the ends of copings, under sills and continuously above projecting trim.

❖ Water that enters the exterior wall can lead to the decay of wood and the degradation of other building materials that are sensitive to moisture. Some of the most common points of moisture penetration in an exterior wall are at intersections of windows, doors, chimneys and roof lines with other framing. These potential points of moisture intrusion must be closed by corrosion-resistant flashings or by other acceptable practices [see Figures 1405.3(1) through (4)]. Materials such as asphalt-saturated felt or building paper are not acceptable for flashing at required locations, in part due to the fact that these materials have a limited ability to resist water penetration. The flashing installation locations stated in this section represent those that are most commonly encountered. As an alternative to the installation of flashing, testing of the specific penetrations, terminations and intersections of the exterior wall finish must be performed in accordance with Section 1403.2 (see commentary, Section 1403.2).

Improperly sloped flashings may enable the accumulation of moisture in the wall assembly, which may result in eventual degradation of the wall finish, building substrate or even the flashing material. Without the extension of the flashing beyond the face of the exterior wall finish, moisture may reenter the wall assembly directly below the flashing through capillary action as indicated in Figure 1405.3(5).

1405.3.1 Exterior wall pockets. In exterior walls of buildings or structures, wall pockets or crevices in which moisture can accumulate shall be avoided or protected with caps or drips, or other approved means shall be provided to prevent water damage.

❖ Changes in building lines and elevations, materials and tops of walls, etc., can produce wall pockets, crevices and other openings that allow the entry of rain, snow, ice and other sources of moisture. Where these openings are present, additional covers, caps, flashings, etc., must be provided to prevent the entry of moisture.

1405.3.2 Masonry. Flashing and weepholes shall be located in the first course of masonry above finished ground level above the foundation wall or slab, and other points of support, including structural floors, shelf angles and lintels where anchored veneers are designed in accordance with Section 1405.5.

❖ Water can penetrate walls constructed of masonry materials from a number of sources. Some of the most common causes of water penetration are: the porous nature of many masonry materials; openings, such as small cracks, that may occur between the masonry unit and mortar joint due to expansion and contraction of the wall over time; and the interface between the masonry and penetrations or dissimilar materials.

Walls constructed of anchored masonry veneer must be designed to accommodate water penetration and to provide a means of both stopping the moisture penetration and for the moisture to leave the wall. This section stipulates the use of flashing to stop moisture penetration into the wall backing and small openings (weepholes) to allow the moisture to exit the wall. In order for these two components to function as intended, the weepholes must be installed immediately above the flashing so that moisture collected by the flashing can exit the wall, preventing the accumulation of water at the flashing location.

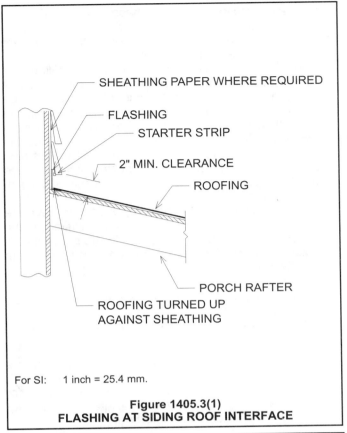

For SI: 1 inch = 25.4 mm.

Figure 1405.3(1)
FLASHING AT SIDING ROOF INTERFACE

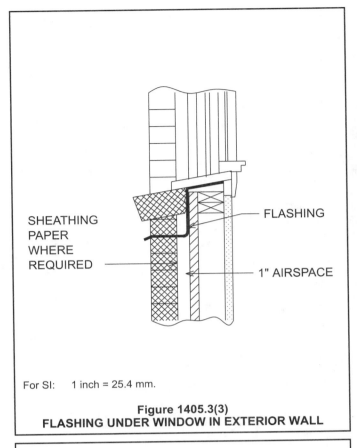

For SI: 1 inch = 25.4 mm.

Figure 1405.3(3)
FLASHING UNDER WINDOW IN EXTERIOR WALL

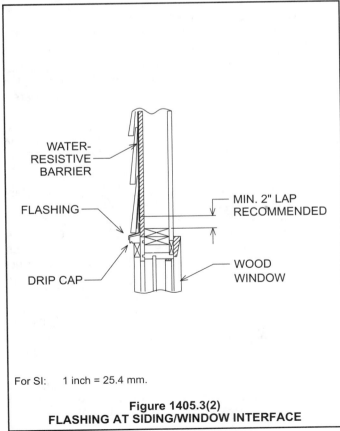

For SI: 1 inch = 25.4 mm.

Figure 1405.3(2)
FLASHING AT SIDING/WINDOW INTERFACE

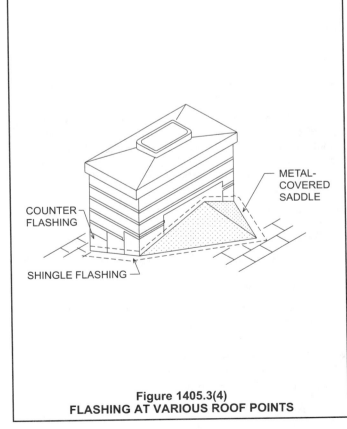

Figure 1405.3(4)
FLASHING AT VARIOUS ROOF POINTS

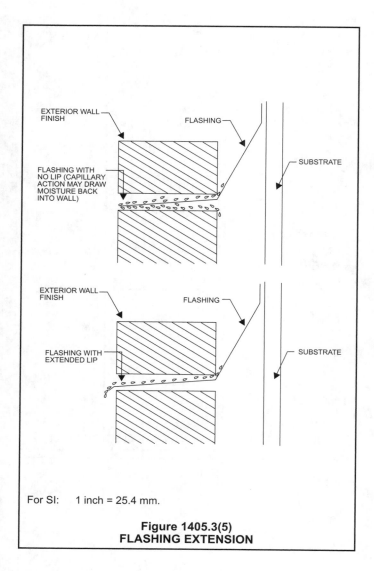

For SI: 1 inch = 25.4 mm.

Figure 1405.3(5)
FLASHING EXTENSION

thickness requirements of this section and three construction criteria are met (see the commentary to Section 1406.2.1 for more discussion on wood veneers as a combustible material on exterior walls).

Item 1 limits the maximum height of the installation of the wood veneer and is coordinated with the provisions of Section 1406.2.2 (see the commentary to Section 1406.2.2 for more discussion on the limitations of use of combustible exterior materials). The increase of one additional story is permitted due to the limited combustible nature of fire-retardant-treated wood (FRTW) siding (see commentary, Section 2303.2). Item 2 requires the backing to be noncombustible, which is consistent with requirements of Chapter 6 for buildings of Type I, II, III and IV construction. The backing must also have the fire-resistance rating required by other provisions of the code, such as in Tables 601 and 602. Item 3 limits the maximum distance that wood veneers may project from the face of the building.

1405.5 Anchored masonry veneer. Anchored masonry veneer shall comply with the provisions of Sections 1405.5, 1405.6, 1405.7 and 1405.8 and Sections 6.1 and 6.2 of ACI 530/ASCE 5/TMS 402.

❖ Sections 6.1 and 6.2 of ACI 530/ASCE 5/TMS 402 are to be used for the design and detailing requirements for anchored masonry veneers. See the commentary for ACI 530/ASCE 5/TMS 402 for an explanation of the anchored masonry veneer requirements.

1405.5.1 Tolerances. Anchored masonry veneers in accordance with Chapter 14 are not required to meet the tolerances in Article 3.3 G1 of ACI 530.1/ASCE 6/TMS 602.

❖ The tolerances in the referenced standard apply to masonry units that are utilized in masonry wall construction and are not intended to be applied to masonry veneers.

1405.5.2 Seismic requirements. Anchored masonry veneer located in Seismic Design Category C, D, E or F shall conform to the requirements of Section 6.2.2.10 of ACI 530/ASCE 5/TMS 402.

❖ This section requires that the seismic design and detailing of anchored masonry veneer be performed in accordance with ACI 530 for veneers located in Seismic Design Categories C, D, E and F (see Chapter 16 for more information on seismic design categories). Section 6.2.2.10 of ACI 530 contains criteria for components such as masonry anchor spacing, joint reinforcement bedding and splicing.

1405.6 Stone veneer. Stone veneer units not exceeding 10 inches (254 mm) in thickness shall be anchored directly to masonry, concrete or to stud construction by one of the following methods:

1. With concrete or masonry backing, anchor ties shall be not less than 0.1055-inch (2.68 mm) corrosion-resistant

1405.4 Wood veneers. Wood veneers on exterior walls of buildings of Type I, II, III and IV construction shall be not less than 1-inch (25 mm) nominal thickness, 0.438-inch (11.1 mm) exterior hardboard siding or 0.375-inch (9.5 mm) exterior-type wood structural panels or particleboard and shall conform to the following:

1. The veneer does not exceed three stories in height, measured from grade, except where fire-retardant-treated wood is used, the height shall not exceed four stories.

2. The veneer is attached to or furred from a noncombustible backing that is fire-resistance rated as required by other provisions of this code.

3. Where open or spaced wood veneers (without concealed spaces) are used, they shall not project more than 24 inches (610 mm) from the building wall.

❖ Wood veneers, which are combustible exterior materials, are permitted for use on buildings of other than combustible (Type V) construction where the minimum wire, or approved equal, formed beyond the base of the backing. The legs of the loops shall be not less than 6 inches (152 mm) in length bent at right angles and laid in the mortar joint, and spaced so that the eyes or loops are 12 inches (305 mm) maximum on center (o.c.) in both directions. There shall be provided not less than a 0.1055-inch (2.68 mm) corrosion-resistant wire tie, or approved equal, threaded through the exposed loops for every 2 square feet (0.2 m²) of stone veneer. This tie shall be a loop having legs not less than 15 inches (381 mm) in length bent so that it will lie in the stone veneer mortar joint. The last 2 inches (51 mm) of each wire leg shall have a right-angle bend. One-inch (25 mm) minimum thickness of cement grout shall be placed between the backing and the stone veneer.

2. With stud backing, a 2-inch by 2-inch (51 by 51 mm) 0.0625-inch (1.59 mm) corrosion-resistant wire mesh with two layers of waterproofed paper backing in accordance with Section 1403.3 shall be applied directly to wood studs spaced a maximum of 16 inches (406 mm) o.c. On studs, the mesh shall be attached with 2-inch-long (51 mm) corrosion-resistant steel wire furring nails at 4 inches (102 mm) o.c. providing a minimum 1.125-inch (29 mm) penetration into each stud and with 8d common nails at 8 inches (203 mm) o.c. into top and bottom plates or with

equivalent wire ties. There shall be not less than a 0.1055-inch (2.68 mm) corrosion-resistant wire, or approved equal, looped through the mesh for every 2 square feet (0.2 m²) of stone veneer. This tie shall be a loop having legs not less than 15 inches (381 mm) in length, so bent that it will lie in the stone veneer mortar joint. The last 2 inches (51 mm) of each wire leg shall have a right-angle bend. One-inch (25 mm) minimum thickness of cement grout shall be placed between the backing and the stone veneer.

❖ This section permits direct anchorage of stone veneers to the wall backing by one of two specified methods. It should be noted that support from below is required to be provided in accordance with Section 1405.5. Since the framing is providing only lateral anchorage, thicknesses of up to 10 inches (254 mm) are permitted.

There are two methods of anchorage prescribed in this section, based upon the construction of the wall backing. The first method specifies the anchor type, size and spacing for concrete and masonry backing (see Figure 1405.6). The second method specifies the anchorage requirements for wood-frame walls.

Both methods require that the anchorage ties be of corrosion-resistant material. Corrosion-resistant ties are necessary because of the potential for corrosion

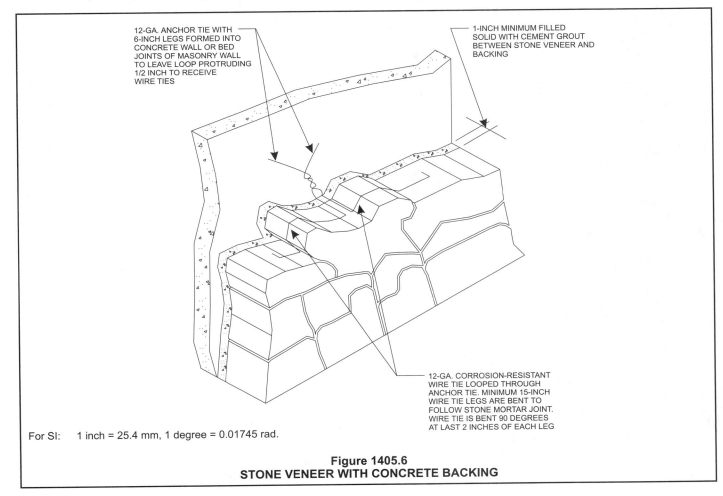

12-GA. ANCHOR TIE WITH 6-INCH LEGS FORMED INTO CONCRETE WALL OR BED JOINTS OF MASONRY WALL TO LEAVE LOOP PROTRUDING 1/2 INCH TO RECEIVE WIRE TIES

1-INCH MINIMUM FILLED SOLID WITH CEMENT GROUT BETWEEN STONE VENEER AND BACKING

12-GA. CORROSION-RESISTANT WIRE TIE LOOPED THROUGH ANCHOR TIE. MINIMUM 15-INCH WIRE TIE LEGS ARE BENT TO FOLLOW STONE MORTAR JOINT. WIRE TIE IS BENT 90 DEGREES AT LAST 2 INCHES OF EACH LEG

For SI: 1 inch = 25.4 mm, 1 degree = 0.01745 rad.

Figure 1405.6
STONE VENEER WITH CONCRETE BACKING

caused by both water penetration and an adverse reaction of the metal with the mortar or other masonry material.

According to the first method, the specified tie spacings, in conjunction with the cement grout, are intended to achieve lateral support points of the veneer at adequate spacings. Where lateral ties are not installed at the required locations, buckling of the masonry veneer under vertical dead loads and lateral loads is possible. It should be noted that the required 1 inch (25 mm) of mortar grout creates walls that are not considered cavity wall construction.

The second method requires that two layers of waterproof paper be installed over the stud construction to provide protection from moisture penetration.

1405.7 Slab-type veneer. Slab-type veneer units not exceeding 2 inches (51 mm) in thickness shall be anchored directly to masonry, concrete or stud construction. For veneer units of marble, travertine, granite or other stone units of slab form ties of corrosion-resistant dowels in drilled holes located in the middle third of the edge of the units spaced a maximum of 24 inches (610 m) apart around the periphery of each unit with not less than four ties per veneer unit. Units shall not exceed 20 square feet (1.9 m²) in area. If the dowels are not tight fitting, the holes shall be drilled not more than 0.063 inch (1.6 mm) larger in diameter than the dowel, with the hole countersunk to a diameter and depth equal to twice the diameter of the dowel in order to provide a tight-fitting key of cement mortar at the dowel locations when the mortar in the joint has set. Veneer ties shall be corrosion-resistant metal capable of resisting, in tension or compression, a force equal to two times the weight of the attached veneer. If made of sheet metal, veneer ties shall be not smaller in area than 0.0336 by 1 inch (0.853 by 25 mm) or, if made of wire, not smaller in diameter than 0.1483-inch (3.76 mm) wire.

❖ Slab-type veneer refers to thin stone masonry veneers that are individually supported. These units do not bear on the successive units below, but are anchored to and supported vertically by the backup construction with corrosion-resistant dowels located at the edges of each unit. This type of veneer anchorage applies to the use of thin stone units that are typically not set in mortar, but are made water tight at the edges through the use of sealant materials. The strength of the veneer ties is required to be adequate to support twice the weight of the veneer in tension to provide for an adequate margin of safety, such that the veneer will remain in place in case it works loose of the grout and backing.

1405.8 Terra cotta. Anchored terra cotta or ceramic units not less than 1.625 inches (41 mm) thick shall be anchored directly to masonry, concrete or stud construction. Tied terra cotta or ceramic veneer units shall be not less than 1.625 inches (41 mm) thick with projecting dovetail webs on the back surface spaced approximately 8 inches (203 mm) o.c. The facing shall be tied to the backing wall with corrosion-resistant metal anchors of not less than No. 8 gage wire installed at the top of each piece in horizontal bed joints not less than 12 inches (305 mm) nor more than 18 inches (457 mm) o.c.; these anchors shall be secured to

0.25-inch (6.4 mm) corrosion-resistant pencil rods that pass through the vertical aligned loop anchors in the backing wall. The veneer ties shall have sufficient strength to support the full weight of the veneer in tension. The facing shall be set with not less than a 2-inch (51 mm) space from the backing wall and the space shall be filled solidly with portland cement grout and pea gravel. Immediately prior to setting, the backing wall and the facing shall be drenched with clean water and shall be distinctly damp when the grout is poured.

❖ The minimum permitted $1^5/_8$-inch (41 mm) thickness for terra cotta units is necessary in order for each unit to have an inherent level of structural integrity. In order to increase the mechanical anchorage of the backing, dovetail webs are required and are embedded in the 2-inch (51 mm) cement grout space. The strength of the veneer ties is required to be adequate to support the full weight of the veneer in tension in order to achieve an adequate margin of safety, such that the veneer will remain in place in case it works loose of the grout and backing (see Figure 1405.8).

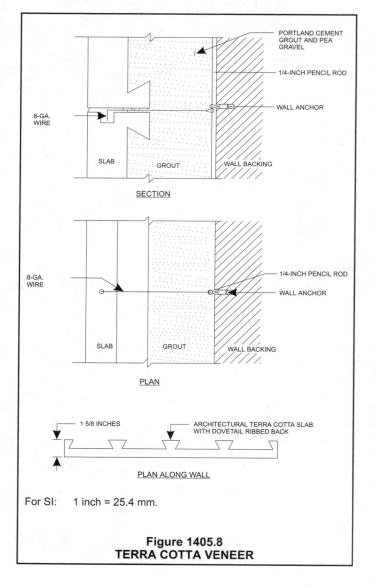

For SI: 1 inch = 25.4 mm.

Figure 1405.8
TERRA COTTA VENEER

1405.9 Adhered masonry veneer. Adhered masonry veneer shall comply with the applicable requirements in Section 1405.9.1 and Sections 6.1 and 6.3 of ACI 530/ASCE 5/TMS 402.

❖ Sections 6.1 and 6.3 of ACI 530/ASCE 5/TMS 402 are to be used for the design and detailing requirements for adhered masonry veneers. See the commentary for ACI 530/ASCE 5/TMS 402 for an explanation of the adhered masonry veneer requirements. Figures 1405.9(1) and 1405.9(2) illustrate two examples of an adhered masonry veneer.

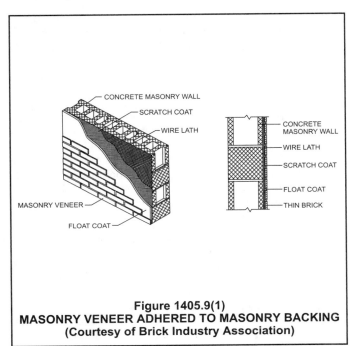

Figure 1405.9(1)
MASONRY VENEER ADHERED TO MASONRY BACKING
(Courtesy of Brick Industry Association)

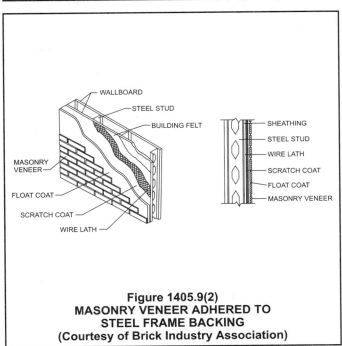

Figure 1405.9(2)
MASONRY VENEER ADHERED TO
STEEL FRAME BACKING
(Courtesy of Brick Industry Association)

1405.9.1 Interior adhered masonry veneers. Interior adhered masonry veneers shall have a maximum weight of 20 psf (0.958 kg/m²) and shall be installed in accordance with Section 1405.9. Where the interior adhered masonry veneer is supported by wood construction, the supporting members shall be designed to limit deflection to 1/600 of the span of the supporting members.

❖ This section addresses masonry veneer that is supported on wood members. The provisions of Section 2104.1.6 are not intended to preclude masonry veneer on wood supports. In fact, Section 2304.12 contains provisions for the support of interior masonry veneer on wood-frame construction. The deflection limit is to minimize cracking of the veneer.

1405.10 Metal veneers. Veneers of metal shall be fabricated from approved corrosion-resistant materials or shall be protected front and back with porcelain enamel, or otherwise be treated to render the metal resistant to corrosion. Such veneers shall not be less than 0.0149-inch (0.378 mm) nominal thickness sheet steel mounted on wood or metal furring strips or approved sheathing on the wood construction.

❖ All metal veneers used as exterior weather coverings are required to be manufactured from materials that are inherently corrosion resistant or to be protected with other materials, such as coatings, to prevent corrosion. For sheet steel, the minimum nominal thickness is 0.0149 inches (0.38 mm), not including the thickness of the corrosion-resistant coating. For aluminum siding, an 0.019-inch-thick (0.48 mm) value is the minimum. Aluminum siding must also conform to AAMA 1402. The minimum thicknesses for other weather coverings are obtained from Table 1405.2.

1405.10.1 Attachment. Exterior metal veneer shall be securely attached to the supporting masonry or framing members with corrosion-resistant fastenings, metal ties or by other approved devices or methods. The spacing of the fastenings or ties shall not exceed 24 inches (610 mm) either vertically or horizontally, but where units exceed 4 square feet (0.4 m²) in area there shall be not less than four attachments per unit. The metal attachments shall have a cross-sectional area not less than provided by W 1.7 wire. Such attachments and their supports shall be capable of resisting a horizontal force in accordance with the wind loads specified in Section 1609, but in no case less than 20 psf (0.958 kg/m²).

❖ This section prescribes minimum requirements for the attachment of metal veneers to the exterior wall construction. The attachment of the veneer may be accomplished through the use of fasteners, metal ties, clips or other means. The means of attaching the panels and supporting members to which the metal veneer is fastened must be able to resist the design wind loads determined in accordance with Chapter 16, but in no case a horizontal load of less than 20 psf (0.958 kg/m²).

1405.10.2 Weather protection. Metal supports for exterior metal veneer shall be protected by painting, galvanizing or by other equivalent coating or treatment. Wood studs, furring strips or other wood supports for exterior metal veneer shall be approved pressure-treated wood or protected as required in Section 1403.2. Joints and edges exposed to the weather shall be caulked with approved durable waterproofing material or by other approved means to prevent penetration of moisture.

❖ Supports for metal veneers are required to be protected from corrosion induced by moisture. The corrosion resistance may be provided through a surface treatment applied to the metal or through the use of materials that are inherently corrosion resistant. Where wood supports are used, they must be protected through pressure treatment or covered with a minimum of one layer of a water-resistive barrier as specified in Section 1403.2. Protection of joints and edges must be provided to prevent the entrance of moisture that may result in damage to the veneer, supports or building framing.

1405.10.3 Backup. Masonry backup shall not be required for metal veneer except as is necessary to meet the fire-resistance requirements of this code.

❖ The type of substrate that metal veneers may be installed over is not limited, except in cases where the wall is required to have a fire-resistance rating in accordance with other sections of the code.

1405.10.4 Grounding. Grounding of metal veneers on buildings shall comply with the requirements of Chapter 27 and the ICC *Electrical Code.*

❖ The need to provide electrical grounding for metal veneers has been widely discussed for more than 30 years. Examples of hazards have been demonstrated where electric services have been violated. Most metal-siding manufacturers have made provisions in their designs for grounding. This section references Chapter 27 and the ICC *Electrical Code*® (ICC EC™), without providing additional specifics regarding grounding of siding.

1405.11 Glass veneer. The area of a single section of thin exterior structural glass veneer shall not exceed 10 square feet (0.93 m²) where it is not more than 15 feet (4572 mm) above the level of the sidewalk or grade level directly below, and shall not exceed 6 square feet (0.56 m²) where it is more than 15 feet (4572 mm) above that level.

❖ Structural glass veneer is produced in the same manner as plate glass (see commentary, Chapter 24), with the addition of metallic oxides that are blended into the glass to provide opacity and color. To accommodate the forces due to expansion and contraction of differing materials in exterior installations, the maximum area of structural glass veneer is limited based upon the height above grade.

1405.11.1 Length and height. The length or height of any section of thin exterior structural glass veneer shall not exceed 48 inches (1219 mm).

❖ Where Section 1405.11 regulates the maximum area of any single section of structural glass veneer, this section limits the maximum dimension of any single section to 48 inches (1219 mm) when used in an exterior application.

1405.11.2 Thickness. The thickness of thin exterior structural glass veneer shall be not less than 0.344 inch (8.7 mm).

❖ Structural glass veneers are produced in a range of thicknesses; however, the minimum thickness permitted for use as an exterior veneer is limited to $^{11}/_{32}$ inches (8.7 mm).

1405.11.3 Application. Thin exterior structural glass veneer shall be set only after backing is thoroughly dry and after application of an approved bond coat uniformly over the entire surface of the backing so as to effectively seal the surface. Glass shall be set in place with an approved mastic cement in sufficient quantity so that at least 50 percent of the area of each glass unit is directly bonded to the backing by mastic not less than 0.25 inch (6.4 mm) thick and not more than 0.625 inch (15.9 mm) thick. The bond coat and mastic shall be evaluated for compatibility and shall bond firmly together.

❖ Structural glass veneer is bonded to a solid backing, such as masonry, through the use of a mastic adhesive. The backing must be provided with a coating that will serve to seal it from moisture absorption, provide a uniform surface to which the mastic can adhere and provide a relatively level surface. The bond coat must be compatible for use with the backing material so that debonding will not occur over time. Some common types of bond coats are a float coat of cement mortar and portland cement stucco.

The glass veneer is attached to the bond coat by applying mastic adhesives. Mastics are characterized by their ability to remain pliable over time, rather than harden. The minimum and maximum thicknesses permitted for the mastic application are specified, since an application that is too thin would not provide an adequate bond and too thick could sag over time. To minimize the possibility of debonding, it is necessary to determine that the materials intended for use as a mastic and as a bond coat are compatible with one another.

1405.11.4 Installation at sidewalk level. Where glass extends to a sidewalk surface, each section shall rest in an approved metal molding, and be set at least 0.25 inch (6.4 mm) above the highest point of the sidewalk. The space between the molding and the sidewalk shall be thoroughly caulked and made water tight.

❖ To protect glass veneer from breakage when installed at sidewalk level, the veneer is required to be supported by a metal molding. The molding must be protected from corrosion and be constructed of a material that is compatible for contact with the glass. The molding is to be located a minimum of $1/4$ inch (6.4 mm) above the sidewalk to minimize the potential for cracking of the veneer due to expansion and contraction.

1405.11.4.1 Installation above sidewalk level. Where thin exterior structural glass veneer is installed above the level of the top of a bulkhead facing, or at a level more than 36 inches (914 mm) above the sidewalk level, the mastic cement binding shall be supplemented with approved nonferrous metal shelf angles located in the horizontal joints in every course. Such shelf angles shall be not less than 0.0478-inch (12 mm) thick and not less than 2 inches (51 mm) long and shall be spaced at approved intervals, with not less than two angles for each glass unit. Shelf angles shall be secured to the wall or backing with expansion bolts, toggle bolts or by other approved methods.

❖ Structural glass veneer installed more than 36 inches (914 mm) above sidewalk level is subject to impact from pedestrians and can pose a hazard if the veneer were to become dislodged from the backing. To minimize the potential for this occurring, it is necessary to provide additional supports that are attached to the backing. The supports are intended to supplement the mastic.

1405.11.5 Joints. Unless otherwise specifically approved by the building official, abutting edges of thin exterior structural glass veneer shall be ground square. Mitered joints shall not be used except where specifically approved for wide angles. Joints shall be uniformly buttered with an approved jointing compound and horizontal joints shall be held to not less than 0.063 inch (1.6 mm) by an approved nonrigid substance or device. Where thin exterior structural glass veneer abuts nonresilient material at sides or top, expansion joints not less than 0.25 inch (6.4 mm) wide shall be provided.

❖ Glass has an expansion coefficient of roughly 4.5, which is less than most metals. To minimize the potential for cracking or chipping of the glass panels and the development of sharp edges due to differential movement, this section stipulates both the requirements for factory finishing of the glass veneer panel edges and the minimum requirements for joints between sections.

1405.11.6 Mechanical fastenings. Thin exterior structural glass veneer installed above the level of the heads of show windows and veneer installed more than 12 feet (3658 mm) above sidewalk level shall, in addition to the mastic cement and shelf angles, be held in place by the use of fastenings at each vertical or horizontal edge, or at the four corners of each glass unit. Fastenings shall be secured to the wall or backing with expansion bolts, toggle bolts or by other methods. Fastenings shall be so designed as to hold the glass veneer in a vertical plane independent of the mastic cement. Shelf angles providing both support and fastenings shall be permitted.

❖ While Section 1405.11.4.1 addresses the requirements for protection of pedestrians for installations of veneer that exceed 3 feet (914 mm) above sidewalk level, this section address installations that exceed 12 feet (3658

mm) above sidewalk level. Installations above 12 feet (3658 mm) are not subject to impact from pedestrians but are subject to wind loads; therefore, each veneer section must be secured in place by mechanical means, in addition to the mastic and shelf angles specified in Sections 1405.11.3 and 1405.11.4, respectively.

1405.11.7 Flashing. Exposed edges of thin exterior structural glass veneer shall be flashed with overlapping corrosion-resistant metal flashing and caulked with a waterproof compound in a manner to effectively prevent the entrance of moisture between the glass veneer and the backing.

❖ The presence of moisture behind the glass veneer can result in loss of strength of either the mastic, the bond coat or both. In addition, corrosion of the supports or fasteners can occur. This can increase the potential for failure of the structural glass veneer. This section addresses the requirements for flashing of structural glass veneers to prevent the penetration of moisture behind the veneers.

1405.12 Exterior windows and doors. Windows and doors installed in exterior walls shall conform to the testing and performance requirements of Section 1714.5.

❖ Windows and doors that are part of the exterior building envelope are to be tested for wind-load resistance in accordance with the methods specified in Section 1714.5 (see commentary, Section 1714.5).

1405.12.1 Installation. Windows and doors shall be installed in accordance with approved manufacturer's instructions. Fastener size and spacing shall be provided in such instructions and shall be calculated based on maximum loads and spacing used in the tests.

❖ Windows and doors that are part of the exterior envelope are to be installed in accordance with the method in which they were tested (see Section 1405.12) and the window or door manufacturer's installation instructions.

1405.13 Vinyl siding. Vinyl siding conforming to the requirements of this section and complying with ASTM D 3679 shall be permitted on exterior walls of buildings of Type V construction located in areas where the basic wind speed specified in Chapter 16 does not exceed 100 miles per hour (161 km/h) and the building height is less than or equal to 40 feet (12 192 mm) in Exposure C. Where construction is located in areas where the basic wind speed exceeds 100 miles per hour (161 km/h), or building heights are in excess of 40 feet (12 192 mm), tests or calculations indicating compliance with Chapter 16 shall be submitted. Vinyl siding shall be secured to the building so as to provide weather protection for the exterior walls of the building.

❖ The installation of vinyl siding on Type V buildings is limited based upon the basic wind speed for a given location. For areas where the basic wind speed is greater than 100 miles per hour (44 m/s), wind-load resistance testing and/or structural calculations supporting the ability of the siding to withstand the basic wind speed must be provided.

1405.13.1 Application. The siding shall be applied over sheathing or materials listed in Section 2304.6. Siding shall be applied to conform with the weather-resistant barrier requirements in Section 1403. Siding and accessories shall be installed in accordance with approved manufacturer's instructions. Unless otherwise specified in the approved manufacturer's instructions, nails used to fasten the siding and accessories shall have a minimum 0.313-inch (7.9 mm) head diameter and 0.125-inch (3.18 mm) shank diameter. The nails shall be corrosion resistant and shall be long enough to penetrate the studs or nailing strip at least 0.75 inch (19 mm). Where the siding is installed horizontally, the fastener spacing shall not exceed 16 inches (406 mm) horizontally and 12 inches (305 mm) vertically. Where the siding is installed vertically, the fastener spacing shall not exceed 12 inches (305 mm) horizontally and 12 inches (305 mm) vertically.

❖ This section prescribes requirements for the installation of vinyl siding. The installation must also comply with ASTM D 4756, which is referenced in ASTM D 3679 (see commentary, Section 1404.9) and the manufacturer's instructions. These instructions are necessary for compliance with the performance requirements of this chapter. The prescriptive fastener requirements of this section correspond with the requirements contained in ASTM D 4756.

1405.14 Cement plaster. Cement plaster applied to exterior walls shall conform to the requirements specified in Chapter 25.

❖ The materials and installation of cement plaster (sometimes referred to as "stucco") are governed by the requirements of Chapter 25. It should be noted that cement plaster is the only plaster type that is permitted for exterior use. Gypsum plaster, which is also governed by Chapter 25, is generally subject to deterioration in the presence of moisture and is, therefore, not permitted for use in exterior applications.

1405.15 Fiber cement siding. Fiber cement siding complying with Section 1404.10 shall be permitted on exterior walls of Type I, II, III, IV and V construction for wind pressure resistance or wind speed exposures as indicated in the manufacturer's compliance report and approved installation instructions. Where specified, the siding shall be installed over sheathing or materials listed in Section 2304.6 and shall be installed to conform to the weather-resistant barrier requirements in Section 1403. Siding and accessories shall be installed in accordance with approved manufacturer's instructions. Unless otherwise specified in the approved manufacturer's instructions, nails used to fasten the siding to wood studs shall be corrosion-resistant round head smooth shank and shall be long enough to penetrate the studs at least 1 inch (25 mm). For metal framing, all-weather screws shall be used and shall penetrate the metal framing at least three full threads.

❖ Fiber cement siding is also a material used for weather-resistant siding. It is manufactured from fiber-reinforced cement, and the code permits its use as either panel siding or horizontal lap siding. Section 1404.10 states that the applicable material standard is ASTM C 1186. Note that the product is required to bear a label, which means that

the manufacturer must have regular inspections by a third-party quality control agency.

1405.16 Fastening. Weather boarding and wall coverings shall be securely fastened with aluminum, copper, zinc, zinc-coated or other approved corrosion-resistant fasteners in accordance with the nailing schedule in Table 2304.9.1 or the approved manufacturer's installation instructions. Shingles and other weather coverings shall be attached with appropriate standard-shingle nails to furring strips securely nailed to studs, or with approved mechanically bonding nails, except where sheathing is of wood not less than 1-inch (25 mm) nominal thickness or of wood structural panels as specified in Table 2308.9.3(3).

❖ The proper attachment of exterior weatherboard or cladding is critical in order to remain durable and stay in place under anticipated weather conditions. This section requires noncorrodible nails and other connectors. The requirements of Table 2304.9.1 address the attachment of wood structural panels and panels of particleboard and fiberboard. Lapped siding products of these materials are to be attached in accordance with the manufacturer's installation instructions.

The relatively narrow and variable widths of wood shingles and shakes make it necessary to provide a continuous nailing base. This is permitted to be wood sheathing, wood structural panels no less than $^5/_{16}$-inch (7.9 mm) thick or horizontal furring strips. Dimension lumber furring strips are required to be of 1-inch (25 mm) minimum nominal thickness. Shingles are attached with special mechanically bonding nails.

Wood shingles and shakes are required to be fastened with at least No. 14 B&S-gage corrosion-resistant nails in accordance with Table 2304.9.1. The nail must be long enough to penetrate the framing, sheathing or furring strip at least $^3/_4$ inch (19.1 mm). In accordance with Sections 1507.8.5 (shingles) and 1507.9.6 (shakes), shingles and shakes must be fastened with no more than two fasteners per shingle or shake.

Wood siding must be attached to the sheathing or framing in accordance with Table 2304.9.1. The method of nailing for the various siding patterns is illustrated in Figure 1405.15. Wood siding, especially plain bevel siding, applied over foam-plastic sheathing, requires special attention to nailing. Cooperative research by the lumber and plastic industries has shown that poor performance may be expected if the nailing is inadequate. Since the foam-plastic sheathing is not an adequate nail base, the siding is required to be nailed through the sheathing to the wood framing members. Condensation can occur on the sheathing and the back of the siding, keeping the back wet. The weather face will remain dry. This differential moisture condition will cause the siding to "curl" or warp and pull improperly installed nails out of the framing members.

1405.17 Fiber cement siding.

❖ Fiber cement siding is also a material used for weather-resistant siding. It is manufactured from fiber-

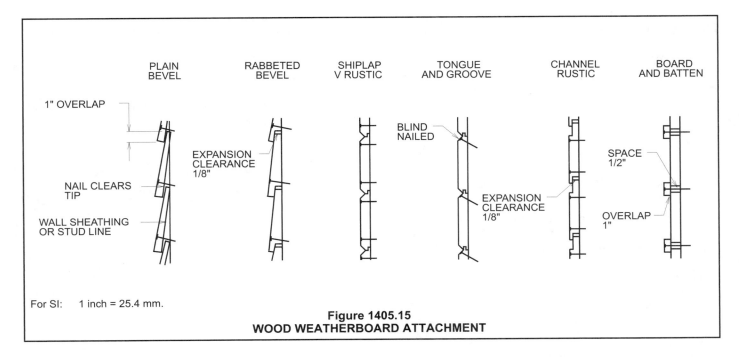

For SI: 1 inch = 25.4 mm.

Figure 1405.15
WOOD WEATHERBOARD ATTACHMENT

reinforced cement, and the code permits its use as either panel siding or horizontal lap siding. The material standard is ASTM C 1186. Note that the product is required to bear a label, which means that the manufacturer must have regular inspections by a third- party quality control agency.

1405.17.1 Panel siding. Panels shall be installed with the long dimension parallel to framing. Vertical joints shall occur over framing members and shall be sealed with caulking or covered with battens. Horizontal joints shall be flashed with Z-flashing and blocked with solid wood framing.

❖ Panel siding of fiber cement is an acceptable exterior wall covering, provided the joints are waterproof. Typical methods for accomplishing the joint treatment include: (1) having the joints occur over framing and protecting such joints with caulking or a continuous batten or (2) flashing and blocking horizontal joints.

1405.17.2 Horizontal lap siding. Lap siding shall be lapped a minimum of $1^1/_4$ inches (32 mm) and shall have the ends sealed with caulking, covered with an H-section joint cover or located over a strip of flashing. Lap siding courses shall be permitted to be installed with the fastener heads exposed or concealed, according to approved manufacturers' instructions.

❖ In order to maintain the necessary weather protection, horizontal fiber cement siding must have a minimum lap of at least 1 $^1/_4$ inches (31.8 mm). The end joints must be treated with one of the following methods: (1) sealed with caulk; (2) covered with an H-section joint cover or (3) located over a strip of flashing. Fastener heads may be exposed or concealed in accordance with approved manufacturers' installation instructions.

SECTION 1406
COMBUSTIBLE MATERIALS ON THE EXTERIOR SIDE OF EXTERIOR WALLS

1406.1 General. This section shall apply to exterior wall coverings, balconies and similar appendages, and bay and oriel windows constructed of combustible materials.

❖ The requirements given in this section apply to exterior elements of combustible construction. The elements included are exterior wall veneers such as vinyl or wood siding; architectural trim; gutters; half-timbering; balconies; bay and oriel windows; plastic panels and foam plastics. The code requires walls to be noncombustible for buildings of construction Types I, II, III and IV in order to limit the fuel load. Many common types of materials used for architectural enhancement of exterior walls are combustible; therefore, the code provides limitations as well as requirements for flame spread and radiant heat exposure testing in order to mitigate the hazards associated with combustible materials as permitted elements of noncombustible construction.

1406.2 Combustible exterior wall coverings. Combustible exterior wall coverings shall comply with this section.

Exception: Plastics complying with Chapter 26.

❖ The specific requirements relating to the combustible exterior wall finish include resistance to ignition through radiant heat exposure testing (see Section 1406.2.1) and limitations on the height and the amount of combustible materials.

The exception is provided for plastic materials that comply with the requirements of Chapter 26 for limitations on flame spread, ignition properties and the amount of plastic.

1406.2.1 Ignition resistance. Combustible exterior wall coverings shall be tested in accordance with NFPA 268.

Exceptions:

1. Wood or wood-based products.

2. Other combustible materials covered with an exterior covering other than vinyl sidings listed in Table 1405.2.

3. Aluminum having a minimum thickness of 0.019 inch (0.48 mm).

4. Exterior wall coverings on exterior walls of Type V construction.

❖ In order to continue to provide the level of safety for adjoining property that is consistent with the fire safety provisions for opening limitations in exterior walls, radiant heat testing is required for combustible materials other than wood, using 1.25 W/cm[2] as a baseline exposure criterion.

This section requires that testing be performed in accordance with the NFPA 268 test method to measure the incident radiant heat that will cause a given exterior finish material to ignite.

There are four exceptions to the test requirements. Exception 1 is given because wood or wood-based products have been chosen as the baseline criterion for evaluation of other combustible materials. Testing in accordance with this section would be redundant since such materials are in compliance by virtue of being the baseline material.

Exception 2 relates to materials covered by one of the veneers listed in Table 1405.2. For example, the exception would apply to the building paper under vinyl siding, not the vinyl siding itself. Again, this is an exception to eliminate any confusion that might result from the presence of combustible components within a wall assembly that are not the exposed exterior finish.

Exception 3 is provided because, although most aluminum alloys are combustible, aluminum has a higher ignition temperature than wood. Although not tested, it is anticipated that the critical radiant heat flux would therefore also be higher.

Exception 4 is provided because its basic intent relates to concerns over flame propagation on exteriors of buildings required to be of noncombustible construction.

1406.2.1.1 Fire separation 5 feet or less. Where installed on exterior walls having a fire separation distance of 5 feet (1524 mm) or less, combustible exterior wall coverings shall not exhibit sustained flaming as defined in NFPA 268.

❖ In order to qualify for use on walls with a fire separation distance as little as 5 feet (1524 mm), the material must not flame when exposed to a 12.5 kW/m[2] exposure.

1406.2.1.2 Fire separation greater than 5 feet. For fire separation distances greater than 5 feet (1524 mm), an assembly shall be permitted that has been exposed to a reduced level of incident radiant heat flux in accordance with the NFPA 268 test method without exhibiting sustained flaming. The minimum fire separa-

tion distance required for the assembly shall be determined from Table 1406.2.1.2 based on the maximum tolerable level of incident radiant heat flux that does not cause sustained flaming of the assembly.

❖ Materials with a lower tolerance to radiant heat (less than 12.5 kW/m[2]) can be used, but the tradeoff is an increased fire separation distance to reduce the potential of ignition from radiant heat.

TABLE 1406.2.1.2
MINIMUM FIRE SEPARATION FOR COMBUSTIBLE VENEERS

FIRE SEPARATION DISTANCE (feet)	TOLERABLE LEVEL INCIDENT RADIANT HEAT ENERGY(kW/m^2)	FIRE SEPARATION DISTANCE (feet)	TOLERABLE LEVEL INCIDENT RADIANT HEAT ENERGY(kW/m^2)
5	12.5	16	5.9
6	11.8	17	5.5
7	11.0	18	5.2
8	10.3	19	4.9
9	9.6	20	4.6
10	8.9	21	4.4
11	8.3	22	4.1
12	7.7	23	3.9
13	7.2	24	3.7
14	6.7	25	3.5
15	6.3		

For SI: 1 foot = 304.8 mm, 1 Btu/H^2 ×°F = .0057 kW/m^2 × K.

❖ The table provides an increased minimum fire separation distance when the resistance of a material to radiant heat is less than 12.5 kW/m[2] required for veneers on walls with a separation distance of 5 feet (1524 mm). The required resistance to radiant heat decreases with increasing separation distances. It is not anticipated that interpolation will be involved, since fire separation distances are not enforced in fractions of feet. For example, if testing indicated a tolerable level of radiant heat energy of 9.5 kW/m[2], the required minimum fire separation distance would be 10 feet (30 480 mm).

1406.2.2 Architectural trim. In buildings of Types I, II, III, and IV construction that do not exceed three stories or 40 feet (12 192 mm) in height above grade plane, exterior wall coverings shall be permitted to be constructed of wood where permitted by Section 1405.4 or other equivalent combustible material. Combustible exterior wall coverings, other than fire-retardant-treated wood complying with Section 2303.2 for exterior installation, shall not exceed 10 percent of an exterior wall surface area where the fire separation distance is 5 feet (1524 mm) or less. Architectural trim that exceeds 40 feet (12 192 mm) in height above grade plane shall be constructed of approved noncombustible materials and shall be secured to the wall with metal or other approved noncombustible brackets.

❖ All architectural trim and exterior veneers on buildings of Type I, II, III and IV construction that are more than 40 feet (12 192 mm) above grade are required to be noncombustible. This is primarily because of the difficulties associated with fire-fighter access to higher portions of a building and to minimize potential injury from flying brands at high elevation. Combustible architectural trim and exterior veneers are permitted below the

40-foot (12 192 mm) height even when the overall building height exceeds 40 feet (12 192 mm). Combustible trim and veneers are also permitted on buildings of Type V construction. It should be noted that although Table 503 limits buildings of Type V construction to 40 feet (12 192 mm) in height, such buildings may be increased in height when protected with an automatic sprinkler system in accordance with Section 504.2.

In order to limit the potential involvement of trim and veneer materials in an exterior exposure fire, combustible exterior trim and veneer (other than code-complying FRTW) is limited to 10 percent of the exterior wall surface area where the fire separation distance is 5 feet (1524 mm) or less. For purposes of determining the 10-percent limitation, the surface area of the wall to which the trim is being attached should be considered, not the aggregate area of all exterior wall surfaces. If the fire separation distance is greater than 5 feet (1524 mm), the amount of combustible trim is not restricted. In all cases, the area of noncombustible and approved FRTW is not limited. This provision applies to all construction types, as indicated by Section 603.1.

1406.2.3 Location. Where combustible exterior wall covering is located along the top of exterior walls, such trim shall be completely backed up by the exterior wall and shall not extend over or above the top of exterior walls.

❖ Combustible trim is not permitted to extend above the exterior wall to which it is attached. This is intended to limit the overall potential involvement of exterior combustible materials in a fire and to reduce the risk of fire spreading to or from such materials from the interior limits of the building. A typical example of architectural trim in this instance is a wood canopy attached to the exterior wall of a strip shopping center, with the canopy fully backed up by the wall.

1406.2.4 Fireblocking. Where the combustible exterior wall covering is furred from the wall and forms a solid surface, the distance between the back of the covering and the wall shall not exceed 1.625 inches (41 mm) and the space thereby created shall be fireblocked in accordance with Section 717 so that there will be no open space exceeding 100 square feet (9.3 m²). Where wood furring strips are used, they shall be of approved wood of natural decay resistance or preservative-treated wood.

Exceptions:

1. Fireblocking of cornices is not required in single-family dwellings.
2. Fireblocking shall not be required where installed on noncombustible framing and the face of the exterior wall finish exposed to the concealed space is covered by one of the following materials:
   2.1. Aluminum having a minimum thickness of 0.019 inch (0.5 mm);
   2.2. Corrosion-resistant steel having a base metal thickness not less than 0.016 inch (0.4 mm) at any point; or

   2.3. Other approved noncombustible materials.

❖ This section limits the area of concealed space behind combustible veneers installed on the exterior of buildings of Type I, II, III or IV construction. The intent of fireblocking is to limit the potential for fire to spread through the concealed spaces formed by combustible veneers or architectural trim. Fireblocking materials and methods are regulated by Section 717.

Exception 1 corresponds to the exception contained in Section 717.2.6. Single-family dwellings are typically constructed entirely of combustible materials and are limited in size and height; therefore, fireblocking in concealed areas behind cornices is not necessary.

Exception 2 recognizes that where the concealed space does not contain combustible framing or furring and the exterior weather covering is noncombustible, fireblocking is not necessary.

1406.3 Balconies and similar projections. Balconies and similar projections of combustible construction, other than fire-retardant-treated wood, shall afford the fire-resistance rating required by Table 601 for floor construction or shall be of Type IV construction as described in Section 602.4, and the aggregate length shall not exceed 50 percent of the building perimeter on each floor.

Exceptions:

1. On buildings of Type I and II construction, three stories or less in height, fire-retardant-treated wood shall be permitted for balconies, porches, decks and exterior stairways not used as required exits.
2. Untreated wood is permitted for pickets and rails, or similar guardrail devices that are limited to 42 inches (1067 mm) in height.
3. Balconies and similar appendages on buildings of Type III, IV and V construction shall be permitted to be of Type V construction, and shall not be required to have a fire-resistance rating where sprinkler protection is extended to these areas.
4. Where sprinkler protection is extended to the balcony areas, the aggregate length of the balcony on each floor shall not be limited.

❖ Because these elements are, in a sense, an extension of floor construction, combustible appendages are required to afford the same required fire-resistance rating as required for floor construction in Table 602, unless the appendage is of FRTW or heavy timber construction (Type IV construction). As an additional safeguard against exterior fire spread, the aggregate length of combustible appendages must not exceed 50 percent of the building perimeter on each floor. Balconies, porches, decks, supplemental exterior stairs and similar appendages in buildings of Types I and II construction are required to be constructed of noncombustible materials in order to prevent fire involvement and fire spread up or along the exterior of a noncombustible building. In buildings of Types III, IV and V construction, the use of combustible materials for these elements is permitted.

Exception 1 permits balconies and similar append-ages to be constructed of FRTW where buildings of Types I and II construction do not exceed three stories in height. This is due to limited combustibility of FRTW. The three-story limitation is similar to the provisions contained in Section 1406.2.2 for combustible trim.

Exception 2 permits the use of combustible guardrails and handrails for all types of construction in an attempt to alleviate the warping and maintenance problems associated with thin lumber members used for pickets and rails. These items need not be constructed of FRTW.

Exception 3 is applicable to buildings of Types III, IV and V construction. Balconies and similar appendages need not be constructed of FRTW nor have a fire-resistance rating when the appendages are protected with an automatic sprinkler system. The presence of sprinkler protection, such as a dry pendent sprinkler, will also serve to limit fire spread from floor to floor.

Balconies, porches, decks and supplemental exterior stairways that are not attached to or supported by the building are separate structures and are to be built accordingly. Although the structural support may be independent, such structures could serve to assist vertical fire spread depending on the construction of the exterior wall and the presence or lack of opening protectives.

1406.4 Bay windows and oriel windows. Bay and oriel windows shall conform to the type of construction required for the building to which they are attached.

Exception: Fire-retardant-treated wood shall be permitted on buildings three stories or less of Type I, II, III and IV construction.

❖ In all buildings of other than Type V construction, bay windows and similar appendages are required to be constructed of noncombustible materials (see Figure 1406.4). This is consistent with the requirements for exterior walls in Type I, II, III and IV construction in Sec-

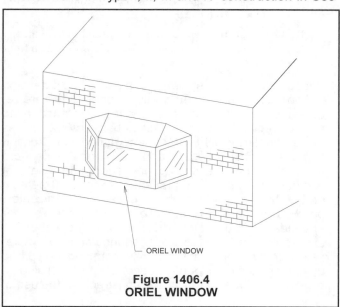

Figure 1406.4
ORIEL WINDOW

tions 602.2 through 602.4. Based upon the limited combustibility of these elements of FRTW, the exception permits their use on buildings of Type I, II, III and IV construction when the elements are constructed of FRTW and the building does not exceed three stories.

SECTION 1407
METAL COMPOSITE MATERIALS (MCM)

1407.1 General. The provisions of this section shall govern the materials, construction and quality of metal composite materials (MCM) for use as exterior wall coverings in addition to other applicable requirements of Chapters 14 and 16.

❖ MCMs are panels with a solid plastic core that is encapsulated with an aluminum facer on both sides. MCMs are typically used in exterior applications as a weather covering; however, they do not provide insulation. Section 1407 regulates MCMs that are used as exterior wall coverings.

1407.2 Exterior wall finish. MCM used as exterior wall finish or as elements of balconies and similar appendages and bay and oriel windows to provide cladding or weather resistance shall comply with Sections 1407.4 through 1407.13.

❖ The plastic core of MCMs is a combustible material. This section provides a correlation with Sections 1406.2, 1406.3 and 1406.4 regulating the use of combustible materials on the exterior of buildings.

1407.3 Architectural trim and embellishments. MCM used as architectural trim or embellishments shall comply with Sections 1407.7 through 1407.13.

❖ In a manner similar to Section 1407.2, the provisions of this section provide a correlation with Section 1406.2.2 regulating the use of combustible trim on the exterior of buildings.

1407.4 Structural design. MCM systems shall be designed and constructed to resist wind loads as required by Chapter 16 for components and cladding.

❖ As previously stated in numerous locations throughout Chapter 14, exterior weather coverings must be designed and installed to resist the design wind loads required by Chapter 16. These provisions apply to the MCM panels, the supporting elements that are part of the MCM system and the attachment of the system to the building frame.

1407.5 Approval. Results of approved tests or an engineering analysis shall be submitted to the building official to verify compliance with the requirements of Chapter 16 for wind loads.

❖ The determination that an MCM system will be able to resist a specific wind load may be accomplished through wind-load testing, structural engineering calculations or a combination of both. In all instances, documentation substantiating the system's ability to resist the design wind loads is required to be submitted to the building official.

1407.6 Weather resistance. MCM systems shall comply with Section 1403 and shall be designed and constructed to resist wind and rain in accordance with this section and the manufacturer's installation instructions.

❖ As with other exterior weather coverings, MCM systems must comply with the requirements of Section 1403 to prevent the entry of moisture into the building or structure. The manufacturer's installation instructions contain details that are specific to a weather-tight installation.

1407.7 Durability. MCM systems shall be constructed of approved materials that maintain the performance characteristics required in Section 1407 for the duration of use.

❖ All materials used as weather coverings are required to exhibit a certain level of durability to be effective in providing protection to the structure and its occupants. MCM panels are a composite construction, and the bond between the metal facing and plastic core must be maintained for the material to maintain the performance characteristics specified in Section 1407. For example, if the bond between the panel and core is not durable, debonding could occur. This could result in detrimental performance, such as possible water intrusion.

1407.8 Fire-resistance rating. Where MCM systems are used on exterior walls required to have a fire-resistance rating in accordance with Section 704, evidence shall be submitted to the building official that the required fire-resistance rating is maintained.

❖ Evidence must be submitted to confirm that the installation of MCM systems will not reduce the fire-resistance rating of exterior walls that are required by Section 704.5 to have a rating. This data may consist of testing performed in accordance with ASTM E 119 of the MCM applied to the fire-resistance-rated wall assembly, as provided for in Section 703.2, or an alternative method, such as engineering analysis, as provided for in Section 703.3 (see commentary, Sections 703.2 and 703.3).

1407.9 Surface-burning characteristics. Unless otherwise specified, MCM shall have a flame spread index of 75 or less and a smoke-developed index of 450 or less when tested as an assembly in the maximum thickness intended for use in accordance with ASTM E84.

❖ MCMs are required to have a basic flame spread rating of 75 and a smoke-developed rating of 450. This is consistent with the requirements for other types of plastics intended for use in building construction, including foam plastics. As indicated in Section 1407.10.1, the flame spread rating required is more restrictive (less than 25) when MCMs are installed on buildings of Type I, II, III and IV construction.

1407.10 Type I, II, III and IV construction. Where installed on buildings of Type I, II, III and IV construction, MCM systems shall comply with Sections 1407.10.1 through 1407.10.4, or 1407.11.

❖ Due to the combustible core of MCM panels, the fire performance of these panels is regulated. Sections 1407.10.1 through 1407.10.4 stipulate the fire testing that is required to permit the use of these materials on buildings of other than Type V construction. It should be noted that the requirements of these sections are similar to the provisions contained in Chapter 26 for foam-plastic materials. Where fire testing is specified, the panels are required to be tested in the maximum thickness intended for use. This is because the maximum thickness represents the greatest potential fuel load in a fire condition.

It should be noted that since MCM panels have an MCM facing with a minimum thickness of 0.019 inches (.48 mm), these materials are not required to undergo the ignition-resistance testing required by Section 1406.2.1.

1407.10.1 Surface-burning characteristics. MCM shall have a flame spread index of not more than 25 and a smoke-developed index of not more than 450 when tested as an assembly in the maximum thickness intended for use in accordance with ASTM E 84.

❖ The limitations on the maximum flame spread and smoke-developed indexes correspond with the requirements of Section 2603.5.4 for foam-plastic materials used on the exterior of buildings of other than Type V construction (see commentary, Section 2603.5.4).

1407.10.2 Thermal barriers. MCM shall be separated from the interior of a building by an approved thermal barrier consisting of 0.5-inch (12.7 mm) gypsum wallboard or equivalent thermal barrier material that will limit the average temperature rise of the unexposed surface to not more than 250°F (121°C) after 15 minutes of fire exposure in accordance with the standard time-temperature curve of ASTM E 119. The thermal barrier shall be installed in such a manner that it will remain in place for not less than 15 minutes based on a test conducted in accordance with UL 1715.

❖ The requirements for the installation of a thermal barrier to separate the MCM from the interior of the building are similar to the requirements contained in Section 2603.5.2. The thermal barrier requirements in this section are identical to the requirements of Section 2603.4 (see the commentary to Section 2603.4 for additional discussion on thermal barriers).

1407.10.3 Thermal barrier not required. The thermal barrier specified for MCM in Section 1407.10.2 is not required where:

1. The MCM system is specifically approved based on tests conducted in accordance with UL 1040 or UL 1715. Such testing shall be performed with the MCM in the maximum thickness intended for use. The MCM system shall include seams, joints and other typical details used in the in-

stallation and shall be tested in the manner intended for use.

2. The MCM is used as elements of balconies and similar appendages, architectural trim or embellishments.

❖ The exemption of the installation of the thermal barrier required by Section 1407.9 may be accomplished in two ways. The first is through the use of large-scale fire testing of the MCM system. The system must be tested in a configuration that incorporates details that are representative of a typical installation. The second provision of this section exempts the installation of a thermal barrier where the MCM is limited to use as a component of balconies and similar appendages or as architectural trim and embellishments.

1407.10.4 Full-scale tests. The MCM exterior wall assembly shall be tested in accordance with, and comply with, the acceptance criteria of NFPA 285. Such testing shall be performed on the MCM system with the MCM in the maximum thickness intended for use.

❖ The final data required to permit the use of MCM systems on buildings of Type I, II, III and IV construction are the result of full-scale testing of the system. This testing is intended to determine that the system will not have the tendency to propagate the spread of fire over the surface of the panels or through the panel core.

1407.11 Alternate conditions. MCM and MCM systems shall not be required to comply with Sections 1407.10.1 through 1407.10.4 provided such systems comply with Section 1407.11.1 or 1407.11.2.

❖ Section 1407.10 provides an alternative to compliance with the requirements of Section 1407.9 based upon a specific set of criteria. These criteria limit the amount of material and height above grade in a manner similar to the provisions of Section 1406 for combustible materials.

1407.11.1 Installations up to 40 feet in height. MCM shall not be installed more than 40 feet (12 190 mm) in height above the grade plane where installed in accordance with Sections 1407.11.1.1 and 1407.11.1.2.

❖ The height limitation of 40 feet (12 192 mm) corresponds to the limitations of Section 1406.2.2 for architectural trim. In addition to the height limitation, the material must comply with the requirements of Sections 1407.11.1 and 1407.11.2.

1407.11.1.1 Fire separation distance of 5 feet or less. Where the fire separation distance is 5 feet (1524 mm) or less, the area of MCM shall not exceed 10 percent of the exterior wall surface.

❖ To limit the potential for the spread of fire between buildings that are in close proximity to one another, the maximum amount of combustible exterior veneer is limited in size. These limitations are identical to those of Section 1406.2.2 for combustible veneers.

1407.11.1.2 Fire separation distance greater than 5 feet. Where the fire separation distance is greater than 5 feet (1524 mm), there shall be no limit on the area of exterior wall surface coverage using MCM.

❖ Because of the height and area limitations of Section 1407.11, the MCM is permitted to have a higher flame spread than that required under Section 1407.10.1. The maximum flame spread of 75 is identical to the maximum flame spread permitted for foam plastics of Section 2603.3.

1407.11.2 Installations up to 50 feet in height. MCM shall not be installed more than 50 feet (15 240 mm) in height above the grade plane where installed in accordance with Sections 1407.11.2.1 and 1407.11.2.2.

❖ The maximum installed height of MCM on buildings of Type I, II, III and IV construction may be increased up to 50 feet (15 240 mm) when the requirements of Sections 1407.11.2.1 and 1407.11.2.2 are complied with.

1407.11.2.1 Self ignition temperature. MCM shall have a self-ignition temperature of 650°F (343°C) or greater when tested in accordance with ASTM D 1929.

❖ Keeping with the requirements in Chapter 26 for other plastic materials, the flame spread of the MCM is not permitted to exceed 75 and the smoke-developed index may not exceed 450. In addition to the limitations on flame spread and smoke development, the MCM must be tested for self-ignition in accordance with ASTM D 1929. These requirements are identical to those contained in Section 2606.4 for light-transmitting plastics (see commentary, Section 2606.4).

1407.11.2.2 Limitations. Sections of MCM shall not exceed 300 square feet (27.9 m²) in area and shall be separated by a minimum of 4 feet (1219 mm) vertically.

❖ The size limitations of this section are consistent with the provisions of Section 2605.2, Item 3 and Section 2606.7.3. The area limitation is intended to minimize the amount of concentrated combustible materials on the exterior of noncombustible buildings. The separation requirement is intended to minimize the potential for extensive flame propagation over the face of the wall.

1407.12 Type V construction. MCM shall be permitted to be installed on buildings of Type V construction.

❖ Because of the combustible nature of Type V construction, the fire testing requirements for MCM used as exterior veneers on these buildings is less severe than for other construction types. This is similar in concept to the provisions of Section 1406.

1407.13 Labeling. MCM shall be labeled in accordance with Section 1703.5.

❖ Because of the composite nature of MCMs and the proprietary nature of the panel cores, the fabrication of the panels must be subject to independent third-party verification. The verification of this inspection is evidenced by labeling of the product by the third-party agency. Refer to Section 1703.5 for more discussion on the labeling of products.

Bibliography

The following resource materials are referenced in this chapter or are relevant to the subject matter addressed in this chapter.

AAMA 1402-86, *Standard Specification for Aluminum Siding, Soffit and Fascia*. Palatine, IL: American Architectural Manufacturers Association, 1986.

ACI 530/ASCE 5/TMS 402-02, *Building Code Requirements for Masonry Structures*. Farmington Hills, MI: American Concrete Institute, 2002.

ACI 530.1/ASCE 6/TMS 602-02, *Specifications for Masonry Structures*. Farmington Hills, MI: American Concrete Institute, 2002.

AHA A135.4-95, *Basic Hardboard*. Palatine, IL: American Hardboard Association, 1995.

AHA A135.6-98, *Hardboard Siding*. Palatine, IL: American Hardboard Association, 1998.

ASTM C 1186—99, *Specification for Flat Nonasbestos Fiber Cement Sheets*. West Conshohocken, PA: ASTM International, 1999.

ASTM D 1929-96, *Test Method for Ignition Properties of Plastics*. West Conshohocken, PA: ASTM International, 1996.

ASTM D 3679-01c, *Specification for Rigid Poly (Vinyl Chloride)(PVC) Siding*. West Conshohocken, PA: ASTM International, 2001.

ASTM D 4756-96, *Standard Practice for Installation of Rigid Poly (Vinyl Chloride)*. West Conshohocken, PA: ASTM International, 1996.

ASTM E 84-01, *Test Methods for Surface-Burning Characteristics of Building Materials*. West Conshohocken, PA: ASTM International, 2001.

ASTM E 96-00, *Standard Test Method for Water Vapor Transmission of Materials*. West Conshohocken, PA: ASTM International, 2000.

ASTM E 119-00, *Test Methods for Fire Tests of Building Construction Materials*. West Conshohocken, PA: ASTM International, 2000.

ASTM E 331-00, *Test Method for Water Penetration of Exterior Walls, Doors by Uniform Static Air Pressure Difference*. West Conshohocken, PA: ASTM International, 2000.

ICC EC-2003, *ICC Electrical Code*. Falls Church, VA: International Code Council, 2003.

IECC-2003, *International Energy Conservation Code*. Falls Church, VA: International Code Council, 2003.

NFPA 268-00, *Standard Test Method for Determining Ignitability of Exterior Wall Assemblies Using a Radiant Heat Energy Source*. Quincy, MA: National Fire Protection Association, 2000.

Chapter 15:
Roof Assemblies And Rooftop Structures

General Comments

Chapter 15 regulates the materials, design, construction and quality of roofs and roof structures for all buildings and structures. Requirements for weather protection, fire classification, flashings and insulation are provided to govern the roof slab or deck and its supporting members, in addition to the covering that is applied to the roof for weather resistance, fire resistance or appearance. Specific installation requirements for certain steep-slope and low-slope roof coverings are provided.

Section 1502 contains definitions of terms associated with the provisions in this chapter.

Section 1503 provides requirements for roof coverings to protect against the effects of weather.

Section 1504 provides performance requirements, including referenced standards, for evaluating the performance of new and innovative roof coverings.

Section 1505 provides a classification system to determine a roof covering's effectiveness against certain fire-test exposures. Specific classifications of roof coverings are required based on the type of construction classification of the structure.

Section 1506 provides requirements for the application of roof covering materials

Section 1507 provides requirements for the installation of roof coverings.

Section 1508 permits combustible roof insulation in all types of construction, subject to applicable code requirements.

Section 1509 establishes specific structures that can be constructed on and above the roof.

Section 1510 provides requirements for reroofing, including adequate structural supports, recovering of existing roof materials and replacement.

The most important requirement of this chapter deals with the fire classification of roof covering materials for their effectiveness against fire test exposure, in accordance with ASTM E 108. This referenced standard provides test methods for measuring the fire performance characteristics of roof coverings from exposure fires. Fire tests on roofing materials were first conducted by Underwriters Laboratories Inc. (UL) and the National Fire Protection Association (NFPA) in the early 1900s. This research was conducted in respon

se to numerous and extensive conflagrations that occurred in that era, which were aided by the low flame spread resistance and high-burning brand characteristics of roofing materials used in the building industry at that time.

Purpose

The provisions in Chapter 15 for roof construction and covering are intended to provide a weather-protective barrier at the roof and, in most circumstances, a fire-retardant barrier to prevent flaming combustible materials, such as flying brands from nearby fires, from penetrating the roof construction. This chapter is essentially prescriptive in nature and is based on decades of experience with various traditional roofing materials. These prescriptive rules are very important for satisfactory performance of the roof covering even though the reason for a particular requirement may be lost. The provisions are based on an attempt to prevent past unsatisfactory performance of the various roofing materials and components.

SECTION 1501
GENERAL

1501.1 Scope. The provisions of this chapter shall govern the design, materials, construction and quality of roof assemblies, and rooftop structures.

❖ This section specifies the scope and applicability of the code for all roofs and roof structures. Requirements are provided to regulate the materials, design and construction of roofs and rooftop structures, such as penthouses, water tanks and other structures. See the definition of "Roof assembly" for the purposes of this chapter. Other roof assemblies addressed elsewhere in the code include membrane structures (see Section 3102) and light-transmitting plastics (see Section 2606).

SECTION 1502
DEFINITIONS

1502.1 General. The following words and terms shall, for the purposes of this chapter and as used elsewhere in this code, have the meanings shown herein.

❖ Definitions of terms can help in the understanding and application of the code requirements. The purpose for including these definitions within this chapter is to provide more convenient access to them without having to refer back to Chapter 2.

For convenience, these terms are also listed in Chapter 2 with a cross reference to this section. Terms that are italicized provide a visual identification throughout the code that a definition exists for that term.

The use and application of all defined terms, including those defined herein, are set forth in Section 201.

BUILT-UP ROOF COVERING. Two or more layers of felt cemented together and surfaced with a cap sheet, mineral aggregate, smooth coating or similar surfacing material.

❖ Because of their low melting points and self-healing characteristics, built-up roofs are typically constructed of coal tar membranes that are commonly installed on lower slopes and dead-level roofs.

INTERLAYMENT. A layer of felt or nonbituminous saturated felt not less than 18 inches (457 mm) wide, shingled between each course of a wood-shake roof covering.

❖ According to Section 1507.9.4, interlayment is required to comply with ASTM D 226, Type I, which is commonly referred to as No. 15 asphalt felt.

MECHANICAL EQUIPMENT SCREEN. A partially enclosed rooftop structure used to aesthetically conceal heating, ventilating and air conditioning (HVAC) electrical or mechanical equipment from view.

❖ These are used mainly for appearance and to conceal the heating, ventilating and air-conditioning (HVAC) equipment from view.

METAL ROOF PANEL. An interlocking metal sheet having a minimum installed weather exposure of 3 square feet (.279 m²) per sheet.

❖ There are two general categories of metal roofing systems: architectural metal roofing and structural metal roofing. Architectural metal roofs are generally water-shedding roof systems and structural metal roofs have hydrostatic (water barrier) characteristics. The difference between a "metal roof panel" and a "metal roof shingle" is the weather exposure (i.e., that portion of the roofing exposed) (see Section 1507.4).

METAL ROOF SHINGLE. An interlocking metal sheet having an installed weather exposure less than 3 square feet (.279 m²) per sheet.

❖ See the definition of "Metal roof panel."

MODIFIED BITUMEN ROOF COVERING. One or more layers of polymer-modified asphalt sheets. The sheet materials shall be fully adhered or mechanically attached to the substrate or held in place with an approved ballast layer.

❖ These are composite sheets consisting of copolymer modified bitumen, often reinforced and sometimes surfaced with various types of films, foils and mats.

PENTHOUSE. An enclosed, unoccupied structure above the roof of a building, other than a tank, tower, spire, dome cupola or bulkhead, occupying not more than one-third of the roof area.

❖ The definition of "Penthouse" is similar to the definition of "Mezzanine," in that specific area limitations are mandated. Any enclosed structure that is located above the surrounding roof surfaces can be considered a penthouse as long as its footprint area is less than or equal to one-third of the roof area. By complying with this area limitation and the space being unoccupied, the penthouse is considered as part of the story below (see Section 1509.2) and, therefore, does not contribute to the height of the building either in number of stories above grade or in feet. If the penthouse exceeds the one-third area limitation, it must be considered as an additional story of the building or structure.

POSITIVE ROOF DRAINAGE. The drainage condition in which consideration has been made for all loading deflections of the roof deck, and additional slope has been provided to ensure drainage of the roof within 48 hours of precipitation.

❖ The primary purpose of positive roof drainage is to allow for prompt roof drainage, thereby preventing the roof from being damaged or adversely impacting the structural support due to the additional load.

REROOFING. The process of recovering or replacing an existing roof covering. See "Roof recover" and "Roof replacement."

❖ This term refers to the process of covering an existing roof system with a new roofing system (see Section 1510).

ROOF ASSEMBLY. A system designed to provide weather protection and resistance to design loads. The system consists of a roof covering and roof deck or a single component serving as both the roof covering and the roof deck. A roof assembly includes the roof deck, vapor retarder, substrate or thermal barrier, insulation, vapor retarder and roof covering.

❖ This is an assembly of interacting roof components (including the roof deck) designed to weatherproof and, normally, to insulate a building's top surface.

ROOF COVERING. The covering applied to the roof deck for weather resistance, fire classification or appearance.

❖ This definition identifies the specific membrane of the entire roof system that provides weather protection and any required resistance to exterior fire exposure. The code has specific performance and prescriptive requirements for this covering to provide a durable, weather-resistant surface for the entire structure. A roof covering is considered the membrane that provides the weather-resistance and fire-performance characteristics required by the code.

ROOF COVERING SYSTEM. See "Roof assembly."

❖ See the definition of "Roof assembly."

ROOF DECK. The flat or sloped surface not including its supporting members or vertical supports.

❖ This term refers to the structural surface to which the roofing and waterproofing system (including insulation) is applied.

ROOF RECOVER. The process of installing an additional roof covering over a prepared existing roof covering without removing the existing roof covering.

❖ This is the process of covering an existing roof system with a new roofing system.

ROOF REPAIR. Reconstruction or renewal of any part of an existing roof for the purposes of its maintenance.

❖ Roofs should be maintained for the purpose of protection. If a section is damaged, then it should be repaired immediately.

ROOF REPLACEMENT. The process of removing the existing roof covering, repairing any damaged substrate and installing a new roof covering.

❖ This is the process of covering an existing roof system with a new roofing system.

ROOF VENTILATION. The natural or mechanical process of supplying conditioned or unconditioned air to, or removing such air from, attics, cathedral ceilings or other enclosed spaces over which a roof assembly is installed.

❖ Ventilation of the attic prevents moisture condensation on cold surfaces, and, therefore, will prevent dry rot on the bottom surface of shingles or wood roof decks.

ROOFTOP STRUCTURE. An enclosed structure on or above the roof of any part of a building.

❖ This definition includes all appurtenances constructed and located above the surrounding roof surfaces. These items (water tanks or cooling towers) or architectural features (spires or cupolas) are regulated by specific code provisions.

SCUPPER. An opening in a wall or parapet that allows water to drain from a roof.

❖ These devices are for the purpose of allowing overflowing water to drain off the roof, and are commonly larger than the roof drain.

SINGLE-PLY MEMBRANE. A roofing membrane that is field applied using one layer of membrane material (either homogeneous or composite) rather than multiple layers.

❖ This is a flexible or semiflexible roof covering or waterproofing layer whose primary function is the exclusion of water.

UNDERLAYMENT. One or more layers of felt, sheathing paper, nonbituminous saturated felt or other approved material over which a steep-slope roof covering is applied.

❖ According to Section 1507.2.3, underlayment is required to comply with ASTM D 266 or 4869, Type I, which is commonly referred to as No. 15 asphalt felt.

SECTION 1503
WEATHER PROTECTION

1503.1 General. Roof decks shall be covered with approved roof coverings secured to the building or structure in accordance with the provisions of this chapter. Roof coverings shall be designed, installed and maintained in accordance with this code and the approved manufacturer's instructions such that the roof covering shall serve to protect the building or structure.

❖ The roof covering provides protection from water intrusion into a building. The roofing system has historically been one of the most problematic areas of a building. Without a properly constructed roof assembly, water may infiltrate the building, causing damage to building materials and contents. A roof assembly, for the purposes of this chapter, is defined as including the roof deck and the components above that make up the roof covering. For a roof covering to perform the way it must to protect the building, it must meet certain requirements. The code provides basic requirements for the construction of roof assemblies. The designer will typically include in his or her roof specification compatibility of materials, deck type, weather conditions, roof slope, structural loads, roof drainage, roof penetrations, energy and future reroofing.

Other code requirements, such as structural and fire resistance, are related to the ability of a roof covering to maintain the weather protection required by this section. For example, it is imperative that the roof covering not be blown off by the wind or it will fail to provide the necessary weather protection.

1503.2 Flashing. Flashing shall be installed in such a manner so as to prevent moisture entering the wall and roof through joints in copings, through moisture-permeable materials and at intersections with parapet walls and other penetrations through the roof plane.

❖ Flashing must be installed in a very specific manner so that moisture does not enter the building construction or get under the roof covering. As indicated, the flashing is installed anyplace the roof covering is interrupted or terminated.

1503.2.1 Locations. Flashing shall be installed at wall and roof intersections, at gutters, wherever there is a change in roof slope or direction and around roof openings. Where flashing is of metal, the metal shall be corrosion resistant with a thickness of not less than 0.019 inch (.483 mm) (No. 26 galvanized sheet).

❖ Flashing consists of components that weatherproof or seal the roof system at discontinuities, such as perimeters, penetrations, walls, expansion joints, valleys, drains and other places where the roof covering is interrupted or terminated. For example, membrane base flashing covers the edge of the field membrane, and cap flashings or counterflashings shield the upper edges of the base flashing.

Metal flashing must be no less than the specified minimum thickness and be corrosion resistant ("Corrosion resistance" is defined in Section 202).

1503.3 Coping. Parapet walls shall be properly coped with noncombustible, weatherproof materials of a width no less than the thickness of the parapet wall.

❖ Coping is the covering piece on top of a wall that is exposed to the weather, usually made of metal, masonry or stone. It is typically sloped to shed water back onto the roof.

[P] 1503.4 Roof drainage. Design and installation of roof drainage systems shall comply with the *International Plumbing Code*.

❖ Positive roof drainage is required for essentially two reasons: to prevent ponding of water and the consequent rapid deterioration of the roof-covering material, including potential for early failure of the roof covering; and to prevent ponding and the resulting failure of the roof structure or, alternatively, the need for increased strength of the roof system to support ponding. Section 1608.3.5 requires the structural design calculations to include the prevention of ponding instability.

1503.4.1 Gutters. Gutters and leaders placed on the outside of buildings, other than Group R-3 as applicable in Section 101.2, private garages and buildings of Type V construction, shall be of noncombustible material or a minimum of Schedule 40 plastic pipe.

❖ All roofs are required to be designed and built with positive drainage (see Section 1507.10.1 for slope requirements of built-up roofs). "Positive roof drainage" is defined in Section 1502. Ponding water has detrimental effects on a roof system, including roof surface and membrane deterioration due to debris accumulation; vegetation and fungal growth; deck deflections that can lead to structural damage; ice formation; tensile splitting and the possibility of water entering the building.

1503.5 Roof ventilation. Intake and exhaust vents shall be provided in accordance with Section 1203.2 and the manufacturer's installation instructions.

❖ During cold weather, condensation is deposited on cold surfaces when, for example, warm, moist air rising from the interior of a building and through the attic comes in contact with the roof deck. Ventilation prevents moisture condensation on the cold surfaces and, therefore, will reduce or prevent dry rot on the bottom surface of shingles or wood roof decks.

SECTION 1504
PERFORMANCE REQUIREMENTS

1504.1 Wind resistance of roofs. Roof decks and roof coverings shall be designed for wind loads in accordance with Chapter 16 and Sections 1504.2, 1504.3 and 1504.4.

❖ It is important that roof coverings and roof decks remain intact and in place when subjected to high winds. Without an intact roof system, the building would be subjected to either water damage, which would reduce its structural stability, or to higher wind pressures for which the building is not designed. The following sections specify requirements for different roof coverings regarding wind loads. The roof decks are included in the requirements of Chapter 16.

1504.1.1 Wind resistance of asphalt shingles. Asphalt shingles shall be designed for wind speeds in accordance with Section 1507.2.7.

❖ The section cites a specific reference in the code, which includes attachment details for asphalt shingles. Roofs with high slopes or in areas subject to higher wind speeds, in accordance with Chapter 16, may require special methods of fastening. Such methods need to be tested in accordance with ASTM D 3161 (see commentary, Section 1507.2.7).

1504.2 Wind resistance of clay and concrete tile. Clay and concrete tile roof coverings shall be connected to the roof deck in accordance with Chapter 16.

❖ Rigid tile roofs, including clay and concrete roof tiles, must meet the wind load provisions of this section and Sections 1609.7.2 and 1609.7.3. Additionally, they must be fastened to the roof in accordance with Section 1507.3.7 based on the wind speed of the area in which they are installed.

1504.3 Wind resistance of nonballasted roofs. Roof coverings installed on roofs in accordance with Section 1507 that are mechanically attached or adhered to the roof deck shall be designed to resist the design wind load pressures for cladding in Chapter 16.

❖ This section addresses wind resistance of steep-slope systems that are mechanically attached by fasteners or adhered to the roof slab or deck by adhesives. Steep-slope roof coverings are those described in Section 1507 with a roof slope of 2:12 or greater. The basic wind speed as obtained from Chapter 16 for the building's location must be modified for the building's overall height above grade and exposure category, wind gust effect and wind importance factor. A structural design is required to demonstrate that the proposed fastening schedule will resist the wind uplift loads.

1504.3.1 Other roof systems. Roof systems with built-up, modified bitumen, fully adhered or mechanically attached single-ply through fastened metal panel roof systems, and other types of membrane roof coverings shall also be tested in accordance with FM 4450, FM 4470, UL 580 or UL 1897.

❖ Minimum requirements addressing wind resistance of low-slope roof systems (typically less than 2:12) are established in this section. These requirements mandate that the roofs and roof coverings be designed and fastened to resist the code-specified design wind load pressures of Chapter 16. This section states the test requirements that other types of membrane roof coverings are to meet.

1504.3.2 Metal panel roof systems. Metal panel roof systems through fastened or standing seam shall be tested in accordance with UL 580 or ASTM E 1592.

❖ This section specifies the test that metal panel roof systems are required to pass before installation. Standards and test methods have been established for the proper installation of metal panel roof systems, whether fastened or standing seamed.

1504.4 Ballasted low-slope roof systems. Ballasted low-slope (roof slope < 2:12) single-ply roof system coverings installed in accordance with Section 1507 shall be designed in accordance with ANSI/SPRI RP-4.

❖ Section 1507 prescribes the installation for ballasted low-slope single-ply roof system coverings. These coverings are to be designed in accordance with ANSI/SPRI RP-4, which is listed in Chapter 35.

1504.5 Edge securement for low-slope roofs. Low-slope membrane roof systems metal edge securement, except gutters, installed in accordance with Section 1507, shall be designed in accordance with ANSI/SPRI ES-1, except the basic wind speed shall be determined from Figure 1609.

❖ This section references the appropriate standard for low-slope roofs, which is ANSI/SPRI ES 1 listed in Chapter 35. This standard is important in preventing failures for low-slope roof systems. Since the standard references ASCE 7, a reference to Figure 1609 is also provided to clarify that the basic wind speeds must be determined using Chapter 16.

1504.6 Physical properties. Roof coverings installed on low-slope roofs (roof slope < 2:12) in accordance with Section 1507 shall demonstrate physical integrity over the working life of the roof based upon 2,000 hours of exposure to accelerated weathering tests conducted in accordance with ASTM G 152, ASTM G 155 or ASTM G 154. Those roof coverings that are subject to cyclical flexural response due to wind loads shall not demonstrate any significant loss of tensile strength for unreinforced membranes or breaking strength for reinforced membranes when tested as herein required.

❖ This section addresses the effects of weather on the performance of the roof covering, requires that weather testing be performed and includes performance criteria for those roof coverings that will have to perform while subjected to wind loads.

The ASTM referenced test methods cover exposure to carbon-arc-type or xenon-arc-type light (with or without water), fluorescent ultraviolet-condensation-type rays and concentrated natural sunlight. In conjunction with these tests, roofing materials must not show any significant signs of failure caused by cyclical wind conditions. The tensile strength in unreinforced membranes and the breaking strength in reinforced membranes must endure such conditions without any significant strength reduction.

1504.7 Impact resistance. Roof coverings installed on low-slope roofs (roof slope < 2:12) in accordance with Section 1507 shall resist impact damage based on the results of tests conducted in accordance with ASTM D 3746, ASTM D 4272, CGSB 37-GP-52M or FM 4470.

❖ All roof-covering materials must withstand impact loads, such as hail and driving rainstorms. The referenced standards set criteria for the testing of certain roofing materials on low-slope roofs, such as plastic film and bituminous roofing systems. Each test uses similar devices to test the material. The tests consist of dropping a dart or missile, which weighs from 2.5 pounds to 5.0 pounds (1.1 kg to 2.3 kg), approximately 4 to 5 feet (1219 to 1524 mm) and then evaluating the material for failure.

SECTION 1505
FIRE CLASSIFICATION

1505.1 General. Roof assemblies shall be divided into the classes defined below. Class A, B and C roof assemblies and roof coverings required to be listed by this section shall be tested in accordance with ASTM E 108 or UL 790. In addition, fire-retardant-treated wood roof coverings shall be tested in accordance with ASTM D 2898. The minimum roof coverings installed on buildings shall comply with Table 1505.1 based on the type of construction of the building.

❖ The code designates the use of any particular classification of roof coverings based on the type of construction of the building. A minimum Class B roof covering is required for all roofs that have a minimum 1-hour fire-resistance rating in accordance with Table 602. Roofs without a required fire-resistance rating require a minimum Class C roof covering.

TABLE 1505.1 – 1505.7   ROOF ASSEMBLIES AND ROOFTOP STRUCTURES

TABLE 1505.1[a,b]
MINIMUM ROOF COVERING CLASSIFICATION
FOR TYPES OF CONSTRUCTION

IA	IB	IIA	IIB	IIIA	IIIB	IV	VA	VB
B	B	B	C[c]	B	C[c]	B	B	C[c]

For SI: 1 foot = 304.8 mm, 1 square foot = 0.0929 m².

a. Unless otherwise required in accordance with the *International Urban Wildland Interface Code* or due to the location of the building within a fire district in accordance with Appendix D.

b. Nonclassified roof coverings shall be permitted on buildings of Group R-3, as applicable in Section 101.2, and Group U occupancies, where there is a minimum fire-separation distance of 6 feet measured from the leading edge of the roof.

c. Buildings that are not more than two stories in height and having not more than 6,000 square feet of projected roof area and where there is a minimum 10-foot fire-separation distance from the leading edge of the roof to a lot line on all sides of the building, except for street fronts or public ways, shall be permitted to have roofs of No. 1 cedar or redwood shakes and No. 1 shingles.

❖ See the commentary to Section 1505.1.

1505.2 Class A roof assemblies. Class A roof assemblies are those that are effective against severe fire test exposure. Class A roof assemblies and roof coverings shall be listed and identified as Class A by an approved testing agency. Class A roof assemblies shall be permitted for use in buildings or structures of all types of construction.

Exception: Class A roof assemblies include those with coverings of brick, masonry, slate, clay or concrete roof tile, exposed concrete roof deck, ferrous or copper shingles or sheets.

❖ Roof coverings that are effective against severe fire-test exposure of ASTM E 108 are to be classified as Class A. Traditional Class A roof coverings include the following noncombustible coverings: masonry, concrete, slate, tile and cement-asbestos. Additionally, any assembly that is tested in accordance with ASTM E 108 requirements by an approved testing agency and is listed and identified by that agency can be used as a Class A roof covering. Classified roof coverings are to be further identified by the manufacturer's designation. The code requirement is for the roof covering to be tested by an approved agency and then addressed as a certain classification. All Class A roof coverings may be used on buildings and structures of any type of construction.

1505.3 Class B roof assemblies. Class B roof assemblies are those that are effective against moderate fire-test exposure. Class B roof assemblies and roof coverings shall be listed and identified as Class B by an approved testing agency.

Exception: Class B roof assemblies include those with coverings of metal sheets and shingles.

❖ Roof coverings that are effective against the moderate fire-test exposure of ASTM E 108 are to be classified as Class B. Traditional Class B roof coverings include metal sheets and shingles. Those roof coverings listed and identified as Class B by an approved testing agency are also allowed (see the commentary to Section 1505.2 for additional information concerning testing

agencies, listing and identification). For buildings and structures with 1-hour fire-resistance-rated roof construction, Class A or B roof coverings are required.

1505.4 Class C roof assemblies. Class C roof assemblies are those that are effective against light fire-test exposure. Class C roof assemblies and roof coverings shall be listed and identified as Class C by an approved testing agency.

❖ Roof coverings that are effective against light fire test exposures of ASTM E 108 are to be classified as Class C. There are no specific materials or products that automatically qualify as Class C roof coverings. Again, those assemblies of roof coverings that are listed and identified as Class C by an approved testing agency may be used. Buildings with nonfire-resistance-rated roofs are required to have a minimum Class C covering.

1505.5 Nonclassified roofing. Nonclassified roofing is approved material that is not listed as a Class A, B or C roof covering.

❖ Nonclassified roofing is a roofing material that is approved by the building official that is not otherwise classified (see the definition of "Approved" in Section 202).

1505.6 Fire-retardant-treated wood shingles and shakes. Fire-retardant-treated wood shakes and shingles shall be treated by impregnation with chemicals by the full-cell vacuum-pressure process, in accordance with AWPA C1. Each bundle shall be marked to identify the manufactured unit and the manufacturer, and shall also be labeled to identify the classification of the material in accordance with the testing required in Section 1505.1, the treating company and the quality control agency.

❖ This section specifically identifies the particular pressure-impregnation method to be used to treat wood shakes and shingles. This method consists of placing bundles of the roofing materials inside a retort, which is a huge cylindrical pressure container. A vacuum is then created inside the retort, which in turn draws the air and moisture from the millions of cells per square inch (psi) in western red cedar, which is commonly used for roofing materials. The fire-retardant chemical mixture is then injected at pressures reaching 150 psi (1034 kPa) into every cell of the shakes and shingles—even the innermost layers. After the products are pressure treated, they are then placed in a special drying kiln, which finishes the materials with a thermal cure at temperatures of up to 2000 °F (1093 °C). This section further identifies the test standard to be followed, the method of labeling and the information required per bundle of shingles in order for the building official to evaluate building compliance. This includes the testing requirements found in Section 1505.1.

1505.7 Special purpose roofs. Special purpose wood shingle or wood shake roofing shall conform with the grading and application requirements of Section 1507.8 or 1507.9. In addition, an underlayment of 0.625-inch (15.9 mm) Type X water-resistant gypsum backing board or gypsum sheathing shall be placed under minimum nominal 0.5-inch-thick (12.7 mm) wood structural panel solid sheathing or 1-inch (25 mm) nominal spaced sheathing.

❖ Informal tests of special-purpose roofs have shown that they provide a significant increase in protection for combustible roof coverings, and again the protection offered is considered appropriate if applied in the manner stipulated.

SECTION 1506
MATERIALS

1506.1 Scope. The requirements set forth in this section shall apply to the application of roof-covering materials specified herein. Roof coverings shall be applied in accordance with this chapter and the manufacturer's installation instructions. Installation of roof coverings shall comply with the applicable provisions of Section 1507.

❖ This section establishes specifications and standards for materials and installation techniques in conjunction with recognized industry standards and acceptable roofing applications. By complying with the specified material requirements, slope limitations, underlayment requirements and fastening schedules, the proposed roof covering can be expected to perform as anticipated.

1506.2 Compatibility of materials. Roofs and roof coverings shall be of materials that are compatible with each other and with the building or structure to which the materials are applied.

❖ Material compatibility requires many considerations. For example, some roof coverings might not be compatible with certain roof deck material, sealants or solvents, and some are not compatible with oils. Additionally, it is necessary to have the correct combinations of materials to adequately respond to moisture, humidity and energy conservation for buildings of specific materials in given climates.

1506.3 Material specifications and physical characteristics. Roof-covering materials shall conform to the applicable standards listed in this chapter. In the absence of applicable standards or where materials are of questionable suitability, testing by an approved agency shall be required by the building official to determine the character, quality and limitations of application of the materials.

❖ This section acknowledges that there may be materials on the market for which the code does not currently reference a test standard. In those cases, the building official will have to require testing by an approved lab to determine the character, quality and application limitations of the materials. The building official must also require testing when he or she questions the suitability of the materials.

1506.4 Product identification. Roof-covering materials shall be delivered in packages bearing the manufacturer's identifying marks and approved testing agency labels required in accordance with Section 1505. Bulk shipments of materials shall be accompanied with the same information issued in the form of a certificate or on a bill of lading by the manufacturer.

❖ Identification of the roofing materials is mandatory in order to verify that they comply with quality standards. In addition to bearing the manufacturer's label or identifying mark, prepared roofing and built-up roofing materials are required by the code to carry a label of an approved agency that inspects the material and finished products during manufacture.

SECTION 1507
REQUIREMENTS FOR ROOF COVERINGS

1507.1 Scope. Roof coverings shall be applied in accordance with the applicable provisions of this section and the manufacturer's installation instructions.

❖ This section requires that the installation be in accordance with the code and the manufacturer's installation instructions. Often, other requirements must be met, such as UL or FM requirements. It is very possible that there will be times that these requirements conflict. Once adopted, the code is law. The other provisions are not law, and compliance with them does not necessarily mean that compliance with the code has been achieved. In cases where conflicts arise, the code requirements govern.

1507.2 Asphalt shingles. The installation of asphalt shingles shall comply with the provisions of this section and Table 1507.2.

❖ This section establishes the criteria for asphalt shingle roofs. Two general types of asphalt shingles specifically apply to this section: strip shingles (e.g., three-tab shingles) and individual interlocking shingles (e.g., t-lock shingles). Asphalt strip shingles, which are the most common, are produced in a variety of finished appearances, including three-tab, random or multitab, no-cutout and laminated architectural. Asphalt shingles are typically classified in two types: cellulose felt reinforced (i.e., organic shingles) and fiberglass mat reinforced (i.e., fiberglass shingles). Although it is not specifically addressed in this chapter, the roofing industry generally recommends the attic space below asphalt shingle roofs be properly ventilated.

Additional information regarding asphalt shingle roofs is provided in the *NRCA Asphalt Shingle Roofing Manual* and the asphalt roofing section of the *NRCA Steep Roofing Manual* published by the National Roofing Contractors Association (NRCA) and the *Residential Asphalt Roofing Manual* published by the Asphalt Roofing Manufacturers Association.

TABLE 1507.2 – 1507.2.3

ROOF ASSEMBLIES AND ROOFTOP STRUCTURES

TABLE 1507.2
ASPHALT SHINGLE APPLICATION

COMPONENT	INSTALLATION REQUIREMENT
1. Roof slope	Asphalt shingles shall only be used on roof slopes of two units vertical in 12 units horizontal (2:12) or greater. For roof slopes from two units vertical in 12 units horizontal (2:12) up to four units vertical in 12 units horizontal (4:12), double underlayment application is required in accordance with Section 1507.2.8.
2. Deck requirement	Asphalt shingles shall be fastened to solidly sheathed roofs.
3. Underlayment	Underlayment shall conform with ASTM D 226, Type 1, or ASTM D 4869, Type 1.
For roof slopes from two units vertical in 12 units horizontal (2:12), up to four units vertical in 12 units horizontal (4:12)	Underlayment shall be two layers applied in the following manner. Apply a minimum 19-inch strip or underlayment felt parallel to and starting at the eaves, fastened sufficiently to hold in place. Starting at the eave, apply 35-inch-wide sheets of underlayment overlapping successive sheets 19 inches and fastened sufficiently to hold in place.
For roof slopes from four units vertical in 12 units horizontal (4:12) or greater	Underlayment shall be one layer applied in the following manner. Underlayment shall be applied shingle fashion, parallel to and starting from the eave and lapped 2 inches, fastened only as necessary to hold in place.
In areas where the average daily temperature in January is 25°F or less or where there is a possibility of ice forming along the eaves causing a backup of water	A membrane that consists of at least two layers of underlayment cemented together or of a self-adhering polymer-modified bitumen sheet shall be used in lieu of normal underlayment and extend from the eave's edge to a point at least 24 inches inside the exterior wall line of the building.
4. Application	—
Attachment	Asphalt shingles shall have the minimum number of fasteners required by the manufacturer and Section 1504.1. Asphalt shingles shall be secured to the roof with not less than four fasteners per strip shingle or two fasteners per individual shingle. Where the roof slope exceeds 20 units vertical in 12 units horizontal (20:12), special methods of fastening are required.
Fasteners	Galvanized, stainless steel, aluminum or copper roofing nails, minimum 12-gage (0.105 inch) shank with a minimum $^3/_8$-inch diameter head. Fasteners shall be long enough to penetrate into the sheathing $^3/_4$ inch or through the thickness of the sheathing.
Flashings	In accordance with Section 1507.2.9.

For SI: 1 inch = 25.4 mm, 1 foot = 304.8 mm, °C = [(°F) - 32]/1.8, 1 mile per hour = 1.609 km/h.

TABLE 1507.2. See above.

❖ Where the roof slope is less than 4:12, water drainage from the roof is slowed down and has a tendency to back up under the roofing and cause leaks. Also, the effect of ice dams at the eaves is more pronounced on "low-slope" roofs and, as a result, special precautions are necessary for satisfactory performance of the roofing materials. The code also specifies double coverage of the shingles. The code requires that underlayment of one layer of Type 15 felt be provided under asphalt shingle roof coverings with a slope at least equal to 4:12.

1507.2.1 Deck requirements. Asphalt shingles shall be fastened to solidly sheathed decks.

❖ Solid sheathed roof decks typically include nominal 1-inch (25 mm) solid lumber or wood structural panels. It is recommended that solid sawn lumber no more than 8 inches (203 mm) in width be used. Wider pieces are more likely to warp and cup. The code is specific about the types of decking allowed depending on the material used. For example, wood roof decking is addressed in Chapter 23.

Manufacturers may also require application of their materials to specific roof deck materials for warranty of their products.

1507.2.2 Slope. Asphalt shingles shall only be used on roof slopes of two units vertical in 12 units horizontal (17-percent slope) or greater. For roof slopes from two units vertical in 12 units horizontal (17-percent slope) up to four units vertical in 12 units horizontal (33-percent slope), double underlayment application is required in accordance with Section 1507.2.8.

❖ Asphalt shingles are intended to be applied to a steep roof. A minimum slope is crucial in the performance of such shingles because it determines the surface drainage.

1507.2.3 Underlayment. Unless otherwise noted, required underlayment shall conform to ASTM D 226, Type I, or ASTM D 4869, Type I.

❖ Asphalt shingle roofs are required to be installed with a felt underlayment. Underlayment serves as secondary protection against wind-driven rain and other moisture penetrations, as well as offering some protection against moisture when individual shingles become damaged or dislodged in high winds.

Underlayment also serves as a separator sheet between the shingles and the deck. As a separator, the underlayment provides some protection to the shingles from uneven roof deck edges and deck fasteners. Underlayment also helps eliminate a rectangular pattern of ridges in the membrane over insulation or deck joints.

Underlayment offers protection from resins and other chemicals contained in wood board sheathing.

Using the appropriate underlayments, decking materials and shingles will help in meeting the fire classification ratings.

Underlayment for asphalt shingle roofs is required to comply with ASTM D 226, Type I, or ASTM D 4869, Type I. Products complying with the Type I designation of these standards are commonly called No. 15 asphalt felt.

1507.2.4 Self-adhering polymer modified bitumen sheet. Self-adhering polymer modified bitumen sheet shall comply with ASTM D 1970.

❖ Self-adhering polymer modified bitumen sheets meeting ASTM D 1970 are intended for use as underlayment for ice dam protection. These underlayments have an adhesive layer that is exposed by the removal of a protective sheet. The top surface of the sheet is suitable to work on during the application of the exposed roofing.

1507.2.5 Asphalt shingles. Asphalt shingles shall have self-seal strips or be interlocking, and comply with ASTM D 225 or ASTM D 3462.

❖ ASTM D 225 covers asphalt roofing in shingle form, composed of single or multiple thicknesses of organic felt saturated and coated on both sides with asphalt and surfaced on the weather side with mineral granules. This standard is well established in the roofing industry. Generally, all organic shingles currently manufactured comply with this standard.

ASTM D 3462 covers asphalt roofing in shingle form, composed of single or multiple thicknesses of glass felt impregnated and coated on both sides with asphalt, and surfaced on the weather side with mineral granules. The shingles may be either locked together during installation, or have a factory-applied self-sealing adhesive. Additionally, asphalt shingles meeting this standard must pass the Class A fire exposure test requirements of ASTM E 108 and the wind resistance test requirements of ASTM D 3161.

Another standard specification, ASTM D 3018, also applies to fiberglass shingles; however, this standard is less stringent than ASTM D 3462 and should not be considered a substitute for ASTM D 3462. It is important to realize that many of the 20-year warranted shingles and some 25-year warranted shingles currently on the market do not comply with ASTM D 3462. Fiberglass shingles complying with ASTM D 3462 will typically be identified as such on the shingle bundle packaging or

the manufacturer will supply a written certification of compliance with the standard.

1507.2.6 Fasteners. Fasteners for asphalt shingles shall be galvanized, stainless steel, aluminum or copper roofing nails, minimum 12 gage [0.105 inch (2.67 mm)] shank with a minimum 0.375 inch-diameter (9.5 mm) head, of a length to penetrate through the roofing materials and a minimum of 0.75 inch (19.1 mm) into the roof sheathing. Where the roof sheathing is less than 0.75 inch (19.1 mm) thick, the nails shall penetrate through the sheathing. Fasteners shall comply with ASTM F 1667.

❖ The fasteners allowed are either of metal that is corrosion resistant or galvanized to be corrosion resistant. It is important to use fasteners that are long enough to penetrate the roof covering, flashing, underlayment and into the deck a minimum of $^3/_4$ inch (19.1 mm). When the deck is less than $^3/_4$ inch (19.1 mm) in thickness, the fastener must penetrate through the deck. It is recommended that the nail be at least $^1/_8$ inch (3.2 mm) through the deck.

1507.2.7 Attachment. Asphalt shingles shall have the minimum number of fasteners required by the manufacturer and Section 1504.1. Asphalt shingles shall be secured to the roof with not less than four fasteners per strip shingle or two fasteners per individual shingle. Where the roof slope exceeds 20 units vertical in 12 units horizontal (166-percent slope), special methods of fastening are required. For roofs located where the basic wind speed in accordance with Figure 1609 is 110 mph or greater, special methods of fastening are required. Special fastening methods shall be tested in accordance with ASTM D 3161, modified to use a wind speed of 110 mph.

❖ This section clarifies that both Section 1504.1 and the manufacturer's instructions must be consulted when installing asphalt shingles. Section 1504.1.1 references Section 1507.2.7 for the installation of asphalt shingles for high wind conditions; therefore, the manufacturer's instructions are the primary requirements. In cases where the manufacturer requires less fasteners, adherence to the code is necessary. Where the manufacturer requires more fasteners, adherence to the manufacturer's recommendations is required in accordance with Section 1503.1.

For roof slopes greater than 20:12, the manufacturer should be consulted for installation requirements. Typically, such an installation will include a 5-inch (127 mm) exposure, with each tab cemented in place using an asphalt roofing cement compatible with the shingle.

Where the basic wind speeds in accordance with Figure 1609 are greater than or equal to 110 mph (48 m/s), special testing is required to determine how such roofing should be fastened. Such fastening methods must specifically be tested to ASTM D 3161, but the test must be modified to accommodate the wind speed of 110 mph (48 m/s). Essentially, ASTM D 3161 deals with several ranges of wind speeds. More basic wind speeds are considered up to 110 mph (48 m/s); therefore, higher wind-prone areas must comply with the modified test.

1507.2.8 Underlayment application. For roof slopes from two units vertical in 12 units horizontal (17-percent slope), up to four units vertical in 12 units horizontal (33-percent slope), underlayment shall be two layers applied in the following manner. Apply a minimum 19-inch-wide (483 mm) strip of underlayment felt parallel with and starting at the eaves, fastened sufficiently to hold in place. Starting at the eave, apply 36-inch-wide (914 mm) sheets of underlayment overlapping successive sheets 19 inches (483 mm) and fastened sufficiently to hold in place. For roof slopes of four units vertical in 12 units horizontal (33-percent slope) or greater, underlayment shall be one layer applied in the following manner. Underlayment shall be applied shingle fashion, parallel to and starting from the eave and lapped 2 inches (51 mm), fastened only as necessary to hold in place.

❖ Roofs of low slope, 2:12 to 4:12, shed water much more slowly than steeper roofs and, therefore, require greater protection from water backing up under the shingles. This is particularly important when considering wind-driven water and ice damming. For such slopes, two layers of underlayment are required. For greater slopes, only one layer of underlayment is necessary, except at eaves where ice dams are potentially a problem. Regardless of the slope, underlayment need only be fastened well enough to stay in place until the shingles are applied. The application of shingles will also serve to fasten the underlayment.

1507.2.8.1 High wind attachment. Underlayment applied in areas subject to high winds (greater than 110 mph in accordance with Figure 1609) shall be applied with corrosion-resistant fasteners in accordance with the manufacturer's instructions. Fasteners are to be applied along the overlap at a maximum spacing of 36 inches (914 mm) on center.

❖ The fasteners must be approved for asphalt shingle installation and should be flat headed and corrosion resistant.

1507.2.8.2 Ice dam membrane. In areas where the average daily temperature in January is 25°F (-4°C) or less or where there is a possibility of ice forming along the eaves causing a backup of water, a membrane that consists of at least two layers of underlayment cemented together or of a self-adhering polymer modified bitumen sheet shall be used in lieu of normal underlayment and extend from the eave's edge to a point at least 24 inches (610 mm) inside the exterior wall line of the building.

Exception: Detached accessory structures that contain no conditioned floor area.

❖ Ice dams form when snow melts over the warmer parts of a roof and refreezes over the colder eaves. This ice formation acts like a dam and causes water to back up beneath the roof covering. The water eventually will leak, causing damage to the structure, including the walls, ceilings and roof [see Figures 1507.8.3(1) and (2)].

There is an exception to this section that exempts accessory buildings from such restrictions, since they are unheated structures where the need for protection against ice dams is unnecessary. The same exception

is found in Sections 1507.5.3, 1507.6.3, 1507.7.3, 1507.8.3 and 1507.9.3

1507.2.9 Flashings. Flashing for asphalt shingles shall comply with this section. Flashing shall be applied in accordance with this section and the asphalt shingle manufacturer's printed instructions.

❖ This section establishes the requirement for roof flashings for asphalt shingles to prevent leakage and further establishes performance criteria.

1507.2.9.1 Base and cap flashing. Base and cap flashing shall be installed in accordance with the manufacturer's instructions. Base flashing shall be of either corrosion-resistant metal of minimum nominal 0.019-inch (0.483 mm) thickness or mineral-surfaced roll roofing weighing a minimum of 77 pounds per 100 square feet (3.76 kg/m²). Cap flashing shall be corrosion-resistant metal of minimum nominal 0.019-inch (0.483 mm) thickness.

❖ Roof system edges must be weatherproofed or sealed where they intersect with other vertical components such as walls or chimneys. Flashing is composed of two parts: the base and the cap. For a general discussion of roof flashings, see the commentary to Section 1503.2.

1507.2.9.2 Valleys. Valley linings shall be installed in accordance with the manufacturer's instructions before applying shingles. Valley linings of the following types shall be permitted:

1. For open valleys (valley lining exposed) lined with metal, the valley lining shall be at least 16 inches (406 mm) wide and of any of the corrosion-resistant metals in Table 1507.2.9.2.

2. For open valleys, valley lining of two plies of mineral-surfaced roll roofing shall be permitted. The bottom layer shall be 18 inches (457 mm) and the top layer a minimum of 36 inches (914 mm) wide.

3. For closed valleys (valleys covered with shingles), valley lining of one ply of smooth roll roofing complying with ASTM D 224 and at least 36 inches (914 mm) wide or types as described in Items 1 and 2 above shall be permitted. Specialty underlayment shall comply with ASTM D 1970.

❖ Valleys are the internal angle formed by the intersection of two sloping roof planes. For asphalt shingle roofs, the valley protection is categorized as open or closed. An open valley is a method of construction in which the steep-slope roofing on both sides is trimmed along each side of the valley, exposing the valley flashing.

Closed valleys are those that are covered with shingles and include methods known as closed-cut valleys and woven valleys.

Closed-cut valleys are a method of valley construction in which shingles from one side of the valley extend across the valley, while shingles from the other side are trimmed back approximately 2 inches (51 mm) from the valley centerline.

Woven valley is a method of valley construction in

which shingles from both sides of the valley extend across the valley and are woven together by overlapping alternate courses as they are applied.

TABLE 1507.2.9.2
VALLEY LINING MATERIAL

MATERIAL	MINIMUM THICKNESS	GAGE	WEIGHT
Copper	—	—	16 oz
Aluminum	0.024 in.	—	—
Stainless steel	—	28	—
Galvanized steel	0.0179 in.	26 (zinc-coated G90)	—
Zinc alloy	0.027 in.	—	—
Lead	—	—	2.5 pounds
Painted terne	—	—	20 pounds

For SI: 1 inch = 25.4 mm, 1 pound = 0.454 kg, 1 ounce = 28.35 g.

❖ These corrosion-resistant metals may be used in open-valley construction as valley lining where the lining is exposed.

1507.2.9.3 Drip edge. Provide drip edge at eaves and gables of shingle roofs. Overlap to be a minimum of 2 inches (51 mm). Eave drip edges shall extend 0.25 inch (6.4 mm) below sheathing and extend back on the roof a minimum of 2 inches (51 mm). Drip edge shall be mechanically fastened a maximum of 12 inches (305 mm) o.c. A cricket or saddle shall be installed on the ridge side of any chimney greater than 30 inches (762 mm) wide. Cricket or saddle coverings shall be sheet metal or of the same material as the roof covering.

❖ Drip edge is a metal flashing or other overhanging component applied at the roof edge intended to control the direction of dripping water and help protect the underlying building components. Drip edge has an outward projecting lower edge to direct the water away from the building. It also is used to break the continuity of contact between the roof perimeter and the wall components to help prevent capillary action.

Metal drip edge is a formed metal flashing that extends back from, and bends down over, the roof edge. Along roof gables, rakes and the eave, the edge drip is applied over the underlayment.

A cricket or saddle is a raised roof substrate or structure constructed to channel or direct surface water around a chimney.

1507.3 Clay and concrete tile. The installation of clay and concrete tile shall comply with the provisions of this section.

❖ Additional information regarding clay and concrete tile roofs is provided in the *NRCA Roofing and Waterproofing Manual*.

1507.3.1 Deck requirements. Concrete and clay tile shall be installed only over solid sheathing or spaced structural sheathing boards.

❖ The choice of deck material is critical to the life of the roof system. Decks must be capable of withstanding loads such as the weight of the tile and associated roof components and accessories. The weight of tile can increase significantly due to water absorption, depending on its porosity. It is important for designers to note that some types of roof tiles may develop increased porosity with age.

1507.3.2 Deck slope. Clay and concrete roof tile shall be installed on roof slopes of $2^1/_2$ units vertical in 12 units horizontal (21-percent slope) or greater. For roof slopes from $2^1/_2$ units vertical in 12 units horizontal (21-percent slope) to four units vertical in 12 units horizontal (33-percent slope), double underlayment application is required in accordance with Section 1507.3.3.

❖ Along with the manufacturer's installation specifications, this section gives specific limitations on the application of concrete and clay tile; thus, tile is not allowed to be installed on slopes less than $2^1/_2$:12 regardless of the manufacturer's literature. Additionally, for low-sloped roofs, double underlayment is required.

1507.3.3 Underlayment. Unless otherwise noted, required underlayment shall conform to: ASTM D 226, Type II; ASTM D 2626 or ASTM D 249 Type I mineral-surfaced roll roofing.

❖ Because of the long service life of many tile roofs, an underlayment should be chosen that will provide protection for a comparable period of time. ASTM D 226 includes the physical requirements for Type II asphalt saturated felt, which is commonly called No. 30 asphalt felt.

ASTM D 2626 covers the felt base sheet with fine mineral surfacing on the top side, with or without perforations. It is intended that nonperforated felt be used in this application. The *NRCA Roofing and Waterproofing Manual* has recommendations regarding the choice of underlayment for specific roof slopes and applications.

1507.3.3.1 Low-slope roofs. For roof slopes from $2^1/_2$ units vertical in 12 units horizontal (21-percent slope), up to four units vertical in 12 units horizontal (33-percent slope), underlayment shall be a minimum of two layers applied as follows:

1. Starting at the eave, a 19-inch (483 mm) strip of underlayment shall be applied parallel with the eave and fastened sufficiently in place.

2. Starting at the eave, 36-inch-wide (914 mm) strips of underlayment felt shall be applied overlapping successive sheets 19 inches (483 mm) and fastened sufficiently in place.

❖ The code requires that the underlayment be two layers, which provides two thicknesses of the underlayment at any point.

1507.3.3.2 High-slope roofs. For roof slopes of four units vertical in 12 units horizontal (33-percent slope) or greater, underlayment shall be a minimum of one layer of underlayment felt applied shingle fashion, parallel to, and starting from the eaves and lapped 2 inches (51 mm), fastened only as necessary to hold in place.

❖ A single layer of underlayment is required on all roof slopes of 4:12 or greater. Reinforced underlayment is required on all roofs with spaced sheathing in order to avoid breakthrough of the tiles.

1507.3.4 Clay tile. Clay roof tile shall comply with ASTM C 1167.

❖ ASTM C 1167 covers clay tiles intended for use as a roof covering where durability and appearance are required to provide a weather-resistant surface of specified design. These tiles are made of clay, shale or other earthy substances and are heat treated at an elevated temperature. The process develops a fired bond between the particulate constituents to provide the strength and durability requirements of ASTM C 1167. The tiles are classified into three grades based on their resistance to weathering. Additionally, the standard covers performance characteristics, including durability; freezing and thawing; strength; efflorescence; permeability and reactive particulates.

1507.3.5 Concrete tile. Concrete roof tiles shall be in accordance with the physical test requirements as follows:

1. The transverse strength of tiles shall be determined according to Section 6.3 of ASTM C 1167 and in accordance with Table 1507.3.5.

2. The absorption of concrete roof tiles shall be according to Section 8 of ASTM C 140. Roof tiles shall absorb not more than 15 percent of the dry weight of the tile during a 24-hour immersion test.

3. Roof tiles shall be tested for freeze/thaw resistance according to Section 8 of ASTM C67. Roof tiles shall show no breakage and not have more than 1 percent loss in dry weight of any individual concrete roof tile.

❖ An ASTM standard for concrete tile is under development. The provisions include many of the same criteria as used in the clay tile standard, such as strength, permeability and freeze and thaw performance criteria.

TABLE 1507.3.5
TRANSVERSE BREAKING STRENGTH
OF CONCRETE ROOF TILE (lbs.)

TILE PROFILE	DRY	
	Average of five tiles	Individual tile
High profile	400	350
Medium profile	300	250
Flat profile	300	250

For SI: 1 pound = 4.45 N.

❖ These tiles are tested after heating in a ventilated oven for 24 hours at a temperature of 230°F to 239°F (110°C to 115°C). The span chosen for the test is required to be 12 inches (305 mm) or two-thirds of the length of the tile, whichever is greater.

1507.3.6 Fasteners. Tile fasteners shall be corrosion resistant and not less than 11 gage, $^5/_{16}$-inch (8.0 mm) head, and of sufficient length to penetrate the deck a minimum of 0.75 inch (19.1 mm) or through the thickness of the deck, whichever is less.

Attaching wire for clay or concrete tile shall not be smaller than 0.083 inch (2.1 mm). Perimeter fastening areas include three tile courses but not less than 36 inches (914 mm) from either side of hips or ridges and edges of eaves and gable rakes.

❖ Fasteners are to be corrosion resistant, such as hot-dipped galvanized steel, aluminum or stainless steel, and needle or diamond pointed with large flat heads.

1507.3.7 Attachment. Clay and concrete roof tiles shall be fastened in accordance with Table 1507.3.7.

❖ The table covers wind speed, mean roof height and roof slope requirements for fastener installation. It also covers interlocking requirements for specific areas and criteria for roofs.

TABLE 1507.3.7. See page 15-13.

❖ The table covers wind speed, mean roof height and roof slope requirements for fastener installation. It also covers interlocking requirements for specific areas and criteria for roofs.

1507.3.8 Application. Tile shall be applied according to the manufacturer's installation instructions, based on the following:

1. Climatic conditions.

2. Roof slope.

3. Underlayment system.

4. Type of tile being installed.

❖ Clay and concrete tile come in two generic forms: roll tile and flat tile. Either one may be interlocking but must be applied according to the manufacturer's instructions.

1507.3.9 Flashing. At the juncture of the roof vertical surfaces, flashing and counterflashing shall be provided in accordance with the manufacturer's installation instructions, and where of metal, shall not be less than 0.019-inch (0.48 mm) (No. 26 galvanized sheet gage) corrosion-resistant. The valley flashing shall extend at least 11 inches (279 mm) from the centerline each way and have a splash diverter rib not less than 1 inch (25 mm) high at the flow line formed as part of the flashing. Sections of flashing shall have an end lap of not less than 4 inches (102 mm). For roof slopes of three units vertical in 12 units horizontal (25-percent slope) and over, the valley flashing shall have a 36-inch-wide (914 mm) underlayment of one layer of Type I underlayment running the full length of the valley, in addition to other required underlayment. In areas where the average daily temperature in January is 25°F (-4°C) or less or where there is a possibility of ice forming along the eaves causing a backup of water, the metal valley flashing underlayment shall be solid cemented to the roofing underlayment for slopes under seven units vertical in 12 units horizontal (58-percent slope) or of self-adhering polymer modified bitumen sheet.

❖ Flashings are required to maintain the integrity of weather-resistant roofs (see commentary, Section 1503.2).

TABLE 1507.3.7
CLAY AND CONCRETE TILE ATTACHMENT[a, b, c]

GENERAL — CLAY OR CONCRETE ROOF TILE			
Maximum basic wind speed (mph)	Mean roof height (feet)	Roof slope up to < 3:12	Roof slope 3:12 and over
85	0-60	One fastener per tile. Flat tile without vertical laps, two fasteners per tile.	Two fasteners per tile. Only one fastener on slopes of 7:12 and less for tiles with installed weight exceeding 7.5 lbs./sq. ft. having a width no greater than 16 inches.
100	0-40		
100	> 40-60	The head of all tiles shall be nailed. The nose of all eave tiles shall be fastened with approved clips. All rake tiles shall be nailed with two nails. The nose of all ridge, hip and rake tiles shall be set in a bead of roofer's mastic.	
110	0-60	The fastening system shall resist the wind forces in Section 1609.7.2.	
120	0-60	The fastening system shall resist the wind forces in Section 1609.7.2.	
130	0-60	The fastening system shall resist the wind forces in Section 1609.7.2.	
All	> 60	The fastening system shall resist the wind forces in Section 1609.7.2.	

INTERLOCKING CLAY OR CONCRETE ROOF TILE WITH PROJECTING ANCHOR LUGS[d, e] (Installations on spaced/solid sheathing with battens or spaced sheathing)				
Maximum basic wind speed (mph)	Mean roof height (feet)	Roof slope up to < 5:12	Roof slope 5:12 < 12:12	Roof slope 12:12 and over
85	0-60	Fasteners are not required. Tiles with installed weight less than 9 lbs./sq. ft. require a minimum of one fastener per tile.	One fastener per tile every other row. All perimeter tiles require one fastener. Tiles with installed weight less than 9 lbs./sq. ft. require a minimum of one fastener per tile.	One fastener required for every tile. Tiles with installed weight less than 9 lbs./sq. ft. require a minimum of one fastener per tile.
100	0-40			
100	> 40-60	The head of all tiles shall be nailed. The nose of all eave tiles shall be fastened with approved clips. All rake tiles shall be nailed with two nails The nose of all ridge, hip and rake tiles shall be set in a bead of roofers's mastic.		
110	0-60	The fastening system shall resist the wind forces in Section 1609.7.2.		
120	0-60	The fastening system shall resist the wind forces in Section 1609.7.2.		
130	0-60	The fastening system shall resist the wind forces in Section 1609.7.2.		
All	> 60	The fastening system shall resist the wind forces in Section 1609.7.2.		

INTERLOCKING CLAY OR CONCRETE ROOF TILE WITH PROJECTING ANCHOR LUGS (Installations on solid sheathing without battens)		
Maximum basic wind speed (mph)	Mean roof height (feet)	All roof slopes
85	0-60	One fastener per tile.
100	0-40	One fastener per tile.
100	> 40-60	The head of all tiles shall be nailed. The nose of all eave tiles shall be fastened with approved clips. All rake tiles shall be nailed with two nails The nose of all ridge, hip and rake tiles shall be set in a bead of roofers's mastic.
110	0-60	The fastening system shall resist the wind forces in Section 1609.7.2.
120	0-60	The fastening system shall resist the wind forces in Section 1609.7.2.
130	0-60	The fastening system shall resist the wind forces in Section 1609.7.2.
All	> 60	The fastening system shall resist the wind forces in Section 1609.7.2.

For SI: 1 inch = 25.4 mm, 1 foot = 304.8 mm, 1 mile per hour = 1.609 km/h, 1 pound per square foot = 0.0478 kn/m^2.

a. Minimum fastener size. Corrosion-resistant nails not less than No. 11 gage with $^5/_{16}$-inch head. Fasteners shall be long enough to penetrate into the sheathing 0.75 inch or through the thickness of the sheathing, whichever is less. Attaching wire for clay and concrete tile shall not be smaller than 0.083 inch.

b. Snow areas. A minimum of two fasteners per tile are required or battens and one fastener.

c. Roof slopes greater than 24:12. The nose of all tiles shall be securely fastened.

d. Horizontal battens. Battens shall be not less than 1inch by 2 inch nominal. Provisions shall be made for drainage by a minimum of $^1/_8$-inch riser at each nail or by 4-foot-long battens with at least a 0.5-inch separation between battens. Horizontal battens are required for slopes over 7:12.

e. Perimeter fastening areas include three tile courses but not less than 36 inches from either side of hips or ridges and edges of eaves and gable rakes.

1507.4 Metal roof panels. The installation of metal roof panels shall comply with the provisions of this section.

❖ There are two general categories of metal roofing systems: architectural metal roofing and structural metal roofing. Architectural metal roofs are generally water-shedding roof systems and structural metal roofs have hydrostatic (water barrier) characteristics.

Architectural metal roof systems are usually characterized by a flat pan with $^3/_4$- to $1^1/_2$-inch (19.1 to 38 mm) ribs on each side. The absence of intermediate ribs and massive side ribs gives a clean appearance, but does not provide the panels the strength to be considered a structural panel. Architectural metal roofing systems are typically designed for steep slopes so that water will shed off the roof panels. This type of design does not require the seams to be water tight; therefore, the slope for such nonwater-tight roof panels is required to be no less than 3:12. One exception is traditional flat seamed, soldered or welded metal roofing that is acceptable on slopes less than 3:12. An example of roofing that has traditionally been used this way is copper. Architectural metal roofing systems require solid decking.

Metal roof panels are defined in Section 1502 as an interlocking metal sheet having a minimum installed weather exposure of 3 square feet (0.28 m²) per sheet. In contrast, metal roof shingles have less than 3 square feet (0.28 m²) of exposure.

1507.4.1 Deck requirements. Metal roof panel roof coverings shall be applied to a solid or closely fitted deck, except where the roof covering is specifically designed to be applied to spaced supports.

❖ The deck for metal roofing is required to be solid or closely fitted, except where the panels are specifically designed to be applied to spaced supports. Structural standing seam metal panel roof systems possess strength characteristics that allow them to span between structural supports, as is commonly used on preengineered metal buildings. The metal panel ribs are not seamed or interlocked like a true standing seam metal roof system.

1507.4.2 Deck slope. The minimum slope for lapped, nonsoldered seam metal roofs without applied lap sealant shall be three units vertical in 12 units horizontal (25-percent slope). The minimum slope for lapped, nonsoldered seam metal roofs with applied lap sealant shall be one-half vertical unit in 12 units horizontal (4-percent slope). The minimum slope for standing seam of roof systems shall be one-quarter unit vertical in 12 units horizontal (2-percent slope).

❖ Typically, structural metal roofing systems are designed to resist water at laps and seams with sealants applied in the seams. It is usually recommended that these types of roof systems have a slope of no less than $^1/_2$:12, although some manufacturers allow slopes as low as $^1/_4$:12. While the code requires a minimum slope of 3:12, a lower slope is permitted on standing-seam-type roof systems, which require only $^1/_4$:12. The term "standing seam" is used to refer to almost any roof panel with a raised vertical rib. Strictly speaking, it only indicates

those metal panels that interlock or are seamed together vertically above the panel's pan.

1507.4.3 Material standards. Metal-sheet roof covering systems that incorporate supporting structural members shall be designed in accordance with Chapter 22. Metal-sheet roof coverings installed over structural decking shall comply with Table 1507.4.3.

❖ The requirements for metal roof coverings that incorporate supporting structural members into their design are contained in Chapter 22. Other metal-sheet roof coverings are required to comply with Table 1507.4.3.

TABLE 1507.4.3
METAL ROOF COVERINGS

ROOF COVERING TYPE	STANDARD APPLICATION RATE/THICKNESS
Aluminum	ASTM B 209, 0.024 inch minimum thickness for roll-formed panels and 0.019 inch minimum thickness for press-formed shingles.
Aluminum-zinc alloy coated steel	ASTM A 792 AZ 50
Copper	16 oz./sq. ft. for metal-sheet roof-covering systems; 12 oz./sq. ft. for preformed metal shingle systems.
Galvanized steel	ASTM A 653 G-90 zinc-coated, 0.013-inch-thick minimum
Lead-coated copper	ASTM B 101
Hard lead	2 lbs./sq. ft.
Soft lead	3 lbs./sq. ft.
Prepainted steel	ASTM A 755
Terne (tin) and terne- coated stainless	Terne coating of 40 lbs. per double base box, field painted where applicable in accordance with manufacturer's installation instructions.

For SI: 1 ounce per square foot = 0.0026 kg/m²,
1 pound per square foot = 4.882 kg/m²,
1 inch = 25.4 mm, 1 pound = 0.454 kg.

❖ This table provides specific guidance on appropriate metal roof coverings. The requirements vary based upon the type of metal used, such as aluminum, steel, copper and lead. Where standards are referenced in other cases, basic criteria is provided.

1507.4.4 Attachment. Metal roofing fastened directly to steel framing shall be attached by approved manufacturers' fasteners. In the absence of manufacturer recommendations, all of the following fasteners shall be used:

1. Galvanized fasteners shall be used for galvanized roofs.

2. 300 series stainless-steel fasteners shall be used for copper roofs.

3. Stainless-steel fasteners are acceptable for all types of metal roofs.

❖ Fasteners used to attach metal roofing to steel framing are to have the manufacturer's approval for that purpose. If such approval is not provided, then galvanized fasteners are required for galvanized roofs, 300 series stainless steel fasteners for copper roofs and stainless-steel fasteners for all types of metal roofs. The purpose of these provisions is to ensure compatibility with roof materials and prevent corrosion.

1507.5 Metal roof shingles. The installation of metal roof shingles shall comply with the provisions of this section.

❖ See the commentary to Section 1507.4 for a general discussion about metal roof systems.

1507.5.1 Deck requirements. Metal roof shingles shall be applied to a solid or closely fitted deck, except where the roof covering is specifically designed to be applied to spaced sheathing.

❖ The manufacturer's instructions must be followed, with any restrictions required by the code.

1507.5.2 Deck slope. Metal roof shingles shall not be installed on roof slopes below three units vertical in 12 units horizontal (25-percent slope).

❖ Metal shingles cannot be installed on roof slopes less than 3:12.

1507.5.3 Underlayment. Underlayment shall conform to ASTM D 226, Type I. In areas where the average daily temperature in January is 25°F (-4°C) or less or where there is a possibility of ice forming along the eaves causing a backup of water, an ice barrier that consists of at least two layers of underlayment cemented together or of a self-adhering polymer-modified bitumen sheet, shall be used in lieu of normal underlayment and extend from the eave's edge to a point at least 24 inches (610 mm) inside the exterior wall line of the building.

Exception: Detached accessory structures that contain no conditioned floor area.

❖ A single layer of underlayment is required under all metal shingles other than flat metal shingles. The underlayment must be in accordance with ASTM D 226, Type I. The method of fastening the covering is to be specified by the manufacturer. In extreme weather conditions, the code requires two layers of felt that provide double coverage at any point [see Figures 1507.8.3(1) and (2)].

The exception is used throughout Chapter 15 for unconditioned accessory buildings (see commentary, Section 1507.2.8.2).

1507.5.4 Material standards. Metal roof shingle roof coverings shall comply with Table 1507.4.3.

❖ ASTM A 653 regulates steel sheet for roofing and siding that is zinc coated on continuous lines and by the cut-length method. Material of this quality is furnished flat, in coils and cut lengths, and is formed in cut lengths. Roofing and siding include corrugated, V-crimp, roll roofing and many special patterns. Corrugated roofing and siding sheet is produced in a number of corrugations, with variations of pitch and depth. ASTM A 755 covers steel sheet that is metallic coated by the hot-dipped process and prepainted by the coil-coating process with organic films for exterior-exposed building products of various qualities. Sheet material of this designation is furnished in coils, cut lengths and formed cut lengths. ASTM B 101 covers lead-coated copper sheets for architectural uses.

1507.5.5 Attachment. Metal roof shingles shall be secured to the roof in accordance with the approved manufacturer's installation instructions.

❖ Fasteners used to attach metal roofing to the roof are to have the manufacturer's approval for that purpose.

1507.5.6 Flashing. Roof valley flashing shall be of corrosion-resistant metal of the same material as the roof covering or shall comply with the standards in Table 1507.4.3. The valley flashing shall extend at least 8 inches (203 mm) from the centerline each way and shall have a splash diverter rib not less than 0.75 inch (19.1 mm) high at the flow line formed as part of the flashing. Sections of flashing shall have an end lap of not less than 4 inches (102 mm). In areas where the average daily temperature in January is 25°F (-4°C) or less or where there is a possibility of ice forming along the eaves causing a backup of water, the metal valley flashing shall have a 36-inch-wide (914 mm) underlayment directly under it consisting of one layer of underlayment running the full length of the valley, in addition to underlayment required for metal roof shingles. The metal valley flashing underlayment shall be solid cemented to the roofing underlayment for roof slopes under seven units vertical in 12 units horizontal (58-percent slope) or of self-adhering polymer-modified bitumen sheet.

❖ See the commentary to Section 1503.2.

1507.6 Mineral-surfaced roll roofing. The installation of mineral-surfaced roll roofing shall comply with this section.

❖ Mineral-surfaced roll roofing is an asphalt roll roofing material, that in some cases, have a selvage edge, which is a specially defined edge (lined for demarcation) designed for some special purpose, such as overlapping or seaming. It is typically 36 inches (914 mm) in width and is surfaced with coarse mineral granules. In mineral-surfaced roll roofing, the selvage is not surfaced with coarse mineral granules to allow better adhesion of the overlapping sheet.

There are two general methods for applying roll roofing: the exposed nail method and the concealed nail method. Depending on the slope and nailing method used, roll roofing can be applied parallel (downslope roof edge) or perpendicular, to the eave (parallel with the rake) (see Section 1507.6.1).

1507.6.1 Deck requirements. Mineral-surfaced roll roofing shall be fastened to solidly sheathed roofs.

❖ Prior to the application of the roll roofing on wood plank or plywood decks, the deck should be inspected for delamination of plywood, warped boards and proper nailing.

For roll roofing applied parallel to the eave, a minimum roof slope of 2:12 is recommended for application by the concealed-nail method, and 6:12 is recommended for application by the exposed-nail method.

For roll roofing applied parallel to the rake, a minimum roof slope of 3:12 is recommended for application by the concealed-nail method, and 4:12 is recommended for application by the exposed-nail method.

1507.6.2 Deck slope. Mineral-surfaced roll roofing shall not be applied on roof slopes below one unit vertical in 12 units horizontal (8-percent slope).

❖ It is not recommended that asphalt roll roofing be applied to deck slopes of less than 2:12, with one exception. Roll roofing with a 19-inch (483 mm) selvage edge (commonly referred to as double coverage or split-sheet) may be applied on a deck with a slope as low as 1:12; however, it should only be used on roofs that will drain by gravity. It is not recommended for decks that will allow ponding and require evaporation to dry the roof.

1507.6.3 Underlayment. Underlayment shall conform to ASTM D 226, Type I. In areas where the average daily temperature in January is 25°F (-4°C) or less or where there is a possibility of ice forming along the eaves causing a backup of water, an ice barrier that consists of at least two layers of underlayment cemented together or of a self-adhering polymer-modified bitumen sheet, shall extend from the eave's edge to a point at least 24 inches (610 mm) inside the exterior wall line of the building.

Exception: Detached accessory structures that contain no conditioned floor area.

❖ Regardless of the type of underlayment required or the slope of the roof, in locations where the January mean temperature is 25°F (-3.8°C) or less, two layers of underlayment should be applied starting from the eaves to a point 24 inches (610 mm) inside the wall line of the building to serve as an ice shield [see Figures 1507.8.3(1) and (2)].

The exception is used throughout Chapter 15 for unconditioned accessory buildings (see commentary, Section 1507.2.8.2).

1507.6.4 Material standards. Mineral-surfaced roll roofing shall conform to ASTM D 224, ASTM D 249, ASTM D 371 or ASTM D 3909.

❖ ASTM D 224 covers asphalt roofing in sheet form. ASTM D 249 covers asphalt roofing in sheet form in widths as agreed upon by the purchaser and the seller. ASTM D 371 covers asphalt roofing in sheet form 36 inches (914 mm) in width or other widths agreed upon between the purchaser and the seller. ASTM D 3909 covers asphalt-impregnated and coated glass felt roll roofing surfaced on the weather side with mineral gran-

ules for use as a cap sheet in the construction of built-up roofs.

1507.7 Slate shingles. The installation of slate shingles shall comply with the provisions of this section.

❖ Slate is a dense, tough, durable natural rock or stone material that is practically nonabsorbent. This natural rock has cleavage planes that allow the slate to be easily split into relatively thin layers. Slate also possesses a natural grain that usually runs perpendicular to the cleavage. Slate is usually split so the length of the shingle runs in the direction of the grain.

Slate roofing has a long service life when consideration is given to all components of the roofing system. Some grades of slate combined with proper roof deck, underlayments, fastening and accessories have a service life in excess of 75 years.

1507.7.1 Deck requirements. Slate shingles shall be fastened to solidly sheathed roofs.

❖ Due to the long service life expected of a slate roof system, careful consideration should be given to the deck material. Deck material that can be expected to last as long as the service life of the roof should be chosen. Additionally, consideration should be given to the weight of the slate when choosing the deck. Slate is available in many different thicknesses and, therefore, the weight will vary.

1507.7.2 Deck slope. Slate shingles shall only be used on slopes of four units vertical in 12 units horizontal (4:12) or greater.

❖ This type of roof covering cannot be installed on roof slopes that are less than 4:12.

1507.7.3 Underlayment. Underlayment shall comply with ASTM D 226, Type II. In areas where the average daily temperature in January is 25°F (-4°C) or less or where there is a possibility of ice forming along the eaves causing a backup of water, an ice barrier that consists of at least two layers of underlayment cemented together or of a self-adhering polymer-modified bitumen sheet, shall extend from the eave's edge to a point at least 24 inches (610 mm) inside the exterior wall line of the building.

Exception: Detached accessory structures that contain no conditioned floor area.

❖ Slate shingle is a watershedding-type roof system and the system itself is not impermeable. The underlayment should be chosen based on the slope and headlap of the shingles [see Figures 1507.8.3(1) and (2)]. Special consideration should be made in more extreme climates for the possibility of ice dam formations.

The exception is used throughout Chapter 15 for unconditioned accessory buildings (see commentary, Section 1507.2.8.2).

1507.7.4 Material standards. Slate shingles shall comply with ASTM C 406.

❖ ASTM C 406 covers the material characteristics, physical requirements and sampling appropriate to selection of slate as a roofing material.

1507.7.5 Application. Minimum headlap for slate shingles shall be in accordance with Table 1507.7.5. Slate shingles shall be secured to the roof with two fasteners per slate.

❖ Slate shingles are drilled or punched with holes for fasteners with consideration for proper headlap. Fastener material should be chosen based on the expected service life. Copper slating nails, stainless steel, aluminum alloy, bronze or cut-brass roofing nails are often specified, depending on the particular project. Unprotected black-iron, electroplated and hot-dipped galvanized fasteners are not recommended.

TABLE 1507.7.5
SLATE SHINGLE HEADLAP

SLOPE	HEADLAP (inches)
4:12 < slope < 8:12	4
8:12 < slope < 20:12	3
slope ≥ 20:12	2

For SI: 1 inch = 25.4 mm.

❖ Headlap is the distance of overlap measured from the uppermost ply or course to the point that it overlaps the undermost ply or course.

1507.7.6 Flashing. Flashing and counterflashing shall be made with sheet metal. Valley flashing shall be a minimum of 15 inches (381 mm) wide. Valley and flashing metal shall be a minimum uncoated thickness of 0.0179-inch (0.455 mm) zinc-coated G90. Chimneys, stucco or brick walls shall have a minimum of two plies of felt for a cap flashing consisting of a 4-inch-wide (102 mm) strip of felt set in plastic cement and extending 1 inch (25 mm) above the first felt and a top coating of plastic cement. The felt shall extend over the base flashing 2 inches (51 mm).

❖ Flashing is required at all intersections between exterior walls and roofs. These are areas where rainwater can easily penetrate the building envelope (see Section 1503.2).

1507.8 Wood shingles. The installation of wood shingles shall comply with the provisions of this section and Table 1507.8.

❖ Wood shingles are defined by the roofing industry as sawed wood products featuring a uniform butt thickness per individual length.

TABLE 1507.8. See page 15-18.

❖ Wood shingles cannot be installed on roof slopes below 3:12. Single-layer underlayment is required at eaves, ridges, hips, valleys and all other changes of roof slope or direction to protect the roof from leakage caused by water backup. Each shingle must be securely fastened to the roof deck with a maximum of two fasteners. Care should be taken so that the fasteners do not cause splitting of the wood shingle. Fasteners are to be as specified in the manufacturer's installation instructions.

1507.8.1 Deck requirements. Wood shingles shall be installed on solid or spaced sheathing. Where spaced sheathing is used, sheathing boards shall not be less than 1-inch by 4-inch (25 mm by 102 mm) nominal dimensions and shall be spaced on centers equal to the weather exposure to coincide with the placement of fasteners.

❖ Spaced sheathing is usually of 1-inch by 4-inch (25 mm by 102 mm) or 1-inch by 6-inch (25 mm by 152 mm) softwood boards. Note that solid sheathing is required for certain locations by Section 1507.8.1.1.

1507.8.1.1 Solid sheathing required. Solid sheathing is required in areas where the average daily temperature in January is 25°F (-4°C) or less or where there is a possibility of ice forming along the eaves causing a backup of water.

❖ Solid sheathing is usually softwood panels, which provide a smooth, even base for the roofing material and help stiffen the entire roof structure. Solid sheathing provides an extra degree of protection where ice damming is a possibility.

1507.8.2 Deck slope. Wood shingles shall be installed on slopes of three units vertical in 12 units horizontal (25-percent slope) or greater.

❖ Wood shingles cannot be installed on roof slopes below 3:12.

1507.8.3 Underlayment. Underlayment shall comply with ASTM D 226, Type I. In areas where the average daily temperature in January is 25°F (-4°C) or less or where there is a possibility of ice forming along the eaves causing a backup of water, an ice barrier that consists of at least two layers of underlayment cemented together or of a self-adhering polymer-modified bitumen sheet shall extend from the eave's edge to a point at least 24 inches (610 mm) inside the exterior wall line of the building.

Exception: Detached accessory structures that contain no conditioned floor area.

❖ ASTM D 226 covers asphalt-saturated organic felts, with or without perforations. This section and Section 1507.8.1.1 address sheathing and underlayment requirements for geographical areas where there is a possibility of ice forming along the eaves causing a backup of water. See Figures 1507.8.3(1) and 1507.8.3(2) for illustrations of ice dams and locations where two layers of underlayment are required.

The exception is used throughout Chapter 15 for unconditioned accessory buildings (see commentary, Section 1507.2.8.2).

TABLE 1507.8   ROOF ASSEMBLIES AND ROOFTOP STRUCTURES

TABLE 1507.8
WOOD SHINGLE AND SHAKE INSTALLATION

ROOF ITEM	WOOD SHINGLES	WOOD SHAKES
1. Roof slope	Wood shingles shall be installed on slopes of three units vertical in 12 units horizontal (3:12) or greater.	Wood shakes shall be installed on slopes of four units vertical in 12 units horizontal (4:12) or greater.
2. Deck requirement	—	—
Temperate climate	Shingles shall be applied to roofs with solid or spaced sheathing. Where spaced sheathing is used, sheathing boards shall not be 4 less than 1″ × 4″ nominal dimensions and shall be spaced on center equal to the weather exposure to coincide with the placement of fasteners.	Shakes shall be applied to roofs with solid or spaced sheathing. Where spaced sheathing is used, sheathing boards shall not be less than 1″ × 4″ nominal dimensions and shall be spaced on center equal to the weather exposure to coincide with the placement of fasteners. When 1″ × 4″ spaced sheathing is installed at 10 inches, boards must be installed between the sheathing boards.
In areas where the average daily temperature in January is 25°F or less or where there is a possibility of ice forming along the eaves causing a backup of water.	Solid sheathing required.	Solid sheathing is required.
3. Interlayment	No requirements.	Interlayment shall comply with ASTM D 226, Type 1.
4. Underlayment	—	—
Temperate climate	Underlayment shall comply with ASTM D 226, Type 1.	Underlayment shall comply with ASTM D 226, Type 1.
In areas where the average daily temperature in January is 25°F or less or where there is a possibility of ice forming along the eaves causing a backup of water.	An ice shield that consists of at least two layers of underlayment cemented together or of a self-adhering polymer-modified bitumen sheet shall extend from the eave's edge to a point at least 24 inches inside the exterior wall line of the building.	An ice shield that consists of at least two layers of underlayment cemented together or of a self-adhering polymer-modified bitumen sheet shall extend from the eave's edge to a point at least 24 inches inside the exterior wall line of the building.
5. Application	—	—
Attachment	Fasteners for wood shingles shall be corrosion resistant with a minimum penetration of 0.75 inch into the sheathing. For sheathing less than 0.5 inch thick, the fasteners shall extend through the sheathing.	Fasteners for wood shakes shall be corrosion resistant with a minimum penetration of 0.75 inch into the sheathing. For sheathing less than 0.5 inch thick, the fasteners shall extend through the sheathing.
No. of fasteners	Two per shingle.	Two per shake.
Exposure	Weather exposures shall not exceed those set forth in Table 1507.8.6	Weather exposures shall not exceed those set forth in Table 1507.9.7
Method	Shingles shall be laid with a side lap of not less than 1.5 inches between joints in courses, and no two joints in any three adjacent courses shall be in direct alignment. Spacing between shingles shall be 0.25 to 0.375 inch.	Shakes shall be laid with a side lap of not less than 1.5 inches between joints in adjacent courses. Spacing between shakes shall not be less than 0.375 inch or more than 0.625 inch for shakes and tapersawn shakes of naturally durable wood and shall be 0.25 to 0.375 inch for preservative taper sawn shakes.
Flashing	In accordance with Section 1507.8.7.	In accordance with Section 1507.9.8.

For SI: 1 inch = 25.4 mm, °C = [(°F) - 32]/1.8.

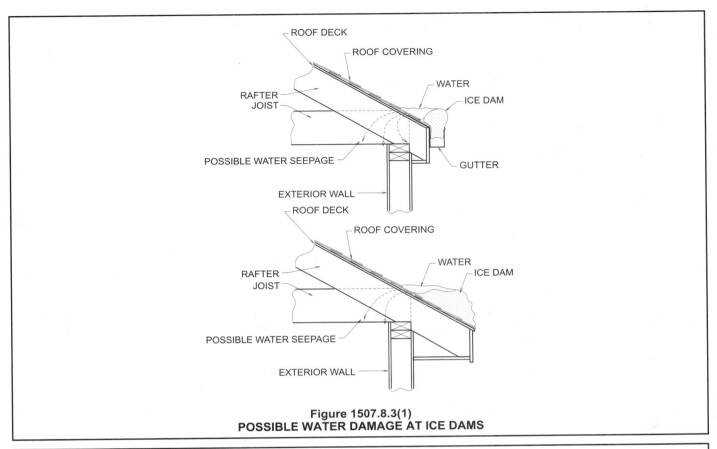

Figure 1507.8.3(1)
POSSIBLE WATER DAMAGE AT ICE DAMS

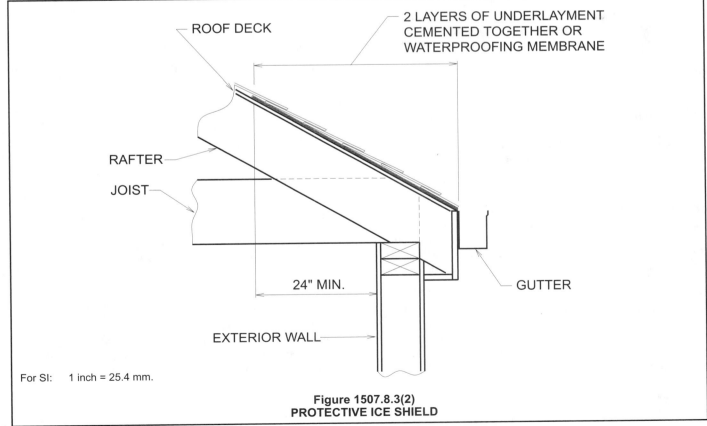

For SI: 1 inch = 25.4 mm.

Figure 1507.8.3(2)
PROTECTIVE ICE SHIELD

1507.8.4 Material standards. Wood shingles shall be naturally durable wood and comply with the requirements of Table 1507.8.4.

❖ Information about the installation of wood shingles is available in the *Design and Application Manual for New Roof Construction* published by the Cedar Shake and Shingle Bureau (CSSB) (see Chapter 35). Additional information regarding wood shingle roofs is provided in the wood roofing section of the *NRCA Roofing and Waterproofing Manual*.

TABLE 1507.8.4
WOOD SHINGLE MATERIAL REQUIREMENTS

MATERIAL	APPLICABLE MINIMUM GRADES	GRADING RULES
Wood shingles of naturally durable wood	1, 2 or 3	CSSB

CSSB = Cedar Shake and Shingle Bureau

❖ Wood shingles cannot be installed on roof slopes below 3:12. Single-layer underlayment is required at eaves, ridges, hips, valleys and all other changes of roof slope or direction to protect the roof from leakage caused by water backup. Each shingle must be securely fastened to the roof deck with a minimum of two fasteners. Fasteners are to be as specified in the manufacturer's installation instructions. Care should be taken so that the fasteners do not cause splitting of the wood shingle.

1507.8.5 Attachment. Fasteners for wood shingles shall be corrosion resistant with a minimum penetration of 0.75 inch (19.1 mm) into the sheathing. For sheathing less than 0.5 inch (12.7 mm) in thickness, the fasteners shall extend through the sheathing. Each shingle shall be attached with a minimum of two fasteners.

❖ Each shingle must be securely fastened to the roof deck with a minimum of two fasteners. Fasteners are to be as specified in the manufacturer's installation instructions. Care should be taken so that the fasteners do not cause splitting of the wood shingle..

1507.8.6 Application. Wood shingles shall be laid with a side lap not less than 1.5 inches (38 mm) between joints in adjacent courses, and not be in direct alignment in alternate courses. Spacing between shingles shall be 0.25 to 0.375 inches (6.4 to 9.5 mm). Weather exposure for wood shingles shall not exceed that set in Table 1507.8.6.

❖ Wood shingle exposure is specified in Table 1507.8.6. Depending on the grade of the material, the total shingle length and the slope of the roof deck, the table specifies the maximum length of exposure for weathering purposes.

TABLE 1507.8.6
WOOD SHINGLE WEATHER EXPOSURE AND ROOF SLOPE

ROOFING MATERIAL	LENGTH (inches)	GRADE	EXPOSURE (inches)	
			3:12 pitch to < 4:12	4:12 pitch or steeper
Shingles of naturally durable wood	16	No. 1	3.75	5
		No. 2	3.5	4
		No. 3	3	3.5
	18	No. 1	4.25	5.5
		No. 2	4	4.5
		No. 3	3.5	4
	24	No. 1	5.75	7.5
		No. 2	5.5	6.5
		No. 3	5	5.5

For SI: 1 inch = 25.4 mm.

❖ The grade of the shingle will be listed in the label required for wood shingles. Each of the three different grades is available in three different lengths. For lower-sloped roofs (3:12 to 4:12), where wind uplift is more significant, the maximum weather exposure is always less than that for steeper-sloped roofs (4:12 and greater).

1507.8.7 Flashing. At the juncture of the roof and vertical surfaces, flashing and counterflashing shall be provided in accordance with the manufacturer's installation instructions, and where of metal, shall not be less than 0.019-inch (0.48 mm) (No. 26 galvanized sheet gage) corrosion-resistant metal. The valley flashing shall extend at least 11 inches (279 mm) from the centerline each way and have a splash diverter rib not less than 1 inch (25 mm) high at the flow line formed as part of the flashing. Sections of flashing shall have an end lap of not less than 4 inches (102 mm). For roof slopes of three units vertical in 12 units horizontal (25-percent slope) and over, the valley flashing shall have a 36-inch-wide (914 mm) underlayment of one layer of Type I underlayment running the full length of the valley, in addition to other required underlayment. In areas where the average daily temperature in January is 25°F (-4°C) or less or where there is a possibility of ice forming along the eaves causing a backup of water, the metal valley flashing underlayment shall be solid cemented to the roofing underlayment for slopes under seven units vertical in 12 units horizontal (58-percent slope).

❖ Improper installation of flashing is the greatest cause of failures of roof covering systems. Whenever one plane of a roof intersects another plane, flashing is required where the planes intersect (see Section 1503.2).

1507.9 Wood shakes. The installation of wood shakes shall comply with the provisions of this section and Table 1507.8.

❖ This section establishes design and installation requirements for wood shakes, which are defined as roofing products split from logs and then shaped as required by the individual manufacturers. Wood shakes must be labeled by an approved third-party inspection agency. A quality control program is required to contain a set of grading rules.

Information about the installation of wood shakes is available in the *Design and Application Manual for New Roof Construction* published by CSSB (see Chapter 35). Additional information is provided in the wood roofing section of the *NRCA Roofing* and *Waterproofing Manual.*

1507.9.1 Deck requirements. Wood shakes shall only be used on solid or spaced sheathing. Where spaced sheathing is used, sheathing boards shall not be less than 1-inch by 4-inch (25 mm by 102 mm) nominal dimensions and shall be spaced on centers equal to the weather exposure to coincide with the placement of fasteners. Where 1-inch by 4-inch (25 mm by 102 mm) spaced sheathing is installed at 10 inches (254 mm) o.c., additional 1-inch by 4-inch (25 mm by 102 mm) boards shall be installed between the sheathing boards.

❖ See the commentary to Section 1507.8.1.

1507.9.1.1 Solid sheathing required. Solid sheathing is required in areas where the average daily temperature in January is 25°F (-4°C) or less or where there is a possibility of ice forming along the eaves causing a backup of water.

❖ See the commentary to Section 1507.8.1.1.

1507.9.2 Deck slope. Wood shakes shall only be used on slopes of four units vertical in 12 units horizontal (33-percent slope) or greater.

❖ Wood shakes cannot be installed on roof slopes less than 4:12 to provide for adequate drainage of the roof.

1507.9.3 Underlayment. Underlayment shall comply with ASTM D 226, Type I. In areas where the average daily temperature in January is 25°F (-4°C) or less or where there is a possibility of ice forming along the eaves causing a backup of water, an ice barrier that consists of at least two layers of underlayment cemented together or a self-adhering polymer-modified bitumen sheet shall extend from the edge of the eave to a point at least 24 inches (610 mm) inside the exterior wall line of the building.

Exception: Detached accessory structures that contain no conditioned floor area.

❖ ASTM D 226 covers asphalt-saturated organic felts, with or without perforations [see Figures 1507.8.3(1) and (2)]. Consideration should be made in more extreme climates for the formation of ice dams in the eaves.

The exception is used throughout Chapter 15 for unconditioned accessory buildings (see commentary, Section 1507.2.8.2).

1507.9.4 Interlayment. Interlayment shall comply with ASTM D 226, Type I.

❖ A single layer of felt interlayment must be shingled between each course on all roof slopes.

1507.9.5 Material standards. Wood shakes shall comply with the requirements of Table 1507.9.5.

❖ Information about the installation of wood shingles is available in the *Design and Application Manual for New Roof Construction* published by CSSB (see Chapter 35). Additional information is provided in the wood roofing section of the *NRCA Roofing* and *Waterproofing Manual.*

TABLE 1507.9.5
WOOD SHAKE MATERIAL REQUIREMENTS

MATERIAL	MINIMUM GRADES	APPLICABLE GRADING RULES
Wood shakes of naturally durable wood	1	CSSB
Taper sawn shakes of naturally durable wood	1 or 2	CSSB
Preservative-treated shakes and shingles of naturally durable wood	1	CSSB
Fire-retardant-treated shakes and shingles of naturally durable wood	1	CSSB
Preservative-treated taper sawn shakes of Southern yellow pine treated in accordance with AWPA Standard C2	1 or 2	TFS

CSSB = Cedar Shake and Shingle Bureau.
TFS = Forest Products Laboratory of the Texas Forest Services.

❖ Wood shakes must be labeled by an approved third-party inspection agency. A quality control program is required to contain a set of grading rules. Care must be taken to follow completely the manufacturer's installation instructions for each particular product. The building official must review and approve all installations with special note of any application limitations.

1507.9.6 Attachment. Fasteners for wood shakes shall be corrosion resistant with a minimum penetration of 0.75 inch (19.1 mm) into the sheathing. For sheathing less than 0.5 inch (12.7 mm) in thickness, the fasteners shall extend through the sheathing. Each shake shall be attached with a minimum of two fasteners.

❖ Each shake must be secured with a minimum of two fasteners. Fasteners are to be as specified in the manufacturer's installation instructions.

1507.9.7 Application. Wood shakes shall be laid with a side lap not less than 1.5 inches (38 mm) between joints in adjacent courses. Spacing between shakes in the same course shall be 0.375 to 0.625 inches (9.5 to 15.9 mm) for shakes and taper sawn shakes of naturally durable wood and shall be 0.25 to 0.375 inch (6.4 to 9.5 mm) for preservative taper sawn shakes. Weather exposure for wood shakes shall not exceed those set in Table 1507.9.7.

❖ Care must be taken to follow completely the manufacturer's installation instructions for each particular product. The building official must review and approve all installations with special note of any application limitations.

TABLE 1507.9.7 – TABLE 1507.10.2

TABLE 1507.9.7
WOOD SHAKE WEATHER EXPOSURE
AND ROOF SLOPE

ROOFING MATERIAL	LENGTH (inches)	GRADE	EXPOSURE (inches) 4:12 PITCH OR STEEPER
Shakes of naturally durable wood	18	No. 1	7.5
	24	No. 1	10a
Preservative-treated taper sawn shakes of Southern yellow pine	18	No. 1	7.5
	24	No. 1	10
	18	No. 2	5.5
	24	No. 2	7.5
Taper sawn shakes of naturally durable wood	18	No. 1	7.5
	24	No. 1	10
	18	No. 2	5.5
	24	No. 2	7.5

For SI: 1 inch = 25.4 mm.

a. For 24-inch by 0.375-inch handsplit shakes, the maximum exposure is 7.5 inches.

❖ The code provides specific weather exposure limitations for two different lengths of shakes. As with wood shingles, the longer the shake, the greater the exposure allowance, since wind uplift resistance is greater for the longer shakes. The exposure limitations are applicable regardless of the roof slope (4:12 or greater). The reduced weather exposure enhances the wind uplift resistance that these wood shakes will be able to provide.

1507.9.8 Flashing. At the juncture of the roof and vertical surfaces, flashing and counterflashing shall be provided in accordance with the manufacturer's installation instructions, and where of metal, shall not be less than 0.019-inch (0.48 mm) (No. 26 galvanized sheet gage) corrosion-resistant metal. The valley flashing shall extend at least 11 inches (279 mm) from the centerline each way and have a splash diverter rib not less than 1 inch (25 mm) high at the flow line formed as part of the flashing. Sections of flashing shall have an end lap of not less than 4 inches (102 mm). For roof slopes of 3 units vertical in 12 units horizontal (25-percent slope) and over, the valley flashing shall have a 36-inch-wide (914 mm) underlayment of one layer of Type I underlayment running the full length of the valley, in addition to other required underlayment. In areas where the average daily temperature in January is 25°F (-4°C) or less or where there is a possibility of ice forming along the eaves causing a backup of water, the metal valley flashing underlayment shall be solid cemented to the roofing underlayment for slopes under seven units vertical in 12 units horizontal (58-percent slope).

❖ Improper installation of flashing is the greatest cause of failure of roof covering systems. Whenever one plane of a roof intersects another plane, flashing is required where the planes intersect [see Section 1503.2 and Figures 1507.8.3(1) and (2)].

1507.10 Built-up roofs. The installation of built-up roofs shall comply with the provisions of this section.

❖ A built-up roof membrane is a continuous, semiflexible multi-ply roof membrane, consisting of plies or layers of saturated felts, coated felts, fabrics or mats between which alternate layers of bitumen are applied. Generally, built-up roof membranes are surfaced with mineral aggregate and bitumen, a liquid-applied coating or a granule-surfaced cap sheet.

1507.10.1 Slope. Built-up roofs shall have a design slope of a minimum of one-fourth unit vertical in 12 units horizontal (2-percent slope) for drainage, except for coal-tar built-up roofs that shall have a design slope of a minimum one-eighth unit vertical in 12 units horizontal (1-percent slope).

❖ Because of their low melting points and self-healing characteristics, coal tar membranes are commonly installed on lower slopes and dead-level roofs. One-quarter inch (6.4 mm) per foot (1 in 48) is the maximum recommended slope for a coal tar built-up roof.

1507.10.2 Material standards. Built-up roof covering materials shall comply with the standards in Table 1507.10.2.

❖ Asphalt and coal tar are the principal bitumens used for roofing purposes.

TABLE 1507.10.2
BUILT-UP ROOFING MATERIAL STANDARDS

MATERIAL STANDARD	STANDARD
Acrylic coatings used in roofing	ASTM D 6083
Aggregate surfacing	ASTM D 1863
Asphalt adhesive used in roofing	ASTM D 3747
Asphalt cements used in roofing	ASTM D 3019; D 2822; D 4586
Asphalt-coated glass fiber base sheet	ASTM D 4601
Asphalt coatings used in roofing	ASTM D1227; D 2823; D 4479
Asphalt glass felt	ASTM D 2178
Asphalt primer used in roofing	ASTM D 41
Asphalt-saturated and asphalt-coated organic felt base sheet	ASTM D 2626
Asphalt-saturated organic felt (perforated)	ASTM D 226
Asphalt used in roofing	ASTM D 312
Coal-tar cements used in roofing	ASTM D 4022; D 5643
Coal-tar saturated organic felt	ASTM D 227
Coal-tar pitch used in roofing	ASTM D 450; Type I or II
Coal-tar primer used in roofing, dampproofing and waterproofing	ASTM D 43
Glass mat, coal tar	ASTM D 4990
Glass mat, venting type	ASTM D 4897
Mineral-surfaced inorganic cap sheet	ASTM D 3909
Thermoplastic fabrics used in roofing	ASTM D 5665, D 5726

❖ The types of materials used in typical built-up roofing must comply with various standards as noted in the table. The following list discusses some of those standards and what they address. Note that the titles and editions of all the standards are located in Chapter 35.

ASTM D 1863 covers the quality and grading of crushed stone, crushed slag and water-worn gravel suitable for use as aggregate surfacing on built-up roofs.

ASTM D 4601 covers asphalt-impregnated and coated glass fiber base sheet, with or without perforations, for use as the first ply of the built-up roofing. When not perforated, this sheet may be used as a vapor retarder under or between roof insulation with a solid top coating of asphaltic material.

ASTM D 2178 covers glass felt impregnated to varying degrees with asphalt, which may be used both with asphalts conforming to the requirements of ASTM D 312 in the construction of built-up roofs, and asphalts conforming to the requirements of ASTM D 449 in the membrane system of waterproofing.

ASTM D 2626 covers the base sheet with fine mineral surfacing on the top side, with or without perforations, for use as the first ply of a built-up roof. When not perforated, this sheet may be used as a vapor retarder under roof insulation.

ASTM D 226 covers asphalt-saturated organic felts, with perforations, that may be used with asphalts conforming to the requirements of ASTM D 312 in the construction of built-up roofs, and with asphalts conforming to the requirements of ASTM D 449 in the membrane system of waterproofing.

ASTM D 312 covers four types of asphalt intended for use in built-up roof construction. The specification is intended for general classification purposes only and does not imply restrictions on the slope at which an asphalt must be used. There are four classifications. Type I includes asphalts that are relatively susceptible to flow at roof temperatures with good adhesive and self-sealing properties. Type II includes asphalts that are moderately susceptible to flow at roof temperatures. Type III includes asphalts that are relatively nonsusceptible to flow at roof temperatures for use in built-up roof construction on slope inclines from 8.3 to 25 percent. Type IV includes asphalts that are generally nonsusceptible to flow at roof temperatures for use in built-up roof construction on slope inclines from approximately 16.7 percent to 50 percent.

ASTM D 227 covers coal-tar saturated organic felt that may be used with coal-tar pitches conforming to the appropriate requirements of ASTM D 450 in the construction of built-up roofs and in the membrane system of waterproofing.

ASTM D 450 covers three types of coal-tar pitch suitable for use in the construction of built-up roofing, dampproofing and membrane waterproofing systems. Only Type I or III can be used in code-recognized roofing systems. Type I is suitable for use in built-up roofing systems with felts conforming to the requirements of ASTM D 227 or as specified by the manufacturer. Type III is suitable for use in built-up roofing systems, but has less volatile components than Type I.

ASTM D 3909 covers asphalt-impregnated and coated glass felt roll roofing surfaced on the weather side with mineral granules for use as a cap sheet in the construction of built-up roofs.

As noted, this list is not all-inclusive when compared to Table 1507.10.2.

1507.11 Modified bitumen roofing. The installation of modified bitumen roofing shall comply with the provisions of this section. Modified bitumen roofing is a membrane-type roofing made up of composite sheets consisting of polymer modified bitumen often reinforced and sometimes surfaced with various types of mats, films, foils and mineral granules. The bitumen is modified through inclusion of one or more polymers. There are two general types of polymer modified bitumen membranes: those with the principal modifier being atactic polypropylene (APP), and those with bitumen modified with styrene butadiene styrene (SBS). They differ in characteristics as well as application. For more information on the application of polymer modified bitumen roofing membranes, refer to the *NRCA Roofing and Waterproofing Manual*.

❖ Modified bitumen roofing is a membrane-type roofing made up of composite sheets consisting of polymer modified bitumen often reinforced and sometimes surfaced with various types of mats, films, foils and mineral granules. The bitumen is modified through inclusion of one or more polymers. There are two general types of polymer modified bitumen membranes: those with the principal modifier being atactic polypropylene (APP) and those with bitumen modified with styrene butadiene styrene (SBS). They differ in characteristics as well as application. For more information on the application of polymer modified bitumen roofing membranes, refer to the *NRCA Roofing and Waterproofing Manual*.

1507.11.1 Slope. Modified bitumen membrane roofs shall have a design slope of a minimum of one-fourth unit vertical in 12 units horizontal (2-percent slope) for drainage.

❖ According to this section, all roofs must have a minimum slope of $^1/_4$:12 so that there is positive drainage of storm water to gutters, roof drains and other components of an approved storm sewer system in order to divert water away from the building or structure.

1507.11.2 Material standards. Modified bitumen roof coverings shall comply with CGSB 37-GP-56M, ASTM D 6162, ASTM D 6163, ASTM D 6164, ASTM D 6222, ASTM D 6223 and ASTM D 6298.

❖ The materials and installation of modified bitumen roof coverings must be in accordance with the listed standards, which regulate modified bituminous roofing membranes, either prefabricated or reinforced.

1507.12 Thermoset single-ply roofing. The installation of thermoset single-ply roofing shall comply with the provisions of this section.

❖ Thermoset is a material that solidifies or "sets" irreversibly when heated. Thinners evaporate during the curing process. Thermoset materials are those whose polymers are chemically cross linked, commonly referred to as being "cured." Once they are fully cured, they can only be bonded to like material with an adhesive, as new molecular links may not be formed. There are four common subcategories of thermoset roof membranes: neoprene, chlorosulfonated polyethylene (CSPE), epichlorohydrin (DCH) and ethylene-propylene-diene monomer (or terpolymer) (EPDM).

1507.12.1 Slope. Thermoset single-ply membrane roofs shall have a design slope of a minimum of one-fourth unit vertical in 12 units horizontal (2-percent slope) for drainage.

❖ All roofs must have a minimum slope of $1/4$:12 so that there is positive drainage of storm water to gutters, roof drains and other components of an approved storm sewer system in order to divert water away from the building or structure.

1507.12.2 Material standards. Thermoset single-ply roof coverings shall comply with RMA RP-1, RP-2 or RP-3, or ASTM D 4637, ASTM D 5019 or CGSB 37-GP-52M.

❖ This section describes the standards to be used for materials and installation. The standards are RMA RP-1, which regulates nonreinforced black EPDM rubber sheets; RMA RP-2, which regulates fabric-reinforced black EPDM rubber sheets; RMA RP-3, which regulates fabric-reinforced black polychloroprene rubber sheets; CGSB 37-GP-52M, which regulates sheet-applied elastomeric roofing membranes; ASTM D 5019, which regulates reinforced nonvulcanized polymeric sheets made from chlorosulfonated polyethylene and polyisobutylene used in single-ply roof membranes and ASTM D 4637, which regulates unreinforced and fabric-reinforced vulcanized rubber sheets made from EPDM or polychloroprene and is intended for use in single-ply roof membranes exposed to weather.

1507.13 Thermoplastic single-ply roofing. The installation of thermoplastic single-ply roofing shall comply with the provisions of this section.

❖ Thermoplastics are materials that soften when heated and harden when cooled. This process can be repeated provided the material is not heated above the point at which decomposition occurs. Thermosplastic materials are distinguished from thermosets in that there is no chemical cross linking. Because of the materials' chemical nature, some thermosplastic membranes may be seamed by either heat (hot air) or solvent welding.

1507.13.1 Slope. Thermoplastic single-ply membrane roofs shall have a design slope of a minimum of one-fourth unit vertical in 12 units horizontal (2-percent slope).

❖ All roofs must have a minimum slope of $1/4$:12, so that there is positive drainage of storm water to gutters, roof drains and other components of an approved storm sewer system in order to divert water away from the building or structure.

1507.13.2 Material standards. Thermoplastic single-ply roof coverings shall comply with ASTM D 4434 or CGSB 37-GP-54M.

❖ This section establishes standards and performance criteria for the installation of thermoplastic single-ply roof coverings. Two standards are recognized: ASTM D 4434 and CGSB 37-GP-54M. These specifications cover flexible sheet made from poly (vinyl chloride) resin intended for use in single-ply roofing membranes exposed to weather. The sheet may be unreinforced or contain nonreinforcing or reinforcing fibers or nonreinforcing or reinforcing fabrics.

1507.14 Sprayed polyurethane foam roofing. The installation of sprayed polyurethane foam roofing shall comply with the provisions of this section.

❖ Sprayed polyurethane foam is a foam-plastic material, constructed by mixing a two-part liquid that is spray applied to form the base of an adhered roof system. The two-part component mixture reacts chemically and immediately expands when applied through special metering equipment to form a closed-cell foam. As the foam rises, it sets into a solid and a skin forms on the surface. The foam provides a thermal insulation and the skin provides a water-resistant surface.

1507.14.1 Slope. Sprayed polyurethane foam roofs shall have a design slope of a minimum of one-fourth unit vertical in 12 units horizontal (2-percent slope) for drainage.

❖ All roofs must have a minimum slope of $1/4$:12, so that there is positive drainage of storm water to gutters, roof drains and other components of an approved storm sewer system in order to divert water away from the building or structure.

1507.14.2 Material standards. Spray-applied polyurethane foam insulation shall comply with ASTM C 1029.

❖ ASTM C 1029 is the standard that regulates spray-applied rigid cellular polyurethane thermally insulated roof coverings. This specification covers the types and physical properties of spray-applied rigid cellular polyurethane intended for use as thermal insulation. The operating temperatures of the surfaces to which the insulation is applied cannot be lower than -22°F(°C) or greater than 225°F (107°C).

1507.14.3 Application. Foamed-in-place roof insulation shall be installed in accordance with the manufacturer's instructions. A liquid-applied protective coating that complies with Section 1507.15 shall be applied no less than 2 hours nor more than 72 hours following the application of the foam.

❖ The roof system must be protected from ultraviolet light and degradation through the use of surfacing. Liquid-applied protected elastomeric coatings are the recommended surfacing for sprayed polyurethane foam roofing systems.

1507.14.4 Foam plastics. Foam plastic materials and installation shall comply with Chapter 26.

❖ Chapter 26 addresses foam-plastic materials and installation.

1507.15 Liquid-applied coatings. The installation of liquid-applied coatings shall comply with the provisions of this section.

❖ Liquid-applied coatings are nonsynthetic roofing materials that are termed "cold applied" because hot bitumen is not utilized in their application.

1507.15.1 Slope. Liquid-applied roofs shall have a design slope of a minimum of one-fourth unit vertical in 12 units horizontal (2-percent slope).

❖ All roofs must have a minimum slope of $^1/_4$:12 so that there is positive drainage of storm water to gutters, roof drains and other components of an approved storm sewer system in order to divert water away from the building or structure.

1507.15.2 Material standards. Liquid-applied roof coatings shall comply with ASTM C 836, ASTM C 957, ASTM D 6083, ASTM D 1227 or ASTM D 3468.

❖ Liquid-applied roof coatings must be installed in accordance with the manufacturer's instructions. The coatings must comply with ASTM C 836, C 957, D 1227 or D 3468. ASTM C 836 describes the required properties and test methods for a cold liquid-applied elastomeric membrane, one or two components, for waterproofing building decks subject to hydrostatic pressure in building areas to be occupied by personnel, vehicles or equipment. This specification only applies to a membrane system above which a separate wearing or traffic course will be applied. ASTM C 957 describes the required properties and test methods for a cold liquid-applied elastomeric membrane for waterproofing building decks not subject to hydrostatic pressure. The specification applies only to a membrane system that has an integral wearing surface. It does not include specific requirements for skid resistance or fire retardance, although both may be important in specific applications. ASTM D 1227 describes emulsified asphalt suitable for use as a protective coating for built-up roofs and other exposed surfaces with inclines of no less than 4 percent. ASTM D 3468 describes liquid-applied neoprene and chlorosulfonated polyethylene synthetic rubber solutions suitable for use in roofing and waterproofing.

SECTION 1508
ROOF INSULATION

1508.1 General. The use of above-deck thermal insulation shall be permitted provided such insulation is covered with an approved roof covering and passes the tests of FM 4450 or UL 1256 when tested as an assembly.

Exception: Foam plastic roof insulation shall conform to the material and installation requirements of Chapter 26.

❖ Insulation may be used as building insulation as well as a substrate to which the roof membrane is applied. In protected roof membrane systems, the insulation is applied over the membrane and is not expected to act as a substrate. The insulation should provide adequate support for the membrane and other associated rooftop materials and permit limited rooftop traffic such as for regular roof inspections and maintenance.

Rigid board roof insulations currently among the most commonly used for low-slope roofs are: cellular glass, glass fiber, mineral fiber, perlite, phenolic foam, polyisocyanurate foam, polystyrene foam, polyurethane foam, wood fiberboard and composite board.

Nonrigid roof insulations are generally used in steep-slope and some metal-roof assembly construction. Commonly made from cellulose, glass fiber and mineral fiber, these nonrigid insulations may be available in batts, blankets and loose-fiber forms.

1508.1.1 Cellulosic fiberboard. Cellulosic fiberboard roof insulation shall conform to the material and installation requirements of Chapter 23.

❖ Chapter 23 addresses cellulosic fiberboard roof installation.

SECTION 1509
ROOFTOP STRUCTURES

1509.1 General. The provisions of this section shall govern the construction of rooftop structures.

❖ This section identifies and establishes the criteria used in evaluating penthouse-type roof structures. Other roof structures, such as water tanks, cooling towers, towers, spires, domes, cupolas, etc., are not to be considered as penthouses since the code has specific provisions for these structures (see Sections 1509.3, 1509.4 and 1509.5).

1509.2 Penthouses. A penthouse or other projection above the roof in structures of other than Type I construction shall not exceed 28 feet (8534 mm) above the roof where used as an enclosure for tanks or for elevators that run to the roof and in all other cases shall not extend more than 18 feet (8534 mm) above the roof. The aggregate area of penthouses and other rooftop structures shall not exceed one-third the area of the supporting roof. A penthouse, bulkhead or any other similar projection above the roof shall not be used for purposes other than shelter of mechanical equipment or shelter of vertical shaft openings in the roof. Provisions such as louvers, louver blades or flashing shall be made to protect the mechanical equipment and the building interior from the elements. Penthouses or bulkheads used for purposes other than permitted by this section shall conform to the requirements of this code for an additional story. The restrictions of this section shall not prohibit the placing of wood flagpoles or similar structures on the roof of any building.

❖ The heights of roof structures are not limited, provided the building is of Type I construction. For other types of construction, the roof structure is limited to 18 feet (5486 mm) in height unless it encloses tanks or elevators that extend to the roof, for which the maximum height is 28 feet (8534 mm). If the height exceeds that permitted by the code, the roof structure should be counted as, and meet all the requirements of, a story. Also, if the area exceeds one-third the roof area, regardless of height, then the penthouse must be considered a story. Regardless of height or area, roof structures used for anything except sheltering mechanical equipment or vertical shaft openings must be considered as an additional story.

1509.2.1 Type of construction. Penthouses shall be constructed with walls, floors and roof as required for the building.

Exceptions:

1. On buildings of Type I and II construction, the exterior walls and roofs of penthouses with a fire separation distance of more than 5 feet (1524 mm) and less than 20 feet (6096 mm) shall be of at least 1-hour fire-resistance-rated noncombustible construction. Walls and roofs with a fire separation distance of 20 feet (6096 mm) or greater shall be of noncombustible construction. Interior framing and walls shall be of noncombustible construction.

2. On buildings of Type III, IV and V construction, the exterior walls of penthouses with a fire separation distance of more than 5 feet (1524 mm) and less than 20 feet (6096 mm) shall be at least 1-hour fire-resistance-rated construction. Walls with a fire separation distance of 20 feet (6096 mm) or greater from a common property line shall be of Type IV or noncombustible construction. Roofs shall be constructed of materials and fire-resistance rated as required in Table 601. Interior framing and walls shall be Type IV or noncombustible construction.

3. Unprotected noncombustible enclosures housing only mechanical equipment and located with a minimum fire separation distance of 20 feet (6096 mm) shall be permitted.

4. On one-story buildings, combustible unroofed mechanical equipment screens, fences or similar enclosures are permitted where located with a fire separation distance of at least 20 feet (6096 mm) from adjacent property lines and where not exceeding 4 feet (1219 mm) in height above the roof surface.

5. Dormers shall be of the same type of construction as the roof on which they are placed, or of the exterior walls of the building.

❖ The general premise is that a penthouse be treated no differently than any other portion of the building as it is constructed of exterior walls, a floor and a roof; however, this section recognizes the reduced exposure

of penthouses when the exterior wall is recessed from the exterior wall of the building.

The exterior walls of the penthouse, which are located within 5 feet (1524 mm) of the building's lot line, are subject to the same requirements as the exterior walls of the building. As such, the exterior walls are required to be of the same fire-resistance rating as the exterior wall of the story immediately below the penthouse.

If a penthouse is set back by a fire separation distance of 5 feet (1524 mm) or more from the lot line, the risk of exposure to adjacent property from a fire within the penthouse, as well as fire exposure to the penthouse from adjacent property, is reduced. Accordingly, if a penthouse is set back 5 feet (1524 mm) or more from the building's lot line, the penthouse construction may be able to take advantage of one of the exceptions, typically based on type of construction, resulting in a reduction in rating of the exterior wall of the penthouse. In many cases, penthouses at least 20 feet (6096 mm) from the lot line are allowed further reductions.

1509.3 Tanks. Tanks having a capacity of more than 500 gallons (2 m³) placed in or on a building shall be supported on masonry, reinforced concrete, steel or Type IV construction provided that, where such supports are located in the building above the lowest story, the support shall be fire-resistance rated as required for Type IA construction.

❖ This section identifies and establishes the criteria used in evaluating rooftop tanks, such as water tanks.

1509.3.1 Valve. Such tanks shall have in the bottom or on the side near the bottom, a pipe or outlet, fitted with a suitable quick opening valve for discharging the contents in an emergency through an adequate drain.

❖ In the event of an emergency, there must be a means to drain the tank. This not only involves a quick-opening valve on the tank but also drains that are capable of handling the discharge. The drains must be sized and installed in accordance with the *International Plumbing Code®* (IPC®).

1509.3.2 Location. Such tanks shall not be placed over or near a line of stairs or an elevator shaft, unless there is a solid roof or floor underneath the tank.

❖ Tanks are not permitted to be located above a stairway or elevator enclosure unless a solid roof or floor deck is provided underneath the tank. This is intended to provide a means of ensurance that a shaft will not be affected by water from a leaking tank.

1509.3.3 Tank cover. Unenclosed roof tanks shall have covers sloping toward the outer edges.

❖ Roof tanks must be covered to prevent the accumulation of rain, ice and snow, which may cause an uncovered tank to overflow.

1509.4 Cooling towers. Cooling towers in excess of 250 square feet (23.2 m²) in base area or in excess of 15 feet (4572 mm) high where located on buildings more than 50 feet (15 240 mm) high shall be of noncombustible construction. Cooling towers shall not exceed one-third of the supporting roof area.

> **Exception:** Drip boards and the enclosing construction of wood not less than 1 inch (25 mm) nominal thickness, provided the wood is covered on the exterior of the tower with noncombustible material.

❖ This section identifies and establishes criteria used in evaluating rooftop cooling towers.

1509.5 Towers, spires, domes and cupolas. Any tower, spire, dome or cupola shall be of a type of construction not less in fire-resistance rating than required for the building to which it is attached except that any such tower, spire, dome or cupola that exceeds 85 feet (25 908 mm) in height above grade, or exceeds 200 square feet (18.6 m²) in horizontal area or is used for any purpose other than a belfry or an architectural embellishment shall be constructed of and supported on Type I or II construction.

❖ This section identifies and establishes criteria used in evaluating rooftop structures other than penthouse-type structures, tanks and cooling towers. Examples of rooftop structures addressed in this section include towers, spires, domes and cupolas. Penthouse-type structures, tanks and cooling towers are addressed in Sections 1509.2, 1509.3 and 1509.4, respectively.

More specifically, when any of the structures addressed in this section exceed either 85 feet (25 908 mm) in height above grade or are greater than 200 square feet (18.6 m²) in area, the construction type is limited to Type I or II. Section 1509.5.1 further restricts construction to noncombustible when the heights become more excessive (see commentary, Section 1509.5.1). These restrictions are related to the difficulty posed in fighting fires in such structures and also the danger to surrounding people, emergency responders and other buildings if such elements should fail during a fire.

1509.5.1 Noncombustible construction required. Any tower, spire, dome or cupola that exceeds 60 feet (18 288) in height above the highest point at which it comes in contact with the roof, or that exceeds 200 square feet (18.6 m²) in area at any horizontal section, or which is intended to be used for any purpose other than a belfry or architectural embellishment, shall be entirely constructed of and supported by noncombustible materials. Such structures shall be separated from the building below by construction having a fire-resistance rating of not less than 1.5 hours with openings protected with a minimum 1.5-hour fire-protection rating. Structures, except aerial supports 12 feet (3658 mm) high or less, flagpoles, water tanks and cooling towers, placed above the roof of any building more than 50 feet (15 240 mm) in height, shall be of noncombustible material and shall be supported by construction of noncombustible material.

❖ These roof structures generally include communication towers, spires, cupolas and similar structures. Such structures must be constructed as required for the building's type of construction classification. Because of the difficulty in fighting fires on tall buildings and structures and generally the hazard associated with locating combustible materials on roofs of taller buildings, when the height of such structures exceeds 60 feet (18 288 mm) above the highest point at which it comes in contact with the roof, or is in excess of 200 square feet (18.6 m²) in area at any horizontal section, the structure and its supports must be of noncombustible construction.

1509.5.2 Towers and spires. Towers and spires where enclosed shall have exterior walls as required for the building to which they are attached. The roof covering of spires shall be of a class of roof covering as required for the main roof of the rest of the structure.

❖ The towers addressed in this section are radio and television antenna towers, church spires and other towers and spires of a similar nature. As with penthouses and roof structures, however, the code intends to obtain construction and fire resistance consistent with that of the building to which they are attached.

SECTION 1510
REROOFING

1510.1 General. Materials and methods of application used for recovering or replacing an existing roof covering shall comply with the requirements of Chapter 15.

> **Exception:** Reroofing shall not be required to meet the minimum design slope requirement of one-quarter unit vertical in 12 units horizontal (2-percent slope) in Section 1507 for roofs that provide positive roof drainage.

❖ This section simply states that when a roof is replaced or recovered it must comply with this chapter for the materials and methods used with only one exception for low-sloped roofs. This section does not mandate that the entire roof be replaced but simply that the portion being replaced complies with Chapter 15.

For low-sloped roofs, the exception indicates that reroofing (i.e., recovering or replacement) is not required to meet the $^{1}/_{4}$:12 minimum slope requirement of Section 1507, provided that the roof has positive drainage. The term "positive drainage" is defined as the drainage condition in which consideration has been made for all loading deflections of the roof deck, and additional roof slope has been provided to ensure drainage of the roof area within 48 hours of rainfall.

1510.2 Structural and construction loads. Structural roof components shall be capable of supporting the roof-covering system and the material and equipment loads that will be encountered during installation of the system.

❖ The structural integrity of the roof must be maintained during reroofing operations, which can significantly contribute to the loading of the roof due to workers and material being present during this period of time. The roof support system must be able to support structurally all additional layers of roof covering material.

1510.3 Recovering versus replacement. New roof coverings shall not be installed without first removing all existing layers of roof coverings where any of the following conditions occur:

1. Where the existing roof or roof covering is water soaked or has deteriorated to the point that the existing roof or roof covering is not adequate as a base for additional roofing.

2. Where the existing roof covering is wood shake, slate, clay, cement or asbestos-cement tile.

3. Where the existing roof has two or more applications of any type of roof covering.

Exceptions:

1. Complete and separate roofing systems, such as standing-seam metal roof systems, that are designed to transmit the roof loads directly to the building's structural system and that do not rely on existing roofs and roof coverings for support, shall not require the removal of existing roof coverings.

2. Metal panel, metal shingle, and concrete and clay tile roof coverings shall be permitted to be installed over existing wood shake roofs when applied in accordance with Section 1510.4.

❖ This section determines when all layers of previously installed roof covering systems must be removed prior to installation of the new roof covering system.

When the existing roof or roof covering is water soaked, it must be allowed to dry completely so as not to trap moisture beneath the new layer of covering. This could cause a rapid deterioration of the new covering material as well as the existing sheathing. The existing covering is required to be removed if it cannot adequately dry out or if its physical properties have been permanently altered.

Wood shake, slate, clay, cement or asbestos-cement tile types of existing roof coverings historically do not make an adequate base for new roof coverings and could prevent the new covering from making a weather-tight seal. They could also allow penetration of water, snow, etc. These types of existing coverings must always be removed.

When the existing roof has two or more layers of any type of covering system, all layers need to be removed to enable the inspector and the contractor to verify that the existing sheathing is not water damaged and is still capable of providing an adequate nailing base.

Exception 1 states that new roofing systems that are designed to transmit all roof loads directly to the structural supports of the building do not necessitate that the existing roofing system be removed. Exception 2 allows certain roof coverings, including metal, concrete panel and clay, to be placed over wood shingle and shake roofs only if any concealed combustible spaces are properly addressed in accordance with Section 1510.4 (see commentary, Section 1510.4).

1510.4 Roof recovering. Where the application of a new roof covering over wood shingle or shake roofs creates a combustible concealed space, the entire existing surface shall be covered with gypsum board, mineral fiber, glass fiber or other approved materials securely fastened in place.

❖ "Roof recovering" is defined as the process of installing an additional roof covering over a prepared existing roof covering without removing the existing roof covering. Where recovering over wood shingles or shakes creates a combustible concealed space, the code requires that the entire surface of the wood shakes and shingles be covered with a material that will reduce the possibility of such materials adding fuel to a fire in such a space.

1510.5 Reinstallation of materials. Existing slate, clay or cement tile shall be permitted for reinstallation, except that damaged, cracked or broken slate or tile shall not be reinstalled. Existing vent flashing, metal edgings, drain outlets, collars and metal counterflashings shall not be reinstalled where rusted, damaged or deteriorated. Aggregate surfacing materials shall not be reinstalled.

❖ This section establishes requirements for materials to be reused for roof coverings. Historically, various types of materials have been removable without substantially damaging the material. Materials such as wood shingles and shakes, roll roofing and asphalt shingles are usually torn or cracked and cannot be reused. Fastener holes also violate the integrity of the material. Before reuse is allowed, materials such as slate, clay or cement tile should be examined thoroughly for cracks and deterioration.

1510.6 Flashings. Flashings shall be reconstructed in accordance with approved manufacturer's installation instructions. Metal flashing to which bituminous materials are to be adhered shall be primed prior to installation.

❖ Flashings to be reused or reconstructed must be in accordance with the manufacturer's installation instructions. Metal flashings that are to be reused for bituminous materials must be primed in accordance with the manufacturer's instructions.

Bibliography

The following resource material is referenced in this chapter or is relevant to the subject matter addressed in this chapter.

ASCE 7-02, *Minimum Design Loads for Buildings and Other Structures.* Reston, VA: American Society of Civil Engineers, 2002.

ASTM A 653/A 653M-01a, *Specification for Steel Sheet, Zinc-coated Galvanized or Zinc-iron Ally-Coated Galvanized by the Hot-dip Process.* West Conshohocken, PA: ASTM International, 2001.

ASTM A 755/A 755M-01, *Specification for Steel Sheet, Metallic-Coated by the Hot-Dip Process and Prepainted by the Coil-Coating Process for Exterior Ex-*

posed Building Products. West Conshohocken, PA: ASTM International, 2001.

ASTM B 101-01, Specification for Lead-Coated Copper Sheets and Strip for Building Construction. West Conshohocken, PA: ASTM International, 2001.

ASTM C 406-00, Specification for Roofing Slate. West Conshohocken, PA: ASTM International, 2000.

ASTM C 836-00, Specification for High-Solids Content, Cold Liquid-Applied Elastomeric Waterproofing Membrane for Use with Separate Wearing Course. West Conshohocken, PA: ASTM International, 2000.

ASTM C 957-98, Specification for High-Solids Content, Cold Liquid-Applied Elastomeric Waterproofing Membrane with Integral Wearing Surface. West Conshohocken, PA: ASTM International, 1998.

ASTM C 1029-96, Specification for Spray-Applied Rigid Cellular Polyurethane Thermal Insulation. West Conshohocken, PA: ASTM International, 1996.

ASTM C 1167-96, Specification for Clay Roof Tiles. West Conshohocken, PA: ASTM International, 1996.

ASTM D 225-01, Specification for Asphalt Shingles (Organic Felt) Surfaced with Mineral Granules. West Conshohocken, PA: ASTM International, 2001.

ASTM D 226-97a, Specification for Asphalt-Saturated Organic Felt Used in Roofing and Waterproofing. West Conshohocken, PA: ASTM International, 1997.

ASTM D 227-97a, Specification for Coal-Tar-Saturated Organic Felt Used in Roofing and Waterproofing. West Conshohocken, PA: ASTM International, 1997.

ASTM D 249-89 (1996), Specification for Asphalt Roll Roofing (Organic Felt) Surfaced with Mineral Granules. West Conshohocken, PA: ASTM International, 1996.

ASTM D 312-00, Specification for Asphalt Used in Roofing. West Conshohocken, PA: ASTM International, 2000.

ASTM D 371-89 (1996), Specification for Asphalt Roll Roofing (Organic Felt) Surfaced with Mineral Granules: Wide-Selvage. West Conshohocken, PA: ASTM International, 1996.

ASTM D 449-89 (1999) e1, Specification for Asphalt Used in Dampproofing and Waterproofing. West Conshohocken, PA: ASTM International, 1999.

ASTM D 450-00, Specification for Coal-Tar Pitch Used in Roofing, Dampproofing and Waterproofing. West Conshohocken, PA: ASTM International, 2000.

ASTM D 1227-00, Specification for Emulsified Asphalt Used as a Protective Coating for Roofing. West Conshohocken, PA: ASTM International, 2000.

ASTM D 1863-93 (2000), Specification for Mineral Aggregate Used on Built-up Roofs. West Conshohocken, PA: ASTM International, 2000.

ASTM D 1970-01, Specification for Self-Adhering Polymer Modified Bituminous Sheet Materials Used as Steep Roof Underlayment for Ice Dam Protection. West Conshohocken, PA: ASTM International, 2001.

ASTM D 2178-97a, Specification for Asphalt Glass Felt Used in Roofing and Waterproofing. West Conshohocken, PA: ASTM International, 1997.

ASTM D 2626-97b, Specification for Asphalt Saturated and Coated Organic Felt Base Sheet Used in Roofing. West Conshohocken, PA: ASTM International, 1997.

ASTM D 2898-94 (1999), Test Methods for Accelerated Weathering for Fire-Retardant-Treated Wood of Fire Testing. West Conshohocken, PA: ASTM International, 1999.

ASTM D 3018-90 (1994) e1, Specification for Class A Asphalt Shingles Surfaced with Mineral Granules. West Conshohocken, PA: ASTM International, 1994.

ASTM D 3161-99a, Test Method for Wind-Resistance of Asphalt Shingles (Fan-Induced Method). West Conshohocken, PA: ASTM International, 1999.

ASTM D 3462-E01, Specification for Asphalt Shingles Made from Glass Felt and Surfaced with Mineral Granules. West Conshohocken, PA: ASTM International, 2001.

ASTM D 3468-99, Specification for Liquid-Applied Neoprene and Chlorosulfonated Polyethylene Used in Roofing and Waterproofing. West Conshohocken, PA: ASTM International, 1999.

ASTM D 3909-97b, Specification for Asphalt Roll Roofing (Glass Felt) Surfaced with Mineral Granules. West Conshohocken, PA: ASTM International, 1997.

ASTM D 4434-96, Specification for Poly (Vinyl Chloride) Sheet Roofing. West Conshohocken, PA: ASTM International, 1996.

ASTM D 4601-98, Specification for Asphalt-Coated Glass Fiber Base Sheet Used in Roofing. West Conshohocken, PA: ASTM International, 1998.

ASTM D 4637-96, Specification for EPDM Sheet Used in Single-Ply Roof Membrane. West Conshohocken, PA: ASTM International, 1996.

ASTM D 4869-93, Specification for Asphalt-Saturated (Organic Felt) Underlayment Used in Steep Slope Roofing. West Conshohocken, PA: ASTM International, 1993.

ASTM E 108-00, Test Methods for Fire Tests of Roof Coverings. West Conshohocken, PA: ASTM International, 2000.

CGSB 37-GP-52M-(1984), Roofing and Waterproofing Membrane, Sheet Applied, Elastomeric. Ottawa, Ontario, Canada: Canadian General Standards Board, 1984.

CGSB 37-GP-56M-80, *Membrane, Modified, Bituminous, Prefabricated, and Reinforced for Roofing—with December 1985 Amendment.* Ottawa, Ontario, Canada: Canadian General Standards Board, 1980.

CGSB 37.54-95, *Polyvinyl Chloride Roofing and Water proffing Membrane.* Ottawa, Ontario, Canada: Canadian General Standards Board,1995.

CSSB, *Design and Application Manual for New Roof Construction.* Sumas, WA: Cedar Shake and Shingle Bureau.

CSSB-97, *Grading and Packing Rules for Western Red Cedar Shakes and Western Red Shingles of the Cedar Shake and Single Bureau.* Sumas, WA: Cedar Shake and Shingle Bureau, 1997.

CSSB-2000, *Grading Rules for Certi-Grade Red Cedar Shingles.* Sumas, WA: Cedar Shake and Shingle Bureau, January 2000.

CSSB-2000, *Grading Rules for Certi-Split Red Cedar Shakes.* Sumas, WA: Cedar Shake and Shingle Bureau, January 2000.

CSSB-2000, *Grading Rules for Certi-Split Taper Sawn Red Cedar Shakes.* Sumas, WA: Cedar Shake and Shingle Bureau, January 2000.

IPC-2003, *International Plumbing Code.* Falls Church, VA: International Code Council, 2003.

New Roof Construction Manual. Sumas, WA: Cedar Shake and Shingle Bureau, January 2001.

NRCA Asphalt Shingle Roofing Manual. Rosemont, IL: National Roofing Contractors Association (NCRA).

NRCA Roofing and Waterproofing Manual, 4th ed. Rosemont, IL: National Roofing Contractors Association.

NRCA Steep Roofing Manual. Rosemont, IL: National Roofing Contractors Association.

Residential Asphalt Roofing Manual, Asphalt Roofing Manufacturers Association.

RMA RP-1-90, *Minimum Requirements for Nonreinforced Black EPDM Rubber Sheets.* Washington, DC: Rubber Manufacturers Association, 1990.

RMA RP-2-90, *Minimum Requirements for Fabric-Reinforced Black EPDM Rubber Sheets.* Washington, DC: Rubber Manufacturers Association, 1990.

RMA RP-3-85, *Minimum Requirements for Fabric-Reinforced Black Polychloroprene Rubber Sheets.* Washington, DC: Rubber Manufacturers Association, 1985.

RMA RP-4-88, *Wind Design Guide for Ballasted Single-Ply Roofing Systems.* Washington, DC: Rubber Manufacturers Association, 1988.

INDEX

Note: This is taken from the index for the 2003 International Building Code. Volume I of the IBC Commentary only in-cludes Chapters 1 through 15.

B

G

Y